TABLE OF ATOMIC MASSES (WEIGHTS) BASED ON CARBON-12

Name	Symbol	Atomic No.	Atomic Mass
Actinium	**Ac**	89	(227)[a]
Aluminum	**Al**	13	26.98154
Americium	**Am**	95	(243)[a]
Antimony	**Sb**	51	121.75
Argon	**Ar**	18	39.948
Arsenic	**As**	33	74.9216
Astatine	**At**	85	(210)[a]
Barium	**Ba**	56	137.34
Berkelium	**Bk**	97	(247)[a]
Beryllium	**Be**	4	9.01218
Bismuth	**Bi**	83	208.9804
Boron	**B**	5	10.81
Bromine	**Br**	35	79.904
Cadmium	**Cd**	48	112.40
Calcium	**Ca**	20	40.08
Californium	**Cf**	98	(251)[a]
Carbon	**C**	6	12.011
Cerium	**Ce**	58	140.12
Cesium	**Cs**	55	132.9054
Chlorine	**Cl**	17	35.453
Chromium	**Cr**	24	51.996
Cobalt	**Co**	27	58.9332
Copper	**Cu**	29	63.546
Curium	**Cm**	96	(247)[a]
Dysprosium	**Dy**	66	162.50
Einsteinium	**Es**	99	(252)[a]
Erbium	**Er**	68	167.26
Europium	**Eu**	63	151.96
Fermium	**Fm**	100	(257)[a]
Fluorine	**F**	9	18.99840
Francium	**Fr**	87	(223)[a]
Gadolinium	**Gd**	64	157.25
Gallium	**Ga**	31	69.72
Germanium	**Ge**	32	72.59
Gold	**Au**	79	196.9665
Hafnium	**Hf**	72	178.49
Helium	**He**	2	4.00260
Holmium	**Ho**	67	164.9304
Hydrogen	**H**	1	1.0079
Indium	**In**	49	114.82
Iodine	**I**	53	126.9045
Iridium	**Ir**	77	192.22
Iron	**Fe**	26	55.847
Krypton	**Kr**	36	83.80
Lanthanum	**La**	57	138.9055
Lawrencium	**Lr**	103	(260)[a]
Lead	**Pb**	82	207.2
Lithium	**Li**	3	6.941
Lutetium	**Lu**	71	174.97
Magnesium	**Mg**	12	24.305
Manganese	**Mn**	25	54.9380
Mendelevium	**Md**	101	(258)[a]
Mercury	**Hg**	80	200.59
Molybdenum	**Mo**	42	95.94
Neodymium	**Nd**	60	144.24
Neon	**Ne**	10	20.179
Neptunium	**Np**	93	237.0482[b]
Nickel	**Ni**	28	58.71
Niobium	**Nb**	41	92.9064
Nitrogen	**N**	7	14.0067
Nobelium	**No**	102	(259)[a]
Osmium	**Os**	76	190.2
Oxygen	**O**	8	15.9994
Palladium	**Pd**	46	106.4
Phosphorus	**P**	15	30.97376
Platinum	**Pt**	78	195.09
Plutonium	**Pu**	94	(244)[a]
Polonium	**Po**	84	(210)[a]
Potassium	**K**	19	39.098
Praseodymium	**Pr**	59	140.9077
Promethium	**Pm**	61	(145)[a]
Protactinium	**Pa**	91	231.0359[b]
Radium	**Ra**	88	226.0254[b]
Radon	**Rn**	86	(222)[a]
Rhenium	**Re**	75	186.2
Rhodium	**Rh**	45	102.9055
Rubidium	**Rb**	37	85.4678
Ruthenium	**Ru**	44	101.07
Samarium	**Sm**	62	150.4
Scandium	**Sc**	21	44.9559
Selenium	**Se**	34	78.96
Silicon	**Si**	14	28.086
Silver	**Ag**	47	107.868
Sodium	**Na**	11	22.98977
Strontium	**Sr**	38	87.62
Sulfur	**S**	16	32.06
Tantalum	**Ta**	73	180.9479
Technetium	**Tc**	43	98.9062[b]
Tellurium	**Te**	52	127.60
Terbium	**Tb**	65	158.9254
Thallium	**Tl**	81	204.37
Thorium	**Th**	90	232.0381[b]
Thulium	**Tm**	69	168.9342
Tin	**Sn**	50	118.69
Titanium	**Ti**	22	47.90
Tungsten	**W**	74	183.85
Unnilhexium	**Unh**	106	(263)[a]
Unnilpentium	**Unp**	105	(262)[a]
Unnilquadium	**Unq**	104	(261)[a]
Uranium	**U**	92	238.029
Vanadium	**V**	23	50.9414
Xenon	**Xe**	54	131.30
Ytterbium	**Yb**	70	173.04
Yttrium	**Y**	39	88.9059
Zinc	**Zn**	30	65.38
Zirconium	**Zr**	40	91.22

[a] Mass number of most stable or best-known isotope

[b] Mass number of the isotope of longest half-life

College Chemistry

Kimburly Zdun
600 Devils Knob Rd
Tiller, OR 97484

College Chemistry

An Introduction to General, Organic, and Biochemistry Fifth Edition

Morris Hein

Leo R. Best

Scott Pattison

Susan Arena

Brooks/Cole Publishing Company
Pacific Grove, California

Brooks/Cole Publishing Company
A Division of Wadsworth, Inc.

Printed in the United States of America

10 9 8 7 6 5 4 3 2

Library of Congress Cataloging-in-Publication Data

College chemistry : an introduction to general, organic, and biochemistry / Morris Hein ... [et al.]. —5th ed.
p. cm.
Includes index.
ISBN 0-534-17526-0
1. Chemistry. I. Hein, Morris.
QD31.2.H43 1992 91-45819
540—dc20 CIP

Sponsoring Editors: Maureen A. Allaire, Harvey C. Pantzis
Editorial Assistant: Beth Wilbur
Print Buyer: Vena M. Dyer
Production: Julie Kranhold, Ex Libris; Joan Marsh
Manuscript Editor: Andrew Alden
Interior Design: Nancy Benedict
Interior Illustration: Nancy Benedict, Lotus, Pat Rogondino
Photo Researcher: Stuart Kenter
Typesetting: Polyglot Pte. Ltd. Compositors
Cover Printing: The Lehigh Press
Printing and Binding: Arcata Graphics, Hawkins County Plant
Cover Design: Vernon T. Boes
Cover Photo: Herb Charles Ohlmeyer/Fran Heyl Associates
Photo credits are listed after the index.

The cover image is a photomicrograph of a human umbilical cord, containing two arteries and one vein that together provide the fetal lifeline. The cross section depicted is magnified 1700 times its actual size.

To Edna, Louise, Joan, and Frank

Preface

To begin the preface of the fifth edition of *College Chemistry*, we wish to express our pleasure at the wide, favorable reception of the previous four editions and to thank our colleagues for the many helpful and constructive suggestions they have made.

We are particularly pleased to introduce our new co-author to this edition, Susan Arena of Mt. San Antonio College. She has contributed many new ideas and new material to this edition, based on her years of experience teaching this course.

The primary purpose of *College Chemistry* is to instruct students in the basic concepts of chemistry. This book is intended for health-science and other students, who, although not majoring in chemistry, require a basic introduction to general, organic chemistry, and biochemistry. The text is written with the assumption that the students have not taken a previous chemistry course and may have a limited mathematical background.

Many students are somewhat apprehensive about beginning the study of chemistry. Accordingly, one of our major goals is to present chemistry in a well-organized and easy to understand fashion. While presenting the technical phases of the subject matter, we strive to develop students' interest, to motivate them to further their scientific knowledge, and to help prepare them for future careers in biology, chemistry, or the health-related occupations.

A certain amount of quantitative reasoning is essential to the study of chemistry. Therefore, another major goal is to provide careful, step-by-step explanations of topics requiring quantitative reasoning. These explanations are generally illustrated by one or more examples. The examples are set up and solved using a dimensional analysis approach. We have taken great care to show each step in the solution of the examples, so that students with only elementary mathematics experience can follow easily. A large number of end-of-chapter exercises are included, to provide sufficient practice in problem-solving for the beginning chemistry student.

Although a number of changes have been made in this edition, the level of the material remains the same. The entire text was carefully reviewed for further improvement in clarity of expression and to provide students with greater assistance in solving problems. Some of the major changes are as follows.

NEW FEATURES

Each new chapter now begins with a **Chapter Preview**, listing sections covered in the chapter and an introduction that relates chemistry to aspects of modern living. Chapter objectives are now located at the end of the chapter in a new section called **Concepts in Review**. End-of-chapter **Exercises** have been expanded and are now numbered consecutively. As an aid to reinforce problem-solving techniques, **Practice Problems** (with answers) have been added immediately after most in-text **Examples**. Throughout the text, we have standardized the use of atomic masses and molar masses to four figures.

Multiple colors have been introduced throughout the text to highlight and identify important features and to assist students in making efficient use of the book. Many full-color illustrations and photos have been added to stimulate student interest and to illustrate chemical principles and applications.

Most chapters feature a new section called **Chemistry in Action**. The topics discussed include a wide range of applications with particular emphasis in the health-science field. Chemistry in Action sections are designed to be supplemental to the main body of the text. However, end-of-chapter exercises do include questions relating to the Chemistry in Action topics.

CHANGES IN ORGANIZATION

Chapter 1 has an added section (1.7) that describes the key pedagogical features of the text and explains their role in the study process. Chapters 3 and 4 have been reorganized so that Chapter 3 covers the classification of matter and Chapter 4 focuses on the properties of matter.

We have observed over the years that students in this course have better success with the practical material than with the theoretical or abstract material. Consequently, a major change in the chapter sequence has been made in this edition. Chapter 5 (Atomic Theory) has been split in two, "Early Atomic Theory and Structure" (Chapter 5) and "Modern Atomic Theory" (Chapter 10). Also moved to follow "Modern Atomic Theory" are the chapters on "The Periodic Table" (Chapter 11) and "Chemical Bonds" (Chapter 12). Because the connection between molecular shape and function is becoming more important in the design and use of chemicals in our lives, we have reorganized Chapter 11 (Periodic Table) and Chapter 12 (Chemical Bonding) in order to clarify and simplify the relationship between structure and reactivity. And, we have introduced two new sections in Chapter 12 to illustrate the use of simple VSEPR theory to predict the shape of molecules. Should any instructor prefer the original sequence of topics, moving Chapters 10, 11, and 12 to follow Chapter 5 will give them the same sequence of topics as in the previous edition.

Some of the benefits of the sequence changes are an earlier treatment of "Nomenclature of Inorganic Compounds" (Chapter 6), "Quantitative Composition of Compounds" (Chapter 7), "Chemical Equations" (Chapter 8), and "Calculations from Chemical Equations" (Chapter 9). Studying these chapters earlier enhances student success with naming compounds both in lecture and in the laboratory. It also increases student confidence in handling chemicals by name, introduces the mole sooner, and provides an earlier understanding and use of chemical equations and calculations based on equations. "Chemistry of Selected Elements" (Chapter 20) has undergone a major revision. The chapter now covers selected elements from each representative family with special emphasis on the use of these elements and their compounds in society.

Two new chapters have been added to the organic and biochemistry portion of the book. Amines and amides (Chapter 26) introduces students to organic nitrogen chemistry. Enzymes have been moved from the chapter on amino acids and proteins to a separate chapter (Chapter 32) to give a more expanded and updated treatment of these important biochemical molecules. A new section covering phosphate esters has been introduced into the chapter on carboxylic acids and esters (Chapter 25).

Learning Aids

Students often feel somewhat apprehensive about undertaking the study of chemistry because they are inexperienced in the use of technical and abstract scientific content. As experienced teachers of these students, we have developed several aids to help students confidently face technical subject matter.

- A list of "Concepts in Review" is given at the end of every chapter to guide students in studying the most important concepts in the chapter.
- Important "Key Terms" are set in boldfaced type where they are defined, and they are also printed in color in the margin.
- The many "Examples" in the text are solved in a step-by-step fashion using the conversion-factor dimensional analysis method.
- Most "Examples" are followed by a "Practice Problem" (with answer) for immediate reinforcement of student learning.
- Answers to mathematical problems are given in Appendix V.
- Complete answers to all end-of-chapter "Exercises" are given in the *Solutions Manual*.
- All chapters contain at least one self-evaluation question. Answers to those questions are also given in Appendix V.
- The Index has been modified to a "Glossary–Index." "Key Terms" are highlighted in the index (with the appropriate page reference) so that students may quickly refer back to the page where the term is defined and read the definition in context.

Supplements to Text

Materials that may be helpful to students and to their instructors have been developed to accompany the text. A short description of them follows.

Study Guide by Peter C. Scott of Linn-Benton Community College includes a self-evaluation section for students to check their understanding of each chapter's objectives, a recap section, and answers to the self-evaluation section.

Solutions Manual includes answers and solutions to all end-of-chapter questions and problems.

Instructor's Supplement includes a set of objective test questions and answers to the test questions.

College Chemistry in the Laboratory, 5th Edition, by Morris Hein, Leo R. Best, Robert L. Miner, and James M. Ritchey includes 42 experiments for a laboratory program that may accompany the lecture course. Also included are Study Aids and Exercises.

Instructor's Manual to accompany the lab manual includes information on the management of the lab, evaluation of experiments, notes for individual experiments, a list of reagents required, and answer keys to each experiment's report form.

EXPTEST, the test generation system, Version 5.0 from Brooks/Cole for IBM PCs or compatibles.

Acknowledgments

It is impossible to thank by name each of the many people who have been involved in this revision of *College Chemistry*. We are grateful for the friendship and many helpful comments and suggestions from our colleagues and students who, over the years, have made this book possible.

No textbook can be completed without the untiring effort of many professionals in publishing. Special thanks to the talented staff at Brooks/Cole, especially Joan Marsh who through her skill, knowledge, and patience directed the production of this colorful book. Much credit also goes to Julie Kranhold of *Ex Libris* for her unfailing attention to detail and persistence in moving the book through production. We also appreciate the guidance of Harvey Pantzis, Executive Editor, and Maureen Allaire, Chemistry Editor.

Chemistry in Action Boxes

Brief Contents

Detailed Contents

Chapter 12
Chemical Bonds: The Formation of Compounds from Atoms 229

Chapter 13
The Gaseous State of Matter 255

Chapter 14
Water and the Properties of Liquids 296

Chapter 15
Solutions 325

Chapter 16
Ionization: Acids, Bases, Salts 362

CHAPTER 22
Unsaturated Hydrocarbons 548

CHAPTER 23
Alcohols, Ethers, Phenols, and Thiols 585

CHAPTER 24
Aldehydes and Ketones 615

CHAPTER 25
Carboxylic Acids and Esters 637

CHAPTER 26
Amides and Amines: Organic Nitrogen Compounds 671

CHAPTER 27
Polymers: Macromolecules 690

CHAPTER 28
Stereoisomerism 707

CHAPTER 29
Carbohydrates 727

CHAPTER 30
Lipids 765

CHAPTER 31
Amino Acids, Polypeptides, and Proteins 788

CHAPTER 32
Enzymes 823

CHAPTER 33
Nucleic Acids and Heredity 841

CHAPTER 34
Nutrition 869

CHAPTER 35
Bioenergetics 895

CHAPTER 36
Carbohydrate Metabolism 916

CHAPTER 37
Metabolism of Lipids and Proteins 936

Appendixes A-1

Glossary/Index I-1

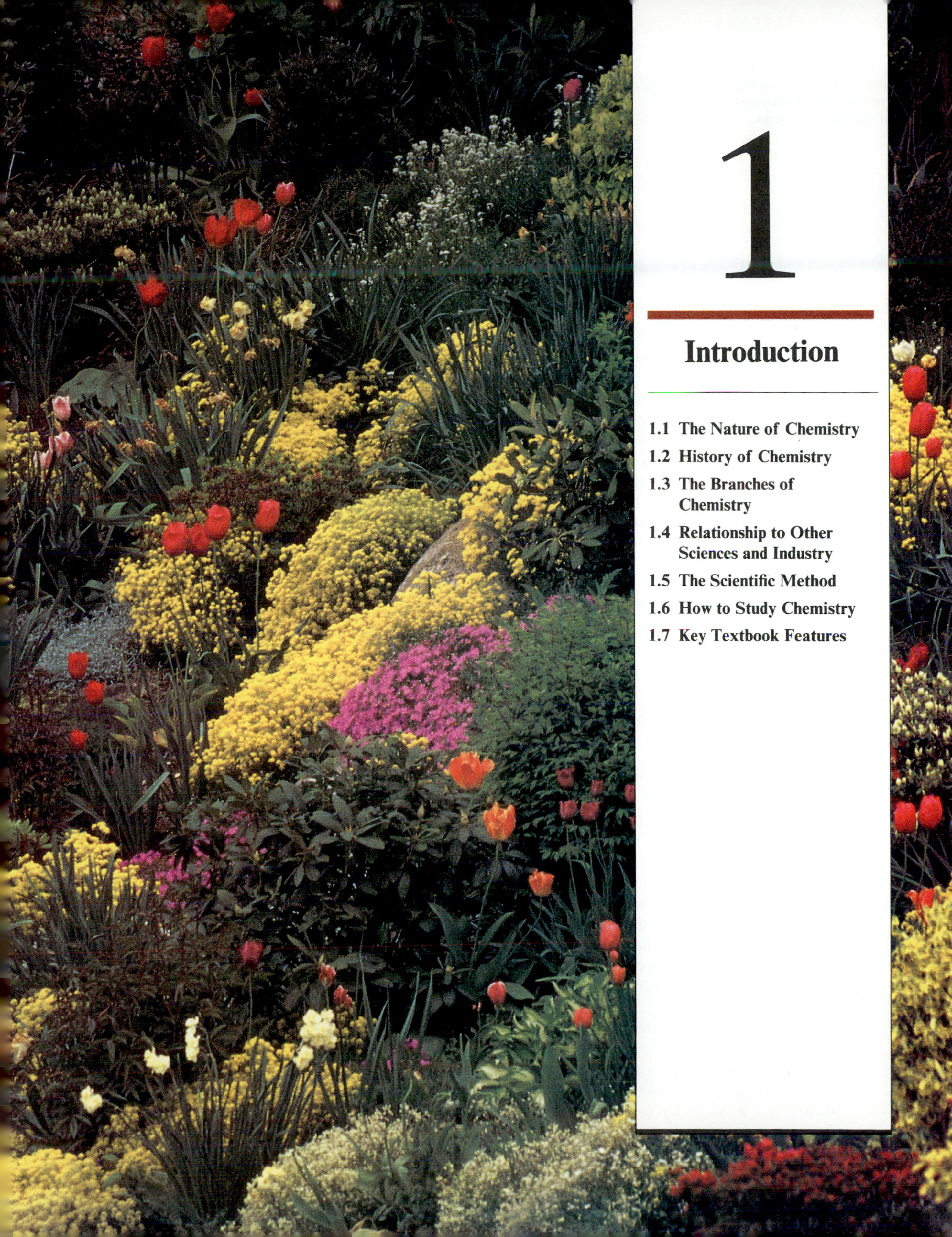

1

Introduction

1.1 The Nature of Chemistry
1.2 History of Chemistry
1.3 The Branches of Chemistry
1.4 Relationship to Other Sciences and Industry
1.5 The Scientific Method
1.6 How to Study Chemistry
1.7 Key Textbook Features

◀ **CHAPTER OPENING PHOTO: The colors of this spring garden result from a series of chemical reactions.**

Have you ever strolled through a spring garden and been amazed at the diversity of colors in the flowers? Or perhaps you have curled up in front of a winter fire and become fascinated watching the flames. And think of those times when you have dropped a beverage container on a hard floor, and were relieved to find that it was plastic instead of glass. All of these phenomena are the result of chemistry—not in the laboratory but rather in our everyday lives. Chemical changes can bring us beautiful colors, warmth and light, or new and exciting products. Chemists seek to understand, explain, and utilize the diversity of materials we find around us.

1.1 THE NATURE OF CHEMISTRY

chemistry

What is chemistry? A popular dictionary gives this definition: **Chemistry** is the science of the composition, structure, properties, and reactions of matter, especially of atomic and molecular systems. Another, somewhat simpler dictionary definition is: Chemistry is the science dealing with the composition of *matter* and the changes in composition that matter undergoes. Neither of these definitions is entirely adequate. Chemistry, along with the closely related science of physics, is a fundamental branch of knowledge. Chemistry is also closely related to biology, not only because living organisms are made of material substances but also because life itself is essentially a complicated system of interrelated chemical processes.

The scope of chemistry is extremely broad: It includes the whole universe and everything, animate and inanimate, in it. Chemistry is concerned not only with the composition and changes in composition of matter, but also with the energy and energy changes associated with matter. Through chemistry we seek to learn and to understand the general principles that govern the behavior of all matter.

The chemist, like other scientists, observes nature and attempts to understand its secrets: What makes a rose red? Why is sugar sweet? What is occurring when iron rusts? Why is carbon monoxide poisonous? Why do people wither with age? Problems such as these—some of which have been solved, some of which are still to be solved—are part of what we call chemistry.

A chemist may interpret natural phenomena, devise experiments that will reveal the composition and structure of complex substances, study methods for

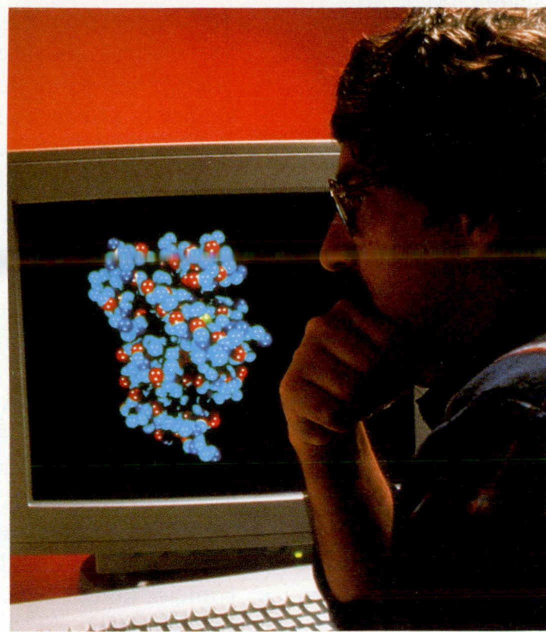

improving natural processes, or, sometimes, synthesize substances unknown in nature. Ultimately, the efforts of successful chemists advance the frontiers of knowledge and at the same time contribute to the well-being of humanity. Chemistry can help us to understand nature; however, one need not be a professional chemist or scientist to enjoy natural phenomena. Nature and its beauty, its simplicity within complexity, are for all to appreciate.

The body of chemical knowledge is so vast that no one can hope to master it all, even in a lifetime of study. However, many of the basic concepts can be learned in a relatively short period of time. These basic concepts have become part of the education required for many professionals, including agriculturists, biologists, dental hygienists, dentists, medical technologists, microbiologists, nurses, nutritionists, pharmacists, physicians, and veterinarians, to name a few.

1.2 The History of Chemistry

People have practiced empirical chemistry from the earliest times. Ancient civilizations were practicing the art of chemistry in such processes as wine-making, glass-making, pottery-making, dyeing, and elementary metallurgy. The early Egyptians, for example, had considerable knowledge of certain chemical processes. Excavations into ancient tombs dated about 3000 B.C. have uncovered workings of gold, silver, copper, and iron, pottery from clay, glass beads, and beautiful dyes and paints, as well as bodies of Egyptian kings in remarkably well-preserved states. Many other cultures made significant developments in chemistry. However, all these developments were empirical; that is, they were achieved by trial and error and did not rest on any valid theory of matter.

Philosophical ideas relating to the properties of matter (chemistry) did not develop as early as those relating to astronomy and mathematics. The Greek philosophers made great strides in philosophical speculation concerning materialistic ideas about chemistry. They led the way to placing chemistry on an intellectual, scientific basis. They introduced the concepts of elements, atoms, shapes of atoms, and chemical combination. They believed that all matter was derived from four elements: earth, air, fire, and water. The Greek philosophers had keen minds and perhaps came very close to establishing chemistry on a sound basis similar to the one that was to develop about 2000 years later. The main shortcoming of the Greek approach to scientific work was a failure to carry out systematic experimentation.

Greek civilization was succeeded by the Roman civilization. The Romans were outstanding in military, political, and economic affairs. They practiced empirical chemical arts such as metallurgy, enameling, glass-making, and pottery-making, but they did very little to advance new and theoretical knowledge. Eventually the Roman civilization was succeeded in Europe by the Dark Ages. During this period European civilization and learning were at a very low ebb.

In the Middle East and in North Africa, knowledge did not decline during the Dark Ages as it did in Western Europe. At this time Arabic cultures made contributions that were of great value to the development of modern chemistry. In particular, the Arabic number system, including the use of zero, gained acceptance; the branch of mathematics known as *algebra* was developed; and alchemy, a sort of pseudochemistry, was practiced extensively.

One of the more interesting periods in the history of chemistry was that of the alchemists (500–1600 A.D.). People have long had a lust for gold, and in those days gold was considered the ultimate, most perfect metal formed in nature. The principal goals of the alchemists were to find a method of prolonging human life indefinitely and to change the base metals, such as iron, zinc, and copper, into gold. They searched for a universal solvent to transmute base metals into gold and for the "philosopher's stone" to rid the body of all diseases and to renew life. In the course of their labors they learned a great deal of chemistry. Unfortunately, much of their work was done secretly because of the mysticism that shrouded their activity, and very few records remain.

Although the alchemists were not guided by sound theoretical reasoning and were clearly not in the intellectual class of the Greek philosophers, they did something that the philosophers had not considered worthwhile. They subjected various materials to prescribed treatments under what might be loosely described as laboratory methods. These manipulations, carried out in alchemical laboratories, not only uncovered many facts of nature but paved the way for the systematic experimentation that is characteristic of modern science.

Alchemy began to decline in the 16th century when Paracelsus (1493–1541), a Swiss physician and outspoken revolutionary leader in chemistry, strongly advocated that the objectives of chemistry be directed toward the needs of medicine and the curing of human ailments. He openly condemned the mercenary efforts of alchemists to convert cheaper metals to gold.

But the real beginning of modern science can be traced to astronomy during the Renaissance. Nicolaus Copernicus (1473–1543), a Polish astronomer, began the downfall of the generally accepted belief in a geocentric universe. Although not all the Greek philosophers had believed that the sun and the stars revolved about the earth, the geocentric concept had come to be generally accepted. The heliocentric (sun-centered) universe concept of Copernicus was based on direct astronomical observation and represented a radical departure from the concepts handed down from Greek and Roman times. The ideas of Copernicus and the invention of the telescope stimulated additional work in astronomy. This work, especially that of Galileo Galilei (1564–1642) and Johannes Kepler (1571–1630), led directly to a rational explanation by Sir Isaac Newton (1642–1727) of the general laws of motion, which he formulated between about 1665 and 1685.

Modern chemistry was slower to develop than astronomy and physics; it began in the 17th and 18th centuries when Joseph Priestley (1733–1804), who discovered oxygen in 1774, and Robert Boyle (1627–1691) began to record and publish the results of their experiments and to discuss their theories openly. Boyle, who has been called the founder of modern chemistry, was one of the first to practice chemistry as a true science. He believed in the experimental method. In his most important book, *The Sceptical Chymist*, he clearly distinguished between an element and a compound or mixture. Boyle is best known today for the gas law that bears his name. A French chemist, Antoine Lavoisier (1743–1794), placed the science on a firm foundation with experiments in which he used a chemical balance to make quantitative measurements of the masses of substances involved in chemical reactions (Figure 1.1).

▲ **FIGURE 1.1** **Antoine Lavoisier (1743–1794).**

The use of the chemical balance by Lavoisier and others later in the 18th century was almost as revolutionary in chemistry as the use of the telescope had been in astronomy. Thereafter, chemistry became a quantitative experimental science. Lavoisier also contributed greatly to the organization of chemical data, to chemical nomenclature, and to the establishment of the Law of Conservation of Mass in chemical changes. During the period from 1803 to 1810, John Dalton (1766–1844), an English schoolteacher, advanced his atomic theory. This theory (see Section 5.2) placed the atomistic concept of matter on a valid rational basis. It remains today as a tremendously important general concept of modern science.

Since the time of Dalton, knowledge of chemistry has advanced in great strides, with the most rapid advancement occurring at the end of the 19th century and during the 20th century. Especially outstanding achievements have been made in determining the structure of the atom, understanding the biochemical fundamentals of life, developing chemical technology, and the mass production of chemicals and related products.

1.3 THE BRANCHES OF CHEMISTRY

Chemistry may be broadly classified into two main branches: *organic* chemistry and *inorganic* chemistry. Organic chemistry is concerned with compounds containing the element carbon. The term *organic* was originally derived from the chemistry of living organisms: plants and animals. Inorganic chemistry deals with all the other elements as well as with some carbon compounds. Substances classified as inorganic are derived mainly from mineral sources rather than from animal or vegetable sources.

Other subdivisions of chemistry, such as analytical chemistry, physical chemistry, biochemistry, electrochemistry, geochemistry, and nuclear chemistry, may be considered specialized fields of, or auxiliary fields to, the two main branches.

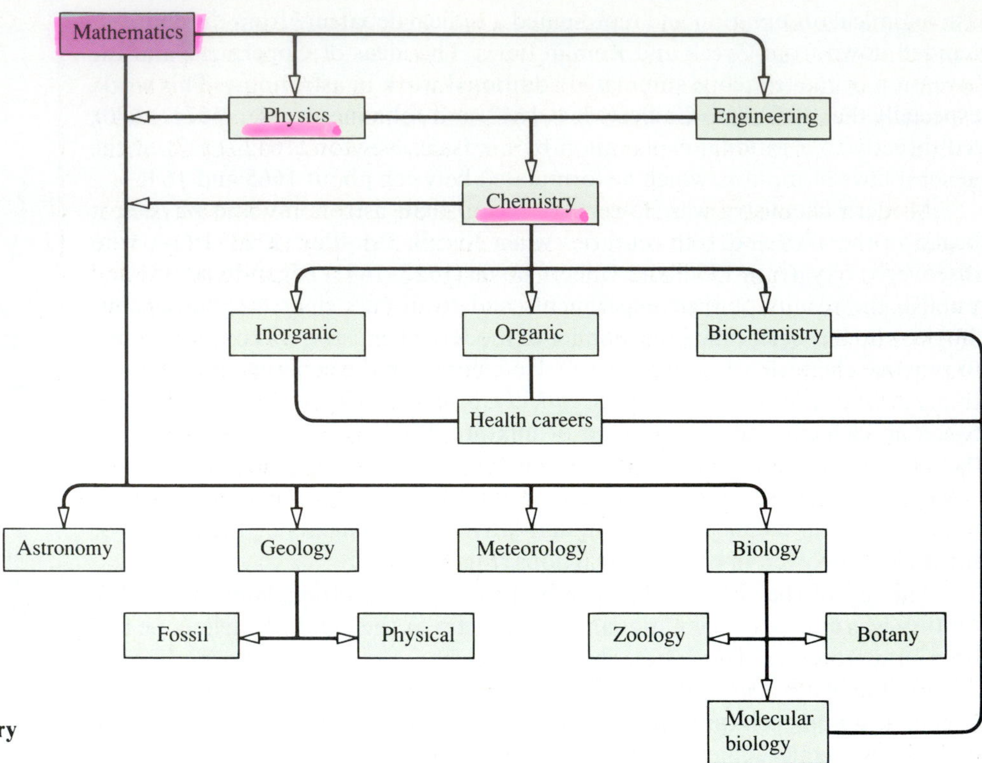

FIGURE 1.2 ▸ The relationship of chemistry to other disciplines.

Chemical engineering is the branch of engineering that deals with the development, design, and operation of chemical processes. A chemical engineer usually begins with a chemist's laboratory-scale process and develops it into an industrial-scale operation.

1.4 Relationship of Chemistry to Other Sciences and Industry

Besides being a science in its own right, chemistry is the servant of other sciences and industry. Chemical principles contribute to the study of physics, biology, agriculture, engineering, medicine, space research, oceanography, and many other sciences (Figure 1.2). Chemistry and physics are overlapping sciences, since both are based on the properties and behavior of matter. Biological processes are chemical in nature. The metabolism of food to provide energy to living organisms is a chemical process. Knowledge of molecular structure of proteins, hormones, enzymes, and the nucleic acids is assisting biologists in their investigations of the composition, development, and reproduction of living cells.

Chemistry is playing an important role in alleviating the growing shortage of food in the world. Agricultural production has been increased with the use of chemical fertilizers, pesticides, and improved varieties of seeds. Chemical refrigerants make possible the frozen food industry, which preserves large amounts of food that might otherwise spoil. Chemistry is also producing synthetic nutrients, but much remains to be done as the world population increases relative to the land available for cultivation. Expanding energy needs have brought about difficult environmental problems in the form of air and water pollution. Chemists and other scientists are working diligently to alleviate these problems.

Advances in medicine and chemotherapy, through the development of new drugs, have contributed to prolonged life and the relief of human suffering. More than 90% of the drugs and pharmaceuticals being used in the United States today have been developed commercially within the past 50 years. The plastics and polymer industry, unknown 60 years ago, has revolutionized the packaging and textile industries and is producing durable and useful construction materials. Energy derived from chemical processes is used for heating, lighting, and transportation. Virtually every industry is dependent on chemicals—for example, the petroleum, steel, rubber, pharmaceutical, electronic, transportation, cosmetic, space, polymer, garment, aircraft, and television industries. (The list could go on.)

1.5 The Scientific Method

Chemistry, as a science or field of knowledge, is concerned with ideas and concepts relating to the behavior of matter. Although these concepts are abstract, their application has had a concrete impact on human culture. This impact is due to modern technology, which may be said to have begun about 200 years ago and which has grown at an accelerating rate ever since.

An important difference between science and technology is that science represents an abstract body of knowledge, and technology represents the physical application of this knowledge to the world in which we live.

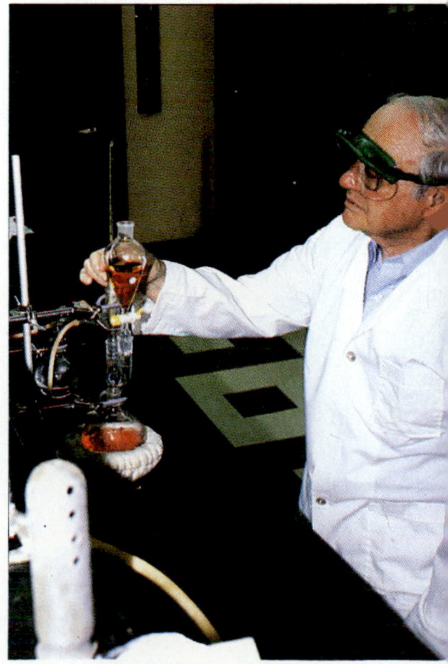

Why has the science of chemistry and its associated technology flourished in the last two centuries? Is it because we are growing more intelligent? No, we have absolutely no reason to believe that the general level of human intelligence is any higher today than it was in the Dark Ages. The use of the scientific method is usually credited with being the most important single factor in the amazing development of chemistry and technology. Although complete agreement is lacking on exactly what is meant by "using the scientific method," the general approach is as follows:

1. Collect facts or data that are relevant to the problem or question at hand, which is usually done by planned experimentation.
2. Analyze the data to find trends (regularities) that are pertinent to the problem. Formulate a hypothesis that will account for the data that have been accumulated and that can be tested by further experimentation.
3. Plan and do additional experiments to test the hypothesis. Such experiments extend beyond the range that is covered in Step 1.
4. Modify the hypothesis as necessary so that it is compatible with all the pertinent experimental data.

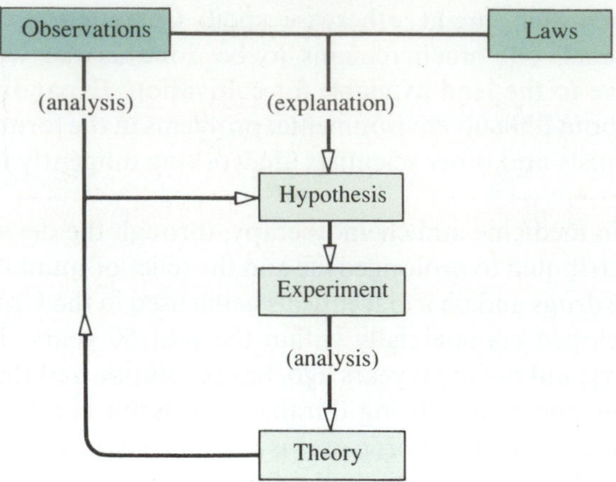

Figure 1.3 ▶
The scientific method.

Confusion sometimes arises regarding the exact meanings of the words *hypothesis*, *theory*, and *law*. A **hypothesis** is a tentative explanation of certain facts that provides a basis for further experimentation. A well-established hypothesis is often called a **theory**. Thus a theory is an explanation of the general principles of certain phenomena with considerable evidence or facts to support it. Hypotheses and theories explain natural phenomena, whereas **scientific laws** are simple statements of natural phenomena to which no exceptions are known under the given conditions.

hypothesis

theory

scientific laws

Although the four steps listed in the preceding paragraph are a broad outline of the general procedure that is followed in much scientific work, they are not a recipe for doing chemistry or any other science (Figure 1.3). But chemistry is an experimental science, and much of its progress has been due to application of the scientific method through systematic research. Occasionally a great discovery is made by accident, but the majority of scientific achievements are accomplished by well-planned experiments.

Many theories and laws are studied in chemistry. They make the study of chemistry or any science easier, because they summarize particular aspects of the science. The student will note that some of the theories advanced by great scientists in the past have since been substantially altered and modified. Such changes do not mean that the discoveries of the past are less significant than those of today. Modification of existing theories in the light of new experimental evidence is essential to the growth and evolution of scientific knowledge.

1.6 How to Study Chemistry

How do you as a student approach a subject such as chemistry with its unfamiliar terminology, symbols, formulas, theories, and laws? All the generally accepted habits of good study are applicable to the study of chemistry. Budget your study time and spend it wisely. In particular, you can spend your time more profitably in regular, relatively short periods of study rather than in one prolonged cram session. Figure 1.4 shows how all the parts of the study process fit together.

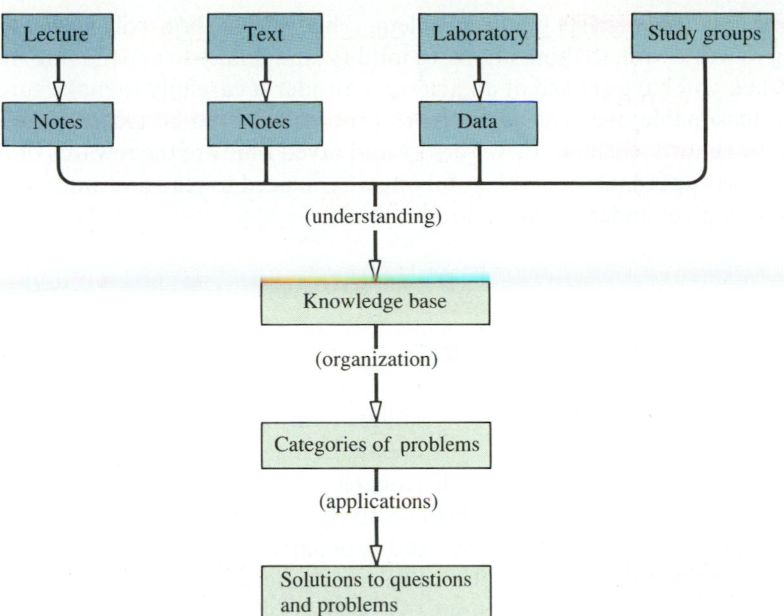

◀ **FIGURE 1.4**
The study process.

Chemistry has its own language, and learning this language is of prime importance to the successful study of chemistry. Chemistry is a subject of many facts. At first you will simply have to memorize some of them. However, you will also learn these facts by referring to them frequently in your studies and by repetitive use. For example, you must learn the symbols of 30 or 40 common elements in order to be able to write chemical formulas and equations. As with the alphabet, repetitive use of these symbols will soon make them part of your vocabulary.

The need for careful reading of assigned material cannot be overemphasized. You should read each chapter at least twice. The first time, read the chapter rapidly, noting especially topic headings, diagrams, and other outstanding features. Then read more thoroughly and deliberately for better understanding. It may be profitable to underline and abstract material during the second reading. Isolated reading may be sufficient for some subjects, but it is not sufficient for learning chemistry. During the lectures, become an active mental participant and try to think along with your instructor, do not just occupy a seat. Lecture and laboratory sessions will be much more meaningful if you have already read the assigned material.

Your studies must include a good deal of *written* chemistry. Chemical symbolism, equations, problem solving, and so on, require much written practice for proficiency. One does not become an accomplished pianist by merely reading or listening to music—it takes practice. One does not become a good baseball player by reading the rules and watching baseball games—it takes practice. So it is with chemistry. One does not become proficient in chemistry by only reading about it—it takes practice.

You will encounter many mathematical problems as you progress through this text. To solve a numerical problem, you should read the problem carefully to determine what is being asked. Then develop a plan for solving the problem. It is a good idea to start by writing down the pertinent material—a formula, a diagram, an equation, the data given in the problem. This information will give you something to work with, to think about, to modify, and finally to expand into an answer. When you have arrived at an answer, consider it carefully to make sure that it is a reasonable one. The solutions to problems should be recorded in a neat, orderly, stepwise fashion. Fewer errors and saved time are the rewards of a neat and orderly approach to problem solving. If you need to read and study still further for complete understanding, do it!

1.7 Key Textbook Features

This new edition of *College Chemistry* includes many tools to help you in the study process. You must be an active participant in the study of chemistry. This requires you to read the text and to ask yourself questions. You must address these questions as you read. To do this effectively, you must read with pencil and paper. There are many calculations and problems in chemistry, hence you will need a scientific calculator to help with the arithmetic, as real numbers can often be cumbersome. To study without the benefits of paper, pencil, and calculator is *inefficient* at best.

The text is designed to assist you in the study process. Each chapter begins with a list of the topics covered, followed by a short description of some practical aspects of the topic in your daily life. This description serves as a bridge to reveal how chemistry is an integral part of our lives.

Most chapters include "Examples" which are worked out in sequence. These will provide you with an illustration of the problem-solving techniques necessary to understand a concept. The "skills" of chemistry are developed in the "Examples." Often, more than one method of solving a problem may be appropriate, thus alternative methods of solution are suggested. Select the method that makes you comfortable and use it to solve the "Practices" that follow the example. Be sure to work the practice problems while you are reading. The answers are provided so you may check them immediately.

At the end of most chapters, you will find a set of "Exercises." Some are less challenging, requiring basic information; while others are more demanding, requiring analysis or application of concepts already learned. The most challenging problems are marked with an asterisk (*). The tables and figures from the chapter, as well as the appendix and endpapers, will assist you with the problems. "Answers to Selected Problems" are found in the appendix. Complete solutions are found in the *Student Solutions Manual.*

If you have difficulty solving a problem or answering a question, refer back to the chapter for a similar situation. Use the examples as a guide. If you still can't solve the problem within a reasonable time, skip it for the moment. Often, leaving a frustrating problem and returning to it later will provide a fresh approach to a solution. Bring troublesome problems to your study group or your instructor for assistance.

"Concepts in Review" is an overview of the important ideas and skills introduced in each chapter. If you accomplish each item in this section, you will understand the fundamental concepts from the chapter. Some chapters also list "Equations in Review." "Marginal Terms" are provided throughout the chapter

Chemistry in Action

Serendipity in Science

Discoveries in the world of chemistry are for the most part made by people who are applying the scientific method in their work. Occasionally, important discoveries are made by chance, or through serendipity. But even when serendipity is involved, a discovery is more likely to be made by someone with a good knowledge of the field. Louis Pasteur summed this up in a statement made long ago: "Chance favors the prepared mind." In chemistry serendipity often can lead to whole new fields and technology.

The synthetic dye industry began in 1856 when William Perkin, an 18-year-old student at the Royal College of Chemistry in London, was attempting to synthesize quinine, a drug used to treat malaria. He reacted two chemicals, aniline sulfate and potassium dichromate, and obtained a black paste. Perkin then extracted the paste with alcohol. Upon evaporating the alcohol, violet crystals appeared which, when dissolved in water, made a beautiful purple solution. He so enjoyed the color he began investigating the solution; he then determined the purple color had a strong affinity for silk. Perkin had discovered the first synthetic aniline dye. Recognizing the commercial possibilities, he immediately left school and, with his father and an older brother, went into the dye manufacturing business. His dye, known as mauve, quickly became a success and inspired other research throughout Europe. By 1870, cloth could be purchased in more and brighter synthetic colors than were ever available with natural dyes.

The Discovery of Artificial Sweeteners

Sweetener	Date	Discoverer
Saccharin	1878	I. Remsen and C. Fahlberg
Cyclamate	1937	M. Sveda
Aspartame	1965	J. Schlatter

Cotton dyeing plant.

A second, more recent, account of chance events in chemistry also led to a multimillion dollar industry (see table). In 1965 James Schlatter was researching anti-ulcer drugs for the pharmaceutical firm G. D. Searle. In the course of his work he accidentally ingested a small amount of his preparation and found, to his surprise, it had an extremely sweet taste. (*Note:* Tasting chemicals of any kind in the laboratory is not a safe procedure.) When purified, the sweet-tasting substance turned out to be aspartame, a molecule consisting of two amino acids joined together. Since only very small quantities are necessary to produce sweetness, it proved to be an excellent low calorie artificial sweetener. Today, under the trade names of "Equal" and "Nutrasweet," aspartame is one of the cornerstones of the artificial sweetener industry.

when important terms are introduced. Chemistry is a new language to you, so the understanding of vocabulary is fundamental to a good foundation. Just as you do not memorize the definition of each new word you encounter, you should not attempt to memorize these terms. Rather, you should seek to understand and incorporate them in your speaking and writing.

The appendix of the text contains valuable information for use in study, including a review of mathematical concepts to assist you in problem solving and a "Glossary/Index" that provides an accessible reference for terms that have been introduced elsewhere in the text. Other tables of useful information are found in the appendix as well. Consider it a handy reference library.

With these tools to assist you in your study of chemistry, you will be prepared to begin a great adventure. Surely, you will feel confusion, frustration, elation, and success along the way. Most certainly you will gain a new appreciation and understanding of the role of chemistry in the world around you.

Exercises

1. Was the concept of a geocentric universe necessarily based on incorrect astronomical observations? Explain.
2. What were the principal goals of the alchemists?
3. What instrument, when first used by chemists, can be considered to be analogous to the use of the telescope by early astronomers? Explain the analogy.
4. Classify the following statements as observation, law, hypothesis, or theory:
   (a) When the pressure remains constant, the volume of a gas is directly proportional to the absolute temperature.
   (b) The water in a closed test tube boiled at 83°C.
   (c) Iron gets heavier when it rusts because it attracts particles of rust from the air.
   (d) All matter is composed of tiny particles called atoms.
   (e) As it approaches its melting point, glass turns a flame yellow.
   (f) Molecules in a gas are always moving.
   (g) When wood burns it decomposes to its elements, which all escape as gas.
5. Which of the following statements are correct? Rewrite the incorrect statements to make them correct.
   (a) Chemistry is the science that deals with the composition of substances and the transformations they undergo.
   (b) Robert Boyle, in the 17th century, clearly distinguished between an element and a compound or mixture.
   (c) Both the knowledge and intellectual capacity of Western European people decreased markedly during the Dark Ages.
   (d) From 1803 to 1810, John Dalton advanced his atomic theory.
   (e) Most of the drugs and pharmaceuticals used in the United States today have been available for at least a century.
   (f) A key feature of the scientific method is to plan and do additional experiments to test a hypothesis.
   (g) Scientific laws are simple statements of natural phenomena to which no exceptions are known.
   (h) Antoine Lavoisier was one of the first chemists to make quantitative measurements using a chemical balance.
   (i) The two main branches of chemistry are organic and biochemistry.
6. List ten examples of matter which you use daily.
7. Distinguish between theory and law.
8. If you perform an experiment and do not get the result you expect based upon your previous knowledge, what would you do next?
9. Design a weekly study plan to use in this course. Include:
   (a) all hours you are enrolled in all classes

(b) study time for each course (plan for 2 hours per class hour)
(c) work hours
(d) relaxation time (life is not all academic)

10. Use the three numbers given and a combination of +, −, ×, and ÷ symbols to yield the boxed number on the right. Remember that trial and error is an acceptable procedure for solution of a problem.

7 5 4 = [16]
6 2 8 = [12]
3 8 6 = [4]

11. Describe the process used by William Perkin which produced the dye mauve in 1856. State the importance of this process.

12. Indicate the important contribution of Remsen to chemistry in 1878.

13. Explain why tasting chemicals is not a safe practice.

2

Standards for Measurement

◀ **CHAPTER OPENING PHOTO: Measuring instruments come in various shapes and sizes, all necessary to allow components to fit together, and permit us to quantify our world.**

Doing an experiment in chemistry is very much like cooking a meal in the kitchen. It is important to know the ingredients *and* the amounts of each in order to have a tasty product. Working on your car requires specific tools, in exact sizes. Buying new carpeting or draperies is an exercise in precise and accurate measurement for a good fit. A small difference in the concentration or amount of medication a pharmacist gives you may have significant effects on your well-being. In all of these cases, a strong foundation in the language, measurement, and use of numbers provides the basis for success. In chemistry, we begin by learning the metric system and the proper units for measuring mass, length, volume, and temperature.

2.1 MASS AND WEIGHT

Chemistry is an experimental science. The results of experiments are usually determined by making measurements. In elementary experiments the quantities that are commonly measured are mass, length, volume, pressure, temperature, and time. Measurements of electrical and optical quantities may also be needed in more sophisticated experimental work.

Although mass and weight are often used interchangeably, the two words have quite different meanings. The **mass** of a body is defined as the amount of matter in that body. The mass of an object is a fixed and unvarying quantity that is independent of the object's location. The mass of an object may be measured on a balance by comparison with other known masses.

mass

The **weight** of a body is the measure of the earth's gravitational attraction for that body. Weight is measured on a device called a scale, which measures force against a spring. Unlike mass, weight varies in relation to (1) the position of an object on or its distance from the earth and (2) whether the rate of motion of the object is changing with respect to the motion of the earth.

weight

Consider an astronaut of mass 70.0 kilograms (154 pounds) who is being shot into orbit. At the instant before blast-off the weight of the astronaut is also 70.0 kilograms. As the distance from the earth increases and the rocket turns into an orbiting course, the gravitational pull on the astronaut's body decreases until a state of weightlessness (zero weight) is attained. However, the mass of the astronaut's body has remained constant at 70.0 kilograms during the entire event.

2.2 Measurement and Significant Figures

To understand certain aspects of chemistry it is necessary to set up and solve problems. Problem solving requires an understanding of the elementary mathematical operations used to manipulate numbers. Numerical values or data are obtained from measurements made in an experiment. A chemist may use these data to calculate the extent of the physical and chemical changes occurring in the substances that are being studied. By appropriate calculations the results of an experiment may be compared with those of other experiments and summarized in ways that are meaningful.

The result of a measurement is expressed by a numerical value together with a unit of that measurement. For example,

numerical value

70.0 kilograms = 154 pounds

unit

Numbers obtained from a measurement are never exact values. They always have some degree of uncertainty due to the limitations of the measuring instrument and the skill of the individual making the measurement. The numerical value recorded for a measurement should give some indication of its reliability (precision). To express maximum precision this number should contain all the digits that are known plus one digit that is estimated. This last estimated digit introduces some uncertainty. Because of this uncertainty every number that expresses a measurement can have only a limited number of digits. These digits, used to express a measured quantity, are known as **significant figures**, or **significant digits**.

significant figures

Suppose we measure temperature on a thermometer calibrated in degrees and we observe that the mercury stops between 21 and 22 (see Figure 2.1a). We then know that the temperature is at least 21 degrees and is less than 22 degrees. To express the temperature with greater precision, we estimate that the mercury is about two-tenths the distance between 21 and 22. The temperature is, therefore, 21.2 degrees. The last digit (2) has some uncertainty, because it is an estimated value. The recorded temperature, 21.2 degrees, is said to have three significant figures. If the mercury stopped exactly on the 22 (Figure 2.1b), the temperature would be recorded as 22.0 degrees. The zero is used to indicate that the temperature was estimated to a precision of one-tenth degree. Finally, look at Figure 2.1c. On this thermometer, the temperature is recorded as 22.11°C (four significant figures). Since the thermometer is calibrated to tenths of a degree, the first estimated digit is the hundredths.

Some numbers are exact and have an infinite number of significant figures. Exact numbers occur in simple counting operations; when you count 25 dollars, you have exactly 25 dollars. Defined numbers, such as 12 inches in 1 foot, 60 minutes in 1 hour, and 100 centimeters in 1 meter, are also considered to be exact numbers. Exact numbers have no uncertainty.

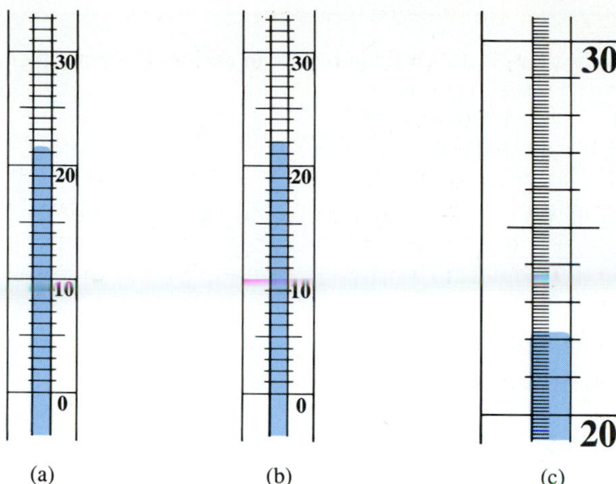

◀ **FIGURE 2.1**
Measuring temperature with various degrees of precision.

Evaluating Zero

In any measurement all nonzero numbers are significant. However, zeros may or may not be significant, depending on their position in the number. Rules for determining when zero is significant in a measurement follow.

1. Zeros between nonzero digits are significant:

   205 has three significant figures.
   2.05 has three significant figures.
   61.09 has four significant figures.

2. Zeros that precede the first nonzero digit are not significant. These zeros are used to locate a decimal point:

   0.0025 has two significant figures (2, 5).
   0.0108 has three significant figures (1, 0, 8).

3. Zeros at the end of a number that include a decimal point are significant:
   0.500 has three significant figures (5, 0, 0).
   25.160 has five significant figures.
   3.00 has three significant figures.
   20. has two significant figures.

4. Zeros at the end of a number without a decimal point are ambiguous and are not considered significant:

   1000 (The zeros may or may not be significant.)
   590 (The zero may or may not be significant.)

   One way of indicating whether these zeros are significant is to write the number using a decimal point and a power of 10. Thus, if the value 1000 has been determined to four significant figures, it is written as 1.000×10^3. If 590 has only two significant figures, it is written as 5.9×10^2 (in this case the zero is not significant).

PRACTICE How many significant figures are in each of these numbers?

(a) 4.5 inches (e) 25.0 grams
(b) 3.025 feet (f) 12.20 liters
(c) 125.0 meters (g) 100,000 people
(d) 0.001 mile (h) 205 birds

Answers: (a) 2 (b) 4 (c) 4 (d) 1 (e) 3 (f) 4 (g) 1 (h) 3

2.3 Rounding off Numbers

rounding off numbers

In calculations we often obtain answers that have more digits than we are justified in using. It is necessary, therefore, to drop the nonsignificant digits in order to express the answer with the proper number of significant figures. When digits are dropped from a number, the value of the last digit retained is determined by a process known as **rounding off numbers**. Two rules will be used in this book for rounding off numbers:

Rule 1 When the first digit after those you want to retain is 4 or less, that digit and all others to its right are dropped. The last digit retained is not changed.

Examples rounded off to four digits:

74.693 = 74.69
This digit is dropped.

1.00629 = 1.006
These two digits are dropped.

Rule 2 When the first digit after those you want to retain is 5 or greater, that digit and all others to the right of it are dropped and the last digit retained is increased by one.

Examples rounded off to four digits:

1.026868 = 1.027
These three digits are dropped.
This digit is changed to 7.

18.02500 = 18.03
These three digits are dropped.
This digit is changed to 3.

12.899 = 12.90
This digit is dropped.
These two digits are changed to 90.

Practice Round off these numbers to the number of significant digits indicated:
(a) 42.246 (four digits) (d) 0.08965 (two digits)
(b) 88.015 (three digits) (e) 225.3 (three digits)
(c) 0.08965 (three digits) (f) 14.150 (three digits)

Answers: (a) 42.25 (Rule 2) (b) 88.0 (Rule 1) (c) 0.0897 (Rule 2) (d) 0.090 (Rule 2) (e) 225 (Rule 1) (f) 14.2 (Rule 2)

2.4 Scientific Notation of Numbers

The age of the earth has been estimated as about 4,500,000,000 (4.5 billion) years. Because this is an estimated value, let us say to the nearest 0.1 billion years, we are justified in using only two significant figures to express it. To express this number with two significant figures we write it using a power of 10 as 4.5×10^9 years.

Very large and very small numbers are often used in chemistry. These numbers can be simplified and conveniently written using a power of 10. Writing a number as a power of 10 is called **scientific notation**.

scientific notation

To write a number in scientific notation, move the decimal point in the original number so that it is located after the first nonzero digit. This new number is multiplied by 10 raised to the proper power (exponent). The power of 10 is equal to the number of places that the decimal point has been moved. If the decimal was moved to the left, the power of 10 will be a positive number. If the decimal was moved to the right, the power of 10 will be a negative number.

The scientific notation of a number is the number written as a factor between 1 and 10 multiplied by 10 raised to a power. For example,

$$2468 = 2.468 \times 10^3$$

number scientific notation of the number

Study the examples that follow.

▲ **Very large numbers, such as the distances between the moon and stars, are often expressed in scientific notation.**

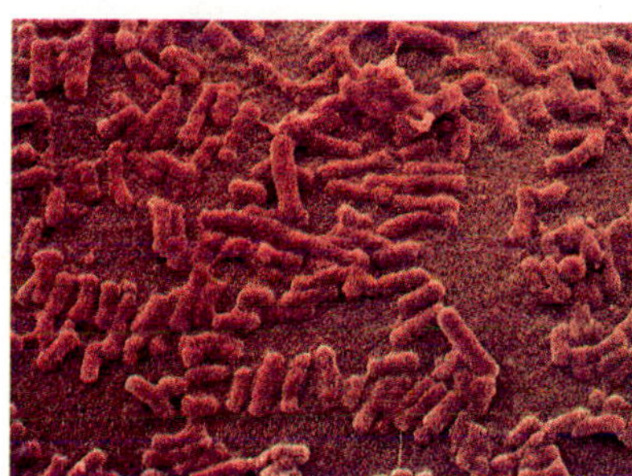

▲ **These tiny *Clostridium perfringens* bacteria have been magnified 2975 times.**

Example 2.1

Write 5283 in scientific notation.

SOLUTION

5283. (3 2 1) Place the decimal between the 5 and the 2. Since the decimal was moved three places to the left, the power of 10 will be 3, and the number 5.283 is multiplied by 10^3.

5.283×10^3 (Correct scientific notation)

Example 2.2

Write 4,500,000,000 in scientific notation (two significant figures).

SOLUTION

4 500 000 000. (9 … 1) Place the decimal between the 4 and the 5. Since the decimal was moved nine places to the left, the power of 10 will be 9, and the number 4.5 is multiplied by 10^9.

4.5×10^9 (Correct scientific notation)

EXAMPLE 2.3

SOLUTION

Write 0.000123 in scientific notation.

0.000123 (4) Place the decimal between the 1 and the 2. Since the decimal was moved four places to the right, the power of 10 will be −4, and the number 1.23 is multiplied by 10^{-4}.

1.23×10^{-4} (Correct scientific notation)

PRACTICE Write the following numbers in scientific notation:
(a) 1200 (four digits) (c) 0.0468
(b) 6,600,000 (two digits) (d) 0.00003

Answers: (a) 1200 = 1.200×10^3 (3) (c) 0.0468 = 4.68×10^{-2} (2)

(b) 6 600 000 = 6.6×10^6 (6) (d) 0.00003 = 3×10^{-5} (5)

2.5 SIGNIFICANT FIGURES IN CALCULATIONS

The results of a calculation based on measurements cannot be more precise than the least precise measurement.

Multiplication or Division

In calculations involving multiplication or division, the answer must contain the same number of significant figures as in the measurement that has the least number of significant figures. Consider the following examples:

EXAMPLE 2.4

SOLUTION

$190.6 \times 2.3 = 438.38$

The value 438.38 was obtained with a hand calculator. The answer should have two significant figures, because 2.3, the number with the fewest significant figures, has only two significant figures. The answer must, therefore, be expressed in scientific notation.

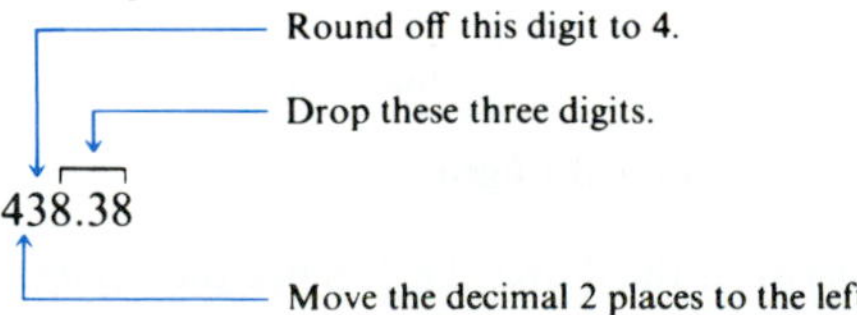
Round off this digit to 4.
Drop these three digits.
438.38
Move the decimal 2 places to the left.

The correct answer is 4.4×10^2.

EXAMPLE 2.5

$$\frac{13.59 \times 6.3}{12} = 7.13475$$

SOLUTION

The value 7.13475 was obtained with a hand calculator. The answer should contain two significant figures because 6.3 and 12 have only two significant figures.

7.1~~3475~~ — Drop these four digits. (3475)

This digit remains the same. (1)

The correct answer is 7.1.

PRACTICE 134 in. × 25 in. = ?

Answer: 3350 in.^{2} = 3.4×10^3 in.^{2}

PRACTICE $\frac{213 \text{ miles}}{4.20 \text{ hours}} = ?$

Answer: 50.7 miles/hour

PRACTICE $\frac{2.2 \times 273}{760} = ?$

Answer: 0.79

Addition or Subtraction

The results of an addition or a subtraction must be expressed to the same precision as the least precise measurement.

EXAMPLE 2.6

Add 125.17, 129.2, and 52.24.

SOLUTION

```
125.17
129.2
 52.24
------
306.61   (306.6)
```

The number with the greatest uncertainty is 129.2. Therefore the answer is rounded off to the nearest tenth. 306.6

Example 2.7

Subtract 14.1 from 132.56.

SOLUTION

$$\begin{array}{r} 132.56 \\ -14.1 \\ \hline 118.46 \end{array} \quad (118.5)$$

14.1 is the number with the least precision. Therefore the answer is rounded off to the nearest tenth. 118.5

Example 2.8

Subtract 120 from 1587.

SOLUTION

$$\begin{array}{r} 1587 \\ -120 \\ \hline 1467 \end{array} \quad (1.47 \times 10^3)$$

120 is the number with greatest uncertainty. The zero is not considered significant, therefore the answer must be rounded to the nearest ten. 1470 or 1.47×10^3.

Example 2.9

Add 5672 and 0.00063.

SOLUTION

$$\begin{array}{r} 5672 \\ +\ 0.00063 \\ \hline 5672.00063 \end{array} \quad (5672)$$

The number with the greatest uncertainty is 5672. Therefore the answer is rounded off to the nearest unit. 5672

Example 2.10

$$\frac{1.039 - 1.020}{1.039} = 0.018286814$$

SOLUTION

The value 0.018286814 was obtained with a hand calculator. When the subtraction in the numerator is done,

$$1.039 - 1.020 = 0.019$$

the number of significant figures changes from four to two. Therefore the answer should contain two significant figures after the division is carried out:

Drop these six digits.

0.018286814

This digit remains the same.

The correct answer is 0.018 or 1.8×10^{-2}.

Practice How many significant figures should the answer contain in each of these calculations?

(a) 14.0×5.2 (e) $119.1 - 3.44$

(b) 0.1682×8.2 (f) $\frac{94.5}{1.2}$

(c) $\frac{160 \times 33}{4}$ (g) $1200 + 6.34$

(d) $8.2 + 0.125$ (h) $1.6 + 23 - 0.005$

Answers: (a) 2 (b) 2 (c) 1 (d) 2 (e) 4 (f) 2 (g) 2 (h) 2

Additional material on mathematical operations is given in the "Mathematical Review" in Appendix I. Review Appendix I and study carefully any portions that are not familiar to you. This study may be done at various times during the course as the need for additional knowledge of mathematical operations arises.

2.6 The Metric System

The **metric system**, or **International System** (**SI**, from *Système International*), is a decimal system of units for measurements of mass, length, time, and other physical quantities. It is built around a set of base units and uses factors of 10 to express larger or smaller numbers of these units. To express quantities that are larger or smaller than the base units, prefixes are added to the names of the units. These prefixes represent multiples of 10, making the metric system a decimal system of measurements. Table 2.1 shows the names, symbols, and numerical values of the prefixes. Some of the more commonly used prefixes are highlighted in color.

metric system or SI

Examples are

1 kilometer = 1000 meters
1 kilogram = 1000 grams
1 millimeter = 0.001 meter
1 microsecond = 0.000001 second

The seven base units in the International System, their abbreviations, and the quantities they measure are given in Table 2.2. Other units are derived from these base units.

The metric system, or International System, is currently used by most of the countries in the world, not only for scientific and technical work, but also in commerce and industry. The United States is currently in the process of changing to the metric system of mass and measurements.

TABLE 2.1 Prefixes and Numerical Values for SI Units*

Prefix	Symbol	Numerical value	Power of 10 equivalent
exa	E	1,000,000,000,000,000,000	10^{18}
peta	P	1,000,000,000,000,000	10^{15}
tera	T	1,000,000,000,000	10^{12}
giga	G	1,000,000,000	10^{9}
mega	M	1,000,000	10^{6}
kilo	k	1,000	10^{3}
hecto	h	100	10^{2}
deka	da	10	10^{1}
—	—	1	10^{0}
deci	d	0.1	10^{-1}
centi	c	0.01	10^{-2}
milli	m	0.001	10^{-3}
micro	μ	0.000001	10^{-6}
nano	n	0.000000001	10^{-9}
pico	p	0.000000000001	10^{-12}
femto	f	0.000000000000001	10^{-15}
atto	a	0.000000000000000001	10^{-18}

* The more commonly used prefixes are in color.

TABLE 2.2 International System Base Units of Measurement

Quantity	Name of unit	Abbreviation
Length	Meter	m
Mass	Kilogram	kg
Temperature	Kelvin	K
Time	Second	s
Amount of substance	Mole	mol
Electric current	Ampere	A
Luminous intensity	Candela	cd

2.7 Measurement of Length

Standards for the measurement of length have an interesting historical development. The Old Testament mentions such units as the *cubit* (the distance from a man's elbow to the tip of his outstretched hand). In ancient Scotland the inch was once defined as a distance equal to the width of a man's thumb.

meter

Reference standards of measurements have undergone continuous improvements in precision. The standard unit of length in the metric system is the **meter**. When the metric system was first introduced in the 1790s, the meter was defined as one ten-millionth of the distance from the equator to the North Pole measured along the meridian passing through Dunkirk, France. In 1889 the meter was

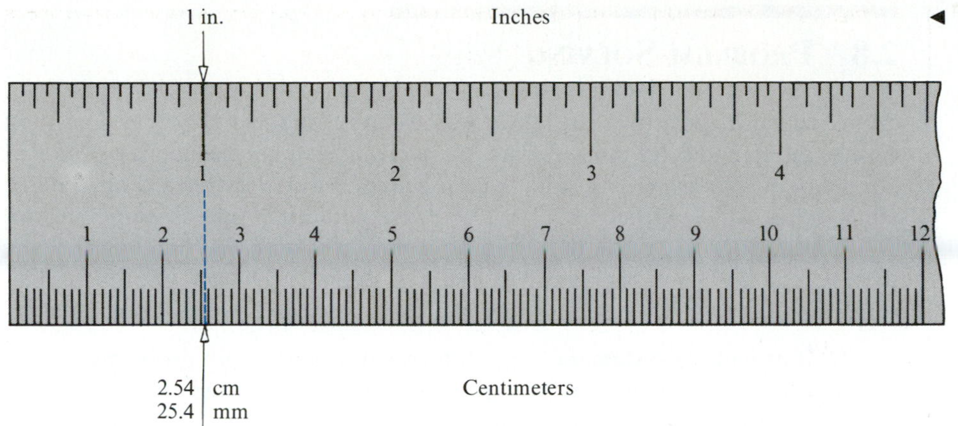

FIGURE 2.2
Comparison of the metric and American systems of length measurement: 2.54 cm = 1 in.

redefined as the distance between two engraved lines on a platinum–iridium alloy bar maintained at 0° Celsius. This international meter bar is stored in a vault at Sèvres near Paris. Duplicate meter bars have been made and are used as standards by many nations.

By the 1950s length could be measured with such precision that a new standard was needed. Accordingly, the length of the meter was redefined in 1960 and again in 1983. The latest definition is: A meter is the distance that light travels in a vacuum during 1/299,792,458 of a second.

A meter is 39.37 inches, a little longer than 1 yard. One meter contains 10 decimeters, 100 centimeters, or 1000 millimeters (see Figure 2.2). A kilometer contains 1000 meters. Table 2.3 shows the relationships of these units.

The nanometer (10^{-9} m) is used extensively in expressing the wavelength of light and in atomic dimensions. See Appendix III for a complete table of common conversions. Other important relationships are:

$$1 \text{ m} = 100 \text{ cm} = 1000 \text{ mm} = 10^6\ \mu\text{m} = 10^{10}\ \text{Å}$$

$$1 \text{ cm} = 10 \text{ mm} = 0.01 \text{ m}$$

$$1 \text{ in.} = 2.54 \text{ cm}$$

$$1 \text{ mile} = 1.61 \text{ km}$$

TABLE 2.3 Units of Length

Unit	Abbreviation	Meter equivalent	Exponential equivalent
Kilometer	km	1000 m	10^3 m
Meter	m	1 m	10^0 m
Decimeter	dm	0.1 m	10^{-1} m
Centimeter	cm	0.01 m	10^{-2} m
Millimeter	mm	0.001 m	10^{-3} m
Micrometer	μm	0.000001 m	10^{-6} m
Nanometer	nm	0.000000001 m	10^{-9} m
Angstrom	Å	0.0000000001 m	10^{-10} m

2.8 Problem Solving

Many chemical principles are illustrated by mathematical concepts. Learning how to set up and solve numerical problems in a systematic fashion is *essential* in the study of chemistry. This skill, once acquired, is also very rewarding in other study areas. An electronic calculator will save you much time in computation.

Usually a problem can be solved by several methods. But in all methods it is best, especially for beginners, to use a systematic, orderly approach. The dimensional analysis, or factor-label, method is stressed in this book because

1. It provides a systematic, straightforward way to set up problems.
2. It gives a clear understanding of the principles involved.
3. It helps in learning to organize and evaluate data.
4. It helps to identify errors because unwanted units are not eliminated if the setup of the problem is incorrect.

The basic steps for solving problems are

1. Read the problem very carefully to determine what is to be solved for, and write it down.
2. Tabulate the data given in the problem. Even in tabulating data it is important to label all factors and measurements with the proper units.
3. Determine which principles are involved and which unit relationships are needed to solve the problem. Sometimes it is necessary to refer to tables for needed data.
4. Set up the problem in a neat, organized, and logical fashion, making sure that unwanted units cancel. Use sample problems in the text as guides for making setups.
5. Proceed with the necessary mathematical operations. Make certain that the answer contains the proper number of significant figures.
6. Check the answer to see if it is reasonable.

Just a few more words about problem solving. Don't allow any formal method of problem solving to limit your use of common sense and intuition. If a problem is clear to you and its solution seems simpler by another method, by all means use it. But in the long run you should be able to solve many otherwise difficult problems by using the dimensional analysis method.

The dimensional analysis method of problem solving converts one unit to another unit by the use of conversion factors.

$$\text{unit}_1 \times \text{conversion factor} = \text{unit}_2$$

If you want to know how many millimeters are in 2.5 meters, you need to convert meters (m) to millimeters (mm). Therefore, you start by writing

$$\text{m} \times \text{conversion factor} = \text{mm}$$

This conversion factor must accomplish two things. It must cancel, or eliminate, meters; and it must introduce millimeters, the unit wanted in the answer. Such a conversion factor will be in fractional form and have meters in the denominator and millimeters in the numerator:

$$\not{m} \times \frac{mm}{\not{m}} = mm$$

We know that 1 m = 1000 mm. From this relationship we can write two factors, 1 m per 1000 mm and 1000 mm per 1 m:

$$\frac{1\ m}{1000\ mm} \quad \text{and} \quad \frac{1000\ mm}{1\ m}$$

Using the factor 1000 mm/1 m, we can set up the calculation for the conversion of 2.5 m to millimeters,

$$2.5\ \not{m} \times \frac{1000\ mm}{1\ \not{m}} = 2500\ mm \quad \text{or} \quad 2.5 \times 10^3\ mm$$

(two significant figures)

Note that, in making this calculation, units are treated as numbers; meters in the numerator are canceled by meters in the denominator.

Now suppose you need to change 215 centimeters to meters. First you must determine that you need to convert centimeters to meters. We start with

$$cm \times \text{conversion factor} = m$$

The conversion factor must have centimeters in the denominator and meters in the numerator:

$$\not{cm} \times \frac{m}{\not{cm}} = m$$

From the relationship 100 cm = 1 m, we can write a factor that will accomplish this conversion:

$$\frac{1\ m}{100\ cm}$$

Now set up the calculation using all the data given.

$$215\ \not{cm} \times \frac{1\ m}{100\ \not{cm}} = \frac{215\ m}{100} = 2.15\ m$$

Some problems may require a series of conversions to reach the correct units in the answer. For example, suppose we want to know the number of seconds in 1 day. We need to go from the unit of days to seconds in this manner:

day ⟶ hours ⟶ minutes ⟶ seconds

This series requires three conversion factors, one for each step. We convert days to hours (hr), hours to minutes (min), and minutes to seconds (s). The conversions can be done individually or in a continuous sequence:

$$\cancel{\text{day}} \times \frac{\text{hr}}{\cancel{\text{day}}} \longrightarrow \cancel{\text{hr}} \times \frac{\text{min}}{\cancel{\text{hr}}} \longrightarrow \cancel{\text{min}} \times \frac{\text{s}}{\cancel{\text{min}}}$$

$$\cancel{\text{day}} \times \frac{\cancel{\text{hr}}}{\cancel{\text{day}}} \times \frac{\cancel{\text{min}}}{\cancel{\text{hr}}} \times \frac{\text{s}}{\cancel{\text{min}}} = \text{s}$$

Inserting the proper factors we calculate the number of seconds in 1 day to be

$$1\,\cancel{\text{day}} \times \frac{24\,\cancel{\text{hr}}}{1\,\cancel{\text{day}}} \times \frac{60\,\cancel{\text{min}}}{1\,\cancel{\text{hr}}} \times \frac{60\ \text{s}}{1\,\cancel{\text{min}}} = 86{,}400.\ \text{s}$$

All five digits in 86,400 are significant, since all the factors in the calculation are exact numbers.

The dimensional analysis, or factor-label, method used in the preceding work shows how unit conversion factors are derived and used in calculations. After you become more proficient with the terms, you can save steps by writing the factors directly in the calculation. The problems that follow give examples of the conversion from American to metric units.

Label all factors with the proper units.

Example 2.11

How many centimeters are in 2.00 ft?

Solution

The stepwise conversion of units from feet to centimeters may be done in this manner: Convert feet to inches; then convert inches to centimeters.

$$\text{ft} \longrightarrow \text{in.} \longrightarrow \text{cm}$$

The conversion factors needed are

$$\frac{12\ \text{in.}}{1\ \text{ft}} \quad \text{and} \quad \frac{2.54\ \text{cm}}{1\ \text{in.}}$$

$$2.00\,\cancel{\text{ft}} \times \frac{12\ \text{in.}}{1\,\cancel{\text{ft}}} = 24.0\ \text{in.}$$

$$24.0\,\cancel{\text{in.}} \times \frac{2.54\ \text{cm}}{1\,\cancel{\text{in.}}} = 61.0\ \text{cm}$$

Since 1 ft and 12 in. are exact numbers, the number of significant figures allowed in the answer is three, based on the number 2.00.

Example 2.12

How many meters are in a 100. yd football field?

Solution

The stepwise conversion of units from yards to meters may be done in this manner, using the proper conversion factors.

$$\text{yd} \longrightarrow \text{ft} \longrightarrow \text{in.} \longrightarrow \text{cm} \longrightarrow \text{m}$$

$$100.\ \cancel{\text{yd}} \times \frac{3\ \text{ft}}{1\ \cancel{\text{yd}}} = 300.\ \text{ft} \qquad (3\ \text{ft/yd})$$

$$300\ \cancel{\text{ft}} \times \frac{12\ \text{in.}}{1\ \cancel{\text{ft}}} = 3600\ \text{in.} \qquad (12\ \text{in./ft})$$

$$3600\ \cancel{\text{in.}} \times \frac{2.54\ \text{cm}}{1\ \cancel{\text{in.}}} = 9144\ \text{cm} \qquad (2.54\ \text{cm/in.})$$

$$9144\ \cancel{\text{cm}} \times \frac{1\ \text{m}}{100\ \cancel{\text{cm}}} = 91.4\ \text{m} \qquad (1\ \text{m}/100\ \text{cm}) \qquad \text{(three significant figures)}$$

Examples 2.11 and 2.12 may be solved using a linear expression and writing down conversion factors in succession. This method often saves one or two calculation steps and allows numerical values to be reduced to simpler terms, leading to simpler calculations. The single linear expressions for Examples 2.11 and 2.12 are

$$2.00\ \cancel{\text{ft}} \times \frac{12\ \cancel{\text{in.}}}{1\ \cancel{\text{ft}}} \times \frac{2.54\ \text{cm}}{1\ \cancel{\text{in.}}} = 61.0\ \text{cm}$$

$$100.\ \cancel{\text{yd}} \times \frac{3\ \cancel{\text{ft}}}{1\ \cancel{\text{yd}}} \times \frac{12\ \cancel{\text{in.}}}{1\ \cancel{\text{ft}}} \times \frac{2.54\ \cancel{\text{cm}}}{1\ \cancel{\text{in.}}} \times \frac{1\ \text{m}}{100\ \cancel{\text{cm}}} = 91.4\ \text{m}$$

Using the units alone (Example 2.12), we see that the stepwise cancellation proceeds in succession until the desired unit is reached.

$$\cancel{\text{yd}} \times \frac{\cancel{\text{ft}}}{\cancel{\text{yd}}} \times \frac{\cancel{\text{in.}}}{\cancel{\text{ft}}} \times \frac{\cancel{\text{cm}}}{\cancel{\text{in.}}} \times \frac{\text{m}}{\cancel{\text{cm}}} = \text{m}$$

PRACTICE How many meters are in 1.00 mile?
Answer: 1.61×10^3 m

EXAMPLE 2.13

How many cubic centimeters (cm^3) are in a box that measures 2.20 in. by 4.00 in. by 6.00 in.?

SOLUTION

First we need to determine the volume of the box in cubic inches ($in.^3$) by multiplying together the length times the width times the height.

$$2.20\ \text{in.} \times 4.00\ \text{in.} \times 6.00\ \text{in.} = 52.8\ \text{in.}^3$$

Now we need to convert $in.^3$ to cm^3, which can be done by using the inches and centimeters relationship three times.

$$\cancel{\text{in.}^3} \times \frac{\text{cm}}{\cancel{\text{in.}}} \times \frac{\text{cm}}{\cancel{\text{in.}}} \times \frac{\text{cm}}{\cancel{\text{in.}}} = \text{cm}^3$$

$$52.8\ \cancel{\text{in.}^3} \times \frac{2.54\ \text{cm}}{1\ \cancel{\text{in.}}} \times \frac{2.54\ \text{cm}}{1\ \cancel{\text{in.}}} \times \frac{2.54\ \text{cm}}{1\ \cancel{\text{in.}}} = 865\ \text{cm}^3$$

Practice How many cubic meters are in a room measuring 8 ft × 10 ft × 12 ft?
Answer: 30 m^3 or 3×10^1 m^3

2.9 Measurement of Mass

kilogram

The gram is used as a unit of mass measurement, but it is a tiny amount of mass; for instance, a nickel has a mass of about 5 grams. Therefore the *standard unit* of mass in the SI system is the **kilogram** (equal to 1000 g). The amount of mass in a kilogram is defined by international agreement as exactly equal to the mass of a platinum–iridium cyclinder (international prototype kilogram) kept in a vault at Sèvres, France. Comparing this unit of mass to 1 lb (16 oz), we find that 1 kg is equal to 2.2 lb. A pound is equal to 454 g (0.454 kg). The same prefixes used in length measurement are used to indicate larger and smaller gram units (see Table 2.4).

A balance is used to measure mass. Some balances will determine the mass of objects to the nearest microgram. The choice of balance depends on the precision required and the amount of material. Several balances are shown in Figure 2.3.

It is convenient to remember that

$$1 \text{ g} = 1000 \text{ mg}$$
$$1 \text{ kg} = 1000 \text{ g}$$
$$1 \text{ kg} = 2.2 \text{ lb}$$
$$1 \text{ lb} = 454 \text{ g}$$

To change grams to milligrams, multiply grams by the conversion factor 1000 mg/g. The setup for converting 25 g to milligrams is

$$25 \cancel{\text{g}} \times \frac{1000 \text{ mg}}{1 \cancel{\text{g}}} = 25{,}000 \text{ mg} \qquad (2.5 \times 10^4 \text{ mg})$$

Note that multiplying a number by 1000 is the same as multiplying the

Table 2.4 Metric Units of Mass

Unit	Abbreviation	Gram equivalent	Exponential equivalent
Kilogram	kg	1000 g	10^3 g
Gram	g	1 g	10^0 g
Decigram	dg	0.1 g	10^{-1} g
Centigram	cg	0.01 g	10^{-2} g
Milligram	mg	0.001 g	10^{-3} g
Microgram	μg	0.000001 g	10^{-6} g

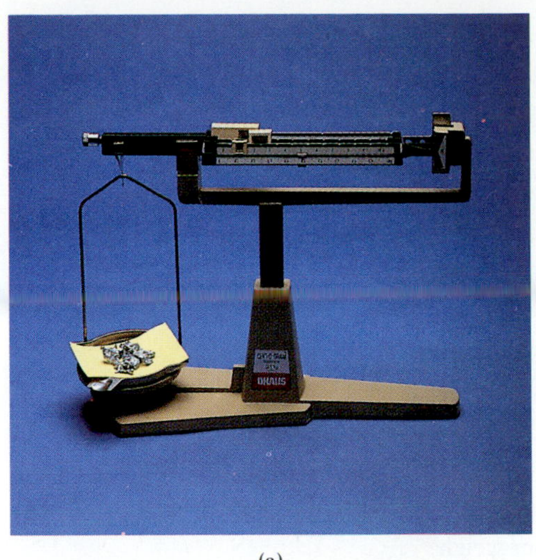

(a)

(b)

(c)

(d)

◀ FIGURE 2.3
(a) A quadruple beam balance with a precision of 0.01 g; (b) a single pan, top-loading balance with a precision of 0.001 g (1 mg); (c) a digital electronic analytical balance with a precision of 0.0001 g; and (d) a digital electronic balance with a precision up to 0.001 g.

number by 10^3 and can be done simply by moving the decimal point three places to the right

$$6.428 \times 1000 = 6428 \qquad (6.428) \; 3$$

To change milligrams to grams, multiply milligrams by the conversion factor 1 g/1000 mg. For example, to convert 155 mg to grams:

$$155\,\cancel{\text{mg}} \times \frac{1\text{ g}}{1000\,\cancel{\text{mg}}} = 0.155\text{ g}$$

Mass conversions from American to metric units are shown in Examples 2.14 and 2.15.

EXAMPLE 2.14

A 1.50 lb package of sodium bicarbonate costs 80 cents. How many grams of this substance are in this package?

SOLUTION

We are solving for the number of grams equivalent to 1.50 lb. Since 1 lb = 454 g, the factor to convert pounds to grams is 454 g/lb.

$$1.50\ \cancel{\text{lb}} \times \frac{454\ \text{g}}{1\ \cancel{\text{lb}}} = 681\ \text{g}$$

Note: The cost of the sodium bicarbonate has no bearing on the question asked in this problem.

EXAMPLE 2.15

Suppose four ostrich feathers weigh 1.00 lb. Assuming that each feather is equal in weight, how many milligrams does a single feather weigh?

SOLUTION

The unit conversion in this problem is from 1 lb/4 feathers to milligrams per feather. Since the unit *feathers* occurs in the denominator of both the starting unit and the desired unit, the unit conversions needed are

$$\text{lb} \longrightarrow \text{g} \longrightarrow \text{mg}$$

$$\frac{1.00\ \cancel{\text{lb}}}{4\ \text{feathers}} \times \frac{454\ \cancel{\text{g}}}{1\ \cancel{\text{lb}}} \times \frac{1000\ \text{mg}}{1\ \cancel{\text{g}}} = \frac{113{,}500\ \text{mg}}{1\ \text{feather}} \qquad (1.14 \times 10^5\ \text{mg/feather})$$

PRACTICE You are traveling in Europe and wake up one morning to find your mass is 75.0 kg. Determine the American equivalent to see whether you need to go on a diet before you return home.

Answer: 165 lb

PRACTICE A tennis ball has a mass of 65 g. Determine the American equivalent in pounds.

Answer: 0.14 lb

2.10 MEASUREMENT OF VOLUME

volume

Volume, as used here, is the amount of space occupied by matter. The SI unit of volume is the *cubic meter* (m^3). However, the liter (pronounced *leeter* and abbreviated L) and the milliliter (abbreviated mL) are the standard units of volume used in most chemical laboratories.

The most common instruments or equipment for measuring liquids are the graduated cylinder, volumetric flask, buret, pipet, and syringe, which are illustrated in Figure 2.4. These pieces are usually made of glass and are available in various sizes.

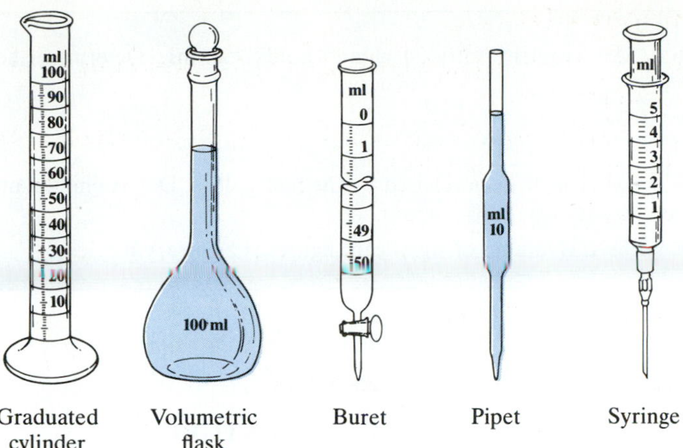

◄ **FIGURE 2.4**
Calibrated glassware for measuring the volume of liquids.

It is convenient to remember that

1 L = 1000 mL = 1000 cm^3

1 mL = 1 cm^3

1 L = 1.057 qt

946 mL = 1 qt

The volume of a cubic or rectangular container can be determined by multiplying its length times width times height. Thus a 10 cm square box has a volume of 10 cm × 10 cm × 10 cm = 1000 cm^3.

The following examples illustrate volume conversions.

EXAMPLE 2.16

How many milliliters are contained in 3.5 liters?

SOLUTION

The conversion factor to change liters to milliliters is 1000 mL/L.

$$3.5 \cancel{L} \times \frac{1000 \text{ mL}}{\cancel{L}} = 3500 \text{ mL} \qquad (3.5 \times 10^3 \text{ mL})$$

Liters may be changed to milliliters by moving the decimal point three places to the right and changing the units to milliliters.

1.500 L = 1500. mL

EXAMPLE 2.17

How many cubic centimeters are in a cube that is 11.1 inches on a side?

SOLUTION

First change inches to centimeters. The conversion factor is 2.54 cm/in.

$$11.1 \cancel{\text{in.}} \times \frac{2.54 \text{ cm}}{1 \cancel{\text{in.}}} = 28.2 \text{ cm on a side}$$

Then change to cubic volume (length × width × height).

$$28.2 \text{ cm} \times 28.2 \text{ cm} \times 28.2 \text{ cm} = 22{,}426 \text{ cm}^3 \qquad (2.24 \times 10^4 \text{ cm}^3)$$

PRACTICE An excellent bottle of chianti holds 750. mL. Determine the volume in quarts.

Answer: 0.793 qt

PRACTICE Milk is often purchased by the half gallon. Determine the number of liters necessary to equal this amount.

Answer: 1.89 L (Number of significant figures is arbitrary.)

2.11 MEASUREMENT OF TEMPERATURE

heat

temperature

Heat is a form of energy associated with the motion of small particles of matter. The term *heat* refers to the quantity of energy within a system or to a quantity of energy added to or taken away from a system. *System* as used here simply refers to the entity that is being heated or cooled. Depending on the amount of heat energy present, a given system is said to be hot or cold. **Temperature** is a measure of the intensity of heat, or how hot a system is, regardless of its size. Heat always flows from a region of higher temperature to one of lower temperature. The SI unit of temperature is the Kelvin. The common laboratory instrument for measuring temperature is a thermometer (see Figure 2.5).

The temperature of a system can be expressed by several different scales. Three commonly used temperature scales are the Celsius scale (pronounced *sell-see-us*), the Kelvin (absolute) scale, and the Fahrenheit scale. The unit of temperature on the Celsius and Fahrenheit scales is called a *degree*, but the size of the Celsius and the Fahrenheit degree is not the same. The symbol for the Celsius and Fahrenheit degrees is °, and it is placed as a superscript after the number and before the symbol for the scales. Thus, 100°C means 100 *degrees Celsius.* The degree sign is not used with Kelvin temperatures.

degrees Celsius = °C

Kelvin (absolute) = K

degrees Fahrenheit = °F

On the Celsius scale the interval between the freezing and boiling temperatures of water is divided into 100 equal parts, or degrees. The freezing point of water is assigned a temperature of 0°C and the boiling point of water a temperature of 100°C. The Kelvin temperature scale is also known as the absolute temperature scale, because 0 K is the lowest temperature theoretically attainable. The Kelvin zero is 273.16 degrees below the Celsius zero. A Kelvin is equal in size to a Celsius degree. The freezing point of water on the Kelvin scale is 273.16 K (usually rounded to 273 K). The Fahrenheit scale has 180 degrees between the freezing and boiling temperatures of water. On this scale the freezing point of water is 32°F and the boiling point is 212°F.

$$0°\text{C} \cong 273\ \text{K} \cong 32°\text{F}$$

The three scales are compared in Figure 2.5. Although absolute zero (0 K) is the lower limit of temperature on these scales, temperature has no known upper

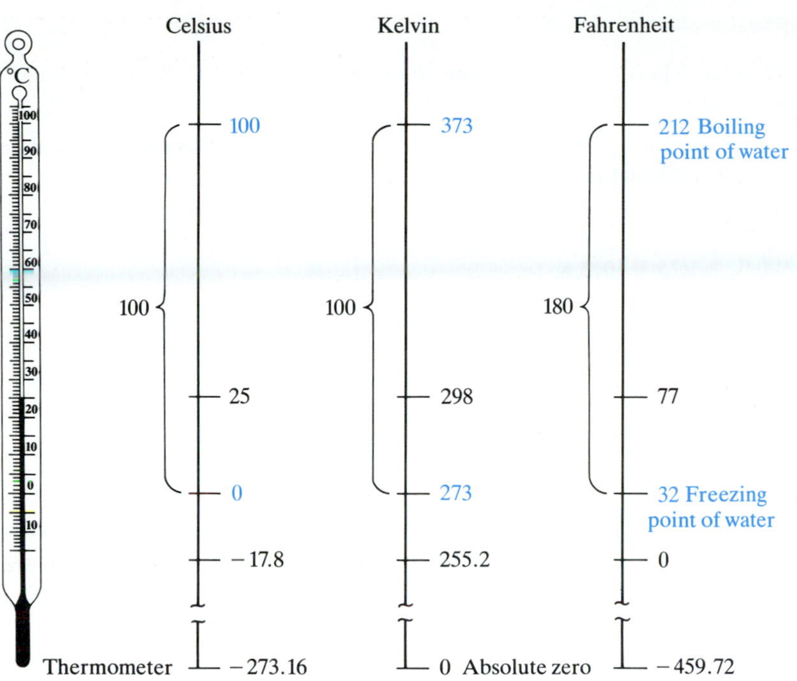

◀ **FIGURE 2.5 Comparison of Celsius, Kelvin, and Fahrenheit temperature scales.**

limit. (Temperatures of several million degrees are known to exist in the sun and in other stars.)

By examining Figure 2.5 we can see that there are 100 Celsius degrees and 100 Kelvins between the freezing and boiling points of water. But there are 180 Fahrenheit degrees between these two temperatures. Hence, the size of a degree on the Celsius scale is the same as the size of one Kelvin, but one Celsius degree corresponds to 1.8 degrees on the Fahrenheit scale.

$$1°\text{C} = 1\ \text{K} = 1.8°\text{F}$$

From these data, mathematical formulas have been derived to convert a temperature on one scale to the corresponding temperature on another scale. These formulas are

$$\text{K} = °\text{C} + 273 \qquad (1)$$

$$°\text{F} = (1.8 \times °\text{C}) + 32 \qquad (2)$$

$$°\text{C} = \frac{(°\text{F} - 32)}{1.8} \qquad (3)$$

Interpretation: Formula (1) states that the addition of 273 to the degrees Celsius converts the temperature to Kelvins. Formula (2) states that to obtain the Fahrenheit temperature corresponding to a given Celsius temperature, we multiply the degrees Celsius by 1.8 and then add 32. Formula (3) states that to obtain the corresponding Celsius temperature, we subtract 32 from the degrees Fahrenheit and then divide this answer by 1.8. Examples of temperature conversions follow.

EXAMPLE 2.18

The temperature at which table salt (sodium chloride) melts is 800.°C. What is this temperature on the Kelvin and Fahrenheit scales?

SOLUTION

We need to calculate K from °C, so we use formula (1) above. We also need to calculate °F from °C; for this calculation we use formula (2).

$$K = {}^{\circ}C + 273$$

$$K = 800.{}^{\circ}C + 273 = 1073\ K$$

$${}^{\circ}F = (1.8 \times {}^{\circ}C) + 32$$

$${}^{\circ}F = (1.8 \times 800.{}^{\circ}C) + 32$$

$${}^{\circ}F = 1440 + 32 = 1472{}^{\circ}F$$

$$800.{}^{\circ}C = 1073\ K = 1472{}^{\circ}F$$

EXAMPLE 2.19

The temperature for December 1 was 110.°F, a new record. Convert this temperature to °C.

SOLUTION

Formula (3) applies here.

$${}^{\circ}C = \frac{({}^{\circ}F - 32)}{1.8}$$

$${}^{\circ}C = \frac{(110. - 32)}{1.8} = \frac{78}{1.8} = 43{}^{\circ}C$$

EXAMPLE 2.20

What temperature on the Fahrenheit scale corresponds to −8.0°C? (Be alert to the presence of the minus sign in this problem.)

SOLUTION

$${}^{\circ}F = (1.8 \times {}^{\circ}C) + 32$$

$${}^{\circ}F = [1.8 \times (-8.0)] + 32 = -14.4 + 32$$

$${}^{\circ}F = 17.6\ {}^{\circ}F$$

Temperatures used in this book are in degrees Celsius (°C) unless specified otherwise. The temperature after conversion should be expressed to the same precision as the original measurement.

PRACTICE Helium boils at 4 K. Convert this temperature to °C and then to °F.
Answer: −269°C; −452°F

PRACTICE "Normal" human body temperature is 98.6°F. Convert this to °C and K.
Answer: 37.0°C; 310.0 K

2.12 Density

Density (d) is the ratio of the mass of a substance to the volume occupied by that mass; it is the mass per unit of volume and is given by the equation

density

$$d = \frac{\text{mass}}{\text{volume}}$$

Density is a physical characteristic of a substance and may be used as an aid to its identification. When the density of a solid or a liquid is given, the mass is usually expressed in grams and the volume in milliliters or cubic centimeters.

$$d = \frac{\text{mass}}{\text{volume}} = \frac{\text{g}}{\text{mL}} \qquad \text{or} \qquad d = \frac{g}{\text{cm}^3}$$

Since the volume of a substance (especially liquids and gases) varies with temperature, it is important to state the temperature along with the density. For example, the volume of 1.0000 g water at 4°C is 1.0000 mL, at 20°C it is 1.0018 mL, and at 80°C it is 1.0290 mL. Density, therefore, also varies with temperature.

The density of water at 4°C is 1.0000 g/mL, but at 80°C the density of water is 0.9718 g/mL.

$$d^{4°\text{C}} = \frac{1.0000\ \text{g}}{1.0000\ \text{mL}} = 1.0000\ \text{g/mL}$$

$$d^{80°\text{C}} = \frac{1.0000\ \text{g}}{1.0290\ \text{mL}} = 0.97182\ \text{g/mL}$$

The density of iron at 20°C is 7.86 g/mL.

$$d^{20°\text{C}} = \frac{7.86\ \text{g}}{1.00\ \text{mL}} = 7.86\ \text{g/mL}$$

The densities of a variety of materials are compared in Figure 2.6.

Densities for liquids and solids are usually represented in terms of grams per milliliter (g/mL) or grams per cubic centimeter (g/cm^3). The density of gases, however, is expressed in terms of grams per liter (g/L). Unless otherwise stated, gas densities are given for 0°C and 1 atmosphere pressure (discussed further in Chapter 13). Table 2.5 lists the densities of a number of common materials.

Suppose that water, carbon tetrachloride, and cottonseed oil are successively poured into a graduated cylinder. The result is a layered three-liquid system (Figure 2.7). Can we predict the order of the liquid layers? Yes, by looking up the densities in Table 2.5. Carbon tetrachloride has the greatest density (1.595 g/mL), and cottonseed oil has the lowest density (0.926 g/mL). Carbon tetrachloride will be the bottom layer, and cottonseed oil will be the top layer. Water, with a density between the other two liquids, will form the middle layer. This information can also be determined by experiment. Add a few milliliters of carbon tetrachloride to a beaker of water. The carbon tetrachloride, being more dense than the water, will sink. Cottonseed oil, being less dense than water, will float when added to the

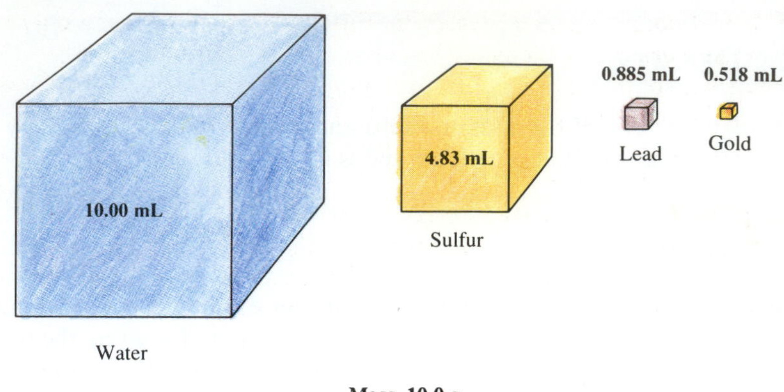

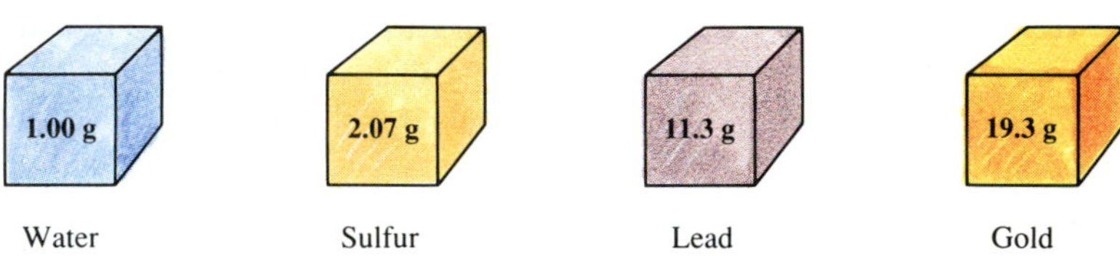

FIGURE 2.6 ▶
(a) Comparison of the volumes of equal masses (10.0 g) of water, sulfur, lead, and gold. (b) Comparison of the masses of equal volumes (1.00 cm^3) of water, sulfur, lead, and gold. (Water is at 4°C; the three solids, at 20°C.)

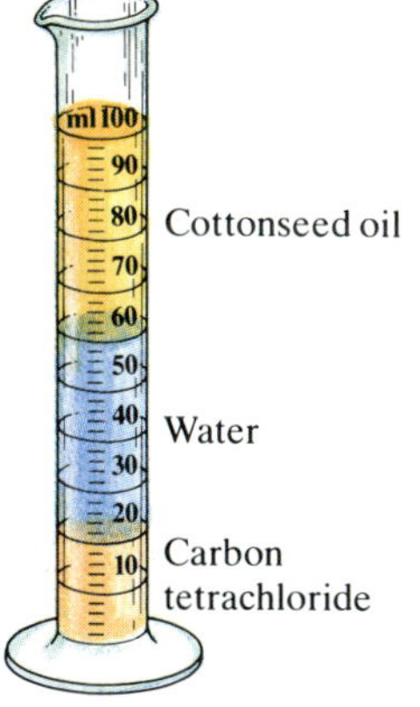

▲
FIGURE 2.7
Relative density of liquids. When three immiscible (not capable of mixing) liquids are poured together, the liquid with the highest density will be the bottom layer. In the case of cottonseed oil, water, and carbon tetrachloride, cottonseed oil is the top layer.

TABLE 2.5 Densities of Some Selected Materials*

Liquids and solids		Gases	
Substance	**Density (g/mL at 20°C)**	**Substance**	**Density (g/L at 0°C)**
Wood (Douglas fir)	0.512	Hydrogen	0.090
Ethyl alcohol	0.789	Helium	0.178
Cottonseed oil	0.926	Methane	0.714
Water (4°C)	**1.000**	Ammonia	0.771
Sugar	1.59	Neon	0.90
Carbon tetrachloride	1.595	Carbon monoxide	1.25
Magnesium	1.74	Nitrogen	1.251
Sulfuric acid	1.84	**Air**	**1.293**
Sulfur	2.07	Oxygen	1.429
Salt	2.16	Hydrogen chloride	1.63
Aluminum	2.70	Argon	1.78
Silver	10.5	Carbon dioxide	1.963
Lead	11.34	Chlorine	3.17
Mercury	13.55		
Gold	19.3		

* For comparing densities the density of water is the reference for solids and liquids; air is the reference for gases.

beaker. Direct comparisons of density in this manner can be made only with liquids that are *immiscible* (do not mix with or dissolve in one another).

The density of air at 0°C is approximately 1.293 g/L. Gases with densities less than this value are said to be "lighter than air." A helium-filled balloon will rise rapidly in air, because the density of helium is only 0.178 g/L.

When an insoluble solid object is dropped into water, it will sink or float, depending on its density. If the object is less dense than water, it will float, displacing a *mass* of water equal to the mass of the object. If the object is more dense than water, it will sink, displacing a *volume* of water equal to the volume of the object. This information can be utilized to determine the volume (and density) of irregularly shaped objects.

The **specific gravity** (sp gr) of a substance is a ratio of the density of that substance to the density of another substance. Water is usually used as the reference standard for solids and liquids. Air is usually used as the reference standard for gases. Specific gravity has no units because the density units cancel. The specific gravity tells us how many times as heavy a liquid, a solid, or a gas is as compared to the reference material.

specific gravity

$$\text{sp gr} = \frac{\text{density of a liquid or solid}}{\text{density of water}} \quad \text{or} \quad \frac{\text{density of a gas}}{\text{density of air}}$$

Sample calculations of density problems follow.

EXAMPLE 2.21

What is the density of a mineral if 427 g of the mineral occupy a volume of 35.0 mL?

SOLUTION

We need to solve for density, so we start by writing the formula for calculating density.

$$d = \frac{\text{mass}}{\text{volume}}$$

Then we substitute the data given in the problem into the equation and solve.

$$\text{mass} = 427\ \text{g} \qquad \text{volume} = 35.0\ \text{mL}$$

$$d = \frac{\text{mass}}{\text{volume}} = \frac{427\ \text{g}}{35.0\ \text{mL}} = 12.2\ \text{g/mL}$$

EXAMPLE 2.22

The density of gold is 19.3 g/mL. What is the mass of 25.0 mL of gold?

SOLUTION

Two ways to solve this problem are: (1) Solve the density equation for mass, then substitute the density and volume data into the new equation and calculate; (2) Solve by dimensional analysis.

Method 1 (a) Solve the density equation for mass:

$$d = \frac{\text{mass}}{\text{volume}} \qquad \text{mass} = d \times \text{volume}$$

(b) Substitute the data and calculate:

$$\text{mass} = \frac{19.3\ \text{g}}{\cancel{\text{mL}}} \times 25.0\ \cancel{\text{mL}} = 482\ \text{g}$$

Method 2 Dimensional analysis: Use density as a conversion factor, converting mL ⟶ g.

The conversion of units is

$$\text{mL} \times \frac{\text{g}}{\text{mL}} = \text{g}$$

$$25.0\,\cancel{\text{mL}} \times \frac{19.3\ \text{g}}{\cancel{\text{mL}}} = 482\ \text{g}$$

EXAMPLE 2.23 Calculate the volume (in mL) of 100. g of ethyl alcohol.

SOLUTION From Table 2.5 we see that the density of ethyl alcohol is 0.789 g/mL. This density also means that 1 mL of the alcohol has a mass of 0.789 g (1 mL/0.789 g).

Method 1 This problem may be done by solving the density equation for volume and then substituting the data in the new equation.

$$d = \frac{\text{mass}}{\text{volume}}$$

$$\text{volume} = \frac{\text{mass}}{d}$$

$$\text{volume} = \frac{100.\,\cancel{\text{g}}}{0.0789\,\cancel{\text{g}}/\text{mL}} = 127\ \text{mL}$$

Method 2 Dimensional analysis. For a conversion factor, we can use either

$$\frac{\text{g}}{\text{mL}} \quad \text{or} \quad \frac{\text{mL}}{\text{g}}$$

In this case the conversion is from g → mL, so we use mL/g. Substituting the data, we get

$$100.\,\cancel{\text{g}} \times \frac{1\ \text{mL}}{0.789\,\cancel{\text{g}}} = 127\ \text{mL of ethyl alcohol}$$

EXAMPLE 2.24 The water level in a graduated cylinder stands at 20.0 mL before and at 26.2 mL after a 16.74 g metal bolt is submerged in the water. (a) What is the volume of the bolt? (b) What is the density of the bolt?

SOLUTION

(a) The bolt will displace a volume of water equal to the volume of the bolt. Thus the increase in volume is the volume of the bolt.

$$\begin{array}{rl} 26.2\ \text{mL} &= \text{volume of water plus bolt} \\ -20.0\ \text{mL} &= \text{volume of water} \\ \hline 6.2\ \text{mL} &= \text{volume of bolt} \end{array}$$

(b) $d = \dfrac{\text{mass of bolt}}{\text{volume of bolt}} = \dfrac{16.74\ \text{g}}{6.2\ \text{mL}} = 2.7\ \text{g/mL}$

Chemistry in Action

Healthy Measurements

Standards for measurement are found throughout our daily lives and have a special impact in the health sciences. The dosage of a drug is often tied directly to body mass. If the recommended dose were 5 mg for each kilogram of body mass, a 50-kg woman would require a 250-mg dose, while an 80-kg man would require 400-mg of the drug. Children's dosages, of course, are much less.

Among the top ten factors in cardiovascular disease is obesity, a known health hazard. Simply measuring the weight of an individual is not a good indicator of leanness or obesity. A person may appear to be thin, yet have a high percentage of body fat. Someone else may appear "overweight" in comparison to published height-weight charts, but actually be especially lean as a result of a large percentage of muscle mass. To assess the body composition of an individual requires a measurement of the percent body fat.

Body fat is defined by health science professionals as the percentage of weight attributable to fat. It is the sum of the essential fat, surrounding and cushioning the internal organs, and the storage fat, which acts as a reservoir for energy in the body. A variety of techniques are currently in use to measure the body composition, including skin-fold tests, bioelectrical impedance, and hydrostatic weighing.

Hydrostatic weighing is considered to be one of the most accurate methods for determining body density. The individual is weighed in air, then seated on a chair suspended from a scale and lowered into a tank of warm water. After exhaling as much as possible the individual submerges in the water and remains under the surface for 5–7 seconds. The underwater weight is recorded during this time. A series of calculations can then be completed resulting in the determination of a percent body fat. The basis for these calculations lies in the variation in density of different tissues shown in the above table. A person with more bone and muscle mass will weigh more in water and have a higher body density.

Density of Various Tissues in the Human Body

Tissue	Density
bone	3.0 g/cm^3
muscle	1.06 g/cm^3
water	1.00 g/cm^3
fat	0.9 g/cm^3

As in all measurements errors in measured values result in variations in calculated results. A 100-g error in underwater weight results in a 1% body fat error. Various values for percent body fat calculations are shown in the table below.

Temperature measurement has long been of interest in the health professions. The traditional glass thermometer is often a nuisance, and is being replaced by newer, more accurate electronic models. Engineers at Johns Hopkins Applied Physics Laboratory have even built a battery-powered transmitting thermometer the size of an aspirin capsule which works after it is swallowed. It is capable of transmitting temperature measurements to within 0.01 degree until it passes out of the body (generally 1–2 days). It can be used to monitor temperature patterns to assist in the treatment of hypothermia, during which the body must be warmed at a slow, constant rate with continual temperature monitoring. The capsule can also be used to prevent hyperthermia in athletes, race drivers, or even in those taking the routine treadmill stress test.

Percent Body Fat

Classification	Male	Female
Lean	<8%	<15%
Healthy	8–15%	15–22%
Plump	16–19%	23–27%
Fat	20–24%	28–33%
Obese	>24%	>33%
Average college-age	15%	25%
Average middle-age	23%	32%
Distance runner	4–9%	6–15%
Tennis player	14–17%	19–22%

Practice Pure silver has a density of 10.5 g/mL. A ring sold as pure silver has a mass of 25.0 g. When placed in a graduated cylinder the water level rises 2.0 mL. Determine whether the ring is actually pure silver or if the customer should see the Better Business Bureau.

Answer: Density is 12.5 g/mL, therefore the ring is *not* pure silver.

Practice The water level in a metric measuring cup is 0.75 L before the addition of 150. g of shortening. The water level after submerging the shortening is 0.92 L. Determine the density of the shortening.

Answer: 0.88 g/mL

Concepts in Review

1. Differentiate between mass and weight; include the instruments used to measure each.
2. Know the metric units of mass, length, and volume.
3. Know the numerical equivalent for the metric prefixes deci, centi, milli, micro, nano, kilo, and mega.
4. Express any number in scientific notation.
5. Express answers to calculations to the proper number of significant figures.
6. Set up and solve problems utilizing the method of dimensional analysis (factor-label method).
7. Convert measurements of mass, length, and volume from American units to metric units, and vice versa.
8. Make temperature conversions among Fahrenheit, Celsius, and Kelvin scales.
9. Differentiate between heat and temperature.
10. Calculate the density, mass, or volume of an object from the appropriate data.

Exercises

An asterisk indicates a more challenging question or problem.

1. Use Table 2.3 to determine how many decimeters make up 1 km.
2. What is the temperature difference in Fahrenheit degrees between 25°C and 100°C? (see Figure 2.5.)
3. Use Figure 2.2 to determine the metric equivalent of 3 in.
4. Why do you suppose the neck of a 100 mL volumetric flask is narrower than the top of a 100 mL graduated cylinder? (See Figure 2.4.)
5. Refer to Table 2.5 and describe the arrangement that would be seen if these three immiscible substances were placed in a 100 mL graduated cylinder: 25 mL mercury, 25 mL carbon tetrachloride, and a cube of magnesium measuring 2.0 cm on an edge.
6. Arrange these materials in order of increasing

density: salt, cottonseed oil, sulfur, aluminum, and ethyl alcohol.

7. Will an argon-filled balloon rise or sink in a methane atmosphere? Explain.
8. Ice floats in cottonseed oil and sinks in ethyl alcohol. The density of ice must therefore lie between what numerical values?
9. What are some of the important advantages of the metric system over the American system of weights and measurements?
10. What are the abbreviations for the following?
    (a) Gram (g) Micrometer
    (b) Kilogram (h) Angstrom
    (c) Milligram (i) Milliliter
    (d) Microgram (j) Microliter
    (e) Centimeter (k) Liter
    (f) Millimeter
11. For the following numbers, tell whether the zeros are significant or are not significant,
    (a) 503 (c) 4200 (e) 100.00
    (b) 0.007 (d) 3.0030 (f) 8.00×10^2
12. State the rules for rounding off numbers.
13. Distinguish between heat and temperature.
14. Distinguish between density and specific gravity.
15. Describe the process of hydrostatic weighing.
16. Why is measuring the weight of a person a poor indication of leanness or obesity?
17. How is body density determined? What is the basis for this determination?
18. Which of the following statements are correct? Rewrite the incorrect statements to make them correct.
    (a) The prefix *micro* indicates one-millionth of the unit expressed.
    (b) The length 10 cm is equal to 1000 mm.
    (c) The number 383.263 rounded to four significant figures becomes 383.3.
    (d) The number of significant figures in the number 29,004 is five.
    (e) The number 0.00723 contains three significant figures.
    (f) The sum of 32.276 + 2.134 should contain four significant figures.
    (g) The product of 18.42 cm × 3.40 cm should contain three significant figures.
    (h) One microsecond is 10^{-6} second.
    (i) One thousand meters is a longer distance than 1000 yards.
    (j) One liter is a larger volume than one quart.
    (k) One centimeter is longer than one inch.
    (l) One cubic centimeter (cm^3) is equal to one milliliter.
    (m) The number 0.0002983 in exponential notation is 2.983×10^{-3}.
    (n) $3.0 \times 10^4 \times 6.0 \times 10^6 = 1.8 \times 10^{11}$
    (o) Temperature is a form of energy.
    (p) The density of water at 4°C is 1.00 g/mL.
    (q) A pipet is a more accurate instrument for measuring 10.0 mL of water than is a graduated cylinder.

Significant Figures, Rounding, Exponential Notation

19. How many significant figures are in each of the following numbers?
    (a) 0.025 (d) 0.0081 (g) 5.50×10^3
    (b) 40.0 (e) 0.0404 (h) 4.090×10^{-3}
    (c) 22.4 (f) 129,042
20. Round off the following numbers to three significant figures:
    (a) 93.246 (e) 4.644
    (b) 8.8726 (f) 129.509
    (c) 0.02854 (g) 34.250
    (d) 21.25 (h) 1.995×10^6
21. Express each of the following numbers in exponential notation:
    (a) 2,900,000 (e) 0.00840
    (b) 0.0456 (f) 40.30
    (c) 0.58 (g) 12,000,000
    (d) 4082.2 (h) 0.0000055
22. Solve the following mathematical problems, stating answers to the proper number of significant figures:
    (a) $12.62 + 1.5 + 0.25 =$
    (b) $4.68 \times 12.5 =$
    (c) $2.25 \times 10^3 \times 4.80 \times 10^4 =$
    (d) $\frac{182.6}{4.6} =$
    (e) $\frac{452 \times 6.2}{14.3} =$
    (f) $1986 + 23.48 + 0.012 =$
    (g) $0.0394 \times 12.8 =$
    (h) $2.92 \times 10^{-3} \times 6.14 \times 10^5 =$
    (i) $\frac{0.4278}{59.6} =$
    (j) $\frac{29.3}{284 \times 415} =$
23. Change these fractions to decimals. Express each answer to three significant figures.
    (a) $\frac{5}{6}$ (b) $\frac{3}{7}$ (c) $\frac{12}{16}$ (d) $\frac{9}{18}$
24. Solve each equation for X:
    (a) $3.42X = 6.5$ (d) $0.298X = 15.3$
    (b) $\frac{X}{12.3} = 7.05$ (e) $\frac{X}{0.819} = 10.9$
    (c) $\frac{0.525}{X} = 0.25$ (f) $\frac{8.4}{X} = 282$

25. Solve each equation for the unknown:

(a) $°C = \frac{212 - 32}{1.8}$ (c) $K = 25 + 273$

(b) $°F = 1.8(22) + 32$ (d) $\frac{8.9 \text{ g}}{\text{mL}} = \frac{40.90 \text{ g}}{\text{volume}}$

Unit Conversions

26. Make the following conversions, showing mathematical setups:

(a) 28.0 cm to m	(l) 12 nm to cm
(b) 1000. m to km	(m) 0.520 km to cm
(c) 9.28 cm to mm	(n) 3.884 Å to nm
(d) 150 mm to km	(o) 42.2 in. to cm
(e) 0.606 cm to km	(p) 0.64 mile to in.
(f) 4.5 cm to Å	(q) 504 miles to km
(g) 6.5×10^{-7} m to Å	(r) 2.00 in.^2 to cm^2
(h) 12.1 m to cm	(s) 35.6 m to ft
(i) 8.0 km to m	(t) 16.5 km to miles
(j) 315 mm to cm	(u) 4.5 in.^3 to mm^3
(k) 25 km to mm	(v) 3.00 mile^3 to mm^3

27. Make the following conversions, showing mathematical setups:

(a) 10.68 g to mg	(e) 164 mg to g
(b) 6.8×10^4 mg to kg	(f) 0.65 kg to mg
(c) 8.54 g to kg	(g) 5.5 kg to g
(d) 42.8 kg to lb	(h) 95 lb to g

28. Make the following conversions, showing mathematical setups:

(a) 25.0 mL to L	(e) 0.468 L to mL
(b) 22.4 L to mL	(f) 35.6 L to gal
(c) 3.5 qt to mL	(g) 9.0 μL to mL
(d) 4.5×10^4 ft^3 to m^3	(h) 20.0 gal to L

29. An automobile traveling at 55 miles per hour is moving at what speed in (a) kilometers per hour and (b) feet per second?

30. Carl Lewis, a sprinter in the 1988 Olympic Games, ran the 100. m dash in 8.90 s. What was his speed in (a) feet per second and (b) miles per hour?

31. A lab experiment requires each student to use 6.55 g of sodium chloride. The instructor opens a new 1.00 lb jar of the salt. If 24 students each take exactly the assigned amount of salt, how much should be left in the bottle at the end of the lab period?

32. When the space probe *Galileo* reaches Jupiter in 1994, it will be traveling at an average speed of 27,000 miles per hour. What will be its speed in (a) miles per second and (b) kilometers per second?

33. How many kilograms does a 170 lb man weigh?

34. The usual aspirin tablet contains 5.0 grains of aspirin. How many grams of aspirin are in one tablet (1 grain = $\frac{1}{7000}$ lb)?

35. The sun is approximately 93 million miles from the earth. How many seconds will it take light to travel from the sun to the earth if the velocity of light is 3.00×10^{10} cm per second?

36. The average mass of the heart of a human baby is about 1 ounce. What is the mass in milligrams?

37. How much would 1.0 kg of potatoes cost if the price is $1.78 for 10 pounds?

38. The price of gold varies greatly and has been as high as $800 per ounce. What is the value of 227 g of gold at $345 per ounce? Gold is priced per troy ounce [1 lb (avoirdupois) = 14.58 oz (troy)].

39. An adult ruby-throated hummingbird has an average mass of 3.2 g, whereas an adult California condor may attain a weight of 21 lb. How many times heavier than the hummingbird is the condor?

40. At 35 cents per liter how much will it cost to fill a 15.8 gal tank with gasoline?

41. How many liters of gasoline will be used to drive 500 miles in a car that averages 34 miles per gallon?

42. Calculate the volume, in liters, of a box 75 cm long by 55 cm wide by 55 cm high.

***43.** Assuming that there are 20 drops in 1.0 mL, how many drops are in one gallon?

44. How many liters of oil are in a 42 gallon barrel of oil?

45. How many milliliters will be delivered by a filled 5.0 μL syringe?

***46.** Calculate the number of milliliters of water in 1.00 cubic foot of water.

***47.** Oil spreads in a thin layer on water and is called an "oil slick." How much area in square meters (m^2) will 200 cm^3 of oil cover if it forms a layer 0.5 nm in thickness?

Temperature Conversions

48. Normal body temperature for humans is 98.6°F. What is this temperature on the Celsius scale?

49. Which is colder, −100°C or −138°F?

50. Make the following conversions, showing mathematical setups:

(a) 162°F to °C	(e) 32°C to °F
(b) 0.0°F to °C	(f) −8.6°F to °C
(c) 0.0°F to K	(g) 273°C to K
(d) −18°C to °F	(h) 212 K to °C

***51.** (a) At what temperature are the Fahrenheit and Celsius temperatures exactly equal?
(b) At what temperature are they numerically equal but opposite in sign?

Density

52. Calculate the density of a liquid if 50.00 mL of the liquid has a mass of 78.26 g.
53. A 12.8 mL sample of bromine has a mass of 39.9 g. What is the density of bromine?
54. When a 32.7 g piece of chromium metal was placed into a graduated cylinder containing 25.0 mL of water, the water level rose to 29.6 mL. Calculate the density of the chromium.
55. Concentrated hydrochloric acid has a density of 1.19 g/mL. Calculate the mass of 500. mL of this acid.
56. An empty graduated cylinder has a mass of 42.817 g. When filled with 50.0 mL of an unknown liquid it has a mass of 106.773 g. What is the density of the liquid?
57. What mass of mercury (density 13.6 g/mL) will occupy a volume of 25.0 mL?
58. A 35.0 mL sample of ethyl alcohol (density 0.789 g/mL) is added to a graduated cylinder that has a mass of 49.28 g. What will be the mass of the cylinder plus the alcohol?
59. You are given three cubes, A, B, and C; one is magnesium, one is aluminum, and the third is silver. All three cubes have the same mass, but cube A has a volume of 25.9 mL, cube B has a volume of 16.7 mL, and cube C has a volume of 4.29 mL. Identify cubes A, B, and C.

*60. A cube of aluminum has a mass of 500. g. What will be the mass of a cube of gold of the same dimensions?

61. A 25.0 mL sample of water at 90°C has a mass of 24.12 g. Calculate the density of water at this temperature.
62. The mass of an empty container is 88.25 g. The mass of the container when filled with a liquid ($d = 1.25$ g/mL) is 150.50 g. What is the volume of the container?
63. Which liquid will occupy the greater volume, 50 g of water or 50 g of ethyl alcohol? Explain.
64. A gold bullion dealer advertised a bar of pure gold for sale. The gold bar had a mass of 3300 g and measured 2.00 cm by 15.0 cm by 6.00 cm. Was the gold bar pure gold? Show evidence for your answer.
65. The largest nugget of gold on record was found in 1872 in New South Wales, Australia, and had a mass of 93.3 kg. Assuming the nugget is pure gold, what is its volume in cubic centimeters? What is it worth by today's standards if gold is $345/oz? (See Exercise 38.)

*66. Forgetful Freddie placed 25.0 mL of a liquid in a graduated cylinder with a mass of 89.450 g when empty. When Freddie placed a metal slug with a mass of 15.434 g into the cylinder, the volume rose to 30.7 mL. Freddie was asked to calculate the density of the liquid and of the metal slug from his data, but he forgot to obtain the mass of the liquid. He was told that if he found the mass of the cylinder containing the liquid and the slug, he would have enough data for the calculations. He did so and found its mass to be 125.934 g. Calculate the density of the liquid and of the metal slug.

3

Classification of Matter

◀ **CHAPTER OPENING PHOTO: The seats in this stadium are classified by color.**

Throughout our lives we seek to bring order into the chaos that surrounds us. To do this, we classify things according to their similarities. In the library, we find books grouped according to the subject, and then by author. Our local department store organizes its merchandise by the size and style of clothing, as well as by the type of customer. The ball park or theater classifies its seats by price and location. The biologist divides the living world into plants and animals; this broad classification is further simplified into various phyla and on to specific genera. In chemistry, this classification process begins with pure substances (such as water or mercury) and mixtures (such as air or vinegar). This process continues and ultimately leads us to the fundamental building blocks of matter—the 109 elements.

▲
FIGURE 3.1
An apparently empty test tube is submerged, mouth downward, in water. Only a small volume of water rises into the tube, which is actually filled with air.

3.1 MATTER DEFINED

The entire universe consists of matter and energy. Every day we come into contact with countless kinds of matter. Air, food, water, rocks, soil, glass, and this book are all different types of matter. Broadly defined, **matter** is *anything* that has mass and occupies space.

matter

Matter may be quite invisible. If an apparently empty test tube is submerged mouth downward in a beaker of water, the water rises only slightly into the tube. The water cannot rise further because the tube is filled with invisible matter: air (see Figure 3.1).

To the eye matter appears to be continuous and unbroken. However, it is actually discontinuous and is composed of discrete, tiny particles called *atoms*. The particulate nature of matter will become evident when we study atomic structure and the properties of gases.

Four states of matter Solid, liquid, gas, and plasma.

3.2 PHYSICAL STATES OF MATTER

Matter exists in three physical states: solid, liquid, and gas. A **solid** has a definite shape and volume, with particles that cohere rigidly to one another. The shape of a solid can be independent of its container. For example, a crystal of sulfur has the same shape and volume whether it is placed in a beaker or simply laid on a glass plate.

solid

Water can exist as a solid ▶ (snow), a liquid (water), and a gas (steam) as shown here at Yellowstone National Park.

Most commonly occurring solids, such as salt, sugar, quartz, and metals, are *crystalline*. Crystalline materials exist in regular, repeating, three-dimensional, geometric patterns. Solids such as plastics, glass, and gels, because they do not have any particular regular internal geometric pattern, are called **amorphous** solids. (*Amorphous* means without shape or form.)

amorphous

liquid

A **liquid** has a definite volume but not a definite shape, with particles that cohere firmly but not rigidly. Although the particles are held together by strong attractive forces and are in close contact with one another, they are able to move freely. Particle mobility gives a liquid fluidity and causes it to take the shape of the container in which it is stored.

gas

A **gas** has indefinite volume and no fixed shape, with particles that are moving independently of one another. Particles in the gaseous state have gained enough energy to overcome the attractive forces that held them together as liquids or solids. A gas presses continuously in all directions on the walls of any container. Because of this quality a gas completely fills a container. The particles of a gas are relatively far apart compared with those of solids and liquids. The actual volume of the gas particles is usually very small in comparison with the volume of the space occupied by the gas. A gas therefore may be compressed into a very small volume or expanded almost indefinitely. Liquids cannot be compressed to any great extent, and solids are even less compressible than liquids.

When a bottle of ammonia solution is opened in one corner of the laboratory, one can soon smell its familiar odor in all parts of the room. The ammonia gas escaping from the solution demonstrates that gaseous particles move freely and rapidly and tend to permeate the entire area into which they are released.

Although matter is discontinuous, attractive forces exist that hold the particles together and give matter its appearance of continuity. These attractive

Table 3.1 Common Materials in the Solid, Liquids and Gaseous States of Matter

Solids	Liquids	Gases
Aluminum	Alcohol	Acetylene
Copper	Blood	Air
Gold	Gasoline	Butane
Polyethylene	Honey	Carbon dioxide
Salt	Mercury	Chlorine
Sand	Oil	Helium
Steel	Vinegar	Methane
Sulfur	Water	Oxygen

Table 3.2 Physical Properties of Solids, Liquids and Gases

State	Shape	Volume	Particles	Compressibility
Solid	Definite	Definite	Rigidly cohering; tightly packed	Very slight
Liquid	Indefinite	Definite	Mobile; cohering	Slight
Gas	Indefinite	Indefinite	Independent of each other and relatively far apart	High

forces are strongest in solids, giving them rigidity; they are weaker in liquids but still strong enough to hold liquids to definite volumes. In gases the attractive forces are so weak that the particles of a gas are practically independent of one another. Table 3.1 lists a number of common materials that exist as solids, liquids, and gases. Table 3.2 summarizes comparative properties of solids, liquids, and gases.

3.3 Substances and Mixtures

The term *matter* refers to all materials or material things that make up the universe. Many thousands of different and distinct kinds of matter or substances exist. A **substance** is a particular kind of matter with a definite, fixed composition. A substance, sometimes known as a *pure substance*, is either an element or a compound. Familiar examples of elements are copper, gold, and oxygen. Familiar compounds are salt, sugar, and water.

substance

We can classify a sample of matter as either *homogeneous* or *heterogeneous* by examining it. **Homogeneous** matter is uniform in appearance and has the same

homogeneous

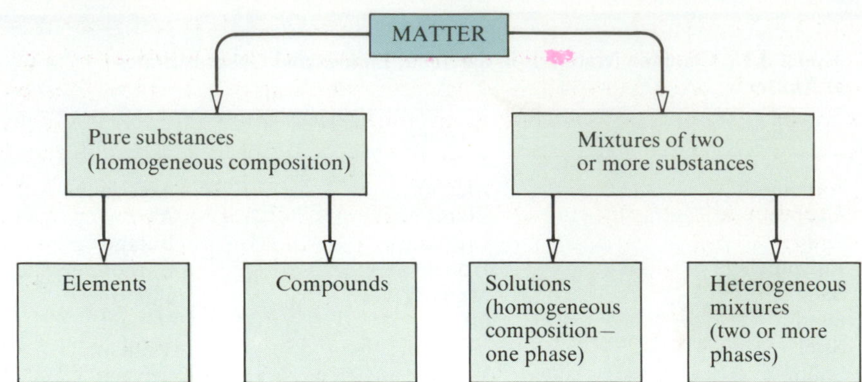

FIGURE 3.2 ▶ Classification of matter. A pure substance is always homogeneous in composition, whereas a mixture always contains two or more substances and may be either homogeneous (a solution) or heterogeneous.

heterogeneous

phase

properties throughout. Matter consisting of two or more physically distinct phases is **heterogeneous**. A **phase** is a homogeneous part of a system separated from other parts by physical boundaries. A system is simply the body of matter under consideration. Whenever we have a system in which visible boundaries exist between the parts or components, that system has more than one phase and is heterogeneous. It does not matter whether the components are in the solid, liquid, or gaseous states.

An important fact to keep in mind is that a pure substance, an element or compound, is always homogeneous in composition. However, a pure substance may exist as different phases in a heterogeneous system. Ice floating in water, for example, is a two-phase system made up of solid water and liquid water. The water in each phase is homogeneous in composition; but because two phases are present, the system is heterogeneous.

mixture

A **mixture** is a material containing two or more substances and can be either heterogeneous or homogeneous. Mixtures are variable in composition. If we add a tablespoonful of sugar to a glass of water, a heterogeneous mixture is formed immediately. The two phases are a solid (sugar) and a liquid (water). But upon stirring the sugar dissolves to form a homogeneous mixture or solution. Both substances are still present: All parts of the solution are sweet and wet. The proportions of sugar and water can be varied simply by adding more sugar and stirring to dissolve.

Many substances do not form homogeneous mixtures. If we mix sugar and fine white sand, a heterogeneous mixture is formed. Careful examination may be needed to decide that the mixture is heterogeneous because the two phases (sugar and sand) are both white solids. Ordinary matter exists mostly as mixtures. If we examine soil, granite, iron ore, or other naturally occurring mineral deposits, we find them to be heterogeneous mixtures. Seawater is a homogeneous mixture (solution) containing many substances. Air is a homogeneous mixture (solution) of several gases. Figure 3.2 illustrates the relationships of substances and mixtures.

3.4 ELEMENTS

All the words in the English dictionary are formed from an alphabet consisting of only 26 letters. All known substances on earth—and most probably in the universe, too—are formed from a sort of "chemical alphabet" consisting of 109 presently known elements. An **element** is a fundamental or elementary substance that cannot be broken down by chemical means to simpler substances. Elements are the building blocks of all substances. The elements are numbered in order of increasing complexity beginning with hydrogen, number 1. Of the first 92 elements, 88 are known to occur in nature. The other four—technetium (43), promethium (61), astatine (85), and francium (87)—either do not occur in nature or have only transitory existences during radioactive decay. With the exception of number 94, plutonium, elements above number 92 are not known to occur naturally but have been synthesized, usually in very small quantities, in laboratories. The discovery of trace amounts of element 94 (plutonium) in nature has been reported recently. The syntheses of elements 107 and 109 were reported in 1981 and 1982. No elements other than those on the earth have been detected on other bodies in the universe.

element

Most substances can be decomposed into two or more simpler substances. Water can be decomposed into hydrogen and oxygen. Sugar can be decomposed into carbon, hydrogen, and oxygen. Table salt is easily decomposed into sodium and chlorine. An element, however, cannot be decomposed into simpler substances by ordinary chemical changes.

If we could take a small piece of an element, say copper, and divide it and subdivide it into smaller and smaller particles, we finally would come to a single unit of copper that we could no longer divide and still have copper. This ultimate particle, the smallest particle of an element that can exist, is called an **atom**. An atom is also the smallest unit of an element that can enter into a chemical reaction. Atoms are made up of still smaller subatomic particles. But these subatomic particles (described in Chapter 5) do not have the properties of elements.

atom

3.5 Distribution of Elements

Elements are distributed very unequally in nature, as shown in Figure 3.3. At normal room temperature two of the elements, bromine and mercury, are liquids; eleven elements, hydrogen, nitrogen, oxygen, fluorine, chlorine, helium, neon, argon, krypton, xenon, and radon, are gases; all the other elements are solids.

Ten elements make up about 99% of the mass of the earth's crust, seawater, and atmosphere. Oxygen, the most abundant of these, constitutes about 50% of this mass. The distribution of the elements shown in Table 3.3 includes the earth's crust to a depth of about 10 miles, the oceans, fresh water, and the atmosphere but does not include the mantle and core of the earth, which are believed to consist of metallic iron and nickel. Because the atmosphere contains relatively little matter, its inclusion has almost no effect on the distribution shown in Table 3.3. But the inclusion of fresh and salt water does have an appreciable effect since water contains about 11.1% hydrogen. Nearly all of the 0.87% hydrogen shown is from water.

The average distribution of the elements in the human body is shown in Table 3.4. Note again the high percentage of oxygen.

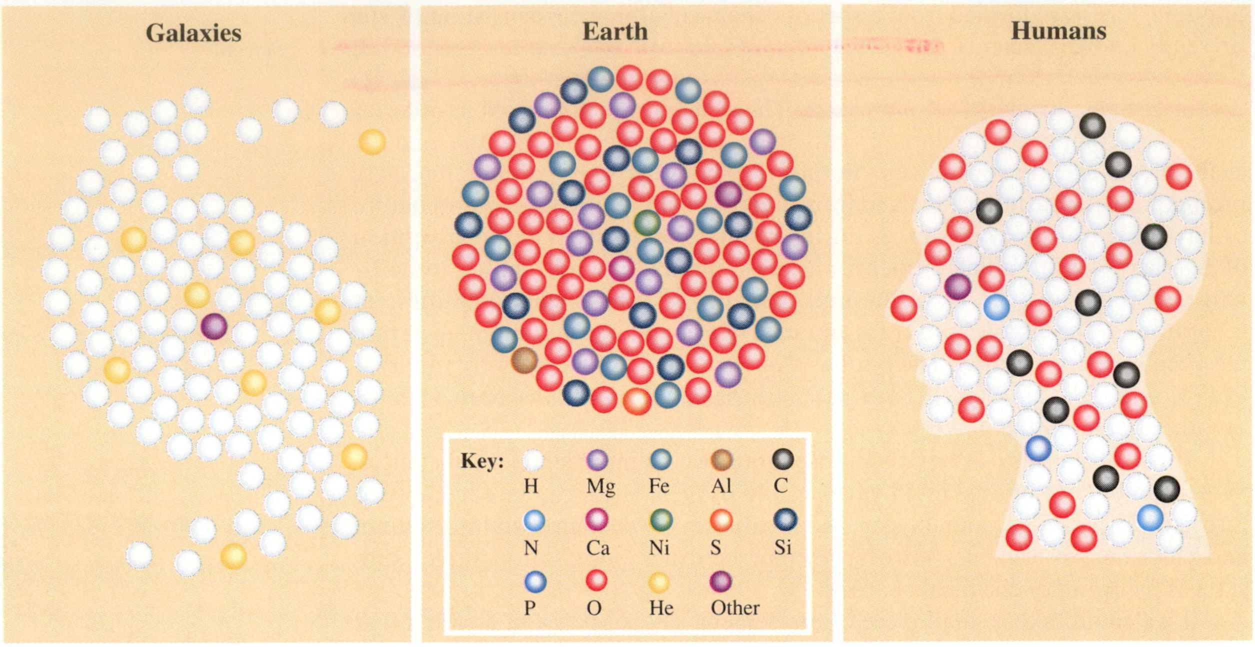

▲
FIGURE 3.3
Distribution of elements in nature

TABLE 3.3 Distribution of the Elements in the Earth's Crust, Seawater, and Atmosphere

Element	Mass percent	Element	Mass percent
Oxygen	49.20	Chlorine	0.19
Silicon	25.67	Phosphorus	0.11
Aluminum	7.50	Manganese	0.09
Iron	4.71	Carbon	0.08
Calcium	3.39	Sulfur	0.06
Sodium	2.63	Barium	0.04
Potassium	2.40	Nitrogen	0.03
Magnesium	1.93	Fluorine	0.03
Hydrogen	0.87		
Titanium	0.58	All others	0.47

TABLE 3.4 Average Elemental Composition of the Human Body

Element	Mass percent
Oxygen	65.0
Carbon	18.0
Hydrogen	10.0
Nitrogen	3.0
Calcium	2.0
Phosphorus	1.0
Traces of several other elements	1.0

3.6 NAMES OF THE ELEMENTS

The names of the elements came to us from various sources. Many are derived from early Greek, Latin, or German words that generally described some property of the element. For example, iodine is taken from the Greek word *iodes*, meaning violetlike. Iodine, indeed, is violet in the vapor state. The name of the metal bismuth had its origin from the German words *weisse masse*, which means white mass. Miners called it *wismat*; it was later changed to *bismat*, and finally to bismuth. Some elements are named for the location of their discovery—for

Table 3.5 Symbols of the Most Common Elements

Element	Symbol	Element	Symbol	Element	Symbol
Aluminum	Al	Fluorine	F	Phosphorus	P
Antimony	Sb	Gold	Au	Platinum	Pt
Argon	Ar	Helium	He	Potassium	K
Arsenic	As	Hydrogen	H	Radium	Ra
Barium	Ba	Iodine	I	Silicon	Si
Bismuth	Bi	Iron	Fe	Silver	Ag
Boron	B	Lead	Pb	Sodium	Na
Bromine	Br	Lithium	Li	Strontium	Sr
Cadmium	Cd	Magnesium	Mg	Sulfur	S
Calcium	Ca	Manganese	Mn	Tin	Sn
Carbon	C	Mercury	Hg	Titanium	Ti
Chlorine	Cl	Neon	Ne	Tungsten	W
Chromium	Cr	Nickel	Ni	Uranium	U
Cobalt	Co	Nitrogen	N	Zinc	Zn
Copper	Cu	Oxygen	O		

example, germanium, discovered in 1886 by Winkler, a German chemist. Others are named in commemoration of famous scientists, such as einsteinium and curium, named for Albert Einstein and Marie Curie, respectively.

3.7 Symbols of the Elements

We all recognize Mr., N.Y., and Ave. as abbreviations for mister, New York, and avenue. In like manner chemists have assigned an abbreviation to each element; these are called **symbols** of the elements. Fourteen of the elements have a single letter as their symbol, six have three-letter symbols, and the rest have two letters. A symbol stands for the element itself, for one atom of the element, and (as we shall see later) for a particular quantity of the element.

symbol

Rules governing symbols of elements are as follows:

1. Symbols are composed of one, two, or three letters.
2. If one letter is used, it is capitalized.
3. If two or three letters are used, the first is capitalized and the others are lowercase letters.

Examples: Sulfur S Barium Ba

The symbols and names of all the elements are given in the table on the inside back cover of this book. Table 3.5 lists the more commonly used symbols. If we examine this table carefully, we note that most of the symbols start with the same letter as the name of the element that is represented. A number of symbols, however, appear to have no connection with the names of the elements they represent (see Table 3.6). These symbols have been carried over from earlier

▲
J. J. Berzelius (1779–1848).

TABLE 3.6 Symbols of the Elements Derived from Early Names*

Present name	Symbol	Former name
Antimony	Sb	Stibium
Copper	Cu	Cuprum
Gold	Au	Aurum
Iron	Fe	Ferrum
Lead	Pb	Plumbum
Mercury	Hg	Hydrargyrum
Potassium	K	Kalium
Silver	Ag	Argentum
Sodium	Na	Natrium
Tin	Sn	Stannum
Tungsten	W	Wolfram

* These symbols are in use today even though they do not correspond to the current name of the element.

names (usually in Latin) of the elements and are so firmly implanted in the literature that their use is continued today.

Special care must be used in writing symbols. Begin each with a capital letter and use a lowercase second letter if needed. For example, consider Co, the symbol for the element cobalt. If through error CO (capital C and capital O) is written, the two elements carbon and oxygen (the *formula* for carbon monoxide) are represented instead of the single element cobalt. Another example of the need for care in writing symbols is the symbol Ca for calcium versus Co for cobalt. The letters must be distinct or else the symbol for the element may be misinterpreted.

Knowledge of symbols is essential for writing chemical formulas and equations. You should begin to learn the symbols immediately because they will be used extensively in the remainder of this book and in any future chemistry courses you may take. One way to learn the symbols is to practice a few minutes a day by making side-by-side lists of names and symbols and then covering each list alternately and writing the corresponding name or symbol. Initially it is a good plan to learn the symbols of the most common elements shown in Table 3.5.

The experiments of alchemists paved the way for the development of chemistry. Alchemists surrounded their work with mysticism, partly by devising a system of symbols known only to practitioners of alchemy (see Figure 3.4). The symbol R (from the Latin *recipe*) is still used in medicine and was established during this time. In the early 1800s the Swedish chemist J. J. Berzelius (1779–1848) made a great contribution to chemistry by devising the present system of symbols using letters of the alphabet.

FIGURE 3.4
Some typical alchemists' symbols.
▼

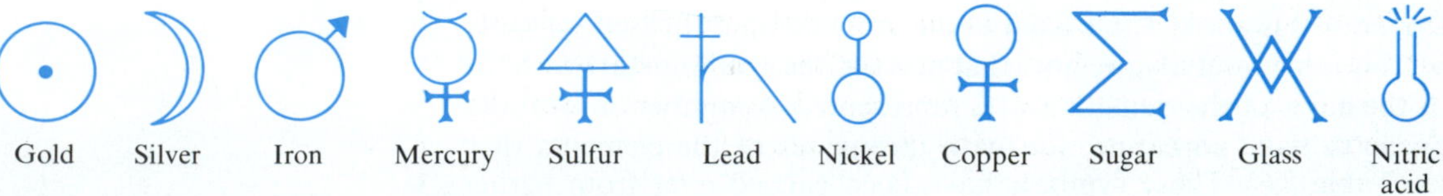

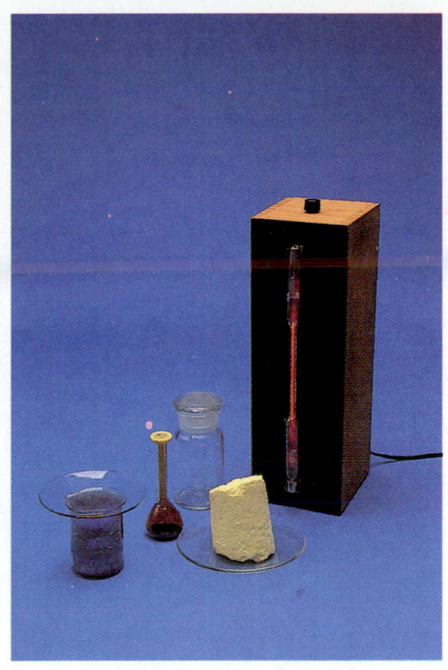

◀ *Left:* Samples of various metals, including aluminum, copper, mercury, titanium, beryllium, cadmium, calcium, and nickel.
Right: Samples of various nonmetals, including iodine, bromine, oxygen, neon, and sulfur.

3.8 Metals, Nonmetals, and Metalloids

metal

nonmetal

metalloid

Three primary classifications of the elements are **metals, nonmetals,** and **metalloids.** Most of the elements are metals. We are familiar with metals because of their widespread use in tools, materials of construction, automobiles, and so on. But nonmetals are equally useful in our everyday life as major components of such items as clothing, food, fuel, glass, plastics, and wood.

The metallic elements are solids at room temperature (mercury is an exception). They have high luster, are good conductors of heat and electricity, are *malleable* (can be rolled or hammered into sheets), and are *ductile* (can be drawn into wires). Most metals have a high melting point and high density. Familiar metals are aluminum, chromium, copper, gold, iron, lead, magnesium, mercury, nickel, platinum, silver, tin, and zinc. Less familiar but still important metals are calcium, cobalt, potassium, sodium, uranium, and titanium.

Metals have little tendency to combine with each other to form compounds. But many metals readily combine with nonmetals such as chlorine, oxygen, and sulfur to form mainly ionic compounds such as metallic chlorides, oxides, and sulfides. In nature the more active metals are found combined with other elements as minerals. A few of the less active ones such as copper, gold, and silver are sometimes found in a native, or free, state.

Nonmetals, unlike metals, are not lustrous, have relatively low melting points and densities, and are generally poor conductors of heat and electricity. Carbon, phosphorus, sulfur, selenium, and iodine are solids; bromine is a liquid; the rest of the nonmetals are gases. Common nonmetals found uncombined in nature are carbon (graphite and diamond), nitrogen, oxygen, sulfur, and the noble gases (helium, neon, argon, krypton, xenon, and radon).

TABLE 3.7 Classification of the Elements into Metals, Metalloids, and Nonmetals

Metals
Metalloids
Nonmetals

1 H																	2 He
3 Li	4 Be											5 B	6 C	7 N	8 O	9 F	10 Ne
11 Na	12 Mg											13 Al	14 Si	15 P	16 S	17 Cl	18 Ar
19 K	20 Ca	21 Sc	22 Ti	23 V	24 Cr	25 Mn	26 Fe	27 Co	28 Ni	29 Cu	30 Zn	31 Ga	32 Ge	33 As	34 Se	35 Br	36 Kr
37 Rb	38 Sr	39 Y	40 Zr	41 Nb	42 Mo	43 Tc	44 Ru	45 Rh	46 Pd	47 Ag	48 Cd	49 In	50 Sn	51 Sb	52 Te	53 I	54 Xe
55 Cs	56 Ba	57 La*	72 Hf	73 Ta	74 W	75 Re	76 Os	77 Ir	78 Pt	79 Au	80 Hg	81 Tl	82 Pb	83 Bi	84 Po	85 At	86 Rn
87 Fr	88 Ra	89 Ac†	104 Unq	105 Unp	106 Unh	107 Uns	108 Uno	109 Une									
			*	58 Ce	59 Pr	60 Nd	61 Pm	62 Sm	63 Eu	64 Gd	65 Tb	66 Dy	67 Ho	68 Er	69 Tm	70 Yb	71 Lu
			†	90 Th	91 Pa	92 U	93 Np	94 Pu	95 Am	96 Cm	97 Bk	98 Cf	99 Es	100 Fm	101 Md	102 No	103 Lr

Nonmetals combine with one another to form molecular compounds such as carbon dioxide (CO_2), methane (CH_4), butane (C_4H_{10}), and sulfur dioxide (SO_2). Fluorine, the most reactive nonmetal, combines readily with almost all the other elements.

Several elements (boron, silicon, germanium, arsenic, antimony, tellurium, and polonium) are classified as *metalloids* and have properties that are intermediate between those of metals and those of nonmetals. The intermediate position of these elements is shown in Table 3.7, which lists and classifies all the elements as metals, nonmetals, or metalloids. Certain metalloids, such as boron, silicon, and germanium, are the raw materials for the semiconductor devices that make our modern electronics industry possible.

3.9 COMPOUNDS

compound

A **compound** is a distinct substance containing two or more elements chemically combined in definite proportions by mass. Compounds, unlike elements, can be decomposed chemically into simpler substances—that is, into simpler compounds and/or elements. Atoms of the elements in a compound are combined in whole-number ratios, never as fractional parts of atoms. Compounds fall into two general types, *molecular* and *ionic*.

molecule

A **molecule** is the smallest uncharged individual unit of a compound formed by the union of two or more atoms. Water is a typical molecular compound. If we divide a drop of water into smaller and smaller particles, we finally obtain a single molecule of water consisting of two hydrogen atoms bonded to one oxygen atom. This molecule is the ultimate particle of water; it cannot be further subdivided without destroying the water molecule and forming hydrogen and oxygen.

ion

cation

anion

An **ion** is a positively or negatively charged atom or group of atoms. An ionic compound is held together by attractive forces that exist between positively and negatively charged ions. A positively charged ion is called a **cation** (pronounced *cat-eye-on*); a negatively charged ion is called an **anion** (pronounced *an-eye-on*).

Compounds

Molecular	*Ionic*
Exist as molecules that consist of two or more elements bonded together	Exist in the form of cations and anions held together by the attractive force of their positive and negative charges

Sodium chloride is a typical ionic compound. The ultimate particles of sodium chloride are positively charged sodium ions and negatively charged chloride ions. Sodium chloride is held together in a crystalline structure by the attractive forces existing between these oppositely charged ions. Although ionic compounds consist of large aggregates of cations and anions, their formulas are normally represented by the simplest possible ratio of the atoms in the compound. For example in sodium chloride the ratio is one sodium ion to one chlorine ion, and the formula is NaCl. The two types of compounds, molecular and ionic, are illustrated in Figure 3.5.

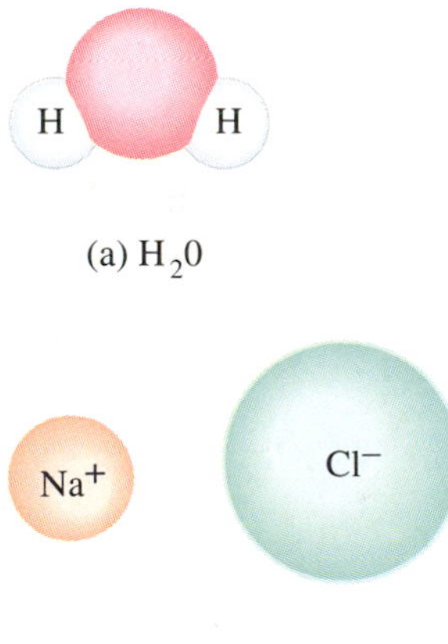

▲
FIGURE 3.5
Representation of molecular and ionic (nonmolecular) compounds. (a) Two hydrogen atoms combined with an oxygen atom to form a molecule of water. (b) A positively charged sodium ion and a negatively charged chloride ion form the compound sodium chloride.

There are more than 9 million known registered compounds, with no end in sight as to the number that will be prepared in the future. Each compound is unique and has characteristic properties. Let us consider two compounds, water and mercury(II) oxide, in some detail. Water is a colorless, odorless, tasteless liquid that can be changed to a solid (ice) at 0°C and to a gas (steam) at 100°C. Composed of two atoms of hydrogen and one atom of oxygen per molecule, water is 11.2% hydrogen and 88.8% oxygen by mass. Water reacts chemically with sodium to produce hydrogen gas and sodium hydroxide, with lime to produce calcium hydroxide, and with sulfur trioxide to produce sulfuric acid. No other compound has all these exact physical and chemical properties; they are characteristic of water alone.

Mercury(II) oxide is a dense, orange-red powder having a ratio of one atom of mercury to one atom of oxygen. Its composition by mass is 92.6% mercury and 7.4% oxygen. When it is heated to temperatures greater than 360°C, a colorless gas, oxygen, and a silvery, liquid metal, mercury, are produced. These specific properties belong to mercury(II) oxide and to no other substance. Thus, a compound may be identified and distinguished from all other compounds by its characteristic properties.

Table 3.8 Elements that Exist as Diatomic Molecules

Element	Symbol	Molecular formula	Normal state
Hydrogen	H	H_2	Colorless gas
Nitrogen	N	N_2	Colorless gas
Oxygen	O	O_2	Colorless gas
Fluorine	F	F_2	Pale yellow gas
Chlorine	Cl	Cl_2	Yellow-green gas
Bromine	Br	Br_2	Reddish-brown liquid
Iodine	I	I_2	Bluish-black solid

3.10 Elements that Exist as Diatomic Molecules

Seven of the elements (all nonmetals) occur as *diatomic molecules*. These elements and their symbols, formulas, and brief descriptions are listed in Table 3.8. Whether found free in nature or prepared in the laboratory, the molecules of these elements always contain two atoms. The formulas of the free elements are therefore always written to show this molecular composition: H_2, N_2, O_2, F_2, Cl_2, Br_2, and I_2.

It is important to see how symbols are used to designate either an atom or a molecule of an element. Consider hydrogen and oxygen. Hydrogen gas is present in volcanic gases and can be prepared by many chemical reactions. Regardless of their source, all samples of free hydrogen gas consist of diatomic molecules. Free hydrogen is designated and its composition is expressed by the formula H_2. Oxygen makes up about 21% by volume of the air that we breathe. This free oxygen is constantly being replenished by photosynthesis; it can also be prepared in the laboratory by several reactions. All free oxygen is diatomic and is designated by the formula O_2. Now consider water, a compound designated by the formula H_2O (sometimes HOH). Water contains neither free hydrogen (H_2) nor free oxygen (O_2). The H_2 part of the formula H_2O simply indicates that two atoms of hydrogen are combined with one atom of oxygen to form water. Thus symbols are used to designate elements, show the composition of molecules of elements, and give the elemental composition of compounds.

3.11 Chemical Formulas

chemical formula

Chemical formulas are used as abbreviations for compounds. A **chemical formula** shows the symbols and the ratio of the atoms of the elements in a compound. Sodium chloride contains one atom of sodium per atom of chlorine; its formula is NaCl. The formula for water is H_2O; it shows that a molecule of water contains two atoms of hydrogen and one atom of oxygen.

The formula of a compound tells us which elements it is composed of and how many atoms of each element are present in a formula unit. For example, a molecule of sulfuric acid is composed of two atoms of hydrogen, one atom of sulfur, and four atoms of oxygen. We could express this compound as HHSOOOO, but the usual formula for writing sulfuric acid is H_2SO_4. The formula may be expressed verbally as "H-two-S-O-four." Numbers that appear partially below the line and to the right of a symbol of an element are called *subscripts*. Thus the 2 and the 4 in H_2SO_4 are subscripts. Characteristics of chemical formulas are

1. The formula of a compound contains the symbols of all the elements in the compound.
2. When the formula contains one atom of an element, the symbol of that element represents that one atom. The number one (1) is not used as a subscript to indicate one atom of an element.
3. When the formula contains more than one atom of an element, the number of atoms is indicated by a subscript written to the right of the symbol of that atom. For example, the two (2) in H_2O indicates two atoms of H in the formula.
4. When the formula contains more than one of a group of atoms that occurs as a unit, parentheses are placed around the group, and the number of units of the group are indicated by a subscript placed to the right of the parentheses. Consider the nitrate group, NO_3^-. The formula for sodium nitrate, $NaNO_3$, has only one nitrate group; therefore no parentheses are needed. Calcium nitrate, $Ca(NO_3)_2$, has two nitrate groups, as indicated by the use of parentheses and the subscript 2. $Ca(NO_3)_2$ has a total of nine atoms: one Ca, two N, and six O atoms. The formula $Ca(NO_3)_2$ is read as "C-A [pause] N-O-three-taken twice."
5. Formulas written as H_2O, H_2SO_4, $Ca(NO_3)_2$, and $C_{12}H_{22}O_{11}$ show only the number and kind of each atom contained in the compound; they do not show the arrangement of the atoms in the compound or how they are chemically bonded to one another.

Figure 3.6 illustrates how symbols and numbers are used in chemical formulas.

EXAMPLE 3.1

Write formulas for the following compounds, the atom composition of which is given. (a) Hydrogen chloride: 1 atom hydrogen + 1 atom chlorine; (b) Methane: 1 atom carbon + 4 atoms hydrogen; (c) Glucose: 6 atoms carbon + 12 atoms hydrogen + 6 atoms oxygen.

SOLUTION

(a) First write the symbols of the atoms in the formula: H Cl Since the ratio of atoms is one to one, we merely bring the symbols together to give the formula for hydrogen chloride as HCl.

(b) Write the symbols of the atoms: C H Now bring the symbols together and place a subscript 4 after the hydrogen atom. The formula is CH_4.

(c) Write the symbols of the atoms: C H O Now write the formula, bringing together the symbols followed by the correct subscripts according to the data given (six C, twelve H, six O). The formula is $C_6H_{12}O_6$.

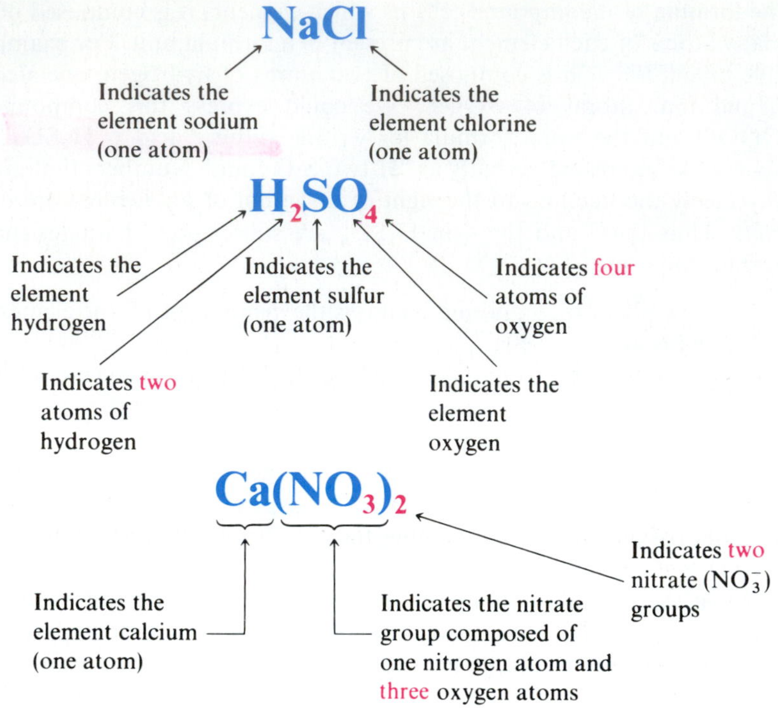

FIGURE 3.6 ▶ Explanation of the formulas NaCl, H_2SO_4, and $Ca(NO_3)_2$.

3.12 MIXTURES

Single substances—elements or compounds—seldom occur naturally in the pure state. Air is a mixture of gases; seawater is a mixture containing a variety of dissolved minerals; ordinary soil is a complex mixture of minerals and various organic materials.

How is a mixture distinguished from a pure substance? A mixture always contains two or more substances that can be present in varying concentrations. Let us consider an example of a homogeneous mixture and an example of a heterogeneous mixture. Homogeneous mixtures (solutions) containing either 5% or 10% salt in water can be prepared simply by mixing the correct amounts of salt and water. These mixtures can be separated by boiling away the water, leaving the salt as a residue. The composition of a heterogeneous mixture of sulfur crystals and iron filings can be varied by merely blending in either more sulfur or more iron filings. This mixture can be separated physically by using a magnet to attract the iron or by adding carbon disulfide to dissolve the sulfur.

Iron(II) sulfide (FeS) contains 63.5% Fe and 36.5% S by mass. If we mix iron and sulfur in this proportion, do we have iron(II) sulfide? No, it is still a mixture; the iron is still attracted by a magnet. But if this mixture is heated strongly, a chemical change (reaction) occurs in which the reacting substances, iron and sulfur, form a new substance, iron(II) sulfide. Iron(II) sulfide, FeS, is a compound of iron and sulfur and has properties that are different from those of

(a)

(b)

(c)

▲
(a) When iron and sulfur exist as pure substances, only the iron is attracted to a magnet. (b) A mixture of iron and sulfur can be separated by using the difference in magnetic attraction. (c) The compound iron (II) sulfide cannot be separated into its elements with a magnet.

either iron or sulfur: It is neither attracted by a magnet nor dissolved by carbon disulfide. The differences between the iron and sulfur *mixture* and the iron(II) sulfide *compound* are as follows:

Mixture of iron and sulfur	Compound of iron and sulfur
Formula: mixture has no definite formula; consists of Fe and S.	Formula: FeS
Composition: mixture contains Fe and S in any proportion by mass	Composition: compound contains 63.5% Fe and 36.5% S by mass.
Separation: Fe and S can be separated by physical means.	Separation: Fe and S can be separated only by chemical change.

The general characteristics of mixtures and compounds are compared in Table 3.9.

Table 3.9 Comparison of Mixtures and Compounds

	Mixture	Compound
Composition	May be composed of elements, compounds, or both in variable composition.	Composed of two or more elements in a definite, fixed proportion by mass.
Separation of components	Separation may be made by physical or mechanical means.	Elements can be separated by chemical changes only.
Identification of components	Components do not lose their identity.	A compound does not resemble the elements from which it is formed.

CHEMISTRY IN ACTION

CARBON—THE CHAMELEON

One of the most diverse elements in the periodic table is carbon. Although it is much less abundant than many other elements it is readily available. Carbon is found free in two forms called **allotropes**—as the mineral graphite and as diamonds. The physical properties of the two allotropes are quite distinct. Diamond consists of transparent, octahedral crystals which are colorless when pure—but may range from pale blue to jet black due to impurities. It is the hardest known substance and an excellent heat conductor. When doped, diamond is an electrical semiconductor. Diamonds for cutting tools are both mined and produced synthetically. The majority of gem diamonds are mined in South Africa but some also come from South America. Graphite, on the other hand, is made from layers or sheets of carbon atoms, which are very soft and the layers easily slip over one another. It is an excellent conductor of electricity. Graphite is mined as massive crystals, or is obtained from heating coal and pitch in very high-temperature furnaces.

Carbon is an essential constituent of plant and animal life. Coal is formed by the gradual decay of plant life enriched in carbon through the loss of carbon dioxide and methane gas during the decaying process. If coal is heated without air present, destructive distillation occurs. This process decomposes the carbon compounds and produces coke, which is 90–95% graphite. Charcoal, consisting of tiny crystals of graphite, is prepared by the destructive distillation of wood. Bone black, also consisting of tiny graphite crystals, is made from the destructive distillation of bones and wastes from packing houses. Carbon black is formed when natural gas is burned with an insufficient quantity of air—a residue of graphite carbon is formed on a cold surface and then scraped off.

Free carbon finds a variety of uses in our society. Diamonds are collected and displayed for their gem quality, used in jewelry, and held as investments. They are also used in cutting and drilling tools and in instruments that require hardness and long lasting quality such as in phonograph needles. Graphite has many uses as a lubricant. Mixtures of clay and graphite are molded into the "lead" used in pencils.—The higher the clay content the harder the "lead". Graphite is also used as electrodes in dry cells and is found in some paints and stove polish.

Charcoal has the quality of absorbing large quantities of gas onto its surface. For this reason it is very useful in water purification systems, in the manufacture of gas masks, and in removing color from solutions (as in the refining of sugar). Coke is an important fuel. It is also used to reduce iron from its ore. Alloys of iron and carbon form the major industrial product we call *steel*. Carbon black is used in making carbon paper, printers ink, and some shoe polish. It is the additive in rubber that makes tires black.

In addition, carbon combines chemically with other elements to form a myriad of useful *compounds*. **Hydrocarbons** containing carbon and hydrogen are commonly found in petroleum and natural gas. There are so many hydrocarbons and their derivatives that an entire branch of chemistry, *organic chemistry*, is dedicated to studying and understanding them (see Chapters 21-37). Carbon is also found as carbon dioxide, CO_2, in our atmosphere. As such, it is one of the greenhouse gases that contributes to global warming and is of great concern to scientists (see Section 8.7). Carbon

▲
Diamonds and graphite.

(*continued*)

(*continued*)
monoxide, CO, is an important fuel gas. It is colorless, odorless, tasteless, and extremely poisonous. Breathing even small quantities of carbon monoxide can be fatal (see Section 17.16). Carbon can combine with chlorine to form chloroform, $CHCl_3$, or carbon tetrachloride, CCl_4. Carbon tetrachloride is used as a cleaning solvent. Carbon also combines with chlorine and fluorine to make a group of compounds known as **chlorofluorocarbons**. These compounds are used widely as refrigerants. They are definite contributors to the destruction of the ozone layer in our atmosphere and are the topic of debate and regulation worldwide (see Section 13.17).

Concepts in Review

1. Identify the three physical states of matter.
2. Distinguish between substances and mixtures.
3. Classify common materials as elements, compounds, or mixtures.
4. Write the symbols when given the names, or write the names when given the symbols, of the common elements listed in Table 3.5.
5. Understand how symbols, including subscripts and parentheses, are used to write chemical formulas.
6. Differentiate among atoms, molecules, and ions.
7. List the characteristics of metals, nonmetals, and metalloids.
8. List the elements that occur as diatomic molecules.

Exercises

1. List four different substances in each of the three states of matter.
2. In terms of the properties of the ultimate particles of a substance, explain
   (a) Why a solid has a definite shape but a liquid does not.
   (b) Why a liquid has a definite volume but a gas does not.
   (c) Why a gas can be compressed rather easily but a solid cannot be compressed appreciably.
3. What evidence can you find in Figure 3.1 that gases occupy space?
4. Which three liquids listed in Table 3.1 are not mixtures?
5. Which of the gases listed in Table 3.1 are not pure substances?
6. When the stopper is removed from a partly filled bottle containing solid and liquid acetic acid at 16.7°C, a strong vinegarlike odor is noticeable immediately. How many acetic acid phases must be present in the bottle? Explain.
7. Is the system enclosed in the bottle of Exercise 6 homogeneous or heterogeneous? Explain.
8. Is a system that contains only one substance necessarily homogeneous? Explain.
9. Is a system that contains two or more substances necessarily heterogeneous? Explain.
10. Are there more atoms of silicon or hydrogen in the earth's crust, seawater, and atmosphere? Use Table 3.3 and the fact that the mass of the silicon atom is about 28 times that of a hydrogen atom.
11. What does the symbol of an element stand for?
12. Write down what you believe to be the symbols for the elements phosphorus, aluminum, hydrogen, potassium, magnesium, sodium, nitrogen, nickel, silver, and plutonium. Now look up the correct symbols and rewrite them, comparing the two sets.

13. Interpret the difference in meanings for each of these pairs:
(a) Si and SI (b) Pb and PB (c) 4 P and P_4
14. List six elements and their symbols in which the first letter of the symbol is different from that of the name.
15. Write the names and symbols for the fourteen elements that have only one letter as their symbol. (See table on inside back cover.)
16. Distinguish between an element and a compound.
17. How many metals are there? Nonmetals? Metalloids? (See Table 3.7)
18. Of the ten most abundant elements in the earth's crust, seawater, and atmosphere, how many are metals? Nonmetals? Metalloids? (Table 3.3)
19. Of the six most abundant elements in the human body, how many are metals? Nonmetals? Metalloids?
20. Why is the symbol for gold Au rather than G or Go?
21. Give the names of (a) the solid diatomic nonmetal and (b) the liquid diatomic nonmetal.
22. Distinguish between a compound and a mixture.
23. What are the two general types of compounds? How do they differ from each other?
24. What is the basis for distinguishing one compound from another?
25. Given the following list of compounds and their formulas, what elements are present in each compound?

(a) Potassium iodide	KI
(b) Sodium carbonate	Na_2CO_3
(c) Aluminum oxide	Al_2O_3
(d) Calcium bromide	$CaBr_2$
(e) Carbon tetrachloride	CCl_4
(f) Magnesium bromide	$MgBr_2$
(g) Nitric acid	HNO_3
(h) Barium sulfate	$BaSO_4$
(i) Aluminum phosphate	$AlPO_4$
(j) Acetic acid	$HC_2H_3O_2$

26. Write the formula for each of the following compounds, the composition of which is given after each name:

(a) Zinc oxide	1 atom Zn, 1 atom O
(b) Potassium chlorate	1 atom K, 1 atom Cl, 3 atoms O
(c) Sodium hydroxide	1 atom Na, 1 atom O, 1 atom H
(d) Aluminum bromide	1 atom Al, 3 atoms Br
(e) Calcium fluoride	1 atom Ca, 2 atoms F
(f) Lead(II) chromate	1 atom Pb, 1 atom Cr, 4 atoms O
(g) Ethyl alcohol	2 atoms C, 6 atoms H, 1 atom O
(h) Benzene	6 atoms C, 6 atoms H

27. Explain the meaning of each symbol and number in the following formulas:
(a) H_2O (e) $C_{12}H_{22}O_{11}$ (sucrose)
(b) $AlBr_3$
(c) Na_2SO_4
(d) $Ni(NO_3)_2$
28. How many atoms are represented in each of these formulas?
(a) KF (f) $NaC_2H_3O_2$
(b) $CaCO_3$ (g) CCl_2F_2 (Freon)
(c) N_2 (h) $Al_2(SO_4)_3$
(d) $Ba(ClO_3)_2$ (i) $(NH_4)_2C_2O_4$
(e) $K_2Cr_2O_7$
29. How many atoms are contained in (a) one molecule of hydrogen, (b) one molecule of water, and (c) one molecule of sulfuric acid?
30. What is the major difference between a cation and an anion?
31. Write the names and formulas of the elements that exist as diatomic molecules.
32. How many atoms of oxygen are represented in each formula?
(a) H_2O (c) H_2O_2 (e) $Al(ClO_3)_3$
(b) $CuSO_4$ (d) $Fe(OH)_3$
33. How many atoms of hydrogen are represented in each formula?
(a) H_2 (c) $C_6H_{12}O_6$
(b) $Ba(C_2H_3O_2)_2$ (d) $HC_2H_3O_2$
34. Distinguish between homogeneous and heterogeneous mixtures.
35. Classify each of the following materials as an element, compound, or mixture:
(a) Air (e) Wine
(b) Oxygen (f) Iodine
(c) Sodium chloride (g) Sulfuric acid
(d) Platinum (h) Crude oil
36. Classify each of the following materials as an element, compound, or mixture:
(a) Paint (e) Sulfuric acid
(b) Salt (f) Silver
(c) Copper (g) Milk
(d) Beer (h) Sodium hydroxide
37. A white solid, on heating, formed a colorless gas and a yellow solid. Assuming that there was no

reaction with the air, is the original solid an element or a compound? Explain.

38. Tabulate the properties that characterize metals and nonmetals.

39. Which of the following are diatomic molecules?
(a) H_2 (c) HCl (e) NO (g) $MgCl_2$
(b) SO_2 (d) H_2O (f) NO_2

40. Name and describe the two allotropes of carbon.

41. List a minimum of three forms of graphite crystals and indicate the source of each.

42. List four uses for graphite in daily life.

43. Indicate four compounds containing carbon, and state their significance in society.

44. Which of the following statements are correct? Rewrite the incorrect statements to make them correct. (Try to answer this question without referring to the text.)
- (a) Liquids are the least compact state of matter.
- (b) Liquids have a definite volume and a definite shape.
- (c) Matter in the solid state is discontinuous; that is, it is made up of discrete particles.
- (d) Wood is homogeneous.
- (e) Wood is a substance.
- (f) Dirt is a mixture.
- (g) Seawater, although homogeneous, is a mixture.
- (h) Any system made up of only one substance is homogeneous.
- (i) Any system containing two or more substances is heterogeneous.
- (j) A solution, although it contains dissolved material, is homogeneous.
- (k) The smallest unit of an element that can exist and enter into a chemical reaction is called a molecule.
- (l) The basic building blocks of all substances, which cannot be decomposed into simpler substances by ordinary chemical change, are compounds.
- (m) The most abundant element in the earth's crust, seawater, and atmosphere by mass is oxygen.
- (n) The most abundant element in the human body, by mass, is carbon.
- (o) Most of the elements are represented by symbols consisting of one or two letters.
- (p) The symbol for copper is Co.
- (q) The symbol for sodium is Na.
- (r) The symbol for potassium is P.
- (s) The symbol for lead is Le.
- (t) Early names for some elements led to unlikely symbols, such as Fe for iron.
- (u) A compound is a distinct substance that contains two or more elements combined in a definite proportion by mass.
- (v) The smallest uncharged individual unit of a compound formed by the union of two or more atoms is called a substance.
- (w) An ion is a positive or negative electrically charged atom or group of atoms.
- (x) Bromine is an element that occurs as a diatomic molecule, Br_2.
- (y) The formula Na_2CO_3 indicates a total of six atoms, including three oxygen atoms.
- (z) A general property of nonmetals is that they are good conductors of heat and electricity.
- (aa) Metals have the properties of ductility and malleability.
- (bb) *Malleable* means that when struck a hard blow the substance will shatter.
- (cc) Elements that have properties intermediate between metals and nonmetals are called mixtures.
- (dd) More of the elements are metals than nonmetals.

***45.** What would be the density of a solution made by mixing 2.50 mL of carbon tetrachloride (CCl_4, $d = 1.595$ g/mL) and 3.50 mL of carbon tetrabromide (CBr_4, $d = 3.420$ g/mL)? Assume that the volume of the mixed liquids is the sum of the two volumes used.

***46.** Only two elements, Br_2 ($d = 3.12$ g/mL) and Hg ($d = 13.6$ g/mL), are liquids at room temperature. How many milliliters of Br_2 will have the same mass as 12.5 mL Hg?

47. Pure gold is too soft a metal for many uses, so it is alloyed to give it more mechanical strength. One particular alloy is made by mixing 60. g of gold, 8.0 g of silver, and 12 g of copper. What carat gold is this alloy if pure gold is considered to be 24 carat?

***48.** Methane, the chief component of natural gas, has the formula CH_4. Each atom of carbon has a mass 12 times greater than an atom of hydrogen. Calculate the mass percent of carbon in methane.

***49.** White gold is a homogeneous solution of 90% gold and 10% palladium. How much gold is present in a bar of white gold with a mass of 8420 g?

***50.** The metal used to make the U.S. nickel coin is an alloy of 75% copper and 25% nickel. What maximum mass of alloy could be produced if only 450 kg of nickel and 1180 kg of copper were on hand?

4

Properties of Matter

◀ **Chapter Opening Photo: Molten iron is poured into a cast and cooled to yield a product.**

The world we live in is a myriad of sights, sounds, smells, and tastes. Our senses help us to describe the objects in our lives. For example, the smell of freshly baked cinnamon rolls along with the sight of a bakery combine to create a mouth-watering desire to gobble down a fresh-baked sample. And so it is with each substance—its own unique properties allow us to identify it and predict its interactions.

These interactions produce both physical and chemical changes. When you eat an apple, the ultimate metabolic result is carbon dioxide and water. These same products are achieved by burning logs. Not only does a chemical change occur in these cases, but an energy change as well. Some reactions release energy (as does the apple or the log) whereas others require energy, such as the production of steel or the melting of ice. Over 90% of our current energy comes from chemical reactions.

4.1 Properties of Substances

properties

physical properties

chemical properties

How do we recognize substances? Each substance has a set of **properties** that is characteristic of that substance and gives it a unique identity. Properties are the personality traits of substances and are classified as either physical or chemical. **Physical properties** are the inherent characteristics of a substance that can be determined without altering its composition; they are associated with its physical existence. Common physical properties are color, taste, odor, state of matter (solid, liquid, or gas), density, melting point, and boiling point. **Chemical properties** describe the ability of a substance to form new substances, either by reaction with other substances or by decomposition.

We can select a few of the physical and chemical properties of chlorine as an example. Physically, chlorine is a gas about 2.4 times heavier than air. It is yellowish-green in color and has a disagreeable odor. Chemically, chlorine will not burn but will support the combustion of certain other substances. It can be used as a bleaching agent, as a disinfectant for water, and in many chlorinated substances such as refrigerants and insecticides. When chlorine combines with the metal sodium, it forms a salt called sodium chloride. These properties, among others, help to characterize and identify chlorine.

Substances, then, are recognized and differentiated by their properties. Table 4.1 lists four substances and tabulates several of their common physical properties. Information about common physical properties, such as that given in

TABLE 4.1 Physical Properties of Chlorine, Water, Sugar, and Acetic Acid

Substance	Color	Odor	Taste	Physical state	Melting point (°C)	Boiling point (°C)
Chlorine	Yellowish-green	Sharp, suffocating	Sharp, sour	Gas	−101.6	−34.6
Water	Colorless	Odorless	Tasteless	Liquid	0.0	100.0
Sugar	White	Odorless	Sweet	Solid	—	Decomposes 170–186
Acetic acid	Colorless	Like vinegar	Sour	Liquid	16.7	118.0

Table 4.1, is available in handbooks of chemistry and physics. Scientists do not pretend to know all the answers or to remember voluminous amounts of data, but it is important for them to know where to look for data in the literature. Handbooks are one of the most widely used resources for scientific data.*

No two substances will have identical physical and chemical properties.

4.2 Physical Changes

physical change

Matter can undergo two types of changes, physical and chemical. **Physical changes** are changes in physical properties (such as size, shape, and density) or changes in the state of matter without an accompanying change in composition. The changing of ice into water and water into steam are physical changes from one state of matter into another. No new substances are formed in these physical changes.

When a clean platinum wire is heated in a burner flame, the appearance of the platinum changes from silvery metallic to glowing red. This change is physical because the platinum can be restored to its original metallic appearance by cooling and, more importantly, because the composition of the platinum is not changed by heating and cooling.

Two such handbooks are David R. Lide, ed., *Handbook of Chemistry and Physics,* 72nd ed. (Cleveland: Chemical Rubber Company, 1992) and Norbert A. Lange, comp., *Handbook of Chemistry,* 13th ed. (New York: McGraw-Hill, 1985).

Before heating
wire is copper-colored
Copper wire: 1.00 g
(100% copper)

Copper and oxygen
from the air combine
chemically on heating

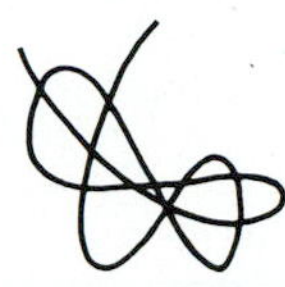
After heating,
wire is black
Copper (II) oxide: 1.251 g
79.9% copper: 1.00 g
20.1% oxygen: 0.251 g

◀ **FIGURE 4.1**
Chemical change: formation of copper(II) oxide from copper and oxygen.

4.3 Chemical Changes

chemical change

In a **chemical change**, new substances are formed that have different properties and composition from the original material. The new substances need not in any way resemble the initial material.

When a clean copper wire is heated in a burner flame, the appearance of the copper changes from coppery metallic to glowing red. Unlike the platinum previously mentioned, the copper is not restored to its original appearance by cooling but has become a black material. This black material is a new substance called copper(II) oxide. It was formed by chemical change when copper combined with oxygen in the air during the heating process. The unheated wire was essentially 100% copper, but the copper(II) oxide is 79.9% copper and 20.1% oxygen. One gram of copper will yield 1.251 g of copper(II) oxide (see Figure 4.1). The platinum was changed only physically when heated, but the copper was changed both physically and chemically when heated.

When 1.00 g of copper reacts with oxygen to yield 1.251 g of copper(II) oxide, the copper must have combined with 0.251 g of oxygen. The percentage of copper and oxygen can be calculated from this data, the copper and oxygen each being a percent of the total mass of copper(II) oxide.

$$1.00 \text{ g copper} + 0.251 \text{ g oxygen} \longrightarrow 1.251 \text{ g copper(II) oxide}$$

$$\frac{1.00 \text{ g copper}}{1.251 \text{ g copper(II) oxide}} \times 100 = 79.9\% \text{ copper}$$

$$\frac{0.251 \text{ g oxygen}}{1.251 \text{ g copper(II) oxide}} \times 100 = 20.1\% \text{ oxygen}$$

Mercury(II) oxide is an orange-red powder that, when subjected to high temperature (500–600°C), decomposes into a colorless gas (oxygen) and a silvery, liquid metal (mercury). The composition and physical appearance of each product are noticeably different from those of the starting compound. When

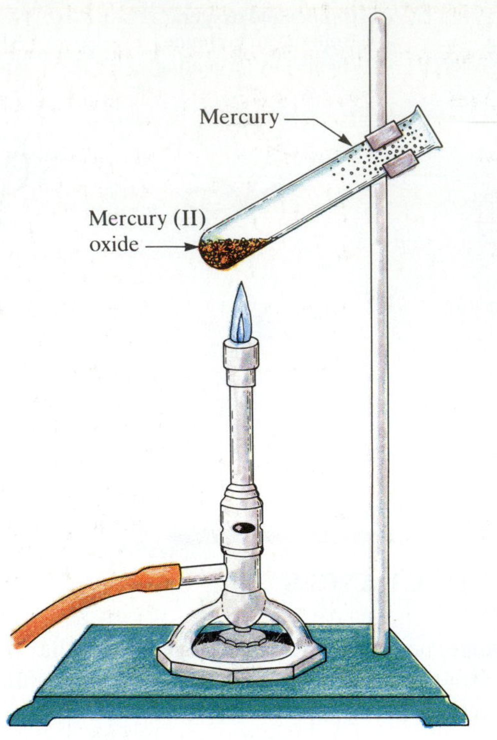

FIGURE 4.2 ▶
Heating mercury(II) oxide causes it to decompose into mercury and oxygen. Observation of the mercury and oxygen, with properties different from those of mercury(II) oxide, is evidence that a chemical change has occurred.

mercury(II) oxide is heated in a test tube (see Figure 4.2), small globules of mercury are observed collecting on the cooler part of the tube. Evidence that oxygen forms is observed when a glowing wood splint, lowered into the tube, bursts into flame. Oxygen supports and intensifies the combustion of the wood. From these observations we conclude that a chemical change has taken place.

Chemists have devised *chemical equations* as a shorthand method for expressing chemical changes. The two previous examples of chemical changes can be represented by the following word equations:

$$\text{copper} + \text{oxygen} \xrightarrow{\Delta} \text{copper(II) oxide} \quad \text{(cupric oxide)} \qquad (1)$$

$$\text{mercury(II) oxide} \xrightarrow{\Delta} \text{mercury} + \text{oxygen} \qquad (2)$$

Equation (1) states: Copper plus oxygen when heated produce copper(II) oxide. Equation (2) states: Mercury(II) oxide when heated produces mercury plus oxygen. The arrow means "produces"; it points to the products. The Greek letter delta (Δ) represents heat. The starting substances (copper, oxygen, and mercury(II) oxide) are called the *reactants*, and the substances produced (copper(II) oxide, mercury, and oxygen) are called the *products*. In later chapters equations are presented in a still more abbreviated form, with symbols to represent substances.

Physical change usually accompanies a chemical change. Table 4.2 lists some common physical and chemical changes. In the examples given in the table, you will note that wherever a chemical change occurs, a physical change occurs also. However, wherever a physical change is listed, only a physical change occurs.

TABLE 4.2 Examples of Processes Involving Physical or Chemical Changes

Process taking place	Type of change	Accompanying observations
Rusting of iron	Chemical	Shiny, bright metal changes to reddish-brown rust.
Boiling of water	Physical	Liquid changes to vapor.
Burning of sulfur in air	Chemical	Yellow solid sulfur changes to gaseous, choking sulfur dioxide.
Boiling an egg	Chemical	Liquid white and yolk change to solids.
Combustion of gasoline	Chemical	Liquid gasoline burns to gaseous carbon monoxide, carbon dioxide, and water
Digesting food	Chemical	Food changes to liquid nutrients and partially solid wastes.
Sawing of wood	Physical	Smaller pieces of wood and sawdust are made from a larger piece of wood.
Burning of wood	Chemical	Wood burns to ashes, gaseous carbon dioxide, and water.
Heating of glass	Physical	Solid becomes pliable during heating, and the glass may change its shape.

4.4 CONSERVATION OF MASS

Law of Conservation of Mass

The **Law of Conservation of Mass** states that no detectable change is observed in the total mass of the substances involved in a chemical change. This law, tested by extensive laboratory experimentation, is the basis for the quantitative mass relationships among reactants and products.

The decomposition of mercury(II) oxide into mercury and oxygen illustrates this law. One hundred grams of mercury(II) oxide decomposes into 92.6 g of mercury and 7.39 g of oxygen.

$$\underset{100.\ g}{\text{mercury(II) oxide}} \longrightarrow \underset{92.6\ g}{\text{mercury}} + \underset{7.39\ g}{\text{oxygen}}$$

100. g Reactant	100. g Products

Operation of the ordinary photographic flashcube also illustrates the law (see Figure 4.3). Sealed within the flashcube are fine wires of magnesium (a metal) and oxygen (a gas). When these reactants are energized, they combine chemically, producing magnesium oxide, a blinding white light, and considerable heat. The

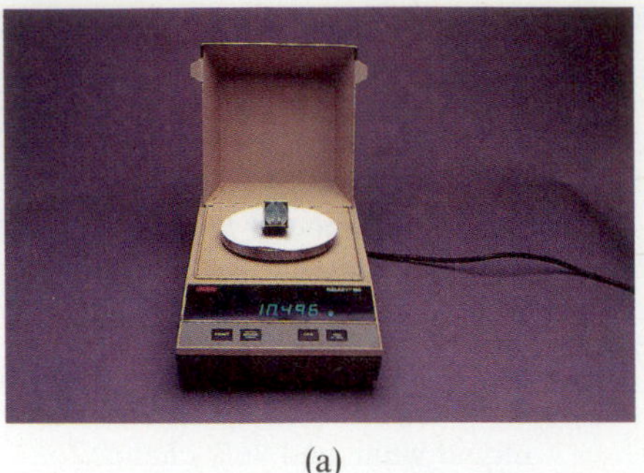

(a)

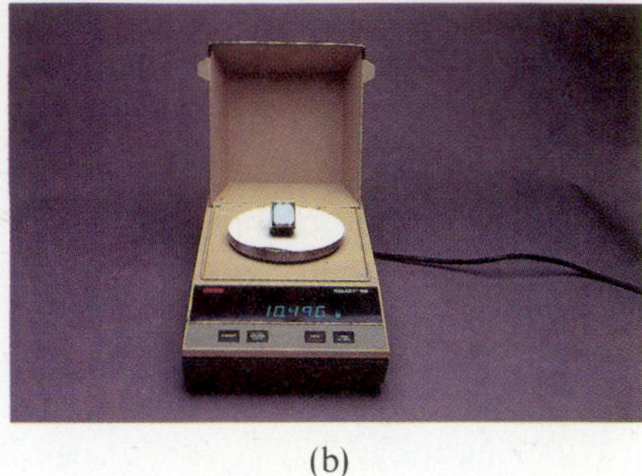

(b)

FIGURE 4.3 ▶ **The flashcube containing magnesium and oxygen, weighs the same (a) before and (b) after it is flashed. When the cube flashes a chemical change occurs, and the original substances change into the white powder magnesium oxide.**

chemical change may be represented by this equation

magnesium + oxygen ⟶ magnesium oxide + heat + light

When the mass of the flashcube is determined before and after the chemical change, as illustrated in Figure 4.3, the cube shows no increase or decrease in mass.

mass of reactants = mass of products

4.5 ENERGY

energy

From the prehistoric discovery that fire could be used to warm shelters and cook food, to the modern-day discovery that nuclear reactors can be used to produce vast amounts of controlled energy, our technical progress has been directed by the ability to produce, harness, and utilize energy. **Energy** is the capacity of matter to do work. Energy exists in several forms; some of the more familiar forms are mechanical, chemical, electrical, heat, nuclear, and radiant or light energy. Matter can have both potential and kinetic energy.

potential energy

Potential energy is stored energy, or energy an object possesses due to its relative position. For example, a ball located 20 ft above the ground has more potential energy than when located 10 ft above the ground and will bounce higher when allowed to fall. Water backed up behind a dam represents potential energy that can be converted into useful work in the form of electrical or mechanical energy. Gasoline is a source of chemical potential energy. When gasoline burns (combines with oxygen), the heat released is associated with a decrease in potential energy. The new substances formed by burning have less chemical potential energy than the gasoline and oxygen did.

kinetic energy

Kinetic energy is the energy that matter possesses due to its motion. When the water behind the dam is released and allowed to flow, its potential energy is changed into kinetic energy, which can be used to drive generators and produce

▲ **The mechanical energy of falling water is converted to electrical energy in a hydroelectric plant.**

electricity. All moving bodies possess kinetic energy. The pressure exerted by a confined gas is due to the kinetic energy of rapidly moving gas particles. We all know the results when two moving vehicles collide: Their kinetic energy is expended in the crash that occurs.

Energy can be converted from one form to another form. Some kinds of energy can be converted to other forms easily and efficiently. For example, mechanical energy can be converted to electrical energy with an electric generator at better than 90% efficiency. On the other hand, solar energy has thus far been directly converted to electrical energy at an efficiency of only about 15%. In chemistry, energy is most frequently expressed as heat.

4.6 Heat: Quantitative Measurement

The SI-derived unit for energy is the joule (pronounced *jool* and abbreviated J). Another unit for heat energy, which has been used for many years, is the calorie (abbreviated cal). The relationship between joules and calories is

$$4.184 \text{ J} = 1 \text{ cal} \qquad \text{(exactly)}$$

To give you some idea of the magnitude of these heat units, 4.184 **joule** or 1 **calorie** is the quantity of heat energy required to change the temperature of 1 g of water by 1°C, usually measured from 14.5°C to 15.5°C.

joule

calorie

Since joule and calorie are rather small units, kilojoules (kJ) and kilocalories (kcal) are used to express heat energy in many chemical processes. The kilocalorie is also known as the nutritional or large Calorie (spelled with a capital *C* and abbreviated Cal). In this book heat energy will be expressed in joules with parenthetical values in calories.

TABLE 4.3 Specific Heat of Selected Substances

Substance	Specific heat J/g°C	Specific heat cal/g°C
Water	4.184	1.00
Ethyl alcohol	2.138	0.511
Ice	2.059	0.492
Aluminum	0.900	0.215
Iron	0.473	0.113
Copper	0.385	0.0921
Gold	0.131	0.0312
Lead	0.128	0.0305

1 kJ = 1000 J

1 kcal = 1000 cal

The difference in the meanings of the terms *heat* and *temperature* can be seen by this example: Visualize two beakers, A and B. Beaker A contains 100 g of water at 20°C, and beaker B contains 200 g of water also at 20°C. The beakers are heated until the temperature of the water in each reaches 30°C. The temperature of the water in the beakers was raised by exactly the same amount, 10°C. But twice as much heat (8368 J or 2000 cal) was required to raise the temperature of the water in beaker B as was required in beaker A (4184 J or 1000 cal).

In the middle of the 18th century Joseph Black, a Scottish chemist, heated and cooled equal masses of iron and lead through the same temperature range. Black noted that much more heat was needed for the iron than for the lead. He had discovered a fundamental property of matter; namely, that every substance has a characteristic heat capacity. Heat capacities may be compared in terms of specific heats. The **specific heat** of a substance is the quantity of heat (lost or gained) required to change the temperature of 1 g of that substance by 1°C. It follows then that the specific heat of water is 4.184 J/g°C (1 cal/g°C). The specific heat of water is high compared with that of most substances. Aluminum and copper, for example, have specific heats of 0.900 and 0.385 J/g°C, respectively (see Table 4.3). The relation of mass, specific heat, temperature change (Δt), and quantity of heat lost or gained by a system is expressed by this general equation:

specific heat

$$\begin{pmatrix}\text{mass of}\\ \text{substance}\end{pmatrix} \times \begin{pmatrix}\text{specific heat}\\ \text{of substance}\end{pmatrix} \times \Delta t = \text{energy (heat)} \qquad (3)$$

Thus, the amount of heat needed to raise the temperature of 200. g of water by 10.0°C can be calculated as follows:

$$200.\ \cancel{\text{g}} \times \frac{4.184\ \text{J}}{\cancel{\text{g}}\cancel{^\circ\text{C}}} \times 10.0\cancel{^\circ\text{C}} = 8.37 \times 10^3\ \text{J}\ (2.00 \times 10^3\ \text{cal})$$

Examples of specific heat problems follow.

EXAMPLE 4.1

Calculate the specific heat of a solid in J/g°C and cal/g°C if 1638 J raise the temperature of 125 g of the solid from 25.0°C to 52.6°C.

SOLUTION

First solve equation (3) to obtain an equation for specific heat.

$$\text{specific heat} = \frac{\text{energy}}{\text{g} \times \Delta t}$$

Now substitute in the data:

$$\text{energy} = 1638\ \text{J} \qquad \text{mass} = 125\ \text{g} \qquad \Delta t = 52.6 - 25.0°\text{C} = 27.6°\text{C}$$

$$\text{specific heat} = \frac{1638\ \text{J}}{125\ \text{g} \times 27.6°\text{C}} = 0.475\ \text{J/g°C}$$

Now convert joules to calories using 1 cal/4.184 J:

$$\text{specific heat} = \frac{0.475\ \cancel{\text{J}}}{\text{g°C}} \times \frac{1\ \text{cal}}{4.184\ \cancel{\text{J}}} = 0.114\ \text{cal/g°C}$$

EXAMPLE 4.2

A sample of a metal with a mass of 212 g is heated to 125.0°C and then dropped into 375 g water at 24.0°C. If the final temperature of the water is 34.2°C, what is the specific heat of the metal? (Assume no heat losses to the surroundings.)

SOLUTION

When the metal enters the water it begins to cool, losing heat to the water. At the same time the temperature of the water rises. This process continues until the temperature of the metal and the temperature of the water are equal, at which point (34.2°C) no net flow of heat occurs.

The heat lost or gained by a system is given by equation (3). We use this equation first to calculate the heat gained by the water and then to calculate the specific heat of the metal.

$$\text{temperature rise of the water } (\Delta t) = 34.2°\text{C} - 24.0°\text{C} = 10.2°\text{C}$$

$$\text{heat gained by the water} = 375\ \text{g} \times \frac{4.184\ \text{J}}{\text{g°C}} \times 10.2°\text{C} = 1.60 \times 10^4\ \text{J}$$

The metal dropped into the water must have a final temperature the same as the water (34.2°C).

$$\text{temperature drop by the metal } (\Delta t) = 125.0°\text{C} - 34.2°\text{C} = 90.8°\text{C}$$

$$\text{heat lost by the metal} = \text{heat gained by the water} = 1.60 \times 10^4\ \text{J}$$

Rearranging equation (3) we get

$$\text{specific heat} = \frac{\text{energy}}{\text{g} \times \Delta t}$$

$$\text{specific heat of the metal} = \frac{1.60 \times 10^4\ \text{J}}{212\ \text{g} \times 90.8°\text{C}} = 0.831\ \text{J/g°C}$$

PRACTICE Calculate the quantity of energy needed to heat 8.0 grams of water from 42°C to 45°C.

Answer: 1.0×10^2 J = 24 cal

PRACTICE A 110.0 g sample of iron at 55.5°C raises the temperature of 150.0 mL of water from 23.0°C to 25.5°C. Determine the specific heat of the iron in calories/g°C.

Answer: 0.114 cal/g°C

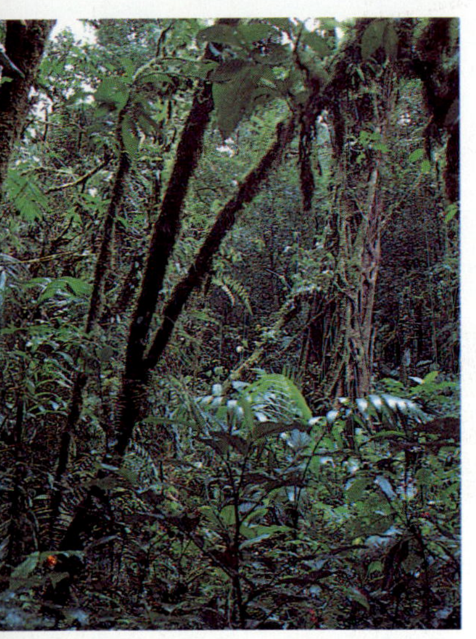

▲
Energy from the sun is used to produce the chemical changes which occur during photosynthesis in the rainforests.

4.7 Energy in Chemical Changes

In all chemical changes matter either absorbs or releases energy. Chemical changes can produce different forms of energy. Electrical energy to start automobiles is produced by chemical changes in the lead storage battery. Light energy for photographic purposes occurs as a flash during the chemical change in the magnesium flashbulb. Heat and light energies are released from the combustion of fuels. All the energy needed for our life processes—breathing, muscle contraction, blood circulation, and so on—is produced by chemical changes occurring within the cells of our bodies.

Conversely, energy is used to cause chemical changes. For example, a chemical change occurs in the electroplating of metals when electrical energy is passed through a salt solution in which the metal is submerged. A chemical change also occurs when radiant energy from the sun is used by green plants in the process of photosynthesis. And, as we saw, a chemical change occurs when heat causes mercury(II) oxide to decompose into mercury and oxygen. Chemical changes are often used primarily to produce energy rather than to produce new substances. The heat or thrust generated by the combustion of fuels is more important than the new substances formed.

4.8 Conservation of Energy

An energy transformation occurs whenever a chemical change occurs. If energy is absorbed during the change, the products will have more chemical potential energy than the reactants. Conversely, if energy is given off in a chemical change, the products will have less chemical potential energy than the reactants. Water, for example, can be decomposed in an electrolytic cell. Electrical energy is absorbed in the decomposition, and the products, hydrogen and oxygen, have a greater chemical potential energy level than that of water. This potential energy is released in the form of heat and light when the hydrogen and oxygen are burned to form water again. Thus, energy can be changed from one form to another or from one substance to another and therefore is not lost.

The energy changes occurring in many systems have been thoroughly studied by many investigators. No system has been found to acquire energy except at the expense of energy possessed by another system. This principle is stated in other words as the **Law of Conservation of Energy**: Energy can be neither created nor destroyed, though it can be transformed from one form to another.

Law of Conservation of Energy

4.9 Interchangeability of Matter and Energy

Sections 4.4–4.8 dealt with matter and energy, which are clearly related; any attempt to deal with one inevitably involves the other. The nature of this relationship eluded the most able scientists until the beginning of the 20th century. Then, in 1905, Albert Einstein (Figure 4.4) presented one of the most original scientific concepts ever devised.

Einstein stated that the quantity of energy E equivalent to the mass m could be calculated by the equation $E = mc^2$. In this equation the energy units of E are joules, m is in grams, and c is the velocity of light (3.0×10^8 m/s). According to Einstein's equation, whenever energy is absorbed or released by a substance, mass is lost or gained. Although the energy changes in chemical reactions are measurable and may appear to be large, the amounts are relatively small. The accompanying difference in mass between reactants and products in chemical changes is so small that it cannot be detected by available measuring instruments. According to Einstein's equation, 9.0×10^7 J of energy are equivalent to 0.0000010 g (1.0 μg) of mass.

$$1\ \mu\text{g mass} = 9.0 \times 10^7\ \text{J energy}$$

In a more practical sense, when 2.8×10^3 g of carbon are burned to carbon dioxide, 9.0×10^7 J of energy are released. Of this very large amount of carbon, only about one-millionth of a gram, which is 3.6×10^{-8}% of the starting mass, is converted to energy. Therefore, in actual practice we may treat the reactants and products of chemical changes as having constant mass. However, because mass and energy are interchangeable, the two laws dealing with the conversation of matter may be combined into a single and generally more accurate statement:

The total amount of mass and energy remains constant during chemical change.

▲
FIGURE 4.4
Albert Einstein (1879–1955).

Chemistry in Action

Energy Balance for Comfort and Convenience

Human beings utilize energy in a variety of ways, but to remain healthy a relatively stable body temperature (37°C) must be maintained. The body tends to lose heat to the surroundings since heat flows from an area of higher temperature to an area of lower temperature. Additional heat energy is used to evaporate moisture and cool our bodies as we perspire. Still more energy is demanded by our daily physical activities. The source for all of this energy is the chemical oxidation of the food we eat.

The energy content of food is determined by burning it in a calorimeter and measuring the heat released. Since the initial substances and the final products of the combustion in the calorimeter are the same as those accomplished through metabolism in the body, a caloric (joule) content can be assigned to each food. Over a long period of time optimum energy requirements (caloric intake) have been determined by classifying people according to age, size, sex, occupation, etc.

If the amount of food taken in contains more energy than we use in heat and daily activities, the excess is stored in the body and weight is gained. If the energy content of the food ingested is less than the heat and activity output, then some of the stored energy is used and weight is lost. We are constantly involved in maintaining the balance between energy lost and energy gained.

Instant hot and cold packs utilize the properties of matter to release or absorb energy as well. When a chemical dissolves in water, energy can be released or absorbed. In a cold pack a small sealed package of ammonium nitrate is placed in a separate sealed pouch containing water. As long as the substances remain separated nothing happens. When the small package is broken, (when the pack is activated) the substances mix and, as the temperature of the solution falls, the solution absorbs heat from the surroundings.

Instant hot packs work in one of two ways. The first type depends on a spontaneous chemical reaction that releases heat energy. In one product an inner paper bag perforated with tiny holes is contained in a plastic envelope. The inner bag contains a mixture of powdered iron, salt, activated charcoal, and dampened sawdust. The heat pack is activated by removing the inner bag, shaking to mix the chemicals, and replacing it in the outer envelope. The heat is the result of a chemical change—that of iron rusting (oxidizing) very rapidly. In the second type of heat pack a physical property of matter is responsible for the production of the heat. The hot pack consists of a tough sealed plastic envelope containing a solution of sodium acetate or sodium thiosulfate. A small crystal is added by squeezing a corner of the pack, or by bending a small metal activator. Crystals then form throughout the solution and release heat to the surroundings. This type of hot pack has the advantages of not being able to overheat and of being reusable. To reuse it is simply heated in boiling water until the crystals dissolve, and then cooled and stored away until needed again.

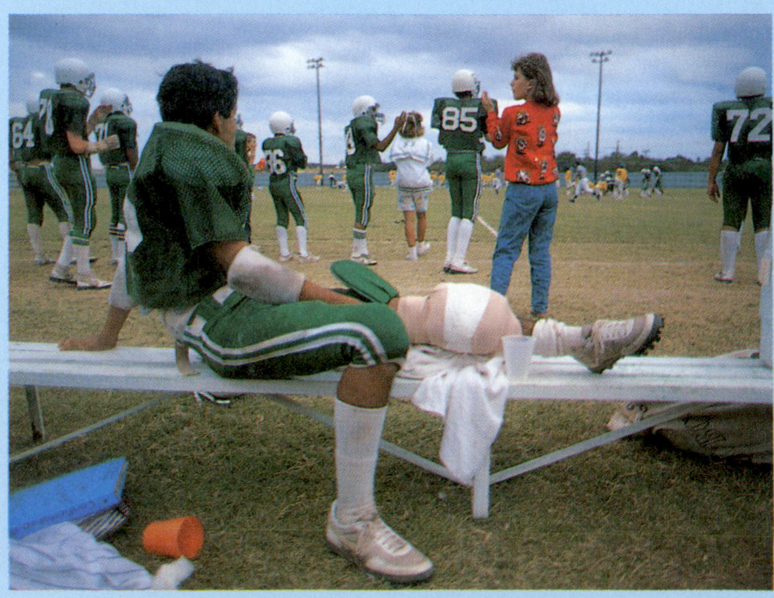

Concepts in Review

1. List the physical properties used to characterize a substance.
2. Distinguish between the physical and chemical properties of matter.
3. Classify changes undergone by matter as either physical or chemical.
4. Distinguish between kinetic and potential energy.
5. State the Law of Conservation of Mass.
6. State the Law of Conservation of Energy.
7. Differentiate clearly between heat and temperature.
8. Make calculations using the equation:

   energy = (mass) × (specific heat) × (Δt)
9. Explain why the laws dealing with the conservation of mass and energy may be combined into a single more accurate general statement.

Exercises

An asterisk indicates a more challenging question or problem.

1. In what physical state does acetic acid exist at 10°C? (See Table 4.1.)
2. In what physical state does chlorine exist at 102 K? (See Table 4.1.)
3. What evidence of chemical change is visible when mercury(II) oxide is heated as shown in Figure 4.2?
4. What chemical changes occur to the matter in the flashcube of Figure 4.3 when it is flashed?
5. What physical changes occur during the electrolysis of water?
6. Distinguish between physical and chemical properties.
7. What is the fundamental difference between a chemical change and a physical change?
8. Classify the following as being primarily physical or primarily chemical changes:
   (a) Formation of a snowflake
   (b) Freezing ice cream
   (c) Boiling water
   (d) Boiling an egg
   (e) Churning cream to make butter
   (f) Souring milk
9. Classify the following as being primarily physical or chemical changes:
   (a) Lighting a candle
   (b) Stirring cake batter
   (c) Dissolving sugar in water
   (d) Decomposition of limestone by heat
   (e) A leaf turning yellow
   (f) Gas escaping from a freshly opened bottle of soda pop
10. Cite the evidence that indicated that only physical changes occurred when a platinum wire was heated in a burner flame.
11. Cite the evidence that indicated that both physical and chemical changes occurred when a copper wire was heated in a burner flame.
12. Cite the evidence that heating mercury(II) oxide brings about a chemical change.
13. Identify the reactants and products for each of the following:
   (a) Heating a copper wire in a burner flame
   (b) Heating mercury(II) oxide as shown in Figure 4.2
   (c) Flashing the flashcube shown in Figure 4.3
14. In a chemical change why can we consider that mass is neither lost nor gained (for practical purposes)?
15. Distinguish between potential and kinetic energy.
16. What happens to the kinetic energy of a speeding automobile when the automobile is braked to a stop?
17. What energy transformation is responsible for the fiery reentry of a returning space vehicle?
18. When the flashcube of Figure 4.3 is flashed, energy is given off to the surroundings. Explain why the mass of the flashcube appears to be the

same after flashing as it was before, even though (according to Einstein) energy is equivalent to mass.

19. Explain how foods are assigned a specific energy (caloric) value.
20. Describe how a "reusable hot pack" works.
21. Which of the following statements are correct? Rewrite the incorrect ones to make them correct.
    (a) An automobile rolling down a hill possesses both kinetic and potential energy.
    (b) When heated in the air, a platinum wire gains mass.
    (c) When heated in the air, a copper wire loses mass.
    (d) 4.184 calories is the equivalent of 1.0 Joule of energy.
    (e) Boiling water represents a chemical change, because a change of state occurs.
    (f) All the following represent chemical changes: baking a cake, frying an egg, leaves changing color, iron changing to rust.
    (g) All of the following represent physical changes: breaking a stick, melting wax, folding a napkin, burning hydrogen to form water.
    (h) Chemical changes can produce electrical energy.
    (i) Electrical energy can produce chemical changes.
    (j) A stretched rubber band possesses kinetic energy.
22. Calculate the boiling point of acetic acid in (a) Kelvins and (b) degrees Fahrenheit. (See Table 4.1)

23. What is the percentage of iron in a mixture that contains 15.0 g of iron, 16.0 g of sulfur, and 18.5 g of sand?
24. How many grams of copper(II) oxide can be obtained from 1.80 g of copper? (See Figure 4.1.)
25. How many grams of copper will combine with 5.50 g of oxygen to form copper(II) oxide?
26. How many grams of mercury can be obtained from 65.0 g of mercury(II) oxide?
27. If 40.0 g of a meat sample contains 3.40 g of fat, what percentage fat is present?
28. When 10.5 g of magnesium was heated in air, 17.4 g of magnesium oxide was produced. Given the chemical reaction

    magnesium + oxygen $\longrightarrow$ magnesium oxide

    (a) What mass of oxygen has combined with the magnesium?
    (b) What percent of the magnesium oxide is magnesium?
29. When aluminum combines with chlorine gas, they produce the substance aluminum chloride. If 4.94 g of aluminum chloride is formed from 1.00 g of aluminum, how many grams of chlorine will combine with 5.50 g of aluminum?

30. How many joules of heat are required to raise the temperature of 80. g of water from 20.°C to 70.°C?

31. How many joules are required to raise the temperature of 80. g of iron from 20.°C to 70.°C? How many calories?
32. A 250. g metal bar requires 5.866 kJ to change its temperature from 22°C to 100.°C. What is the specific heat of the metal in J/g°C?
*33. A 20.0 g piece of a metal at 203°C is dropped into 100. g of water at 25.0°C. The water temperature rises to 29.0°C. Calculate the specific heat (J/g°C) of the metal. Assume that all the heat lost by the metal is transferred to the water and no heat is lost to the surroundings.
*34. Assuming no heat losses by the system, what will be the final temperature when 50. g of water at 10.°C are mixed with 10. g of water at 50.°C?
*35. A 325 g piece of gold at 427°C is dropped into 200. mL of water at 22.0°C. The specific heat of gold is 0.131 J/g°C. Calculate the final temperature of the mixture. (Assume no heat losses to the surroundings.)
*36. The specific heat of zinc is 0.096 cal/g°C. Determine the energy required to raise the temperature of 250. g of zinc from room temperature (24.°C) to 150.°C
*37. If 40,000. J were absorbed by 500. g H_2O at 10.0°C, what would be the final temperature of the H_2O?
*38. Antimony (1.00 kg) absorbed 30.7 kJ, thus raising the temperature of the antimony from 20.°C to its melting point of 630.°C. Calculate the specific heat of antimony.
39. Three 500. g pans of iron, aluminum, and copper were used to fry an egg. Which pan would fry the egg (105°C) the quickest? Explain.
*40. A 500. g iron bar at 212°C is placed in 2.0 L of water at 24.0°C. What will be the change in the temperature of the water?
*41. The heat of combustion of a sample of coal is 5500 cal/g. What quantity of this coal must be burned to heat 500. g of water from 20.°C to 90.°C?
42. The mass of a U.S. 25-cent coin is about 5.5 g.
    (a) How many joules would be released by the complete conversion of a 25-cent coin to energy? How many calories?
    (b) The energy calculated in (a) could heat how many gallons of water from room temperature to the boiling point if 1.27×10^6 J are needed to heat a gallon of water from room temperature (20.°C) to boiling (100.°C)?

5

Early Atomic Theory and Structure

◀ **CHAPTER OPENING PHOTO: A computer model of this complex molecule allows chemists to look at interactions between enzymes and substrates.**

Since ancient times, we have sought to turn substances into gold. The Rumpelstiltskin fairy tale and many other legends surround "alchemy." Yet, the alchemists discovered many elements, and founded the sciences of medicine, pharmacology, and metallurgy.

Pure substances are classified into elements and compounds. But just what makes a substance possess its unique properties? Salt tastes salty, but how small a piece of salt will retain this property? Carbon dioxide puts out fires, is used by plants to produce oxygen, and forms "dry" ice when solidified. But how small a mass of this material still behaves like carbon dioxide?

When substances finally reach the atomic, ionic, or molecular level they are in simplest identifiable form. Further division produces a loss of characteristic properties. What particles lie within an atom or ion? How are these tiny particles alike? How do they differ? How far can we continue to divide them? Alchemists began the quest, early chemists laid the foundation, and the modern chemist continues to build and expand on models of the atom.

5.1 EARLY THOUGHTS

The structure of matter has long intrigued and engaged the minds of people. The seed of modern atomic theory was sown during the time of the ancient Greek philosophers. About 440 B.C. Empedocles stated that all matter was composed of four "elements"—earth, air, water, and fire. Democritus (about 470–370 B.C.), one of the early atomistic philosophers, thought that all forms of matter were finitely divisible into invisible particles, which he called atoms. He held that atoms were in constant motion and that they combined with one another in various ways. This purely speculative hypothesis was not based on scientific observations. Shortly thereafter, Aristotle (384–322 B.C.) opposed the theory of Democritus and endorsed and advanced the Empedoclean theory. So strong was the influence of Aristotle that his theory dominated the thinking of scientists and philosophers until the beginning of the 17th century. The term *atom* is derived from the Greek word *atomos*, meaning indivisible.

5.2 Dalton's Atomic Theory

More than 2000 years after Democritus, the English schoolmaster John Dalton (1766–1844) revived the concept of atoms and proposed an atomic theory based on facts and experimental evidence. This theory, described in a series of papers published during the period 1803–1810, rested on the idea of a different kind of atom for each element. The essence of **Dalton's atomic theory** may be summed up as follows:

Dalton's atomic theory

1. Elements are composed of minute, indivisible particles called atoms.
2. Atoms of the same element are alike in mass and size.
3. Atoms of different elements have different masses and sizes.
4. Chemical compounds are formed by the union of two or more atoms of different elements.
5. Atoms combine to form compounds in simple numerical ratios such as one to one, two to one, two to three, and so on.
6. Atoms of two elements may combine in different ratios to form more than one compound.

Dalton's atomic theory stands as a landmark in the development of chemistry. The major premises of his theory are still valid today. However, some of the statements must be modified or qualified because investigations since Dalton's time have shown that (1) atoms are composed of subatomic particles; (2) not all the atoms of a specific element have the same mass; and (3) atoms, under special circumstances, can be decomposed.

5.3 Composition of Compounds

A large number of experiments extending over a long period of time have established the fact that a particular compound always contains the same elements in the same proportions by mass. For example, water will always contain 11.2% hydrogen and 88.8% oxygen by mass. The fact that water contains hydrogen and oxygen in this particular ratio does not mean that hydrogen and oxygen cannot combine in some other ratio. However, a compound with a different ratio would not be water. In fact, hydrogen peroxide is made up of two atoms of hydrogen and two atoms of oxygen per molecule and contains 5.9% hydrogen and 94.1% oxygen by mass; its properties are markedly different from those of water.

	Water	Hydrogen peroxide
Percent H	11.2	5.9
Percent O	88.8	94.1
Atomic composition	2 H + 1 O	2 H + 2 O

▲ John Dalton (1766–1844)

Law of Definite Composition

The **Law of Definite Composition** states: A compound always contains two or more elements combined in a definite proportion by mass.

Let us consider two elements, oxygen and hydrogen, that form more than one compound. In water there are 8.0 g of oxygen for each gram of hydrogen. In hydrogen peroxide there are 16.0 g of oxygen for each gram of hydrogen. The masses of oxygen are in the ratio of small whole numbers, 16:8 or 2:1. Hydrogen peroxide has twice as much hydrogen (by mass) as does water. Using Dalton's atomic theory, we deduce that hydrogen peroxide has twice as many oxygens per hydrogen as water. In fact, we now write the formulas for water as H_2O and for hydrogen peroxide as H_2O_2.

Law of Multiple Proportions

The **Law of Multiple Proportions** states: Atoms of two or more elements may combine in different ratios to produce more than one compound. The reliability of this law and the law of Definite Composition is the cornerstone of the science of chemistry. In essence these laws state that (1) the composition of a particular substance will always be the same no matter what its origin or how it is formed, and (2) the composition of different compounds formed from the same elements will always be unique.

5.4 The Nature of Electric Charge

Many of us have received a shock after walking across a carpeted area on a dry day. We have also experienced the static associated with combing our hair, and have had our clothing cling to us. All of these phenomena result from an accumulation of *electric charge*. This charge may be transferred from one object to another. The properties of electric charge follow:

1. Charge may be of two types, positive and negative.
2. Unlike charges attract (positive attracts negative) and like charges repel (negative repels negative and positive repels positive).
3. Charge may be transferred from one object to another, by contact or induction.
4. The less distance between two charges the greater the force of attraction between unlike charges (or repulsion between identical charges).

5.5 Discovery of Ions

▲ Michael Faraday (1791–1867)

The great English scientist Michael Faraday (1791–1867) made the discovery that certain substances when dissolved in water could conduct an electric current. He also noticed that certain compounds could be decomposed into their elements by passing an electric current through the compound. Atoms of some elements were attracted to the positive electrode, while atoms of other elements were attracted to the negative electrode. Faraday concluded that these atoms were electrically charged. He called them *ions* after the Greek word, meaning "wanderer".

Any moving charge is an electric current. The electrical charge must travel through a substance known as a conducting medium. The most familiar conducting media are metals that are used in the form of wires.

The Swedish scientist Svante Arrhenius (1859–1927) extended Faraday's work. Arrhenius reasoned that an ion was an atom carrying a positive or negative charge. When a compound, such as sodium chloride (NaCl), was melted it conducted electricity. Water was unnecessary. Arrhenius' explanation of this conductivity was that upon melting the sodium chloride dissociated, or broke up, into charged ions, Na^+ and Cl^-. The Na^+ ions moved toward the negative electrode (cathode), whereas the Cl^- migrated toward the positive electrode (anode). Thus, positive ions are cations, and negative ions are anions.

From Faraday's and Arrhenius' work with ions, Irish physicist G. J. Stoney (1826–1911) realized there must be some fundamental unit of electricity associated with atoms. He named this unit the *electron* in 1891. Unfortunately he had no means of supporting his idea with experimental proof. Evidence remained elusive until 1897 when English physicist J. J. Thomson was able to show experimentally the existence of the electron.

5.6 Subatomic Parts of the Atom

The concept of the atom—a particle so small that, until recently, it could not be seen even with the most powerful microscope—and the subsequent determination of its structure stand among the very greatest creative intellectual human achievements.

Any visible quantity of an element contains a vast number of identical atoms. But when we refer to an atom of an element, we isolate a single atom from the multitude in order to present the element in its simplest form. Figure 5.1 illustrates the hypothetical isolation of a single copper atom from its crystalline lattice.

Let us examine this tiny particle we call the atom. The diameter of a single atom ranges from 0.1 to 0.5 nanometers (1 nm = 1×10^{-9} m). Hydrogen, the smallest atom, has a diameter of about 0.1 nm. To arrive at some idea of how small an atom is, consider this dot (•), which has a diameter of about 1 mm, or 1×10^6 nm. It would take 10 million hydrogen atoms to form a line of atoms across this dot. As inconceivably small as atoms are, they contain smaller particles, the **subatomic particles**, such as electrons, protons, neutrons.

subatomic particles

The development of atomic theory was helped in large part by the invention of new instruments. The Crookes tube, developed by Sir William Crookes in 1875, opened the door to the subatomic structure of the atom (Figure 5.2). The emissions generated in a Crookes tube are called *cathode rays.* Thomson (1856–1940) demonstrated in 1897 that cathode rays (1) travel in straight lines, (2) are negative in charge, (3) are deflected by electric and magnetic fields, (4) produce sharp shadows, and (5) are capable of moving a small paddle wheel. This was the experimental discovery of the fundamental unit of charge, the electron.

electron

The **electron** (e^-) is a particle with a negative electrical charge and a mass of 9.110×10^{-28} g. This mass is 1/1837 the mass of a hydrogen atom and

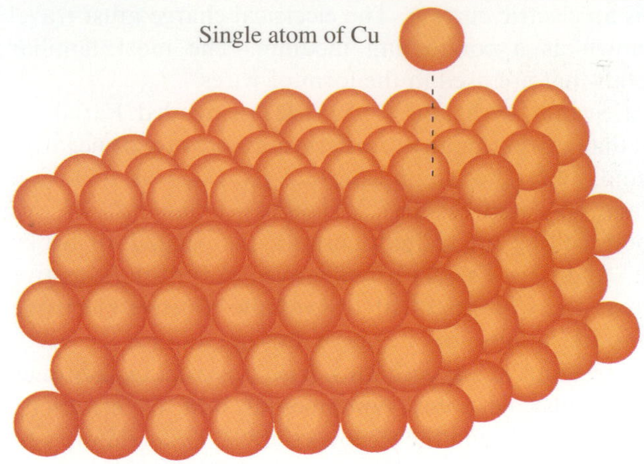

FIGURE 5.1 ▶
Left: A single atom of copper compared with copper as it occurs in its regular crystalline lattice structure. *Right:* A scanning tunneling microscope shows an array of copper atoms.

corresponds to 0.0005486 atomic mass unit (amu) (defined in Section 5.11). One atomic mass unit has a mass of 1.6606×10^{-24} g. Although the actual charge of an electron is known, its value is too cumbersome for practical use and therefore has been assigned a relative electrical charge of -1. The size of an electron has not been determined exactly, but its diameter is believed to be less than 10^{-12} cm.

Protons were first observed by German physicist E. Goldstein (1850–1930) in 1886. However, it was Thomson who discovered the nature of the proton. He showed that the proton was a particle, and he calculated its mass to be about 1837 times that of an electron. The **proton** (p) is a particle with a relative mass of 1 amu and an actual mass of 1.673×10^{-24} g. Its relative charge $(+1)$ is equal in magnitude, but of opposite sign, to the charge on the electron. The mass of a proton is only very slightly less than that of a hydrogen atom.

proton

Thomson had shown that atoms contained both negatively and positively charged particles. Clearly, the Dalton model of the atom was no longer acceptable. Atoms were not indivisible. They were composed of smaller parts. Thomson proposed a new model of the atom.

Thomson model of the atom

In the **Thomson model of the atom** the electrons were negatively charged particles imbedded in the atomic sphere. Since atoms were electrically neutral the sphere also contained an equal number of protons, or positive charges. A neutral atom could become an ion by gaining or losing electrons.

Positive ions were explained by assuming that the neutral atom had lost electrons. An atom with a net charge of $+1$ (for example, H^+ or Li^+) would have lost one electron. An atom with a net charge of $+3$ (for example, Al^{3+}) would have lost three electrons.

Negative ions were explained by assuming that extra electrons had been added to the atoms. A net charge of -1 (for example, Cl^- or F^-) would be produced by the addition of one electron. A net charge of -2 (for example, O^{2-} or S^{2-}) would be explained as the addition of two electrons.

neutron

The third major subatomic particle was discovered in 1932 by James Chadwick (1891–1974). This particle, the **neutron** (n), bears neither a positive

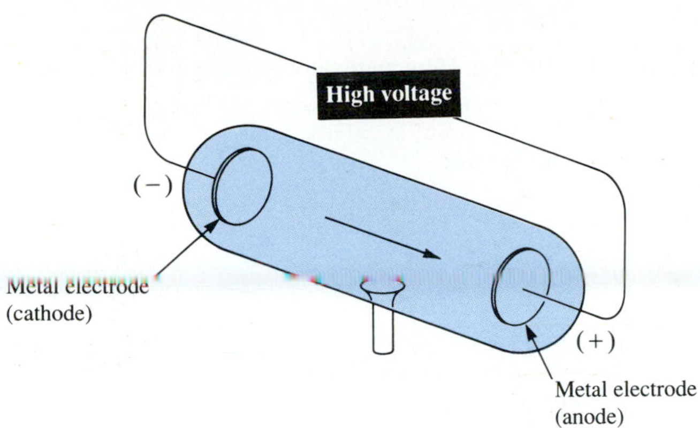

▲
FIGURE 5.2
Crookes tube. In a Crookes tube a gas is contained at very low pressure. When a current is applied across electrodes within the tube a stream of electrons travels from the cathode toward the anode.

TABLE 5.1 Electrical Charge and Relative Mass of Electrons, Protons, and Neutrons

Particle	Symbol	Relative electrical charge	Relative mass (amu)	Actual mass (g)
Electron	e^-	-1	$\frac{1}{1837}$	9.110×10^{-28}
Proton	p	$+1$	1	1.673×10^{-24}
Neutron	n	0	1	1.675×10^{-24}

nor a negative charge and has a relative mass of about 1 amu. Its actual mass (1.675×10^{-24} g) is only very slightly greater than that of a proton. The properties of these three subatomic particles are summarized in Table 5.1.

Nearly all the ordinary chemical properties of matter can be explained in terms of atoms consisting of electrons, protons, and neutrons. The discussion of atomic structure that follows is based on the assumption that atoms contain only these principal subatomic particles. Many other subatomic particles such as mesons, positrons, neutrinos, and antiprotons have been discovered. At this time it is not clear whether all these particles are actually present in the atom or whether they are produced by reactions occurring within the nucleus. The fields of atomic and particle or high-energy physics are fascinating and have attracted many young scientists in recent years. This interest has resulted in a great deal of research that is producing a long list of subatomic particles. Descriptions of the properties of many of these particles are to be found in recent physics textbooks and in various articles appearing in current issues of *Scientific American.*

▲
Joseph Thomson (1856–1940)

EXAMPLE 5.1

The mass of a helium atom is 6.65×10^{-24} g. How many atoms are in a 4.0 g sample of helium?

SOLUTION

$$4.0 \text{ g} \times \frac{1 \text{ atom}}{6.65 \times 10^{-24} \text{ g}} = 6.0 \times 10^{23} \text{ atoms}$$

PRACTICE The mass of an atom of hydrogen is 1.673×10^{-24} g. How many atoms are in a 10.0 g sample of hydrogen?

Answer: 5.98×10^{24} atoms

5.7 THE NUCLEAR ATOM

The discovery that positively charged particles were present in atoms came soon after the discovery of radioactivity by Henri Becquerel in 1896. Radioactive elements spontaneously emit alpha, beta, and gamma rays from their nuclei (see Chapter 19).

Ernest Rutherford (Figure 5.3) had, by 1907, established that the positively charged alpha particles emitted by certain radioactive elements were ions of the element helium. Rutherford used these alpha particles to establish the nuclear nature of atoms. In experiments performed in 1911, he directed a stream of positively charged helium ions (alpha particles) at a very thin sheet of gold foil (about 1000 atoms thick). See Figure 5.4(a). He observed that most of the alpha particles passed through the foil with little or no deflection; but a few of the particles were deflected at large angles, and occasionally one even bounced back from the foil (Figure 5.4b). It was known that like charges repel each other and that an electron with a mass of 1/1837 amu could not possibly have an appreciable effect on the path of a 4 amu alpha particle, which is about 7350 times more massive than an electron. Rutherford therefore reasoned that each gold atom must contain a positively charged mass occupying a relatively tiny volume and that, when an alpha particle approached close enough to this positive mass, it was deflected. Rutherford spoke of this positively charged mass as the *nucleus* of the atom. Because alpha particles have relatively high masses, the extent of the deflections (some actually bounced back) indicated to Rutherford that the nucleus is relatively very heavy and dense. (The density of the nucleus of a hydrogen atom is about 10^{12} g/cm^3, about one trillion times the density of water.) Because most of the alpha particles passed through the thousand or so gold atoms without any apparent deflection, he further concluded that most of an atom consists of empty space.

▲
FIGURE 5.3
Ernest Rutherford (1871–1937).

When we speak of the mass of an atom, we are, for practical purposes, referring primarily to the mass of the nucleus. The nucleus contains all the

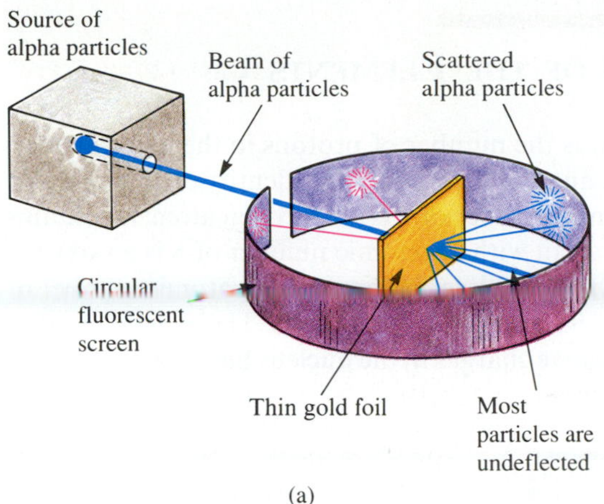

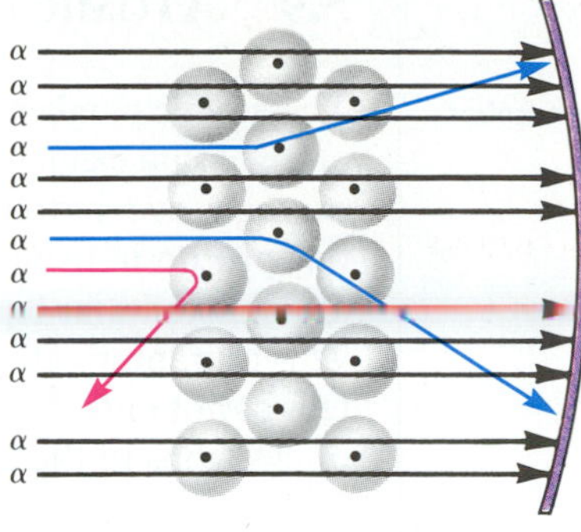

▲ **FIGURE 5.4**
(a) Diagram representing Rutherford's experiment on alpha particle scattering. (b) Positive alpha particles (α), emanating from a radioactive source, were directed at a thin gold foil. Diagram illustrates the deflection and repulsion of the positive alpha particles by the positive nuclei of the gold atoms.

protons and neutrons, which represent more than 99.9% of the total mass of any atom (see Table 5.1). By way of illustration, the largest number of electrons known to exist in an atom is 109. The mass of even 109 electrons is only about 1/17 of the mass of a single proton or neutron. The mass of an atom, therefore, is primarily determined by the combined masses of its protons and neutrons.

5.8 GENERAL ARRANGEMENT OF SUBATOMIC PARTICLES

The alpha particle scattering experiments of Rutherford established that the atom contains a dense, positively charged nucleus. The later work of Chadwick demonstrated that the atom contains neutrons, which are particles with mass but no charge. He also noted that light, negatively charged electrons are present and offset the positive charges in the nucleus. Based on this experimental evidence, a general description of the atom and the location of its subatomic particles was devised. Each atom consists of a **nucleus** surrounded by electrons. The nucleus contains protons and neutrons but not electrons. In a neutral atom the positive charge of the nucleus (due to protons) is exactly offset by the negative electrons. Because the charge of an electron is equal but of opposite sign to the charge of a proton, a neutral atom must contain exactly the same number of electrons as protons. However, this generalized picture of atomic structure provides no information on the arrangement of electrons within the atom. Electron configuration is discussed in Chapter 10.

nucleus

A neutral atom contains the same number of protons and electrons.

5.9 Atomic Numbers of the Elements

atomic number

The **atomic number** of an element is the number of protons in the nucleus of an atom of that element. The atomic number determines the identity of an atom. For example, every atom with an atomic number of 1 is a hydrogen atom; it contains one proton in its nucleus. Every atom with an atomic number of 8 is an oxygen atom; it contains 8 protons in its nucleus. Every atom with an atomic number of 92 is a uranium atom; it contains 92 protons in its nucleus. The atomic number tells us not only the number of positive charges in the nucleus but also the number of electrons in the neutral atom.

atomic number = number of protons in the nucleus

There is no need to memorize the atomic numbers of the elements as a periodic table is commonly provided in texts, laboratories, and on examinations. The atomic numbers of all elements are shown in the Periodic Table on the inside front cover of this book and are also listed in the Table of Atomic Masses on the inside front cover.

5.10 Isotopes of the Elements

Shortly after Rutherford's conception of the nuclear atom, experiments were performed to determine the masses of individual atoms. These experiments showed that the masses of nearly all atoms were greater than could be accounted for by simply adding up the masses of all the protons and electrons that were known to be present in an atom. This fact led to the concept of the neutron, a particle with no charge but with a mass about the same as that of a proton. Because this particle has no charge, it was very difficult to detect, and the existence of the neutron was not proven experimentally until 1932. All atomic nuclei except that of the simplest hydrogen atom are now believed to contain neutrons.

All atoms of a given element have the same number of protons, but experimental evidence has shown that, in most cases, all atoms of a given element do not have identical masses because atoms of the same element may have different numbers of neutrons in their nuclei.

isotopes

Atoms of an element having the same atomic number but different atomic masses are called **isotopes** of that element. Atoms of the various isotopes of an element, therefore, have the same number of protons and electrons but different numbers of neutrons.

Three isotopes of hydrogen (atomic number 1) are known. Each has one proton in the nucleus and one electron. The first isotope (protium), without a neutron, has a mass of 1; the second isotope (deuterium), with one neutron in the nucleus, has a mass of 2; the third isotope (tritium), with two neutrons, has a mass of 3 (see Figure 5.5).

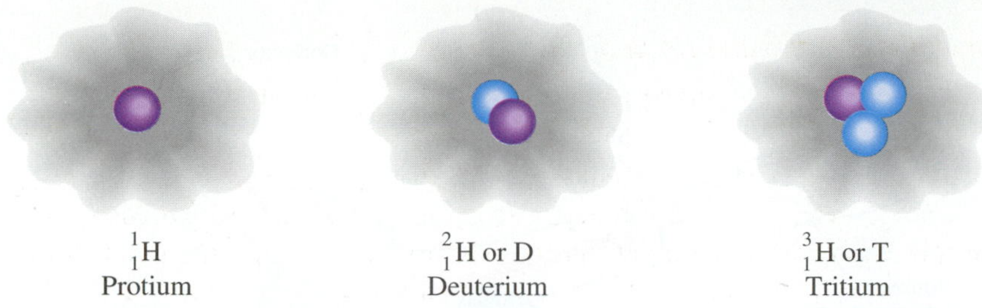

◄ **FIGURE 5.5** **Schematic diagram of the isotopes of hydrogen. The number of protons (purple) and neutrons (blue) are shown within the nucleus. The electron (e^-) exists outside the nucleus (gray).**

The three isotopes of hydrogen may be represented by the symbols ^1_1H, ^2_1H, ^3_1H, indicating an atomic number of 1 and mass numbers of 1, 2, and 3, respectively. This method of representing atoms is called *isotopic notation.* The subscript (Z) is the atomic number; the superscript (A) is the **mass number**, which is the sum of the number of protons and the number of neutrons in the nucleus.

mass number

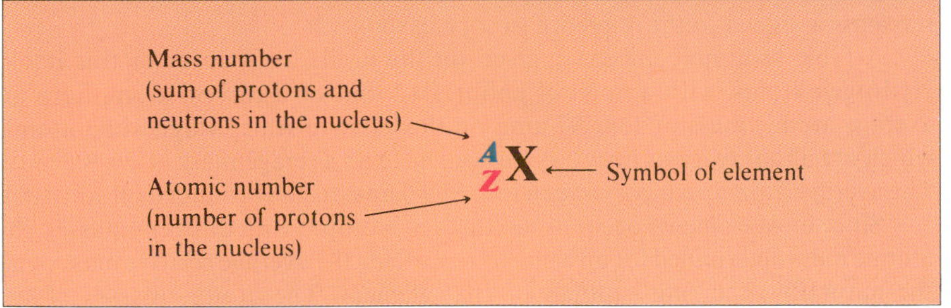

The hydrogen isotopes may also be referred to as hydrogen-1, hydrogen-2, and hydrogen-3.

Most of the elements occur in nature as mixtures of isotopes. However, not all isotopes are stable; some are radioactive and are continuously decomposing to form other elements. For example, of the seven known isotopes of carbon, only two, carbon-12 and carbon-13, are stable. Of the seven known isotopes of oxygen, only three, $^{16}_8O$, $^{17}_8O$, and $^{18}_8O$, are stable. Of the fifteen known isotopes of arsenic, $^{75}_{33}As$ is the only one that is stable.

5.11 Atomic Mass (Atomic Weight)

The mass of a single atom is far too small to measure individually on a balance. But fairly precise determinations of the masses of individual atoms can be made with an instrument called a *mass spectrometer* (see Figure 5.6). The mass of a single hydrogen atom is 1.6736×10^{-24} g. However, it is neither convenient nor practical to compare the actual masses of atoms expressed in grams; therefore a table of relative atomic masses using *atomic mass units* was devised. (The term *atomic weight* is often used instead of atomic mass.) The carbon isotope, having six protons and six neutrons and designated carbon-12, or $^{12}_6C$, was chosen as the standard for atomic masses. This reference isotope was assigned a value of exactly 12 atomic mass units (amu). Thus one **atomic mass unit** is defined as equal

atomic mass unit

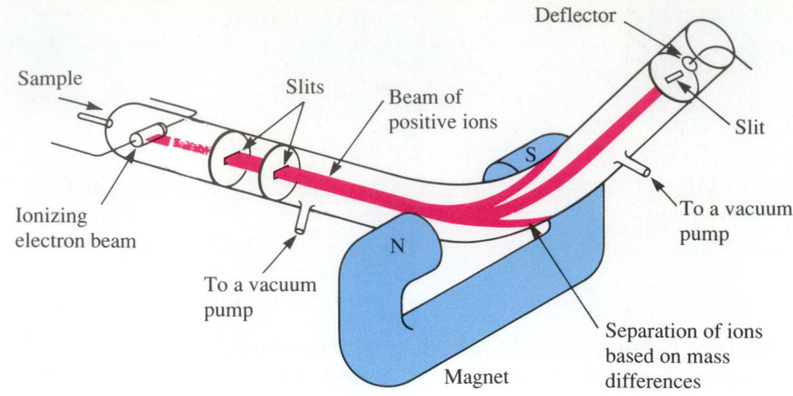

FIGURE 5.6 ► Schematic diagram of a modern mass spectrometer. A beam of positive ions is produced from the sample as it enters the chamber. The positive ions are then accelerated as they pass through slits in an electric field. When the ions enter the magnetic field, they are deflected differently, depending on mass and charge. The ions are then detected at the end of the tube. From intensity and position of the lines on the mass spectrogram, the different isotopes of the elements and their relative amounts can be determined.

to exactly 1/12 of the mass of a carbon-12 atom. The actual mass of a carbon-12 atom is 1.9927×10^{-23} g, and that of one atomic mass unit is 1.6606×10^{-24} g. In the table of atomic masses all elements then have values that are relative to the mass assigned to the reference isotope carbon-12.

A table of atomic masses is given on the inside front cover of this book. Hydrogen atoms, with a mass of about 1/12 that of a carbon atom, have an average atomic mass of 1.00797 amu on this relative scale. Magnesium atoms, which are about twice as heavy as carbon, have an average mass of 24.305 amu. The average atomic mass of oxygen is 15.9994 amu (usually rounded off to 16.00).

Since most elements occur as mixtures of isotopes with different masses, the atomic mass determined for an element represents the average relative mass of all the naturally occurring isotopes of that element. The atomic masses of the individual isotopes are approximately whole numbers, because the relative masses of the protons and neutrons are approximately 1.0 amu each. Yet we find that the atomic masses given for many of the elements deviate considerably from whole numbers. For example, the atomic mass of rubidium is 85.4678 amu, that of copper is 63.546 amu, and that of magnesium is 24.305 amu. The deviation of an atomic mass from a whole number is due mainly to the unequal occurrence of the various isotopes of an element. It is also due partly to the difference between the mass of a free proton or neutron and the mass of these same particles in the nucleus. For example, the two principal isotopes of copper are $^{63}_{29}Cu$ and $^{65}_{29}Cu$. It is apparent that copper-63 atoms are the more abundant isotope, since the atomic mass of copper, 63.546 amu, is closer to 63 than to 65 amu (see Figure 5.7). The actual values of the copper isotopes observed by mass spectra determination are shown here:

Isotope	Isotopic mass (amu)	Abundance (%)	Average atomic mass (amu)
$^{63}_{29}Cu$	62.9298	69.09	63.55
$^{65}_{29}Cu$	64.9278	30.91	

The average atomic mass can be calculated by multiplying the atomic mass of each isotope by the fraction of each isotope present and adding the results. The

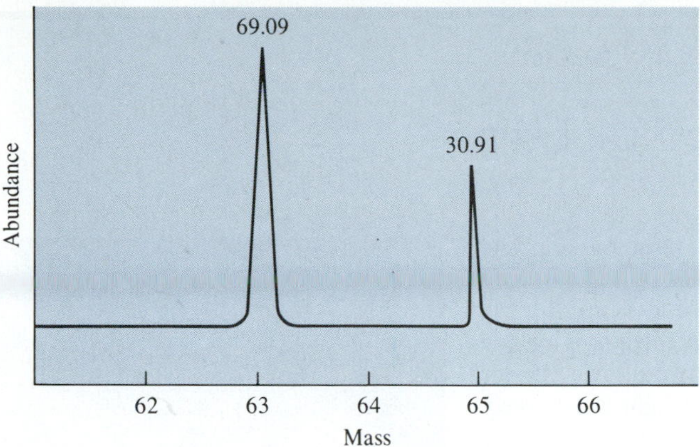

◀ **FIGURE 5.7**
A typical reading from a mass spectrometer. The two principal isotopes of copper are shown with the abundance (%) given.

calculation for copper is

$$62.9298 \text{ amu} \times 0.6909 = 43.48 \text{ amu}$$
$$64.9278 \text{ amu} \times 0.3091 = \underline{20.07} \text{ amu}$$
$$63.55 \text{ amu}$$

atomic mass

The **atomic mass** of an element is the average relative mass of the isotopes of that element referred to the atomic mass of carbon-12 (exactly 12.0000 amu).

The relationship between mass number and atomic number is such that, if we subtract the atomic number from the mass number of a given isotope, we obtain the number of neutrons in the nucleus of an atom of that isotope. Table 5.2 shows the application of this method of determining the number of neutrons. For example, the fluorine atom ($^{19}_{9}F$), atomic number 9, having a mass of 19 amu, contains 10 neutrons:

$$\text{Mass number} - \text{Atomic number} = \text{Number of neutrons}$$
$$19 - 9 = 10$$

The atomic masses given in the table on the inside front cover of this book are values accepted by international agreement. You need not memorize atomic masses. In most of the calculations needed in this book, use of atomic masses to four significant figures will give results of sufficient accuracy.

TABLE 5.2 Determination of the Number of Neutrons in an Atom by Subtracting Atomic Number from Mass Number

	Hydrogen ($^{1}_{1}H$)	Oxygen ($^{16}_{8}O$)	Sulfur ($^{32}_{16}S$)	Fluorine ($^{19}_{9}F$)	Iron ($^{56}_{26}Fe$)
Mass number	1	16	32	19	56
Atomic number	(−)1	(−)8	(−)16	(−)9	(−)26
Number of neutrons	0	8	16	10	30

Chemistry in Action

Triboluminescence

Some substances give off light when they are rubbed, crushed, or broken in a phenomenon called triboluminescence. Examples of substances exhibiting this property include crystals of quartz, sugar cubes, adhesive tape torn off certain surfaces, and Wintergreen Lifesavers.

Linda M. Sweeting, a chemist at Towson State University in Maryland, investigated the sparks created by a Wintergreen Lifesaver when crushed inside the mouth in a dark room. When a sugar crystal is fractured separate areas of positive and negative charge form on opposite sides of the crack. Electrons tend to leap the gap and neutralize the charge. When the electrons collide with nitrogen molecules in the air small amounts of light are emitted. (Lightning is a somewhat similar phenomenon but on a much grander scale.) Addition of wintergreen molecules changes the outcome though because it absorbs some of the light energy from the electron leap and re-emits it as bright blue-green flashes. These are the sparks we see as we crush a Lifesaver in the dark.

Triboluminescence of Wintergreen Lifesavers.

Example 5.2 How many protons, neutrons, and electrons are found in an atom of ${}^{14}_{6}C$?

SOLUTION The element is carbon, atomic number 6. The number of protons or electrons equals the atomic number and is 6. The number of neutrons is determined by subtracting the atomic number from the mass number: $14 - 6 = 8$.

Practice How many protons, neutrons, and electrons are in each of the following?

(a) ${}^{16}_{8}O$ (b) ${}^{80}_{35}Br$ (c) ${}^{235}_{92}U$ (d) ${}^{64}_{29}Cu$

Answers:		protons	neutrons	electrons
	(a)	8	8	8
	(b)	35	45	35
	(c)	92	143	92
	(d)	29	35	29

Chemistry in Action

Chemical Fraud

The bottle of vanilla flavoring found in most kitchens may be labeled either as "vanilla extract" or as "imitation vanilla extract." Vanilla extract is an alcohol and water solution of materials extracted from vanilla beans. Imitation vanilla extract is made from lignin, a waste product in the wood pulp industry, which is converted to vanillin—the same molecule found in natural vanilla extract. The advantage of this process lies in monetary savings. The cost of producing natural vanilla is approximately $125 per gallon, whereas production of synthetic vanillin is approximately $60 per gallon.

One of the roles of a government chemist is to prevent fraudulent mislabeling of products. Since the compounds in both products are chemically the same, another method must be used to detect the source of the molecule. This is done by inspecting the carbon atoms in the vanillin, using a technique called stable isotope ratio analysis (SIRA). Two isotopes in organic compounds are carbon-12 and carbon-13. The ratio of these isotopes is slightly different for natural vanillin and imitation vanillin. It is from these slight differences that government scientists distinguish whether the vanillin is from the vanilla bean or from the lignin.

Example 5.3

Chlorine is found in nature as two isotopes, $^{37}_{17}Cl$ (24.47%) and $^{35}_{17}Cl$ (75.53%). The atomic masses are 36.96590 amu and 34.96885 amu, respectively. Determine the average atomic mass of chlorine.

Solution

Multiply each mass by its percentage and add the results to find the average.

$$(0.2447) \times (36.96590 \text{ amu}) + (0.7553) \times (34.96885 \text{ amu})$$
$$= 35.4575 \text{ amu}$$
$$= 35.46 \text{ amu (4 significant figures)}$$

Practice Silver is found in two isotopes with atomic masses of 106.9041 and 108.9047 amu, respectively. The first isotope represents 51.82% and the second 48.18%. Determine the average atomic mass of silver.

Answer: 107.9 amu

Concepts in Review

1. State the major provisions of Dalton's atomic theory.
2. State the Law of Definite Composition and indicate its significance.
3. State the Law of Multiple Proportions and indicate its significance.
4. Give the names, symbols, and relative masses of the three principal subatomic particles.
5. Describe the Thomson model of the atom.
6. Describe the atom as conceived by Ernest Rutherford after his alpha-scattering experiment.
7. Determine the atomic number, mass number, or number of neutrons of an isotope when given the values of any two of these three items.
8. Name and distinguish among the three isotopes of hydrogen.
9. Calculate the average atomic mass of an element, given the isotopic masses and the abundance of its isotopes.
10. Determine the number of protons, neutrons, and electrons from the atomic number and atomic mass of an atom.

Exercises

An asterisk indicates a more challenging question or problem.

1. What are the atomic numbers of (a) copper, (b) nitrogen, (c) phosphorus, (d) radium, and (e) zinc?
2. A neutron is approximately how many times heavier than an electron?
3. Explain why, in Rutherford's experiments, some alpha particles were scattered at large angles by the gold foil or even bounced back from it?

Atomic Structure

4. From the point of view of a chemist what are the essential differences among a proton, a neutron, and an electron?
5. Describe the general arrangement of subatomic particles in the atom.
6. What part of the atom contains practically all its mass?
7. What experimental evidence led Rutherford to conclude each of the following?
   (a) The nucleus of the atom contains most of the atomic mass.
   (b) The nucleus of the atom is positively charged.
   (c) The atom consists of mostly empty space.
8. What contribution did each of the following scientists make to the atomic theory?
   (a) Dalton
   (b) Thomson
   (c) Rutherford
9. Which of the following statements are correct? Rewrite the incorrect statements to make them correct.
   (a) John Dalton developed an important atomic theory in the early 1800s.
   (b) Dalton said that elements are composed of minute indivisible particles called atoms.
   (c) Dalton said that when atoms combine to form compounds, they do so in simple numerical ratios.
   (d) Dalton said that atoms are composed of protons, neutrons, and electrons.
   (e) All of Dalton's theory is still considered valid today.
   (f) Hydrogen is the smallest atom.
   (g) A proton is about 1837 times as heavy as an electron.
   (h) The nucleus of an atom contains protons, neutrons, and electrons.
10. What are the numbers of protons, neutrons, and electrons in each of the following?
    (a) $^{79}_{35}Br$ (b) $^{131}_{56}Ba$ (c) $^{238}_{92}U$ (d) $^{56}_{26}Fe$

11. What letters are used to designate atomic number and mass number in isotopic notation of atoms?
12. Write isotopic notation symbols for the following:
    (a) $Z = 26, A = 55$ (d) $Z = 14, A = 29$
    (b) $Z = 12, A = 26$ (e) $Z = 79, A = 188$
    (c) $Z = 3, A = 6$
13. Give the isotopic notation ($^{73}_{32}Ge$, for example) for:
    (a) An atom containing 27 protons, 32 neutrons, and 27 electrons.
    (b) An atom containing 110 neutrons, 74 electrons, and 74 protons.
14. Which of the following statements are correct? Rewrite the incorrect statements to make them correct.
    (a) An element with an atomic number of 29 has 29 protons, 29 neutrons, and 29 electrons.
    (b) An atom of the isotope $^{60}_{26}Fe$ has 34 neutrons in its nucleus.
    (c) $^{2}_{1}H$ is a symbol for the isotope deuterium.
    (d) An atom of $^{31}_{15}P$ contains 15 protons, 16 neutrons, and 31 electrons.
    (e) In the isotope $^{6}_{3}Li$, $Z = 3$ and $A = 3$.
    (f) Isotopes of a given element have the same number of protons but differ in the number of neutrons.
    (g) The three isotopes of hydrogen are called protium, deuterium, and tritium.
    (h) $^{23}_{11}Na$ and $^{24}_{11}Na$ are isotopes.
    (i) $^{24}_{11}Na$ has one more electron than $^{23}_{11}Na$.
    (j) $^{24}_{11}Na$ has one more proton than $^{23}_{11}Na$.
    (k) $^{24}_{11}Na$ has one more neutron than $^{23}_{11}Na$.
    (l) Only a few of the elements exist in nature as mixtures of isotopes.

Atomic mass

15. Explain why the atomic masses of elements are not whole numbers.
16. Is the isotopic mass of a given isotope ever an exact whole number? Is it always? In answering, consider the masses of $^{12}_{6}C$ and $^{63}_{29}Cu$.
17. Which of the isotopes of calcium in Exercise 23 is the most abundant isotope? Can you be sure? Explain your choice.
18. In what ways are isotopes alike? In what ways are they different?
19. What special names are given to the isotopes of hydrogen?
20. List the similarities and differences in the three isotopes of hydrogen.
21. What is the symbol and name of the element that has an atomic number of 24 and a mass number of 52.
22. An atom of an element has a mass number of 201 and has 121 neutrons in its nucleus.
    (a) What is the electrical charge of the nucleus?
    (b) What is the symbol and name of the element?
23. What is the nuclear composition of the six naturally occurring isotopes of calcium having mass numbers of 40, 42, 43, 44, 46, and 48?
24. Lanthanum exists in four naturally occurring isotopes. Explain how these isotopes affect the atomic mass of lanthanum.
25. Distinguish between an atom and an ion.
26. Indicate the appropriate atomic mass of an element with 42 protons, 54 neutrons, and 42 electrons.
27. Define an isotope and state two examples.
28. Which of the following statements are correct? Rewrite the incorrect statements to make them correct.
    (a) One atomic mass unit has a mass 12 times as much as one carbon-12 atom.
    (b) The atomic masses of protium and deuterium differ by about 100%.
    (c) $^{23}_{11}Na$ and $^{24}_{11}Na$ have the same atomic masses.
    (d) The atomic mass of an element represents the average relative atomic mass of all the naturally occurring isotopes of that element.
29. Indicate the relationship between triboluminescence and wintergreen lifesavers.
30. Differentiate between synthetic and natural vanilla extract. Explain why this is important in our society.
31. Describe and indicate the value of the SIRA technique.
32. Using the periodic table at the front of the book, determine which of the first twenty elements have the same number of protons, neutrons, and electrons.
33. Complete the following table with the appropriate data for each isotope given:

Atomic number	Mass number	Symbol of element	Number of protons	Number of neutrons
(a) 8	16			
(b)		Ni		30
(c)	199		80	

*34. The actual mass of one atom of an unknown isotope is 2.18×10^{-22} amu. Calculate the atomic mass of this isotope.

35. Naturally occurring silver exists as two stable isotopes, ^{107}Ag with a mass of 106.9041 amu (51.82%) and ^{109}Ag with a mass of 108.9047 amu (48.18%). Calculate the average atomic mass of silver.

36. Naturally occurring magnesium consists of three stable isotopes: ^{24}Mg, 23.985 amu (78.99%); ^{25}Mg, 24.986 amu (10.00%); and ^{26}Mg, 25.983 amu (11.01%). Calculate the average atomic mass of magnesium

37. The mass of an atom of argon is 6.63×10^{-24} g. How many atoms are in a 40.0-g sample of argon?

38. 68.9257 amu is the mass of 60.4% of the atoms of an element with only two naturally occurring isotopes. The atomic mass of the other isotope is 70.9249 amu. Determine the average atomic mass of the element. Identify the element.

39. Complete the following table:

Element	Symbol	Atomic number	Number of protons	Number of neutrons	Number of electrons
(a) Platinum					
(b)	^{30}P				
(c)		53			
(d)			36		
(e)				45	34
(f)	^{40}Ca				

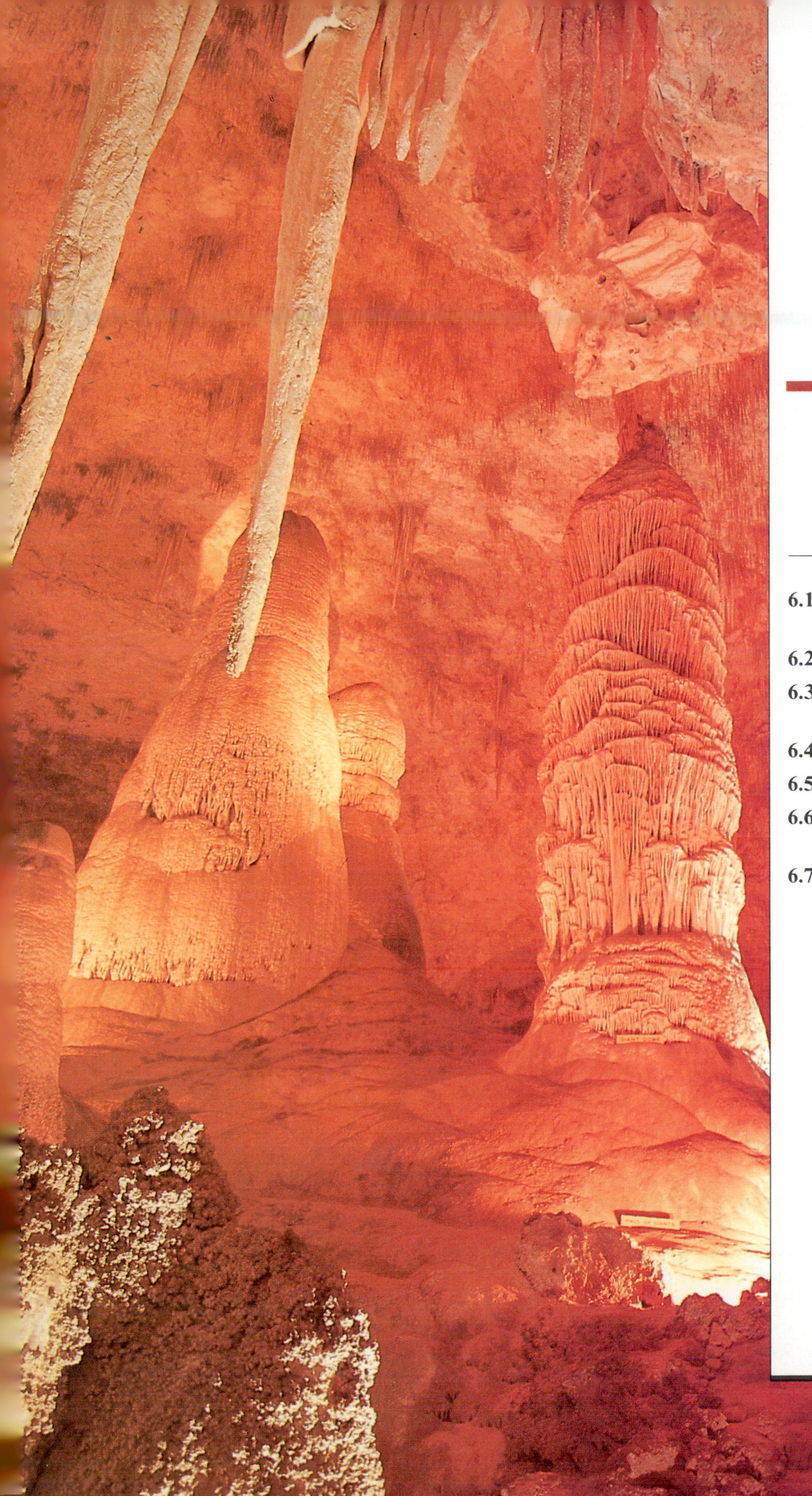

6

Nomenclature of Inorganic Compounds

◀ **Chapter Opening Photo: Stalactites and stalagmites are formed from calcium carbonate, commonly called limestone.**

As children, we begin to communicate with other people in our lives by learning the names of objects around us. As we continue to develop, we learn to speak and use language to complete a wide variety of tasks. As we enter school we begin to learn of other languages—the languages of mathematics, of other cultures, of computers. In each case, we begin by learning the names of the building blocks, and then proceed to more abstract concepts. In chemistry, a new language beckons also—a whole new way of describing the objects so familiar to us in our daily lives—the nomenclature of compounds. Only after learning the language are we able to understand the complexities of the modern model of the atom and its applications to the various fields we have chosen as our careers.

6.1 Common and Systematic Names

Chemical nomenclature is the system of names that chemists use to identify compounds. When a new substance is formulated, it must be named in order to distinguish it from all other substances (see Figure 6.1). In this chapter, we will restrict our discussion to the nomenclature of inorganic compounds, compounds which do not generally contain carbon. The naming of organic compounds, carbon-containing compounds, will be covered separately in Chapter 20.

Common names are arbitrary names that are not based on the chemical composition of compounds. Before chemistry was systematized, a substance was given a name that generally associated it with one of its outstanding physical or chemical properties. For example, *quicksilver* was a common name for mercury, and nitrous oxide (N_2O), used as an anesthetic in dentistry, has been called *laughing* gas because it induces laughter when inhaled. Water and ammonia also are common names because neither provides any information about the chemical composition of the compounds. If every substance were assigned a common name, the amount of memorization required to learn over nine million names would be astronomical.

Common names have distinct limitations, but they remain in frequent use. Often common names continue to be used in industry because the systematic name is too long or too technical for everyday use. For example, calcium oxide (CaO) is called lime by plasterers; photographers refer to *hypo* rather than sodium thiosulfate ($Na_2S_2O_3$); and nutritionists use *vitamin D_3*, instead of *9,10-secocholesta-5,7,10(19)-trien-3- β-ol* ($C_{27}H_{44}O$). Table 6.1 lists the common names, formulas, and systematic names of some familiar substances.

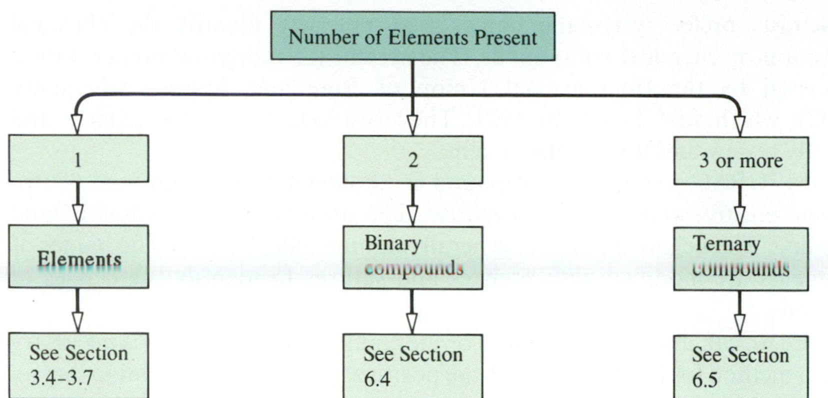

FIGURE 6.1
Flow chart summarizing the locations of rules for naming inorganic substances.

TABLE 6.1 Common Names, Formulas, and Chemical Names of Familiar Substances

Common name	Formula	Chemical name
Acetylene	C_2H_2	Ethyne
Lime	CaO	Calcium oxide
Slaked lime	$Ca(OH)_2$	Calcium hydroxide
Water	H_2O	Water
Galena	PbS	Lead(II) sulfide
Alumina	Al_2O_3	Aluminum oxide
Baking soda	$NaHCO_3$	Sodium hydrogen carbonate
Cane or beet sugar	$C_{12}H_{22}O_{11}$	Sucrose
Blue stone, blue vitriol	$CuSO_4 \cdot 5\ H_2O$	Copper(II) sulfate pentahydrate
Borax	$Na_2B_4O_7 \cdot 10\ H_2O$	Sodium tetraborate decahydrate
Brimstone	S	Sulfur
Calcite, marble, limestone	$CaCO_3$	Calcium carbonate
Cream of tartar	$KHC_4H_4O_6$	Potassium hydrogen tartrate
Epsom salts	$MgSO_4 \cdot 7\ H_2O$	Magnesium sulfate heptahydrate
Gypsum	$CaSO_4 \cdot 2\ H_2O$	Calcium sulfate dihydrate
Grain alcohol	C_2H_5OH	Ethanol, ethyl alcohol
Hypo	$Na_2S_2O_3$	Sodium thiosulfate
Laughing gas	N_2O	Dinitrogen monoxide
Litharge	PbO	Lead(II) oxide
Lye, caustic soda	$NaOH$	Sodium hydroxide
Milk of magnesia	$Mg(OH)_2$	Magnesium hydroxide
Muriatic acid	HCl	Hydrochloric acid
Oil of vitriol	H_2SO_4	Sulfuric acid
Plaster of paris	$CaSO_4 \cdot \frac{1}{2}\ H_2O$	Calcium sulfate hemihydrate
Potash	K_2CO_3	Potassium carbonate
Pyrite (fool's gold)	FeS_2	Iron disulfide
Quicksilver	Hg	Mercury
Sal ammoniac	NH_4Cl	Ammonium chloride
Saltpeter (chile)	$NaNO_3$	Sodium nitrate
Table salt	$NaCl$	Sodium chloride
Vinegar	$HC_2H_3O_2$	Acetic acid
Washing soda	$Na_2CO_3 \cdot 10\ H_2O$	Sodium carbonate decahydrate
Wood alcohol	CH_3OH	Methanol, methyl alcohol

Chemists prefer systematic names that precisely identify the chemical composition of chemical compounds. The system for inorganic nomenclature was devised by the International Union of Pure and Applied Chemistry (IUPAC), which first began in 1921. They continue to meet regularly and constantly review and update the system.

In the IUPAC system, the compound is considered to be composed of two parts, one positive and the other negative. The positive portion is named and written first. The negative portion, generally nonmetallic, follows. The names of the elements are modified with prefixes and suffixes to identify the classes of compounds.

Before we can discuss the specifics of inorganic nomenclature, it is necessary to learn a method for keeping track of the positive, negative, or zero value that an atom has within a compound or ion.

6.2 Oxidation Numbers

oxidation number

oxidation state

The **oxidation number** or **oxidation state** of an element is an integer value assigned to each element in a compound or ion. This value allows us to keep track of electrons associated with each atom. Oxidation numbers have a variety of uses in chemistry—from writing formulas, to predicting properties of compounds, and assisting in the balancing of oxidation-reduction reactions in which electrons are transferred (Chapter 19).

As a starting point, the oxidation number of an uncombined element, regardless of whether it is monatomic or diatomic, is zero. Other oxidation numbers are assigned by the following somewhat arbitrary set of rules:

1. Any element in its free state has an oxidation number of zero (Examples: Na, Mg, H_2, O_2, Cl_2).
2. Metals generally have positive oxidation numbers.
3. The oxidation number of hydrogen in a compound or an ion is generally $+1$. An exception is a metal hydride when hydrogen is second in the formula and has an oxidation number of -1 (Examples: NaH, H is -1; HCl, H is $+1$).
4. The oxidation number of oxygen in a compound or an ion is generally -2, with the exception of peroxides where it is -1 (Examples: H_2O, O is -2; H_2O_2, O is -1).
5. The oxidation number of a monatomic ion is the same as the charge on the ion (Examples: Cl^-, Mg^{2+}).
6. The algebraic sum of the oxidation numbers for all the atoms in a compound must equal zero.
7. The algebraic sum of the oxidation numbers for all the atoms in a polyatomic ion (ions containing more than one atom) must equal the charge on the ion.

The oxidation numbers of many elements are predictable from their position in the periodic table. In Figure 6.2, groups of elements are labeled with their oxidation number at the top of selected columns. This figure also shows the oxidation numbers for selected common ions.

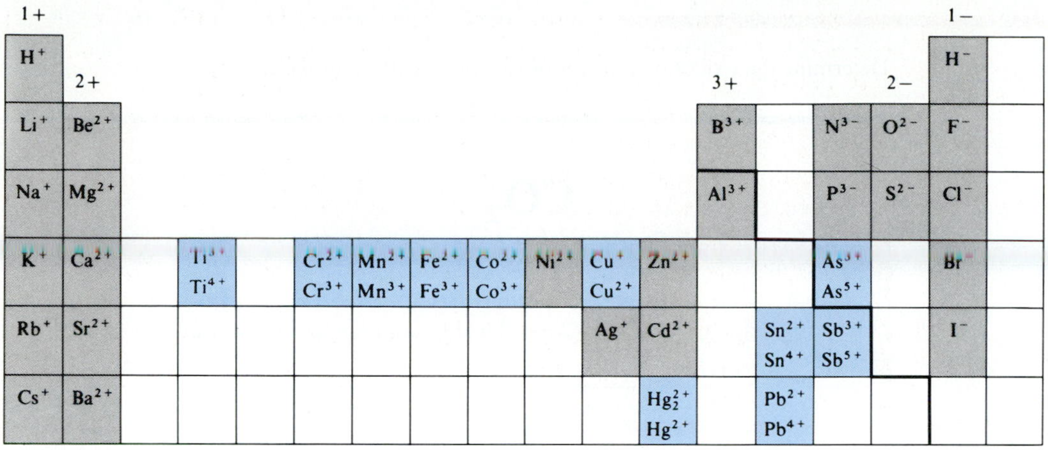

1+	2+											3+			2−	1−	
H^+																H^-	
Li^+	Be^{2+}											B^{3+}		N^{3-}	O^{2-}	F^-	
Na^+	Mg^{2+}											Al^{3+}		P^{3-}	S^{2-}	Cl^-	
K^+	Ca^{2+}		Ti^{3+} Ti^{4+}		Cr^{2+} Cr^{3+}	Mn^{2+} Mn^{3+}	Fe^{2+} Fe^{3+}	Co^{2+} Co^{3+}	Ni^{2+}	Cu^+ Cu^{2+}	Zn^{2+}			As^{3+} As^{5+}		Br^-	
Rb^+	Sr^{2+}									Ag^+	Cd^{2+}		Sn^{2+} Sn^{4+}	Sb^{3+} Sb^{5+}		I^-	
Cs^+	Ba^{2+}										Hg_2^{2+} Hg^{2+}		Pb^{2+} Pb^{4+}				

▲
FIGURE 6.2
Oxidation numbers of selected elements in the periodic table. The elements with multiple oxidation numbers are highlighted in blue. Hg_2^{2+} is a diatomic ion, indicating a + 1 charge from each atom in the ion.

TABLE 6.2 Names, Formulas, and Charges of Some Common Polyatomic Ions

Name	Formula	Charge	Name	Formula	Charge
Acetate	$C_2H_3O_2^-$	−1	Cyanide	CN^-	−1
Ammonium	NH_4^+	+1	Dichromate	$Cr_2O_7^{2-}$	−2
Arsenate	AsO_4^{3-}	−3	Hydroxide	OH^-	−1
Bicarbonate	HCO_3^-	−1	Nitrate	NO_3^-	−1
Bisulfate	HSO_4^-	−1	Nitrite	NO_2^-	−1
Bromate	BrO_3^-	−1	Permanganate	MnO_4^-	−1
Carbonate	CO_3^{2-}	−2	Phosphate	PO_4^{3-}	−3
Chlorate	ClO_3^-	−1	Sulfate	SO_4^{2-}	−2
Chromate	CrO_4^{2-}	−2	Sulfite	SO_3^{2-}	−2

The names, formulas, and ionic charges of some common polyatomic ions are given in Table 6.2. A list of both monatomic and polyatomic ions is given on the inside back cover of this book. Writing formulas of compounds and chemical equations is facilitated by a knowledge of oxidation numbers and ionic charges.

Use the following steps to find the oxidation number for an element within a compound.

Step 1 Write the oxidation number of each known atom below the atom in the formula.

Step 2 Multiply each oxidation number by the number of atoms of that element in the compound.

Step 3 Write an equation indicating the sum of all the oxidation numbers in the compound. Remember that the sum of all the oxidation numbers in a compound must equal zero.

EXAMPLE 6.1

Determine the oxidation number of carbon in carbon dioxide, CO_2.

CO_2

Step 1 -2
Step 2 $(-2)2$
Step 3 $C + (-4) = 0$
$C = +4$ (oxidation number for carbon)

EXAMPLE 6.2

Determine the oxidation number for sulfur in sulfuric acid, H_2SO_4:

$H_2S\ O_4$

Step 1 $+1$ -2
Step 2 $2(+1) = +2$ $4(-2) = -8$
Step 3 $+2 + S + (-8) = 0$
$S = +6$

PRACTICE Determine the oxidation number of (a) S in Na_2SO_4, (b) As in K_3AsO_4, (c) C in $CaCO_3$.
Answers: (a) $S = +6$ (b) $As = +5$ (c) $C = +4$

Oxidation numbers in a polyatomic ion (ions containing more than one atom) are determined in a similar fashion, remembering that in a polyatomic ion the sum of the oxidation numbers must equal the charge on the ion instead of zero.

EXAMPLE 6.3

Determine the oxidation number of manganese in the permanganate ion MnO_4^-.

MnO_4^-

Step 1 -2
Step 2 $(-2)4$
Step 3 $Mn + (-8) = -1$ (the charge on the atom)
$Mn = +7$ (oxidation number for manganese)

Example 6.4

Determine the oxidation number of carbon in the oxalate ion $C_2O_4^{2-}$

$$C_2O_4^{2-}$$

-2

Step 1 $(-2)4$
Step 2 $2C + (-8) = -2$ (the charge on the ion)
Step 3 $2C = +6$
$C = +3$ (oxidation number for C)

Practice Determine the oxidation numbers of (a) N in NH_4^+, (b) Cr in $Cr_2O_7^{2-}$, (c) P in PO_4^{3-}.
Answers: (a) $N = -3$ (b) $Cr = +6$ (c) $P = +5$
(Note: H is $+1$ in (a) even though it comes second in the formula. N is not a metal.)

6.3 Using Ions to Write Formulas of Compounds

The sum of the oxidation numbers of all the atoms in a compound is zero. This statement applies to all substances. For ionic compounds the sum of the charges of all the ions in the compound must be zero. Hence the formulas of ionic substances can easily be determined and written by simply combining the ions in the simplest proportion that makes the sum of the ionic charges add up to zero.

To illustrate: Sodium chloride consists of Na^+ and Cl^- ions. Since $(+1) + (-1) = 0$, these ions combine in a one-to-one ratio, and the formula is written NaCl. Calcium fluoride is made up of Ca^{2+} and F^- ions; one Ca^{2+} ion $(+2)$ and two F^- ions (-2) are needed to make zero, so the formula is CaF_2. Aluminum oxide is a bit more complicated, because it consists of Al^{3+} and O^{2-} ions. Since 6 is the lowest common multiple of 3 and 2, we have $2(+3) + 3(-2) = 0$; that is, two Al^{3+} ions $(+6)$ and three O^{2-} ions (-6) are needed; therefore, the formula is Al_2O_3.

The foregoing compounds all are made up of monatomic ions. The same procedure is used for polyatomic ions. Consider calcium hydroxide, which is made up of Ca^{2+} and OH^- ions. Since $(+2) + 2(-1) = 0$, one Ca^{2+} and two OH^- ions are needed, so the formula is $Ca(OH)_2$. The parentheses are used to enclose the OH^- so that two hydroxide ions can be shown. It is not correct to write CaO_2H_2 in place of $Ca(OH)_2$ because the identity of the compound would be lost by so doing. Note that the positive ion is written first in formulas. The following table provides examples of formula writing for ionic compounds.

1. Cobalt (II) chloride, $CoCl_2$
2. Sodium chloride, NaCl
3. Lead sulfide, PbS
4. Sulfur, S
5. Zinc, Zn
6. Marble chips, $CaCO_3$
7. Logwood chips
8. Charcoal, C
9. Mercury (II) iodide, HgI_2
10. Pyrite, FeS_2
11. Chromium (III) oxide, Cr_2O_3
12. Iron (II) sulfate, $FeSO_4$
13. Sodium sulfite, Na_2SO_3
14. Rosin
15. Sodium thiosulfate, $Na_2S_2O_3$
16. Iron, Fe
17. Aluminum, Al
18. Potassium hexacyanoferrate, $K_3Fe(CN)_6$
19. Potassium chromium sulfate, $KCr(SO_4)_2$
20. Menthol, $C_{10}H_{19}OH$
21. Potassium permanganate, $KMnO_4$
22. Ammonium nickel sulfate, $(NH_4)_2Ni(SO_4)_2$
23. Copper (II) sulfate pentahydrate, $CuSO_4 \cdot 5H_2O$
24. Sodium chromate, Na_2CrO_4
25. Trilead tetraoxide, Pb_3O_4
26. Hydroquinone, $C_6H_4(OH)_2$
27. Copper, Cu

▲ Compounds of transition elements are typically very colorful and are useful as paint pigments. Compounds of elements on the left of the periodic table reflect all wavelengths of light, which gives them a white color. Carbon compounds take a variety of colors.

Name of compound	Ions	Lowest common multiple	Sum of charges on ions	Formula
Sodium bromide	Na^+, Br^-	1	$(+1) + (-1) = 0$	NaBr
Potassium sulfide	K^+, S^{2-}	2	$2(+1) + (-2) = 0$	K_2S
Zinc sulfate	Zn^{2+}, SO_4^{2-}	2	$(+2) + (-2) = 0$	$ZnSO_4$
Ammonium phosphate	NH_4^+, PO_4^{3-}	3	$3(+1) + (-3) = 0$	$(NH_4)_3PO_4$
Aluminum chromate	Al^{3+}, CrO_4^{2-}	6	$2(+3) + 3(-2) = 0$	$Al_2(CrO_4)_3$

The sum of the charges on the ions of an ionic compound must equal zero. The sum of the oridation number for all atoms in a polyatomic ion must equal the charge on the ion.

EXAMPLE 6.5

Write formulas for (a) calcium chloride; (b) iron (III) sulfide; (c) aluminum sulfate.

SOLUTION

(a) Use the following steps for calcium chloride.

Step 1 From the name we know that calcium chloride is composed of calcium and chloride ions. First write down the formulas of these ions.

Ca^{2+} and Cl^-

Step 2 To write the formula of the compound, combine the smallest numbers of Ca^{2+} and Cl^- ions to give a charge sum equal to zero. In this case the lowest common multiple of the charges is 2:

$$(Ca^{2+}) + 2(Cl^-) = 0$$
$$(+2) + 2(-1) = 0$$

Therefore, the formula is $CaCl_2$.

(b) Use the same procedure for iron(III) sulfide.

Step 1 Write down the formulas for the iron(III) and sulfide ions.

Fe^{3+} and S^{2-}

Step 2 Use the smallest numbers of these ions required to give a charge sum equal to zero. The lowest common multiple of the charges is 6:

$$2(Fe^{3+}) + 3(S^{2-}) = 0$$
$$2(+3) + 3(-2) = 0$$

Therefore, the formula is Fe_2S_3.

(c) Use the same procedure for aluminum sulfate.

Step 1 Write down the formulas for the aluminum and sulfate ions.

Al^{3+} and SO_4^{2-}

Step 2 Use the smallest numbers of these ions required to give a charge sum equal to zero. The lowest common multiple of the charges is 6:

$$2(Al^{3+}) + 3(SO_4^{2-}) = 0$$
$$2(+3) + 3(-2) = 0$$

Therefore, the formula is $Al_2(SO_4)_3$. Note the use of parentheses around the SO_4^{2-} ion.

6.4 BINARY COMPOUNDS

Binary compounds contain only two different elements. Their names have two parts: the name of the more positive element followed by the name of the more negative element modified to end in *ide*. [Some nonbinary compounds have names ending in *ide*, but they are exceptions to the rule and are discussed in part (d) of this section.]

(a) Binary Compounds in Which the Positive Element has a Fixed Oxidation State Most of these compounds contain a metal and a nonmetal. The chemical name is composed of the name of the metal followed by the name of the nonmetal, which has been modified to an identifying stem plus the suffix *ide*.

For example, sodium chloride, NaCl, is composed of one atom each of sodium and chlorine. The name of the metal, sodium, is written first and is not modified. The second part of the name is derived from the nonmetal, chlorine, by using the stem *chlor* and adding the ending *ide*; it is named *chloride*. The compound name is sodium chloride.

NaCl

Elements:	Sodium (metal)
	Chlorine (nonmetal)
	name modified to the stem *chlor* + *ide*
Name of compound:	Sodium chloride

Stems of the more common negative-ion forming elements are shown in Table 6.3. Table 6.4 shows some compounds with names ending in *ide*.

Compounds may contain more than one atom of the same element, but as long as they contain only two different elements and if only one compound of

TABLE 6.3 Examples of Elements Forming Negative Ions

Symbol	Element	Stem	Binary name ending
B	Boron	Bor	Boride
Br	Bromine	Brom	Bromide
Cl	Chlorine	Chlor	Chloride
F	Fluorine	Fluor	Fluoride
H	Hydrogen	Hydr	Hydride
I	Iodine	Iod	Iodide
N	Nitrogen	Nitr	Nitride
O	Oxygen	Ox	Oxide
P	Phosphorus	Phosph	Phosphide
S	Sulfur	Sulf	Sulfide

TABLE 6.4 Examples of Compounds with Names Ending in *ide*

Formula	Name	Formula	Name
$AlCl_3$	Aluminum chloride	ZnS	Zinc sulfide
Al_2O_3	Aluminum oxide	LiI	Lithium iodide
CaC_2	Calcium carbide	$MgBr_2$	Magnesium bromide
HCl	Hydrogen chloride	NaH	Sodium hydride
HI	Hydrogen iodide	Na_2O	Sodium oxide

these two elements exists, the name follows the rule for binary compounds:

Examples: $CaBr_2$ Calcium bromide; Mg_3N_2 Magnesium nitride; Ag_2O Silver oxide

(b) Binary Compounds Containing Metals of Variable Oxidation Numbers Two systems are commonly used for compounds in this category. The official system, designated by the International Union of Pure and Applied Chemistry (IUPAC), is known as the *Stock System*. In the Stock System, when a compound contains a metal that can have more than one oxidation number, the oxidation number of the metal is designated by a Roman numeral placed in parentheses immediately following the name of the metal. The negative element is treated in the usual manner for binary compounds.

Oxidation number	+1	+2	+3	+4	+5	+6
Roman numeral	(I)	(II)	(III)	(IV)	(V)	(VI)

Examples:

$FeCl_2$	Iron(II) chloride	Fe^{2+}
$FeCl_3$	Iron(III) chloride	Fe^{3+}
CuCl	Copper(I) chloride	Cu^{+}
$CuCl_2$	Copper(II) chloride	Cu^{2+}

The fact that $FeCl_2$ has two chloride ions, each with a −1 charge, establishes that the oxidation number of Fe is +2. To distinguish between the two iron chlorides, $FeCl_2$ is named iron(II) chloride and $FeCl_3$ is named iron(III) chloride.

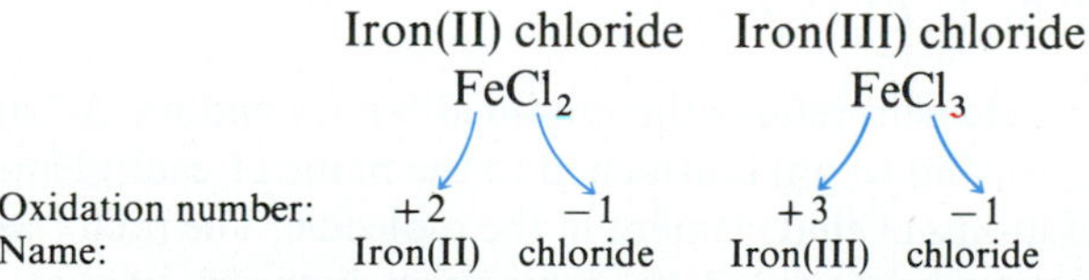

When a metal has only one possible oxidation state, we need not distinguish one oxidation state from another, so Roman numerals are not needed. Thus we do not say calcium(II) chloride for $CaCl_2$, but rather calcium chloride, since the oxidation number of calcium is understood to be +2.

In classical nomenclature, when the metallic ion has only two oxidation numbers, the name of the metal (usually the Latin name) is modified with the

TABLE 6.5 Names and Oxidation Numbers of Some Common Metal Ions that Have More than One Oxidation Number

Formula	Stock system name	Classical name
Cu^{1+}	Copper(I)	Cuprous
Cu^{2+}	Copper(II)	Cupric
Hg^{1+} $(Hg_2)^{2+}$	Mercury(I)	Mercurous
Hg^{2+}	Mercury(II)	Mercuric
Fe^{2+}	Iron(II)	Ferrous
Fe^{3+}	Iron(III)	Ferric
Sn^{2+}	Tin(II)	Stannous
Sn^{4+}	Tin(IV)	Stannic
Pb^{2+}	Lead(II)	Plumbous
Pb^{4+}	Lead(IV)	Plumbic
As^{3+}	Arsenic(III)	Arsenous
As^{5+}	Arsenic(V)	Arsenic
Ti^{3+}	Titanium(III)	Titanous
Ti^{4+}	Titanium(IV)	Titanic

suffixes *ous* and *ic* to distinguish between the two. The lower oxidation state is given the *ous* ending, and the higher one, the *ic* ending.

Examples:	$FeCl_2$	Ferrous chloride	Fe^{2+}	(lower oxidation state)
	$FeCl_3$	Ferric chloride	Fe^{3+}	(higher oxidation state)
	CuCl	Cuprous chloride	Cu^{+}	(lower oxidation state)
	$CuCl_2$	Cupric chloride	Cu^{2+}	(higher oxidation state)

Table 6.5 lists some common metals that have more than one oxidation number.

Notice that the *ous–ic* naming system does not give the oxidation state of an element but merely indicates that at least two oxidation states exist. The Stock System avoids any possible uncertainty by clearly stating the oxidation number.

(c) Binary Compounds Containing Two Nonmetals In a compound between two nonmetals, the element that occurs first in the series is written and named first.

Si, B, P, H, C, S, I, Br, N, Cl, O, F

The name of the second element retains the modified binary ending. A Latin or Greek prefix (*mono, di, tri,* and so on) is attached to the name of each element to indicate the number of atoms of that element in the molecule. The prefix *mono* is generally omitted except when needed to distinguish between two or more compounds, such as carbon monoxide, CO, and carbon dioxide, CO_2. Some common prefixes and their numerical equivalences follow.

Mono = 1	*Tetra* = 4	*Hepta* = 7	*Nona* = 9
Di = 2	*Penta* = 5	*Octa* = 8	*Deca* = 10
Tri = 3	*Hexa* = 6		

Here are some examples of compounds that illustrate this system:

CO	Carbon monoxide	N_2O	Dinitrogen monoxide
CO_2	Carbon dioxide	N_2O_4	Dinitrogen tetroxide
PCl_3	Phosphorus trichloride	NO	Nitrogen oxide
PCl_5	Phosphorus pentachloride	N_2O_3	Dinitrogen trioxide
P_2O_5	Diphosphorus pentoxide	S_2Cl_2	Disulfur dichloride
CCl_4	Carbon tetrachloride	S_2F_{10}	Disulfur decafluoride

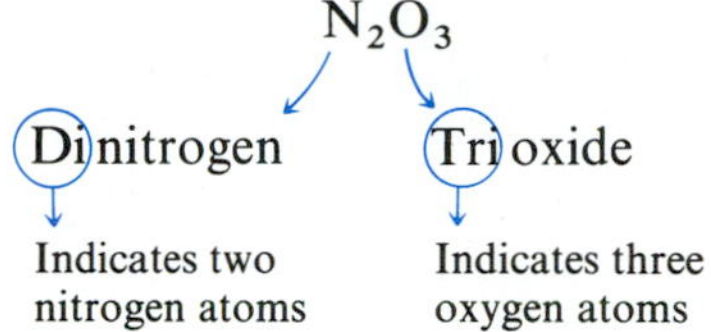

Some common names, such as ammonia for NH_3, are exceptions to the system.

(d) Exceptions That Use *ide* Endings Three notable exceptions that use the *ide* ending are hydroxides (OH^-), cyanides (CN^-), and ammonium (NH_4^+) compounds. These polyatomic ions, when combined with another element, take the ending *ide*, even though more than two elements are present in the compound.

NH_4I	Ammonium iodide
$Ca(OH)_2$	Calcium hydroxide
KCN	Potassium cyanide

(e) Acids Derived From Binary Compounds Certain binary hydrogen compounds, when dissolved in water, form solutions that have *acid* properties. Because of this property, these compounds are given acid names in addition to their regular *ide* names. For example, HCl is a gas and is called *hydrogen chloride*, but its water solution is known as *hydrochloric acid*. Binary acids are composed of hydrogen and one other nonmetallic element. However, not all binary hydrogen compounds are acids. To express the formula of a binary acid it is customary to write the symbol of hydrogen first, followed by the symbol of the second element (for example, HCl, HBr, H_2S). When we see formulas such as CH_4 or NH_3, we understand that these compounds are not normally considered to be acids.

To name a binary acid, place the prefix *hydro* in front of, and the suffix *ic* after, the stem of the nonmetal name. Then add the word *acid*.

	HCl	H_2S
Examples:	*Hydro* chlor/*ic acid* (hydrochloric acid)	*Hydro* sulfur/*ic acid* (hydrosulfuric acid)

Acids are hydrogen-containing substances that liberate hydrogen ions when dissolved in water. The same formula is often used to express binary hydrogen compounds such as HCl, regardless of whether they are dissolved in water. Table 6.6 shows several examples of binary acids.

TABLE 6.6 Names and Formulas of Selected Binary Acids

Formula	Acid name	Formula	Acid name
HF	Hydrofluoric acid	HI	Hydriodic acid
HCl	Hydrochloric acid	H_2S	Hydrosulfuric acid
HBr	Hydrobromic acid	H_2Se	Hydroselenic acid

EXAMPLE 6.6

Name the compound CaS.

Step 1 From the formula it is a two-element compound and follows the rules for binary compounds.

Step 2 The compound is composed of Ca, a metal, and S, a nonmetal. Elements in the periodic table under the +2 column have only one oxidation state (Figure 6.1). Thus, we name the positive part of the compound *calcium.*

Step 3 Modify the name of the second element to the identifying stem *sulf* and add the binary ending *ide* to form the name of the negative part, *sulfide.*

Step 4 The name of the compound, therefore, is *calcium sulfide.*

EXAMPLE 6.7

Name the compound FeS.

Step 1 This compound follows the rules for a binary compound and, like CaS, must be a sulfide.

Step 2 It is a compound of Fe, a metal, and S, a nonmetal. In the oxidation number tables we see that Fe has two oxidation numbers. In sulfides, the oxidation number of S is −2. Therefore, the oxidation number of Fe must be +2, and the name of the positive part of the compound is *iron(II)*, or *ferrous.*

Step 3 We have already determined that the name of the negative part of the compound will be *sulfide.*

Step 4 The name of FeS is *iron(II) sulfide*, or *ferrous sulfide.*

EXAMPLE 6.8

Name the compound PCl_5

Step 1 Phosphorus and chlorine are nonmetals, so the rules for naming binary compounds containing two nonmetals apply. Phosphorus is named first (see Section 6.4c).Therefore the compound is a chloride.

Step 2 No prefix is needed for phosphorus because each molecule has only one atom of phosphorus. The prefix *penta* is used with chloride to indicate the five chlorine atoms. (PCl_3 is also a known compound.)

Step 3 The name for PCl_5 is *phosphorus pentachloride.*

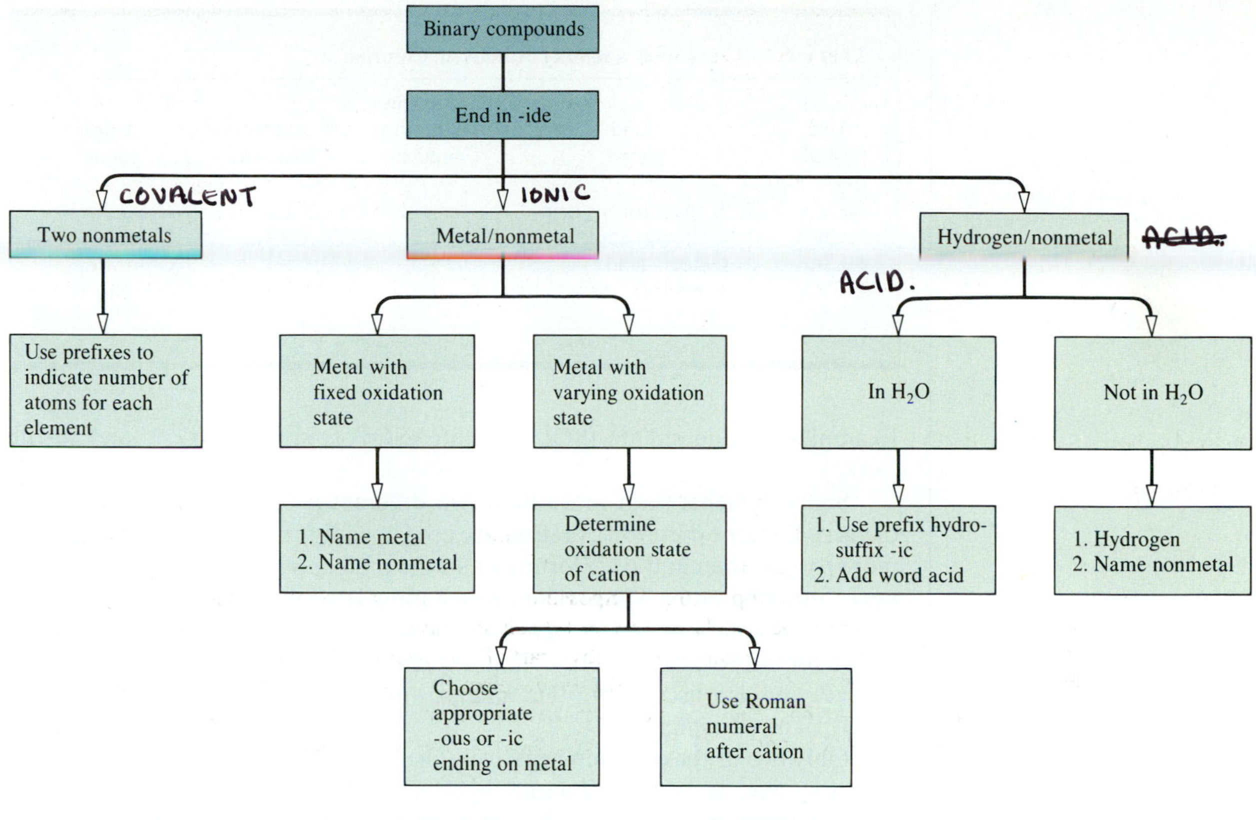

▲ **FIGURE 6.3** **Flow diagram for naming binary compounds.**

PRACTICE Name each of the following binary compounds: (a) KBr, (b) Ca_3N_2, (c) SO_3, (d) SnF_2, (e) $CuCl_2$, (f) N_2O_4.

Answers: (a) potassium bromide (d) tin(II) fluoride or stannous fluoride
(b) calcium nitride (e) copper(II) chloride or cupric chloride
(c) sulfur trioxide (f) dinitrogen tetraoxide

A summary of the approach to naming binary compounds is shown in Figure 6.3.

6.5 Ternary Compounds

Ternary compounds contain three elements and usually consist of a cation (either hydrogen or a metal), combined with a negative polyatomic ion. In general, when naming ternary compounds, the positive group is given first, followed by the name of the negative ion.

In order to name ternary compounds the names of the polyatomic ions must be known. Many polyatomic ions contain oxygen and generally have the suffix *ate* or *ite*. Unfortunately, the suffix does not indicate the number of oxygen atoms present. The *ate* form contains more oxygen atoms than the *ite* form.

TABLE 6.7 Oxy-Acids and Oxy-Anions of Chlorine

Acid formula	Acid name	Chlorine oxidation number	Anion formula	Anion name
$HClO$	*Hypo*chlor*ous* acid	+1	ClO^-	*Hypo*chlor*ite*
$HClO_2$	Chlor*ous* acid	+3	ClO_2^-	Chlor*ite*
$HClO_3$	Chlor*ic* acid	+5	ClO_3^-	Chlor*ate*
$HClO_4$	*Per*chloric acid	+7	ClO_4^-	*Per*chlor*ate*

Examples include sulfate (SO_4^{2-}), sulfite (SO_3^{2-}), nitrate (NO_3^-), and nitrite (NO_2^-).

Some elements form more than two different polyatomic ions containing oxygen. To name these ions, prefixes are used in addition to the suffix. To indicate more oxygen than in the *ate* form we add the prefix *per*, which is a short form of *hyper*, meaning more. The prefix *hypo*, meaning less (oxygen in this case), is used for the ion containing less oxygen than the *ite* form. An example of this system is shown for the polyatomic ions containing chlorine and oxygen in Table 6.7. The prefixes are also used with other similar ions, such as iodate (IO_3^-), bromate (BrO_3^-), and phosphate (PO_4^{3-}).

Only two of the common negatively charged polyatomic ions do not use the *ate/ite* system. These exceptions are hydroxide (OH^-) and cyanide (CN^-). Care must be taken with these, as their endings can easily be confused with the *ide* ending for binary compounds (Section 6.4d).

There are two common positively charged polyatomic ions as well—the ammonium and the hydronium ions. The ammonium ion (NH_4^+) is frequently found in ternary compounds (Section 6.4), whereas the hydronium ion (H_3O^+) is usually seen in aqueous solutions of acids (Chapter 16).

(a) Ternary Oxy-Acids Inorganic ternary compounds containing hydrogen, oxygen, and one other element are called *oxy-acids*. The element other than hydrogen or oxygen in these acids is often a nonmetal, but can be a metal.

The first step in naming these acids is to determine that the compound in question is really an oxy-acid. The keys to identification are that (1) hydrogen is the first element in the compound's formula; and (2) the second portion of the formula consists of a polyatomic ion containing oxygen.

Hydrogen in a ternary oxy-acid is not specifically designated in the acid name. The presence of hydrogen in the compound is indicated by the use of the word *acid* in the name of the substance. To determine the particular type of acid the polyatomic ion following hydrogen must be examined. The name of the polyatomic ion is modified in the following manner: (1) *ate* changes to an *ic* ending; (2) *ite* changes to an *ous* ending. Note also that this rule is similar to the earlier rule used in the Stock system for naming binary compounds. The *ic* ending represents a higher oxidation state, whereas the *ous* ending represents a lower oxidation state. Consider the following examples:

H_2SO_4 sulf*ate* ⟶ sulfur*ic* acid (S is +6)
H_2SO_3 sulf*ite* ⟶ sulfur*ous* acid (S is +4)

TABLE 6.8 Formulas and Names of Selected Ternary Oxy-Acids

Formula	Acid name	Formula	Acid name
H_2SO_3	Sulfurous acid	$HC_2H_3O_2$	Acetic acid
H_2SO_4	Sulfuric acid	$H_2C_2O_4$	Oxalic acid
HNO_2	Nitrous acid	H_2CO_3	Carbonic acid
HNO_3	Nitric acid	$HBrO_3$	Bromic acid
H_3PO_2	Hypophosphorous acid	HIO_3	Iodic acid
H_3PO_3	Phosphorous acid	H_3BO_3	Boric acid
H_3PO_4	Phosphoric acid		

HNO_3 nitr*ate* ⟶ nitr*ic* acid (N is +5)

HNO_2 nitr*ite* ⟶ nitr*ous* acid (N is +3)

The complete system for naming ternary oxy-acids is shown for the various acids containing chlorine in Table 6.7.

Examples of other ternary oxy-acids and their names are shown in Table 6.8.

(b) Salts of Ternary Acids A *salt* is formed when the hydrogen of an acid is replaced by a metal or ammonium ion. The name of the cation is given first, followed by the name of the polyatomic anion. The names of the anions are derived from the names of the corresponding ternary acids by changing the *ous* and *ic* endings to *ite* and *ate*, respectively. The stem portion of the acid name is not changed.

Ternary oxy-acid		*Ternary oxy-salt*
ous ending of acid	becomes	*ite* ending in salt
ic ending of acid	becomes	*ate* ending in salt

Thus the sulf*ite* ion (SO_3^{2-}) is derived from sulfur*ous* acid (H_2SO_3) and the sulf*ate* ion (SO_4^{2-}), from sulfur*ic* acid (H_2SO_4). The names and formulas of the sodium, calcium, and aluminum salts of sulfurous and sulfuric acids are as follows:

Na_2SO_3	$CaSO_3$	$Al_2(SO_3)_3$
Sodium sulfite	Calcium sulfite	Aluminum sulfite
Na_2SO_4	$CaSO_4$	$Al_2(SO_4)_3$
Sodium sulfate	Calcium sulfate	Aluminum sulfate

The complete system for naming ternary acids and their anions, using the oxy-acids of chlorine as a model, is shown in Table 6.7. A comparison of the acid, anion, and salt names for a number of ternary oxy-compounds is presented in Table 6.9.

The endings *ous*, *ic*, *ite*, and *ate* are part of classical nomenclature; they are not used in the Stock System to indicate different oxidation states of the elements. These endings are still used, however, in naming many common compounds. The Stock name for H_2SO_4 is tetraoxosulfuric(VI) acid, and that for H_2SO_3 is trioxosulfuric(IV) acid. These Stock names are not commonly used.

TABLE 6.9 Comparison of Acid, Anion and Salt Names for Selected Ternary Oxy-Compounds

Acid	Anion	Formulas and names of representative salts	
H_2SO_4 Sulfuric acid	SO_4^{2-} Sulfate ion	$CaSO_4$ $Fe_2(SO_4)_3$	Calcium sulfate Iron(III) sulfate or ferric sulfate
H_2SO_3 Sulfurous acid	SO_3^{2-} Sulfite ion	Na_2SO_3 Ag_2SO_3	Sodium sulfite Silver sulfite
HNO_3 Nitric acid	NO_3^- Nitrate ion	KNO_3 $Hg(NO_3)_2$	Potassium nitrate Mercury(II) nitrate or mercuric nitrate
HNO_2 Nitrous acid	NO_2^- Nitrite ion	KNO_2 $Co(NO_2)_2$	Potassium nitrite Cobalt(II) nitrite or cobaltous nitrite
H_2CO_3 Carbonic acid	CO_3^{2-} Carbonate ion	Li_2CO_3 $BaCO_3$	Lithium carbonate Barium carbonate
H_3PO_4 Phosphoric acid	PO_4^{3-} Phosphate ion	$AlPO_4$ $Zn_3(PO_4)_2$	Aluminum phosphate Zinc phosphate
H_3PO_3 Phosphorous acid	PO_3^{3-} Phosphite ion	Na_3PO_3 $Zn_3(PO_3)_2$	Sodium phosphite Zinc phosphite
HIO_3 Iodic acid	IO_3^- Iodate ion	$AgIO_3$ $Cu(IO_3)_2$	Silver iodate Copper(II) iodate or cupric iodate
$HC_2H_3O_2$ Acetic acid	$C_2H_3O_2^-$ Acetate ion	$Pb(C_2H_3O_2)_2$ $NH_4C_2H_3O_2$	Lead(II) acetate Ammonium acetate
$H_2C_2O_4$ Oxalic acid	$C_2O_4^{2-}$ Oxalate ion	CaC_2O_4 $(NH_4)_2C_2O_4$	Calcium oxalate Ammonium oxalate

EXAMPLE 6.9 (a) Name the salt KNO_3 and (b) name the acid HNO_3 from which this salt can be derived.

SOLUTION

(a) **Step 1** The compound contains three elements and follows the rules for ternary compounds.

Step 2 The salt is composed of a K^+ ion and a NO_3^- ion. The name of the positive part of the compound is *potassium.*

Step 3 Since it is a ternary salt, the name will end in *ite* or *ate.* In the oxidation number tables, we see that the name of the NO_3^- ion is *nitrate.*

Step 4 The name of the compound is *potassium nitrate.*

(b) The name of the acid follows the rules for ternary oxy-acids. Because the name of the salt KNO_3 ends in *ate*, the name of the corresponding acid will end in *ic acid.* Change the *ate* ending of nitrate to *ic.* Thus *nitrate* becomes *nitric*, and the name of the acid is *nitric acid.*

PRACTICE Name the following salts and the acids from which they are derived. (a) $CaSO_4$, H_2SO_4; (b) Mg_3PO_4, H_3PO_4; (c) Na_2CO_3, H_2CO_3.

Answers: (a) calcium sulfate, sulfuric acid
(b) magnesium phosphate, phosphoric acid
(c) sodium carbonate, carbonic acid

6.6 Salts with More than One Kind of Positive Ion

Salts can be formed from acids that contain two or more acid hydrogen atoms by replacing one or more of the hydrogen atoms with a metal. Each positive group is named first, and then the appropriate salt ending is added.

Acid	Salt	Name of salt
H_2CO_3	$NaHCO_3$	Sodium hydrogen carbonate or sodium bicarbonate
H_2S	$NaHS$	Sodium hydrogen sulfide or sodium bisulfide
H_3PO_4	$MgNH_4PO_4$	Magnesium ammonium phosphate
H_2SO_4	$NaKSO_4$	Sodium potassium sulfate

Note the name *sodium bicarbonate* given in the table. The prefix *bi* is commonly used to indicate a compound in which one of two acid hydrogen atoms has been replaced by a metal. The HCO_3^- ion is known as the hydrogen carbonate ion or the bicarbonate ion. Another example is sodium bisulfate, which has the formula $NaHSO_4$. Table 6.10 shows examples of other salts that contain more than one kind of positive ion.

Note that prefixes are also used in chemical nomenclature to give special clarity or emphasis to certain compounds as well as to distinguish between two or more compounds.

Examples:
Na_3PO_4 Trisodium phosphate
Na_2HPO_4 Disodium hydrogen phosphate
NaH_2PO_4 Sodium dihydrogen phosphate

6.7 Bases

Inorganic bases contain the hydroxide ion, OH^-, in chemical combination with a metal ion. These compounds are called *hydroxides*. The OH^- group is named as a single ion and is given the ending *ide*. Several common bases are listed below:

$NaOH$ Sodium hydroxide
KOH Potassium hydroxide
NH_4OH Ammonium hydroxide
$Ca(OH)_2$ Calcium hydroxide
$Ba(OH)_2$ Barium hydroxide

A summary of the approach to naming ternary compounds is shown in Figure 6.4. We have now looked at ways of naming a variety of inorganic compounds—binary compounds consisting of a metal and a nonmetal and of two nonmetals, and binary and ternary acids, salts, and bases. These

TABLE 6.10 Names of Selected Salts that Contain More than One Kind of Positive Ion

Acid	Salt	Name of salt
H_2SO_4	$KHSO_4$	Potassium hydrogen sulfate or potassium bisulfate
H_2SO_3	$Ca(HSO_3)_2$	Calcium hydrogen sulfite or calcium bisulfite
H_2S	NH_4HS	Ammonium hydrogen sulfide or ammonium bisulfide
H_3PO_4	$MgNH_4PO_4$	Magnesium ammonium phosphate
H_3PO_4	NaH_2PO_4	Sodium dihydrogen phosphate
H_3PO_4	Na_2HPO_4	Disodium hydrogen phosphate
$H_2C_2O_4$	KHC_2O_4	Potassium hydrogen oxalate or potassium binoxalate
H_2SO_4	$KAl(SO_4)_2$	Potassium aluminum sulfate
H_2CO_3	$Al(HCO_3)_3$	Aluminum hydrogen carbonate or aluminum bicarbonate

compounds are just a small part of the classified chemical compounds. Most of the remaining classes are in the broad field of organic chemistry under such categories as hydrocarbons, alcohols, ethers, aldehydes, ketones, phenols, and carboxylic acids.

FIGURE 6.4 ▶ Flow diagram for naming ternary compounds.

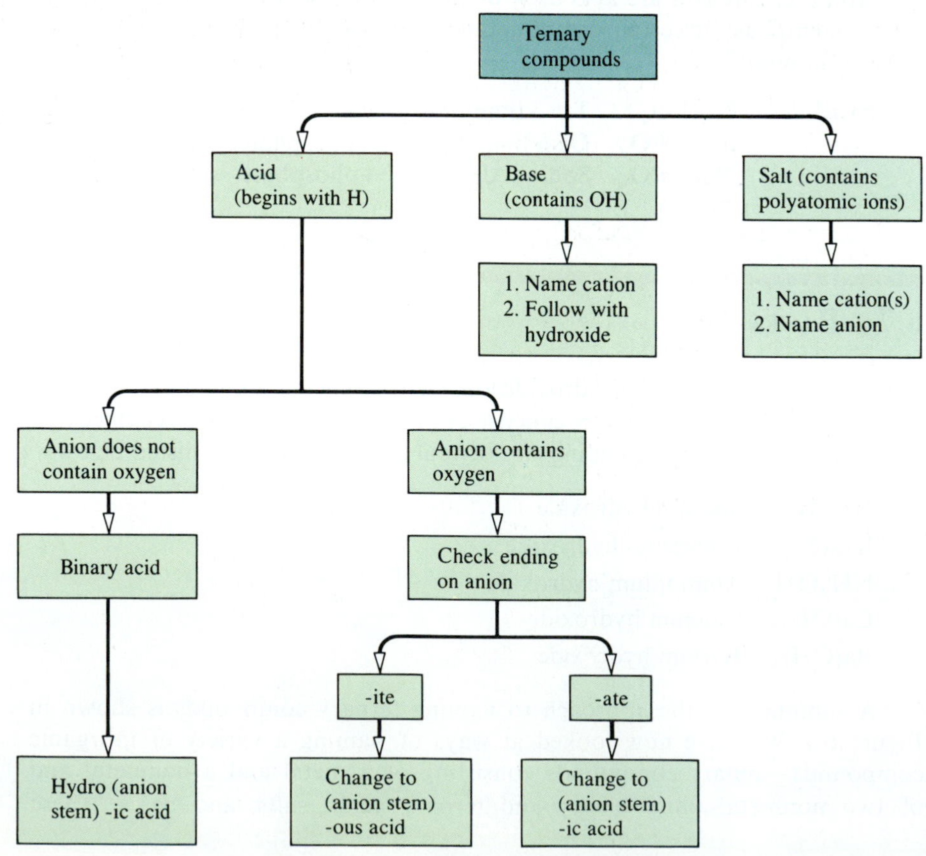

CHEMISTRY IN ACTION

CHARGES IN YOUR LIFE

Ions are used in living organisms to perform many important functions. For example, electrical neutrality must be maintained both inside and outside the body cells. Within the cell, neutrality is maintained by potassium ions (K^+) and hydrogen phosphate ions (HPO_4^{2-}). Outside the cell in the intercellular fluid the ions responsible for neutrality are sodium (Na^+) and chloride (Cl^-).

Another important ion within organisms is magnesium (Mg^{2+}), found in chlorophyll and used during nerve and muscle activity, as well as in conjunction with certain enzymes. Iron ions (Fe^{2+}) are incorporated within the hemoglobin molecule and are an integral part of the oxygen transport system within the body. Calcium ions (Ca^{2+}) are part of the matrix of both bones and teeth and play a significant role in the clotting of blood.

Ions also have a significant role in the detergent industry. Water is said to be "hard" when it contains relatively high concentrations of Ca^{2+} and Mg^{2+} ions. In solution, soaps combine with these ions to form an insoluble scum. This material forms the common bathtub ring and, in laundry, settles out on the clothes leaving them gray and dingy. Although soaps and detergents are similar in their cleansing actions detergents are less likely to form this scum (see Chapter 25). For this reason many people who live in areas with "hard" water use detergents instead of soaps.

Phosphate ions are used with detergents for two purposes. They help to make the cleaning agents function better by changing the acidity of the wash solution. Secondly, the phosphate ion attaches to the Ca^{2+} and Mg^{2+} preventing the formation of scum. However, the use of phosphates in detergent formulations has, in some instances, led to an environmental problem known as *eutrophication*. Phosphates are excellent plant fertilizers so when waste water containing detergents reaches streams or lakes algae begin to multiply very rapidly. The subsequent death and decay of excess algae leads to eutrophic conditions in which dissolved oxygen disappears from the water. The absence of dissolved oxygen then causes the death of fish and other aquatic life forms. This problem has become so serious that many states have banned the sale of detergents containing phosphates. For liquid detergents the problem has, in many cases, been solved by replacing phosphates with another chemical (EDTA). But the search continues to find an adequate phosphate replacement for use in powdered detergents.

Concepts in Review

1. State the rules for assigning oxidation numbers.
2. Write the formulas of compounds formed by combining the ions from Figure 6.1 and Table 6.2 (or from the inside back cover of this book) in the correct ratios.
3. Assign the oxidation number to each element in a compound or ion.
4. Write the names or formulas for inorganic binary compounds in which the metal has only one common oxidation state.
5. Write the names or formulas for inorganic binary compounds that contain metals of variable oxidation state, using either the Stock System or classical nomenclature.
6. Write the names or formulas for inorganic binary compounds that contain two nonmetals.
7. Write the names or formulas for binary acids.
8. Write the names or formulas for ternary oxy-acids.
9. Write the names or formulas for ternary salts.
10. Given the formula of a salt, write the name and formula of the acid from which the salt may be derived.
11. Write the names or formulas for salts that contain more than one kind of positive ion.
12. Write the names or formulas for inorganic bases.

Exercises

1. Using the oxidation numbers shown in Figure 6.1, write formulas for:
   (a) The oxygen compounds of Ca, Ni, Ag, Al, Na
   (b) The fluorine compounds of Li, Al, B, Sr, Mg
2. Use the oxidation number table on the inside back cover of your book to determine the formulas for compounds composed of the following ions:
   (a) Sodium and chlorate
   (b) Hydrogen and sulfate
   (c) Tin(II) and acetate
   (d) Copper(I) and oxide
   (e) Zinc and bicarbonate
   (f) Iron(III) and carbonate
3. Determine the oxidation number of the element underlined in each formula:
   (a) $\underline{Mn}CO_3$ (d) $K\underline{Mn}O_4$ (g) $\underline{W}Cl_5$
   (b) $\underline{Sn}F_4$ (e) $Ba\underline{C}O_3$ (h) $K_2\underline{Cr}_2O_7$
   (c) $K\underline{N}O_3$ (f) $\underline{P}Cl_3$

4. Determine the oxidation number of the element underlined in each formula:
   (a) $\underline{In}I_3$ (d) $\underline{C}_2H_5OH$ (g) $\underline{Fe}_2(CO_3)_3$
   (b) $K\underline{Cl}O_3$ (e) $Mg(\underline{N}O_3)_2$ (h) $Na\underline{Cl}O_4$
   (c) $Na_2\underline{S}O_4$ (f) $\underline{Sn}O_2$
5. Write the formula of the compound that would be formed between the given elements:
   (a) Na and I (c) Al and O (e) Cs and Cl
   (b) Ba and F (d) K and S (f) Sr and Br
6. Write the formula of the compound that would be formed between the given elements:
   (a) Ba and O (c) Al and Cl (e) Li and Si
   (b) H and S (d) Be and Br (f) Mg and P
7. Does the fact that two elements combine in a one-to-one atomic ratio mean that their oxidation numbers are both 1? Explain.
8. Write formulas for the following cations (do not forget to include the charges): sodium, magnesium, aluminum, copper(II), iron(II), ferric, lead(II), silver, cobalt(II), barium, hydrogen,

mercury(II), tin(II), chromium(III), stannic, manganese(II) bismuth(III).

9. Write formulas for the following anions (do not forget to include the charges): chloride, bromide, fluoride, iodide, cyanide, oxide, hydroxide, sulfide, sulfate, bisulfate, bisulfite, chromate, carbonate, bicarbonate, acetate, chlorate, permanganate, oxalate.

10. Complete the table, filling in each box with the proper formula.

Anions

Cations	Br^-	O^{2-}	NO_3^-	PO_4^{3-}	CO_3^{2-}
K^+	KBr				
Mg^{2+}					
Al^{3+}					
Zn^{2+}				$Zn_3(PO_4)_2$	
H^+					

11. Complete the table, filling in each box with the proper formula.

Anions

Cations	SO_4^{2-}	Cl^-	AsO_4^{3-}	$C_2H_3O_2^-$	CrO_4^{2-}
NH_4^+			$(NH_4)_3AsO_4$		
Ca^{2+}					
Fe^{3+}	$Fe_2(SO_4)_3$				
Ag^+					
Cu^{2+}					

12. State how each of the following is used in naming inorganic compounds: *ide, ous, ic, hypo, per, ite, ate*, Roman numerals.

13. Write formulas for the following binary compounds, all of which are composed of nonmetals:
(a) Carbon monoxide
(b) Sulfur trioxide
(c) Carbon tetrabromide
(d) Phosphorus trichloride
(e) Nitrogen dioxide
(f) Dinitrogen pentoxide

14. Name the following binary compounds, all of which are composed of nonmetals:
(a) CO_2 (e) SO_2 (i) NF_3
(b) N_2O (f) N_2O_4 (j) CS_2
(c) PCl_5 (g) P_2O_5
(d) CCl_4 (h) OF_2

15. Name the following compounds:
(a) K_2O (d) $BaCO_3$ (g) $Zn(NO_3)_2$
(b) NH_4Br (e) Na_3PO_4 (h) Ag_2SO_4
(c) CaI_2 (f) Al_2O_3

16. Write formulas for the following compounds:
(a) Sodium nitrate
(b) Magnesium fluoride
(c) Barium hydroxide
(d) Ammonium sulfate
(e) Silver carbonate
(f) Calcium phosphate

17. Name each of the following compounds by both the Stock (IUPAC) System, and the *ous–ic* system:
(a) $CuCl_2$ (d) $FeCl_3$ (g) $As(C_2H_3O_2)_3$
(b) $CuBr$ (e) SnF_2 (h) TiI_3
(c) $Fe(NO_3)_2$ (f) $HgCO_3$

18. Write formulas for the following compounds:
    (a) Tin(IV) bromide (d) Mercuric nitrite
    (b) Copper(I) sulfate (e) Titanic sulfide
    (c) Ferric carbonate (f) Iron(II) acetate
19. Write formulas for the following acids:
    (a) Hydrochloric acid (d) Carbonic acid
    (b) Chloric acid (e) Sulfurous acid
    (c) Nitric acid (f) Phosphoric acid
20. Name the following acids:
    (a) HNO_2 (d) HBr (g) HF
    (b) H_2SO_4 (e) H_3PO_3 (h) $HBrO_3$
    (c) $H_2C_2O_4$ (f) $HC_2H_3O_2$
21. Write formulas for the following acids:
    (a) Acetic acid (d) Boric acid
    (b) Hydrofluoric acid (e) Nitrous acid
    (c) Hypochlorous acid (f) Hydrosulfuric acid
22. Name the following acids:
    (a) H_3PO_4 (c) HIO_3 (e) HClO (g) HI
    (b) H_2CO_3 (d) HCl (f) HNO_3 (h) $HClO_4$
23. Name the following compounds:
    (a) $Ba(NO_3)_2$ (d) $MgSO_4$ (g) NiS
    (b) $NaC_2H_3O_2$ (e) $CdCrO_4$ (h) $Sn(NO_3)_4$
    (c) PbI_2 (f) $BiCl_3$ (i) $Ca(OH)_2$
24. Write formulas for the following compounds:
    (a) Silver sulfite
    (b) Cobalt(II) bromide
    (c) Tin(II) hydroxide
    (d) Aluminum sulfate
    (e) Manganese(II) fluoride
    (f) Ammonium carbonate
    (g) Chromium(III) oxide
    (h) Cupric chloride
    (i) Potassium permanganate
    (j) Barium nitrite
    (k) Sodium peroxide
    (l) Ferrous sulfate
    (m) Potassium dichromate
    (n) Bismuth(III) chromate
25. Write formulas for the following compounds:
    (a) Sodium chromate
    (b) Magnesium hydride
    (c) Nickel(II) acetate
    (d) Calcium chlorate
    (e) Lead(II) nitrate
    (f) Potassium dihydrogen phosphate
    (g) Manganese(II) hydroxide
    (h) Cobalt(II) bicarbonate
    (i) Sodium hypochlorite
    (j) Arsenic(V) carbonate
    (k) Chromium(III) sulfite
    (l) Antimony(III) sulfate
    (m) Sodium oxalate
    (n) Potassium thiocyanate
26. Write the name of each salt and the formula and name of the acid from which the salt may be derived. [*Example*: NiC_2O_4, nickel(II) oxalate; $H_2C_2O_4$, oxalic acid.]
    (a) $ZnSO_4$ (k) $Ca(HSO_4)_2$
    (b) $HgCl_2$ (l) $As_2(SO_3)_3$
    (c) $CuCO_3$ (m) $Sn(NO_2)_2$
    (d) $Cd(NO_3)_2$ (n) $FeBr_3$
    (e) $Al(C_2H_3O_2)_3$ (o) $KHCO_3$
    (f) CoF_2 (p) $BiAsO_4$
    (g) $Cr(ClO_3)_3$ (q) $Fe(BrO_3)_2$
    (h) Ag_3PO_4 (r) $(NH_4)_2HPO_4$
    (i) NiS (s) NaClO
    (j) $BaCrO_4$ (t) $KMnO_4$
27. Write the chemical formula for each of the following substances:
    (a) Baking soda (i) Fool's gold
    (b) Lime (j) Saltpeter
    (c) Epsom salts (k) Limestone
    (d) Muriatic acid (l) Cane sugar
    (e) Vinegar (m) Milk of magnesia
    (f) Potash (n) Washing soda
    (g) Lye (o) Grain alcohol
    (h) Quicksilver
28. Explain how "hard water" may cause white clothing to become gray and dingy.
29. Why are phosphates added to detergents? What are the potential problems associated with this action?
30. Give the name and formula of three salts with *ide* endings that are not binary compounds. Do not use the exact salts used as examples in the chapter.
31. Which of these statements are correct? Rewrite each incorrect statement to make it correct.
    (a) An oxidation number can be positive, negative, or zero.
    (b) The oxidation number of Li, Na, and K in compounds is -1.
    (c) The sum of the oxidation numbers for all the atoms in a polyatomic ion is zero.
    (d) The formula for calcium hydride is CaH_2.
    (e) All the following compounds are acids: H_2SO_4, HCl, HNO_3, $NaC_2H_3O_2$.
    (f) The ions of all the following metals have an oxidation number of $+2$: Ca, Ba, Sr, Cd, Zn.
    (g) The formulas for nitrous and sulfurous acids are HNO_2 and H_2SO_3.
    (h) The formula for the compound between Fe^{3+} and O^{2-} is Fe_3O_2.
    (i) The oxidation number of Cr in K_2CrO_4 is $+6$.

(j) The oxidation number of Sn in $SnCl_4$ is +4.
(k) The oxidation number of Co in $CoCl_2$ is +4.
(l) The name for $NaNO_2$ is sodium nitrite.
(m) The name for $Ca(ClO_3)_2$ is calcium chlorate.
(n) The name for CuO is copper(I) oxide.
(o) The name for SO_4^{2-} is sulfate ion.
(p) The name for N_2O_4 is dinitrogen tetroxide.
(q) The name for Na_2O is disodium oxide.
(r) If the name of an anion ends with *ide*, the name of the corresponding acid will start with *hydro*.
(s) If the name of an anion ends with *ite*, the corresponding acid name will end with *ic*.
(t) If the name of an acid ends with *ous*, the corresponding salt name will end with *ate*.
(u) In FeI_2, the iron is iron(II) because it is combined with two I^- ions.
(v) In Cu_2SO_4, the copper is copper(II) because there are two copper ions.
(w) In Ru_2O_3, we can deduce the oxidation state of Ru as +3 because two Ru ions are combined with three oxide ions.
(x) When two nonmetals combine, prefixes of *di*, *tri*, *tetra*, and so on are used to specify how many atoms of each element are in a molecule.
(y) N_2O_3 is called dinitrogen trioxide.
(z) $Sn(CrO_4)_2$ is called tin dichromate.
(aa) In the Stock System of nomenclature, when a compound contains a metal that can have more than one oxidation number, the oxidation number of the metal is designated by a Roman numeral written immediately after the name of the metal.

7

Quantitative Composition of Compounds

◄ **CHAPTER OPENING PHOTO: The taste of chocolate truffles is enhanced by careful control of the quantities of each ingredient.**

Knowing the substances contained in a product is not sufficient information to produce the product. An artist can create an incredible array of colors from a limited number of pigments. A pharmacist can combine the same drugs in various amounts to produce different effects in patients. Cosmetics, cereals, cleaning products, and pain remedies all show a list of ingredients on the labels of the packages.

In each of these products the key to the production lies in the amount of each ingredient. The pharmaceutical industry maintains strict regulations on the amounts of the ingredients in the medicines we purchase. The formulas for soft drinks and most cosmetics are trade secrets. Small deviations in the composition of these products can result in large losses or lawsuits for these organizations.

The composition of compounds is an important concept in chemistry. The numerical relationships among elements within compounds and the measurement of exact quantities of particles are fundamental tasks for the chemist.

7.1 The Mole

In the laboratory we normally determine the mass of substances on a balance in units of grams. But, when we run a chemical reaction, the reaction occurs between atoms and molecules. For example, in the reaction between magnesium and sulfur, one atom of sulfur reacts with one atom of magnesium.

$$Mg + S \longrightarrow MgS$$

However, when we measure the masses of these two elements that react, we find that 24.31 g of Mg are required to react with 32.06 g of S. Because magnesium and sulfur react in a 1:1 atom ratio, we can conclude from this experiment that 24.31 g of Mg contain the same number of atoms as 32.06 g of S. How many atoms are in 24.31 g of Mg or 32.06 g of S? These two amounts each contain one mole of atoms.

The mole (abbreviated mol) is one of the seven base units in the International System and is the unit for an amount of substance. The mole is a counting unit as in other things that we count, such as a dozen (12) eggs or a gross (144) of pencils. But a mole is a much larger number of things, namely 6.022×10^{23}. Thus one mole contains 6.022×10^{23} entities of anything. In reference to our reaction between magnesium and sulfur, 1 mol Mg (24.31 g) contains 6.022×10^{23} atoms of magnesium and 1 mol S (32.06 g) contains 6.022×10^{23} atoms of sulfur.

Avogadro's number

The number 6.022×10^{23} is known as **Avogadro's number** in honor of Amedeo Avogadro (1776–1856), an Italian physicist. Avogadro's number is an important constant in chemistry and physics and has been experimentally determined by several independent methods.

▲
Amedeo Avogadro (1776–1856).

Avogadro's number = 6.022×10^{23}

It is difficult to imagine how large Avogadro's number really is, but perhaps the following analogy will help express it: If 10,000 people started to count Avogadro's number and each counted at the rate of 100 numbers per minute each minute of the day, it would take them over 1 trillion (10^{12}) years to count the total number. So, even the most minute amount of matter contains extremely large numbers of atoms. For example, 1 mg (0.001 g) of sulfur contains 2×10^{19} atoms of sulfur.

mole

Avogadro's number is the basis for the amount of substance that is used to express a particular number of chemical species, such as atoms, molecules, formula units, ions, or electrons. This amount of substance is the mole. We define a **mole** as an amount of a substance containing the same number of formula units as there are atoms in exactly 12 g of carbon-12. (Recall that carbon-12 is the reference isotope for atomic masses.) Other definitions are used, but they all relate to a mole being Avogadro's number of formula units of a substance. A **formula unit** is the atom or molecule indicated by the formula of the substance under consideration—for example, Mg, MgS, H_2O, O_2, ${}^{75}_{33}As$.

formula unit

From the above definition we can say that the atomic mass in grams of any element contains one mole of atoms.

The term *mole* is so commonplace in chemical jargon that chemists use it as freely as the words *atom* and *molecule*. The mole is used in conjunction with many different particles, such as atoms, molecules, ions, and electrons, to represent Avogadro's number of these particles. If we can speak of a mole of atoms, we can also speak of a mole of molecules, a mole of electrons, a mole of ions, understanding that in each case we mean 6.022×10^{23} formula units of these particles.

1 mole of atoms = 6.022×10^{23} atoms
1 mole of molecules = 6.022×10^{23} molecules
1 mole of ions = 6.022×10^{23} ions

molar mass

The atomic mass in grams of an element contains Avogadro's number of atoms and is defined as the **molar mass** of that element. To determine the molar mass of an element, change the units of the atomic mass found in the periodic table from amu to grams. Sulfur has an atomic mass of 32.06 amu. One molar mass of sulfur has a mass of 32.06 grams and contains 6.022×10^{23} atoms of sulfur. See the following table.

Element	Atomic mass	Molar mass	Number of atoms
H	1.008 amu	1.008 g	6.022×10^{23}
Mg	24.31 amu	24.31 g	6.022×10^{23}
Na	22.99 amu	22.99 g	6.022×10^{23}

1 molar mass (g) = 1 mole of atoms
= Avogadro's number (6.022×10^{23}) of atoms

We frequently encounter problems that require conversions involving quantities of mass, numbers, and moles of atoms of an element. Conversion factors that can be used for this purpose are

(a) Grams to atoms: $\dfrac{6.022 \times 10^{23} \text{ atoms of the element}}{1 \text{ molar mass of the element}}$

(b) Atoms to grams: $\dfrac{1 \text{ molar mass of the element}}{6.022 \times 10^{23} \text{ atoms of the element}}$

(c) Grams to moles: $\dfrac{1 \text{ mole of the element}}{1 \text{ molar mass of the element}}$

(d) Moles to grams: $\dfrac{1 \text{ molar mass of the element}}{1 \text{ mole of the element}}$

EXAMPLE 7.1

How many moles of iron does 25.0 g of Fe represent?

SOLUTION

The problem requires that we change grams of Fe to moles of Fe. We look up the atomic mass of Fe in the table of atomic masses on the periodic table or table of atomic masses and find it to be 55.85. Then we use the proper conversion factor to obtain moles. The conversion factor is (c).

grams Fe ⟶ moles Fe $\qquad \text{grams Fe} \times \dfrac{1 \text{ mole Fe}}{1 \text{ molar mass Fe}}$

$$25.0 \cancel{\text{g Fe}} \times \frac{1 \text{ mol Fe}}{55.85 \cancel{\text{g Fe}}} = 0.448 \text{ mol Fe}$$

EXAMPLE 7.2

How many magnesium atoms are contained in 5.00 g of Mg?

SOLUTION

The problem requires that we change grams of magnesium to atoms of magnesium.

grams Mg ⟶ atoms Mg

We find the atomic mass of magnesium to be 24.31 and set up the calculation using conversion factor (a).

$$\text{grams Mg} \times \frac{6.022 \times 10^{23} \text{ atoms Mg}}{1 \text{ molar mass Mg}}$$

$$5.00 \cancel{\text{g Mg}} \times \frac{6.022 \times 10^{23} \text{ atoms Mg}}{24.31 \cancel{\text{g Mg}}} = 1.24 \times 10^{23} \text{ atoms Mg}$$

An alternative solution is first to convert grams of magnesium to moles of magnesium, which are then changed to atoms of magnesium.

$$\text{grams Mg} \longrightarrow \text{moles Mg} \longrightarrow \text{atoms Mg}$$

Use conversion factor (c) followed by (a). The calculation setup is

$$5.00 \cancel{\text{g Mg}} \times \frac{\cancel{1 \text{ mol Mg}}}{24.31 \cancel{\text{g Mg}}} \times \frac{6.022 \times 10^{23} \text{ atoms Mg}}{\cancel{1 \text{ mol Mg}}} = 1.24 \times 10^{23} \text{ atoms Mg}$$

Thus 1.24×10^{23} atoms of Mg are contained in 5.00 g of Mg.

EXAMPLE 7.3 What is the mass, in grams, of one atom of carbon?

SOLUTION The molar mass of carbon is 12.01 g. The factor needed to convert atoms to grams is conversion factor (b).

$$\text{atoms C} \longrightarrow \text{grams C}$$

$$\text{atoms C} \times \frac{1 \text{ molar mass}}{6.022 \times 10^{23} \text{ atoms C}}$$

$$1 \cancel{\text{atom C}} \times \frac{12.01 \text{ g C}}{6.022 \times 10^{23} \cancel{\text{atoms C}}} = 1.994 \times 10^{-23} \text{ g C}$$

EXAMPLE 7.4 What is the mass of 3.01×10^{23} atoms of sodium?

SOLUTION The information needed to solve this problem is the molar mass of Na (22.99 g) and conversion factor (b).

$$\text{atoms Na} \longrightarrow \text{grams Na}$$

$$\text{atoms Na} \times \frac{1 \text{ molar mass Na}}{6.022 \times 10^{23} \text{ atoms Na}}$$

$$3.01 \times 10^{23} \cancel{\text{atoms Na}} \times \frac{22.99 \text{ g Na}}{6.022 \times 10^{23} \cancel{\text{atoms Na}}} = 11.5 \text{ g Na}$$

EXAMPLE 7.5 What is the mass of 0.252 mol of Cu?

SOLUTION The information needed to solve this problem is the molar mass of Cu (63.55 g) and conversion factor (d).

$$\text{moles Cu} \longrightarrow \text{grams Cu}$$

$$\text{moles Cu} \times \frac{1 \text{ molar mass Cu}}{1 \text{ mole Cu}}$$

$$0.252 \cancel{\text{mol Cu}} \times \frac{63.55 \text{ g Cu}}{1 \cancel{\text{mol Cu}}} = 16.0 \text{ g Cu}$$

EXAMPLE 7.6

How many oxygen atoms are present in 1.00 mol of oxygen molecules?

SOLUTION

Oxygen is a diatomic molecule with the formula O_2. Therefore a molecule of oxygen contains two atoms of oxygen.

$$\frac{2 \text{ atoms O}}{1 \text{ molecule } O_2}$$

The sequence of conversions is

$$\text{moles } O_2 \longrightarrow \text{molecules } O_2 \longrightarrow \text{atoms O}$$

Two conversion factors are needed; they are

$$\frac{6.022 \times 10^{23} \text{ molecules } O_2}{1 \text{ mol } O_2} \quad \text{and} \quad \frac{2 \text{ atoms O}}{1 \text{ molecule } O_2}$$

The calculation is

$$1.00 \cancel{\text{mol } O_2} \times \frac{6.022 \times 10^{23} \cancel{\text{molecules } O_2}}{1 \cancel{\text{mol } O_2}} \times \frac{2 \text{ atoms O}}{1 \cancel{\text{molecule } O_2}} = 1.20 \times 10^{24} \text{ atoms O}$$

PRACTICE What is the mass of 2.50 mol of helium?
Answer: 10.0 g of helium

PRACTICE How many atoms are present in 0.025 mol of iron?
Answer: 1.51×10^{22} atoms

7.2 Molar Mass of Compounds

One mole of a compound contains Avogadro's number of formula units of that compound. The terms *molecular weight* and *formula weight* have been used in the past to refer to the mass of one mole of a compound. However, the term *molar mass* is more inclusive, since it can be used for all types of compounds.

If the formula of a compound is known, its molar mass may be determined by adding together the molar masses of all the atoms in the formula. If more than one atom of any element is present, its mass must be added as many times as it is used.

One-mole samples of various ▶ substances. *Clockwise from lower left:* magnesium, carbon, copper(II) sulfate, copper, mercury, potassium permanganate, cadmium, and sodium chloride (*center*).

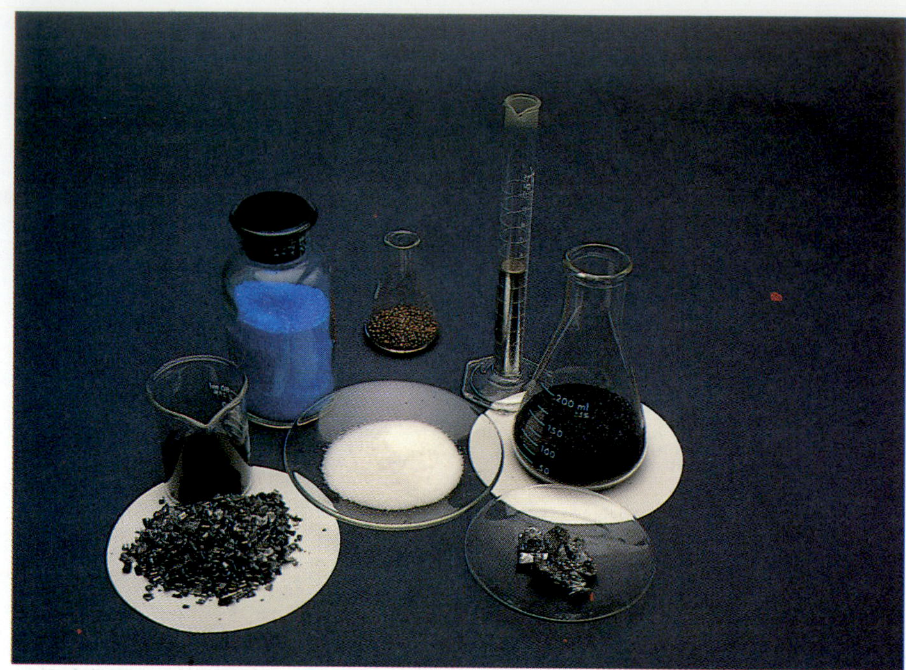

EXAMPLE 7.7

The formula for water is H_2O. What is its molar mass?

SOLUTION

Proceed by looking up the molar masses of H (1.008 g) and O (16.00 g) and adding together the masses of all the atoms in the formula unit. Water contains two atoms of H and one atom of O. Thus,

$$\begin{aligned} 2\ \text{H} &= 2 \times 1.008\ \text{g} = 2.016\ \text{g} \\ 1\ \text{O} &= 1 \times 16.00\ \text{g} = \underline{16.00\ \text{g}} \\ & \qquad\qquad\qquad 18.02\ \text{g} = \text{molar mass} \end{aligned}$$

EXAMPLE 7.8

Calculate the molar mass of calcium hydroxide, $Ca(OH)_2$.

SOLUTION

The formula of this substance contains one atom of Ca and two atoms each of O and H. Proceed as in Example 7.7. Thus

$$\begin{aligned} 1\ \text{Ca} &= 1 \times 40.08\ \text{g} = 40.08\ \text{g} \\ 2\ \text{O} &= 2 \times 16.00\ \text{g} = 32.00\ \text{g} \\ 2\ \text{H} &= 2 \times 1.008\ \text{g} = \underline{2.016\ \text{g}} \\ & \qquad\qquad\qquad 74.10\ \text{g} = \text{molar mass} \end{aligned}$$

The molar masses of elements in this text are rounded off to four significant figures to simplify calculations. (However, this simplification cannot be made in more exacting chemical work.)

Practice Calculate the molar mass of KNO_3.
Answer: 101.1 g KNO_3

A one molar mass of a compound is one mole of that compound and therefore contains Avogadro's number of formula units or molecules. Consider the compound hydrogen chloride, HCl. One atom of H combines with one atom of Cl to form one molecule of HCl. When 1 molar mass of H (1.008 g of H representing 1 mol or 6.022×10^{23} H atoms) combines with 1 molar mass of Cl (35.45 g of Cl representing 1 mol or 6.022×10^{23} Cl atoms), 1 molar mass of HCl (36.46 g of HCl representing 1 mol or 6.022×10^{23} HCl molecules) is produced. These relationships are summarized in the following table.

H	Cl	HCl
6.022×10^{23} H *atoms*	6.022×10^{23} Cl *atoms*	6.022×10^{23} HCl *molecules*
1 mol H *atoms*	1 mol Cl *atoms*	1 mol HCl *molecules*
1.008 g H	35.45 g Cl	36.46 g HCl
1 molar mass H	1 molar mass Cl	1 molar mass HCl

In dealing with diatomic elements (H_2, O_2, N_2, F_2, Cl_2, Br_2, and I_2), special care must be taken to distinguish between a mole of atoms and a mole of molecules. For example consider *one* mole of oxygen molecules, which has a mass of 32.00 g. This quantity is equal to *two* moles of oxygen atoms. The key concept is that one mole represents Avogadro's number of the particular chemical entity—atoms, molecules, formula units and so forth—that is under consideration.

$$1 \text{ mol } H_2O = 18.02 \text{ g } H_2O = 6.022 \times 10^{23} \text{ molecules}$$
$$1 \text{ mol NaCl} = 58.44 \text{ g NaCl} = 6.022 \times 10^{23} \text{ formula units}$$
$$1 \text{ mol } H_2 = 2.016 \text{ g } H_2 = 6.022 \times 10^{23} \text{ molecules}$$
$$1 \text{ mol } HNO_3 = 63.02 \text{ g } HNO_3 = 6.022 \times 10^{23} \text{ molecules}$$
$$1 \text{ mol } K_2SO_4 = 174.3 \text{ g } K_2SO_4 = 6.022 \times 10^{23} \text{ formula units}$$

1 mol = 6.022×10^{23} formula units or molecules
= 1 molar mass of a compound

We often need to convert moles of a compound to grams, and grams of a compound to moles. The factors for these conversions are

$$\text{grams to moles: } \frac{1 \text{ mol of a substance}}{1 \text{ molar mass of the substance}}$$

$$\text{moles to grams: } \frac{1 \text{ molar mass of a substance}}{1 \text{ mol of the substance}}$$

EXAMPLE 7.9 What is the mass of 1 mol of sulfuric acid, H_2SO_4?

SOLUTION One mole of H_2SO_4 is one molar mass of H_2SO_4. The problem, therefore, is solved in a similar manner to Examples 7.7 and 7.8. Look up the molar masses of H, S, and O, and solve.

$$\begin{aligned} 2\text{ H} &= 2 \times 1.008 \text{ g} = 2.016 \text{ g} \\ 1\text{ S} &= 1 \times 32.06 \text{ g} = 32.06 \text{ g} \\ 4\text{ O} &= 4 \times 16.00 \text{ g} = \underline{64.00 \text{ g}} \\ & \qquad\qquad\qquad 98.08 \text{ g} = \text{mass of 1 mol of } H_2SO_4 \end{aligned}$$

EXAMPLE 7.10 How many moles of NaOH are there in 1.00 kg of sodium hydroxide?

SOLUTION First we know that

$$1 \text{ molar mass} = (22.99 \text{ g} + 16.00 \text{ g} + 1.008 \text{ g}) \text{ or } 40.00 \text{ g NaOH}$$

To convert grams to moles we use the conversion factor

$$\frac{1 \text{ mol}}{1 \text{ molar mass}} \quad \text{or} \quad \frac{1 \text{ mol NaOH}}{40.00 \text{ g NaOH}}$$

Use this conversion sequence:

kg NaOH ⟶ g NaOH ⟶ mol NaOH

The calculation is

$$1.00 \cancel{\text{kg NaOH}} \times \frac{1000 \cancel{\text{g NaOH}}}{\cancel{\text{kg NaOH}}} \times \frac{1 \text{ mol NaOH}}{40.00 \text{ g NaOH}} = 25.0 \text{ mol NaOH}$$

$$1.00 \text{ kg NaOH} = 25.0 \text{ mol NaOH}$$

EXAMPLE 7.11 What is the mass of 5.00 mol of water?

SOLUTION First we know that

$$1 \text{ mol } H_2O = 18.02 \text{ g} \qquad \text{(Example 7.7)}$$

The conversion is

mol H_2O ⟶ g H_2O

To convert moles to grams use the conversion factor

$$\frac{1 \text{ molar mass } H_2O}{1 \text{ mol } H_2O} \quad \text{or} \quad \frac{18.02 \text{ g } H_2O}{1 \text{ mol } H_2O}$$

The calculation is

$$5.00 \cancel{\text{mol } H_2O} \times \frac{18.02 \text{ g } H_2O}{1 \cancel{\text{mol } H_2O}} = 90.1 \text{ g } H_2O$$

EXAMPLE 7.12

How many molecules of HCl are there in 25.0 g of hydrogen chloride?

SOLUTION

From the formula we find that the molar mass of HCl is 36.46 g (1.008 g + 35.45 g). The sequence of conversions is

$$\text{g HCl} \longrightarrow \text{mol HCl} \longrightarrow \text{molecules HCl}$$

using the conversion factors

$$\frac{1\text{ mol HCl}}{36.46\text{ g HCl}} \quad \text{and} \quad \frac{6.022 \times 10^{23}\text{ molecules HCl}}{1\text{ mol HCl}}$$

$$25.0\,\cancel{\text{g HCl}} \times \frac{1\,\cancel{\text{mol HCl}}}{36.46\,\cancel{\text{g HCl}}} \times \frac{6.022 \times 10^{23}\text{ molecules HCl}}{1\,\cancel{\text{mol HCl}}} = 4.13 \times 10^{23}\text{ molecules HCl}$$

PRACTICE What is the mass of 0.150 mol of Na_2SO_4?
Answer: 21.3 g Na_2SO_4

PRACTICE How many moles are there in 500.0 g of $AlCl_3$?
Answer: 3.751 mol of $AlCl_3$

7.3 PERCENT COMPOSITION OF COMPOUNDS

Percent means parts per one hundred parts. Just as each piece of pie is a percent of the whole pie, each element in a compound is a percent of the whole compound. The **percent composition of a compound** is the *mass percent* of each element in the compound. The molar mass represents the total mass, or 100%, of the compound. Thus the percent composition of water, H_2O, is 11.19% H and 88.79% O by mass. According to the Law of Definite Composition, the percent composition must be the same no matter what size sample is taken.

percent composition of a compound

The percent composition of a compound can be determined if its formula is known or if the masses of two or more elements that have combined with each other are known or are experimentally determined.

If the formula is known, it is essentially a two-step process to determine the percent composition.

Step 1 Calculate the molar mass as was done in Section 7.2.
Step 2 Divide the total mass of each element in the formula by the molar mass and multiply by 100. This gives the percent composition.

$$\frac{\text{total mass of the element}}{\text{molar mass}} \times 100 = \text{percent of the element}$$

EXAMPLE 7.13

Calculate the percent composition of sodium chloride, NaCl

SOLUTION

Step 1 Calculate the molar mass of NaCl:

$$1\ \text{Na} = 1 \times 22.99\ \text{g} = 22.99\ \text{g}$$
$$1\ \text{Cl} = 1 \times 35.45\ \text{g} = \underline{35.45\ \text{g}}$$
$$58.44\ \text{g (molar mass)}$$

Step 2 Now calculate the percent composition. We know there are 22.99 g Na and 35.45 g Cl in 58.44 g NaCl.

$$\text{Na:}\quad \frac{22.99\ \text{g Na}}{58.44\ \text{g NaCl}} \times 100 = 39.34\%\ \text{Na}$$

$$\text{Cl:}\quad \frac{35.45\ \text{g Cl}}{58.44\ \text{g NaCl}} \times 100 = \underline{60.66\%\ \text{Cl}}$$
$$100.00\%\ \text{total}$$

In any two-component system, if one percent is known, the other is automatically defined by the difference; that is, if Na = 39.34%, then Cl = 100% − 39.34% = 60.66%. However, the calculation of the percent of each component should be carried out, since this provides a check against possible error. The percent composition data should add up to 100 ± 0.2%.

EXAMPLE 7.14

Calculate the percent composition of potassium chloride, KCl.

SOLUTION

Step 1 Calculate the molar mass of KCl:

$$1\ \text{K} = 1 \times 39.10\ \text{g} = 39.10\ \text{g}$$
$$1\ \text{Cl} = 1 \times 35.45\ \text{g} = \underline{35.45\ \text{g}}$$
$$74.55\ \text{g}\quad \text{(molar mass)}$$

Step 2 Now calculate the percent composition. We know there are 39.10 g K and 35.45 g Cl in 74.55 g KCl.

$$\text{K:}\quad \frac{39.10\ \text{g K}}{74.55\ \text{g KCl}} \times 100 = 52.45\%\ \text{K}$$

$$\text{Cl:}\quad \frac{35.45\ \text{g Cl}}{74.55\ \text{g KCl}} \times 100 = \underline{47.55\%\ \text{Cl}}$$
$$100.00\%\ \text{total}$$

Comparing the data calculated for NaCl and for KCl, we see that NaCl contains a higher percentage of Cl by mass, although each compound has a one-to-one atom ratio of Cl to Na and Cl to K. The reason for this mass percent difference is that Na and K do not have the same atomic masses.

It is important to realize that, when we compare 1 mol of NaCl with 1 mol of KCl, each quantity contains the same number of Cl atoms—namely, 1 mol of Cl atoms. However, if we compare equal masses of NaCl and KCl, there will be more Cl atoms in the mass of NaCl since NaCl has a higher mass percent of Cl.

1 mole NaCl contains	100 g NaCl contains
1 mol Na 1 mol Cl 60.66% Cl	39.34 g Na 60.66 g Cl

1 mole KCl contains	100 g KCl contains
1 mol K 1 mol Cl 47.55% Cl	52.43 g K 47.55 g Cl

EXAMPLE 7.15

Calculate the percent composition of potassium sulfate, K_2SO_4.

SOLUTION

Step 1 Calculate the molar mass of K_2SO_4:

$$\begin{aligned} 2\,K &= 2 \times 39.10\text{ g} = 78.20\text{ g} \\ 1\,S &= 1 \times 32.06\text{ g} = 32.06\text{ g} \\ 4\,O &= 4 \times 16.00\text{ g} = \underline{64.00\text{ g}} \\ &\qquad\qquad\qquad 174.3\text{ g} \quad \text{(molar mass)} \end{aligned}$$

Step 2 Now calculate the percent composition. We know there are 78.20 g of K, 32.06 g of S, and 64.00 g of O in 174.26 g of K_2SO_4.

$$\text{K:} \quad \frac{78.20\text{ g K}}{174.26\text{ g } K_2SO_4} \times 100 = 44.88\%\text{ K}$$

$$\text{S:} \quad \frac{32.06\text{ g S}}{174.26\text{ g } K_2SO_4} \times 100 = 18.40\%\text{ S}$$

$$\text{O:} \quad \frac{64.00\text{ g O}}{174.26\text{ g } K_2SO_4} \times 100 = \underline{36.73\%\text{ O}}$$

$$100.01\%\text{ total}$$

PRACTICE Calculate the percent composition of $Ca(NO_3)_2$
Answer: Ca = 24.42%, N = 17.07%, O = 58.50%

PRACTICE Calculate the percent composition of K_2CrO_4.
Answer: K = 40.27%, Cr = 26.78%, O = 32.96%

The percent composition can be determined from experimental data without knowing the formula of a compound. This determination is done by calculating the mass of each element in a compound as a percentage of the total mass of the compound formed.

EXAMPLE 7.16

When heated in the air, 1.63 g of zinc, Zn, combine with 0.40 g of oxygen, O_2, to form zinc oxide. Calculate the percent composition of the compound formed.

SOLUTION

First, calculate the total mass of the compound formed.

$$\begin{array}{ll} 1.63\text{ g Zn} & \\ 0.40\text{ g } O_2 & \\ \hline 2.03\text{ g} & = \text{total mass of product} \end{array}$$

Then divide the mass of each element by the total mass (2.03 g) and multiply by 100.

$$\frac{1.63\ \cancel{g}}{2.03\ \cancel{g}} \times 100 = 80.3\%\ \text{Zn}$$

$$\frac{0.40\ \cancel{g}}{2.03\ \cancel{g}} \times 100 = \frac{19.7\%\ \text{O}}{100.0\%\ \text{total}}$$

The compound formed contains 80.3% Zn and 19.7% O.

PRACTICE Aluminum chloride is formed by reacting 13.43 g aluminum with 53.18 g chlorine. What is the percent composition of the compound?

Answer: Al = 20.16%, Cl = 79.84%

7.4 EMPIRICAL FORMULA VERSUS MOLECULAR FORMULA

empirical formula

molecular formula

The **empirical formula**, or *simplest formula*, gives the smallest whole-number ratio of the atoms that are present in a compound. This formula gives the relative number of atoms of each element in the compound.

The **molecular formula** is the true formula, representing the total number of atoms of each element present in one molecule of a compound. It is entirely possible that two or more substances will have the same percent composition, yet be distinctly different compounds. For example, acetylene, C_2H_2, is a common gas used in welding; benzene, C_6H_6, is an important solvent obtained from coal tar and is used in the synthesis of styrene and nylon. Both acetylene and benzene contain 92.3% C and 7.7% H. The smallest ratio of C and H corresponding to these percents is CH (1:1). Therefore the empirical formula for both acetylene and benzene is CH, even though it is known that the molecular formulas are C_2H_2 and C_6H_6, respectively. It is not uncommon for the molecular formula to be the same as the empirical formula. If the molecular formula is not the same, it will be an integral (whole number) multiple of the empirical formula.

TABLE 7.1 Molecular Formulas of Two Compounds Having an Empirical Formula with a 1:1 Ratio of Carbon and Hydrogen Atoms

Formula	Composition % C	Composition % H	Molar mass
CH (empirical)	92.3	7.7	13.02 (empirical)
C_2H_2 (acetylene)	92.3	7.7	26.04 (2 × 13.02)
C_6H_6 (benzene)	92.3	7.7	78.06 (6 × 13.02)

TABLE 7.2 Some Empirical and Molecular Formulas

Compound	Empirical formula	Molecular formula	Compound	Empirical formula	Molecular formula
Acetylene	CH	C_2H_2	Diborane	BH_3	B_2H_6
Benzene	CH	C_6H_6	Hydrazine	NH_2	N_2H_4
Ethylene	CH_2	C_2H_4	Hydrogen	H	H_2
Formaldehyde	CH_2O	CH_2O	Chlorine	Cl	Cl_2
Acetic acid	CH_2O	$C_2H_4O_2$	Bromine	Br	Br_2
Glucose	CH_2O	$C_6H_{12}O_6$	Oxygen	O	O_2
Hydrogen chlorine	HCl	HCl	Nitrogen	N	N_2
Carbon dioxide	CO_2	CO_2			

$$CH = \text{empirical formula}$$
$$(CH)_2 = C_2H_2 = \text{acetylene} \quad \text{(molecular formula)}$$
$$(CH)_6 = C_6H_6 = \text{benzene} \quad \text{(molecular formula)}$$

Table 7.1 summarizes the data concerning these CH formulas. Table 7.2 shows empirical and molecular formula relationships of other compounds.

7.5 Calculation of Empirical Formulas

It is possible to establish an empirical formula because (1) the individual atoms in a compound are combined in whole-number ratios and (2) each element has a specific atomic mass.

In order to calculate the empirical formula we need to know (1) the elements that are combined, (2) their atomic masses, and (3) the ratio by mass or percentage in which they are combined. If elements A and B form a compound, we may represent the empirical formula as A_xB_y, where x and y are small whole numbers that represent the number of atoms of A and B. To write the empirical formula we must determine x and y.

The solution to this problem requires three or four steps.

Step 1 Assume a definite starting quantity (usually 100.0 g) of the compound, if not given, and express the mass of each element in grams.

Step 2 Multiply the mass (in grams) of each element by the factor 1 mol/1 molar mass to convert grams to moles. This conversion gives the number of moles of atoms of each element in the quantity assumed. At this point these numbers will usually not be whole numbers.

Step 3 Divide each of the values obtained in Step 2 by the smallest of these values. If the numbers obtained by this procedure are whole numbers, use them as subscripts in writing the empirical formula. If the numbers obtained are not whole numbers, go on to Step 4.

Step 4 Multiply the values obtained in Step 3 by the smallest number that will convert them to whole numbers. Use these whole numbers as the subscripts in the empirical formula. For example, if the ratio of A to B is 1.0:1.5, multiply both numbers by 2 to obtain a ratio of 2:3. The empirical formula then is A_2B_3.

EXAMPLE 7.17

Calculate the empirical formula of a compound containing 11.19% hydrogen, H, and 88.79% oxygen, O.

SOLUTION

Step 1 Express each element in grams. If we assume that there are 100.0 g of material, then the percent of each element is equal to the grams of each element in 100 g, and the percent sign can be omitted.

H = 11.19 g

O = 88.79 g

Step 2 Multiply the grams of each element by the proper mol/molar mass factor to obtain the relative number of moles of atoms:

$$\text{H:}\quad 11.19\,\cancel{\text{g H}} \times \frac{1\text{ mol H atoms}}{1.008\,\cancel{\text{g H}}} = 11.10\text{ mol H atoms}$$

$$\text{O:}\quad 88.79\,\cancel{\text{g O}} \times \frac{1\text{ mol O atoms}}{16.00\,\cancel{\text{g O}}} = 5.549\text{ mol O atoms}$$

The formula could be expressed as $H_{11.10}O_{5.556}$. However, it is customary to use the smallest whole-number ratio of atoms. This ratio is calculated in Step 3.

Step 3 Change these numbers to whole numbers by dividing each of them by the smaller number.

$$H = \frac{11.10\text{ mol}}{5.549\text{ mol}} = 2 \qquad O = \frac{5.549\text{ mol}}{5.549\text{ mol}} = 1$$

In this step the ratio of atoms has not changed, because we divided the number of moles of each element by the same number.

The simplest ratio of H to O is 2:1.

Empirical formula = H_2O

EXAMPLE 7.18

The analysis of a salt showed that it contained 56.58% potassium, K, 8.68% carbon, C, and 34.73% oxygen, O. Calculate the empirical formula for this substance.

SOLUTION

Steps 1 and 2 After changing the percent of each element to grams, find the relative number of moles of each element by multiplying by the proper mol/molar mass factor.

$$\text{K:}\quad 56.58\ \cancel{\text{g K}} \times \frac{1\ \text{mol K atoms}}{39.10\ \cancel{\text{g K}}} = 1.447\ \text{mol K atoms}$$

$$\text{C:}\quad 8.68\ \cancel{\text{g C}} \times \frac{1\ \text{mol C atoms}}{12.01\ \cancel{\text{g C}}} = 0.723\ \text{mol C atoms}$$

$$\text{O:}\quad 34.73\ \cancel{\text{g O}} \times \frac{1\ \text{mol O atoms}}{16.00\ \cancel{\text{g O}}} = 2.171\ \text{mol O atoms}$$

Step 3 Divide each number of moles by the smallest value.

$$\text{K} = \frac{1.447\ \text{mol}}{0.723\ \text{mol}} = 2.00$$

$$\text{C} = \frac{0.723\ \text{mol}}{0.723\ \text{mol}} = 1.00$$

$$\text{O} = \frac{2.171\ \text{mol}}{0.723\ \text{mol}} = 3.00$$

The simplest ratio of K:C:O is 2:1:3.

Empirical formula = K_2CO_3

EXAMPLE 7.19

A sulfide of iron was formed by combining 2.233 g of iron, Fe, with 1.926 g of sulfur, S. What is the empirical formula of the compound?

SOLUTION

Steps 1 and 2 The grams of each element are given, so we use them directly in our calculations. Calculate the relative number of moles of each element by multiplying grams of each element by the proper mol/molar mass factor.

$$\text{Fe:}\quad 2.233\ \cancel{\text{g Fe}} \times \frac{1\ \text{mol Fe atoms}}{55.85\ \cancel{\text{g Fe}}} = 0.03998\ \text{mol Fe atoms}$$

$$\text{S:}\quad 1.926\ \cancel{\text{g S}} \times \frac{1\ \text{mol S atoms}}{32.06\ \cancel{\text{g S}}} = 0.06007\ \text{mol S atoms}$$

Step 3 Divide each number of moles by the smaller of the two numbers.

$$\text{Fe} = \frac{0.03998\ \text{mol}}{0.03998\ \text{mol}} = 1.000$$

$$\text{S} = \frac{0.06007\ \text{mol}}{0.03998\ \text{mol}} = 1.503$$

Step 4 We still have not reached a ratio that will give a formula containing whole numbers of atoms, so we must double each value to obtain a ratio of 2.000 atoms of Fe to 3.000 atoms of S. Doubling both values does not change the ratio of Fe and S atoms.

Fe: $1.000 \times 2 = 2.000$

S: $1.503 \times 2 = 3.006$

Empirical formula = Fe_2S_3

PRACTICE Calculate the empirical formula of a compound containing 53.33% C, 11.11% H, and 35.53% O.
Answer: C_2H_5O

PRACTICE Calculate the empirical formula of a compound that contains 43.7% phosphorus and 56.3% O by mass.
Answer: P_2O_5

In many of these calculations results may vary somewhat from an exact whole number, which can be due to experimental errors in obtaining the data or from rounding off numbers. Calculations that vary by no more than ±0.1 from a whole number can usually be rounded off to the nearest whole number. Deviations greater than about 0.1 unit usually mean that the calculated ratios need to be multiplied by a factor to make them all whole numbers. For example, an atom ratio of 1:1.33 should be multiplied by 3 to make the ratio 3:4.

7.6 Calculation of the Molecular Formula from the Empirical Formula

The molecular formula can be calculated from the empirical formula if the molar mass, in addition to data for calculating the empirical formula, is known. The molecular formula, as stated in Section 7.4, will be equal to or some multiple of the empirical formula. For example, if the empirical formula of a compound of hydrogen and fluorine is HF, the molecular formula can be expressed as $(HF)_n$, where $n = 1, 2, 3, 4, \ldots$ This n means that the molecular formula could be HF, H_2F_2, H_3F_3, H_4F_4, and so on. To determine the molecular formula, we must evaluate n.

$$n = \frac{\text{molar mass}}{\text{mass of empirical formula}} = \text{number of empirical formula units}$$

What we actually calculate is the number of units of the empirical formula that is contained in the molecular formula.

EXAMPLE 7.20 A compound of nitrogen and oxygen with a molar mass of 92.00 g was found to have an empirical formula of NO_2. What is its molecular formula?

SOLUTION

Step 1 Let n be the number of (NO_2) units in a molecule; then the molecular formula is $(NO_2)_n$.

Step 2 Each (NO_2) unit has a mass of [14.01 g + (2 × 16.00 g)] or 46.01 g. The molar mass of $(NO_2)_n$ is 92.00 g and the number of 46.01 units in 92.00 is 2.

$$n = \frac{92.00\ \text{g}}{46.01\ \text{g}} = 2 \quad \text{(empirical formula units)}$$

Step 3 The molecular formula is $(NO_2)_2$, or N_2O_4.

EXAMPLE 7.21

The hydrocarbon propylene has a molar mass of 42.00 g and contains 14.3% H and 85.7% C. What is its molecular formula?

SOLUTION

Step 1 First find the empirical formula:

$$\text{C:} \quad 85.7\ \cancel{\text{g C}} \times \frac{1\ \text{mol C atoms}}{12.01\ \cancel{\text{g C}}} = 7.14\ \text{mol C atoms}$$

$$\text{H:} \quad 14.3\ \cancel{\text{g H}} \times \frac{1\ \text{mol H atoms}}{1.008\ \cancel{\text{g H}}} = 14.2\ \text{mol H atoms}$$

Divide each value by the smaller number of moles.

$$\text{C} = \frac{7.14\ \text{mol}}{7.14\ \text{mol}} = 1.00$$

$$\text{H} = \frac{14.2\ \text{mol}}{7.14\ \text{mol}} = 1.99$$

Empirical formula = CH_2

Step 2 Determine the molecular formula from the empirical formula and the molar mass.

Molecular formula = $(CH_2)_n$

Molar mass = 42.00 g

Each CH_2 unit has a mass of (12.01 g + 2.016 g) or 14.03 g. The number of CH_2 units in 42.00 g is 3.

$$n = \frac{42.0\ \text{g}}{14.03\ \text{g}} = 3 \quad \text{(empirical formula units)}$$

The molecular formula is $(CH_2)_3$, or C_3H_6.

PRACTICE Calculate the empirical and molecular formulas of a compound that contains 80.0% C, 20.0% H, and has a molar mass of 30.00 g.

Answers: The empirical formula is CH_3
The molecular formula is C_2H_6

CHEMISTRY IN ACTION

THE TASTE OF CHEMISTRY

Flavorings, seasonings and preservatives—notably salt—have been added to foods since ancient civilizations. Spices were originally added as preservatives when refrigeration was unavailable. These spices contained mild antiseptics and antioxidants and were effective in prolonging the time during which food could be eaten. Over the course of time a great variety of substances (food additives) came to be used in foods—preservatives, colorings, flavorings, antioxidants, sweetners, etc. Many of these substances are now regarded as virtual necessities for processing foods. However, there is genuine concern that some additives, particularly those which are not naturally present in foods, may be detrimental to health when consumed. This concern has led to the passage of federal and state laws regulating the food industry.

The United States Food and Drug Administration (FDA) is the principle agency charged with enforcing federal laws concerning food and must approve the use of all food additives. The FDA divides food additives into several categories. These include a classification known as GRAS (Generally Regarded As Safe). The GRAS designation includes substances that were in use in 1958 and that met certain specifications for safety. All substances introduced after 1958 have been approved on an individual basis.

Before commercial use of a new food additive, a company must provide the FDA with satisfactory evidence that the chemical is safe for the proposed usage. Providing this evidence of safety is complex and expensive. Although mainly concerned with chemistry, this task requires the services of many people trained in a variety of disciplines—biochemistry, microbiology, medicine, physiology, etc. Research and testing may have to be done in several laboratories before final FDA approval (or rejection) is obtained.

The amount of an additive which is considered to be safe in foods is determined by the maximum tolerable daily intake (MTDI). This is the amount of the food additive that can be eaten daily for a lifetime without adverse effects. It is calculated on the basis of body mass (mg/kg/day). If there is any doubt regarding the safety of an additive a time limit is determined and a conditional MTDI is issued and reviewed with further testing at the end of the initial time period. To establish the MTDI, experiments are run on animals, increasing the quantities of the additive in successive experiments until acute and chronic toxicity occurs. A minimum of two species of animals must be tested, the most sensitive species forming the basis for the appropriate level for the additive. This quantity is then divided by 100 (for additional safety) to set the MTDI.

Once the MDTI is established all foods in which the use of the additive is proposed must be considered. An estimate is made of the maximum amount of the additive that could be ingested. On the basis of this information the use may then be restricted to only certain foods (as in the exclusion of sulfites from most meats) or broadened to include additional foods.

Unfortunately, even with all the quantitative testing it is extremely difficult to determine safe levels of additives. The majority of toxicological data is from animal testing. The results are often not the same as in humans—chemicals toxic to one species may not be so for humans. Children add a further complication.

(*continued*)

(continued)

They cannot simply be considered as small adults. A child has a much greater energy demand per kilogram than an adult. Often childrens' chemical defense mechanisms are considerably different than those of an adult.

Flavorings are the largest class of food additives. There are 1100–1400 natural or synthetic flavorings used in foods. The task of checking for effects of these additives is herculean. Flavorings are complex mixtures of chemicals used in very small amounts making them difficult to analyze. Distinctions between artificial and natural flavorings must also be monitored (see Chemistry in Action, Chapter 5). Chemists rely on quantitative analysis techniques (such as gas-liquid chromatography) to separate and identify components in these mixtures. It is possible to detect compounds in amounts as low as 10 $\mu g/kg$ (parts per billion)

Concepts in Review

1. Explain the meaning of the mole.
2. Discuss the relationship between a mole and Avogadro's number.
3. Convert grams, atoms, molecules, and molar masses to moles, and vice versa.
4. Determine the molar mass of a compound from the formula.
5. Calculate the percent composition of a compound from its formula.
6. Calculate the percent composition of a compound from experimental data on combining masses.
7. Explain the relationship between an empirical formula and a molecular formula.
8. Determine the empirical formula for a compound from its percent composition.
9. Calculate the molecular formula of a compound from its percent composition and molar mass.

Exercises

An asterisk indicates a more challenging question or problem.

1. What is a mole?
2. Which would have a higher mass: a mole of potassium atoms, or a mole of gold atoms?
3. Which would contain more atoms: a mole of potassium atoms, or a mole of gold atoms?
4. Which would contain more electrons: a mole of potassium atoms, or a mole of gold atoms?

*5. If the atomic mass scale had been defined differently, with an atom of $^{12}_{6}C$ being defined as a mass of 50 amu, would this have any effect on the value of Avogadro's number? Explain.

6. What is the numerical value of Avogadro's number?
7. What is the relationship between Avogadro's number and the mole?
8. Complete the following statements, supplying the proper quantity.
   (a) A mole of oxygen atoms (O) contains ________ atoms.
   (b) A mole of oxygen molecules (O_2) contains ________ molecules.
   (c) A mole of oxygen molecules (O_2) contains ________ atoms.
   (d) A mole of oxygen atoms (O) has a mass of ________ grams.

(e) A mole of oxygen molecules (O_2) has a mass of ________ grams.

9. Which of the following statements are correct? Rewrite the incorrect statements to make them correct.
   (a) One atomic mass of any element contains 6.022×10^{23} atoms.
   (b) The mass of one atom of chlorine is $\frac{35.45 \text{ g}}{6.022 \times 10^{23} \text{ atoms}}$.
   (c) A mole of magnesium atoms (24.31 g) contains the same number of atoms as a mole of sodium atoms (22.99 g).
   (d) A mole of bromine atoms contains 6.022×10^{23} atoms of bromine.
   (e) A mole of chlorine molecules (Cl_2) contains 6.022×10^{23} atoms of chlorine.
   (f) A mole of aluminum atoms has the same mass as a mole of tin atoms.
   (g) A mole of H_2O contains 6.022×10^{23} atoms.
   (h) A mole of hydrogen molecules (H_2) contains 1.204×10^{24} electrons.
10. How many molecules are present in 1 molar mass of sulfuric acid, H_2SO_4? How many atoms are present?
11. In calculating the empirical formula of a compound from its percent composition, why do we choose to start with 100.0 g of the compound?
12. List four characteristics of food additives.
13. Explain how the maximum tolerable daily intake of an additive is determined.
14. Why is it difficult to determine safe levels of food additives?
15. Which of the following statements are correct? Rewrite the incorrect statements to make them correct.
   (a) A mole of sodium and a mole of sodium chloride contain the same number of sodium atoms.
   (b) One mole of nitrogen gas (N_2) has a mass of 14.01 g.
   (c) The percent of oxygen is higher in K_2CrO_4 than it is in Na_2CrO_4.
   (d) The number of Cr atoms is the same in a mole of K_2CrO_4 as it is in a mole of Na_2CrO_4.
   (e) Both K_2CrO_4 and Na_2CrO_4 contain the same percent by mass of Cr.
   (f) A molar mass of sucrose, $C_{12}H_{22}O_{11}$, contains 1 mol of sucrose molecules.
   (g) Two moles of nitric acid, HNO_3, contain 6 moles of oxygen atoms.
   (h) The empirical formula of sucrose, $C_{12}H_{22}O_{11}$, is CH_2O.
   (i) A hydrocarbon that has a molar mass of 280 and an empirical formula of CH_2 has a molecular formula of $C_{22}H_{44}$.
   (j) The empirical formula is often called the simplest formula.
   (k) The empirical formula of a compound gives the smallest whole-number ratio of the atoms that are present in a compound.
   (l) If the molecular formula and the empirical formula of a compound are not the same, the empirical formula will be an integral multiple of the molecular formula.
   (m) The empirical formula of benzene, C_6H_6, is CH.
   (n) A compound having an empirical formula of CH_2O, and a molar mass of 60, has a molecular formula of $C_3H_6O_3$.
16. Determine the molar masses of the following compounds:
   (a) KBr (f) Fe_3O_4
   (b) Na_2SO_4 (g) $C_{12}H_{22}O_{11}$
   (c) $Pb(NO_3)_2$ (h) $Al_2(SO_4)_3$
   (d) C_2H_5OH (i) $(NH_4)_2HPO_4$
   (e) $HC_2H_3O_2$
17. Determine the molar mass of the following compounds:
   (a) NaOH (f) C_6H_5COOH
   (b) Ag_2CO_3 (g) $C_6H_{12}O_6$
   (c) Cr_2O_3 (h) $K_4Fe(CN)_6$
   (d) $(NH_4)_2CO_3$ (i) $BaCl_2 \cdot 2\,H_2O$
   (e) $Mg(HCO_3)_2$

Moles and Avogadro's number

18. How many moles of atoms are contained in the following?
   (a) 22.5 g Zn
   (b) 0.688 g Mg
   (c) 4.5×10^{22} atoms Cu
   (d) 382 g Co
   (e) 0.055 g Sn
   (f) 8.5×10^{24} molecules N_2
19. How many moles are contained in the following?
   (a) 25.0 g NaOH (d) 14.8 g CH_3OH
   (b) 44.0 g Br_2 (e) 2.88 g Na_2SO_4
   (c) 0.684 g $MgCl_2$ (f) 4.20 lb ZnI_2
20. Calculate the number of grams in each of the following:
   (a) 0.550 mol Au
   (b) 15.8 mol H_2O
   (c) 12.5 mol Cl_2
   (d) 3.15 mol NH_4NO_3
   (e) 4.25×10^{-4} mol H_2SO_4
   (f) 4.5×10^{22} molecules CCl_4

(g) 0.00255 mol Ti
(h) 1.5×10^{16} atoms S

21. How many molecules are contained in each of the following:
(a) 1.26 mol O_2
(b) 0.56 mol C_6H_6
(c) 16.0 g CH_4
(d) 1000. g HCl

22. Calculate the mass in grams of each of the following:
(a) 1 atom Pb
(b) 1 atom Ag
(c) 1 molecule H_2O
(d) 1 molecule $C_3H_5(NO_3)_3$

23. Make the following conversions:
(a) 8.66 mol Cu to grams Cu
(b) 125 mol Au to kilograms Au
(c) 10 atoms C to moles C
(d) 5000 molecules CO_2 to moles CO_2
(e) 28.4 g S to moles S
(f) 2.50 kg NaCl to moles NaCl
(g) 42.4 g Mg to atoms Mg
(h) 485 mL Br_2 ($d = 3.12$ g/mL) to moles Br_2

24. One mole of carbon disulfide (CS_2) contains
(a) How many carbon disulfide molecules?
(b) How many carbon atoms?
(c) How many sulfur atoms?
(d) How many total atoms of all kinds?

25. White phosphorus is one of several forms of phosphorus and exists as a waxy solid consisting of P_4 molecules. How many atoms are present in 0.350 mol of P_4?

26. How many grams of sodium contain the same number of atoms as 10.0 g of potassium?

27. One atom of an unknown element is found to have a mass of 1.79×10^{-23} g. What is the molar mass of this element?

*28. If a stack of 500 sheets of paper is 4.60 cm high, what will be the height, in meters, of a stack of Avogadro's number of sheets of paper?

29. There are about 5.0 billion (5.0×10^9) people on earth. If 1 mole of dollars were distributed equally among these people, how many dollars would each person receive?

*30. If 20 drops of water equal 1.0 mL (1.0 cm^3),
(a) How many drops of water are there in a cubic mile of water?
(b) What would be the volume in cubic miles of a mole of drops of water?

*31. Silver has a density of 10.5 g/cm^3. If 1.00 mol of silver were shaped into a cube,
(a) What would be the volume of the cube?
(b) What would be the length of one side of the cube?

32. How many atoms of oxygen are contained in each of the following?
(a) 16.0 g O_2
(b) 0.622 mol MgO
(c) 6.00×10^{22} molecules $C_6H_{12}O_6$
(d) 5.0 mol MnO_2
(e) 250 g $MgCO_3$
(f) 5.0×10^{18} molecules H_2O

33. Calculate the number of:
(a) Grams of silver in 25.0 g AgBr
(b) Grams of chlorine in 5.00 g $PbCl_2$
(c) Grams of nitrogen in 6.34 mol $(NH_4)_3PO_4$
(d) Grams of oxygen in 8.45×10^{22} molecules SO_3
(e) Grams of hydrogen in 45.0 g C_3H_8O

*34. A sulfuric acid solution contains 65.0% H_2SO_4 by mass and has a density of 1.55 g/mL. How many moles of the acid are present in 1.00 L of the solution?

*35. A nitric acid solution containing 72.0% HNO_3 by mass has a density of 1.42 g/mL. How many moles of HNO_3 are present in 100 mL of the solution?

36. Given 1.00 g samples of each of the following compounds, CO_2, O_2, H_2O, and CH_3OH,
(a) Which sample will contain the largest number of molecules?
(b) Which sample will contain the largest number of atoms?
Show proof for your answers.

37. How many grams of Fe_2S_3 will contain a total number of atoms equal to Avogadro's number?

Percent composition

38. Calculate the percent composition by mass of the following compounds:
(a) NaBr (c) $FeCl_3$ (e) $Al_2(SO_4)_3$
(b) $KHCO_3$ (d) $SiCl_4$ (f) $AgNO_3$

39. Calculate the percent composition of the following compounds:
(a) $ZnCl_2$ (d) $(NH_4)_2SO_4$
(b) $NH_4C_2H_3O_2$ (e) $Fe(NO_3)_3$
(c) MgP_2O_7 (f) ICl_3

40. Calculate the percent of iron, Fe, in the following compounds:
(a) FeO (c) Fe_3O_4
(b) Fe_2O_3 (d) $K_4Fe(CN)_6$

41. Which of the following chlorides has the highest and which has the lowest percentage of chlorine, Cl, by mass, in its formula?
(a) KCl (c) $SiCl_4$
(b) $BaCl_2$ (d) LiCl

42. A 6.20 g sample of phosphorus was reacted with oxygen to form an oxide with a mass of 14.20 g. Calculate the percent composition of the compound.

43. A sample of ethylene chloride was analyzed to contain 6.00 g of C, 1.00 g of H, and 17.75 g of Cl. Calculate the percent composition of ethylene chloride.

44. How many grams of lithium will combine with 20.0 grams of sulfur to form the compound Li_2S?

45. Calculate the percentage of
(a) Mercury in $HgCO_3$
(b) Oxygen in $Ca(ClO_3)_2$
(c) Nitrogen in $C_{10}H_{14}N_2$ (nicotine)
(d) Mg in $C_{55}H_{72}MgN_4O_5$ (chlorophyll)

46. Answer the following by examination of the formulas. Check your answers by calculations if you wish. Which compound has the:
(a) Higher percent by mass of hydrogen, H_2O or H_2O_2?
(b) Lower percent by mass of nitrogen, NO or N_2O_3?
(c) Higher percent by mass of oxygen, NO_2 or N_2O_4?
(d) Lower percent by mass of chlorine, $NaClO_3$ or $KClO_3$?
(e) Higher percent by mass of sulfur, $KHSO_4$ or K_2SO_4?
(f) Lower percent by mass of chromium, Na_2CrO_4 or $Na_2Cr_2O_7$?

Empirical and molecular formulas

47. Calculate the empirical formula of each compound from the percent compositions given.
(a) 63.6% N, 36.4% O
(b) 46.7% N, 53.3% O
(c) 25.9% N, 74.1% O
(d) 43.4% Na, 11.3% C, 45.3% O
(e) 18.8% Na, 29.0% Cl, 52.3% O
(f) 72.02% Mn, 27.98% O

48. Calculate the empirical formula of each compound from the percent compositions given.
(a) 64.1% Cu, 35.9% Cl
(b) 47.2% Cu, 52.8% Cl
(c) 51.9% Cr, 48.1% S
(d) 55.3% K, 14.6% P, 30.1% O
(e) 38.9% Ba, 29.4% Cr, 31.7% O
(f) 3.99% P, 82.3% Br, 13.7% Cl

49. A sample of tin (Sn) having a mass of 3.996 g was oxidized and found to have combined with 1.077 g of oxygen. Calculate the empirical formula of this oxide of tin.

50. A 3.054 g sample of vanadium (V) combined with oxygen to form 5.454 g of product. Calculate the empirical formula for this compound.

***51.** Zinc and sulfur react to form zinc sulfide, ZnS. If we mix 19.5 g of zinc and 9.40 g of sulfur, have we added sufficient sulfur to fully react all the zinc? Show evidence for your answer.

52. Hydroquinone is an organic compound commonly used as a photographic developer. It has a molar mass of 110.1 g/mol and a composition of 65.45% C, 5.45% H, and 29.09% O. Calculate the molecular formula of hydroquinone.

53. Fructose is a very sweet natural sugar that is present in honey, fruits, and fruit juices. It has a molar mass of 180.1 g/mol and a composition of 40.0% C, 6.7% H, and 53.3% O. Calculate the molecular formula of fructose.

54. Aspirin is well known as a pain reliever (analgesic) and as a fever reducer (anti-pyretic). It has a molar mass of 180.2 g/mol and a composition of 60.0% C, 4.48% H, and 35.5% O. Calculate the molecular formula of aspirin.

55. How many grams of oxygen are contained in 8.50 g $Al_2(SO_4)_3$?

56. Gallium arsenide is one of the newer materials used to make semiconductor chips for use in supercomputers. Its composition is 48.2% Ga and 51.8% As. What is the empirical formula?

57. Listed below are the compositions of four different compounds of carbon and chlorine. Determine both the empirical formula and the molecular formula for each.

Percentage C	Percentage Cl	Molar mass
(a) 7.79	92.21	153.8
(b) 10.13	89.87	236.7
(c) 25.26	74.74	284.8
(d) 11.25	88.75	319.6

8

Chemical Equations

◀ Chapter Opening Photo: **The thermite reaction releases enough energy to melt iron, and so can be used to weld railroad rails.**

In today's world much of our energy is directed toward expressing information in a concise, useful manner. From our earliest days in childhood, we are taught to translate ideas and desires into sentences. In mathematics, we learn to translate numerical relationships and situations into mathematical expressions and equations. An historian translates a thousand years of history into a 500-page textbook. A secretary translates an entire letter or document into a few lines of shorthand. A film maker translates an entire event, such as the Olympics, into several hours of entertainment.

And so it is in chemistry. A chemist uses a chemical equation to translate the reactions that are observed over widely varying timeframes in the laboratory or in nature. Chemical equations provide the necessary means (1) to summarize the reaction, (2) to determine the substances that are reacting, (3) to predict the products, and (4) to indicate the amounts of all component substances in the reaction.

8.1 The Chemical Equation

In a chemical reaction the substances entering the reaction are called reactants and the substances formed are called the products. During a chemical reaction atoms, molecules, or ions interact and rearrange themselves to form the products. During this process chemical bonds are broken and new bonds are formed. The reactants and products may be in the solid, liquid, or gaseous state, or in solution.

chemical equation

word equation

A **chemical equation** is a shorthand expression for a chemical change or reaction. It shows, among other things, the rearrangement of the atoms that are involved in the reaction. A **word equation** states in words, in equation form, the substances involved in a chemical reaction. For example, when mercury(II) oxide is heated, it decomposes to form mercury and oxygen. The word equation for this decomposition is

$$\text{mercury(II) oxide} + \text{heat} \longrightarrow \text{mercury} + \text{oxygen}$$

From the chemist's point of view this method of describing a chemical reaction is inadequate. It is bulky and cumbersome to use and does not give quantitative information. The chemical equation, using symbols and formulas, is a far better way to describe the decomposition of mercury(II) oxide:

$$2\ HgO \xrightarrow{\Delta} 2\ Hg + O_2\uparrow$$

This equation gives all the information from the word equation plus formulas, composition, reactive amounts of all the substances involved in the reaction, and much additional information (see Section 8.4). Even though a chemical equation provides much quantitative information, it is still not a complete description; it does not tell us how much energy is needed to cause decomposition, what we observe during the reaction, or anything about the rate of reaction. This information must be obtained from other sources or from experimentation.

8.2 Format for Writing Chemical Equations

A chemical equation uses the chemical symbols and formulas of the reactants and products and other symbolic terms to represent a chemical reaction. Equations are written according to this general format:

1. The reactants are separated from the products by an arrow ($\rightarrow$) that indicates the direction of the reaction. A double arrow ($\rightleftarrows$) indicates that the reaction goes in both directions and establishes an equilibrium between the reactants and the products.
2. The reactants are placed to the left and the products to the right of the arrow. A plus sign ($+$) is placed between reactants and between products when needed.
3. Conditions required to carry out the reaction may, if desired, be placed above or below the arrow or equality sign. For example, a delta sign placed over the arrow ($\xrightarrow{\Delta}$) indicates that heat is supplied to the reaction.
4. Coefficients (whole numbers) are placed in front of substances (for example, $2\ H_2O$) to balance the equation and to indicate the number of formula units (atoms, molecules, moles, ions) of each substance reacting or being produced. When no number is shown, it is understood that one formula unit of the substance is indicated.
5. The physical state of a substance is indicated by the following symbols: (*s*) for solid state; (*l*) for liquid state; (*g*) for gaseous state; and (*aq*) for substances in aqueous solution.

Symbols commonly used in equations are given in Table 8.1.

8.3 Writing and Balancing Equations

To represent the quantitative relationships of a reaction, the chemical equation must be balanced. A **balanced equation** contains the same number of each kind of atom on each side of the equation. The balanced equation, therefore, obeys the Law of Conservation of Mass.

balanced equation

TABLE 8.1 Symbols Commonly Used in Chemical Equations

Symbol	Meaning
$\rightarrow$	Yields; produces (points to products)
$\rightleftarrows$	Reversible reaction; equilibrium between reactants and products
$\uparrow$	Gas evolved (written after a substance)
$\downarrow$	Solid or precipitate formed (written after a substance)
(s)	Solid state (written after a substance)
(l)	Liquid state (written after a substance)
(g)	Gaseous state (written after a substance)
(aq)	Aqueous solution (substance dissolved in water)
Δ	Heat
$+$	Plus or added to (placed between substances)

The ability to balance equations must be acquired by every chemistry student. Simple equations are easy to balance, but some care and attention to detail are required. The way to balance an equation is to adjust the number of atoms of each element so that it is the same on each side of the equation. But we must not change a correct formula in order to achieve a balanced equation. Each equation must be treated on its own merits; we have no simple "plug in" formula for balancing equations. The following outline gives a general procedure for balancing equations. Study this outline and refer to it as needed when working examples. There is no substitute for practice in learning to write and balance chemical equations.

1. Identify the Reaction for Which the Equation is to be Written Formulate a description or word equation for the reaction if needed (e.g., mercury(II) oxide decomposes yielding mercury and oxygen).

2. Write the Unbalanced, or Skeleton, Equation Make sure that the formula for each substance is correct and that the reactants are written to the left and the products to the right of the arrow (e.g., $HgO \rightarrow Hg + O_2$). The correct formulas must be known or ascertained from the periodic table, oxidation numbers, lists of ions, or experimental data.

3. Balance the Equation Use the following steps as necessary:

(a) Count and compare the number of atoms of each element on each side of the equation and determine those that must be balanced.

(b) Balance each element, one at a time, by placing whole numbers (coefficients) in front of the formulas containing the unbalanced element. It is usually best to balance metals first, then nonmetals, then hydrogen and oxygen. Select the smallest coefficients that will give the same number of atoms of the element on each side. A coefficient placed before a formula multiplies every atom in the formula by that number (for example, 2 H_2SO_4 means two molecules of sulfuric acid and also means four H atoms, two S atoms, and eight O atoms.)

(c) Check all other elements after each individual element is balanced to see whether, in balancing one element, other elements have become unbalanced. Make adjustments as needed.

(d) Balance polyatomic ions such as SO_4^{2-}, which remain unchanged from one side of the equation to the other, in the same way as individual atoms.

(e) Do a final check, making sure that each element and/or polyatomic ion is balanced and that the smallest possible set of whole-number coefficients has been used.

$$4\ HgO \longrightarrow 4\ Hg + 2\ O_2 \quad \text{(incorrect form)}$$
$$2\ HgO \longrightarrow 2\ Hg + O_2 \quad \text{(correct form)}$$

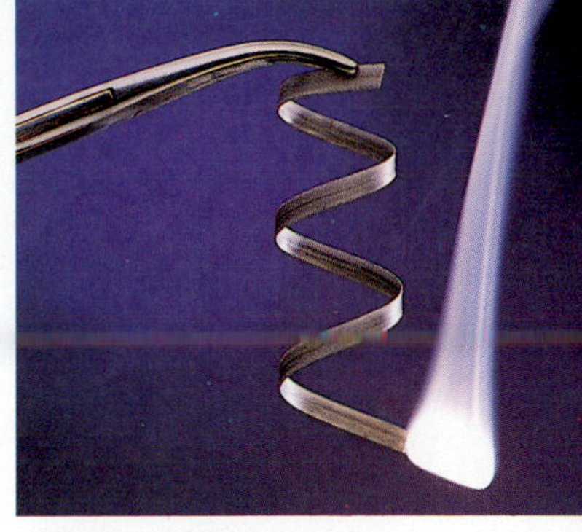

▲ **A burst of UV light is emitted as a ribbon of metallic magnesium combusts in air, producing magnesium oxide.**

Not all chemical equations can be balanced by the simple method of inspection just described. The balancing of more complex equations such as oxidation-reduction equations is described in Chapter 18.

The following examples show stepwise sequences leading to balanced equations. Study each example carefully.

EXAMPLE 8.1

Write the balanced equation for the reaction that takes place when magnesium metal is burned in air to produce magnesium oxide.

SOLUTION

1. *Word equation*

magnesium + oxygen ⟶ magnesium oxide

2. *Skeleton equation*

$$Mg + O_2 \longrightarrow MgO \quad \text{(unbalanced)}$$

3. *Balance*

(a) Oxygen is not balanced. Two O atoms appear on the left side and one on the right side.

(b) Place the coefficient 2 before MgO.

$$Mg + O_2 \longrightarrow 2\ MgO \quad \text{(unbalanced).}$$

(c) Now Mg is not balanced. One Mg atom appears on the left side and two on the right side. Place a 2 before Mg.

$$2\ Mg + O_2 \longrightarrow 2\ MgO \quad \text{(balanced)}$$

(d) *Check*: Each side has two Mg and two O atoms.

EXAMPLE 8.2

When methane, CH_4, undergoes complete combustion, it reacts with oxygen to produce carbon dioxide and water. Write the balanced equation for this reaction.

SOLUTION

1. *Word equation*

methane + oxygen ⟶ carbon dioxide + water

2. *Skeleton equation*

$$CH_4 + O_2 \longrightarrow CO_2 + H_2O \quad \text{(unbalanced)}$$

3. *Balance*
   (a) Carbon is balanced. Hydrogen and oxygen are not balanced.
   (b) Balance H atoms by placing a 2 before H_2O.

   $$CH_4 + O_2 \longrightarrow CO_2 + 2\ H_2O \quad \text{(unbalanced)}$$

   Each side of the equation has four H atoms; oxygen is still not balanced. Place a 2 before O_2 to balance the oxygen atoms.

   $$CH_4 + 2\ O_2 \longrightarrow CO_2 + 2\ H_2O \quad \text{(balanced)}$$

   (c) *Check*: The equation is correctly balanced; it has one C, four O, and four H atoms on each side.

EXAMPLE 8.3 Oxygen and potassium chloride are formed by heating potassium chlorate. Write a balanced equation for this reaction.

SOLUTION

1. *Word equation*

   $$\text{potassium chlorate} \xrightarrow{\Delta} \text{potassium chloride} + \text{oxygen}$$

2. *Skeleton equation*

   $$KClO_3 \xrightarrow{\Delta} KCl + O_2 \quad \text{(unbalanced)}$$

3. *Balance*
   (a) Oxygen is unbalanced (three O atoms on the left and two on the right side).
   (b) How many oxygen atoms are needed? The subscripts of oxygen (3 and 2) in $KClO_3$ and O_2 have a least common multiple of 6. Therefore coefficients for $KClO_3$ and O_2 are needed to give six oxygen atoms on each side. Place a 2 before $KClO_3$ and a 3 before O_2 to give six O atoms on each side.

   $$2\ KClO_3 \xrightarrow{\Delta} KCl + 3\ O_2 \quad \text{(unbalanced)}$$

   Now K and Cl are not balanced. Place a 2 before KCl, which balances both K and Cl at the same time.

   $$2\ KClO_3 \xrightarrow{\Delta} 2\ KCl + 3\ O_2 \quad \text{(balanced)}$$

   (c) *Check*: Each side now contains two K, two Cl, and six O atoms.

EXAMPLE 8.4 Balance by starting with the word equation given.

SOLUTION

1. *Word equation*

   $$\text{silver nitrate} + \text{hydrogen sulfide} \longrightarrow \text{silver sulfide} + \text{nitric acid}$$

2. *Skeleton equation*

   $$AgNO_3 + H_2S \longrightarrow Ag_2S + HNO_3 \quad \text{(unbalanced)}$$

3. *Balance*
   (a) Ag and H are unbalanced.

(b) Place a 2 in front of $AgNO_3$ to balance Ag.

$$2\ AgNO_3 + H_2S \longrightarrow Ag_2S + HNO_3 \quad \text{(unbalanced)}$$

(c) H and NO_3^- are still unbalanced. Balance by placing a 2 in front of HNO_3.

$$2\ AgNO_3 + H_2S \longrightarrow Ag_2S + 2\ HNO_3 \quad \text{(balanced)}$$

(d) In this example N and O atoms are balanced by balancing the NO_3^- ion as a unit.

(e) *Check*: Each side has two Ag, two H, and one S atom. Also, each side has two NO_3^- ions.

EXAMPLE 8.5

Balance by starting with the word equation given.

SOLUTION

1. *Word equation*

aluminum hydroxide + sulfuric acid ⟶ aluminum sulfate + water

2. *Skeleton equation*

$$Al(OH)_3 + H_2SO_4 \longrightarrow Al_2(SO_4)_3 + H_2O \quad \text{(unbalanced)}$$

3. *Balance*

(a) All elements are unbalanced.

(b) Balance Al by placing a 2 in front of $Al(OH)_3$. Treat the unbalanced SO_4^{2-} ion as a unit and balance by placing a 3 before H_2SO_4.

$$2\ Al(OH)_3 + 3\ H_2SO_4 \longrightarrow Al_2(SO_4)_3 + H_2O \quad \text{(unbalanced)}$$

Balance the unbalanced H and O by placing a 6 in front of H_2O.

$$2\ Al(OH)_3 + 3\ H_2SO_4 \longrightarrow Al_2(SO_4)_3 + 6\ H_2O \quad \text{(balanced)}$$

(c) *Check*: Each side has two Al, twelve H, three S, and eighteen O atoms.

EXAMPLE 8.6

When the fuel in a butane gas stove undergoes complete combustion, it reacts with oxygen to form carbon dioxide and water. Write the balanced equation for this reaction.

SOLUTION

1. *Word equation*

butane + oxygen ⟶ carbon dioxide + water

2. *Skeleton equation*

$$C_4H_{10} + O_2 \longrightarrow CO_2 + H_2O \quad \text{(unbalanced)}$$

3. *Balance*

(a) All elements are unbalanced.

(b) Balance C by placing a 4 in front of CO_2.

$$C_4H_{10} + O_2 \longrightarrow 4\ CO_2 + H_2O \quad \text{(unbalanced)}$$

Balance H by placing a 5 in front of H_2O.

$$C_4H_{10} + O_2 \longrightarrow 4\,CO_2 + 5\,H_2O \quad \text{(unbalanced)}$$

Oxygen remains unbalanced. The oxygen atoms on the right side are fixed, because 4 CO_2 and 5 H_2O are derived from the single C_4H_{10} molecule on the left. When we try to balance the O atoms, we find that there is no whole number that can be placed in front of O_2 to bring about a balance. The equation can be balanced if we use $6\frac{1}{2}\,O_2$ and then double the coefficients of each substance, including the $6\frac{1}{2}\,O_2$, to obtain the balanced equation.

$$C_4H_{10} + 6\tfrac{1}{2}\,O_2 \longrightarrow 4\,CO_2 + 5\,H_2O \quad \text{(balanced—incorrect form)}$$

$$2\,C_4H_{10} + 13\,O_2 \longrightarrow 8\,CO_2 + 10\,H_2O \quad \text{(balanced)}$$

(c) *Check*: Each side now has eight C, twenty H, and twenty-six O atoms.

Practice Balance the following word equation.

Aluminum + oxygen ⟶ Aluminum oxide

Answer: the coefficients are 4, 2 ⟶ 2

Practice Balance the following word equation.

Magnesium hydroxide + phosphoric acid ⟶ magnesium phosphate + water

Answer: the coefficients are 3, 2 ⟶ 1, 6

8.4 What Information Does an Equation Tell Us?

Depending on the particular context in which it is used, a formula can have different meanings. The meanings refer either to an individual chemical entity (atom, ion, molecule, or formula unit) or to a mole of that chemical entity. For example, the formula H_2O can be used to indicate any of the following:

1. 2 H atoms and 1 O atom
2. 1 molecule of water
3. 1 mol of water
4. 6.022×10^{23} molecules of water
5. 1 molar mass of water
6. 18.02 g of water

Formulas used in equations can be expressed in units of individual chemical entities or as moles, the latter being more commonly used. For example, in the reaction of hydrogen and oxygen to form water,

$$2\ H_2 + O_2 \longrightarrow 2\ H_2O$$

the 2 H_2 can represent 2 molecules or 2 moles of hydrogen; the O_2, 1 molecule or 1 mole of oxygen; and the 2 H_2O, 2 molecules or 2 moles of water. In terms of moles, this equation is stated: 2 moles of H_2 react with 1 mole of O_2 to give 2 moles of H_2O.

As indicated earlier, a chemical equation is a shorthand description of a chemical reaction. Interpretation of the balanced equation gives us the following information:

1. What the reactants are and what the products are
2. The formulas of the reactants and products
3. The number of molecules or formula units of reactants and products in the reaction
4. The number of atoms of each element involved in the reaction
5. The number of molar masses of each substance used or produced
6. The number of moles of each substance
7. The number of grams of each substance used or produced

Consider the equation

$$H_2(g) + Cl_2(g) \longrightarrow 2\ HCl(g)$$

This equation states that hydrogen gas reacts with chlorine gas to produce hydrogen chloride, also a gas. Let us summarize all the information relating to the equation. The information that can be stated about the relative amount of each substance, with respect to all other substances in the balanced equation, is written below its formula in the following equation:

$$H_2(g) \quad + \quad Cl_2(g) \quad \longrightarrow \quad 2\ HCl(g)$$

Hydrogen	Chlorine	Hydrogen Chloride
1 molecule	1 molecule	2 molecules
2 atoms	2 atoms	2 atoms H + 2 atoms Cl
1 molar mass	1 molar mass	2 molar masses
1 mole	1 mole	2 moles
2.016 g	70.90 g	2 × 36.46 g or 72.92 g

These data are very useful in calculating quantitative relationships that exist among substances in a chemical reaction. For example, if we react 2 moles of hydrogen (twice as much as is indicated by the equation) with 2 moles of chlorine, we can expect to obtain 4 moles, or 145.8 g, of hydrogen chloride as a product. We will study this phase of using equations in more detail in the next chapter.

Let us try another equation. When propane gas is burned in air, the products are carbon dioxide, CO_2, and water, H_2O. The balanced equation and its interpretation are as follows:

$$C_3H_8(g) \quad + \quad 5\,O_2(g) \quad \longrightarrow \quad 3\,CO_2(g) \quad + \quad 4\,H_2O(g)$$

Propane	Oxygen	Carbon dioxide	Water
1 molecule	5 molecules	3 molecules	4 molecules
3 atoms C	10 atoms O	3 atoms C	8 atoms H
8 atoms H		6 atoms O	4 atoms O
1 molar mass	5 molar masses	3 molar masses	4 molar masses
1 mole	5 moles	3 moles	4 moles
44.09 g	5 × 32.00 g (160.0 g)	3 × 44.01 g (132.0 g)	4 × 18.02 g (72.08 g)

8.5 Types of Chemical Equations

Chemical equations represent chemical changes or reactions. To be of any significance an equation must represent an actual or possible reaction. Part of the problem of writing equations is determining the products formed. We have no sure method of predicting products, nor do we have time to carry out experimentally all the reactions we may wish to consider. Therefore we must use data reported in the writings of other workers, certain rules to aid in our predictions, and the atomic structure and combining capacities of the elements to help us predict the formulas of the products of a chemical reaction. The final proof of the existence of any reaction, of course, is in the actual observation of the reaction in the laboratory (or elsewhere).

Reactions are classified into types to assist in writing equations and to aid in predicting other reactions. Many chemical reactions fit one or another of the four principal reaction types that are discussed in the following paragraphs. Reactions are also classified as oxidation–reduction. Special methods are used to balance complex oxidation–reduction equations (see Chapter 18).

combination or synthesis reaction

1. Combination or Synthesis Reaction In a **combination reaction**, two reactants combine to give one product. The general form of the equation is

$$A + B \longrightarrow AB$$

in which A and B are either elements or compounds and AB is a compound. The formula of the compound in many cases can be determined from a knowledge of the oxidation numbers of the reactants in their combined states. Some reactions that fall into this category are the following:

(a) metal + oxygen ⟶ metal oxide

$$2\,Mg(s) + O_2(g) \xrightarrow{\Delta} 2\,MgO(s)$$

$$4\,Al(s) + 3\,O_2(g) \xrightarrow{\Delta} 2\,Al_2O_3(s)$$

(b) nonmetal + oxygen ⟶ nonmetal oxide

$$S(s) + O_2(g) \xrightarrow{\Delta} SO_2(g)$$

$$N_2(g) + O_2(g) \xrightarrow{\Delta} 2\ NO(g)$$

(c) metal + nonmetal ⟶ salt

$$2\ Na(s) + Cl_2(g) \longrightarrow 2\ NaCl(s)$$

$$2\ Al(s) + 3\ Br_2(l) \longrightarrow 2\ AlBr_3(s)$$

(d) metal oxide + water ⟶ base (metal hydroxide)

$$Na_2O(s) + H_2O(l) \longrightarrow 2\ NaOH(aq)$$

$$CaO(s) + H_2O(l) \longrightarrow Ca(OH)_2(aq)$$

(e) nonmetal oxide + water ⟶ oxy-acid

$$SO_3(g) + H_2O(l) \longrightarrow H_2SO_4(aq)$$

$$N_2O_5(s) + H_2O(l) \longrightarrow 2\ HNO_3(aq)$$

2. Decomposition Reaction In a **decomposition reaction** a single substance is decomposed or broken down to give two or more different substances. The reaction may be considered the reverse of combination. The starting material must be a compound, and the products may be elements or compounds. The general form of the equation is

decomposition reaction

$$AB \longrightarrow A + B$$

Predicting the products of a decomposition reaction can be difficult and requires an understanding of each individual reaction. Heating oxygen-containing compounds often results in decomposition. Some reactions that fall into this category are the following:

(a) Metal oxides. Some metal oxides decompose to yield the free metal plus oxygen, others give a lower oxide, and some are very stable, resisting decomposition by heating.

$$2\ HgO(s) \xrightarrow{\Delta} 2\ Hg(l) + O_2(g)$$

$$2\ PbO_2(s) \xrightarrow{\Delta} 2\ PbO(s) + O_2(g)$$

(b) Carbonates and bicarbonates decompose to yield CO_2 when heated.

$$CaCO_3(s) \xrightarrow{\Delta} CaO(s) + CO_2(g)$$

$$2\ NaHCO_3(s) \xrightarrow{\Delta} Na_2CO_3(s) + H_2O(g) + CO_2(g)$$

(c) Miscellaneous.

$$2\ KClO_3(s) \xrightarrow{\Delta} 2\ KCl(s) + 3\ O_2(g)$$

$$2\ NaNO_3(s) \xrightarrow{\Delta} 2\ NaNO_2(s) + O_2(g)$$

$$2\ H_2O_2(l) \xrightarrow{\Delta} 2\ H_2O(l) + O_2(g)$$

single-displacement reaction

3. Single-Displacement Reaction In a **single-displacement reaction** one element reacts with a compound to take the place of one of the elements of that compound. A different element and a different compound are formed. The general form of the equation is

$$\underset{\text{metal}}{A} + BC \longrightarrow B + AC \quad \text{or} \quad \underset{\text{halogen}}{A} + BC \longrightarrow C + BA$$

▲ **A single displacement reaction occurs as elemental potassium reacts vigorously with water, producing aqueous potassium hydroxide and hydrogen gas.**

If A is a metal, A will replace B to form AC, providing A is a more reactive metal than B. If A is a halogen, it will replace C to form BA, providing A is a more reactive halogen than C.

A brief activity series of selected metals (and hydrogen) and halogens are shown in Table 8.2. The series are listed in descending order of chemical reactivity, with the most active metals and halogens at the top. From such series it is possible to predict many chemcial reactions. Any metal on the list will replace the ions of those metals that appear anywhere underneath it on the list. For example, zinc metal will replace hydrogen from a hydrochloric acid solution. But copper metal, which is underneath hydrogen on the list and thus less reactive than hydrogen, will not replace hydrogen from a hydrochloric acid solution. Some reactions that fall into this category follow.

(a) metal + acid ⟶ hydrogen + salt

$$Zn(s) + 2\ HCl(aq) \longrightarrow H_2(g) + ZnCl_2(aq)$$
$$2\ Al(s) + 3\ H_2SO_4(aq) \longrightarrow 3\ H_2(g) + Al_2(SO_4)_3(aq)$$

(b) metal + water ⟶ hydrogen + metal hydroxide or metal oxide

$$2\ Na(s) + 2\ H_2O \longrightarrow H_2(g) + 2\ NaOH(aq)$$
$$Ca(s) + 2\ H_2O \longrightarrow H_2(g) + Ca(OH)_2(aq)$$
$$3\ Fe(s) + 4\ \underset{\text{Steam}}{H_2O(g)} \longrightarrow 4\ H_2(g) + Fe_3O_4(s)$$

(c) metal + salt ⟶ metal + salt

$$Fe(s) + CuSO_4(aq) \longrightarrow Cu(s) + FeSO_4(aq)$$
$$Cu(s) + 2\ AgNO_3(aq) \longrightarrow 2\ Ag(s) + Cu(NO_3)_2(aq)$$

(d) halogen + halogen salt ⟶ halogen + halogen salt

$$Cl_2(g) + 2\ NaBr(aq) \longrightarrow Br_2(l) + 2\ NaCl(aq)$$
$$Cl_2(g) + 2\ KI(aq) \longrightarrow I_2(s) + 2\ KCl(aq)$$

A common chemical reaction is the displacement of hydrogen from water or acids. This reaction is a good illustration of the relative reactivity of metals and the use of the activity series.

K, Ca, and Na displace hydrogen from cold water, steam, and acids.
Mg, Al, Zn, and Fe displace hydrogen from steam and acids.
Ni, Sn, and Pb displace hydrogen only from acids.
Cu, Ag, Hg, and Au do not displace hydrogen.

TABLE 8.2 Activity Series

Metals	Halogens
K	F_2
Ca	Cl_2
Na	Br_2
Mg	I_2
Al	
Zn	
Fe	
Ni	
Sn	
Pb	
H	
Cu	
Ag	
Hg	
Au	

(↑ increasing activity)

EXAMPLE 8.7

Will a reaction occur between (a) nickel metal and hydrochloric acid and (b) tin metal and a solution of aluminum chloride? Write balanced equations for the reactions.

SOLUTION

(a) Nickel is more reactive than hydrogen so it will displace hydrogen from hydrochloric acid. The products are hydrogen gas and a salt of Ni^{2+} and Cl^- ions.

$$Ni(s) + 2\,HCl(aq) \longrightarrow H_2(g) + NiCl_2(aq)$$

(b) According to the activity series, tin is less reactive than aluminum, so no reaction will occur.

$$Sn(s) + AlCl_3(aq) \longrightarrow \text{no reaction}$$

PRACTICE Write balanced equations for the reactions:
(a) iron metal and a solution of magnesium chloride
(b) zinc metal and a solution of lead (II) nitrate

Answers: (a) $Fe + MgCl_2 \longrightarrow$ no reaction
(b) $Zn + Pb(NO_3)_2 \longrightarrow Pb\downarrow + Zn(NO_3)_2$

4. Double-Displacement or Metathesis Reaction In a **double-displacement reaction**, two compounds exchange partners with each other to produce two different compounds. The general form of the equation is

double-displacement or metathesis reaction

$$AB + CD \longrightarrow AD + CB$$

This reaction may be thought of as an exchange of positive and negative groups, in which A combines with D, and C combines with B. In writing the formulas of the products, we must take into account the oxidation numbers or charges of the combining groups.

It is possible to write an equation in the form of a double displacement reaction when a reaction has not occurred. For example, when solutions of sodium chloride and potassium nitrate are mixed, the following equation can be written:

$$NaCl(aq) + KNO_3(aq) \longrightarrow NaNO_3(aq) + KCl(aq)$$

When the procedure is carried out, no physical changes are observed, indicating that no chemcial reaction has taken place.

A double displacement reaction is accompanied by evidence of such reactions as the evolution of heat, the formation of an insoluble precipitate, or the production of bubbles of a gas. Some of the reactions that fall into these categories follow.

▲ A double displacement reaction results from mixing silver nitrate with sodium chromate, which yields silver chromate precipitate.

(a) Neutralization of an acid and a base The production of a molecule of water from an H^+ and an OH^- ion will be accompanied by a release of heat, which can be detected by touching the reaction container.

$$\text{acid} + \text{base} \longrightarrow \text{salt} + \text{water}$$
$$HCl(aq) + NaOH(aq) \longrightarrow NaCl(aq) + H_2O(l)$$
$$H_2SO_4(aq) + Ba(OH)_2(aq) \longrightarrow BaSO_4\downarrow + 2\,H_2O(l)$$

(b) Formation of an insoluble precipitate The solubilities of the products can be determined by consulting the Solubility Table in Appendix IV. One or both of the products may be insoluble.

$$BaCl_2(aq) + 2\,AgNO_3(aq) \longrightarrow 2\,AgCl\downarrow + Ba(NO_3)_2(aq)$$
$$FeCl_3(aq) + 3\,NH_4OH(aq) \longrightarrow Fe(OH)_3\downarrow + 3\,NH_4Cl(aq)$$

(c) Metal oxide + acid Heat is released by the production of a molecule of water.

$$\text{metal oxide} + \text{acid} \longrightarrow \text{salt} + \text{water}$$
$$CuO(s) + 2\,HNO_3(aq) \longrightarrow Cu(NO_3)_2(aq) + H_2O(l)$$
$$CaO(s) + 2\,HCl(aq) \longrightarrow CaCl_2(aq) + H_2O(l)$$

(d) Formation of a gas A gas such as HCl or H_2S may be produced directly as in these two examples:

$$H_2SO_4(l) + NaCl(s) \longrightarrow NaHSO_4(s) + HCl\uparrow$$
$$2\,HCl(aq) + ZnS(s) \longrightarrow ZnCl_2(aq) + H_2S\uparrow$$

A gas can be produced indirectly. Some unstable compounds formed in a double displacement reaction, such as H_2CO_3, H_2SO_3, and NH_4OH, will decompose to form water and a gas:

$$2\,HCl(aq) + Na_2CO_3(aq) \longrightarrow 2\,NaCl(aq) + H_2CO_3 \longrightarrow 2\,NaCl(aq) + H_2O(l) + CO_2\uparrow$$
$$2\,HNO_3(aq) + K_2SO_3(aq) \longrightarrow 2\,KNO_3(aq) + H_2SO_3(aq) \longrightarrow 2\,KNO_3(aq) + H_2O(l) + SO_2\uparrow$$
$$NH_4Cl(aq) + NaOH(aq) \longrightarrow NaCl(aq) + NH_4OH(aq) \longrightarrow NaCl(aq) + H_2O(l) + NH_3\uparrow$$

EXAMPLE 8.8

Write the equation for the reaction between aqueous solutions of hydrobromic acid and potassium hydroxide.

SOLUTION

First write the formulas for the reactants. They are HBr and KOH. Then classify the type of reaction that would occur between them. Because the reactants are compounds and one is an acid and the other is a base, the reaction will be of the neutralization type:

$$\text{acid} + \text{base} \longrightarrow \text{salt} + \text{water}$$

Now rewrite the equation by putting down the formulas for the known substances:

$$HBr(aq) + KOH(aq) \longrightarrow \text{salt} + H_2O$$

In this reaction, which is a double-displacement type, the H^+ from the acid combines with the OH^- from the base to form water. The salt must be composed of the other two ions,

K^+ and Br^-. We determine the formula of the salt to be KBr from the fact that K is a +1 cation and Br is a −1 anion. The final balanced equation is

$$HBr(aq) + KOH(aq) \longrightarrow KBr(aq) + H_2O(l)$$

EXAMPLE 8.9

Complete and balance the equation for the reaction between aqueous solutions of barium chloride and sodium sulfate.

SOLUTION

First determine the formulas for the reactants. They are $BaCl_2$ and Na_2SO_4. Then classify these substances as acids, bases, or salts. Both substances are salts. Since both substances are compounds, the reaction looks as though it will be of the double-displacement type. Start writing the equation with the reactants:

$$BaCl_2(aq) + Na_2SO_4(aq) \longrightarrow$$

If the reaction is double-displacement, Ba^{2+} will be written combined with SO_4^{2-}, and Na^+ with Cl^- as the products. The balanced equation is

$$BaCl_2(aq) + Na_2SO_4(aq) \longrightarrow BaSO_4 + 2\ NaCl$$

The final step is to determine the nature of the products, which controls whether or not the reaction will take place. If both products are soluble, we may merely have a mixture of all the ions in solution. But if an insoluble precipitate is formed, the reaction will definitely occur. We know from experience that NaCl is fairly soluble in water, but what about $BaSO_4$? The Solubility Table in Appendix IV can give us this information. From this table we see that $BaSO_4$ is insoluble in water, so it will be a precipitate in the reaction. Thus the reaction will occur, forming a white precipitate. The equation is

$$BaCl_2(aq) + Na_2SO_4(aq) \longrightarrow BaSO_4\downarrow + 2\ NaCl(aq)$$

PRACTICE Complete and balance the equations for the reactions:
(a) potassium phosphate + barium chloride
(b) hydrochloric acid + nickel carbonate
(c) ammonium chloride + sodium nitrate

Answers: (a) $2\ K_3PO_4(aq) + 3\ BaCl_2(aq) \longrightarrow Ba_3(PO_4)_2\downarrow + 6\ KCl(aq)$
(b) $2HCl(aq) + NiCO_3 \longrightarrow NiCl_2(aq) + H_2O + CO_2\uparrow$
(c) $NH_4Cl(aq) + NaNO_3(aq) \longrightarrow$ no reaction

Some reactions we attempt may fail because the substances are not reactive, or the proper conditions for reaction may not be present. For example, mercury(II) oxide does not decompose until it is heated; magnesium does not burn in air or oxygen until the temperature is raised to the point at which it begins to react. When silver is placed in a solution of copper(II) sulfate, no reaction takes place; however, when a strip of copper is placed in a solution of silver nitrate, the single-displacement reaction (3c on p. 158) takes place because copper is a more reactive metal than silver. The successful prediction of the products of a reaction is not always easy. The ability to predict products correctly comes with knowledge and experience. Although you may not be able to predict many reactions at this point, as you continue you will find that reactions can be categorized, and that prediction of the products thereby becomes easier, if not always certain.

We have a great deal yet to learn about which substances react with each other, how they react, and what conditions are necessary to bring about their reaction. It is possible to make accurate predictions concerning the occurrence of proposed reactions. Such predictions require, in addition to appropriate data, a good knowledge of thermodynamics, a subject usually reserved for advanced courses in chemistry and physics. But even without the formal use of thermodynamics your knowledge of such generalities as the four reaction types just cited, the periodic table, atomic structure, oxidation numbers, and so on, can be put to good use in predicting reactions and in writing equations. Indeed, such applications serve to make chemistry an interesting and fascinating study.

8.6 Heat in Chemical Reactions

Energy changes always accompany chemical reactions. One reason why reactions occur is that the products attain a lower, more stable energy state than the reactants. For the products to attain this more stable state, energy must be liberated and given off to the surroundings as heat (or as heat and work). When a solution of a base is neutralized by the addition of an acid, the liberation of heat energy is signaled by an immediate rise in the temperature of the solution. When an automobile engine burns gasoline, heat is certainly liberated; at the same time, part of the liberated energy does the work of moving the automobile.

exothermic reaction

endothermic reaction

Reactions are either exothermic or endothermic. **Exothermic reactions** liberate heat; **endothermic reactions** absorb heat. In an exothermic reaction heat is a product and may be written on the right side of the equation for the reaction. In an endothermic reaction heat can be regarded as a reactant and is written on the left side of the equation. Examples indicating heat in an exothermic and an endothermic reaction follow.

$$H_2(g) + Cl_2(g) \longrightarrow 2\ HCl(g) + 185\ \text{kJ}\ (44.2\ \text{kcal}) \quad \text{(exothermic)}$$
$$N_2(g) + O_2(g) + 181\ \text{kJ}\ (43.2\ \text{kcal}) \longrightarrow 2\ NO(g) \quad \text{(endothermic)}$$

heat of reaction

The quantity of heat produced by a reaction is known as the **heat of reaction**. The units used can be kilojoules or kilocalories. Consider the reaction represented by this equation

$$C(s) + O_2(g) \longrightarrow CO_2(g) + 393\ \text{kJ}\ (94.0\ \text{kcal})$$

When the heat liberated is expressed as part of the equation, the substances are expressed in units of moles. Thus, when 1 mol (12.01 g) of C combines with 1 mol

(32.00 g) of O_2, 1 mol (44.01 g) of CO_2 is formed and 393 kJ (94.0 kcal) of heat are liberated. In this reaction, as in many others, the heat energy is more useful than the chemical products.

Aside from relatively small amounts of energy from nuclear processes, the sun is the major provider of energy for life on earth. The sun maintains the temperature necessary for life and also supplies light energy for the endothermic photosynthetic reactions carried on by green plants. In photosynthesis carbon dioxide and water are converted to free oxygen and glucose.

$$6\ CO_2 + 6\ H_2O + 2519\ \text{kJ}\ (673\ \text{kcal}) \longrightarrow \underset{\text{Glucose}}{C_6H_{12}O_6} + 6\ O_2$$

Nearly all of the chemical energy used by living organisms is obtained from glucose or compounds derived from glucose. Modern technology depends for energy on fossil fuels—coal, petroleum, and natural gas. The energy is obtained from the combustion (burning) of these fuels, which are converted to carbon dioxide and water. **Combustion** is the term for a chemical reaction in which heat and light are given off.

combustion

Fossil fuels constitute a huge energy reservoir. Some coal is about 90% carbon. Since 393 kJ are obtained from the combustion of 1 mol (12.01 g) of carbon, the combustion of a single ton of coal yields about 2.68×10^{10} J (6.40×10^9 cal) of energy. At 4.184 J/g°C, this energy is enough to heat about 21,000 gallons of water from room temperature to the boiling point (20° to 100°C).

Natural gas is primarily methane, CH_4. Petroleum is a mixture of hydrocarbons (compounds of carbon and hydrogen). Liquefied petroleum gas (LPG) is a mixture of propane (C_3H_8) and butane (C_4H_{10}).

The combustion of these fuels releases a tremendous amount of energy, but reactions won't occur to a significant extent at ordinary temperatures. A spark or a flame must be present before methane will ignite. The amount of energy that must be supplied to start a chemical reaction is called the **activation energy**. Once the activation energy has been provided, enough energy is generated to keep the reaction going.

activation energy

$$CH_4(g) + 2\ O_2(g) \longrightarrow CO_2(g) + 2\ H_2O(g) + 890\ \text{kJ}\ (213\ \text{kcal})$$
$$C_3H_8(g) + 5\ O_2(g) \longrightarrow 3\ CO_2(g) + 4\ H_2O(g) + 2200\ \text{kJ}\ (526\ \text{kcal})$$
$$2\ C_8H_{18}(l) + 25\ O_2(g) \longrightarrow 16\ CO_2(g) + 18\ H_2O(g) + 10{,}900\ \text{kJ}\ (2606\ \text{kcal})$$

Be careful not to confuse an exothermic reaction that merely requires heat (activation energy) to get it started with a truly endothermic process. The combustion of magnesium is highly exothermic, yet magnesium must be heated to a fairly high temperature in air before combustion begins. Once started, however, the combustion reaction goes very vigorously until either the magnesium or the available supply of oxygen is exhausted. The electrolytic decomposition of water to hydrogen and oxygen is highly endothermic. If the electric current is shut off when this process is going on, the reaction stops instantly. The relative energy levels of reactants and products in exothermic and in endothermic processes are presented graphically in Figure 8.1.

FIGURE 8.1 ► Energy changes in exothermic and endothermic reactions.

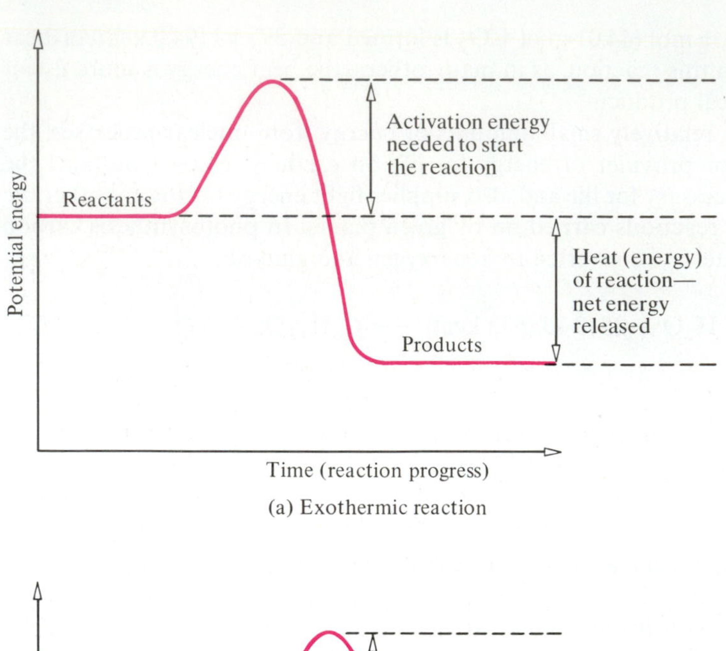

(a) Exothermic reaction

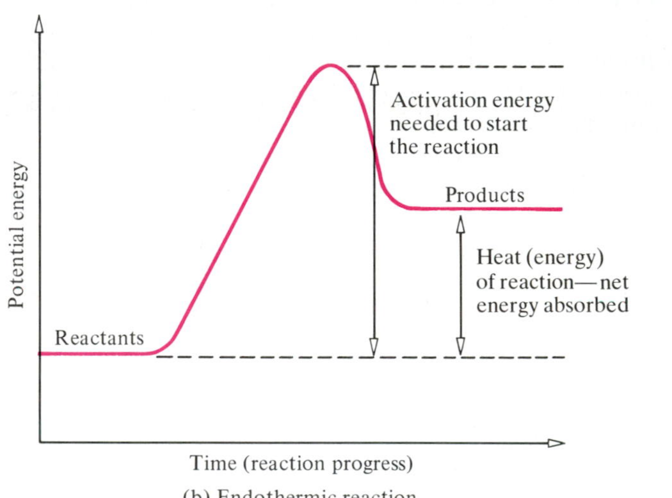

(b) Endothermic reaction

In reaction (a) of Figure 8.1, the products are at a lower energy level than the reactants. Energy (heat) is given off, and the reaction is exothermic. In reaction (b) the products are at a higher energy level than the reactants. Energy has therefore been absorbed, and the reaction is endothermic.

Examples of endothermic and exothermic processes that can be demonstrated easily in the laboratory are shown in Figure 8.2. When dissolving ammonium chloride, NH_4Cl, in water, we observe an endothermic process. The temperature changes from 24.5°C to 18.1°C when 10 g of NH_4Cl are added to 100 mL of water. This energy, in the form of heat, is taken from the immediate surroundings, the water, causing the salt solution to become colder. In the second example, we observe a temperature change from 24.8°C to 47.0°C when 10 mL of concentrated sulfuric acid, H_2SO_4, are dissolved in 100 mL of water. This reaction is an exothermic process. In both examples the temperature changes are large enough to be detected by touching the containers.

▲
FIGURE 8.2
Left: In the endothermic process $Ba(OH)_2$ and NH_4SCN are mixed in a beaker in a puddle of water. The water freezes causing the reaction beaker to stick to the board. *Right:* In the exothermic process a mixture of sugar and potassium chlorate are ignited with a drop of concentrated sulfuric acid.

8.7 Global Warming: The Greenhouse Effect

Fossil fuels, derived from coal and petroleum, provide the energy we use to power our industries, heat and light our homes and workplaces, and run our cars. These fuels are made primarily of hydrocarbons—compounds containing carbon and hydrogen. As these fuels are burned they produce carbon dioxide and water. Over 50 billion tons of carbon dioxide are released into our atmosphere each year.

The concentration of carbon dioxide has been monitored by scientists since 1958. Analysis of the air trapped in a core sample of snow from Antartica provides data on carbon dioxide levels for the past 160,000 years. The results of this study show that as the carbon dioxide increased, the global temperature increased as well. The levels of carbon dioxide remained reasonably constant from the last ice age, 100,000 years ago, until the Industrial Revolution. Since then the concentration of carbon dioxide in our atmosphere has risen 15% to an all-time high.

Carbon dioxide is a minor component in our atmosphere and is not usually considered to be a pollutant. The concern expressed by scientists arises from the dramatic increase occuring in the Earth's atmosphere. Without the influence of man in the environment the exchange of carbon dioxide between plants and animals would be relatively balanced. Our continued use of fossil fuels has led to an increase of 7.4% carbon dioxide between 1900 and 1970, and has additionally increased 3.5% during the 1980s. See Figure 8.3.

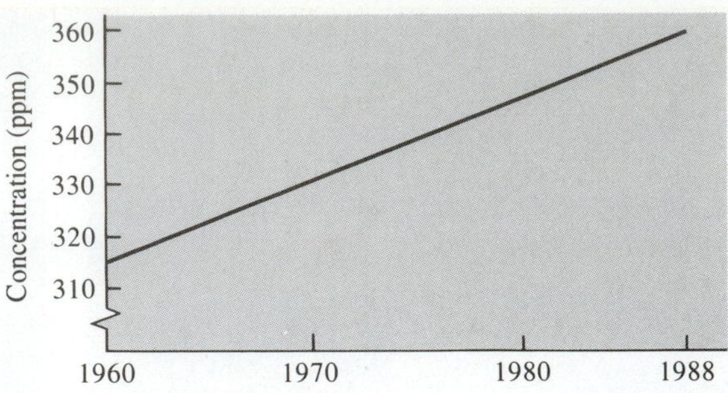

FIGURE 8.3 ▶ Concentration of carbon dioxide in the atmosphere.

In addition to the larger consumption of fossil fuels, there are still other factors that are increasing carbon dioxide in the atmosphere: Rain forests in Brazil are being destroyed by cutting and burning to make room for increased population and agricultural needs. Carbon dioxide is added to the atmosphere during the burning and the loss of trees diminishes the uptake of carbon dioxide by plants.

About half of all the carbon dioxide released into our atmosphere annually remains there, thus increasing the concentration. The other half of the carbon dioxide is absorbed by plants during photosynthesis and is dissolved in the ocean forming bicarbonates and carbonates.

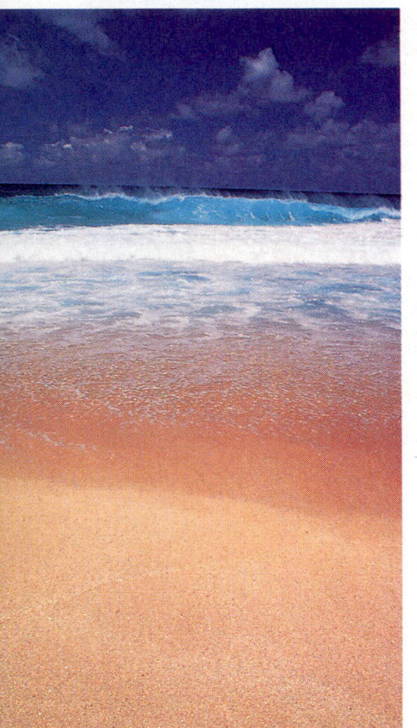

▲ Carbon dioxide dissolves in the ocean, forming carbonates and bicarbonates.

Carbon dioxide and other greenhouse gases, such as methane and water, act to warm our atmosphere by trapping heat near the surface of the Earth. Solar radiation strikes the Earth and warms the surface. The warmed surface reradiates this energy as heat. The greenhouse gases absorb some of this heat energy from the surface, which warms our atmosphere. This is the same principle found in a greenhouse where sunlight comes through the glass yet heat cannot escape. The air in the greenhouse warms, producing a climate considerably different than in nature. In the atmosphere these greenhouse gases are acting to warm our air and produce changes in our climate.

Long-term effects of global warming are still a matter of speculation and debate. Considerations include the melting of the polar icecaps, which would cause a rise in sea level and lead to major flooding on the coasts of our continents. Further effects could include shifts in rainfall patterns, producing droughts and extreme seasonal changes in such major agricultural regions as California.

To reverse these trends in releasing carbon dioxide into the atmosphere will require the development of new energy sources to cut our dependence on fossil fuels, an end to deforestation worldwide, and intense efforts to improve conservation. On an individual basis each of us can play a significant role. For example, the simple conversion of a 100-watt incandescent light bulb to a fluorescent bulb can reduce electrical consumption for that light 20% and the bulb can last 10 times longer. Recycling, switching to more fuel-efficient cars, and energy efficient appliances, heaters, and air-conditioners all result in decreased energy consumption and less carbon dioxide in our atmosphere.

Chemistry in Action

Autumn Leaf Color

As well as contributing to the metabolic cycles in plants, chemical reactions can be the source of natural beauty. The wondrous display of color seen in the leaf display of autumn is the result of chemical reactions.

Chlorophylls are responsible for the common green color in plants and are necessary for the plant to produce food, in the process called *photosynthesis*. Photosynthesis involves many chemical reactions but can be summarized by the following equation:

$$\underset{\text{carbon dioxide}}{6\,CO_2} + \underset{\text{water}}{6\,H_2O} \xrightarrow[\text{sunlight}]{\text{chlorophyll}} \underset{\text{glucose}}{C_6H_{12}O_6} + \underset{\text{oxygen}}{6\,O_2}$$

There are several types of chlorophyll, including chlorophyll a, chlorophyll b, and chlorophyll c. All photosynthetic plants contain chlorophyll a, and some contain chlorophyll b and chlorophyll c, which are called *accessory pigments*. The additional pigments extend the range of colored light that plants can use for photosynthesis. In addition, plants also contain carotenoids (orange, yellow, and red pigments), which protect the plant from destructive potential of chlorophyll. The carotenoids absorb the high-energy oxygen that is released when the chlorophyll absorbs light energy, and they release the oxygen when it can be used constructively. The least prevalent pigments are *anthocyanins* (red or blue). These pigments are responsible for the great diversity of combinations and concentrations of colors in the leaves. The carotenoids and anthocyanins are masked by the chlorophylls and can only be seen as the chlorophyll disintegrates in the autumn.

To produce a brilliant autumn display, the proper set of conditions must exist. The trees need a long, vigorous growing season with plenty of water for photosynthesis. The autumn days should be bright to provide a longer photosynthetic period with plenty of sugar production. When less intense sunlight and cold nights signal the leaves to stop photosynthesis, the protein bound to the chlorophyll begins to release from it. The protein breaks down into amino acids that go to the roots for storage. The chlorophyll decomposes and the green color fades away. The other pigments, especially the yellows and reds of the carotenoids, can now be seen. Sugar, trapped in the leaves during the very chilly nights, is changed through a set of complex reactions into anthocyanins to produce a variety of colors, mostly reds. The more productive the leaf has been the greater the concentration of the other pigments and the more brilliant the color.

Concepts in Review

1. Know the format used in setting up chemical equations.
2. Recognize the various symbols commonly used in writing chemical equations.
3. Balance simple chemical equations.
4. Interpret a balanced equation in terms of the relative numbers or amounts of molecules, atoms, grams, or moles of each substance represented.

5. Classify equations as combination, decomposition, single-displacement, or double-displacement reactions.
6. Use the activity series to predict whether a single-displacement reaction will occur.
7. Complete and balance equations for simple combination, decomposition, single-displacement, and double-displacement reactions when given the reactants.
8. Distinguish between exothermic and endothermic reactions, and relate the quantity of heat to the amounts of substances involved in the reaction.
9. Identify the major sources of chemical energy and their uses.

EXERCISES

1. Balance the following equations:
   (a) $H_2 + O_2 \longrightarrow H_2O$
   (b) $H_2 + Br_2 \longrightarrow HBr$
   (c) $C + Fe_2O_3 \longrightarrow Fe + CO$
   (d) $H_2O_2 \longrightarrow H_2O + O_2$
   (e) $Ba(ClO_3)_2 \xrightarrow{\Delta} BaCl_2 + O_2$
   (f) $H_2SO_4 + NaOH \longrightarrow H_2O + Na_2SO_4$
   (g) $NH_4I + Cl_2 \longrightarrow NH_4Cl + I_2$
   (h) $CrCl_3 + AgNO_3 \longrightarrow Cr(NO_3)_3 + AgCl$
   (i) $Al_2(CO_3)_3 \xrightarrow{\Delta} Al_2O_3 + CO_2$
   (j) $Al + C \xrightarrow{\Delta} Al_4C_3$
2. Classify the reactions in Exercise 1. as combination, decomposition, single displacement, or double displacement.
3. Balance the following equations:
   (a) $SO_2 + O_2 \longrightarrow SO_3$
   (b) $Al + MnO_2 \xrightarrow{\Delta} Mn + Al_2O_3$
   (c) $Na + H_2O \longrightarrow NaOH + H_2$
   (d) $AgNO_3 + Ni \longrightarrow Ni(NO_3)_2 + Ag$
   (e) $Bi_2S_3 + HCl \longrightarrow BiCl_3 + H_2S$
   (f) $PbO_2 \xrightarrow{\Delta} PbO + O_2$
   (g) $LiAlH_4 \xrightarrow{\Delta} LiH + Al + H_2$
   (h) $KI + Br_2 \longrightarrow KBr + I_2$
   (i) $K_3PO_4 + BaCl_2 \longrightarrow KCl + Ba_3(PO_4)_2$
4. Balance the following equations.
   (a) $MnO_2 + CO \longrightarrow Mn_2O_3 + CO_2$
   (b) $Mg_3N_2 + H_2O \longrightarrow Mg(OH)_2 + NH_3$
   (c) $C_3H_5(NO_3)_3 \longrightarrow CO_2 + H_2O + N_2 + O_2$
   (d) $FeS + O_2 \longrightarrow Fe_2O_3 + SO_2$
   (e) $Cu(NO_3)_2 \longrightarrow CuO + NO_2 + O_2$
   (f) $NO_2 + H_2O \longrightarrow HNO_3 + NO$
   (g) $Al + H_2SO_4 \longrightarrow Al_2(SO_4)_3 + H_2$
   (h) $HCN + O_2 \longrightarrow N_2 + CO_2 + H_2O$
   (i) $B_5H_9 + O_2 \longrightarrow B_2O_3 + H_2O$
   (j) $NH_3 + O_2 \longrightarrow NO + H_2O$
5. Change the following word equations into formula equations and balance them:
   (a) Copper + Sulfur $\xrightarrow{\Delta}$ Copper(I) sulfide
   (b) Phosphoric acid + Calcium hydroxide $\longrightarrow$ Calcium phosphate + Water
   (c) Silver oxide $\xrightarrow{\Delta}$ Silver + Oxygen
   (d) Iron(III) chloride + Sodium hydroxide $\longrightarrow$ Iron(III) hydroxide + Sodium chloride
   (e) Nickel(II) phosphate + Sulfuric acid $\longrightarrow$ Nickel(II) sulfate + Phosphoric acid
   (f) Zinc carbonate + Hydrochloric acid $\longrightarrow$ Zinc chloride + Water + Carbon dioxide
   (g) Silver nitrate + Aluminum chloride $\longrightarrow$ Silver chloride + Aluminum nitrate
6. Change the following word equations into formula equations and balance them:
   (a) Water $\longrightarrow$ Hydrogen + Oxygen
   (b) Acetic acid + Potassium hydroxide $\longrightarrow$ Potassium acetate + Water
   (c) Phosphorus + Iodine $\longrightarrow$ Phosphorus triiodide
   (d) Aluminum + Copper(II) sulfate $\longrightarrow$ Copper + Aluminum sulfate
   (e) Ammonium sulfate + Barium chloride $\longrightarrow$ Ammonium chloride + Barium sulfate
   (f) Sulfur tetrafluoride + Water $\longrightarrow$ Sulfur dioxide + Hydrogen fluoride
   (g) Chromium(III) carbonate $\xrightarrow{\Delta}$ Chromium(III) oxide + Carbon dioxide
7. Complete and balance the equations for these combination reactions:
   (a) $K + O_2 \longrightarrow$
   (b) $Al + Cl_2 \longrightarrow$
   (c) $CO_2 + H_2O \longrightarrow$
   (d) $CaO + H_2O \longrightarrow$

8. Complete and balance the equations for these decomposition reactions:
   (a) $HgO \xrightarrow{\Delta}$ (c) $MgCO_3 \xrightarrow{\Delta}$
   (b) $NaClO_3 \xrightarrow{\Delta}$ (d) $PbO_2 \xrightarrow{\Delta} PbO +$
9. Complete and balance the equations for these single-displacement reactions:
   (a) $Zn + H_2SO_4 \longrightarrow$
   (b) $AlI_3 + Cl_2 \longrightarrow$
   (c) $Mg + AgNO_3 \longrightarrow$
   (d) $Al + CoSO_4 \longrightarrow$
10. Use the activity series to predict which of the following reactions will occur. Complete and balance the equations. Where no reaction will occur, write "no reaction" as the product.
   (a) $Ag + H_2SO_4(aq) \longrightarrow$
   (b) $Cl_2 + NaBr(aq) \longrightarrow$
   (c) $Mg + ZnCl_2(aq) \longrightarrow$
   (d) $Pb + AgNO_3(aq) \longrightarrow$
   (e) $Cu + FeCl_3(aq) \longrightarrow$
   (f) $H_2 + Al_2O_3(s) \xrightarrow{\Delta}$
   (g) $Al + HBr(aq) \longrightarrow$
   (h) $I_2 + HCl(aq) \longrightarrow$
11. Complete and balance the equations for these double-displacement reactions:
   (a) $ZnCl_2 + KOH \longrightarrow$
   (b) $CuSO_4 + H_2S \longrightarrow$
   (c) $Ca(OH)_2 + H_3PO_4 \longrightarrow$
   (d) $(NH_4)_3PO_4 + Ni(NO_3)_2 \longrightarrow$
   (e) $Ba(OH)_2 + HNO_3 \longrightarrow$
   (f) $(NH_4)_2S + HCl \longrightarrow$
12. Predict which of the following double-displacement reactions will occur. Complete and balance the equations. Where no reaction will occur, write "no reaction" as the product.
   (a) $AgNO_3(aq) + KCl(aq) \longrightarrow$
   (b) $Ba(NO_3)_2(aq) + MgSO_4(aq) \longrightarrow$
   (c) $H_2SO_4(aq) + Mg(OH)_2(aq) \longrightarrow$
   (d) $MgO(s) + H_2SO_4(aq) \longrightarrow$
   (e) $Na_2CO_3(aq) + NH_4Cl(aq) \longrightarrow$
13. Complete and balance the equations for these reactions. All reactions yield products:
   (a) $H_2 + I_2 \longrightarrow$
   (b) $CaCO_3 \xrightarrow{\Delta}$
   (c) $Mg + H_2SO_4 \longrightarrow$
   (d) $FeCl_2 + NaOH \longrightarrow$
   (e) $SO_2 + H_2O \longrightarrow$
   (f) $SO_3 + H_2O \longrightarrow$
   (g) $Ca + H_2O \longrightarrow$
   (h) $Bi(NO_3)_3 + H_2S \longrightarrow$
14. Complete and balance the equations for the following reactions. All reactions yield products:
   (a) $Ba + O_2 \longrightarrow$
   (b) $NaHCO_3 \xrightarrow{\Delta} Na_2CO_3 +$
   (c) $Ni + CuSO_4 \longrightarrow$
   (d) $MgO + HCl \longrightarrow$
   (e) $H_3PO_4 + KOH \longrightarrow$
   (f) $C + O_2 \longrightarrow$
   (g) $Al(ClO_3)_3 \xrightarrow{\Delta} O_2 +$
   (h) $CuBr_2 + Cl_2 \longrightarrow$
   (i) $SbCl_3 + (NH_4)_2S \longrightarrow$
   (j) $NaNO_3 \xrightarrow{\Delta} NaNO_2 +$
15. What is the purpose of balancing equations?
16. What is represented by the numbers (coefficients) that are placed in front of the formulas in a balanced equation?
17. Interpret the following chemical reactions in terms of the number of moles of each reactant and product:
   (a) $MgBr_2 + 2\ AgNO_3 \longrightarrow Mg(NO_3)_2 + 2\ AgBr$
   (b) $N_2 + 3\ H_2 \longrightarrow 2\ NH_3$
   (c) $2\ C_3H_7OH + 9\ O_2 \longrightarrow 6\ CO_2 + 8\ H_2O$
18. Interpret each of the following equations in terms of the relative number of moles of each substance involved and indicate whether the reaction is exothermic or endothermic:
   (a) $2\ Na + Cl_2 \longrightarrow 2\ NaCl + 822\ kJ\ (196.4\ kcal)$
   (b) $PCl_5 + 92.9\ kJ\ (22.2\ kcal) \longrightarrow PCl_3 + Cl_2$
19. Write balanced equations for each of these reactions, including the heat term:
   (a) Lime, CaO, is converted to slaked lime, $Ca(OH)_2$, by reaction with water. The reaction liberates 65.3 kJ (15.6 kcal) of heat for each mole of lime reacted.
   (b) The industrial production of aluminum metal from aluminum oxide is an endothermic electrolytic process requiring 1630 kJ per mole of Al_2O_3. Oxygen is also a product.
20. Write balanced equations for the combustion of the following hydrocarbons.
   (a) ethane, C_2H_6 (c) heptane, C_7H_{16}
   (b) benzene, C_6H_6
21. Which of the following statements are correct? Rewrite the incorrect ones to make them correct.
   (a) The coefficients in front of the formulas in a balanced chemical equation give the relative number of moles of the reactants and products in the reaction.
   (b) A balanced chemical equation is one that has the same number of moles on each side of the equation.
   (c) In a chemical equation, the symbol $\xrightarrow{\Delta}$ indicates that the reaction is exothermic.

(d) A chemical change that absorbs heat energy is said to be endothermic.
(e) In the reaction $H_2 + Cl_2 \longrightarrow 2\ HCl$, 100 molecules of HCl are produced for every 50 molecules of H_2 reacted.
(f) The symbol (*aq*) after a substance in an equation means that the substance is in a water solution.
(g) The equation $H_2O \longrightarrow H_2 + O_2$ can be balanced by placing a 2 in front of H_2O.
(h) In the equation $3\ H_2 + N_2 \longrightarrow 2\ NH_3$ there are fewer moles of product than there are moles of reactants.
(i) The total number of moles of reactants and products represented by this equation is 5 moles:
$Mg + 2\ HCl \longrightarrow MgCl_2 + H_2$
(j) One mole of glucose, $C_6H_{12}O_6$, contains 6 moles of carbon atoms.
(k) The reactants are the substances produced by the chemical reaction.
(l) In a balanced equation, each side of the equation contains the same number of atoms of each element.
(m) When a precipitate is formed in a chemical reaction, it can be indicated in the equation with the symbol ↓ or (*s*) immediately before the formula of the substance precipitated.
(n) When a gas is evolved in a chemical reaction, it can be indicated in the equation with the symbol ↑ or (*g*) immediately following the formula of the gas.
(o) According to the equation $3\ H_2 + N_2 \longrightarrow 2\ NH_3$, 4 mol of NH_3 will be formed when 6 mol of H_2 and 2 mol of N_2 react.
(p) The products of an exothermic reaction are at a lower energy state than the reactants.
(q) The combustion of hydrocarbons produces carbon dioxide and water as products.

22. List the various pigments found in plants and state the function of each.

23. State the four necessary requirements for successful photosynthesis.

24. Draw a flowchart illustrating how leaves "change color" in the autumn of the year.

25. List the factors that contribute to an increase in carbon dioxide in our atmosphere.

26. List three gases considered to be greenhouse gases. Explain why they are given this name.

27. How can the effects of global warming be reduced?

28. What happens to carbon dioxide released into our atmosphere?

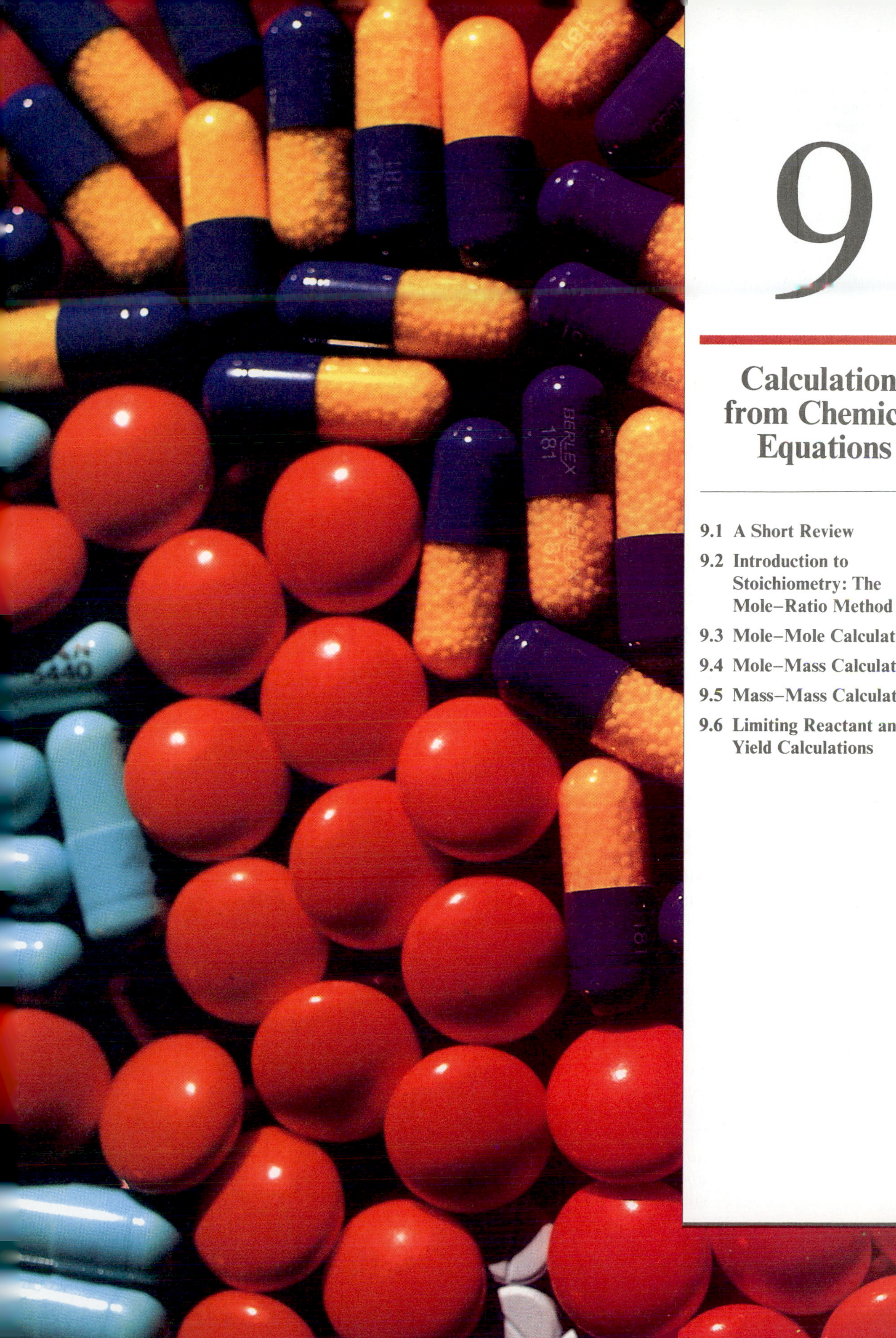

9

Calculations from Chemical Equations

◄ CHAPTER OPENING PHOTO: **The precise and necessary amount of a drug must be calculated prior to manufacturing pills.**

Often, the old adage "waste not, want not" is appropriate in our daily life. For example, a hostess determines the quantity of food and beverage necessary to serve all her guests. These amounts are defined by specific recipes and a knowledge of the particular likes and dislikes of the guests. A seamstress determines the amount of material, lining, and trim necessary to produce a gown for her client by relying on a pattern or her own experience to guide the selection. A carpet layer determines the correct amount of carpet and padding necessary to recarpet a customer's house by calculating the floor area. The IRS determines the correct deduction for federal income taxes from your paycheck based on your expected annual income.

The chemist also finds it necessary to calculate amounts of products or reactants using a balanced chemical equation. With these calculations, the chemist can control the amount of product by scaling the reaction up or down to fit the needs of the laboratory, and thereby minimize waste or excess materials formed during the reaction.

9.1 A Short Review

Molar Mass The molar mass is the sum of the atomic masses of all the atoms in a molecule. The molar mass also applies to the mass of any formula unit—atoms, molecules, or ions; it is the atomic mass of an atom, or the sum of the atomic masses in a molecule or an ion.

Relationship Between Molecule and Mole A molecule is the smallest unit of a molecular substance (for example, Cl_2), and a mole is Avogadro's number, 6.022×10^{23}, of molecules of that substance. A mole of chlorine (Cl_2) has the same number of molecules as a mole of carbon dioxide, a mole of water, or a mole of any other molecular substance. When we relate molecules to molar mass 1 molar mass = 1 mol, or 6.022×10^{23} molecules.

In addition to referring to molecular substances, the term *mole* may refer to any chemical species. It represents a quantity in grams equal to the molar mass, and may be applied to atoms, ions, electrons, and formula units of nonmolecular substances.

$$1 \text{ mole} = \begin{cases} 6.022 \times 10^{23} \text{ molecules} \\ 6.022 \times 10^{23} \text{ formula units} \\ 6.022 \times 10^{23} \text{ atoms} \\ 6.022 \times 10^{23} \text{ ions} \end{cases}$$

Other useful mole relationships are

$$\text{number of moles} = \frac{\text{grams of a substance}}{\text{molar mass of the substance}}$$

$$\text{number of moles} = \frac{\text{grams of a monatomic element}}{\text{molar mass of the element}}$$

$$\text{number of moles} = \frac{\text{number of molecules}}{6.022 \times 10^{23} \text{ molecules/mole}}$$

Two other useful equalities can be derived algebraically from each of these mole relationships. What are they?

Balanced Equations When using chemical equations for calculations of mole–mass–volume relationships between reactants and products, the equations must be balanced. Remember that the number in front of a formula in a balanced chemical equation can represent the number of moles of that substance in the chemical reaction.

9.2 Introduction to Stoichiometry: The Mole–Ratio Method

It is often necessary to calculate the amount of a substance that is produced from, or needed to react with, a given quantity of another substance. The area of chemistry that deals with the quantitative relationships among reactants and products is known as **stoichiometry** (*stoy-key-ah-meh-tree*). Although several methods are known, we firmly believe that the *mole* or *mole-ratio* method is generally best for solving problems in stoichiometry. This method is straightforward and, in our opinion, makes it easy to see and understand the relationships of the reacting species.

stoichiometry

A **mole ratio** is a ratio between the number of moles of any two species involved in a chemical reaction. For example, in the reaction

mole ratio

$$\underset{2\text{ mol}}{2\,H_2} + \underset{1\text{ mol}}{O_2} \longrightarrow \underset{2\text{ mol}}{2\,H_2O}$$

six mole ratios apply only to this reaction:

$$\frac{2\text{ mol } H_2}{1\text{ mol } O_2} \qquad \frac{2\text{ mol } H_2}{2\text{ mol } H_2O} \qquad \frac{1\text{ mol } O_2}{2\text{ mol } H_2}$$

$$\frac{1\text{ mol } O_2}{2\text{ mol } H_2O} \qquad \frac{2\text{ mol } H_2O}{2\text{ mol } H_2} \qquad \frac{2\text{ mol } H_2O}{1\text{ mol } O_2}$$

The mole ratio is a conversion factor used to convert the number of moles of one substance to the corresponding number of moles of another substance in a chemical reaction. For example, if we want to calculate the number of moles of H_2O that can be obtained from 4.0 mol of O_2, we use the mole ratio 2 mol H_2O/1 mol O_2.

$$4.0\ \cancel{\text{mol } O_2} \times \frac{2\text{ mol } H_2O}{1\ \cancel{\text{mol } O_2}} = 8.0\text{ mol } H_2O$$

Since stoichiometric problems are encountered throughout the entire field of chemistry, it is profitable to master this general method for their solution. The mole-ratio method makes use of three simple basic operations.

1. Convert the quantity of starting substance to moles (if it is not given in moles).
2. Convert the moles of starting substance to moles of desired substance.
3. Convert the moles of desired substance to the units specified in the problem.

Like learning to balance chemical equations, learning to make stoichiometric calculations requires practice. A detailed step-by-step description of the general method, together with a variety of worked examples, is given in the following paragraphs. Study this material and apply the method to the problems at the end of this chapter.

Use a balanced equation. Write a balanced equation for the chemical reaction in question or check to see that the equation given is balanced.

Step 1 Determine the number of moles of starting substance. Identify the starting substance from the data given in the statement of the problem. When the starting substance is given in moles, use it in that form. If it is not in moles, convert the quantity of the starting substance to moles. For example, if grams of a starting substance are given,

$$\begin{pmatrix}\text{moles of starting}\\ \text{substance}\end{pmatrix} = \begin{pmatrix}\text{grams of starting}\\ \text{substance}\end{pmatrix} \times \frac{\begin{pmatrix}\text{1 mol of}\\ \text{starting substance}\end{pmatrix}}{\begin{pmatrix}\text{molar mass of}\\ \text{starting substance}\end{pmatrix}}$$

Step 2 Determine the mole ratio of the desired substance to the starting substance. The number of moles of each substance in the balanced equation is indicated by the coefficient in front of each substance. Use these coefficients to set up the mole ratio:

$$\text{mole ratio} = \frac{\text{moles of desired substance in the equation}}{\text{moles of starting substance in the equation}}$$

Calculate the number of moles of the desired substance. Multiply the number of moles of starting substance (from Step 2) by the mole ratio (from Step 3) to obtain the number of moles of desired substance:

$$\begin{pmatrix}\text{moles of desired}\\ \text{substance}\end{pmatrix} = \underset{\text{From Step 2}}{\begin{pmatrix}\text{moles of starting}\\ \text{substance}\end{pmatrix}} \times \underset{\text{Mole ratio from Step 3}}{\frac{\begin{pmatrix}\text{moles of desired substance}\\ \text{in the equation}\end{pmatrix}}{\begin{pmatrix}\text{moles of starting substance}\\ \text{in the equation}\end{pmatrix}}}$$

Note that the units of moles of starting substance cancel out in the numerator and the denominator.

Step 3 Calculate the desired substance in the units specified in the problem. If the answer is to be in moles, the problem is finished in Step 4. If units other than moles are wanted, multiply the moles of the desired substance (from Step 4) by the appropriate factor to convert moles to the units required.
For example, if grams of the desired substance are wanted,

$$\begin{pmatrix}\text{grams of desired}\\ \text{substance}\end{pmatrix} = \underset{\text{From Step 4}}{\begin{pmatrix}\text{moles of desired}\\ \text{substance}\end{pmatrix}} \times \frac{\begin{pmatrix}\text{molar mass of}\\ \text{desired substance}\end{pmatrix}}{\begin{pmatrix}\text{one mole of}\\ \text{desired substance}\end{pmatrix}}$$

Use the conversion factors 6.022×10^{23} atoms/mol or 6.022×10^{23} molecules/mol when the problem asks for the answer in atoms or molecules.

The steps for converting the mass of starting substance A to either mass, atoms, or molecules of desired substance B are summarized in the flow chart below:

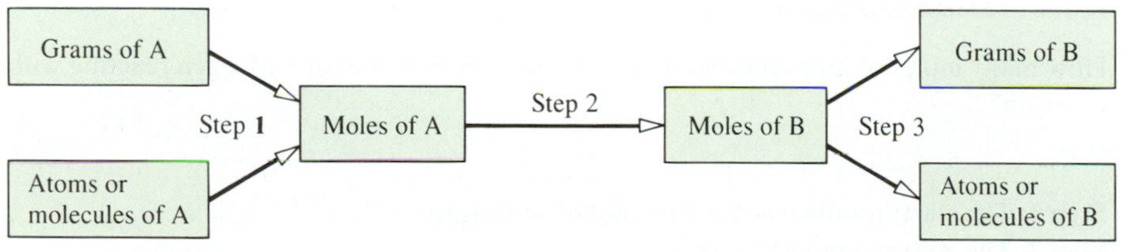

9.3 Mole–Mole Calculations

The first application of the mole-ratio method of solving stoichiometric problems is that of mole-mole calculations.

The quantity of starting substance is given in moles and the quantity of desired substance is requested in moles. Steps 2 and 5 of the mole-ratio method (Section 9.2) are not required. Understanding the use of mole ratios will be very helpful in solving later problems. Some illustrative problems follow.

EXAMPLE 9.1

How many moles of carbon dioxide will be produced by the complete oxidation of 2.0 mol of glucose ($C_6H_{12}O_6$), according to the following reaction?

$$\underset{1\text{ mol}}{C_6H_{12}O_6} + \underset{6\text{ mol}}{6\,O_2} \longrightarrow \underset{6\text{ mol}}{6\,CO_2} + \underset{6\text{ mol}}{6\,H_2O}$$

SOLUTION

The balanced equation states that 6 mol of CO_2 will be produced from 1 mol of $C_6H_{12}O_6$. Even though we can readily see that 12 mol of CO_2 will be formed from 2.0 mol of $C_6H_{12}O_6$, the mole-ratio method of solving the problem is shown here.

Step 1 The number of moles of starting substance is 2.0 mol $C_6H_{12}O_6$.
Step 2 The conversion needed is

$$\text{moles } C_6H_{12}O_6 \longrightarrow \text{moles } CO_2$$

Step 3 Multiply 2.0 mol of glucose (given in the problem) by this mole ratio.

$$2.0 \cancel{\text{mol } C_6H_{12}O_6} \times \frac{6 \text{ mol } CO_2}{1 \cancel{\text{mol } C_6H_{12}O_6}} = 12 \text{ mol } CO_2$$

Again note the use of units. The moles of $C_6H_{12}O_6$ cancel, leaving the answer in units of moles of CO_2.

EXAMPLE 9.2

How many moles of ammonia can be produced from 8.00 mol of hydrogen reacting with nitrogen?

SOLUTION

Step 1 The starting substance is 8.00 mol of hydrogen.
Step 2 The conversion needed is

$$\text{moles } H_2 \longrightarrow \text{moles } NH_3$$

The balanced equation states that we get 2 mol of NH_3 for every 3 mol of H_2 that react. Set up the mole ratio of desired substance (NH_3) to starting substance (H_2):

$$\text{mole ratio} = \frac{2 \text{ mol } NH_3}{3 \text{ mol } H_2} \quad \text{(from equation)}$$

EXAMPLE 9.3

Given the balanced equation

$$\underset{1\ \text{mol}}{K_2Cr_2O_7} + \underset{6\ \text{mol}}{6\ KI} + 7\ H_2SO_4 \longrightarrow Cr_2(SO_4)_3 + 4\ K_2SO_4 + \underset{3\ \text{mol}}{3\ I_2} + 7\ H_2O$$

calculate (a) the number of moles of potassium dichromate ($K_2Cr_2O_7$) that will react with 2.0 mol of potassium iodide (KI); (b) the number of moles of iodine (I_2) that will be produced from 2.0 mol of potassium iodide.

SOLUTION

After the equation is balanced, we are concerned only with $K_2Cr_2O_7$, KI, and I_2, and we can ignore all the other substances. The equation states that 1 mol of $K_2Cr_2O_7$ will react with 6 mol of KI to produce 3 mol of I_2.

(a) Calculate the number of moles of $K_2Cr_2O_7$.

Step 1 The starting substance is 2.0 mol of KI.
Step 2 The conversion needed is

$$\text{moles KI} \longrightarrow \text{moles } K_2Cr_2O_7$$

Set up the mole ratio of desired substance to starting substance:

$$\text{mole ratio} = \frac{1\ \text{mol}\ K_2Cr_2O_7}{6\ \text{mol KI}} \quad \text{(from equation)}$$

(b) Calculate the number of moles of I_2.

Step 1 The equation given is balanced and the moles of starting substance are 2.0 mol KI as in part (a).
Step 2 The conversion needed is

$$\text{moles KI} \longrightarrow \text{moles } I_2$$

Set up the mole ratio of desired substance to starting substance:

$$\text{mole ratio} = \frac{3\ \text{mol}\ I_2}{6\ \text{mol KI}} \quad \text{(from equation)}$$

EXAMPLE 9.4

How many molecules of water can be produced by reacting 0.010 mol of oxygen with hydrogen?

SOLUTION

The balanced equation is $2H_2 + O_2 \rightarrow 2H_2O$

The sequence of conversions needed in the calculation is

$$\text{moles } O_2 \longrightarrow \text{moles } H_2O \longrightarrow \text{molecules } H_2O$$

▲
The space shuttle is powered by H_2 and O_2 reacting to H_2O.

Step 1 The starting substance is 0.010 mol O_2.

Step 2 The conversion needed is moles $O_2 \longrightarrow$ moles H_2O. Set up the mole ratio of desired substance to starting substance:

$$\text{mole ratio} = \frac{2 \text{ mol } H_2O}{1 \text{ mol } O_2} \quad \text{(from equation)}$$

Step 3 Since the problem asks for molecules instead of moles of H_2O, we must convert moles to molecules. Use the conversion factor $(6.022 \times 10^{23}$ molecules)/mole.

$$0.020 \cancel{\text{mol}}\ H_2O \times \frac{6.022 \times 10^{23} \text{ molecules}}{1 \cancel{\text{mol}}} = 1.2 \times 10^{22} \text{ molecules } H_2O$$

Note that 0.020 mol is still quite a large number of water molecules.

PRACTICE How many moles of aluminum oxide will be produced from 0.50 mol of oxygen?

$$4\ Al + 3\ O_2 \longrightarrow 2\ Al_2O_3$$

Answer: 0.33 mol Al_2O_3

PRACTICE How many moles of aluminum hydroxide are required to produce 22.0 mol of water?

$$2\ Al(OH)_3 + 3\ H_2SO_4 \longrightarrow Al_2(SO_4)_3 + 6\ H_2O$$

Answer: 7.33 mol $Al(OH)_3$

9.4 Mole–Mass Calculations

The object of this type of problem is to calculate the mass of one substance that reacts with or is produced from a given number of moles of another substance in a chemical reaction. If the mass of the starting substance is given, it is necessary to convert it to moles. The mole ratio is used to convert from moles of starting substance to moles of desired substance. Moles of desired substance can then be changed to mass if required.

As you are now familiar with the mole-ratio method, Steps 3 and 4 (Section 9.2) are combined in the examples. Each example is solved in a continuous calculation sequence, after the step-by-step method is demonstrated.

EXAMPLE 9.5

What mass of hydrogen can be produced by reacting 6.0 mol of aluminum with hydrochloric acid?

SOLUTION

First calculate the moles of hydrogen produced, using the mole-ratio method, and then calculate the mass of hydrogen by multiplying the moles of hydrogen by its grams per mole. The sequence of conversions in the calculation is

moles Al ⟶ moles H_2 ⟶ grams H_2

The balanced equation is $2Al\ (s) + 6HCl\ (aq) \rightarrow 2AlCl_3\ (aq) + 3H_2\ (g)$

Step 1 The starting substance is 6.0 mol of aluminum.
Step 2 Calculate moles of H_2 by the mole-ratio method.

$$6.0\ \cancel{\text{mol Al}} \times \frac{3\ \text{mol H}_2}{2\ \cancel{\text{mol Al}}} = 9.0\ \text{mol H}_2$$

Step 3 Convert moles of H_2 to grams [g = mol × (g/mol)]:

$$9.0\ \cancel{\text{mol H}_2} \times \frac{2.016\ \text{g H}_2}{1\ \cancel{\text{mol H}_2}} = 18\ \text{g H}_2 \quad \text{(Answer)}$$

We see that 18 g of H_2 can be produced by reacting 6.0 mol of Al with HCl. The following setup combines all the above steps into one continuous calculation:

$$6.0\ \cancel{\text{mol Al}} \times \frac{3\ \cancel{\text{mol H}_2}}{2\ \cancel{\text{mol Al}}} \times \frac{2.016\ \text{g H}_2}{1\ \cancel{\text{mol H}_2}} = 18\ \text{g H}_2$$

EXAMPLE 9.6

How many moles of water can be produced by burning 325 g of octane (C_8H_{18})?

The balanced equation is $2C_8H_{18}\ (g) + 25O_2\ (g) \rightarrow 16CO_2\ (g) + 18H_2O\ (g)$

SOLUTION

The sequence of conversions in the calculation is

grams C_8H_{18} ⟶ moles C_8H_{18} ⟶ moles H_2O

Step 1 The starting substance is 325 g C_8H_{18}. Convert 325 g of C_8H_{18} to moles:

$$325\ \cancel{\text{g C}_8\text{H}_{18}} \times \frac{1\ \text{mol C}_8\text{H}_{18}}{114.2\ \cancel{\text{g C}_8\text{H}_{18}}} \longrightarrow 2.85\ \text{mol C}_8\text{H}_{18}$$

Step 2 Calculate the moles of water by the mole-ratio method:

$$2.85\ \cancel{\text{mol C}_8\text{H}_{18}} \times \frac{18\ \text{mol H}_2\text{O}}{2\ \cancel{\text{mol C}_8\text{H}_{18}}} = 25.7\ \text{mol H}_2\text{O} \quad \text{(Answer)}$$

The calculation in a continuous sequence is

$$325\ \cancel{\text{g C}_8\text{H}_{18}} \times \frac{1\ \cancel{\text{mol C}_8\text{H}_{18}}}{114.2\ \cancel{\text{g C}_8\text{H}_{18}}} \times \frac{18\ \text{mol H}_2\text{O}}{2\ \cancel{\text{mol C}_8\text{H}_{18}}} = 25.6\ \text{mol H}_2\text{O}$$

Practice How many moles of potassium chloride can be produced from 100.0 g of potassium chlorate?

$$2\ \text{KClO}_3 \longrightarrow 2\ \text{KCl} + 3\ \text{O}_2$$

Answer: 0.8157 mol KCl

Practice How many grams of silver nitrate are required to produce 0.25 mol of silver sulfide?

$$2\ \text{AgNO}_3 + \text{H}_2\text{S} \longrightarrow \text{Ag}_2\text{S} + 2\ \text{HNO}_3$$

Answer: 85 g $AgNO_3$

9.5 Mass–Mass Calculations

Solving mass-mass stoichiometry problems requires all steps of the mole-ratio method. The mass of starting substance is converted to moles. The mole ratio is then used to determine moles of desired substance which in turn is converted to mass.

Steps 3 and 4 (Section 9.2) are combined in the examples. At the end of each example the problem is solved in a continuous calculation sequence.

Example 9.7 What mass of carbon dioxide is produced by the complete combustion of 100. g of the hydrocarbon pentane, C_5H_{12}?

Solution The sequence of conversions in the calculation is

grams C_5H_{12} ⟶ moles C_5H_{12} ⟶ moles CO_2 ⟶ grams CO_2

The balanced equation is $C_5H_{12} + 8O_2 \rightarrow 5CO_2 + 6H_2O$

Step 1 The starting substance is 100. g of C_5H_{12}. Convert 100. g of C_5H_{12} to moles:

$$100.\ \cancel{g\ C_5H_{12}} \times \frac{1\ mol\ C_5H_{12}}{72.15\ \cancel{g\ C_5H_{12}}} = 1.39\ mol\ C_5H_{12}$$

Step 2 Calculate the moles of CO_2 by the mole-ratio method:

$$1.39\ \cancel{mol\ C_5H_{12}} \times \frac{5\ mol\ CO_2}{1\ \cancel{mol\ C_5H_{12}}} = 6.95\ mol\ CO_2$$

Step 3 Convert moles of CO_2 to grams:

$$mol\ CO_2 \times \frac{molar\ mass\ CO_2}{1\ mol\ CO_2} = grams\ CO_2$$

$$6.95\ \cancel{mol\ CO_2} \times \frac{44.01\ g\ CO_2}{1\ \cancel{mol\ CO_2}} = 306\ g\ CO_2$$

The calculation in a continuous sequence is

$$100.\ \cancel{g\ C_5H_{12}} \times \frac{1\ \cancel{mol\ C_5H_{12}}}{72.15\ \cancel{g\ C_5H_{12}}} \times \frac{5\ \cancel{mol\ CO_2}}{1\ \cancel{mol\ C_5H_{12}}} \times \frac{44.01\ g\ CO_2}{1\ \cancel{mol\ CO_2}} = 305\ g\ CO_2$$

EXAMPLE 9.8

How many grams of nitric acid, HNO_3, are required to produce 8.75 g of dinitrogen monoxide, N_2O, according to the following equation?

$$\underset{10\ mol}{4\ Zn(s) + 10\ HNO_3(aq)} \longrightarrow 4\ Zn(NO_3)_2(aq) + \underset{1\ mol}{N_2O(g)} + 5\ H_2O(l)$$

SOLUTION

The sequence of conversions in the calculation is

grams N_2O ⟶ moles N_2O ⟶ moles HNO_2 ⟶ grams HNO_3

Step 1 The starting substance is 8.75 g of N_2O. Convert 8.75 g of N_2O to moles:

$$8.75\ \cancel{g\ N_2O} \times \frac{1\ mol\ N_2O}{44.02\ \cancel{g\ N_2O}} = 0.199\ mol\ N_2O$$

Step 2 Calculate the moles of CO_2 by the mole-ratio method:

$$0.199\ \cancel{mol\ N_2O} \times \frac{10\ mol\ HNO_3}{1\ \cancel{mol\ N_2O}} = 1.99\ mol\ HNO_3$$

Step 3 Convert moles of HNO_3 to grams:

$$1.99\ \cancel{mol\ HNO_3} \times \frac{63.02\ g\ HNO_3}{1\ \cancel{mol\ HNO_3}} = 125\ g\ HNO_3$$

The calculation in a continuous sequence is

$$8.75\ \cancel{g\ N_2O} \times \frac{1\ \cancel{mol\ N_2O}}{44.02\ \cancel{g\ N_2O}} \times \frac{10\ \cancel{mol\ HNO_3}}{1\ \cancel{mol\ N_2O}} \times \frac{63.02\ g\ HNO_3}{1\ \cancel{mol\ HNO_3}} = 125\ g\ HNO_3$$

PRACTICE How many grams of chromium (III) chloride are required to produce 75.0 g of silver chloride?

$$CrCl_3 + 3\ AgNO_3 \longrightarrow Cr(NO_3)_3 + 3\ AgCl$$

Answer: 27.6 g $CrCl_3$

PRACTICE What mass of water is produced by the complete combustion of 225.0 g of butane (C_4H_{10})?

$$2\ C_4H_{10} + 13\ O_2 \longrightarrow 8\ CO_2 + 10\ H_2O$$

Answer: 348.8 g H_2O

9.6 Limiting Reactant and Yield Calculations

In many chemical processes the quantities of the reactants used are such that the amount of one reactant is in excess of the amount of a second reactant in the reaction. The amount of the product(s) formed in such a case will depend on the reactant that is not in excess. Thus the reactant that is not in excess is known as the **limiting reactant** (sometimes called the limiting reagent), because it limits the amount of product that can be formed.

limiting reactant

▲ The amount of product formed into each pill is determined by the limiting reactant.

As an example, consider the case in which solutions containing 1.0 mol of sodium hydroxide and 1.5 mol of hydrochloric acid are mixed:

$$\underset{1\ \text{mol}}{NaOH} + \underset{1\ \text{mol}}{HCl} \longrightarrow \underset{1\ \text{mol}}{NaCl} + \underset{1\ \text{mol}}{H_2O}$$

According to the equation it is possible to obtain 1.0 mol of NaCl from 1.0 mol of NaOH, and 1.5 mol of NaCl from 1.5 mol of HCl. However, we cannot have two different yields of NaCl from the reaction. When 1.0 mol of NaOH and 1.5 mol of HCl are mixed, there is insufficient NaOH to react with all of the HCl. Therefore, HCl is the reactant in excess and NaOH is the limiting reactant. Since the amount of NaCl formed is dependent on the limiting reactant, only 1.0 mol of NaCl will be formed. Because 1.0 mol of NaOH reacts with 1.0 mol of HCl, 0.5 mol of HCl remains unreacted.

$$\left.\begin{array}{l}1.0\ \text{mol NaOH}\\ 1.5\ \text{mol HCl}\end{array}\right\} \longrightarrow \begin{array}{l}1.0\ \text{mol NaCl}\\ 1.0\ \text{mol } H_2O\end{array} + 0.5\ \text{mol HCl unreacted}$$

Problems giving the amounts of two reactants are generally of the limiting-reactant type, and there are several methods used to identity them in a chemical reaction. In the most direct method two steps are needed to determine the limiting reactant and the amount of product formed.

1. Calculate the amount of product (moles or grams, as needed) that can be formed from each reactant.

2. Determine which reactant is limiting. (The reactant that gives the least amount of product is the limiting reactant and the other reactant is in excess. The limiting reactant will determine the amount of product formed in the reaction.)

Sometimes it is necessary to calculate the amount of excess reactant.

3. This can be done by first calculating the amount of excess reactant required to react with the limiting reactant. Then subtract the amount that reacts from the starting quantity of the reactant in excess. This result is the amount of that substance that remains unreacted.

EXAMPLE 9.9

How many moles of Fe_3O_4 can be obtained by reacting 16.8 g Fe with 10.0 g H_2O? Which substance is the limiting reactant? Which substance is in excess?

SOLUTION

$$3\,Fe(s) + 4\,H_2O(g) \xrightarrow{\Delta} Fe_3O_4(s) + 4\,H_2(g)$$

Calculate the moles of Fe_3O_4 that can be formed from each reactant.

g reactant ⟶ mol reactant ⟶ mol Fe_3O_4

1. Calculate the moles of Fe_3O_4 that can be formed from each reactant.

$$16.8\,\text{g Fe} \times \frac{1\,\text{mol Fe}}{55.85\,\text{g Fe}} \times \frac{1\,\text{mol } Fe_3O_4}{3\,\text{mol Fe}} = 0.100\,\text{mol } Fe_3O_4$$

$$10.0\,\text{g } H_2O \times \frac{1\,\text{mol } H_2O}{18.02\,\text{g } H_2O} \times \frac{1\,\text{mol } Fe_3O_4}{4\,\text{mol } H_2O} = 0.139\,\text{mol } Fe_3O_4$$

2. Determine the limiting reactant. The limiting reactant is Fe because it produces less Fe_3O_4; the H_2O is in excess. The yield of product is 0.100 mol of Fe_3O_4.

EXAMPLE 9.10

How many grams of silver bromide, AgBr, can be formed when solutions containing 50.0 g of $MgBr_2$ and 100.0 g of $AgNO_3$ are mixed together? How many grams of the excess reactant remain unreacted?

SOLUTION

$$MgBr_2(aq) + 2\,AgNO_3(aq) \longrightarrow 2\,AgBr\downarrow + Mg(NO_3)_2(aq)$$

1. Calculate the grams of AgBr that can be formed from each reactant.

g reactant ⟶ mol reactant ⟶ mol AgBr ⟶ g AgBr

$$50.0\,\text{g } MgBr_2 \times \frac{1\,\text{mol } MgBr_2}{184.1\,\text{g } MgBr_2} \times \frac{2\,\text{mol AgBr}}{1\,\text{mol } MgBr_2} \times \frac{187.8\,\text{g AgBr}}{1\,\text{mol AgBr}} = 102\,\text{g AgBr}$$

$$100.0\,\text{g } AgNO_3 \times \frac{1\,\text{mol } AgNO_3}{169.9\,\text{g } AgNO_3} \times \frac{2\,\text{mol AgBr}}{2\,\text{mol } AgNO_3} \times \frac{187.8\,\text{g AgBr}}{1\,\text{mol AgBr}} = 110.5\,\text{g AgBr}$$

2. Determine the limiting reactant. The limiting reactant is $MgBr_2$ because it gives less AgBr. $AgNO_3$ is in excess. The yield is 102 g AgBr.
3. Calculate the grams of unreacted $AgNO_3$.
   Calculate the grams of $AgNO_3$ that will react with 50.0 g of $MgBr_2$.

g $MgBr_2$ ⟶ mol $MgBr_2$ ⟶ mol $AgNO_3$ ⟶ g $AgNO_3$

$$50.0\,\text{g } MgBr_2 \times \frac{1\,\text{mol } MgBr_2}{184.1\,\text{g } MgBr_2} \times \frac{2\,\text{mol } AgNO_3}{1\,\text{mol } MgBr_2} \times \frac{169.9\,\text{g } AgNO_3}{1\,\text{mol } AgNO_3} = 92.3\,\text{g } AgNO_3$$

Thus 92.3 g of $AgNO_3$ react with 50.0 g of $MgBr_2$. The amount of $AgNO_3$ that remains unreacted is

$$100.0 \text{ g } AgNO_3 - 92.3 \text{ g } AgNO_3 = 7.7 \text{ g } AgNO_3 \text{ unreacted}$$

The final mixture will contain 102 g AgBr(*s*), 7.7 g $AgNO_3$ and $Mg(NO_3)_2$ in solution.

Practice How many grams of hydrogen chloride can be produced from 0.490 g of hydrogen and 50.0 g of chlorine?

$$H_2(g) + Cl_2(g) \longrightarrow 2\,HCl(g)$$

Answer: 17.7 g HCl

Practice How many grams of barium sulfate will be formed from 200.0 g of barium nitrate and 100.0 grams of sodium sulfate?

$$Ba(NO_3)_2 + Na_2SO_4 \longrightarrow BaSO_4 + 2\,NaNO_3$$

Answer: 164.4 g $BaSO_4$

The quantities of the products that we have been calculating from equations represent the maximum yield (100%) of product according to the reaction represented by the equation. Many reactions, especially those involving organic substances, fail to give a 100% yield of product. The main reasons for this failure are the side reactions that give products other than the main product and the fact that many reactions are reversible. In addition, some product may be lost in handling and transferring from one vessel to another. The **theoretical yield** of a reaction is the calculated amount of product that can be obtained from a given amount of reactant, according to the chemical equation. The **actual yield** is the amount of product that we finally obtain.

theoretical yield

actual yield

percent yield

The **percent yield** is the ratio of the actual yield to the theoretical yield multiplied by 100. Both yields must have the same units.

$$\frac{\text{actual yield}}{\text{theoretical yield}} \times 100 = \text{percent yield}$$

For example, if the theoretical yield calculated for a reaction is 14.8 g, and the amount of product obtained is 9.25 g, the percent yield is

$$\text{percent yield} = \frac{9.25 \text{ g}}{14.8 \text{ g}} \times 100 = 62.5\%$$

Example 9.11

Carbon tetrachloride was prepared by reacting 100. g of carbon disulfide and 100. g of chlorine. Calculate the percent yield if 65.0 g of CCl_4 was obtained from the reaction.

$$CS_2 + 3\,Cl_2 \longrightarrow CCl_4 + S_2Cl_2$$

SOLUTION

In this problem we need to determine the limiting reactant in order to calculate the quantity of CCl_4 (theoretical yield) that can be formed. Then we can compare that amount with the 65.0 g of CCl_4 actual yield to calculate the percent yield.

1. Determine the theoretical yield. Calculate the grams of CCl_4 that can be formed from each reactant.

g reactant ⟶ mol reactant ⟶ mol CCl_4 ⟶ g CCl_4

$$100.\ \text{g } CS_2 \times \frac{1\ \text{mol } CS_2}{76.13\ \text{g } CS_2} \times \frac{1\ \text{mol } CCl_4}{1\ \text{mol } CS_2} \times \frac{153.8\ \text{g } CCl_4}{1\ \text{mol } CCl_4} = 202\ \text{g } CCl_4$$

$$100.\ \text{g } Cl_2 \times \frac{1\ \text{mol } Cl_2}{70.90\ \text{g } Cl_2} \times \frac{1\ \text{mol } CCl_4}{3\ \text{mol } Cl_2} \times \frac{153.8\ \text{g } CCl_4}{1\ \text{mol } CCl_4} = 72.3\ \text{g } CCl_4$$

2. Determine the limiting reactant. The limiting reactant is Cl_2 because it gives less CCl_4. The CS_2 is in excess. The theoretical yield is 72.3 g CCl_4.
3. Calculate the percent yield.

   According to the equation, 72.3 g of CCl_4 is the maximum amount or theoretical yield of CCl_4 possible from 100. g of Cl_2. Actual yield is 65.0 g of CCl_4.

$$\text{percent yield} = \frac{65.0\ \text{g}}{72.3\ \text{g}} \times 100 = 89.9\%$$

EXAMPLE 9.12

Silver bromide was prepared by reacting 200.0 g of magnesium bromide and an adequate amount of silver nitrate. Calculate the percent yield if 375.0 g of silver bromide was obtained from the reaction.

SOLUTION

$$MgBr_2 + 2\ AgNO_3 \longrightarrow Mg(NO_3)_3 + 2\ AgBr$$

1. Determine the theoretical yield. Calculate the grams of AgBr that can be formed.

$$200.0\ \text{g } MgBr_2 \times \frac{1\ \text{mol } MgBr_2}{184.1\ \text{g } MgBr_2} \times \frac{2\ \text{mol AgBr}}{1\ \text{mol } MgBr_2} \times \frac{187.8\ \text{g AgBr}}{1\ \text{mol AgBr}} = 408.0\ \text{g AgBr}$$

   The theoretical yield is 408.0 g AgBr
2. Calculate the percent yield. According to the equation, 408.0 g AgBr is the maximum amount of AgBr possible from 200.0 g $MgBr_2$. Actual yield is 375.0 g AgBr.

$$\text{Percent yield} = \frac{375.0\ \text{g AgBr}}{408.0\ \text{g AgBr}} \times 100 = 91.91\%$$

PRACTICE Aluminum oxide was prepared by heating 225 g of chromium(II) oxide with 125 g of aluminum. Calculate the percent yield if 100.0 g of aluminum oxide was obtained.

$$2\ Al + 3\ CrO \longrightarrow Al_2O_3 + 3\ Cr$$

Answer: 88.5% yield

Chemistry in Action

A Shrinking Technology

In the high tech world of computers the microchip has miniaturized the field of electronics. In order to produce ever smaller computers, calculators, and even microbots (micro-sized robots) precise quantities of chemicals in exact proportions are required.

Engineers at Bell Laboratories, Massachusetts Institute of Technology, University of California, and Stanford University are busy trying to produce parts for tiny machines and robots. New techniques are allowing them to produce gears smaller than a grain of sand, and motors lighter than a speck of dust.

The secret behind these tiny circuits is to print the entire circuit or blueprint at one time. Computers are used to draw a design of how the chip will look. This image is then transferred into a pattern, or mask, with details finer than a human hair. Light is then shined through the mask onto a silicon coated surface. The process is similar to photography. The areas created on the silicon exhibit high or low resistance to chemical etching. Next chemicals are applied to etch away the silicon.

Micromachinery is produced in the same way. First a thin layer of silicon dioxide is applied (sacrificial material). Then a layer of polysilicon is carefully applied (structural material). A mask is then applied and the whole structure is covered with plasma (excited gas). The plasma acts as a tiny sandblaster removing everything the mask doesn't protect. This process is repeated constructing the entire machine. When the entire assembly is complete the whole machine is placed in hydrofluoric acid, which dissolves all the sacrificial material and permits the various parts of the machine to move.

More development is necessary to turn these micromachines into true microbots, but work is currently underway to design methods of locomotion and sensing imaging systems. Possible uses for these microbots include "smart" pills which could contain sensors (see Chemistry in Action, Chapter 2), or drug reservoirs (currently done for birth control). Tiny pumps, once inside the body, will be able to dispense the proper amount of medication at precisely the correct site. These microbots are currently in production for the treatment of diabetes (to release insulin).

A cog in a micromachine.

When solving problems, you will achieve better results if at first you do not try to take shortcuts. Write the data and numbers in a logical, orderly manner. Make certain that the equations are balanced and that the computations are accurate and expressed to the correct number of significant figures. Remember that units are very important; a number without units has little meaning. Finally, a scientific calculator can save you many hours of tedious computations.

Concepts in Review

1. Write mole ratios for any two substances involved in chemical reaction.
2. Outline the mole or mole-ratio method for making stoichiometric calculations.

3. Calculate the number of moles of a desired substance obtainable from a given number of moles of a starting substance in a chemical reaction (mole-to-mole calculations).
4. Calculate the mass of a desired substance obtainable from a given number of moles of a starting substance in a chemical reaction, and vice versa (mole-to-mass and mass-to-mole calculations).
5. Calculate the mass of a desired substance involved in a chemical reaction from a given mass of a starting substance (mass-to-mass calculation).
6. Deduce the limiting reactant or reagent when given the amounts of starting substances, and then calculate the moles or mass of desired substance obtainable from a given chemical reaction (limiting reactant calculation).
7. Apply theoretical yield or actual yield to any of the foregoing types of problems, or calculate theoretical and actual yields of a chemical reaction.

EXERCISES

An asterisk indicates a more challenging question or problem.

Mole Review Problems

1. Calculate the number of moles in each of the following quantities:
   (a) 25.0 g KNO_3
   (b) 10.8 g $Ca(NO_3)_2$
   (c) 5.4×10^2 g $(NH_4)_2C_2O_4$
   (d) 2.10 kg $NaHCO_3$
   (e) 525 mg $ZnCl_2$
   (f) 56 millimol NaOH
   (g) 9.8×10^{24} molecules CO_2
   (h) 250 mL ethyl alcohol, C_2H_5OH ($d = 0.789$ g/mL)
   *(i) 16.8 mL H_2SO_4 solution ($d = 1.727$ g/mL, 80.0% H_2SO_4 by mass)
2. Calculate the number of grams in each of the following quantities:
   (a) 2.55 mol $Fe(OH)_3$
   (b) 0.00844 mol $NiSO_4$
   (c) 125 kg $CaCO_3$
   (d) 0.0600 mol $HC_2H_3O_2$
   (e) 10.5 mol NH_3
   (f) 0.725 mol Bi_2S_3
   (g) 72 millimol HCl
   (h) 4.50×10^{21} molecules glucose, $C_6H_{12}O_6$
   (i) 500 mL of liquid Br_2 ($d = 3.119$ g/mL)
   *(j) 75 mL K_2CrO_4 solution ($d = 1.175$ g/mL, 20.0% K_2CrO_4 by mass)
3. Which contains the larger number of molecules?
   (a) 10.0 g H_2O or 10.0 g H_2O_2
   (b) 25.0 g HCl or 85.0 g $C_6H_{12}O_6$

Mole-Ratio Problems

4. Given the equation for the combustion of isopropyl alcohol:

   $$2\,C_3H_7OH + 9\,O_2 \longrightarrow 6\,CO_2 + 8\,H_2O$$

   what is the mole ratio of:
   (a) CO_2 to C_3H_7OH
   (b) C_3H_7OH to O_2
   (c) O_2 to CO_2
   (d) H_2O to C_3H_7OH
   (e) CO_2 to H_2O
   (f) H_2O to O_2
5. For the reaction

   $$3\,CaCl_2 + 2\,H_3PO_4 \longrightarrow Ca_3(PO_4)_2 + 6\,HCl$$

   set up the mole ratio of
   (a) $CaCl_2$ to $Ca_3(PO_4)_2$
   (b) HCl to H_3PO_4
   (c) $CaCl_2$ to H_3PO_4
   (d) $Ca_3(PO_4)_2$ to H_3PO_4
   (e) HCl to $Ca_3(PO_4)_2$
   (f) H_3PO_4 to HCl
6. How many moles of Cl_2 can be produced from 5.60 mol HCl?

   $$4\,HCl + O_2 \longrightarrow 2\,Cl_2 + 2\,H_2O$$

7. How many grams of sodium hydroxide can be produced from 500 g of calcium hydroxide according to this equation?

   $$Ca(OH)_2 + Na_2CO_3 \longrightarrow 2\,NaOH + CaCO_3$$

8. Given the equation

   $$Al_4C_3 + 12\,H_2O \longrightarrow 4\,Al(OH)_3 + 3\,CH_4$$

(a) How many moles of water are needed to react with 100. g of Al_4C_3?
(b) How many moles of $Al(OH)_3$ will be produced when 0.600 mol of CH_4 is formed?

9. How many grams of zinc phosphate, $Zn_3(PO_4)_2$, are formed when 10.0 g of Zn are reacted with phosphoric acid?

10. Given the equation

$$4\ FeS_2 + 11\ O_2 \longrightarrow 2\ Fe_2O_3 + 8\ SO_2$$

(a) How many moles of Fe_2O_3 can be made from 1.00 mol of FeS_2?
(b) How many moles of O_2 are required to react with 4.50 mol of FeS_2?
(c) If the reaction produces 1.55 mol of Fe_2O_3, how many moles of SO_2 are produced?
(d) How many grams of SO_2 can be formed from 0.512 mol of FeS_2?
(e) If the reaction produces 40.6 g of SO_2, how many moles of O_2 were reacted?
(f) How many grams of FeS_2 are needed to produce 221 g of Fe_2O_3?

11. An early method of producing chlorine was by the reaction of pyrolusite, MnO_2, and hydrochloric acid. How many moles of HCl will react with 1.05 mol of MnO_2? (Balance the equation first.)

$$MnO_2(s) + HCl(aq) \longrightarrow Cl_2(g) + MnCl_2(aq) + H_2O$$

12. 180 g of zinc were dropped into a beaker of hydrochloric acid. After the reaction ceased, 35 g of unreacted zinc remained in the beaker:

$$Zn + HCl \longrightarrow ZnCl_2 + H_2$$

(a) How many moles of hydrogen gas were produced?
(b) How many grams of HCl were reacted?

13. In a blast furnace, iron(III) oxide reacts with coke (carbon) to produce molten iron and carbon monoxide:

$$Fe_2O_3 + 3\ C \longrightarrow 2\ Fe + 3\ CO$$

How many kilograms of iron would be formed from 125 kg of Fe_2O_3?

14. How many grams of steam and iron must react to produce 375 g of magnetic iron oxide, Fe_3O_4?

$$3\ Fe(s) + 4\ H_2O(g) \longrightarrow Fe_3O_4(s) + 4\ H_2(g)$$

15. Ethane gas, C_2H_6, burns in air (i.e., reacts with the oxygen in air) to form carbon dioxide and water:

$$2\ C_2H_6 + 7\ O_2 \longrightarrow 4\ CO_2 + 6\ H_2O$$

(a) How many moles of O_2 are needed for the complete combustion of 15.0 mol of ethane?
(b) How many grams of CO_2 are produced for each 8.00 g of H_2O produced?
(c) How many grams of CO_2 will be produced by the combustion of 75.0 g of C_2H_6?

Limiting-Reactant and Percent-Yield Problems

16. In the following equations, determine which reactant is the limiting reactant and which reactant is in excess. The amounts mixed together are shown below each reactant. Show evidence for your answers.

(a) $KOH + HNO_3 \longrightarrow KNO_3 + H_2O$
16.0 g   12.0 g
(b) $2\ NaOH + H_2SO_4 \longrightarrow Na_2SO_4 + 2\ H_2O$
10.0 g   10.0 g
(c) $2\ Bi(NO_3)_3 + 3\ H_2S \longrightarrow Bi_2S_3 + 6\ HNO_3$
50.0 g   6.00 g
(d) $3\ Fe + 4\ H_2O \longrightarrow Fe_3O_4 + 4\ H_2$
40.0 g   16.0 g

17. Given the equation

$$Fe(s) + CuSO_4(aq) \longrightarrow Cu(s) + FeSO_4(aq)$$

(a) When 2.0 mol of Fe and 3.0 mol of $CuSO_4$ are reacted, what substances will be present when the reaction is over? How many moles of each substance are present?
(b) When 20.0 g of Fe and 40.0 g of $CuSO_4$ are reacted, what substances will be present when the reaction is over? How many grams of each substance are present?

***18.** The reaction for the combustion of propane, C_3H_8, is:

$$C_3H_8 + 5\ O_2 \longrightarrow 3\ CO_2 + 4\ H_2O$$

(a) If 5.0 mol of C_3H_8 and 5.0 mol of O_2 are reacted, how many moles of CO_2 can be produced?
(b) If 3.0 mol of C_3H_8 and 20.0 mol of O_2 are reacted, how many moles of CO_2 can be produced?
(c) If 20.0 mol of C_3H_8 and 3.0 mol of O_2 are reacted, how many moles of CO_2 can be produced?
*(d) If 2.0 mol of C_3H_8 and 14.0 mol of O_2 are placed in a closed container, and they react

to completion (until one reactant is completely used up), what compounds are present in the container after the reaction, and how many moles of each compound are present?

(e) If 20.0 g of C_3H_8 and 20.0 g of O_2 are reacted, how many grams of CO_2 can be produced?

(f) If 20.0 g of C_3H_8 and 80.0 g of O_2 are reacted, how many grams of CO_2 can be produced?

(g) If 20.0 g C_3H_8 and 200 g of O_2 are reacted, how many grams of CO_2 can be produced?

19. Aluminum reacts with bromine to form aluminum bromide:

$$2\,Al + 3\,Br_2 \longrightarrow 2\,AlBr_3$$

If 25.0 g of Al and 100. g of Br_2 are reacted, and 64.2 g of $AlBr_3$ product are recovered, what is the percent yield for the reaction?

20. Methyl alcohol, CH_3OH, is made by reacting carbon monoxide and hydrogen in the presence of certain metal oxide catalysts. How much alcohol can be obtained by reacting 40.0 g of CO and 10.0 g of H_2? How many grams of excess reactant remain unreacted?

$$CO(g) + 2\,H_2(g) \longrightarrow CH_3OH(l)$$

21. Iron was reacted with a solution containing 400. g of copper(II) sulfate. The reaction was stopped after 1 hour, and 151 g of copper were obtained. Calculate the percent yield of copper obtained.

$$Fe(s) + CuSO_4(aq) \longrightarrow Cu(s) + FeSO_4(aq)$$

***22.** Ethyl alcohol, C_2H_5OH, also called grain alcohol, can be made by the fermentation of sugar, which often comes from starch in grain:

$$\underset{\text{Glucose}}{C_6H_{12}O_6} \longrightarrow \underset{\text{Ethyl alcohol}}{2\,C_2H_5OH} + 2\,CO_2$$

If an 84.6% yield of ethyl alcohol is obtained,

(a) What mass of ethyl alcohol will be produced from 750 g of glucose?

(b) What mass of glucose should be used to produce 475 g of C_2H_5OH?

***23.** Carbon disulfide, CS_2, can be made from coke, C, and sulfur dioxide, SO_2:

$$3\,C + 2\,SO_2 \longrightarrow CS_2 + 2\,CO_2$$

If the actual yield of CS_2 is 86.0% of the theoretical yield, what mass of coke is needed to produce 950 g of CS_2?

***24.** Acetylene, C_2H_2, can be manufactured by the reaction of water and calcium carbide, CaC_2:

$$CaC_2 + 2\,H_2O \longrightarrow C_2H_2(g) + Ca(OH)_2$$

When 44.5 g of commercial grade (impure) calcium carbide were reacted, 0.540 mol of C_2H_2 was produced. Assuming that all of the CaC_2 was reacted to C_2H_2, what is the percent of CaC_2 in the commercial grade material?

Additional Problems

***25.** Both $CaCl_2$ and $MgCl_2$ react with $AgNO_3$ to precipitate AgCl. When solutions containing equal masses of $CaCl_2$ and $MgCl_2$ are reacted, which salt will produce the larger amount of AgCl? Show proof.

***26.** An astronaut excretes about 2500 g of water a day. If lithium oxide, Li_2O, is used in the spaceship to absorb this water, how many kilograms of Li_2O must be carried for a 30-day space trip for three astronauts?

$$Li_2O + H_2O \longrightarrow 2\,LiOH$$

***27.** Much commerical hydrochloric acid is prepared by the reaction of concentrated sulfuric acid with sodium chloride:

$$H_2SO_4 + 2\,NaCl \longrightarrow Na_2SO_4 + 2\,HCl$$

How many kilograms of concentrated H_2SO_4, 96% H_2SO_4 by mass, are required to produce 20.0 L of concentrated hydrochloric acid ($d = 1.20$ g/mL, 42.0% HCl by mass)?

***28.** Gastric juice contains about 3.0 g of HCl per liter. If a person produces about 2.5 L of gastric juice per day, how many antacid tablets, each containing 400. mg of $Al(OH)_3$, are needed to neutralize all the HCl produced in one day?

$$Al(OH)_3(s) + 3\,HCl(aq) \longrightarrow AlCl_3(aq) + 6\,H_2O(l)$$

***29.** 12.82 g of a mixture of $KClO_3$ and NaCl are heated strongly. The $KClO_3$ reacts according to the following equation:

$$2\,KClO_3(s) \longrightarrow 2\,KCl(s) + 3\,O_2(g)$$

The NaCl does not undergo any reaction. After

the heating, the mass of the residue (KCl and NaCl) is 9.45g. Assuming that all the loss of mass represents loss of oxygen gas, calculate the percent of $KClO_3$ in the original mixture.

30. Phosphine, PH_3, can be prepared by the hydrolysis of calcium phosphide, Ca_3P_2:

$$Ca_3P_2 + 6\,H_2O \longrightarrow 3\,Ca(OH)_2 + 2\,PH_3$$

Based on this equation, which of the following statements are correct? Show evidence to support your answer.

(a) One mole of Ca_3P_2 produces 2 mol of PH_3.
(b) One gram of Ca_3P_2 produces 2 g of PH_3.
(c) Three moles of $Ca(OH)_2$ are produced for each 2 mol of PH_3 produced.
(d) The mole ratio between phosphine and calcium phosphide is

$$\frac{2 \text{ mol } PH_3}{1 \text{ mol } Ca_3P_2}$$

(e) When 2.0 mol of Ca_3P_2 and 3.0 mol of H_2O react, 4.0 mol of PH_3 can be formed.
(f) When 2.0 mol of Ca_3P_2 and 15.0 mol of H_2O react, 6.0 mol of $Ca(OH)_2$ can be formed.
(g) When 200. g of Ca_3P_2 and 100. g of H_2O react, Ca_3P_2 is the limiting reactant.
(h) When 200. g of Ca_3P_2 and 100. g of H_2O react, the theoretical yield of PH_3 is 57.4 g.

31. The equation representing the reaction used for the commercial preparation of hydrogen cyanide is

$$2\,CH_4 + 3\,O_2 + 2\,NH_3 \longrightarrow 2\,HCN + 6\,H_2O$$

Based on this equation, which of the statements below are correct? Rewrite incorrect statements to make them correct.

(a) Three moles of O_2 are required for 2 mol of NH_3.
(b) Twelve moles of HCN are produced for every 16 mol of O_2 that react.
(c) The mole ratio between H_2O and CH_4 is

$$\frac{6 \text{ mol } H_2O}{2 \text{ mol } CH_4}$$

(d) When 12 mol of HCN are produced, 4 mol of H_2O will be formed.
(e) When 10 mol of CH_4, 10 mol of O_2, and 10 mol of NH_3 are mixed and reacted, O_2 is the limiting reactant.
(f) When 3 mol each of CH_4, O_2, and NH_3 are mixed and reacted, 3 mol of HCN will be produced.

32. What is the purpose of a mask in the manufacture of tiny circuits and machines?

33. Why is silicon dioxide called a sacrificial material in the production of micromachinery?

34. What are some possible uses for micromachines in our society? Indicate some of the difficulties that must be overcome to implement these machines.

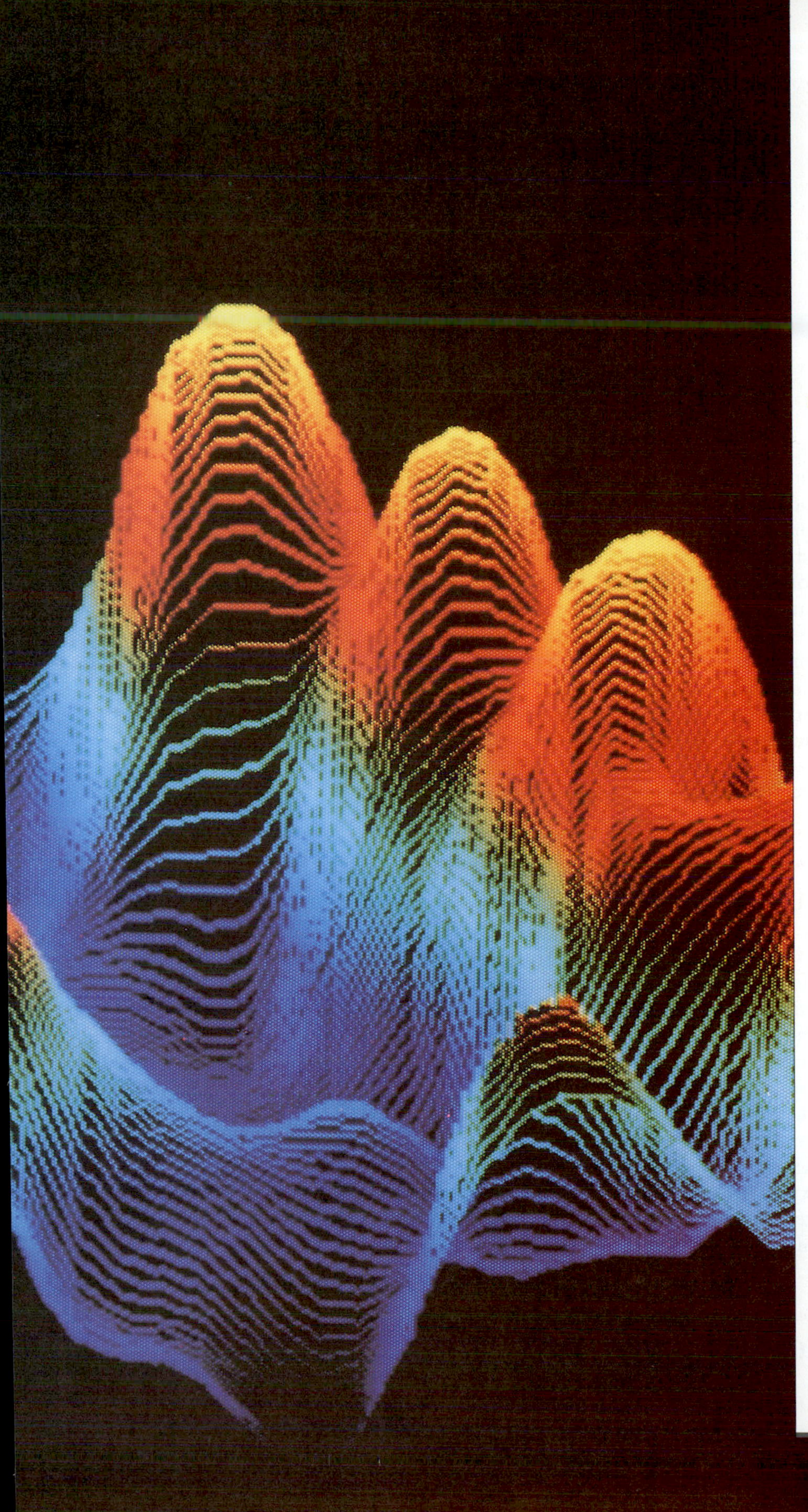

10

Modern Atomic Theory

◀ CHAPTER OPENING PHOTO: A scanning tunneling microscope reveals a molecule of DNA.

How do we go about studying an object too small to see? Think back to that birthday when you had a present you could look at but not yet open. Simply judging from the wrapping and size of the box was not very useful. Shaking, turning, and lifting the package all gave clues, indirectly, to the contents. After all the experiments were done on the package, a fairly good hypothesis of the contents could be made. But was it correct? The only means of absolute verification would be opening the package.

The same is true for chemists in their study of the atom. Atoms are so very small that it is not possible to use the normal senses and rules to describe them. Chemists are working within a package essentially in the dark. Yet, as such instruments as x-ray machines and scanning-tunneling microscopes and such measuring devices as spectrophotometers and magnetic resonance imaging (mri) have progressed along with man's skill in mathematics and probability, the secrets of the atom are beginning to be unraveled.

10.1 INTRODUCTION

In the last 200 years, vast amounts of data have been accumulated to support atomic theory. When atoms were originally suggested by the early Greeks, no evidence existed to physically support their ideas. Lavoisier, Faraday, Arrhenius, and others did a variety of experiments, which culminated in the Dalton theory of the atom. Because of the limitations of Dalton's model, modifications were proposed by Thomson and then by Rutherford to evolve eventually into the nuclear atom. These early models of the atom work reasonably well—in fact, we continue to use them today to visualize a variety of chemical concepts. There remain, however, questions that these models cannot answer, including an explanation of how atomic structure relates to the periodic table. In this chapter, we will look at our modern model of the atom to see how it varies from and improves upon the earlier atomic theories.

10.2 THE BOHR ATOM

At high temperatures, or when subjected to high voltages, elements in the gaseous state give off colored light. Brightly colored neon signs illustrate this prop-

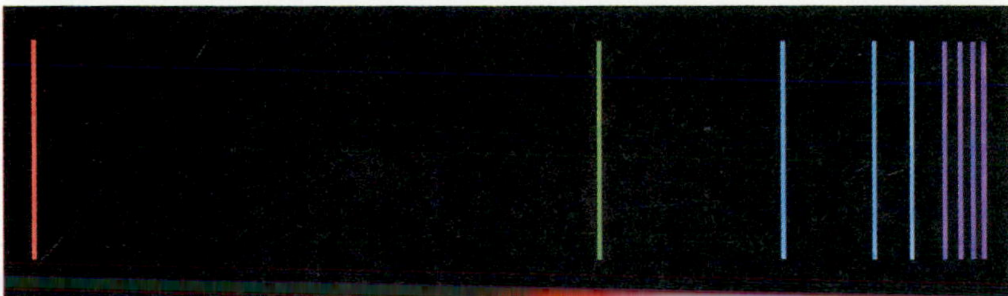

◀ FIGURE 10.1
Line spectrum of hydrogen. Each line corresponds to the wavelength of the energy emitted when the electron of a hydrogen atom, which has absorbed energy, falls back to a lower energy level.

erty of matter very well. When passed through the prism or the grating of a spectroscope, the light emitted by a gas appears as a set of brightly colored lines and is called a line spectrum. These colored lines indicate that the light is being emitted only at certain wavelengths, or frequencies, that correspond to specific colors. Each element possesses a unique set of these spectral lines that is different from the sets of all the other elements. This is illustrated in Figure 10.1.

In 1912–1913, while studying the line spectra of hydrogen, Niels Bohr (1885–1962), a Danish physicist, made a significant contribution to the rapidly growing knowledge of atomic structure. His research led him to believe that electrons in an atom exist in specific regions at various distances from the nucleus. He also visualized the electrons as rotating in orbits around the nucleus, like planets rotating around the sun.

Bohr's first paper in this field dealt with the hydrogen atom, which he described as a single electron rotating in an orbit about a relatively heavy nucleus. He applied the concept of energy quanta, proposed in 1900 by the German physicist Max Planck (1858–1947), to the observed line spectra of hydrogen. Planck stated that energy is never emitted in a continuous stream but only in small discrete packets called quanta (Latin, *quantus*, how much). Bohr theorized that electrons have several possible orbits at different distances from the nucleus and that an electron had to be in one specific orbit (energy level) or another; it could not exist between orbits. In other words, the energy of the electron is said to be quantized. Bohr also stated that, when a hydrogen atom absorbed one or more quanta of energy, its electron would "jump" to an orbit at a greater distance from the nucleus.

Bohr was able to account for spectral lines this way. A number of orbits are available, each corresponding to a different energy level. The orbit closest to the nucleus is the lowest, or ground state, energy level; orbits at increasing distances are the second, third, fourth, etc., energy levels. When an electron falls from a high-energy orbit to a lower one (say, from the fourth to the second), a quantum of energy is emitted as light at a specific frequency, or wavelength. This light corresponds to one of the lines visible in the hydrogen spectrum (see Figure 10.1). Several lines are visible in this spectrum. Each line corresponds to a specific electron energy-level shift within the hydrogen atom.

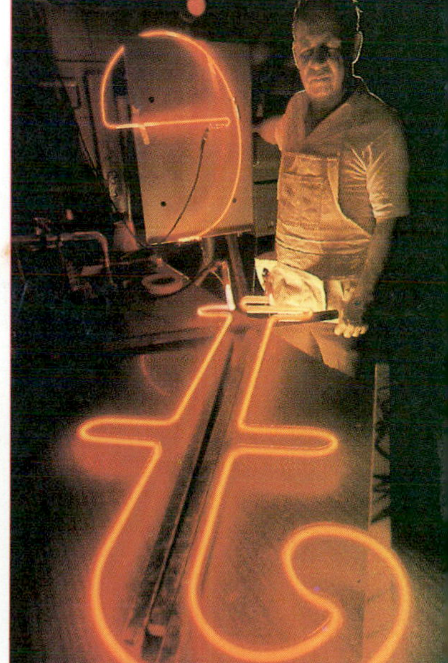

▲
The light in this neon sign is the result of electrons falling from one orbit to another.

The chemical properties of an element and its position in the periodic table (Chapter 11) depend on electron behavior within the atoms. In turn, much of our knowledge of the behavior of electrons within atoms is based on spectroscopy. Niels Bohr contributed a great deal to our knowledge of atomic structure by (1) suggesting quantized energy levels for electrons and (2) showing that spectral lines result from the radiation of small increments of energy (Planck's

▲
From left to right:
Max Planck (1858–1947), Erwin Schrödinger (1887–1961); Niels Bohr (1885–1962).

quanta) when electrons shift from one energy level to another. Bohr's calculations succeeded very well in correlating the experimentally observed spectral lines with electron energy levels for the hydrogen atom. However, Bohr's methods of calculation did not succeed for heavier atoms. More theoretical work on atomic structure was needed.

In 1924 the French physicist, Louis de Broglie, suggested that moving electrons had properties of waves as well as mass. In 1926 Erwin Schrödinger, an Austrian physicist, introduced a new method of calculation—quantum mechanics, or wave mechanics. By Schrödinger's method electrons are described in mathematical terms as having dual characteristics; that is, some electron properties are best described in terms of waves (like light) and others in terms of particles having mass.

orbital

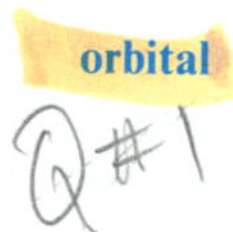

Since the late 1920's quantum mechanical concepts have generally replaced Bohr's ideas in theoretical considerations of atomic structure. One important difference between the Bohr and the quantum mechanics concepts is in the treatment of electrons. Quantum mechanics retains the concept of electrons being in specific energy levels. But the electrons are treated not as revolving about the nucleus in *orbits* but as being located in *orbitals*. An **orbital** is simply a region in space about the nucleus where there is a high probability of finding a given electron.

10.3 Energy Levels of Electrons

energy levels

electron shells

Not all the electrons in an atom are located the same distance from the nucleus. As pointed out by both the Bohr theory and quantum mechanics, the probability of finding electrons is greatest at certain specified distances from the nucleus, called **energy levels**. Energy levels are also referred to as **electron shells** and may contain only a limited number of electrons. The main or principal energy levels (**n**) are numbered, starting with **n** = 1 as the energy level nearest to the nucleus and going to **n** = 7, for the known elements. (Theoretically, the number of energy

TABLE 10.1 Maximum Number of Electrons that Can Occupy Each Principal Energy Level

Principal energy level, n	Maximum number of electrons in each energy level, $2n^2$
1	$2 \times 1^2 = 2$
2	$2 \times 2^2 = 8$
3	$2 \times 3^2 = 18$
4	$2 \times 4^2 = 32$
5	$2 \times 5^2 = 50^a$

[a] The theoretical value of 50 electrons in energy level 5 has never been attained in any element known to date.

levels is infinite.) Each succeeding energy level is located farther from the nucleus. The electrons in energy levels at increasing distances from the nucleus have increasingly higher energies. The order of energy for the principal energy levels **n** is

Principal energy levels: $1 < 2 < 3 < 4 < 5 < 6 < 7$

The number of electrons that can exist in each energy level is limited. The maximum number of electrons for a specific energy level can be calculated from the formula $2\mathbf{n}^2$, where **n** is the number of the principal energy level. For example, energy level 1 ($\mathbf{n} = 1$) can have a maximum of two electrons ($2 \times 1^2 = 2$); energy level 2 ($\mathbf{n} = 2$) can have a maximum of eight electrons ($2 \times 2^2 = 8$), and so on. Table 10.1 shows the maximum number of electrons that can exist in each of the first five energy levels.

10.4 ENERGY SUBLEVELS

The principal energy levels contain sublevels designated by the letters *s*, *p*, *d*, and *f*. These orbitals are the ones in which electrons are located. The *s* sublevel consists of one orbital; the *p* sublevel consists of three orbitals; the *d* sublevel consists of five orbitals; and the *f* sublevel consists of seven orbitals. An electron spins on its own axis in one of only two directions, clockwise or counterclockwise. As a result, only two electrons can occupy the same orbital, one spinning clockwise and the other spinning counterclockwise. When an orbital contains two electrons, the electrons are said to be paired. Because no more than two electrons can exist in an orbital, the maximum numbers of electrons that can exist in the sublevels are 2 in the *s* orbital, 6 in the three *p* orbitals, 10 in the five *d* orbitals, and 14 in the seven *f* orbitals.

Type of sublevel	Number of orbitals possible	Number of electrons possible
s	1	2
p	3	6
d	5	10
f	7	14

The order of energy of the sublevels within a principal energy level is the following: *s* electrons are lower in energy than *p* electrons, which are lower than *d* electrons, which are lower than *f* electrons. This order may be expressed in the following manner:

Sublevel energy: $s < p < d < f$

Not all principal energy levels contain each and every type of sublevel. To determine what types of sublevels occur in any given energy level, we need to know the maximum number of electrons possible in that energy level (see Table 10.1), and we need to use these three rules:

1. No more than two electrons can occupy one orbital.
2. Electrons occupy the lowest possible energy sublevels; they enter a higher sublevel only when the lower sublevels are filled.
3. Orbitals in a given sublevel of equal energy are each occupied by a single electron before a second electron enters them. For example, all three *p* orbitals must contain one electron before a second electron enters a *p* orbital.

The maximum number of electrons in the first energy level is two; both are *s* orbital electrons, designated as $1s^2$. (The *s* orbital in the second energy level ($\mathbf{n} = 2$) is written as 2*s*, in the third energy level as 3*s*, and so on.) The second energy level, with a maximum of eight electrons, contains only *s* and *p* electrons—namely, a maximum of two *s* and six *p* electrons, designated as $2s^22p^6$. The following diagram shows how to read these electron designations:

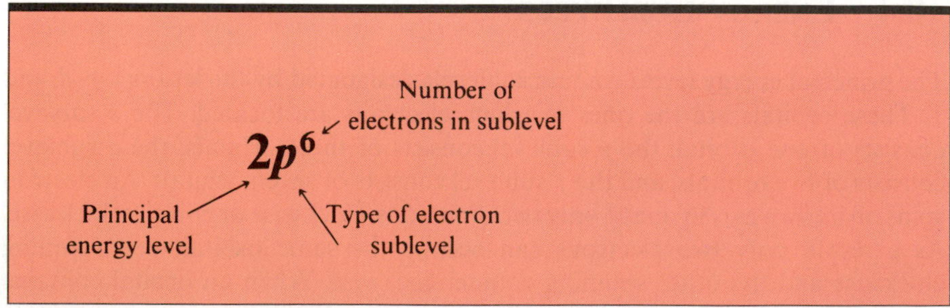

If each orbital contains two electrons, the second energy level can have four orbitals (8 electrons): one *s* orbital and three individual *p* orbitals. These three *p*

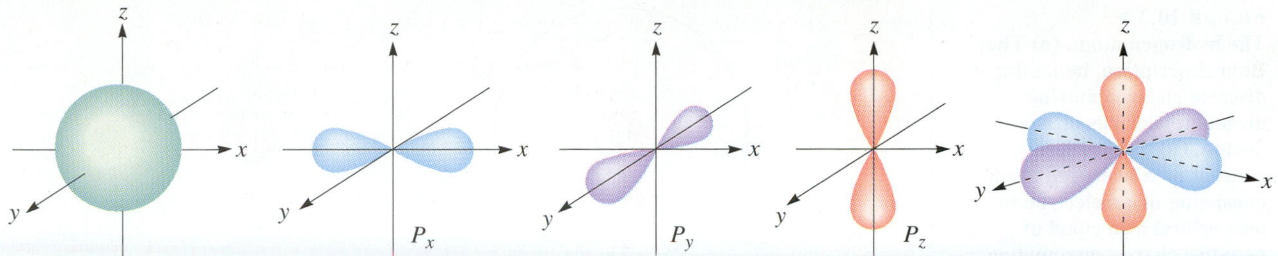

▲
FIGURE 10.2
Perspective representation of the s, p_x, p_y, and p_z atomic orbitals.

orbitals are energetically equivalent to each other and are labeled $2p_x$, $2p_y$, and $2p_z$ to indicate their orientation in space (see Figure 10.2). The symbols $3s^2$, $3p^6$, and $3d^{10}$ show the sublevel breakdown of electrons in the third energy level. From this line of reasoning, we can see that, if an atom has sufficient electrons, f electrons first appear in the fourth energy level. Table 10.2 shows the type of sublevel electron orbitals and the maximum number of orbitals and electrons in each energy level. No elements in the fifth, sixth and seventh energy levels contain the calculated maximum number of electrons.

Since the *spdf* atomic orbitals have definite orientations in space, they are represented by particular spatial shapes. At this time we will consider only the s and p orbitals. The s orbitals are spherically symmetrical about the nucleus, as illustrated in Figure 10.2. A $2s$ orbital is a larger sphere than a $1s$ orbital. The p orbitals (p_x, p_y, p_z) are dumbbell-shaped and are oriented at right angles to each other along the x, y, and z axes in space. An electron has an equal probability of being located in either lobe of the p orbital. In illustrations such as Figure 10.2, the boundaries of the orbitals enclose the region of the greatest probability (about a 90% chance) of finding an electron. In the ground state, or lowest energy level, of a hydrogen atom, this region is a sphere having a radius of 0.053 nm.

TABLE 10.2 Sublevel Electron Orbitals in Each Principal Energy Level and the Maximum Number of Orbitals and Electrons in Each Energy Level

Principal energy level	Sublevels	Maximum number of orbitals	Maximum number of electrons
1	*s*	1	2
2	*s, p*	4	8
3	*s, p, d*	9	18
4	*s, p, d, f*	16	32
5	*s, p, d, f*	Incomplete[a]	(50)[a]
6	*s, p, d*	Incomplete[a]	(72)[a]
7	*s*	Incomplete[a]	(98)[a]

[a] Insufficient electrons to complete the shell.

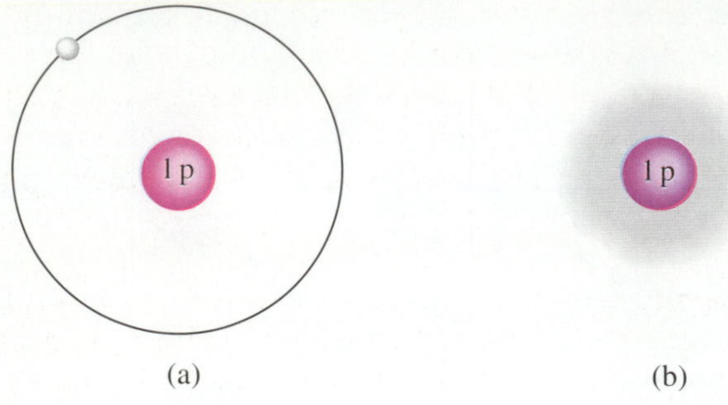

FIGURE 10.3 ▶
The hydrogen atom. (a) The Bohr description, indicating a discrete electron moving around its nucleus of one proton. (b) The modern concept of a hydrogen atom consisting of an electron in an *s* orbital as a cloud of negative charge surrounding the proton in the nucleus.

10.5 The Simplest Atom: Hydrogen

The common hydrogen atom, consisting of a nucleus containing one proton and an electron orbital containing one electron, is the simplest known atom. (Some hydrogen atoms are known to contain one or two neutrons in their nucleus. (See Section 5.10 on isotopes.) The electron in hydrogen occupies an *s* orbital in the first energy level. This electron does not move in any definite path but rather in a rapid random motion throughout its entire orbital, forming an electron "cloud" about the nucleus. The diameter of the nucleus is believed to be about 10^{-13} cm, and the diameter of the electron orbital to be about 10^{-8} cm. Hence, the diameter of the electron orbital of a hydrogen atom is about 100,000 times greater than the diameter of the nucleus.

What we have, then, is a positive nucleus surrounded by an electron cloud formed by an electron in an *s* orbital. The net electrical charge on the hydrogen atom is zero; it is called a *neutral atom.* Figure 10.3 shows two methods of representing a hydrogen atom.

10.6 Atomic Structures of the First Twenty Elements

Starting with hydrogen and progressing in order of increasing atomic number to helium, lithium, beryllium, and so on, the atoms of each successive element contain one more proton and one more electron than do the atoms of the preceding element. This sequence continues, without exception, through the entire list of known elements. It is one of the most impressive examples of order in nature.

The number of neutrons also increases as we progress through the list of elements. But this number, unlike the number of protons and electrons, does not increase in a perfectly uniform manner. Furthermore, atoms of the same element may contain different numbers of neutrons; such atoms are called *isotopes.* For example, three different hydrogen isotopes are described in Section 5.10. The most abundant helium isotope (element number 2) contains two neutrons, but two other helium isotopes exist, containing one and four neutrons, respectively.

The ground-state electron structures of the first 20 elements fall into a regular pattern. The one hydrogen electron is in the first energy level as are both helium electrons. The electron structures for hydrogen and helium are $1s^1$ and $1s^2$. The maximum number of electrons in the first energy level is two ($2\mathbf{n}^2 = 2 \times 1^2 = 2$; see Section 10.3), so the two electrons fill the first energy level of helium.

An atom with three electrons will have its third electron in the second energy level because the first level can contain only two electrons. Thus in lithium (atomic number 3) the third electron is in the $2s$ sublevel of the second energy level. Lithium has the electron structure $1s^22s^1$.

In succession, the atoms of beryllium (4), boron (5), carbon (6), nitrogen (7), oxygen (8), fluorine (9), and neon (10) have one more proton and one more electron than the preceding element until, in neon, both the first and second energy levels are filled to capacity, with 2 and 8 electrons, respectively. The electron structures of these ten elements are shown below:

H $1s^1$
He $1s^2$
Li $1s^22s^1$
Be $1s^22s^2$
B $1s^22s^22p^1$
C $1s^22s^22p^2$
N $1s^22s^22p^3$
O $1s^22s^22p^4$
F $1s^22s^22p^5$
Ne $1s^22s^22p^6$

Element 11, sodium (Na), has two electrons in the first energy level and eight electrons in the second energy level, with the remaining electron occupying the $3s$ orbital in the third energy level. The electron structure of sodium is $1s^22s^22p^63s^1$. Magnesium (12), aluminum (13), silicon (14), phosphorus (15), sulfur (16), chlorine (17), and argon (18) follow in order. Each of these elements will add one electron to the third energy level up to argon, which has eight electrons in the third energy level.

Up through the $3p$ level the sequence of filling the sublevels is exactly as expected, based on the increasing principal and sublevel energy levels. However, after the $3p$ level is filled, variations occur. The third energy level might logically be expected to fill to its capacity of 18 electrons with $3d$ electrons before electrons enter the $4s$ sublevel. However, this order of filling the third energy level does not occur because the $4s$ sublevel is at a lower energy than the $3d$ sublevel (see Figure 10.4). Consequently, because the sublevels fill in order of increasing energy, the last electron in potassium (19) and the last two electrons in calcium (20) are in the $4s$ sublevel. The electron structures for potassium and calcium are

K $1s^22s^22p^63s^23p^64s^1$
Ca $1s^22s^22p^63s^23p^64s^2$

This break in sequence does not invalidate the formula $2\mathbf{n}^2$, which prescribes the maximum number of electrons that each shell can contain but not the order in which the shells are filled. Table 10.3 shows the electron structure of the first 20 elements.

TABLE 10.3 Electron Structure of the First Twenty Elements

Element	Number of protons (atomic number)	Number of electrons	Electron structure
H	1	1	$1s^1$
He	2	2	$1s^2$
Li	3	3	$1s^22s^1$
Be	4	4	$1s^22s^2$
B	5	5	$1s^22s^22p^1$
C	6	6	$1s^22s^22p^2$
N	7	7	$1s^22s^22p^3$
O	8	8	$1s^22s^22p^4$
F	9	9	$1s^22s^22p^5$
Ne	10	10	$1s^22s^22p^6$
Na	11	11	$1s^22s^22p^63s^1$
Mg	12	12	$1s^22s^22p^63s^2$
Al	13	13	$1s^22s^22p^63s^23p^1$
Si	14	14	$1s^22s^22p^63s^23p^2$
P	15	15	$1s^22s^22p^63s^23p^3$
S	16	16	$1s^22s^22p^63s^23p^4$
Cl	17	17	$1s^22s^22p^63s^23p^5$
Ar	18	18	$1s^22s^22p^63s^23p^6$
K	19	19	$1s^22s^22p^63s^23p^64s^1$
Ca	20	20	$1s^22s^22p^63s^23p^64s^2$

The relative energies of the electron orbitals are shown in Figure 10.4. The order given can be used to determine the electron distribution in the atoms of the elements, although some exceptions to the pattern are known. Suppose we wish to determine the electron structure of a chlorine atom (atomic number 17), which has 17 electrons. Following the order in Figure 10.4, we begin by placing two electrons in the $1s$ orbital, then two electrons in the $2s$ orbital, and then six electrons in the $2p$ orbitals. We now have used ten electrons.

$1s^22s^22p^6$

Finally we place the next two electrons in the $3s$ orbital and the remaining five electrons in the $3p$ orbitals, which uses all 17 electrons, giving the electron structure for a chlorine atom as $1s^22s^22p^63s^23p^5$. The sum of the superscripts equals 17, the number of electrons in the atom. This procedure is summarized below.

Order of orbitals to be filled:	$1s2s2p3s3p$
Distribution of the 17 electrons in a chlorine atom:	$1s^22s^22p^63s^23p^5$

EXAMPLE 10.1 What is the electron distribution in a phosphorus atom?

SOLUTION First determine the number of electrons contained in a phosphorus atom. The atomic number of phosphorus is 15; therefore, each atom contains 15 protons and 15 electrons.

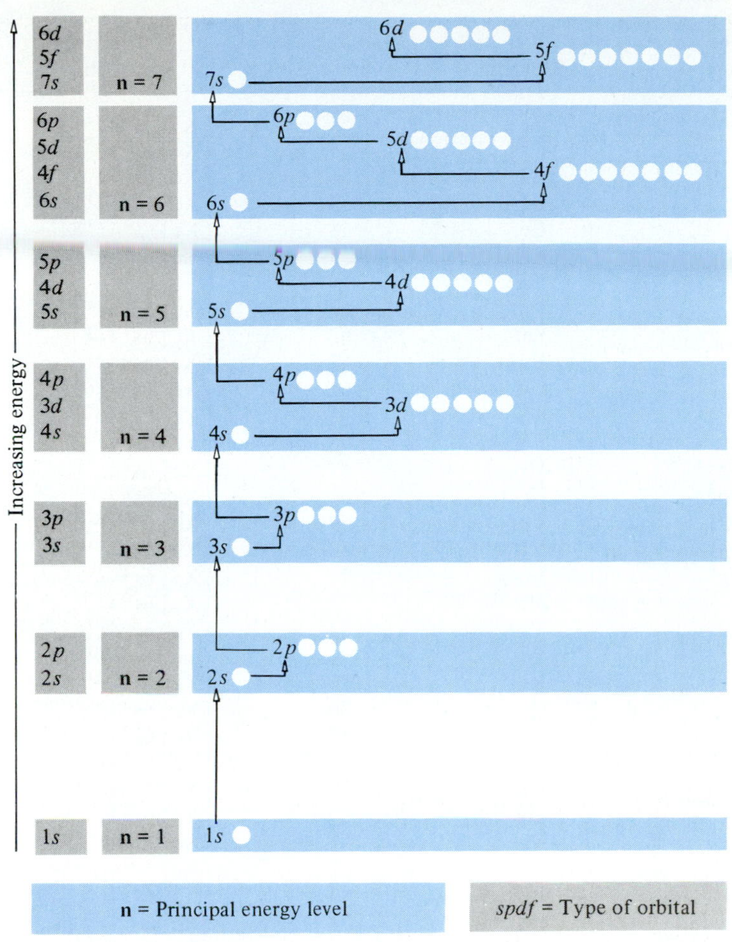

◀ **FIGURE 10.4**
Order of filling electron orbitals. Each circle represents an orbital, which can contain two electrons. The progression of electrons filling the energy sublevels is shown, starting with 1*s* and going to 6*d*. All lower sublevels must be filled before an electron enters the next higher sublevel. For example, all three 2*p* orbitals must be filled before an electron enters the 3*s* sublevel. (Some exceptions to this order are known.)

Now tabulate the number of electrons in each principal and subenergy level until all 15 electrons are assigned.

Sublevel	Number of e^-	Total e^-
1*s* orbital	$2e^-$	2
2*s* orbital	$2e^-$	4
2*p* orbital	$6e^-$	10
3*s* orbital	$2e^-$	12
3*p* orbital	$3e^-$	15

Therefore the electron distribution in phosphorus is $1s^22s^22p^63s^23p^3$.

PRACTICE Determine the electron configuration of each of the following elements:
(a) Al (b) Si (c) Be
Answers: (a) $1s^22s^22p^63s^23p^1$ (b) $1s^22s^22p^63s^23p^2$ (c) $1s^22s^2$

Table 10.4 Electron Structures of the Elements

Element	Atomic number	Electron structure
H	1	$1s^1$
He	2	$1s^2$
Li	3	$1s^2 2s^1$
Be	4	$1s^2 2s^2$
B	5	$1s^2 2s^2 2p^1$
C	6	$1s^2 2s^2 2p^2$
N	7	$1s^2 2s^2 2p^3$
O	8	$1s^2 2s^2 2p^4$
F	9	$1s^2 2s^2 2p^5$
Ne	10	$1s^2 2s^2 2p^6$
Na	11	[Ne] $3s^1$
Mg	12	[Ne] $3s^2$
Al	13	[Ne] $3s^2 3p^1$
Si	14	[Ne] $3s^2 3p^2$
P	15	[Ne] $3s^2 3p^3$
S	16	[Ne] $3s^2 3p^4$
Cl	17	[Ne] $3s^2 3p^5$
Ar	18	[Ne] $3s^2 3p^6$
K	19	[Ar] $4s^1$
Ca	20	[Ar] $4s^2$
Sc	21	[Ar] $4s^2 3d^1$
Ti	22	[Ar] $4s^2 3d^2$
V	23	[Ar] $4s^2 3d^3$
Cr	24	[Ar] $4s^1 3d^5$
Mn	25	[Ar] $4s^2 3d^5$
Fe	26	[Ar] $4s^2 3d^6$
Co	27	[Ar] $4s^2 3d^7$
Ni	28	[Ar] $4s^2 3d^8$
Cu	29	[Ar] $4s^1 3d^{10}$
Zn	30	[Ar] $4s^2 3d^{10}$
Ga	31	[Ar] $4s^2 3d^{10} 4p^1$
Ge	32	[Ar] $4s^2 3d^{10} 4p^2$
As	33	[Ar] $4s^2 3d^{10} 4p^3$
Se	34	[Ar] $4s^2 3d^{10} 4p^4$
Br	35	[Ar] $4s^2 3d^{10} 4p^5$
Kr	36	[Ar] $4s^2 3d^{10} 4p^6$
Rb	37	[Kr] $5s^1$
Sr	38	[Kr] $5s^2$
Y	39	[Kr] $5s^2 4d^1$
Zr	40	[Kr] $5s^2 4d^2$
Nb	41	[Kr] $5s^1 4d^4$
Mo	42	[Kr] $5s^1 4d^5$
Tc	43	[Kr] $5s^2 4d^5$
Ru	44	[Kr] $5s^1 4d^7$
Rh	45	[Kr] $5s^1 4d^8$
Pd	46	[Kr] $4d^{10}$
Ag	47	[Kr] $5s^1 4d^{10}$
Cd	48	[Kr] $5s^2 4d^{10}$
In	49	[Kr] $5s^2 4d^{10} 5p^1$
Sn	50	[Kr] $5s^2 4d^{10} 5p^2$
Sb	51	[Kr] $5s^2 4d^{10} 5p^3$
Te	52	[Kr] $5s^2 4d^{10} 5p^4$
I	53	[Kr] $5s^2 4d^{10} 5p^5$
Xe	54	[Kr] $5s^2 4d^{10} 5p^6$
Cs	55	[Xe] $6s^1$
Ba	56	[Xe] $6s^2$
La	57	[Xe] $6s^2 5d^1$
Ce	58	[Xe] $6s^2 4f^1 5d^1$
Pr	59	[Xe] $6s^2 4f^3$
Nd	60	[Xe] $6s^2 4f^4$
Pm	61	[Xe] $6s^2 4f^5$
Sm	62	[Xe] $6s^2 4f^6$
Eu	63	[Xe] $6s^2 4f^7$
Gd	64	[Xe] $6s^2 4f^7 5d^1$
Tb	65	[Xe] $6s^2 4f^9$
Dy	66	[Xe] $6s^2 4f^{10}$
Ho	67	[Xe] $6s^2 4f^{11}$
Er	68	[Xe] $6s^2 4f^{12}$
Tm	69	[Xe] $6s^2 4f^{13}$
Yb	70	[Xe] $6s^2 4f^{14}$
Lu	71	[Xe] $6s^2 4f^{14} 5d^1$
Hf	72	[Xe] $6s^2 4f^{14} 5d^2$
Ta	73	[Xe] $6s^2 4f^{14} 5d^3$
W	74	[Xe] $6s^2 4f^{14} 5d^4$
Re	75	[Xe] $6s^2 4f^{14} 5d^5$
Os	76	[Xe] $6s^2 4f^{14} 5d^6$
Ir	77	[Xe] $6s^2 4f^{14} 5d^7$
Pt	78	[Xe] $6s^1 4f^{14} 5d^9$
Au	79	[Xe] $6s^1 4f^{14} 5d^{10}$
Hg	80	[Xe] $6s^2 4f^{14} 5d^{10}$
Tl	81	[Xe] $6s^2 4f^{14} 5d^{10} 6p^1$
Pb	82	[Xe] $6s^2 4f^{14} 5d^{10} 6p^2$
Bi	83	[Xe] $6s^2 4f^{14} 5d^{10} 6p^3$
Po	84	[Xe] $6s^2 4f^{14} 5d^{10} 6p^4$
At	85	[Xe] $6s^2 4f^{14} 5d^{10} 6p^5$
Rn	86	[Xe] $6s^2 4f^{14} 5d^{10} 6p^6$
Fr	87	[Rn] $7s^1$
Ra	88	[Rn] $7s^2$
Ac	89	[Rn] $7s^2 6d^1$
Th	90	[Rn] $7s^2 6d^2$
Pa	91	[Rn] $7s^2 5f^2 6d^1$
U	92	[Rn] $7s^2 5f^3 6d^1$
Np	93	[Rn] $7s^2 5f^4 6d^1$
Pu	94	[Rn] $7s^2 5f^6$
Am	95	[Rn] $7s^2 5f^7$
Cm	96	[Rn] $7s^2 5f^7 6d^1$
Bk	97	[Rn] $7s^2 5f^9$
Cf	98	[Rn] $7s^2 5f^{10}$
Es	99	[Rn] $7s^2 5f^{11}$
Fm	100	[Rn] $7s^2 5f^{12}$
Md	101	[Rn] $7s^2 5f^{13}$
No	102	[Rn] $7s^2 5f^{14}$
Lr	103	[Rn] $7s^2 5f^{14} 6d^1$
Unq	104	[Rn] $7s^2 5f^{14} 6d^2$
Unp	105	[Rn] $7s^2 5f^{14} 6d^3$
Unh	106	[Rn] $7s^2 5f^{14} 6d^4$

Note: For simplicity of expression, symbols of the chemically stable noble gases are used as a portion of the electron structure for the elements beyond neon. For example, the electron structure of a sodium atom, Na, consists of ten electrons, as in neon [Ne], plus a $3s^1$ electron. Detailed electron structures for the noble gases are given in Table 10.5.

10.7 Electron Structures of the Elements Beyond Calcium

The elements following calcium have a less regular pattern of adding electrons. The lowest energy level available for the 21st electron is the 3*d* sublevel. Scandium (21) has one more electron than calcium (20). Its electron structure will be the same as calcium plus one electron in the 3*d* sublevel. The electron structure for scandium is $1s^2 2s^2 2p^6 3s^2 3p^6 4s^2 3d^1$. The elements following scandium—titanium (22) through copper (29)—continue to add *d* electrons until the third energy level has its maximum of 18. Two exceptions in the orderly electron addition are chromium (24) and copper (29), the structures of which are given in Table 10.4. The third energy level of electrons is first completed in the element copper. Table 10.4 shows the order of filling of the electron orbitals and the electron configuration for the elements.

Example 10.2

What is the electron structure for a sulfur atom (atomic number 16) and for an iron atom (atomic number 26)?

Solution

Look in Table 10.4 for the element with atomic number 16 and write down its structure, [Ne] $3s^2 3p^4$. [Ne] is an abbreviated structure for neon, which is $1s^2 2s^2 2p^6$. Therefore, the electron structure for a sulfur atom is $1s^2 2s^2 2p^6 3s^2 3p^4$.

For an iron atom, Table 10.4 shows a structure of [Ar] $4s^2 3d^6$. This notation means that the electron structure of iron consists of the electron structure for argon plus $4s^2 3d^6$. The table shows that the structure for [Ar] is [Ne] $3s^2 3p^6$, which is equal to $1s^2 2s^2 2p^6 3s^2 3p^6$. Therefore, the electron structure for iron is $1s^2 2s^2 2p^6 3s^2 3p^6 4s^2 3d^6$.

If Table 10.4 is not available, the structure can be determined by tabulating electrons as in Example 10.1 and using Figure 10.4. Structures for all the noble gases, He, Ne, Ar, and so on, are also given in Table 10.5 (p. 207).

Practice Write the electron configuration for each of the following elements, using Table 10.4.

(a) Co (b) Cd (c) Pt

Answers: (a) [Ar]$4s^2 3d^7$ (b) [Kr]$5s^2 4d^{10}$ (c) [Xe]$6s^1 4f^{14} 5d^9$

10.8 Diagramming Atomic Structures

We can use several methods to diagram the atomic structures of atoms, depending on what we are trying to illustrate. When we want to show both the nuclear makeup and the electron structure of each energy level (without orbital detail), we can use a diagram such as Figure 10.5.

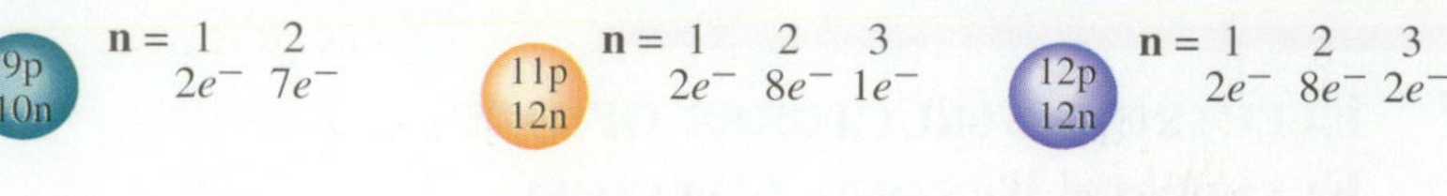

Fluorine atom Sodium atom Magnesium atom

FIGURE 10.5 ▶ Atomic structure diagrams of fluorine, sodium, and magnesium atoms. The numbers of protons and neutrons are shown in the nucleus. The number of electrons is shown in each principal energy level outside the nucleus.

A method of diagramming subenergy levels is shown in Figure 10.6. Each orbital is represented by a square □. When the orbital contains one electron, an arrow (↑) is placed in the square. A second arrow, pointing downward (↓), indicates the second electron in that orbital.

The diagram for hydrogen is [↑]. Helium, with two electrons, is drawn as [↑↓]; both electrons are 1*s* electrons. The diagram for lithium shows three electrons in two energy levels, $1s^22s^1$. All four electrons of beryllium are *s* electrons, $1s^22s^2$. Boron has the first *p* electron, which is located in the $2p_x$ orbital. Because it is energetically more difficult for the next *p* electron to pair up with the electron in the p_x orbital than to occupy a second *p* orbital, the second *p* electron in carbon is located in the $2p_y$ orbital. The third *p* electron in nitrogen is still unpaired and is found in the $2p_z$ orbital. The next three electrons pair with each of the 2*p* electrons through the element neon. Also shown in Figure 10.6 are the equivalent linear expressions for these orbital electron structures.

The electrons in successive elements are found in sublevels of increasing energy. The general sequence of increasing energy of sublevels and the order of filling sublevels with electrons is

1*s* 2*s* 2*p* 3*s* 3*p* 4*s* 3*d* 4*p* 5*s* 4*d* 5*p* 6*s* 4*f* 5*d* 6*p* 7*s* 5*f* 6*d*

Minor variations from the electron structure predicted by the foregoing general sequence are found in a number of atoms. Table 10.4 shows the accepted ground-state electron structure for the elements.

FIGURE 10.6 ▶ Subenergy level electron structure of hydrogen through sodium atoms. Each electron is indicated by an arrow placed in the square, which represents the orbital.

Element	Orbital electron structure						Linear expression of electron structure
	1*s*	2*s*	$2p_x$	$2p_y$	$2p_z$	3*s*	
H	↑						$1s^1$
He	↑↓						$1s^2$
Li	↑↓	↑	□	□	□		$1s^22s^1$
Be	↑↓	↑↓	□	□	□		$1s^22s^2$
B	↑↓	↑↓	↑	□	□		$1s^22s^22p_x^1$
C	↑↓	↑↓	↑	↑	□		$1s^22s^22p_x^12p_y^1$
N	↑↓	↑↓	↑	↑	↑		$1s^22s^22p_x^12p_y^12p_z^1$
O	↑↓	↑↓	↑↓	↑	↑		$1s^22s^22p_x^22p_y^12p_z^1$
F	↑↓	↑↓	↑↓	↑↓	↑		$1s^22s^22p_x^22p_y^22p_z^1$
Ne	↑↓	↑↓	↑↓	↑↓	↑↓		$1s^22s^22p_x^22p_y^22p_z^2$
Na	↑↓	↑↓	↑↓	↑↓	↑↓	↑	$1s^22s^22p_x^22p_y^22p_z^23s^1$

EXAMPLE 10.3

Diagram the electron structure of a zinc atom and a rubidium atom. Use the $1s^2 2s^2 2p^6$, etc., method.

SOLUTION

The atomic number of zinc is 30; therefore it has 30 protons and 30 electrons in a neutral atom. Using Figure 10.4 tabulate the 30 electrons as follows:

Orbital	Number of e^-	Total e^-
$1s$	$2e^-$	2
$2s$	$2e^-$	4
$2p$	$6e^-$	10
$3s$	$2e^-$	12
$3p$	$6e^-$	18
$4s$	$2e^-$	20
$3d$	$10e^-$	30

The electron structure of a zinc atom is $1s^2 2s^2 2p^6 3s^2 3p^6 4s^2 3d^{10}$. Check by adding the superscripts, which should equal 30.

The atomic number of rubidium is 37; therefore it has 37 protons and 37 electrons in a neutral atom. With a little practice, and using Figure 10.4, the electron structure may be written directly in the linear form. The electron structure of a rubidium atom is $1s^2 2s^2 2p^6 3s^2 3p^6 4s^2 3d^{10} 4p^6 5s^1$. Check by adding the superscripts, which should equal 37.

10.9 Lewis-Dot Representation of Atoms

The Lewis-dot (or electron-dot) method of representing atoms, proposed by the American chemist G. N. Lewis (1875–1946), uses the symbol for the element and dots for electrons. The number of dots placed around the symbol equals the number of *s* and *p* electrons in the outermost energy level of the atom. Paired dots represent paired electrons; unpaired dots represent unpaired electrons. For example, **H·** is the Lewis symbol for a hydrogen atom, $1s^1$; **:Ḃ** is the Lewis symbol for a boron atom, $1s^2 2s^2 2p^1$; **:Ï·** is an iodine atom, which has seven electrons in its outermost energy level. In the case of boron, the symbol B represents the boron nucleus and the $1s^2$ electrons; the dots represent only the $2s^2 2p^1$ electrons.

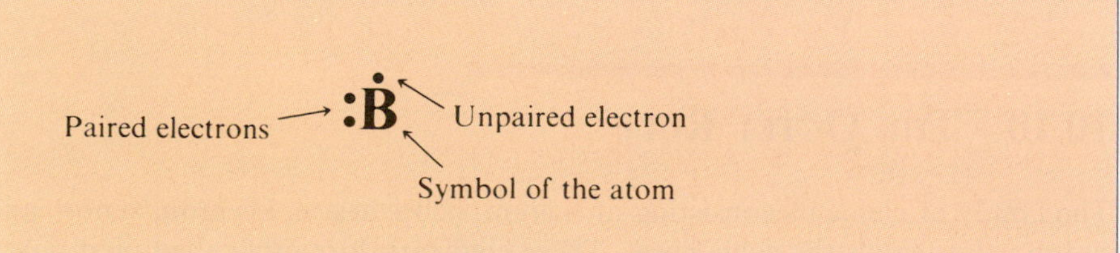

The Lewis-dot method is often used, not only because of its simplicity of expression, but also because much of the chemistry of the atom is directly

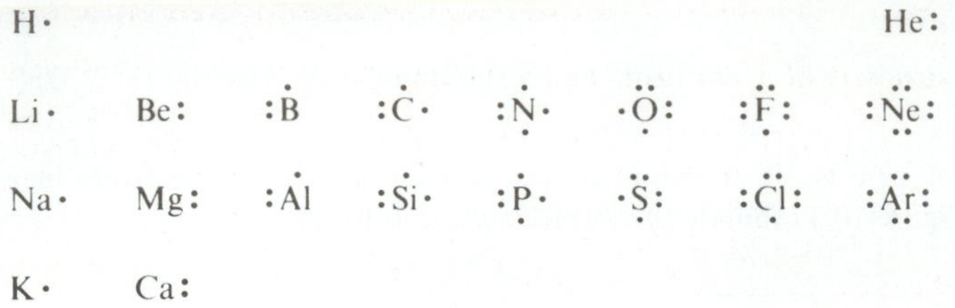

FIGURE 10.7 ►
Lewis-dot diagrams of the first twenty elements. Dots represent electrons in the outermost energy level only.

H·							He:
Li·	Be:	:B	:C·	:N·	·O:	:F:	:Ne:
Na·	Mg:	:Al	:Si·	:P·	·S:	:Cl:	:Ar:
K·	Ca:						

associated with the electrons in the outermost energy level. This association is especially true for the first 20 elements and the remaining Group A elements of the periodic table (see Chapter 11). Figure 10.7 shows Lewis-dot diagrams for the elements hydrogen through calcium.

EXAMPLE 10.4

Write the Lewis-dot structure for a phosphorus atom.

SOLUTION

First establish the electron structure for a phosphorus atom. It is $1s^2 2s^2 2p^6 3s^2 3p^3$. Note that there are five electrons in the outermost principal energy level; they are $3s^2 3p^3$. Write the symbol for phosphorus and place the five electrons as dots around it:

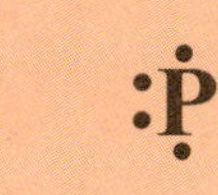

The $3s^2$ electrons are paired and are represented by the paired dots. The $3p^3$ electrons, which are unpaired, are represented by the single dots.

PRACTICE Write the Lewis-dot structure for the following elements:
(a) N (b) Al (c) Sr (d) Br
Answers: (a) :N· (b) :Al (c) Sr: (d) :Br:

10.10 THE OCTET RULE

The family of elements consisting of helium, neon, argon, krypton, xenon, and radon is known as the **noble gases**. These elements, formerly called *inert gases*, have almost no chemical reactivity; in fact, no compounds of any of them were known before 1962.

noble gases

TABLE 10.5 Arrangement of Electrons in the Noble Gases[a]

Noble gas	Symbol	Electron structure					
		n = 1	2	3	4	5	6
Helium	He	$1s^2$					
Neon	Ne	$1s^2$	$2s^22p^6$				
Argon	Ar	$1s^2$	$2s^22p^6$	$3s^23p^6$			
Krypton	Kr	$1s^2$	$2s^22p^6$	$3s^23p^63d^{10}$	$4s^24p^6$		
Xenon	Xe	$1s^2$	$2s^22p^6$	$3s^23p^63d^{10}$	$4s^24p^64d^{10}$	$5s^25p^6$	
Radon	Rn	$1s^2$	$2s^22p^6$	$3s^23p^63d^{10}$	$4s^24p^64d^{10}4f^{14}$	$5s^25p^65d^{10}$	$6s^26p^6$

[a] Each gas except helium has eight electrons in its outermost energy level.

Each of these elements, except helium, has an outer shell of eight electrons, two *s* and six *p* (see Table 10.5). The only shell of helium is an *s* orbital filled with two electrons. The electron structure of the noble gases is such that the outer shell *s* and *p* orbitals are filled with paired electrons. This arrangement is very stable and makes the atoms of the noble gases chemically unreactive. Recognition of the extraordinary stability of this structure led to the **octet rule**: Through chemical changes many of the elements tend to attain an electron structure of eight electrons in their outermost energy level, identical to that of the chemically stable noble gases. Although the octet rule is useful and applies to the behavior of many elements and compounds, it is not universally applicable; some elements do not obey this rule. Applications of the octet rule are given in Chapter 12.

octet rule

Concepts in Review

1. Describe the atom as conceived by Niels Bohr.
2. Discuss the contributions to atomic theory made by Dalton, Thomson, Rutherford, Bohr, Chadwick, and Schrödinger.
3. Describe what is meant by an electron orbital.
4. Determine the maximum number of electrons that can exist in the principal energy levels and sublevels.
5. Write the electron configuration for any of the first 56 elements.
6. Explain what is represented by the Lewis-dot (electron-dot) structure of an element.
7. Write the Lewis-dot (electron-dot) symbols for the first twenty elements.
8. Understand the basis for the octet rule.

Chemistry in Action

Yes, We Can See Atoms!

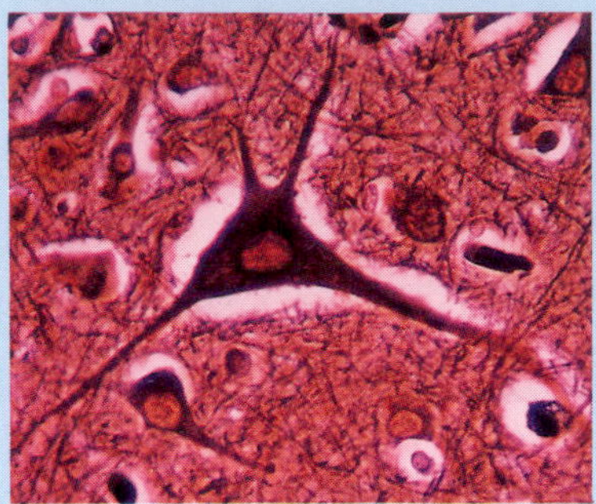

▲ **Neurons, optical microscope**

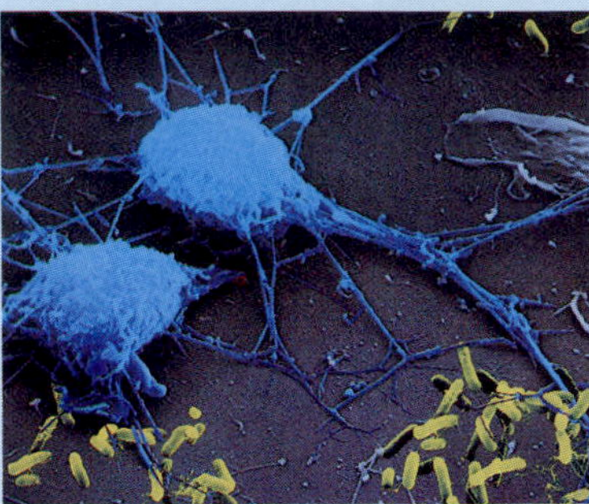

▲ **Neurons, electron microscope**

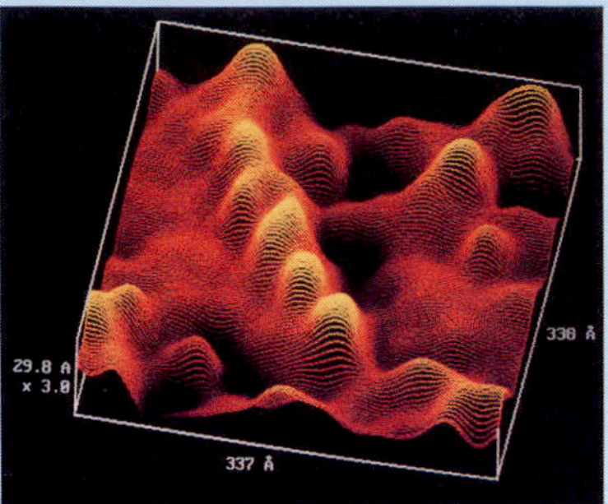

▲ **DNA molecule, scanning-tunneling microscope**

For centuries scientists have argued and theorized over the nature and existence of atoms. Today physicists and chemists can produce pictures of atoms and even move them individually from place to place. This new-found ability to see atoms, molecules, and even watch chemical reactions occur is the direct result of the evolution of the microscope.

An optical microscope is capable of viewing objects as small as the size of a cell. To see smaller objects an electron microscope is necessary. Since the eye cannot respond to a beam of electrons the image is produced on a fluorescent screen or on film. These microscopes have been used for some time to photograph large molecules. Unfortunately, however, the objects must be placed in a vacuum and the electrons must be high energy to see tiny objects. If the sample is fragile, as are most molecules, it can be destroyed before the image is formed.

In 1981, Gerd Binnig and Heinrich Rohrer, two scientists from IBM, invented the first scanning-probe microscope. These instruments are fundamentally different from previous microscopes. In a scanning-probe microscope a probe is placed near the surface of a sample and a parameter of some sort (voltage, magnetic field, etc.) is measured. As the probe is moved across the surface, an image is produced—in the same manner a child would determine the identity of an object sealed in an opaque bag. The first of the scanning-probe instruments was called a scanning-tunneling microscope. It produced the first clear pictures of silicon atoms in January 1983. The greatest limitation for the scanning-tunneling microscope is that, in order to be viewed, organic molecules must be given a thin coating of metal so that electrons are free to jump from the surface to the probe.

In 1985, a team of physicists from Stanford University and IBM found a solution to this problem. In a new instrument, known as an atomic-force microscope, the probe measures tiny electric forces between electrons instead of the actual movement of electrons from surface to probe. The greatest advantage of this approach is that the probe is so gentle that even very fragile molecules remain intact. The probe is a tiny shard of a diamond attached to a tiny piece of silicon, and works like a phonograph needle. At University of California, Santa Barbara a group of scientists have even succeeded in making a movie of the formation of a blood clot on the molecular level.

In industry another type of scanning-probe instrument has been developed to check the quality of microelectronic equipment. Researchers at IBM have developed a laser-force microscope in which a tiny wire probe measures small attractive forces (surface tension of water on the sample) to show imperfections as small as 25 atoms across.

Exercises

1. What is an electron orbital?
2. Under which conditions can a second electron enter an orbital already containing one electron?
3. What is meant when we say that the electron structure of an atom is in its ground state?
4. How do 1*s* and 2*s* orbitals differ? How are they alike?
5. What letters are used to designate the energy sublevels?
6. List the following electron sublevels in order of increasing energy: 2*s*, 2*p*, 4*s*, 1*s*, 3*d*, 3*p*, 4*p*, 3*s*.
7. How many *s* electrons, *p* electrons, and *d* electrons are possible in any electron shell?
8. How many protons are in the nucleus of an atom of each of these elements: H, B, F, Sc, Ag, U, Br, Sb, and Pb?
9. Give the electron structure ($1s^22s^22p^6\ldots$) for B, Ti, Zn, Br, and Sr.
10. Why is the eleventh electron of the sodium atom located in the third energy level rather than in the second energy level?
11. What is the major difference between an orbital and a Bohr orbit?
12. Explain how the spectral lines of hydrogen occur.
13. Explain how Bohr used the data of the hydrogen bright-line spectrum to support his solar system model of atomic structure.
14. Explain why and how Bohr's solar system model of the atom was modified to the cloud model of the atom.
15. Using the formula $2\mathbf{n}^2$, calculate the number of electrons that can exist in principal energy levels: $\mathbf{n} = 1, 2, 3, 4, 5$, and 6.
16. How many electrons can be present in the fourth energy level?
17. How many orbitals can exist in the third energy level? What are they?
18. Sketch the *s*, p_x, p_y, and p_z orbitals.
19. Diagram the atomic structures of the following atoms:
    (a) $^{14}_{7}N$ (d) $^{91}_{40}Zr$
    (b) $^{35}_{17}Cl$ (e) $^{127}_{53}I$
    (c) $^{65}_{30}Zn$
20. Write out the electron configuration (long form) for each of the following elements:
    (a) Chlorine (d) Iron
    (b) Silver (e) Iodine
    (c) Lithium
21. Show the Lewis-dot structures for C, Mg, Al, Cl, and K.
22. In the designation $3d^7$, give the significance of the 3, the *d*, and the 7.
23. Using the method shown in Figure 10.6, show the orbital electron structure for an atom of: (a) Si (b) S (c) Ar (d) V
24. What electron structure do the noble gases have in common?
25. Why is the last electron in potassium located in the fourth energy level rather than in the third energy level?
26. Which atoms have the following electron structures?
    (a) $1s^22s^22p^63s^2$
    (b) $1s^22s^22p^5$
    (c) $1s^22s^22p^63s^23p^64s^23d^8$
    (d) $1s^22s^22p^63s^23p^64s^23d^5$
    (e) $1s^22s^22p^63s^23p^64s^23d^{10}4p^65s^14d^5$
27. Show the electron structures ($1s^22s^22p^6\ldots$) for elements of atomic numbers:
    (a) 8 (b) 11 (c) 17 (d) 23 (e) 28 (f) 34
28. Using only Table 10.4 show the electron structures ($1s^22s^22p^6\ldots$) for the elements having the following numbers of electrons:
    (a) 9 (b) 26 (c) 31 (d) 39 (e) 52
29. Which elements have the following electron structures?
    (a) $[Ar]4s^23d^1$ (c) $[Kr]5s^24d^{10}5p^2$
    (b) $[Ar]4s^23d^{10}4p^6$ (d) $[Xe]6s^1$
30. Identify these atoms from their atomic structure diagrams:
    (a) (16p 16n) $2e^-$ $8e^-$ $6e^-$
    (b)
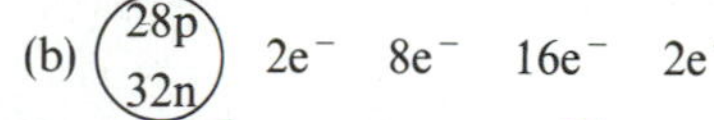

31. Explain how scanning-probe microscopes are different from previous microscopes.
32. Describe the functioning of a scanning-tunneling microscope.
33. Differentiate between the scanning-tunneling microscope and an atomic force microscope.
34. Diagram the atomic structures (as in Exercise 30) for these atoms:
    (a) $^{27}_{13}Al$ (b) $^{51}_{23}V$
35. State the octet rule and its relationship to the noble gases.

36. Write Lewis-dot symbols for these atoms: He, B, O, Na, Si, Ar, Ga, Ca, Br, and Kr.
37. Which of the following statements are correct? Rewrite each incorrect statement to make it correct.
   (a) In the ground state, electrons tend to occupy orbitals having the lowest possible energy.
   (b) The maximum number of *p* electrons in the first energy level is six.
   (c) A 2*s* electron is in a lower energy state than a 2*p* electron.
   (d) The electron structure for a carbon atom is $1s^2 2s^2 2p^2$.
   (e) The $2p_x$, $2p_y$, and $2p_z$ electron orbitals are all in the same energy state.
   (f) The energy level of a 3*d* electron is higher than that of a 4*s* electron.
   (g) The electron structure for a calcium atom is $1s^2 2s^2 2p^6 3s^2 3p^6 3d^2$.
   (h) There are seven principal energy levels for the known elements.
   (i) The third energy level can have a maximum of 18 electrons.
   (j) The number of possible *d* electrons in the third energy level is ten.
   (k) The first *f* electron occurs in the fourth principal energy level.
   (l) The Lewis-dot symbol for nitrogen is **:Ṅ·** (with a dot below)
   (m) The Lewis-dot symbol for potassium is **P·**
   (n) Atoms of all the noble gases (except helium) have eight electrons in their outermost energy level.
   (o) A *p* orbital is spherically symmetrical around the nucleus.
   (p) An atom of nitrogen has two electrons in a 1*s* orbital, two electrons in a 2*s* orbital, and one electron in each of three different 2*p* orbitals.
   (q) The maximum number of electrons that can occupy a specific energy level **n** is given by $2\mathbf{n}^2$.
   (r) ${}^{12}_{6}C$ is an isotope of carbon that is used as the reference standard for the atomic mass system.
   (s) The Lewis-dot symbol for the noble gas helium is **:He:** (with dots above and below)
   (t) When an orbital contains two electrons, the electrons have parallel spins.
   (u) The Bohr theory proposed that electrons move around the nucleus in circular orbits.
   (v) Bohr concluded from his experiment that the positive charge and almost all the mass were concentrated in a very small nucleus.

11

The Periodic Table

◀ CHAPTER OPENING PHOTO: The aurora borealis is produced when charged particles interact with atoms in the atmosphere.

Organizing and classifying large volumes of information for easy use and convenience is a necessary task in modern society. In the hardware store, paint colors are sorted into color chips—first by major color group, then by shades or hues within the specific color. In the local winery, wines are classified as red, white, or rosé—then further grouped by type of grape, origin and year. The use of personal computers has soared in recent years, and allows us to classify many things, such as addresses, birthdays, or household finances. Even the calendar itself is broken into rows and columns to permit us to organize our time and activities by day, week, and month.

11.1 THE ORIGIN OF THE PERIODIC TABLE

The periodic table represents the efforts of a number of chemists. These researchers were trying to find some logical way in which to organize the elements. Chemists of the early 19th century had sufficient knowledge of the properties of elements to recognize similarities among groups of elements. As early as 1817 J. W. Döbereiner (1780–1849), professor at University of Jena in Germany, observed the existence of *triads* of similarly behaving elements, in which the middle element had an atomic mass approximating the average of the other two elements. He also noted that for many other properties the value for the central element was approximately the average of the values of the other two elements.

▲ Dimitri Mendeleev (1834–1907)

In 1864, J. A. R. Newlands (1837–1898), an English chemist, reported his *Law of Octaves.* In his studies Newlands observed that, when the elements were arranged according to increasing atomic masses, every eighth element had similar properties. (The noble gases were not yet discovered at that time.) Newland's theory was ridiculed by his contemporaries in the Royal Chemical Society, and they refused to publish his work. Many years later, however, Newlands was awarded the highest honor of the society for this important contribution to the development of the periodic law.

In 1869 Dmitri Ivanovitch Mendeleev (1834–1907) of Russia and Lothar Meyer (1830–1895) of Germany independently published their periodic arrangements of the elements that were based on increasing atomic masses. Because his arrangement was published slightly earlier and was in a somewhat more useful form than that of Meyer, Mendeleev's name is usually associated with the modern periodic table.

TABLE 11.1 Comparison of Properties of Eka-Silicon (predicted by Mendeleev) with Germanium

Property	Eka-silicon, Es (predicted)	Germanium, Ge (observed)
Atomic mass	72	72.6
Color of metal	Dirty gray	Grayish-white
Density	5.5 g/mL	5.35 g/mL
Oxide formula	EsO_2	GeO_2
Oxide density	4.7 g/mL	4.70 g/mL
Chloride formula	$EsCl_4$	$GeCl_4$
Chloride density	1.9 g/mL	1.87 g/mL
Boiling temperature of chloride	Under 100°C	86°C

Only about 63 elements were known when Mendeleev constructed his table. He arranged these elements so that those with similar properties fit into columns to form family groups. The arrangement left many gaps between elements, and Mendeleev predicted that these spaces would be filled as new elements were discovered. For example, spaces for undiscovered elements were left after calcium, under aluminum, and under silicon. He called these unknown elements *eka*-boron, *eka*-aluminum, and *eka*-silicon. The term *eka* comes from Sanskrit meaning "one" and was used to indicate that the missing element was one place away in the table from the element indicated. Mendeleev even predicted the physical and chemical properties of these undiscovered elements. The three elements scandium (atomic number 21), gallium (31), and germanium (32), were actually discovered during Mendeleev's lifetime and were found to have properties which agree very closely with his predictions. The amazing way in which Mendeleev's predictions were fulfilled is illustrated in Table 11.1.

The term *periodic* means recurring at regular intervals. The original periodic tables were based on the premise that the properties of the elements are periodic functions of their atomic masses. However, if we look carefully there are some exceptions to this rule. For example, the atomic mass for argon is greater than that for potassium, yet argon must be placed with the other noble gases before potassium. When sorted by its chemical properties potassium belongs with the alkali metals. These small discrepancies were resolved by the work of British physicist H. G. Moseley (1887–1915) and by the discovery of isotopes. The elements in the periodic table are now arranged by increasing atomic number, resolving the irregularities of Mendeleev's original table.

11.2 ARRANGEMENT OF THE PERIODIC TABLE

The modern **periodic table** is shown in Table 11.2 and on the inside front cover of this book. In this table the elements are arranged horizontally in numerical order according to atomic numbers. The resulting seven rows are called **periods**.

periodic table

periods

TABLE 11.2 Periodic Table of the Elements

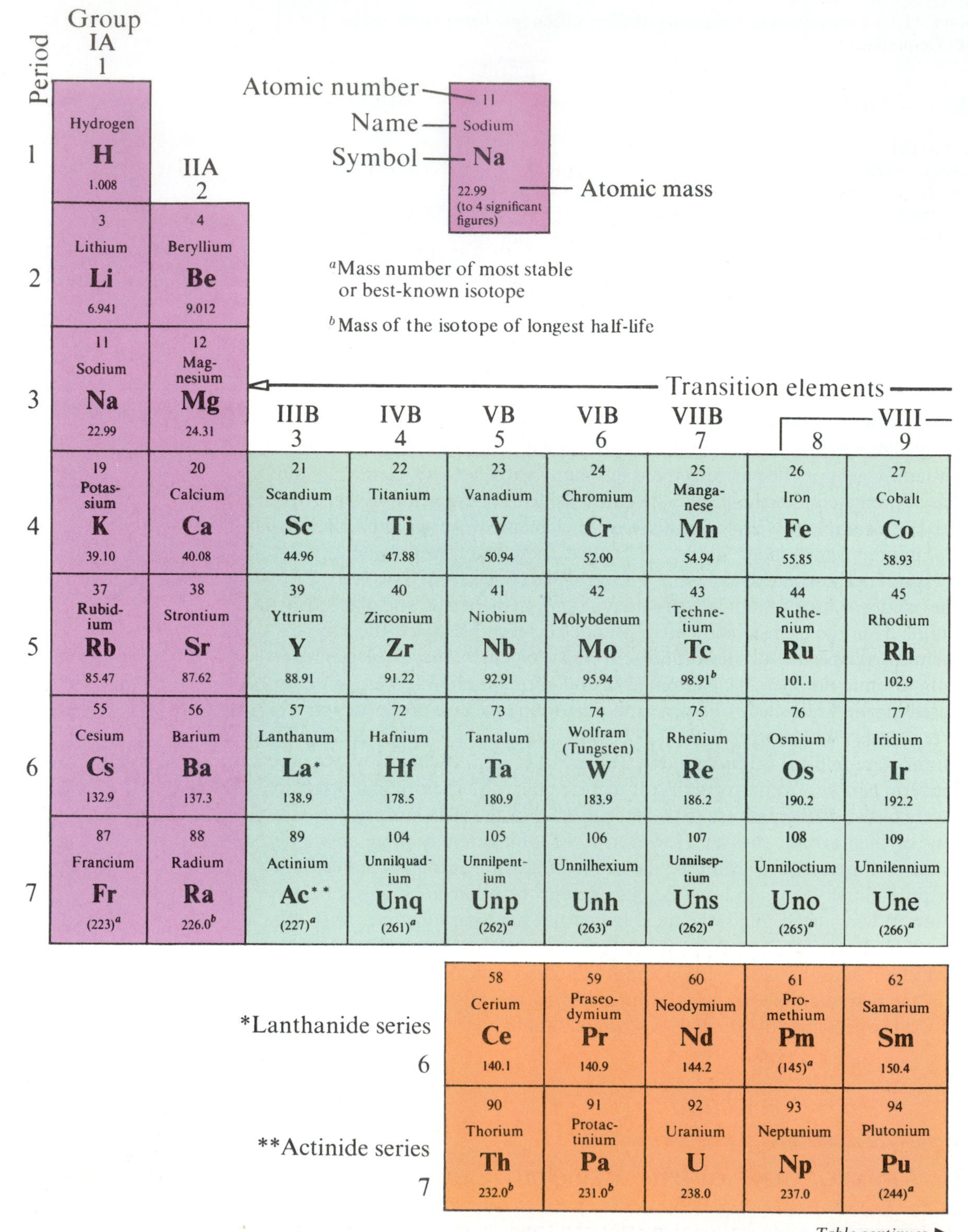

[a]Mass number of most stable or best-known isotope

[b]Mass of the isotope of longest half-life

Period	Group IA 1	IIA 2	IIIB 3	IVB 4	VB 5	VIB 6	VIIB 7	VIII 8	VIII 9
1	1 Hydrogen **H** 1.008								
2	3 Lithium **Li** 6.941	4 Beryllium **Be** 9.012							
3	11 Sodium **Na** 22.99	12 Mag-nesium **Mg** 24.31							
4	19 Potas-sium **K** 39.10	20 Calcium **Ca** 40.08	21 Scandium **Sc** 44.96	22 Titanium **Ti** 47.88	23 Vanadium **V** 50.94	24 Chromium **Cr** 52.00	25 Manga-nese **Mn** 54.94	26 Iron **Fe** 55.85	27 Cobalt **Co** 58.93
5	37 Rubid-ium **Rb** 85.47	38 Strontium **Sr** 87.62	39 Yttrium **Y** 88.91	40 Zirconium **Zr** 91.22	41 Niobium **Nb** 92.91	42 Molybdenum **Mo** 95.94	43 Techne-tium **Tc** 98.91[b]	44 Ruthe-nium **Ru** 101.1	45 Rhodium **Rh** 102.9
6	55 Cesium **Cs** 132.9	56 Barium **Ba** 137.3	57 Lanthanum **La*** 138.9	72 Hafnium **Hf** 178.5	73 Tantalum **Ta** 180.9	74 Wolfram (Tungsten) **W** 183.9	75 Rhenium **Re** 186.2	76 Osmium **Os** 190.2	77 Iridium **Ir** 192.2
7	87 Francium **Fr** (223)[a]	88 Radium **Ra** 226.0[b]	89 Actinium **Ac**** (227)[a]	104 Unnilquad-ium **Unq** (261)[a]	105 Unnilpent-ium **Unp** (262)[a]	106 Unnilhexium **Unh** (263)[a]	107 Unnilsep-tium **Uns** (262)[a]	108 Unniloctium **Uno** (265)[a]	109 Unnilennium **Une** (266)[a]

Series					
*Lanthanide series 6	58 Cerium **Ce** 140.1	59 Praseo-dymium **Pr** 140.9	60 Neodymium **Nd** 144.2	61 Pro-methium **Pm** (145)[a]	62 Samarium **Sm** 150.4
**Actinide series 7	90 Thorium **Th** 232.0[b]	91 Protac-tinium **Pa** 231.0[b]	92 Uranium **U** 238.0	93 Neptunium **Np** 237.0	94 Plutonium **Pu** (244)[a]

Table continues ▶

Atomic masses are based on carbon-12. Atomic masses in parentheses indicate the most stable or best-known isotope. Slight disagreement exists as to the exact electronic configuration of several of the high-atomic-number elements. Names and symbols for elements 104 to 109 are unofficial.

10	IB 11	IIB 12	IIIA 13	IVA 14	VA 15	VIA 16	VIIA 17	Noble gases 18
								2 Helium **He** 4.003
			5 Boron **B** 10.81	6 Carbon **C** 12.01	7 Nitrogen **N** 14.01	8 Oxygen **O** 16.00	9 Fluorine **F** 19.00	10 Neon **Ne** 20.18
			13 Aluminum **Al** 26.98	14 Silicon **Si** 28.09	15 Phosphorus **P** 30.97	16 Sulfur **S** 32.06	17 Chlorine **Cl** 35.45	18 Argon **Ar** 39.95
28 Nickel **Ni** 58.71	29 Copper **Cu** 63.55	30 Zinc **Zn** 65.38	31 Gallium **Ga** 69.72	32 Germanium **Ge** 72.59	33 Arsenic **As** 74.92	34 Selenium **Se** 78.96	35 Bromine **Br** 79.90	36 Krypton **Kr** 83.80
46 Palladium **Pd** 106.4	47 Silver **Ag** 107.9	48 Cadmium **Cd** 112.4	49 Indium **In** 114.8	50 Tin **Sn** 118.7	51 Antimony **Sb** 121.8	52 Telurium **Te** 127.6	53 Iodine **I** 126.9	54 Xenon **Xe** 131.3
78 Platinum **Pt** 195.1	79 Gold **Au** 197.0	80 Mercury **Hg** 200.6	81 Thallium **Tl** 204.4	82 Lead **Pb** 207.2	83 Bismuth **Bi** 209.0	84 Polonium **Po** (210)[a]	85 Astatine **At** (210)[a]	86 Radon **Rn** (222)[a]

Inner transition elements

63 Europium **Eu** 152.0	64 Gadolinium **Gd** 157.3	65 Terbium **Tb** 158.9	66 Dysprosium **Dy** 162.5	67 Holmium **Ho** 164.9	68 Erbium **Er** 167.3	69 Thulium **Tm** 168.9	70 Ytterbium **Yb** 173.0	71 Lutetium **Lu** 175.0
95 Americium **Am** (243)[a]	96 Curium **Cm** (247)[a]	97 Berkelium **Bk** (247)[a]	98 Californium **Cf** (251)[a]	99 Einsteinium **Es** (252)[a]	100 Fermium **Fm** (257)[a]	101 Mendelevium **Md** (258)[a]	102 Nobelium **No** (259)[a]	103 Lawrencium **Lr** (260)[a]

groups

families of elements

Each period with the exception of the first begins with an alkali metal and ends with a noble gas. This arrangement forms vertical columns of elements which have similar chemical properties. These columns are known as **groups** or **families of elements**.

The number of each period corresponds with the number of the outermost energy level that contains electrons for elements in that period. Those in row 1 contain electrons only on level 1, while those in row 2 contain electrons in levels 1 and 2. In row 3, electrons are found on levels 1, 2 and 3, and so on.

representative elements

transition elements

The groups or families of elements form the columns on the periodic table. There are several systems in use today for numbering the groups. In one system the columns are numbered from left to right using the numbers 1–18. We will use a system that numbers the columns by Roman numerals and the letters A and B. The A groups are known as the **representative elements**. The numbers I through VII indicate the number of outermost energy level electrons for the elements in that group. For example, Be is in Group IIA and contains two electrons in its outermost energy level. Oxygen is in Group VIA and contains six electrons in the outermost level. The B groups and Group VIII elements form the **transition elements**. The behavior of these elements is more complex than that of the representative elements. Since this is an introductory course we will concentrate on the representative elements.

Other families of elements beside the noble gases have family names. Group IA elements are called the *alkali metals*; Group IIA, the alkaline earth metals; and Group VII A the halogens.

11.3 Electron Configurations and the Periodic Table

In Chapter 10 we learned to write electron configurations for the elements. The filling of the energy levels within the atom is a periodic property of the elements. To understand the connection between orbital filling and the periodic table examine Table 11.3. The *s* and *p* blocks are composed of Group A elements and the noble gases. The *s* block consists of elements in Group IA, IIA, and helium. Each element in this block has one or two *s* electrons in its outer energy level. The *p* block contains Group IIIA through VIIA elements and the noble gases. The elements each contain two *s* electrons and from one to six *p* electrons in the outermost energy level. The *d* block includes the transistion elements consisting of Groups IB through VIIB and Group VIII. The *f* block of elements are shown below the main body of the periodic table. These elements are known as the inner transition elements. They are shown below the table to conserve space.

A periodic table is almost always available to you. If you understand the relationship between electron configuration and the position of the elements on the table you can write the electron configuration of almost any atom.

Example 11.1

Use the periodic table to write the electron configuration for phosphorus and tin.

Solution

Phosphorus is element 15 and is located in Period 3, Group VA. The electron configuration must have a full first and second energy level.

$$\text{P} \quad 1s^22s^22p^63s^23p^3$$

TABLE 11.3 Arrangement of Elements According to the Sublevel of Electrons Being Filled in Their Atomic Structure

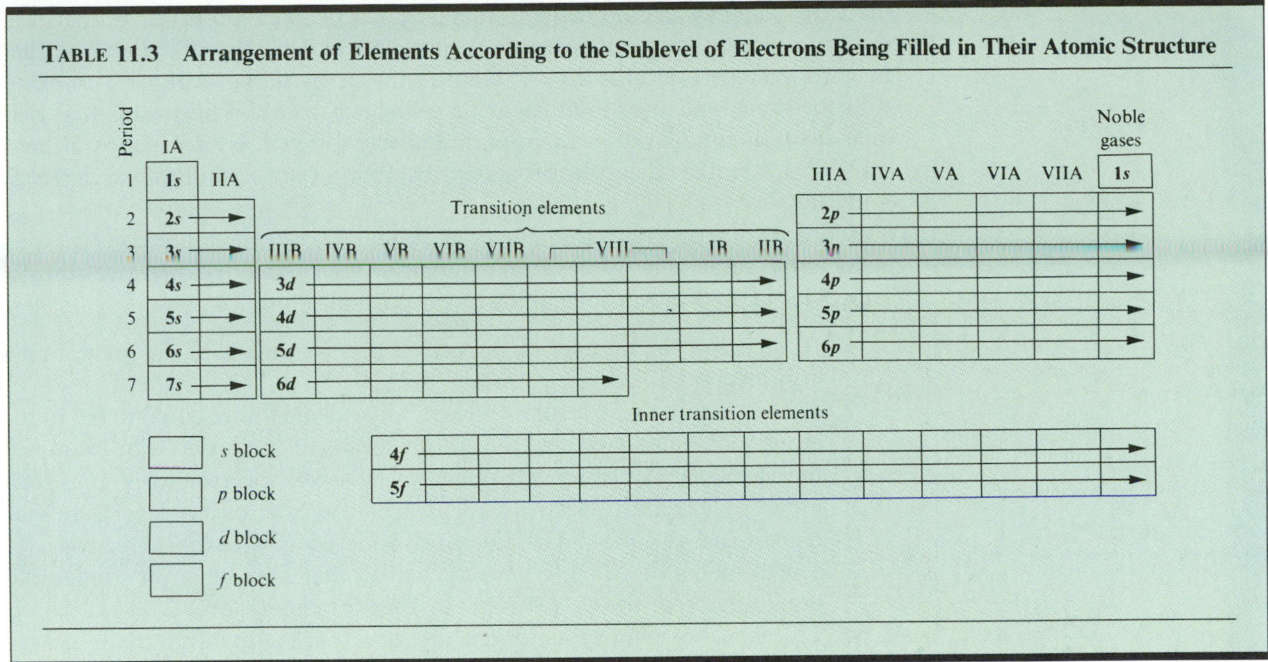

Just by looking across each row and counting the element blocks the electron configuration is determined.

Period	IA	IIA	Transition elements (IIIB IVB VB VIB VIIB VIII IB IIB)	IIIA	IVA	VA	VIA	VIIA	Noble gases
1	1s								1s
2	2s 1	2		2p 1	2	3	4	5	6
3	3s 1	2		3p 1	2 P	3			
4	4s		3d	4p					
5	5s		4d	5p					
6	6s		5d	6p					
7	7s		6d						

Tin is element 50 in Period 5, Group IVA, two places after the transition metals. It must have 2 electrons in the 5*p* series. Its electron configuration is

$$\text{Sn} \quad 1s^2 2s^2 2p^6 3s^2 3p^6 4s^2 3d^{10} 4p^6 5s^2 4d^{10} 5p^2$$

Notice that the *d* series of electrons is always one energy level behind the period number.

PRACTICE Use the periodic table to write the electron configuration for 0, Ca, Zn.

Answer: O $1s^2 2s^2 2p^4$

Ca $1s^2 2s^2 2p^6 3s^2 3p^6 4s^2$

Zn $1s^2 2s^2 2p^6 3s^2 3p^6 4s^2 3d^{10}$

The early chemists classified the elements based on their observed properties. Modern atomic theory gives us an explanation of why the properties of the elements vary periodically. As we build up atoms by filling orbitals with electrons the same type of orbitals occur on each energy level. This means that the same electron configuration reappears regularly for each level. Groups of elements show similar chemical properties because of these outermost electron similarities.

Consider the periodic table shown in Table 11.4. For the representative elements in this table only the electron configuration of the outer shell electrons is given. This table illustrates the following important points:

1. In each row the number of the period corresponds with the highest energy level occupied by electrons.
2. The Group numbers for the representative elements are equal to the total number of outer shell electrons in the atoms of the group. For example the elements in Group VIIA always have the electron configuration ns^2np^5. The *d* and *f* electrons are always in a lower energy level than the highest energy level and so are not considered as outermost electrons.
3. The elements of a family have the same outer shell electron configuration except that the electrons are in different energy levels.
4. The elements within each of the *s, p, d, f* blocks are filling the *s, p, d, f* orbitals as shown in Table 11.3.
5. Within the transition elements some discrepancies in the order of filling occur. Explanation of these discrepancies and the similar ones which occur in the inner transition elements are beyond the scope of this book.

EXAMPLE 11.2 Write Lewis-dot symbols for the representative elements Na, K, Se, and Br.

SOLUTION Locate the element on the periodic table. The Group A number is the same as the number of valence electrons. Place this number of dots around the symbol of the element to show the correct Lewis symbol.

Na, K	Group IA	$(1e^-)$	Na· K·
Sr	Group IIA	$(2e^-)$	Sr:
Se	Group VIA	$(6e^-)$	·S̤̈e·
Br	Group VIIA	$(7e^-)$	:B̤̈r·

11.4 Periodic Trends in Atomic Properties

Although atomic theory and electron configuration can help us to understand the arrangement and behavior of the elements it is important to remember that the design of the periodic table is based on the observation of properties of the elements. Before we proceed to use the concept of atomic structure to explain how and why atoms combine to form compounds it is important to understand the characteristic properties of the elements and the trends that occur in these

TABLE 11.4 Outermost Electron Configurations of the Elements

Period	IA ← Group number	IIA											IIIA	IVA	VA	VIA	VIIA	Noble gases
1	1 H $1s^1$																	2 He $1s^2$
2	3 Li $2s^1$	4 Be $2s^2$											5 B $2s^22p^1$	6 C $2s^22p^2$	7 N $2s^22p^3$	8 O $2s^22p^4$	9 F $2s^22p^5$	10 Ne $2s^22p^6$
3	11 Na $3s^1$	12 Mg $3s^2$											13 Al $3s^23p^1$	14 Si $3s^23p^2$	15 P $3s^23p^3$	16 S $3s^23p^4$	17 Cl $3s^23p^5$	18 Ar $3s^23p^6$
4	19 K $4s^1$	20 Ca $4s^2$	21 Sc $4s^23d^1$	22 Ti $4s^23d^2$	23 V $4s^23d^3$	24 Cr $4s^13d^5$	25 Mn $4s^23d^5$	26 Fe $4s^23d^6$	27 Co $4s^23d^7$	28 Ni $4s^23d^8$	29 Cu $4s^13d^{10}$	30 Zn $4s^23d^{10}$	31 Ga $4s^24p^1$	32 Ge $4s^24p^2$	33 As $4s^24p^3$	34 Se $4s^24p^4$	35 Br $4s^24p^5$	36 Kr $4s^24p^6$
5	37 Rb $5s^1$	38 Sr $5s^2$	39 Y $5s^24d^1$	40 Zr $5s^24d^2$	41 Nb $5s^14d^4$	42 Mo $5s^14d^5$	43 Tc $5s^14d^6$	44 Ru $5s^14d^7$	45 Rh $5s^14d^8$	46 Pd $5s^24d^{10}$	47 Ag $5s^24d^{10}$	48 Cd $5s^24d^{10}$	49 In $5s^25p^1$	50 Sn $5s^15p^2$	51 Sb $5s^15p^3$	52 Te $5s^15p^4$	53 I $5s^15p^5$	54 Xe $5s^25p^6$
6	55 Cs $6s^1$	56 Ba $6s^2$	57 La $6s^25d^1$	72 Hf $6s^25d^2$	73 Ta $6s^25d^3$	74 W $6s^25d^5$	75 Re $6s^25d^5$	76 Os $6s^25d^6$	77 Ir $6s^25d^7$	78 Pt $6s^15d^9$	79 Au $6s^15d^{10}$	80 Hg $6s^25d^{10}$	81 Tl $6s^26p^1$	82 Pb $6s^26p^2$	83 Bi $6s^26p^3$	84 Po $6s^26p^4$	85 At $6s^26p^5$	86 Rn $6s^26p^6$
7	87 Fr $7s^1$	88 Ra $7s^2$	89 Ac $7s^26d^1$	104 Unq $7s^26d^2$	105 Unp $7s^26d^3$	106 Unh $7s^26d^4$	107 Uns $7s^26d^5$	108 Uno $7s^26d^6$	109 Une $7s^26d^7$									

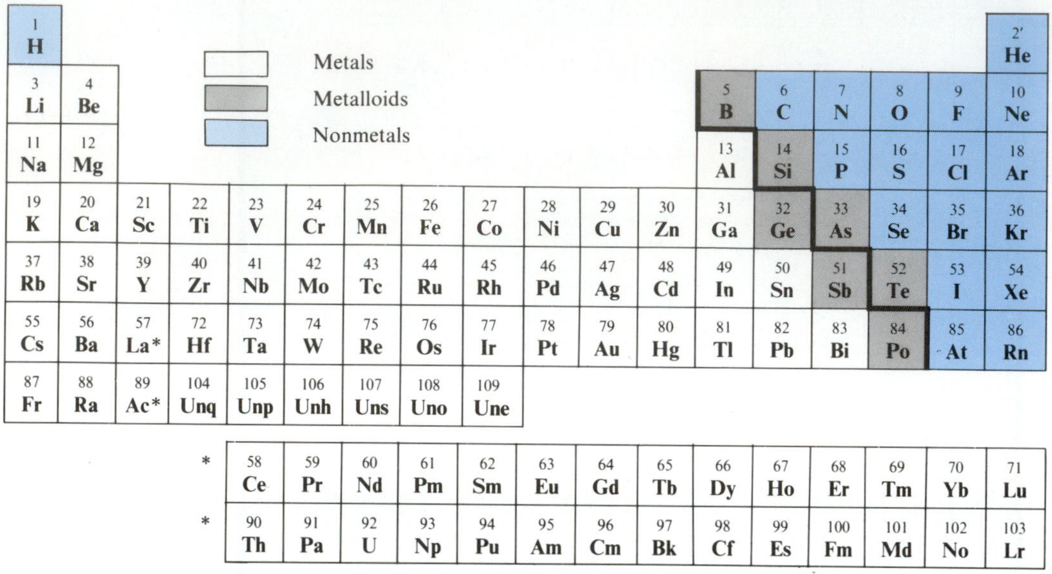

Metals

Metalloids

Nonmetals

1 H																	2 He
3 Li	4 Be											5 B	6 C	7 N	8 O	9 F	10 Ne
11 Na	12 Mg											13 Al	14 Si	15 P	16 S	17 Cl	18 Ar
19 K	20 Ca	21 Sc	22 Ti	23 V	24 Cr	25 Mn	26 Fe	27 Co	28 Ni	29 Cu	30 Zn	31 Ga	32 Ge	33 As	34 Se	35 Br	36 Kr
37 Rb	38 Sr	39 Y	40 Zr	41 Nb	42 Mo	43 Tc	44 Ru	45 Rh	46 Pd	47 Ag	48 Cd	49 In	50 Sn	51 Sb	52 Te	53 I	54 Xe
55 Cs	56 Ba	57 La*	72 Hf	73 Ta	74 W	75 Re	76 Os	77 Ir	78 Pt	79 Au	80 Hg	81 Tl	82 Pb	83 Bi	84 Po	85 At	86 Rn
87 Fr	88 Ra	89 Ac*	104 Unq	105 Unp	106 Unh	107 Uns	108 Uno	109 Une									

*	58 Ce	59 Pr	60 Nd	61 Pm	62 Sm	63 Eu	64 Gd	65 Tb	66 Dy	67 Ho	68 Er	69 Tm	70 Yb	71 Lu
*	90 Th	91 Pa	92 U	93 Np	94 Pu	95 Am	96 Cm	97 Bk	98 Cf	99 Es	100 Fm	101 Md	102 No	103 Lr

▲
FIGURE 11.1
Classification of the elements into metals, nonmetals, and metalloids.

properties on the periodic table. These trends in observed properties allow us to use the periodic table to accurately predict properties and reactions of a diversity of substances without needing to possess the substance or complete the reaction.

Metals and Nonmetals

In Section 3.8 we began our study of the periodic table by classifying elements as metals, nonmetals, or metalloids. The heavy stair-step line beginning at boron and running diagonally down the periodic table separates the elements into metals and nonmetals. Metals are usually lustrous, malleable, and good conductors of heat and electricity. Nonmetals are just the opposite—nonlustrous, brittle, and poor conductors. Metalloids are found bordering the heavy diagonal line and may have properties of both metals and nonmetals.

Most elements are classified as metals as shown in Figure 11.1. Metals are found on the left side of the stair-step line, while the nonmetals are located toward the upper right of the table. Note that hydrogen does not fit into the division of metals and nonmetals. It displays nonmetallic properties under normal conditions, even though it has only one outermost electron like the alkali metals. Hydrogen is considered to be a unique element.

It is the chemical properties of metals and nonmetals that are most interesting. Metals tend to lose electrons forming positive ions, while nonmetals tend to gain electrons forming negative ions. When a metal reacts with a nonmetal a transfer of electrons from the metal to the nonmetal frequently occurs.

Atomic Radius

The radii of the representative elements are shown in Figure 11.2. Notice that the radii of the atoms tends to increase down each group and that they tend to decrease from left to right across a period.

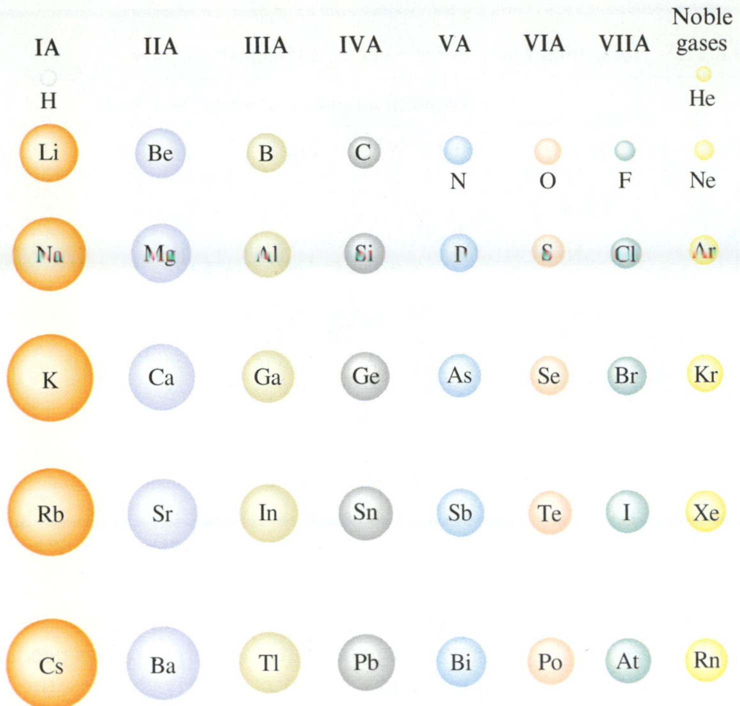

◀ **FIGURE 11.2**
Relative atomic radii for representative elements. Atomic radius decreases across a period and increases down a group.

The increase in radius down a column can be understood if we consider the electron structure of the atoms. For each step down a column an additional energy level is added to the atom. The average distance from the nucleus to the outside edge of the atom must increase as each new energy level is added. Atoms get bigger as electrons are placed in these new higher energy levels.

Understanding the decrease in atomic radius across a row requires more thought, however. As we move from left to right across a period electrons within the same block are being added to the same energy level. Since these orbitals are all approximately the same size it would seem that the atoms should be about the same size, but each time an electron is added a proton is also added to the nucleus. The increase in positive charge in the nucleus pulls the electrons closer to the nucleus, resulting in a gradual decrease in atomic radius across a row.

Ionization Energy

ionization energy

The **ionization energy** of an atom is the energy required to remove an electron from the atom. For example

$$\text{Na} + \text{ionization energy} \longrightarrow \text{Na}^+ + e^-$$

The first ionization energy is the amount of energy required to remove the first electron from an atom, the second is the amount required to remove the second electron from that atom, and so on.

Table 11.5 gives the ionization energies for the removal of one to five electrons from several elements. The table shows that even higher amounts of energy are needed to remove the second, third, fourth, and fifth electrons. This makes sense because removing electrons does not decrease the positive charge

TABLE 11.5 Ionization Energies for Selected Elements*

Element	Required amounts of energy (kJ/mol)				
	1st e^-	2nd e^-	3rd e^-	4th e^-	5th e^-
H	1,314				
He	2,372	5,247			
Li	520	7,297	11,810		
Be	900	1,757	14,845	21,000	
B	800	2,430	3,659	25,020	32,810
C	1,088	2,352	4,619	6,222	37,800
Ne	2,080	3,962	6,276	9,376	12,190
Na	496	4,565	6,912	9,540	13,355

* Values are expressed in kilojoules per mole, showing energies required to remove 1 to 5 electrons per atom. Color indicates the energy needed to remove an electron from a noble-gas electron structure.

in the nucleus; the remaining electrons are held even more tightly. The data in Table 11.5 also show that an extra large ionization energy (red) is needed when an electron is removed from a noble gas structure, clearly showing the stability of this configuration.

First ionization energies have been experimentally determined for most elements. Figure 11.3 is a graphic plot of these ionization energies for representative elements in the first four periods. Note these important points:

1. Ionization energy in Group A elements decreases from top to bottom in a group. For example in Group IA the ionization energy changes from 520 kJ/mol for Li to 419 kJ/mol for K.

2. From left to right across a period the ionization energy gradually increases. Noble gases have a relatively high value confirming the nonreactive nature of these elements and the stability of an eight-electron structure for the outermost electrons.

Why does ionization energy increase down each column? The size of the atoms increase by the addition of new outer energy levels. The distance between outermost electrons and the nucleus is therefore increases, and the electron is easier to remove. In addition to this distance there is an increasing number of completely filled energy levels that shield the outermost electron from the full attraction of the nucleus.

Moving across a period we see an increased nuclear charge as protons are added. As the electrons are pulled in more tightly it becomes more difficult to remove an electron. Ionization energy increases.

To relate ionization energy to the metals and nonmetals we must consider that a metal tends to lose electrons to form a cation. This means that metals have relatively low ionization energies. Thus, the most chemically active metals are located on the lower left of the periodic table since they are the largest elements in each group. Nonmetals behave just the opposite of metals. Nonmetals tend to gain electrons and thus have relatively high ionization energies. The most active nonmetals are found in the upper right corner of the table.

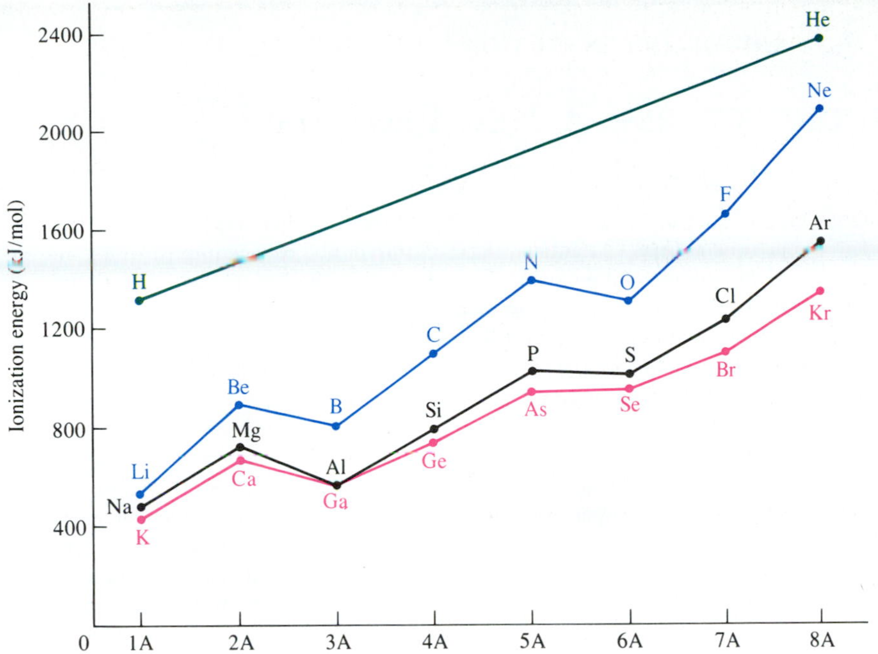

◀ **FIGURE 11.3**
Periodic relationship of the first ionization energy for representative elements in the first four periods.

11.5 VALUE OF THE PERIODIC TABLE

The arrangement of the elements into a modern periodic table has been and continues to be of great value to chemists and chemistry students. This value increases with one's knowledge of chemistry. For a given element the following data can usually be obtained directly from the table: name, symbol, atomic number, atomic mass, electron configuration, group number, period number, and whether or not the element is a metal, nonmetal, or metalloid. From the location of the element in the periodic table, such properties as ionization energy, density, atomic size, atomic volume, and oxidation states can be estimated and compared with other elements.

The periodic table is still used as guide in predicting the synthesis of possible new elements. It presents a very large amount of chemical information in compact form and correlates the properties and relationships of all the elements. The table is so useful that a copy hangs in nearly every chemistry lecture hall and laboratory in the world.

Chemistry in Action

Superconductors—A New Frontier

When electric current flows through a wire, resistance slows the current and heats the wire. In order to keep the current flowing this electrical "friction" must be overcome by adding energy to the system. In fact the existence of electrical resistance limits the efficiency of all electrical devices.

In 1911, a Dutch scientist, Heike Kamerlingh Onnes, discovered at very cold temperatures (near 0 K), electrical resistance disappears. Onnes named this phenomenon *superconductivity*. Scientists have been fascinated by it ever since. Unfortunately, because such low temperatures were required liquid helium was necessary to cool the wires. At a cost of $7 per liter, commercial applications were far too expensive to be considered.

When cooled below the critical temperature, the cube of superconductor levitates over the disk in a phenomenon known as the "Meissner effect."
▼

For many years scientists were convinced that superconductivity was not possible at higher temperatures (not even as high as 77 K, the boiling point of liquid nitrogen, a bargain at $0.17 per liter). The first high-temperature superconductor, developed in 1986, was superconducting at 30 K. This material is a complex metal oxide that has a sandwich-like crystal structure with copper and oxygen atoms on the inside and barium and lanthanum atoms on the outside.

Scientists immediately tried to develop materials that would be superconducting at even higher temperatures. To do this they relied on their knowledge of the periodic table and the properties of chemical families. Paul Chu, University of Houston, Texas, found the critical temperature could be raised by compressing the superconducting oxide. The pressure was too intense to be useful commercially, so Chu looked for another way to bring the layers closer together. He recognized that this could be accomplished by replacing the barium with strontium, an element in the same family with similar chemical properties and a smaller ionic radius. The idea was successful, the critical temperature changed from 30 K to 40 K. He then tried replacing the strontium with calcium (same family, smaller still) but to no avail. The new material had a lower critical temperature! But Chu persisted, and on January 12, 1987, by substituting yttrium for lanthanum (same family, smaller radius) he produced a new superconductor having a critical temperature of 95 K, well above the 77 K boiling point of liquid nitrogen. This material has the formula $YBa_2Cu_3O_7$ and is a good candidate for commercial applications.

There remain several barriers to cross before superconductors are in wide use. The material is brittle and easily broken, nonmalleable, and does not carry as high a current per unit cross section as conventional conductors. Many researchers are currently working to surmount these problems and develop potential uses for superconductors including high-speed levitation trains, tiny efficient electric motors, and smaller faster computers.

Concepts in Review

1. Describe briefly the contributions of Döbereiner, Newlands, Mendeleev, Meyer, and Moseley to the development of the periodic table.
2. Indicate the locations of the metals, nonmetals, metalloids, and noble gases in the periodic table.
3. Indicate in the periodic table the areas where the *s*, *p*, *d*, and *f* orbitals are being filled.
4. Describe how atomic radii vary (a) from left to right in a period, and (b) from top to bottom in a group.
5. Describe how the ionization energies of the elements vary with respect to (a) the position in the periodic table and (b) the removal of successive electrons.
6. Determine the number of valence electrons in any atom for any Group A element.
7. Distinguish between representative elements and transition elements.
8. Identify groups of elements by their special family names.
9. Describe the changes in outer-shell electron structure (a) when moving from left to right in a period, and (b) when going from top to bottom in a group.
10. Explain the relationship between group number and the number of outer-shell electrons for the representative elements.
11. Write Lewis-dot symbols for the representative elements from their position in the periodic table.

Exercises

1. From the standpoint of electron structure, what do the elements in the *s* block have in common?
2. Write the symbols for the elements having atomic numbers 8, 16, 34, 52, and 84. What do these elements have in common?
3. Mendeleev described a then-undiscovered element as eka-silicon. When this element was discovered, (a) what was it named and (b) how did the actual density agree with that predicted by Mendeleev?
4. How does the size of the atoms of the elements in the third period vary from left to right?
5. Write the symbols of the family of elements that have seven electrons in their outer energy level.
6. Write the symbols of the alkali metal family in order of increasing atomic size.
7. What is the greatest number of elements to be found in any period? Which periods have this number?
8. From the standpoint of energy level, how does the placement of the last electron differ in Group A elements from that of Group B elements?
9. What were Döbereiner's triads? In what way did they lead to later developments in periodicity?
10. What do you think is the basis for Newlands' Law of Octaves?
11. Given that the noble gases were not discovered at the time of Newlands' Law of Octaves, could his law be extended as far as the element bromine? Explain.
12. What is meant by the term *periodicity* as applied to the elements?
13. Why are some missing elements in Mendeleev's periodic table considered a victory for his table rather than a defeat?
14. How does our modern periodic table differ from Mendeleev's?

15. What additional understanding of periodic properties was added by the work of H. G. J. Moseley?
16. Explain why much more ionization energy is required to remove the first electron from neon than from sodium.
17. Explain the large increase in ionization energy needed to remove the third electron from beryllium compared with that needed for the second electron.
18. Does the first ionization energy increase or decrease from top to bottom in the periodic table for the alkali metal family? Explain.
19. Does the first ionization energy increase or decrease from top to bottom in the periodic table for the noble gas family? Explain.
20. Why does barium (Ba) have a lower ionization energy than beryllium (Be)?
21. Why is there such a large increase in the ionization energy required to remove the second electron from a sodium atom as opposed to the first?
22. Find the places in the modern periodic table where elements are not in proper sequence according to atomic mass.
23. How are elements in a period related to one another?
24. How are elements in a group related to one another?
25. What is common about the electron structures of the alkali metals?
26. Why would you expect the elements zinc, cadmium, and mercury to be in the same chemical family?
27. Draw the Lewis-dot symbols for Cs, Ba, Tl, Pb, Po, At, and Rn. How do these structures correlate with the group in which each element occurs?
28. Pick the electron structures below that represent elements in the same chemical family:
    (a) $1s^22s^1$
    (b) $1s^22s^22p^4$
    (c) $1s^22s^22p^2$
    (d) $1s^22s^22p^63s^23p^4$
    (e) $1s^22s^22p^63s^23p^6$
    (f) $1s^22s^22p^63s^23p^64s^2$
    (g) $1s^22s^22p^63s^23p^64s^1$
    (h) $1s^22s^22p^63s^23p^64s^23d^1$
29. Pick the electron structures below that represent elements in the same chemical family:
    (a) $[He]2s^22p^6$
    (b) $[Ne]3s^1$
    (c) $[Ne]3s^2$
    (d) $[Ne]3s^23p^3$
    (e) $[Ar]4s^13d^{10}$
    (f) $[Ar]4s^23d^{10}4p^6$
    (g) $[Ar]4s^23d^5$
    (h) $[Kr]5s^14d^{10}$
30. In the periodic table, calcium, element 20, is surrounded by elements 12, 19, 21, and 38. Which of these have physical and chemical properties most resembling calcium?
31. Oxygen is a gas. Sulfur is a solid. What is it about their electron structures that causes them to be grouped in the same chemical family?
32. Classify each of the following elements as metals, nonmetals, or metalloids:
    (a) Potassium
    (b) Plutonium
    (c) Sulfur
    (d) Antimony
    (e) Iodine
    (f) Tungsten
    (g) Molybdenum
    (h) Germanium
33. In which period and group does an electron first appear in a *d* orbital?
34. How many electrons occur in the outer shell of Group IIIA and IIIB elements? Why are they different?
35. In which groups are transition elements located?
36. How do the electron structures of the transition elements differ from representative elements?
37. Which element in each of the following pairs has the larger atomic radius?
    (a) Na or K (c) O or F (e) Ti or Zr
    (b) Na or Mg (d) Br or I
38. Which element in each of Groups IA–VIIA has the smallest atomic radius?
39. Why does the atomic size increase in going down any family of the periodic table?
40. All the atoms within each Group A family of elements can be represented by the same Lewis-dot symbol. Complete the table, expressing the Lewis-dot symbol for each group. Use E to represent the elements.

Group	IA	IIA	IIIA	IVA	VA	VIA	VIIA
	E·						

41. Why should the discovery of the existence of isotopes have any bearing on the fact that the periodicity of the elements is a function of their atomic numbers and not their atomic masses?

Try to answer Exercises 42–44 without referring to the periodic table.

42. The atomic numbers of the noble gases are 2, 10, 18, 36, 54, and 86. What are the atomic numbers for the elements with six electrons in their outer electron shells?

43. Element number 87 is in Group IA, period 7. Describe its outermost energy level. How many energy levels of electrons does it have?

44. If element 36 is a noble gas, in which groups would you expect elements 35 and 37 to occur?

45. Write a paragraph describing the general features of the periodic table.

46. Rank the following five elements according to the radii of their atoms, from smallest to largest: Na, Mg, Cl, K, and Rb.

47. Which of the following statements are correct? Rewrite the incorrect statements to make them correct.

(a) Properties of the elements are periodic functions of their atomic numbers.

(b) There are more nonmetallic elements than metallic elements.

(c) Metallic properties of the elements increase from left to right across a period.

(d) Metallic properties of the elements increase from top to bottom in a family of elements.

(e) Calcium is a member of the alkaline earth metal family.

(f) Iron belongs to the alkali metal family.

(g) Bromine belongs to the halogen family.

(h) Neon is a noble gas.

(i) Group A elements do not contain partially filled d or f sublevels.

(j) An atom of oxygen is larger than an atom of lithium.

(k) An atom of sulfur is larger than an atom of oxygen.

(l) An atom of aluminum (Group IIIA) has five electrons in its outer shell.

(m) The amount of energy required to remove one electron from an atom is known as the ionization energy.

(n) Elements with a high ionization energy tend to have very metallic properties.

(o) Uranium is an inner transition element.

(p) The element $[Ar]4s^2 3d^{10} 4p^5$ is a halogen.

(q) The element $[Kr]5s^2$ is a nonmetal.

(r) The element with $Z = 12$ forms compounds similar to the element with $Z = 37$.

(s) Nitrogen, fluorine, neon, gallium, and bromine are all nonmetals.

(t) The Lewis-dot symbol for tin is $:\dot{Sn}\cdot$

(u) The atom having an outer-shell electron structure of $5s^2 5p^2$ would be in period 6, Group IVA.

(v) The yet-to-be-discovered element with an atomic number of 118 would be a noble gas.

(w) The most metallic element of Group VIIA is iodine.

48. What is meant by the term "superconductor"?

49. What is the important relationship between 77 K and superconductivity?

50. What was the relationship on the periodic table that enabled Chu to find a high temperature superconductor?

51. Indicate the limitations of current superconducting material.

12

Chemical Bonds: The Formation of Compounds from Atoms

◄ **Chapter Opening Photo: A variety of types of chemical bonds are necessary to produce the beautiful coral found in reefs.**

For centuries, we have been aware that certain metals cling to a magnet. Balloons may stick to a wall. Why? Recently, the activity of a superconductor floating in air has been the subject of television commercials. High-speed levitation trains are heralded to be the wave of the future. How do they function? In each case, forces of attraction and repulsion are at work.

Interestingly, human interactions suggest that "opposites attract" and "likes repel." Attractions draw us into friendships and significant relationships, whereas repulsive forces may produce debate and antagonism. We form and break apart interpersonal bonds throughout our lives.

In chemistry, we also see this phenomenon, Substances form chemical bonds as a result of electrical attractions. Chemical bonds provide the tremendous diversity of compounds found in nature.

12.1 Chemical Bonds

Except in very rare instances matter does not fly apart spontaneously. It is prevented from doing so by forces acting at the ionic and molecular levels. Through chemical reactions atoms tend to attain more stable states at lower chemical potential energy levels. Atoms react chemically by losing, gaining, or sharing electrons. Forces arise from electron transferring and electron sharing interactions. Those forces that hold oppositely charged ions together or that bind atoms together in molecules are called **chemical bonds**. The two principal types of bonds are the ionic bond and the covalent bond.

chemical bonds

12.2 Electrons in the Outer Shell: Valence Electrons

One outstanding property of the elements is their tendency to form a stable outer-shell electron structure. For many elements this stable outer shell contains eight electrons (two *s* and six *p*) identical to the outer-shell electron structure of the noble gases. Atoms undergo rearrangements of electron structure to attain a state of greater stability. These rearrangements are accomplished by losing, gaining, or sharing electrons with other atoms. For example, a hydrogen atom has a tendency to accept another electron and thus attain an electron structure like that of the stable noble gas helium; a fluorine atom can acquire one more electron to attain a stable electron structure like neon; a sodium atom tends to lose one electron to attain a stable electron structure like neon.

$$Na + energy \longrightarrow Na^+ + 1\ e^-$$

valence electrons

The electrons in the outermost shell of an atom are responsible for most of this electron activity and are called the **valence electrons**. In Lewis-dot symbols of atoms, the dots represent the outer-shell electrons and thus also represent the valence electrons. For example, hydrogen has one valence electron; sodium, one; aluminum, three; and oxygen, six. When a rearrangement of these electrons takes place between atoms, a chemical change occurs.

H·	$\ddot{\text{Al}}$·	·$\ddot{\text{O}}$:
One valence electron	Three valence electrons	Six valence electrons

12.3 The Ionic Bond: Transfer of Electrons from One Atom to Another

The chemistry of the elements, especially the representative ones, is to attain an outer-shell electron structure like that of the chemically stable noble gases. With the exception of helium, this stable structure consists of eight electrons in the outer shell (see Table 10.5).

Let us look at the electron structures of sodium and chlorine to see how each element can attain a structure of 8 electrons in its outer shell. A sodium atom has 11 electrons: 2 in the first energy level, 8 in the second energy level, and 1 in the third energy level. A chlorine atom has 17 electrons: 2 in the first energy level, 8 in the second energy level, and 7 in the third energy level. If a sodium atom transfers or loses its 3*s* electron, its third energy level becomes vacant, and it becomes a sodium ion with an electron configuration identical to that of the noble gas neon. This process absorbs energy.

11+ $2e^-$ 8^- $1e^-$ ⟶ [11+ $2e^-$ 8^-]$^{1+}$ + $1e^-$

Na atom ($1s^22s^22p^63s^1$) Na$^+$ ion ($1s^22s^22p^6$)

An atom that has lost or gained electrons will have a plus or minus electric charge, depending on which charged particles, protons or electrons, are in excess. Remember that a charged atom or group of atoms is called an *ion*.

By losing a negatively charged electron, the sodium atom becomes a positively charged particle known as a sodium ion. The charge, +1, results because the nucleus still contains 11 positively charged protons, and the electron orbitals contain only 10 negatively charged electrons. The charge is indicated by a plus sign (+) and is written as a superscript after the symbol of the element (Na^+).

A chlorine atom with 7 electrons in the third energy level needs 1 electron to pair up with its one unpaired 3*p* electron to attain the stable outer-shell electron structure of argon. By gaining 1 electron the chlorine atom becomes a chloride ion (Cl^-), a negatively charged particle containing 17 protons and 18 electrons. This process releases energy.

$$17^+\ 2e^-\ 8e^-\ 7e^- + 1e^- \longrightarrow \left[17^+\ 2e^-\ 8e^-\ 8e^-\right]^{1-}$$

Cl atom ($1s^22s^22p^63s^23p^5$) Cl^- ion ($1s^22s^22p^63s^23p^6$)

Consider the case in which sodium and chlorine atoms react with each other. The $3s$ electron from the sodium atom transfers to the half-filled $3p$ orbital in the chlorine atom to form a positive sodium ion and a negative chloride ion. The compound sodium chloride results because the Na^+ and Cl^- ions are strongly attracted to each other by their opposite electrostatic charges. The force holding the oppositely charged ions together is an ionic bond.

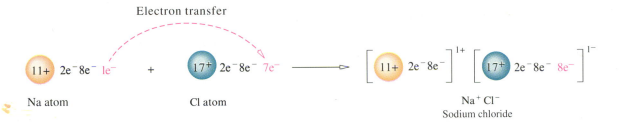

Electron transfer

$$11+\ 2e^-\ 8e^-\ 1e^- + 17^+\ 2e^-\ 8e^-\ 7e^- \longrightarrow \left[11+\ 2e^-\ 8e^-\right]^{1+}\left[17^+\ 2e^-\ 8e^-\ 8e^-\right]^{1-}$$

Na atom Cl atom Na^+Cl^- Sodium chloride

The Lewis-dot representation of sodium chloride formation is shown below.

$$Na\cdot + \cdot\ddot{\underset{..}{Cl}}: \longrightarrow [Na]^+\left[:\ddot{\underset{..}{Cl}}:\right]^-$$

The chemical reaction between sodium and chlorine is a very vigorous one, producing considerable heat in addition to the salt formed. When energy is released in a chemical reaction, the products are more stable than the reactants. Note that in NaCl both atoms attained a noble-gas electron structure.

Sodium chloride is made up of cubic crystals in which each sodium ion is surrounded by six chloride ions and each chloride ion by six sodium ions, except at the crystal surface. It is helpful to remember that a cation is always smaller than its parent atom whereas an anion is always larger than its parent atom. A visible crystal is a regularly arranged aggregate of millions of these ions, but the ratio of sodium to chloride ions is one-to-one, hence the formula NaCl. The cubic crystalline lattice arrangement of sodium chloride is shown in Figure 12.1.

Figure 12.2 contrasts the relative sizes of sodium and chlorine atoms with those of their ions. The sodium ion is smaller than the atom due primarily to two factors: (1) The sodium atom has lost its outer shell of 1 electron, thereby reducing its size; and (2) the 10 remaining electrons are now attracted by 11 protons and are thus drawn closer to the nucleus. Conversely the chloride ion is larger than the atom because it has 18 electrons but only 17 protons; the nuclear attraction on each electron is thereby decreased, allowing the chlorine atom to expand as it forms an ion.

We have seen that when sodium reacts with chlorine, each atom becomes an electrically charged ion. Sodium chloride, like all ionic substances, is held together by the attraction existing between positive and negative charges. An **ionic bond** is the attraction between oppositely charged ions.

ionic bond

Ionic bonds are formed whenever one or more electrons are transferred from one atom to another. The metals, which have relatively little attraction for their valence electrons, tend to form ionic bonds when they combine with nonmetals.

It is important to recognize that substances with ionic bonds do not exist as molecules. In sodium chloride, for example, the bond does not exist solely between a single sodium ion and a single chloride ion. Each sodium ion in the crystal attracts six near-neighbor negative chloride ions; in turn, each negative chloride ion attracts six near-neighbor positive sodium ions (see Figure 12.1).

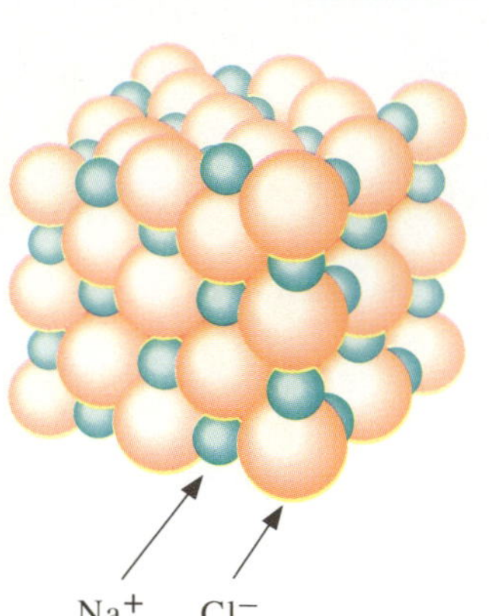

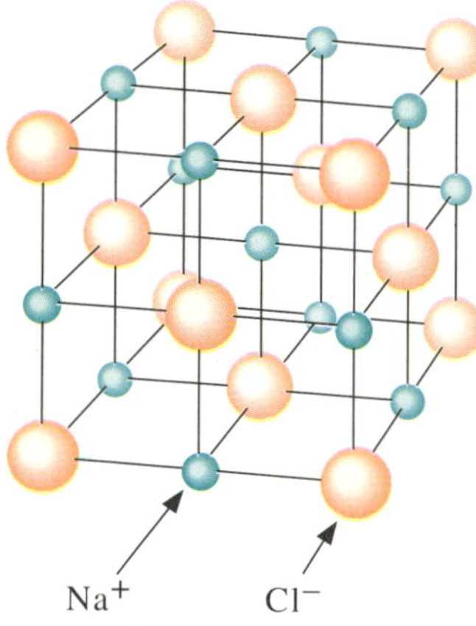

▲ FIGURE 12.1
Sodium chloride crystal. Diagram represents a small fragment of sodium chloride, which forms cubic crystals. Each sodium ion is surrounded by six chloride ions, and each chloride ion is surrounded by six sodium ions.

A metal will usually have one, two, or three electrons in its outer energy level. In reacting, metal atoms characteristically lose these electrons, attain the electron structure of a noble gas, and become positive ions. A nonmetal, on the other hand, is only a few electrons short of having a complete octet in its outer energy level and thus has a tendency to gain electrons. In reacting with metals, nonmetal atoms characteristically gain one, two or three electrons, attain the electron structure of a noble gas, and become negative ions. The ions formed by loss of electrons are much smaller than the corresponding metal atoms; the ions formed by gaining electrons are larger than the corresponding nonmetal atoms. The actual dimensions of the atomic and ionic radii of several metals and nonmetals are given in Table 12.1.

Study the following examples. Note the loss and gain of electrons between atoms; also note that the ions in each compound have a noble-gas electron structure.

TABLE 12.1 Change in Atomic Radii of Selected Metals and Nonmetals*

Atomic radius (nm)	Ionic radius (nm)	Atomic radius (nm)	Ionic radius (nm)
Li 0.152	Li^+ 0.060	F 0.071	F^- 0.136
Na 0.186	Na^+ 0.095	Cl 0.099	Cl^- 0.181
K 0.227	K^+ 0.133	Br 0.114	Br^- 0.195
Mg 0.160	Mg^{2+} 0.065	O 0.074	O^{2-} 0.140
Al 0.143	Al^{3+} 0.050	S 0.103	S^{2-} 0.184

* The metals shown lose electrons to become positive ions. The nonmetals gain electrons to become negative ions.

FIGURE 12.2 ▶
Relative radii of sodium and chlorine atoms and their ions.

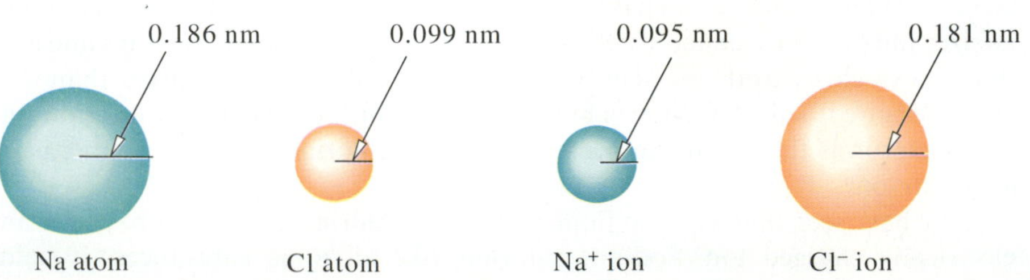

EXAMPLE 12.1

Explain how magnesium and chlorine combine to form magnesium chloride, $MgCl_2$.

SOLUTION

A magnesium atom of electron structure $1s^2 2s^2 2p^6 3s^2$ must lose two electrons or gain six electrons to reach a stable electron structure. If magnesium reacts with chlorine and each chlorine atom can accept only one electron, two chlorine atoms will be needed for the two electrons from one magnesium atom. The compound formed will contain one magnesium ion and two chloride ions. The magnesium atom, having lost two electrons, becomes a magnesium ion with a +2 charge. Each chloride ion will have a −1 charge. The transfer of electrons from a magnesium atom to two chlorine atoms is shown in the following illustration.

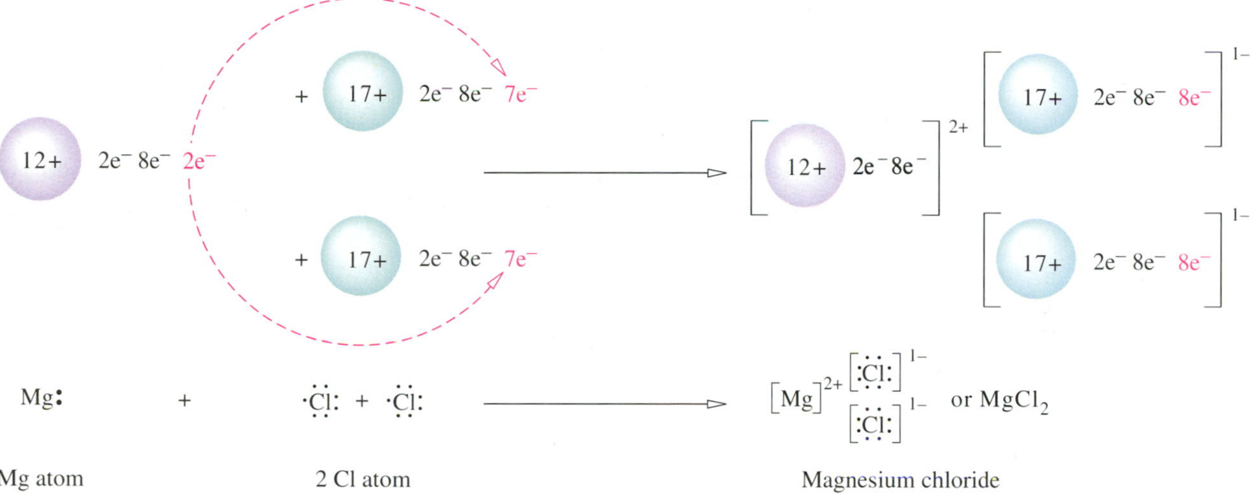

EXAMPLE 12.2

Explain the formation of sodium fluoride, NaF, from its elements.

SOLUTION

11+ 2e⁻ 8e⁻ 1e⁻ + 9+ 2e⁻ 7e⁻ ⟶ [11+ 2e⁻ 8e⁻]$^{1+}$ [9+ 2e⁻ 8e⁻]$^{1-}$

Na·	+	·F:	⟶	$[Na]^+ [:F:]^-$
Sodium atom		Fluorine atom		Sodium fluoride

The fluorine atom, with seven electrons in its outer shell, behaves similarly to the chlorine atom.

EXAMPLE 12.3

Explain the formation of aluminum chloride, $AlCl_3$, from its elements.

SOLUTION

$$\dot{\text{Al}}\cdot + 3\,\cdot\text{Cl}: \longrightarrow [\text{Al}]^{3+}\,[:\text{Cl}:]^-\,[:\text{Cl}:]^-\,[:\text{Cl}:]^- \quad \text{or} \quad \text{AlCl}_3$$

Aluminum atom	Chlorine atoms	Aluminum chloride

Each chlorine atom can accept only one electron. Therefore three chlorine atoms are needed to combine with the three outer-shell electrons of one aluminum atom. The aluminum atom has lost three electrons to become an aluminum ion, Al^{3+}, with a +3 charge.

EXAMPLE 12.4 Explain the formation of magnesium oxide, MgO, from its elements.

SOLUTION

12+ 2e⁻ 8e⁻ 2e⁻ + 8+ 2e⁻ 6e⁻ ⟶ [12+ 2e⁻ 8e⁻]$^{2+}$ [8+ 2e⁻ 6e⁻]$^{2-}$

Mg: + ·Ö: ⟶ [Mg]$^{2+}$[:Ö:]$^{2-}$

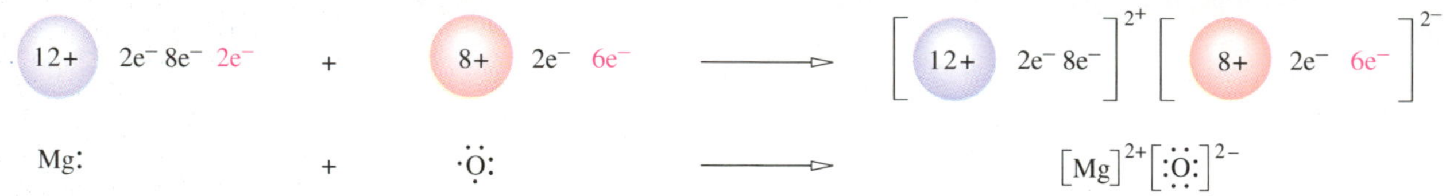

Magnesium atom Oxygen atom Magnesium oxide

The magnesium atom, with two electrons in the outer energy level, exactly fills the need of one oxygen atom for two electrons. The resulting compound has a ratio of one atom of magnesium to one atom of oxygen. The oxygen (oxide) ion has a −2 charge, having gained two electrons. In combining with oxygen, magnesium behaves the same way as when combining with chlorine; it loses two electrons.

EXAMPLE 12.5 Explain the formation of sodium sulfide, Na_2S, from its elements.

SOLUTION

Na· + Na· + ·S̈: ⟶ [Na]$^+$ [Na]$^+$ [:S̈:]$^{2-}$ or Na_2S

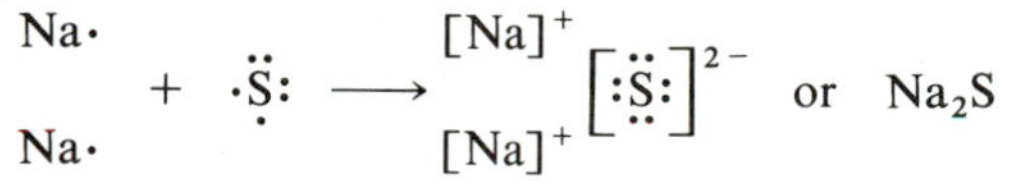

Sodium atoms Sulfur atom Sodium sulfide

Two sodium atoms supply the electrons that one sulfur atom needs to make eight in its outer shell.

EXAMPLE 12.6 Explain the formation of aluminum oxide, Al_2O_3, from its elements.

SOLUTION

Äl· Äl· + ·Ö: ·Ö: ·Ö: ⟶ [Al]$^{3+}$ [Al]$^{3+}$ [:Ö:]$^{2-}$ [:Ö:]$^{2-}$ [:Ö:]$^{2-}$ or Al_2O_3

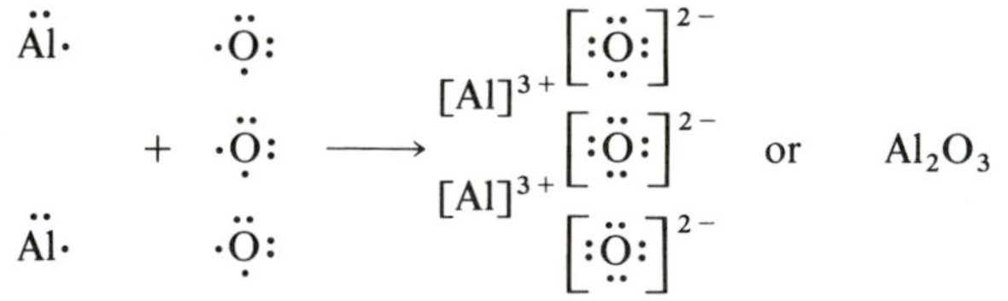

Aluminum atoms Oxygen atoms Aluminum oxide

One oxygen atom, needing two electrons, cannot accommodate the three electrons from one aluminum atom. One aluminum atom falls one electron short of the four electrons needed by two oxygen atoms. A ratio of two atoms of aluminum to three atoms of oxygen, involving the transfer of six electrons (two to each oxygen atom), gives each atom a stable electron configuration.

Note that in each of the examples above, outer shells containing eight electrons were formed in all the negative ions. This formation resulted from the pairing of all the *s* and *p* electrons in these outer shells.

12.4 Predicting Formulas of Ionic Compounds

In the previous examples we have seen that when a metal and a nonmetal react to form an ionic compound, the metal loses one or more electrons to the nonmetal. In Chapter 6, where we learned to name compounds and write formulas, we saw that Group IA metals always formed +1 cations, whereas Group IIA formed +2 cations. Group VIIA elements formed −1 anions and Group VIA formed −2 anions.

We now realize that this pattern is directly related to the stability of the noble gas configuration. Metals lose electrons to attain the electron configuration of a noble gas (the previous one on the periodic table). Nonmetals form ions by gaining enough electrons to achieve the electron configuration of the noble gas following it on the periodic table. These observations lead us to an important chemical principle: In almost all stable chemical compounds of representative elements, each atom attains a noble gas electron configuration. This is the central concept forming the basis for our understanding of chemical bonding.

We can apply this principle in predicting the formulas of ionic compounds. To predict the formula of an ionic compound we must recognize that chemical compounds are always electrically neutral. In addition, the metal will lose electrons to achieve noble gas configuration and the nonmetal will gain electrons to a noble gas configuration. Consider the compound formed between magnesium and oxygen. Magnesium has two valence electrons whereas oxygen has six valence electrons:

Mg [Ne] $3s^2$
O [He] $2s^2 2p^4$

If magnesium loses two electrons it will achieve the configuration of Ne. By gaining two electrons oxygen achieves the configuration of Ne. Consequently a pair of electrons is transferred between atoms. Now we have Mg^{2+} and O^{2-}. Since compounds are electrically neutral there must be a ratio of one Mg to one O and the empirical formula is MgO.

The same principle works for many other cases. Since the key to the principle lies in the electron configuration, the periodic table can be used to extend the prediction even further. Because of similar electron structures, the elements in a family generally form compounds with the same atomic ratios. In general, if we know the atomic ratio of a particular compound, say sodium chloride (NaCl), we can predict the atomic ratios and formulas of the other alkali metal chlorides. These formulas are LiCl, KCl, RbCl, CsCl, and FrCl (see Table 12.2).

In a similar way, if we know that the formula of the oxide of hydrogen is H_2O, we can predict that the formula of the sulfide will be H_2S, because sulfur has the same outer-shell electron structure as oxygen. It must be recognized, however, that these are only predictions; it does not necessarily follow that every element in a group will behave like the others or even that a predicted compound will actually exist. Knowing the formulas for potassium chlorate, bromate, and iodate to be $KClO_3$, $KBrO_3$, and KIO_3, we can correctly predict the corresponding

TABLE 12.2 Formulas of Compounds Formed by Alkali Metals

Lewis-dot structure	Monoxides	Chlorides	Bromides	Sulfates
Li·	Li_2O	LiCl	LiBr	Li_2SO_4
Na·	Na_2O	NaCl	NaBr	Na_2SO_4
K·	K_2O	KCl	KBr	K_2SO_4
Rb·	Rb_2O	RbCl	RbBr	Rb_2SO_4
Cs·	Cs_2O	CsCl	CsBr	Cs_2SO_4

sodium compounds to have the formulas $NaClO_3$, $NaBrO_3$, and $NaIO_3$. Fluorine belongs to the same family of elements (Group VIIA) as chlorine, bromine, and iodine. So we can predict that potassium and sodium fluorates will have the formulas KFO_3 and $NaFO_3$. But this prediction would not be correct, because potassium and sodium fluorates are not known to exist. However, if they did exist, the formulas could very well be correct, for these predictions are based on comparisons with known formulas and similar electron structures.

In our discussion we refer only to representative metals (Groups IA, IIA, IIIA). The transition metals (Group B) show more complicated behavior (they form multiple ions) and their formulas are not as easily predicted.

EXAMPLE 12.7

The formula for calcium sulfide is CaS and that for lithium phosphide is Li_3P. Predict formulas for (a) magnesium sulfide, (b) potassium phosphide and (c) magnesium selinide.

SOLUTION

(a) Look in the periodic table for calcium and magnesium. They are both in Group IIA. Since the formula for calcium sulfide is CaS, it is reasonable to predict that the formula for magnesium sulfide is MgS.

(b) Find lithium and potassium in the periodic table. They are in Group IA. Since the formula for lithium phosphide is Li_3P, it is reasonable to predict that K_3P is the formula for potassium phosphide.

(c) Find selenium in the periodic table. It is in Group VIA just below sulfur. Therefore it is reasonable to assume that selenium forms selenide in the same way that sulfur forms sulfide. Since MgS_4 was the predicted formula for magnesium sulfide in part (a), it is reasonable to assume that the formula for magnesium selenide is $MgSe_4$.

12.5 THE COVALENT BOND: SHARING ELECTRONS

Some atoms do not transfer electrons from one atom to another to form ions. Instead they form a chemical bond by sharing pairs of electrons between them.

covalent bond

A **covalent bond** consists of a pair of electrons shared between two atoms. This bonding concept was introduced in 1916 by G. N. Lewis of the University of California at Berkeley. In the millions of compounds that are known, the covalent bond is the predominant chemical bond.

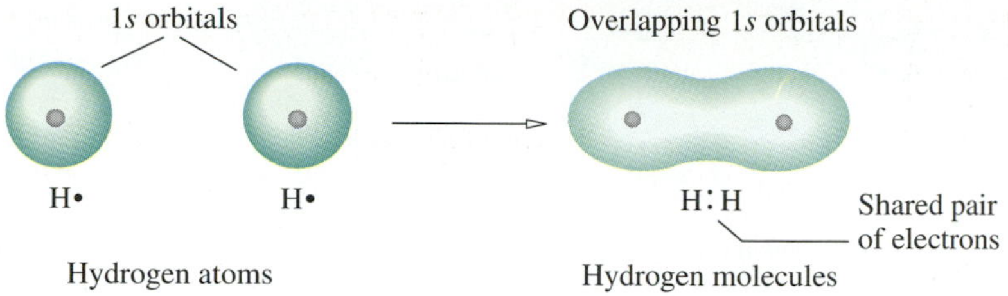

◀ **FIGURE 12.3**
The formation of a hydrogen molecule from two hydrogen atoms. The two 1s orbitals overlap, forming the H_2 molecule. In this molecule the two electrons are shared between the atoms, forming a covalent bond.

True molecules exist in substances in which the atoms are covalently bonded. It is proper to refer to molecules of such substances as hydrogen, chlorine, hydrogen chloride, carbon dioxide, water, or sugar. These substances contain only covalent bonds and exist as aggregates of molecules. We do not use the term *molecule* when talking about ionically bonded compounds such as sodium chloride, because such substances exist as large aggregates of positive and negative ions, not as molecules.

A study of the hydrogen molecule will give an insight into the nature of the covalent bond and its formation. The formation of a hydrogen molecule, H_2, involves the overlapping and pairing of 1*s* electron orbitals from two hydrogen atoms. This overlapping and pairing is shown in Figure 12.3. Each atom contributes one electron of the pair that is shared jointly by two hydrogen nuclei. The orbital of the electrons now includes both hydrogen nuclei, but probability factors show that the most likely place to find the electrons (the point of highest electron density) is between the two nuclei. The two nuclei are shielded from each other by the pair of electrons, allowing the two nuclei to be drawn very close to each other. (The average bond length between the hydrogen nuclei is 7.4×10^{-9} cm.)

The tendency for hydrogen atoms to form a molecule is very strong. In the molecule, each electron is attracted by two positive nuclei. This attraction gives the hydrogen molecule a more stable structure than the individual hydrogen atoms had. Energy is released when a bond forms between two atoms. Consequently, the same amount of energy is needed to break that bond. Experimental evidence of stability is shown by the fact that 436 kJ are needed to break the bonds between the hydrogen atoms in 1 mole of hydrogen molecules. The strength of a bond may be determined by the energy required to break it. The energy required to break a covalent bond is known as the *bond dissociation energy*. The following bond dissociation energies illustrate relative bond strengths. (All substances are considered to be in the gaseous state and to form neutral atoms.)

Reaction	Bond dissociation energy (kJ/mol)
$H_2 \longrightarrow 2\,H$	436
$N_2 \longrightarrow 2\,N$	946
$O_2 \longrightarrow 2\,O$	595
$F_2 \longrightarrow 2\,F$	153
$Cl_2 \longrightarrow 2\,Cl$	243
$Br_2 \longrightarrow 2\,Br$	193
$I_2 \longrightarrow 2\,I$	151

▲
Gilbert N. Lewis (1875–1946)

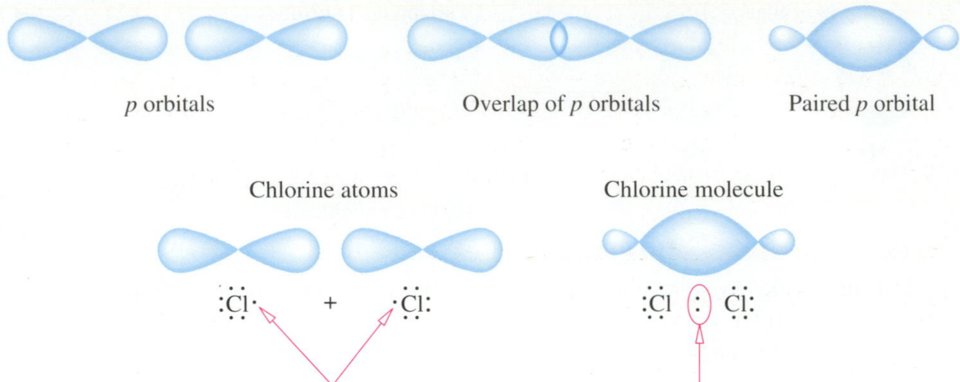

FIGURE 12.4 ▶ Pairing of *p* electrons in the formation of a chlorine molecule.

The formula for chlorine gas is Cl_2. When the two atoms of chlorine combine to form this molecule, the electrons must interact in a manner that is similar to that shown in the preceding example. Each chlorine atom would be more stable with eight electrons in its outer shell. But chlorine atoms are identical, and neither is able to pull an electron away from the other. What happens is this: The unpaired 3*p* electron orbital of one chlorine atom overlaps the unpaired 3*p* electron orbital of the other atom, resulting in a pair of electrons that are mutually shared between the two atoms. Each atom furnishes one of the pair of shared electrons. Thus, each atom attains a stable structure of eight electrons by sharing an electron pair with the other atom. The pairing of the *p* electrons and formation of a chlorine molecule are illustrated in Figure 12.4. Neither chlorine atom has a positive or negative charge, since both contain the same number of protons and have equal attraction for the pair of electrons being shared. Other examples of molecules in which electrons are equally shared between two atoms are hydrogen, H_2; oxygen, O_2; nitrogen, N_2; fluorine, F_2; bromine, Br_2; and iodine, I_2. Note that more than one pair of electrons may be shared between atoms.

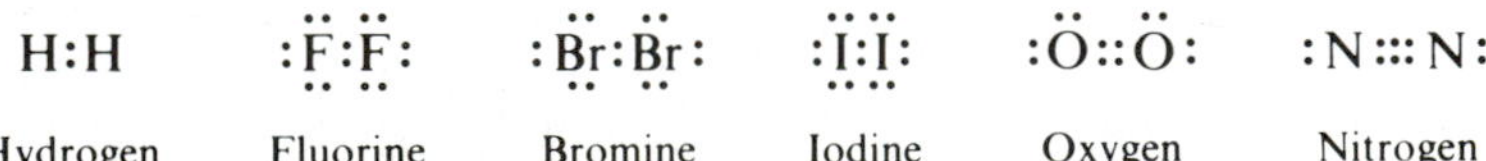

The Lewis structure given for oxygen does not adequately account for all the properties of the oxygen molecule. Other theories explaining the bonding in oxygen molecules have been advanced; however, they are complex and beyond the scope of this book.

A common practice in writing structures is to replace the pair of dots used to represent a shared pair of electrons by a dash (—). But you must remember that a dash represents a shared pair of electrons. One dash represents a single bond; two dashes, a double bond; and three dashes, a triple bond. The six structures just shown may be written thus:

H—H :F̤̈—F̤̈: :B̤̈r—B̤̈r: :Ï̤—Ï̤: :Ö=Ö: :N≡N:

The ionic bond and the covalent bond represent two extremes. In ionic bonding the atoms are so different that electrons are transferred between them forming a charged pair of ions. In covalent bonding two identical atoms share electrons equally. The bond is the mutual attraction of the two nuclei for the shared electrons. Between these extremes lie many cases in which the electronegativity of the atoms are not different enough for a transfer of electrons, but are different enough so the electrons cannot be shared equally. This unequal sharing of electrons results in the formation of a **polar covalent bond**.

polar covalent bond

12.6 Electronegativity

When two different kinds of atoms share a pair of electrons, one atom assumes a partial positive charge and the other a partial negative charge with respect to each other. This difference in charge occurs because the two atoms exert unequal attraction for the pair of shared electrons. The attractive force that an atom of an element has for shared electrons in a molecule is known as its **electronegativity**. Elements differ in their electronegativities. For example, both hydrogen and chlorine need one electron to form stable electron configurations. They share a pair of electrons in hydrogen chloride, HCl. Chlorine is more electronegative and therefore has a greater attraction for the shared electrons than does hydrogen. As a result the pair of electrons is displaced toward the chlorine atom, giving it a partial negative charge and leaving the hydrogen atom with a partial positive charge. It should be understood that the electron is not transferred entirely to the chlorine atom, as in the case of sodium chloride, and no ions are formed. The entire molecule, HCl, is electrically neutral. A partial charge is usually indicated by the Greek letter delta, δ. Thus, a partial positive charge is represented by $\delta+$ and a partial negative charge by $\delta-$.

electronegativity

Hydrogen chloride

The pair of shared electrons in HCl is closer to the more electronegative chlorine atom than to the hydrogen atom, giving chlorine a partial negative charge with respect to the hydrogen atom.

The electronegativity, or ability of an atom to attract an electron, depends on several factors: (1) the charge on the nucleus, (2) the distance of the outer electrons from the nucleus, and (3) the amount the nucleus is shielded from the outer-shell electrons by intervening shells of electrons. A scale of relative electronegativities, in which the most electronegative element, fluorine, is assigned a value of 4.0, was developed by the Nobel laureate (1954 and 1962) Linus Pauling (b. 1901). Table 12.3 shows that the relative electronegativity of the nonmetals is high and that of the metals is low. These electronegativities indicate that atoms of metals have a greater tendency to lose electrons than do atoms of nonmetals and that nonmetals have a greater tendency to gain electrons than do metals. The higher the electronegativity value, the greater the attraction for electrons. Notice that electronegativity generally increases from left to right across a period and decreases down a group for the representative elements. The highest electronegativity is 4.0 for fluorine, and the lowest is 0.7 for francium and cesium. It is

▲ Nobel laureate Linus Pauling (1901–   ).

TABLE 12.3 Relative Electronegativity of the Elements*

Key: 9 — Atomic number; F — Symbol; 4.0 — Electronegativity

1 **H** 2.1																	2 **He**
3 **Li** 1.0	4 **Be** 1.5											5 **B** 2.0	6 **C** 2.5	7 **N** 3.0	8 **O** 3.5	9 **F** 4.0	10 **Ne**
11 **Na** 0.9	12 **Mg** 1.2											13 **Al** 1.5	14 **Si** 1.8	15 **P** 2.1	16 **S** 2.5	17 **Cl** 3.0	18 **Ar**
19 **K** 0.8	20 **Ca** 1.0	21 **Sc** 1.3	22 **Ti** 1.4	23 **V** 1.6	24 **Cr** 1.6	25 **Mn** 1.5	26 **Fe** 1.8	27 **Co** 1.8	28 **Ni** 1.8	29 **Cu** 1.9	30 **Zn** 1.6	31 **Ga** 1.6	32 **Ge** 1.8	33 **As** 2.0	34 **Se** 2.4	35 **Br** 2.8	36 **Kr**
37 **Rb** 0.8	38 **Sr** 1.0	39 **Y** 1.2	40 **Zr** 1.4	41 **Nb** 1.6	42 **Mo** 1.8	43 **Tc** 1.9	44 **Ru** 2.2	45 **Rh** 2.2	46 **Pd** 2.2	47 **Ag** 1.9	48 **Cd** 1.7	49 **In** 1.7	50 **Sn** 1.8	51 **Sb** 1.9	52 **Te** 2.1	53 **I** 2.5	54 **Xe**
55 **Cs** 0.7	56 **Ba** 0.9	57–71 **La–Lu** 1.1–1.2	72 **Hf** 1.3	73 **Ta** 1.5	74 **W** 1.7	75 **Re** 1.9	76 **Os** 2.2	77 **Ir** 2.2	78 **Pt** 2.2	79 **Au** 2.4	80 **Hg** 1.9	81 **Tl** 1.8	82 **Pb** 1.8	83 **Bi** 1.9	84 **Po** 2.0	85 **At** 2.2	86 **Rn**
87 **Fr** 0.7	88 **Ra** 0.9	89–103 **Ac–Lr** 1.1–1.7	104 **Unq**	105 **Unp**	106 **Unh**	107 **Uns**	108 **Uno**	109 **Une**									

* The electronegativity value is given below the symbol of each element.

important to remember that the higher the electronegativity the stronger an atom holds electrons.

nonpolar

The polarity of a bond is determined by the difference in electronegativity values of the atoms forming the bond. See Figure 12.5. If the electronegativities are the same the bond is **nonpolar** and the electrons are shared equally. If the atoms have greatly different electronegativities the bond is very polar. At the extreme one or more electrons are actually transferred and an ionic bond results. In general, if the difference in electronegativity values is greater than 1.7–1.9 the bond is classified as ionic.

dipole

A **dipole** is a molecule that is electrically asymmetrical, causing it to be oppositely charged at two points. A dipole is often written as (+ −). A hydrogen chloride molecule is polar and behaves as a small dipole. The HCl dipole may be written as H +⟶ Cl. The arrow points toward the negative end of the dipole. Molecules of H_2O, HBr, and ICl are polar.

$$\overset{\delta+}{\text{H}} \;{+\!\!\longrightarrow}\; \overset{\delta-}{\text{Cl}} \qquad \overset{\delta+}{\text{H}} \;{+\!\!\longrightarrow}\; \overset{\delta-}{\text{Br}} \qquad \overset{\delta+}{\text{I}} \;{+\!\!\longrightarrow}\; \overset{\delta-}{\text{Cl}}$$

$$\overset{\delta-}{\text{O}} \quad \overset{\delta+}{\text{H}} \nearrow \text{O} \nwarrow \overset{\delta+}{\text{H}}$$

How do we know whether a bond between two atoms is ionic or covalent? The difference in electronegativity between two atoms determines the character of the bond formed between them. As the difference in electronegativity increases, the polarity of the bond (or percent ionic character) increases. As a rule, if the electronegativity difference between two bonded atoms is greater than 1.7–1.9, the bond will be more ionic than covalent. If the electronegativity difference is greater than 2.0, the bond is strongly ionic. If the electronegativity difference is less than 1.5, the bond is strongly covalent.

Care must be taken to distinguish between polar bonds and polar molecules. A covalent bond between different kinds of atoms is always polar. But a molecule containing different kinds of atoms may or may not be polar, depending on its shape or geometry. Molecules of HF, HCl, HBr, HI, and ICl are all polar because each contains a single polar bond. However, CO_2, CH_4, and CCl_4 are nonpolar molecules despite the fact that all three contain polar bonds. The carbon dioxide molecule, O=C=O, is nonpolar because the carbon-oxygen dipoles cancel each other by acting in opposite directions.

$$\overleftarrow{\;+\;}\;\overrightarrow{+\;}$$
$$O{=}C{=}O$$

Dipoles in opposite directions

Methane (CH_4) and carbon tetrachloride (CCl_4) are nonpolar because the four C—H and C—Cl polar bonds are identical, and, because these bonds emanate from the center to the corners of a tetrahedron in the molecule, the effect of their polarities cancel one another (Figure 12.7, page 249).

We have said that water is a polar molecule. If the atoms in water were linear like those in carbon dioxide, the two O—H dipoles would cancel each other, and the molecule would be nonpolar. However, water is definitely polar and has a nonlinear (bent) structure with an angle of 105° between the two O—H bonds.

The relationships among types of bonds are summarized in Figure 12.6. It is important to realize that bonding is a continuum, the difference between ionic and covalent is a gradual change.

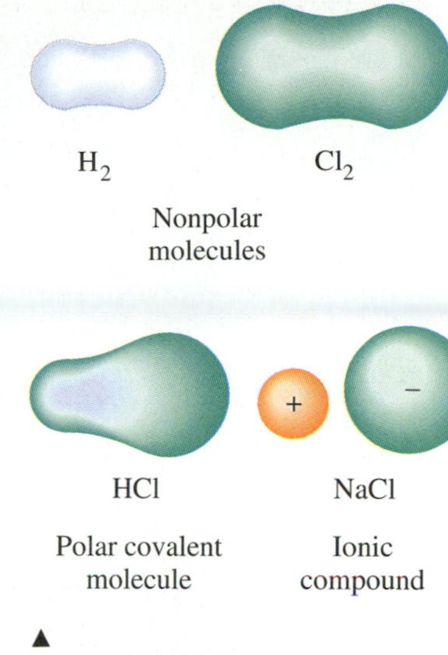

▲ **FIGURE 12.5**
Nonpolar, polar covalent, and ionic compounds.

Bond type	Covalent	Polar covalent	Ionic
Electronegativity difference	0	Intermediate	3.3

◀ **FIGURE 12.6**
Relating bond type to electronegativity difference between atoms.

12.7 Writing Lewis Structures

As we have seen, Lewis structures are a convenient way of showing the covalent bonds in many molecules or ions of the representative elements. In writing Lewis structures the object is to connect the atoms in a molecule with covalent bonds by rearranging the valence electrons of the atoms so that each atom has eight outer-shell electrons around it. Exceptions to this rule are hydrogen, which requires only two electrons, and several other elements such as lithium, beryllium, and boron.

The most difficult part of writing Lewis structures is determining the arrangement of the atoms in a molecule or an ion. In simple molecules with more than two atoms, one atom will be the central atom surrounded by the other atoms. Thus Cl_2O has two possible arrangements, Cl—Cl—O or Cl—O—Cl. Usually, but not always, the single atom in the formula (except H) will be the central atom.

Although Lewis structures for many molecules and ions can be written by inspection, the following procedure will be helpful for learning to write these structures:

1. Obtain the total number of valence electrons to be used in the structure by adding the number of valence electrons in all of the atoms in the molecule or ion. If you are writing the structure of an ion, add one electron for each negative charge or subtract one electron for each positive charge on the ion. Remember, the number of valence electrons of Group A elements is the same as their group number in the periodic table.
2. Write down the skeletal arrangement of the atoms and connect them with a single covalent bond (two dots or one dash). Hydrogen, which contains only one bonding electron, can form only one covalent bond. Oxygen atoms are not normally bonded to each other, except in compounds known to be peroxides. Oxygen atoms normally have a maximum of two covalent bonds, two single bonds or one double bond.
3. Subtract two electrons for each single bond you used in Step 2 from the total number of electrons calculated in Step 1. This calculation gives you the net number of electrons available for completing the structure.
4. Distribute pairs of electrons (pairs of dots) around each atom (except hydrogen) to give each atom a total of eight electrons around it.
5. If there are not enough electrons to give these atoms eight electrons, change single bonds between atoms to double or triple bonds by shifting unbonded pairs of electrons as needed. Check to see that each atom (except H) has eight electrons around it. A double bond counts as four electrons for each atom to which it is bonded.

EXAMPLE 12.8 How many valence electrons are in each of these atoms: Cl, H, C, O, N, S, P, I?

SOLUTION You can look in Table 10.4 for the electron structure, or, if the element is in an A Group of the periodic table, the number of valence electrons is equal to the group number.

Atom	Periodic group	Valence electrons
Cl	VIIA	7
H	IA	1
C	IVA	4
O	VIA	6
N	VA	5
S	VIA	6
P	VA	5
I	VIIA	7

EXAMPLE 12.9 Write the Lewis structure for water, H_2O.

SOLUTION

Step 1 The total number of valence electrons is 8, 2 from the two hydrogen atoms and 6 from the oxygen atom.

Step 2 The two hydrogen atoms are connected to the oxygen atom. Write the skeletal structure:

```
H O
  H
```

Place two dots between the hydrogen and oxygen atoms to form the covalent bonds:

```
H:O
  ..
  H
```

Step 3 Subtract the 4 electrons used in Step 2 from 8 to obtain 4 electrons yet to be used.
Step 4 Distribute the four electrons around the oxygen atom. Hydrogen atoms cannot accommodate any more electrons.

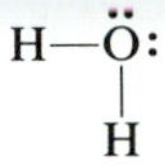

```
   ..
H—O:
   |
   H
```

This arrangement is the Lewis structure. Each atom has a noble-gas electron structure.

EXAMPLE 12.10

Write Lewis structures for a molecule of (a) methane, CH_4, and (b) carbon tetrachloride, CCl_4.

SOLUTION

Part (a)

Step 1 The total number of valence electrons is 8, 1 from each hydrogen atom and 4 from the carbon atom.
Step 2 The skeletal structure contains four H atoms around a central C atom. Place 2 electrons between the C and each H.

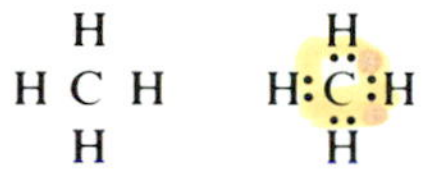

```
  H          H
                ..
H C H      H:C:H
                ..
  H          H
```

Step 3 Subtract the 8 electrons used in Step 2 from 8 to obtain zero electrons yet to be placed. Therefore, the Lewis structure must be as written in Step 2.

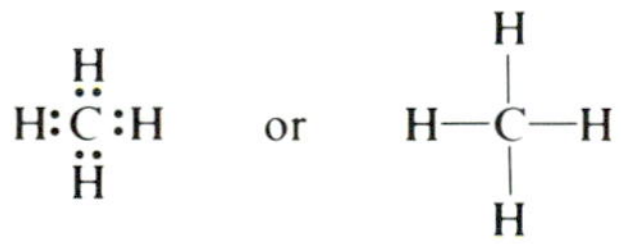

```
                       H
   H                   |
   ..
H:C:H      or      H—C—H
   ..
   H                   |
                       H
```

Part (b)

Step 1 The total number of valence electrons to be used is 32, 4 from the carbon atom and 7 from each of the four chlorine atoms.
Step 2 The skeletal structure contains the four Cl atoms around a central C atom. Place 2 electrons between the C and each Cl.

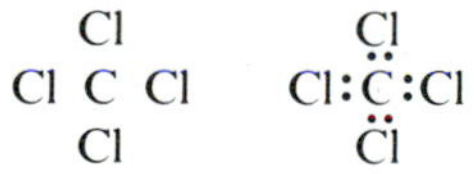

```
   Cl            Cl
                 ..
Cl C Cl      Cl:C:Cl
                 ..
   Cl            Cl
```

Step 3 Subtract the 8 electrons used in Step 2 from 32, to obtain 24 electrons yet to be placed.
Step 4 Distribute the 24 electrons (12 pairs) around the Cl atoms so that each Cl atom has 8 electrons around it.

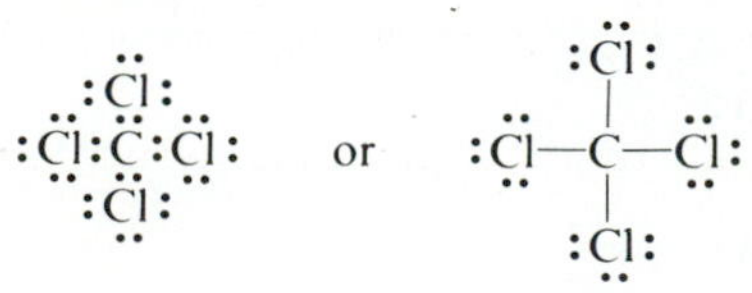

```
     ..                       ..
    :Cl:                     :Cl:
 ..  ..  ..                    |
:Cl:C:Cl:      or       :Cl—C—Cl:
 ..  ..  ..              ..    |    ..
    :Cl:                     :Cl:
     ..                       ..
```

This arrangement is the Lewis structure; CCl_4 contains four covalent bonds.

EXAMPLE 12.11 Write Lewis structures for (a) carbon dioxide, CO_2, and (b) nitric acid, HNO_3.

SOLUTION **Part (a)**

Step 1 The total number of valence electrons is 16, 4 from the carbon atom and 6 from each oxygen atom.

Step 2 The two O atoms are bonded to a central C atom. Write the skeletal structure and place 2 electrons between the C and each O atom.

O:C:O

Step 3 Subtract the 4 electrons used in Step 2 from 16 to obtain 12 electrons yet to be placed.

Step 4 Distribute the 12 electrons around the C and O atoms. Several possibilities exist:

:Ö:C:Ö: :Ö:C:Ö: :Ö:C:O:

I II III

Step 5 All the atoms do not have 8 electrons around them. Move one pair of unbonded electrons from each O atom in structure I and place one additional pair of electrons between each C and O atom, forming two double bonds:

:Ö::C::Ö: or :Ö=C=Ö:

Each atom now has 8 electrons around it. Carbon is sharing four pairs of electrons, and each oxygen is sharing two pairs. These bonds are known as double bonds since each involves sharing two pairs of electrons.

Part (b)

Step 1 The total number of valence electrons is 24, 1 from the hydrogen atom, 5 from the nitrogen atom, and 6 from each oxygen atom.

Step 2 The three O atoms are bonded to a central N atom. The H atom is bonded to one of the O atoms. Write the skeletal structure and place 2 electrons between each pair of atoms:

O
H:O:N:O

Step 3 Subtract the 8 electrons used in Step 2 from 24 to obtain 16 electrons yet to be placed.

Step 4 Distribute the 16 electrons around the N and O atoms:

:Ö ← electron deficient
H:Ö:N:Ö:

Step 5 One pair of electrons is still needed to give all the N and O atoms 8 electrons. Move the unbonded pair of electrons from the N atom and place it between the N and the electron-deficient O atom, making a double bond:

:Ö
H:Ö:N:Ö: or H—Ö—N—Ö: (with :Ö double-bonded to N)

This arrangement is the Lewis structure.

Example 12.12

Write the Lewis structure for a sulfate ion, SO_4^{2-}.

SOLUTION

Step 1 These five atoms have 30 valence electrons (6 in each atom) plus 2 electrons from the -2 charge, which makes 32 electrons to be placed.

Step 2 In the sulfate ion the sulfur is the central atom surrounded by the four oxygen atoms. Write the skeletal structure and place two electrons between each pair of atoms:

```
   O
   ··
O:S:O
   ··
   O
```

Step 3 Subtract the 8 electrons used in Step 2 from 32 to give 24 electrons yet to be placed.

Step 4 Distribute the 24 electrons around the four O atoms and indicate that the sulfate ion has a -2 charge:

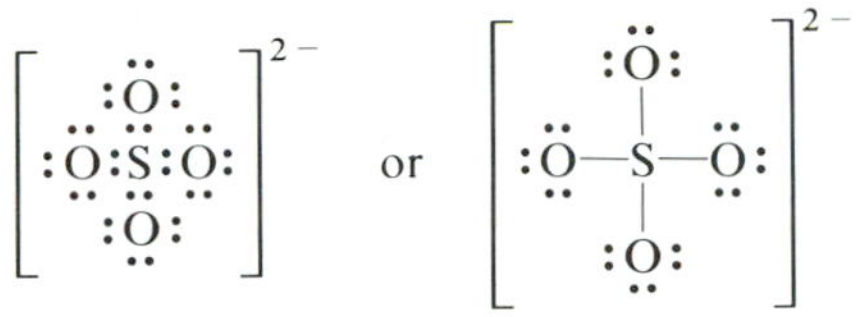

```
[   ··   ]2−          [      ··      ]2−
[  :O:   ]            [     :O:      ]
[ ·· ·· ··]     or    [  ··  |   ··  ]
[:O:S:O: ]            [ :O — S — O:  ]
[ ·· ·· ··]           [  ··  |   ··  ]
[  :O:   ]            [     :O:      ]
[   ··   ]            [      ··      ]
```

This arrangement is the Lewis structure. Each atom has 8 electrons around it.

Practice Write the Lewis structures for each of the following: (a) NH_3 (b) SiO_2 (c) SO_3 (d) ClO_3^- (e) H_3O^+

Answers:

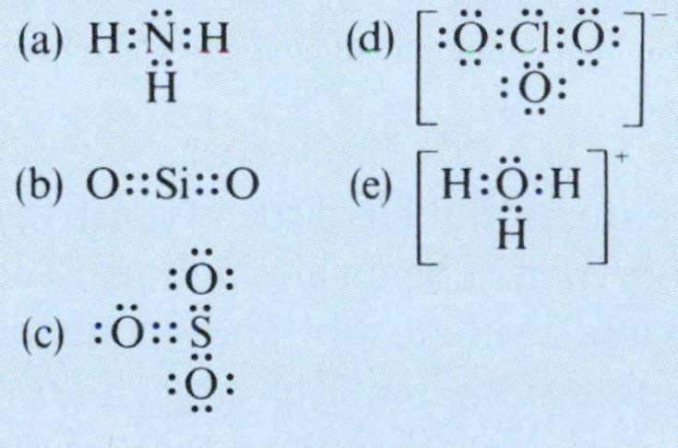

```
(a) H:N:H          (d) [:O:Cl:O:]−
      H                [   :O:   ]

(b) O::Si::O       (e) [H:O:H]+
                       [  H  ]
         :O:
(c) :O::S
         :O:
```

Although many compounds follow the octet rule for covalent bonding, the rule has numerous exceptions. Sometimes it is impossible to write a structure in which each atom has eight electrons around it. For example, in BF_3 the boron atom has only 6 electrons around it, and in SF_6 the sulfur atom has 12 electrons around it.

Even though there are exceptions, most molecules can be described using Lewis structures where each atom has a noble gas electron configuration. This is a useful model for understanding chemistry.

12.8 Coordinate Covalent Bonds

In Section 12.5 we saw that a covalent bond was formed by the overlapping of electron orbitals between two atoms. The two atoms each furnish an electron to make a pair that is shared between them.

coordinate covalent bond

Covalent bonds can also be formed by one atom furnishing both electrons that are shared between the two atoms. The bond so formed is called a **coordinate covalent bond**. This bond is often designated by an arrow pointing away from the electron-pair donor (for example, A $\longrightarrow$ B). Once formed, a coordinate covalent bond cannot be distinguished from any other covalent bond; it is simply a pair of electrons shared between two atoms.

The Lewis structures of sulfurous and sulfuric acids show coordinate covalent bonds between the sulfur and the oxygen atoms that are not bonded to hydrogen atoms. The colored dots indicate the electrons of the sulfur atom.

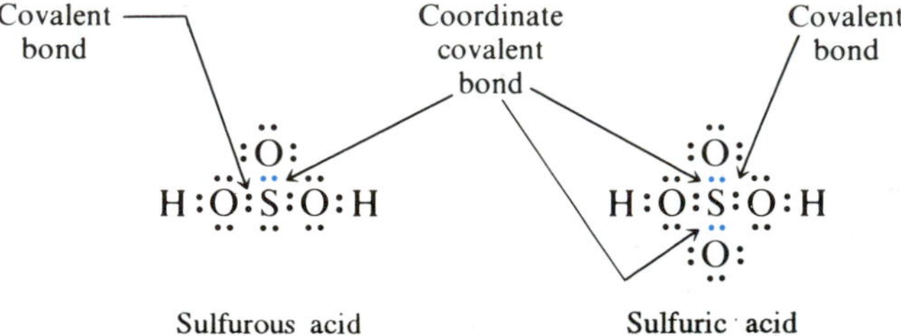

Sulfurous acid   Sulfuric acid

The lone (unbonded) pair of electrons on the sulfur atom in sulfurous acid allows room for another oxygen atom with six valence electrons to fit perfectly into its structure to form sulfuric acid. Other atoms with six valence electrons in their outer shell, such as sulfur, could also fit into this pattern. The coordinate covalent bond explains the formation of many complex molecules.

12.9 Polyatomic Ions

A polyatomic ion is a stable group of atoms that has either a positive or a negative charge and behaves as a single unit in many chemical reactions. Sodium sulfate, Na_2SO_4, contains two sodium ions and a sulfate ion. The sulfate ion, SO_4^{2-}, is a polyatomic ion composed of one sulfur atom and four oxygen atoms and has a charge of -2. One sulfur and four oxygen atoms have a total of 30 electrons in their outer shells. The sulfate ion contains 32 outer-shell electrons and therefore has a charge of -2. In this case the two additional electrons come from the two sodium atoms, which are now sodium ions.

$$[\mathrm{Na}]^+ \left[\begin{array}{c} :\ddot{\mathrm{O}}: \\ :\ddot{\mathrm{O}}:\ddot{\mathrm{S}}:\ddot{\mathrm{O}}: \\ :\ddot{\mathrm{O}}: \end{array} \right]^{2-} [\mathrm{Na}]^+ \qquad \left[\begin{array}{c} :\ddot{\mathrm{O}}: \\ :\ddot{\mathrm{O}}:\ddot{\mathrm{S}}:\ddot{\mathrm{O}}: \\ :\ddot{\mathrm{O}}: \end{array} \right]^{2-}$$

Sodium sulfate   Sulfate ion

Sodium sulfate has both ionic and covalent bonds. Ionic bonds exist between each of the sodium ions and the sulfate ion. Covalent bonds are present between the sulfur and oxygen atoms within the sulfate ion. One important difference between the ionic and covalent bonds in this compound can be demonstrated by dissolving sodium sulfate in water. It dissolves in water forming three charged particles, two sodium ions and one sulfate ion, per formula unit of sodium sulfate:

$$\underset{\text{Sodium sulfate}}{Na_2SO_4} \xrightarrow{\text{Water}} \underset{\text{Sodium ions}}{2\,Na^+} + \underset{\text{Sulfate ion}}{SO_4^{2-}}$$

The SO_4^{2-} ion remains as a unit, held together by covalent bonds; whereas, where the bonds were ionic, dissociation of the ions took place. Do not think, however, that polyatomic ions are so stable that they cannot be altered. Chemical reactions by which polyatomic ions can be changed to other substances do exist.

12.10 Molecular Structure

So far in our discussion of bonding we have used Lewis structures to represent valence electrons in molecules and ions. Lewis structures do not indicate anything regarding the molecular or geometric structure of a molecule. The three-dimensional arrangement of the atoms within a molecule is a significant feature in understanding molecular interactions.

Consider several examples. Water is known to have the geometric structure:

```
H     H
 \   /
   O
```

known as "bent" or "V-shaped." Carbon dioxide exhibits a linear shape:

O=C=O

whereas $AlCl_3$ forms a third molecular structure

```
Cl      Cl
  \    /
    Al
    |
    Cl
```

This last structure is called *trigonal planar* since all the atoms lie in one plane in a triangle arrangement. One of the more common molecular structures is the tetrahedron illustrated by the molecule methane, CH_4 shown in the margin. How can chemists predict the geometric structure of a molecule? We will look at a model (VSEPR model) developed to assist in making predictions from the Lewis structure of a molecule.

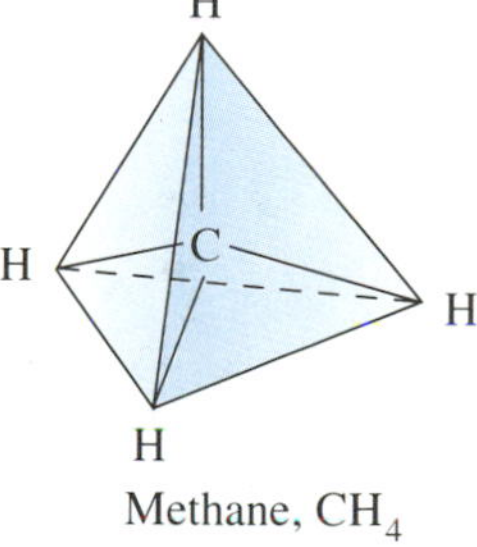

Methane, CH_4

12.11 The VSEPR Model for Molecular Structure

The chemical properties of a substance are closely related to the structure of its molecules. A change in a single site on a large biomolecule can make a difference in whether or not a particular reaction occurs.

Instrumental analysis can be used to determine exact spatial arrangements of atoms. Quite often, though, we only need to be able to predict the approximate structure of a molecule. A relatively simple model has been developed by chemists to allow us to make predictions of shape from Lewis structures.

The Valence Shell Electron Pair Repulsion (VSEPR) model is based on the idea that electron pairs will repel each other electrically and will seek to minimize this repulsion. In order to accomplish this minimization, the electron pairs will be arranged around a central atom as far apart as possible. Consider $BeCl_2$, a molecule with only two pairs of electrons surrounding the central atom. These pairs of electrons are arranged 180° apart for maximum separation.

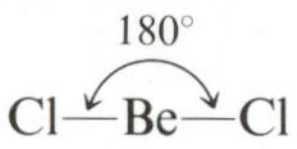

180°
Cl—Be—Cl

linear structure

The molecular structure can now be labeled as a **linear structure**. When only two pairs of electrons surround an atom they should be placed 180° apart to give a linear structure.

What occurs when there are three pairs of electrons on the central atom? Consider the BF_3 molecule. The greatest separation of electron pairs occurs when the angles between atoms are 120°:

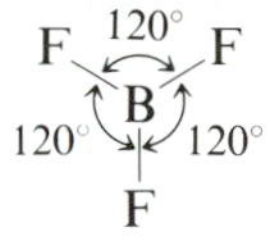

F 120° F
B
120° 120°
F

trigonal planar

This arrangement of atoms is flat (planar) and is usually called **trigonal planar**. When three pairs of electrons surround an atom they should be placed 120° apart to give trigonal planar structure.

Now consider the most common situation, CH_4, with four pairs of electrons on the central carbon atom. In this case, the central atom exhibits a noble gas electron structure. What arrangement will best minimize the electron pair repulsions? At first it seems that an obvious choice is a 90° angle with all the atoms in a single plane:

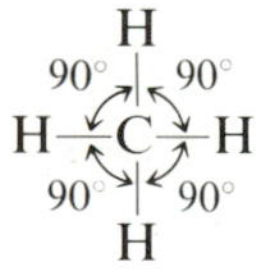

H
90° 90°
H—C—H
90° 90°
H

However, we must consider that molecules are three-dimensional. This concept results in a structure in which the electron pairs are 109.5° apart.

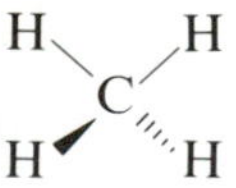

H H
C
H H

tetrahedral structure

In this diagram the wedged line protrudes from the page whereas the dashed line recedes behind the paper. Two examples showing representations of this arrangement know as **tetrahedral structure** are given in Figure 12.7. When four pairs of electrons surround an atom they should be placed 109.5° apart to give a tetrahedral structure.

The VSEPR model is based upon the premise that we are counting electron pairs. It is quite possible that one or more of these electron pairs may be non-

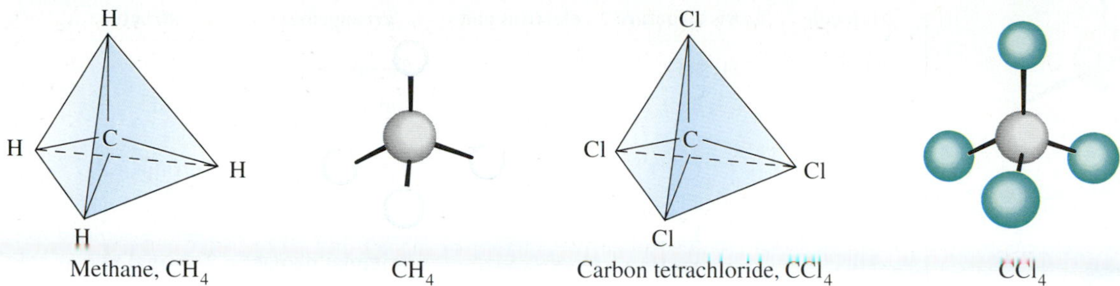

▲
FIGURE 12.7
Ball-and-stick models of methane and carbon tetrachloride. Methane and carbon tetrachloride are nonpolar molecules because their polar bonds cancel each other in tetrahedral arrangement of their atoms. The carbon atoms are located in the centers of the tetrahedrons.

bonding or lone pairs. What happens to the molecular structure in these cases? Consider the ammonia molecule. First draw the Lewis structure to determine the number of electron pairs around the central atom.

$$\begin{array}{c} \mathrm{H:\ddot{\underset{\cdot\cdot}{N}}:H} \\ \mathrm{H} \end{array}$$

Since there are four pairs of electrons the arrangement will be tetrahedral. However, only three on the pairs are bonded to another atom so the shape of the molecule itself is pyramidal. It is important to understand that the placement of the electron pairs determines the structure but the name of the shape of the molecule is determined by the position of the atoms themselves. Therefore, ammonia has a pyramidal shape, not a tetrahedral shape. See Figure 12.8.

Now consider the effect of two lone pairs in the water molecule. The Lewis structure for water is

$$\begin{array}{l} \mathrm{H-\ddot{O}:} \\ \quad\ \ | \\ \quad\ \mathrm{H} \end{array}$$

The four electron pairs indicate a tetrahedral arrangement is necessary (see Figure 12.7). The molecule is not tetrahedral because two of the electron pairs are unbonded pairs. The atoms in the water molecule form a "bent" shape as shown in Figure 12.9. Using the VSEPR model helps to explain some of the unique properties of the water molecule. Because it is bent and not linear we can see that the molecule is polar.

The properties of water which cause it to be involved in so many interesting and important roles are largely a function of its shape and polarity. We will consider water in greater detail in Chapter 14.

(a)

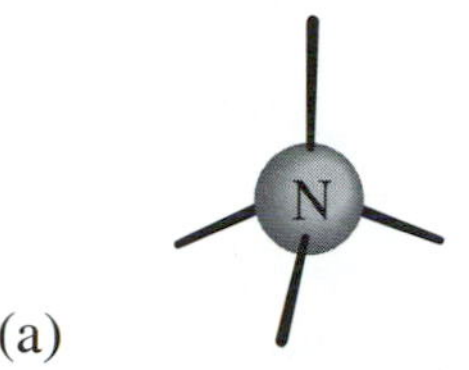

(b)

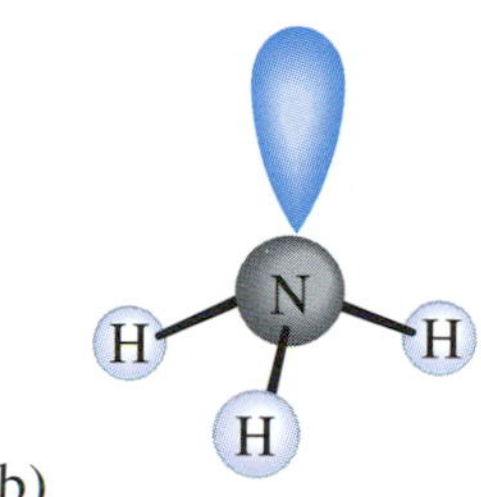

(c)

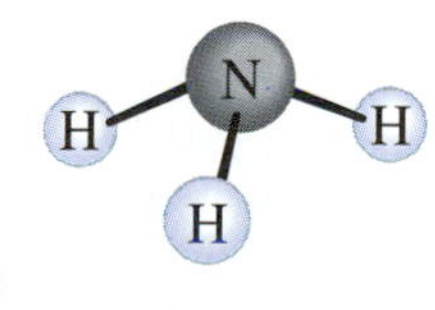

▲
FIGURE 12.8
(a) The tetrahedral arrangement of electron pairs around the N atom in the NH_3 molecule. (b) Three pairs are shared and one is unshared. (c) The NH_3 molecule has a trigonal pyramidal structure.

EXAMPLE 12.13

Predict the molecular structure for each molecule: H_2S, $BeCl_2$, CCl_4, $AlBr_3$.

SOLUTION

1. Draw the Lewis structure.
2. Count the electron pairs and determine the arrangement that will minimize repulsions.
3. Determine the positions of the atoms and name the structure.

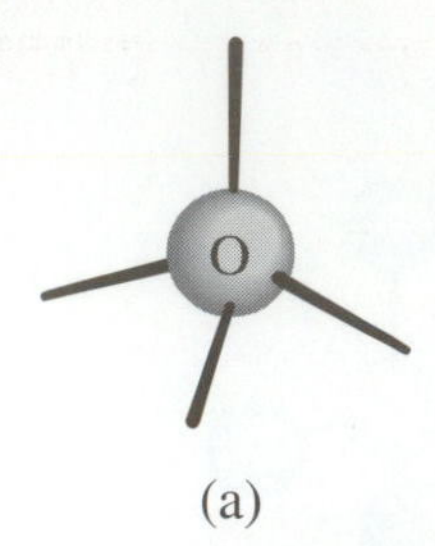

(a)

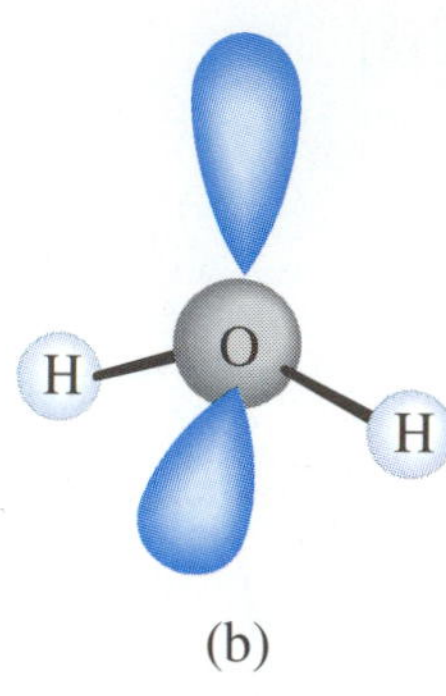

(b)

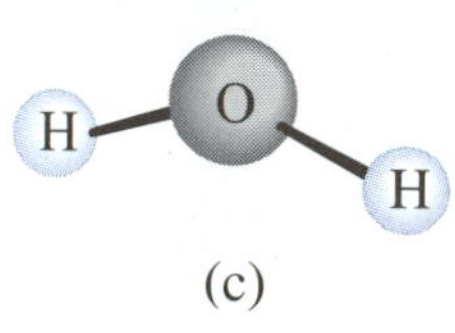

(c)

▲
Figure 12.9
(a) The tetrahedral arrangement of the four electron pairs around oxygen in the H_2O molecule. (b) Two of the pairs are shared and two are unshared. (c) The H_2O molecule has a bent molecular structure.

Molecule	Lewis structure	Number of electron pairs	Electron pair arrangement	Molecular structure
H_2S	H:S:H	4	Tetrahedral	Bent
$MgCl_2$	Cl:Be:Cl	2	Linear	Linear
CCl_4	Cl above; Cl:C:Cl; Cl below	4	Tetrahedral	Tetrahedral
AlF_3	F above; F:Al:F	3	Trigonal planar	Trigonal planar

Practice Predict the shape for CF_4, NF_3, and $AlCl_4^-$.
Answer: Tetrahedral, pyramidal, tetrahedral.

Concepts in Review

1. Describe (a) the formation of ions by electron transfer and (b) the nature of the chemical bond formed by electron transfer.
2. Show by means of Lewis-dot structures the formation of an ionic compound from atoms.
3. Describe a crystal of sodium chloride.
4. Predict the relative sizes of an atom and a monatomic ion for a given element.
5. Describe the covalent bond and predict whether a given covalent bond would be polar or nonpolar.
6. Draw Lewis-dot structures for the diatomic elements.
7. Identify single, double, and triple covalent bonds.
8. Describe the changes in electronegativity in (1) moving across a period and (2) moving down a group in the periodic table.
9. Predict formulas of simple compounds formed between the representative (Group A) elements using the periodic table.
10. Describe the effect of electronegativity on the type of chemical bonds in a compound.
11. Draw Lewis-dot structures for (a) the molecules of covalent compounds and (b) polyatomic ions.
12. Describe the difference between polar and nonpolar bonds.
13. Distinguish clearly between ionic and molecular substances.
14. Predict whether the bonding in a compound will be primarily ionic or covalent.
15. Distinguish coordinate-covalent from covalent bonds in a Lewis-dot structure.
16. Describe the VSEPR model for molecular shape.
17. Use the VSEPR model to determine molecular structure from Lewis structure for given compounds.

LIQUID CRYSTALS

What do mood rings, bullet-resistant vests, wristwatches, calculators, and changing color thermometers have in common? The chemicals that make each of them work are liquid crystals. In a normal crystal the molecules have an orderly arrangement. In a liquid crystal the molecules can flow and maintain an orderly arrangement at the same time.

The molecules in all types of liquid crystals are linear and polar. The atoms of linear molecules tend to lie in a relatively straight line and the molecules are generally much longer than they are wide. Polar molecules are attracted to each other, and certain linear ones are able to line up in an orderly fashion, without solidifying, to form liquid crystals. Such a substance can be used as a liquid crystal display (LCD), to change color with changing temperature, or to make a super-strong synthetic fiber.

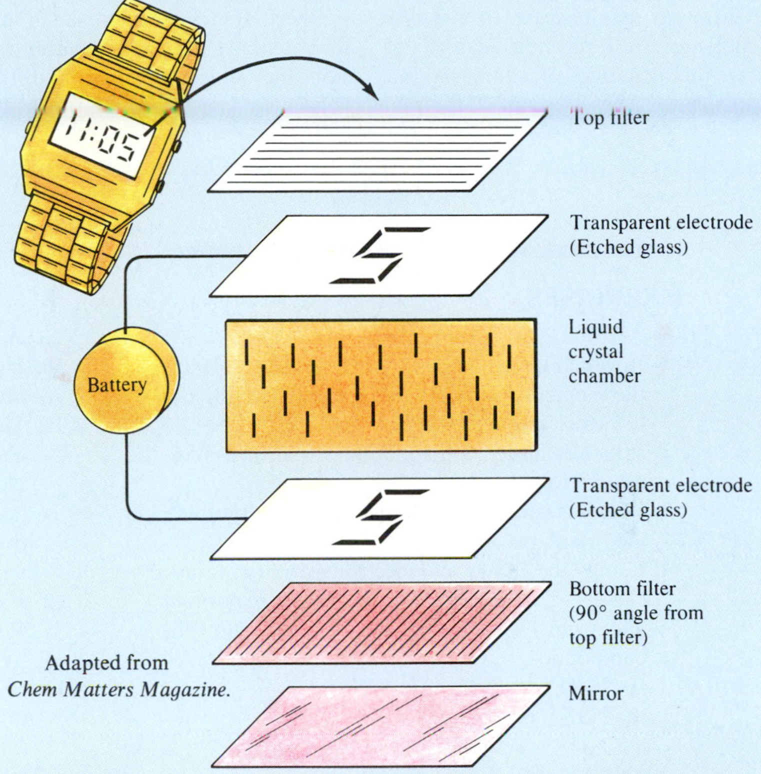

Adapted from *Chem Matters Magazine.*

The key to these products that change color with temperature is in the twisted arrangement of the molecules. The linear molecules form a generally flat surface. In the liquid crystal the molecules lie side by side in a nearly flat layer. The next layer is similar but at an angle to the one below. These closely packed flat layers have a special effect on light. As the light strikes the surface, some of it is reflected from the top layer and more from lower layers. When the same wavelength is reflected from many layers a color is observed. This is similar to the rainbow of colors formed by oil in a puddle on the street or the film of a soap bubble. As the temperature increases the molecules move faster, causing a change in the angle and the space between the layers. These changes result in a color change in the reflected light. Different compounds change color within different temperature ranges, allowing a variety of practical and amusing applications.

In liquid crystal displays (LCDs) in watches and calculators the process is similar. Normally the LCD acts as a mirror reflecting the light which enters it. The display is created by a series of layers, consisting of a filter, a glass etched with tiny lines, a liquid crystal chamber, a second etched glass, a bottom filter, and a mirror (see diagram). The molecules at the top of the liquid crystal chamber align with the lines on the top layer of glass whereas those at the bottom align with the grooves on that glass (90-degree turn from the top). In between the molecules line up as closely as possible with neighboring molecules to form a twisted spiral. To display a number a tiny current is sent to the proper SnO_2 segments on the etched piece of glass. The plates become charged, and the polar molecules of the liquid crystal are attracted to the charged segments thus destroying the carefully arranged spirals. The pattern of reflected light is changed and a numeral appears.

Another type of liquid crystal (nematic) contains molecules that all point in the same direction. These liquid crystals are used to manufacture very strong synthetic fibers. The molecules in nematic crystals all line up in the same direction but are free to slide past each other.

Perhaps the best example of these liquid crystals is the manufacture of Kevlar, a synthetic fiber used in

(*continued*)

bullet resistant vests, canoes, and parts of the space shuttle. Kevlar is a synthetic polymer, like nylon or polyester, that gains strength by passing through a liquid crystal state during its manufacture. In a typical polymer the long molecular chains are jumbled together, somewhat like spaghetti. The strength of the material is limited by the disorderly arrangement.

The trick is to get the molecules to line up parallel to each other. Once the giant molecules have been synthesized they are dissolved in sulfuric acid. At the proper concentration the molecules align and the solution is forced through tiny holes in a nozzle and further aligned. The sulfuric acid is removed in a water bath forming solid fibers in near perfect alignment. One strand of Kevlar is stronger than an equal-sized strand of steel. It has a much lower density as well, making it a material of choice in bullet-resistant vests.

EXERCISES

1. In which general areas of the periodic table are the elements with (a) the highest and (b) the lowest electronegativities found?

2. Which is larger, a magnesium atom or a magnesium ion? Explain.

3. Which is smaller, a bromine atom or a bromide ion? Explain.

4. Using the table of electronegativity values (Table 12.3), indicate which element is positive and which is negative in the following compounds:

(a) H_2O (e) NO (i) CCl_4
(b) NaF (f) CH_4 (j) IBr
(c) NH_3 (g) HCl (k) MgH_2
(d) PbS (h) LiH (l) OF_2

5. Classify the bond between the following pairs of elements as principally ionic or principally covalent (use Table 12.3):

(a) Sodium and chlorine
(b) Carbon and hydrogen
(c) Chlorine and carbon
(d) Calcium and oxygen
(e) Hydrogen and sulfur
(f) Barium and oxygen
(g) Fluorine and fluorine
(h) Potassium and fluorine

6. Explain what happens to the electron structures of Mg and Cl atoms when they react to form $MgCl_2$.

7. Write an equation representing (a) the change of a fluorine atom to a fluoride ion (F^-) and (b) the change of a calcium atom to a calcium ion (Ca^{2+}).

8. Use Lewis-dot symbols to show the electron transfer for the formation of the following ionic compounds from the atoms.

(a) MgF_2 (b) K_2O (c) CaO (d) NaBr

9. What are valence electrons?

10. How many valence electrons are in each of these atoms? H, K, Mg, He, Al, Si, N, P, O, Cl

11. How many electrons must be gained or lost for each of the following to achieve a noble-gas electron structure?

(a) A calcium atom (d) A chloride ion
(b) A sulfur atom (e) A nitrogen atom
(c) A helium atom (f) A potassium atom

12. Explain why potassium forms a K^+ ion but not a K^{2+} ion.

13. Why does an aluminum ion have a +3 charge?

14. Which is larger? Explain.

(a) A potassium atom or a potassium ion
(b) A bromine atom or a bromide ion
(c) A magnesium ion or an aluminum ion
(d) Fe^{2+} or Fe^{3+}

15. Let E be any representative element. Following the pattern in the table, write the formulas for the hydrogen and oxygen compounds of

(a) Na (c) Al (e) Sb (g) Cl
(b) Ca (d) Sn (f) Se

Group IA	IIA	IIIA	IVA	VA	VIA	VIIA
EH	EH_2	EH_3	EH_4	EH_3	H_2E	HE
E_2O	EO	E_2O_3	EO_2	E_2O_5	EO_3	E_2O_7

16. Group IB elements have one electron in their outer shell, as do Group IA elements. Would you expect them to form compounds such as CuCl, AgCl, and AuCl? Explain.

17. The formula for lead(II) bromide is $PbBr_2$; predict formulas for tin(II) and germanium(II) bromides. (If needed, see Section 6.4, part (b), for the use of Roman numerals in naming compounds.)

18. The formula for sodium sulfate is Na_2SO_4. Write the names and formulas for the other alkali metal sulfates.
19. The formula for calcium bromide is $CaBr_2$. Write formulas for magnesium bromide, strontium bromide, and barium bromide.
20. Write Lewis-dot structures for
 (a) Na (b) Br^- (c) O^{2-} (d) Ga (e) Ga^{3+}
21. Why is it not proper to speak of sodium chloride molecules?
22. What is a covalent bond? How does it differ from an ionic bond?
23. What is different about the formation of a coordinate covalent bond from that of an ordinary covalent bond?
24. Classify the bonding in each compound as ionic or covalent:
 (a) H_2O (c) MgO (e) HCl (g) NH_3
 (b) NaCl (d) Br_2 (f) $BaCl_2$ (h) SO_2
25. Predict the type of bond that would be formed between each of the following pairs of atoms:
 (a) Na and N (d) H and Si
 (b) N and S (e) O and F
 (c) Br and I
26. Draw Lewis structures for:
 (a) H_2 (b) N_2 (c) Cl_2 (d) O_2 (e) Br_2
27. Draw Lewis structures for:
 (a) NCl_3 (c) H_3PO_4 (e) H_2S
 (b) H_2CO_3 (d) C_2H_6 (f) CS_2
28. Briefly comment on the structure $Na{:}\ddot{\underset{\cdot\cdot}{O}}{:}Na$ for the compound Na_2O.
29. What are the four most electronegative elements?
30. Draw Lewis structures for:
 (a) Ba^{2+} (e) SO_4^{2-} (h) CO_3^{2-}
 (b) Al^{3+} (f) SO_3^{2-} (i) ClO_3^-
 (c) I^- (g) CN^- (j) NO_3^-
 (d) S^{2-}
31. The Lewis structure for chloric acid is

```
   ..  ..  ..
 H:O:Cl:O:
   ..  ..  ..
      :O:
       ..
```

 Point out the covalent and coordinate covalent bonds in this structure.
32. Draw the Lewis structure for ammonia (NH_3). What type of bond is present? Can this molecule form coordinate covalent bonds? If so, how many?
33. Rank these elements from highest electronegativity to lowest: Mg, S, F, H, O, Cs.
34. Classify the following molecules as polar or nonpolar:
 (a) H_2O (c) CF_4 (e) CO_2
 (b) HBr (d) F_2 (f) NH_3
35. Is it possible for a molecule to be nonpolar even though it contains polar covalent bonds? Explain.
36. Why is CO_2 a nonpolar molecule, whereas CO is a polar molecule?
37. Which of these statements are correct? (Try to answer this exercise using only the periodic table.) Rewrite each incorrect statement to make it correct.
 (a) If the formula for calcium iodide is CaI_2, then the formula for cesium iodide is CsI_2.
 (b) Metallic elements tend to have relatively low electronegativities.
 (c) If the formula for aluminum oxide is Al_2O_3, then the formula for gallium oxide is Ga_2O_3.
 (d) Sodium and chlorine react to form molecules of NaCl.
 (e) A chlorine atom has fewer electrons than a chloride ion.
 (f) The noble gases have a tendency to lose one electron to become positively charged ions.
 (g) The chemical bonds in a water molecule are ionic.
 (h) The chemical bonds in a water molecule are polar.
 (i) Valence electrons are those electrons in the outermost shell of an atom.
 (j) An atom with eight electrons in its outer shell has all its *s* and *p* orbitals filled.
 (k) Fluorine has the lowest electronegativity of all the elements.
 (l) Oxygen has a greater electronegativity than carbon.
 (m) A cation is larger than its corresponding neutral atom.
 (n) Cl_2 is more ionic in character than HCl.
 (o) A neutral atom with eight electrons in its valence shell will likely be an atom of a noble gas.
 (p) A nitrogen atom has four valence electrons.
 (q) An aluminum atom must lose three electrons to become an aluminum ion, Al^{3+}.
 (r) A stable group of atoms that has either a positive or a negative charge and behaves as a single unit in many chemical reactions is called a polyatomic ion.
 (s) Sodium sulfate, Na_2SO_4, has covalent bonds between sulfur and the oxygen atoms, and ionic bonds between the sodium ions and the sulfate ion.
 (t) The water molecule is a dipole.
 (u) In an ethylene molecule, C_2H_4,

```
H       H
 \     /
  C═C
 /     \
H       H
```

two pairs of electrons are shared between the carbon atoms.

(v) The octet rule is mainly useful for atoms where only *s* and *p* electrons enter into bonding.

(w) When electrons are transferred from one atom to another, the resulting compound contains ionic bonds.

(x) A phosphorus atom, $\cdot\ddot{P}\cdot$, needs three additional electrons to attain a stable octet of electrons.

(y) The simplest compound between oxygen, $\cdot\ddot{O}\cdot$, and fluorine, $:\ddot{F}\cdot$, atoms is FO_2.

(z) The $H:\ddot{Cl}:$ molecule has three unshared pairs of electrons.

(aa) The smaller the difference in electronegativity between two atoms, the more ionic the bond between them will be.

(bb) Lewis structures are mainly useful for the representative elements.

(cc) The correct Lewis structure for NH_3 is

H:N̈:H
  H

(dd) The correct Lewis structure for CO_2 is

$:\ddot{O}:C:\ddot{O}:$

(ee) The correct Lewis structure for SO_4^{2-} is

$\left[:\ddot{O}:\ddot{O}:\ddot{S}:\ddot{O}:\ddot{O}:\right]^{2-}$

(ff) In period 5 of the periodic table, the element having the lowest ionization energy is Xe.

(gg) An atom having an electron structure of $1s^22s^22p^63s^23p^2$ has four valence electrons.

(hh) When an atom of bromine becomes a bromide ion, its size increases.

(ii) The structures that show that H_2O is a dipole and that CO_2 is not a dipole are

$H{-}\ddot{O}{-}H$ and $:\ddot{O}{=}C{=}\ddot{O}:$

(jj) The Cl^- and S^{2-} ions have the same electron structure.

(kk) A molecule with the central atom surrounded by two pairs of electrons has a linear shape.

(ll) A molecule with a central atom having three bonding pairs and one nonbonding electron pair will have a tetrahedral electron structure and a tetrahedral shape.

(mm) A molecule in which the central atom is surrounded by three bonding pairs of electrons will have a trigonal planar shape.

(nn) A molecule with two bonding and two nonbonding pairs of electrons will have a bent shape.

38. Use VSEPR theory to predict the shape of the following molecules.
(a) SiH_4
(b) PH_3
(c) SeF_2

39. Give the number and arrangement of the electron pairs around the central atom.
(a) Si in SiO_2
(b) S in H_2S
(c) Al in AlH_3

40. Estimate the bond angle between atoms in each of these molecules:
(a) H_2S (c) NH_4^+
(b) NH_3 (d) $SiCl_4$

41. Use VSEPR theory to predict the structure of the following polyatomic ions:
(a) Sulfate ion
(b) Chlorate ion
(c) Periodate ion

42. Indicate three uses for liquid crystals.

43. Explain how a liquid crystal display works.

44. What is Kevlar? How does it attain the property of super strength?

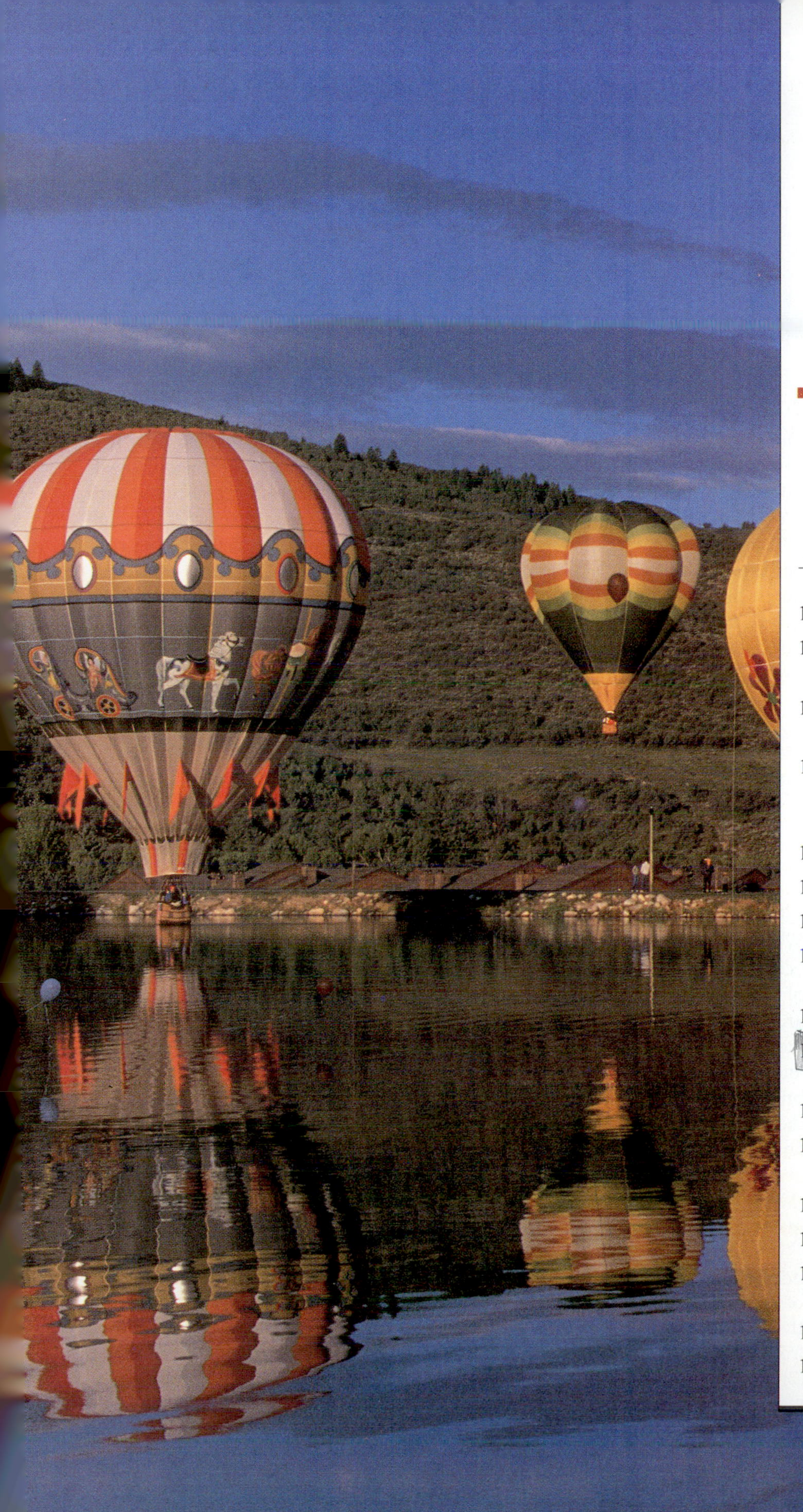

13

The Gaseous State of Matter

◄ Chapter Opening Photo: Hot air balloons illustrate the properties of gases.

Our atmosphere is composed of a mixture of gases, including nitrogen, oxygen, carbon dioxide, ozone, and trace amounts of other gases. Carbon dioxide is taken in by plants and converted to carbohydrates. Yet it also is associated with the potentially hazardous greenhouse effect. Ozone surrounds the earth at high altitudes and protects us from harmful ultraviolet rays, but it also destroys rubber and plastics. We require air to live, yet scuba divers must be concerned about oxygen poisoning, nitrogen narcosis, and the bends.

In chemistry, the study of the behavior of gases, allows us to lay a foundation for understanding our atmosphere and the effects that gases have on our lives.

13.1 General Properties

In Chapter 3, solids, liquids, and gases are described in a brief outline. In this chapter we shall consider the behavior of gases in greater detail.

Gases are the least compact and most mobile of the three states of matter. A solid has a rigid structure, and its particles remain in essentially fixed positions. When a solid absorbs sufficient heat, it melts and changes into a liquid. Melting occurs because the molecules (or ions) have absorbed enough energy to break out of the rigid crystal lattice structure of the solid. The molecules or ions in the liquid are more energetic than they were in the solid, as shown by their increased mobility. Molecules in the liquid state are *coherent*; that is, they cling to one another. When the liquid absorbs additional heat, the more energetic molecules break away from the liquid surface and go into the gaseous state. Gases represent the most mobile state of matter. Gas molecules move with very high velocities and have high kinetic energy (KE). The average velocity of hydrogen molecules at 0°C is over 1600 meters (1 mile) per second. Because of the high velocities of their molecules, mixtures of gases are uniformly distributed within the container in which they are confined.

A quantity of a substance occupies a much greater volume as a gas than it does as a liquid or a solid. For example, 1 mole of water (18.02 g) has a volume of 18 mL at 4°C. This same amount of water would occupy about 22,400 mL in the gaseous state—more than a 1200-fold increase in volume. We may assume from this difference in volume that (1) gas molecules are relatively far apart, (2) gases can be greatly compressed, and (3) the volume occupied by a gas is mostly empty space.

13.2 The Kinetic-Molecular Theory

Careful scientific studies of the behavior and properties of gases were begun in the 17th century by Robert Boyle (1627–1691). This work was carried forward by many investigators after Boyle. The accumulated data were used in the second half of the 19th century to formulate a general theory to explain the behavior and properties of gases. This theory is called the **Kinetic-Molecular Theory (KMT)**. The KMT has since been extended to cover, in part, the behavior of liquids and solids. It ranks today with the atomic theory as one of the greatest generalizations of modern science.

Kinetic-Molecular Theory (KMT)

The KMT is based on the motion of particles, particularly gas molecules. A gas that behaves exactly as outlined by the theory is known as an **ideal**, or **perfect**, **gas**. Actually no ideal gases exist, but under certain conditions of temperature and pressure gases approach ideal behavior, or at least show only small deviations from it. Under extreme conditions, such as very high pressure and low temperature, real gases may deviate greatly from ideal behavior. For example, at low temperature and high pressure many gases become liquids.

ideal, or perfect, gas

The principal assumptions of the Kinetic-Molecular Theory are

1. Gases consist of tiny (submicroscopic) molecules.
2. The distance between molecules is large compared with the size of the molecules themselves. The volume occupied by a gas consists mostly of empty space.
3. Gas molecules have no attraction for one another.
4. Gas molecules move in straight lines in all directions, colliding frequently with one another and with the walls of the container.
5. No energy is lost by the collision of a gas molecule with another gas molecule or with the walls of the container. All collisions are perfectly elastic.
6. The average kinetic energy for molecules is the same for all gases at the same temperature, and its value is directly proportional to the Kelvin temperature.

The kinetic energy (KE) of a molecule is one-half its mass times its velocity squared. It is expressed by the equation

$$KE = \frac{1}{2}mv^2$$

where m is the mass and v is the velocity of the molecule.

All gases have the same kinetic energy at the same temperature. Therefore, from the kinetic energy equation we can see that, if we compare the velocities of the molecules of two gases, the lighter molecules will have a greater velocity than the heavier ones. For example, calculations show that the velocity of a hydrogen molecule is four times the velocity of an oxygen moelcule.

Due to their molecular motion gases have the property of **diffusion**, the ability of two or more gases to mix spontaneously until they form a uniform mixture. The diffusion of gases may be illustrated by use of the apparatus shown in Figure 13.1. Two large flasks, one containing reddish-brown bromine vapors and the other dry air, are connected by a side tube. When the stopcock between

diffusion

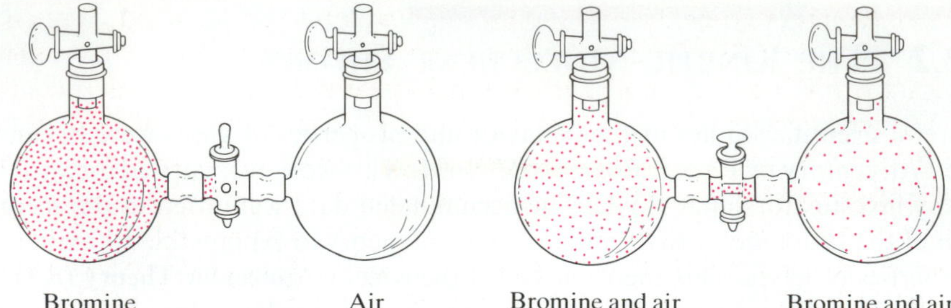

FIGURE 13.1 ▶ **Diffusion of gases. When the stopcock between the two flasks is opened, colored bromine molecules can be seen diffusing into the flask containing air.**

the flasks is opened, the bromine and air will diffuse into each other. After standing awhile, both flasks will contain bromine and air.

effusion

If we put a pinhole in a balloon, the gas inside will effuse or flow out of the balloon. **Effusion** is a process by which gas molecules pass through a very small orifice (opening) from a container at higher pressure to one at lower pressure.

Graham's Law of Effusion

Thomas Graham (1805–1869), a Scottish chemist, observed that the rate of effusion was dependent on the density of a gas. This observation led to **Graham's Law of Effusion**: The rates of effusion of two gases at the same temperature and pressure are inversely proportional to the square roots of their densities or molar masses,

$$\frac{\text{rate of effusion of gas A}}{\text{rate of effusion of gas B}} = \sqrt{\frac{d\text{B}}{d\text{A}}} = \sqrt{\frac{\text{molar mass B}}{\text{molar mass A}}}$$

A major application of Graham's law occurred during World War II with the separation of the isotopes of uranium-235 (U-235) and uranium-238 (U-238). Naturally occurring uranium consists of 0.7% U-235, 99.3% U-238, and a trace of U-234. However, only U-235 is useful as fuel for nuclear reactors and atomic bombs, so the concentration of U-235 in the mixture of isotopes had to be increased.

Uranium was first changed to uranium hexafluoride, UF_6, a white solid that readily goes into the gaseous state. The gaseous mixture of $^{235}UF_6$ and $^{238}UF_6$ was then allowed to effuse through porous walls. Although the effusion rate of the lighter gas is only slightly faster than that of the heavier one,

$$\frac{\text{effusion rate } ^{235}\text{UF}_6}{\text{effusion rate } ^{238}\text{UF}_6} = \sqrt{\frac{\text{molar mass } ^{238}\text{UF}_6}{\text{molar mass } ^{235}\text{UF}_6}} = \sqrt{\frac{352}{349}} = 1.0043$$

the separation and enrichment of U-235 was accomplished by subjecting the gaseous mixture to several thousand stages of effusion.

The properties of an ideal gas are independent of the molecular constitution of the gas. Mixtures of gases also obey the Kinetic-Molecular Theory if the gases in the mixture do not enter into a chemical reaction with one another.

13.3 Measurement of Pressure of Gases

pressure

Pressure is defined as force per unit area. Do gases exert pressure? Yes. When a rubber balloon is inflated with air, it stretches and maintains its larger size because the pressure on the inside is greater than that on the outside. Pressure results from the collisions of gas molecules with the walls of the balloon (see Figure 13.2). When the gas is released, the force or pressure of the air escaping from the small neck propels the balloon in a rapid, irregular flight. If the balloon is inflated until it bursts, the gas escaping all at once causes an explosive noise. This pressure that gases display can be measured; it can also be transformed into useful work. Steam under pressure, used in the steam locomotive, played an important role in the early development of the United States. Compressed steam is used today to generate at least part of the electricity for many cities. Compressed air is used to operate many different kinds of mechanical equipment.

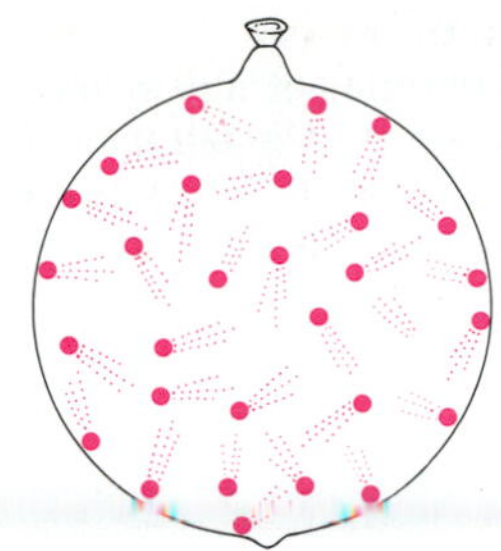

▲
FIGURE 13.2
Here we see the pressure resulting from the collisions of gas molecules with the walls of the balloon. This pressure keeps the balloon inflated.

The mass of air surrounding the earth is called the *atmosphere*. It is composed of about 78% nitrogen, 21% oxygen, and 1% argon, and other minor constituents by volume (see Table 13.1). The outer boundary of the atmosphere is not known precisely, but more than 99% of the atmosphere is below an altitude of 20 miles (32 km). Thus, the concentration of gas molecules in the atmosphere decreases with altitude, and at about 4 miles the amount of oxygen is insufficient to sustain human life. The gases in the atmosphere exert a pressure known as **atmospheric pressure**. The pressure exerted by a gas depends on the number of molecules of gas present, the temperature, and the volume in which the gas is confined. Gravitational forces hold the atmosphere relatively close to the earth and prevent air molecules from flying off into outer space. Thus, the atmospheric pressure at any point is due to the mass of the atmosphere pressing downward at that point.

atmospheric pressure

The pressure of a gas can be measured with a pressure gauge, a manometer, or a **barometer**. A mercury barometer is commonly used in the laboratory to measure atmospheric pressure. A simple barometer of this type may be prepared by completely filling a long tube with pure, dry mercury and inverting the open end into an open dish of mercury. If the tube is longer than 760 mm, the mercury level will drop to a point at which the column of mercury in the tube is just supported by the pressure of the atmosphere. If the tube is properly prepared, a vacuum will exist above the mercury column. The mass of mercury, per unit area, is equal to the pressure of the atmosphere. The column of mercury is supported by the pressure of the atmosphere, and the height of the column is a measure of this pressure (see Figure 13.3). The mercury barometer was invented in 1643 by the Italian physicist E. Torricelli (1608–1647), for whom the unit of pressure *torr* was named.

barometer

TABLE 13.1 Average Composition of Normal Dry Air

Gas	Percent by volume	Gas	Percent by volume
N_2	78.08	He	0.0005
O_2	20.95	CH_4	0.0002
Ar	0.93	Kr	0.0001
CO_2	0.033	Xe, H_2, and N_2O	Trace
Ne	0.0018		

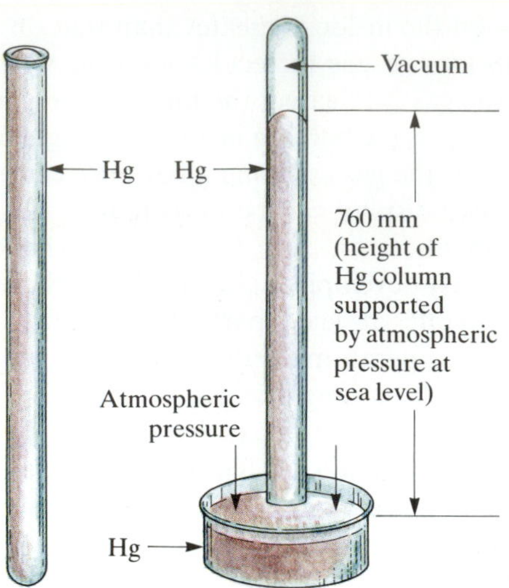

FIGURE 13.3 ▶ Preparation of a mercury barometer. The full tube of mercury at the left is inverted and placed in a dish of mercury.

1 atmosphere

Air pressure is measured and expressed in many units. The standard atmospheric pressure, or simply **1 atmosphere**, is the pressure exerted by a column of mercury 760 mm high at a temperature of 0°C. The abbreviation for atmosphere is atm. The normal pressure of the atmosphere at sea level is 1 atm or 760 torr or 760 mm Hg. The SI unit for pressure is the pascal (Pa), where 1 atm = 101,325 Pa or 101.3 kPa. Other units for expressing pressure are inches of mercury, centimeters of mercury, the millibar (mbar), and pounds per square inch ($lb/in.^2$ or psi). The meteorologist uses inches of mercury in reporting atmospheric pressure. The values of these units equivalent to 1 atm are summarized in Table 13.2 (1 atm ≡ 760 torr ≡ 760 mm Hg ≡ 76 cm Hg ≡ 101,325 Pa ≡ 1013 mbar ≡ 29.9 in. Hg ≡ 14.7 $lb/in.^2$). (The symbol ≡ means *identical with.*)

TABLE 13.2 Pressure Units Equivalent to 1 Atmosphere

1 atm
760 torr
760 mm Hg
76 cm Hg
101,325 Pa
1013 mbar
29.9 in. Hg
14.7 $lb/in.^2$

Atmospheric pressure varies with altitude. The average pressure at Denver, Colorado, 1.61 km (1 mile) above sea level, is 630 torr (0.83 atm). Atmospheric pressure is 0.5 atm at about 5.5 km (3.4 miles) altitude.

Other liquids besides mercury may be employed for barometers, but they are not as useful as mercury because of the difficulty of maintaining a vacuum above the liquid and because of impractical heights of the liquid column. For example, a pressure of 1 atm will support a column of water about 10,336 mm (33.9 ft) high.

Pressure is often measured by reading the heights of mercury columns in millimeters on barometers and manometers. Thus pressure may be recorded as mm Hg. But in many applications the torr is superseding mm Hg as a unit of pressure. In problems dealing with gases it is necessary to make interconversions among the various pressure units. Since atm, torr, and mm Hg are common pressure units, we use illustrative problems involving all three of these units.

1 atm = 760 torr = 760 mm Hg

EXAMPLE 13.1 The average atmospheric pressure at Walnut, California, is 740. mm Hg. Calculate this pressure in (a) torr and (b) atmospheres.

SOLUTION This problem can be solved using conversion factors relating one unit of pressure to another.

(a) To convert mm Hg to torr, use the conversion factor 760 torr/760 mm Hg (1 torr/1 mm Hg):

$$740.\ \cancel{\text{mm Hg}} \times \frac{1\ \text{torr}}{1\ \cancel{\text{mm Hg}}} = 740.\ \text{torr}$$

(b) To convert mm Hg to atm, use the conversion factor 1 atm/760. mm Hg:

$$740.\ \cancel{\text{mm Hg}} \times \frac{1\ \text{atm}}{760.\ \cancel{\text{mm Hg}}} = 0.934\ \text{atm}$$

PRACTICE A barometer reads 1.12 atm. Calculate the corresponding pressure in (a) torr and (b) mm Hg.

Answer: (a) 851 torr (b) 851 mm Hg

13.4 Dependence of Pressure on Number of Molecules and Temperature Relationship

Pressure is produced by gas molecules colliding with the walls of a container. At a specific temperature and volume, the number of collisions depends on the number of gas molecules present. The number of collisions can be increased by increasing the number of gas molecules present. If we double the number of molecules, the frequency of collisions and the pressure should double. We find, for an ideal gas, that this doubling is actually what happens: When the temperature and volume are kept constant, the pressure is directly proportional to the number of moles or molecules of gas present. Figure 13.4 illustrates this concept.

A good example of this molecule-pressure relationship may be observed in an ordinary cylinder of compressed gas that is equipped with a pressure gauge. When the valve is opened, gas escapes from the cylinder. The volume of the cylinder is constant, and the decrease in quantity (moles) of gas is registered by a drop in pressure indicated on the gauge.

The pressure of a gas in a fixed volume also varies with temperature. When the temperature is increased, the kinetic energy of the molecules increases, causing more frequent collisions of the molecules with the walls of the container. This increase in collision frequency results in a pressure increase (see Figure 13.5).

13.5 Boyle's Law

Robert Boyle demonstrated experimentally that, at constant temperature (T), the volume (V) of a fixed mass of a gas is inversely proportional to the pressure (P). This relationship of P and V is known as **Boyle's law**. Mathematically, Boyle's

Boyle's law

law may be expressed

$$V \propto \frac{1}{P} \quad \text{(mass and temperature are constant)}$$

This equation says that the volume varies ($\propto$) inversely with the pressure, at

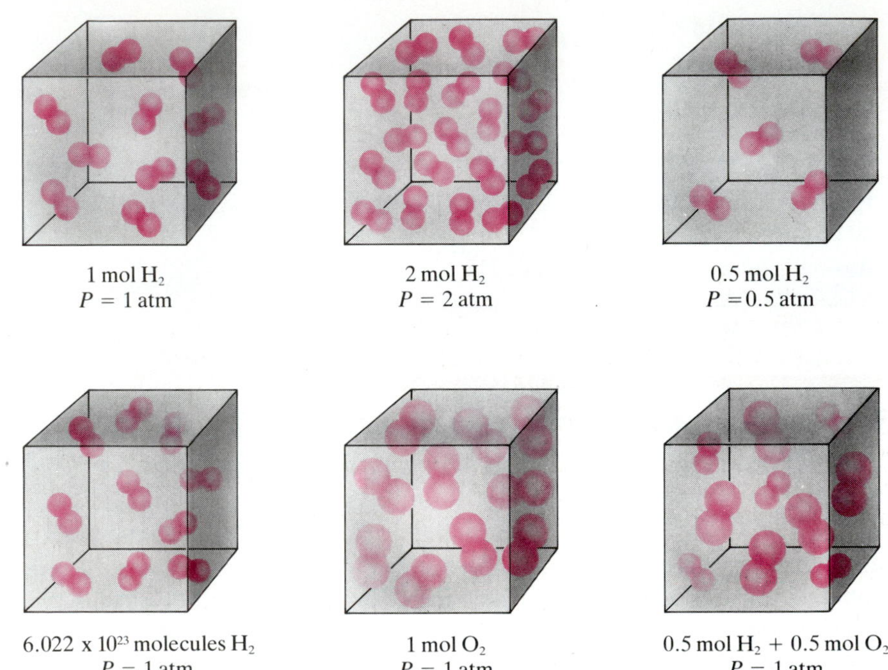

FIGURE 13.4 ▶ The pressure exerted by a gas is directly proportional to the number of molecules present. In each case shown, the volume is 22.4 L and the temperature is 0°C.

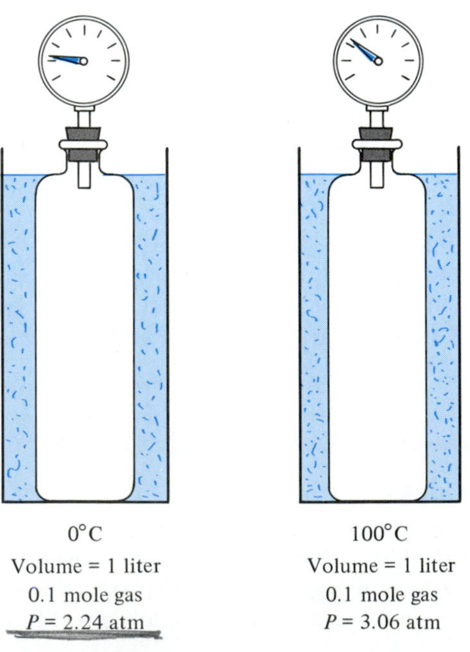

FIGURE 13.5 ▶ The pressure of a gas in a fixed volume increases with increasing temperature. The increased pressure is due to more frequent collisions of the gas molecules with the walls of the container at the higher temperature.

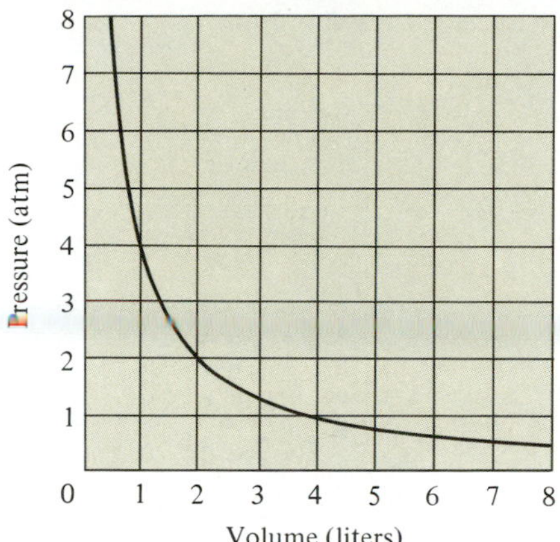

◀ **FIGURE 13.6**
Graph of pressure versus volume showing the inverse *PV* relationship of an ideal gas

constant mass and temperature. When the pressure on a gas is increased, its volume will decrease, and vice versa. The inverse relationship of pressure and volume is shown graphically in Figure 13.6.

Boyle demonstrated that, when he doubled the pressure on a specific quantity of a gas, keeping the temperature constant, the volume was reduced to one-half the original volume; when he tripled the pressure on the system, the new volume was one-third the original volume; and so on. His demonstration showed that the product of volume and pressure is constant if the temperature is not changed:

$$\text{PV} = \text{constant} \qquad \text{or} \qquad PV = k \quad \text{(mass and temperature are constant)}$$

Let us demonstrate this law by using a cylinder with a movable piston so that the volume of gas inside the cylinder may be varied by changing the external pressure (see Figure 13.7). We assume that the temperature and the number of gas molecules do not change. Let us start with a volume of 1000 mL and a pressure of 1 atm. When we change the pressure to 2 atm, the gas molecules are crowded closer together, and the volume is reduced to 500 mL. When we increase the pressure to 4 atm, the volume becomes 250 mL.

Note that the product of the pressure times the volume is the same number in each case, substantiating Boyle's law. We may then say that

$$P_1V_1 = P_2V_2$$

where P_1V_1 is the pressure–volume product at one set of conditions, and P_2V_2 is the product at another set of conditions. In each case the new volume may be calculated by multiplying the starting volume by a ratio of the two pressures involved. Of course, the ratio of pressures used must reflect the direction in which the volume should change. When the pressure is changed from 1 atm to 2 atm, the ratio to be used is 1 atm/2 atm. Now we can verify the results given in Figure 13.7:

(a) Starting volume, 1000 mL; pressure change, 1 atm ⟶ 2 atm

$$1000\text{ mL} \times \frac{1\,\cancel{\text{atm}}}{2\,\cancel{\text{atm}}} = 500\text{ mL}$$

▲
Robert Boyle (1627–1691)

FIGURE 13.7 ▶
The effect of pressure on the volume of a gas.

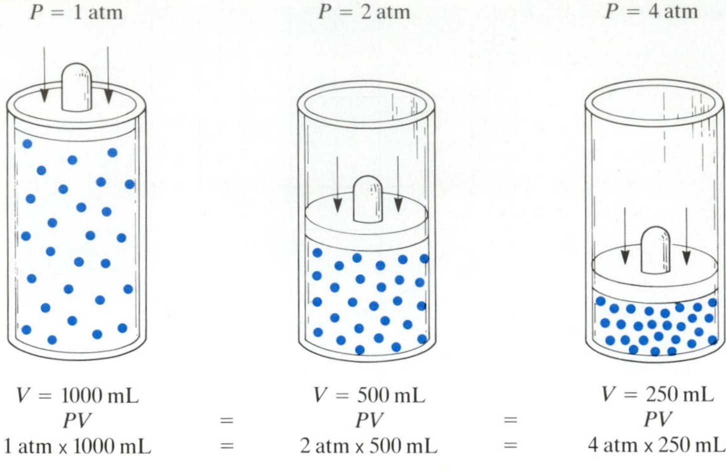

(b) Starting volume, 1000 mL; pressure change, 1 atm ⟶ 4 atm

$$1000 \text{ mL} \times \frac{1 \cancel{\text{atm}}}{4 \cancel{\text{atm}}} = 250 \text{ mL}$$

(c) Starting volume, 500 mL; pressure change, 2 atm ⟶ 4 atm

$$500 \text{ mL} \times \frac{2 \cancel{\text{atm}}}{4 \cancel{\text{atm}}} = 250 \text{ mL}$$

In summary, a change in the volume of a gas due to a change in pressure can be calculated by multiplying the original volume by a ratio of the two pressures. If the pressure is increased, the ratio should have the smaller pressure in the numerator and the larger pressure in the denominator. If the pressure is decreased, the larger pressure should be in the numerator and the smaller pressure in the denominator.

new volume = original volume × ratio of pressures

Examples of problems based on Boyle's law follow. If no mention is made of temperature, assume that it remains constant.

EXAMPLE 13.2

What volume will 2.50 L of a gas occupy if the pressure is changed from 760. mm Hg to 630. mm Hg?

SOLUTION

Method (a): Conversion Factors

Step 1 Determine whether pressure is being increased or decreased.

pressure decreases ⟶ volume increases

Step 2 Multiply the original volume by a ratio of pressures that will result in an increase in volume.

$$V = 2.50 \text{ L} \times \frac{760. \cancel{\text{mm Hg}}}{630. \cancel{\text{mm Hg}}} = 3.02 \text{ L (new volume)}$$

Method (b): Algebraic Equation

Step 1 Organize given information.

$$P_1 = 760.\ \text{mm Hg} \qquad V_1 = 2.50\ \text{L}$$
$$P_2 = 630.\ \text{mm Hg} \qquad V_2 = ?$$

Step 2 Write and solve equation for the unknown.

$$P_1V_1 = P_2V_2 \qquad V_2 = V_1 \times \frac{P_1}{P_2}$$

Step 3 Put given information into equation and calculate.

$$V_2 = 2.50\ \text{L} \times \frac{760.\ \cancel{\text{mm Hg}}}{630.\ \cancel{\text{mm Hg}}} = 3.02\ \text{L}$$

EXAMPLE 13.3

A given mass of hydrogen occupies 40.0 L at 700. torr. What volume will it occupy at 5.0 atm pressure?

SOLUTION

Method (a): Conversion Factors

Step 1 Determine whether the pressure is being increased or decreased. Notice that in order to compare the values the units must be the same.

$$700.\ \cancel{\text{torr}} \times \frac{1\ \text{atm}}{760\ \cancel{\text{torr}}} = 0.921\ \text{atm}$$

pressure increases ⟶ volume decreases

Step 2 Multiply the original volume by a ratio of pressures that will result in a decrease in volume.

$$V = 40.0\ \text{L} \times \frac{0.921\ \cancel{\text{atm}}}{5.00\ \cancel{\text{atm}}} = 7.37\ \text{L}$$

Method (b): Algebraic Equation

Step 1 Organize the given information. Remember to make units the same.

$$P_1 = 700.\ \text{torr} = 0.921\ \text{atm} \qquad V_1 = 40.0\ \text{L}$$
$$P_2 = 5.00\ \text{atm} \qquad V_2 = ?$$

Step 2 Write and solve equation for the unknown.

$$P_1V_1 = P_2V_2 \qquad V_2 = V_1 \times \frac{P_1}{P_2}$$

Step 3 Put given information into equation and calculate.

$$V_2 = 40.0\ \text{L} \times \frac{0.921\ \cancel{\text{atm}}}{5.00\ \cancel{\text{atm}}} = 7.37\ \text{L}$$

EXAMPLE 13.4 A gas occupies a volume of 200. mL at 400. torr pressure. To what pressure must the gas be subjected in order to change the volume to 75.0 mL?

SOLUTION

Method (a): Conversion Factors

Step 1 Determine whether volume is being increased or decreased.

volume decreases ⟶ pressure increases

Step 2 Multiply the original pressure by a ratio of volumes that will result in an increase in pressure.

new pressure = original pressure × ratio of volumes

$$P = 400.\text{ torr} \times \frac{200.\ \cancel{\text{mL}}}{75.0\ \cancel{\text{mL}}} = 1067\text{ torr} \quad \text{or} \quad 1.07 \times 10^3\text{ torr} \quad \text{(new pressure)}$$

Method (b): Algebraic Equation

Step 1 Organize given information. Remember to make units the same.

$P_1 = 400.$ torr $\quad V_1 = 200.$ mL

$P_2 = ?$ $\quad V_2 = 75.0$ mL

Step 2 Write and solve equation for the unknown.

$$P_1V_1 = P_2V_2 \qquad P_2 = P_1 \times \frac{V_1}{V_2}$$

Step 3 Put given information into equation and calculate.

$$P_2 = P_1 \times \frac{V_1}{V_2} = 400.\text{ torr} \times \frac{200.\ \cancel{\text{mL}}}{75.0\ \cancel{\text{mL}}} = 1.07 \times 10^3\text{ torr}$$

PRACTICE A gas occupies a volume of 3.86 L at 0.750 atm. What must the pressure be changed to in order for the volume to change to 4.86 L?
Answer: 0.600 atm

13.6 CHARLES' LAW

The effect of temperature on the volume of a gas was observed in about 1787 by the French physicist J. A. C. Charles (1746–1823). Charles found that various gases expanded by the same fractional amount when heated through the same temperature interval. Later it was found that if a given volume of any gas initially at 0°C was cooled by 1°C, the volume decreased by $\frac{1}{273}$; if cooled by 2°C, by $\frac{2}{273}$;

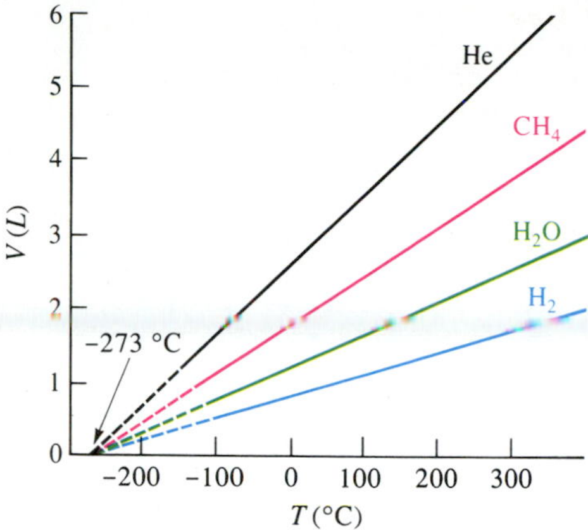

◀ **FIGURE 13.8 Volume-temperature relationship of gases. Extrapolated portion of the graph is shown by the broken line.**

if cooled by 20°C, by $\frac{20}{273}$; and so on. Since each degree of cooling reduced the volume by $\frac{1}{273}$, it was apparent that any quantity of any gas would have zero volume, if it could be cooled to −273°C. Of course no real gas can be cooled to −273°C for the simple reason that it liquefies before that temperature is reached. However, −273°C (more precisely −273.16°C) is referred to as *absolute zero*; this temperature is the zero point on the Kelvin (absolute) temperature scale. It is the temperature at which the volume of an ideal, or perfect, gas would become zero.

The volume–temperature relationship for gases is shown graphically in Figure 13.8. Experimental data show the graph to be a straight line that, when extrapolated, crosses the temperature axis at −273.16°C, or absolute zero.

Charles' law

In modern form, **Charles' law** states that at *constant pressure* the volume of a fixed mass of any gas is directly proportional to the absolute temperature. Mathematically, Charles' law may be expressed as

$$V \propto T \quad (P \text{ is constant})$$

which means that the volume of a gas varies directly with the absolute temperature when the pressure remains constant. In equation form Charles' law may be written as

$$V = kT \qquad \text{or} \qquad \frac{V}{T} = k \quad \text{(at constant pressure)}$$

where k is a constant for a fixed mass of the gas. If the absolute temperature of a gas is doubled, the volume will double. (A capital T is usually used for absolute temperature, K, and a small t for °C.)

To illustrate, let us return to the gas cylinder with the movable or free-floating piston, (see Figure 13.9). Assume that the cylinder labeled (a) contains a quantity of gas and the pressure on it is 1 atm. When the gas is heated, the molecules move faster, and their kinetic energy increases. This action should increase the number of collisions per unit of time and therefore increase the pressure. However, the increased internal pressure will cause the piston to rise to a level at which the internal and external pressures again equal 1 atm, as we see in cylinder (b). The net result is an increase in volume due to an increase in temperature.

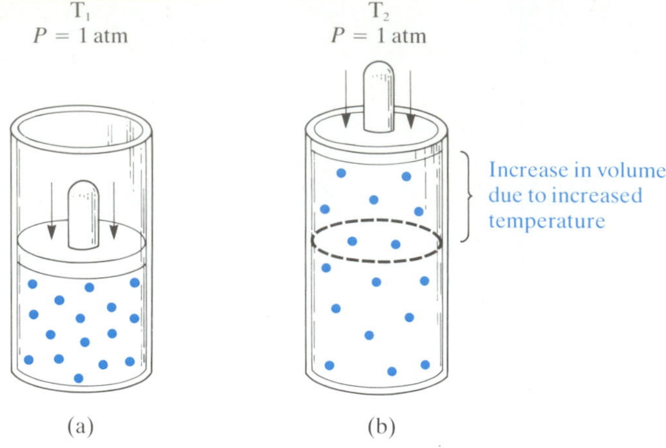

FIGURE 13.9 ▶
The effect of temperature on the volume of a gas. The gas in cylinder (a) is heated from T_1 to T_2. With the external pressure constant at 1 atm, the free-floating piston rises, resulting in an increased volume, as shown in cylinder (b).

Another equation relating the volume of a gas at two different temperatures is

$$\frac{V_1}{T_1} = \frac{V_2}{T_2} \quad \text{(constant } P\text{)}$$

where V_1 and T_1 are one set of conditions and V_2 and T_2 are another set of conditions.

A simple experiment showing the variation of the volume of a gas with temperature is illustrated in Figure 13.10. A flask to which a balloon is attached is immersed in either ice water or hot water. In ice water the volume is reduced, as shown by the collapse of the balloon; in hot water the gas expands and the balloon increases in size.

(a)

(b)

(c)

▲
FIGURE 13.10
The air-filled balloons in (a) are placed in liquid nitrogen (b). The volume of the air decreases tremendously at this temperature. In (c) the balloons are removed from the beaker and return to their original volume as they warm.

EXAMPLE 13.5

Three liters of hydrogen at −20.°C are allowed to warm to a room temperature of 27°C. What is the volume at room temperature if the pressure remains constant?

SOLUTION

Method (a): Conversion Factors

Step 1 Determine whether temperature is being increased or decreased. Remember temperature must be changed to Kelvin in gas problems.

$$-20.°\text{C} + 273 = 253\text{ K}$$

$$27°\text{C} + 273 = 300.\text{ K}$$

temperature increases ⟶ volume increases

Step 2 Multiply the original volume by a ratio of temperatures that will result in an increase in volume.

$$V = 3.00\text{ L} \times \frac{300.\,\cancel{\text{K}}}{253\,\cancel{\text{K}}} = 3.56\text{ L} \quad \text{(new volume)}$$

Method (b): Algebraic Equation

Step 1 Organize given information. Remember to make units the same.

$V_1 = 3.00$ L $\quad T_1 = -20.°\text{C} = 253$ K

$V_2 = ?$ $\quad T_2 = 27°\text{C} = 300.$ K

Step 2 Write and solve equation for the unknown.

$$\frac{V_1}{T_1} = \frac{V_2}{T_2} \qquad V_2 = V_1 \times \frac{T_2}{T_1}$$

Step 3 Put given information into equation and calculate.

$$V_2 = V_1 \times \frac{T_2}{T_1} = 3.00\text{ L} \times \frac{300.\,\cancel{\text{K}}}{253\,\cancel{\text{K}}} = 3.56\text{ L}$$

EXAMPLE 13.6

If 20.0 L of oxygen are cooled from 100°C to 0°C, what is the new volume?

SOLUTION

Since no mention is made of pressure, assume that pressure does not change.

Method (a): Conversion Factors

Step 1 Change °C to K:

$$100°\text{C} + 273 = 373\text{ K}$$

$$0°\text{C} + 273 = 273\text{ K}$$

Step 2 The ratio of temperature to be used is 273 K/373 K, because the final volume should be smaller than the original volume. The calculation is

$$V = 20.0\text{ L} \times \frac{273\,\cancel{\text{K}}}{373\,\cancel{\text{K}}} = 14.6\text{ L} \quad \text{(new volume)}$$

Method (b): Algebraic Equation

Step 1 Organize the given information. Remember to make units coincide.

$V_1 = 20.0\ \text{L} \qquad T_1 = 100°\text{C} = 373\ \text{K}$

$V_2 = ? \qquad T_2 = 0°\text{C} = 273\ \text{K}$

Step 2 Write and solve equation for the unknown.

$$\frac{V_1}{T_1} = \frac{V_2}{T_2} \qquad V_2 = V_1 \times \frac{T_2}{T_1}$$

Step 3 Put given information into equation and calculate.

$$V_2 = V_1 \times \frac{T_2}{T_1} = 20.0\ \text{L} \times \frac{273\ \text{K}}{373\ \text{K}} = 14.6\ \text{L}$$

PRACTICE A 4.50 L container of nitrogen at 28.0°C is heated to 56.0°C. Assuming the volume of the container can vary, what is the new volume of the gas?
Answer: 4.92 L

▲ **Joseph Louis Gay-Lussac (1778–1850).**

13.7 GAY-LUSSAC'S LAW

Three variables—pressure, P; volume, V; and temperature, T—are needed to describe a fixed amount of a gas. Boyle's law, $PV = k$, relates pressure and volume at constant temperature; Charles' law, $V = kT$, relates volume and temperature at constant pressure. A third relationship involving pressure and temperature at constant volume is stated: The pressure of a fixed mass of a gas, at constant volume, is directly proportional to the Kelvin temperature. In equation form the relationship is

$$P = kT \quad \text{(at constant volume)} \qquad \frac{P_1}{T_1} = \frac{P_2}{T_2}$$

Gay-Lussac's law

This relationship is a modification of Charles' law and is sometimes called **Gay-Lussac's law**.

EXAMPLE 13.7

The pressure of a container of helium is 650. torr at 25°C. If the sealed container is cooled to 0°C, what will the pressure be?

SOLUTION

Method (a): Conversion Factors

Step 1 Determine whether temperature is being increased or decreased.

temperature decreases ⟶ pressure decreases

Step 2 Multiply the original pressure by a ratio of Kelvin temperatures which will result in a decrease in pressure.

$$650.\text{ torr} \times \frac{273\text{ K}}{298\text{ K}} = 595\text{ torr}$$

Method (b): Algebraic Equation

Step 1 Organize given information. Remember to make units the same.

$P_1 = 650.$ torr $\quad T_1 = 25°\text{C} = 298$ K

$P_2 = ?$ $\quad T_2 = 0°\text{C} = 273$ K

Step 2 Write and solve equation for the unknown.

$$\frac{P_1}{T_1} = \frac{P_2}{T_2} \qquad P_2 = P_1 \times \frac{T_2}{T_1}$$

Step 3 Put given information into equation and calculate.

$$P_2 = 650.\text{ torr} \times \frac{273\text{ K}}{298\text{ K}} = 595\text{ torr}$$

PRACTICE A gas cylinder contains 40.0 L of gas at 45.0°C and has a pressure of 650. torr. What will the pressure be if the temperature is changed to 100.°C?
Answer: 762 torr

We may summarize the effects of changes in pressure, temperature, and quantity of a gas as follows:

1. In the case of a fixed or constant volume,
   (a) when the temperature is increased, the pressure increases.
   (b) when the quantity of a gas is increased, the pressure increases (T remaining constant).
2. In the case of a variable volume,
   (a) when the external pressure is increased, the volume decreases (T remaining constant).
   (b) when the temperature of a gas is increased, the volume increases (P remaining constant).
   (c) when the quantity of a gas is increased, the volume increases (P and T remaining constant).

13.8 Standard Temperature and Pressure

standard conditions
standard temperature and pressure (STP)

In order to compare volumes of gases, common reference points of temperature and pressure were selected and called **standard conditions** or **standard temperature and pressure** (abbreviated STP). Standard temperature is 273 K (0°C), and standard pressure is 1 atm or 760 torr or 760 mm Hg or 101.325 kPa. For purposes of comparison, volumes of gases are usually changed to STP conditions.

standard temperature = 273 K or 0°C
standard pressure = 1 atm or 760 torr or 760 mm Hg or 101.325 kPa

13.9 Combined Gas Laws: Simultaneous Changes in Pressure, Volume, and Temperature

When temperature and pressure change at the same time, the new volume may be calculated by multiplying the initial volume by the correct ratios of both pressure and temperature, as follows:

$$\text{final volume} = \text{initial volume} \times \begin{pmatrix}\text{ratio of} \\ \text{pressures}\end{pmatrix} \times \begin{pmatrix}\text{ratio of} \\ \text{temperatures}\end{pmatrix}$$

This equation combines Boyle's and Charles' laws, and the same considerations for the pressure and the temperature ratios should be used in the calculation. The four possible variations are:

1. Both T and P cause an increase in volume.
2. Both T and P cause a decrease in volume.
3. T causes an increase and P causes a decrease in volume.
4. T causes a decrease and P causes an increase in volume.

The P, V, and T relationships for a given mass of any gas, in fact, may be expressed as a single equation, $PV/T = k$. For problem solving this equation is usually written

$$\frac{P_1V_1}{T_1} = \frac{P_2V_2}{T_2}$$

where P_1, V_1, and T_1 are the initial conditions and P_2, V_2, and T_2 are the final conditions.

This equation can be solved for any one of the six variables and is useful in dealing with the pressure–volume–temperature relationships of gases. Note that when T is constant ($T_1 = T_2$), Boyle's law is represented; when P is constant ($P_1 = P_2$), Charles' law is represented; and when V is constant ($V_1 = V_2$), the modified Charles' or Gay-Lussac's law is represented.

Example 13.8

Given 20.0 L of ammonia gas at 5°C and 730. torr, calculate the volume at 50.°C and 800. torr.

SOLUTION

Step 1 Organize the given information, putting temperatures in Kelvin.

$P_1 = 730.\ \text{torr}$ $\quad P_2 = 800.\ \text{torr}$

$V_1 = 20.0\ \text{L}$ $\quad V_2 = ?$

$T_1 = 5°\text{C} = 278\ \text{K}$ $\quad T_2 = 50.°\text{C} = 323\ \text{K}$

Method (a): Conversion Factors

Step 2 Set up ratios of T and P.

$$T\ \text{ratio} = \frac{323\ \text{K}}{278\ \text{K}} \quad \text{(increase in } T \text{ should increase } V\text{)}$$

$$P\ \text{ratio} = \frac{730.\ \text{torr}}{800.\ \text{torr}} \quad \text{(increase in } P \text{ should decrease } V\text{)}$$

Step 3 Multiply the original pressure by the ratios.

$$V_2 = 20.0\ \text{L} \times \frac{730.\ \cancel{\text{torr}}}{800.\ \cancel{\text{torr}}} \times \frac{323\ \cancel{\text{K}}}{278\ \cancel{\text{K}}} = 21.2\ \text{L}$$

Method (b): Algebraic Equation

Step 2 Write and solve equation for the unknown.

Solve $\dfrac{P_1V_1}{T_1} = \dfrac{P_2V_2}{T_2}$ for V_2 by multiplying both sides of the equation by T_2/P_2 and rearranging to obtain

$$V_2 = \frac{V_1P_1T_2}{P_2T_1}$$

Step 3 Put given information into equation and calculate.

$$V_2 = \frac{20.0\ \text{L} \times 730.\ \cancel{\text{torr}} \times 323\ \cancel{\text{K}}}{800.\ \cancel{\text{torr}} \times 278\ \cancel{\text{K}}} = 21.2\ \text{L}$$

EXAMPLE 13.9

To what temperature (°C) must 10.0 L of nitrogen at 25°C and 700. torr be heated in order to have a volume of 15.0 L and a pressure of 760. torr?

SOLUTION

Step 1 Organize the given information, putting temperatures in Kelvin.

$P_1 = 700.\ \text{torr}$ $\quad P_2 = 760.\ \text{torr}$

$V_1 = 10.0\ \text{L}$ $\quad V_2 = 15.0\ \text{L}$

$T_1 = 25°\text{C} = 298\ \text{K}$ $\quad T_2 = ?$

Method (a): Conversion Factors

Step 2 Set up ratios of V and P.

$$P\ \text{ratio} = \frac{760.\ \text{torr}}{700.\ \text{torr}} \quad \text{(increase in } P \text{ should increase } T\text{)}$$

$$V\ \text{ratio} = \frac{15.0\ \text{L}}{10.0\ \text{L}} \quad \text{(increase in } V \text{ should increase } T\text{)}$$

Step 3 Multiply the original pressure by the ratios.

$$T_2 = 298 \text{ K} \times \frac{760. \cancel{\text{torr}} \times 15.0 \cancel{\text{L}}}{700. \cancel{\text{torr}} \times 10.0 \cancel{\text{L}}} = 485 \text{ K}$$

Method (b): Algebraic Equation

Step 2 Write and solve equation for unknown.

$$\frac{P_1V_1}{T_1} = \frac{P_2V_2}{T_2} \qquad T_2 = \frac{T_1P_2V_2}{P_1V_1}$$

Step 3 Put given information into equation and calculate.

$$T_2 = \frac{298 \text{ K} \times 760. \cancel{\text{torr}} \times 15.0 \cancel{\text{L}}}{700. \cancel{\text{torr}} \times 10.0 \cancel{\text{L}}} = 485 \text{ K}$$

In either method since the problem asks for °C, we must subtract 273 from the Kelvin answer:

$$485 \text{ K} - 273 = 212°\text{C}$$

EXAMPLE 13.10

The volume of a gas-filled balloon is 50.0 L at 20.°C and 742 torr. What volume will it occupy at standard temperature and pressure (STP)?

SOLUTION

Step 1 Organize the given information, putting temperatures in Kelvin.

$P_1 = 742.$ torr $\qquad P_2 = 760.$ torr (standard pressure)

$V_1 = 50.0$ L $\qquad V_2 = ?$

$T_1 = 20.°\text{C} = 293$ K $\qquad T_2 = 273$ K (standard pressure)

Method (a): Conversion Factors

Step 2 Set up ratios of T and P.

$$T \text{ ratio} = \frac{273 \text{ K}}{293 \text{ K}} \quad [\text{decrease in } T \text{ should decrease } V]$$

$$P \text{ ratio} = \frac{742 \text{ torr}}{760. \text{ torr}} \quad [\text{increase in } P \text{ should decrease } V]$$

Step 3 Multiply the original volume by the ratios.

$$V_2 = 50.0 \text{ L} \times \frac{273 \cancel{\text{K}}}{293 \cancel{\text{K}}} \times \frac{742 \cancel{\text{torr}}}{760. \cancel{\text{torr}}} = 45.5 \text{ L}$$

Method (b): Algebraic Equation

Step 2 Write and solve equation for the unknown.

$$\frac{P_1V_1}{T_1} = \frac{P_2V_2}{T_2} \qquad V_2 = \frac{P_1V_1T_2}{P_2T_1}$$

Step 3 Put given information into equation and calculate.

$$V_2 = \frac{742 \text{ torr} \times 50.0 \text{ L} \times 273 \text{ K}}{760. \text{ torr} \times 293 \text{ K}} = 45.5 \text{ L}$$

PRACTICE 15.00 L of gas at 45.0°C and 800. torr is heated to 400.°C, and the pressure changed to 300. torr. What is the new volume?

Answer: 84.7 L

PRACTICE To what temperature must 5.0 L of oxygen at 50.°C and 600. torr be heated in order to have a volume of 10.0 L and a pressure of 800. torr?

Answer: 861 K (588°C)

13.10 Dalton's Law of Partial Pressures

If gases behave according to the Kinetic-Molecular Theory, there should be no difference in the pressure–volume–temperature relationships whether the gas molecules are all the same or different. This similarity in the behavior of gases is the basis for an understanding of **Dalton's Law of Partial Pressures**, which states that the total pressure of a mixture of gases is the sum of the partial pressures exerted by each of the gases in the mixture. Each gas in the mixture exerts a pressure that is independent of the other gases present. These pressures are called **partial pressures**. Thus, if we have a mixture of three gases, A, B, and C, exerting partial pressures of 50 torr, 150 torr, and 400 torr, respectively, the total pressure will be 600 torr.

Dalton's Law of Partial Pressures

partial pressure

$$P_{\text{Total}} = p_A + p_B + p_C$$
$$P_{\text{Total}} = 50 \text{ torr} + 150 \text{ torr} + 400 \text{ torr} = 600 \text{ torr}$$

We can see an application of Dalton's law in the collection of insoluble gases over water. When prepared in the laboratory, oxygen is commonly collected by the downward displacement of water. Thus the oxygen is not pure but is mixed with water vapor (see Figure 13.11). When the water levels are adjusted to the same height inside and outside the bottle, the pressure of the oxygen plus water vapor inside the bottle is equal to the atmospheric pressure:

$$P_{\text{atm}} = p_{O_2} + p_{H_2O}$$
$$p_{O_2} = P_{\text{atm}} - p_{H_2O}$$

To determine the amount of O_2 or any other gas collected over water, we must subtract the pressure of the water vapor from the total pressure of the gas. The vapor pressure of water at various temperatures is tabulated in Appendix II.

FIGURE 13.11 ▶ Oxygen collected over water.

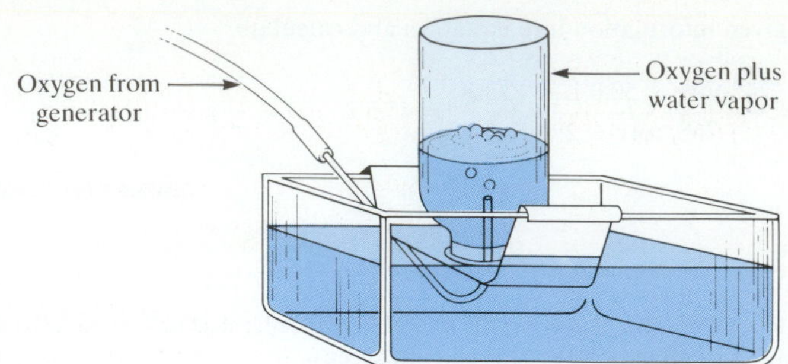

EXAMPLE 13.11

A 500. mL sample of oxygen was collected over water at 23°C and 760. torr. What volume will the dry O_2 occupy at 23°C and 760. torr? The vapor pressure of water at 23°C is 21.2 torr.

SOLUTION

To solve this problem, we must first determine the pressure of the oxygen alone, by subtracting the pressure of the water vapor present.

Step 1 Determine the pressure of dry O_2.

$$P_{\text{Total}} = 760.\ \text{torr} = p_{O_2} + p_{H_2O}$$
$$p_{O_2} = 760.\ \text{torr} - 21.2\ \text{torr} = 739\ \text{torr} \quad (\text{dry } O_2)$$

Step 2 Organize the given information.

$P_1 = 739$ torr $\quad P_2 = 760.$ torr

$V_1 = 500.$ mL $\quad V_2 = ?$

T is constant

Step 3 Solve as a Boyle's law problem:

$$V = 500.\ \text{mL} \times \frac{739\ \text{torr}}{760.\ \text{torr}} = 486\ \text{mL dry } O_2$$

PRACTICE Hydrogen gas was collected by downward displacement of water. A volume of 600.0 mL of gas was collected at 25.0°C and 740.0 torr. What volume will the dry hydrogen occupy at STP?

Answer: 518 mL

13.11 AVOGADRO'S LAW

Gay-Lussac's Law of Combining Volumes of Gases

Early in the 19th century J. L. Gay-Lussac (1778–1850) of France studied the volume relationships of reacting gases. His results, published in 1809, were summarized in a statement known as **Gay-Lussac's Law of Combining Volumes of Gases**: *When measured at the same temperature and pressure, the ratios of the*

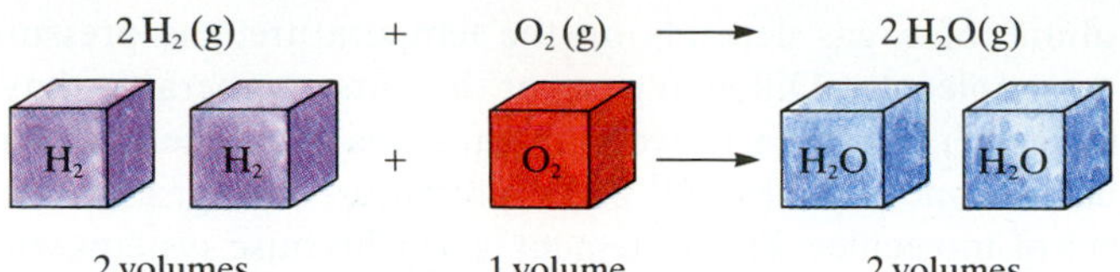

◀ **FIGURE 13.12 Gay-Lussac's Law of Combining Volumes of Gases applied to the reaction of hydrogen and oxygen. When measured at the same temperature and pressure, hydrogen and oxygen react in a volume ratio of 2 to 1.**

volumes of reacting gases are small whole numbers. Thus, H_2 and O_2 combine to form water vapor in a volume ratio of 2 to 1 (Figure 13.12); H_2 and Cl_2 react to form HCl in a volume ratio of 1 to 1; and H_2 and N_2 react to form NH_3 in a volume ratio of 3 to 1.

Two years later, in 1811, Amedeo Avogadro used the Law of Combining Volumes of Gases to make a simple but significant and far-reaching generalization concerning gases. **Avogadro's Law** states:

Avogadro's Law

Equal volumes of different gases at the same temperature and pressure contain the same number of molecules.

This law was a real breakthrough in understanding the nature of gases. (1) It offered a rational explanation of Gay-Lussac's Law of Combining Volumes of Gases and indicated the diatomic nature of such elemental gases as hydrogen, chlorine, and oxygen; (2) it provided a method for determining the molar masses of gases and for comparing the densities of gases of known molar mass (see Sections 13.12 and 13.13); and (3) it afforded a firm foundation for the development of the Kinetic-Molecular Theory.

On a volume basis hydrogen and chlorine react thus:

hydrogen +	chlorine ⟶	hydrogen chloride
1 volume	1 volume	2 volumes

By Avogadro's law, equal volumes of hydrogen and chlorine contain the same number of molecules. Therefore, hydrogen molecules react with chlorine molecules in a 1:1 ratio. Since two volumes of hydrogen chloride are produced, one molecule of hydrogen and one molecule of chlorine must produce two molecules of hydrogen chloride. Therefore, each hydrogen molecule and each chlorine molecule must be made up of two atoms. The coefficients of the balanced equation for the reaction give the correct ratios for volumes, molecules, and moles of reactants and products:

H_2 +	Cl_2 ⟶	2 HCl
1 volume	1 volume	2 volumes
1 molecule	1 molecule	2 molecules
1 mol	1 mol	2 mol

By like reasoning oxygen molecules also must contain at least two atoms because one volume of oxygen reacts with two volumes of hydrogen to produce two volumes of water vapor.

The volume of a gas depends on the temperature, the pressure, and the number of gas molecules. Different gases at the same temperature have the same average kinetic energy. Hence, if two different gases are at the same temperature, occupy equal volumes, and exhibit equal pressures, each gas must contain the same number of molecules. This statement is true because systems with identical *PVT* properties can be produced only by equal numbers of molecules having the same average kinetic energy.

13.12 Mole-Mass-Volume Relationships of Gases

molar volume

Because a mole contains 6.022×10^{23} molecules (Avogadro's number), a mole of any gas will have the same volume as a mole of any other gas at the same temperature and pressure. It has been experimentally determined that the volume occupied by a mole of any gas is 22.4 L at standard temperature and pressure. This volume, 22.4 L, is known as the **molar volume** of a gas. The molar volume is a cube about 28.2 cm (11.1 in.) on a side (see Figure 13.13). The molar masses of several gases, each occupying 22.4 L at STP, are also shown in Figure 13.13.

One mole of a gas occupies 22.4 liters at STP.

The molar volume is useful for determining the molar mass of a gas or of substances that can be easily vaporized. If the mass and the volume of a gas at STP are known, we can calculate its molar mass. For example, 1 liter of pure oxygen at STP has a mass of 1.429 g. The molar mass of oxygen may be calculated by multiplying the mass of 1 liter by 22.4 L/mol.

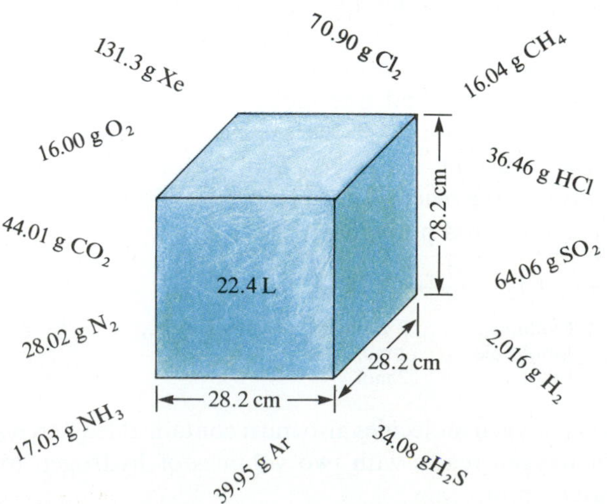

Figure 13.13 ▶ One mole of a gas occupies 22.4 L at STP. The mass given for each gas is the mass of 1 mol.

$$\frac{1.429 \text{ g}}{1 \not{L}} \times \frac{22.4 \not{L}}{1 \text{ mol}} = 32.00 \text{ g/mol} \quad \text{(molar mass)}$$

If the mass and volume are at other than standard conditions, we change the volume to STP and then calculate the molar mass. Note that we do not correct the mass to standard conditions—only the volume.

The molar volume, 22.4 L/mol, is used as a conversion factor to convert grams per liter to grams per mole (molar mass) and also to convert liters to moles. The two conversion factors are

$$\frac{22.4 \text{ L}}{1 \text{ mol}} \quad \text{and} \quad \frac{1 \text{ mol}}{22.4 \text{ L}}$$

These conversions must be done at STP except under certain special circumstances. Examples follow.

EXAMPLE 13.12

If 2.00 L of a gas measured at STP have a mass of 3.23 g, what is the molar mass of the gas?

SOLUTION

The unit of molar mass is g/mol; the conversion is from

$$\frac{\text{g}}{\text{L}} \longrightarrow \frac{\text{g}}{\text{mol}}$$

The starting amount is $\frac{3.23 \text{ g}}{2.00 \text{ L}}$. The conversion factor is $\frac{22.4 \text{ L}}{1 \text{ mol}}$.

The calculation is $\frac{3.23 \text{ g}}{2.00 \not{L}} \times \frac{22.4 \not{L}}{1 \text{ mol}} = 36.2 \text{ g/mol}$ (molar mass)

EXAMPLE 13.13

Measured at 40°C and 630. torr, the mass of 691 mL of ethyl ether is 1.65 g. Calculate the molar mass of ethyl ether.

SOLUTION

Step 1 Organize the given information, converting temperatures to Kelvin. Note that we must change to STP in order to determine molar mass.

$P_1 = 630.$ torr $\qquad P_2 = 760.$ torr

$V_1 = 691$ mL $\qquad V_2 = ?$

$T_1 = 313$ K(40.°C) $\qquad T_2 = 273$ K

Step 2 Use either the conversion factor method or the algebraic method and the combined gas law to correct the volume (V_2) to STP:

$$V_2 = 691 \text{ mL} \times \frac{273 \not{K}}{313 \not{K}} \times \frac{630. \not{\text{torr}}}{760. \not{\text{torr}}} = 500. \text{ mL} = 0.500 \text{ L} \quad \text{(at STP)}$$

Step 3 In the example V_2 is the volume for 1.65 g of the gas, so we can now find the molar mass by converting g/L to g/mol.

$$\frac{1.65 \text{ g}}{0.500 \not{L}} \times \frac{22.4 \not{L}}{\text{mol}} = 73.9 \text{ g/mol}$$

PRACTICE A gas with a mass of 86 g occupies 5.00 L at 25°C and 3.00 atm pressure. What is the molar mass of the gas?
Answer: 1.4×10^2 g/mol

13.13 Density of Gases

The density, *d*, of a gas is its mass per unit volume, which is generally expressed in grams per liter (g/L) as follows:

$$d = \frac{\text{mass}}{\text{volume}} = \frac{\text{g}}{\text{L}}$$

Because the volume of a gas depends on temperature and pressure, both should be given when stating the density of a gas. The volume of a solid or liquid is hardly affected by changes in pressure and is changed only slightly when the temperature is varied. Increasing the temperature from 0°C to 50°C will reduce the density of a gas by about 18% if the gas is allowed to expand, whereas a 50°C rise in the temperature of water (0°C $\longrightarrow$ 50°C) will change its density by less than 0.2%.

The density of a gas at any temperature and pressure can be determined by calculating the mass of gas present in 1 L. At STP, in particular, the density can be calculated by multiplying the molar mass of the gas by 1 mol/22.4 L.

$$d\text{ (at STP)} = \text{molar mass} \times \frac{1\text{ mol}}{22.4\text{ L}}$$

$$\text{molar mass} = d\text{ (at STP)} \times \frac{22.4\text{ L}}{1\text{ mol}}$$

Table 13.3 lists the densities of some common gases.

TABLE 13.3 Density of Common Gases at STP

Gas	Molar mass	Density (g/L at STP)	Gas	Molar mass	Density (g/L at STP)
H_2	2.016	0.0900	H_2S	34.08	1.52
CH_4	16.04	0.716	HCl	36.45	1.63
NH_3	17.03	0.760	F_2	38.00	1.70
C_2H_2	26.04	1.16	CO_2	44.01	1.96
HCN	27.03	1.21	C_3H_8	44.09	1.97
CO	28.01	1.25	O_3	48.00	2.14
N_2	28.02	1.25	SO_2	64.06	2.86
air	**(28.9)**	**(1.29)**	Cl_2	70.90	3.17
O_2	32.00	1.43			

EXAMPLE 13.14

Calculate the density of Cl_2 at STP.

SOLUTION

First calculate the molar mass of Cl_2. It is 70.90 g/mol. Since d = g/L, the conversion is

$$\frac{\text{g}}{\text{mol}} \longrightarrow \frac{\text{g}}{\text{L}}$$

The conversion factor is $\frac{1\text{ mol}}{22.4\text{ L}}$.

$$d = \frac{70.90\text{ g}}{1\text{ }\cancel{\text{mol}}} \times \frac{1\text{ }\cancel{\text{mol}}}{22.4\text{ L}} = 3.165\text{ g/L}$$

PRACTICE The molar mass of a gas is 20. g. Calculate the density of the gas at STP.

Answer: 0.89 g/L

13.14 Ideal Gas Equation

We have used four variables in calculations involving gases: the volume, V; the pressure, P; the absolute temperature, T; and the number of molecules or moles, which is abbreviated n. Combining these variables into a single expression, we obtain

$$V \propto \frac{nT}{P} \qquad \text{or} \qquad V = \frac{nRT}{P}$$

where R is a proportionality constant known as the *ideal gas constant*. The equation is commonly written as

$$PV = nRT$$

and is known as the **ideal gas equation**. This equation states in a single expression what we have considered in our earlier discussions: The volume of a gas varies directly with the number of gas molecules and the absolute temperature, and varies inversely with the pressure. The value and units of R depend on the units of P, V, and T. We can calculate one value of R by taking 1 mol of a gas at STP conditions. Solve the equation for R:

ideal gas equation

$$R = \frac{PV}{nT} = \frac{1\text{ atm} \times 22.4\text{ L}}{1\text{ mol} \times 273\text{ K}} = 0.0821\,\frac{\text{L-atm}}{\text{mol-K}}$$

The units of R in this case are liter-atmospheres (L-atm) per mol-K. When the value of R = 0.0821 L-atm/mol-K, P is in atmospheres, n is in moles, V is in liters, and T is in Kelvin.

The ideal gas equation can be used to calculate any one of the four variables if the other three are known.

EXAMPLE 13.15

What pressure will be exerted by 0.400 mol of a gas in a 5.00 L container at 17.0°C?

SOLUTION

Step 1 Organize the given information, putting temperatures in Kelvin

$P = ?$

$V = 5.00 \text{ L}$

$T = 290. \text{ K}$

$n = 0.400 \text{ mol}$

Step 2 Write and solve equation for the unknown.

$$PV = nRT \qquad \text{or} \qquad P = \frac{nRT}{V}$$

Step 3 Put given information into equation and calculate.

$$P = \frac{0.400 \cancel{\text{mol}} \times 0.0821 \cancel{\text{L}}\text{-atm}/\cancel{\text{mol-K}} \times 290. \cancel{\text{K}}}{5.00 \cancel{\text{L}}} = 1.90 \text{ atm}$$

EXAMPLE 13.16

How many moles of oxygen gas are in a 50.0 L tank at 22.0°C if the pressure gauge reads 2000. lb/in.^{2}?

SOLUTION

Step 1 Organize the given information, putting the temperature in Kelvin and changing pressure to atmospheres.

$$P = \frac{2000 \cancel{\text{lb}}}{\cancel{\text{in.}^2}} \times \frac{1 \text{ atm}}{14.7 \cancel{\text{lb/in.}^2}} = 136 \text{ atm}$$

$V = 50.0 \text{ L}$

$T = 295 \text{ K}$

$n = ?$

Step 2 Write and solve equation for the unknown.

$$PV = nRT \qquad \text{or} \qquad n = \frac{PV}{RT}$$

Step 3 Put given information into equation and calculate.

$$n = \frac{136 \cancel{\text{atm}} \times 50.0 \cancel{\text{L}}}{(0.0821 \cancel{\text{L-atm}}/\text{mol-}\cancel{\text{K}}) \times 295 \cancel{\text{K}}} = 281 \text{ mol } O_2$$

PRACTICE A 23.8 L cylinder contains oxygen gas at 20.0°C and 732 torr. How many moles of oxygen are in the cylinder?

Answer: 0.953 mol

The molar mass of a gaseous substance can be determined using the ideal gas equation. Since molar mass = g/mol, then mol = g/molar mass. Using M for molar mass we can substitute g/M for n (moles) in the ideal gas equation to get

$$PV = \frac{g}{M}RT \qquad \text{or} \qquad M = \frac{gRT}{PV} \quad \text{(modified ideal gas equation)}$$

which will allow us to calculate the molar mass, M, for any substance in the gaseous state.

EXAMPLE 13.17

Calculate the molar mass of butane gas, if 3.69 g occupy 1.53 L at 20°C and 1 atm pressure.

SOLUTION

Change 20°C to 293 K and substitute the data into the modified ideal gas equation:

$$M = \frac{gRT}{PV} = \frac{3.69\text{ g} \times 0.0821\ \cancel{\text{L-atm}} \times 293\ \cancel{\text{K}}}{1\ \cancel{\text{atm}} \times 1.53\ \cancel{\text{L}} \times \text{mol-}\cancel{\text{K}}} = 58.0\text{ g/mol}$$

PRACTICE A sample of 0.286 g of a certain gas occupies 50.0 mL at standard temperature and 76.0 cm Hg pressure. Determine the molar mass of the gas.

Answer: 128 g/mol

13.15 STOICHIOMETRY INVOLVING GASES

Mole–Volume (Gas) and Mass–Volume (Gas) Calculations Stoichiometric problems involving gas volumes can be solved by the general mole-ratio method outlined in Chapter 9. The factors 1 mol/22.4 L and 22.4 L/1 mol are used for converting volume to moles and moles to volume, respectively. These conversion factors are used under the assumption that the gases are at STP and behave as ideal gases. In practice, gases are measured at other than STP conditions, and the volumes are converted to STP for stoichiometric calculations.

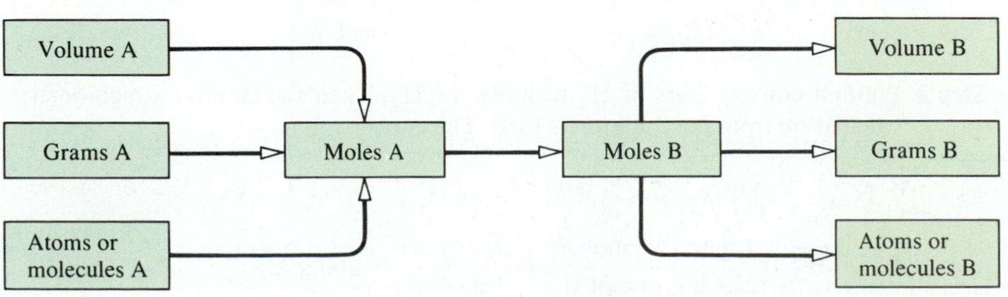

In a balanced equation, the number preceding the formula of a gaseous substance represents the number of moles or molar volumes (22.4 L at STP) of that substance.

The following are examples of typical problems involving gases and chemical equations.

EXAMPLE 13.18

What volume of oxygen (at STP) can be formed from 0.500 mol of potassium chlorate?

SOLUTION

Step 1 Write the balanced equation:

$$\underset{2\text{ mol}}{2\ KClO_3} \longrightarrow 2\ KCl + \underset{3\text{ mol}}{3\ O_2}\uparrow$$

Step 2 The starting amount is 0.500 mol $KClO_3$. The conversion is from

moles $KClO_3 \longrightarrow$ moles $O_2 \longrightarrow$ liters O_2

Step 3 Calculate the moles of O_2, using the mole-ratio method:

$$0.500\ \cancel{\text{mol KClO}_3} \times \frac{3\text{ mol } O_2}{2\ \cancel{\text{mol KClO}_3}} = 0.750\text{ mol } O_2$$

Step 4 Convert moles of O_2 to liters of O_2. The moles of a gas at STP are converted to liters by multiplying by the molar volume, 22.4 L/mole:

$$0.750\ \cancel{\text{mol}}\ O_2 \times \frac{22.4\text{ L}}{1\ \cancel{\text{mol}}} = 16.8\text{ L } O_2$$

Setting up a continuous calculation, we obtain

$$0.500\ \cancel{\text{mol KClO}_3} \times \frac{3\ \cancel{\text{mol}}\ O_2}{2\ \cancel{\text{mol KClO}_3}} \times \frac{22.4\text{ L}}{1\ \cancel{\text{mol}}} = 16.8\text{ L } O_2$$

EXAMPLE 13.19

How many grams of aluminum must react with sulfuric acid to produce 1.25 L of hydrogen gas at STP?

SOLUTION

Step 1 The balanced equation is

$$\underset{2\text{ mol}}{2\ Al(s)} + 3\ H_2SO_4(aq) \longrightarrow Al_2(SO_4)_3(aq) + \underset{3\text{ mol}}{3\ H_2(g)}$$

Step 2 We first convert liters of H_2 to moles of H_2. Then the familiar stoichiometric calculation from the equation is used. The conversion is

L $H_2 \longrightarrow$ mol $H_2 \longrightarrow$ mol Al $\longrightarrow$ g Al

$$1.25\ \cancel{\text{L}\ H_2} \times \frac{1\ \cancel{\text{mol}}}{22.4\ \cancel{\text{L}}} \times \frac{2\ \cancel{\text{mol Al}}}{3\ \cancel{\text{mol}\ H_2}} \times \frac{26.98\text{ g Al}}{1\ \cancel{\text{mol Al}}} = 1.00\text{ g Al}$$

EXAMPLE 13.20

What volume of hydrogen, collected at 30°C and 700. torr pressure, will be formed by reacting 50.0 g of aluminum with hydrochloric acid?

$$\underset{2\ \text{mol}}{2\ \text{Al}(s)} + 6\ \text{HCl}(aq) \longrightarrow 2\ \text{AlCl}_3(aq) + \underset{3\ \text{mol}}{3\ \text{H}_2(g)}$$

SOLUTION

In this problem the conditions are not at STP, so we cannot use the method shown in Example 13.18. Either we need to calculate the volume at STP from the equation and then convert this volume to the conditions given in the problem, or we can use the ideal gas equation. Let's use the ideal gas equation.

First calculate the moles of H_2 obtained from 50.0 g of Al. Then, using the ideal gas equation, calculate the volume of H_2 at the conditions given in the problem.

Step 1 Moles of H_2: The conversion is

grams Al ⟶ moles Al ⟶ moles H_2

$$50.0\ \text{g Al} \times \frac{1\ \text{mol Al}}{26.98\ \text{g Al}} \times \frac{3\ \text{mol H}_2}{2\ \text{mol Al}} = 2.78\ \text{mol H}_2$$

Step 2 Liters of H_2: Solve $PV = nRT$ for V and substitute the data into the equation.
Convert °C to K: 30°C + 273 = 303 K.
Convert torr to atm: 700. torr × 1 atm/760. torr = 0.921 atm.

$$V = \frac{nRT}{P} = \frac{2.78\ \text{mol H}_2 \times 0.0821\ \text{L-atm} \times 303\ \text{K}}{0.921\ \text{atm} \times \text{mol-K}} = 75.1\ \text{L H}_2$$

Note: The volume at STP is 62.3 L H_2.

PRACTICE If 10.0 g of sodium peroxide (Na_2O_2) react with water to produce sodium hydroxide and oxygen, how many liters of oxygen will be produced at 20°C and 750. torr?

$$2\ \text{Na}_2\text{O}_2 + 2\ \text{H}_2\text{O} \longrightarrow 4\ \text{NaOH} + \text{O}_2$$

Answer: 1.56 L O_2

Volume–Volume Calculations When all substances in a reaction are in the gaseous state, simplifications in the calculation can be made that are based on Avogadro's law that gases under identical conditions of temperature and pressure contain the same number of molecules and occupy the same volume. Using this same law, we can also state that, under the same conditions of temperature and pressure, the volumes of gases reacting are proportional to the numbers of moles of the gases in the balanced equation. Consider the reaction

$H_2(g)$	+ $Cl_2(g)$	⟶ $2\ HCl(g)$
1 mol	1 mol	2 mol
22.4 L	22.4 L	2 × 22.4 L
1 volume	1 volume	2 volumes
Y volume	*Y* volume	2 *Y* volumes

In this reaction 22.4 L of hydrogen will react with 22.4 L of chlorine to give $2 \times 22.4 = 44.8$ L of hydrogen chloride gas. This statement is true because these volumes are equivalent to the number of reacting moles in the equation. Therefore, *Y* volume of H_2 will combine with *Y* volume of Cl_2 to give 2 *Y* volumes of HCl. For example, 100 L of H_2 react with 100 L of Cl_2 to give 200 L of HCl; if the 100 L of H_2 and of Cl_2 are at 50°C, they will give 200 L of HCl at 50°C. When the temperature and pressure before and after a reaction are the same, volumes can be calculated without changing the volumes to STP.

For reacting gases: Volume–volume relationships are the same as mole–mole relationships.

EXAMPLE 13.21

What volume of oxygen will react with 150 L of hydrogen to form water vapor? What volume of water vapor will be formed?

SOLUTION

Assume that both reactants and products are measured at the same conditions. Calculation by reacting volumes:

$$2\,H_2(g) + O_2(g) \longrightarrow 2\,H_2O(g)$$

$2\,H_2(g)$	$O_2(g)$	$2\,H_2O(g)$
2 mol	1 mol	2 mol
2×22.4 L	22.4 L	2×22.4 L
2 volumes	1 volume	2 volumes
150 L	75 L	150 L

For every two volumes of H_2 that react, one volume of O_2 reacts and two volumes of $H_2O(g)$ are produced.

$$150\text{ L } H_2 \times \frac{1\text{ volume } O_2}{2\text{ volumes } H_2} = 75\text{ L } O_2$$

$$150\text{ L } H_2 \times \frac{1\text{ volume } O_2}{1\text{ volume } H_2} = 150\text{ L } H_2$$

EXAMPLE 13.22

The equation for the preparation of ammonia is

$$3H_2 + N_2 \xrightarrow{400^\circ C} 2\,NH_3$$

Assuming that the reaction goes to completion,

(a) What volume of H_2 will react with 50 L of N_2?
(b) What volume of NH_3 will be formed from 50 L of N_2?
(c) What volume of N_2 will react with 100 mL of H_2?
(d) What volume of NH_3 will be produced from 100 mL of H_2?
(e) If 600 mL of H_2 and 400 mL of N_2 are sealed in a flask and allowed to react, what amounts of H_2, N_2, and NH_3 are in the flask at the end of the reaction?

SOLUTION

The answers to parts (a)–(d) are shown in the boxes and can be determined from the equation by inspection, using the principle of reacting volumes.

$3\,H_2$	+ N_2	$\longrightarrow$ $2\,NH_3$
3 volumes	1 volume	2 volumes

	H_2	N_2	NH_3
(a)	**150 L**	50 L	
(b)		50 L	**100 L**
(c)	100 mL	**33.3 mL**	
(d)	100 mL		**66.7 mL**

(e) Volume ratio from the equation $= \dfrac{3 \text{ volumes } H_2}{1 \text{ volume } N_2}$

$$\text{Volume ratio used} = \frac{600 \text{ mL } H_2}{400 \text{ mL } N_2} = \frac{3 \text{ volumes } H_2}{2 \text{ volumes } N_2}$$

Comparing these two ratios, we see that an excess of N_2 is present in the gas mixture. Therefore, the reactant limiting the amount of NH_3 that can be formed is H_2:

$$3\,H_2 + N_2 \longrightarrow 2\,NH_3$$

600 mL 200 mL 400 mL

In order to have a 3:1 ratio of volumes reacting, 600 mL of H_2 will react with 200 mL of N_2 to produce 400 mL of NH_3, leaving 200 mL of N_2 unreacted. At the end of the reaction the flask will contain 400 mL of NH_3 and 200 mL of N_2.

Practice What volume of oxygen will react with 15.0 L of propane (C_3H_8) to form carbon dioxide and water? How much carbon dioxide will be formed? How much water?

$$C_3H_8(g) + 5\,O_2(g) \longrightarrow 3\,CO_2(g) + 4\,H_2O(g)$$

Answers: 75.0 L O_2 45.0 L CO_2 60.0 L H_2O

13.16 Real Gases

All the gas laws are based on the behavior of an ideal gas—that is, a gas with a behavior that is described exactly by the gas laws for all possible values of *P*, *V*, and *T*. Most real gases actually do behave very nearly as predicted by the gas laws over a fairly wide range of temperatures and pressures. However, when conditions are such that the gas molecules are crowded closely together (high pressure and/or low temperature), they show marked deviations from ideal behavior. Deviations occur because molecules have finite volumes and also have intermolecular attractions, which result in less compressibility at high pressures and greater compressibility at low temperatures than predicted by the gas laws. Many gases become liquids at high pressure and low temperature.

13.17 Air Pollution

Chemical reactions occur among the gases that are emitted into our atmosphere. In recent years, concern has been growing regarding the effect these reactions have upon our environment and our lives.

The outer portion of the atmosphere plays a significant role in determining the conditions for life at the surface of the earth. This stratosphere protects the surface from the intense radiation and particles bombarding our planet. Some of the high-energy radiation from the sun acts upon oxygen molecules in the stratosphere, converting them into ozone, O_3. Different molecular forms of an element are called **allotropes** of that element. Thus oxygen and ozone are allotropic forms of oxygen.

allotrope

$$O_2 \xrightarrow{\text{Sunlight}} \underset{\text{Oxygen atoms}}{O + O}$$

$$O_2 + O \longrightarrow O_3$$

Ultraviolet radiation from the sun is highly damaging to living tissues of plants and animals. The ozone layer, however, shields the earth by absorbing ultraviolet radiation and thus prevents most of this lethal radiation from reaching the earth's surface. The reaction that occurs is the reverse of the preceding one:

$$O_3 \xrightarrow[\text{radiation}]{\text{Ultraviolet}} O_2 + O + \text{heat}$$

Scientists have become concerned about a growing hazard to the ozone layer. Chlorofluorocarbon propellants, such as the Freons, CCl_3F and CCl_2F_2, which were used in aerosol spray cans and are used in refrigeration and air conditioning units, are stable compounds and remain unchanged in the lower atmosphere. When these chlorofluorocarbons are carried by convection currents to the stratosphere, they absorb ultraviolet radiation and produce chlorine atoms (chlorine free radicals) that in turn react with ozone. The following reaction sequence involving free radicals (see Section 17.15) has been proposed to explain the partial destruction of the ozone layer by chlorofluorocarbons.

$$\underset{\text{Fluorocarbon molecule}}{CCl_3F} \xrightarrow[\text{radiation}]{\text{Ultraviolet}} \underset{\text{Fluorocarbon free radical}}{\cdot CCl_2F} + \underset{\text{Chlorine free radical (atom)}}{Cl\cdot}$$

$$Cl\cdot + O_3 \longrightarrow ClO\cdot + O_2$$

$$ClO\cdot + O \longrightarrow O_2 + Cl\cdot$$

Because a chlorine atom is generated for each ozone molecule that is destroyed, a single chlorofluorocarbon molecule can be responsible for the destruction of many ozone molecules. During the past decade, scientists have discovered an annual thinning in the ozone layer over Antarctica. This is what we call the "hole" in the ozone layer. If this hole were to occur over populated regions of the world, severe effects would result, including a rise in the cancer rate, increased climatic temperatures, and vision problems.

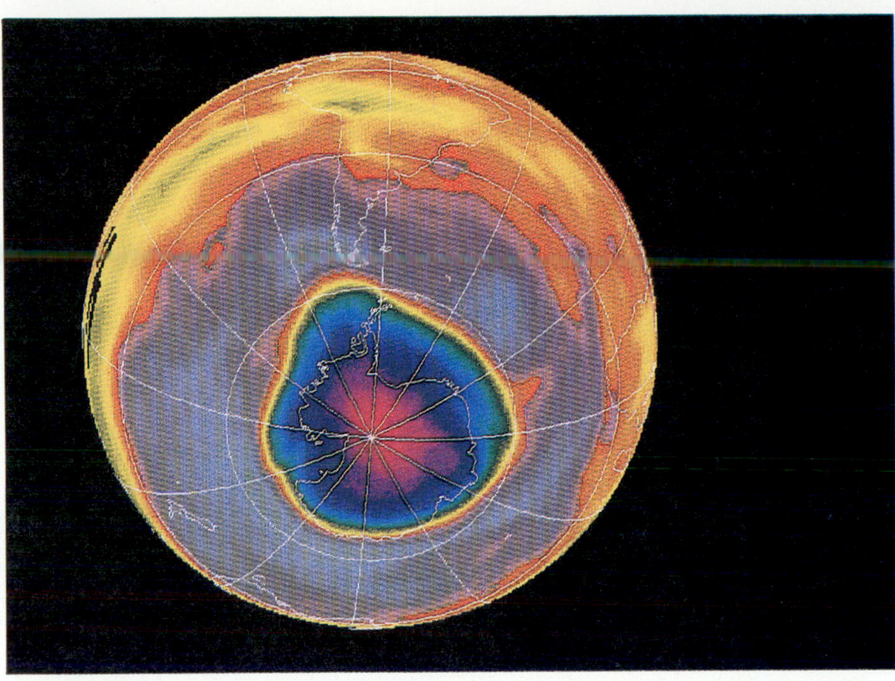

◀ **Satellite map showing a severe depletion or "hole" in the ozone layer over Antarctica in October, 1990. The hole is believed to be due to pollution of the atmosphere by chlorofluorocarbons used in aerosols and refrigerants.**

Ozone can be prepared by passing air or oxygen through an electrical discharge:

$$3\ O_2(g) + 286\ \text{kJ}\ (68.4\ \text{kcal}) \xrightarrow[\text{discharge}]{\text{Electrical}} 2\ O_3(g)$$

The characteristic pungent odor of ozone is noticeable in the vicinity of electrical machines and power transmission lines. Ozone is formed in the atmosphere during electrical storms and by the photochemical action of ultraviolet radiation on a mixture of nitrogen dioxide and oxygen. Areas with high air pollution are subject to high atmospheric ozone concentrations.

Ozone is not a desirable low-altitude constituent of the atmosphere, because it is known to cause extensive plant damage, cracking of rubber, and the formation of eye-irritating substances. Concentrations of ozone greater than 0.1 part per million (ppm) of air cause coughing, choking, headache, fatigue, and reduced resistance to respiratory infection. Concentrations between 10 and 20 ppm are fatal to humans.

In addition to ozone, the air in urban areas contains nitrogen oxides, substances which are components of smog. The term *smog* refers to air pollution in urban environments. Often the chemical reactions occur as part of a *photochemical process.* Nitric oxide (NO) is oxidized in the air or in automobile engines to produce nitrogen dioxide (NO_2). In the presence of light,

$$NO_2 \longrightarrow NO + O$$

In addition to nitrogen oxides, combustion of fossil fuels releases CO_2, CO, and sulfur oxides. Incomplete combustion releases unburned and partially burned hydrocarbons.

Chemistry in Action

Physiological Effects of Pressure Changes

The human body has a variety of methods for coping with the changes in atmospheric pressure. As we travel to the mountains, fly in an airplane, or take a high-speed elevator to the top of a skyscraper the pressure around us decreases. Our ears are sensitive to this because the eardrum (tympanic membrane) has air on both sides of it. The difference in pressure is relieved by yawning or moving the jaw to open the tubes (Eustacian) that connect the middle ear and throat and allow the pressure inside the eardrum to equalize with the outside.

Divers must also contend with the effects of pressure, most notably in body cavities containing air, such as the lungs, ears, and sinuses. Scuba divers do not experience a crushing effect of pressure at increased depths because the tank regulators deliver air at the same pressure as that of the surroundings. The diver must always breathe out regularly while ascending to the surface. Failure to do so may cause the lungs to expand, thus rupturing some of the alveoli, and resulting in loss of consciousness, brain damage, or heart attack. This is a clear application of Boyle's law.

Divers are also affected by consequences of Henry's law, which states that the amount of gas that will dissolve in a liquid varies directly with the pressure above the liquid. This means that during a dive the gases entering the lungs are absorbed into the blood to a greater extent than at the water's surface. If the diver returns too rapidly to the surface, the swift pressure reduction may cause dissolved gases to produce bubbles in the blood resulting in a condition known as *decompression sickness* or "the bends." The only successful method of treatment for this involves the use of a decompression chamber to increase the pressure once again and slowly decompress the diver back to normal pressure.

In the medical field, *hyperbaric units* are used to treat patients who have cells starved for oxygen. In these units the whole room may be placed at high pressure (2 or 3 atm) and the entire staff as well as the patient undergo gradual compression and, following treatment, decompression. These units are widely used to treat carbon monoxide poisoning. Oxygen is dissolved directly into the plasma giving the tissues temporary relief from oxygen deprivation. Hyperbaric units are also effective in treating other problems such as skin grafts, severe thermal burns, and radiation tissue damage.

Society is continually attempting to discover, understand, and control emissions that contribute to this sort of atmospheric chemistry. It is a problem that each one of us faces as we look to the future if we want to continue to support life as we know it on our planet.

Concepts in Review

1. State the principal assumptions of the Kinetic Molecular Theory.
2. Estimate the relative rates of effusion of two gases of known molar mass.
3. Sketch and explain the operation of a mercury barometer.
4. List two factors that determine gas pressure in a vessel of fixed volume.
5. State Boyle's, Charles' and Gay-Lussac's laws. Use all of them in problems.
6. State the combined gas law. Indicate when it is used.
7. Use Dalton's Law of Partial Pressures and the combined gas law to determine the dry STP volume of a gas collected over water.
8. State Avogadro's law.
9. Understand the mole-mass-volume relationship of gases.
10. Determine the density of any gas at STP.
11. Determine the molar mass of a gas from its density at a known temperature and pressure.
12. Solve problems involving the ideal gas equation.
13. Make mole-volume, mass-volume, and volume-volume stoichiometric calculations from balanced chemical equations.
14. State two reasons why real gases may deviate from the behavior predicted for an ideal gas.

Exercises

An asterisk indicates a more challenging question or problem

1. What evidence is used to show diffusion in Figure 13.1? If hydrogen, H_2, and oxygen, O_2, were in the two flasks, how could we prove that diffusion had taken place?
2. How does the air pressure inside the balloon shown in Figure 13.2 compare with the air pressure outside the balloon? Explain.
3. According to Table 13.1, what two gases are the major constituents of dry air?
4. How does the pressure represented by 1 torr compare in magnitude to the pressure represented by 1 mm Hg? See Table 13.2.
5. In which container illustrated in Figure 13.5 are the molecules of gas moving faster? Assume both gases to be hydrogen.
6. In Figure 13.6, what gas pressure corresponds to a volume of 4 L?
7. How do the data illustrated in Figure 13.6 substantiate Boyle's law?
8. What effect would you observe in Figure 13.9 if T_2 were lower than T_1?
9. In the diagram shown in Figure 13.11, is the pressure of the oxygen plus water vapor inside the bottle equal to, greater than, or less than the atmospheric pressure outside the bottle? Explain.
10. List five gases in Table 13.3 that are more dense than air. Explain the basis for your selections.
11. What are the basic assumptions of the Kinetic-Molecular Theory?
12. Arrange the following gases, all at standard temperature, in order of increasing relative molecular velocities: H_2, CH_4, Rn, N_2, F_2, He. What is your basis for determining the order?
13. List, in descending order, the average kinetic energies of the molecules in Exercise 12.
14. What are the four parameters used to describe the behavior of a gas?

15. What are the characteristics of an ideal gas?
16. Under what condition of temperature, high or low, is a gas least likely to exhibit ideal behavior? Explain.
17. Under what conditions of pressure, high or low, is a gas least likely to exhibit ideal behavior? Explain.
18. Compare, at the same temperature and pressure, equal volumes of H_2 and O_2 as to:
   (a) Number of molecules
   (b) Mass
   (c) Number of moles
   (d) Average kinetic energy of the molecules
   (e) Rate of effusion
   (f) Density
19. How does the Kinetic-Molecular Theory account for the behavior of gases as described by:
   (a) Boyle's law
   (b) Charles' law
   (c) Dalton's Law of Partial Pressures
20. Explain how the reaction $N_2(g) + O_2(g) \xrightarrow{\Delta} 2\ NO(g)$ proves that nitrogen and oxygen are diatomic molecules.
21. What is the reason for referring gases to STP?
22. Is the conversion of oxygen to ozone an exothermic or endothermic reaction? How do you know?
23. Write formulas for an oxygen atom, an oxygen molecule, and an ozone molecule. How many electrons are in an oxygen molecule?
24. When constant pressure is maintained, what effect does heating a mole of N_2 gas have on the following?
   (a) Its density
   (b) Its mass
   (c) The average kinetic energy of its molecules
   (d) The average velocity of its molecules
   (e) The number of N_2 molecules in the sample
25. What causes "the bends"?
26. State Henry's Law and indicate its significance to a scuba diver.
27. Assuming ideal gas behavior, which of the following statements are correct? (Try to answer without referring to your text.) Rewrite each incorrect statement to make it correct.
   (a) The pressure exerted by a gas at constant volume is independent of the temperature of the gas.
   (b) At constant temperature, increasing the pressure exerted on a gas sample will cause a decrease in the volume of the gas sample.
   (c) At constant pressure, the volume of a gas is inversely proportional to the absolute temperature.
   (d) At constant temperature, doubling the pressure on a gas sample will cause the volume of the gas sample to decrease to one-half its original volume.
   (e) Compressing a gas at constant temperature will cause its density and mass to increase.
   (f) Equal volumes of CO_2 and CH_4 gases at the same temperature and pressure contain
      (1) The same number of molecules
      (2) The same mass
      (3) The same densities
      (4) The same number of moles
      (5) The same number of atoms
   (g) At constant temperature, the average kinetic energy of O_2 molecules at 200 atm pressure is greater than the average kinetic energy of O_2 molecules at 100 atm pressure.
   (h) According to Charles' law, the volume of a gas becomes zero at $-273°C$.
   (i) One liter of O_2 gas at STP has the same mass as 1 L of O_2 gas at 273°C and 2 atm pressure.
   (j) The volume occupied by a gas depends only on its temperature and pressure.
   (k) In a mixture containing O_2 molecules and N_2 molecules, the O_2 molecules, on the average, are moving faster than the N_2 molecules.
   (l) $PV = k$ is a statement of Charles' law.
   (m) If the temperature of a sample of gas is increased from 25°C to 50°C, the volume of the gas will increase by 100%.
   (n) One mole of chlorine, Cl_2, at 20°C and 600 torr pressure contains 6.022×10^{23} molecules.
   (o) One mole of H_2 plus 1 mole of O_2 in an 11.2 L container exert a pressure of 4 atm at 0°C.
   (p) When the pressure on a sample of gas is halved, with the temperature remaining constant, the density of the gas is also halved.
   (q) When the temperature of a sample of gas is increased at constant pressure, the density of the gas will decrease.
   (r) According to the equation

$$2\ KClO_3(s) \xrightarrow{\Delta} 2\ KCl(s) + 3\ O_2(g),$$

      1 mol of $KClO_3$ will produce 67.2 L of O_2 at STP.
   (s) $PV = nRT$ is a statement of Avogadro's law.
   (t) STP conditions are 1 atm and 0°C.

Pressure Units

28. The barometer reads 715 mm Hg. Calculate the corresponding pressure in
(a) atmospheres (d) torrs
(b) inches of Hg (e) millibars
(c) lb/in.^{2} (f) kilopascals

29. Express the following pressures in atmospheres:
(a) 28 mm Hg (c) 795 torr
(b) 6000. cm Hg (d) 5.00 kPa

Boyle's, Charles', and Gay-Lussac's Laws

Boyles

30. A gas occupies a volume of 400. mL at 500. mm Hg pressure. What will be its volume, at constant temperature, if the pressure is changed to (a) 760 mm Hg; (b) 250 torr; (c) 2.00 atm?

31. A 500. mL sample of a gas is at a pressure of 640. mm Hg. What must be the pressure, at constant temperature, if the volume is changed to (a) 855 mL and (b) 450 mL?

32. At constant temperature, what pressure would be required to compress 2500 L of hydrogen gas at 1.0 atm pressure into a 25 L tank?

33. Given 6.00 L of N_2 gas at $-25°C$, what volume will the nitrogen occupy at (a) 0.0°C; (b) 0.0°F; (c) 100.K; (d) 345 K? (Assume constant pressure.)

34. Given a sample of a gas at 27°C, at what temperature would the volume of the gas sample be doubled, the pressure remaining constant?

***35.** A gas sample at 22°C and 740 torr pressure is heated until its volume is doubled. What pressure would restore the sample to its original volume?

36. A gas occupies 250 mL at 700. torr and 22°C. When the pressure is changed to 500. torr, what temperature (°C) is needed to maintain the same volume?

37. Hydrogen stored in a metal cylinder has a pressure of 252 atm at 25°C. What will be the pressure in the cylinder when the cylinder is lowered into liquid nitrogen at $-196°C$?

38. The tires on an automobile were filled with air to 30. psi at 71.0°F. When driving at high speeds, the tires become hot. If the tires have a bursting pressure of 44 psi, at what temperature (°F) will the tires "blow out"?

Combined Gas Laws

39. A gas occupies a volume of 410 mL at 27°C and 740 mm Hg pressure. Calculate the volume the gas would occupy at (a) STP; (b) 250.°C and 680 mm Hg pressure.

40. What volume would 5.30 L of H_2 gas at STP occupy at 70°C and 830 torr pressure?

41. What pressure will 800. mL of a gas at STP exert when its volume is 250. mL at 30°C?

42. An expandable balloon contains 1400. L of He at 0.950 atm pressure and 18°C. At an altitude of 22 miles (temperature 2.0°C and pressure 4.0 torr), what will be the volume of the balloon?

43. A gas occupies 22.4 L at 2.50 atm and 27°C. What will be its volume at 1.50 atm and $-5.00°C$?

44. How many gas molecules are present in 600. mL of N_2O at 40°C and 400. torr pressure? How many atoms are present? What would be the volume of the sample at STP?

Dalton's Law of Partial Pressures

45. What would be the partial pressure of O_2 gas collected over water at 20°C and 720. torr pressure? (Check Appendix II for the vapor pressure of water.)

46. An equilibrium mixture contains H_2 at 600. torr pressure, N_2 at 200 torr pressure, and O_2 at 300. torr pressure. What is the total pressure of the gases in the system?

47. A sample of methane gas, CH_4, was collected over water at 25.0°C and 720. torr. The volume of the wet gas is 2.50 L. What will be the volume of the dry methane at standard pressure?

***48.** 5.00 L of CO_2 at 500. torr and 3.00 L of CH_4 at 400. torr are put into a 10.0 L container. What is the pressure exerted by the gases in the container?

Mole-Mass-Volume Relationships

49. What volume will 2.5 mol of Cl_2 occupy at STP?

50. A steel cylinder contains 60. mol of H_2 at a pressure of 1500 lb/in.^{2}. (a) How many moles of H_2 are in the cylinder when the pressure reads 850 lb/in.^{2}? (b) How many grams of H_2 were initially in the cylinder?

51. How many grams of CO_2 are present in 2500 mL of CO_2 at STP?

52. At STP, 560 mL of a gas have a mass of 1.08 g. What is the molar mass of the gas?

53. What volume will each of the following occupy at STP?
(a) 1.0 mol of NO_2
(b) 17.05 g of NO_2
(c) 1.20×10^{24} molecules of NO_2

54. How many molecules of NH_3 gas are present in a 1.00 L flask at STP?

*55. How many moles of Cl_2 are in one cubic meter ($1.00\ m^3$) at STP?

Density of Gases

56. A gas has a density at STP of 1.78 g/L. What is its molar mass?
57. Calculate the density of the following gases at STP:
(a) Kr (b) He (c) SO_3 (d) C_4H_8
58. Calculate the density of:
(a) F_2 gas at STP
(b) F_2 gas at 27°C and 1 atm pressure
*59. At what temperature (°C) will the density of methane, CH_4, be 1.0 g/L at 1.0 atm pressure?

Ideal Gas Equation and Stoichiometry

60. Using the ideal gas equation, $PV = nRT$, calculate:
(a) The volume of 0.510 mole of H_2 at 47°C and 1.6 atm pressure
(b) The number of grams in 16.0 L of CH_4 at 27°C and 600. torr pressure
*(c) The density of CO_2 at 4.00 atm pressure and −20°C
*(d) The molar mass of a gas having a density of 2.58 g/L at 27°C and 1.00 atm pressure.
Hints for (c) and (d): n = moles = m/molar mass, and $d = m/V$.
61. What is the molar mass of a gas if 1.15 g occupy 0.215 L at 0.813 atm and 30°C?
62. At 27°C and 750 torr pressure, what will be the volume of 2.3 mol of Ne?
63. What volume will a mixture of 5.00 mol of H_2 and 0.500 mol of CO_2 occupy at STP?
64. 4.50 mol of a gas occupy 0.250 L at 4.15 atm. What is the Kelvin temperature of the system?
65. How many moles of N_2 gas occupy 5.20 L at 250 K and 0.500 atm?
66. What volume of hydrogen at STP can be produced by reacting 8.30 mol of Al with sulfuric acid? The equation is

$$2\ Al(s) + 3\ H_2SO_4(aq) \longrightarrow Al_2(SO_4)_3(aq) + 3\ H_2(g)$$

67. Given the equation

$$4\ NH_3(g) + 5\ O_2(g) \longrightarrow 4\ NO(g) + 6\ H_2O(g)$$

(a) How many moles of NH_3 are required to produce 5.5 mol of NO?
(b) How many moles of NH_3 will react with 7.0 mol of O_2?
(c) How many liters of NO can be made from 12 L of O_2 and 10 L of NH_3 at STP?
(d) At constant temperature and pressure how many liters of NO can be made by the reaction of 800. mL of O_2?
(e) At constant temperature and pressure, what is the maximum volume, in liters, of NO that can be made from 3.0 L of NH_3 and 3.0 L of O_2?
(f) How many grams of O_2 must react to produce 60. L of NO measured at STP?
*(g) How many grams of NH_3 must react to produce a total of 32 L of products, NO plus H_2O, measured at STP?
68. Given the equation

$$4\ FeS(s) + 7\ O_2(g) \xrightarrow{\Delta} 2\ Fe_2O_3(s) + 4\ SO_2(g)$$

(a) How many liters of O_2, measured at STP, will react with 0.600 kg of FeS?
(b) How many liters of SO_2, measured at STP, will be produced from 0.600 kg of FeS?
*69. Acetylene, C_2H_2, and hydrogen fluoride react to give difluoroethane.

$$C_2H_2(g) + 2\ HF(g) \longrightarrow C_2H_4F_2(g)$$

When 1.0 mol of C_2H_2 and 5.0 mol of HF are reacted in a 10.0 L flask, what will be the pressure in the flask at 0°C when the reaction is complete?

Additional Problems

70. What are the relative rates of effusion of N_2 and He?
*71. (a) What are the relative rates of effusion of methane, CH_4, and helium, He?
(b) If these two gases are simultaneously introduced into opposite ends of a 100. cm tube and allowed to diffuse toward each other, at what distance from the helium end will molecules of the two gases meet?
*72. A gas has a percent composition by mass of 85.7% carbon and 14.3% hydrogen. At STP the density of the gas is 2.50 g/L. What is the molecular formula of the gas?
*73. Assume that the reaction $2\ CO(g) + O_2(g) \longrightarrow 2\ CO_2(g)$ goes to completion. When 10. mol of CO and 8.0 mol of O_2 react in a closed 10. L vessel,
(a) How many moles of CO, O_2, and CO_2 are present at the end of the reaction?

(b) What will be the total pressure in the flask at 0°C?

*74. 250 mL of O_2, measured at STP, were obtained by the decomposition of the $KClO_3$ in a 1.20 g mixture of KCl and $KClO_3$:

$$2\ KClO_3(s) \longrightarrow 2\ KCl(s) + 3\ O_2(g)$$

What is the percent by mass of $KClO_3$ in the mixture?

*75. Look at the apparatus below. When a small amount of water is squirted into the flask containing ammonia gas (by squeezing the bulb of the medicine dropper), water from the beaker fills the flask through the long glass tubing. Explain this phenomenon. (Remember that ammonia dissolves in water.)

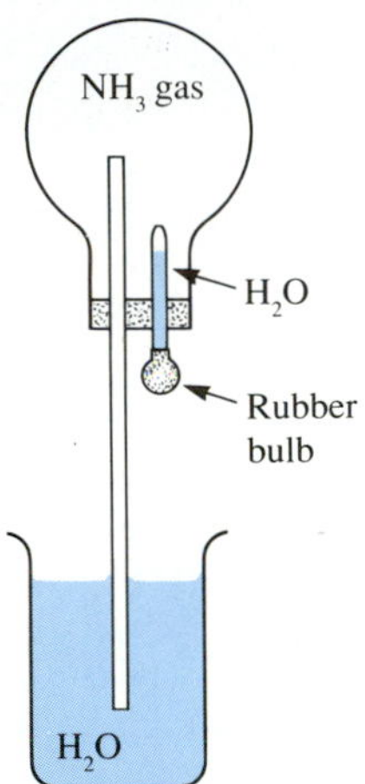

*76. Determine the pressure of the gas in each of the figures below:

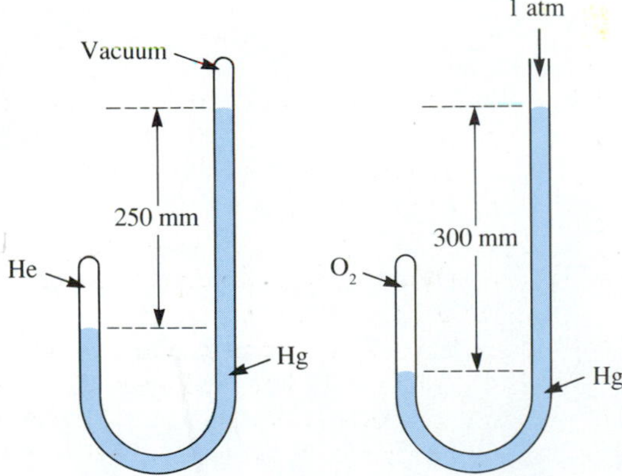

*77. Consider the arrangement of gases shown below. If the valve between the gases is opened and the temperature is held constant:
(a) Determine the pressure of each gas.
(b) Determine the total pressure in the system.

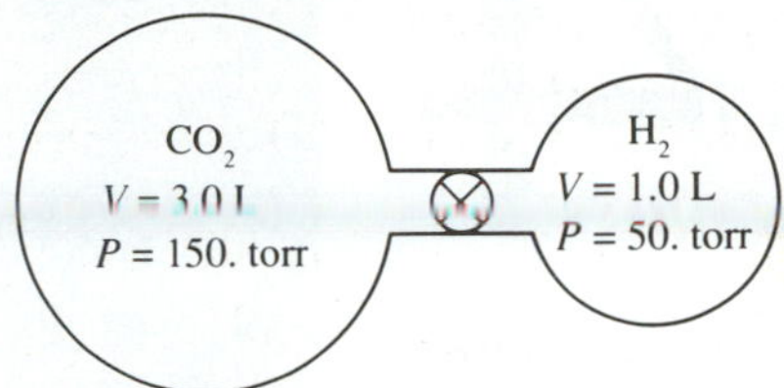

*78. Air has a mass of 1.29 g/L at STP. Calculate the density of air on Pikes Peak, where the pressure is 450 torr and the temperature is 17°C.

*79. A room is 16 ft × 12 ft × 12 ft. Would air enter or leave the room and how much, if the temperature in the room changed from 27°C to −3°C with the pressure remaining constant? Show evidence.

*80. A steel cylinder contained 50.0 L of oxygen gas under a pressure of 40.0 atm and a temperature of 25°C. What was the pressure in the cylinder during a storeroom fire that caused the temperature to rise 152°C? (Be careful!)

14

Water and the Properties of Liquids

◀ **CHAPTER OPENING PHOTO: Water on planet Earth as viewed from an Apollo spacecraft.**

Planet Earth, that magnificent blue sphere we enjoy viewing from the space shuttle, is spectacular. Over 75% of the earth is covered with water. We are born from it, drink it, bathe in it, cook with it, enjoy its beauty in waterfalls and rainbows, and stand in awe of the majesty of icebergs. Water supports and enhances life.

In chemistry water provides the medium for numerous reactions. The shape of the water molecule is the basis for hydrogen bonds. These bonds determine the unique properties and reactions of water. The tiny water molecule holds the answers to many of the mysteries of chemical reactions.

14.1 LIQUIDS AND SOLIDS

In the last chapter, we found that a gas can be a substance containing particles that are far apart, in rapid random motion, and independent of each other. The Kinetic Molecular Theory, along with the ideal gas equation summarizes the behavior of most gases at relatively high temperatures and low pressures.

Solids are obviously very different from gases. A solid contains particles very close together, has a high density, compresses negligibly, and maintains its shape regardless of container. These characteristics indicate large attractive forces between particles. The model for solids is very different from the one for gases.

Liquids, on the other hand, lie somewhere between the extremes of gases and solids. A liquid contains particles that are close together, is essentially incompressible, and has a definite volume. These properties are very similar to solids. However, a liquid also takes the shape of its container, which is closer to the model of a gas.

Although liquids and solids show similar properties, they differ tremendously from gases. No simple mathematical relationship, like the ideal gas equation, works well for liquids or solids. Instead, these models are directly related to the forces of attraction between molecules. With these general statements in mind, let us consider some of the specific properties of liquids.

14.2 Evaporation

When beakers of water, ethyl ether, and ethyl alcohol are allowed to stand uncovered in an open room, the volumes of these liquids gradually decrease. The process by which this change takes place is called *evaporation.*

Attractive forces exist between molecules in the liquid state. Not all of these molecules, however, have the same kinetic energy. Molecules that have greater than average kinetic energy can overcome the attractive forces and break away from the surface of the liquid to become a gas. **Evaporation** or **vaporization** is the escape of molecules from the liquid state to the gas or vapor state.

evaporation

vaporization

In evaporation, molecules of higher than average kinetic energy escape from a liquid, leaving it cooler than it was before they escaped. For this reason, evaporation of perspiration is one way the human body cools itself and keeps its temperature constant. When volatile liquids such as ethyl chloride (C_2H_5Cl) are sprayed on the skin, they evaporate rapidly, cooling the area by removing heat. The numbing effect of the low temperature produced by evaporation of ethyl chloride allows it to be used as a local anesthetic for minor surgery.

Solids such as iodine, camphor, naphthalene (moth balls), and, to a small extent, even ice will go directly from the solid to the gaseous state, bypassing the liquid state. This change is a form of evaporation and is called **sublimation**:

sublimation

$$\text{liquid} \xrightarrow{\text{Evaporation}} \text{vapor}$$

$$\text{solid} \xrightarrow{\text{Sublimation}} \text{vapor}$$

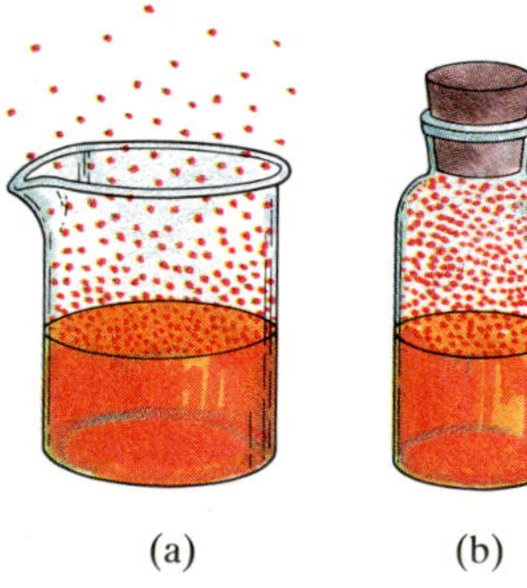

▲
Figure 14.1
(a) Molecules in an open beaker evaporate from the liquid and disperse into the atmosphere. Evaporation will continue until all the liquid is gone. (b) Molecules leaving the liquid are confined to a limited space. With time, the concentration in the vapor phase will increase until an equilibrium between liquid and vapor is established.

14.3 Vapor Pressure

When a liquid vaporizes in a closed system as shown in Figure 14.1, part (b), some of the molecules in the vapor or gaseous state strike the surface and return to the liquid state by the process of *condensation.* The rate of condensation increases until it is equal to the rate of vaporization. At this point, the space above the liquid is said to be saturated with vapor, and an equilibrium, or steady state, exists between the liquid and the vapor. The equilibrium equation is

$$\text{liquid} \underset{\text{Condensation}}{\overset{\text{Vaporization}}{\rightleftarrows}} \text{vapor}$$

This equilibrium is dynamic; both processes—vaporization and condensation—are taking place, even though one cannot see or measure a change. The number of molecules leaving the liquid in a given time interval is equal to the number of molecules returning to the liquid.

At equilibrium the molecules in the vapor exert a pressure like any other gas. The pressure exerted by a vapor in equilibrium with its liquid is known as the **vapor pressure** of the liquid. The vapor pressure may be thought of as an internal pressure, a measure of the "escaping" tendency of molecules to go from the liquid to the vapor state. The vapor pressure of a liquid is independent of the amount of

vapor pressure

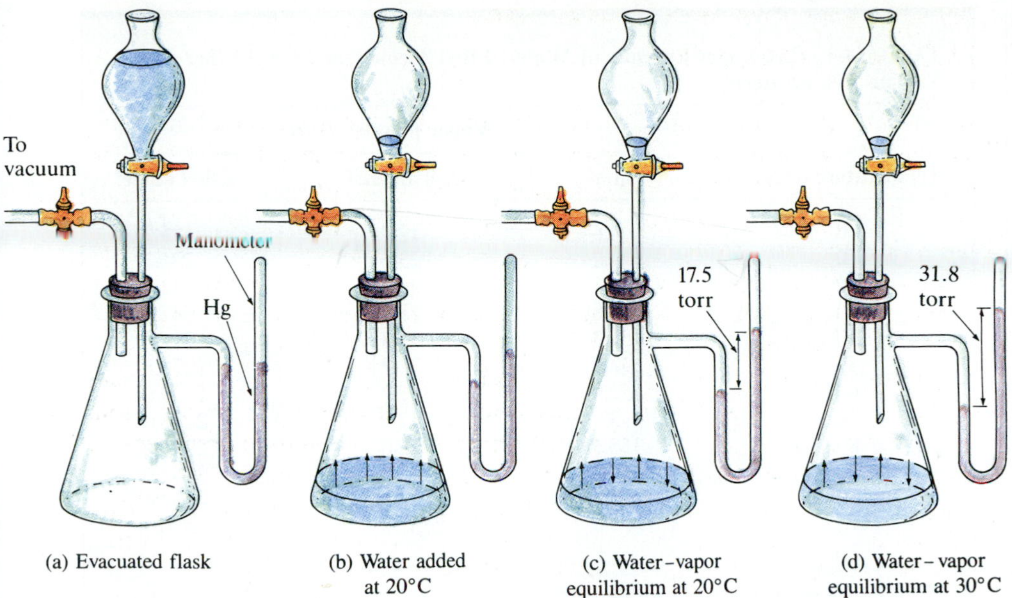

◀ **FIGURE 14.2** **Measurement of the vapor pressure of water at 20°C and 30°C. In flask (a) the system is evacuated. The mercury manometer attached to the flask shows equal pressure in both legs. In (b) water has been added to the flask and begins to evaporate, exerting pressure as indicated by the manometer. In (c), when equilibrium is established, the pressure inside the flask remains constant at 17.5 torr. In (d) the temperature is changed to 30°C, and equilibrium is reestablished with the vapor pressure at 31.8 torr.**

liquid and vapor present, but it increases as the temperature rises (see Table 14.1). Figure 14.2 illustrates a liquid–vapor equilibrium and the measurement of vapor pressure.

When equal volumes of water, ethyl ether, and ethyl alcohol are placed in separate beakers and allowed to evaporate at the same temperature, we observe that the ether evaporates faster than the alcohol, which evaporates faster than the water. This order of evaporation is consistent with the fact that ether has a higher vapor pressure at any particular temperature than ethyl alcohol or water. One reason for this higher vapor pressure is that the attraction is less between ether molecules than between alcohol molecules or between water molecules. The vapor pressures of these three compounds at various temperatures are compared in Table 14.1.

volatile

Substances that evaporate readily are said to be **volatile**. A volatile liquid has a relatively high vapor pressure at room temperature. Ethyl ether is a very volatile liquid; water is not too volatile; mercury, which has a vapor pressure of 0.0012 torr at 20°C, is essentially a nonvolatile liquid. Most substances that are normally solids are nonvolatile (solids that sublime are exceptions).

14.4 SURFACE TENSION

Have you ever observed water and mercury in the form of small drops? These liquids occur as drops due to *surface tension* of liquids. A droplet of liquid that is not falling or under the influence of gravity (as on the space shuttle) will form a sphere. Minimum surface area is found in the geometrical form of the sphere. The molecules within the liquid are attracted to the surrounding liquid molecules. However, at the surface of the liquid, the attraction is nearly all inward, pulling the surface into a spherical shape. The resistance of a liquid to an increase in its

TABLE 14.1 The Vapor Pressure of Water, Ethyl Alcohol, and Ethyl Ether at Various Temperatures

	Vapor pressure (torr)		
Temperature (°C)	Water	Ethyl alcohol	Ethyl ether[a]
0	4.6	12.2	185.3
10	9.2	23.6	291.7
20	17.5	43.9	442.2
30	31.8	78.8	647.3
40	55.3	135.3	921.3
50	92.5	222.2	1276.8
60	152.9	352.7	1729.0
70	233.7	542.5	2296.0
80	355.1	812.6	2993.6
90	525.8	1187.1	3841.0
100	760.0	1693.3	4859.4
110	1074.6	2361.3	6070.1

[a] Note that the vapor pressure of ethyl ether at temperatures of 40°C and higher exceeds standard pressure, 760 torr, which indicates that the substance has a low boiling point and therefore should be stored in a cool place in a tightly sealed container.

surface tension

surface area is called the **surface tension** of the liquid. Substances with large attractive forces between molecules have high surface tensions. The effect of surface tension in water is illustrated by the phenomenon of floating a needle on the surface of still water. Other examples include the movement of the water strider insect across a calm pond, or the beading of water on a freshly waxed car.

capillary action

Liquids also exhibit a phenomenon known as **capillary action**, which is the spontaneous rising of a liquid in a narrow tube. This action results from the *cohesive forces* within the liquid, and the *adhesive forces* between the liquid and the walls of the container. If the forces between the liquid and the container are greater than those within the liquid itself, the liquid will climb the walls of the container. For example, consider the California sequoia tree, which can be over 200 feet in height. Under atmospheric pressure, water will only rise 33 feet, but capillary action will cause water to rise from the roots to all parts of the tree.

meniscus

The meniscus in liquids is further evidence of these cohesive and adhesive forces. When a liquid is placed in a glass cylinder the surface of the liquid shows a curve called the **meniscus** (see Figure 14.3). The concave shape of the meniscus shows that the adhesive forces between the glass and the liquid (water) are stronger than the cohesive forces within the liquid. In a nonpolar substance such

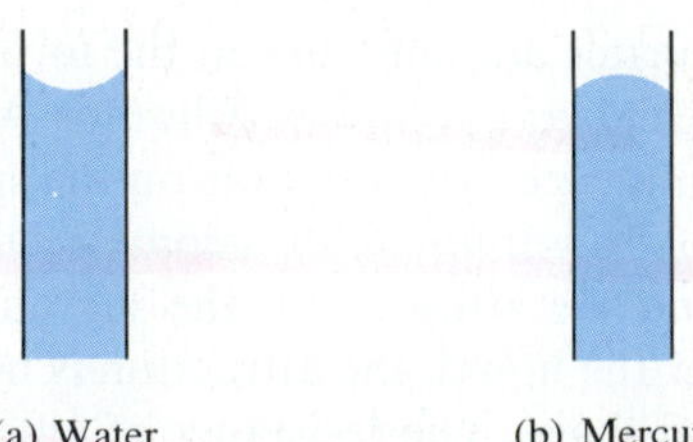

FIGURE 14.3 ▶ The meniscus is the characteristic curve of the surface of a liquid in a narrow capillary tube.

as mercury, the meniscus is convex, indicating that the cohesive forces within the mercury are greater than the adhesive forces between the glass wall and the mercury.

14.5 Boiling Point

The boiling temperature of a liquid is associated with its vapor pressure. We have seen that the vapor pressure increases as the temperature increases. When the internal or vapor pressure of a liquid becomes equal to the external pressure, the liquid boils. (By external pressure we mean the pressure of the atmosphere above the liquid.) The boiling temperature of a pure liquid remains constant as long as the external pressure does not vary.

The boiling point (bp) of water is 100°C at 1 atm pressure. Table 14.1 shows that the vapor pressure of water at 100°C is 760 torr, a figure we have seen many times before. The significant fact here is that the boiling point is the temperature at which the vapor pressure of the water or other liquid is equal to standard, or atmospheric, pressure at sea level. These relationships lead to the following definition: The **boiling point** is the temperature at which the vapor pressure of a liquid is equal to the external pressure above the liquid.

boiling point

We can readily see that a liquid has an infinite number of boiling points. When we give the boiling point of a liquid, we should also state the pressure. When we express the boiling point without stating the pressure, we mean it to be the **standard** or **normal boiling point** at standard pressure (760 torr). Using Table 14.1 again, we see that the normal boiling point of ethyl ether is between 30°C and 40°C, and for ethyl alcohol it is between 70°C and 80°C, because, for each compound, 760 torr pressure lies within these stated temperature ranges. At the normal boiling point, 1 g of a liquid changing to a vapor (gas) absorbs an amount of energy equal to its heat of vaporization (see Table 14.2).

standard or normal boiling point

The boiling point at various pressures may be evaluated by plotting the data of Table 14.2 on the graph in Figure 14.4, where temperature is plotted horizontally along the x axis and vapor pressure is plotted vertically along the y axis. The resulting curves are known as **vapor pressure curves**. Any point on these curves represents a vapor–liquid equilibrium at a particular temperature and pressure. We may find the boiling point at any pressure by tracing a horizontal

vapor pressure curves

Table 14.2 Physical Properties of Ethyl Chloride, Ethyl Ether, Ethyl Alcohol, and Water

	Boiling point (°C)	Melting point (°C)	Heat of vaporization J/g (cal/g)	Heat of fusion J/g (cal/g)
Ethyl chloride	13	−139	387 (92.5)	—
Ethyl ether	34.6	−116	351 (83.9)	—
Ethyl alcohol	78.4	−112	855 (204.3)	104 (24.9)
Water	100.0	0	2259 (540)	335 (80)

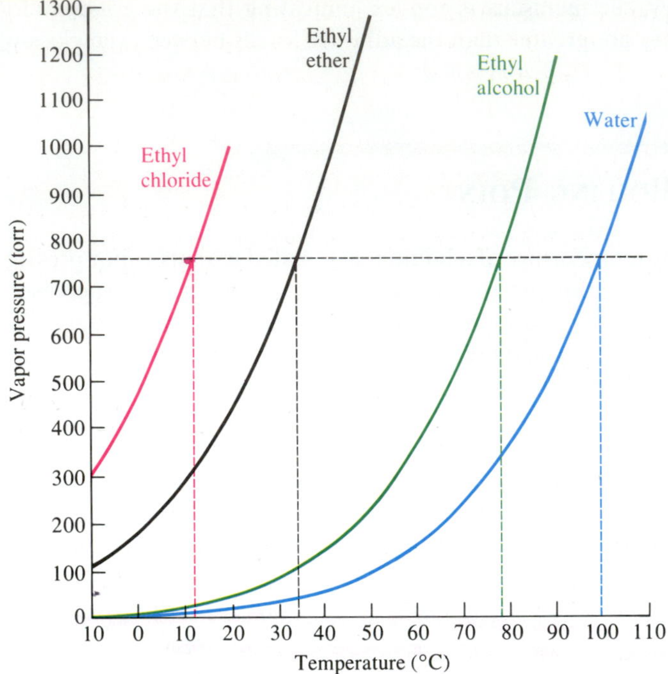

FIGURE 14.4 ▶
Vapor pressure–temperature curves for ethyl chloride, ethyl ether, ethyl alcohol, and water.

line from the designated pressure to a point on the vapor pressure curve. From this point we draw a vertical line to obtain the boiling point on the temperature axis. Four such points are shown in Figure 14.4; they represent the normal boiling points of the four compounds at 760 torr pressure. By reversing this process, you can ascertain at what pressure a substance will boil at a specific temperature. The boiling point is one of the most commonly used physical properties for characterizing and identifying substances.

PRACTICE Use the graph in Figure 14.4 to determine the boiling points of ethyl chloride, ethyl ether, ethyl alcohol, and water at 600 torr.

Answers: 8.5°C, 28°C, 73°C, 93°C

PRACTICE The average atmospheric pressure in Denver is 0.83 atm. What is the boiling point of water in Denver?

Answer: approximately 93°C

14.6 FREEZING POINT OR MELTING POINT

As heat is removed from a liquid, the liquid becomes colder and colder, until a temperature is reached at which it begins to solidify. A liquid that is changing into a solid is said to be *freezing*, or *solidifying*. When a solid is heated continuously, a temperature is reached at which the solid begins to liquefy. A solid that is changing into a liquid is said to be *melting*. The temperature at which the solid phase of a substance is in equilibrium with its liquid phase is known as the **freezing point** or **melting point** of that substance. The equilibrium equation is

freezing or melting point

$$\text{solid} \underset{\text{Freezing}}{\overset{\text{Melting}}{\rightleftarrows}} \text{liquid}$$

When a solid is slowly and carefully heated so that a solid–liquid equilibrium is achieved and then maintained, the temperature will remain constant as long as both phases are present. The energy is used solely to change the solid to the liquid. The melting point is another physical property that is commonly used for characterizing substances.

The most common example of a solid–liquid equilibrium is ice and water. In a well-stirred system of ice and water, the temperature remains at 0°C as long as both phases are present. The melting point changes with pressure but only slightly unless the pressure change is very large.

14.7 Changes of State

The majority of solids undergo two changes of state upon heating. A solid changes to a liquid at its melting point, and a liquid changes to a gas at its boiling point. This warming process can be represented by a graph called a *heating curve* (Figure 14.5). This figure shows ice being heated at a constant rate. As energy flows into the ice, the vibrations within the crystal increase and the temperature rises ($A \longrightarrow B$). Eventually, the molecules begin to break free from the crystal and melting occurs ($B \longrightarrow C$). During the melting process all energy goes into breaking down the crystal structure; the temperature remains constant.

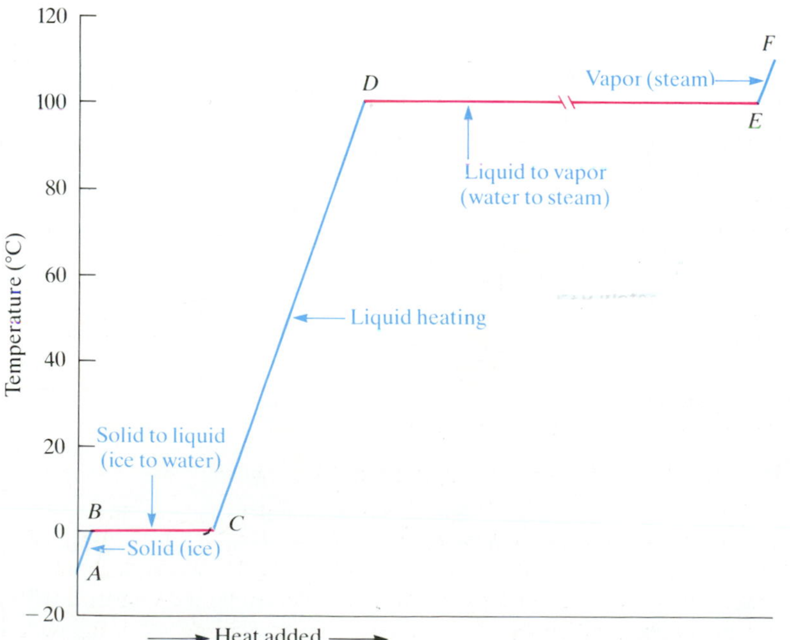

◀ **FIGURE 14.5**
Heating curve—the absorption of heat by a substance from the solid state to the vapor state. Using water as an example, the interval *AB* represents the ice phase; *BC* interval, the melting of ice to water; *CD* interval, the elevation of the temperature of water from 0°C to 100°C; *DE* interval, the boiling of water to steam; and *EF* interval, the heating of steam.

heat of fusion

heat of vaporization

The energy required to change 1 g of a solid at its melting point into a liquid is called the **heat of fusion**. When the solid has completely melted, the temperature once again rises ($C \longrightarrow D$); the energy input is increasing the molecular motion within the water. At 100°C, the water reaches its boiling point; the temperature remains constant while the added energy is used to vaporize the water to steam ($D \longrightarrow E$). The **heat of vaporization** is the energy required to change 1 g of liquid to vapor at its normal boiling point. The attractive forces between the liquid molecules are overcome during vaporization. Beyond this temperature, all the water exists as steam and is being further heated ($E \longrightarrow F$).

EXAMPLE 14.1

How many joules of energy are needed to change 10.0 g of ice at 0.00°C to water at 20.0°C?

SOLUTION

Ice will absorb 335 J/g (heat of fusion) in going from a solid at 0°C to a liquid at 0°C. An additional 4.184 J/g°C (specific heat of water) are needed to raise the temperature of the water for each 1°C.

Joules needed to melt the ice:

$$10.0\ \cancel{g} \times \frac{335\ \text{J}}{1\ \cancel{g}} = 3.35 \times 10^3\ \text{J (801 cal)}$$

Joules needed to heat the water from 0.00°C to 20.0°C:

$$10.0\ \cancel{g} \times \frac{4.184\ \text{J}}{1\ \cancel{g}\cancel{°C}} \times 20.0\cancel{°C} = 837\ \text{J (200. cal)}$$

Thus, 3350 J + 837 J = 4.19×10^3 J (1.00×10^3 cal) are needed.

EXAMPLE 14.2

How many kilojoules of energy are needed to change 20.0 g of water at 20.°C to steam at 100.°C?

SOLUTION

Kilojoules needed to heat the water from 20.°C to 100.°C:

$$20.0\ \text{g} \times \frac{4.184\ \text{J}}{1\ \text{g}\cancel{°C}} \times \frac{1\ \text{kJ}}{1000\ \text{J}} \times 80.\cancel{°C} = 7.0 \times 10^3\ \text{kJ}\ (1.6 \times 10^3\ \text{kcal})$$

Kilojoules needed to change water at 100.°C to steam at 100.°C:

$$20.0\ \cancel{g} \times \frac{2.26\ \text{kJ}}{1\ \cancel{g}} = 4.52 \times 10^4\ \text{kJ}\ (1.08 \times 10^4\ \text{kcal})$$

Thus, 6.7×10^3 kJ + 4.52×10^4 kJ = 5.2×10^4 kJ (1.2×10^3 kcal) are needed.

PRACTICE How many kilojoules of energy are required to change 50.0 g of ethyl alcohol from 60.0°C to vapor at 78.4°C? The specific heat of ethyl alcohol is 2.138 J/g°C.

Answer: 44.8 kJ

14.8 Occurrence of Water

Water is our most common natural resource; it covers about 75% of the earth's surface. Not only is it found in the oceans and seas, in lakes, rivers, streams, and in glacial ice deposits, it is also always present in the atmosphere and in cloud formations.

About 97% of the earth's water is in the oceans. This *saline* water contains vast amounts of dissolved minerals. More than 70 elements have been detected in the mineral content of seawater. Only four of these—chlorine, sodium, magnesium, and bromine—are now commercially obtained from the sea. The world's *fresh* water comprises the other 3%, of which about two-thirds is locked up in polar ice caps and glaciers. The remaining fresh water is found in ground water, lakes, and the atmosphere.

Water is an essential constituent of all living matter. It is the most abundant compound in the human body, making up about 70% of total body mass. About 92% of blood plasma is water; about 80% of muscle tissue is water; and about 60% of a red blood cell is water. Water is more important than food in the sense that a person can survive much longer without food than without water.

14.9 Physical Properties of Water

Water is a colorless, odorless, tasteless liquid with a melting point of 0°C and a boiling point of 100°C at 1 atm pressure. The heat of fusion of water is 335 J/g (80 cal/g). The heat of vaporization of water is 2.26 kJ/g (540 cal/g). The values for water for both the heat of fusion and the heat of vaporization are high compared with those for other substances; these high values indicate that strong attractive forces are acting between the molecules.

Ice and water exist together in equilibrium at 0°C, as shown in Figure 14.6. When ice at 0°C melts, it absorbs 335 J/g (80 cal/g) in changing into a liquid; the temperature remains at 0°C. In order to refreeze the water we have to remove 335 J/g (80 cal/g) from the liquid at 0°C.

In Figure 14.6 both boiling water and steam are shown to have a temperature of 100°C. It takes 418 J (100 cal) to heat 1 g of water from 0°C to 100°C, but water at its boiling point absorbs 2.26 kJ/g (540 cal/g) in changing to steam. Although boiling water and steam are both at the same temperature, steam contains considerably more heat per gram and can cause more severe burns than hot water. In Table 14.3 the physical properties of water are tabulated and compared with those of other hydrogen compounds of Group VIA elements.

The maximum density of water is 1.000 g/mL at 4°C. Water has the unusual property of contracting in volume as it is cooled to 4°C and then expanding when cooled from 4°C to 0°C. Therefore 1 g of water occupies a volume greater than 1 mL at all temperatures except 4°C. Although most liquids contract in volume all the way down to the point at which they solidify, a large increase (about 9%) in volume occurs when water changes from a liquid at 0°C to a solid (ice) at 0°C. The density of ice at 0°C is 0.917 g/mL, which means that ice, being less dense than water, will float in water.

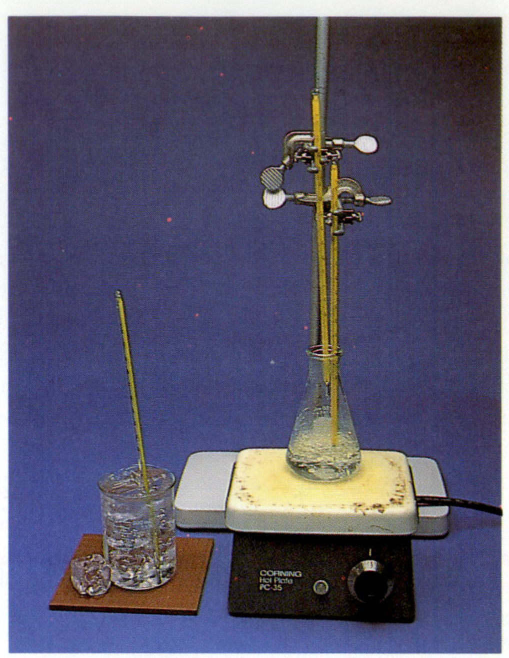

FIGURE 14.6 ▶
Water equilibrium systems. In the beaker on the left, ice and water are in equilibrium at 0°C; in the flask on the right, boiling water and steam are in equilibrium at 100 C.

TABLE 14.3 Physical Properties of Water and Other Hydrogen Compounds of Group VIA Elements

Formula	Color	Molar mass	Melting point (°C)	Boiling point, 1 atm (°C)	Heat of fusion J/g (cal/g)	Heat of vaporization J/g (cal/g)
H_2O	Colorless	18.02	0.00	100.0	335 (80.0)	2.26×10^3 (540)
H_2S	Colorless	34.08	−85.5	−60.3	69.9 (16.7)	548 (131)
H_2Se	Colorless	80.98	−65.7	−41.3	31 (7.4)	238 (57.0)
H_2Te	Colorless	129.6	−49	−2	—	179 (42.8)

14.10 STRUCTURE OF THE WATER MOLECULE

A single water molecule consists of two hydrogen atoms and one oxygen atom. Each H atom is attached to the O atom by a single covalent bond. This bond is formed by the overlap of the 1*s* orbital of hydrogen with an unpaired 2*p* orbital of oxygen. The average distance between the two nuclei is known as the *bond length.* The O—H bond length in water is 0.096 nm. The water molecule is nonlinear and has a V-shaped structure with an angle of about 105° between the two bonds (see Figure 14.7).

Oxygen is the second most electronegative element. As a result, the two covalent OH bonds in water are polar. If the three atoms in a water molecule were aligned in a linear structure such as H ⟶ O ⟵ H, the two polar bonds would be acting in equal and opposite directions and the molecule would be nonpolar.

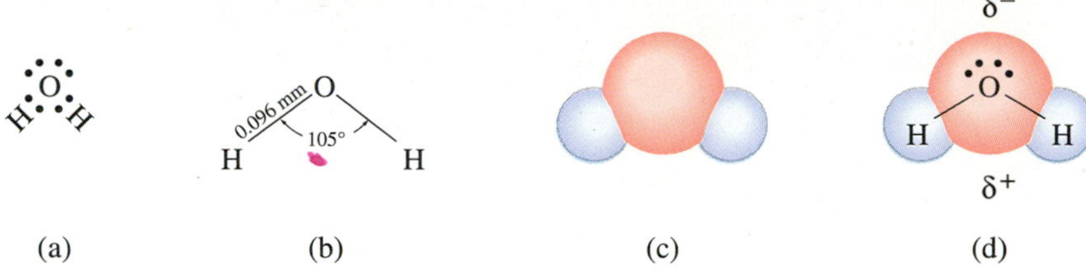

◀ **FIGURE 14.7 Diagrams of a water molecule: (a) electron distribution, (b) bond angle and O—H bond length, (c) molecular orbital structure, and (d) dipole representation.**

However, water is a highly polar molecule. Therefore, it does not have a linear structure. When atoms are bonded together in a nonlinear fashion, the angle formed by the bonds is called the *bond angle*. In water the HOH bond angle is 105°. The two polar covalent bonds and the bent structure result in a partial negative charge on the oxygen atom and a partial positive charge on each hydrogen atom. The polar nature of water is responsible for many of its properties, including its behavior as a solvent.

14.11 The Hydrogen Bond

Table 14.3 compares the physical properties of H_2O, H_2S, H_2Se, and H_2Te. From this comparison it is apparent that four physical properties of water—melting point, boiling point, heat of fusion, and heat of vaporization—are extremely high and do not fit the trend relative to the molar masses of the four compounds. If the properties of water followed the progression shown by the other three compounds, we would expect the melting point of water to be below −85°C and the boiling point to be below −60°C.

Why does water have these anomalous physical properties? The answer is that liquid water molecules are linked together by hydrogen bonds. A **hydrogen bond** is a chemical bond that is formed between polar molecules that contain hydrogen covalently bonded to a small, highly electronegative atom such as fluorine, oxygen, or nitrogen (F—H, O—H, N—H). The bond is actually the dipole–dipole attraction of polar molecules containing these three types of polar bonds.

hydrogen bond

Elements that have significant hydrogen bonding ability are F, O, and N.

What is a hydrogen bond, or H-bond? Because a hydrogen atom has only one electron, it can form only one covalent bond. When it is attached to a strong electronegative atom such as oxygen, a hydrogen atom will also be attracted to an oxygen atom of another molecule, forming a bond (or bridge) between the two molecules. Water has two types of bonds: covalent bonds that exist between hydrogen and oxygen atoms within a molecule and hydrogen bonds that exist between hydrogen and oxygen atoms in different water molecules.

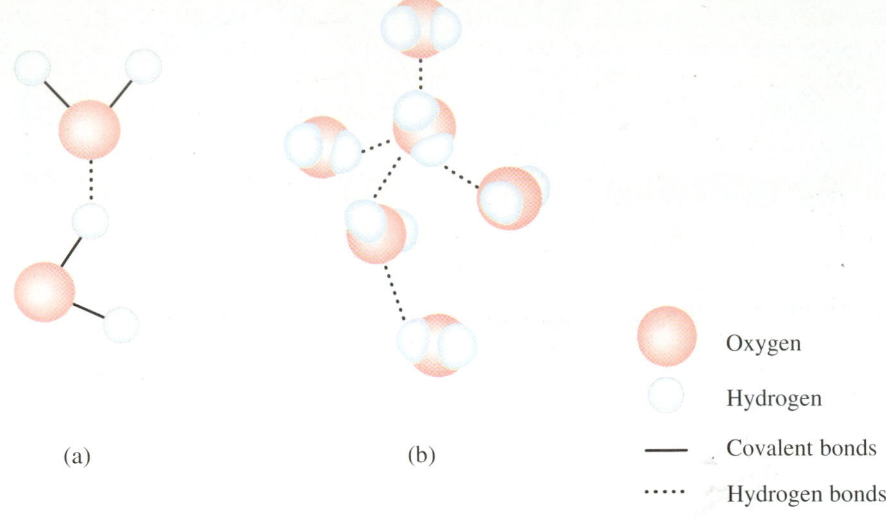

FIGURE 14.8 ▶
Hydrogen bonding. Water in the liquid and solid states exists as aggregates in which the water molecules are linked together by hydrogen bonds.

Hydrogen bonds are *intermolecular* bonds; that is, they are formed between atoms in different molecules. They are somewhat ionic in character because they are formed by electrostatic attraction. Hydrogen bonds are much weaker than the ionic or covalent bonds that unite atoms to atoms to form compounds. Despite their weakness, hydrogen bonds are of great chemical importance.

The oxygen atom in water can form two hydrogen bonds—one through each of the unbonded pairs of electrons. Figure 14.8 shows (a) two water molecules linked by a hydrogen bond and (b) six water molecules linked by hydrogen bonds. A dash (—) is used for the covalent bond and a dotted line (····) for the hydrogen bond. In water each molecule is linked to others through hydrogen bonds to form a three-dimensional aggregate of water molecules. This intermolecular hydrogen bonding effectively gives water the properties of a much larger, heavier molecule, explaining in part its relatively high melting point, boiling point, heat of fusion, and heat of vaporization. As water is heated and energy is absorbed, hydrogen bonds are continually being broken until at 100°C, with the absorption of an additional 2.26 kJ/g (540 cal/g), water separates into individual molecules, going into the gaseous state. Sulfur, selenium, and tellurium are not sufficiently electronegative for their hydrogen compounds to behave like water. As a result, H-bonding in H_2S, H_2Se, and H_2Te is only of small consequence (if any) to their physical properties. For example, the lack of hydrogen bonding is one reason why H_2S is a gas and not a liquid at room temperature.

Fluorine, the most electronegative element, forms the strongest hydrogen bonds. This bonding is strong enough to link hydrogen fluoride molecules together as *dimers*, H_2F_2, or as larger, $(HF)_n$, molecular units. The dimer structure may be represented in this way:

F
H H
F
H—bond

The existence of salts, such as KHF_2 and NH_4HF_2, verifies the hydrogen fluoride (bifluoride) structure, HF_2^- (F—H···· F)$^-$, where one H atom is bonded to two F atoms through one covalent bond and one hydrogen bond.

Hydrogen bonding can occur between two different atoms that are capable of forming H-bonds. Thus we may have an O···· H—N or O—H···· N linkage in which the hydrogen atom forming the H-bond is between an oxygen and a nitrogen atom. This form of the H-bond exists in certain types of protein molecules and many biologically active substances.

EXAMPLE 14.3

Would you expect hydrogen bonding to occur between molecules of the following substances?

```
        H  H                     H     H
        |  |                     |     |
(a)  H—C—C—O—H         (b)  H—C—O—C—H
        |  |                     |     |
        H  H                     H     H
```

Ethyl alcohol — Dimethyl ether

SOLUTION

(a) Hydrogen bonding should occur in ethyl alcohol because one hydrogen atom is bonded to an oxygen atom.

```
    H  H           H  H  H
    |  |           |  |  |
 H—C—C—O—H···O—C—C—H
    |  |       ↑      |  |
    H  H    H-bond    H  H
```

(b) There is no hydrogen bonding in dimethyl ether because all the hydrogen atoms are bonded only to carbon atoms.

Both ethyl alcohol and dimethyl ether have the same molar mass (46.07). Although both compounds have the same molecular formula, C_2H_6O, ethyl alcohol has a much higher boiling point (78.4°C) than dimethyl ether (−23.7°C), because of hydrogen bonding between the alcohol molecules.

PRACTICE Would you expect hydrogen bonding to occur between molecules of the following substances?

```
                                                       H
                                                       |
       H  H  H            H  H  H          H  H  H—C—H
       |  |  |            |  |  |          |  |     |
(a) H—C—C—N—H  (b) H—C—N—C—H  (c) H—C—C———N
       |  |               |     |          |  |     |
       H  H               H     H          H  H  H—C—H
                                                       |
                                                       H
```

Answers: (a) yes (b) yes (c) no

14.12 Formation and Chemical Properties of Water

Water is very stable to heat; it decomposes to the extent of only about 1% at temperatures up to 2000°C. Pure water is a nonconductor of electricity. But when a small amount of sulfuric acid or sodium hydroxide is added, the solution is readily decomposed into hydrogen and oxygen by an electric current. Two volumes of hydrogen are produced for each volume of oxygen.

$$2\,H_2O(l) \xrightarrow[H_2SO_4 \text{ or } NaOH]{\text{Electrical energy}} 2\,H_2(g) + O_2(g)$$

Formation Water is formed when hydrogen burns in air. Pure hydrogen burns very smoothly in air, but mixtures of hydrogen and air or oxygen explode when ignited. The reaction is strongly exothermic.

$$2\,H_2(g) + O_2(g) \longrightarrow 2\,H_2O(g) + 484 \text{ kJ } (116 \text{ kcal})$$

Water is produced by a variety of other reactions, especially by (1) acid–base neutralizations, (2) combustion of hydrogen-containing materials, and (3) metabolic oxidation in living cells:

1. $HCl(aq) + NaOH(aq) \longrightarrow NaCl(aq) + H_2O(l)$
2. $\underset{\text{Acetylene}}{2\,C_2H_2(g)} + 5\,O_2(g) \longrightarrow 4\,CO_2(g) + 2\,H_2O(g) + 1212 \text{ kJ } (289.6 \text{ kcal})$

   $\underset{\text{Methane}}{CH_4} + 2\,O_2 \longrightarrow CO_2 + 2\,H_2O + 803 \text{ kJ } (192 \text{ kcal})$
3. $\underset{\text{Glucose}}{C_6H_{12}O_6} + 6\,O_2 \xrightarrow{\text{Enzymes}} 6\,CO_2 + 6\,H_2O + 2519 \text{ kJ } (673 \text{ kcal})$

The combustion of acetylene shown in (2) is strongly exothermic and is capable of producing very high temperatures. It is used in oxygen–acetylene torches to cut and weld steel and other metals. Methane is known as natural gas and is commonly used as fuel for heating and cooking. The reaction of glucose with oxygen shown in (3) is the reverse of photosynthesis. It is the overall reaction by which living cells obtain needed energy by metabolizing glucose to carbon dioxide and water.

Reactions of Water with Metals and Nonmetals The reactions of metals with water at different temperatures show that these elements vary greatly in their reactivity. Metals such as sodium, potassium, and calcium react with cold water to produce hydrogen and a metal hydroxide. A small piece of sodium added to water melts from the heat produced by the reaction, forming a silvery metal ball, which rapidly flits back and forth on the surface of the water. Caution must be used when experimenting with this reaction, because the hydrogen produced is frequently ignited by the sparking of the sodium, and it will explode, spattering sodium. Potassium reacts even more vigorously than sodium. Calcium sinks in water and liberates a gentle stream of hydrogen. The equations for these reactions are

$$2\,Na(s) + 2\,H_2O(l) \longrightarrow H_2\uparrow + 2\,NaOH(aq)$$
$$2\,K(s) + 2\,H_2O(l) \longrightarrow H_2\uparrow + 2\,KOH(aq)$$
$$Ca(s) + 2\,H_2O(l) \longrightarrow H_2\uparrow + Ca(OH)_2(aq)$$

Zinc, aluminum, and iron do not react with cold water but will react with steam at high temperatures, forming hydrogen and a metallic oxide. The equations are

$$Zn(s) + H_2O(steam) \longrightarrow H_2\uparrow + ZnO(s)$$
$$2\,Al(s) + 3\,H_2O(steam) \longrightarrow 3\,H_2\uparrow + Al_2O_3(s)$$
$$3\,Fe(s) + 4\,H_2O(steam) \longrightarrow 4\,H_2\uparrow + Fe_3O_4(s)$$

Copper, silver, and mercury are examples of metals that do not react with cold water or steam to produce hydrogen. We conclude that sodium, potassium, and calcium are chemically more reactive than zinc, aluminum, and iron, which are more reactive than copper, silver, and mercury.

Certain nonmetals react with water under various conditions. For example, fluorine reacts violently with cold water, producing hydrogen fluoride and free oxygen. The reactions of chlorine and bromine are much milder, producing what is commonly known as "chlorine water" and "bromine water," respectively. Chlorine water contains HCl, HOCl, and dissolved Cl_2; the free chlorine gives it a yellow-green color. Bromine water contains HBr, HOBr, and dissolved Br_2; the free bromine gives it a reddish-brown color. Steam passed over hot coke (carbon) produces a mixture of carbon monoxide and hydrogen that is known as "water gas." Since water gas is combustible, it is useful as a fuel. It is also the starting material for the commercial production of several alcohols. The equations for these reactions are

$$2\,F_2(g) + 2\,H_2O(l) \longrightarrow 4\,HF(aq) + O_2(g)$$
$$Cl_2(g) + H_2O(l) \longrightarrow HCl(aq) + HOCl(aq)$$
$$Br_2(l) + H_2O(l) \longrightarrow HBr(aq) + HOBr(aq)$$
$$C(s) + H_2O(g) \xrightarrow{1000^\circ C} CO(g) + H_2(g)$$

Reactions of Water with Metal and Nonmetal Oxides Metal oxides that react with water to form bases are known as **basic anhydrides.** Examples are

basic anhydride

$$CaO(s) + H_2O(l) \longrightarrow \underset{\text{Calcium hydroxide}}{Ca(OH)_2(aq)}$$

$$Na_2O(s) + H_2O(l) \longrightarrow \underset{\text{Sodium hydroxide}}{2\,NaOH(aq)}$$

Certain metal oxides, such as CuO and Al_2O_3, do not form basic solutions because the oxides are insoluble in water.

Nonmetal oxides that react with water to form acids are known as **acid anhydrides**. Examples are

acid anhydride

$$CO_2(g) + H_2O(l) \rightleftharpoons \underset{\text{Carbonic acid}}{H_2CO_3(aq)}$$

$$SO_2(g) + H_2O(l) \rightleftharpoons H_2SO_3(aq)$$

Sulfurous acid

$$N_2O_5(s) + H_2O(l) \longrightarrow 2\ HNO_3(aq)$$

Nitric acid

The word *anhydrous* means "without water." An anhydride is a metal oxide or a nonmetal oxide derived from a base or an oxy-acid by the removal of water. To determine the formula of an anhydride, the elements of water, H_2O, are removed from an acid or base formula until all the hydrogen is removed. Sometimes more than one formula unit is needed to remove all the hydrogen as water. The formula of the anhydride then consists of the remaining metal or nonmetal and the remaining oxygen atoms. In calcium hydroxide, removal of water as indicated leaves CaO as the anhydride:

$$Ca\begin{matrix} O\,\boxed{H} \\ \boxed{OH} \end{matrix} \xrightarrow{\Delta} CaO + H_2O$$

In sodium hydroxide, H_2O cannot be removed from one formula unit, so two formula units of NaOH must be used, leaving Na_2O as the formula of the anhydride:

$$\begin{matrix} NaO\,\boxed{H} \\ Na\,\boxed{OH} \end{matrix} \xrightarrow{\Delta} Na_2O + H_2O$$

The removal of water from sulfuric acid, H_2SO_4, gives the acid anhydride SO_3:

$$H_2SO_4 \xrightarrow{\Delta} SO_3 + H_2O$$

The foregoing are examples of typical reactions of water but are by no means a complete list of the known reactions of water.

14.13 Hydrates

hydrate

water of hydration

water of crystallization

When certain salt solutions are allowed to evaporate, some water molecules remain as part of the crystalline salt that is left after evaporation is complete. Solids that contain water molecules as part of their crystalline structure are known as **hydrates**. Water in a hydrate is known as **water of hydration**, or **water of crystallization**.

Formulas for hydrates are expressed by first writing the usual anhydrous (without water) formula for the compound and then adding a dot followed by the number of water molecules present. An example is $BaCl_2 \cdot 2\ H_2O$. This formula tells us that each formula unit of this salt contains one barium ion, two chloride ions, and two water molecules. A crystal of the salt contains many of these units in its crystalline lattice.

In naming hydrates, we first name the compound exclusive of the water and then add the term *hydrate*, with the proper prefix representing the number of water molecules in the formula. For example, $BaCl_2 \cdot 2\ H_2O$ is called *barium chloride dihydrate*. Hydrates are true compounds and follow the Law of Definite Composition. The molar mass of $BaCl_2 \cdot 2\ H_2O$ is 244.2 g/mol; it contains 56.22% barium, 29.03% chlorine, and 14.76% water.

Water molecules in hydrates are bonded by electrostatic forces between polar water molecules and the positive or negative ions of the compound. These forces are not as strong as covalent or ionic chemical bonds. As a result, water of crystallization can be removed by moderate heating of the compound. A partially dehydrated or completely anhydrous compound may result. When $BaCl_2 \cdot 2\ H_2O$ is heated, it loses its water at about 100°C:

$$BaCl_2 \cdot 2\ H_2O \xrightarrow{100^\circ C} BaCl_2 + 2\ H_2O\uparrow$$

When a solution of copper(II) sulfate ($CuSO_4$) is allowed to evaporate, beautiful blue crystals containing 5 moles of water per mole of $CuSO_4$ are formed. The formula for this hydrate is $CuSO_4 \cdot 5\ H_2O$; it is called copper(II) sulfate pentahydrate, or cupric sulfate pentahydrate. When $CuSO_4 \cdot 5\ H_2O$ is heated, water is lost, and a pale green-white powder, anhydrous $CuSO_4$, is formed.

$$CuSO_4 \cdot 5\ H_2O \xrightarrow{250^\circ C} CuSO_4 + 5\ H_2O\uparrow$$

When water is added to anhydrous copper(II) sulfate, the foregoing reaction is reversed, and the salt turns blue again. Because of this outstanding color change, anhydrous copper(II) sulfate has been used as an indicator to detect small amounts of water. The formation of the hydrate is noticeably exothermic.

The formula for plaster of paris is $(CaSO_4)_2 \cdot H_2O$. When mixed with the proper quantity of water, plaster of paris forms a dihydrate and sets to a hard mass. It is, therefore, useful for making patterns for the reproduction of art objects, molds, and surgical casts. The chemical reaction is

$$(CaSO_4)_2 \cdot H_2O(s) + 3\ H_2O(l) \longrightarrow 2\ CaSO_4 \cdot 2\ H_2O(s)$$

The occurrence of hydrates is commonplace in salts. Table 14.4 lists a number of common hydrates.

TABLE 14.4 Selected Hydrates

Hydrate	Name	Hydrate	Name
$CaCl_2 \cdot 2\ H_2O$	Calcium chloride dihydrate	$Na_2CO_3 \cdot 10\ H_2O$	Sodium carbonate decahydrate
$Ba(OH)_2 \cdot 8\ H_2O$	Barium hydroxide octahydrate	$(NH_4)_2C_2O_4 \cdot H_2O$	Ammonium oxalate monohydrate
$MgSO_4 \cdot 7\ H_2O$	Magnesium sulfate heptahydrate	$NaC_2H_3O_2 \cdot 3\ H_2O$	Sodium acetate trihydrate
$SnCl_2 \cdot 2\ H_2O$	Tin(II) chloride dihydrate	$Na_2B_4O_7 \cdot 10\ H_2O$	Sodium tetraborate decahydrate
$CoCl_2 \cdot 6\ H_2O$	Cobalt(II) chloride hexahydrate	$Na_2S_2O_3 \cdot 5\ H_2O$	Sodium thiosulfate pentahydrate

14.14 Hygroscopic Substances: Deliquescence; Efflorescence

hygroscopic substances

Many anhydrous salts and other substances readily absorb water from the atmosphere. Such substances are said to be **hygroscopic**. This property can be observed in the following simple experiment: Spread a 10–20 g sample of anhydrous copper(II) sulfate on a watch glass and set it aside so that the salt is exposed to the air. Then determine the mass of the sample periodically for 24 hours, noting the increase in mass and the change in color. Water is absorbed from the atmosphere, forming the blue pentahydrate $CuSO_4 \cdot 5\ H_2O$.

deliquescence

dessicant

Some compounds continue to absorb water beyond the hydrate stage to form solutions. A substance that absorbs water from the air until it forms a solution is said to be **deliquescent**. A few granules of anhydrous calcium chloride or pellets of sodium hydroxide exposed to the air will appear moist in a few minutes, and within an hour will absorb enough water to form a puddle of solution. Diphosphorus pentoxide (P_2O_5) picks up water so rapidly that its mass cannot be determined accurately except in an anhydrous atmosphere.

Compounds that absorb water are useful as drying agents (desiccants). Refrigeration systems must be kept dry with such agents or the moisture will freeze and clog the tiny orifices in the mechanism. Bags of drying agents are often enclosed in packages containing iron or steel parts to absorb moisture and prevent rusting. Anhydrous calcium chloride, magnesium sulfate, sodium sulfate, calcium sulfate, silica gel, and diphosphorus pentoxide are some of the compounds commonly used for drying liquids and gases that contain small amounts of moisture.

efflorescence

The process by which crystalline materials spontaneously lose water when exposed to the air is known as **efflorescence**. Glauber's salt ($Na_2SO_4 \cdot 10\ H_2O$), a transparent crystalline salt, loses water when exposed to the air. One can actually observe these well-defined, large crystals crumbling away as they lose water and form a white, noncrystalline-appearing powder. From our discussion of the decomposition of hydrates, we can predict that heat will increase the rate of efflorescence. The rate also depends on the concentration of moisture in the air. A dry atmosphere will allow the process to take place more rapidly.

14.15 Natural Waters

Natural fresh waters are not pure, but contain dissolved minerals, suspended matter, and sometimes harmful bacteria. The water supplies of large cities are usually drawn from rivers or lakes. Such water is generally unsafe to drink without treatment. To make such water potable (that is, safe to drink), it is treated by some or all of the following processes (See Figure 14.9).

1. **Screening** Removal of relatively large objects, such as trash, fish, and so on.

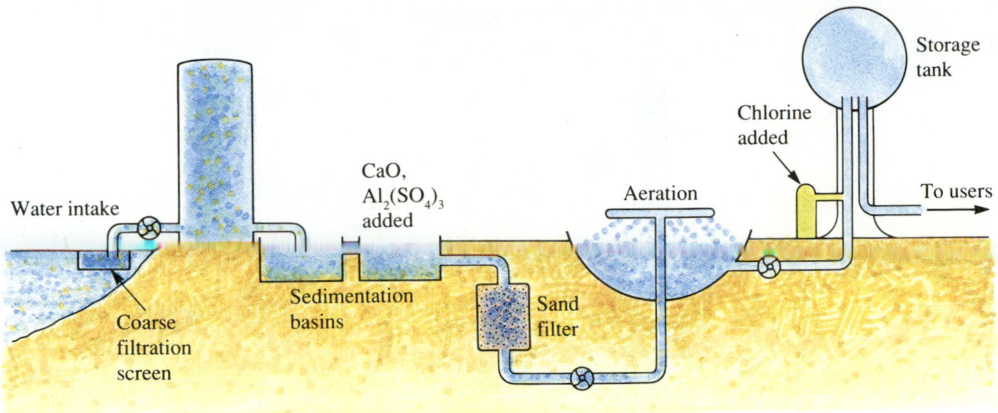

▲ **FIGURE 14.9 Typical municipal water treatment plant.**

2. **Flocculation and sedimentation** Chemicals, usually lime (CaO) and alum (aluminum sulfate), are added to form a flocculent jellylike precipitate of aluminum hydroxide. This precipitate traps most of the fine suspended matter in the water and carries it to the bottom of the sedimentation basin.
3. **Sand filtration** Water is drawn from the top of the sedimentation basin and passed downward through fine sand filters. Nearly all the remaining suspended matter and bacteria are removed by the sand filters.
4. **Aeration** Water is drawn from the bottom of the sand filters and is aerated by spraying. The purpose of this process is to remove objectionable odors and tastes.
5. **Disinfection** In the final stage chlorine gas is injected into the water to kill harmful bacteria before the water is distributed to the public. Ozone is also used in some countries to disinfect water. In emergencies water may be disinfected by simply boiling for a few minutes.

If the drinking water of children contains an optimum amount of fluoride ion, their teeth will be more resistant to decay. Therefore, in many communities NaF or Na_2SiF_6 is added to the water supply to bring the fluoride ion concentration up to the optimum level of about 1.0 ppm. Excessively high concentrations of fluoride ion can cause mottling of the teeth.

Water that contains dissolved calcium and magnesium salts is called *hard water*. One drawback of hard water is that ordinary soap does not lather well in it; the soap reacts with the calcium and magnesium ions to form an insoluble greasy scum. However, synthetic soaps, known as detergents or syndets, are available; they have excellent cleaning qualities and do not form precipitates with hard water. Hard water is also undesirable because it causes "boiler scale" to form on the walls of water heaters and steam boilers, which greatly reduces their efficiency.

Four techniques used to "soften" hard water are distillation, chemical precipitation, ion exchange, and demineralization. In distillation the water is boiled, and the steam thus formed is condensed to a liquid again, leaving the minerals behind in the distilling vessel. Figure 14.10 illustrates a simple laboratory distillation apparatus. Commercial stills are available that are capable of producing hundreds of liters of distilled water per hour.

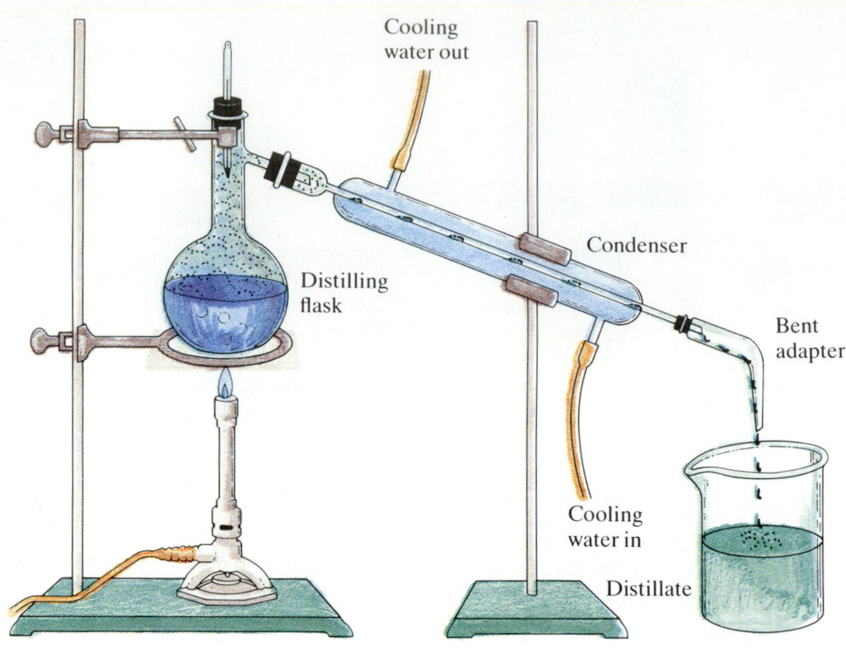

FIGURE 14.10 ▶ Simple laboratory setup for distillation of liquids.

Calcium and magnesium ions are precipitated from hard water by adding sodium carbonate and lime. Insoluble calcium carbonate and magnesium hydroxide are precipitated and are removed by filtration or sedimentation.

In the ion-exchange method, used in many households, hard water is effectively softened as it is passed through a bed or tank of zeolite. Zeolite is a complex sodium aluminum silicate. In this process sodium ions replace objectionable calcium and magnesium ions, and the water is thereby softened:

$$Na_2Zeolite(s) + Ca^{2+}(aq) \longrightarrow CaZeolite(s) + 2\ Na^{+}(aq)$$

The zeolite is regenerated by back-flushing with concentrated sodium chloride solution, reversing the foregoing reaction.

The sodium ions that are present in water softened either by chemical precipitation or by the zeolite process are not objectionable to most users of soft water.

In demineralization both cations and anions are removed by a two-stage ion-exchange system. Special synthetic organic resins are used in the ion-exchange beds. In the first stage metal cations are replaced by hydrogen ions. In the second stage anions are replaced by hydroxide ions. The hydrogen and hydroxide ions react, and essentially pure, mineral-free water leaves the second stage.

The oceans are an inexhaustible source of water; however, seawater contains about 3.5 lb of salts per 100 lb of water. This 35,000 ppm of dissolved salts makes seawater unfit for agricultural and domestic uses. Water that contains less than 1000 ppm of salts is considered reasonably good for drinking, and potable (safe to drink) water is already being obtained from the sea in many parts of the world. Continuous research is being done in an effort to make usable water from the oceans more abundant and economical. See Figure 14.11.

◄ FIGURE 14.11
This desalination plant in Jubail, Saudi Arabia, supplies much of the water for that area.

14.16 WATER POLLUTION

Polluted water was formerly thought of as water that was unclear, had a bad odor or taste, and contained disease-causing bacteria. However, such factors as increased population, industrial requirements for water, atmospheric pollution, toxic waste dumps, and use of pesticides have greatly modified the problem of water pollution.

Many of the newer pollutants are not removed or destroyed by the usual water-treatment processes. For example, among the 66 organic compounds found in the drinking water of a major city on the Mississippi River, 3 are labeled slightly toxic, 17 moderately toxic, 15 very toxic, 1 extremely toxic, and 1 supertoxic. Two are known carcinogens (cancer-producing agents), 11 are suspect, and 3 are metabolized to carcinogens. The United States Public Health Service classifies water pollutants under eight broad categories. These categories are shown in Table 14.5.

Many outbreaks of disease or poisoning, such as typhoid, dysentery, and cholera have been attributed directly to drinking water. Rivers and streams are an easy means for municipalities to dispose of their domestic and industrial waste products. Much of this water is used again by people downstream, and then discharged back into the water source. Then another community still farther downstream draws the same water and discharges its own wastes. Thus, along waterways such as the Mississippi and Delaware rivers, water is withdrawn and discharged many times. If this water is not properly treated, harmful pollutants will build up, causing epidemics of various diseases.

Hazardous waste products are unavoidable in the manufacture of many products that we use in everyday life. One common way to dispose of these wastes is to place them in toxic waste dumps. What has been found after many years of

TABLE 14.5 Classification of Water Pollutants

Type of pollutant	Examples
Oxygen-demanding wastes	Decomposable organic wastes from domestic sewage and industrial wastes of plant and animal origin
Infectious agents	Bacteria, viruses, and other organisms from domestic sewage, animal wastes, and animal process wastes
Plant nutrients	Principally compounds of nitrogen and phosphorus
Organic chemicals	Large numbers of chemicals synthesized by industry, pesticides, chlorinated organic compounds
Other minerals and chemicals	Inorganic chemicals from industrial operations, mining, oil field operations, and agriculture
Radioactive substances	Waste products from mining and processing of radioactive materials, airborne radioactive fallout, increased use of radioactive materials in hospitals and research
Heat from industry	Large quantities of heated water returned to water bodies from power plants and manufacturing facilities after use for cooling
Sediment from land erosion	Solid matter washed into streams and oceans by erosion, rain, and water run off

▲ **Researchers sample lake water to determine the level of organic contamination.**

disposing of wastes in this manner is that toxic substances have seeped into the ground-water deposits. As a result many people have become ill, and water wells have been closed until satisfactory methods of detoxifying this water are found. This problem is serious, because one-half the United States population gets its drinking water from ground water. To clean up the thousands of industrial dumps and to find and implement new and safe methods of disposing of wastes will be very costly.

Mercury and its compounds have long been known to be highly toxic. Mercury gets into the body primarily in the foods we eat. Although it is not an essential mineral for the body, mercury accumulates in the blood, kidneys, liver, and brain tissues. Mercury in the brain causes serious damage to the central nervous system.

The sequence of events that have led to incidents of mercury poisoning is as follows: Mercury and its compounds are used in many industries and in agriculture, primarily as a fungicide in the treatment of seeds. One of the largest uses is in the electrochemical conversion of sodium chloride brines to chlorine and sodium hydroxide, as represented by this equation:

$$2\ NaCl + 2\ H_2O \xrightarrow{\text{Electrolysis}} Cl_2 + 2\ NaOH + H_2$$

Although no mercury is shown in the chemical equation, it is used in the process

Chemistry in Action

Moisturizers and Sweetness

Moisturizers have long been used to protect and rehydrate the skin. These products contain humectants and emollients, which increase the water content of the skin in different ways. The emollients cover the skin with a layer of material which is immiscible with water to prevent the water from within the skin from evaporating. In contrast humectants add water to the skin by attracting water vapor from the air.

The most common humectants are sorbitol, glycerin, and polypropylene glycol. Each molecule is polar, containing multiple —OH groups (see diagram).

Structures of some common humectants.

```
    H              H
    |              |
H—C—OH         H—C—OH
    |              |
H—C—OH         H—C—OH
    |              |
H—C—OH         H—C—H
    |              |
    H              H

 Glycerin      Propylene
                 glycol

    H
    |
H—C—OH
    |
H—C—OH
    |
HO—C—H
    |
H—C—OH
    |
H—C—OH
    |
H—C—OH
    |
    H

 Sorbitol
```

The oxygen atom in each —OH group is considerably more electronegative than the hydrogen atom. This electronegativity difference results in a partial negative charge on the oxygen, whereas the hydrogen carries a partial positive charge. This polarity is the basis for the attraction between the humectant and the water molecules (also polar).

Emollients are composed of hydrophobic (water insoluble) molecules. These products are made of nonpolar molecules. There is a great diversity of compounds in this category, including animal oils, vegetable oils, exotic oils (such as jojoba and aloe vera), and synthetic oils. In each case the molecules form a water-insoluble layer on the skin, which traps the skin's own moisture and feels smooth to the touch.

Intermolecular forces also provide the chemical basis for the sweet taste of many consumer products. There are more than 50 different molecules that have a sweet taste and all of them have similar molecular shapes. The triangle of sweetness theory, developed by Lamont Kier (Massachusetts College of Pharmacy) indicates that these molecules contain three sites that produce the proper structure to attach to the taste bud and trigger the response that registers "sweet" in our brains.

Taste buds are composed of proteins that can form hydrogen bonds with other molecules. The proteins contain —N—H and —OH groups (with hydrogen available to bond) as well as C═O groups (providing oxygen for hydrogen bonding). Molecules that are sweet also contain H-bonding groups including —OH, $—NH_2$, and oxygen or nitrogen. The molecules must not only have the proper atoms to form hydrogen bonds, but must also contain a region that is hydrophobic (repels H_2O). The triangle in the diagram below actually shows the three necessary sites along with the distances between them. The molecule must contain the three sites located at just the proper distances.

The search for new and better sweeteners is continuing. The perfect sweetener would have the following qualities: (1) be as sweet or sweeter than sucrose (table sugar), (2) be nontoxic, (3) be quick to register sweet on the taste buds, (4) be easy to release so the taste doesn't linger, (5) have no calories, (6) be stable when cooked or dissolved, and (7) of course, be inexpensive. Scientists are continuing to search for different molecules that possess these qualities.

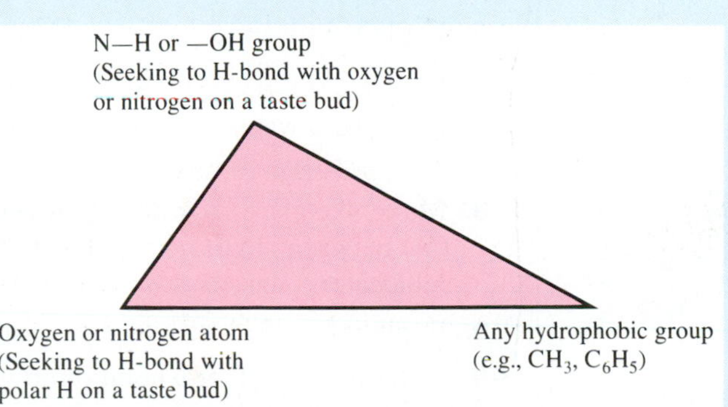

◄ **Triangle of sweetness**

for electrical contact, and small amounts are discharged along with spent brine solutions. Thus considerable quantities of mercury, in low concentrations, have been discharged into lakes and other surface waters from the effluents of these manufacturing plants. The mercury compounds discharged into the water are converted by bacterial action and other organic compounds to methyl mercury, $(CH_3)_2Hg$, which then accumulates in the bodies of fish. Several major episodes of mercury poisoning that have occurred in the past years were the result of eating mercury-contaminated fish. The best way to control this contaminant is at the source, and much has been done since 1970 to eliminate the discharge of mercury in industrial wastes. In 1976 the Environmental Protection Agency banned the use of all mercury-containing insecticides and fungicides.

Many other major water pollutants have been recognized and steps have been taken to eliminate them. Three that pose serious problems are lead, detergents, and chlorine-containing organic compounds. Lead poisoning, for example, has been responsible for many deaths in past years. One major toxic action of lead in the body is the inhibition of the enzyme necessary for the production of hemoglobin in the blood. The usual intake of lead into the body is through food. However, extraordinary amounts of lead can be ingested from water running through lead pipes and by using lead-containing ceramic containers for storage of food and beverages.

It has been clearly demonstrated that waterways rendered so polluted that the water is neither fit for human use nor able to sustain marine life can be successfully restored. However, keeping our lakes and rivers free from pollution is a very costly and complicated process.

Concepts in Review

1. List the common properties of liquids and solids. Explain how they are different from gases.
2. Explain the process of evaporation from the standpoint of kinetic energy.
3. Relate vapor pressure data or vapor pressure curves of different substances to their relative rates of evaporation and to their relative boiling points.
4. Explain the forces involved in surface tension of a liquid. Give two common examples.
5. Explain why a meniscus forms on the surface of liquids in a container.
6. Explain what is occurring throughout the heating curve for water.
7. Describe a water molecule with respect to the Lewis structure, bond angle, and polarity.
8. Make sketches showing hydrogen bonding (a) between water molecules, (b) between hydrogen fluoride molecules, and (c) between ammonia molecules.
9. Explain the effect of hydrogen bonding on the physical properties of water.
10. Determine whether a compound will or will not form hydrogen bonds.
11. Identify metal oxides as basic anhydrides and write balanced equations for their reactions with water.

12. Identify nonmetal oxides as acid anhydrides and write balanced equations for their reactions with water.
13. Deduce the formula of the acid anhydride or basic anhydride when given the formula of the corresponding acid or base.
14. Identify, name, and write equations for the complete dehydration of hydrates.
15. Outline the processes necessary to prepare potable water from a contaminated river source.
16. Describe how water may be softened by distillation, chemical precipitation, ion exchange, and demineralization.
17. Complete and balance equations for (a) the reactions of water with Na, K, and Ca; (b) the reaction of steam with Zn, Al, Fe, and C; and (c) the reaction of water with halogens.

Exercises

An asterisk indicates a more challenging question or problem.

1. Compare the potential energy of the two states of water shown in Figure 14.6.
2. In what state (solid, liquid, or gas) would H_2S, H_2Se, and H_2Te be at 0°C? (See Table 14.3.)
3. The two thermometers in Figure 14.6 read 100°C. What is the pressure of the atmosphere?
4. Draw a diagram of a water molecule and point out the areas that are the negative and positive ends of the dipole.
5. If the water molecule were linear, with all three atoms in a straight line rather than in the shape of a V, as shown in Figure 14.7, what effect would this have on the physical properties of water?
6. Based on Table 14.4, how do we specify 1, 2, 3, 4, 5, 6, 7, and 8 molecules of water in the formulas of hydrates?
7. Would the distillation setup in Figure 14.10 be satisfactory for separating salt and water? Ethyl alcohol and water? Explain.
8. If the liquid in the flask in Figure 14.10 is ethyl alcohol and the atmospheric pressure is 543 torr, what temperature will show on the thermometer? (Use Figure 14.4.)
9. If water were placed in both containers in Figure 14.1, would both have the same vapor pressure at the same temperature? Explain.
10. In Figure 14.1, in which case, (a) or (b), will the atmosphere above the liquid reach a point of saturation?
11. Suppose that a solution of ethyl ether and ethyl alcohol were placed in the closed bottle in Figure 14.1. Use Figure 14.4 for information on the substances.
    (a) Would both substances be present in the vapor?
    (b) If the answer to part (a) is yes, which would have more molecules in the vapor?
12. In Figure 14.2, if 50% more water had been added in part (b), what equilibrium vapor pressure would have been observed in (c)?
13. At approximately what temperature would each of the substances listed in Table 14.2 boil when the pressure is 30 torr? (See Figure 14.4.)
14. Use the graph in Figure 14.4 to find the following:
    (a) The boiling point of water at 500 torr pressure
    (b) The normal boiling point of ethyl alcohol
    (c) The boiling point of ethyl ether at 0.50 atm
15. Consider Figure 14.5.
    (a) Why is line *BC* horizontal? What is happening in this interval?
    (b) What phases are present in the interval *BC*?
    (c) When heating is continued after point *C*, another horizontal line, *DE*, is reached at a higher temperature. What does this line represent?
16. List six physical properties of water.
17. What condition is necessary for water to have its maximum density? What is its maximum density?
18. Account for the fact that an ice–water mixture remains at 0°C until all the ice is melted, even though heat is applied to it.

19. Which contains less heat, ice at 0°C or water at 0°C? Explain.
20. Why does ice float in water? Would ice float in ethyl alcohol ($d = 0.789$ g/mL)? Explain.
21. If water molecules were linear instead of bent, would the heat of vaporization be higher or lower? Explain.
22. The heat of vaporization for ethyl ether is 351 J/g (83.9 cal/g) and that for ethyl alcohol is 855 J/g (204.3 cal/g). Which of these compounds has hydrogen bonding? Explain.
23. Would there be more or less H-bonding if water molecules were linear instead of bent? Explain.
24. Which would show hydrogen bonding, ammonia, NH_3, or methane, CH_4? Explain.
25. In which condition are there fewer hydrogen bonds between molecules: water at 40°C or water at 80°C?
26. Which compound, $H_2NCH_2CH_2NH_2$ or $CH_3CH_2CH_2NH_2$, would you expect to have the higher boiling point? Explain your answer. (Both compounds have similar molar masses.)
27. Explain why rubbing alcohol, which has been warmed to body temperature, still feels cold when applied to your skin.
28. The vapor pressure at 20°C is given for the following compounds:

Methyl alcohol	96	torr
Acetic acid	11.7	torr
Benzene	74.7	torr
Bromine	173	torr
Water	17.5	torr
Carbon tetrachloride	91	torr
Mercury	0.0012	torr
Toluene	23	torr

(a) Arrange these compounds in their order of increasing rate of evaporation.
(b) Which substance listed would have the highest boiling point, and which would have the lowest?

29. Suggest a method whereby water could be made to boil at 50°C.
30. If a dish of water initially at 20°C is placed in a living room maintained at 20°C, the water temperature will fall below 20°C. Explain.
31. Explain why a higher temperature is obtained in a pressure cooker than in an ordinary cooking pot.
32. What is the relationship between vapor pressure and boiling point?
33. On the basis of the Kinetic-Molecular Theory, explain why vapor pressure increases with temperature.
34. Why does water have such a relatively high boiling point?
35. The boiling point of ammonia, NH_3, is −33.4°C and that of sulfur dioxide, SO_2, is −10.0°C. Which has the higher vapor pressure at −40°C?
36. Explain what is occurring physically when a substance is boiling.
37. Explain why HF (bp = 19.4°C) has a higher boiling point than HCl (bp = −85°C), whereas F_2 (bp = −188°C) has a lower boiling point than Cl_2 (bp = −34°C).
38. Can ice be colder than 0°C? Explain.
39. Why does a boiling liquid maintain a constant temperature when heat is continuously being added?
40. At what specific temperature will copper have a vapor pressure of 760 torr?
41. Why does a lake freeze from the top down?
42. What water temperature would you theoretically expect to find at the bottom of a very deep lake? Explain.
43. Write equations to show how the following metals react with water: aluminum, calcium, iron, sodium, zinc. State the conditions for each reaction.
44. Is the formation of hydrogen and oxygen from water an exothermic or an endothermic reaction? How do you know?
45. (a) What is an anhydride?
(b) What type of compound will be an acid anhydride?
(c) What type of compound will be a basic anhydride?
46. (a) Write the formulas for the anhydrides of the following acids:
H_2SO_3, H_2SO_4, HNO_3, $HClO_4$, H_2CO_3, H_3PO_4
(b) Write the formulas for the anhydrides of the following bases:
NaOH, KOH, $Ba(OH)_2$, $Ca(OH)_2$, $Mg(OH)_2$
47. Complete and balance the following equations:
(a) $Ba(OH)_2 \xrightarrow{\Delta}$
(b) $CH_3OH + O_2 \longrightarrow$
Methyl alcohol
(c) $Rb + H_2O \longrightarrow$
(d) $SnCl_2 \cdot 2\,H_2O \xrightarrow{\Delta}$
(e) $HNO_3 + NaOH \longrightarrow$
(f) $Li_2O + H_2O \longrightarrow$
(g) $KOH \xrightarrow{\Delta}$
(h) $Ba + H_2O \longrightarrow$
(i) $Cl_2 + H_2O \longrightarrow$
(j) $SO_3 + H_2O \longrightarrow$

(k) $H_2SO_3 + KOH \longrightarrow$
(l) $CO_2 + H_2O \longrightarrow$

48. Name each of the following hydrates:
(a) $BaBr_2 \cdot 2\,H_2O$ (d) $MgNH_4PO_4 \cdot 6\,H_2O$
(b) $AlCl_3 \cdot 6\,H_2O$ (e) $FeSO_4 \cdot 7\,H_2O$
(c) $FePO_4 \cdot 4\,H_2O$ (f) $SnCl_4 \cdot 5\,H_2O$

49. Explain how anhydrous copper(II) sulfate ($CuSO_4$) can act as an indicator for moisture.

50. Write formulas for magnesium sulfate heptahydrate and disodium hydrogen phosphate dodecahydrate.

51. Distinguish between deionized water and:
(a) Hard water (c) Distilled water
(b) Soft water

52. How can soap function to make soft water from hard water? What objections are there to using soap for this purpose?

53. What substance is commonly used to destroy bacteria in water?

54. What chemical, other than chlorine or chlorine compounds, can be used to disinfect water for domestic use?

55. Some organic pollutants in water can be oxidized by dissolved molecular oxygen. What harmful effect can result from this depletion of oxygen in the water?

56. Why should you not drink liquids that are stored in ceramic containers, especially unglazed ones?

57. Write the chemical equation showing how magnesium ions are removed by a zeolite water softener.

58. Write an equation to show how hard water containing calcium chloride ($CaCl_2$) is softened by using sodium carbonate (Na_2CO_3).

59. Which of the following statements are correct? Rewrite each incorrect statement to make it correct.
(a) The process of a substance changing directly from a solid to a gas is called sublimation.
(b) When water is decomposed, the volume ratio of H_2 to O_2 is 2:1, but the mass ratio of H_2 to O_2 is 1:8.
(c) Hydrogen sulfide is a larger molecule than water.
(d) The changing of ice into water is an exothermic process.
(e) Water and hydrogen fluoride are both nonpolar molecules.
(f) Hydrogen bonding is stronger in H_2O than in H_2S because oxygen is more electronegative than sulfur.
(g) $H_2O_2 \longrightarrow 2\,H_2O + O_2$ represents a balanced equation for the decomposition of hydrogen peroxide.
(h) Steam at 100°C can cause more severe burns than liquid water at 100°C.
(i) The density of water is independent of temperature.
(j) Liquid A boils at a lower temperature than liquid B. This fact indicates that liquid A has a lower vapor pressure than liquid B at any particular temperature.
(k) Water boils at a higher temperature in the mountains than at sea level.
(l) No matter how much heat you put under an open pot of pure water on a stove, you cannot heat the water above its boiling point.
(m) The vapor pressure of a liquid at its boiling point is equal to the prevailing atmospheric pressure.
(n) The normal boiling temperature of water is 273°C.
(o) The pressure exerted by a vapor in equilibrium with its liquid is known as the vapor pressure of the liquid.
(p) Sodium, potassium, and calcium each react with water to form hydrogen gas and a metal hydroxide.
(q) Calcium oxide reacts with water to form calcium hydroxide and hydrogen gas.
(r) Carbon dioxide is the hydride of carbonic acid.
(s) Water in a hydrate is known as water of hydration or water of crystallization.
(t) A substance that spontaneously loses its water of hydration when exposed to the air is said to be efflorescent.
(u) A substance that absorbs water from the air until it forms a solution is deliquescent.
(v) Distillation is effective for softening water because the minerals boil away, leaving soft water behind.
(w) The original source of mercury-contaminated fish is industrial pollution.
(x) Disposal of toxic industrial wastes in toxic waste dumps has been found to be a very satisfactory long-term solution to the problem of what to do with these wastes.
(y) The amount of heat needed to change 1 mole of ice at 0°C to a liquid at 0°C is 6.02 kJ (1.44 kcal).
(z) $BaCl_2 \cdot 2\,H_2O$ has a higher percentage of water than does $CaCl_2 \cdot 2\,H_2O$.

60. Explain how humectants and emollients help to moisturize the skin.

61. Explain the triangle theory of sweetness.

62. What are the characteristics of a good sweetener?

63. How is the chemical structure of a humectant different from that of an emollient?
64. How many moles of compound are in 100. g of each of these hydrates?
(a) $CoCl_2 \cdot 6\,H_2O$ (b) $FeI_2 \cdot 4\,H_2O$
65. How many moles of water can be obtained from 100. g of each of these hydrates?
(a) $CoCl_2 \cdot 6\,H_2O$ (b) $FeI_2 \cdot 4\,H_2O$
66. When a person purchases epsom salts, $MgSO_4 \cdot 7\,H_2O$, what percent of the compound is water?
67. Calculate the mass percent of water in the hydrate $Al_2(SO_4)_3 \cdot 18\,H_2O$.
68. Sugar of lead, a hydrate of lead acetate, $Pb(C_2H_3O_2)_2$, contains 14.2% H_2O. What is the formula for the hydrate?
69. A 25.0 g sample of a hydrate of $FePO_4$ was heated until no more water was driven off. The mass of anhydrous sample is 16.9 g. What is the formula of the hydrate?
70. How many joules are needed to change 120. g of water at 20.°C to steam at 100.°C?
71. How many joules of energy must be removed from 126 g of water at 24°C to form ice at 0°C?
72. How many calories are required to change 225 g of ice at 0°C to steam at 100.°C?
73. The molar heat of vaporization is the number of joules required to change 1 mole of a substance from liquid to vapor at its boiling point. What is the molar heat of vaporization of water?
*74. Suppose 100. g of ice at 0°C are added to 300. g of water at 25°C. Is this sufficient ice to lower the temperature of the system to 0°C and still have ice remaining? Show evidence for your answer.
*75. If 75 g of $H_2O(s)$ at 0.0°C were added to 1.5 L of water at 75°C, what would be the final temperature of the mixture?
*76. The specific heat of zinc is 0.096 cal/g°C. Determine the energy required to raise the temperature of 250. g of zinc from room temperature to 150.°C.
*77. If 9560 J of energy were absorbed by 500. g of ice at 0.0°C, what would be the final temperature?
*78. Suppose 150. g of ice at 0.0°C is added to 0.120 L of water at 45°C. If the mixture is stirred and allowed to cool to 0.0°C, how many grams of ice remain?
*79. Suppose 35.0 g of steam at 100.°C are added to 300. g of water at 25°C. Is there sufficient steam to heat all the water to 100.°C and still have steam remaining? Show evidence for your answer.
80. How many joules of energy would be liberated by condensing 50.0 mol of steam at 100.0°C and allowing the liquid to cool to 30.0°C?
81. How many kilojoules of energy are needed to convert 100. g of ice at −10.0°C to water at 20.0°C. (The specific heat of ice at −10.0°C is 2.01 J/g°C.)
82. What mass of water must be decomposed to produce 25.0 L of oxygen at STP?
83. Compare the volume occupied by 1.00 mol of liquid water at 0°C and 1.00 mol of water vapor at STP.
84. How many grams of water will react with each of the following?
(a) 1.00 mol K (d) 1.00 mole SO_3
(b) 1.00 mol Ca (e) 1.00 g MgO
(c) 1.00 g Na (f) 1.00 g N_2O_5
*85. Suppose 1.00 mol of water evaporates in 1.00 day. How many water molecules, on the average, leave the liquid each second?
*86. A quantity of sulfuric acid is added to 100. mL of water. The final volume of the solution is 122 mL and it has a density of 1.26 g/mL. What mass of acid was added? Assume the density of the water is 1.00 g/mL.
87. A mixture of 80.0 mL of hydrogen and 60.0 mL of oxygen is ignited by a spark to form water.
(a) Does any gas remain unreacted? Which one, H_2 or O_2?
(b) What volume of which gas (if any) remains unreacted? (Assume the same conditions before and after the reaction.)

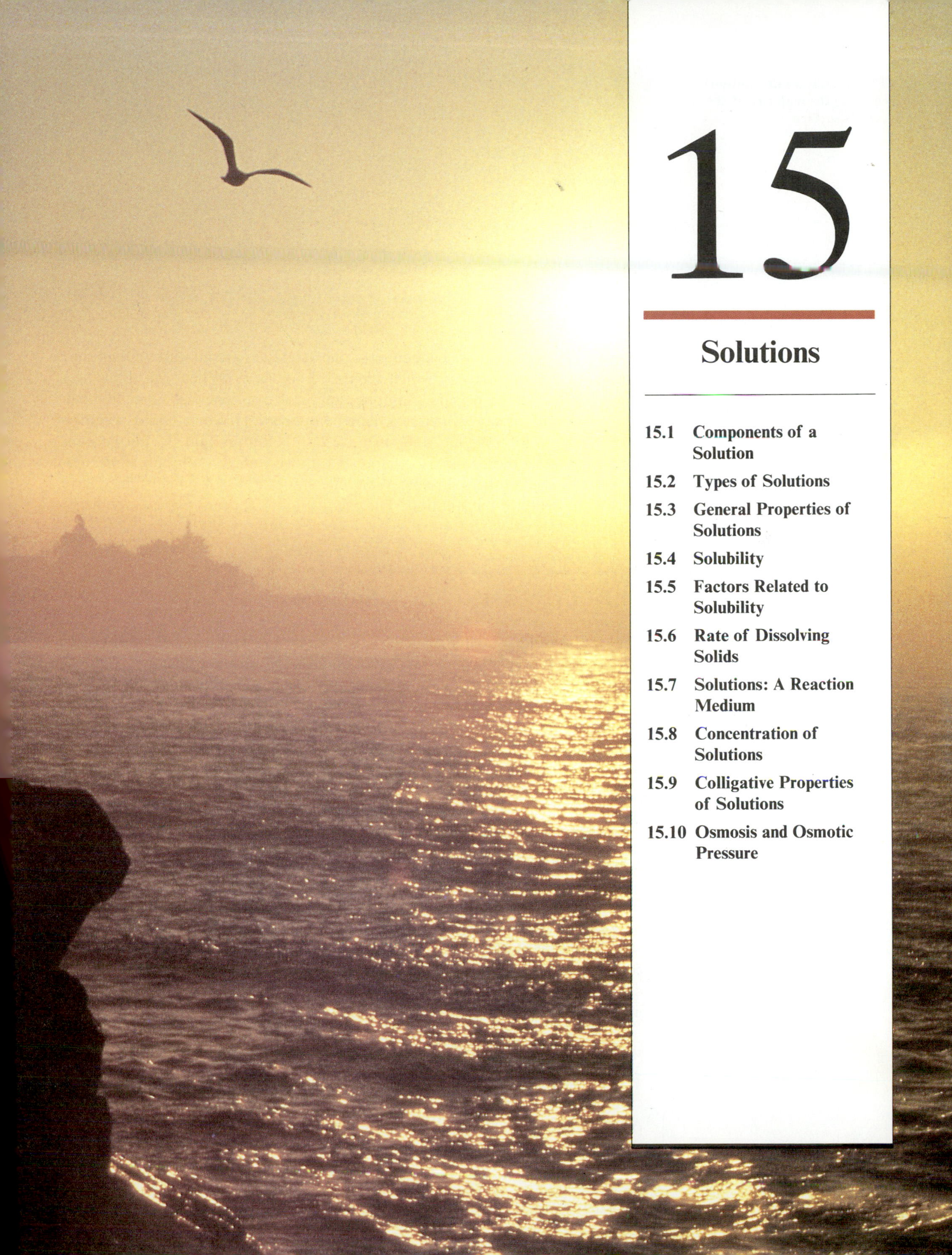

15

Solutions

◀ CHAPTER OPENING PHOTO: The ocean is a salt solution covering the majority of the earth's surface.

Most of the substances we encounter in our daily lives are mixtures. Often they are homogeneous mixtures, which are called solutions. When you think of a solution, juices, blood plasma, shampoo, soft drinks, or wine may come to mind. These solutions all have water as a main component. However, many common items, such as air, gasoline, and steel are also solutions that do not contain water. What are the necessary components in a solution? Why do some substances mix while others do not? What effect does a dissolved substance have on the properties of the solution? The chemistry of solutions plays a significant role in our everyday lives.

15.1 Components of a Solution

solution

solute

solvent

The term **solution** is used in chemistry to describe a system in which one or more substances are homogeneously mixed or dissolved in another substance. A simple solution has two components, a solute and a solvent. The **solute** is the component that is dissolved or the least abundant component in the solution. The **solvent** is the dissolving agent or the most abundant component in the solution. For example, when salt is dissolved in water to form a solution, salt is the solute and water is the solvent. Complex solutions containing more than one solute and/or more than one solvent are common.

15.2 Types of Solutions

From the three states of matter—solid, liquid, and gas—it is possible to have nine different types of solutions: solid dissolved in solid, solid dissolved in liquid, solid dissolved in gas, liquid dissolved in liquid, and so on. Of these, the most common solutions are solid dissolved in liquid, liquid dissolved in liquid, gas dissolved in liquid, and gas dissolved in gas. Some common types of solutions are listed in Table 15.1.

15.3 General Properties of Solutions

A true solution is one in which the particles of dissolved solute are molecular or ionic in size, generally in the range of 0.1 to 1 nm (10^{-8} to 10^{-7} cm). The

TABLE 15.1 Common Types of Solutions

Phase of solution	Solute	Solvent	Example
Gas	Gas	Gas	Air
Liquid	Gas	Liquid	Soft drinks
Liquid	Liquid	Liquid	Antifreeze
Liquid	Solid	Liquid	Salt water
Solid	Gas	Solid	H_2 in Pt.
Solid	Solid	Solid	Brass

properties of a true solution are as follows:

1. It is a homogeneous mixture of two or more components, solute and solvent.
2. It has a variable composition; that is, the ratio of solute to solvent may be varied.
3. The dissolved solute is molecular or ionic in size.
4. It may be either colored or colorless but is usually transparent.
5. The solute remains uniformly distributed throughout the solution and will not settle out with time.
6. The solute generally can be separated from the solvent by purely physical means (for example, by evaporation).

These properties are illustrated by water solutions of sugar and of potassium permanganate. Suppose that we prepare two sugar solutions, the first containing 10 g of sugar added to 100 mL of water and the second containing 20 g of sugar added to 100 mL of water. Each solution is stirred until all the solute dissolves, demonstrating that we can vary the composition of a solution. Every portion of the solution has the same sweet taste because the sugar molecules are uniformly distributed throughout. If confined so that no solvent is lost, the solution will taste and appear the same a week or a month later. The properties of the solution are unaltered after the solution is passed through filter paper. But by carefully evaporating the water, we can recover the sugar from the solution.

To observe the dissolving of potassium permanganate ($KMnO_4$) we affix a few crystals of $KMnO_4$ to paraffin wax or rubber cement at the end of a glass rod and submerge the entire rod, with the wax-permanganate end up, in a cylinder of water. Almost at once the beautiful purple color of dissolved permanganate ions (MnO_4^-) appears at the top of the rod and streams to the bottom of the cylinder as the crystals dissolve. The purple color at first is mostly at the bottom of the cylinder because potassium permanganate is denser than water. But after a while the purple color disperses until it is evenly distributed throughout the solution. This dispersal demonstrates that molecules and ions move about freely and spontaneously (diffuse) in a liquid or solution.

Once formed, a solution is permanent; the solute particles do not settle out. Solution permanency is explained in terms of the Kinetic-Molecular Theory (see Section 13.2). According to the KMT both the solute and solvent particles (molecules and/or ions) are in constant random thermal motion. This motion is energetic enough to prevent the solute particles from settling out under the influence of gravity. This same ceaseless, random thermal motion is also responsible for diffusion in liquids as well as in gases.

15.4 Solubility

solubility

The term **solubility** describes the amount of one substance (solute) that will dissolve in a specified amount of another substance (solvent) under stated conditions. For example, 36.0 g of sodium chloride (NaCl) will dissolve in 100 g of water at 20°C. We say, then, that the solubility of NaCl in water is 36.0 g per 100 g of water at 20°C.

Solubility is often used in a relative way. We say that a substance is very soluble, moderately soluble, slightly soluble, or insoluble. Although these terms do not accurately indicate how much solute will dissolve, they are frequently used to describe the solubility of a substance qualitatively.

miscible

immiscible

Two other terms often used to describe solubility are miscible and immiscible. Liquids that are capable of mixing and forming a solution are **miscible**; those that do not form solutions or are generally insoluble in each other are **immiscible**. Methyl alcohol and water are miscible in each other in all proportions. Carbon tetrachloride and water are immiscible, forming two separate layers when they are mixed. Miscible and immiscible systems are illustrated in Figure 15.1.

The general rules for the solubility of common salts and hydroxides are given in Table 15.2. The solubilities of over 200 compounds are given in the Solubility Table in Appendix IV. Solubility data for thousands of compounds can be found by consulting standard reference sources.*

concentration of a solution

The quantitative expression of the amount of dissolved solute in a particular quantity of solvent is known as the **concentration of a solution.** Several methods of expressing concentration will be described in Section 15.8.

15.5 Factors Related to Solubility

Predicting solubilities is complex and difficult. Many variables, such as size of ions, charge on ions, interaction between ions, interaction between solute and solvent, and temperature, bear upon the problem. Because of the factors involved, the general rules of solubility given in Table 15.2 have many exceptions. However, the rules are very useful, because they do apply to many of the more common compounds that we encounter in the study of chemistry. Keep in mind that these are rules, not laws, and are therefore subject to exceptions. Fortunately

* Two commonly used handbooks are *Lange's Handbook of Chemistry*, 13th ed. (New York: McGraw-Hill, 1985), and *Handbook of Chemistry and Physics*, 72nd ed. (Cleveland: Chemical Rubber Co., 1991).

the solubility of a solute is relatively easy to determine experimentally. Factors related to solubility are discussed in the following paragraphs.

The Nature of the Solute and Solvent The old adage "like dissolves like" has merit, in a general way. Polar or ionic substances tend to be more miscible, or soluble, with other polar substances. Nonpolar substances tend to be miscible with other nonpolar substances and less miscible with polar substances. Thus, mineral acids, bases, and salts, which are polar, tend to be much more soluble in water, which is polar, than in solvents such as ether, carbon tetrachloride, or benzene, which are essentially nonpolar. Sodium chloride, an ionic substance, is soluble in water, slightly soluble in ethyl alcohol (less polar than water), and insoluble in ether and benzene. Pentane (C_5H_{12}), a nonpolar substance, is only slightly soluble in water but is very soluble in benzene and ether.

At the molecular level the formation of a solution from two nonpolar substances, such as carbon tetrachloride and benzene, can be visualized as a process of simple mixing. The nonpolar molecules, having little tendency to either attract or repel one another, easily intermingle to form a homogeneous mixture.

Solution formation between polar substances is much more complex. For example, the process by which sodium chloride dissolves in water is illustrated in

▲
FIGURE 15.1
Miscible and immiscible systems. On the left is a immiscible mixture of iodine water and CCl_4. The mixture is heterogeneous, and two liquid layers are observed. On the right is a miscible mixture of $CuSO_4(aq)$ and CH_3OH. The solution consists of a single phase uniformly dispersed.

TABLE 15.2 General Solubility Rules for Common Salts and Hydroxides[a]

Class	Solubility in cold water[b]
Nitrates	Most nitrates are soluble.
Acetates	Most acetates are soluble.
Chlorides, Bromides, Iodides	Most chlorides, bromides, and iodides are soluble, except those of Ag, Hg(I), and Pb(II); $PbCl_2$ and $PbBr_2$ are slightly soluble in hot water.
Sulfates	Most sulfates are soluble except those of Ba, Sr and Pb; Ca and Ag sulfates are slightly soluble.
Carbonates, Phosphates	Most carbonates and phosphates are insoluble except those of Na, K, and NH_4^+. Many bicarbonates and acid phosphates are soluble.
Hydroxides	Most hydroxides are insoluble except those of the alkali metals and NH_4OH; $Ba(OH)_2$ and $Ca(OH)_2$ are slightly soluble.
Sodium salts, Potassium salts, Ammonium salts	Most common salts of these ions are soluble.
Sulfides	Most sulfides are insoluble except those of the alkali metals, ammonium, and the alkaline earth metals (Ca, Mg, Ba).

[a] When we say a substance is soluble, we mean that the substance is reasonably soluble. All substances have some solubility in water, although the amount of solubility may be very small; the solubility of silver iodide, for example, is about 1×10^{-8} mol AgI/liter H_2O.
[b] These rules have exceptions.

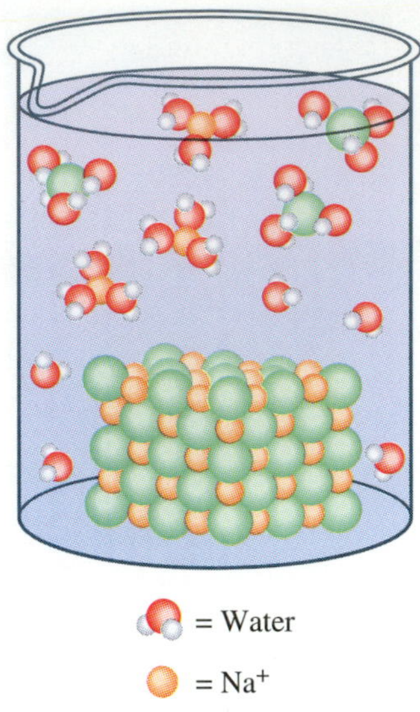

▲ **FIGURE 15.2**
Dissolution of sodium chloride in water. Polar water molecules are attracted to Na^+ and Cl^- ions in the salt crystal, weakening the attraction between the ions. As the attraction between the ions weakens, the ions move apart and become surrounded by water dipoles. The hydrated ions slowly diffuse away from the crystal to become dissolved in solution.

Figure 15.2. Water molecules are very polar and are attracted to other polar molecules or ions. When salt crystals (NaCl) are put into water, polar water molecules become attracted to the sodium and chloride ions on the crystal surfaces and weaken the attraction between Na^+ and Cl^- ions. The positive end of the water dipole is attracted to the Cl^- ions, and the negative end of the water dipole to the Na^+ ions. The weakened attraction permits the ions to move apart, making room for more water dipoles. Thus, the surface ions are surrounded by water molecules, becoming hydrated ions, $Na^+(aq)$ and $Cl^-(aq)$, and slowly diffuse away from the crystals and dissolve in solution.

$$\text{NaCl(crystal)} \xrightarrow{H_2O} Na^+(aq) + Cl^-(aq)$$

Examination of the data in Table 15.3 reveals some of the complex questions relating to solubility. For example: Why are lithium halides, except for lithium fluoride (LiF), more soluble than sodium and potassium halides? Why are the solubilites of LiF and sodium fluoride (NaF) so low in comparison with those of the other salts? Why does not the solubility of LiF, NaF, and NaCl increase proportionately with temperature, as do the solubilities of the other salts? Sodium chloride is appreciably soluble in water but is insoluble in concentrated hydrochloric acid (HCl) solution. On the other hand, LiF and NaF are not very soluble in water but are quite soluble in hydrofluoric acid (HF) solution—why? These questions will not be answered directly here, but it is hoped that your curiosity will be aroused to the point that you will do some reading and research on the properties of solutions.

The Effect of Temperature on Solubility Temperature has major effects on the solubility of most substances, and most solutes have a limited solubility in a

TABLE 15.3 Solubility of Alkali Metal Halides in Water

Salt	Solubility (g salt/100 g H_2O) 0°C	100°C
LiF	0.12	0.14 (at 35°C)
LiCl	67	127.5
LiBr	143	266
LiI	151	481
NaF	4	5
NaCl	35.7	39.8
NaBr	79.5	121
NaI	158.7	302
KF	92.3 (at 18°C)	Very soluble
KCl	27.6	57.6
KBr	53.5	104
KI	127.5	208

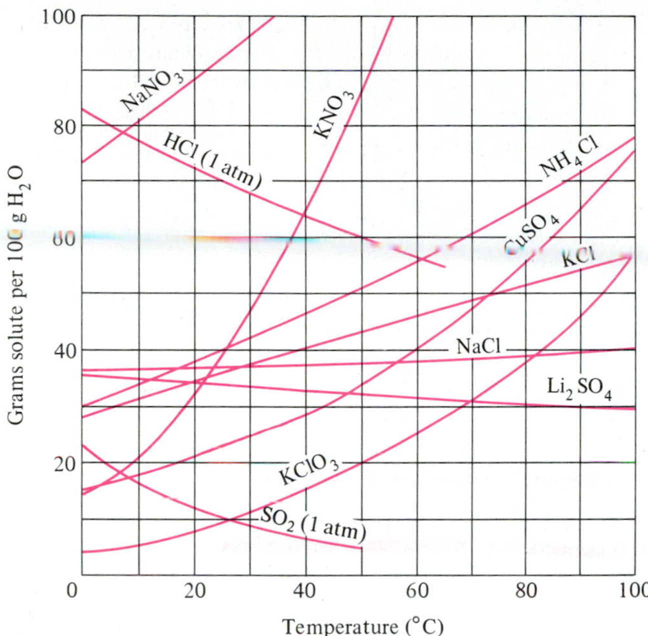

◀ **FIGURE 15.3 Solubility of various compounds in water.**

specific solvent at a fixed temperature. For most solids dissolved in a liquid, an increase in temperature results in increased solubility (see Figure 15.3). However, no completely valid general rule governs the solubility of solids in liquids with change in temperature. Some solids increase in solubility only slightly with increasing temperature (see NaCl in Figure 15.3); other solids decrease in solubility with increasing temperature (see Li_2SO_4 in Figure 15.3).

On the other hand, the solubility of a gas in water usually decreases with increasing temperature (see HCl and SO_2 in Figure 15.3). The tiny bubbles that are formed when water is heated are due to the decreased solubility of air at higher temperatures. The decreased solubility of gases at higher temperatures is explained in terms of the KMT by assuming that, in order to dissolve, the gas molecules must form "bonds" of some sort with the molecules of the liquid. An increase in temperature decreases the solubility of the gas because it increases the kinetic energy (speed) of the gas molecules and thereby decreases their ability to form "bonds" with the liquid molecules.

The Effect of Pressure on Solubility Small changes in pressure have little effect on the solubility of solids in liquids or liquids in liquids but have a marked effect on the solubility of gases in liquids. The solubility of a gas in a liquid is directly proportional to the pressure of that gas above the solution. Thus, the amount of a gas that is dissolved in solution will double if the pressure of that gas over the solution is doubled. For example, carbonated beverages contain dissolved carbon dioxide at pressures greater than atmospheric pressure. When a bottle of carbonated soda is opened, the pressure is immediately reduced to the atmospheric pressure, and the excess dissolved carbon dioxide bubbles out of the solution. ↑P = ↑ Solubility.

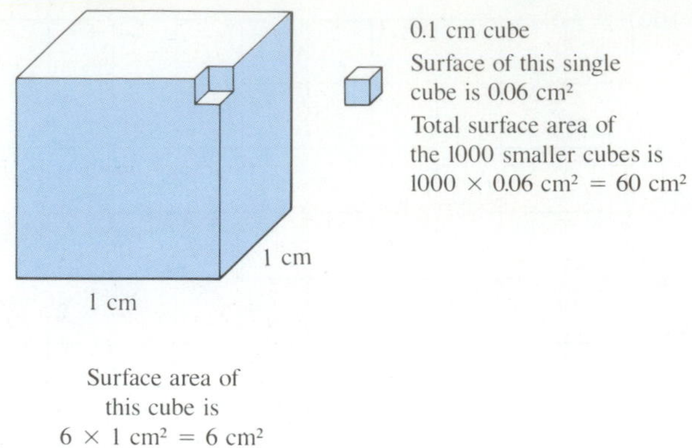
0.1 cm cube
Surface of this single cube is 0.06 cm²
Total surface area of the 1000 smaller cubes is 1000 × 0.06 cm² = 60 cm²
1 cm
1 cm
Surface area of this cube is 6 × 1 cm² = 6 cm²

FIGURE 15.4 ▶
Surface area of crystals. A crystal 1 cm on a side has a surface area of 6 cm^2. Subdivided into 1000 smaller crystals, each 0.1 cm on a side, the total surface area is increased to 60 cm^2.

15.6 RATE OF DISSOLVING SOLIDS

The rate at which a solid dissolves is governed by (1) the size of the solute particles, (2) the temperature, (3) the concentration of the solution, and (4) agitation or stirring.

Particle Size A solid can dissolve only at the surface that is in contact with the solvent. Because the surface to volume ratio increases as size decreases, smaller crystals dissolve faster than large ones. For example, if a salt crystal 1 cm on a side (6 cm^2 surface area) is divided into 1000 cubes, each 0.1 cm on a side, the total surface of the smaller cubes is 60 cm^2—a tenfold increase in surface area (see Figure 15.4).

Temperature In most cases the rate of dissolving of a solid increases with temperature. This increase is due to kinetic effects. The solvent molecules move more rapidly at higher temperatures and strike the solid surfaces more often and harder, causing the rate of dissolving to increase.

Concentration of the Solution When the solute and solvent are first mixed, the rate of dissolving is at its maximum. As the concentration of the solution increases and the solution becomes more nearly saturated with the solute, the rate of dissolving decreases greatly. The rate of dissolving is pictured graphically in Figure 15.5. Note that about 17 g dissolve in the first 5 minute interval, but only about 1 g dissolves in the fourth 5 minute interval. Although different solutes show different rates, the rate of dissolving always becomes very slow as the concentration approaches the saturation point.

Agitation or Stirring The effect of agitation or stirring is kinetic. When a solid is first put into water, the only solvent with which it comes in contact is in the immediate vicinity. As the solid dissolves, the amount of dissolved solute around the solid becomes more and more concentrated, and the rate of dissolving slows down. If the mixture is not stirred, the dissolved solute diffuses very slowly through the solution; weeks may pass before the solid is entirely dissolved. Stirring distributes the dissolved solute rapidly through the solution, and more solvent is brought into contact with the solid, causing it to dissolve more rapidly.

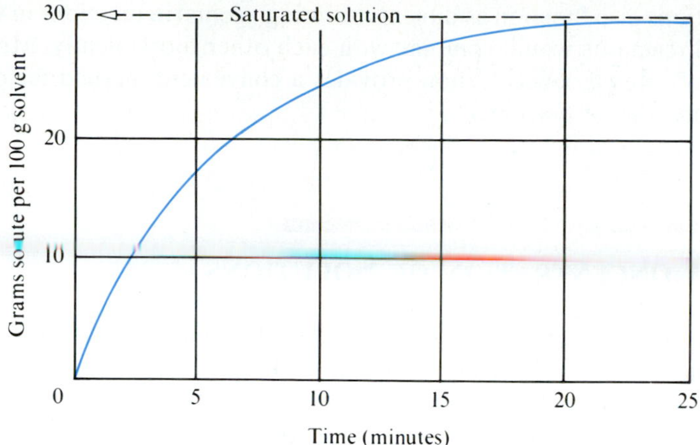

◄ FIGURE 15.5
Rate of dissolution of a solid solute in a solvent. The rate is maximum at the beginning and decreases as the concentration approaches saturation.

15.7 Solutions: A Reaction Medium

Many solids must be put into solution in order to undergo appreciable chemical reaction. We can write the equation for the double-displacement reaction between sodium chloride and silver nitrate:

$$NaCl + AgNO_3 \longrightarrow AgCl + NaNO_3$$

But suppose we mix solid NaCl and solid $AgNO_3$ and look for a chemical change. If any reaction occurs, it is slow and virtually undetectable. In fact, the crystalline structures of NaCl and $AgNO_3$ are so different that we could separate them by tediously picking out each kind of crystal from the mixture. But if we dissolve the sodium chloride and silver nitrate separately in water and mix the two solutions, we observe the immediate formation of a white, curdy precipitate of silver chloride.

Molecules or ions must come into intimate contact or collide with one another in order to react. In the foregoing example, the two solids did not react because the ions were securely locked within their crystal structures. But when the sodium chloride and silver nitrate are dissolved, their crystal lattices are broken down and the ions become mobile. When the two solutions are mixed, the mobile Ag^+ and Cl^- ions come into contact and react to form insoluble AgCl, which precipitates out of solution. The soluble Na^+ and NO_3^- ions remain mobile in solution but form the crystalline salt $NaNO_3$ when the water is evaporated:

$$NaCl(aq) + AgNO_3(aq) \longrightarrow AgCl\downarrow + NaNO_3(aq)$$

$$\underset{\text{Sodium chloride solution}}{Na^+(aq) + Cl^-(aq)} + \underset{\text{Silver nitrate solution}}{Ag^+(aq) + NO_3^-(aq)} \xrightarrow{H_2O} \underset{\text{Silver chloride}}{AgCl\downarrow} + \underset{\text{Sodium nitrate in solution}}{Na^+(aq) + NO_3^-(aq)}$$

The mixture of the two solutions provides a medium or space in which the Ag^+ and Cl^- ions can react. (See Chapter 16 for further discussion of ionic reactions.)

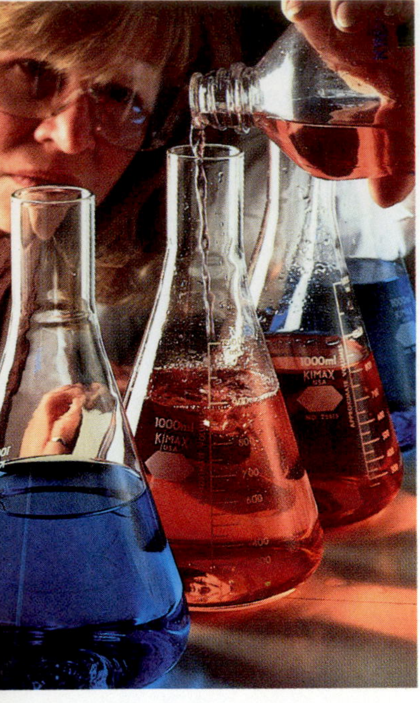

Solutions also function as diluents (diluting agents) in reactions in which the undiluted reactants would combine with each other too violently. Moreover, a solution of known concentration provides a convenient method for delivering specific amounts of reactants.

15.8 Concentration of Solutions

The concentration of a solution expresses the amount of solute dissolved in a given quantity of solvent or solution. Because reactions are often conducted in solution, it is important to understand the methods of expressing concentration and to know how to prepare solutions of particular concentrations. The concentration of a solution may be expressed qualitatively or quantitatively. Let's begin with a look at the qualitative methods of expressing concentration.

Dilute and Concentrated Solutions

When we say that a solution is *dilute* or *concentrated*, we are expressing, in a relative way, the amount of solute present. One gram of salt and 2 g of salt in solution are both dilute solutions when compared with the same volume of a solution containing 20 g of salt. Ordinary concentrated hydrochloric acid (HCl) contains 12 moles of HCl per liter of solution. In some laboratories the dilute acid is made by mixing equal volumes of water and the concentrated acid. In other laboratories the concentrated acid is diluted with two or three volumes of water, depending on its use. The term **dilute solution**, then, describes a solution that contains a relatively small amount of dissolved solute. Conversely, a **concentrated solution** contains a relatively large amount of dissolved solute.

dilute solution

concentrated solution

Saturated, Unsaturated, and Supersaturated Solutions

At a specific temperature there is a limit to the amount of solute that will dissolve in a given amount of solvent. When this limit is reached, the resulting solution is said to be *saturated*. For example, when we put 40.0 g of KCl into 100 g of H_2O at 20°C, we find that 34.0 g of KCl dissolve and 6.0 g of KCl remain undissolved. The solution formed is a saturated solution of KCl.

Two processes are occurring simultaneously in a saturated solution. The solid is dissolving into solution, and at the same time the dissolved solute is crystallizing out of solution. This may be expressed as

$$\text{solute (undissolved)} \rightleftharpoons \text{solute (dissolved)}$$

When these two opposing processes are occurring at the same rate, the amount of solute in solution is constant, and a condition of equilibrium is established between dissolved and undissolved solute. **A saturated solution** contains dissolved solute in equilibrium with undissolved solute.

saturated solution

It is important to state the temperature of a saturated solution, because a solution that is saturated at one temperature may not be saturated at another. If

TABLE 15.4 Saturated Solutions at 20°C and 50°C

Solute	Solubility (g solute/100 g H_2O) 20°C	50°C	Solute	Solubility (g solute/100 g H_2O) 20°C	50°C
NaCl	36.0	37.0	$KClO_3$	7.4	19.3
KCl	34.0	42.6	$AgNO_3$	222.0	455.0
$NaNO_3$	88.0	114.0	$C_{12}H_{22}O_{11}$	203.9	260.4

the temperature of a saturated solution is changed, the equilibrium is disturbed, and the amount of dissolved solute will change to reestablish equilibrium.

A saturated solution may be either dilute or concentrated, depending on the solubility of the solute. A saturated solution can be conveniently prepared by dissolving a little more than the saturated amount of solute at a temperature somewhat higher than room temperature. Then the amount of solute in solution will be in excess of its solubility at room temperature; and, when the solution cools, the excess solute will crystallize, leaving the solution saturated. (In this case, the solute must be more soluble at higher temperatures and must not form a supersaturated solution.) Examples expressing the solubility of saturated solutions at two different temperatures are given in Table 15.4.

unsaturated solution

An **unsaturated solution** contains less solute per unit of volume than does its corresponding saturated solution. In other words, additional solute can be dissolved in an unsaturated solution without altering any other conditions. Consider a solution made by adding 40 g of KCl to 100 g of H_2O at 20°C (see Table 15.4). The solution formed will be saturated and will contain about 6 g of undissolved salt, because the maximum amount of KCl that can dissolve in 100 g of H_2O at 20°C is 34 g. If the solution is now heated and maintained at 50°C, all the salt will dissolve and, in fact, even more can be dissolved. Thus the solution at 50°C is unsaturated.

supersaturated solution

In some circumstances, solutions can be prepared that contain more solute than that needed for a saturated solution at a particular temperature. Such solutions are said to be **supersaturated.** However, we must qualify this definition by noting that a supersaturated solution is unstable. Disturbances such as jarring, stirring, scratching the walls of the container, or dropping in a "seed" crystal cause the supersaturation to break. When a supersaturated solution is disturbed, the excess solute crystallizes out rapidly, returning the solution to a saturated state.

Supersaturated solutions are not easy to prepare but may be made from certain substances by dissolving, in warm solvent, an amount of solute greater than that needed for a saturated solution at room temperature. The warm solution is then allowed to cool very slowly. With the proper solute and careful work, a supersaturated solution will result. Two substances commonly used to demonstrate this property are sodium thiosulfate pentahydrate, $Na_2S_2O_3 \cdot 5\ H_2O$, and sodium sulfate, Na_2SO_4 (from a saturated solution at 30°C).

EXAMPLE 15.1

Will a solution made by adding 2.5 g of $CuSO_4$ to 10 g of H_2O be saturated or unsaturated at 20°C?

SOLUTION

To answer this question we first need to know the solubility of $CuSO_4$ at 20°C. From Figure 15.3 we see that the solubility of $CuSO_4$ at 20°C is about 21 g per 100 g of H_2O. This amount is equivalent to 2.1 g of $CuSO_4$ per 10 g of H_2O.

Since 2.5 g per 10 g of H_2O is greater than 2.1 g per 10 g of H_2O, the solution will be saturated and 0.4 g of $CuSO_4$ will be undissolved.

PRACTICE Will a solution made by adding 9.0 g NH_4Cl to 20 g of H_2O be saturated or unsaturated at 50°C?

Answer: unsaturated

Each of these methods for expressing concentration is useful on a qualitative basis. But suppose we wish to compare two solutions with concentrations that are nearly the same. We would need to introduce quantities into our concentration system. A variety of concentration units are quantitative. Let's examine several of them.

Mass Percent Solution

This method expresses concentration as the percent of solute in a given mass of solution. It says that for a given mass of solution a certain percent of that mass is solute. Suppose that we take a bottle from the reagent shelf that reads "sodium hydroxide, NaOH, 10%." This statement means that for every 100 g of this solution, 10 g will be NaOH and 90 g will be water. (Note that this amount of solution is 100 g and not 100 mL.) We could also make this same concentration of solution by dissolving 2.0 g of NaOH in 18 g of water. Mass percent concentrations are most generally used for solids dissolved in liquids.

$$\text{mass percent} = \frac{\text{g solute}}{\text{g solute} + \text{g solvent}} \times 100 = \frac{\text{g solute}}{\text{g solution}} \times 100$$

As instrumentation advances in chemistry, our ability to measure the concentration of dilute solutions is increasing as well. Instead of mass percent, chemists now commonly use **parts per million (ppm).**

parts per million (ppm)

$$\text{parts per million} = \frac{\text{g solute}}{\text{g solute} + \text{g solvent}} \times 1{,}000{,}000$$

Currently, air and water contaminants, drugs in the human body, and pesticide residues are some substances measured in parts per million.

EXAMPLE 15.2

What is the mass percent of sodium hydroxide in a solution that is made by dissolving 8.00 g of NaOH in 50.0 g of H_2O?

SOLUTION

grams of solute (NaOH) = 8.00 g

grams of solvent (H_2O) = 50.0 g

$$\frac{8.00 \text{ g NaOH}}{8.00 \text{ g NaOH} + 50.0 \text{ g } H_2O} \times 100 = 13.8\% \text{ NaOH solution}$$

EXAMPLE 15.3

What masses of potassium chloride (KCl) and water are needed to make 250. g of 5.00% solution?

SOLUTION

The percent expresses the mass of the solute.

250. g = total mass of solution

5.00% of 250. g = 0.0500 × 250. g = 12.5 g KCl (solute)

250. g − 12.5 g = 237.5 g H_2O

Dissolving 12.5 g of KCl in 237.5 g of H_2O gives a 5.00% KCl solution.

EXAMPLE 15.4

A 34.0% sulfuric acid solution has a density of 1.25 g/mL. How many grams of H_2SO_4 are contained in 1.00 L of this solution?

SOLUTION

Since H_2SO_4 is the solute, we first solve the mass percent equation for grams of solute:

$$\text{mass percent} = \frac{\text{g solute}}{\text{g solution}} \times 100$$

$$\text{g solute} = \frac{\text{mass percent} \times \text{g solution}}{100}$$

The mass percent is given in the problem. We need to determine the grams of solution. The mass of the solution can be calculated from the density data.

Convert density (g/mL) to grams:

$$1.00 \text{ L} = 1.00 \times 10^3 \text{ mL}$$

$$\frac{1.25 \text{ g}}{\text{mL}} \times 1.00 \times 10^3 \text{ mL} = 1250 \text{ g} \quad \text{(mass of solution)}$$

Now we have all the figures to calculate the grams of solute.

$$\text{g solute} = \frac{34.0 \times 1250 \text{ g}}{100} = 425 \text{ g } H_2SO_4$$

1.00 L of 34.0% H_2SO_4 solution contains 425 g of H_2SO_4.

PRACTICE What is the mass percent of Na_2SO_4 in a solution that is made by dissolving 25.0 g of Na_2SO_4 in 225.0 g of H_2O?

Answer: 10.0% Na_2SO_4 solution

The student should note that the concentration expressed as mass percent is independent of the formula of the solute.

Mass/Volume Percent (m/v)

This method expresses concentration as grams of solute per 100 mL of solution. With this system, a 10.0% (m/v) glucose solution is made by dissolving 10.0 g of glucose in water, diluting to 100 mL, and mixing. The 10.0% (m/v) solution could also be made by diluting 20.0 g to 200 mL, 50.0 g to 500 mL, and so on. Of course, any other appropriate dilution ratio may be used.

$$\text{mass/volume percent} = \frac{\text{g solute}}{\text{mL solution}} \times 100$$

Volume Percent

Solutions that are formulated from two liquids are often expressed as *volume percent* with respect to the solute. The volume percent is the volume of a liquid in 100 mL of solution. The label on a bottle of ordinary rubbing alcohol reads "isopropyl alcohol, 70% by volume." Such a solution could be made by mixing 70 mL of alcohol with water to make a total volume of 100 mL. We cannot use 30 mL of water because the two volumes are not necessarily additive.

$$\text{volume percent} = \frac{\text{volume of liquid in question}}{\text{total volume of solution}} \times 100$$

Volume percent is used to express the concentration of alcohol in beverages. Wines generally contain 12% alcohol by volume. This translates into 12 mL of alcohol in each 100 mL of wine. The beverage industry also uses the concentration unit of *proof* (twice the volume percent). Pure alcohol is 100%, therefore 200 proof. Scotch is 86 proof or 43% alcohol.

Molarity

Mass percent solutions do not equate or express the molar masses of the solute in solution. For example, 1000. g of 10.0% NaOH solution contain 100. g of NaOH; 1000. g 10.0% KOH solution contain 100. g of KOH. In terms of moles of NaOH and KOH, these solutions contain

$$\text{moles NaOH} = 100.\text{ g NaOH} \times \frac{1\text{ mol NaOH}}{40.00\text{ g NaOH}} = 2.50\text{ mol NaOH}$$

$$\text{moles KOH} = 100.\text{ g KOH} \times \frac{1\text{ mol KOH}}{56.11\text{ g KOH}} = 1.78\text{ mol KOH}$$

From these figures we see that the two 10.0% solutions do not contain the same number of moles of NaOH and KOH. Yet one mole of each of these bases will neutralize the same amount of acid. As a result we find that a 10.0% NaOH solution has more reactive alkali than a 10.0% KOH solution.

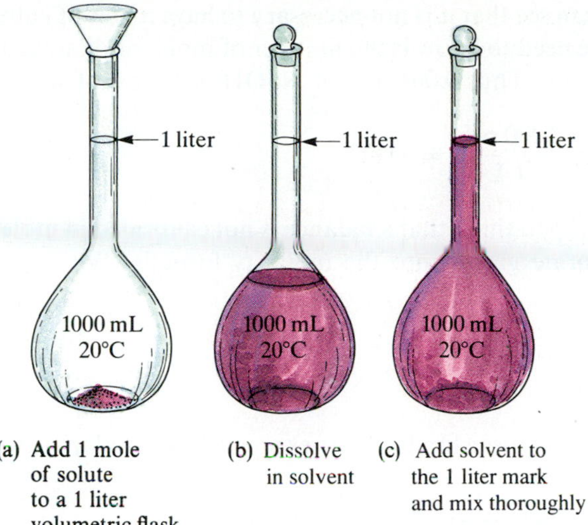

(a) Add 1 mole of solute to a 1 liter volumetric flask (b) Dissolve in solvent (c) Add solvent to the 1 liter mark and mix thoroughly

◄ **FIGURE 15.6 Preparation of a 1 *M* solution.**

We need a method of expressing concentration that will easily indicate how many moles of solute are present per unit volume of solution. For this purpose the molar method of expressing concentration is used.

A 1 molar solution contains 1 mol, or 1 molar mass of solute per liter of solution. For example, to make a 1 molar solution of sodium hydroxide (NaOH), we dissolve 40.00 g of NaOH (1 mol) in water and dilute the solution with more water to a volume of 1 liter. The solution contains 1 mol of the solute in 1 L of solution and is said to be 1 molar (1 *M*) in concentration. Figure 15.6 illustrates the preparation of a 1 molar solution. Note that the volume of the solute and the solvent together is 1 liter.

The concentration of a solution can, of course, be varied by using more or less solute or solvent; but in any case the **molarity** of a solution is the number of moles of solute per liter of solution. A capital *M* is the abbreviation for molarity. The units of molarity are moles per liter. The expression "2.0 *M* NaOH" means a 2.0 molar solution of NaOH (2.0 mol, or 80. g, of NaOH dissolved in 1 L of solution).

molarity

$$\text{molarity} = M = \frac{\text{number of moles of solute}}{\text{liter of solution}} = \frac{\text{moles}}{\text{liter}}$$

Flasks that are calibrated to contain specific volumes at a particular temperature are used to prepare solutions of a desired concentration. These *volumetric flasks* have a calibration mark on the neck to indicate accurately the measured volume. Molarity is based on a specific volume of solution and therefore will vary slightly with temperature because volume varies with temperature (1000 mL of H_2O at 20°C = 1001 mL at 25°C).

Suppose we want to make 500 mL of 1 *M* solution. This solution can be prepared by determining the mass of 0.5 mol of the solute and diluting with water in a 500 mL (0.5 L) volumetric flask. The molarity will be

$$M = \frac{0.5 \text{ mol solute}}{0.5 \text{ L solution}} = 1 \text{ molar}$$

Thus you can see that it is not necessary to have a liter of solution to express molarity. All we need to know is the number of moles of dissolved solute and the volume of solution. Thus 0.001 mol of NaOH in 10 mL of solution is 0.1 *M*:

$$\frac{0.001 \text{ mol}}{10 \cancel{\text{mL}}} \times \frac{1000 \cancel{\text{mL}}}{1 \text{ L}} = 0.1\ M$$

When we stop to think that a balance is not calibrated in moles but in grams, we can incorporate grams into the molarity formula. We do so by using the relationship

$$\text{moles} = \frac{\text{grams of solute}}{\text{molar mass}}$$

Substituting this relationship into our expression for molarity, we get

$$M = \frac{\text{mol}}{\text{L}} = \frac{\text{g solute}}{\text{molar mass solute} \times \text{L solution}}$$

$$= \frac{\text{g}}{\text{molar mass} \times \text{L}}$$

We can now determine the mass of any amount of a solute that has a known formula, dilute it to any volume, and calculate the molarity of the solution using this formula.

The molarities of the concentrated acids commonly used in the laboratory are

HCl 12 *M* $HC_2H_3O_2$ 17 *M* HNO_3 16 *M* H_2SO_4 18 *M*

Example 15.5

What is the molarity of a solution containing 1.4 mol of acetic acid ($HC_2H_3O_2$) in 250. mL of solution?

Solution

Substitute the data, 1.4 mol and 250. mL (0.250 L), directly into the equation for molarity.

$$M = \frac{\text{mol}}{\text{L}} = \frac{1.4 \text{ mol}}{0.250 \text{ L}} = \frac{5.6 \text{ mol}}{\text{L}} = 5.6\ M$$

By the unit conversion method we note that the concentration given in the problem statement is 1.4 mol per 250. mL (mol/mL). Since molarity = mol/L, the needed conversion is

$$\frac{\text{mol}}{\text{mL}} \longrightarrow \frac{\text{mol}}{\text{L}} = M$$

$$\frac{1.40 \text{ mol}}{250 \cancel{\text{mL}}} \times \frac{1000 \cancel{\text{mL}}}{1 \text{ L}} = \frac{5.6 \text{ mol}}{1 \text{ L}} = 5.6\ M$$

Example 15.6

What is the molarity of a solution made by dissolving 2.00 g of potassium chlorate ($KClO_3$) in enough water to make 150. mL of solution?

SOLUTION

This problem can be solved using the unit conversion method. The steps in the conversions must lead to units of moles/liter.

$$\frac{\text{g KClO}_3}{\text{mL}} \longrightarrow \frac{\text{g KClO}_3}{\text{L}} \longrightarrow \frac{\text{mol KClO}_3}{\text{L}} = M$$

The data are

g = 2.00 g molar mass $KClO_3$ = 122.6 g/mol volume = 150. mL

$$\frac{2.00\ \text{g KClO}_3}{150.\ \text{mL}} \times \frac{1000\ \text{mL}}{1\ \text{L}} \times \frac{1\ \text{mol KClO}_3}{122.6\ \text{g KClO}_3} = \frac{0.109\ \text{mol}}{1\ \text{L}} = 0.109\ M$$

EXAMPLE 15.7

How many grams of potassium hydroxide are required to prepare 600. mL of 0.450 *M* KOH solution?

SOLUTION

The conversion is

milliliters ⟶ liters ⟶ moles ⟶ grams

The data are

$$\text{volume} = 600.\ \text{mL} \qquad M = \frac{0.450\ \text{mol}}{\text{L}} \qquad \text{molar mass KOH} = \frac{56.11\ \text{g KOH}}{\text{mol}}$$

The calculation is

$$600.\ \text{mL} \times \frac{1\ \text{L}}{1000\ \text{mL}} \times \frac{0.450\ \text{mol}}{\text{L}} \times \frac{56.11\ \text{g KOH}}{\text{mol}} = 15.1\ \text{g KOH}$$

PRACTICE What is the molarity of a solution made by dissolving 7.50 g of magnesium nitrate, $Mg(NO_3)_2$, in enough water to make 25.0 mL of solution?
Answer: 2.02 *M*

EXAMPLE 15.8

How many milliliters of 2.00 *M* HCl will react with 28.0 g of NaOH?

SOLUTION

Step 1 Write and balance the equation for the reaction:

$$\text{HCl}(aq) + \text{NaOH}(aq) \longrightarrow \text{NaCl}(aq) + \text{H}_2\text{O}(aq)$$

The equation states that 1 mol of HCl reacts with 1 mol of NaOH.

Step 2 Find the number of moles of NaOH in 28.0 g of NaOH:

g NaOH ⟶ mol NaOH

$$28.0\ \text{g NaOH} \times \frac{1\ \text{mol}}{40.00\ \text{g}} = 0.700\ \text{mol NaOH}$$

$$28.0\ \text{g NaOH} = 0.700\ \text{mol NaOH}$$

Step 3 Solve for moles and volume of HCl needed. From Steps 1 and 2 we see that 0.700 mol of HCl will react with 0.700 mol of NaOH, because the ratio of moles reacting is 1:1. We know that 2.00 M HCl contains 2.00 mol of HCl per liter; therefore, the volume that contains 0.700 mol of HCl will be less than 1 L.

$$\text{mol NaOH} \longrightarrow \text{mol HCl} \longrightarrow \text{L HCl} \longrightarrow \text{mL HCl}$$

$$0.700\,\cancel{\text{mol NaOH}} \times \frac{1\,\cancel{\text{mol HCl}}}{1\,\cancel{\text{mol NaOH}}} \times \frac{1\text{ L HCl}}{2.00\,\cancel{\text{mol HCl}}} = 0.350\text{ L HCl}$$

$$0.350\,\cancel{\text{L}}\text{ HCl} \times \frac{1000\text{ mL}}{1\,\cancel{\text{L}}} = 350.\text{ mL HCl}$$

Therefore, 350. mL of 2.00 M HCl contain 0.700 mol HCl and will react with 0.700 mol, or 28.0 g, of NaOH.

Example 15.9

What volume of 0.250 M solution can be prepared from 16.0 g of potassium carbonate (K_2CO_3)?

SOLUTION

We are starting with 16.0 g of K_2CO_3 and need to find the volume of 0.250 M solution that can be prepared from this K_2CO_3.

The conversion therefore is

$$\text{g } K_2CO_3 \longrightarrow \text{mol } K_2CO_3 \longrightarrow \text{L solution}$$

The data are

$$16.0\text{ g } K_2CO_3 \qquad M = \frac{0.250\text{ mol}}{1\text{ L}} \qquad \text{molar mass } K_2CO_3 = \frac{138.2\text{ g } K_2CO_3}{1\text{ mol}}$$

$$16.0\,\cancel{\text{g } K_2CO_3} \times \frac{1\,\cancel{\text{mol } K_2CO_3}}{138.2\,\cancel{\text{g } K_2CO_3}} \times \frac{1\text{ L}}{0.250\,\cancel{\text{mol } K_2CO_3}} = 0.463\text{ L (463 mL)}$$

Thus, 463 mL of 0.250 M solution can be made from 16.0 g K_2CO_3.

Example 15.10

Calculate the number of moles of nitric acid in 325 mL of 16 M HNO_3 solution.

SOLUTION

Use the equation

$$\text{moles} = \text{liters} \times M$$

Substitute the data given in the problem and solve:

$$\text{moles} = 0.325\,\cancel{\text{L}} \times \frac{16\text{ mol } HNO_3}{1\,\cancel{\text{L}}} = 5.2\text{ mol } HNO_3$$

Practice What volume of 0.035 M $AgNO_3$ can be made from 5.0 g of $AgNO_3$?
Answer: 0.84 L (840 mL)

Dilution Problems

Chemists often find it necessary to dilute solutions from one concentration to another by adding more solvent to the solution. If a solution is diluted by adding pure solvent, the volume of the solution increases, but the number of moles of solute in the solution remains the same. Thus, the moles/liter (molarity) of the solution decreases. It is important to read a problem carefully to distinguish between (1) how much solvent must be added to dilute a solution to a particular concentration and (2) to what volume a solution must be diluted to prepare a solution of a particular concentration.

EXAMPLE 15.11

Calculate the molarity of a sodium hydroxide solution that is prepared by mixing 100. mL of 0.20 *M* NaOH with 150. mL of water.

SOLUTION

This problem is a dilution problem. If we double the volume of a solution by adding water, we cut the concentration in half. Therefore, the concentration of the above solution should be less than 0.10 *M*. In the dilution, the moles of NaOH remain constant; the molarity and volume change. The final volume is (100. mL + 150. mL) or 250. mL.

To solve this problem, (1) calculate the moles of NaOH in the original solution, and (2) divide the moles of NaOH by the final volume of the solution to obtain the new molarity.

Step 1 Calculate the moles of NaOH in the original solution.

$$M = \frac{\text{mol}}{\text{L}} \qquad \text{mol} = \text{L} \times M$$

$$0.100\,\cancel{\text{L}} \times \frac{0.20 \text{ mol NaOH}}{1\,\cancel{\text{L}}} = 0.020 \text{ mol NaOH}$$

Step 2 Solve for the new molarity, taking into account that the total volume of the solution after dilution is 250. mL (0.250 L).

$$M = \frac{0.020 \text{ mol NaOH}}{0.250 \text{ L}} = 0.080\ M \text{ NaOH}$$

Alternate Solution When the moles of solute in a solution before and after dilution are the same, then the moles before and after dilution may be set equal to each other:

$$\text{mol}_1 = \text{mol}_2$$

where mol_1 = moles before dilution, and mol_2 = moles after dilution. Then

$$\text{mol}_1 = \text{L}_1 \times M_1 \qquad \text{mol}_2 = \text{L}_2 \times M_2$$

$$\text{L}_1 \times M_1 = \text{L}_2 \times M_2$$

When both volumes are in the same units, a more general statement can be made:

$$V_1 \times M_1 = V_2 \times M_2$$

For this problem

$$V_1 = 100.\text{ mL} \qquad M_1 = 0.20\ M$$

$$V_2 = 250.\text{ mL} \qquad M_2 = M_2 \quad \text{(unknown)}$$

Then

$$100.\ \text{mL} \times 0.20\ M = 250.\ \text{mL} \times M_2$$

Solving for M_2, we get

$$M_2 = \frac{100.\ \cancel{\text{mL}} \times 0.20\ M}{250.\ \cancel{\text{mL}}} = 0.080\ M\ \text{NaOH}$$

Practice Calculate the molarity of a solution prepared by diluting 125 mL of 0.400 M $K_2Cr_2O_7$ with 875 mL of water.

Answer: $5.00 \times 10^{-2}\ M$

Example 15.12

How many grams of silver chloride, AgCl, will be precipitated by adding sufficient silver nitrate, $AgNO_3$, to react with 1500. mL of 0.400 M $BaCl_2$ (barium chloride) solution?

$$2\ AgNO_3(aq) + \underset{1\ \text{mol}}{BaCl_2(aq)} \longrightarrow \underset{2\ \text{mol}}{2\ AgCl\downarrow} + Ba(NO_3)_2(aq)$$

Solution

This problem is a stoichiometry problem. The fact that $BaCl_2$ is in solution means that we need to consider the volume and concentration of the solution in order to know the number of moles of $BaCl_2$ reacting.

Step 1 Determine the number of moles of $BaCl_2$ in 1500. mL of 0.400 M solution:

$$M = \frac{\text{mol}}{\text{L}} \qquad \text{mol} = \text{L} \times M \qquad 1500.\ \text{mL} = 1.500\ \text{L}$$

$$1.500\ \cancel{\text{L}} \times \frac{0.400\ \text{mol}\ BaCl_2}{\cancel{\text{L}}} = 0.600\ \text{mol}\ BaCl_2$$

Step 2 Use the mole-ratio method to calculate the moles and grams of AgCl:

mol $BaCl_2$ ⟶ mol AgCl ⟶ g AgCl

$$0.600\ \cancel{\text{mol}\ BaCl_2} \times \frac{2\ \cancel{\text{mol AgCl}}}{1\ \cancel{\text{mol}\ BaCl_2}} \times \frac{143.4\ \text{g AgCl}}{\cancel{\text{mol AgCl}}} = 172\ \text{g AgCl}$$

Practice How many grams of lead(II) iodide will be precipitated by adding sufficient $Pb(NO_3)_2$ to react with 750 mL of 0.250 M KI solution?

$$2\ KI + Pb(NO_3)_2 \longrightarrow PbI_2\downarrow + 2\ KNO_3$$

Answer: 31 g

Normality

Normality is another way of expressing the concentration of a solution. It is based on an alternate chemical unit of mass called the *equivalent mass*. The **normality** of a solution is the concentration expressed as the number of equivalent masses (equivalents, abbreviated equiv) of solute per liter of solution. A 1 normal (1 *N*) solution contains 1 equivalent mass of solute per liter of solution. Normality is widely used in analytical chemistry because it simplifies many of the calculations involving solution concentration.

normality

$$\text{normality} = N = \frac{\text{number of equivalents of solute}}{\text{1 liter of solution}} = \frac{\text{equivalents}}{\text{liter}}$$

where

$$\text{number of equivalents of solute} = \frac{\text{grams of solute}}{\text{equivalent mass of solute}}$$

Every substance may be assigned an equivalent mass. The equivalent mass may be equal either to the molar mass of the substance or to an integral fraction of the molar mass (that is, the molar mass divided by 2, 3, 4, and so on). To gain an understanding of the meaning of equivalent mass, let us start by considering these two reactions:

$$\underset{\substack{\text{1 mole}\\(36.46\text{ g})}}{HCl(aq)} + \underset{\substack{\text{1 mol}\\(40.00\text{ g})}}{NaOH(aq)} \longrightarrow NaCl(aq) + H_2O$$

$$\underset{\substack{\text{1 mol}\\(98.08\text{ g})}}{H_2SO_4(aq)} + \underset{\substack{\text{2 mol}\\(80.00\text{ g})}}{2\,NaOH(aq)} \longrightarrow Na_2SO_4(aq) + 2\,H_2O$$

We note first that 1 mol of hydrochloric acid (HCl) reacts with 1 mol of sodium hydroxide (NaOH) and 1 mol of sulfuric acid (H_2SO_4) reacts with 2 mol of NaOH. If we make 1 molar solutions of these substances, 1 L of 1 *M* HCl will react with 1 L of 1 *M* NaOH, and 1 L of 1 *M* H_2SO_4 will react with 2 L of 1 *M* NaOH. From this reaction, we can see that H_2SO_4 has twice the chemical capacity of HCl when reacting with NaOH. We can, however, adjust these acid solutions to be equivalent in reactivity by dissolving only 0.5 mol of H_2SO_4 per liter of solution. By doing so, we find that we are required to use 49.04 g of H_2SO_4 per liter (instead of 98.08 g of H_2SO_4 per liter) to make a solution that is equivalent to one made from 36.46 g of HCl per liter. These masses, 49.04 g of H_2SO_4 and 36.46 g of HCl, are chemically equivalent and are known as the equivalent masses of these substances, because each will react with the same amount of NaOH (40.00 g). The equivalent mass of HCl is equal to its molar mass, but that of H_2SO_4 is one-half its molar mass. Table 15.5 summarizes these relationships.

Thus, 1 L of solution containing 36.46 g of HCl would be 1 *N*, and 1 L of solution containing 49.04 g of H_2SO_4 would also be 1 *N*. A solution containing 98.08 g of H_2SO_4 (1 mol) per liter would be 2 *N* when reacting with NaOH in the given equation.

TABLE 15.5 Comparison of Molar and Normal Solutions of HCl and H_2SO_4 reacting with NaOH

	Molar mass	Concentration	Volumes that react	Equivalent mass	Concentration	Volumes that react
HCl	36.45	1 *M*	1 liter	36.46	1 *N*	1 liter
NaOH	40.00	1 *M*	1 liter	40.00	1 *N*	1 liter
H_2SO_4	98.08	1 *M*	1 liter	49.04	1 *N*	1 liter
NaOH	40.00	1 *M*	2 liters	40.00	1 *N*	1 liter

equivalent mass

The **equivalent mass** is the mass of a substance that will react with, combine with, contain, replace, or in any other way be equivalent to 1 mole of hydrogen atoms or hydrogen ions.

Normality and molarity can be interconverted in the following manner.

$$N = \frac{\text{equiv}}{\text{L}} \qquad M = \frac{\text{mol}}{\text{L}}$$

$$N = M \times \frac{\text{equiv}}{\text{mol}} = \frac{\cancel{\text{mol}}}{\text{L}} \times \frac{\text{equiv}}{\cancel{\text{mol}}} = \frac{\text{equiv}}{\text{L}}$$

$$M = N \times \frac{\text{mol}}{\text{equiv}} = \frac{\cancel{\text{equiv}}}{\text{L}} \times \frac{\text{mol}}{\cancel{\text{equiv}}} = \frac{\text{mol}}{\text{L}}$$

Thus a 2.0 *N* H_2SO_4 solution is 1.0 *M*.

One application of normality and equivalents is in acid–base neutralization reactions. An equivalent of an acid is that mass of the acid that will furnish 1 mol of H^+ ions. An equivalent of a base is that mass of base that will furnish 1 mol of OH^- ions. Using concentrations in normality, one equivalent of acid (A) will react with one equivalent of base (B).

$$N_A = \frac{\text{equiv}_A}{\text{L}_A} \quad \text{and} \quad N_B = \frac{\text{equiv}_B}{\text{L}_B}$$

$$\text{equiv}_A = \text{L}_A \times N_A \quad \text{and} \quad \text{equiv}_B = \text{L}_B \times N_B$$

Since $\text{equiv}_A = \text{equiv}_B$,

$$\text{L}_A \times N_A = \text{L}_B \times N_B$$

When both volumes are in the same units, we can write a more general equation:

$$V_A N_A = V_B N_B$$

which states that the volume of acid times the normality of the acid equals the volume of base times the normality of the base.

EXAMPLE 15.13

(a) What is the normality of an H_2SO_4 solution if 25.00 mL of the solution requires 22.48 mL of 0.2018 *N* NaOH for complete neutralization? (b) What is the molarity of the H_2SO_4 solution?

SOLUTION

(a) Solve for N_A by substituting the data into

$$V_A N_A = V_B N_B$$

$$25.00 \text{ mL} \times N_A = 22.48 \text{ mL} \times 0.2018\ N$$

$$N_A = \frac{22.48 \text{ mL} \times 0.2018\ N}{25.00 \text{ mL}} = 0.1815\ N\ H_2SO_4$$

(b) When H_2SO_4 is completely neutralized it furnishes 2 equivalents of H^+ ions per mole of H_2SO_4. The conversion from N to M is

$$\frac{\text{equiv}}{\text{L}} \longrightarrow \frac{\text{mol}}{\text{L}} \qquad H_2SO_4 = 0.1815\ N$$

$$\frac{0.1815 \text{ equiv}}{1 \text{ L}} \times \frac{1 \text{ mol}}{2 \text{ equiv}} = 0.09075 \text{ mol/L}$$

The H_2SO_4 solution is 0.09075 M.

Practice What is the normality of a NaOH solution if 50.0 mL of the solution requires 23.72 mL of 0.0250 N H_2SO_4? What is the molarity of the NaOH?
Answers: $1.19 \times 10^{-2}\ N$ $\quad$ $1.19 \times 10^{-2}\ M$

The equivalent mass of a substance may be variable; its value is dependent on the reaction that the substance is undergoing. Consider the reactions represented by these equations:

$$NaOH + H_2SO_4 \longrightarrow NaHSO_4 + H_2O$$

$$2\ NaOH + H_2SO_4 \longrightarrow Na_2SO_4 + 2\ H_2O$$

In the first reaction 1 mol of sulfuric acid furnishes 1 mol of hydrogen atoms. Therefore the equivalent mass of sulfuric acid is the molar mass, namely 98.08 g. But in the second reaction 1 mol of H_2SO_4 furnishes 2 mol of hydrogen atoms. Therefore, the equivalent mass of the sulfuric acid is one-half the molar mass, or 49.04 g. A summary of the quantitative concentration units is found in Table 15.6.

15.9 Colligative Properties of Solutions

Two solutions, one containing 1 mol (60.06 g) of urea (NH_2CONH_2) and the other containing 1 mol (342.3 g) of sucrose ($C_{12}H_{22}O_{11}$) in 1 kg of water, both have a freezing point of $-1.86°C$, not 0°C as for pure water. Urea and sucrose are distinctly different substances, yet they lower the freezing point of the water by the same amount. The only thing apparently common to these two solutions is that each contains 1 mol (6.022×10^{23} molecules) of solute and 1 kg of solvent.

TABLE 15.6 Concentration Units for Solutions

Units	Symbol	Definition
Mass percent	%m/m	$\frac{\text{Mass solute}}{\text{Mass solution}} \times 100$
Parts per million	ppm	$\frac{\text{Mass solute}}{\text{Mass solution}} \times 1{,}000{,}000$
Mass/volume percent	%m/v	$\frac{\text{Mass solute}}{\text{mL solution}} \times 100$
Volume percent	%v/v	$\frac{\text{mL solute}}{\text{mL solution}} \times 100$
Molarity	M	$\frac{\text{Moles solute}}{\text{L solution}}$
Normality	N	$\frac{\text{Equivalents solute}}{\text{L solution}}$
Molality	m	$\frac{\text{Moles solute}}{\text{kg solvent}}$

In fact, if we dissolve one mole of any nonionizable solute in 1 kg of water, the freezing point of the resulting solution will be $-1.86°C$.

These results lead us to conclude that the freezing point depression for a solution containing 6.022×10^{23} solute molecules (particles) and 1 kg of water is a constant, namely, 1.86°C. Freezing point depression is a general property of solutions. Furthermore the amount by which the freezing point is depressed is the same for all solutions made with a given solvent; that is, each solvent shows a characteristic *freezing point depression constant*. Freezing point depression constants for several solvents are given in Table 15.7.

The solution formed by the addition of a nonvolatile solute to a solvent has a lower freezing point, a higher boiling point, and a lower vapor pressure than that

TABLE 15.7 Freezing Point Depression and Boiling Point Elevation Constants of Selected Solvents

Solvent	Freezing point of pure solvent (°C)	Freezing point depression constant, K_f $\left(\frac{°\text{C kg solvent}}{\text{mol solute}}\right)$	Boiling point of pure solvent (°C)	Boiling point elevation constant, K_b $\left(\frac{°\text{C kg solvent}}{\text{mol solute}}\right)$
Water	0.00	1.86	100.0	0.512
Acetic acid	16.6	3.90	118.5	3.07
Benzene	5.5	5.1	80.1	2.53
Camphor	178	40	208.2	5.95

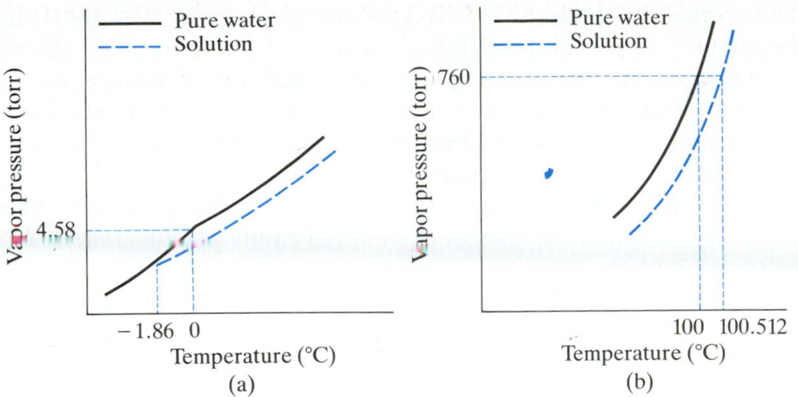

◀ **FIGURE 15.7**
Vapor pressure curves of pure water and water solutions, showing (a) freezing point depression and (b) boiling point elevation effects. (Concentration: 1 mol solute/1 kg water.)

of the pure solvent. All these effects are related and are known as colligative properties. The **colligative properties** are properties that depend only upon the number of solute particles in a solution and not on the nature of those particles. Freezing point depression, boiling point elevation, and vapor pressure lowering are colligative properties of solutions.

colligative properties

The colligative properties of a solution can be considered in terms of vapor pressure. The vapor pressure of a pure liquid depends on the tendency of molecules to escape from its surface. Thus, if 10% of the molecules in a solution are nonvolatile solute molecules, the vapor pressure of the solution is 10% lower than that of the pure solvent. The vapor pressure is lower because the surface of the solution contains 10% nonvolatile molecules and 90% of the volatile solvent molecules. A liquid boils when its vapor pressure equals the pressure of the atmosphere. Thus, we can see that the solution just described as having a lower vapor pressure will have a higher boiling point than the pure solvent. The solution with a lowered vapor pressure does not boil until it has been heated above the boiling point of the solvent (see Figure 15.7). Each solvent has its own characteristic boiling point elevation constant (see Table 15.7). The boiling point elevation constant is based on a solution that contains 1 mole of solute particles per kilogram of solvent. For example, the boiling point elevation constant for a solution containing 1 mole of solute particles per kilogram of water is 0.512°C, which means that this water solution will boil at 100.512°C.

The freezing behavior of a solution can also be considered in terms of lowered vapor pressure. Figure 15.7 shows the vapor pressure relationships of ice, water, and a solution containing 1 mole of solute per kilogram of water. The freezing point of water is at the intersection of the water and ice vapor pressure curves—that is, at the point where water and ice have the same vapor pressure. Because the vapor pressure of water is lowered by the solute, the vapor pressure curve of the solution does not interesect the vapor pressure curve of ice until the solution has been cooled below the freezing point of pure water. Thus it is necessary to cool the solution below 0°C in order to freeze out ice.

The foregoing discussion dealing with freezing point depressions is restricted to *un-ionized* substances. The discussion of boiling point elevations is restricted to *nonvolatile* and un-ionized substances. The colligative properties of ionized substances (Electrolytes, Chapter 16) are not under consideration at this point.

Some practical applications involving colligative properties are (1) use of salt-ice mixtures to provide low freezing temperatures for homemade ice creams, (2) use of sodium chloride or calcium chloride to melt ice from streets, and (3) use of ethylene glycol and water mixtures as antifreeze in automobile radiators (ethylene glycol also raises the boiling point of radiator fluid and thus allows the engine to operate at a higher temperature).

Both the freezing point depression and the boiling point elevation are directly proportional to the number of moles of solute per kilogram of solvent. When we deal with the colligative properties of solutions, another concentration expression, *molality*, is used. The **molality** (m) of a solute is the number of moles of solute per kilogram of solvent:

molality

$$m = \frac{\text{mol solute}}{\text{kg solvent}}$$

Note that a lowercase m is used for molality concentrations and a capital M for molarity. The difference between molality and molarity is that molality refers to moles of solute *per kilogram of solvent*, whereas molarity refers to moles of solute *per liter of solution*. For un-ionized substances, the colligative properties of a solution are directly proportional to its molality.

Molality is independent of volume. It is a mass-to-mass relationship of solute to solvent and allows for experiments, such as freezing point depression and boiling point elevation, to be conducted at variable temperatures.

The following equations are used in calculations involving colligative properties and molality.

$$\Delta t_f = mK_f \qquad \Delta t_b = mK_b \qquad m = \frac{\text{mol solute}}{\text{kg solvent}}$$

m = molality; mol solute/kg solvent

Δt_f = freezing point depression; °C

Δt_b = boiling point elevation; °C

K_f = freezing point depression constant; °C kg solvent/mol solute

K_b = boiling point elevation constant; °C kg solvent/mol solute

EXAMPLE 15.14

What is the molality (m) of a solution prepared by dissolving 2.70 g of CH_3OH in 25.0 g of H_2O?

SOLUTION

Since $m = \frac{\text{mol solute}}{\text{kg solvent}}$ the conversion is

$$\frac{2.70\text{ g } CH_3OH}{25.0\text{ g } H_2O} \longrightarrow \frac{\text{mol } CH_3OH}{25.0\text{ g } H_2O} \longrightarrow \frac{\text{mol } CH_3OH}{1\text{ kg } H_2O}$$

The molar mass of CH_3OH is (12.01 + 4.032 + 16.00) or 32.04 g/mol.

$$\frac{2.70\ \cancel{\text{g } CH_3OH}}{25.0\ \cancel{\text{g } H_2O}} \times \frac{1\text{ mol } CH_3OH}{32.04\ \cancel{\text{g } CH_3OH}} \times \frac{1000\ \cancel{\text{g } H_2O}}{1\text{ kg } H_2O} = \frac{3.38\text{ mol } CH_3OH}{1\text{ kg } H_2O}$$

The molality is 3.38 m.

PRACTICE What is the molality of a solution prepared by dissolving 150.0 g of $C_6H_{12}O_6$ in 600.0 g of H_2O?

Answer: 1.39 *m*

EXAMPLE 15.15

A solution is made by dissolving 100. g of ethylene glycol ($C_2H_6O_2$) in 200. g of water. What is the freezing point of this solution?

SOLUTION

To calculate the freezing point of the solution, we first need to calculate Δt_f, the change in freezing point. Use the equation

$$\Delta t_f = mK_f = \frac{\text{mol solute}}{\text{kg solvent}} \times K_f$$

K_f (for water): $\dfrac{1.86°\text{C kg solvent}}{\text{mol solute}}$ (from Table 15.7)

mol solute: $100.\ \cancel{\text{g } C_2H_6O_2} \times \dfrac{1 \text{ mol } C_2H_6O_2}{62.07\ \cancel{\text{g } C_2H_6O_2}} = 1.61 \text{ mol } C_2H_6O_2$

kg solvent: $200.\ \cancel{\text{g}}\ H_2O \times \dfrac{1 \text{ kg}}{1000\ \cancel{\text{g}}} = 0.200 \text{ kg } H_2O$

$$\Delta t_f = \frac{1.61\ \cancel{\text{mol } C_2H_6O_2}}{0.200\ \cancel{\text{kg } H_2O}} \times \frac{1.86°\text{C}\ \cancel{\text{kg } H_2O}}{1\ \cancel{\text{mol } C_2H_6O_2}} = 15.0°\text{C}$$

The freezing point depression, 15.0°C, must be subtracted from 0°C, the freezing point of the pure solvent.

$$\text{freezing point of solution} = \text{freezing point of solvent} - \Delta t_f$$
$$= 0.0°\text{C} - 15.0°\text{C} = -15.0°\text{C}.$$

Therefore, the freezing point of the solution is −15.0°C.

The calculation can also be done using the equation

$$\Delta t_f = K_f \times \frac{\text{g solute}}{\text{molar mass solute}} \times \frac{1}{\text{kg solvent}}$$

EXAMPLE 15.16

A solution made by dissolving 4.71 g of a compound of unknown molar mass in 100.0 g of water has a freezing point of −1.46°C. What is the molar mass of the compound?

SOLUTION

First substitute the data in $\Delta t_f = mK_f$ and solve for *m*.

$\Delta t_f = +1.46$ since the solvent, water, freezes at 0°C.

$$K_f = \frac{1.86°\text{C kg } H_2O}{\text{mol solute}}$$

$$1.46°\text{C} = mK_f = m \times \frac{1.86°\text{C kg } H_2O}{\text{mol solute}}$$

$$m = \frac{1.46\ \cancel{°\text{C}} \times \text{mol solute}}{1.86\ \cancel{°\text{C}} \times \text{kg } H_2O} = \frac{0.785 \text{ mol solute}}{\text{kg } H_2O}$$

Now convert the data, 4.71 g solute/100.0 g H_2O, to g/mol.

$$\frac{4.71\ \text{g solute}}{100.0\ \text{g } H_2O} \times \frac{1000\ \text{g } H_2O}{1\ \text{kg } H_2O} \times \frac{1\ \text{kg } H_2O}{0.785\ \text{mol solute}} = 60.0\ \text{g/mol}$$

The molar mass of the compound is 60.0 g/mol.

PRACTICE What is the freezing point of the solution in the previous practice problem? What is the boiling point?

Answers: freezing point = −2.59°C boiling point = 100.72°C

15.10 OSMOSIS AND OSMOTIC PRESSURE

semipermeable membrane

osmosis

When red blood cells are put into distilled water, they gradually swell and, in time, may burst. If red blood cells are put in a 5% urea (or a 5% salt) solution, they gradually shrink and take on a wrinkled appearance. The cells behave in this fashion because they are enclosed in semipermeable membranes. A **semipermeable membrane** allows the passage of water (solvent) molecules through it in either direction but prevents the passage of solute molecules or ions. When two solutions of different concentrations (or water and a water solution) are separated by a semipermeable membrane, water diffuses through the membrane from the solution of lower concentration into the solution of higher concentration. The diffusion of water, either from a dilute solution or from pure water, through a semipermeable membrane into a solution of higher concentration is called **osmosis**.

All solutions exhibit *osmotic pressure.* Osmotic pressure is another colligative property; it is dependent only on the concentration of the solute particles and is independent of their nature. The osmotic pressure of a solution can be measured by determining the amount of counterpressure needed to prevent

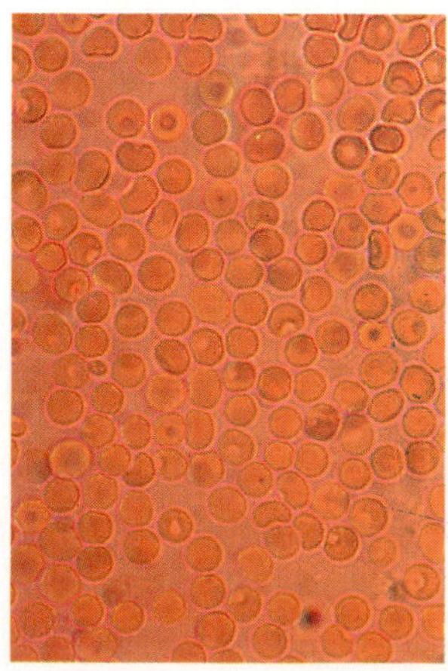
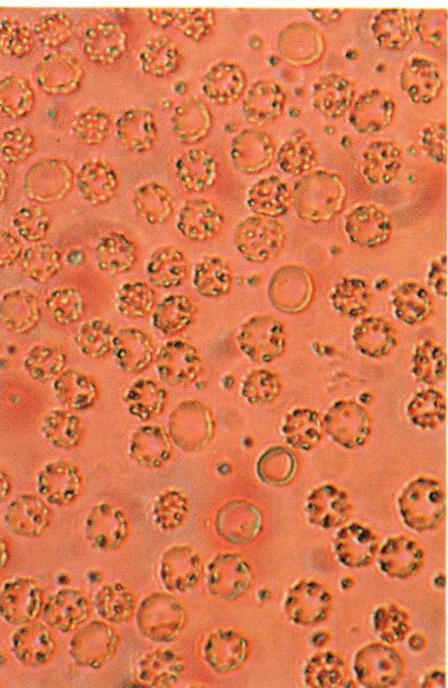
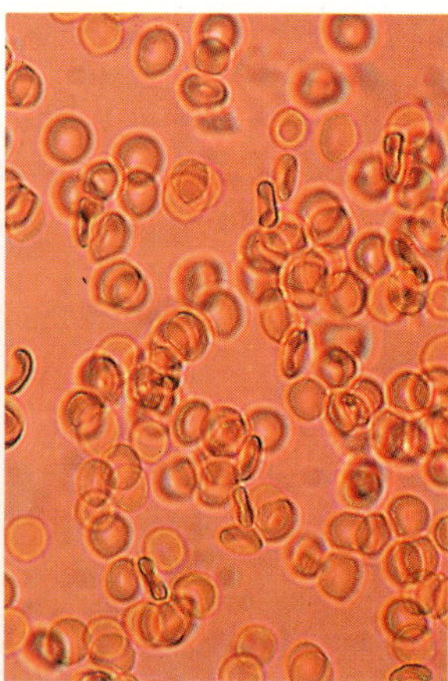

Human red blood cells. *Left:* In a hypotonic solution (0.2% saline) the cells swell as water moves into the cell center. *Center:* In a hypertonic solution (1.6% saline) water leaves the cells causing them to crenate (shrink). *Right:* In an isotonic solution the concentration is the same inside and outside the cell (0.9%) saline). Cells remain in original size.

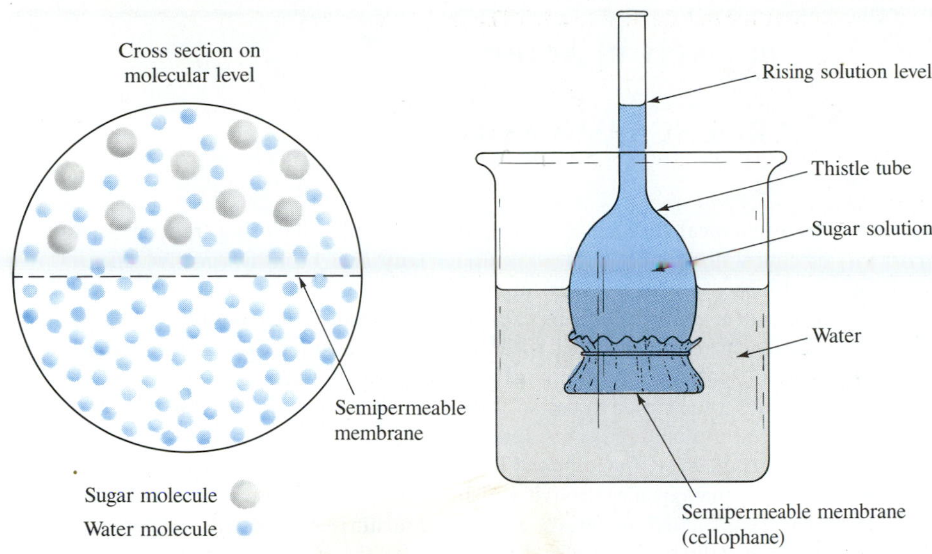

◄ **FIGURE 15.8**
Laboratory demonstration of osmosis: As a result of osmosis, water passes through the membrane causing the solution to rise in the thistle tube.

osmosis; this pressure can be very large. The osmotic pressure of a solution containing 1 mol of solute particles in 1 kg of water is about 22.4 atm, which is about the same as the pressure exerted by 1 mol of a gas confined in a volume of 1 L at 0°C.

Osmosis has a role in many biological processes, and semipermeable membranes occur commonly in living organisms. For example, the roots of plants are covered with tiny structures called root hairs; soil water enters the plant by osmosis, passing through the semipermeable membranes covering the root hairs. Artificial or synthetic membranes can also be made. Cellophane that has been treated to remove the waterproof coating is a good semipermeable membrane.

Osmosis can be demonstrated with the simple laboratory setup shown in Figure 15.8. As a result of osmotic pressure, water passes through the cellophane membrane into the thistle tube, causing the solution level to rise. In osmosis the net transfer of water is always from a less concentrated to a more concentrated solution; that is, the effect is toward equalization of the concentration on both sides of the membrane. It should also be noted that the effective movement of water in osmosis is always from the region of *higher water concentration* to the region of *lower water concentration.*

Osmosis can be explained by assuming that a semipermeable membrane has passages that permit water molecules—but no other molecules or ions—to pass in either direction. Both sides of the membrane are constantly being struck by water molecules in random motion. The number of water molecules crossing the membrane is proportional to the number of water molecule–membrane impacts per unit of time. Because the solute molecules or ions reduce the concentration of water, there are more water molecules, and more water molecule impacts, on the side with the lower solute concentration (more dilute solution). The greater number of water molecule–membrane impacts on the dilute side thus causes a net transfer of water to the more concentrated solution. Again note that the overall process involves the net transfer, by diffusion through the membrane, of water molecules from a region of higher water concentration (dilute solution) to one of lower water concentration (more concentrated solution).

CHEMISTRY IN ACTION

MICROENCAPSULATION

Producing chemical reactions that will occur at precisely the correct moment is one of the tasks facing the chemist in industry. For this to happen, one or more of the reactants must be stored separately and released under controlled conditions precisely when the reaction is desired. A technique developed to accomplish this is microencapsulation in which reactive chemicals—solids, liquids or gases—are sealed inside tiny capsules. The material forming the wall of the capsule is carefully chosen so that the encapsulated chemicals can be released, at the appropriate time, by one of several methods. This release can be accomplished in a variety of ways which include dissolving the capsules, diffusion through the capsule walls, and by mechanical, thermal, electrical, or chemical disruption of the capsules.

In one type of microencapsulation, water diffuses into the capsule and forms a solution which then diffuses out into the surroundings at a constant rate. Some types of capsules contain materials that dissolve at a certain level of acidity and form pores in the capsule through which the encapsulated materials escape. Still other types of capsules dissolve completely over a given period of time, releasing their contents into the system.

Diverse applications of microencapsulation are found in our daily lives. Carbonless paper, often used in credit-card receipts makes use of pressure-sensitive microcapsules containing colorless dye precursors. Another reactive substance is present and converts the precursor to the colored form when pressure is applied by a pen or printer.

Adhesives are frequently encapsulated to prevent them from becoming tacky too soon. The active surfaces on pressure sensitive labels and certain self-sealing envelopes are coated with encapsulated adhesives that are released by pressure. Heat sensitive encapsulation is used for the adhesives in iron-on patches for clothing.

Many microencapsulated products are to be found in the kitchen. Flavorings are encapsulated to make them easier to store in a powdered state, cut evaporation, and reduce reactions with the air. These advantages increase the shelf life of products. The flavoring microcapsules may also be heat sensitive and release their contents during cooking or pressure sensitive (as in chewing gum) and release their contents upon chewing.

Still other encapsulated products are found in our bathrooms. Time-release microencapsulation is currently being used in deodorants, moisturizers, colognes and perfumes. The encapsulation process prevents evaporation, decomposition, and unwanted reactions with the air and other ingredients. Drugs and medications are frequently encapsulated to dissolve slowly over a long period of time in the body. These medications generally work in the intestinal tract but some may also be given by injection to work within other tissues.

Fragrances have undergone microencapsulation in such products as cosmetics, health care products, detergents, and even foods. Gas companies use encapsulated propyl mercaptan, $CH_3CH_2CH_2SH$, to teach children how to detect a gas leak (natural gas without this substance is odorless). Encapsulated fragrances are responsible for the ever-present scratch-and-sniff labels found in childrens books and fashion magazines. When the paper is scratched or pulled open the fragrance is released into the air.

Other applications of microencapsulation include time-release pesticides and neutralizer for contact lenses, as well as special additives in detergents, cleaners, and paints.

This explanation is a simplified picture of osmosis. No one has ever seen the hypothetical passages that allow water molecules, and no other kinds of molecules or ions, to pass through them. Alternative explanations have been proposed. Our discussion has been confined to water solutions, but osmotic pressure is a general colligative property and osmosis is known to occur in non-aqueous systems.

A 0.90% (0.15 M) sodium chloride solution is known as a **physiological saline solution** because it is *isotonic* with blood plasma; that is, it has the same osmotic pressure as blood plasma. Because each mole of NaCl yields about 2 moles of ions when in solution, the solute particle concentration in physiological saline solution is nearly 0.30 M. Five percent glucose solution (0.28 molar) is also approximately isotonic with blood plasma. Blood cells neither swell nor shrink in an isotonic solution. The cells described in the first paragraph of this section swelled in water because the water was *hypotonic* to cell plasma. The cells shrank in 5% urea solution because the urea solution was *hypertonic* to the cell plasma. In order to prevent possible injury to blood cells by osmosis, fluids for intravenous use are usually made up at approximately isotonic concentration.

physiological saline solution

Concepts in Review

1. Describe the types of solutions.
2. List the general properties of solutions.
3. Describe and illustrate the process by which an ionic substance dissolves in water.
4. Indicate the effects of temperature and pressure on the solubility of solids and gases in liquids.
5. Identify and explain the factors affecting the rate at which a solid dissolves in a liquid.
6. Use a solubility table or graph to determine whether a solution is saturated, unsaturated, or supersaturated at a given temperature.
7. Calculate the mass percent or volume percent for a solution.
8. Calculate the amount of solute in a given quantity of a solution when given the mass percent or volume percent of a solution.
9. Calculate the molarity of a solution from the volume and the mass, or moles, of solute.
10. Calculate the mass of a substance necessary to prepare a solution of specified volume and molarity.
11. Determine the resulting molarity in a typical dilution problem.
12. Apply stoichiometry to chemical reactions involving solutions.
13. Understand the concepts of equivalent mass and normality and use them in calculations.
14. Explain the effect of a solute on the vapor pressure of a solvent.
15. Explain the effect of a solute on boiling point and freezing point of a solution.

16. Calculate the boiling and freezing points of a solution from concentration data.
17. Calculate molality and molar mass of a solute from boiling or freezing point data.

EXERCISES

An asterisk indicates a more challenging question or problem.

1. Make a sketch indicating the orientation of water molecules (a) about a single sodium ion and (b) about a single chloride ion in solution.
2. Which of the substances listed below are reasonably soluble and which are insoluble in water?
   (a) KOH (f) PbI_2
   (b) $NiCl_2$ (g) $MgCO_3$
   (c) ZnS (h) $CaCl_2$
   (d) $AgC_2H_3O_2$ (i) $Fe(NO_3)_3$
   (e) Na_2CrO_4 (j) $BaSO_4$
3. Estimate the number of grams of sodium fluoride that would dissolve in 100 g of water at 50°C. (See Table 15.3.)
4. What is the solubility at 25°C of each of the substances listed below? (See Figure 15.3.)
   (a) Potassium chloride (c) Potassium nitrate
   (b) Potassium chlorate
5. What is different in the solubility trend of the potassium halides compared with that of the lithium halides and the sodium halides? (See Table 15.3.)
6. What is the solubility, in grams of solute per 100 g of H_2O, of (a) $KClO_3$ at 60°C, (b) HCl at 20°C, (c) Li_2SO_4 at 80°C, and (d) KNO_3 at 0°C? (See Figure 15.3.)
7. Which substance, KNO_3 or NH_4Cl, shows the greater increase in solubility with increased temperature? (See Figure 15.3.)
8. Does a 2 molal solution in benzene or a 1 molal solution in camphor show the greater freezing point depression? (See Table 15.7)
9. What would be the total surface area if the 1 cm cube in Figure 15.4 were cut into cubes 0.01 cm on a side?
10. At which temperatures—10°C, 20°C, 30°C, 40°C, or 50°C—would you expect a solution made from 63 g of ammonium chloride and 150 g of water to be unsaturated? (See Figure 15.3.)
11. Explain why the rate of dissolving decreases as shown in Figure 15.5.
12. Would the volumetric flasks in Figure 15.6 be satisfactory for preparing normal solutions? Explain.
13. Assume that the thistle tube in Figure 15.8 contains 1.0 *M* sugar solution and that the water in the beaker has just been replaced by a 2.0 *M* solution of urea. Would the solution level in the thistle tube continue to rise, remain constant, or fall? Explain.
14. Name and distinguish between the two components of a solution.
15. Is it always apparent in a solution which component is the solute, for example, in a solution of a liquid in a liquid?
16. Explain why the solute does not settle out of a solution.
17. Is it possible to have one solid dissolved in another? Explain.
18. An aqueous solution of KCl is colorless, $KMnO_4$ is purple, and $K_2Cr_2O_7$ is orange. What color would you expect of an aqueous solution of $Na_2Cr_2O_7$? Explain.
19. Explain why carbon tetrachloride will dissolve benzene but will not dissolve sodium chloride.
20. Some drinks like tea are consumed either hot or cold, whereas others like Coca Cola are drunk only cold. Why?
21. Why is air considered to be a solution?
22. In which will a teaspoonful of sugar dissolve more rapidly, 200 mL of iced tea or 200 mL of hot coffee? Explain in terms of the KMT.
23. What is the effect of pressure on the solubility of gases in liquids? Solids in liquids?
24. Why do smaller particles dissolve faster than large ones?
25. In a saturated solution containing undissolved solute, solute is continuously dissolving, but the concentration of the solution remains unchanged. Explain.
26. Explain why there is no apparent reaction when crystals of $AgNO_3$ and NaCl are mixed, but a reaction is apparent immediately when solutions of $AgNO_3$ and NaCl are mixed.

27. What do we mean when we say that concentrated nitric acid (HNO_3) is 16 molar?
28. Will 1 liter of 1 molar NaCl contain more chloride ions than 0.5 liter of 1 molar $MgCl_2$? Explain.
29. Champagne is usually cooled in a refrigerator prior to opening. It is also opened very carefully. What would happen if a warm bottle of champagne is shaken and opened quickly and forcefully?
30. Explain how a supersaturated solution of $NaC_2H_3O_2$ can be prepared and proven to be supersaturated.
31. Explain in terms of the KMT how a semipermeable membrane functions when placed between pure water and a 10% sugar solution.
32. Which has the higher osmotic pressure, a solution containing 100 g of urea (NH_2CONH_2) in 1 kg of H_2O or a solution containing 150 g of glucose ($C_6H_{12}O_6$) in 1 kg of H_2O?
33. Explain why a lettuce leaf in contact with salad dressing containing salt and vinegar soon becomes wilted and limp whereas another lettuce leaf in contact with plain water remains crisp.
34. A group of shipwreck survivors floated for several days on a life raft before being rescued. Those who had drunk some seawater were found to be suffering the most from dehydration. Explain.
35. Which of the following statements are correct? Rewrite the incorrect statements to make them correct.
   (a) A solution is a homogeneous mixture.
   (b) It is possible for the same substance to be the solvent in one solution and the solute in another.
   (c) A solute can be removed from a solution by filtration.
   (d) Saturated solutions are always concentrated solutions.
   (e) If a solution of sugar in water is allowed to stand undisturbed for a long time, the sugar will gradually settle to the bottom of the container.
   (f) It is not possible to prepare an aqueous 1.0 M AgCl solution.
   (g) Gases are generally more soluble in hot water than in cold water.
   (h) It is impossible to prepare a two-phase liquid mixture from two liquids that are miscible with each other in all proportions.
   (i) A solution that is 10% NaCl by mass always contains 10 g of NaCl.
   (j) Small changes in pressure have little effect on the solubility of solids in liquids but a marked effect on the solubility of gases in liquids.
   (k) How fast a solute dissolves depends mainly on the size of the solute particles, the temperature of the solvent, and the degree of agitation or stirring taking place.
   (l) In order to have a 1 molar solution you must have 1 mol of solute dissolved in sufficient solvent to give 1 L of solution.
   (m) Dissolving 1 mole of NaCl in 1 liter of water will give a 1 molar solution.
   (n) One mole of solute in 1 L of solution has the same concentration as 0.1 mol of solute in 100 mL of solution.
   (o) When 100 mL of 0.200 M HCl is diluted to 200 mL volume by the addition of water, the resulting solution is 0.100 M and contains one-half the number of moles of HCl as were in the original solution.
   (p) Fifty milliliters of 0.1 M H_2SO_4 will neutralize the same volume of 0.1 M NaOH as 100 mL of 0.1 M HCl.
   (q) Fifty milliliters of 0.1 N H_2SO_4 will neutralize the same volume of 0.1 M NaOH as 100 mL of 0.1 M HCl.
   (r) The molarity of a solution will vary slightly with temperature.
   (s) The equivalent mass of $Ca(OH)_2$ is one-half its molar mass.
   (t) Gram for gram methyl alochol, CH_3OH, is more effective than ethyl alcohol, C_2H_5OH, in lowering the freezing point of water.
   (u) An aqueous solution that freezes below 0°C will have a normal boiling point below 100°C.
   (v) The colligative properties of a solution depend on the number of solute particles dissolved in solution.
   (w) A solution of 1.00 mol of a nonionizable solute and 1000 g of water will freeze at −1.86°C and will boil at 99.5°C at atmospheric pressure.
   (x) Water will diffuse from a 0.1 M sugar solution to a 0.2 M sugar solution when these two solutions are separated by a semipermeable membrane.
   (y) An isotonic salt solution has the same osmotic pressure as blood plasma.
   (z) Red blood cells will neither swell nor shrink when placed into an isotonic salt solution.
36. What disadvantages are there in expressing the concentration of solutions as dilute or concentrated?

37. Explain how concentrated H_2SO_4 can be both 18 molar and 36 normal in concentration.

38. Describe how you would prepare 750 mL of 5 molar NaCl solution.

39. Arrange the following bases (in descending order) according to the volume of each that will react with 1 liter of 1 *M* HCl: (a) 1 *M* NaOH, (b) 1.5 *M* $Ca(OH)_2$, (c) 2 *M* KOH, and (d) 0.6 *M* $Ba(OH)_2$.

40. Explain in terms of vapor pressure why the boiling point of a solution containing a nonvolatile solute is higher than that of the pure solvent.

41. Explain why the freezing point of a solution is lower than the freezing point of the pure solvent.

42. Which would be colder, a glass of water and crushed ice or a glass of Seven-Up and crushed ice? Explain.

43. When water and ice are mixed, the temperature of the mixture is 0°C. But, if methyl alcohol and ice are mixed, a temperature of −10°C is readily attained. Explain why the two mixtures show such different temperature behavior.

44. Which would be more effective in lowering the freezing point of 500. g of water?
(a) 100. g of sucrose ($C_{12}H_{22}O_{11}$) or 100. g of ethyl alcohol (C_2H_5OH)
(b) 100. g of sucrose or 20.0 g of ethyl alcohol
(c) 20.0 g of ethyl alcohol or 20.0 g of methyl alcohol (CH_3OH)

45. Is the molarity of a 5 molal aqueous solution of NaCl greater or less than 5 molar? Explain.

46. Express, in terms of its molarity, the normality of an H_2SO_4 solution used to titrate NaOH solutions. (Assume both hydrogens react.)

47. What is microencapsulation?

48. Explain how a scratch-and-sniff label works.

49. What are the purposes for timed-release microencapsulation.

50. State three types of microencapsulation systems and give a practical application of each.

Percent Solutions

51. Calculate the mass percent of the following solutions.
(a) 25.0 g NaBr + 100. g H_2O
(b) 1.20 g K_2SO_4 + 10.0 g H_2O
(c) 40.0 g $Mg(NO_3)_2$ + 500. g H_2O

52. How many grams of a solution that is 12.5% by mass $AgNO_3$ would contain the following?
(a) 30.0 g of $AgNO_3$
(b) 0.400 mol of $AgNO_3$

53. Calculate the mass percent of the following solutions:
(a) 60.0 g NaCl + 200.0 g H_2O
(b) 0.25 mol $HC_2H_3O_2$ + 3.0 mol H_2O
(c) 1.0 molal solution of $C_6H_{12}O_6$ in water

54. How much solute is present in each of the following?
(a) 65 g of 5.0% KCl solution
(b) 250. g of 15.0% K_2CrO_4 solution
*(c) A solution that contains 100.0 g of water and is 6.0% by mass sodium bicarbonate, $NaHCO_3$.

55. What mass of 5.50% solution can be prepared from 25.0 g of KCl?

56. Physiological saline (NaCl) solutions used in intravenous injections have a concentration of 0.90% NaCl by mass.
(a) How many grams of NaCl are needed to prepare 500. g of this solution?
*(b) How much water must evaporate from this solution to give a solution that is 9.0% NaCl by mass?

57. A solution is made from 50.0 g of KNO_3 and 175 g of H_2O. How many grams of water must evaporate to give a saturated solution of KNO_3 in water at 20°C? (See Figure 15.3.)

58. Calculate the mass/volume percent of a solution made by dissolving
(a) 22.0 g of CH_3OH dissolved in C_2H_5OH to make 100. mL of solution
(b) 4.20 g of NaCl dissolved in H_2O to make 12.5 mL of solution

59. What is the volume percent of these solutions?
(a) 10.0 mL of CH_3OH dissolved in water to a volume of 40.0 mL.
(b) 2.0 mL of CCl_4 dissolved in benzene to a volume of 9.0 mL.

60. What volume of 70.% rubbing alcohol can you prepare if you have only 150 mL of pure isopropyl alcohol on hand?

61. At 20°C an aqueous solution of HNO_3 that is 35.0% HNO_3 by mass has a density of 1.21 g/mL.
(a) How many grams of HNO_3 are present in 1.00 L of this solution?
(b) What volume of this solution will contain 500. g of HNO_3?

Molarity Problems

62. Calculate the molarity of the following solutions:
(a) 0.10 mol of solute in 250 mL of solution
(b) 2.5 mol of NaCl in 0.650 L of solution.

(c) 0.025 mol of HCl in 10. mL of solution
(d) 0.35 mol $BaCl_2 \cdot 2\ H_2O$ in 593 mL of solution

63. Calculate the molarity of the following solutions:
(a) 53.0 g of Na_2CrO_4 in 1.00 L of solution
(b) 260 g of $C_6H_{12}O_6$ in 800. mL of solution
(c) 1.50 g of $Al_2(SO_4)_3$ in 2.00 L of solution
(d) 0.0282 g of $Ca(NO_3)_2$ in 1.00 mL of solution

64. Calculate the number of moles of solute in each of the following solutions:
(a) 40.0 L of 1.0 *M* LiCl
(b) 25.0 mL of 3.00 *M* H_2SO_4
(c) 349 mL of 0.0010 *M* NaOH
(d) 5000. mL of 3.1 *M* $CoCl_2$

65. Calculate the grams of solute in each of the following solutions:
(a) 150 L of 1.0 *M* NaCl
(b) 0.035 L of 10.0 *M* HCl
(c) 260 mL of 18 *M* H_2SO_4
(d) 8.00 mL of 8.00 *M* $Na_2C_2O_4$

66. How many milliters of 0.256 *M* KCl solution will contain the following?
(a) 0.430 mol of KCl
(b) 10.0 mol of KCl
(c) 20.0 g of KCl
*(d) 71.0 g of chloride ion, Cl^-

*67. What is the molarity of a nitric acid solution, if the solution is 35.0% HNO_3 by mass and has a density of 1.21 g/mL?

Dilution Problems

68. What will be the molarity of the resulting solutions made by mixing the following (Assume volumes are additive.)
(a) 200. mL 12 *M* HCl + 200. mL H_2O
(b) 60.0 mL 0.60 *M* $ZnSO_4$ + 500. mL H_2O
(c) 100. mL 1.0 *M* HCl + 150 mL 2.0 *M* HCl

69. Calculate the volume of concentrated reagent required to prepare the diluted solutions indicated:
(a) 12 *M* HCl to prepare 400. mL of 6.0 *M* HCl
(b) 15 *M* NH_3 to prepare 50. mL of 6.0 *M* NH_3
(c) 16 *M* HNO_3 to prepare 100. mL of 2.5 *M* HNO_3
(d) 18 *M* H_2SO_4 to prepare 250 mL of 10.0 *N* H_2SO_4

70. To what volume must a solution of 80.0 g of H_2SO_4 in 500. mL of solution be diluted to give a 0.10 *M* solution?

71. What will be the molarity of each of the solutions made by mixing 250 mL of 0.75 *M* H_2SO_4 with (a) 150 mL of H_2O, (b) 250 mL of 0.70 *M* H_2SO_4, and (c) 400. mL of 2.50 *M* H_2SO_4?

72. How many milliliters of water must be added to 300. mL of 1.40 *M* HCl to make a solution that is 0.500 *M* HCl?

73. A 10.0 mL sample of 16 *M* HNO_3 solution is diluted to 500. mL. What is the molarity of the final solution?

74. Given a 5.00 *M* KOH solution, how would you prepare 250. mL of 0.625 *M* KOH?

Stoichiometry Problems

75. 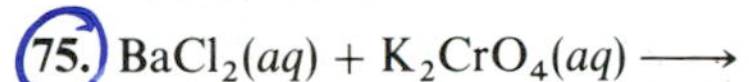

$$BaCl_2(aq) + K_2CrO_4(aq) \longrightarrow BaCrO_4(s) + 2\ KCl(aq)$$

Using the above equation, calculate the following:
(a) The grams of $BaCrO_4$ that can be obtained from 100. mL of 0.300 *M* $BaCl_2$
(b) The volume of 1.0 *M* $BaCl_2$ solution needed to react with 50.0 mL of 0.300 *M* K_2CrO_4 solution

76. $$3\ MgCl_2(aq) + 2\ Na_3PO_4(aq) \longrightarrow Mg_3(PO_4)_2(s) + 6\ NaCl(aq)$$

Using the above equation, calculate:
(a) The milliliters of 0.250 *M* Na_3PO_4 that will react with 50.0 mL of 0.250 *M* $MgCl_2$.
(b) The grams of $Mg_3(PO_4)_2$ that will be formed from 50.0 mL of 0.250 M $MgCl_2$.

77. (a) How many moles of hydrogen will be liberated from 200. mL of 3.00 *M* HCl reacting with an excess of magnesium? The equation is

$$Mg(s) + 2\ HCl(aq) \longrightarrow MgCl_2(aq) + H_2(g)$$

(b) How many liters of hydrogen gas, H_2, measured at 27°C and 720 torr, will be obtained? (*Hint*: Use the ideal gas equation.]

*78. What is the molarity of an HCl solution, 150. mL of which, when treated with excess magnesium, liberates 3.50 L of H_2 gas measured at STP?

79. Given the balanced equation

$$6\ FeCl_2 + K_2Cr_2O_7 + 14\ HCl \longrightarrow 6\ FeCl_3 + 2\ CrCl_3 + 2\ KCl + 7\ H_2O$$

(a) How many moles of KCl will be produced from 2.0 mol of $FeCl_2$?
(b) How many moles of $CrCl_3$ will be produced from 1.0 mol of $FeCl_2$?

(c) How many moles of $FeCl_2$ will react with 0.050 mol of $K_2Cr_2O_7$?
(d) How many milliliters of 0.060 M $K_2Cr_2O_7$ will react with 0.025 mol of $FeCl_2$?
(e) How many milliliters of 6.0 M HCl will react with 15.0 mL of 6.0 M $FeCl_2$?

80. $2\ KMnO_4 + 16\ HCl \longrightarrow 2\ MnCl_2 + 5\ Cl_2 + 8\ H_2O + 2\ KCl$

Calculate the following using the above equation:
(a) The moles of Cl_2 produced from 0.50 mol of $KMnO_4$
(b) The moles of HCl required to react with 1.0 L of 2.0 M $KMnO_4$
(c) The milliliters of 6.0 M HCl required to react with 200. mL of 0.50 M $KMnO_4$
(d) The liters of Cl_2 gas at STP produced by the reaction of 75.0 mL of 6.0 M HCl

Equivalent Mass and Normality Problems

81. Calculate the equivalent mass of the acid and base in each of the following reactions:
(a) $HCl + NaOH \longrightarrow NaCl + H_2O$
(b) $2\ HCl + Ba(OH)_2 \longrightarrow BaCl_2 + 2\ H_2O$
(c) $H_2SO_4 + Ca(OH)_2 \longrightarrow CaSO_4 + 2\ H_2O$
(d) $H_2SO_4 + KOH \longrightarrow KHSO_4 + H_2O$
(e) $H_3PO_4 + 2\ LiOH \longrightarrow Li_2HPO_4 + 2\ H_2O$

82. What is the normality of the following solutions? Assume complete neutralization.
(a) 4.0 M HCl
(b) 0.243 M HNO_3
(c) 3.0 M H_2SO_4
(d) 1.85 M H_3PO_4
(e) 0.250 M $HC_2H_3O_2$

83. What is the normality of an H_2SO_4 solution if 36.26 mL are required to neutralize 2.50 g of $Ca(OH)_2$?

84. Which will be more effective in neutralizing stomach acid (HCl) a tablet containing 12.0 g of $Mg(OH)_2$ or a tablet containing 10.0 g of $Al(OH)_3$? Show evidence for your answer.

85. What volume of 0.2550 N NaOH is required to neutralize
(a) 20.22 mL of 0.1254 N HCl
(b) 14.86 mL of 0.1246 N H_2SO_4
(c) 18.00 mL of 0.1430 M H_2SO_4

Molality and Colligative Properties Problems

86. Calculate the molality of these solutions.
(a) 14.0 g of CH_3OH in 100. g of H_2O
(b) 2.50 mol of benzene (C_6H_6) in 250 g of CCl_4
(c) 1.0 g of $C_6H_{12}O_6$ in 1.0 g of H_2O

87. Which would be more effective as an antifreeze in an automobile radiator? (a) 10 kg of methyl alcohol, CH_3OH, or 10 kg of ethyl alcohol, C_2H_5OH (b) 10 m solution of methyl alcohol or 10 m solution of ethyl alcohol?

88. Automobile battery acid is 38% H_2SO_4 and has a density of 1.29 g/mL. Calculate the molality and the molarity of this solution.

***89.** A sugar solution made to feed humming birds contains 1.00 lb of sugar to 4.00 lb of water. Can this solution be put outside, without freezing, where the temperature falls to 20.0°F at night? Show evidence for your answer.

***90.** What would be (a) the boiling point and (b) the molality of an aqueous sugar ($C_{12}H_{22}O_{11}$) solution that freezes at −5.4°C?

91. (a) What is the molality of a solution containing 100.0 g of ethylene glycol, $C_2H_6O_2$, in 150.0 g of water?
(b) What is the boiling point of this solution?
(c) What is the freezing point of this solution?

92. What is (a) the molality, (b) the freezing point, and (c) the boiling point of a solution containing 2.68 g of naphthalene, $C_{10}H_8$, in 38.4 g of benzene (C_6H_6)?

***93.** The freezing point of a solution of 8.00 g of an unknown compound dissolved in 60.0 g of acetic acid is 13.2°C. Calculate the molar mass of the compound.

94. What is the molar mass of a compound if 4.80 g of the compound dissolved in 22.0 g of H_2O gives a solution that freezes at −2.50°C?

95. A solution of 6.20 g of $C_2H_6O_2$ in water has a freezing point of −0.372°C. How many grams of H_2O are in the solution?

96. What (a) mass and (b) volume of ethylene glycol ($C_2H_6O_2$, density = 1.11 g/mL) should be added to 12.0 L of water in an automobile radiator to protect it from freezing at −20.0°C? (c) To what temperature Fahrenheit will the radiator be protected?

Additional Problems

97. How many grams of solution, 10.% NaOH by mass, are required to neutralize 150 mL of a 1.0 M HCl solution?

***98.** How many grams of solution, 10.% NaOH by mass, are required to neutralize 250. g of a 1.0 molal solution of HCl?

***99.** A sugar syrup solution contains 15.0% sugar, $C_{12}H_{22}O_{11}$, by mass, and has a density of 1.06 g/mL.
(a) How many grams of sugar are in 1.0 L of this syrup?

(b) What is the molarity of this solution?

(c) What is the molality of this solution?

***100.** A solution of 3.84 g of C_4H_2N (empirical formula) in 250. g of benzene depresses the freezing point of benzene 0.614C. What is the molecular formula of the compound?

***101.** Hydrochloric acid (HCl) is sold as a concentrated aqueous solution at 12.0 mol/L. If the density of the solution is 1.18 g/mL, determine the molality of the solution.

***102.** How many grams of KNO_3 must be used to make 450 mL of a solution that is to contain 5.5 mg/mL of the potassium ion? Calculate the molarity of the solution.

16

Ionization: Acids, Bases, Salts

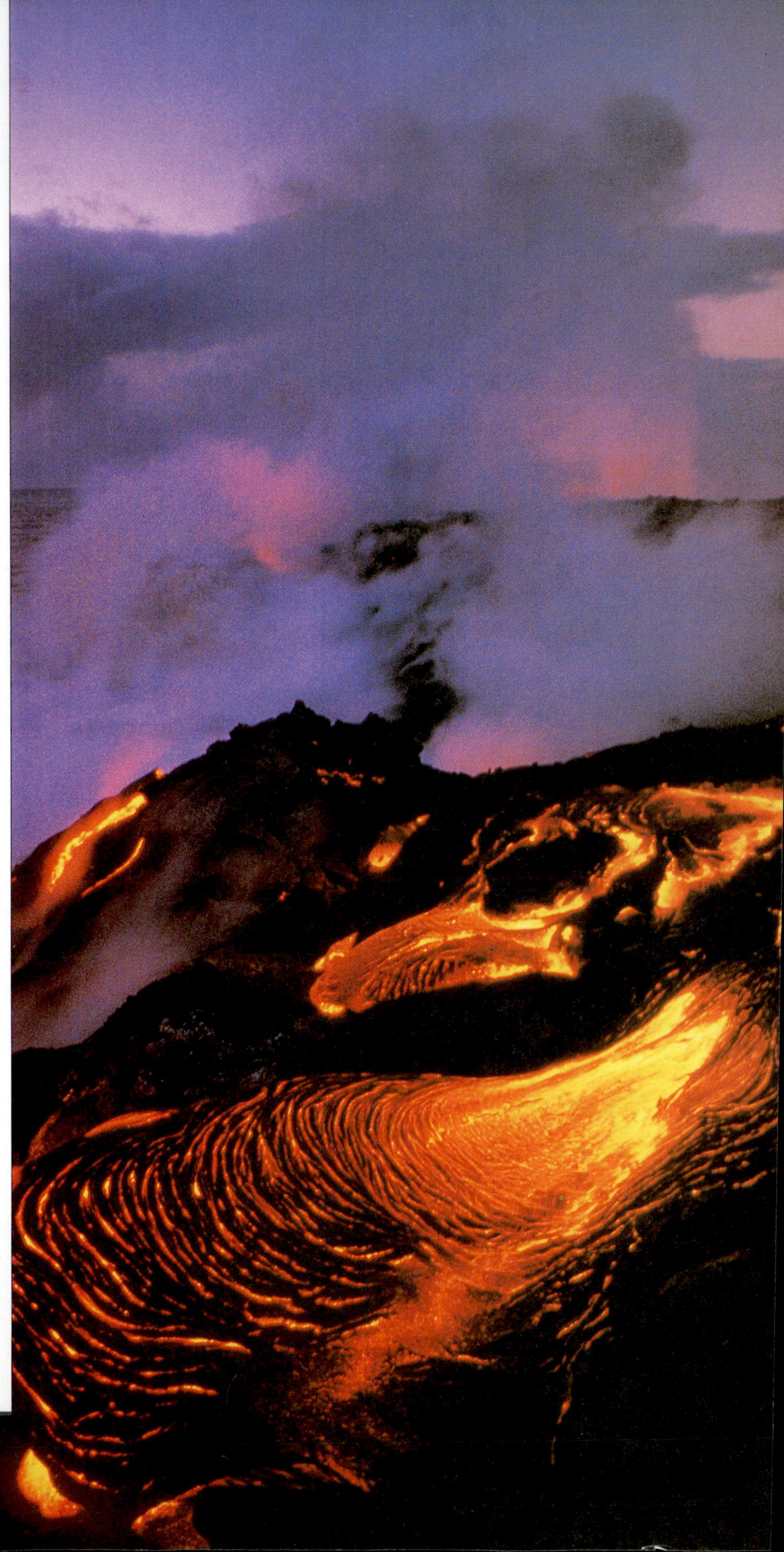

◀ **CHAPTER OPENING PHOTO: As lava flows into the sea, a slightly acidic solution is formed.**

Many significant chemical reactions occur in aqueous solutions. Water is the medium for reactions that provide the foundation for successful life on this planet. Volcanos and hot springs yield acidic solutions formed from hydrochloric acid and sulfur dioxide. These acids mix with bases, such as calcium carbonate, dissolved from rocks and microscopic animals to form the salts of our oceans and the beauty of stalactites and stalagmites in limestone caves. Photosynthesis and respiration reactions cannot occur without an adequate balance between acids and bases. Even small deviations in this balance can be detrimental. In recent years, we have become very concerned about the effects of acid rain upon our environment. Consumers are rapidly becoming more informed about household products, such as shampoos, cleaning supplies, antacids, and paints, because of their effects upon our daily lives and on our planet.

16.1 Acids and Bases

The word *acid* is derived from the Latin *acidus*, meaning "sour" or "tart," and is also related to the Latin word *acetum*, meaning "vinegar." Vinegar has been known since antiquity as the product of the fermentation of wine and apple cider. The sour constituent of vinegar is acetic acid ($HC_2H_3O_2$).

Some of the characteristic properties commonly associated with acids are the following: Water solutions of acids taste sour and change the color of litmus, a vegetable dye, from blue to red. Water solutions of nearly all acids react with (1) metals such as zinc and magnesium to produce hydrogen gas, (2) bases to produce water and a salt, and (3) carbonates to produce carbon dioxide. These properties are due to the hydrogen ions, H^+, released by acids in a water solution.

Classically, a *base* is a substance capable of liberating hydroxide ions, OH^-, in water solution. Hydroxides of the alkali metals (Group IA) and alkaline earth metals (Group IIA), such as LiOH, NaOH, KOH, $Ca(OH)_2$, and $Ba(OH)_2$, are the most common inorganic bases. Water solutions of bases are called *alkaline solutions* or *basic solutions.* They have a bitter or caustic taste, a slippery, soapy feeling, the ability to change litmus from red to blue, and the ability to interact with acids to form a salt and water.

Several theories have been proposed to answer the question "What is an acid and a base?" One of the earliest, most significant of these theories was advanced in a doctoral thesis in 1884 by Svante Arrhenius (1859–1927), a Swedish scientist, who stated that an acid is a hydrogen-containing substance that dissociates to produce hydrogen ions, and that a base is a hydroxide-containing substance that dissociates to produce hydroxide ions in aqueous solutions. Arrhenius postulated that the hydrogen ions were produced by the dissociation of acids in water, and that the hydroxide ions were produced by the dissociation of bases in water:

$$\underset{\text{Acid}}{HA} \longrightarrow H^+ + A^-$$

$$\underset{\text{Base}}{MOH} \longrightarrow M^+ + OH^-$$

Thus, an acid solution contains an excess of hydrogen ions and a base an excess of hydroxide ions.

In 1923 the Brønsted–Lowry proton transfer theory was introduced by J. N. Brønsted (1897–1947), a Danish chemist, and T. M. Lowry (1847–1936), an English chemist. This theory states that an acid is a proton donor and a base is a proton acceptor.

Consider the reaction of hydrogen chloride gas with water to form hydrochloric acid:

$$HCl(g) + H_2O(l) \longrightarrow H_3O^+(aq) + Cl^-(aq) \qquad (1)$$

In the course of the reaction, HCl donates, or gives up, a proton to form a Cl^- ion, and H_2O accepts a proton to form the H_3O^+ ion. Thus, HCl is an acid and H_2O is a base, according to the Brønsted–Lowry theory.

A hydrogen ion, H^+, is nothing more than a bare proton and does not exist by itself in an aqueous solution. In water a proton combines with a polar water molecule to form a hydrated hydrogen ion, H_3O^+ [that is, $H(H_2O)^+$], commonly called a **hydronium ion**. The proton is attracted to a polar water molecule, forming a coordinate-covalent bond with one of the two pairs of unshared electrons:

hydronium ion

$$H^+ + H:\underset{\underset{H}{\cdot\cdot}}{\overset{\cdot\cdot}{O}}: \longrightarrow \left[H:\underset{\underset{H}{\cdot\cdot}}{\overset{\cdot\cdot}{O}}:H\right]^+$$

Hydronium ion

Note the electron structure of the hydronium ion. For simplicity of expression in equations, we often use H^+ instead of H_3O^+, with the explicit understanding that H^+ is always hydrated in solution.

Whereas the Arrhenius theory is restricted to aqueous solutions, the Brønsted–Lowry approach has applications in all media and has become the more important theory when the chemistry of substances in solutions other than water is studied. Ammonium chloride (NH_4Cl) is a salt, yet its water solution has an acidic reaction. From this test we must conclude that NH_4Cl has acidic properties. The Brønsted–Lowry explanation shows that the ammonium ion, NH_4^+, is a proton donor, and water is the proton acceptor:

$$\underset{\text{Acid}}{NH_4^+} \rightleftharpoons \underset{\text{Base}}{NH_3} + \underset{\text{Acid}}{H^+} \qquad (2)$$

$$\underset{\text{Acid}}{NH_4^+} + \underset{\text{Base}}{H_2O} \longrightarrow \underset{\text{Acid}}{H_3O^+} + \underset{\text{Base}}{NH_3} \qquad (3)$$

A Brønsted–Lowry acid is a proton (H^+) donor.
A Brønsted–Lowry base is a proton (H^+) acceptor.

The Brønsted–Lowry theory also applies to certain cases where no solution is involved. For example, in the reaction of hydrogen chloride and ammonia gases, HCl is the proton donor and NH_3 is the base. [Remember that (g) after a formula in equations stands for a gas.]

$$\underset{\text{Acid}}{HCl(g)} + \underset{\text{Base}}{NH_3(g)} \longrightarrow \underset{\text{Acid}}{NH_4^+} + \underset{\text{Base}}{Cl^-} \tag{4}$$

When a Brønsted–Lowry acid donates a proton, it forms the conjugate base of that acid. When a base accepts a proton, it forms the conjugate acid of that base.

In equations (1), (3), and (4) a conjugate acid and base are produced as products. The formulas of a conjugate acid–base pair differ by one proton (H^+). In equation (1) the conjugate acid–base pairs are HCl–Cl^- and H_3O^+–H_2O. Cl^- is the conjugate base of HCl, and HCl is the conjugate acid of Cl^-. H_2O is the conjugate base of H_3O^+, and H_3O^+ is the conjugate acid of H_2O.

conjugate acid–base pair

$$\underset{\text{acid}}{HCl(g)} + \underset{\text{base}}{H_2O(l)} \longrightarrow \underset{\text{base}}{Cl^-(aq)} + \underset{\text{acid}}{H_3O^+(aq)}$$

conjugate acid–base pair

In equation (3) the conjugate acid–base pairs are NH_4^+–NH_3 and H_3O^+–H_2O; in equation (4) they are HCl–Cl^- and NH_4^+–NH_3.

EXAMPLE 16.1

Write the formula for (a) the conjugate base of H_2O and of HNO_3, and (b) the conjugate acid of SO_4^{2-} and of $C_2H_3O_2^-$. The difference between an acid or a base and its conjugate is one proton, H^+.

SOLUTION

(a) To write the conjugate base of an acid, remove one proton from the acid formula. Thus,

$$H_2O \xrightarrow{-H^+} OH^- \quad \text{(Conjugate base)}$$

$$HNO_3 \xrightarrow{-H^+} NO_3^- \quad \text{(Conjugate base)}$$

Note that, by removing an H^+, the conjugate base becomes more negative than the acid by one minus charge.

(b) To write the conjugate acid of a base, add one proton to the formula of the base. Thus,

$$SO_4^{2-} \xrightarrow{+H^+} HSO_4^- \quad \text{(Conjugate acid)}$$

$$C_2H_3O_2^- \xrightarrow{+H^+} HC_2H_3O_2 \quad \text{(Conjugate acid)}$$

In each case the conjugate acid becomes more positive than the base by one positive charge due to the addition of H^+.

PRACTICE Indicate the conjugate base for each of the following acids:
(a) H_2CO_3 (b) HNO_2 (c) $HC_2H_3O_2$
Answers: (a) HCO_3^- (b) NO_2^- (c) $C_2H_3O_2^-$

PRACTICE Indicate the conjugate acids for each of the following bases:
(a) HSO_4^- (b) Cl^- (c) OH^-
Answers: (a) H_2SO_4 (b) HCl (c) H_2O

A more general concept of acids and bases was introduced by Gilbert N. Lewis. The Lewis theory deals with the way in which a substance with an unshared pair of electrons reacts in an acid–base type of reaction. According to this theory a base is any substance that has an unshared pair of electrons (electron-pair donor), and an acid is any substance that will attach itself to or accept a pair of electrons. In the reaction

$$\underset{\text{Acid}}{H^+} + \underset{\text{Base}}{\begin{matrix} & H & \\ : & \ddot{N} & :H \\ & \overset{..}{H} & \end{matrix}} \longrightarrow \left[\begin{matrix} & H & \\ H: & \ddot{H} & :H \\ & \overset{..}{H} & \end{matrix}\right]^+$$

H^+ is a Lewis acid and $:NH_3$ is a Lewis base. According to the Lewis theory, substances other than proton donors (for example, BF_3) behave as acids:

$$\underset{\text{Acid}}{\begin{matrix} & F \\ F: & \ddot{B} \\ & \overset{..}{F} \end{matrix}} + \underset{\text{Base}}{\begin{matrix} & H & \\ : & \ddot{N} & :H \\ & \overset{..}{H} & \end{matrix}} \longrightarrow \begin{matrix} & F & H & \\ F: & \ddot{B}: & \ddot{N} & :H \\ & \overset{..}{F} & \overset{..}{H} & \end{matrix}$$

The Lewis and Brønsted–Lowry bases are identical because, to accept a proton, a base must have an unshared pair of electrons.

The three theories are summarized in Table 16.1. These theories explain how acid–base reactions occur. We will generally use the theory that best explains the reaction that is under consideration. Most of our examples will refer to aqueous solutions. It is important to realize that in an aqueous acidic solution the H^+ ion concentration is always greater than OH^- ion concentration. And, vice versa, in an aqueous basic solution the OH^- ion concentration is always greater than the H^+ ion concentration. When the H^+ and OH^- ion concentrations in a solution are equal, the solution is neutral; that is, it is neither acidic nor basic.

16.2 Reactions of Acids

In aqueous solutions the H^+ or H_3O^+ ions are responsible for the characteristic reactions of acids. All the following reactions are in an aqueous medium.

Table 16.1 Summary of Acid–Base Definitions

Theory	Acid	Base
Arrhenius	A hydrogen-containing substance that produces hydrogen ions in aqueous solution	A hydroxide-containing substance that produces hydroxide ions in aqueous solution
Brønsted–Lowry	A proton (H^+) donor	A proton (H^+) acceptor
Lewis	Any species that will bond to an unshared pair of electrons (electron-pair acceptor)	Any species that has an unshared pair of electrons (electron-pair donor)

Reaction with Metals Acids react with metals that lie above hydrogen in the activity series of elements to produce hydrogen and a salt (see Section 18.5).

$$\text{acid} + \text{metal} \longrightarrow \text{hydrogen} + \text{salt}$$
$$2\,HCl(aq) + Ca(s) \longrightarrow H_2\uparrow + CaCl_2(aq)$$
$$H_2SO_4(aq) + Mg(s) \longrightarrow H_2\uparrow + MgSO_4(aq)$$
$$6\,HC_2H_3O_2(aq) + 2\,Al(s) \longrightarrow 3\,H_2\uparrow + 2\,Al(C_2H_3O_2)_3(aq)$$

Acids such as nitric acid (HNO_3) are oxidizing substances (see Chapter 18) and react with metals to produce water instead of hydrogen. For example,

$$3\,Zn(s) + 8\,HNO_3(\text{dilute}) \longrightarrow 3\,Zn(NO_3)_2(aq) + 2\,NO(g) + 4\,H_2O$$

Reaction with Bases The interaction of an acid and a base is called a *neutralization reaction.* In aqueous solutions, the products of this reaction are a salt and water.

$$\text{acid} + \text{base} \longrightarrow \text{salt} + \text{water}$$
$$HBr(aq) + KOH(aq) \longrightarrow KBr(aq) + H_2O(l)$$
$$2\,HNO_3(aq) + Ca(OH)_2(aq) \longrightarrow Ca(NO_3)_2(aq) + 2\,H_2O(l)$$
$$2\,H_3PO_4(aq) + 3\,Ba(OH)_2(aq) \longrightarrow Ba_3(PO_4)_2\downarrow + 6\,H_2O(l)$$

Reaction with Metal Oxides This reaction is closely related to that of an acid with a base. With an aqueous acid, the products are a salt and water.

$$\text{acid} + \text{metal oxide} \longrightarrow \text{salt} + \text{water}$$
$$2\,HCl(aq) + Na_2O(s) \longrightarrow 2\,NaCl(aq) + H_2O(l)$$
$$H_2SO_4(aq) + MgO(s) \longrightarrow MgSO_4(aq) + H_2O(l)$$
$$6\,HCl(aq) + Fe_2O_3(s) \longrightarrow 2\,FeCl_3(aq) + 3\,H_2O(l)$$

Reaction with Carbonates Many acids react with carbonates to produce carbon dioxide, water, and a salt. Carbonic acid (H_2CO_3) is not the product, because it is unstable and decomposes into water and carbon dioxide.

acid + carbonate ⟶ salt + water + carbon dioxide

$$2\,HCl(aq) + Na_2CO_3(aq) \longrightarrow 2\,NaCl(aq) + H_2O(l) + CO_2\uparrow$$

$$H_2SO_4(aq) + MgCO_3(s) \longrightarrow MgSO_4(aq) + H_2O(l) + CO_2\uparrow$$

16.3 Reactions of Bases

High pH

The OH^- ions are responsible for the characteristic reactions of bases. All the following reactions are in an aqueous medium.

Reaction with Acids Bases react with acids to produce a salt and water. See reaction of acids with bases in Section 16.2.

amphoteric

Amphoteric Hydroxides Hydroxides of certain metals, such as zinc, aluminum, and chromium, are **amphoteric**; that is, they are capable of reacting as either an acid or a base. When treated with a strong acid, they behave like bases; when reacted with a strong base, they behave like acids.

$$Zn(OH)_2(s) + 2\,HCl(aq) \longrightarrow ZnCl_2(aq) + 2\,H_2O(l)$$

$$Zn(OH)_2(s) + 2\,NaOH(aq) \longrightarrow Na_2Zn(OH)_4(aq)$$

Reaction of NaOH and KOH with Certain Metals Some amphoteric metals react directly with the strong bases, sodium hydroxide and potassium hydroxide, to produce hydrogen.

base + metal + water ⟶ salt + hydrogen

$$2\,NaOH(aq) + Zn(s) + 2\,H_2O(l) \longrightarrow Na_2Zn(OH)_4(aq) + H_2\uparrow$$

$$2\,KOH(aq) + 2\,Al(s) + 6\,H_2O(l) \longrightarrow 2\,KAl(OH)_4(aq) + 3\,H_2\uparrow$$

Reaction with Salts Bases will react with many salts in solution due to the formation of insoluble metal hydroxides.

base + salt ⟶ metal hydroxide↓ + salt

$$2\,NaOH(aq) + MnCl_2(aq) \longrightarrow Mn(OH)_2\downarrow + 2\,NaCl(aq)$$

$$3\,Ca(OH)_2(aq) + 2\,FeCl_3(aq) \longrightarrow 2\,Fe(OH)_3\downarrow + 3\,CaCl_2(aq)$$

$$2\,KOH(aq) + CuSO_4(aq) \longrightarrow Cu(OH)_2\downarrow + K_2SO_4(aq)$$

16.4 Salts

Salts are very abundant in nature. Most of the rocks and minerals of the earth's mantle are salts of one kind or another. Huge quantities of dissolved salts also exist in the oceans. Salts may be considered to be compounds that have been derived from acids and bases. They consist of positive metal or ammonium ions (H^+ excluded) combined with negative nonmetal ions (OH^- and O^{2-} excluded).

◀ These strange mineral formations called "tufa" exist at Mono Lake in California. Tufa are formed by water bubbling through sand saturated with NaCl, Na_2CO_3 and Na_2SO_4.

The positive ion is the base counterpart and the nonmetal ion is the acid counterpart:

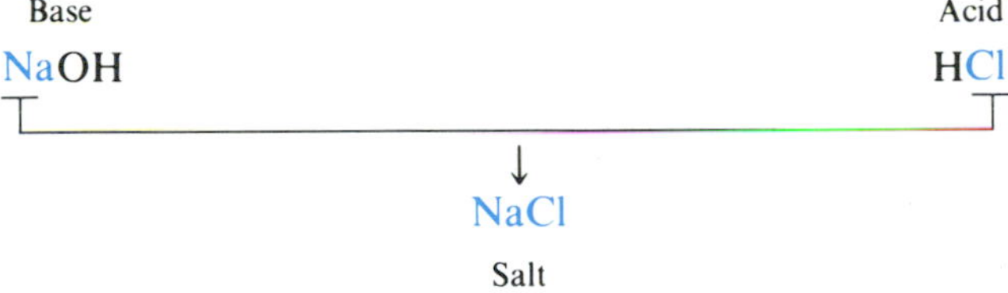

Salts are usually crystalline and have high melting and boiling points.

From a single acid such as hydrochloric acid (HCl), we can produce many chloride salts by replacing the hydrogen with metal ions (for example, NaCl, KCl, RbCl, $CaCl_2$, $NiCl_2$). Hence, the number of known salts greatly exceeds the number of known acids and bases. Salts are ionic compounds. If the hydrogen atoms of a binary acid are replaced by a nonmetal, the resulting compound has covalent bonding and is therefore not considered to be a salt (for example, PCl_3, S_2Cl_2, Cl_2O, NCl_3, ICl).

A review of Chapter 6 on the nomenclature of acids, bases, and salts may be beneficial at this point.

16.5 Electrolytes and Nonelectrolytes

Some of the most convincing evidence as to the nature of chemical bonding within a substance is the ability (or lack of ability) of a water solution of the substance to conduct electricity.

FIGURE 16.1 ► A simple conductivity apparatus for testing electrolytes and nonelectrolytes in solution. The light bulb glows because the solution contains an electrolyte (acetic acid solution).

We can show that solutions of certain substances are conductors of electricity by using a simple conductivity apparatus consisting of a pair of electrodes connected to a voltage source through a light bulb and switch (see Figure 16.1). If the medium between the electrodes is a conductor of electricity, the light bulb will glow when the switch is closed. When chemically pure water is placed in the beaker and the switch is closed, the light does not glow, indicating that water is a virtual nonconductor. When we dissolve a small amount of sugar in the water and test the solution, the light still does not glow, showing that a sugar solution is also a nonconductor. But, when a small amount of salt, NaCl, is dissolved in water and this solution is tested, the light glows brightly. Thus, the salt solution conducts electricity. A fundamental difference exists between the chemical bonding of sugar and that of salt. Sugar is a covalently bonded (molecular) substance; salt is a substance with ionic bonds.

electrolyte

nonelectrolyte

Substances whose aqueous solutions are conductors of electricity are called **electrolytes**. Substances whose solutions are nonconductors are known as **nonelectrolytes**. The classes of compounds that are electrolytes are acids, bases, and salts. Solutions of certain oxides also are conductors because the oxides form an acid or a base when dissolved in water. One major difference between electrolytes and nonelectrolytes is that electrolytes are capable of producing ions in solution, whereas nonelectrolytes do not have this property. Solutions that contain a sufficient number of ions will conduct an electric current. Although pure water is essentially a nonconductor, many city water supplies contain enough dissolved ionic matter to cause the light to glow dimly when the water is tested in a conductivity apparatus. Table 16.2 lists some common electrolytes and nonelectrolytes.

Acids, bases, and salts are electrolytes.

TABLE 16.2 Representative Electrolytes and Nonelectrolytes

Electrolytes		Nonelectrolytes	
H_2SO_4	$HC_2H_3O_2$	$C_{12}H_{22}O_{11}$ (sugar)	CH_3OH (methyl alcohol)
HCl	NH_4OH	C_2H_5OH (ethyl alcohol)	$CO(NH_2)_2$ (urea)
HNO_3	K_2SO_4	$C_2H_4(OH)_2$ (ethylene glycol)	O_2
NaOH	$NaNO_3$	$C_3H_5(OH)_3$ (glycerol)	H_2O

16.6 DISSOCIATION AND IONIZATION OF ELECTROLYTES

Arrhenius received the 1903 Nobel Prize in chemistry for his work on electrolytes. He stated that a solution conducts electricity because the solute dissociates immediately upon dissolving into electrically charged particles called *ions*. The movement of these ions toward oppositely charged electrodes causes the solution to be a conductor. According to his theory, solutions that are relatively poor conductors contain electrolytes that are only partly dissociated. Arrhenius also believed that ions exist in solution whether or not an electric current is present. In other words, the electric current does not cause the formation of ions. Positive ions, attracted to the cathode, are cations; negative ions, attracted to the anode, are anions.

dissociation

We have seen that sodium chloride crystals consist of sodium and chloride ions held together by ionic bonds. **Dissociation** is the process by which the ions of a salt separate as the salt dissolves. When placed in water, the sodium and chloride ions are attracted by the polar water molecules, which surround each ion as it dissolves. In water, the salt dissociates, forming hydrated sodium and chloride ions (see Figure 16.2). The sodium and chloride ions in solution are bonded to a specific number of water dipoles and have less attraction for each other than they had in the crystalline state. The equation representing this dissociation is

$$NaCl(s) + (x + y)\,H_2O \longrightarrow Na^+(H_2O)_x + Cl^-(H_2O)_y$$

A simplified dissociation equation in which the water is omitted but understood to be present is

$$NaCl \longrightarrow Na^+ + Cl^-$$

It is important to remember that sodium chloride exists in an aqueous solution as hydrated ions and not as NaCl units, even though the formula NaCl or $Na^+ + Cl^-$ is often used in equations.

The chemical reactions of salts in solution are the reactions of their ions. For example, when sodium chloride and silver nitrate react and form a precipitate of silver chloride, only the Ag^+ and Cl^- ions participate in the reaction. The Na^+ and NO_3^- remain as ions in solution.

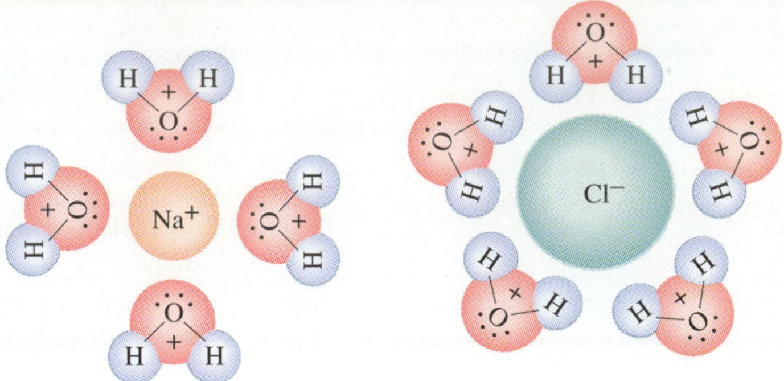

FIGURE 16.2 ▶ Hydrated sodium and chloride ions. When sodium chloride dissolves in water, each Na^+ and Cl^- ion becomes surrounded by water molecules. The negative end of the water dipole is attracted to the Na^+ ion, and the positive end is attracted to the Cl^- ion.

$$Ag^+ + Cl^- \longrightarrow AgCl\downarrow$$

In many cases the number of molecules of water associated with a particular ion is known. For example, the blue color of the copper(II) ion is due to the hydrated ion $Cu(H_2O)_4^{2+}$. The hydration of ions can be demonstrated in a striking way with cobalt(II) chloride. When cobalt(II) chloride hexahydrate is dissolved in water, a pink solution forms due to the $Co(H_2O)_6^{2+}$ ions. If concentrated hydrochloric acid is added to this pink solution, the color gradually changes to blue. If water is then added to the blue solution, the color changes to pink again. These color changes are due to the exchange of water molecules and chloride ions on the cobalt ion. The $CoCl_4^{2-}$ ion is blue. Thus, the hydration of the cobalt ion is a reversible or equilibrium reaction (see Chapter 17). The equilibrium equation representing these changes is

$$\underset{\text{Pink}}{Co(H_2O)_6^{2+}} + 4\,Cl^- \rightleftharpoons \underset{\text{Blue}}{CoCl_4^{2-}} + 6\,H_2O$$

ionization

Ionization is the formation of ions; it occurs as a result of chemical reaction of certain substances with water. Glacial acetic acid (100% $HC_2H_3O_2$) is a liquid that behaves as a nonelectrolyte when tested by the method described in Section 16.5. But a water solution of acetic acid conducts an electric current (as indicated by the dull-glowing light of the conductivity apparatus). The equation for the reaction with water, which forms hydronium and acetate ions, is

$$\underset{\text{Acid}}{HC_2H_3O_2} + \underset{\text{Base}}{H_2O} \rightleftharpoons \underset{\text{Acid}}{H_3O^+} + \underset{\text{Base}}{C_2H_3O_2^-}$$

or, in the simplified equation,

$$HC_2H_3O_2 \rightleftharpoons H^+ + C_2H_3O_2^-$$

In this ionization reaction, water serves not only as a solvent but also as a base according to the Brønsted–Lowry theory.

Hydrogen chloride is predominantly covalently bonded, but when dissolved in water it reacts to form hydronium and chloride ions:

$$HCl(g) + H_2O(l) \longrightarrow H_3O^+(aq) + Cl^-(aq)$$

When a hydrogen chloride solution is tested for conductivity, the light glows brilliantly, indicating many ions in the solution.

Ionization occurs in each of the above two reactions with water, producing ions in solution. The necessity for water in the ionization process can be demonstrated by dissolving hydrogen chloride in a nonpolar solvent such as benzene, and testing the solution for conductivity. The solution fails to conduct electricity, indicating that no ions are produced.

The terms *dissociation* and *ionization* are often used interchangeably to describe processes taking place in water. But, strictly speaking, the two are different. In the dissociation of a salt, the salt already exists as ions; when it dissolves in water, the ions separate, or dissociate, and increase in mobility. In the ionization process, ions are produced by the reaction of a compound with water.

16.7 Strong and Weak Electrolytes

Electrolytes are classified as strong or weak depending on the degree, or extent, of dissociation or ionization. **Strong electrolytes** are essentially 100% ionized in solution; **weak electrolytes** are much less ionized (based on comparing 0.1 *M* solutions). Most electrolytes are either strong or weak, with a few classified as moderately strong or weak. Most salts are strong electrolytes. Acids and bases that are strong electrolytes (highly ionized) are called *strong acids* and *strong bases.* Acids and bases that are weak electrolytes (slightly ionized) are called *weak acids* and *weak bases.*

strong electrolyte

Strong Acids
Strong Bases

weak electrolyte

For equivalent concentrations, solutions of strong electrolytes contain many more ions than do solutions of weak electrolytes. As a result, solutions of strong electrolytes are better conductors of electricity. Consider the two solutions, 1 *M* HCl and 1 *M* $HC_2H_3O_2$. Hydrochloric acid is almost 100% ionized; acetic acid is about 1% ionized. Thus HCl is a strong acid, and $HC_2H_3O_2$ is a weak acid. Hydrochloric acid has about 100 times as many hydronium ions in solution as acetic acid, making the HCl solution much more acidic.

One can distinguish between strong and weak electrolytes experimentally using the apparatus described in Section 16.5. A 1 *M* HCl solution causes the light to glow brilliantly, but a 1 *M* $HC_2H_3O_2$ solution causes only a dim glow. In a similar fashion the strong base sodium hydroxide (NaOH) may be distinguished from the weak base ammonium hydroxide (NH_4OH). The ionization of a weak electrolyte in water is represented by an equilibrium equation showing that both the un-ionized and ionized forms are present in solution. In the equilibrium equation of $HC_2H_3O_2$ and its ions, we say that the equilibrium lies "far to the left" because relatively few hydrogen and acetate ions are present in solution:

$$HC_2H_3O_2(aq) \rightleftharpoons H^+ + C_2H_3O_2^-$$

We have previously used a double arrow in an equation to represent reversible processes in the equilibrium between dissolved and undissolved solute in a saturated solution. A double arrow ($\rightleftarrows$) is also used in the ionization equation of soluble weak electrolytes to indicate that the solution contains a considerable amount of the un-ionized compound in equilibrium with its ions in solution. (See Section 17.1 for a discussion of reversible reactions). A single arrow is used to indicate that the electrolyte is essentially all in the ionic form in the solution. For

TABLE 16.3 Strong and Weak Electrolytes

Strong electrolytes		Weak electrolytes	
Most soluble salts	$HClO_4$	$HC_2H_3O_2$	$H_2C_2O_4$
H_2SO_4	NaOH	H_2CO_3	H_3BO_3
HNO_3	KOH	HNO_2	HClO
HCl	$Ca(OH)_2$	H_2SO_3	NH_4OH
HBr	$Ba(OH)_2$	H_2S	HF

example, nitric acid is a strong acid; nitrous acid is a weak acid. Their ionization equations in water may be indicated as

$$HNO_3(aq) \xrightarrow{H_2O} H^+ + NO_3^-$$

$$HNO_2(aq) \overset{H_2O}{\rightleftharpoons} H^+ + NO_2^-$$

Practically all soluble salts; acids such as sulfuric, nitric, and hydrochloric acids; and bases such as sodium, potassium, calcium, and barium hydroxides are strong electrolytes. Weak electrolytes include numerous other acids and bases such as acetic acid, nitrous acid, carbonic acid, and ammonium hydroxide. The terms *strong acid*, *strong base*, *weak acid*, and *weak base* refer to whether an acid or base is a strong or weak electrolyte. A brief list of strong and weak electrolytes is given in Table 16.3.

Electrolytes yield two or more ions per formula unit upon dissociation, the actual number being dependent on the compound. Dissociation is complete or nearly complete for nearly all soluble salts and for certain other strong electrolytes such as those given in Table 16.3. The following are dissociation equations for several strong electrolytes. In all cases the ions are actually hydrated.

$$NaOH \xrightarrow{H_2O} Na^+ + OH^- \qquad \text{2 ions in solution per formula unit}$$

$$Na_2SO_4 \xrightarrow{H_2O} 2\ Na^+ + SO_4^{2-} \qquad \text{3 ions in solution per formula unit}$$

$$AlCl_3 \xrightarrow{H_2O} Al^{3+} + 3\ Cl^- \qquad \text{4 ions in solution per formula unit}$$

$$Fe_2(SO_4)_3 \xrightarrow{H_2O} 2\ Fe^{3+} + 3\ SO_4^{2-} \qquad \text{5 ions in solution per formula unit}$$

One mole of NaCl will give 1 mol of Na^+ ions and 1 mol of Cl^- ions in solution, assuming complete dissociation of the salt. One mole of $CaCl_2$ will give 1 mol of Ca^{2+} ions and 2 mol of Cl^- ions in solution.

$$\underset{1\ \text{mol}}{NaCl} \xrightarrow{H_2O} \underset{1\ \text{mol}}{Na^+} + \underset{1\ \text{mol}}{Cl^-}$$

$$\underset{1\ \text{mol}}{CaCl_2} \xrightarrow{H_2O} \underset{1\ \text{mol}}{Ca^{2+}} + \underset{2\ \text{mol}}{2\ Cl^-}$$

EXAMPLE 16.2

What is the molarity of each ion in a solution of (a) 2.0 M NaCl, and (b) 0.40 M K_2SO_4? (Assume complete dissociation.)

SOLUTION

(a) According to the dissociation equation,

$$\underset{1\ \text{mol}}{NaCl} \xrightarrow{H_2O} \underset{1\ \text{mol}}{Na^+} + \underset{1\ \text{mol}}{Cl^-}$$

the concentration of Na^+ is equal to that of NaCl (1 mol NaCl $\longrightarrow$ 1 mol Na^+) and the concentration of Cl^- is also equal to that of NaCl. Therefore, the concentrations of the ions in 2.0 M NaCl are 2.0 M Na^+ and 2.0 M Cl^-.

(b) According to the dissociation equation,

$$\underset{1\ \text{mol}}{K_2SO_4} \xrightarrow{H_2O} \underset{2\ \text{mol}}{2\,K^+} + \underset{1\ \text{mol}}{SO_4^{2-}}$$

The concentration of K^+ is twice that of K_2SO_4 and the concentration of SO_4^{2-} is equal to that of K_2SO_4. Therefore, the concentrations of the ions in 0.40 M K_2SO_4 are 0.80 M K^+ and 0.40 M SO_4^{2-}.

PRACTICE What is the molarity of each ion in a solution of (a) 0.050 M of $MgCl_2$, and (b) 0.070 M of $AlCl_3$?
Answers: (a) 0.050 M Mg^{2+}, 0.10 M Cl^- (b) 0.070 M Al^{3+}, 0.21 M Cl^-

Colligative Properties of Electrolyte Solutions

Depends on # of ions
moles of ions

We have learned that when 1 mol of sucrose, a nonelectrolyte, is dissolved in 1000 g of water, the solution freezes at $-1.86°C$. When 1 mol of NaCl is dissolved in 1000 g of water, the freezing point of the solution is not $-1.86°C$, as might be expected, but is closer to $-3.72°C$ (-1.86×2). The reason for the lower freezing point is that 1 mol of NaCl in solution produces 2 mol of particles ($2 \times 6.022 \times 10^{23}$ ions) in solution. Thus, the freezing point depression produced by 1 mol of NaCl is essentially equivalent to that produced by 2 mol of a nonelectrolyte. An electrolyte such as $CaCl_2$, which yields three ions in water, gives a freezing point depression of about three times that of a nonelectrolyte. These freezing point data provide additional evidence that electrolytes dissociate when dissolved in water. The other colligative properties are similarly affected by substances that yield ions in aqueous solutions.

16.8 Ionization of Water

The more we study chemistry, the more intriguing the little molecule of water becomes. Two equations commonly used to show how water ionizes are

$$\underset{\text{Acid}}{H_2O} + \underset{\text{Base}}{H_2O} \rightleftharpoons \underset{\text{Acid}}{H_3O^+} + \underset{\text{Base}}{OH^-}$$

and

$$H_2O \rightleftharpoons H^+ + OH^-$$

The first equation represents the Brønsted–Lowry concept, with water reacting as both an acid and a base, forming a hydronium ion and a hydroxide ion. The second equation is a simplified version, indicating that water ionizes to give a hydrogen and a hydroxide ion. Actually, the proton, H^+, is hydrated and exists as a hydronium ion. In either case equal molar amounts of acid and base are produced so that water is neutral, having neither H^+ nor OH^- ions in excess. The ionization of water at 25°C produces an H^+ ion concentration of 1.0×10^{-7} mole per liter and an OH^- ion concentration of 1.0×10^{-7} mole per liter. These concentrations are usually expressed as

$$[H^+] \text{ or } [H_3O^+] = 1.0 \times 10^{-7}\ \text{mol/L}$$
$$[OH^-] = 1.0 \times 10^{-7}\ \text{mol/L}$$

These figures mean that about two out of every billion water molecules are ionized. This amount of ionization, small as it is, is a significant factor in the behavior of water in many chemical reactions.

The square brackets, [ ], indicate that the concentration is in moles per liter. Thus $[H^+]$ means the concentration of H^+ in moles per liter.

16.9 Introduction to pH

The acidity of an aqueous solution depends on the concentration of hydrogen or hydronium ions. The acidity of solutions involved in a chemical reaction is often critically important, especially for biochemical reactions. The pH scale of acidity was devised to fill the need for a simple, convenient numerical way to state the acidity of a solution. Values on the pH scale are obtained by mathematical conversion of H^+ ion concentrations to pH by these expressions:

$$\text{pH} = \log \frac{1}{[H^+]} \qquad \text{or} \qquad \text{pH} = -\log[H^+]$$

pH

where $[H^+] = H^+$ or H_3O^+ ion concentration in moles per liter. The **pH** is defined as the logarithm (log) of the reciprocal of the H^+ or H_3O^+ ion concentration in moles per liter. The scale itself is based on the H^+ concentration in water at 25°C. At this temperature, water has an H^+ concentration of 1×10^{-7} mole per liter and is calculated to have a pH of 7.

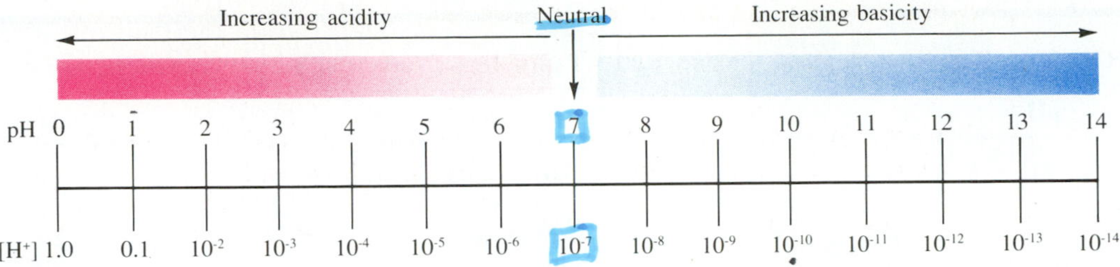

▲
FIGURE 16.3
The pH scale of acidity and basicity.

$$\text{pH} = \log\frac{1}{[\text{H}^+]} = \log\frac{1}{[1 \times 10^{-7}]} = \log 1 \times 10^7 = 7$$

By an alternate and mathematically equivalent definition, pH is the *negative* logarithm of the H^+ or H_3O^+ concentration in moles per liter:

$$\text{pH} = -\log[\text{H}^+] = -\log[1 \times 10^{-7}] = -(-7) = 7$$

The pH of pure water at 25°C is 7 and is said to be neutral; that is, it is neither acidic nor basic, because the concentrations of H^+ and OH^- are equal. Solutions that contain more H^+ ions than OH^- ions have pH values less than 7, and solutions that contain less H^+ ions than OH^- ions have values greater than 7.

pH < 7.00 Acidic solution
pH = 7.00 Neutral solution
pH > 7.00 Basic solution

When $[\text{H}^+] = 1 \times 10^{-5}$ mol/L, pH = 5 (acidic)
When $[\text{H}^+] = 1 \times 10^{-9}$ mol/L, pH = 9 (basic)

Instead of saying that the hydrogen ion concentration in the solution is 1×10^{-5} mole per liter, it is customary to say that the pH of the solution is 5. The smaller the pH value, the more acidic the solution (see Figure 16.3).

At a given molarity a strong acid is more acidic (has a higher H^+ concentration) than a weak acid. For example, in a 0.100 *M* concentration, the pH of HCl is 1.00 and that of $HC_2H_3O_2$ is 2.87. As a weak acid is made more dilute, greater percentages of its molecules ionize, and the pH tends to approach that of a strong acid at comparable dilutions. This behavior is illustrated by the following comparative data for hydrochloric acid (100% ionized) and acetic acid.

HCl solution			**$HC_2H_3O_2$ solution**		
M	**pH**	**% ionized**	**M**	**pH**	**% ionized**
0.100	1.00	100	0.100	2.87	1.35
0.0100	2.00	100	0.0100	3.37	4.27
0.00100	3.00	100	0.00100	3.90	12.6

TABLE 16.4 The pH Scale for Expressing Acidity

$[H^+]$ (mol/L)	pH	
1×10^{-14}	14	↑
1×10^{-13}	13	
1×10^{-12}	12	
1×10^{-11}	11	Increasing basicity
1×10^{-10}	10	
1×10^{-9}	9	
1×10^{-8}	8	
1×10^{-7}	7	Neutral
1×10^{-6}	6	
1×10^{-5}	5	
1×10^{-4}	4	
1×10^{-3}	3	Increasing acidity
1×10^{-2}	2	
1×10^{-1}	1	
1×10^{0}	0	↓

TABLE 16.5 The pH of Some Common Solutions

Solution	pH
Gastric juice	1.0
0.1 *M* HCl	1.0
Lemon juice	2.3
Vinegar	2.8
0.1 *M* $HC_2H_3O_2$	2.9
Orange juice	3.7
Tomato juice	4.1
Coffee, black	5.0
Urine	6.0
Milk	6.6
Pure water (25°C)	7.0
Blood	7.4
Household ammonia	11.0
1 *M* NaOH	14.0

The pH scale, along with its interpretation, is given in Table 16.4, and Table 16.5 lists the pH of some common solutions. Note that a change of only 1 pH unit means a tenfold increase or decrease in H^+ ion concentration. For example, a solution with a pH of 3.0 is ten times more acidic than a solution with a pH of 4.0. A simplified method of determining pH from $[H^+]$ follows:

$$[H^+] = 1 \times 10^{-5}$$

When this number is exactly 1 (the 1) — pH = this number (5) (the exponent)
pH = 5

$$[H^+] = 2 \times 10^{-5}$$

When this number is between 1 and 10 (the 2) — pH is between this number and next lower number (4 and 5) (the exponent)
pH = 4.7

logarithm

Calculation of the pH value corresponding to any H^+ ion concentration requires the use of logarithms. Logarithms are exponents. The **logarithm** (log) of a number is simply the power to which 10 must be raised to give that number. Thus the log of 100 is 2 ($100 = 10^2$), and the log of 1000 is 3 ($1000 = 10^3$). The log of 500 is 2.70, but you cannot easily determine this value without a scientific calculator. Determining a logarithm used to be a tedious task requiring a published standard log table, but with the availability of affordable calculators with log capability, the log table method is virtually obsolete.

Let us determine the pH of a solution with $[H^+] = 2 \times 10^{-5}$. The exponent (-5) indicates that the pH is between 4 and 5. Enter 2×10^{-5} into your calculator and press the log key. The number $-4.69\ldots$ will be displayed. The pH is then

$$\text{pH} = -\log[H^+] = -(-4.69\ldots) = 4.7$$

EXAMPLE 16.3

What is the pH of a solution with an $[H^+]$ of (a) 1.0×10^{-11} and (b) 6.0×10^{-4}?

SOLUTION

(a) $[H^+] = 1.0 \times 10^{-11}$
$pH = -\log(1.0 \times 10^{-11})$
$pH = 11$

(b) $[H^+] = 6.0 \times 10^{-4}$
$\log 6.0 \times 10^{-4} = -3.22$
$pH = -\log[H^+]$
$pH = -(-3.22) = 3.22$

PRACTICE What is the pH of a solution with $[H^+]$ of (a) 3.9×10^{-12}, (b) 1.3×10^{-3}, and (c) 3.72×10^{-6}?
Answers: (a) 11.41 (b) 2.89 (c) 5.43

The measurement and control of pH is extremely important in many fields of science and technology. The proper soil pH is necessary to grow certain types of plants successfully. The pH of certain foods is too acidic for some diets. Many biological processes are delicately controlled pH systems. The pH of human blood is regulated to very close tolerances through the uptake or release of H^+ by mineral ions such as HCO_3^-, HPO_4^{2-}, and $H_2PO_4^-$. Changes in the pH of the blood by as little as 0.4 pH unit result in death.

Compounds with colors that change at particular pH values are used as indicators in acid–base reactions. For example, phenolphthalein, an organic compound, is colorless in acid solution and changes to pink at a pH of 8.3. When a solution of sodium hydroxide is added to a hydrochloric acid solution containing phenolphthalein, the change in color (from colorless to pink) indicates that all the acid is neutralized. Commercially available pH test paper, such as shown in Figure 16.4, contains chemical indicators. The indicator in the paper takes on different colors when wetted with solutions of different pH. Thus the pH of a solution can be estimated by placing a drop on the test paper and comparing the color of the test paper with a color chart calibrated at different pH values. Electronic pH meters of the type shown in Figure 16.5 are used for making rapid and precise pH determinations.

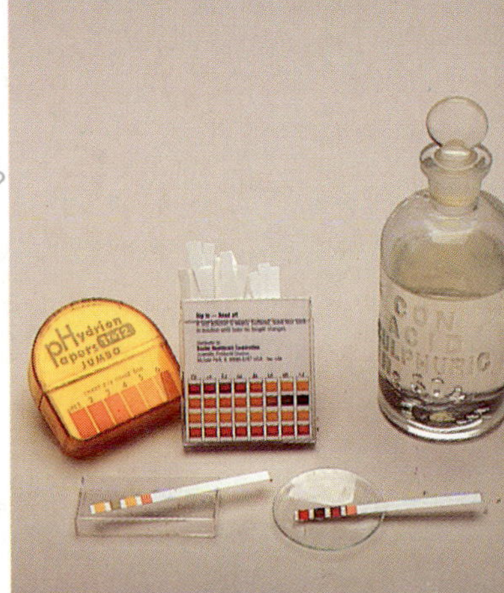

FIGURE 16.4
pH test paper for determining the approximate acidity of solutions.

16.10 NEUTRALIZATION

neutralization

The reaction of an acid and a base to form a salt and water is known as **neutralization**. We have seen this reaction before; but now, in the light of what we have learned about ions and ionization, let us reexamine the process of neutralization.

Consider the reaction that occurs when solutions of sodium hydroxide and hydrochloric acid are mixed. The ions present initially are Na^+ and OH^- from the base and H^+ and Cl^- from the acid. The products, sodium chloride and water, exist as Na^+ and Cl^- ions and H_2O molecules. A chemical equation representing this reaction is

$$HCl(aq) + NaOH(aq) \longrightarrow NaCl(aq) + H_2O(l) \quad (5)$$

This equation, however, does not show that HCl, NaOH, and NaCl exist as ions in solution. The following total ionic equation gives a better representation of the reaction:

$$(H^+ + Cl^-) + (Na^+ + OH^-) \longrightarrow Na^+ + Cl^- + H_2O(l) \quad (6)$$

spectator ion

Equation (6) shows that the Na^+ and Cl^- ions did not react. These ions are called **spectator ions** because they were present but did not take part in the reaction. The only reaction that occurred was that between the H^+ and OH^- ions. Therefore, the equation for the neutralization can be written as this net ionic equation:

$$\underset{\text{Acid}}{H^+} + \underset{\text{Base}}{OH^-} \longrightarrow \underset{\text{Water}}{H_2O(l)} \quad (7)$$

This simple net ionic equation (7) represents not only the reaction of sodium hydroxide and hydrochloric acid but also the reaction of any acid with any base in an aqueous solution. The driving force of a neutralization reaction is the ability of an H^+ ion and an OH^- ion to react and form a molecule of un-ionized water.

titration

The amount of acid, base, or other species in a sample may be determined by titration. **Titration** is the process of measuring the volume of one reagent that is required to react with a measured mass or volume of another reagent. Let us consider the titration of an acid with a base. A measured volume of acid of unknown concentration is placed in a flask, and a few drops of an indicator solution are added. Base solution of known concentration is slowly added from a buret to the acid until the indicator changes color (see Figure 16.6). The indicator selected is one that changes color when the stoichiometric quantity (according to the equation) of base has been added to the acid. At this point, known as the *end point of the titration*, the titration is complete, and the volume of base used to

▲ **FIGURE 16.5**
An electronic pH meter. Accurate measurements may be made by meters of this type.

FIGURE 16.6 ▶
pH meters are often used with computers and burets to perform titrations. The data is gathered and graphed on the computer at the same time the titration is being done.

neutralize the acid is read from the buret. The concentration or amount of acid in solution can be calculated from the titration data and the chemical equation for the reaction. Illustrative problems follow.

EXAMPLE 16.4

Suppose that 42.00 mL of 0.150 *M* NaOH solution is required to titrate 50.00 mL of hydrochloric acid solution. What is the molarity of the acid solution?

SOLUTION

The equation for the reaction is

$$NaOH(aq) + HCl(aq) \longrightarrow NaCl(aq) + H_2O(l)$$

In this neutralization NaOH and HCl react in a 1:1 mole ratio. Therefore, the moles of HCl in solution are equal to the moles of NaOH required to react with it. First we calculate the moles of NaOH used, and from this value we determine the moles of HCl.

Data: 42.00 mL of 0.150 *M* NaOH   50.00 mL of HCl
Molarity of acid = *M* (unknown)

Moles of NaOH:

$$M = \text{mol/L} \qquad 42.00\ \text{mL} = 0.04200\ \text{L}$$

$$0.04200\,\cancel{\text{L}} \times \frac{0.150\ \text{mol NaOH}}{1\,\cancel{\text{L}}} = 0.00630\ \text{mol NaOH}$$

Since NaOH and HCl react in a 1:1 mole ratio, 0.00630 mol of HCl was present in the 50.00 mL of HCl solution. Therefore the molarity of the HCl is

$$M = \frac{\text{mol}}{\text{L}} = \frac{0.00630\ \text{mol HCl}}{0.05000\ \text{L}} = 0.126\ M\ \text{HCl}$$

EXAMPLE 16.5

Suppose that 42.00 mL of 0.150 *M* NaOH solution is required to titrate 50.00 mL of sulfuric acid (H_2SO_4) solution. What is the molarity of the acid solution?

SOLUTION

The equation for the reaction is

$$2\ NaOH(aq) + H_2SO_4(aq) \longrightarrow Na_2SO_4(aq) + 2\ H_2O(l)$$

The same amount of base (0.00630 mol of NaOH) is used in this titration as in Example 16.4, but the mole ratio of acid to base in the reaction is 1:2. The moles of H_2SO_4 reacted can be calculated by using the mole-ratio method.

Data: 42.00 mL of 0.150 *M* NaOH = 0.00630 mol NaOH

$$0.00630\,\cancel{\text{mol NaOH}} \times \frac{1\ \text{mol}\ H_2SO_4}{2\,\cancel{\text{mol NaOH}}} = 0.00315\ \text{mol}\ H_2SO_4$$

Therefore 0.00315 mol of H_2SO_4 was present in 50.00 mL of H_2SO_4 solution. The molarity of the H_2SO_4 is

$$M = \frac{\text{mol}}{\text{L}} = \frac{0.00315\ \text{mol}\ H_2SO_4}{0.05000\ \text{L}} = 0.0630\ M\ H_2SO_4$$

Example 16.6 A 25.00 mL sample of H_2SO_4 solution required 14.26 mL of 0.2240 *N* NaOH for complete neutralization. What is the normality and the molarity of the sulfuric acid?

SOLUTION The equation for the reaction is

$$2\ NaOH(aq) + H_2SO_4(aq) \longrightarrow Na_2SO_4(aq) + 2\ H_2O(l)$$

The normality of the acid can be calculated from

$$V_A N_A = V_B N_B$$

Substitute the data in the problem and solve for N_A.

$$25.00\ mL \times N_A = 14.26\ mL \times 0.2240\ N$$

$$N_A = \frac{14.26\ mL \times 0.2240\ N}{25.00\ mL} = 0.1278\ N\ H_2SO_4$$

The normality of the acid is 0.1278 *N*.

Because H_2SO_4 furnishes 2 equivalents of H^+ per mole, the conversion to molarity is

$$\frac{equiv.}{L} \times \frac{mol}{equiv.}$$

$$\frac{0.1278\ \cancel{equiv.\ H_2SO_4}}{1\ L} \times \frac{1\ mol\ H_2SO_4}{2\ \cancel{equiv.\ H_2SO_4}} = 0.06390\ mol/L$$

The H_2SO_4 solution is 0.06390 *M*.

Practice A 50.0 mL sample of HCl required 24.81 mL of 0.1250 *M* NaOH for neutralization. What is the molarity of the acid?
Answer: 0.06203 *M* HCl

16.11 Acid Rain

Acid rain is defined as any precipitation which is more acidic than normal. The increase in acidity may be from natural or man-made sources. Rain acidity varies throughout the world and across the United States. The pH of rain is generally lower in the eastern United States and higher in the west. Unpolluted rain has a pH of 5.6, and so is slightly acidic. This acidity results from the dissolution of carbon dioxide in the water producing carbonic acid:

$$CO_2(g) + H_2O(l) \longrightarrow H_2CO_3(aq) \rightleftharpoons H^+(aq) + HCO_3^-(aq)$$

Rain is considered acidic when the pH falls below 5.6. Although the details of acid rain formation are not yet fully understood, chemists know the general process involves the following steps:

1. Emission of nitrogen and sulfur oxides into the air
2. Transportation of these oxides throughout the atmosphere
3. Chemical reactions between the oxides and water forming sulfuric acid (H_2SO_4) and nitric acid (HNO_3)
4. Rain or snow, which carries the acids to the surface

The oxides may also be deposited directly on a dry surface and become acidic when normal rain falls on them.

Acid rain is not a new phenomenon. Rain was probably acidic in the early days of our planet as volcanic eruptions, fires, and decomposition of organic matter released large volumes of nitrogen and sulfur oxides into the atmosphere. Use of fossil fuels, especially since the Industrial Revolution about 250 years ago, has made significant changes in the amounts of atmospheric pollutants. As increasing amounts of fossil fuels have been burned, more and more sulfur and nitrogen oxides have poured into the atmosphere thus increasing the acidity of rain.

Acid rain affects a variety of factors in our environment. Fresh water plants and animals have declined significantly when rain is acidic. Large numbers of fish and plants die when acidic water from spring thaws enters lakes. Aluminum is leached from the soil by acidic rain water; the aluminum compounds adversely affect the gills of fish. In addition to leaching aluminum from the soil acid rain also causes other valuable minerals, such as magnesium and calcium, to dissolve and leave the soil. It can also dissolve the waxy protective coat on plant leaves making them vulnerable to attack by bacteria and fungi.

In our cities, acid rain is responsible for extensive and continuing damage to buildings, monuments, and statues. It may also reduce the durability of paint and promote the deterioration of paper, leather, and cloth. In short we are just beginning to explore the effects of acid rain on human beings and on our food chain.

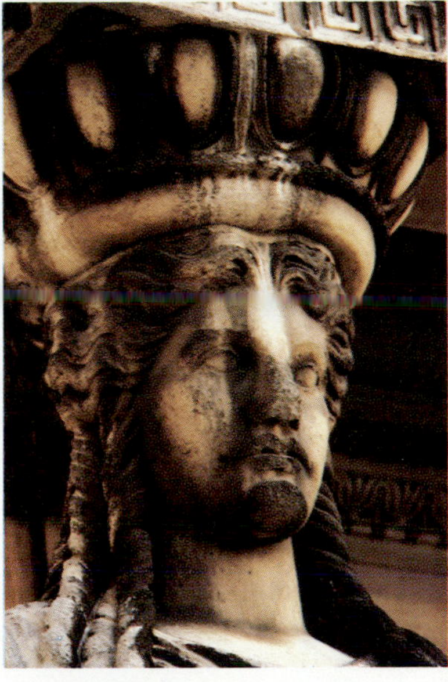

▲ **Marble masterpieces, sculpted to last forever, are slowly disappearing as acid rain dissolves the calcium carbonate ($CaCO_3$).**

16.12 Writing Ionic Equations

In Section 16.10 we wrote the reaction of hydrochloric acid and sodium hydroxide in three different equations. Equation (5) was the un-ionized equation; equation (6) was the total ionic equation; and equation (7) was the net ionic equation. In the **un-ionized equation**, compounds are written in their molecular, or normal, formula expressions. In the **total ionic equation**, compounds are written to show the form in which they are predominantly present: strong electrolytes as ions in solution; and nonelectrolytes, weak electrolytes, precipitates, and gases in their molecular (or un-ionized) forms. In the **net ionic equation**, only those molecules or ions that have changed are included in the equation; ions or molecules that do not change (spectators) are omitted.

un-ionized equation

total ionic equation

net ionic equation

Up to this point, when balancing an equation we have been concerned only with the atoms of the individual elements. Because ions are electrically charged, ionic equations often end up with a net electrical charge. A balanced equation must have the same net charge on each side, whether that charge is positive, negative, or zero. Therefore, when balancing an ionic equation, we must make sure that both the same number of each kind of atom and the same net electrical charge are present on each side.

Following is a list of rules for writing ionic equations:

1. Strong electrolytes in solution are written in their ionic form.
2. Weak electrolytes are written in their molecular (un-ionized) from.
3. Nonelectrolytes are written in their molecular form.
4. Insoluble substances, precipitates, and gases are written in their molecular forms.
5. The net ionic equation should include only those substances that have undergone a chemical change. Spectator ions are omitted from the net ionic equation.
6. Equations must be balanced, both in atoms and in electrical charge.

Study the examples below. Note that all reactions are in solution.

EXAMPLE 16.7

$HNO_3(aq) + KOH(aq) \longrightarrow KNO_3(aq) + H_2O(l)$ Un-ionized equation

$(H^+ + NO_3^-) + (K^+ + OH^-) \longrightarrow (K^+ + NO_3^-) + H_2O$ Total ionic equation

$H^+ + OH^- \longrightarrow H_2O$ Net ionic equation

HNO_3, KOH, and KNO_3 are soluble, strong electrolytes. K^+ and NO_3^- are spectator ions, have not changed, and are not included in the net ionic equation. Water is a nonelectrolyte and is written in the molecular form.

EXAMPLE 16.8

$2\ AgNO_3(aq) + BaCl_2(aq) \longrightarrow 2\ AgCl\downarrow + Ba(NO_3)_2(aq)$

$(2\ Ag^+ + 2\ NO_3^-) + (Ba^{2+} + 2\ Cl^-) \longrightarrow 2\ AgCl\downarrow + (Ba^{2+} + 2\ NO_3^-)$

$Ag^+ + Cl^- \longrightarrow AgCl\downarrow$ Net ionic equation

Although silver chloride (AgCl) is an ionic salt, it is written in the un-ionized form on the right side of the ionic equations because most of the Ag^+ and Cl^- ions are no longer in solution but have formed a precipitate of AgCl. Ba^{2+} and NO_3^- are spectator ions.

EXAMPLE 16.9

$Na_2CO_3(aq) + H_2SO_4(aq) \longrightarrow Na_2SO_4(aq) + H_2O(l) + CO_2\uparrow$

$(2\ Na^+ + CO_3^{2-}) + (2\ H^+ + SO_4^{2-}) \longrightarrow (2\ Na^+ + SO_4^{2-}) + H_2O + CO_2\uparrow$

$CO_3^{2-} + 2\ H^+ \longrightarrow H_2O + CO_2\uparrow$ Net ionic equation

Carbon dioxide (CO_2) is a gas and evolves from the solution; Na^+ and SO_4^{2-} are spectator ions.

EXAMPLE 16.10

$HC_2H_3O_2(aq) + NaOH(aq) \longrightarrow NaC_2H_3O_2(aq) + H_2O(l)$

$HC_2H_3O_2 + (Na^+ + OH^-) \longrightarrow (Na^+ + C_2H_3O_2^-) + H_2O$

$HC_2H_3O_2 + OH^- \longrightarrow C_2H_3O_2^- + H_2O$ Net ionic equation

Acetic acid ($HC_2H_3O_2$), a weak acid, is written in the molecular form, but sodium acetate ($NaC_2H_3O_2$), a soluble salt, is written in the ionic form. The Na^+ ion is the only spectator ion in this reaction. Both sides of the net ionic equation have a -1 electrical charge.

Example 16.11

$$Mg(s) + 2\,HCl(aq) \longrightarrow MgCl_2(aq) + H_2\uparrow$$
$$Mg + (2\,H^+ + 2\,Cl^-) \longrightarrow (Mg^{2+} + 2\,Cl^-) + H_2\uparrow$$
$$Mg + 2\,H^+ \longrightarrow Mg^{2+} + H_2\uparrow \qquad \text{Net ionic equation}$$

The net electrical charge on both sides of the equation is $+2$.

Example 16.12

$$H_2SO_4(aq) + Ba(OH)_2(aq) \longrightarrow BaSO_4\downarrow + 2\,H_2O(l)$$
$$(2\,H^+ + SO_4^{2-}) + (Ba^{2+} + 2\,OH^-) \longrightarrow BaSO_4\downarrow + 2\,H_2O(l)$$
$$2\,H^+ + SO_4^{2-} + Ba^{2+} + 2\,OH^- \longrightarrow BaSO_4\downarrow + 2\,H_2O(l) \qquad \text{Net ionic equation}$$

Barium sulfate ($BaSO_4$) is a highly insoluble salt. If we conduct this reaction using the conductivity apparatus described in Section 16.5, the light glows brightly at first but goes out when the reaction is complete, because almost no ions are left in solution. The $BaSO_4$ precipitates out of solution, and water is a nonconductor of electricity.

16.13 Colloids: An Introduction

When we add sugar to a flask of water and shake it, the sugar dissolves and forms a clear homogeneous *solution*. When we do the same experiment with very fine sand and water, the sand particles form a *suspension*, which settles when the shaking stops. When we repeat the experiment again using ordinary corn starch, we find that the starch does not dissolve in cold water; but if the mixture is heated and stirred, the starch forms a cloudy, opalescent *dispersion*. This dispersion does not appear to be clear and homogeneous like the sugar solution. Yet it is not obviously heterogeneous and does not settle like the sand suspension. In short, its properties are intermediate between those of the sugar solution and those of the sand suspension. The starch dispersion is actually a colloid. The name *colloid* is derived from the Greek *kolla*, meaning *glue*, and was coined by the English scientist Thomas Graham in 1861. Graham classified solutes as crystalloids if they diffused through a parchment membrane and as colloids if they did not diffuse through the membrane.

As it is now used, the word **colloid** means a dispersion in which the dispersed particles are larger than the solute ions or molecules of a true solution and smaller than the particles of a mechanical suspension. The term does not imply a gluelike quality, although most glues are colloidal materials. The size of colloidal particles ranges from a lower limit of about 1 nm (10^{-7} cm) to an upper limit of about 1000 nm (10^{-4} cm). **colloid**

The fundamental difference between a colloidal dispersion and a true solution is the size, not the nature, of the particles. The solute particles in a solution are usually single ions or molecules that may be hydrated to varying degrees.

TABLE 16.6 Types of Colloidal Dispersions

Type	Name	Examples
Gas in liquid	Foam	Whipped cream, soap suds
Gas in solid	Solid foam	Styrofoam, foam rubber, pumice
Liquid in gas	Liquid aerosol	Fog, clouds
Liquid in liquid	Emulsion	Milk, vinegar in oil salad dressing, mayonnaise
Liquid in solid	Solid emulsion	Cheese, opals, jellies
Solid in gas	Solid aerosol	Smoke, dust in air
Solid in liquid	Sol	India ink, gold sol
Solid in solid	Solid sol	Tire rubber, certain gems (for example, rubies)

Colloidal particles are usually aggregations of ions or molecules. However, the molecules of some polymers such as proteins are large enough to be classified as colloidal particles when in solution. To appreciate fully the differences in relative sizes, the volumes, not just the linear dimensions, of colloidal particles and solute particles must be compared. The difference in volumes can be approximated by assuming that the particles are spheres. A large colloidal particle has a diameter of about 500 nm, whereas a fair-sized ion or molecule has a diameter of about 0.5 nm. Thus, the diameter of the colloidal particle is about 1000 times that of the solute particle. Because the volumes of spheres are proportional to the cubes of their diameters, we can calculate that the volume of a colloidal particle can be up to a billion ($10^3 \times 10^3 \times 10^3 = 10^9$) times greater than that of a solution particle.

Colloids are mixtures in which one component, the *dispersed phase*, exists as discrete particles in the other component, the *dispersing phase* or *dispersing medium*. The components of a colloidal dispersion are also sometimes called the *discontinuous phase* and the *continuous phase*. Each component, or phase, can exist as a solid, a liquid, or a gas. The components cannot be mutually soluble, nor can both phases of the dispersion be gases, because such conditions would result in an ordinary solution. Hence only eight types of colloidal dispersions, based on the possible physical states of the phases, are known. The eight types, with specific examples, are listed in Table 16.6.

16.14 Preparation of Colloids

Colloidal dispersions can be prepared by two methods: (1) *dispersion*, the breaking down of larger particles to colloidal size, and (2) *condensation*, the formation of colloidal particles from solutions.

Homogenized milk is a good example of a colloid prepared by dispersion. Milk, as drawn from the cow, is an unstable emulsion of fat in water. The fat globules are so large that they rise and form a cream layer in a few hours. To avoid separation of the cream, the milk is homogenized by pumping it through very small holes, or orifices, at high pressure. The violent shearing action of this treatment breaks the fat globules into particles well within the colloidal size range. The butterfat in homogenized milk remains dispersed indefinitely. Colloid mills, which reduce particles to colloidal size by grinding or shearing, are used in preparing many commercial products such as paints, cosmetics, and salad dressings.

The preparation of colloids by condensation frequently involves a precipitation reaction in a dilute solution. For example, a good colloidal sulfur sol can be made by bubbling hydrogen sulfide into a solution of sulfur dioxide. Solid sulfur is formed and dispersed as a colloid:

$$SO_2 + 2\,H_2S \longrightarrow \underset{\text{Colloidal sulfur}}{3\,S} + 2\,H_2O$$

A colloidal dispersion is also easily made by adding iron(III) chloride solution to boiling water. The reddish-brown colloidal dispersion that is formed probably consists of iron(III) hydroxide and hydrated iron(III) oxide:

$$FeCl_3 + 3\,HOH \xrightarrow{H_2O} Fe(OH)_3 + 3\,HCl$$

$$Fe(OH)_3 \xrightarrow{H_2O} Fe_2O_3 \cdot x\,H_2O$$

A great many products for home use (insecticides, insect repellents, and deodorants, to name a few) are packaged as aerosols. The active ingredient, either a liquid or a solid, is dissolved in a liquefied gas and sealed under pressure in a container fitted with a release valve. When this valve is opened, the pressurized solution is ejected. The liquefied gas vaporizes, and the active ingredient is converted to a colloidal aerosol almost instantaneously.

16.15 Properties of Colloids

In 1827 Robert Brown (1773–1858), while observing a strongly illuminated aqueous suspension of pollen under a high-powered microscope, noted that the pollen grains appeared to have a trembling, erratic motion. He determined later that this erratic motion was not confined to pollen but was characteristic of colloidal particles in general. The random motion of colloidal particles, first reported by Brown, is called **Brownian movement**. It is readily observed in cigarette smoke. The smoke is confined in a small transparent chamber and is illuminated with a strong beam of light at right angles to the optical axis of the microscope. The smoke particles appear as tiny randomly moving lights, because the light is reflected from their surfaces. This motion is due to the continual bombardment of the smoke particles by air molecules. Since Brownian movement can be seen when colloidal particles are dispersed in either a gaseous or a liquid medium, it affords nearly direct visual proof that matter at the molecular level actually is moving randomly, as postulated by the Kinetic-Molecular Theory.

Brownian movement

When an intense beam of light is passed through an ordinary solution and is viewed at an angle, the beam passing through the solution is hardly visible. A

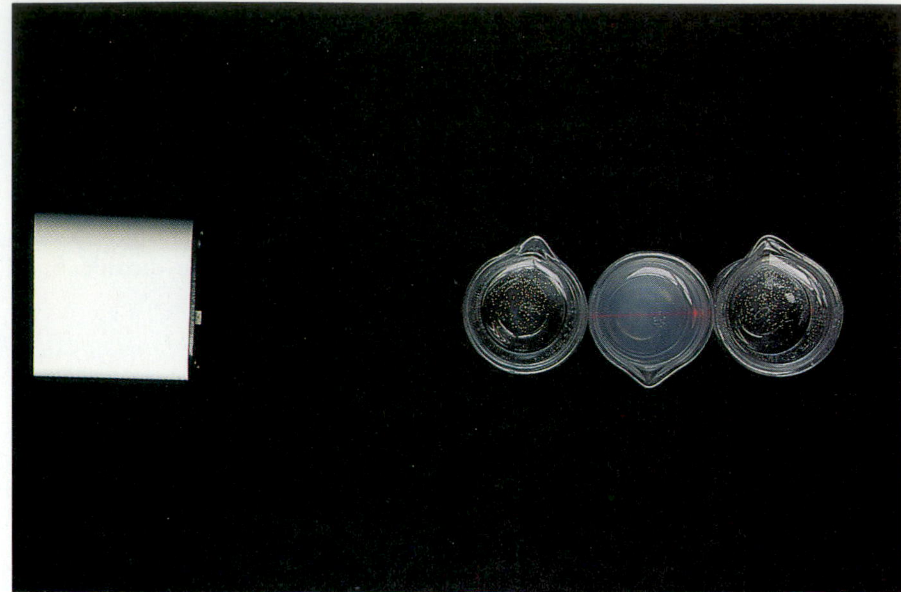

FIGURE 16.7 ▶ **The Tyndall effect. The beakers on the left and right each contain a true solution, whereas the beaker in the center contains a collodial starch solution. A laser beam, emitted by the laser on the far left, scatters in the colloid but is invisible in the true solution.**

beam of light, however, is clearly visible and sharply outlined when it is passed through a colloidal dispersion (see Figure 16.7). This phenomenon is known as the **Tyndall effect**. It was first described by Michael Faraday in 1857 and later amplified by John Tyndall. The Tyndall effect, like the Brownian movement, can be observed in nearly all colloidal dispersions. It occurs because the colloidal particles are large enough to scatter the rays of visible light. The ions or molecules of true solutions are too small to scatter light and, therefore, do not exhibit a noticeable Tyndall effect.

Tyndall effect

Another important characteristic of colloids is that the particles have relatively huge surface areas. We saw in Section 15.6 that the surface area is increased tenfold when a 1 cm cube is divided into 1000 cubes with sides of 0.1 cm. When a 1 cm cube is divided into colloidal-size cubes measuring 10^{-6} cm, the combined surface area of all the particles becomes a million times greater than that of the original cube.

Colloidal particles become electrically charged when they adsorb ions on their surfaces. This property is directly related to the large surface area presented by the many tiny particles. Adsorption occurs because the atoms or ions at the surface of a particle are not completely surrounded by other atoms or ions as are those in the interior. Consequently these surface atoms or ions attract and adsorb ions or polar molecules from the dispersion medium onto the surfaces of the colloidal particles. The particles in a given dispersion tend to adsorb only ions having one kind of charge. For example, cations are primarily adsorbed on iron(III) hydroxide sol, resulting in positively charged colloidal particles. On the other hand, the particles of an arsenic(III) sulfide (As_2S_3) sol primarily adsorb anions, resulting in negatively charged colloidal particles. The properties of true solutions, colloidal dispersions, and mechanical suspensions are summarized and compared in Table 16.7.

Adsorption should not be confused with *absorption*. Adsorption refers to the adhesion of molecules or ions to a surface, whereas absorption refers to the taking in of one material by another material.

TABLE 16.7 Comparison of the Properties of True Solutions, Colloidal Dispersions, and Suspensions

	Particle size (nm)	Ability to pass through filter paper	Ability to pass through parchment	Exhibits Tyndall effect	Exhibits Brownian movement	Settles out on standing	Appearance
True solution	<1	Yes	Yes	No	No	No	Transparent, homogeneous
Colloidal dispersion	1–1000	Yes	No	Yes	Yes	Generally does not	Usually not transparent, but may appear to be homogeneous
Suspension	>1000	No	No	—	No	Yes	Not transparent, heterogeneous

16.16 STABILITY OF COLLOIDS

The stability of different dispersions varies with the properties of the dispersed and dispersing phases. We have noted that nonhomogenized milk is a colloid, yet it will separate after standing for a few hours. However, the particles of a good colloidal dispersion will remain in suspension indefinitely. As a case in point, a ruby-red gold sol has been kept in the British Museum for more than a century without noticeable settling. (This specimen is kept for historical interest; it was prepared by Michael Faraday.) The particles of a specific colloid remain dispersed for two reasons: (1) They are bombarded by the molecules of the dispersing phase, which keeps the particles in motion (Brownian movement) so that gravity does not cause them to settle out; (2) since the colloidal particles have the same kind of electrical charge, they repel each other. This mutual repulsion prevents the dispersed particles from coalescing to larger particles, which would settle out of suspension.

In certain types of colloids, the presence of a material known as a protective colloid is necessary for stability. Egg yolk, for example, acts as a stabilizer, or protective colloid, in mayonnaise. The yolk adsorbs on the surfaces of the oil particles and prevents them from coalescing.

16.17 APPLICATIONS OF COLLOIDAL PROPERTIES

Activated charcoal has an enormous surface area, approximately 1 million square centimeters per gram in some samples. Hence, charcoal is very effective in selectively adsorbing the polar molecules of some poisonous gases and is used

Chemistry In Action

Hair Care and pH—A Delicate Balance

Hair shampoo advertisements often proclaim the proper pH for their products, but does controlling the pH of hair care products really make hair cleaner, shiny, or stronger?

Each strand of hair is composed of many long chains of amino acid linked together as polymers called proteins. The individual chains can connect with other chains in one of three ways: (1) hydrogen bonds,

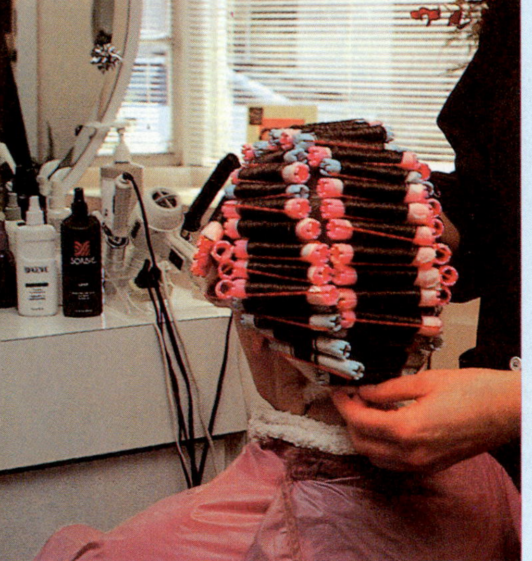

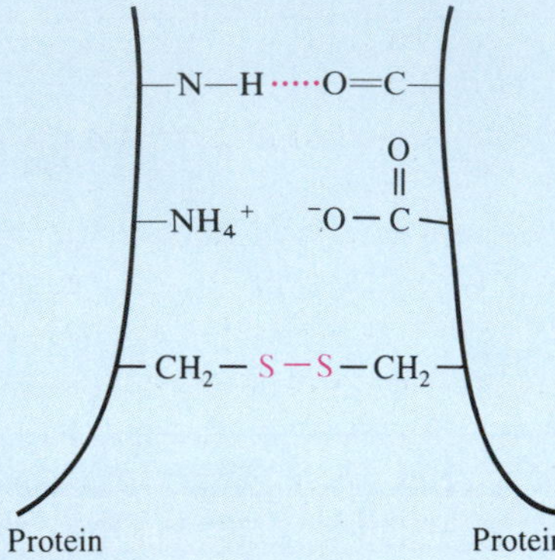

Interactions between strands of hair protein

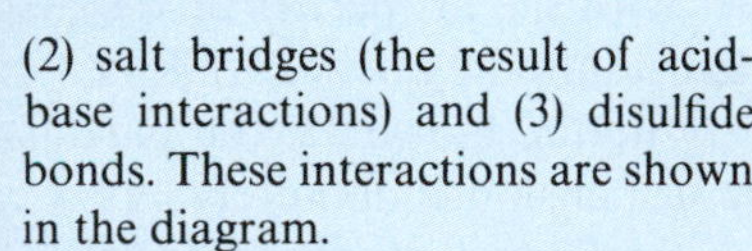

(2) salt bridges (the result of acid-base interactions) and (3) disulfide bonds. These interactions are shown in the diagram.

When hair is wet with water, the hydrogen bonds are broken. As the wet hair is shaped, set, or dried, the hydrogen bonds form at new positions and hold the hair in the style desired. If an acidic solution (pH 1.0–2.0) is used on the hair, the hydrogen bonds and the salt bridges are both broken, leaving only the disulfide bonds to hold the chains together. In a mildly alkaline solution (pH 8.5) some of the disulfide bonds are also broken. The outer surface of the hair becomes rough and light does not reflect evenly from the surface, making the hair look dull. Using an alkaline shampoo will cause damage by continued breakage of the disulfide bonds resulting in a condition called "split ends." If the pH is increased further to approximately 12.0 the hair dissolves as all types of bonds break. This is the working basis for depilatories (hair removers), such as Neet(TM) and Nair(TM).

Hair has its maximum strength at pH 4.0–5.0. Shampooing tends to leave the hair slightly alkaline so an acid rinse is sometimes used to bring the pH back into the normal range. Lemon juice or vinegar are common household products that are used for this purpose. The shampoo may also be "acid-balanced," containing a weak acid (such as citric acid) to counteract the alkalinity of the solution formed when the detergent interacts with water.

in gas masks. Charcoal can be used to adsorb impurities from liquids as well as from gases, and large amounts are used to remove substances that have objectionable tastes and odors from water supplies. In sugar refineries activated charcoal is used to adsorb colored impurities from the raw sugar solutions.

The Cottrell process is widely used for dust and smoke control in many urban and industrial areas. This process, devised by an American, Frederick Cottrell (1877–1948), takes advantage of the fact that the particulate matter in dust and smoke is electrically charged. Air to be cleaned of dust or smoke is passed between electrode plates charged with a high voltage. Positively charged particles are attracted to, neutralized, and thereby precipitated at the negative electrodes. Negatively charged particles are removed in the same fashion at the positive electrodes. Large Cottrell units are fitted with devices for automatic removal of precipitated material. In some installations, particularly at cement mills and smelters, the value of the dust collected may be sufficient to pay for the precipitation equipment. Small units, designed for removing dust and pollen from air in the home, are now on the market. Unfortunately, Cottrell units remove only particulate matter; they cannot remove gaseous pollutants such as carbon monoxide, sulfur dioxide, and nitrogen oxides.

Thomas Graham found that a parchment membrane would allow the passage of true solutions but would prevent the passage of colloidal dispersions. Dissolved solutes can be removed from colloidal dispersions through the use of such a membrane by a process called **dialysis**. The membrane itself is called a *dialyzing membrane*. Many animal membranes can act as dialyzing membranes. Artificial membranes are made from such materials as parchment paper, collodion, or certain kinds of cellophane. Dialysis can be demonstrated by putting a colloidal starch dispersion and some copper(II) sulfate solution in a parchment paper bag and suspending it in running water. In a few hours the blue color of the copper(II) sulfate has disappeared, and only the starch dispersion remains in the bag.

dialysis

An interesting application of dialysis has been the development of artificial kidneys. These devices are dialyzing units that are able to act as kidneys by removing soluble waste products from the blood. The blood of a patient suffering from partial kidney failure is passed through the artificial kidney machine for several hours. During passage through the machine, the soluble waste products are removed by dialysis.

Concepts in Review

1. State the general characteristics of acids and bases.
2. Define an acid and base in terms of the Arrhenius, Brønsted-Lowry, and Lewis theories.
3. Identify acid-base conjugate pairs in a reaction.
4. When given the reactants, complete and balance equations for the reactions of acids with bases, metals, metal oxides, and carbonates.
5. When given the reactants, complete and balance equations of the reaction of an amphoteric hydroxide with either a strong acid or a strong base.
6. Write balanced equations for the reaction of sodium hydroxide or potassium hydroxide with zinc and with aluminum.
7. Classify common compounds as electrolytes or nonelectrolytes.
8. Distinguish between strong and weak electrolytes.
9. Understand the process of dissociation and ionization.
10. Write equations for the dissociation of acids, bases, and salts in water.
11. Describe and write equations for the ionization of water.
12. Understand pH as an expression of hydrogen ion concentration or hydronium ion concentration.
13. Given pH as an integer, indicate the H^+ molarity, and vice versa.
14. Use a calculator to estimate pH values from corresponding H^+ molarities.
15. Understand the process of acid-base neutralization.

16. Calculate the molarity, normality, or volume of an acid or base solution from appropriate titration data.
17. Write un-ionized, total ionic, and net ionic equations for neutralization equations.
18. Discuss colloids and describe methods for their preparation.
19. Describe the characteristics that distinguish true solutions, colloidal dispersions, and mechanical suspensions.
20. Explain (a) how colloidal dispersion can be stabilized and (b) how they can be precipitated.

Exercises

An asterisk indicates a more challenging question or problem.

1. Since a hydrogen ion and a proton are identical, what differences exist between the Arrhenius and Brønsted–Lowry definitions of an acid? (See Table 16.1.)
2. According to Figure 16.1, what type of substance must be in solution in order for the bulb to light?
3. Which of the following classes of compounds are electrolytes: acids, alcohols, bases, salts? (See Table 16.2.)
4. What two differences are apparent in the arrangement of water molecules about the hydrated ions as depicted in Figure 16.2?
5. The pH of a solution with a hydrogen ion concentration of 0.003 *M* is between what two whole numbers? (See Table 16.4.)
6. Which is the more acidic, tomato juice or blood? (See Table 16.5.)
7. Using each of the three acid–base theories (Arrhenius, Brønsted–Lowry, and Lewis), define an acid and a base.
8. For each of the acid–base theories referred to in Exercise 7, write an equation illustrating the neutralization of an acid with a base.
9. Identify the conjugate acid–base pairs in the following equations:
   (a) $HCl + NH_3 \longrightarrow NH_4^+ + Cl^-$
   (b) $HCO_3^- + OH^- \rightleftharpoons CO_3^{2-} + H_2O$
   (c) $HCO_3^- + H_3O^+ \rightleftharpoons H_2CO_3 + H_2O$
   (d) $HC_2H_3O_2 + H_2O \rightleftharpoons H_3O^+ + C_2H_3O_2^-$
   (e) $HC_2H_3O_2 + H_2SO_4 \rightleftharpoons H_2C_2H_3O_2^+ + HSO_4^-$
   (f) The two-step ionization of sulfuric acid,
   $H_2SO_4 + H_2O \longrightarrow H_3O^+ + HSO_4^-$
   $HSO_4^- + H_2O \rightleftharpoons H_3O^+ + SO_4^{2-}$
   (g) $HClO_4 + H_2O \longrightarrow H_3O^+ + ClO_4^-$
   (h) $CH_3O^- + H_3O^+ \longrightarrow CH_3OH + H_2O$
10. Write the Lewis structure for (a) bromide ion, (b) hydroxide ion, and (c) cyanide ion. Why are these ions considered to be bases according to the Brønsted–Lowry and Lewis acid–base theories?
11. Complete and balance the following equations:
   (a) $Mg(s) + HCl(aq) \longrightarrow$
   (b) $BaO(s) + HBr(aq) \longrightarrow$
   (c) $Al(s) + H_2SO_4(aq) \longrightarrow$
   (d) $Na_2CO_3(aq) + HCl(aq) \longrightarrow$
   (e) $Fe_2O_3(s) + HBr(aq) \longrightarrow$
   (f) $Ca(OH)_2(aq) + H_2CO_3(aq) \longrightarrow$
   (g) $NaOH(aq) + HBr(aq) \longrightarrow$
   (h) $KOH(aq) + HCl(aq) \longrightarrow$
   (i) $Ca(OH)_2(aq) + HI(aq) \longrightarrow$
   (j) $Al(OH)_3(s) + HBr(aq) \longrightarrow$
   (k) $Na_2O(s) + HClO_4(aq) \longrightarrow$
   (l) $LiOH(aq) + FeCl_3(aq) \longrightarrow$
   (m) $NH_4OH(aq) + FeCl_2(aq) \longrightarrow$
12. Into what three classes of compounds do electrolytes generally fall?
13. Which of the following compounds are electrolytes? Consider each substance to be mixed with water.
   (a) HCl
   (b) CO_2
   (c) $CaCl_2$
   (d) $C_{12}H_{22}O_{11}$ (sugar)
   (e) C_3H_7OH (rubbing alcohol)
   (f) CCl_4 (insoluble)
   (g) $NaHCO_3$ (baking soda)
   (h) N_2 (insoluble gas)
   (i) $AgNO_3$
   (j) $HCOOH$ (formic acid)
   (k) $RbOH$
   (l) K_2CrO_4
14. Name each compound listed in Table 16.3.

15. A solution of HCl in water conducts an electric current, but a solution of HCl in benzene does not. Explain this behavior in terms of ionization and chemical bonding.

16. How do salts exist in their crystalline structure? What occurs when they are dissolved in water?

17. An aqueous methyl alcohol, CH_3OH, solution does not conduct an electric current, but a solution of sodium hydroxide, NaOH, does. What does this information tell us about the OH group in the alcohol?

18. Why does molten sodium chloride conduct electricity?

19. Explain the difference between dissociation of ionic compounds and ionization of molecular compounds.

20. Distinguish between strong and weak electrolytes.

21. Explain why ions are hydrated in aqueous solutions.

22. Indicate, by simple equations, how the following substances dissociate or ionize in water:

(a) $Cu(NO_3)_2$	(e) NH_4Br
(b) $HC_2H_3O_2$	(f) K_2SO_4
(c) HNO_2	(g) $NaClO_3$
(d) LiOH	(h) K_3PO_4

23. What is the main distinction between water solutions of strong and weak electrolytes?

24. What are the relative concentrations of $H^+(aq)$ and $OH^-(aq)$ in (a) a neutral solution, (b) an acid solution, and (c) a basic solution?

25. Write the net ionic equation for the reaction of an acid with a base in an aqueous solution.

26. The solubility of hydrogen chloride gas in water, a polar solvent, is much greater than its solubility in benzene, a nonpolar solvent. How can you account for this difference?

27. Pure water, containing both acid and base ions, is neutral. Why?

28. Which of the following statements are correct? Rewrite each incorrect statement to make it correct.

(a) The Arrhenius theory of acids and bases is restricted to aqueous solutions.

(b) The Brønsted–Lowry theory of acids and bases is restricted to solutions other than aqueous solutions.

(c) All substances that are acids according to the Brønsted–Lowry theory will also be acids by the Lewis theory.

(d) All substances that are acids according to the Lewis theory will also be acids by the Brønsted–Lowry theory.

(e) An electron-pair donor is a Lewis acid.

(f) All Arrhenius acid–base neutralization reactions can be represented by a single net ionic equation.

(g) When an ionic compound dissolves in water, the ions separate; this process is called ionization.

(h) In the autoionization of water

$$2\,H_2O \rightleftharpoons H_3O^+ + OH^-$$

the H_3O^+ and the OH^- constitute a conjugate acid–base pair.

(i) In the reaction in part (h), H_2O is both the acid and the base.

(j) Most common Na^+, K^+, and NH_4^+ salts are soluble in water.

(k) A solution of pH 3 is 100 times more acidic than a solution of pH 5.

(l) In general, ionic substances when placed in water will give a solution capable of conducting an electric current.

(m) The terms *dissociation* and *ionization* are synonymous.

(n) A solution of $Mg(NO_3)_2$ will produce three ions per formula unit in solution.

(o) The terms *strong acid*, *strong base*, *weak acid*, and *weak base* refer to whether an acid or base solution is concentrated or dilute.

(p) pH is defined as the negative logarithm of the molar concentration of H^+ ions (or H_3O^+ ions).

(q) All reactions may be represented by net ionic equations.

(r) One mole of $CaCl_2$ contains more anions than cations.

(s) It is possible to boil seawater at a lower temperature than that required to boil pure water (both at the same pressure).

(t) It is possible to have a neutral aqueous solution whose pH is not 7.

(u) The size of colloidal particles ranges from 1 mm to 1000 mm.

(v) The Tyndall effect is observable in both colloidal and true solutions.

29. Indicate the fundamental difference between a colloidal dispersion and a true solution.

30. List the methods used to prepare colloidal dispersions and briefly explain how each is accomplished.

31. Explain the Tyndall effect and how it may be used to distinguish between a colloidal dispersion and a true solution.

32. Distinguish between adsorption and absorption.

33. Why do particles in a specific colloid remain dispersed?

34. Explain the process of dialysis, giving a practical application in society.

35. Diagram and label the types of interactions occurring between strands of protein in the hair.

36. What is the difference between a shampoo and a depilatory? Include a discussion of the types of bonds broken in your answer.

37. Why should the pH of hair be maintained between 4 and 5? How can this be accomplished?

Review problems

38. Calculate the molarity of the ions present in each of the following salt solutions. Assume each salt to be 100% dissociated.

(a) 0.015 *M* NaCl
(b) 4.25 *M* $NaKSO_4$
(c) 0.75 *M* $ZnBr_2$
(d) 1.65 *M* $Al_2(SO_4)_3$
(e) 0.20 *M* $CaCl_2$
(f) 22.0 g KI in 500 mL of solution
(g) 900 g $(NH_4)_2SO_4$ in 20.0 L of solution
(h) 0.0120 g $Mg(ClO_3)_2$ in 1.00 mL of solution

39. In Exercise 29, how many grams of each ion would be present in 100. mL of each solution?

40. What is the concentration of Ca^{2+} ions in a solution of CaI_2 having an I^- ion concentration of 0.520 *M*?

41. What is the molar concentration of all ions present in a solution prepared by mixing the following? Neglect the concentration of H^+ and OH^- from water. Also, assume volumes of solutions are additive.

(a) 30.0 mL of 1.0 *M* NaCl and 40.0 mL of 1.0 *M* NaCl
(b) 30.0 mL of 1.0 *M* HCl and 30.0 mL of 1.0 *M* NaOH
(c) 100.0 mL of 2.0 *M* KCl and 100.0 mL of 1.0 *M* $CaCl_2$
*(d) 100.0 mL of 0.40 *M* KOH and 100.0 mL of 0.80 *M* HCl
(e) 35.0 mL of 0.20 *M* $Ba(OH)_2$ and 35.0 mL of 0.20 *M* H_2SO_4
(f) 1.00 L of 1.0 *M* $AgNO_3$ and 500 mL of 2.0 *M* NaCl.

42. How many milliliters of 0.40 *M* HCl can be made by diluting 100. mL of 12 *M* HCl with water?

43. Given the data for the following six titrations, calculate the molarity of the HCl in titrations (a), (b), and (c), and the molarity of the NaOH in titrations (d), (e), and (f).

	mL HCl	Molarity HCl	mL NaOH	Molarity NaOH
(a)	40.13	*M* HCl	37.70	0.728
(b)	19.00	*M* HCl	33.66	0.306
(c)	27.25	*M* HCl	18.00	0.555
(d)	37.19	0.126	31.91	*M* NaOH
(e)	48.04	0.482	24.02	*M* NaOH
(f)	13.13	1.425	39.39	*M* NaOH

44. If 29.26 mL of 0.430 *M* HCl neutralizes 20.40 mL of $Ba(OH)_2$ solution, what is the molarity of the $Ba(OH)_2$ solution? The reaction is

$$Ba(OH)_2(aq) + 2\ HCl(aq) \longrightarrow BaCl_2(aq) + 2\ H_2O$$

45. A 1 molal solution of acetic acid in water freezes at a lower temperature than a 1 molal solution of ethyl alcohol, C_2H_5OH, in water. Explain.

46. At the same cost per pound, which alcohol, CH_3OH or C_2H_5OH, would be more economical to purchase as an antifreeze for your car?

47. How does a hydronium ion differ from a hydrogen ion?

48. Arrange, in decreasing order of freezing points, 1 molal aqueous solutions of HCl, $HC_2H_3O_2$, $C_{12}H_{22}O_{11}$ (sucrose), and $CaCl_2$. (List the one with the highest freezing point first.)

49. At 100°C the H^+ concentration in water is about 1×10^{-6} mol/L, about 10 times that of water at 25°C. At which of these temperatures is (a) the pH of water the greater, (b) the hydrogen ion (hydronium ion) concentration the higher, and (c) the water neutral?

50. What is the relative difference in H^+ concentration in solutions that differ by 1 pH unit?

51. Rewrite the following unbalanced equations, changing them into balanced net ionic equations. All reactions are in water solution.

(a) $K_2SO_4(aq) + Ba(NO_3)_2(aq) \longrightarrow KNO_3(aq) + BaSO_4(s)$
(b) $CaCO_3(s) + HCl(aq) \longrightarrow CaCl_2(aq) + CO_2(g) + H_2O(l)$
(c) $Mg(s) + HC_2H_3O_2(aq) \longrightarrow Mg(C_2H_3O_2)_2(aq) + H_2(g)$
(d) $H_2S(g) + CdCl_2(aq) \longrightarrow CdS(s) + HCl(aq)$
(e) $Zn(s) + H_2SO_4(aq) \longrightarrow ZnSO_4(aq) + H_2(g)$
(f) $AlCl_3(aq) + Na_3PO_4(aq) \longrightarrow AlPO_4(s) + NaCl(aq)$

52. In each of the following pairs which solution is more acidic? (All are water solutions.)
    (a) 1 molar HCl or 1 molar H_2SO_4?
    (b) 1 molar HCl or 1 molar $HC_2H_3O_2$?
    (c) 1 molar HCl or 2 molar HCl?
    (d) 1 normal H_2SO_4 or 1 molar H_2SO_4?

53. What volume (in milliliters) of 0.245 *M* HCl will neutralize (a) 50.0 mL of 0.100 *M* $Ca(OH)_2$ and (b) 10.0 g of $Al(OH)_3$? The equations are
    (a) $2\ HCl(aq) + Ca(OH)_2(aq) \longrightarrow CaCl_2(aq) + 2\ H_2O(l)$
    (b) $3\ HCl(aq) + Al(OH)_3(s) \longrightarrow AlCl_3(aq) + 3\ H_2O(l)$

54. A sample of pure sodium carbonate with a mass of 0.452 g was dissolved in water and neutralized with 42.4 mL of hydrochloric acid. Calculate the molarity of the acid:

$$Na_2CO_3(aq) + 2\ HCl(aq) \longrightarrow 2\ NaCl(aq) + CO_2(g) + H_2O(l)$$

55. What volume (mL) of 0.1234 *M* HCl is needed to neutralize 2.00 g $Ca(OH)_2$?

56. How many grams of KOH are required to neutralize 50.00 mL of 0.240 *M* HNO_3?

*57. A 0.200 g sample of impure NaOH requires 18.25 mL of 0.2406 *M* HCl for neutralization. What is the mass percent of NaOH in the sample?

*58. A batch of sodium hydroxide was found to contain sodium chloride as an impurity. To determine the amount of impurity, a 1.00 g sample was analyzed and found to require 49.90 mL of 0.466 *M* HCl for neutralization. What is the percent of NaCl in the sample?

*59. What volume of H_2 gas, measured at 27°C and 700. torr pressure, can be obtained by reacting 5.00 g of zinc metal with (a) 100. mL of 0.350 *M* HCl and (b) 200. mL of 0.350 *M* HCl? The equation is

$$Zn(s) + 2\ HCl(aq) \longrightarrow ZnCl_2(aq) + H_2(g)$$

60. Calculate the pH of solutions having the following H^+ ion concentrations:
    (a) 0.01 *M*
    (b) 1.0 *M*
    (c) 6.5×10^{-9} *M*
    (d) 1×10^{-7} *M*
    (e) 0.50 *M*
    (f) 0.00010 *M*

61. Calculate the pH of the following:
    (a) Orange juice, 3.7×10^{-4} *M* H^+
    (b) Vinegar, 2.8×10^{-3} *M* H^+
    (c) Black coffee, 5.0×10^{-5} *M* H^+
    (d) Limewater, 3.4×10^{-11} *M* H^+

62. Two drops (0.1 mL) of 1.0 *M* HCl are added to water to make 1.0 L of solution. What is the pH of this solution if the HCl is 100% ionized?

63. What volume of concentrated (18.0 *M*) sulfuric acid must be used to prepare 50.0 L of 5.00 *M* solution?

64. Three (3.0) grams of NaOH are added to 500 mL of 0.10 *M* HCl. Will the resulting solution be acidic or basic? Show evidence for your answer.

65. A 10.00 mL sample of base solution requires 28.92 mL of 0.1240 *N* H_2SO_4 for neutralization. What is the normality of the base?

66. How many milliliters of 0.325 *N* HNO_3 are required to neutralize 32.8 mL of 0.225 *N* NaOH?

67. How many milliliters of 0.325 *N* H_2SO_4 are required to neutralize 32.8 mL of 0.225 *N* NaOH?

68. What is the normality and the molarity of a 25.00 mL sample of H_3PO_4 solution that requires 22.68 mL of 0.5000 *N* NaOH for complete neutralization?

69. How many milliliters of 0.10 *N* NaOH is required to neutralize 60. mL of 0.20 N H_2SO_4?

*70. A 25 mL solution of H_2SO_4 requires 40. mL of 0.20 NaOH for neutralization.
    (a) What is the normality of the sulfuric acid solution?
    (b) How many grams of sulfuric acid are contained in the 25 mL?

*71. A solution of 40.0 mL of HCl is neutralized by 20.0 mL of NaOH solution. The resulting neutral solution is evaporated to dryness, and the residue is found to have a mass of 0.117 g. Calculate the normality of HCl and NaOH solutions.

*72. The equivalent masses of organic acids are often determined by titration with a standard solution of a base. Determine the equivalent mass of benzoic acid, if 0.305 g of it require 25 mL of 0.10 *N* NaOH for neutralization.

*73. Determine the equivalent mass of succinic acid, if 0.738 g is required to neutralize 125 mL of 0.10 *N* base.

17

Chemical Equilibrium

◀ **Chapter Opening Photo: A nautilus shell is an example of chemical equilibrium in action.**

Thus far, we have considered chemical change to proceed from reactants to products. Does that mean that reactions stop? No. A solute dissolves until the solution becomes saturated. Once a solid remains undissolved in a container, the system appears to be at rest. The human body is a marvelous chemical factory, yet from day to day it appears to be quite the same. For example, the blood remains at a constant pH, even though all sorts of chemical reactions are taking place. A terrarium can be watered and sealed for long periods of time with no ill effects. An antacid is advertised to absorb excess stomach acid and *not* change the pH of the stomach. In all of these cases, reactions are proceeding, even though visible signs of chemical change are absent. When the system is at equilibrium, chemical reactions are dynamic. As the Frenchman Alphonse Karr so succinctly stated in 1849, "The more things change the more they stay the same."

17.1 Reversible Reactions

In the preceding chapters we have treated chemical reactions mainly as reactants changing to products. However, many reactions do not go to completion. Some reactions do not go to completion because they are reversible; that is, when the products are formed, they react to produce the starting reactants.

We have encountered reversible systems before. One is the vaporization of a liquid by heating and its subsequent condensation by cooling:

$$\text{liquid} + \text{heat} \longrightarrow \text{vapor}$$
$$\text{vapor} + \text{cooling} \longrightarrow \text{liquid}$$

The interconversion of nitrogen dioxide (NO_2) and dinitrogen tetroxide (N_2O_4) offers visible evidence of the reversibility of a reaction. NO_2 is a reddish-brown gas that changes, with cooling, to N_2O_4, a yellow liquid that boils at 21.2°C, and then to a colorless solid (N_2O_4) that melts at −11.2°C. The reaction is reversible by heating N_2O_4.

$$2\ NO_2(g) \xrightarrow{\text{Cooling}} N_2O_4(l)$$
$$N_2O_4(l) \xrightarrow{\text{Heating}} 2\ NO_2(g)$$

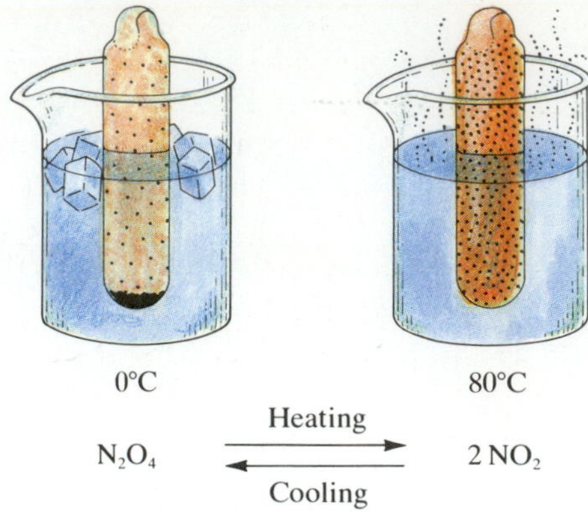

FIGURE 17.1 ▶ Reversible reaction of nitrogen dioxide (NO_2) and dinitrogen tetroxide (N_2O_4). More of the reddish-brown NO_2 molecules are visible in the tube that is heated than in the tube that is cooled.

These two reactions may be represented by a single equation with a double arrow, $\rightleftarrows$, to indicate that the reactions are taking place in both directions at the same time.

$$2\ NO_2(g) \rightleftharpoons N_2O_4(l)$$

This reversible reaction can be demonstrated by sealing samples of NO_2 in two tubes and placing one tube in warm water and the other in ice water (see Figure 17.1). Heating promotes disorder or randomness in a system, so we would expect more NO_2, a gas, to be present at higher temperatures.

reversible chemical reaction

A **reversible chemical reaction** is one in which the products formed react to produce the original reactants. Both the forward and reverse reactions occur simultaneously. The forward reaction is called *the reaction to the right*, and the reverse reaction is called *the reaction to the left*. A double arrow is used in the equation to indicate that the reaction is reversible.

17.2 RATES OF REACTION

chemical kinetics

Every reaction has a rate, or speed, at which it proceeds. Some are fast and some are extremely slow. The study of reaction rates and reaction mechanisms is known as **chemical kinetics**.

The rate of a reaction is variable and depends on the concentration of the reacting species, the temperature, the presence or absence of catalytic agents, and the nature of the reactants. Consider the hypothetical reaction

$$A + B \longrightarrow C + D \quad \text{(forward reaction)}$$
$$C + D \longrightarrow A + B \quad \text{(reverse reaction)}$$

in which a collision between A and B is necessary for a reaction to occur. The rate at which A and B react depends on the concentration or the number of A and B molecules present; it will be fastest, for a fixed set of conditions, when they are first mixed. As the reaction proceeds, the number of A and B molecules available for reaction decreases, and the rate of reaction slows down. If the reaction is reversible, the speed of the reverse reaction is zero at first and gradually increases as the concentrations of C and D increase. As the number of A and B molecules decreases, the forward rate slows down because A and B cannot find one another as often in order to accomplish a reaction. To counteract this diminishing rate of reaction, an excess of one reagent is often used to keep the reaction from becoming impractically slow. Collisions between molecules may be likened to the space games at the video arcades. When many objects are on the screen, collisions occur frequently; but if only a few objects are present, collisions can usually be avoided.

17.3 Chemical Equilibrium

equilibrium

Any system at **equilibrium** represents a dynamic state in which two or more opposing processes are taking place at the same time and at the same rate. A chemical equilibrium is a dynamic system in which two or more opposing chemical reactions are going on at the same time and at the same rate. When the rate of the forward reaction is exactly equal to the rate of the reverse reaction, a condition of **chemical equilibrium** exists (see Figure 17.2). The concentrations of the products and the reactants are not changing, and the system appears to be at a standstill because the products are reacting at the same rate at which they are being formed.

chemical equilibrium

Chemical equilibrium:
rate of forward reaction = rate of reverse reaction

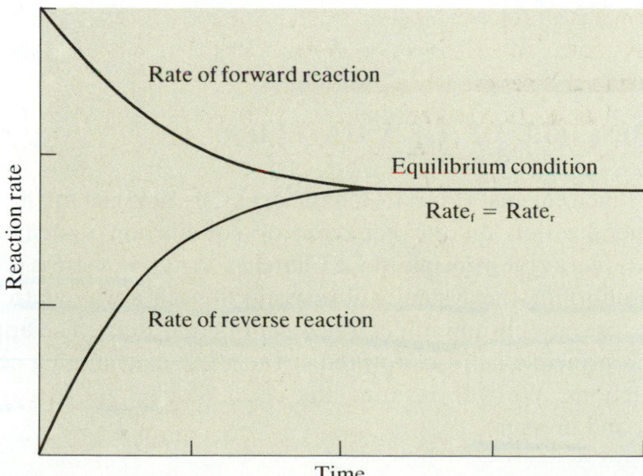

◄ FIGURE 17.2
The graph illustrates that the rates of the forward and reverse reactions become equal at some point in time. The forward reaction rate decreases as a result of decreasing amounts of reactants. The reverse reaction rate starts at zero and increases as the amount of product increases. When the two rates become equal, a state of chemical equilibrium has been reached.

A saturated salt solution is in a condition of equilibrium:

$$NaCl(s) \rightleftharpoons Na^+(aq) + Cl^-(aq)$$

At equilibrium, salt crystals are continuously dissolving, and Na^+ and Cl^- ions are continuously crystallizing. Both processes are occurring at the same rate.

The ionization of weak electrolytes is another chemical equilibrium system:

$$HC_2H_3O_2(aq) + H_2O(l) \rightleftharpoons H_3O^+(aq) + C_2H_3O_2^-(aq)$$

In this reaction, the equilibrium is established in a 1 *M* solution when the forward reaction has gone about 1%; that is, when only 1% of the acetic acid molecules in solution have ionized. Therefore, only a relatively few ions are present, and the acid behaves as a weak electrolyte. In any acid–base equilibrium system the position of equilibrium is toward the weaker conjugate acid and base. In the ionization of acetic acid, $HC_2H_3O_2$ is a weaker acid than H_3O^+, and H_2O is a weaker base than $C_2H_3O_2^-$.

The reactions represented by

$$H_2(g) + I_2(g) \rightleftharpoons 2\,HI(g)$$

provide another example of a chemical equilibrium. Theoretically, 1.00 mol of hydrogen should react with 1.00 mol of iodine to yield 2.00 mol of hydrogen iodide. Actually, when 1.00 mol of H_2 and 1.00 mol of I_2 are reacted at 700 K, only 1.58 mol of HI are present when equilibrium is attained. Since 1.58 is 79% of the theoretical yield of 2.00 mol of HI, the forward reaction is only 79% complete at equilibrium. The equilibrium mixture will also contain 0.21 mol each of unreacted H_2 and I_2 (1.00 mol − 0.79 mol = 0.21 mol).

$$\underset{1.00\ \text{mol}}{H_2} + \underset{1.00\ \text{mol}}{I_2} \xrightarrow{700\ \text{K}} \underset{2.00\ \text{mol}}{2\,HI}$$

(This equation would represent the condition if the reaction were 100% complete; 2.00 mol of HI would be formed and no H_2 and I_2 would be left unreacted.)

$$\underset{0.21\ \text{mol}}{H_2} + \underset{0.21\ \text{mol}}{I_2} \xrightleftharpoons{700\ \text{K}} \underset{1.58\ \text{mol}}{2\,HI}$$

(This equation represents the actual equilibrium attained starting with 1.00 mol each of H_2 and I_2. It shows that the forward reaction is only 79% complete.)

17.4 Principle of Le Chatelier

In 1888 the French chemist Henri Le Chatelier (1850–1936) set forth a simple, far-reaching generalization on the behavior of equilibrium systems. This generalization, known as the **principle of Le Chatelier**, states: If a stress is applied to a system in equilibrium, the system will respond in such a way as to relieve that stress and restore equilibrium under a new set of conditions. The application of Le Chatelier's principle helps us to predict the effect of changing conditions in chemical reactions. We will examine the effect of changes in concentration, temperature, and pressure.

principle of Le Chatelier

17.5 Effect of Concentration on Reaction Rate and Equilibrium

The way in which the rate of a chemical reaction depends on the concentration of the reactants must be determined experimentally. Many simple, one-step reactions result from a collision between two molecules or ions. The rate of such one-step reactions can be altered by changing the concentration of the reactants or products. An increase in concentration of the reactants provides more individual reacting species for collisions and results in an increase in the rate of reaction.

An equilibrium is disturbed when the concentration of one or more of its components is changed. As a result the concentration of all the species will change, and a new equilibrium mixture will be established. Consider the hypothetical equilibrium represented by the equation

$$A + B \rightleftharpoons C + D$$

where A and B react in one step to form C and D. When the concentration of B is increased, the following occurs:

1. The rate of the reaction to the right (forward) increases. This rate is proportional to the concentration of A times the concentration of B.
2. The rate to the right becomes greater than the rate to the left.
3. Reactants A and B are used faster than they are produced; C and D are produced faster than they are used.
4. After a period of time, rates to the right and left become equal, and the system is again in equilibrium.
5. In the new equilibrium the concentration of A is less and the concentrations of B, C, and D are greater than in the original equilibrium.

Conclusion: The equilibrium has shifted to the right.

Applying this change in concentration to the equilibrium mixture of 1.00 mol of hydrogen and 1.00 mol of iodine from Section 17.3, we find that, when an additional 0.20 mol of I_2 is added, the yield of HI (based on H_2) is 85% (1.70 mol) instead of 79%. A comparison of the two systems, after the new equilibrium mixture is reached, follows.

Original equilibrium	New equilibrium
1.00 mol H_2 + 1.00 mol I_2	1.00 mol H_2 + 1.20 mol I_2
Yield: 79% HI	Yield: 85% HI (based on H_2)
Equilibrium mixture contains:	Equilibrium mixture contains:
1.58 mol HI	1.70 mol HI
0.21 mol H_2	0.15 mol H_2
0.21 mol I_2	0.35 mol I_2

Analyzing this new system, we see that, when the 0.20 mol I_2 was added, the equilibrium shifted to the right in order to counteract the increase in I_2 concentration. Some of the H_2 reacted with added I_2 and produced more HI, until an equilibrium mixture was established again. When I_2 was added, the concentration of I_2 increased, the concentration of H_2 decreased, and the concentration of HI increased. What do you think would be the effects of adding (a) more H_2 or (b) more HI?

The equation

$$\underset{\text{Pale yellow}}{Fe^{3+}(aq)} + \underset{\text{Colorless}}{SCN^-(aq)} \rightleftharpoons \underset{\text{Red}}{Fe(SCN)^{2+}(aq)}$$

represents an equilibrium that is used in certain analytical procedures as an indicator because of the readily visible, intense red color of the complex $Fe(SCN)^{2+}$ ion. A very dilute solution of iron(III), Fe^{3+}, and thiocyanate, SCN^-, is light red. When the concentration of either Fe^{3+} or SCN^- is increased, the equilibrium shift to the right is observed by an increase in the intensity of the color, resulting from the formation of additional $Fe(SCN)^{2+}$.

If either Fe^{3+} or SCN^- is removed from solution, the equilibrium will shift to the left, and the solution will become lighter in color. When Ag^+ is added to the solution, a white precipitate of silver thiocyanate (AgSCN) is formed, thus removing SCN^- ion from the equilibrium.

$$Ag^+(aq) + SCN^-(aq) \longrightarrow AgSCN\downarrow$$

The system accordingly responds to counteract the change in SCN^- concentration by shifting the equilibrium to the left. This shift is evident by a decrease in the intensity of the red color due to a decreased concentration of $Fe(SCN)^{2+}$.

Let us now consider the effect of changing the concentrations in the equilibrium mixture of chlorine water. The equilibrium equation is

$$Cl_2(aq) + 2\,H_2O \rightleftharpoons HOCl(aq) + H_3O^+(aq) + Cl^-(aq)$$

The variation in concentrations and the equilibrium shifts are tabulated in the following table. An X in the second or third column indicates the reagent that is increased or decreased. The fourth column indicates the direction of the equilibrium shift.

	Concentration		
Reagent	**Increase**	**Decrease**	**Equilibrium shift**
Cl_2	—	X	Left
H_2O	X	—	Right
HOCl	X	—	Left
H_3O^+	—	X	Right
Cl^-	X	—	Left

Consider the equilibrium in a 0.100 *M* acetic acid solution:

$$HC_2H_3O_2 + H_2O \rightleftharpoons H_3O^+ + C_2H_3O_2^-$$

In this solution the concentration of the hydronium ion (H_3O^+), which is a measure of the acidity, is 1.34×10^{-3} mol/L, corresponding to a pH of 2.87. What will happen to the acidity when 0.100 mol of sodium acetate ($NaC_2H_3O_2$) is added to 1 L of 0.100 *M* $HC_2H_3O_2$? When $NaC_2H_3O_2$ dissolves, it dissociates into sodium ions (Na^+) and acetate ions ($C_2H_3O_2^-$). The acetate ion from the salt is a common ion to the acetic acid equilibrium system and increases the total acetate ion concentration in the solution. As a result the equilibrium shifts to the left, decreasing the hydronium ion concentration and lowering the acidity of the solution. Evidence of this decrease in acidity is shown by the fact that the pH of a solution that is 0.100 *M* in $HC_2H_3O_2$ and 0.100 *M* in $NaC_2H_3O_2$ is 4.74. The pH of several different solutions of $HC_2H_3O_2$ and $NaC_2H_3O_2$ is shown in the table that follows. Each time the acetate ion is increased, the pH increases, indicating a further shift in the equilibrium toward un-ionized acetic acid.

Solution	pH
1 L 0.100 *M* $HC_2H_3O_2$	2.87
1 L 0.100 *M* $HC_2H_3O_2$ + 0.100 mol $NaC_2H_3O_2$	4.74
1 L 0.100 *M* $HC_2H_3O_2$ + 0.200 mol $NaC_2H_3O_2$	5.05
1 L 0.100 *M* $HC_2H_3O_2$ + 0.300 mol $NaC_2H_3O_2$	5.23

In summary, we can say that, when the concentration of a reagent on the left side of an equation is increased, the equilibrium shifts to the right. When the concentration of a reagent on the right side of an equation is increased, the equilibrium shifts to the left. In accordance with Le Chatelier's principle the equilibrium always shifts in the direction that tends to reduce the concentration of the added reactant.

Practice Aqueous chromate ion (CrO_4^{2-}) exists in equilibrium with aqueous dichromate ion ($Cr_2O_7^{2-}$) in an acidic solution. What effect will (a) increasing the dichromate ion and (b) adding HCl have on the equilibrium?

$$2\,CrO_4^{2-} + 2\,H^+ \rightleftharpoons Cr_2O_7^{2-} + H_2O$$

Answers: (a) Equilibrium shifts left (b) Equilibrium shifts right

17.6 Effect of Pressure on Reaction Rate and Equilibrium

Changes in pressure significantly affect the reaction rate only when one or more of the reactants or products is a gas. In these cases the effect of increasing the pressure of the reacting gases is equivalent to increasing their concentrations. In the reaction

$$CaCO_3(s) \overset{\Delta}{\rightleftharpoons} CaO(s) + CO_2(g)$$

calcium carbonate decomposes into calcium oxide, and carbon dioxide when heated above 825°C. Increasing the pressure of the equilibrium system by adding CO_2 or by decreasing the volume, speeds up the reverse reaction and causes the equilibrium to shift to the left. The increased pressure gives the same effect as that caused by increasing the concentration of CO_2, the only gaseous substance in the reaction.

When the pressure on a gas is increased, its volume is decreased. In a system composed entirely of gases, an increase in pressure will cause the reaction and the equilibrium to shift to the side that contains the smaller volume or smaller number of moles. This shift occurs because the increase in pressure is partially relieved by the system's shifting its equilibrium toward the side in which the substances occupy the smaller volume.

Prior to World War I, Fritz Haber (1868–1934) in Germany invented the first major process for the fixation of nitrogen. In the Haber process nitrogen and hydrogen are reacted together in the presence of a catalyst at moderately high temperature and pressure to produce ammonia. The catalyst consists of iron and iron oxide with small amounts of potassium and aluminum oxides. For this process, Haber received the Nobel Prize in chemistry in 1918.

$$N_2(g) + 3\,H_2(g) \rightleftharpoons 2\,NH_3(g) + 92.5\text{ kJ}(22.1\text{ kcal}) \qquad (\text{at } 25°C)$$

$N_2(g)$	$3\,H_2(g)$	$2\,NH_3(g)$
1 mol	3 mol	2 mol
1 volume	3 volumes	2 volumes

The left side of the equation in the Haber process represents four volumes of gas combining to give two volumes of gas on the right side of the equation. An increase in the total pressure on the system shifts the equilibrium to the right. This increase in pressure results in a higher concentration of both reactants and products. The equilibrium shifts to the right when the pressure is increased, because fewer moles of NH_3 than moles of N_2 and H_2 are in the equilibrium reaction.

Ideal conditions for the Haber process are 200°C and 1000 atm pressure. However, at 200°C the rate of reaction is very slow, and at 1000 atm extraordinarily heavy equipment is required. As a compromise the reaction is run at 400–600°C and 200–350 atm pressure, which gives a reasonable yield at a reasonable rate. The effect of pressure on the yield of ammonia at one particular temperature is shown in Table 17.1.

When the total number of gaseous molecules on both sides of an equation is the same, a change in pressure does not cause an equilibrium shift. The following reaction is an example.

$$N_2(g) + O_2(g) \rightleftharpoons 2\,NO(g)$$

$N_2(g)$	$O_2(g)$	$2\,NO(g)$
1 mol	1 mol	2 mol
1 volume	1 volume	2 volumes
6.022×10^{23} molecules	6.022×10^{23} molecules	$2 \times 6.022 \times 10^{23}$ molecules

When the pressure on this system is increased, the rate of both the forward and the reverse reactions will increase because of the higher concentrations of N_2,

TABLE 17.1 The Effect of Pressure on the Conversion of H_2 and N_2 to NH_3 at 450°C[a]

Pressure (atm)	Yield of NH_3(%)	Pressure (atm)	Yield of NH_3(%)
10	2.04	300	35.5
30	5.8	600	53.4
50	9.17	1000	69.4
100	16.4		

[a] The starting ratio of H_2 to N_2 is 3 moles to 1 mole.

O_2, and NO. But the equilibrium will not shift, because the increase in concentration of molecules is the same on both sides of the equation and the decrease in volume is the same on both sides of the equation.

EXAMPLE 17.1

What effect would an increase in pressure have on the position of equilibrium in the following reactions?

(a) $2\ SO_2(g) + O_2(g) \rightleftharpoons 2\ SO_3(g)$
(b) $H_2(g) + Cl_2(g) \rightleftharpoons 2\ HCl(g)$
(c) $N_2O_4(l) \rightleftharpoons 2\ NO_2(g)$

SOLUTION

(a) The equilibrium will shift to the right because the substance on the right has a smaller volume than those on the left.
(b) The equilibrium position will be unaffected because the volumes (or moles) of gases on both sides of the equation are the same.
(c) The equilibrium will shift to the left because $N_2O_4(l)$ occupies a much smaller volume than does $2\ NO_2(g)$.

PRACTICE What effect would an increase in pressure have on the position of the equilibrium in the following reactions?
(a) $2\ NO(g) + Cl_2(g) \rightleftharpoons 2\ NOCl(g)$ (b) $COBr_2(g) \rightleftharpoons CO(g) + Br_2(g)$
Answers: (a) Equilibrium shifts right (b) Equilibrium shifts left

17.7 EFFECT OF TEMPERATURE ON REACTION RATE AND EQUILIBRIUM

An increase in temperature generally increases the rate of reaction. Molecules at elevated temperatures are more energetic and have more kinetic energy; thus, their collisions are more likely to result in a reaction. However, we cannot assume that the rate of a desired reaction will increase indefinitely as the temperature is raised. High temperatures may cause the destruction or decomposition of the

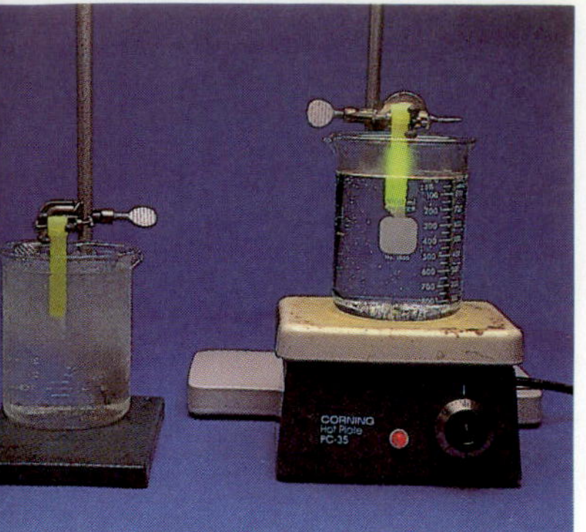

▲ **Light sticks. The chemical reaction that produces light in these light sticks is endothermic. Placing the light stick in hot water (right) favors this reaction, producing a brighter light than when the light stick is in ice water (left).**

reactants and products or may initiate reactions other than the one desired. For example, when calcium oxalate (CaC_2O_4) is heated to 500°C, it decomposes into calcium carbonate and carbon monoxide:

$$CaC_2O_4(s) \xrightarrow{500°C} CaCO_3(s) + CO(g)$$

If calcium oxalate is heated to 850°C, the products are calcium oxide, carbon monoxide, and carbon dioxide:

$$CaC_2O_4(s) \xrightarrow{850°C} CaO(s) + CO(g) + CO_2(g)$$

When heat is applied to a system in equilibrium, the reaction that absorbs heat is favored. When the process, as written, is endothermic, the forward reaction is increased. When the reaction is exothermic, the reverse reaction is favored. In this sense heat may be treated as a reactant in endothermic reactions or as a product in exothermic reactions. Therefore, temperature is analogous to concentration when applying Le Chatelier's principle to heat effects on a chemical reaction.

Hot coke (C) is a very reactive element. In the reaction

$$C(s) + CO_2(g) + \text{heat} \rightleftharpoons 2\,CO(g)$$

very little, if any, CO is formed at room temperature. At 1000°C the equilibrium mixture contains about an equal number of moles of CO and CO_2. At higher temperatures the equilibrium shifts to the right, increasing the yield of CO. The reaction is endothermic, and, as can be seen, the equilibrium is shifted to the right at higher temperatures.

Phosphorus trichloride reacts with dry chlorine gas to form phosphorus pentachloride. The reaction is exothermic:

$$PCl_3(l) + Cl_2(g) \rightleftharpoons PCl_5(s) + 88\ \text{kJ}(21\ \text{kcal})$$

Heat must continuously be removed during the reaction to obtain a good yield of the product. According to the principle of Le Chatelier, heat will cause the product, PCl_5, to decompose, re-forming PCl_3 and Cl_2. The equilibrium mixture at 200°C contains 52% PCl_5, and at 300°C it contains 3% PCl_5, verifying that heat causes the equilibrium to shift to the left.

When the temperature of a system is raised, the rate of reaction increases because of increased kinetic energy and more frequent collisions of the reacting species. In a reversible reaction the rate of both the forward and the reverse reactions is increased by an increase in temperature; however, the reaction that absorbs heat increases to a greater extent, and the equilibrium shifts to favor that reaction. The following example illustrates these effects.

EXAMPLE 17.2

What effect would an increase in temperature have on the position of equilibrium in the following reactions?

$$4\,HCl(g) + O_2(g) \rightleftharpoons 2\,H_2O(g) + 2\,Cl_2(g) + 95.4\ \text{kJ}\ (28.4\ \text{kcal}) \quad (1)$$

$$H_2(g) + Cl_2(g) \rightleftharpoons 2\,HCl(g) + 185\ \text{kJ}\ (44.2\ \text{kcal}) \quad (2)$$

$$CH_4(g) + 2\,O_2(g) \rightleftharpoons CO_2(g) + 2\,H_2O(g) + 890\ \text{kJ}\ (212.8\ \text{kcal}) \quad (3)$$

$$N_2O_4(l) + 58.6\ kJ\ (14\ kcal) \rightleftharpoons 2\ NO_2(g) \quad (4)$$

$$2\ CO_2(g) + 566\ kJ\ (135.2\ kcal) \rightleftharpoons 2\ CO(g) + O_2(g) \quad (5)$$

$$H_2(g) + I_2(g) + 51.9\ kJ\ (12.4\ kcal) \rightleftharpoons 2\ HI(g) \quad (6)$$

SOLUTION

Reactions (1), (2), and (3) are exothermic; an increase in temperature will cause the equilibrium to shift to the left. Reactions (4), (5), and (6) are endothermic; an increase in temperature will cause the equilibrium to shift to the right.

PRACTICE What effect would an increase in temperature have on the position of the equilibrium in the following reactions?
(a) $2\ SO_2(g) + O_2(g) \rightleftharpoons 2\ SO_3(g) + 198\ kJ$
(b) $H_2(g) + CO_2(g) + 41\ kJ \rightleftharpoons H_2O(g) + CO(g)$
Answers: (a) Equilibrium shifts left (b) Equilibrium shifts right

17.8 Effect of Catalysts on Reaction Rate and Equilibrium

catalyst

A **catalyst** is a substance that influences the rate of a reaction and can be recovered essentially unchanged at the end of the reaction. A catalyst does not shift the equilibrium of a reaction; it affects only the speed at which the equilibrium is reached. If a catalyst does not affect the equilibrium, then it follows that it must affect the rate of both the forward and the reverse reactions equally.

The reaction between phosphorus trichloride (PCl_3) and sulfur is highly exothermic, but it is so slow that very little product, thiophosphoryl chloride ($PSCl_3$), is obtained, even after prolonged heating. When a catalyst, such as aluminum chloride ($AlCl_3$), is added, the reaction is complete in a few seconds:

$$PCl_3(l) + S(s) \xrightarrow{AlCl_3} PSCl_3(l)$$

In the laboratory preparation of oxygen, manganese dioxide is used as a catalyst to increase the rates of decomposition of both potassium chlorate and hydrogen peroxide.

$$2\ KClO_3(s) \xrightarrow[\Delta]{MnO_2} 2\ KCl(s) + 3\ O_2(g)$$

$$2\ H_2O_2(aq) \xrightarrow{MnO_2} 2\ H_2O(l) + O_2(g)$$

Catalysts are extremely important to industrial chemistry. Hundreds of chemical reactions that are otherwise too slow to be of practical value have been put to commercial use once a suitable catalyst was found. But in the area of biochemistry catalysts are of supreme importance because nearly all chemical reactions in all forms of life are completely dependent on biochemical catalysts known as *enzymes*.

17.9 Equilibrium Constants

In a reversible chemical reaction at equilibrium, the concentrations of the reactants and products are constant; that is, they are not changing. The rates of the forward and reverse reactions are constant, and an equilibrium constant expression can be written relating the products to the reactants. For the general reaction

$$\mathrm{a\,A + b\,B \rightleftharpoons c\,C + d\,D}$$

at constant temperature, the following equilibrium constant expression can be written:

$$K_{eq} = \frac{[\mathrm{C}]^c[\mathrm{D}]^d}{[\mathrm{A}]^a[\mathrm{B}]^b}$$

equilibrium constant, K_{eq}

where K_{eq} is constant at a particular temperature and is known as the **equilibrium constant**. The quantities in brackets are the concentrations of each substance in moles per liter. The superscript letters a, b, c, and d are the coefficients of the substances in the balanced equation. The units for K_{eq} are not the same for every equilibrium reaction; however, the units are generally omitted. Observe that the concentration of each substance is raised to a power that is the same as the substance's numerical coefficient in the balanced equation. The convention is to place the concentrations of the products (the substances on the right side of the equation as written) in the numerator and the concentrations of the reactants in the denominator.

Example 17.3

Write equilibrium constant expressions for

(a) $3\,H_2(g) + N_2(g) \rightleftharpoons 2\,NH_3(g)$
(b) $CO(g) + 2\,H_2(g) \rightleftharpoons CH_3OH(g)$

Solution

(a) The only product, NH_3, has a coefficient of 2. Therefore the numerator of the equilibrium constant will be $[NH_3]^2$. Two reactants are present, H_2 with a coefficient of 3 and N_2 with a coefficient of 1. Therefore the denominator will be $[H_2]^3[N_2]$. The equilibrium constant expression is

$$K_{eq} = \frac{[NH_3]^2}{[H_2]^3[N_2]}$$

(b) For this equation the numerator is $[CH_3OH]$ and the denominator is $[CO][H_2]^2$. The equilibrium constant expression is

$$K_{eq} = \frac{[CH_3OH]}{[CO][H_2]^2}$$

PRACTICE Write the equilibrium constant expressions for
(a) $2\ N_2O_5(g) \rightleftharpoons 4\ NO_2(g) + O_2(g)$
(b) $4\ NH_3(g) + 3\ O_2(g) \rightleftharpoons 2\ N_2(g) + 6\ H_2O(g)$

Answers: (a) $K_{eq} = \dfrac{[NO_2]^4[O_2]}{[N_2O_5]^2}$ (b) $K_{eq} = \dfrac{[N_2]^2[H_2O]^6}{[NH_3]^4[O_2]^3}$

The magnitude of an equilibrium constant indicates the extent to which the forward and reverse reactions take place. When K_{eq} is greater than 1, the amount of the products at equilibrium is greater than the amount of the reactants. When K_{eq} is less than 1, the amount of reactants at equilibrium is greater than the amount of the products. A very large value for K_{eq} indicates that the forward reaction goes essentially to completion. A very small K_{eq} means that the reverse reaction goes nearly to completion and that the equilibrium is far to the left (toward the reactants). Two examples follow:

$$H_2(g) + I_2(g) \rightleftharpoons 2\ HI(g) \qquad K_{eq} = 54.8 \text{ at } 425^\circ C$$

This K_{eq} indicates that considerably more product than reactants is present at equilibrium.

$$COCl_2(g) \rightleftharpoons CO(g) + Cl_2(g) \qquad K_{eq} = 7.6 \times 10^{-4} \text{ at } 400^\circ C$$

This K_{eq} indicates that $COCl_2$ is stable and that very little decomposition to CO and Cl_2 occurs at 400°C. The equilibrium is far to the left.

When the molar concentrations of all the species in an equilibrium reaction are known, the K_{eq} can be calculated by substituting the concentrations into the equilibrium constant expression.

EXAMPLE 17.4

Calculate the K_{eq} for the following reaction based on concentrations of: $PCl_5 = 0.030$ mol/L; $PCl_3 = 0.97$ mol/L; $Cl_2 = 0.97$ mol/L at 300°C.

$$PCl_5(g) \rightleftharpoons PCl_3(g) + Cl_2(g)$$

SOLUTION

First write the K_{eq} expression; then substitute the respective concentrations into this equation and solve:

$$K_{eq} = \frac{[PCl_3][Cl_2]}{[PCl_5]} = \frac{[0.97][0.97]}{[0.030]} = 31$$

This K_{eq} is considered to be a fairly large value, indicating that at 300°C the decomposition of PCl_5 proceeds far to the right.

PRACTICE Calculate the K_{eq} for the following reaction. Is the forward or the reverse reaction favored?

$$2\ NO(g) + O_2(g) \rightleftharpoons 2\ NO_2(g)$$

when $[NO] = 0.050\ M$, $[O_2] = 0.75\ M$, $[NO_2] = 0.25\ M$
Answer: $K_{eq} = 33$; The forward reaction is favored

17.10 Ionization Constants

acid ionization constant, K_a

As a first application of an equilibrium constant, let us consider the constant for acetic acid in solution. Because it is a weak acid, an equilibrium is established between molecular $HC_2H_3O_2$ and its ions in solution. The constant is called the **acid ionization constant, K_a**, a special type of equilibrium constant. The concentration of water in the solution is large compared to the other concentrations and does not change appreciably, so we may use the following simplified equation to set up the constant:

$$HC_2H_3O_2 \rightleftharpoons H^+ + C_2H_3O_2^-$$

The ionization constant expression is the concentration of the products divided by the concentration of the reactants:

$$K_a = \frac{[H^+][C_2H_3O_2^-]}{[HC_2H_3O_2]}$$

It states that the ionization constant, K_a, is equal to the product of the hydrogen ion $[H^+]$ concentration and the acetate ion $[C_2H_3O_2^-]$ concentration divided by the concentration of the un-ionized acetic acid $[HC_2H_3O_2]$.

At 25°C a 0.100 M $HC_2H_3O_2$ solution is 1.34% ionized and has a hydrogen ion concentration of 1.34×10^{-3} mol/L. From this information we can calculate the ionization constant for acetic acid.

A 0.100 M solution initially contains 0.100 mol of acetic acid per liter. Of this 0.100 mol, only 1.34%, or 1.34×10^{-3} mol, is ionized, which gives an H^+ ion concentration of 1.34×10^{-3} mol/L. Because each molecule of acid that ionizes yields one H^+ and one $C_2H_3O_2^-$, the concentration of $C_2H_3O_2^-$ ions is also 1.34×10^{-3} mol/L. This ionization leaves $0.100 - 0.00134 = 0.0987$ mol/L of un-ionized acetic acid.

	Initial concentration (mol/L)	Equilibrium concentration (mol/L)
$[HC_2H_3O_2]$	0.100	0.0987
$[H^+]$	0	0.00134
$[C_2H_3O_2^-]$	0	0.00134

TABLE 17.2 Ionization Constants (K_a) of Weak Acids at 25°C

Acid	Formula	K_a	Acid	Formula	K_a
Acetic	$HC_2H_3O_2$	1.8×10^{-5}	Hydrocyanic	HCN	4.0×10^{-10}
Benzoic	$HC_7H_5O_2$	6.3×10^{-5}	Hypochlorous	HClO	3.5×10^{-8}
Carbolic (phenol)	HC_6H_5O	1.3×10^{-10}	Nitrous	HNO_2	4.5×10^{-4}
Cyanic	HCNO	2.0×10^{-4}	Hydrofluoric	HF	6.5×10^{-4}
Formic	$HCHO_2$	1.8×10^{-4}			

Substituting these concentrations in the equilibrium expression, we obtain the value for K_a:

$$K_a = \frac{[H^+][C_2H_3O_2^-]}{[HC_2H_3O_2]} = \frac{[1.34 \times 10^{-3}][1.34 \times 10^{-3}]}{[0.0987]} = 1.82 \times 10^{-5}$$

The K_a for acetic acid, 1.82×10^{-5}, is small and indicates that the position of the equilibrium is far toward the un-ionized acetic acid. In fact, a 0.100 *M* acetic acid solution is 98.7% un-ionized.

Once the K_a for acetic acid is established, it can be used to describe other systems containing H^+, $C_2H_3O_2^-$, and $HC_2H_3O_2$ in equilibrium at 25°C. The ionization constants for several other weak acids are listed in Table 17.2.

EXAMPLE 17.5

What is the H^+ ion concentration in a 0.50 *M* $HC_2H_3O_2$ solution? The ionization constant, K_a, for $HC_2H_3O_2$ is 1.8×10^{-5}.

SOLUTION

To solve this problem, first write the equilibrium equation and the K_a expression:

$$HC_2H_3O_2 \rightleftharpoons H^+ + C_2H_3O_2^- \qquad K_a = \frac{[H^+][C_2H_3O_2^-]}{[HC_2H_3O_2]} = 1.8 \times 10^{-5}$$

We know that the initial concentration of $HC_2H_3O_2$ is 0.50 *M*. We also know from the ionization equation that one $C_2H_3O_2^-$ is produced for every H^+ produced; that is, the $[H^+]$ and the $[C_2H_3O_2^-]$ are equal. To solve, let $Y = [H^+]$, which also equals the $[C_2H_3O_2^-]$. The un-ionized $[HC_2H_3O_2]$ remaining will then be $0.50 - Y$, the starting concentration minus the amount that ionized.

$$[H^+] = [C_2H_3O_2^-] = Y \qquad [HC_2H_3O_2] = 0.50 - Y$$

Substituting these values into the K_a expression, we obtain

$$K_a = \frac{(Y)(Y)}{0.50 - Y} = \frac{Y^2}{0.50 - Y} = 1.8 \times 10^{-5}$$

An exact solution of this equation for Y requires the use of a mathematical equation known as the quadratic equation. However, an approximate solution is obtained if we assume that Y is small and can be neglected compared with 0.50. Then $0.50 - Y$ will be equal to approximately 0.50. The equation now becomes

$$\frac{Y^2}{0.50} = 1.8 \times 10^{-5}$$

$$Y^2 = 0.50 \times 1.8 \times 10^{-5} = 0.90 \times 10^{-5} = 9.0 \times 10^{-6}$$

Taking the square root of both sides of the equation, we obtain

$$Y = \sqrt{9.0 \times 10^{-6}} = 3.0 \times 10^{-3} \text{ mol/L}$$

Thus, $[H^+]$ is approximately 3.0×10^{-3} mol/L in a 0.50 M $HC_2H_3O_2$ solution. The exact solution to this problem, using the quadratic equation, gives a value of 2.99×10^{-3} mol/L for $[H^+]$, showing that we were justified in neglecting Y compared with 0.50.

PRACTICE Calculate the hydrogen ion concentration in (a) 0.100 M hydrocyanic acid (HCN) solution and (b) 0.0250 M carbolic acid (HC_6H_5O) solution.
Answers: (a) 6.32×10^{-6} (b) 1.80×10^{-6}

EXAMPLE 17.6 Calculate the percent ionization in a 0.50 M $HC_2H_3O_2$ solution.

SOLUTION The percent ionization of a weak acid, $HA(aq) \rightleftharpoons H^+(aq) + A^-(aq)$, is found by dividing the concentration of the H^+ or A^- ions at equilibrium by the initial concentration of HA. For acetic acid

$$\frac{\text{concentration of } [H^+] \text{ or } [C_2H_3O_2^-]}{\text{initial concentration of } [HC_2H_3O_2]} \times 100 = \text{percent ionized}$$

To solve this problem we first need to calculate $[H^+]$. This calculation has already been done in Example 17.5 for a 0.50 M solution.

$$[H^+] = 3.0 \times 10^{-3} \text{ mol/L in a 0.50 } M \text{ solution} \quad \text{(from Example 17.5)}$$

This $[H^+]$ represents a fractional amount of the initial 0.50 M $HC_2H_3O_2$. Therefore

$$\frac{3.0 \times 10^{-3} \text{ mol/L}}{0.50 \text{ mol/L}} \times 100 = 0.60\% \text{ ionized}$$

A 0.50 M $HC_2H_3O_2$ solution is 0.60% ionized.

PRACTICE Calculate the percent ionization for the solution in the previous practice problem.
Answers: (a) 6.32×10^{-3}% ionized (b) 7.20×10^{-3}% ionized

TABLE 17.3 Relationship of H^+ and OH^- Concentrations in Water Solutions

$[H^+]$	$[OH^-]$	K_w	pH	pOH
1.00×10^{-2}	1.00×10^{-12}	1.00×10^{-14}	2.00	12.0
1.00×10^{-4}	1.00×10^{-10}	1.00×10^{-14}	4.00	10.0
2.00×10^{-6}	5.00×10^{-9}	1.00×10^{-14}	5.70	8.30
1.00×10^{-7}	1.00×10^{-7}	1.00×10^{-14}	7.00	7.00
1.00×10^{-9}	1.00×10^{-5}	1.00×10^{-14}	9.00	5.00

17.11 ION PRODUCT CONSTANT FOR WATER

We have seen that water ionizes to a slight degree. This ionization is represented by these equilibrium equations:

$$H_2O + H_2O \rightleftharpoons H_3O^+ + OH^- \quad (7)$$

$$H_2O \rightleftharpoons H^+ + OH^- \quad (8)$$

Equation (7) is the more accurate representation of the equilibrium, since free protons (H^+) do not exist in water. Equation (8) is a simplified and often-used representation of the water equilibrium. The actual concentration of H^+ produced in pure water is very minute and amounts to only 1.00×10^{-7} mol/L at 25°C. In pure water

$$[H^+] = [OH^-] = 1.00 \times 10^{-7} \text{ mol/L}$$

since both ions are produced in equal molar amounts, as shown in equation (8).

The $H_2O \rightleftharpoons H^+ + OH^-$ equilibrium exists in water and in all water solutions. A special equilibrium constant called the **ion product constant for water**, $\boldsymbol{K_w}$, applies to this equilibrium. The constant K_w is defined as the product of the H^+ ion concentration and the OH^- ion concentration, each in moles per liter:

ion product constant for water, K_w

$$K_w = [H^+][OH^-]$$

The numerical value of K_w is 1.00×10^{-14}, since for pure water at 25°C

$$K_w = [H^+][OH^-] = [1.00 \times 10^{-7}][1.00 \times 10^{-7}] = 1.00 \times 10^{-14}$$

The value of K_w for all water solutions at 25°C is the constant 1.00×10^{-14}. It is important to realize that, as the concentration of one of these ions, H^+ or OH^-, increases, the other decreases. However, the product of $[H^+]$ and $[OH^-]$ always equals the constant 1.00×10^{-14}. This relationship can be seen in the examples shown in Table 17.3. If the concentration of one ion is known, the concentration of the other can be calculated from the K_w expression.

$$K_w = [H^+][OH^-] \qquad [H^+] = \frac{K_w}{[OH^-]} \qquad [OH^-] = \frac{K_w}{[H^+]}$$

EXAMPLE 17.7

What is the concentration of (a) H^+ and (b) OH^- in a 0.001 *M* HCl solution? Assume that HCl is 100% ionized.

SOLUTION

(a) Since all the HCl is ionized, $H^+ = 0.001$ mol/L or 1×10^{-3} mol/L.

$$\underset{0.00\,M}{HCl} \longrightarrow \underset{0.001\,M}{H^+} + \underset{0.001\,M}{Cl^-}$$

$[H^+] = 1 \times 10^{-3}$ mol/L

(b) To calculate the $[OH^-]$ in this solution, use the following equation and substitute the values for K_w and $[H^+]$:

$$[OH^-] = \frac{K_w}{[H^+]}$$

$$[OH^-] = \frac{1.00 \times 10^{-14}}{1 \times 10^{-3}} = 1 \times 10^{-11} \text{ mol/L}$$

PRACTICE Determine the $[H^+]$ and $[OH^-]$ in
(a) 5.0×10^{-5} *M* HNO_3 (b) 2.0×10^{-6} *M* KOH

Answers: (a) $[H^+] = 5.0 \times 10^{-5}$ $[OH^-] = 2.0 \times 10^{-10}$
(b) $[H^+] = 5.0 \times 10^{-9}$ $[OH^-] = 2.0 \times 10^{-6}$

EXAMPLE 17.8

What is the pH of a 0.010 *M* NaOH solution? Assume that NaOH is 100% ionized.

SOLUTION

Since all the NaOH is ionized, $OH^- = 0.010$ mol/L or 1.0×10^{-2} mol/L.

$$\underset{0.00\,M}{NaOH} \longrightarrow \underset{0.010\,M}{Na^+} + \underset{0.010\,M}{OH^-}$$

To find the pH of the solution we first calculate the H^+ concentration. Use the following equation and substitute the values for K_w and $[OH^-]$.

$$[H^+] = \frac{K_w}{[OH^-]} = \frac{1.00 \times 10^{-14}}{1.0 \times 10^{-2}} = 1.0 \times 10^{-12} \text{ mol/L}$$

$$pH = -\log[H^+] = -\log(1.0 \times 10^{-12}) = 12$$

PRACTICE Determine the pH for the solutions following:
(a) 5.0×10^{-5} *M* HNO_3 (b) 2.0×10^{-6} *M* KOH.
Answers: (a) 4.30 (b) 8.30

Just as pH is used to express the acidity of a solution, pOH is used to express the basicity of an aqueous solution. The pOH is related to the OH^- ion concentration in the same way that the pH is related to the H^+ ion concentration:

$$\text{pOH} = \log \frac{1}{[OH^-]} \qquad \text{or} \qquad \text{pOH} = -\log[OH^-]$$

Thus, a solution in which $[OH^-] = 1.0 \times 10^{-2}$, as in Example 17.8, will have pOH = 2.0.

In pure water, where $[H^+] = 1 \times 10^{-7}$ and $[OH^-] = 1 \times 10^{-7}$, the pH is 7, and the pOH is 7. The sum of the pH and pOH is always 14.

$$\text{pH} + \text{pOH} = 14$$

In Example 17.8 the pH can also be found by first calculating the pOH from the OH^- ion concentration and then subtracting from 14.

$$\text{pH} = 14 - \text{pOH} = 14 - 2 = 12$$

17.12 Solubility Product Constant

The **solubility product constant**, abbreviated K_{sp}, is another application of the equilibrium constant. It is the equilibrium constant of a slightly soluble salt. The following example illustrates how K_{sp} is evaluated.

solubility product constant, K_{sp}

The solubility of silver chloride (AgCl) in water is 1.3×10^{-5} mol/L at 25°C. The equation for the equilibrium between AgCl and its ions in solution is

$$AgCl(s) \rightleftharpoons Ag^+ + Cl^-$$

The equilibrium constant expression is

$$K_{eq} = \frac{[Ag^+][Cl^-]}{[AgCl(s)]}$$

The amount of solid AgCl does not affect the equilibrium system provided that some is present. In other words, the concentration of solid silver chloride is constant whether 1 mg or 10 g of the salt are present. Therefore, the product obtained by multiplying the two constants K_{eq} and $[AgCl(s)]$ is also a constant. This constant is the solubility product constant, K_{sp}.

$$K_{eq} \times [AgCl(s)] = [Ag^+][Cl^-] = K_{sp}$$
$$K_{sp} = [Ag^+][Cl^-]$$

The K_{sp} is equal to the product of the Ag^+ ion and the Cl^- ion concentrations, each in moles per liter. When 1.3×10^{-5} mol/L of AgCl dissolves, it produces

TABLE 17.4 Solubility Product Constants (K_{sp}) at 25°C

Compound	K_{sp}	Compound	K_{sp}
$AgCl$	1.7×10^{-10}	CaF_2	3.9×10^{-11}
$AgBr$	5×10^{-13}	CuS	9×10^{-45}
AgI	8.5×10^{-17}	$Fe(OH)_3$	6×10^{-38}
$AgC_2H_3O_2$	2×10^{-3}	PbS	7×10^{-29}
Ag_2CrO_4	1.9×10^{-12}	$PbSO_4$	1.3×10^{-8}
$BaCrO_4$	8.5×10^{-11}	$Mn(OH)_2$	2.0×10^{-13}
$BaSO_4$	1.5×10^{-9}		

1.3×10^{-5} mol/L each of Ag^+ and Cl^-. From these concentrations the K_{sp} can be evaluated.

$$[Ag^+] = 1.3 \times 10^{-5}\ \text{mol/L} \qquad [Cl^-] = 1.3 \times 10^{-5}\ \text{mol/L}$$
$$K_{sp} = [Ag^+][Cl^-] = [1.3 \times 10^{-5}][1.3 \times 10^{-5}] = 1.7 \times 10^{-10}$$

Once the K_{sp} value for AgCl is established, it can be used to describe other systems containing Ag^+ and Cl^-.

The K_{sp} expression does not have a denominator. It consists only of the concentrations (mol/L) of the ions in solution. As in other equilibria expressions, each of these concentrations is raised to a power that is the same number as its coefficient in the balanced equation. The equilibrium equations and the K_{sp} expressions for several other substances follow.

$$AgBr(s) \rightleftharpoons Ag^+ + Br^- \qquad K_{sp} = [Ag^+][Br^-]$$
$$BaSO_4(s) \rightleftharpoons Ba^{2+} + SO_4^{2-} \qquad K_{sp} = [Ba^{2+}][SO_4^{2-}]$$
$$Ag_2CrO_4(s) \rightleftharpoons 2\ Ag^+ + CrO_4^{2-} \qquad K_{sp} = [Ag^+]^2[CrO_4^{2-}]$$
$$CuS(s) \rightleftharpoons Cu^{2+} + S^{2-} \qquad K_{sp} = [Cu^{2+}][S^{2-}]$$
$$Mn(OH)_2(s) \rightleftharpoons Mn^{2+} + 2\ OH^- \qquad K_{sp} = [Mn^{2+}][OH^-]^2$$
$$Fe(OH)_3(s) \rightleftharpoons Fe^{3+} + 3\ OH^- \qquad K_{sp} = [Fe^{3+}][OH^-]^3$$

Table 17.4 lists K_{sp} values for these and several other substances.

When the product of the molar concentration of the ions in solution, each raised to its proper power, is greater than the K_{sp} for that substance, precipitation should occur. If the ion product is less than the K_{sp} value, no precipitation will occur.

EXAMPLE 17.9 Write K_{sp} expressions for AgI and PbI_2, both of which are slightly soluble salts.

SOLUTION First write the equilibrium equations:

$$AgI(s) \rightleftharpoons Ag^+ + I^-$$
$$PbI_2(s) \rightleftharpoons Pb^{2+} + 2\ I^-$$

Since the concentration of the solid crystals is constant, the K_{sp} equals the product of the molar concentrations of the ions in solution. In the case of PbI_2, the $[I^-]$ must be squared.

$$K_{sp} = [Ag^+][I^-]$$
$$K_{sp} = [Pb^{2+}][I^-]^2$$

PRACTICE Write the K_{sp} expression for
(a) $Cr(OH)_3$ (b) $Cu_3(PO_4)_2$
Answers: (a) $K_{sp} = [Cr^{3+}][OH^-]^3$ (b) $K_{sp} = [Cu^{2+}]^3[PO_4^{3-}]^2$

EXAMPLE 17.10

The K_{sp} value for lead sulfate is 1.3×10^{-8}. Calculate the solubility of $PbSO_4$ in grams per liter.

SOLUTION

First write the equilibrium equation and the K_{sp} expression:

$$PbSO_4 \rightleftharpoons Pb^{2+} + SO_4^{2-}$$
$$K_{sp} = [Pb^{2+}][SO_4^{2-}] = 1.3 \times 10^{-8}$$

Since the lead sulfate that is in solution is completely dissociated, the concentration of $[Pb^{2+}]$ or $[SO_4^{2-}]$ is equal to the solubility of $PbSO_4$ in moles per liter.

Let

$$Y = [Pb^{2+}] = [SO_4^{2-}]$$

Substitute Y into the K_{sp} equation and solve.

$$[Pb^{2+}][SO_4^{2-}] = [Y][Y] = 1.3 \times 10^{-8}$$
$$Y^2 = 1.3 \times 10^{-8}$$
$$Y = 1.1 \times 10^{-4} \text{ mol/L}$$

The solubility of $PbSO_4$, therefore, is 1.1×10^{-4} mol/L. Now convert mol/L to g/L:

1 mol of $PbSO_4$ has a mass of (207.2 g + 32.06 g + 64.00 g) or 303.3 g

$$\frac{1.1 \times 10^{-4} \cancel{\text{mol}}}{\text{L}} \times \frac{303.3 \text{ g}}{\cancel{\text{mol}}} = 3.3 \times 10^{-2} \text{ g/L}$$

The solubility of $PbSO_4$ is 3.3×10^{-2} g/L.

PRACTICE The K_{sp} value for CuS is 9.0×10^{-45}. Calculate the solubility in grams per liter.
Answer: 9.1×10^{-21} g/L

An ion added to a solution that already contains that ion is called a *common ion*. When a common ion is added to an equilibrium solution of a weak electrolyte or a slightly soluble salt, the equilibrium shifts according to Le Chatelier's principle. For example, when silver nitrate, $AgNO_3$, is added to a saturated solution of silver chloride, AgCl ($AgCl(s) \rightleftharpoons Ag^+ + Cl^-$), the equilibrium shifts to the left due to the increase in the Ag^+ concentration. As a result, the Cl^- concentration and the solubility of AgCl decreases. AgCl and $AgNO_3$ have the common ion Ag^+. A shift in the equilibrium position upon addition of an ion already contained in the solution is known as the **common ion effect**.

common ion effect

EXAMPLE 17.11 Silver nitrate, $AgNO_3$, is added to a saturated AgCl solution until the Ag^+ concentration is 0.10 *M*. What will be the Cl^- concentration remaining in solution?

SOLUTION This problem is an example of the common ion effect. The addition of $AgNO_3$ puts more Ag^+ in solution; the Ag^+ combines with Cl^- and causes the equilibrium to shift to the left, reducing the Cl^- concentration in solution.

We use the K_{sp} to calculate the Cl^- ion concentration remaining in solution. The K_{sp} is constant at a particular temperature and remains the same no matter how we change the concentration of the species involved.

$$K_{sp} = [Ag^+][Cl^-] = 1.7 \times 10^{-10} \qquad [Ag^+] = 0.10 \text{ mol/L}$$

We then substitute the concentration of Ag^+ ion into the K_{sp} expression and calculate the Cl^- concentration.

$$[0.10][Cl^-] = 1.7 \times 10^{-10}$$

$$[Cl^-] = \frac{1.7 \times 10^{-10}}{0.10} = 1.7 \times 10^{-9} \text{ mol/L}$$

This calculation shows a 10,000-fold reduction of Cl^- ions in solution. It illustrates that Cl^- ions may be quantitatively removed from solution with an excess of Ag^+ ions.

PRACTICE Sodium sulfate (Na_2SO_4) is added to a saturated solution of $BaSO_4$ until the concentration of sulfate ion is 2.0×10^{-2} *M*. What will be the concentration of the Ba^{2+} ions remaining in solution?

Answer: 7.5×10^{-8} mol/L

17.13 HYDROLYSIS

hydrolysis

Hydrolysis is the term used for the general reaction in which a water molecule is split. For example, the net ionic hydrolysis reaction for a sodium acetate solution is

$$C_2H_3O_2^- + H_2O(l) \rightleftharpoons HC_2H_3O_2(aq) + OH^-$$

Table 17.5 Ionic Composition of Salts and the Nature of the Aqueous Solutions They Form

Type of salt	Nature of aqueous solution	Examples
Weak base–strong acid	Acidic	NH_4Cl, NH_4NO_3
Strong base–weak acid	Basic	$NaC_2H_3O_2$, K_2CO_3
Weak base–weak acid	Depends on the salt	$NH_4C_2H_3O_2$, NH_4NO_2
Strong base–strong acid	Neutral	NaCl, KBr

In the reaction above the water molecule is split, with the H^+ combining with $C_2H_3O_2^-$ to give the weak acid $HC_2H_3O_2$ and the OH^- going into solution, making the solution more basic.

Salts that contain an ion of a weak acid or a weak base undergo hydrolysis. For example, a 0.10 *M* NH_4Cl solution has a pH of 5.1, and a 0.10 *M* NaCN solution has a pH of 11.1. The hydrolysis reactions that cause these solutions to become acidic or basic are

$$NH_4^+ + H_2O(l) \rightleftharpoons NH_4OH(aq) + H^+$$
$$CN^- + H_2O(l) \rightleftharpoons HCN(aq) + OH^-$$

The ions of a salt derived from a strong acid and a strong base, such as NaCl, do not undergo hydrolysis and thus form neutral solutions. Table 17.5 lists the ionic composition of various salts and the nature of the aqueous solutions that they form.

Practice Indicate whether each of the following salts would produce acidic, basic, or neutral aqueous solution.
(a) KCN (b) $NaNO_3$ (c) NH_4Br
Answers: (a) basic (b) neutral (c) acidic

17.14 Buffer Solutions: The Control of pH

The control of pH within narrow limits is critically important in many chemical applications and vitally important in many biological systems. For example, human blood must be maintained between pH 7.35 and 7.45 for the efficient transport of oxygen from the lungs to the cells. This narrow pH range is maintained by buffer systems in the blood.

A **buffer solution** resists changes in pH when diluted or when small amounts of acid or base are added. Two common types of buffer solutions are (1) a weak acid mixed with a salt of that weak acid and (2) a weak base mixed with a salt of that weak base.

buffer solution

A salt water aquarium is a ▶ buffer system.

The action of a buffer system can be understood by considering a solution of acetic acid and sodium acetate. The weak acid, $HC_2H_3O_2$, is mostly un-ionized and is in equilibrium with its ions in solution. The salt, $NaC_2H_3O_2$, is completely ionized.

$$HC_2H_3O_2(aq) \rightleftharpoons H^+(aq) + C_2H_3O_2^-(aq)$$
$$NaC_2H_3O_2(aq) \longrightarrow Na^+(aq) + C_2H_3O_2^-(aq)$$

Because the salt is completely ionized, the solution contains a much higher concentration of acetate ions than would be present if only acetic acid were in solution. The acetate ion represses the ionization of acetic acid and also reacts with water, causing the solution to have a higher pH (be more basic) than an acetic acid solution (see Section 17.5). Thus, a 0.1 *M* acetic acid solution has a pH of 2.87, but a solution that is 0.1 *M* in acetic acid and 0.1 *M* in sodium acetate has a pH of 4.74. This difference in pH is the result of the common ion effect.

A buffer solution has a built-in mechanism that counteracts the effect of adding acid or base. Consider the effect of adding HCl or NaOH to an acetic acid–sodium acetate buffer. When a small amount of HCl is added, the acetate ions of the buffer combine with the H^+ ions from HCl to form un-ionized acetic acid, thus neutralizing the added acid and maintaining the approximate pH of the solution. When NaOH is added, the OH^- ions react with acetic acid to neutralize the added base and thus maintain the approximate pH. The equations for these reactions are

$$H^+ + C_2H_3O_2^- \rightleftharpoons HC_2H_3O_2(aq)$$
$$OH^- + HC_2H_3O_2(aq) \rightleftharpoons H_2O(l) + C_2H_3O_2^-$$

Data comparing the changes in pH caused by adding HCl and NaOH to pure water and to an acetic acid–sodium acetate buffer solution are shown in Table 17.6.

TABLE 17.6 Changes in pH Caused by the Addition of HCl and NaOH

Solution	pH	Change in pH
H_2O (1000 mL)	7	—
H_2O + 0.010 mol HCl	2	5
H_2O + 0.010 mol NaOH	12	5
Buffer solution (1000 mL)		
0.10 *M* $HC_2H_3O_2$ + 0.10 *M* $NaC_2H_3O_2$	4.74	—
Buffer + 0.010 mol HCl	4.66	0.08
Buffer + 0.010 mol NaOH	4.83	0.09

The human body has a number of buffer systems. One of these, the bicarbonate–carbonic acid buffer, HCO_3^-–H_2CO_3, maintains the blood plasma at a pH of 7.4. The phosphate system, HPO_4^{2-}–$H_2PO_4^-$, is an important buffer in the red blood cells as well as in other places in the body.

17.15 Mechanism of Reactions

How a reaction occurs—that is, the manner in which it proceeds—is known as the **mechanism of the reaction**. The mechanism is the path, or route, the atoms and molecules take to arrive at the products. Our aim here is not to study the mechanisms themselves but to show that chemical reactions occur by specific routes.

mechanism of a reaction

When hydrogen and iodine are mixed at room temperature, we observe no appreciable reaction. In this case, the reaction takes place as a result of collisions between H_2 and I_2 molecules, but at room temperature the collisions do not result in reaction because the molecules lack sufficient energy to react. We say that an energy barrier to reaction exists. If heat is added, the kinetic energy of the molecules increases. When molecules of H_2 and I_2 with sufficient energy collide, an intermediate product, known as the **activated complex**, is formed. The amount of energy needed to form the activated complex is known as the **activation energy**. The activated complex, H_2I_2, is in a metastable form and has an energy level higher than that of the reactants or the product. It can decompose to form either the reactants or the product. Three steps constitute the mechanism of the reaction: (1) collision of an H_2 and an I_2 molecule; (2) formation of the activated complex, H_2I_2; and (3) decomposition to the product, HI. The various steps in the formation of HI are shown in Figure 17.3. Figure 17.4 illustrates the energy relationships in this reaction.

activated complex

activation energy

The reaction of hydrogen and chlorine proceeds by a different mechanism. When H_2 and Cl_2 are mixed and kept in the dark, essentially no product is formed. But, if the mixture is exposed to sunlight or ultraviolet radiation, it reacts very rapidly. The overall reaction is

$$H_2(g) + Cl_2(g) \longrightarrow 2\,HCl(g)$$

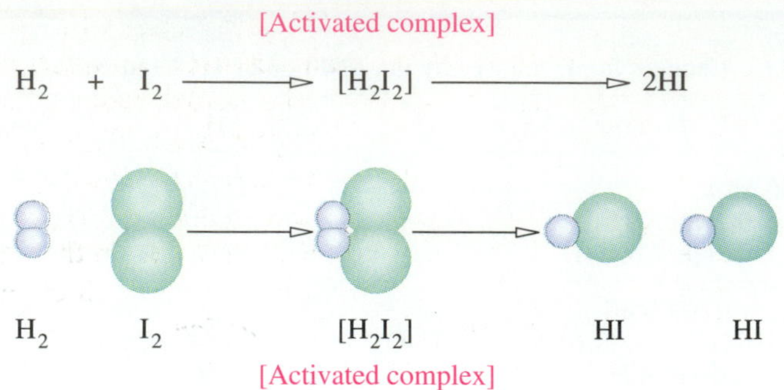

FIGURE 17.3
Mechanism of the reaction between hydrogen and iodine. H_2 and I_2 molecules of sufficient energy unite, forming the intermediate activated complex that decomposes to the product, hydrogen iodide.

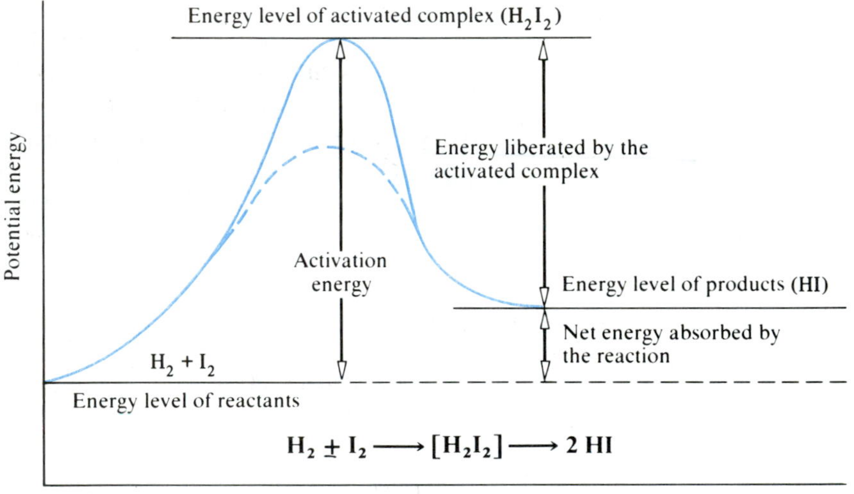

FIGURE 17.4 ▶
Relative energy diagram for the reaction between hydrogen and iodine. Energy equal to the activation energy is put into the system to form the activated complex, H_2I_2. When this complex decomposes, it liberates energy, forming the product. In this case, the product is at a higher energy level than the reactants, indicating that the reaction is endothermic and that energy is absorbed during the reaction. The dotted line represents the effect that a catalyst would have on the reaction. The catalyst lowers the activation energy, thereby increasing the rate of the reaction.

free radical

This reaction proceeds by what is known as a *free radical mechanism.* A **free radical** is a neutral atom or group of atoms containing one or more unpaired electrons. Both atomic chlorine ($:\ddot{\underset{..}{Cl}}\cdot$) and atomic hydrogen (H·) have an unpaired electron and are free radicals. The reaction occurs in three steps:

Step 1 *Initiation*

$$:\ddot{\underset{..}{Cl}}:\ddot{\underset{..}{Cl}}: + h\nu \longrightarrow :\ddot{\underset{..}{Cl}}\cdot + :\ddot{\underset{..}{Cl}}\cdot$$

Chlorine free radicals

In this step a chlorine molecule absorbs energy in the form of a photon, *hv*, of light or ultraviolet radiation. The energized chlorine molecule then splits into two chlorine free radicals.

Step 2 *Propagation*

$$:\ddot{\underset{..}{Cl}}\cdot + H:H \longrightarrow HCl + H\cdot$$

Hydrogen free radical

$$H\cdot + :\ddot{\underset{..}{Cl}}:\ddot{\underset{..}{Cl}}: \longrightarrow HCl + :\ddot{\underset{..}{Cl}}\cdot$$

This step begins when a chlorine free radical reacts with a hydrogen molecule to produce a molecule of hydrogen chloride and a hydrogen free radical. The hydrogen radical then reacts with another chlorine molecule to form hydrogen chloride and another chlorine free radical. This chlorine radical can repeat the process by reacting with another hydrogen molecule, and the reaction continues to propagate itself in this manner until one or both of the reactants are used up. Almost all of the product is formed in this step.

Step 3 *Termination*

$$:\ddot{\underset{..}{Cl}}\cdot + :\ddot{\underset{..}{Cl}}\cdot \longrightarrow Cl_2$$

$$H\cdot + H\cdot \longrightarrow H_2$$

$$H\cdot + :\ddot{\underset{..}{Cl}}\cdot \longrightarrow HCl$$

Hydrogen and chlorine free radicals can react in any of the three ways shown. Unless further activation occurs, the formation of hydrogen chloride will terminate when the radicals form molecules. In an exothermic reaction such as that between hydrogen and chlorine, usually enough heat and light energy is available to maintain the supply of free radicals, and the reaction will continue until at least one reactant is exhausted.

Concepts in Review

1. Describe a reversible reaction.
2. Explain why the rate of the forward reaction decreases and the rate of the reverse reaction increases as a chemical reaction approaches equilibrium.
3. State and understand the qualitative effect of Le Chatlier's principle.
4. Predict how the rate of a chemical reaction is affected by (a) changes in the concentration of reactants, (b) changes in pressure of gaseous reactants, (c) changes in temperature, and (d) the presence of a catalyst.
5. Write the equilibrium constant expression for a chemical reaction from a balanced chemical equation.
6. Explain the meaning of the numerical constant, K_{eq}, when given the concentration of the reactants and products in equilibrium.
7. Calculate the concentration of one substance in an equilibrium when given the equilibrium constant and the concentrations of all the other substances.
8. Calculate the equilibrium constant, K_{eq}, when given the concentration of reactants and products in equilibrium.
9. Calculate the ionization constant for a weak acid from appropriate data.
10. Calculate the concentrations of all the chemical species in a solution of a weak acid when given the percent ionization or the ionization constant.
11. Compare the relative strengths of acids by using their ionization constants.
12. Use the ion product constant for water, K_w, to calculate $[H^+]$, $[OH^-]$, pH, and pOH when given any one of these quantities.

Chemistry in Action

Exchange of Oxygen and Carbon Dioxide in the Blood

The transport of oxygen and carbon dioxide between the lungs and tissues is a complex process that involves several reversible reactions, each of which behaves in accordance with Le Chatelier's principle.

The binding of oxygen to hemoglobin is a reversible reaction. The oxygen molecule must attach to the hemoglobin (Hb) and later detach again. The equilibrium equation for this reaction can be written:

$$Hb + O_2 \rightleftharpoons HbO_2$$

In the lungs, the concentration of oxygen is high and favors the forward reaction. Oxygen quickly binds to the hemoglobin until it is saturated with oxygen.

In the tissues the concentration of oxygen is lower and in accordance with Le Chatelier's principle the equilibrium position shifts to the left, and the hemoglobin releases oxygen to the tissues. Approximately 45% of the oxygen diffuses out of the capillaries into the tissues where it may be picked up by *myoglobin*, another carrier molecule.

Myglobin functions as an oxygen storage molecule, holding the oxygen until it is required in the energy-producing portions of the cell. The reaction between myoglobin (Mb) and oxygen can be written as an equilibrium reaction:

$$Mb + O_2 \rightleftharpoons MbO_2$$

Since both the hemoglobin and myoglobin equations are so similar, what accounts for the transfer of the oxygen from the hemoglobin to the myoglobin? Although both equilibria involve similar interactions the affinity between oxygen and hemoglobin is different than the affinity between myoglobin and oxygen. In the tissues the position of the hemoglobin equilibrium is such that the hemoglobin is 55% saturated with oxygen, whereas the myoglobin is at 90% oxygen saturation. Under these conditions hemoglobin will release oxygen while myoglobin will bind oxygen. Thus oxygen is loaded onto hemoglobin in the lungs and unloaded in the tissues cells.

Carbon dioxide produced by oxidation in the cells must be removed from the tissues. Oxygen-depleted hemoglobin molecules accomplish this by becoming carriers of carbon dioxide. The carbon dioxide does not bind at the heme site as the oxygen does, but rather at one end of the protein chain. When carbon dioxide dissolves in water some of the CO_2 reacts to release hydrogen ions:

$$CO_2 + H_2O \rightleftharpoons HCO_3^- + H^+$$

In order to facilitate the removal of CO_2 from the tissues this equilibrium needs to be moved toward the right. This shift is accomplished by the removal of H^+ from the tissues by the hemoglobin molecule. The deoxygenated hemoglobin molecule can bind H^+ ions as well as CO_2. In the lungs this whole process is reversed so the CO_2 is removed from the hemoglobin and exhaled.

Molecules that are similar in structure to the oxygen molecule can become involved in competing equilibria. Hemoglogin is capable of binding with carbon monoxide (CO), nitric oxide (NO), and cyanide (CN^-). The extent of the competition depends upon affinity. Since these molecules have a greater affinity for hemoglobin than oxygen they will effectively displace oxygen from hemoglobin. For example

$$HbO_2 + CO \rightleftharpoons HbCO + O_2$$

Since the affinity of hemoglobin for CO is 150 times stronger than its affinity for oxygen, the equilibrium position lies far to the right. This explains why CO is a poisonous substance and why oxygen is administered to victims of CO poisoning. The hemoglobin molecules can only transport oxygen if the CO is released and the oxygen shifts the equilibrium toward the left.

Capillary and red blood cells
▼

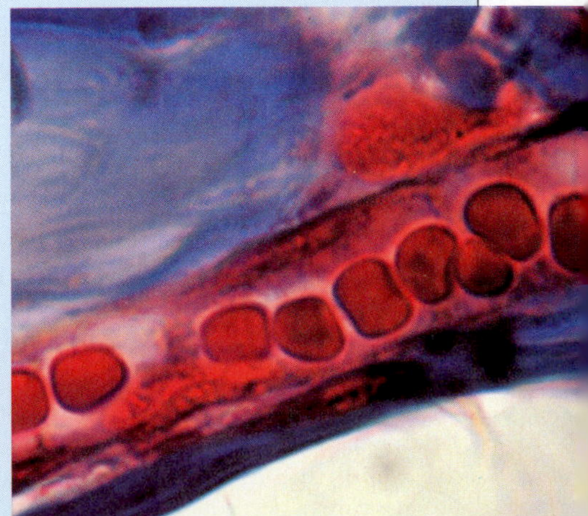

13. Calculate the solubility product constant, K_{sp}, of a slightly soluble salt when given its solubility, or vice versa.
14. Compare relative solubilities of salts if solubility products are known.
15. Discuss the common ion effect on a system at equilibrium.
16. Explain hydrolysis and why some salts form acidic or basic aqueous solutions.
17. Explain how a buffer solution is able to counteract the addition of small amounts of either H^+ or OH^- ions.
18. Draw the relative energy diagram of an exothermic or endothermic reaction. Label the activation energy, and show the effect of a catalyst.

Exercises

An asterisk indicates a more challenging question or problem.

1. How would you expect the two tubes in Figure 17.1 to appear if both are at 25°C?
2. Is the reaction $N_2O_4 \rightleftharpoons 2\ NO_2$ (Figure 17.1) exothermic or endothermic?
3. At equilibrium how do the forward and reverse reaction rates compare? (See Figure 17.2.)
4. Would the reaction of 30 mol of H_2 and 10 mol of N_2 produce a greater yield of NH_3 if carried out in a 1 L or a 2 L vessel? (See Table 17.1.)
5. Of the acids listed in Table 17.2, which ones are stronger than acetic acid and which are weaker?
6. For each of the solutions in Table 17.3, what is the sum of the pH plus the pOH? What would be the pOH of a solution whose pH was -1?
7. Tabulate the relative order of molar solubilities of AgCl, AgBr, AgI, $AgC_2H_3O_2$, $PbSO_4$, $BaSO_4$, $BaCrO_4$, and PbS. (Use Table 17.4.) List the most soluble first.
8. Which compound in each of the following pairs has the greater molar solubility? (See Table 17.4.)
   (a) $Mn(OH)_2$ or Ag_2CrO_4
   (b) $BaCrO_4$ or Ag_2CrO_4
9. Using Table 17.6, explain how the acetic acid–sodium acetate buffer system maintains its pH when 0.010 mol of HCl is added to 1 L of the buffer solution.
10. How would Figure 17.4 be altered if the reaction were exothermic?
11. Express the following reversible systems in equation form:
    (a) A mixture of ice and liquid water at 0°C
    (b) Liquid water and vapor at 100°C in a pressure cooker
    (c) Crystals of Na_2SO_4 in a saturated aqueous solution of Na_2SO_4
    (d) A closed system containing boiling sulfur dioxide, SO_2
12. Explain why a precipitate of NaCl forms when hydrogen chloride gas is passed into a saturated aqueous solution of NaCl.
13. Why does the rate of a reaction usually increase when the concentration of one of the reactants is increased?
14. Consider the following system at equilibrium:

$$4\ NH_3(g) + 3\ O_2(g) \rightleftharpoons 2\ N_2(g) + 6\ H_2O(g) + 1531\ kJ$$

    (a) Is the reaction exothermic or endothermic?
    (b) If the system's state of equilibrium is disturbed by the addition of O_2, in which direction, left or right, must reaction occur to reestablish equilibrium? After the new equilibrium has been established, how will the final molar concentrations of NH_3, O_2, N_2, and H_2O compare (increase or decrease) with their concentrations before the addition of the O_2?
    (c) If the system's state of equilibrium is disturbed by the addition of heat, in which direction will reaction occur, left or right, to reestablish equilibrium?
15. Consider the following system at equilibrium:

$$N_2(g) + 3\ H_2(g) \rightleftharpoons 2\ NH_3(g) + 92.5\ kJ$$

Complete the following table. Indicate changes in moles by entering I, D, N, or ? in the table. (I = increase, D = decrease, N = no change, ? = insufficient information to determine.)

Change or stress imposed on the system at equilibrium	Direction of reaction, left or right, to reestablish equilibrium	Change in number of moles		
		N_2	H_2	NH_3
(a) Add N_2	Right	I	D	I
(b) Remove H_2	Left	I	D	
(c) Decrease volume of reaction vessel				
(d) Increase volume of reaction vessel				
(e) Increase temperature				
(f) Add catalyst				
(g) Add both H_2 and NH_3				

16. If pure hydrogen iodide, HI, is placed in a vessel at 700 K, will it decompose? Explain.

17. For each of the equations that follow, tell in which direction, left or right, the equilibrium will shift when the following changes are made: The temperature is increased; the pressure is increased by decreasing the volume of the reaction vessel; a catalyst is added.
(a) $3\,O_2(g) + 271\text{ kJ} \rightleftharpoons 2\,O_3(g)$
(b) $CH_4(g) + Cl_2(g) \rightleftharpoons CH_3Cl(g) + HCl(g) + 110\text{ kJ}$
(c) $2\,NO(g) + 2\,H_2(g) \rightleftharpoons N_2(g) + 2\,H_2O(g) + 665\text{ kJ}$
(d) $2\,SO_3(g) + 197\text{ kJ} \rightleftharpoons 2\,SO_2(g) + O_2(g)$
(e) $4\,NH_3(g) + 3\,O_2(g) \rightleftharpoons 2\,N_2(g) + 6\,H_2O(g) + 1531\text{ kJ}$

18. Explain why an increase in temperature causes the rate of reaction to increase.

19. Give a word description of how equilibrium is reached when the substances A and B are first mixed and react as

$$A + B \rightleftharpoons C + D$$

20. Utilizing Le Chatelier's principle, indicate the shift (if any) that would occur to

$$C_2H_6(g) + \text{heat} \rightleftharpoons C_2H_4(g) + H_2(g)$$

(a) If the concentration of hydrogen gas is decreased
(b) If the temperature is lowered
(c) If a catalyst is added
(d) If C_2H_6 is removed from the system
(e) If the volume of the container is increased

21. With dilution, aqueous solutions of acetic acid, $HC_2H_3O_2$, show increased ionization. For example, a 1.0 *M* solution of acetic acid is 0.42% ionized, whereas a 0.10 *M* solution is 1.34% ionized. Explain this behavior using the ionization equation and equilibrium principles.

22. A 1.0 *M* solution of acetic acid ionizes less and has a higher concentration of H^+ ions than a 0.10 *M* acetic acid solution. Explain this behavior. (See Exercise 21 for data.)

23. Write the equilibrium constant expression for each of the following reactions:
(a) $4\,HCl(g) + O_2(g) \rightleftharpoons 2\,Cl_2(g) + 2\,H_2O(g)$
(b) $N_2(g) + 3\,H_2(g) \rightleftharpoons 2\,NH_3(g)$
(c) $PCl_5(g) \rightleftharpoons PCl_3(g) + Cl_2(g)$
(d) $HClO_2(aq) \rightleftharpoons H^+(aq) + ClO_2^-(aq)$
(e) $NH_4OH(aq) \rightleftharpoons NH_4^+(aq) + OH^-(aq)$
(f) $4\,NH_3(g) + 5\,O_2(g) \rightleftharpoons 4\,NO(g) + 6\,H_2O(g)$

24. What would cause two separate samples of pure water to have slightly different pH values?

25. Why are the pH and pOH equal in pure water?

26. What effect will increasing the H^+ ion concentration of a solution have on (a) pH, (b) pOH, (c) $[OH^-]$, and (d) K_w?

27. Write the solubility product expression (K_{sp}) for each of the following substances:
(a) CuS (e) $Fe(OH)_3$
(b) $BaSO_4$ (f) Sb_2S_5
(c) $PbBr_2$ (g) CaF_2
(d) Ag_3AsO_4 (h) $Ba_3(PO_4)_2$

28. Explain why silver acetate, $AgC_2H_3O_2$, is more soluble in nitric acid than in water. [*Hint:* Write the equilibrium equation first and then consider the effect of the acid on the acetate ion.] What would happen if hydrochloric acid, HCl, were used in place of nitric acid?

29. Decide whether each of the following salts forms an acidic, a basic, or a neutral aqueous solution.
(a) KCl (d) $(NH_4)_2SO_4$ (g) $NaNO_2$
(b) Na_2CO_3 (e) $Ca(CN)_2$ (h) NaF
(c) K_2SO_4 (f) $BaBr_2$

30. Dissolution of sodium acetate, $NaC_2H_3O_2$, in pure water gives a basic solution. Why? [*Hint:* A small amount of $HC_2H_3O_2$ is formed.]

31. Write hydrolysis equations for aqueous solutions of these salts.
(a) KNO_2 (c) NH_4NO_3
(b) $Mg(C_2H_3O_2)_2$ (d) Na_2SO_3

32. Write hydrolysis equations for the following ions.
(a) HCO_3^- (b) NH_4^+ (c) OCl^- (d) ClO_2^-

33. Describe why the pH of a buffer solution remains almost constant when a small amount of acid or base is added to it.

34. One of the important pH-regulating systems in the blood consists of a carbonic acid–sodium bicarbonate buffer.

$$H_2CO_3(aq) \rightleftharpoons H^+(aq) + HCO_3^-(aq)$$
$$NaHCO_3(aq) \longrightarrow Na^+(aq) + HCO_3^-(aq)$$

Explain how this buffer resists changes in pH when (a) excess acid and (b) excess base get into the bloodstream.

35. How does Le Chatelier's principle explain the transport of oxygen from the lungs to the tissues?

36. State the similarities and differences between hemoglobin and myoglobin.

37. Explain how the removal of carbon dioxide from the tissues is facilitated and accomplished by the hemoglobin molecule.

38. Which of the following statements are correct? Rewrite the incorrect statements to make them correct.

(a) In a reaction at equilibrium the concentrations of reactants and products are equal.
(b) A catalyst increases the concentrations of products present at equilibrium.
(c) Enzymes are the catalysts in living systems.
(d) A catalyst lowers the activation energy of a reaction by equal amounts for both the forward and the reverse reactions.
(e) If an increase in temperature causes an increase in the concentration of products present at equilibrium, the reaction is exothermic.
(f) The magnitude of an equilibrium constant is independent of the reaction temperature.
(g) A large equilibrium constant for a reaction indicates that the reaction, at equilibrium, favors products over reactants.
(h) The amount of product obtained at equilibrium is proportional to how fast equilibrium is attained.
(i) The study of reaction rates and reaction mechanisms is known as chemical kinetics.
(j) For the reaction

$$CaCO_3(s) \overset{\Delta}{\rightleftharpoons} CaO(s) + CO_2(g)$$

increasing the pressure of CO_2 present at equilibrium will cause the reaction to shift left.

(k) The larger the value of the equilibrium constant, the greater the proportion of products present at equilibrium.
(l) At chemical equilibrium the rate of the reverse reaction is equal to the rate of the forward reaction.

Statements (m)–(s) pertain to the equilibrium system.

$$2\,NO(g) + O_2(g) \rightleftharpoons 2\,NO_2(g) + \text{heat}$$

(m) The reaction as shown is endothermic.
(n) Increasing the temperature will cause the equilibrium to shift left.
(o) Increasing the temperature will increase the magnitude of the equilibrium constant, K_{eq}.
(p) Decreasing the volume of the reaction vessel will shift the equilibrium to the right and decrease the concentrations of the reactants.
(q) Removal of some of the O_2 will cause an increase in the concentration of NO.
(r) High temperatures and pressures favor increased yields of NO_2.
(s) The equilibrium constant expression for the reaction is

$$K_{eq} = \frac{[NO_2]^2}{[NO]^2[O_2]}$$

(t) A solution with an H^+ ion concentration of 1×10^{-5} mol/L has a pOH of 9.
(u) An aqueous solution that has an OH^- ion concentration of 1×10^{-4} mol/L has an H^+ ion concentration of 1×10^{-10} mol/L.
(v) $K_w = [H^+][OH^-] = 1.00 \times 10^{-14}$, and pH + pOH = 14.
(w) As solid $BaSO_4$ is added to a saturated solution of $BaSO_4$, the magnitude of its K_{sp} increases.
(x) A solution of pOH 10 is basic.
(y) The pH increases as the $[H^+]$ increases.
(z) The pH of 0.050 *M* $Ca(OH)_2$ is 13.

Equilibrium Constants

39. What is the maximum number of moles of HI that can be obtained from a reaction mixture containing 2.30 mol of I_2 and 2.10 mol of H_2?

40. (a) How many moles of hydrogen iodide, HI, will be produced when 2.00 mol of H_2 and 2.00 mol of I_2 are reacted at 700 K? (Reaction is 79% complete.)
(b) Addition of 0.27 mol of I_2 to the system increases the yield of HI to 85%. How many moles of H_2, I_2, and HI are now present?

(c) From the data in part (a), calculate K_{eq} for the reaction at 700 K.

*41. 6.00 g of hydrogen, H_2, and 200. g of iodine, I_2, are reacted at 500. K. After equilibrium is reached, analysis shows that the flask contains 64.0 g of HI. How many moles of H_2, I_2, and HI are present in this equilibrium mixture?

42. What is the equilibrium constant of the reaction shown if a 20. L flask contains 0.10 mol of PCl_3, 1.50 mol of Cl_2, and 0.22 mol of PCl_5?

$$PCl_3(g) + Cl_2(g) \rightleftharpoons PCl_5$$

43. If the rate of a reaction doubles for every 10° rise in temperature, how much faster will the reaction go at 100°C than at 30°C?

Ionization Constants

44. Calculate the ionization constant for each of the following monoprotic acids. Each acid ionizes as follows: $HA \rightleftharpoons H^+ + A^-$.

Acid	Acid concentration	$[H^+]$
Hypochlorous, HOCl	0.10 *M*	5.9×10^{-5} mol/L
Propanoic, $HC_3H_5O_2$	0.15 *M*	1.4×10^{-3} mol/L
Hydrocyanic, HCN	0.20 *M*	8.9×10^{-6} mol/L

45. Calculate (a) the H^+ ion concentration, (b) the pH, and (c) the percent ionization of a 0.25 *M* solution of $HC_2H_3O_2$. ($K_a = 1.8 \times 10^{-5}$.)

46. A 1.0 *M* solution of a weak acid, HA, is 0.52% ionized. Calculate the ionization constant, K_a, for the acid.

47. A 0.15 *M* solution of a weak acid, HA, has a pH of 5. Calculate the ionization constant, K_a, for the acid.

48. Calculate the percent ionization and pH of solutions of $HC_2H_3O_2$ having the following molarities: (a) 1.0 *M*, (b) 0.10 *M*, and (c) 0.010 *M*. ($K_a = 1.8 \times 10^{-5}$.)

49. A 0.37 *M* solution of a weak acid, HA, has a pH of 3.7. What is the K_a for this acid?

50. A 0.23 *M* solution of a weak acid, HA, has a pH of 2.89. What is the K_a for this acid?

51. A common laboratory reagent is 6.0 *M* HCl. Calculate the $[H^+]$, $[OH^-]$, pH, and pOH of this solution.

52. Calculate the pH and the pOH of the following solutions:
(a) 0.00010 *M* HCl
(b) 0.010 *M* NaOH
(c) 0.0025 *M* NaOH
(d) 0.10 *M* HClO ($K_a = 3.5 \times 10^{-8}$)
*(e) Saturated $Fe(OH)_2$ solution ($K_{sp} = 8.0 \times 10^{-16}$)

53. Calculate the $[OH^-]$ in each of these solutions:
(a) $[H^+] = 1.0 \times 10^{-4}$
(b) $[H^+] = 2.8 \times 10^{-6}$
(c) $[H^+] = 4.0 \times 10^{-9}$

54. Calculate the $[H^+]$ in each of these solutions:
(a) $[OH^-] = 6.0 \times 10^{-7}$
(b) $[OH^-] = 1 \times 10^{-8}$
(c) $[OH^-] = 4.5 \times 10^{-6}$

Solubility Product Constant

55. Given the following solubility data, calculate the solubility product constant for each substance.
(a) $BaSO_4$, 3.9×10^{-5} mol/L
(b) Ag_2CrO_4, 7.8×10^{-5} mol/L
(c) ZnS, 3.5×10^{-12} mol/L
(d) $Pb(IO_3)_2$, 4.0×10^{-5} mol/L
(e) Bi_2S_3, 4.9×10^{-15} mol/L
(f) AgCl, 0.0019 g/L
(g) $CaSO_4$, 0.67 g/L
(h) $Zn(OH)_2$, 2.33×10^{-4} g/L
(i) Ag_3PO_4, 6.73×10^{-3} g/L

56. Calculate the molar solubility for each of the following substances:
(a) $BaCO_3$, $K_{sp} = 2.0 \times 10^{-9}$
(b) $AlPO_4$, $K_{sp} = 5.8 \times 10^{-19}$
(c) Ag_2SO_4, $K_{sp} = 1.5 \times 10^{-5}$
(d) $Mg(OH)_2$, $K_{sp} = 7.1 \times 10^{-12}$

57. Calculate, for each of the substances in Question 56, the solubility in grams per 100 mL of water.

58. The K_{sp} of CaF_2 is 3.9×10^{-11}. Calculate (a) the molar concentrations of Ca^{2+} and F^- in a saturated solution, and (b) the grams of CaF_2 that will dissolve in 500 mL of water.

*59. The following pairs of solutions are mixed. Show by calculation whether or not a precipitate will form.
(a) 100 mL of 0.010 *M* Na_2SO_4 and 100 mL of 0.001 *M* $Pb(NO_3)_2$
(b) 50.0 mL of 1.0×10^{-4} *M* $AgNO_3$ and 100. mL of 1.0×10^{-4} *M* NaCl
(c) 1.0 g $Ca(NO_3)_2$ in 150 mL H_2O and 250 mL of 0.01 *M* NaOH

K_{sp} $PbSO_4 = 1.3 \times 10^{-8}$
K_{sp} $AgCl = 1.7 \times 10^{-10}$
K_{sp} $Ca(OH)_2 = 1.3 \times 10^{-6}$

60. $BaCl_2$ is added to a saturated $BaSO_4$ solution until the Ba^{2+} concentration is 0.050 *M*. (a) What concentration of SO_4^{2-} remains in solution? (b) How many grams of $BaSO_4$ remain dissolved in 100. mL of the solution? ($K_{sp} = 1.5 \times 10^{-9}$ for $BaSO_4$.)

61. The concentration of a solution is 0.10 *M* Ba^{2+} and 0.10 *M* Sr^{2+}. Which sulfate, $BaSO_4$ or $SrSO_4$, will precipitate first when a dilute solution of H_2SO_4 is added dropwise to the solution? Show evidence for your answer. ($K_{sp} = 1.5 \times 10^{-9}$ for $BaSO_4$ and $K_{sp} = 3.5 \times 10^{-7}$ for $SrSO_4$.)

62. How many moles of AgBr will dissolve in 1.0 L of (a) 0.10 *M* NaBr and (b) 0.10 *M* $MgBr_2$? ($K_{sp} = 5.0 \times 10^{-13}$ for AgBr.)

63. The K_{sp} for $PbCl_2$ is 2.0×10^{-5}. Will a precipitate form when 0.050 mol of $Pb(NO_3)_2$ and 0.010 mol of NaCl are dissolved in 1.0 L H_2O? Show evidence for your answer.

64. Calculate the K_{eq} for the reaction $SO_2(g) + O_2(g) \rightleftharpoons SO_3(g)$ when the equilibrium concentrations of the gases are at 530°C. $SO_3 = 11.0$ *M*, $SO_2 = 4.20$ *M*, and $O_2 = 0.6 \times 10^{-3}$ *M*.

65. If it takes 0.048 g of BaF_2 to saturate 15.0 mL of water, what is the K_{sp} of BaF_2?

66. The K_{eq} for the formation of ammonia gas from its elements is 4.0. If the equilibrium concentrations of nitrogen gas and hydrogen gas are both 2.0 *M*, what is the equilibrium concentration of the ammonia gas?

67. The K_{sp} of $SrSO_4$ is 7.6×10^{-7}. Should precipitation occur when 25.0 mL of 1.0×10^{-3} *M* of $SrCl_2$ solution is mixed with 15.0 mL of 2.0×10^{-3} *M* Na_2SO_4? Prove.

***68.** The solubility of Hg_2I_2 in H_2O is 3.04×10^{-7} g/L. The reaction $Hg_2I_2 \rightleftharpoons Hg_2^{2+} + 2I^-$ represents equilibrium. Calculate the K_{sp}.

69. Calculate the H^+ ion concentration and the pH of buffer solutions that are 0.20 *M* in $HC_2H_3O_2$ and contain sufficient sodium acetate to make the $C_2H_3O_2^-$ ion concentration equal to (a) 0.10 *M* and (b) 0.20 *M*. (K_a $HC_2H_3O_2 = 1.8 \times 10^{-5}$.)

70. (a) When 1.0 mL of 1.0 *M* HCl is added to 50 mL of 1.0 *M* NaCl, the H^+ ion concentration changes from 1×10^{-7} *M* to 2.0×10^{-2} *M*.

(b) When 1.0 mL of 1.0 *M* HCl is added to 50. mL of a buffer solution that is 1.0 *M* in $HC_2H_3O_2$ and 1.0 *M* in $NaC_2H_3O_2$, the H^+ ion concentration changes from 1.8×10^{-5} *M* to 1.9×10^{-5} *M*.

Calculate the initial pH and the pH change in each solution (log 1.8 = 0.26; log 1.9 = 0.28; log 2.0 = 0.30).

18

Oxidation–Reduction

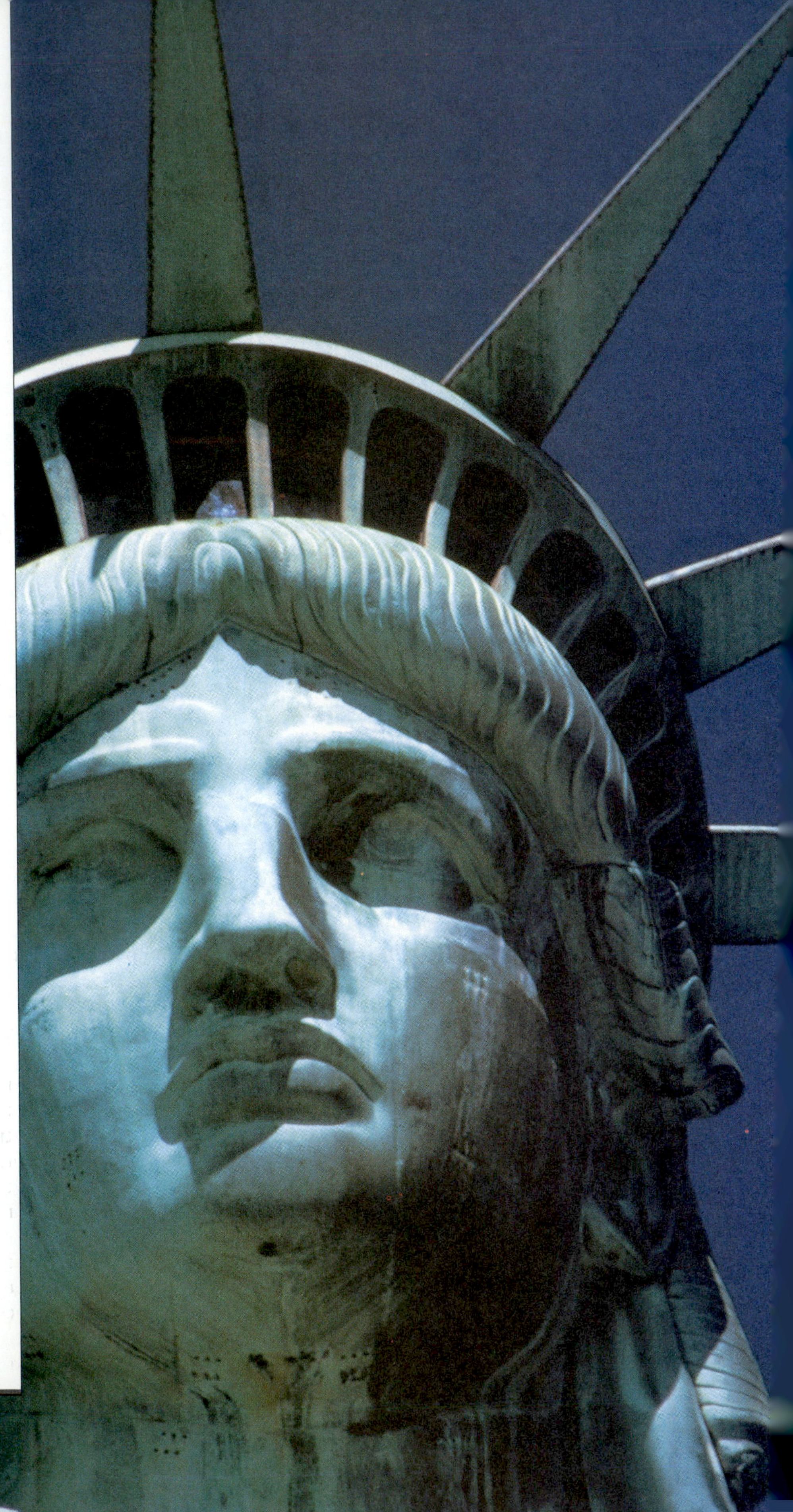

◄ **CHAPTER OPENING PHOTO: The green patina covering the Statue of Liberty is the result of oxidation-reduction reactions.**

The variety of chemical reactions that occur in our daily lives is amazing. Our society seems to run on batteries—in calculators, cars, toys, lights, thermostats, radios, televisions, and more. We polish sterling silver, paint iron railings, and galvanize nails to combat corrosion. Jewelry and computer chips are electroplated with very thin coatings of gold or silver. Clothes are bleached, and photographs are developed in solutions using chemical reactions that involve electron transfer. Tests for glucose in urine, or alcohol in the breath show vivid color changes. Plants turn energy into chemical compounds through a series of reactions called the *electron transport chain.* All of these reactions involve the transfer of electrons between substances in a chemical process called *oxidation-reduction.*

18.1 OXIDATION NUMBER

The oxidation number of an atom (sometimes called its oxidation state) can be considered to represent the number of electrons lost, gained, or unequally shared by the atom. Oxidation numbers can be zero, positive, or negative. When the oxidation number of an atom is zero, the atom has the same number of electrons assigned to it as there are in the free neutral atom. When the oxidation number is positive, the atom has fewer electrons assigned to it than there are in the neutral atom. When the oxidation number is negative, the atom has more electrons assigned to it than there are in the neutral atom.

The oxidation number of an atom that has lost or gained electrons to form an ion is the same as the plus or minus charge of the ion. In the ionic compound NaCl the oxidation numbers are clearly established to be $+1$ for the Na^+ ion and -1 for the Cl^- ion. The Na^+ ion has one less electron than the neutral Na atom; and the Cl^- ion has one more electron than the neutral Cl atom. In $MgCl_2$ two electrons have transferred from the Mg atom to the Cl atoms; thus, the oxidation number of Mg is $+2$.

In covalently bonded substances, where electrons are shared between two atoms, oxidation numbers are assigned by a somewhat arbitrary system based on relative electronegativities. For symmetrical covalent molecules, such as H_2 and Cl_2, each atom is assigned an oxidation number of zero because the bonding pair of electrons is shared equally between two like atoms, neither of which is more electronegative than the other.

H:H   :C̤̈l:C̤̈l:

When the covalent bond is between two unlike atoms, the bonding electrons are shared unequally because the more electronegative element has a greater attraction for them. In this case the oxidation numbers are determined by assigning both electrons to the more electronegative element.

Thus in compounds with covalent bonds, such as NH_3 and H_2O,

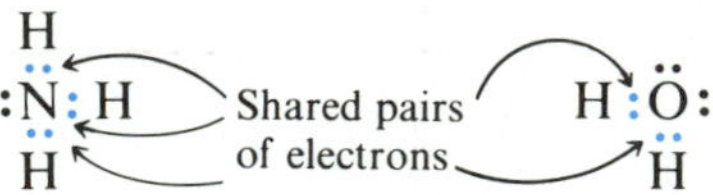

the pairs of electrons are unequally shared between the atoms and are attracted toward the more electronegative elements, N and O. This unequal sharing causes the N and O atoms to be relatively negative with respect to the H atoms. At the same time it causes the H atoms to be relatively positive with respect to the N and O atoms. In H_2O both pairs of shared electrons are assigned to the O atom, giving it two electrons more than the neutral O atom. At the same time, each H atom is assigned one electron less than the neutral H atom. Therefore, the O atom is assigned an oxidation number of -2, and each H atom is assigned an oxidation number of $+1$. In NH_3 the three pairs of shared electrons are assigned to the N atom, giving it three electrons more than the neutral N atom. At the same time, each H atom has one electron less than the neutral atom. Therefore, the N atom is assigned an oxidation number of -3, and each H atom is assigned an oxidation number of $+1$.

The assignment of correct oxidation numbers to elements is essential for balancing oxidation–reduction equations. Review Section 6.2 regarding oxidation numbers, oxidation number tables, and the determination of oxidation numbers from formulas. Table 12.3 lists relative electronegativities of the elements. Rules for assigning oxidation numbers are given in Section 6.2 and are summarized in Table 18.1. Examples showing oxidation numbers in compounds and ions are given in Table 18.2.

Higher electronegativity controls electrons

TABLE 18.1 Rules for Assigning Oxidation Number

1. All elements in their free state (uncombined with other elements) have an oxidation number of zero (for example, Na, Cu, Mg, H_2, O_2, Cl_2, N_2).
2. H is $+1$, except in metal hydrides, where it is -1 (for example, NaH, CaH_2).
3. O is -2, except in peroxides, where it is -1, and in OF_2, where it is $+2$.
4. The metallic element in an ionic compound has a positive oxidation number.
5. In covalent compounds the negative oxidation number is assigned to the most electronegative atom.
6. The algebraic sum of the oxidation numbers of the elements in a compound is zero.
7. The algebraic sum of the oxidation numbers of the elements in a polyatomic ion is equal to the charge of the ion.

TABLE 18.2 Oxidation Numbers of Atoms in Selected Compounds

Ion or compound	Oxidation number	Ion or compound	Oxidation number
H_2O	H, +1; O, −2	$Al_2(SO_4)_3$	Al, +3; S, +6; O, −2
SO_2	S, +4; O, −2	NO	N, +2; O, −2
CH_4	C, −4; H, +1	BCl_3	B, +3; Cl, −1
CO_2	C, +4; O, −2	SO_4^{2-}	S, +6; O, −2
$KMnO_4$	K, +1; Mn, +7; O, −2	NO_3^-	N, +5; O, −2
Na_3PO_4	Na, +1; P, +5; O, −2	CO_3^{2-}	C, +4; O, −2

EXAMPLE 18.1

Determine the oxidation number of each element in (a) KNO_3 and (b) SO_4^{2-}.

SOLUTION

(a) K is a Group IA metal; therefore it has an oxidation number of +1. The oxidation number of each O atom is −2 (Table 18.1, Rule 3). Using these values and the fact that the sum of the oxidation numbers of all the atoms in a compound is zero, we can determine the oxidation number of N.

$$KNO_3$$

$$+1 + N + 3(-2) = 0$$

$$N = +6 - 1 = +5$$

The oxidation numbers are K, +1; N, +5; O, −2.

(b) SO_4^{2-} is an ion; therefore, the sum of oxidation numbers of the S and the O atoms must be −2, the charge of the ion. The oxidation number of each O atom is −2 (Table 18.1, Rule 3). Then

$$SO_4^{2-}$$

$$S + 4(-2) = -2, \qquad S - 8 = -2$$

$$S = -2 + 8 = +6$$

The oxidation numbers are S, +6; O, −2.

PRACTICE Determine the oxidation number of each element in the following compounds:

(a) $BeCl_2$ (b) HClO (c) H_2O_2 (d) NH_4^+ (e) BrO_3^- (f) CrO_4^{2-}

Answers: (a) Be = +2; Cl = −1
(b) H = +1; Cl = +1; O = −2
(c) H = +1; O = −1
(d) N = −3; H = +1
(e) Br = +5; O = −2
(f) Cr = +6; O = −2

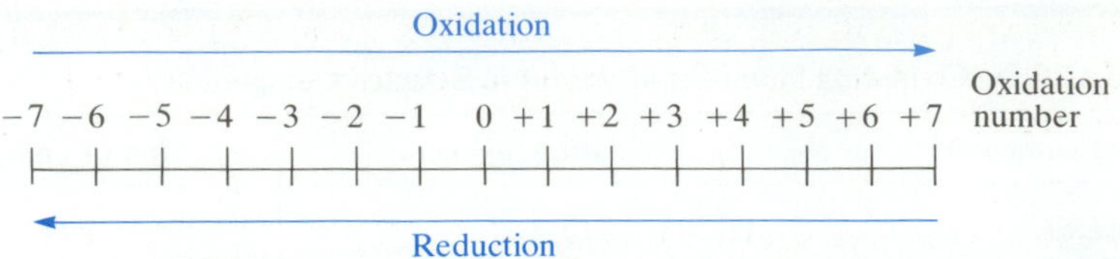

FIGURE 18.1 ▶ Oxidation and reduction. Oxidation results in an increase in the oxidation number, and reduction results in a decrease in the oxidation number.

18.2 Oxidation–Reduction

oxidation–reduction

redox

Oxidation–reduction, also known as **redox**, is a chemical process in which the oxidation number of an element is changed. The process may involve the complete transfer of electrons to form ionic bonds or only a partial transfer or shift of electrons to form covalent bonds.

oxidation

reduction

Oxidation occurs whenever the oxidation number of an element increases as a result of losing electrons. Conversely, **reduction** occurs whenever the oxidation number of an element decreases as a result of gaining electrons. For example, a change in oxidation number from +2 to +3 or from −1 to 0 is oxidation; a change from +5 to +2 or from −2 to −4 is reduction (see Figure 18.1). Oxidation and reduction occur simultaneously in a chemical reaction; one cannot take place without the other.

Many combination, decomposition, and single-displacement reactions involve oxidation–reduction. Let us examine the combustion of hydrogen and oxygen from this point of view:

$$2\,H_2 + O_2 \longrightarrow 2\,H_2O$$

oxidizing agent

reducing agent

Both reactants, hydrogen and oxygen, are elements in the free state and have an oxidation number of zero. In the product, water, hydrogen has been oxidized to +1 and oxygen reduced to −2. The substance that causes an increase in the oxidation state of another substance is called an **oxidizing agent**. The substance that causes a decrease in the oxidation state of another substance is called a **reducing agent**. In this reaction the oxidizing agent is free oxygen, and the reducing agent is free hydrogen. In the reaction

$$Zn(s) + H_2SO_4(aq) \longrightarrow ZnSO_4(aq) + H_2\uparrow$$

metallic zinc is oxidized, and hydrogen ions are reduced. Zinc is the reducing agent, and hydrogen ions, the oxidizing agent. Electrons are transferred from the zinc metal to the hydrogen ions. The reaction is better expressed as

$$Zn^0 + 2\,H^+ + SO_4^{2-} \longrightarrow Zn^{2+} + SO_4^{2-} + H_2^0\uparrow$$

Oxidation:	**Increase in oxidation number**
	Loss of electrons
Reduction:	**Decrease in oxidation number**
	Gain of electrons

The oxidizing agent is reduced and gains electrons. The reducing agent is oxidized and loses electrons. The transfer of electrons is characteristic of all redox reactions.

18.3 Balancing Oxidation–Reduction Equations

Many simple redox equations can be balanced readily by inspection, or trial and error.

$$Na + Cl_2 \longrightarrow NaCl \quad \text{(unbalanced)}$$
$$2\,Na + Cl_2 \longrightarrow 2\,NaCl \quad \text{(balanced)}$$

Balancing this equation is certainly not complicated. But as we study more complex reactions and equations, such as

$$P + HNO_3 + H_2O \longrightarrow NO + H_3PO_4 \quad \text{(unbalanced)}$$
$$3\,P + 5\,HNO_3 + 2\,H_2O \longrightarrow 5\,NO + 3\,H_3PO_4 \quad \text{(balanced)}$$

the trial-and-error method of finding the proper numbers to balance the equation would take an unnecessarily long time.

One systematic method for balancing oxidation–reduction equations is based on the transfer of electrons between the oxidizing and reducing agents. Consider the first equation again.

$$Na^0 + Cl_2^0 \longrightarrow Na^+Cl^- \quad \text{(unbalanced)}$$

In this reaction sodium metal loses one electron per atom when it changes to a sodium ion. At the same time chlorine gains one electron per atom. Because chlorine is diatomic, two electrons per molecule are needed to form a chloride ion from each atom. These electrons are furnished by two sodium atoms. Stepwise, the reaction may be written as two half-reactions, the oxidation half-reaction and the reduction half-reaction:

Oxidation half-reaction	$2\,Na^0 \longrightarrow 2\,Na^+ + 2\,e^-$
Reduction half-reaction	$Cl_2^0 + 2\,e^- \longrightarrow 2\,Cl^-$
	$2\,Na^0 + Cl_2^0 \longrightarrow 2\,Na^+Cl^-$

When the two half-reactions, each containing the same number of electrons, are added together algebraically, the electrons cancel out. In this reaction there are no excess electrons; the two electrons lost by the two sodium atoms are utilized by chlorine. In all redox reactions the loss of electrons by the reducing agent must equal the gain of electrons by the oxidizing agent. Sodium is oxidized; chlorine is reduced. Chlorine is the oxidizing agent; sodium is the reducing agent.

The following examples illustrate a systematic method of balancing more complicated redox equations by the change-in-oxidation-number method.

Example 18.2

Balance the equation

$$Sn + HNO_3 \longrightarrow SnO_2 + NO_2 + H_2O \quad \text{(unbalanced)}$$

SOLUTION

Step 1 Assign oxidation numbers to each element to identify the elements that are being oxidized and those that are being reduced. Write the oxidation numbers below each element in order to avoid confusing them with the charge on an ion.

$$\underset{0}{Sn} + \underset{+1}{H}\underset{+5}{N}\underset{-2}{O_3} \longrightarrow \underset{+4}{Sn}\underset{-2}{O_2} + \underset{+4}{N}\underset{-2}{O_2} + \underset{+1}{H_2}\underset{-2}{O}$$

Note that the oxidation numbers of Sn and N have changed.

Step 2 Now write two new equations, using only the elements that change in oxidation number. Then add electrons to bring the equations into electrical balance. One equation represents the oxidation step; the other represents the reduction step. The oxidation step produces electrons; the reduction step uses electrons.

oxidation $\quad Sn^0 \longrightarrow Sn^{4+} + 4\,e^-$ $\quad$ (Sn^0 loses 4 electrons)

reduction $\quad N^{5+} + 1\,e^- \longrightarrow N^{4+}$ $\quad$ (N^{5+} gains 1 electron)

Step 3 Now multiply the two equations by the smallest whole numbers that will make the loss of electrons by the oxidation step equal to the number of electrons gained in the reduction step. In this reaction the oxidation step is multiplied by 1 and the reduction step by 4. The equations become

oxidation $\quad Sn^0 \longrightarrow Sn^{4+} + 4\,e^-$ $\quad$ (Sn^0 loses 4 electrons)

reduction $\quad 4\,N^{5+} + 4\,e^- \longrightarrow 4\,N^{4+}$ $\quad$ (4 N^{5+} gain 4 electrons)

We have now established the ratio of the oxidizing to the reducing agent as being four atoms of N to one atom of Sn.

Step 4 Now transfer the coefficient that appears in front of each substance in the balanced oxidation–reduction equations to the corresponding substance in the original equation. We need to use 1 Sn, 1 SnO_2, 4 HNO_3, and 4 NO_2:

$$Sn + 4\,HNO_3 \longrightarrow SnO_2 + 4\,NO_2 + H_2O \quad \text{(unbalanced)}$$

Step 5 In the usual manner, balance the remaining elements that are not oxidized or reduced to give the final balanced equation:

$$Sn + 4\,HNO_3 \longrightarrow SnO_2 + 4\,NO_2 + 2\,H_2O \quad \text{(balanced)}$$

In balancing the final elements, we must not change the ratio of the elements that were oxidized and reduced. We should make a final check to ensure that both sides of the equation have the same number of atoms of each element. The final balanced equation contains 1 atom of Sn, 4 atoms of N, 4 atoms of H, and 12 atoms of O on each side.

Because each new equation may present a slightly different problem and because proficiency in balancing equations requires practice, we will work through a few more problems.

EXAMPLE 18.3

Balance the equation

$$I_2 + Cl_2 + H_2O \longrightarrow HIO_3 + HCl \quad \text{(unbalanced)}$$

SOLUTION

Step 1 Assign oxidation numbers:

$$\underset{0}{I_2} + \underset{0}{Cl_2} + \underset{+1\,-2}{H_2O} \longrightarrow \underset{+1\,+5\,-2}{H\ I\ O_3} + \underset{+1\,-1}{HCl}$$

The oxidation numbers of I_2 and Cl_2 have changed, I_2 from 0 to +5, and Cl_2 from 0 to −1.

Step 2 Write oxidation and reduction steps. Balance the number of atoms and then balance the electrical charge using electrons.

oxidation $\quad I_2 \longrightarrow 2\,I^{5+} + 10\,e^-$ $\quad$ (I_2 loses 10 electrons)

reduction $\quad Cl_2 + 2\,e^- \longrightarrow 2\,Cl^-$ $\quad$ (Cl_2 gains 2 electrons)

Step 3 Adjust loss and gain of electrons so that they are equal. Multiply the oxidation step by 1 and the reduction step by 5.

oxidation $\quad I_2 \longrightarrow 2\,I^{5+} + 10\,e^-$ $\quad$ (I_2 loses 10 electrons)

reduction $\quad 5\,Cl_2 + 10\,e^- \longrightarrow 10\,Cl^-$ $\quad$ (5 Cl_2 gain 10 electrons)

Step 4 Transfer the coefficients from the balanced redox equations into the original equation. We need to use 1 I_2, 2 HIO_3, 5 Cl_2, and 10 HCl.

$$I_2 + 5\,Cl_2 + H_2O \longrightarrow 2\,HIO_3 + 10\,HCl \quad \text{(unbalanced)}$$

Step 5 Balance the remaining elements, H and O:

$$I_2 + 5\,Cl_2 + 6\,H_2O \longrightarrow 2\,HIO_3 + 10\,HCl \quad \text{(balanced)}$$

Check: The final balanced equation contains 2 atoms of I, 10 atoms of Cl, 12 atoms of H, and 6 atoms of O on each side.

EXAMPLE 18.4

Balance the equation

$$K_2Cr_2O_7 + FeCl_2 + HCl \longrightarrow CrCl_3 + KCl + FeCl_3 + H_2O \quad \text{(unbalanced)}$$

SOLUTION

Step 1 Assign oxidation numbers (Cr and Fe have changed):

$$\underset{+1\,+6\,-2}{K_2Cr_2O_7} + \underset{+2\,-1}{FeCl_2} + \underset{+1\,-1}{HCl} \longrightarrow \underset{+3\,-1}{CrCl_3} + \underset{+1\,-1}{KCl} + \underset{+3\,-1}{FeCl_3} + \underset{+1\,-2}{H_2O}$$

Step 2 Write the oxidation and reduction steps. Balance the number of atoms and then balance the electrical charge using electrons.

oxidation $\quad Fe^{2+} \longrightarrow Fe^{3+} + 1\,e^-$ $\quad$ (Fe^{2+} loses 1 electron)

reduction $\quad 2\,Cr^{6+} + 6\,e^- \longrightarrow 2\,Cr^{3+}$ $\quad$ (2 Cr^{6+} gain 6 electrons)

Step 3 Balance the loss and gain of electrons. Multiply the oxidation step by 6 and the reduction step by 1 to equalize the transfer of electrons.

oxidation $6\ Fe^{2+} \longrightarrow 6\ Fe^{3+} + 6\ e^-$ ($6\ Fe^{2+}$ lose 6 electrons)

reduction $2\ Cr^{6+} + 6\ e^- \longrightarrow 2\ Cr^{3+}$ ($2\ Cr^{6+}$ gain 6 electrons)

Step 4 Transfer the coefficients from the balanced redox equations into the original equation. (Note that one formula unit of $K_2Cr_2O_7$ contains two Cr atoms.) We need to use 1 $K_2Cr_2O_7$, 2 $CrCl_3$, 6 $FeCl_2$, and 6 $FeCl_3$.

$$K_2Cr_2O_7 + 6\ FeCl_2 + HCl \longrightarrow 2\ CrCl_3 + KCl + 6\ FeCl_3 + H_2O \quad \text{(unbalanced)}$$

Step 5 Balance the remaining elements in this order: K, Cl, H, O.

$$K_2Cr_2O_7 + 6\ FeCl_2 + 14\ HCl \longrightarrow 2\ CrCl_3 + 2\ KCl + 6\ FeCl_3 + 7\ H_2O \quad \text{(balanced)}$$

Check: The final balanced equation contains 2 K atoms, 2 Cr atoms, 7 O atoms, 6 Fe atoms, 26 Cl atoms, and 14 H atoms on each side.

Practice Balance the following equations using the change in oxidation number method:

(a) $HNO_3 + S \longrightarrow NO_2 + H_2SO_4 + H_2O$
(b) $CrCl_3 + MnO_2 + H_2O \longrightarrow MnCl_2 + H_2CrO_4$
(c) $KMnO_4 + HCl + H_2S \longrightarrow KCl + MnCl_2 + S + H_2O$

Answers: The coefficients for the balanced equations from left to right are
(a) 6, 1, 6, 1, 2 (b) 2, 3, 2, 3, 2 (c) 2, 6, 5, 2, 2, 5, 8

18.4 Balancing Ionic Redox Equations

The main difference between balancing ionic and balancing molecular redox equations is in the handling of ions. In addition to having the same number of each kind of element on both sides of the final equation, the net charges must also be equal. In assigning oxidation numbers we must be careful to consider the charge on the ions. In many respects, balancing ionic equations is much simpler than balancing molecular equations.

Several methods can be used to balance ionic redox equations. These methods include, with slight modification, the oxidation-number method just shown for molecular equations. But the most popular is probably the ion–electron method, which is explained in the following paragraphs.

The ion–electron method uses ionic charges and electrons to balance ionic redox equations. Oxidation numbers are not formally used, but it is necessary to determine what is being oxidized and what is being reduced. The method is as follows:

1. Write the two half-reactions that contain the elements being oxidized and reduced.
2. Balance the elements other than oxygen and hydrogen.
3. Balance oxygen and hydrogen: (a) acidic solutions; (b) basic solutions.
   (a) For reactions that occur in acidic solution, use H^+ and H_2O to balance oxygen and hydrogen. For each oxygen needed, use one H_2O. Then add H^+ as needed to balance the hydrogen atoms.
   (b) Balancing equations that occur in alkaline solutions is a bit more complicated. For reactions that occur in alkaline solutions, first balance as though the reaction were in an acidic solution, using Steps 1, 2, and 3(a). Then add as many OH^- ions to each side of the equation as there are H^+ ions in the equation. Now combine the H^+ and OH^- ions into water (for example, 4 H^+ and 4 OH^- give 4 H_2O). Rewrite the equation, canceling equal numbers of water molecules that appear on opposite sides of the equation.
4. Add electrons (e^-) to each half-reaction to bring them into electrical balance.
5. Since the loss and gain of electrons must be equal, multiply each half-reaction by the appropriate number to make the number of electrons the same in each half-reaction.
6. Add the two half-reactions together, canceling electrons and any other identical substances that appear on opposite sides of the equation.

EXAMPLE 18.5

Balance this equation using the ion–electron method:

$$MnO_4^- + S^{2-} \longrightarrow Mn^{2+} + S^0 \quad \text{(acidic solution)}$$

SOLUTION

Step 1 Write two half-reactions, one containing the element being oxidized and the other, the element being reduced. (Use the entire molecule or ion.)

oxidation $S^{2-} \longrightarrow S^0$

reduction $MnO_4^- \longrightarrow Mn^{2+}$

Step 2 Balance elements other than oxygen and hydrogen (accomplished in Step 1 in this example—1 S and 1 Mn on each side).

Step 3 Balance O and H. The oxidation requires neither O nor H, but the reduction equation needs 4 H_2O on the right and 8 H^+ on the left [Step 3(a)].

$$S^{2-} \longrightarrow S^0$$

$$8\,H^+ + MnO_4^- \longrightarrow Mn^{2+} + 4\,H_2O$$

Step 4 Balance electrically with electrons.

$$S^{2-} \longrightarrow S^0 + 2\,e^- \qquad \text{(net charge} = -2 \text{ on each side)}$$

$$5\,e^- + 8\,H^+ + MnO_4^- \longrightarrow Mn^{2+} + 4\,H_2O \qquad \text{(net charge} = +2 \text{ on each side)}$$

Step 5 Equalize loss and gain of electrons. In this case multiply the oxidation equation by 5 and the reduction equation by 2.

$$5\ S^{2-} \longrightarrow 5\ S^0 + 10\ e^-$$

$$10\ e^- + 16\ H^+ + 2\ MnO_4^- \longrightarrow 2\ Mn^{2+} + 8\ H_2O$$

Step 6 Add the two half-reactions together, cancelling the 10 e⁻ from each side, to obtain the balanced equation.

$$5\ S^{2-} \longrightarrow 5\ S^0 + \cancel{10\ e^-}$$

$$\cancel{10\ e^-} + 16\ H^+ + 2\ MnO_4^- \longrightarrow 2\ Mn^{2+} + 8\ H_2O$$

$$16\ H^+ + 2\ MnO_4^- + 5\ S^{2-} \longrightarrow 2\ Mn^{2+} + 5\ S^0 + 8\ H_2O \quad \text{(balanced)}$$

Check: Both sides of the equation have a charge of +4 and contain the same number of atoms of each element.

EXAMPLE 18.6

Balance the following equation.

$$CrO_4^{2-} + Fe(OH)_2 \longrightarrow Cr(OH)_3 + Fe(OH)_3 \quad \text{(basic solution)}$$

SOLUTION

Step 1 Write the two half-reactions.

oxidation $Fe(OH)_2 \longrightarrow Fe(OH)_3$

reduction $CrO_4^{2-} \longrightarrow Cr(OH)_3$

Step 2 Balance elements other than H and O (accomplished in Step 1).

Step 3 Balance O and H as though the solution were acidic (Step 3a). Use H_2O and H^+. To balance O and H in the oxidation equation, add 1 H_2O on the left and 1 H^+ on the right side of the equation.

$$Fe(OH)_2 + H_2O \longrightarrow Fe(OH)_3 + H^+$$

Add 1 OH^- to each side.

$$Fe(OH)_2 + H_2O + OH^- \longrightarrow Fe(OH)_3 + H^+ + OH^-$$

Combine H^+ and OH^- as H_2O and rewrite, cancelling H_2O on each side [Step 3(b)].

$$Fe(OH)_2 + \cancel{H_2O} + OH^- \longrightarrow Fe(OH)_3 + \cancel{H_2O}$$

$$\boxed{Fe(OH)_2 + OH^- \longrightarrow Fe(OH)_3}$$

To balance O and H in the reduction equation, add 1 H_2O on the right and 5 H^+ on the left.

$$CrO_4^{2-} + 5\ H^+ \longrightarrow Cr(OH)_3 + H_2O$$

Add 5 OH^- to each side.

$$CrO_4^{2-} + 5\ H^+ + 5\ OH^- \longrightarrow Cr(OH)_3 + H_2O + 5\ OH^-$$

Combine $5\ H^+ + 5\ OH^- \longrightarrow 5\ H_2O$.

$CrO_4^{2-} + 5\,H_2O \longrightarrow Cr(OH)_3 + H_2O + 5\,OH^-$

Rewrite, cancelling 1 H_2O from each side.

$$CrO_4^{2-} + 4\,H_2O \longrightarrow Cr(OH)_3 + 5\,OH^-$$

Step 4 Balance electrically with electrons.

$Fe(OH)_2 + OH^- \longrightarrow Fe(OH)_3 + e^-$ (balanced oxidation equation)

$CrO_4^{2-} + 4\,H_2O + 3\,e^- \longrightarrow Cr(OH)_3 + 5\,OH^-$ (balanced reduction equation)

Step 5 Equalize the loss and gain of electrons. Multiply the oxidation reaction by 3.

$3\,Fe(OH)_2 + 3\,OH^- \longrightarrow 3\,Fe(OH)_3 + 3\,e^-$

$CrO_4^{2-} + 4\,H_2O + 3\,e^- \longrightarrow Cr(OH)_3 + 5\,OH^-$

Step 6 Add the two half-reactions together, cancelling the 3 e^- and 3 OH^- from each side of the equation.

$$3\,Fe(OH)_2 + 3\,OH^- \longrightarrow 3\,Fe(OH)_3 + \cancel{3e^-}$$
$$CrO_4^{2-} + 4\,H_2O + \cancel{3e^-} \longrightarrow Cr(OH)_3 + 5\,OH^-$$

$CrO_4^{2-} + 3\,Fe(OH)_2 + 4\,H_2O \longrightarrow Cr(OH)_3 + 3\,Fe(OH)_3 + 2\,OH^-$ (balanced)

Check: Each side of the equation has a charge of -2 and contains the same number of atoms of each element.

PRACTICE Balance the following equations using the ion-electron method:
(a) $I^- + NO_2^- \longrightarrow I_2 + NO$ (acidic solution)
(b) $Cl_2 + IO_3^- \longrightarrow IO_4^- + Cl^-$ (basic solution)
(c) $AuCl_4^- + Sn^{2+} \longrightarrow Sn^{4+} + AuCl + Cl^-$

Answers: (a) $4H^+ + 2I^- + 2NO_2^- \longrightarrow I_2 + 2NO + 2H_2O$
(b) $2OH^- + Cl_2 + IO_3^- \longrightarrow IO_4^- + H_2O + 2Cl^-$
(c) $AuCl_4^- + Sn^{2+} \longrightarrow Sn^{4+} + AuCl + 3Cl^-$

Ionic equations can also be balanced by using the change in oxidation number method shown in Section 18.3. Steps 1, 2, 3, and 4 are the same. In Step 5 the electrical charges are balanced with H^+ in acidic solutions and with OH^- in basic solutions. H_2O is then used to complete the balancing of the equation. Let us use the same equation as in Example 18.6 as an example of this method.

EXAMPLE 18.7

Balance the following equation using the change in oxidation number method.

$CrO_4^{2-} + Fe(OH)_2 \longrightarrow Cr(OH)_3 + Fe(OH)_3$ (basic solution)

SOLUTION

Steps 1 and 2 Assign oxidation numbers and balance the charges with electrons.

reduction $Cr^{6+} + 3\ e^- \longrightarrow Cr^{3+}$ (Cr^{6+} gains 3 e^-)

oxidation $Fe^{2+} \longrightarrow Fe^{3+} + e^-$ (Fe^{2+} loses 1 e^-)

Step 3 Equalize the loss and gain of electrons. Multiply the oxidation step by 3.

$3\ Fe^{2+} \longrightarrow 3\ Fe^{3+} + 3\ e^-$ (3 Fe^{2+} lose 3 e^-)

$Cr^{6+} + 3\ e^- \longrightarrow Cr^{3+}$ (Cr^{6+} gains 3 e^-)

Step 4 Transfer coefficients back to the original equation.

$$CrO_4^{2-} + 3\ Fe(OH)_2 \longrightarrow Cr(OH)_3 + 3\ Fe(OH)_3$$

Step 5 Balance electrically. Because the solution is basic, use OH^- to balance charges. The charge on the left side is -2, and on the right side is 0. Add 2 OH^- ions to the right side of the equation.

$$CrO_4^{2-} + 3\ Fe(OH)_2 \longrightarrow Cr(OH)_3 + 3\ Fe(OH)_3 + 2\ OH^-$$

Adding 4 H_2O to the left side balances the equation.

$$CrO_4^{2-} + 3\ Fe(OH)_2 + 4\ H_2O \longrightarrow Cr(OH)_3 + 3\ Fe(OH)_3 + 2\ OH^- \quad \text{(balanced)}$$

Check: Each side of the equation has a charge of -2 and contains the same number of atoms of each element.

Practice Balance each of the following equations using the change in oxidation number method:

(a) $Zn \longrightarrow Zn(OH)_4^{2-} + H_2$ (basic solution)
(b) $H_2O_2 + Sn^{2+} \longrightarrow Sn^{4+}$ (acidic solution)
(c) $Cu + Cu^{2+} \longrightarrow Cu_2O$ (basic solution)

Answers: (a) $Zn + 2\ H_2O + 2\ OH^- \longrightarrow Zn(OH)_4^{2-} + H_2$
(b) $H_2O_2 + Sn^{2+} + 2\ H^+ \longrightarrow Sn^{4+} + 2\ H_2O$
(c) $Cu + Cu^{2+} + 2\ OH^- \longrightarrow Cu_2O + H_2O$

18.5 Activity Series of Metals

Knowledge of the relative chemical reactivities of the elements helps us predict the course of many chemical reactions.

Calcium reacts with cold water to produce hydrogen, and magnesium reacts with steam to produce hydrogen. Therefore, calcium is considered to be a more reactive metal than magnesium.

$$Ca(s) + 2\ H_2O(l) \longrightarrow Ca(OH)_2(aq) + H_2(g)$$

$$Mg(s) + \underset{\text{Steam}}{H_2O(g)} \longrightarrow MgO(s) + H_2(g)$$

◀ **A strip of copper has been placed in silver nitrate solution. Silver crystals are formed on the wire, and the pale blue of the solution indicates the presence of copper ions.**

The difference in their activity is attributed to the fact that calcium loses its two valence electrons more easily than does magnesium and is therefore more reactive and/or more readily oxidized than magnesium.

When a strip of copper is placed in a solution of silver nitrate ($AgNO_3$), free silver begins to plate out on the copper. After the reaction has continued for some time, we can observe a blue color in the solution, indicating the presence of copper(II) ions. If a strip of silver is placed in a solution of copper(II) nitrate, $Cu(NO_3)_2$, no reaction is visible. The equations are

$$Cu^0 + 2\,AgNO_3(aq) \longrightarrow 2\,Ag^0 + Cu(NO_3)_2(aq)$$

$$Cu^0 + 2\,Ag^+ \longrightarrow 2\,Ag^0 + Cu^{2+} \qquad \text{net ionic equation}$$

$$Cu^0 \longrightarrow Cu^{2+} + 2\,e^- \qquad \text{oxidation of } Cu^0$$

$$Ag^+ + e^- \longrightarrow Ag^0 \qquad \text{reduction of } Ag^+$$

$$Ag^0 + Cu(NO_3)_2(aq) \longrightarrow \text{no reaction}$$

In the reaction between Cu and $AgNO_3$, electrons are transferred from Cu^0 atoms to Ag^+ ions in solution. Copper has a greater tendency than silver to lose electrons, so an electrochemical force is exerted upon silver ions to accept electrons from copper atoms. When a Ag^+ ion accepts an electron, it is reduced to a Ag^0 atom and is no longer soluble in solution. At the same time, Cu^0 is oxidized and goes into solution as Cu^{2+} ions. From this reaction we can conclude that copper is more reactive than silver.

Metals such as sodium, magnesium, zinc, and iron, which react with solutions of acids to liberate hydrogen, are more reactive than hydrogen. Metals such as copper, silver, and mercury, which do not react with solutions of acids to liberate hydrogen, are less reactive than hydrogen. By studying a series of reactions such as those given above, we can list metals according to their chemical activity, placing the most active at the top and the least active at the bottom. This list is called the **activity series of metals**. Table 18.3 shows some of the common

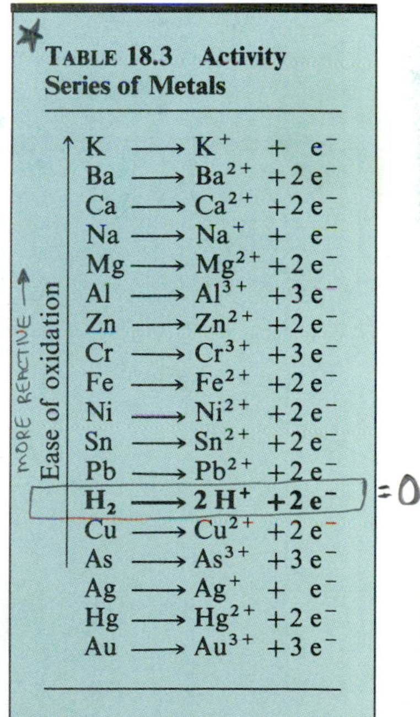

TABLE 18.3 Activity Series of Metals

Ease of oxidation ↑			
K	⟶	K^+	$+\ e^-$
Ba	⟶	Ba^{2+}	$+2\,e^-$
Ca	⟶	Ca^{2+}	$+2\,e^-$
Na	⟶	Na^+	$+\ e^-$
Mg	⟶	Mg^{2+}	$+2\,e^-$
Al	⟶	Al^{3+}	$+3\,e^-$
Zn	⟶	Zn^{2+}	$+2\,e^-$
Cr	⟶	Cr^{3+}	$+3\,e^-$
Fe	⟶	Fe^{2+}	$+2\,e^-$
Ni	⟶	Ni^{2+}	$+2\,e^-$
Sn	⟶	Sn^{2+}	$+2\,e^-$
Pb	⟶	Pb^{2+}	$+2\,e^-$
H_2	⟶	**$2\,H^+$**	**$+2\,e^-$**
Cu	⟶	Cu^{2+}	$+2\,e^-$
As	⟶	As^{3+}	$+3\,e^-$
Ag	⟶	Ag^+	$+\ e^-$
Hg	⟶	Hg^{2+}	$+2\,e^-$
Au	⟶	Au^{3+}	$+3\,e^-$

activity series of metals

metals in the series. The arrangement corresponds to the ease with which the elements listed are oxidized or lose electrons, with the most easily oxidizable element listed first. More extensive tables are available in chemistry reference books.

The general principles governing the arrangement and use of the activity series are as follows:

1. The reactivity of the metals listed decreases from top to bottom.
2. A free metal can displace the ion of a second metal from solution provided that the free metal is above the second metal in the Activity Series.
3. Free metals above hydrogen react with nonoxidizing acids in solution to liberate hydrogen gas.
4. Free metals below hydrogen do not liberate hydrogen from acids.
5. Conditions such as temperature and concentration may affect the relative position of some of these elements.

Two examples of the application of the activity series are given in the following problems.

EXAMPLE 18.8 Will zinc metal react with dilute sulfuric acid?

SOLUTION From Table 18.3 we see that zinc is above hydrogen; therefore zinc atoms will lose electrons more readily than hydrogen atoms. Hence, zinc atoms will reduce hydrogen ions from the acid to form hydrogen gas and zinc ions. In fact, these reagents are commonly used for the laboratory preparation of hydrogen. The equation is

$$Zn(s) + H_2SO_4(aq) \longrightarrow ZnSO_4(aq) + H_2(g)$$

$$Zn + 2\,H^+ \longrightarrow Zn^{2+} + H_2(g) \qquad \text{net ionic equation}$$

EXAMPLE 18.9 Will a reaction occur when copper metal is placed in an iron(II) sulfate solution?

SOLUTION No, copper lies below iron in the series, loses electrons less easily than iron, and therefore will not displace iron(II) ions from solution. In fact, the reverse is true. When an iron nail is dipped into a copper(II) sulfate solution, it becomes coated with free copper. The equations are

$$Cu(s) + FeSO_4(aq) \longrightarrow \text{no reaction}$$

$$Fe(s) + CuSO_4(aq) \longrightarrow FeSO_4(aq) + Cu\downarrow$$

From Table 18.3 we may abstract the following pair in their relative position to each other.

$$Fe \longrightarrow Fe^{2+} + 2\,e^-$$

$$Cu \longrightarrow Cu^{2+} + 2\,e^-$$

According to the second principle listed above on the use of the activity series, we can predict that free iron will react with copper(II) ions in solution to form free copper metal and iron(II) ions in solution.

$$Fe(s) + Cu^{2+}(aq) \longrightarrow Fe^{2+}(aq) + Cu(s) \qquad \text{net ionic equation}$$

PRACTICE Indicate whether the following reactions will occur:
(a) Sodium metal is placed in dilute hydrochloric acid.
(b) A piece of lead is placed in magnesium nitrate solution.
(c) Mercury is placed in a solution of silver nitrate.
Answers: (a) yes (b) no (c) no

18.6 Electrolytic and Voltaic Cells

The process in which electrical energy is used to bring about chemical change is known as **electrolysis.** An **electrolytic cell** uses electrical energy to produce a nonspontaneous chemical reaction. The use of electrical energy has many applications in the chemical industry—for example, in the production of sodium, sodium hydroxide, chlorine, fluorine, magnesium, aluminum, and pure hydrogen and oxygen, and in the purification and electroplating of metals.

electrolysis

electrolytic cell

What happens when an electric current is passed through a solution? Let us consider a hydrochloric acid solution in a simple electrolytic cell, as shown in Figure 18.2. The cell consists of a source of direct current (a battery) connected to two electrodes that are immersed in a solution of hydrochloric acid. The negative electrode is called the **cathode** because cations are attracted to it. The positive electrode is called the **anode** because anions are attracted to it. The cathode is attached to the negative pole and the anode to the positive pole of the battery. The battery supplies electrons to the cathode.

cathode

anode

When the electric circuit is completed; positive hydronium ions (H_3O^+) migrate to the cathode, where they pick up electrons and evolve hydrogen gas. At the same time the negative chloride ions (Cl^-) migrate to the anode, where they lose electrons and evolve chlorine gas.

Reaction at the cathode

(reduction) $H_3O^+ + 1\,e^- \longrightarrow H^0 + H_2O$

$H^0 + H^0 \longrightarrow H_2(g)$

Reaction at the anode

(oxidation) $Cl^- \longrightarrow Cl^0 + 1\,e^-$

$Cl^0 + Cl^0 \longrightarrow Cl_2(g)$

Net reaction $2\,HCl(aq) \xrightarrow{\text{Electrolysis}} H_2(g) + Cl_2(g)$

$2Cl^- + 2H_2O \rightarrow Cl_2 + 2OH^- + H_2$

Note that oxidation–reduction has taken place. Chloride ions lose electrons (are oxidized) at the anode, and hydronium ions gain electrons (are reduced) at the cathode.

Oxidation always occurs at the anode and reduction at the cathode.

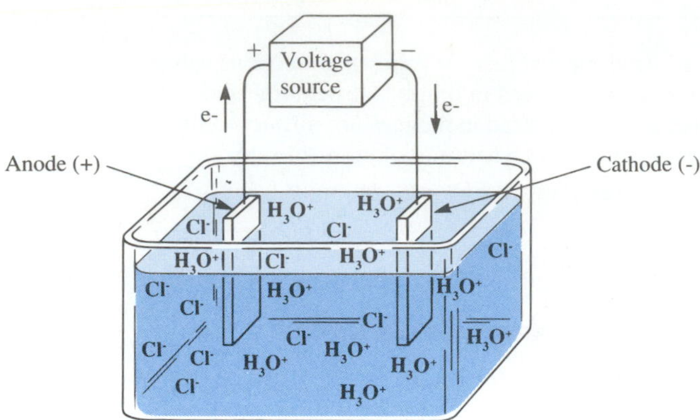

FIGURE 18.2 ▶
Electrolysis. During the electrolysis of a hydrochloric acid solution, positive hydronium ions are attracted to the cathode, where they gain electrons and form hydrogen gas. Chloride ions migrate to the anode, where they lose electrons and form chlorine gas. The equation for this process is
$2\,HCl(aq) \rightarrow H_2(g) + Cl_2(g)$

When concentrated sodium chloride solutions (brines) are electrolyzed, the products are sodium hydroxide, hydrogen, and chlorine. The overall reaction is

$$2\,Na^+ + 2\,Cl^- + 2\,H_2O \xrightarrow{\text{Electrolysis}} 2\,Na^+ + 2\,OH^- + H_2(g) + Cl_2(g)$$

The net ionic equation is

$$2\,Cl^-(aq) + 2\,H_2O(l) \longrightarrow 2\,OH^-(aq) + H_2(g) + Cl_2(g)$$

During the electrolysis, Na^+ ions move toward the cathode and Cl^- ions move toward the anode. The anode reaction is similar to that of hydrochloric acid; chlorine is liberated.

$$2\,Cl^-(aq) \longrightarrow Cl_2(g) + 2\,e^-$$

Even though Na^+ ions are attracted by the cathode, the facts show that hydrogen is liberated there. No evidence of metallic sodium is found, but the area around the cathode tests alkaline from accumulated OH^- ions. The reaction at the cathode is

$$2\,H_2O(l) + 2\,e^- \longrightarrow H_2(g) + 2\,OH^-(aq)$$

If the electrolysis is allowed to continue until all the chloride is reacted, the solution remaining will contain only sodium hydroxide, which on evaporation yields solid NaOH. Large tonnages of sodium hydroxide and chlorine are made by this process.

When molten sodium chloride (without water) is subjected to electrolysis, metallic sodium and chlorine gas are formed:

$$2\,Na^+(l) + 2\,Cl^-(l) \xrightarrow{\text{Electrolysis}} 2\,Na(l) + Cl_2(g)$$

▲
Electroplating is a process involving oxidation reduction reactions. It can be used to protect an object or for decorative effect.

An important electrochemical application is the electroplating of metals. Electroplating is the art of covering a surface or an object with a thin adherent electrodeposited metal coating. Electroplating is done for protection of the surface of the base metal or for a purely decorative effect. The layer deposited is surprisingly thin, varying from as little as 5×10^{-5} cm to 2×10^{-3} cm, depending on the metal and the intended use. The object to be plated is set up as the cathode and is immersed in a solution containing ions of the metal to be plated. When an electric current passes through the solution, metal ions that

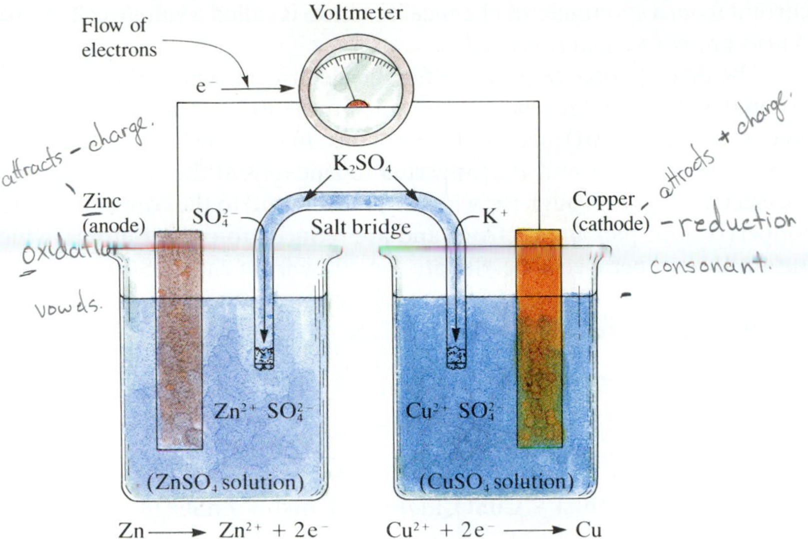

◄ **FIGURE 18.3**
Zinc–copper voltaic cell. The cell has a potential of 1.1 volts when $ZnSO_4$ and $CuSO_4$ solutions are 1.0 *M*. The salt bridge provides electrical contact between the two half-cells.

migrate to the cathode are reduced, depositing on the object as the free metal. In most cases the metal deposited on the object is replaced in the solution by using an anode of the same metal. The following equations show the chemical changes in the electroplating of nickel:

Reaction at the cathode	$Ni^{2+}(aq) + 2\,e^- \longrightarrow Ni(s)$	Ni plated out on an object
Reaction at the anode	$Ni(s) \longrightarrow Ni^{2+}(aq) + 2\,e^-$	Ni replenished in solution

Metals commonly used in commercial electroplating are copper, nickel, zinc, lead, cadmium, chromium, tin, gold, and silver.

In the electrolytic cell shown in Figure 18.2, electrical energy from the voltage source is used to bring about nonspontaneous redox reactions. The hydrogen and chlorine produced have more potential energy than was present in the hydrochloric acid before electrolysis.

Conversely, some spontaneous redox reactions can be made to supply useful amounts of electrical energy. When a piece of zinc is put in a copper(II) sulfate solution, the zinc quickly becomes coated with metallic copper. We expect this coating to happen because zinc is above copper in the activity series; copper(II) ions are therefore reduced by zinc atoms:

$$Zn^0(s) + Cu^{2+}(aq) \longrightarrow Zn^{2+}(aq) + Cu^0(s)$$

This reaction is clearly a spontaneous redox reaction, but simply dipping a zinc rod into a copper(II) sulfate solution will not produce useful electric current. However, when we carry out this reaction in the cell shown in Figure 18.3, an electric current is produced. The cell consists of a piece of zinc immersed in a zinc sulfate solution and connected by a wire through a voltmeter to a piece of copper immersed in copper(II) sulfate solution. The two solutions are connected by a salt bridge. Such a cell produces an electric current and a potential of about 1.1 volts when both solutions are 1.0 *M* in concentration. A cell that produces electric

voltaic cell

current from a spontaneous chemical reaction is called a **voltaic cell**. A voltaic cell is also known as a *galvanic cell.*

The driving force responsible for the electric current in the zinc–copper cell originates in the great tendency of zinc atoms to lose electrons relative to the tendency of copper(II) ions to gain electrons. In the cell shown in Figure 18.3 zinc atoms lose electrons and are converted to zinc ions at the zinc electrode surface; the electrons flow through the wire (external circuit) to the copper electrode. Here copper(II) ions pick up electrons and are reduced to copper atoms, which plate out on the copper electrode. Sulfate ions flow from the $CuSO_4$ solution via the salt bridge into the $ZnSO_4$ solution (internal circuit) to complete the circuit. The equations for the reactions of this cell are

anode	$Zn^0(s) \longrightarrow Zn^{2+}(aq) + 2\,e^-$	(oxidation)
cathode	$Cu^{2+}(aq) + 2\,e^- \longrightarrow Cu^0(s)$	(reduction)
net ionic	$Zn^0(s) + Cu^{2+}(aq) \longrightarrow Zn^{2+}(aq) + Cu^0(s)$	
overall	$Zn(s) + CuSO_4(aq) \longrightarrow Cu(s) + ZnSO_4(aq)$	

The redox reaction, the movement of electrons in the metallic or external part of the circuit, and the movement of ions in the solution or internal part of the circuit of the copper–zinc cell are very similar to the actions that occur in the electrolytic cell of Figure 18.2. The only important difference is that the reactions of the zinc–copper cell are spontaneous. This spontaneity is the crucial difference between all voltaic and electrolytic cells.

Voltaic cells use chemical reactions to produce electrical energy, and electrolytic cells use electrical energy to produce chemical reactions.

Although the zinc–copper voltaic cell is no longer used commercially, it was used to energize the first transcontinental telegraph lines. Such cells were the direct ancestors of the many different kinds of "dry" cells that operate portable radio and television sets, automatic cameras, tape recorders, and so on.

One such "dry" cell, the alkaline zinc–mercury cell, is shown diagrammatically in Figure 18.4. The reactions occurring in this cell are

anode	$Zn^0 + 2\,OH^- \longrightarrow ZnO + H_2O + 2\,e^-$	(oxidation)
cathode	$HgO + H_2O + 2\,e^- \longrightarrow Hg^0 + 2\,OH^-$	(reduction)
net ionic	$Zn^0 + Hg^{2+} \longrightarrow Zn^{2+} + Hg^0$	
overall	$Zn^0 + HgO \longrightarrow ZnO + Hg^0$	

To offset the relatively high initial cost, this cell (a) provides current at a very steady potential of about 1.5 volts, (b) has an exceptionally long service life—that is, high energy output to weight ratio, (c) is completely self-contained, and (d) can be stored for relatively long periods of time when not in use.

An automobile storage battery is an energy reservoir. The charged battery acts as a voltaic cell and through chemical reactions furnishes electrical energy to operate the starter, lights, radio, and so on. When the engine is running, a

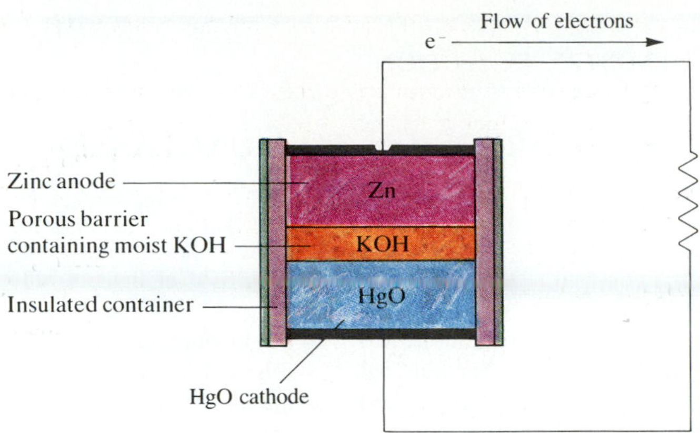

◄ **FIGURE 18.4** **Diagram of an alkaline zinc–mercury cell.**

generator, or alternator, produces and forces an electric current through the battery and, by electrolytic chemical action, restores it to the charged condition.

The cell unit consists of a lead plate filled with spongy lead and a lead dioxide plate, both immersed in dilute sulfuric acid solution, which serves as the electrolyte (see Figure 18.5). When the cell is discharging, or acting as a voltaic cell, these reactions occur:

Pb plate (anode)	$Pb^0 \longrightarrow Pb^{2+} + 2\,e^-$	(oxidation)
PbO_2 plate (cathode)	$PbO_2 + 4\,H^+ + 2\,e^- \longrightarrow Pb^{2+} + 2\,H_2O$	(reduction)
Net ionic redox reaction	$Pb^0 + PbO_2 + 4\,H^+ \longrightarrow 2\,Pb^{2+} + 2\,H_2O$	
Precipitation reaction on plates	$Pb^{2+} + SO_4^{2-} \longrightarrow PbSO_4(s)$	

Because lead(II) sulfate is insoluble, the Pb^{2+} ions combine with SO_4^{2-} ions to form a coating of $PbSO_4$ on each plate. The overall chemical reaction of the cell is

$$Pb(s) + PbO_2(s) + 2\,H_2SO_4(aq) \xrightarrow[\text{cycle}]{\text{Discharge}} 2\,PbSO_4(s) + 2\,H_2O(l)$$

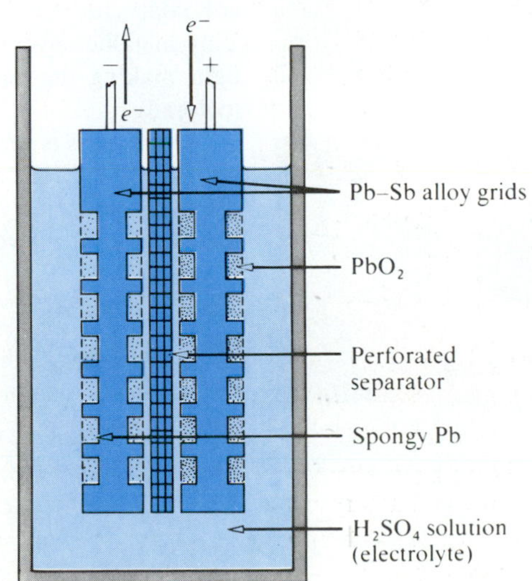

◄ **FIGURE 18.5** **Cross-sectional diagram of a lead storage battery cell.**

Chemistry in Action

Trading Electrons—Corrosion to Sunglasses

Corrosion has been a problem for the world since the earliest use of metals. One familiar example of the complexities of corrosion has been the renovation of our famous Statue of Liberty. The beautiful lady was given to the United States by France by 1886. She was constructed of a skeleton of iron bars (ribs) which were bent to precisely conform to her shape and covered with a thin skin of copper. Her skin was connected to the skeleton with special copper "saddles" or bands which overlapped the iron bars and were riveted in place. Iron will corrode (oxidize) in moist air, and if in contact with copper and an electrolyte (sea water), the corrosion rate can increase one-hundred-fold. The harbor environment of New York fosters accelerated corrosion; and within less than a century the iron ribs had corroded to a point where the lady was in serious danger of collapse.

In order to repair and reinforce the statue each of the iron bars (over 1300 in all) were replaced one at a time with stainless steel, which corrodes much less easily. At the same time the copper saddles were coated with Teflon to insulate them from contact with the iron and further reduce future corrosion.

The external appearance of the statue was left essentially unchanged. The bluish green patina is a natural product formed as copper undergoes atmospheric corrosion. Patina may take several forms and exists as a protective film that adheres tightly to the surface of the copper and reduces the rate of corrosion underneath. The patina on the Statue of Liberty is primarily $CuSO_4 \cdot 3Cu(OH)_2$. It has protected the lady well—during her hundred years the copper skin has thinned only about 4%. Unfortunately, acid rain appears to be changing the patina to $CuSO_4 \cdot 2Cu(OH)_2$, which doesn't bond as closely to the copper surface. Chemists are concerned that this change will result in the loss of patina and increased corrosion of the copper skin.

Oxidation–reduction reactions are also the basis for many interesting and useful applications in technology. One such application is photochromic glass used for the lenses in light sensitive glasses. Lenses manufactured by the Corning Glass Company can change from transmitting 85% of light to only transmitting 22% of light when exposed to bright sunlight.

Photochromic glass is composed of linked tetrahedrons of silicon and oxygen atoms jumbled together in a disorderly array, with crystals of silver chloride caught in between the silica tetrahedrons. When the glass is clear the visible light passes right through the molecules. The glass absorbs ultraviolet light, however, and this energy triggers an oxidation–reduction reaction between Ag^+ and Cl^-.

$$Ag^+ + Cl^- \xrightarrow{\text{uv light}} Ag^0 + Cl^0$$

To prevent the reaction from reversing itself immediately a few ions of Cu^+ are incorporated into the silver chloride crystal. These Cu^+ ions react with the newly formed chlorine atoms:

$$Cu^+ + Cl^0 \longrightarrow Cu^{2+} + Cl^-$$

The silver atoms move to the surface of the crystal and form small colloidal clusters of silver metal. This metallic silver absorbs visible light making the lens appear dark (colored).

As the glass is removed from the light the Cu^{2+} ions slowly move to the surface of the crystal where they interact with the silver metal

$$Cu^{2+} + Ag^0 \longrightarrow Cu^+ + Ag^+$$

The glass clears as the silver ions rejoin chloride ions in the crystals.

The cell can be recharged by reversing the chemical reaction. This reversal is accomplished by forcing an electric current through the cell in the opposite direction. Lead sulfate and water are reconverted to lead, lead dioxide, and sulfuric acid:

$$2\ PbSO_4(s) + 2\ H_2O(l) \xrightarrow[\text{cycle}]{\text{Charge}} Pb(s) + PbO_2(s) + 2\ H_2SO_4(aq)$$

The electrolyte in a lead storage battery is a 38% by mass sulfuric acid solution having a density of 1.29 g/mL. As the battery is discharged, sulfuric acid is removed, thereby decreasing the density of the electrolyte solution. The state of charge or discharge of the battery can be estimated by measuring the density (or specific gravity) of the electrolyte solution with a hydrometer. When the density has dropped to about 1.05 g/mL, the battery needs recharging.

In a commercial battery, each cell consists of a series of cell units of alternating lead–lead dioxide plates separated and supported by wood, glass wool, or fiberglass. The energy storage capacity of a single cell is limited, and its electrical potential is only about 2 volts. Therefore, a bank of six cells is connected in series to provide the 12 volt output of the usual automobile battery.

Concepts in Review

1. Assign oxidation numbers to all the elements in a given compound or ion.
2. Determine which element is being oxidized and which element is being reduced in an oxidation–reduction reaction.
3. Identify the oxidizing agent and the reducing agent in an oxidation–reduction reaction.
4. Balance oxidation–reduction equations in molecular and ionic forms.
5. Outline the general principles concerning the activity series of the metals.
6. Use the activity series to determine whether a proposed single-displacement reaction will occur.
7. Distinguish between an electrolytic and a voltaic cell.
8. Draw a voltaic cell that will produce electric current from an oxidation–reduction reaction involving two metals and their salts.
9. Identify the anode reaction and the cathode reaction in a given electrolytic or voltaic cell.

EXERCISES

An asterisk indicates a more challenging question or problem.

1. In the equation

$$I_2 + 5\,Cl_2 + 6\,H_2O \longrightarrow 2\,HIO_3 + 10\,HCl$$

(a) Has iodine been oxidized or has it been reduced ?
(b) Has chlorine been oxidized or has it been reduced? (See Figure 18.1.)

2. Based on Table 18.3, which element of each pair is more active?
(a) Ag or Al (b) Na or Ba (c) Ni or Cu

3. Based on Table 18.3, will the following combinations react in aqueous solution?
(a) $Zn + Cu^{2+}$ (e) $Ba + FeCl_2$
(b) $Ag + H^+$ (f) $Pb + NaCl$
(c) $Sn + Ag^+$ (g) $Ni + Hg(NO_3)_2$
(d) $As + Mg^{2+}$ (h) $Al + CuSO_4$

4. The reaction between powdered aluminum and iron(III) oxide (in the Thermite process), producing molten iron, is very exothermic.
(a) Write the equation for the chemical reaction that occurs.
(b) Explain in terms of Table 18.3 why a reaction occurs.
(c) Would you expect a reaction between powdered iron and aluminum oxide?
(d) Would you expect a reaction between powdered aluminum and chromium(III) oxide?

5. Write equations for the chemical reaction of each of the following metals with dilute solutions of (a) hydrochloric acid and (b) sulfuric acid: aluminum, chromium, gold, iron, copper, magnesium, mercury, and zinc. If a reaction will not occur, write "no reaction" as the product. (See Table 18.3.)

6. A $NiCl_2$ solution is placed in the apparatus shown in Figure 18.2, instead of the HCl solution shown. Write equations for:
(a) The anode reaction
(b) The cathode reaction
(c) The net electrochemical reaction

7. What is the major distinction between the reactions occurring in Figure 18.2 and those in Figure 18.3?

8. In the cell shown in Figure 18.3:
(a) What would be the effect of removing the voltmeter and connecting the wires shown coming to the voltmeter?
(b) What would be the effect of removing the salt bridge?

9. What is the oxidation number of the underlined element in each compound?
(a) $\underline{Na}Cl$ (e) $H_2\underline{S}O_3$ (i) $\underline{N}H_3$
(b) $Fe\underline{Cl}_3$ (f) $\underline{N}H_4Cl$ (j) $K\underline{Cl}O_3$
(c) $\underline{Pb}O_2$ (g) $K\underline{Mn}O_4$ (k) $K_2\underline{Cr}O_4$
(d) $Na\underline{N}O_3$ (h) $\underline{I}_2$ (l) $K_2\underline{Cr}_2O_7$

10. What is the oxidation number of the underlined elements?
(a) $\underline{S}^{2-}$ (d) $\underline{Mn}O_4^-$ (g) $\underline{As}O_4^{3-}$
(b) $\underline{N}O_2^-$ (e) $\underline{Bi}^{3+}$ (h) $Fe(\underline{O}H)_3$
(c) $Na_2\underline{O}_2$ (f) $\underline{O}_2$ (i) $\underline{I}O_3^-$

11. In the following half-reactions, which element is changing oxidation state? Is the half-reaction an oxidation or a reduction? Supply the proper number of electrons to the proper side to balance each equation.
(a) $Zn^{2+} \longrightarrow Zn$
(b) $2\,Br^- \longrightarrow Br_2$
(c) $MnO_4^- + 8\,H^+ \longrightarrow Mn^{2+} + 4\,H_2O$
(d) $Ni \longrightarrow Ni^{2+}$
(e) $SO_3^{2-} + H_2O \longrightarrow SO_4^{2-} + 2\,H^+$
(f) $NO_3^- + 4\,H^+ \longrightarrow NO + 2\,H_2O$
(g) $S_2O_4^{2-} + 2\,H_2O \longrightarrow 2\,SO_3^{2-} + 4\,H^+$
(h) $Fe^{2+} \longrightarrow Fe^{3+}$

12. In the following unbalanced equations,
(a) Identify the element that is oxidized and the element that is reduced.
(b) Identify the oxidizing agent and the reducing agent.
(1) $Cr + HCl \longrightarrow CrCl_3 + H_2$
(2) $SO_4^{2-} + I^- + H^+ \longrightarrow H_2S + I_2 + H_2O$
(3) $AsH_3 + Ag^+ + H_2O \longrightarrow H_3AsO_4 + Ag + H^+$
(4) $Cl_2 + NaBr \longrightarrow NaCl + Br_2$

13. Balance these equations by the change-in-oxidation-number method.
(a) $Zn + S \longrightarrow ZnS$
(b) $AgNO_3 + Pb \longrightarrow Pb(NO_3)_2 + Ag$
(c) $Fe_2O_3 + CO \longrightarrow Fe + CO_2$
(d) $H_2S + HNO_3 \longrightarrow S + NO + H_2O$
(e) $MnO_2 + HBr \longrightarrow MnBr_2 + Br_2 + H_2O$
(f) $Cl_2 + KOH \longrightarrow KCl + KClO_3 + H_2O$
(g) $Ag + HNO_3 \longrightarrow AgNO_3 + NO + H_2O$
(h) $CuO + NH_3 \longrightarrow N_2 + Cu + H_2O$
(i) $PbO_2 + Sb + NaOH \longrightarrow PbO + NaSbO_2 + H_2O$
(j) $H_2O_2 + KMnO_4 + H_2SO_4 \longrightarrow O_2 + MnSO_4 + K_2SO_4 + H_2O$

14. Balance the following ionic redox equations using the ion–electron method. All reactions occur in acidic solution.
 (a) $Zn + NO_3^- \longrightarrow Zn^{2+} + NH_4^+$
 (b) $NO_3^- + S \longrightarrow NO_2 + SO_4^{2-}$
 (c) $PH_3 + I_2 \longrightarrow H_3PO_2 + I^-$
 (d) $Cu + NO_3^- \longrightarrow Cu^{2+} + NO$
 (e) $ClO_3^- + I^- \longrightarrow I_2 + Cl^-$
 (f) $Cr_2O_7^{2-} + Fe^{2+} \longrightarrow Cr^{3+} + Fe^{3+}$
 (g) $MnO_4^- + SO_2 \longrightarrow Mn^{2+} + SO_4^{2-}$
 (h) $H_3AsO_3 + MnO_4^- \longrightarrow H_3AsO_4 + Mn^{2+}$
 *(i) $ClO_3^- + Cl^- \longrightarrow Cl_2$
 *(j) $Cr_2O_7^{2-} + H_3AsO_3 \longrightarrow Cr^{3+} + H_3AsO_4$
15. Balance the following ionic redox equations using the ion–electron method. All reactions occur in basic solutions.
 (a) $Cl_2 + IO_3^- \longrightarrow Cl^- + IO_4^-$
 (b) $MnO_4^- + ClO_2^- \longrightarrow MnO_2 + ClO_4^-$
 (c) $Se \longrightarrow Se^{2-} + SeO_3^{2-}$
 (d) $MnO_4^- + SO_3^{2-} \longrightarrow MnO_2 + SO_4^{2-}$
 (e) $ClO_2(g) + SbO_2^- \longrightarrow ClO_2^- + Sb(OH)_6^-$
 *(f) $Fe_3O_4 + MnO_4^- \longrightarrow Fe_2O_3 + MnO_2$
 *(g) $BrO^- + Cr(OH)_4^- \longrightarrow Br^- + CrO_4^{2-}$
 *(h) $P_4 \longrightarrow HPO_3^{2-} + PH_3$
 *(i) $Al + OH^- \longrightarrow Al(OH)_4^- + H_2$
 (j) $Al + NO_3^- \longrightarrow NH_3 + Al(OH)_4^-$
16. Why are oxidation and reduction said to be complementary processes?
17. When molten $CaBr_2$ is electrolyzed, calcium metal and bromine are produced. Write equations for the two half-reactions that occur at the electrodes. Label the anode half-reaction and the cathode half-reaction.
18. Why is direct current used instead of alternating current in the electroplating of metals?
19. The chemical reactions taking place during discharge in a lead storage battery are

$$Pb + SO_4^{2-} \longrightarrow PbSO_4$$

$$PbO_2 + SO_4^{2-} + 4\,H^+ \longrightarrow PbSO_4 + 2\,H_2O$$

 (a) Complete each half-reaction by supplying electrons.
 (b) Which reaction is oxidation and which is reduction?
 (c) Which reaction occurs at the anode of the battery?
20. What property of lead dioxide and lead(II) sulfate makes it unnecessary to have salt bridges in the cells of a lead storage battery?
21. Explain why the density of the electrolyte in a lead storage battery decreases during the discharge cycle.
22. In one type of alkaline cell used to power devices such as portable radios, Hg^{2+} ions are reduced to metallic mercury when the cell is being discharged. Does this reduction occur at the anode or the cathode? Explain.
23. Differentiate between an electrolytic cell and a voltaic cell.
24. Why is a porous barrier or a salt bridge necessary in some voltaic cells?
25. Use the following redox equation to indicate:

$$KMnO_4 + HCl \longrightarrow KCl + MnCl_2 + H_2O + Cl_2$$

 (a) The oxidizing agent
 (b) The reducing agent
 (c) The number of electrons that are transferred per mole of oxidizing agent
26. Explain the chemical basis for the corrosion and deterioration of the Statue of Liberty.
27. What is patina? How does it act to protect the object from further corrosion?
28. What is photochromic glass? How does it work?
29. Which of the following statements are correct? Rewrite the incorrect statements to make them correct.
 (a) An atom of an element in the uncombined state has an oxidation number of zero.
 (b) The oxidation number of molybdenum in Na_2MoO_4 is +4.
 (c) The oxidation number of an ion is the same as the electrical charge on the ion.
 (d) The process in which an atom or an ion loses electrons is called reduction.
 (e) The reaction $Fe^{3+} + e^- \longrightarrow Fe^{2+}$ is a reduction reaction.
 (f) In the reaction

$$2\,Al + 3\,CuCl_2 \longrightarrow 2\,AlCl_3 + 3\,Cu$$

 aluminum is the oxidizing agent.
 (g) In a redox reaction the oxidizing agent is reduced, and the reducing agent is oxidized.
 (h) $Cu^0 \longrightarrow Cu^{2+}$ is a balanced oxidation half-reaction.
 (i) In the electrolysis of sodium chloride brine (solution), Cl_2 gas is formed at the cathode, and hydroxide ions are formed at the anode.
 (j) In any cell, electrolytic or voltaic, reduction takes place at the cathode, and oxidation occurs at the anode.
 (k) In the Zn–Cu voltaic cell, the reaction at the anode is $Zn \longrightarrow Zn^{2+} + 2\,e^-$.

The statements in (l) through (o) pertain to this activity series:

Ba Mg Zn Fe H Cu Ag

(l) The reaction $Zn + MgCl_2 \rightarrow Mg + ZnCl_2$ is a spontaneous reaction.
(m) Barium is more active than copper.
(n) Silver metal will react with acids to liberate hydrogen gas.
(o) Iron is a better reducing agent than zinc.
(p) Oxidation and reduction occur simultaneously in a chemical reaction; one cannot take place without the other.
(q) A free metal can displace from solution the ions of a metal that lies below the free metal in the activity series.
(r) In electroplating, the piecc to be electroplated with a metal is attached to the cathode.
(s) In an automobile lead storage battery, the density of the sulfuric acid solution decreases as the battery discharges.
(t) In an electrolytic cell, chemical energy is used to produce electrical energy.

30. How many moles of NO gas will be formed by the reaction of 25.0 g of silver with nitric acid? [See the equation given in Exercise 13(g).]

31. What volume of chlorine gas, measured at STP, is required to react with excess KOH to form 0.300 mol of $KClO_3$? [See the equation given in Exercise 13(f).]

32. What mass of $KMnO_4$ would be needed to react with 100 mL of H_2O_2 solution ($d = 1.031$ g/mL, 9.0% H_2O_2 by mass)? [See the equation given in Exercise 13(j).]

***33.** What volume of 0.200 M $K_2Cr_2O_7$ will be required to oxidize 5.00 g of H_3AsO_3? [See the equation given in Exercise 14(j).]

***34.** What volume of 0.200 M $K_2Cr_2O_7$ will be required to oxidize the Fe^{2+} ion in 60.0 mL of 0.200 M $FeSO_4$ solution? [See the equation given in Exercise 14(f).]

35. A sample of crude potassium iodide was analyzed using the following reaction (not balanced):

$$I^- + SO_4^{2-} \longrightarrow I_2 + H_2S \quad \text{(acid solution)}$$

If a 4.00 g sample of crude KI produced 2.79 g of iodine, what is the percent purity of the KI?

What mass of copper is formed when 35.0 L of ammonia gas, measured at STP, reacts with copper(II) oxide? [See the equation given in Exercise 13(h).]

***37.** What volume of NO gas, measured at 28°C and 744 torr, will be formed by the reaction of 0.500 mol of Ag reacting with excess nitric acid? [See the equation given in Exercise 13(g).]

38. How many moles of H_2 can be produced from 100. g of Al according to the following reaction?

$$Al + OH^- \longrightarrow Al(OH)_4^- + H_2 \quad \text{(basic solution)}$$

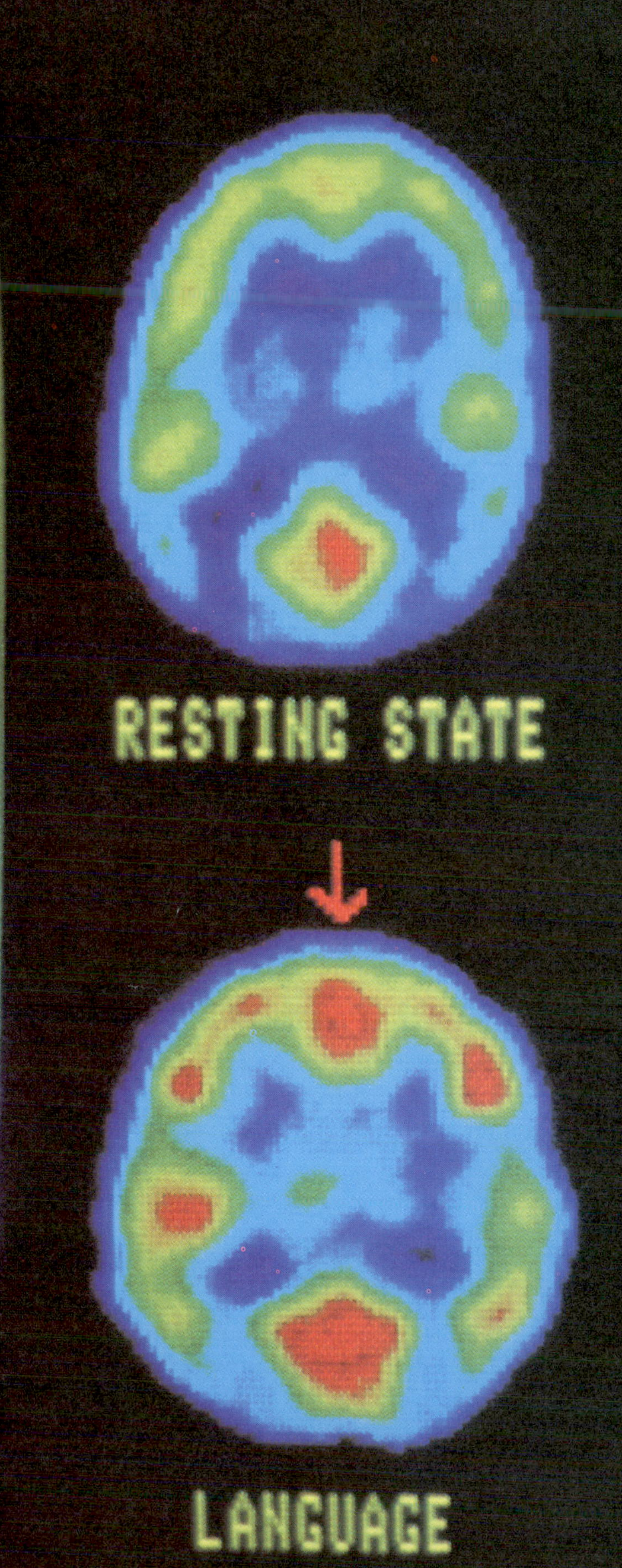

19

Nuclear Chemistry

◀ **Chapter Opening Photo: A positron emission tomography (PET) scan of the human brain reveals the areas that are active during various activities.**

An unusual telephone conversation originated from the University of Chicago in December of 1942. The caller, American physicist Arthur Compton, said, "Jim, the Italian navigator has just landed in the new world." The reference was to the great physicist Enrico Fermi; the landing was the first controlled atomic chain reaction; and the new world was the dawning of the nuclear age.

A new world indeed! Locked within the nucleus of an atom is a source of tremendous energy. Its power can be devastating, as seen in the effects of nuclear weapons or accidents, such as Three Mile Island or Chernobyl. Nuclear energy can be harnassed to perform such useful tasks as generating power, treating cancer, and preserving food. Isotopes are used in medicine to diagnose illness and detect minute quantities of drugs or hormones. Researchers are using radioactive tracers to sequence the human genome. The applications of nuclear chemistry are important in medicine, industry, art, and research. Its impact both threatens and enhances our lives and our future.

19.1 Discovery of Radioactivity

One of the most important steps leading to the discovery of radioactivity was made in 1895 by Wilhelm Konrad Roentgen (1845–1923). Roentgen discovered X rays when he observed that a vacuum discharge tube, enclosed in a thin, black cardboard box, caused a nearby piece of paper coated with barium platinocyanide to glow with a brilliant phosphorescence. From this and other experiments he concluded that certain rays, which he called X rays, were emitted from the discharge tube, penetrated the box, and caused the salt to glow. Roentgen also showed that X rays could penetrate other bodies and affect photographic plates. This observation led to the development of X-ray photography.

Shortly after this discovery Antoine Henri Becquerel (1852–1908) attempted to show a relationship between X rays and the phosphorescence of uranium salts. In one of his experiments he wrapped a photographic plate in black paper, placed a sample of uranium salt on it, and exposed it to sunlight. The developed photographic plate showed that rays emitted from the salt had penetrated the paper. Later Becquerel prepared to repeat the experiment, but, because the sunlight was intermittent, he placed the entire setup in a drawer. Several days later he developed the photographic plate, expecting to find it only slightly affected. He was amazed to observe an intense image on the plate. He repeated the experiment in total darkness and obtained the same results, proving that the uranium salt emitted rays that affected the photographic plate without its being exposed to sunlight. In this way Becquerel discovered radioactivity. The name radioactivity was given to this phenomenon two years later (in 1898) by Marie Curie. **Radioactivity** is the spontaneous emission of particles and/or rays from the nucleus of an atom. Elements having this property are said to be radioactive. Becquerel later showed that the rays coming from uranium were able to ionize air and were also capable of penetrating thin sheets of metal.

radioactivity

In 1898 Marie Sklodowska Curie (1867–1934) and her husband Pierre Curie (1859–1906) turned their research interests to radioactivity. In a short time the Curies discovered two new elements, polonium and radium, both of which are radioactive. To confirm their work on radium they processed 1 ton of pitchblende residue ore to obtain 0.1 g of pure radium chloride, which they used to make further studies on the properties of radium and to determine its atomic mass.

Marie Curie (1867–1934).

In 1899 Ernest Rutherford began to investigate the nature of the rays emitted from uranium. He found two rays, which he called *alpha* and *beta rays.* Soon he realized that uranium, while emitting these rays, was changing into another element. By 1912, over 30 radioactive isotopes were known, and many more are known today. The *gamma ray*, a third ray emitted from radioactive materials and similar to an X ray, was discovered by Paul Villard in 1900. After the description of the nuclear atom by Rutherford, the phenomenon of radioactivity was attributed to reactions taking place in the nuclei of atoms.

The symbolism and notation described for isotopes in Chapter 5 is very useful in nuclear chemistry and is briefly reviewed here.

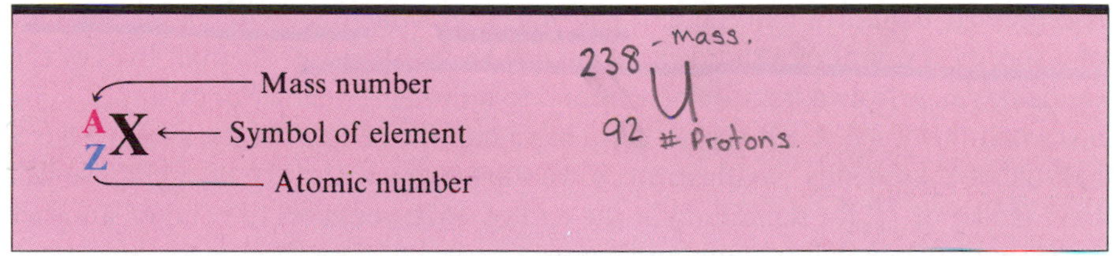

For example, $^{238}_{92}U$ represents a uranium isotope with an atomic number of 92 and a mass number of 238. This isotope is also designated as U-238 or uranium-238 and contains 92 protons and 146 neutrons. The protons and neutrons collectively are known as **nucleons**. The mass number is the total number of nucleons in the nucleus. Table 19.1 shows the iostopic notations for several particles associated with nuclear chemistry.

nucleon

When we speak of isotopes we generally mean atoms of the same element with different masses, such as $^{16}_{8}O$, $^{17}_{8}O$, $^{18}_{8}O$. In nuclear chemistry we use the term **nuclide** to mean any isotope of any atom. Thus $^{16}_{8}O$ and $^{235}_{92}U$ are referred to as nuclides. Nuclides that spontaneously emit radiation are referred to as *radionuclides.*

nuclide

TABLE 19.1 Symbols in Isotopic Notation for Several Particles (and Small Isotopes) Associated with Nuclear Chemistry

Particle	Symbol	Atomic number *Z*	Mass number *A*
Neutron	${}^{1}_{0}n$	0	1
Proton	${}^{1}_{1}H$	1	1
Beta particle (electron)	${}^{0}_{-1}e$	−1	0
Positron (positive electron)	${}^{0}_{+1}e$	1	0
Alpha particle (helium nucleus)	${}^{4}_{2}He$	2	4
Deuteron (heavy hydrogen nucleus)	${}^{2}_{1}H$	1	2

19.2 NATURAL RADIOACTIVITY

radioactive decay

Radioactive elements continuously undergo **radioactive decay**, or disintegration, to form different elements. The chemical properties of an element are associated with its electronic structure, but radioactivity is a property of the nucleus. Therefore, neither ordinary changes of temperature and pressure nor the chemical or physical state of an element has any effect on its radioactivity.

The principal emissions from the nuclei of radionuclides are known as alpha rays (or particles), beta rays (or particles), and gamma rays. Upon losing an alpha or beta particle, the radioactive element changes into a different element. This process will be explained in detail later.

half-life

Each radioactive nuclide disintegrates at a specific and constant rate, which is expressed in units of half-life. The **half-life** ($t_{1/2}$) is the time required for one-half of a specific amount of a radioactive nuclide to disintegrate. The half-lives of the elements range from a fraction of a second to billions of years. For example, ${}^{238}_{92}U$ has a half-life of 4.5×10^9 years, ${}^{226}_{88}Ra$ has a half-life of 1620 years, and ${}^{15}_{6}C$ has a half-life of 2.4 seconds. To illustrate, if we start today with 1.0 g of ${}^{226}_{88}Ra$, we will have 0.50 g of ${}^{226}_{88}Ra$ remaining at the end of 1620 years; at the end of another 1620 years, 0.25 g will remain; and so on.

$$1.0 \text{ g } {}^{226}_{88}Ra \xrightarrow[\text{1620 years}]{t_{1/2}} 0.50 \text{ g } {}^{226}_{88}Ra \xrightarrow[\text{1620 years}]{t_{1/2}} 0.25 \text{ g } {}^{226}_{88}Ra$$

TABLE 19.2 Half-Lives for Radium, Carbon, and Uranium Isotopes

Isotope	Half-life	Isotope	Half-life
Ra-223	11.7 days	C-14	5668 years
Ra-224	3.64 days	C-15	2.4 seconds
Ra-225	14.8 days	U-235	7.1×10^8 years
Ra-226	1620 years	U-238	4.5×10^9 years
Ra-228	6.7 years		

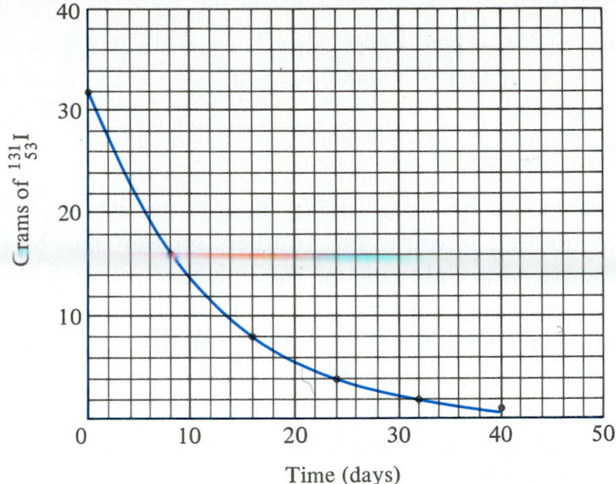

FIGURE 19.1 Radioactive decay curve for $^{131}_{53}I$, which has a half-life of 8 days:

$$^{131}_{53}I \longrightarrow \, ^{131}_{54}Xe + \, ^{0}_{-1}e$$

The half-lives of the various radioisotopes of the same element are different from one another. Half-lives for some isotopes of radium, carbon, and uranium are listed in Table 19.2. A radioactive decay curve is illustrated in Figure 19.1.

EXAMPLE 19.1

The half-life of $^{131}_{53}I$ is 8 days. How much ^{131}I will be left of a 32-g sample after five half-lives?

SOLUTION

This problem can be solved by using the graph in Figure 19.1. To find the number of grams of ^{131}I left after one half-life, trace a perpendicular line from 8 days on the x axis to the line on the graph. Now trace a horizontal line from this point on the plotted line to the y axis and read the corresponding grams of ^{131}I. This process continues for each half-life, adding 8 days to the previous value on the x axis.

Half-lives	0	1	2	3	4	5
Number of days		8	16	24	32	40
Amount remaining	32 g	16 g	8 g	4 g	2 g	1 g

Starting with 32 g, 1 g of ^{131}I will be left after five half-lives (40 days).

EXAMPLE 19.2

In how many half-lives will 10.0 g of a radioactive nuclide decay to less than 10% of its original value?

SOLUTION

Ten percent of the original amount is 1.0 g. After the first half-life, half of the original material remains and half has decayed. After the second half-life, one fourth of the original material remains, which is one-half of the starting amount at the end of the first half-life. This progression continues, reducing the quantity remaining by half for each half-life that passes.

Half-lives	0	1	2	3	4
Percent remaining	100%	50%	25%	12.5%	6.25%
Amount remaining	10.0 g	5.00 g	2.50 g	1.25 g	0.625 g

Therefore, the amount remaining will be less than 10% sometime between the third and the fourth half-lives.

PRACTICE The half-life of $^{14}_{6}C$ is 5668 years. How much $^{14}_{6}C$ will remain after six half-lives in a sample that initially contains 25.0 g?
Answer: 0.391 g

Nuclides are said to be either *stable* (nonradioactive) or *unstable* (radioactive). All elements that have atomic numbers greater than 83 (bismuth) are naturally radioactive, although some of the nuclides have extremely long half-lives. Some of the naturally occurring nuclides of elements 81, 82, and 83 are radioactive, and some are stable. Only a few naturally occurring elements that have atomic numbers less than 81 are radioactive. However, no stable isotopes of element 43 (technetium) or of element 61 (promethium) are known.

Radioactivity is believed to be a result of an unstable ratio of neutrons to protons in the nucleus. Stable nuclides of elements up to about atomic number 20 generally have about a 1:1 neutron to proton ratio. In elements above number 20 the neutron to proton ratio in the stable nuclides gradually increases to about 1.5:1 in element number 83 (bismuth). When the neutron to proton ratio is too high or too low, alpha, beta, or other particles are emitted to achieve a more stable nucleus.

19.3 Properties of Alpha Particles, Beta Particles, and Gamma Rays

The classical experiment proving that alpha and beta particles are oppositely charged was performed by Marie Curie (see Figure 19.2). She placed a radioactive source in a hole in a lead block and positioned two poles of a strong electromagnet so that the radiations that were given off passed between them. The paths of three different kinds of radiation were detected by means of a photographic plate placed some distance beyond the electromagnet. The lighter beta particles were strongly deflected toward the positive pole of the electromagnet; the heavier alpha particles were less strongly deflected and in the opposite direction. The uncharged gamma rays were not affected by the electromagnet and struck the photographic plates after traveling along a path straight out of the lead block.

alpha particle

Alpha Particle An **alpha particle** consists of two protons and two neutrons, has a mass of about 4 amu, a charge of +2, and is considered to be a doubly charged helium atom. It is usually given one of the following symbols: α, He^{2+}, or $^{4}_{2}He$. When an alpha particle is emitted from the nucleus, a different element is formed. The atomic number of the new element is 2 less and the mass is 4 amu less than that of the starting element

> Loss of an alpha particle from the nucleus results in
> loss of 4 in the mass number (A)
> loss of 2 in the atomic number (Z)

For example, when $^{238}_{92}U$ loses an alpha particle, $^{234}_{90}Th$ is formed, because two

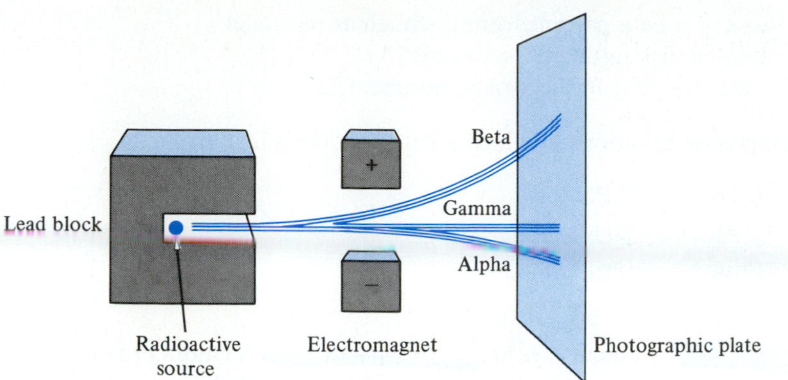

◀ FIGURE 19.2 The effect of an electromagnetic field on alpha, beta, and gamma rays. Lighter beta particles are deflected considerably more than alpha particles. Alpha and beta particles are deflected in opposite directions. Gamma radiation is not affected by the electromagnetic field.

neutrons and two protons are lost from the uranium nucleus. This disintegration may be written as an equation:

$${}^{238}_{92}\mathrm{U} \longrightarrow {}^{234}_{90}\mathrm{Th} + \alpha \qquad \text{or} \qquad {}^{238}_{92}\mathrm{U} \longrightarrow {}^{234}_{90}\mathrm{Th} + {}^{4}_{2}\mathrm{He}$$

For the loss of an alpha particle from ${}^{226}_{88}\mathrm{Ra}$, the equation is

$${}^{226}_{88}\mathrm{Ra} \longrightarrow {}^{222}_{86}\mathrm{Rn} + {}^{4}_{2}\mathrm{He} \qquad \text{or} \qquad {}^{226}_{88}\mathrm{Ra} \longrightarrow {}^{222}_{86}\mathrm{Rn} + \alpha$$

A nuclear equation, like a chemical equation, consists of reactants and products and must be balanced. To have a balanced nuclear equation the sum of the mass numbers (superscripts) on both sides of the equation must be equal, and the sum of the atomic numbers (subscripts) on both sides of the equation must be equal.

Sum of mass numbers equals 226

$${}^{226}_{88}\mathrm{Ra} \longrightarrow {}^{222}_{86}\mathrm{Rn} + {}^{4}_{2}\mathrm{He}$$

Sum of atomic numbers equals 88

What new nuclide will be formed when ${}^{230}_{90}\mathrm{Th}$ loses an alpha particle? This loss is equivalent to two protons and two neutrons. The new nuclide will have a mass of (230 − 4) or 226 amu and will contain (90 − 2) or 88 protons, so its atomic number is 88. Locate the corresponding element on the periodic chart. It is ${}^{226}_{88}\mathrm{Ra}$ or radium-226.

beta particle

Beta Particle The **beta particle** is identical in mass and charge to an electron; its charge is −1. Both a beta particle and a proton are produced by the decomposition of a neutron.

$${}^{1}_{0}\mathrm{n} \longrightarrow {}^{1}_{1}\mathrm{p} + {}^{\;0}_{-1}\mathrm{e}$$

The beta particle leaves, and the proton remains in the nucleus. When an atom loses a beta particle from its nucleus, a different element is formed that has essentially the same mass but an atomic number that is 1 greater than that of the starting element. The beta particle is written as β or ${}^{\;0}_{-1}\mathrm{e}$.

Loss of a beta particle from the nucleus results in
no change in the mass number (A)
increase of 1 in the atomic number (Z)

Examples of equations in which a beta particle is lost are

$$^{234}_{90}\text{Th} \longrightarrow {}^{234}_{91}\text{Pa} + \beta$$
$$^{234}_{91}\text{Pa} \longrightarrow {}^{234}_{92}\text{U} + {}^{\;0}_{-1}\text{e}$$
$$^{210}_{82}\text{Pb} \longrightarrow {}^{210}_{83}\text{Bi} + \beta$$

γ

gamma ray

Gamma Ray **Gamma rays** are photons of energy. A gamma ray is similar to an X ray, but is more energetic. They have no electrical charge and no measurable mass. Gamma rays emanate from the nucleus in many radioactive changes along with either alpha or beta particles. The designation for a gamma ray is γ. Gamma radiation does not result in a change of atomic number or the mass of an element.

Loss of a gamma ray from the nucleus results in
no change in mass number (A) or atomic number (Z)

Example 19.3

(a) Write an equation for the loss of an alpha particle from the nuclide $^{194}_{78}\text{Pt}$.
(b) What nuclide is formed when $^{228}_{88}\text{Ra}$ loses a beta particle from its nucleus?

SOLUTION

(a) Loss of an alpha particle, $^{4}_{2}\text{He}$, means the loss of two neutrons and two protons. This change results in a decrease of 4 in the mass number and a decrease of 2 in the atomic number.

Mass of new nuclide: $A - 4$ or $194 - 4 = 190$

Atomic number of new nuclide: $Z - 2$ or $78 - 2 = 76$

Looking up element number 76 on the periodic table, we find it to be osmium, Os. The equation, then, is

$$^{194}_{78}\text{Pt} \longrightarrow {}^{190}_{76}\text{Os} + {}^{4}_{2}\text{He}$$

(b) The loss of a beta particle from a $^{228}_{88}\text{Ra}$ nucleus means a gain of 1 in the atomic number with no essential change in mass.
The new nuclide will have an atomic number of $(Z + 1)$ or 89, which is actinium, Ac. The nuclide formed is $^{228}_{89}\text{Ac}$.

$$^{228}_{88}\text{Ra} \longrightarrow {}^{228}_{89}\text{Ac} + {}^{\;0}_{-1}\text{e}$$

Example 19.4

What nuclide will be formed when $^{214}_{82}\text{Pb}$ successively emits β, β, and α partices from its nucleus? Write successive equations showing these changes.

SOLUTION

The changes brought about in the three-steps outlined are as follows:

β loss: Increase of 1 in the atomic number; no change in mass

β loss: Increase of 1 in the atomic number; no change in mass

α loss: Decrease of 2 in the atomic number; decrease of 4 in the mass

The equations are

$$^{214}_{82}Pb \xrightarrow{-\beta} {}^{214}_{83}X \xrightarrow{-\beta} {}^{214}_{84}X \xrightarrow{-\alpha} {}^{210}_{82}X$$

where X stands for the new nuclide formed. Looking up each of these elements by their atomic numbers, we rewrite the equations

$$^{214}_{82}Pb \xrightarrow{-\beta} {}^{214}_{83}Bi \xrightarrow{-\beta} {}^{214}_{84}Po \xrightarrow{-\alpha} {}^{210}_{82}Pb$$

Practice What nuclide will be formed when $^{230}_{90}Th$ emits an alpha particle?
Answer: $^{226}_{88}Ra$

The ability of radioactive rays to pass through various objects is in proportion to the speed at which they leave the nucleus. Gamma rays travel at the velocity of light (186,000 miles per second) and are capable of penetrating several inches of lead. The velocities of beta particles are variable, the fastest being about nine-tenths the velocity of light. Alpha particles have velocities less than one-tenth the velocity of light. Figure 19.3 illustrates the relative penetrating power of these rays. A few sheets of paper will stop alpha particles; a thin sheet of aluminum will stop both alpha and beta particles; and a 5 cm block of lead will reduce, but not completely stop, gamma radiation. In fact it is difficult to stop all gamma radiation. Table 19.3 summarizes the properties of alpha, beta, and gamma radiation.

19.4 Radioactive Disintegration Series

The naturally occurring radioactive elements with a higher atomic number than lead (Pb) fall into three orderly disintegration series. Each series proceeds from one element to the next by the loss of either an alpha or a beta particle, finally ending in a nonradioactive nuclide. The uranium series starts with $^{238}_{92}U$ and ends with $^{206}_{82}Pb$. The thorium series starts with $^{232}_{90}Th$ and ends with $^{208}_{82}Pb$. The actinium series starts with $^{235}_{92}U$ and ends with $^{207}_{82}Pb$. A fourth series, the

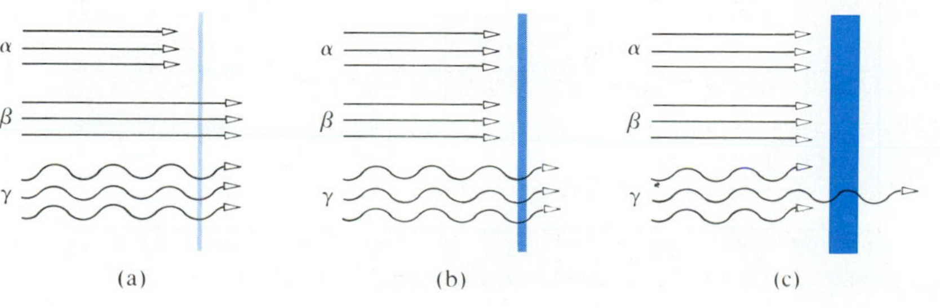

◄ **Figure 19.3** Relative penetrating ability of alpha, beta, and gamma radiation. (a) Thin sheet of paper; (b) thin sheet of aluminum; (c) 5 cm lead block.

TABLE 19.3 Characteristics of Nuclear Radiation

Radiation	Symbol	Mass (amu)	Electrical charge	Velocity	Composition	Ionizing power
Alpha	α $^{4}_{2}He$	4	+2	Variable, less than 10% the speed of light	Identical to He^{2+}	High
Beta	β $^{0}_{-1}e$	$\frac{1}{1837}$	−1	Variable, up to 90% the speed of light	Identical to an electron	Moderate
Gamma	γ	0	0	Speed of light	Photons or electromagnetic waves of energy	Almost none

neptunium series, starts with the synthetic element $^{241}_{94}Pu$ and ends with the stable bismuth nuclide $^{209}_{83}Bi$. The uranium series is shown in Figure 19.4; gamma radiation, which accompanies alpha and beta radiation, is not shown in the figure.

By using such a series and the half-lives of its members, scientists have been able to approximate the age of certain geologic deposits. This approximation was done by comparing the amount of $^{238}_{92}U$ with the amount of $^{206}_{82}Pb$ and other nuclides in the series that are present in a particular geologic formation. Rocks found in Canada and Finland have been calculated to be about 3.0×10^9 (3 billion) years old. Some meteorites have been determined to be 4.5×10^9 years old.

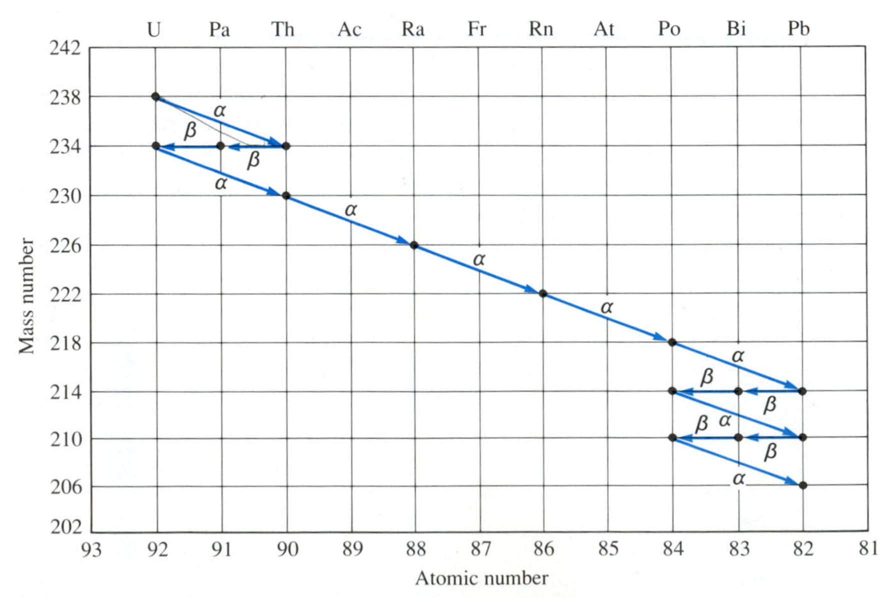

FIGURE 19.4 ▶ The uranium disintegration series. $^{238}_{92}U$ decays by a series of alpha (α) and beta (β) emissions to the stable nuclide $^{206}_{82}Pb$.

Practice What nuclides are formed when $^{238}_{92}U$ undergoes the following decays?
(a) an alpha particle and a beta particle
(b) 3 alpha particles and 2 beta particles
Answers: (a) $^{234}_{91}Pa$ (b) $^{226}_{88}Ra$

Chemistry in Action

Artifacts and Geologic Formations

An interesting outgrowth of the use of the radionuclide techniques is *radiocarbon dating*. The method is based on the decay rate of C-14 and was devised by the American chemist W. F. Libby, who received the Nobel Prize in chemistry in 1960 for this work. The principle of radiocarbon dating is as follows: Carbon dioxide in the atmosphere contains a fixed ratio of radioactive C-14 to ordinary C-12, because C-14 is produced at a steady rate in the atmosphere by bombardment of N-14 by neutrons from cosmic ray sources.

$$^{14}_{7}N + ^{1}_{0}n \longrightarrow ^{14}_{6}C + ^{1}_{1}H$$

Plants that consume carbon dioxide during photosynthesis and animals that eat the plants contain the same proportion of C-14 to C-12 as long as they are alive. When an organism dies, the amount of C-12 remains fixed, but the C-14 content diminishes according to its half-life (5668 years). By comparing the ratio of C-14 to C-12 in an object to the same ratio in living plants, one can estimate the age of the object being evaluated. In 5668 years, one-half of the radiocarbon initially present will have undergone decomposition. In 11,336 years, one-fourth of the original C-14 will be left. The age of fossil material, archaeological specimens, and old wood can be determined by this method. The age of specimens from ancient Egyptian tombs calculated by radiocarbon dating correlates closely with the chronological age established by Egyptologists. Charcoal samples obtained at Darrington Walls, a wood-henge in Great Britain, were determined to be about 4000 years old. Radiocarbon dating instruments currently in use enable researchers to date specimens back as far as 70,000 years. This technique was used recently to estimate the age of the shroud of Turin.

Radioactive decay has been used to date samples other than those containing carbon. For example, the age of rock formations containing uranium has been approximated by determining the ratio of U-238 to Pb-206. Lead-206 is the last isotope formed in the U-238 disintegration series. Thus, a geologic deposit containing a 1:1 ratio of U-238 to Pb-206 would correspond to a time lapse of one half-life of U-238, which is 4.5×10^9 years, assuming that all the lead came from the decay of U-238. The age of moon rocks returned to earth by the Apollo missions were calculated by similar techniques.

Close up of a moon rock taken from Apollo 12. ▼

19.5 Transmutation of Elements

transmutation

Transmutation is the conversion of one element into another by either natural or artificial means. Transmutation occurs spontaneously in natural radioactive disintegrations. Alchemists tried for centuries to convert lead and mercury into gold by artificial means. But transmutation by artifical means was not achieved until 1919, when Ernest Rutherford succeeded in bombarding the nuclei of nitrogen atoms with alpha particles and produced oxygen nuclides and protons. The nuclear equation for this transmutation can be written as

$$^{14}_{7}N + \alpha \longrightarrow {}^{17}_{8}O + {}^{1}_{1}H \qquad \text{or} \qquad {}^{14}_{7}N + {}^{4}_{2}He \longrightarrow {}^{17}_{8}O + {}^{1}_{1}H$$

It is believed that the alpha particle enters the nitrogen nucleus, forming $^{18}_{9}F$ as an intermediate, which then decomposes into the products

Rutherford's experiments opened the door to nuclear transmutations of all kinds. Atoms were bombarded by alpha particles, neutrons, protons, deuterons ($^{2}_{1}H$), electrons, and so forth. Massive instruments were developed for accelerating these particles to very high speeds and energies to aid their penetration of the nucleus. Some of these instruments are the famous cyclotron, developed by E. O. Lawrence at the University of California; the Van de Graaf electrostatic generator; the betatron; and the electron and proton synchrotrons. With these instruments many nuclear transmutations became possible. Equations for a few of these follows.

$$^{7}_{3}Li + {}^{1}_{1}H \longrightarrow 2\,{}^{4}_{2}He$$
$$^{40}_{18}Ar + {}^{1}_{1}H \longrightarrow {}^{40}_{19}K + {}^{1}_{0}n$$
$$^{23}_{11}Na + {}^{1}_{1}H \longrightarrow {}^{23}_{12}Mg + {}^{1}_{0}n$$
$$^{114}_{48}Cd + {}^{2}_{1}H \longrightarrow {}^{115}_{48}Cd + {}^{1}_{1}H$$
$$^{2}_{1}H + {}^{2}_{1}H \longrightarrow {}^{3}_{1}H + {}^{1}_{1}H$$
$$^{209}_{83}Bi + {}^{2}_{1}H \longrightarrow {}^{210}_{84}Po + {}^{1}_{0}n$$
$$^{16}_{8}O + {}^{1}_{0}n \longrightarrow {}^{13}_{6}C + {}^{4}_{2}He$$
$$^{238}_{92}U + {}^{12}_{6}C \longrightarrow {}^{244}_{98}Cf + 6\,{}^{1}_{0}n$$

A shorthand notation for nuclear bombardment reactions includes, in order, the nuclide being bombarded, followed by the bombarding particle, a comma, and the particle released all in parentheses, and finally the nuclide produced. The shorthand notation for the last two equations above are

$$^{16}_{8}O(n, \alpha)^{13}_{6}C \qquad {}^{238}_{92}U({}^{12}_{6}C, 6n)^{244}_{98}Cf$$

Note that the reactants are to the left and products to the right of the comma.

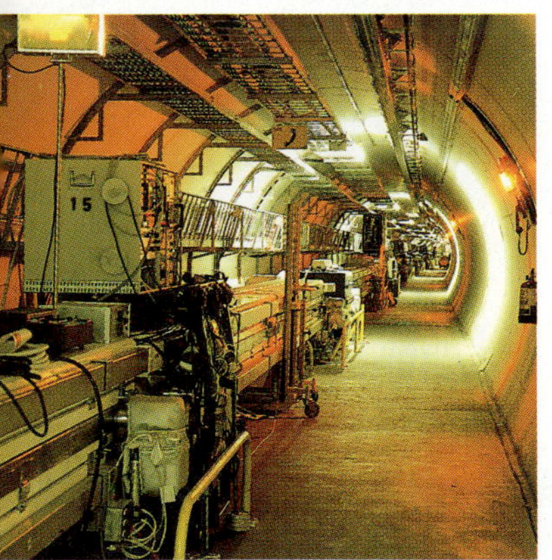

▲ **This particle accelerator is located underground near Geneva, Switzerland. It is used to accelerate and collide electrons and positrons.**

Practice Use shorthand notation to represent the following nuclear equation:

$$^{209}_{83}Bi + {}^{2}_{1}H \longrightarrow {}^{210}_{84}Po + {}^{1}_{0}n$$

Answer: $^{209}_{83}Bi({}^{2}_{1}H, n)^{210}_{84}Po$

19.6 Artificial Radioactivity

Irene Joliot-Curie, a daughter of Pierre and Marie Curie, and her husband Frederic Joliot-Curie observed that when aluminum-27 was bombarded with alpha particles, neutrons and positrons (positive electrons) were emitted as part of the products. When the source of alpha particles was removed, neutrons ceased to be produced, but positrons continued to be emitted. This obervation suggested that the neutrons and positrons were coming from two separate reactions. It also indicated that one of the products of the first reaction was radioactive. After further investigation they discovered that, when aluminum-27 is bombarded with alpha particles, phosphorus-30 and neutrons are produced. Phosphorus-30 is radioactive, has a half-life of 2.5 minutes, and decays to silicon-30 with the emission of a positron. The equations for these reactions follow.

$$^{27}_{13}\text{Al} + ^{4}_{2}\text{He} \longrightarrow ^{30}_{15}\text{P} + ^{1}_{0}\text{n}$$

$$^{30}_{15}\text{P} \longrightarrow ^{30}_{14}\text{Si} + ^{0}_{+1}\text{e}$$

The radioactivity of nuclides produced in this manner is known as **artificial radioactivity** or **induced radioactivity**. Artificial radionuclides behave like natural radioactive elements in two ways: they disintegrate in a definite fashion and they have a specific half-life. The Joliot-Curies received the Nobel Prize in chemistry in 1935 for the discovery of artificial, or induced, radioactivity.

artificial radioactivity

induced radioactivity

19.7 Measurement of Radioactivity

Radiation from radioactive sources is so energetic that it is called *ionizing radiation*. When it strikes an atom or a molecule, one or more electrons are knocked off, and an ion is created. One of the common instruments used to detect and measure radioactivity, the Geiger counter, depends on this fact. The instrument consists of a Geiger–Müller detecting tube and a counting device. The detector tube is a pair of oppositely charged electrodes in an argon gas–filled chamber fitted with a thin window. When radiation, such as a beta particle, passes through the window into the tube, some argon is ionized, and a momentary pulse of current (discharge) flows between the electrodes. These current pulses are electronically amplified in the counter and appear as signals in the form of audible clicks, flashing lights, meter defections, or numerical readouts (Figure 19.5a).

The amount of radiation that an individual encounters can be measured by a film badge (Figure 19.5(b)). A piece of photographic film in a light proof holder is worn in areas where radiation might be encountered. The silver grains in the film will darken when exposed to radiation. The badges are processed after a predetermined time interval to determine the amount of radiation the wearer has been exposed to.

A scintillation counter is used to measure radioactivity for biomedical applications. A scintillator is composed of molecules that emit light when exposed

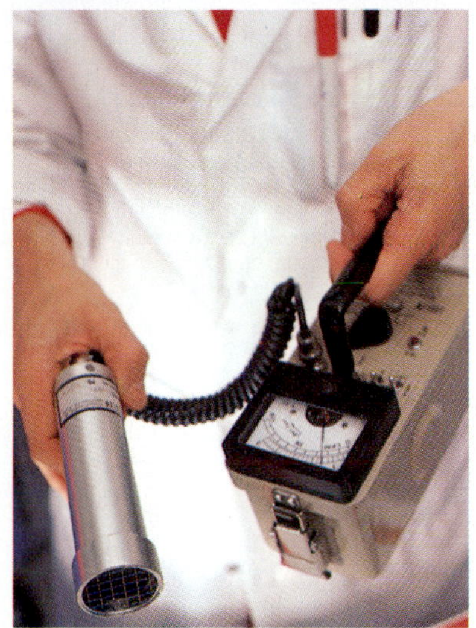

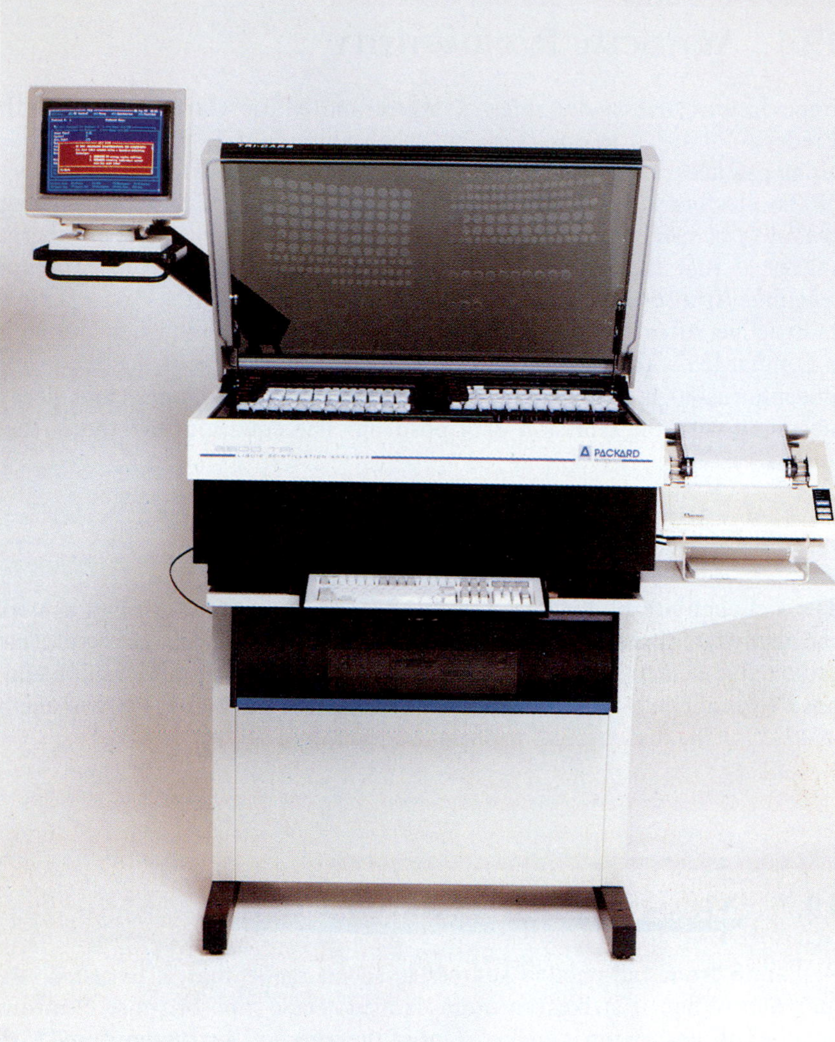

▲
FIGURE 19.5
***Top left*: Film badge to measure radioactivity. *Bottom left*: Geiger-Mueller survey meter. *Right*: Liquid scintillation counter.**

to ionizing radiation. A light sensitive detector counts the flashes and converts them to a numerical readout (Figure 19.5(c)).

curie

The *curie* is the unit used to express the amount of radioactivity produced by an element. One **curie** is defined as the quantity of radioactive material giving 3.7×10^{10} disintegrations per second. The basis for this figure is pure radium, which has an activity of 1 curie per gram. Because the curie is such a large quantity, the millicurie and microcurie, representing one-thousandth and one-millionth of a curie, respectively, are more practical and more commonly used.

The curie only measures radioactivity emitted by a radionuclide. Different units are required to measure exposure to radiation. The Roentgen (R) is the unit that quantifies exposure to gamma or X rays. One Roentgen is defined as the amount of radiation required to produce 2.1×10^{9} ions per cm^3 of dry air. The rad (*r*adiation *a*bsorbed *d*ose) is defined as the amount of radiation that provides 0.01 J of energy per kilogram of matter. The amount of radiation absorbed will change depending on the type of matter. The Roentgen and the rad are

CHEMISTRY IN ACTION

ISOTOPES IN AGRICULTURE

Agricultural research scientists use gamma radiation from Co-60 or other sources to develop disease-resistant and highly productive grains and other crops. The seeds are exposed to gamma radiation to induce mutations. The most healthy and vigorous plants grown from the irradiated seed are then selected and propagated to obtain new and improved varieties for commercial use. Preservation of foodstuffs by radiation is another beneficial application. Food is exposed to either gamma radiation or a beam of beta particles supplied by Co-60 or Cs-137. Microorganisms that can cause food spoilage are destroyed, but the temperature of the food is raised only slightly. The food does not become radioactive as a result of this process, but the shelf-life is extended significantly.

Radioactivity has been used to control and, in some areas, to eliminate the screw-worm fly. The larvae of this obnoxious insect pest burrow into wounds in livestock. The female fly, like a queen bee, mates only once. When large numbers of gamma ray–sterilized male flies are released at the proper time in an area infested with screw-worm flies, the majority of the females mate with sterile males. As a consequence, the flies fail to reproduce sufficiently to maintain their numbers. This technique has also been used to eradicate the Mediterranean fruit fly in some areas.

numerically similar. One Roentgen of gamma radiation provides 0.92 rad in bone tissue.

Neither the rad nor the Roentgen indicates the biological damage caused by radiation. One rad of alpha particles has the ability to cause ten times more damage than one rad of gamma rays or beta particles. Another unit, rem (*r*oentgen *e*quivalent to *m*an) takes into account the degree of biological effect caused by the type of radiation exposure. One rem is equal to the dose in rads multiplied by a factor specific to the form of radiation. The factor is 10 for alpha particles and 1 for both beta particles and gamma rays. Units of radiation are summarized in Table 19.4.

TABLE 19.4 Radiation Units

Curie (Ci)	A unit of radioactivity indicating the rate of decay of a radioactive substance. 1 Ci = 3.7×10^{10} disintegrations/sec
Roentgen (R)	A unit of exposure of gamma or X radiation based on the quantity of ionization produced in air 1 R = 2.1×10^9 ions/cm^3
Rad	A unit of absorbed dose of radiation indicating the energy absorbed from any ionizing radiation. 1 rad = 0.01 J/kg matter
Rem	A unit of radiation dose equivalent. 1 rem is equal to the dose in rads multiplied by a factor dependent on the particular type of radiation. 1 rem = 1 rad × factor
Gray (Gy) (SI unit)	Energy absorbed by tissue. 1 Gy = 1 J/kg tissue (1 Gy = 100 rad)

19.8 Nuclear Fission

nuclear fission

In **nuclear fission** a heavy nuclide splits into two or more intermediate-sized fragments when struck in a particular way by a neutron. The fragments are called *fission products*. As the atom splits, it releases energy and two or three neutrons, each of which can cause another nuclear fission. The first instance of nuclear fission was reported in January 1939 by the German scientists Otto Hahn and F. Strassmann. Detecting isotopes of barium, krypton, cerium, and lanthanum after bombarding uranium with neutrons, the scientists were led to believe that the uranium nucleus had been split.

Characteristics of nuclear fission are

1. Upon absorption of a neutron, a heavy nuclide splits into two or more smaller nuclides (fission products).
2. The mass of the nuclides formed range from about 70 to 160 amu.
3. Two or more neutrons are produced from the fission of each atom.
4. Large quantities of energy are produced as a result of the conversion of a small amount of mass into energy.
5. Most nuclides produced are radioactive and continue to decay until they reach a stable nucleus.

One suggested process by which this fission takes place is illustrated in Figure 19.6. When a heavy nucleus captures a neutron, the energy increase may be sufficient to cause deformation of the nucleus until the mass finally splits into two fragments, releasing energy and usually two or more neutrons.

chain reaction

critical mass

In a typical fission reaction, a $^{235}_{92}U$ nucleus captures a neutron and forms unstable $^{236}_{92}U$. This $^{236}_{92}U$ nucleus undergoes fission, quickly disintegrating into two fragments, such as $^{139}_{56}Ba$ and $^{94}_{36}Kr$, and three neutrons. The three neutrons in turn may be captured by three other $^{235}_{92}U$ atoms, each of which undergoes fission, producing nine neutrons, and so on. A reaction of this kind, in which the products cause the reaction to continue or magnify, is known as a **chain reaction**. For a chain reaction to continue, enough fissionable material must be present so that each atomic fission causes, on the average, at least one additional fission. The minimum quantity of an element needed to support a self-sustaining chain reaction is called the **critical mass**. Since energy is released in each atomic fission, chain reactions constitute a possible source of a steady supply of energy. A chain reaction is illustrated in Figure 19.7. Two of the many possible ways in which $^{235}_{92}U$ may fission are shown by the following equations.

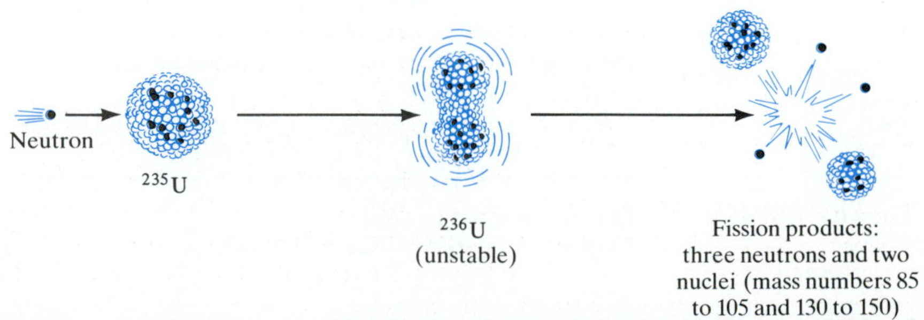

FIGURE 19.6
The fission process. When a neutron is captured by a heavy nucleus, the nucleus becomes more unstable. The more energetic nucleus begins to deform, resulting in fission. Two nuclear fragments and three neutrons are produced by this fission process.

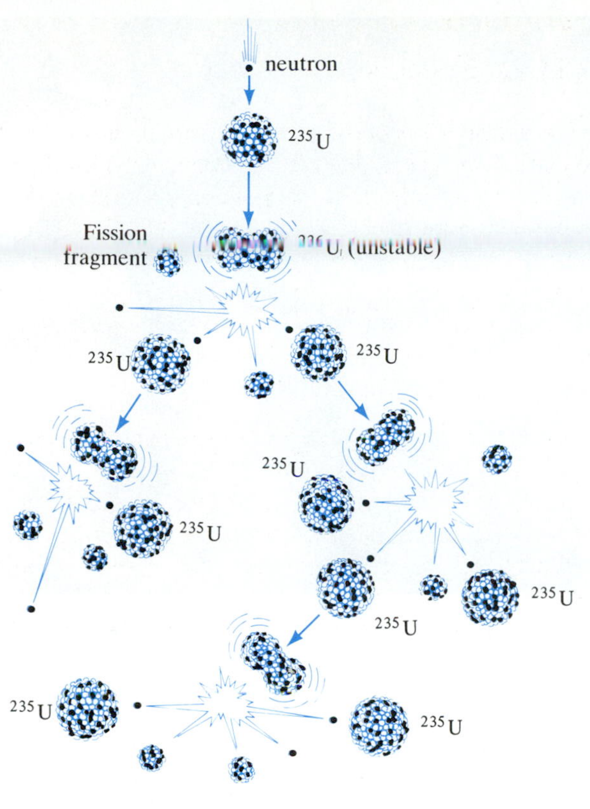

◄ FIGURE 19.7
Fission and chain reaction of $^{235}_{92}U$. Each fission produces two major fission fragments and three neutrons, which may be captured by other $^{235}_{92}U$ nuclei, continuing the chain reaction.

$$^{235}_{92}U + ^{1}_{0}n \longrightarrow ^{139}_{56}Ba + ^{94}_{36}Kr + 3\,^{1}_{0}n$$

$$^{235}_{92}U + ^{1}_{0}n \longrightarrow ^{144}_{54}Xe + ^{90}_{38}Sr + 2\,^{1}_{0}n$$

19.9 NUCLEAR POWER

Nearly all electricity for commercial use is produced by machines consisting of a turbine linked by a drive shaft to an electrical generator. The energy required to run the turbine can be supplied by falling water, as in hydroelectric power plants, or by steam generated by heat from fuel, as in thermal power plants. Thermal power plants burn fossil fuel—coal, oil, or natural gas.

The world's demand for energy, largely from fossil fuels, has continued to grow at an ever-increasing rate for about 250 years. Even at present rates of consumption, the estimated world supply of fossil fuels is sufficient for only a few centuries. Although the United States has large coal and oil shale deposits, it is currently importing about 40% of its oil supply. We clearly need to develop alternative energy sources. At present uranium is the most productive alternative energy source, and about 17% of the electrical energy used in the United States is generated in power plants using uranium fuel.

A technician carefully ▶ manipulates a fuel rod containing enriched uranium-235. The blue glow is characteristic of a nuclear reactor in operation.

A nuclear power plant is a thermal power plant in which heat is produced by a nuclear reactor instead of by combustion of fossil fuel. The major components of a nuclear reactor are (1) an arrangement of nuclear fuel, called the reactor core; (2) a control system, which regulates the rate of fission and thereby the rate of heat generation; and (3) a cooling system, which removes the heat from the reactor and also keeps the core at the proper temperature. One type of reactor uses metal slugs containing uranium enriched from the normal 0.7% U-235 to about 3% U-235. The self-sustaining fission reaction is moderated, or controlled, by adjustable control rods containing substances that slow down and capture some of the neutrons produced. Ordinary water, heavy water, and molten sodium are typical coolants used. Energy obtained from nuclear reactions in the form of heat is used in the production of steam to drive turbines for generating electricity. (See Figure 19.8.)

Two events that demonstrate the potential dangers of nuclear power are the accidents at Three Mile Island, Pennsylvania (1979) and Chernobyl, U.S.S.R. (1986). Both of these nuclear accidents resulted from the loss of coolant to the reactor core. The reactors at Three Mile Island are covered by concrete containment buildings and therefore released a relatively small amount of radioactive material into the atmosphere. Because the Soviet Union did not require containment structures on nuclear power plants, the Chernobyl accident resulted in 31 deaths and the resettlement of 135,000 people. The release of large quantities of I-131, Cs-134 and Cs-137 may cause long-term health problems in that exposed population.

Another major disadvantage of nuclear-fueled power plants is that they produce highly radioactive waste products, some of which have half-lives of thousands of years. As yet, no agreement has been reached on how to safely dispose of these dangerous wastes.

In the United States reactors designed for commercial power production

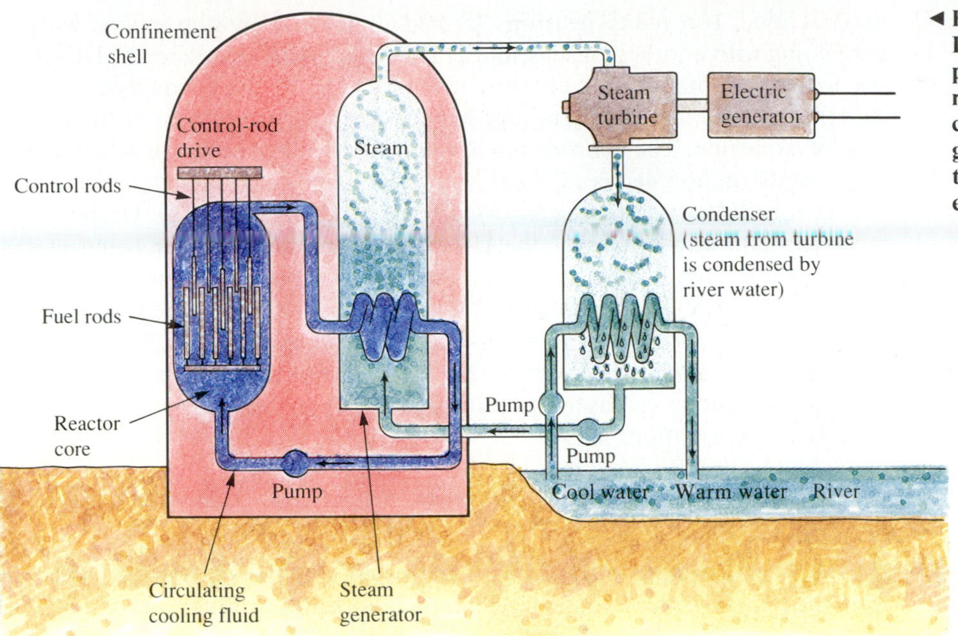

◀ **FIGURE 19.8**
Diagram of a nuclear power plant. Heat produced by the reactor core is carried by a cooling fluid to a steam generator, which turns a turbine that produces electricity.

use uranium oxide, U_3O_8, that is enriched with the relatively scarce fissionable U-235 isotope. Because the supply of U-235 is limited, a new type of reactor known as the *breeder reactor* has been developed. Breeder reactors are designed to produce additional fissionable material at the same time that the fission reaction is occurring. In a breeder reactor, excess neutrons convert nonfissionable isotopes, such as U-238 or Th-232, to the fissionable isotopes, Pu-239 or U-233.

$$^{238}_{92}U + ^{1}_{0}n \longrightarrow ^{239}_{92}U \xrightarrow{-\beta} ^{239}_{93}Np \xrightarrow{-\beta} ^{239}_{94}Pu$$

$$^{232}_{90}Th + ^{1}_{0}n \longrightarrow ^{233}_{90}Th \xrightarrow{-\beta} ^{233}_{91}Pa \xrightarrow{-\beta} ^{233}_{92}U$$

These transmutations make it possible to greatly extend the supply of fuel for nuclear reactors. No breeder reactors are presently in commercial operation in the United States, but several of them are being operated in Europe and Great Britain.

19.10 The Atomic Bomb

The atomic bomb is a fission bomb; it operates on the principle of a very fast chain reaction that releases a tremendous amount of energy. An atomic bomb and a nuclear reactor both depend on self-sustaining nuclear fission chain reactions. The essential difference is that in a bomb the fission is "wild," or uncontrolled, whereas in a nuclear reactor the fission is moderated and carefully controlled. A minimum critical mass of fissionable material is needed for a bomb, or a major explosion will not occur. When a quantity smaller than the critical mass is used, too many neutrons formed in the fission step escape without combining with another nucleus, and a chain reaction does not occur. Therefore, the fissionable material of an atomic bomb must be stored as two or more subcritical masses and brought together to form the critical mass at the desired time of explosion. The temperature developed in an atomic bomb is believed to be about 10 million degrees Celsius.

The nuclides used in atomic bombs are U-235 and Pu-239. Uranium deposits contain about 0.7% of the U-235 isotope, the remainder being U-238. Uranium-238 does not undergo fission except with very high energy neutrons. It was discovered, however, that U-238 captures a low-energy neutron without undergoing fission and that the product, U-239 changes to Pu-239 (plutonium) by a beta decay process. Plutonium-239 readily undergoes fission upon capture of a neutron and is therefore useful for nuclear weapons. The equations for the nuclear transformations are

$$^{238}_{92}U + ^{1}_{0}n \longrightarrow ^{239}_{92}U \xrightarrow{-\beta} ^{239}_{93}Np \xrightarrow{-\beta} ^{239}_{94}Pu$$

The hazards of an atomic bomb explosion include not only shock waves from the explosive pressure and tremendous heat, but also intense radiation in the form of alpha, beta, gamma, and ultraviolet rays. Gamma rays and X rays can penetrate deeply into the body, causing burns, sterilization, and mutation of the genes, which may have an adverse effect on future generations. Both radioactive fission products and unfissioned material are present after the explosion. If the bomb explodes near the ground, many tons of dust are lifted into the air. Radioactive material adhering to this dust, known as *fallout*, is spread by air currents over wide areas of the land and constitutes a lingering source of radiation hazard.

Today nuclear war is probably the most awesome threat facing civilization. Only two rather primitive fission-type atom bombs were used to destroy the Japanese cities of Hiroshima and Nagasaki and bring World War II to an early end. Now tens of thousands of nuclear weapons are in existence, and hundreds more are being manufactured each year. Many of these weapons are at least several hundred times more powerful than the Hiroshima and Nagasaki bombs. The United States and the Soviet Union (now the Commonwealth of Independent States), the so-called superpowers, have made considerable progress toward reaching an effective agreement on the control and reduction of nuclear armaments. The threat of nuclear war is increased by the fact that the number of nations possessing nuclear weapons is steadily increasing.

19.11 Nuclear Fusion

nuclear fusion

The process of uniting the nuclei of two light elements to form one heavier nucleus is known as **nuclear fusion**. Such reactions can be used for producing energy, because the mass of the two individual nuclei that are fused into a single nucleus is greater than the mass of the nucleus formed by their fusion. The mass differential is liberated in the form of energy. Fusion reactions are responsible for the tremendous energy output of the sun. Thus, aside from relatively small amounts from nuclear fission and radioactivity, fusion reactions are the ultimate source of our energy, even the energy from fossil fuels. They are also responsible for the devastating power of the thermonuclear, or hydrogen, bomb.

Fusion reactions require temperatures on the order of tens of millions of degrees for initiation. Such temperatures are present in the sun but have been produced only momentarily on earth. For example, the hydrogen, or fusion, bomb is triggered by the temperature of an exploding fission bomb.

Two typical fusion reactions are

$$\underset{\text{Tritium}}{{}^{3}_{1}\text{H}} + \underset{\text{Deuterium}}{{}^{2}_{1}\text{H}} \longrightarrow {}^{4}_{2}\text{He} + {}^{1}_{0}\text{n} + \text{energy}$$

$$\underset{\substack{3.0150\\ \text{amu}}}{{}^{3}_{1}\text{H}} + \underset{\substack{1.0079\\ \text{amu}}}{{}^{1}_{1}\text{H}} \longrightarrow \underset{\substack{4.0026\\ \text{amu}}}{{}^{4}_{2}\text{He}} + \text{energy}$$

The total mass of the reactants in the second equation is 4.0229 amu, which is 0.0203 amu greater than the mass of the product. This difference in mass is manifested in the great amount of energy liberated.

During the past 40 to 45 years, a large amount of research on controlled nuclear fusion reactions has been done in the United States and in other countries, especially the Soviet Union. So far, the goal of controlled nuclear fusion has not been attained, although the required ignition temperature has been reached in several devices. Evidence to date leads us to believe that it is possible to develop a practical fusion power reactor. Fusion power, if it can be developed, will be far superior to fission power because: (1) Virtually infinite amounts of energy are to be had from fusion. Uranium supplies for fission power are limited, but heavy hydrogen, or deuterium (the most likely fusion fuel), is relatively abundant. It is estimated that the deuterium in a cubic mile of seawater is an energy resource (as fusion fuel) greater than the petroleum reserves of the entire world. (2) From an environmental viewpoint, fusion power is much "cleaner" than fission power because fusion reactions, in contrast to uranium and plutonium fission reactions, do not produce large amounts of long-lived and dangerously radioactive isotopes.

19.12 Mass-Energy Relationship in Nuclear Reactions

Because large amounts of energy are released in nuclear reactions, significant amounts of mass are converted to energy. We stated earlier that the amount of mass converted to energy in chemical changes is considered insignificant. In fission reactions about 0.1% of the mass is converted into energy. In fusion reactions as much as 0.5% of the mass may be changed into energy. The Einstein equation, $E = mc^2$, can be used to calculate the energy liberated, or available, when the mass loss is known. For example, in the reaction

$$\underset{7.016\text{ g}}{{}^{7}_{3}\text{Li}} + \underset{1.008\text{ g}}{{}^{1}_{1}\text{H}} \longrightarrow \underset{4.003\text{ g}}{{}^{4}_{2}\text{He}} + \underset{4.003\text{ g}}{{}^{4}_{2}\text{He}} + \text{energy}$$

the mass difference between the reactants and products (8.024 g − 8.006 g) is 0.018 g. The energy equivalent to this amount of mass is 1.62×10^{12} J (3.8×10^{11} cal). By comparison this amount is more than 4 million times greater than the 3.9×10^{5} J (9.4×10^{4} cal) of energy obtained from the complete combustion of 12.01 g (1 mol) of carbon.

mass defect

nuclear binding energy

The mass of a nucleus is actually less than the sum of the masses of the protons and neutrons that make up that nucleus. The difference between the mass of the protons and the neutrons in a nucleus and the mass of the nucleus is known as the **mass defect**. The energy equivalent to this difference in mass is known as the **nuclear binding energy**. This energy is the amount that would be required to break up a nucleus into its individual protons and neutrons. The higher the binding energy, the more stable is the nucleus. Elements of intermediate atomic masses have high binding energies. For example, iron (element number 26) has a very high binding energy and therefore has a very stable nucleus. Just as electrons attain less energetic and more stable arrangements through ordinary chemical reactions, neutrons and protons attain less energetic and more stable arrangements through nuclear fission or fusion reactions. Thus, when uranium undergoes fission, the products have less mass (and greater binding energy) than the original uranium. In like manner, when hydrogen and lithium fuse to form helium, the helium has less mass (and greater binding energy) than the hydrogen and lithium. It is this conversion of mass to energy that accounts for the very large amounts of energy associated with both nuclear fission and fusion reactions.

19.13 Transuranium Elements

transuranium elements

The elements following uranium on the periodic chart and having atomic numbers greater than 92 are known as the **transuranium elements**. All of them are synthetic radioactive elements; none of them occur naturally.

The first transuranium element, number 93, was discovered in 1939 by Edwin M. McMillan (1907–1991) at the University of California while he was investigating the fission of uranium. He named it neptunium for the planet Neptune. In 1941 element 94, plutonium, was identified as a beta decay product of neptunium:

$$^{238}_{93}\text{Np} \longrightarrow \, ^{238}_{94}\text{Pu} + \, ^{0}_{-1}\text{e}$$
$$^{239}_{93}\text{Np} \longrightarrow \, ^{239}_{94}\text{Pu} + \, ^{0}_{-1}\text{e}$$

Plutonium is one of the most important fissionable elements known today.

Since 1964 the discoveries of six new transuranium elements, numbers 104–109, have been announced. All of these elements were produced in minute quantities by high energy particle accelerators. In the case of element 109, the discovery was based on the detection of a single atom of $^{266}_{109}\text{X}$, which existed for about five milliseconds! The reported synthesis was achieved by bombarding a $^{209}_{83}\text{Bi}$ target with highly accelerated iron-58 nuclei.

$$^{209}_{83}\text{Bi} + \, ^{58}_{26}\text{Fe} \longrightarrow \, ^{266}_{109}\text{X} + \, ^{1}_{0}\text{n}$$

TABLE 19.5 Transuranium Elements

Element	Symbol	Atomic number	Discovery date
Neptunium	Np	93	1939
Plutonium	Pu	94	1941
Americium	Am	95	1944
Curium	Cm	96	1944
Berkelium	Bk	97	1949
Californium	Cf	98	1950
Einsteinium	Es	99	1953
Fermium	Fm	100	1953
Mendelevium	Md	101	1955
Nobelium	No	102	1957
Lawrencium	Lr	103	1961
Unnilquadium	Unq	104	1964
Unnilpentium	Unp	105	1970
Unnilhexium	Unh	106	1974
Unnilseptium	Uns	107	1981
Unniloctium	Uno	108	1988
Unnilennium	Une	109	1982

Groups of scientists in the United States and the Soviet Union involved in high energy particle research to produce new elements disagree on the names for elements 104, 105 and 106. Consequently the International Union of Pure and Applied Chemistry (IUPAC) has suggested a systematic method for naming all elements beyond 103. The method combines letters from the Greek and Latin words for the numbers plus the ending *ium*. For example, *un* for 1, *nil* for 0, and *quad* for 4 plus the ending *ium* forms unnilquadium, element 104. Thus unnilpentium is 105 and unnilhexium is 106. Table 19.5 lists all the presently known transuranium elements.

19.14 Biological Effects of Radiation

Radiation that has enough energy to dislocate bonding electrons and create ions when passing through matter is classified as *ionizing radiation*. Alpha, beta, and gamma rays, along with X rays, fall into this classification. Ionizing radiation can damage or kill living cells. This damage is particularly devastating when it occurs in the cell nuclei and affects molecules involved in cell reproduction. The overall effects of radiation on living organisms fall into these general categories: (1) acute, or short-term, effects; (2) long-term effects, and (3) genetic effects.

Acute Radiation Damage High levels of radiation, especially of gamma or X rays, produce nausea, vomiting, and diarrhea. The effect has been likened to a sunburn throughout the body. If the dosage is high enough, death will occur in a few days. The damaging effects of radiation appear to be centered in the nuclei of the cells, and cells that are undergoing rapid cell division are most susceptible to damage. It is for this reason that cancers are often treated with gamma radiation from a Co-60 source. Cancerous cells are multiplying rapidly and are destroyed by a level of radiation that does not seriously damage normal cells.

Long-Term Radiation Damage Protracted exposure to low levels of any form of ionizing radiation can weaken the organism and lead to the onset of malignant tumors, even after fairly long time delays. The largest exposure to man-made sources of radiation is from X rays. Evidence suggests that a number of early workers in radioactivity and X-ray technology may have had their lives shortened by long-term radiation damage.

A number of women who had been employed in the early 1920s to paint luminous numbers on watch dials died some years later from the effects of radiation. These women had ingested radium by using their lips to point the brushes used on the job. Radium was retained in their bodies and, as an alpha emitter with a half-life of about 1620 years, continued to inflict radiation damage.

Strontium-90 isotopes are present in the fallout from atmospheric testing of nuclear weapons. Strontium is in the same periodic-table group as calcium, and its chemical behavior is similar to that of calcium. Hence, when foods contaminated with Sr-90 are eaten, Sr-90 ions are laid down in the bone tissue along with ordinary calcium ions. Strontium-90 is a beta emitter with a half-life of 28 years. Blood cells that are manufactured in bone marrow are affected by the radiation from Sr-90. Hence, there is concern that Sr-90 accumulation in the environment may cause an increase in the incidence of leukemia and bone cancers. Fortunately, the United States and the Soviet Union agreed to stop atmospheric testing of nuclear weapons several years ago; however, some countries are still doing testing in the atmosphere.

Genetic Effects All the information needed to create an individual of a particular species, be it a bacterial cell or a human being, is contained within the nucleus of a cell. This genetic information is encoded in the structure of DNA (deoxyribonucleic acid) molecules, which make up genes. The DNA molecules form precise duplicates of themselves when cells divide, thus passing genetic information from one generation to the next. Radiation can damage DNA molecules. If the damage is not severe enough to prevent the individual from reproducing, a mutation (a heritable variation in the offspring) may result. Most mutation-induced traits are undesirable. Unfortunately, if the bearer of the altered genes survives to reproduce, these traits are passed along to succeeding generations. In other words, the genetic effects of increased radiation exposure are found in future generations, not in the present generation.

Because radioactive rays are hazardous to health and living tissue, special precautions must be taken in designing laboratories and nuclear reactors, in disposing of waste materials, and in monitoring the radiation exposure of people working in this field. For example, personnel working in areas of hazardous radiation wear film badges or pocket dosimeters to provide them with an accurate indication of cumulative radiation exposure.

Chemistry in Action

Tracers, Diagnosis, and Therapy

Compounds containing a radionuclide are described as being *labeled* or *tagged*. These compounds undergo their normal chemical reactions, but their location can be detected because of their radioactivity. When such compounds are given to a plant or an animal, the movement of the nuclide can be traced through the organism by the use of a Geiger counter or other detecting device.

One important use of the tracer technique was determining the pathway by which CO_2 becomes fixed into carbohydrate ($C_6H_{12}O_6$) during photosynthesis. The net equation for the photosynthesis is

$$6\,CO_2 + 6\,H_2O \longrightarrow C_6H_{12}O_6 + 6O_2$$

Radioactive $^{14}CO_2$ was injected into a colony of green algae. The algae was then placed in the dark, killed at selected time intervals, and the radioactive compounds separated by paper chromatography and analyzed. From these results a series of light independent photosynthetic reactions were elucidated.

Some other examples of biological research in which tracer techniques have been employed are (1) in determining the rate of phosphate uptake by plants, using radiophosphorus; (2) the flow of nutrients in the digestive tract using radioactive barium compounds; (3) the accumulation of iodine in the thyroid gland, using radioactive iodine; and (4) the absorption of iron by the hemoglobin of the blood, using radioactive iron. In chemistry, the uses are unlimited. The study of reaction mechanisms, the measurement of the rates of chemical reactions, and the determination of physical constants are just a few of the areas of application.

Radioactive tracers are commonly used in medical diagnosis. Because radiation must be detected outside of the body, radionuclides that emit gamma rays are usually chosen. The radionuclide must also be effective at a low concentration and have a short half-life to reduce the possibility of damage to the patient.

Radioactive iodine (I-131) is used to determine thyroid function, where the body concentrates iodine. In this process, a small amount of radioactive potassium or sodium iodide is ingested. A detector is focused on the thyroid gland and measures the amount of iodine in the gland. This picture can be compared to that of a normal thyroid to detect any differences.

Doctors can examine the heart's pumping performance and check for evidence of obstruction in the coronary arteries by *nuclear scanning*. The radionuclide Tl-201, when injected into the bloodstream, lodges in healthy heart muscle. Thallium-201 emits gamma radiation, which is detected by a special imaging device called a *scintillation camera*. The data obtained are simultaneously translated into pictures by a computer. With this technique doctors can observe whether heart tissue has died after a heart attack and whether blood is flowing freely through the coronary passages.

One of the newest applications of nuclear chemistry is the use of positron emission tomography (PET) in the measurement of dynamic processes in the body, such as oxygen use or blood flow. In this application a compound is made that contains a positron-emitting nuclide such as C-11, O-15, or N-13. The compound is injected into the body, and the patient is placed in an instrument that detects the positron emissions. A computer produces a three-dimensional image of the area.

PET scans have been used to locate the areas of the brain involved with epileptic seizures. The brain uses glucose almost exclusively for energy. Glucose that has been tagged with C-11 is injected, and an image of the brain is produced. Diseased areas that use glucose at a rate different than normal tissue can then be visualized.

For many years radium has been used in the treatment of cancer. Co-60 and Cs-137 are now extensively used for radiation therapy. The effectiveness of this therapy is dependent on the fact that rapidly growing or dividing malignant cells are more susceptible to radiation damage than normal cells. Cobalt-60 emits both beta particles and gamma rays. The radiation is focused on the area where the tumor is located, but it is very difficult to limit exposure only to malignant cells. Many patients suffer from radiation sickness following this type of therapy.

Iodine-131 can be used for the treatment of hyperthyroidism. The therapeutic dose is larger than that used for diagnosis. The thyroid gland selectively concentrates the I-131. The section of the gland that is hyperactive will be exposed to a large dose of the isotope and be specifically destroyed. First Lady Barbara Bush went through this procedure in 1989.

Concepts in Review

1. Outline the historical development of nuclear chemistry, including the major contributions of Henri Becquerel, Marie Curie, Ernest Rutherford, Irene Joliet-Curie, Otto Hahn, Fritz Strassmann, and Edwin McMillen.
2. Write balanced nuclear chemical equations using isotopic notation.
3. Determine the amount of radionuclide remaining after a given period of time when the starting amount and half-life are given.
4. List the characteristics that distinguish alpha, beta, and gamma rays from the standpoint of mass, charge, relative velocities, and penetrating power.
5. Describe the effect of a magnetic field on alpha particles, beta particles, and gamma rays.
6. Describe a radioactive disintegration series, and predict which isotope would be formed by the loss of specified numbers of alpha and beta particles from a given radionuclide.
7. Discuss the transmutation of elements.
8. Indicate the methods used for the detection of radiation.
9. Distinguish between radioactive disintegration and nuclear fission reactions.
10. Explain how the fission of U-235 can lead to a chain reaction and why a critical mass is necessary.
11. Explain how the energy from nuclear fission is converted to electrical energy.
12. Explain the difference between fission reactions in a nuclear reactor and those of an atomic bomb.
13. Explain what is meant by the term *nuclear fusion* and why a massive effort to develop controlled nuclear fusion is in progress.
14. Indicate the significance of mass defect and nuclear binding energy.
15. Indicate the major effects of radiation on living organisms.
16. Explain how the age of objects can be determined using radioactivity.
17. Indicate several current uses for radioactive tracers.

Exercises

An asterisk indicates a more challenging question or problem.

1. To afford protection from radiation injury, which kind of radiation requires (a) the most shielding? (b) the least shielding?
2. Why is an alpha particle deflected less than a beta particle in passing through an electromagnetic field?
3. Name three pairs of nuclides that might be obtained by fissioning U-235 atoms.
4. Identify the following people and their associations with the early history of radioactivity.
   (a) Antoine Henri Becquerel
   (b) Marie and Pierre Curie
   (c) Wilhelm Roentgen
   (d) Ernest Rutherford
   (e) Otto Hohn and Fritz Strassmann
5. Why is the radioactivity of an element unaffected by the usual factors that affect the rate of chemical reactions, such as ordinary changes of temperature and concentration?
6. Distinguish between the terms *isotope* and *nuclide*.

7. Indicate the number of protons, neutrons, and nucleons in each of the following nuclei.
(a) ${}^{35}_{17}Cl$ (b) ${}^{226}_{88}Ra$ (c) ${}^{235}_{92}U$ (d) ${}^{82}_{35}Br$
8. Explain the phenomenon of half-life for a radionuclide.
9. The half-life of Pu-244 is 76 million years. If the age of the earth is about 5 billion years, discuss the feasibility of this nuclide's being found as a naturally occurring nuclide.
10. Tell how alpha, beta, and gamma radiation are distinguished from the standpoint of:
(a) Charge
(b) Relative mass
(c) Nature of particle or ray
(d) Relative penetrating power
11. How are the mass and the atomic number of a nucleus affected by the loss of the following:
(a) An alpha particle
(b) A beta particle
12. Distinguish between natural and artificial radioactivity.
13. What is a radioactive disintegration series?
14. Briefly discuss the transmutation of elements.
15. Write nuclear equations for the alpha decay of:
(a) ${}^{218}_{85}At$ (b) ${}^{221}_{87}Fr$ (c) ${}^{192}_{78}Pt$ (d) ${}^{210}_{84}Po$
16. Write nuclear equations for the beta decay of:
(a) ${}^{14}_{6}C$ (b) ${}^{137}_{55}Cs$ (c) ${}^{239}_{93}Np$ (d) ${}^{90}_{38}Sr$
17. Stable Pb-208 is formed from Th-232 in the thorium disintegration series by successive α, β, β, α, α, α, α, β, β, α particle emissions. Write the symbol (including mass and atomic number) for each nuclide formed in this series.
18. The nuclide Np-237 loses a total of seven alpha particles and four beta particles. What nuclide remains after these losses?
19. Bismuth-211 decays by alpha emission to give a nuclide that in turn decays by beta emission to yield a stable nuclide. Show these two steps with nuclear equations.
20. Write nuclear equations for the following:
(a) Conversion of ${}^{13}_{6}C$ to ${}^{14}_{6}C$
(b) Conversion of ${}^{30}_{15}P$ to ${}^{30}_{14}Si$
21. Complete and balance the following nuclear equations by supplying the missing particles.
(a) ${}^{27}_{13}Al + {}^{4}_{2}He \longrightarrow {}^{30}_{15}P + ?$
(b) ${}^{27}_{14}Si \longrightarrow {}^{0}_{+1}e + ?$
(c) $? + {}^{2}_{1}H \longrightarrow {}^{13}_{7}N + {}^{1}_{0}n$
(d) $? \longrightarrow {}^{82}_{36}Kr + {}^{0}_{-1}e$
(e) ${}^{66}_{29}Cu \longrightarrow {}^{66}_{30}Zn + ?$
(f) ${}^{0}_{-1}e + ? \longrightarrow {}^{7}_{3}Li$
(g) ${}^{27}_{13}Al + {}^{4}_{2}He \longrightarrow {}^{30}_{14}Si + ?$
(h) ${}^{85}_{37}Rb + ? \longrightarrow {}^{82}_{35}Br + {}^{4}_{2}He$
(i) ${}^{214}_{83}Bi \longrightarrow {}^{4}_{2}He + ?$
22. Write shorthand notations for the following nuclear equations.
(a) ${}^{27}_{13}Al + {}^{4}_{2}He \longrightarrow {}^{30}_{15}P + {}^{1}_{0}n$
(b) ${}^{14}_{7}N + {}^{4}_{2}He \longrightarrow {}^{17}_{8}O + {}^{1}_{1}H$
(c) ${}^{59}_{27}Co + {}^{1}_{0}n \longrightarrow {}^{60}_{27}Co + \gamma$
(d) ${}^{249}_{98}Cf + {}^{12}_{6}C \longrightarrow {}^{257}_{104}Unq + 4\,{}^{1}_{0}n$
23. Write nuclear equations from these shorthand notations.
(a) ${}^{23}_{11}Na(p, n){}^{23}_{12}Mg$
(b) ${}^{209}_{83}Bi(d, n){}^{210}_{84}P_o$ (d = deuteron, ${}^{2}_{1}H$)
(c) ${}^{238}_{92}U({}^{16}_{8}O, 5\ n){}^{249}_{100}Fm$
(d) ${}^{9}_{4}Be(\alpha, n){}^{12}_{6}C$
24. Describe the use of a Geiger counter in measuring radioactivity.
25. What does each of these units represent?
(a) Curie (c) Rad or Gray
(b) Roentgen (d) Rem
26. What was the contribution to nuclear physics of Otto Hahn and Fritz Strassmann?
27. What is a breeder reactor? Explain how it accomplishes the "breeding."
28. What is the essential difference between the nuclear reactions in a nuclear reactor and those in an atomic bomb?
29. Why must a certain minimum amount of fissionable material be present before a self-supporting chain reaction can occur?
30. Are the terms *atomic bomb* and *hydrogen bomb* synonymous? If not, what is the major distinction between them?
31. What is mass defect and nuclear binding energy?
32. Explain why radioactive rays are classified as ionizing radiation.
33. Give a brief description of the biological hazards associated with radioactivity.
34. Strontium-90 has been found to occur in radioactive fallout. Why is there so much concern about this radionuclide being found in cow's milk? (Half-life of Sr-90 is 28 years.)
35. What is a radioactive tracer? How is it used?
36. Describe the radiocarbon method for dating archaeological artifacts.
37. How might radioactivity be used to locate a leak in an underground pipe?
38. Anthropologists have found bones whose age suggests that the human line may have emerged in Africa as much as 4 million years ago. If wood or charcoal were found with such bones, would C-14 dating be useful in dating the bones? Explain.
39. Strontium-90 has a half-life of 28 years. If a

1.00 mg sample were stored for 112 years, what mass of Sr-90 would remain?

40. If radium costs \$50,000 a gram, how much will 0.0100 g of $^{226}RaCl_2$ cost if the price is based only on the radium content?

41. Strontium-90 has a half-life of 28 years. If a sample was tested in 1980 and found to be emitting 240 counts/minute, in what year would the same sample be found to be emitting 30 counts/minute? How much of the original Sr-90 would be left?

42. An archaeological specimen was analyzed and found to be emitting only 25% as much carbon-14 radiation per gram of carbon as newly cut wood. How old is this specimen?

43. Barium-141 is a beta emitter. What is the half-life if a 16.0 g sample of the nuclide decays to 0.500 g in 90 minutes?

***44.** Calculate (a) the mass defect and (b) the binding energy of $^{7}_{3}Li$. Mass data: $^{7}_{3}Li = 7.0160$ g; n = 1.0087 g; p = 1.0073 g; $e^- = 0.00055$ g; 1.0 g = 9.0×10^{13} J (from $E = mc^2$).

***45.** For the fission reaction

$$^{235}_{92}U + ^{1}_{0}n \longrightarrow ^{94}_{38}Sr + ^{139}_{54}Xe + 3\,^{1}_{0}n + \text{energy}$$

Mass data: U-235 = 235.0439 amu; Sr-94 = 93.9154 amu; Xe-139 = 138.9179 amu; n = 1.0087 amu; 1.0 g = 9.0×10^{13} J. Calculate:

(a) The energy released in joules for a single event (one uranium atom splitting)

(b) The energy released in joules per mole of uranium splitting

(c) The percentage of the mass that is lost in the reaction

***46.** For the fusion reaction

$$^{1}_{1}H + ^{2}_{1}H \longrightarrow ^{3}_{2}He + \text{energy}$$

Mass data: $^{1}_{1}H = 1.00794$; $^{2}_{1}H = 2.01410$; $^{3}_{2}He = 3.01603$; 1.0 g = 9.0×10^{13} J. Calculate:

(a) The energy released in joules per mole of He-3 formed

(b) The percentage of the mass that is lost in the reaction

***47.** In the disintegration series $^{235}_{92}U \longrightarrow ^{207}_{82}Pb$, how many alpha and beta particles are emitted?

48. List three devices used for radiation detection and explain their operation.

49. The half-life of I-123 is 13 hours. If 10 mg of I-123 is given a patient, how much I-123 remains after 3 days and 6 hours?

50. Clearly distinguish between fission and fusion. Give an example of each.

51. Which of the following statements are correct? Rewrite the incorrect statements to make them correct.

(a) Radioactivity was dicovered by Marie Curie.

(b) An atom of $^{59}_{28}Ni$ has 59 neutrons.

(c) The loss of a beta particle by an atom of $^{75}_{33}As$ forms an atom of increased atomic number.

(d) The emission of an alpha particle from the nucleus of an atom lowers its atomic number by 4 and lowers its mass number by 2.

(e) Emission of gamma radiation from the nucleus of an atom leaves both the atomic number and mass number unchanged.

(f) Relatively few naturally occurring radioactive nuclides have atomic numbers below 81.

(g) The longer the half-life of a radionuclide, the more slowly it decays.

(h) The beta ray has the greatest penetrating power of all the rays emitted from the nucleus of an atom.

(i) Radioactivity is due to an unstable ratio of neutrons to nucleons in an atom.

(j) The half-life of different nuclides can vary from a fraction of a second to millions of years.

(k) The symbol $_{+1}^{0}e$ is used to indicate a positron, which is a positively charged particle with the mass of an electron.

(l) The disintegration of $^{226}_{88}Ra$ into $^{214}_{83}Po$ involves the loss of three alpha particles and two beta particles.

(m) If 1.0 g of a radionuclide has a half-life of 7.2 days, the half-life of 0.50 g of that nuclide is 3.6 days.

(n) If the mass of a radionuclide is reduced by radioactive decay from 12 g to 0.75 g in 22 hours, the half-life of the isotope is 5.5 hours.

(o) A very high temperature is required to initiate nuclear fusion reactions.

(p) Radiocarbon dating of archaeological artifacts is based on an increase in the ratio of C-14 to C-12 in the object.

(q) Cancers are often treated by radiation from Co-60, which destroys the rapidly growing cancer cells.

(r) High levels of radiation produce nausea, vomiting, and diarrhea.

(s) Carbon-14, used in radiocarbon dating, is produced in living matter by the beta decay of N-14.

(t) Radioactive tracers are small amounts of radioactive nuclides of selected elements whose progress in chemical or biological processes can be followed using a radiation detection instrument.

20

Chemistry of Selected Elements

◄ Chapter Opening Photo: **The beauty and noise of a fireworks display are the result of the chemistry of specific elements.**

It is amazing to consider the vast number of substances and products that surround us. Even more astounding is the fact that all forms of matter are combinations of only slightly more than 100 elements. The human body is a complex of billions of cells, but all the cells are primarily combinations of carbon, hydrogen, oxygen, and nitrogen. This is typical of many substances, either man-made or in the natural state. Each contains countless numbers of atoms and molecules, but is limited to very few elements. In an earlier chapter, the design of the periodic table was considered. Each element belongs to a particular family, whose members are similar in electronic structure. The chemical behavior of elements in each family is similar as well. In this chapter we will take a brief tour of the periodic table to discuss the chemistry of selected elements and discover some of the uses and characteristics of these elements, which make them an integral part of our lives.

20.1 Periodic Trends: A Quick Review

A number of important properties of elements correspond directly to the position of the elements on the periodic table (Chapter 11). Several of these properties are summarized in Figure 20.1.

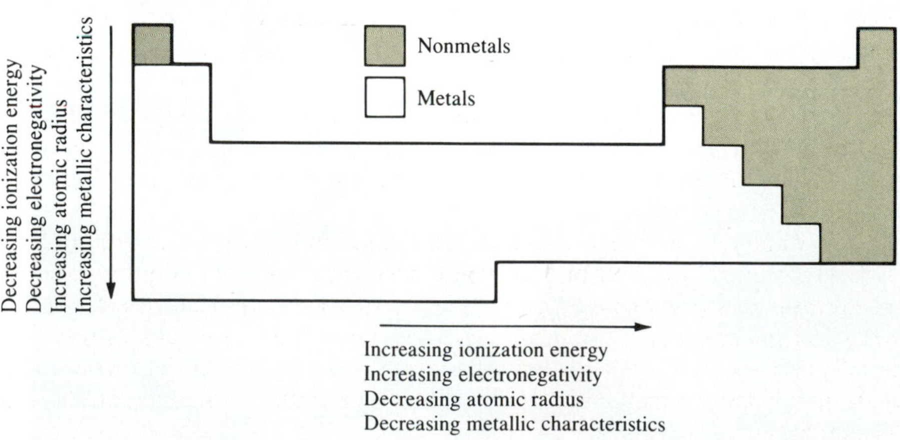

Figure 20.1 ► **A summary of periodic trends**

Table 20.1 Physical and Chemical Properties of Metals and Nonmetals

Metals	Nonmetals
Lustrous	Nonlustrous
Good thermal conductors	Poor thermal conductors
Good electrical conductors	Poor electrical conductors
Malleable	Brittle
Ductile	Shatter easily
Form positive ions in compounds	Form negative ions in compounds
Form basic oxides	Form acidic oxides

One of the most useful of these characteristics is the division of the elements into metals and nonmetals. More than three-fourths of the elements are metals. The outstanding properties of metals and nonmetals are summarized in Table 20.1.

The first member of a family often exhibits chemistry that is significantly different from that of other members of the same family. The differences are partially the result of the smaller size and greater electronegativity of the first member. Hydrogen is so small and unique that it is often classified as a family of one.

20.2 Preparation of the Elements

The distribution of the elements in the earth's crust, oceans, and atmosphere is given in Table 3.3. Oxygen, by far the most abundant, is found as O_2 in the air, H_2O in the sea, and in silicates and oxides in the crust of the earth. The most common metals are Al and Fe. They usually occur combined with nonmetals as ores.

In the world of living organisms the distribution of elements is dramatically different (Table 3.4). Only carbon, hydrogen, nitrogen, and oxygen are found in significant amounts. Many other elements are present in trace amounts, but play a vital role in metabolism.

Only about a quarter of the elements are found in nature in the free state. Most are found in compounds. **Metallurgy** is the science of obtaining and refining metals from their ores and the study of the properties and applications of metals. An **ore** is the commercial mineral source of a metal. Since metals in ores are usually in the form of positive ions (cations), the processing of an ore involves the *reduction* of the ion to the free metal (oxidation number 0). Reduction may be accomplished by the following methods:

metallurgy

ore

Method	Example
Carbon reduction	$2\ PbO + C \xrightarrow{\Delta} 2\ Pb + CO_2$
Active metal (Mg, Zn) reduction	$TiCl_4 + 2\ Mg \xrightarrow{\Delta} Ti + 2\ MgCl_2$
Hydrogen reduction	$WO_3 + 3\ H_2 \xrightarrow{\Delta} W + 3\ H_2O$
Electrolysis	$MgCl_2 \xrightarrow[\text{current}]{\text{Electric}} Mg + Cl_2$
Heating	$2\ Ag_2O \xrightarrow{\Delta} 4\ Ag + O_2$

The nonmetals are prepared in a variety of ways. Nitrogen and oxygen are most frequently obtained from the *liquefication* of air. Hydrogen can be obtained by electrolysis of water or the decomposition of methane (natural gas). Sulfur is mined in its elemental form, and halogens are produced by the oxidation of halide salt solutions.

Alloys

alloy

An **alloy** is a mixture (usually a solid) that contains two or more elements and has the characteristics of a metal. Making an alloy is one of the important ways that the properties of a pure metallic element can be modified. Alloys are made in order to improve the tensile strength, wearability, corrosion resistance, or conductivity of pure metals. **Steel**, for example, is an alloy of iron containing small amounts of carbon that is much stronger and harder than pure iron. Pure gold is too soft to use in jewelry, but alloys of gold and copper or silver form beautiful decorative pieces.

steel

solution alloy

intermetallic compounds

amalgam

Alloys can be classified as *solution alloys* or *intermetallic compounds.* In a **solution alloy** the metals are combined in the molten state and the components mix randomly and uniformly. Examples of this type of alloy include brass, pewter, and nitinol (a nickel–titanium alloy that "remembers" its shape). **Intermetallic compounds** are homogeneous alloys with definite composition. Dental **amalgam** (alloys of Hg) dates back to early China. In modern society these intermetallic compounds are playing an increasing role in diverse fields. Razor blades are advertised with edges of Cr_3Pt. Stereo headphones contain permanent lightweight magnets of Co_5Sm (a cobalt–samarium alloy). Most recently, intermetallic compounds have played an important role in developing superconductors. These materials are utilized as superconducting magnets in magnetic resonance imaging (MRI), a noninvasive technique for observing tissues in living organisms. The number of alloys and their uses are almost limitless. Table 20.2 gives several more examples of the thousands of alloys now available.

20.3 Hydrogen: A Unique Element

Hydrogen is the most abundant element in the universe. In the early days of chemistry, hydrogen was considered to be the precursor to all the other elements. Hydrogen combines chemically with most of the elements. The history of hydrogen includes many of the scientists we've encountered before. Jacques

Table 20.2 Composition of Selected Alloys

Alloy	Percent composition
Stainless steel	74 Fe, 18 Cr, 8 Ni, 0.18 C (Many others are known.)
Storage battery plates	94 Pb, 6 Sb
Coinage silver (U.S.)	90 Ag, 10 Cu
Plumber's solder	67 Pb, 33 Sn
Babbitt metal	90 Sn, 7 Sb, 3 Cu
Yellow brass	67 Cu, 33 Zn
10 carat gold	42 Au, 38–46 Cu, 12–20 Ag
18 carat gold	75 Au, 10–20 Ag, 5–15 Cu
Nichrome	60 Ni, 40 Cr
Stellite	55 Co, 15 W, 25 Cr, 5 Mo
Spring steel	98.6 Fe, 1 Cr, 0.4 C

Charles, in 1783, first used hydrogen to float his balloon above France. By 1928 a dirigible designed to carry passengers was built in Germany, soon to be followed by the infamous *Hindenburg*. Unfortunately the flammable nature of hydrogen resulted in the explosion and destruction of the *Hindenburg* in 1939, killing many passengers. This accident caused hydrogen to be classified by many people as a dangerous substance. Today the use of hydrogen is once again expanding.

Chemically hydrogen is used commercially in the production of ammonia, NH_3, by the Haber process:

$$N_2 + 3\,H_2 \rightleftarrows 2\,NH_3$$

Most of the ammonia manufactured this way is converted to fertilizers. Hydrogen is also used to produce methanol, CH_3OH, by reaction with carbon monoxide.

$$2\,H_2 + CO \xrightarrow[\Delta]{\text{catalyst}} CH_3OH$$

The methanol produced this way can be used as a gasoline additive or converted itself to gasoline. Methanol is also considered to be an alternative fuel, and it is used in the manufacture of adhesives and plastics.

A more evident use of hydrogen is found in the kitchen. Fats and oils, from peanut butter to solid shortening or margarine, are hardened by the addition of hydrogen to carbon–carbon double bonds in the molecules. Hydrogenation converts inexpensive oils to popular cooking products.

Hydrogen has also been proposed to be the fuel of the future. It is currently in use as a fuel for the space shuttle. The large external tanks are filled with hydrogen and oxygen which combine to form water vapor, providing energy to lift the shuttle into orbit. On board the shuttle are hydrogen–oxygen fuel cells to provide power for the shuttle during flight. Cars propelled by hydrogen have been suggested and prototypes exist. There are several drawbacks to these models, including a convenient method for storing hydrogen and a need for an inexpensive supply of hydrogen.

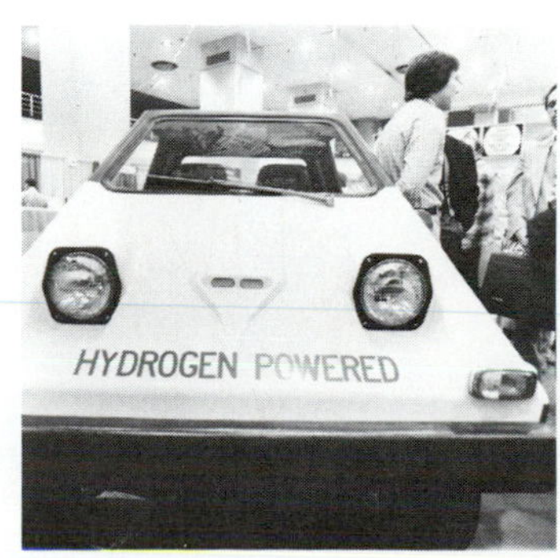

TABLE 20.3 Physical Properties of the Alkali Metals

Element	Atomic number	Atomic mass	Electron configuration	Metallic atomic radius (nm)	Radius of M^+ ion (nm)	Density (20°C, g/mL)	Melting point (°C)	Boiling point (°C)
Li	3	6.941	$[He]2s^1$	0.158	0.060	0.53	186.0	1336
Na	11	22.99	$[Ne]3s^1$	0.186	0.095	0.97	97.7	880
K	19	39.10	$[Ar]4s^1$	0.238	0.133	0.86	63.6	760
Rb	37	85.47	$[Kr]5s^1$	0.253	0.148	1.53	39.0	700
Cs	55	132.9	$[Xe]6s^1$	0.272	0.169	1.90	28.4	670

20.4 The Alkali Metals: Group IA

Physical Properties

alkali metals

The elements of Group IA in the periodic table are known as the **alkali metals**. They are named alkali metals because all of them form strong basic aqueous solutions. All Group IA metals are silvery-white in color when freshly cut, are soft and ductile, and are good conductors of electricity. Their physical properties are summarized in Table 20.3.

The alkali metals are among the most reactive elements and are never found in the free state in nature. Their chemical activity increases with size and atomic number. Because these metals react readily with oxygen and moisture, they are stored in a vacuum or under an inert liquid such as kerosene, benzene, or toluene. The similarity in chemical properties of this family is due to a similar electron configuration. An atom of an alkali metal has only one valence electron. Therefore, these elements form ions generally having a +1 oxidation number.

Chemical Properties

The chemical properties of alkali metals depend on the ease with which they each lose one outer shell electron. The ion formed when an electron is lost from an alkali metal has a noble gas electron configuration and, therefore, has great stability. The energy required to remove an electron from an alkali metal (ionization energy, see Chapter 11) is lower than that for other metals. The alkali metals are excellent reducing agents. A number of these oxidation–reduction reactions are summarized in Table 20.4.

Applications of Alkali Metal Compounds

Uses of the alkali metals come primarily from their ability to lose electrons. The formation of basic solutions in reactions with water is a sometimes spectacular reaction. Sodium hydroxide, or lye (NaOH), is widely used in the manufacturing of other chemicals and in processing pulp in the paper industry. Sodium

TABLE 20.4 Chemical Reactions of Alkali Metals

Reacts with	Products	Example
H_2O	Hydroxide, basic solution	$2\ Na + 2\ H_2O \longrightarrow 2\ NaOH + H_2\uparrow$
H_2	Hydride	$2\ Rb + H_2 \longrightarrow 2\ RbH$
Halogens	Metal halides (salts)	$2\ Na + Cl_2 \longrightarrow 2\ NaCl$
Oxygen	Oxide	$4\ Li + O_2 \longrightarrow 2\ Li_2O$
	Peroxide	$2\ Na + O_2 \longrightarrow Na_2O_2$
	Superoxide	$K + O_2 \longrightarrow KO_2$

hydroxide is also used to treat cotton, producing mercerized cotton, often used in thread. Many household cleaners contain lye. Soap is formed when sodium hydroxide is reacted with fats.

In reaction with oxygen, the alkali metals produce a diversity of useful compounds. Sodium peroxide can be reacted with water to produce hydrogen peroxide:

$$Na_2O_2 + 2\ H_2O \longrightarrow 2\ NaOH + H_2O_2$$

Hydrogen peroxide is a good oxidizing agent often used as a bleach for hair or as a disinfectant. K, Rb, and Cs react with oxygen to form superoxides. These compounds (especially KO_2) are used as a source of oxygen in emergency breathing equipment. The carbon dioxide and water vapor exhaled by the wearer react with the superoxide to form oxygen.

$$4\ KO_2 + 4\ CO_2 + 2\ H_2O \longrightarrow 4\ KHCO_3 + 3\ O_2$$

The salts formed by the alkali metals also find many uses. Sodium nitrate (Chile saltpeter) is found in large deposits formed by bacteria acting on saltwater organisms in shallow water. Sodium nitrate can be converted into potassium nitrate in a metathesis reaction:

$$NaNO_3 + KCl \longrightarrow KNO_3 + NaCl$$

Potassium nitrate has long been used in gunpowder and as the charge for pyrotechnics. Sodium bicarbonate ($NaHCO_3$) is commonly called baking soda, and is a component of baking powder and antacids. Lithium carbonate is currently prescribed as a treatment for manic depression.

Biological Applications

In living organisms sodium and potassium ions play a vital role in the functioning of biochemical systems. Inside living cells the concentration of potassium ions is very high compared to the concentration of sodium ions, while outside the cell the reverse is true. In order to maintain these levels an elaborate mechanism is required to transport these ions through the cell membrane. Potassium ions help the cell to maintain intercellular osmotic pressure and regulate pH. Potassium ions promote metabolic reactions, as well as playing an important role in membrane polarization. This polarization is necessary for proper nerve impulse conduction and muscle contraction. Sources of potassium in the diet include avocados, peanut butter, potatoes, and bananas.

Sodium ions are absorbed from foods by a transport system. The blood concentration of sodium ions is regulated by the kidneys. The major role of sodium ions in the extracellular fluid is to maintain osmotic pressure and regulate water balance. Sodium ions are also significant in the regulation of pH, the maintenance of membrane polarization, and the transportation of substances through cell membranes.

20.5 Alkaline Earth Metals: Group IIA

Physical Properties

alkaline earth metals

The **alkaline earth metals** were named for the alkaline properties of their oxides and because early chemists called all insoluble nonmetallic substances that were not affected by fire "earth." Group IIA metals are too reactive to be found free in nature but are generally harder and less reactive than the Group IA metals. At ordinary temperatures beryllium and magnesium are stable in air, but barium and strontium are so reactive that they are usually stored under an inert solvent such as kerosene. Radium is highly radioactive. Some of the physical properties of the alkaline earth metals are shown in Table 20.5.

Magnesium is the only metal that is obtained commercially from seawater (Figure 20.2). The process involves the following steps:

1. Precipitation of magnesium ions from seawater (which contains approximately 0.13% Mg^{2+}) as magnesium hydroxide by adding calcium oxide

$$CaO(s) + H_2O(l) \longrightarrow Ca^{2+}(aq) + 2\,OH^-(aq)$$
$$Mg^{2+}(aq) + 2\,OH^-(aq) \longrightarrow Mg(OH)_2(s)$$

2. Filtering out the $Mg(OH)_2$ and converting it to magnesium chloride by reaction with hydrochloric acid

$$Mg(OH)_2(s) + 2\,HCl(aq) \longrightarrow MgCl_2(aq) + 2\,H_2O$$

3. Drying and electrolyzing the $MgCl_2$ to obtain magnesium and chlorine

$$MgCl_2(l) \xrightarrow{\text{Electrolysis}} Mg(s) + Cl_2(g)$$

Chemical Properties and Applications

All the Group IIA elements react by losing the outermost pair of *s* electrons to form +2 ions. Hence, the formulas of their compounds are similar ($BeSO_4$, $MgSO_4$, $CaSO_4$, and so on). However, these elements differ widely in reactivity.

Beryllium is the least reactive of the alkaline earth metals. It is resistant to oxidation and stable in air at ordinary temperatures, but oxidizes rapidly at about 800°C.

$$2\,Be + O_2 \longrightarrow 2\,BeO$$

TABLE 20.5 Physical Properties of the Alkaline Earth Metals

Element	Atomic number	Atomic mass	Electron configuration	Metallic atomic radius (nm)	Radius of M^{2+} ion (nm)	Density (20°C, g/mL)	Melting point (°C)	Boiling point (°C)
Be	4	9.012	$[He]2s^2$	0.112	0.031	1.85	1278	2970
Mg	12	24.31	$[Ne]3s^2$	0.116	0.065	1.74	651	1107
Ca	20	40.08	$[Ar]4s^2$	0.197	0.099	1.55	843	1487
Sr	38	87.62	$[Kr]5s^2$	0.215	0.113	2.6	769	1384
Ba	56	137.3	$[Xe]6s^2$	0.222	0.135	3.6	725	1140
Ra	88	226.0	$[Rn]7s^2$	0.246	—	5.5	700	<1737

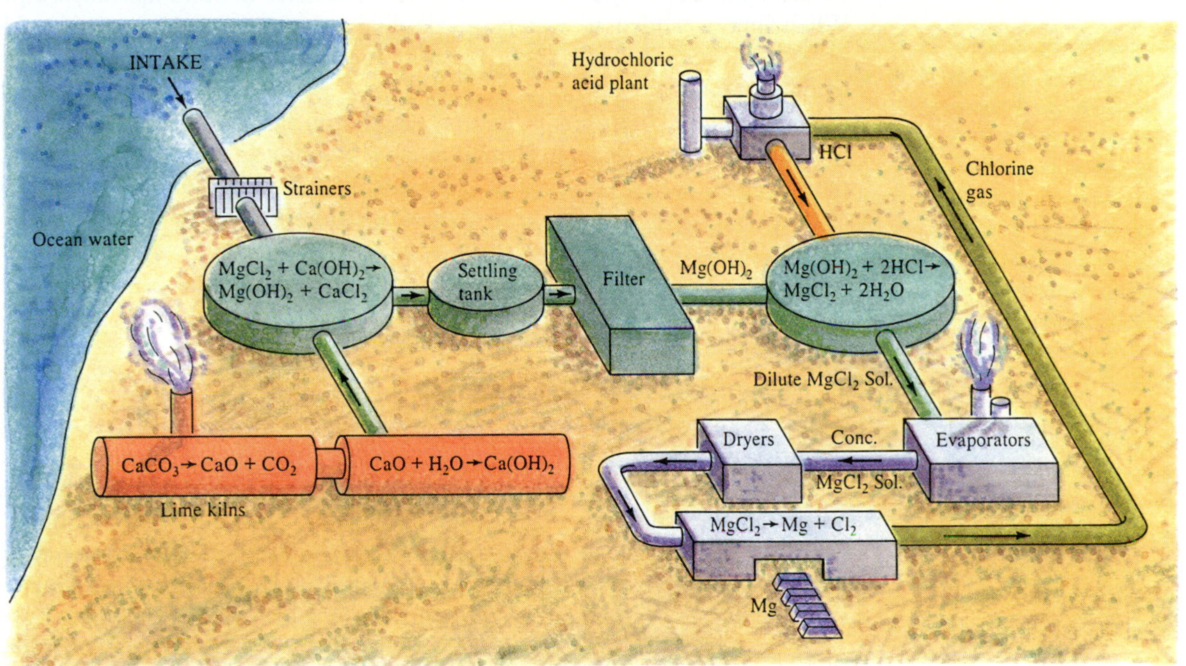

▲
FIGURE 20.2
Plant for processing Mg^{2+} ions from seawater.

Because of its many special properties, beryllium is useful in technology. For example, it is highly permeable to X rays and is therefore used in X-ray tube windows. In nuclear technology beryllium serves as a moderator to slow down neutrons; and when bombarded with alpha particles, it is a good source of neutrons.

Alloys of beryllium and copper form a metal that conducts electricity almost as well as copper. Because of its strength it does not generate sparks, and this alloy can be made into tools for use in areas where explosive gases are present. Its light weight and strength also have made it a choice (an expensive one) for golf clubs.

The use of magnesium is also based on its property of high strength-to-weight ratios. When exposed to air, magnesium quickly becomes coated with a thin, transparent layer of MgO. This layer protects the surface from further corrosion at ordinary temperatures. The majority of magnesium produced is made into lightweight alloys. These are used in aircraft, automotive parts, lightweight tools, and luggage trim.

The oxides and carbonates of calcium are compounds of economic value and special interest. Calcium carbonate decomposes to CaO (lime) in a reaction that ranks among the oldest chemistry known to us. The primary use of lime until 1900 was as a building material in the form of mortar (a sand, lime, and water mixture). Mortar was used by the Greeks in building temples. The chemistry involved in the mortar hardening process is unique: The mortar slowly absorbs CO_2 from the air and the lime reverts back to calcium carbonate.

$$Ca(OH)_2 + CO_2 \longrightarrow CaCO_3 + H_2O$$

The sand simply remains stuck in the particles, forming a hard substance resistant to wear. Calcium carbonate is found as beautiful stalactites and stalagmites in caverns, and also forms marble, which we see so often in buildings and sculpture. When lime is added to water, an alkaline solution of calcium hydroxide is produced. This solution has been used in the removal of hair from hides in the leather tanning process.

The mineral gypsum is another common calcium compound, hydrous calcium sulfate. Although gypsum is much less common than calcium carbonate, the several hundred square miles of sand in the White Sands Desert in New Mexico are composed of finely divided gypsum. Plaster of Paris is made by heating gypsum:

$$\underset{\text{Gypsum}}{2\,CaSO_4\cdot 2\,H_2O} \xrightarrow{\Delta} \underset{\text{Plaster of Paris}}{(CaSO_4)_2\cdot H_2O} + 3\,H_2O$$

To make surgical casts or plaster molds and impressions, the foregoing reaction is reversed by wetting powdered plaster of Paris to form a solid cast of calcium sulfate dihydrate.

Calcium and magnesium salts present problems when dissolved in our water. Water supplies containing Ca^{2+}, Mg^{2+}, and Fe^{3+} are called "hard water." The walls of pipes and the insides of tea kettles and coffee pots often become coated with insoluble carbonates of these ions. To prevent this scaly build-up the water is "softened" by removing the Ca^{2+} and Mg^{2+} ions (see Section 14.15).

Biological Applications

Calcium and magnesium are required for many essential biological processes in both animal and plant life. Magnesium is present in the chlorophyll of green plants and is required for the activity of some enzymes involved in animal metabolism. It is involved in a variety of metabolic functions within the mitochondria and is associated with the production and use of ATP, the central energy transfer molecule of the cell. Calcium is necessary for nitrogen utilization by plants.

Ninety-nine percent of the calcium in humans is found as inorganic salts in the bones and teeth. In dissolved form calcium ions play a role in nerve impulses, muscle contraction, hormone regulation, and blood coagulation. The absorption of calcium by the body is dependent on a variety of factors, including the need for calcium. The presence of vitamin D and a high protein intake increase absorption of calcium, while excessive fat seems to lower absorption. The role of calcium in the prevention of osteoporosis has been well documented. Good sources of calcium in the diet include milk and leafy green vegetables.

The remaining Group IIA elements are toxic to living systems. Interestingly, physicians often use barium sulfate (a barium "cocktail") to observe the digestive tract by X rays. This does not mean the doctor is poisoning the patient; rather, the $BaSO_4$ is *very* insoluble and passes through the digestive system without absorption. Strontium is believed to be a biologically nonessential element; however, the isotope Sr-90, present in nuclear fallout, can be absorbed by the body and concentrates in bone tissue creating a radiation hazard. Strontium chloride is used as a dental desensitizer in some toothpastes.

20.6 Flame Tests and Spectroscopy

All Group IA elements and calcium, strontium, and barium from Group IIA can be detected by simple flame tests. A Bunsen burner and a piece of clean platinum wire are the only essential pieces of apparatus needed to make flame tests. When salts of the elements are volatilized in the burner flame, the characteristic colors are visible.

Element	Color
Li	Red
Na	Bright yellow (persistent)
K	Violet (short duration)
Rb	Violet
Ca	Brick red (short duration)
Sr	Crimson
Ba	Light green

Flame color has been exploited by spectroscopy both to identify elements and to provide detailed information on the structure of atoms. Electrons can exist only at certain definite quantized energy levels within atoms (Section 5.6). At ordinary temperatures the electrons in atoms are at their lowest, or ground-state, energy levels. When heated in a Bunsen flame (or electric arc for some elements), some ground-state electrons are promoted to higher excited-state energy levels. In falling back to the ground state, each electron releases the same amount of energy (as radiation) that was absorbed when it was promoted to the excited state. If the wavelength of this radiation falls in the visible spectrum (400–700 nm), it is seen as colored light corresponding to a particular wavelength. When such light is passed through a prism (or ruled grating) in a spectroscope, it is found to consist of narrow colored bands called *spectral lines.* Each band (or line) corresponds to the energy emitted when an electron falls from a specific excited state to a specific lower energy level. Since the number of electrons in each element is different, each element has a characteristic set of spectral lines that is different from that of every other element. This set of spectral lines identifies the element and provides a means of determining the energy levels that are occupied by electrons within the atoms of that element.

A spectacular example of the colors produced by the nitrate salts of Groups IA and IIA can be seen in fireworks displays. The colors of flames in treated pine cones or commercial pressed sawdust logs also result from these salts.

20.7 Group IIIA Elements

Physical Properties

The elements in Group IIIA vary from the most abundant metal on earth to the first metalloid (boron) we've encountered in our survey. The elements do not occur free in nature but are generally found as oxides in many global locations. The physical properties of these elements are summarized in Table 20.6.

Boron is relatively rare; it is usually found combined with oxygen in borates. Large deposits of borax, $Na_2B_4O_7 \cdot 10\ H_2O$, are mined in the Mojave Desert near Boron, California. In the 1800s deposits of borate were mined in Death Valley using the famous 20-mule teams to haul the ore. Another source of boron is boric acid, H_3BO_3, found in the volcanic regions of Italy.

Applications

Although boron is not a common element it finds many interesting uses. Diboron trioxide can be reduced with carbon to form tetraboron carbide.

$$2\ B_2O_3 + 4\ C \longrightarrow B_4C + 3\ CO_2$$

This material is extremely hard and has a low density. One of its major uses is in bulletproof armor. Borax is probably the most important boron compound. It is mildly alkaline and works well as a cleansing agent or water softener. Borax can be reacted with sulfuric acid to form boric acid.

$$Na_2B_4O_7 \cdot 10\ H_2O + H_2SO_4 \longrightarrow 4\ H_3BO_3 + Na_2SO_4 + 5\ H_2O$$

Table 20.6 Physical Properties of Group IIIA Elements

Element	Atomic number	Atomic mass	Electron configuration	Density (20°C, g/mL)	Melting point (°C)	Boiling point (°C)
B	5	10.81	$[He]2s^22p^1$	2.34	2300	2550
Al	13	26.98	$[Ne]3s^23p^1$	2.70	660	2327
Ga	31	69.72	$[Ar]4s^23d^{10}4p^1$	5.91	29.78	2403
In	49	114.8	$[Kr]5s^24d^{10}5p^1$	7.31	156.6	2000
Tl	81	204.4	$[Xe]6s^24f^{14}5d^{10}6p^1$	11.85	303.5	1457

◀ **Borax mining in California.**

Boric acid is a weak acid that is slightly toxic. It is used as a mild antiseptic in the home and can be used in dilute solution as an eyewash. Boric acid also finds use as a flame retardant in insulating materials made from cellulose. The dehydration of boric acid yields boric anhydride (diboron trioxide).

$$2\,H_3BO_3 \longrightarrow B_2O_3 + 3\,H_2O$$

This oxide is used in the manufacture of borosilicate glass, commonly known as Pyrex. This type of glass has a higher softening temperature and expands less on heating than other types of glass, so it can be used for freezer-to-microwave (or stove) cookware.

Aluminum (generally spelled *aluminium* in countries other than the United States) is the most abundant metal and the third most abundant element in the earth's crust, being exceeded only by oxygen and silicon. Aluminum occurs mainly in clays, feldspars, and granite as complex aluminum silicates and in bauxite ores as hydrated aluminum oxide. Pure aluminum oxide (Al_2O_3) is known as *alumina.*

Alumina occurs naturally as the mineral corundum. Corundum is a very hard material, standing next to diamond on the hardness scale, and it is used as an abrasive in sandpaper and toothpaste. Emery, also widely used as an abrasive, is an impure corundum consisting of Al_2O_3 and Fe_3O_4. Rubies and sapphires are much-sought-after precious gemstones. They consist of corundum colored by small amounts of other metallic oxides. Artificial rubies and sapphires are produced that are practically indistinguishable from the natural stones. In addition to their use as gemstones and bearings in watches and other instruments, artificial rubies and sapphires are used in lasers for the production of intense beams of coherent light.

Although aluminum is the most abundant metal in the earth's crust, free aluminum metal was not known until 1825. In that year the Danish scientist Hans Christian Oersted succeeded in making some impure samples of aluminum metal by heating aluminum chloride with potassium-mercury amalgam. However, the reduction of aluminum by an alkali metal was an inherently costly process, and aluminum remained nearly as expensive as gold and platinum during most of the 19th century. Finally in 1886 a relatively cheap method for the electrolytic production of aluminum was devised. Since then, aluminum has been used in ever-increasing amounts. Aluminum has a density of only 2.7 g/mL, a value about one-third that of other common metals such as iron, copper, and zinc. Because it is a lightweight metal and has the ability to form a variety of useful alloys, aluminum has many important commercial applications.

Aluminum is an excellent conductor of heat. As an electrical conductor, it is surpassed only by silver and copper. However, when a comparison is made with wires having the same mass per unit of length, aluminum wire, being larger, surpasses both copper and silver as an electrical conductor. For this reason—and because of the increasing scarcity of copper—high-voltage transmission cables are now usually made of aluminum wires reinforced with a steel core for additional strength.

Aluminum metal has numerous uses for structural materials as well as in packaging. It is most commonly found as an alloy, since alone aluminum is soft and weak. When combined with copper and smaller amounts of silicon and magnesium, the alloy is used in large quantities to build airplanes. In combination with manganese a softer, more corrosion-resistant alloy is formed for use in kitchen utensils, lawn furniture, highway and directional signs, aluminum foil, and soda cans.

▲ **The vibrant colors in fabrics are produced by adhering dye molecules to the cloth with a mordant such as aluminum hydroxide.**

Bauxite or clay can be treated with sulfuric acid to produce aluminum sulfate. Double salts, such as $(KAl(SO_4)_2 \cdot 12\ H_2O$ form a class of substances known as alums. These aluminum compounds are used in the paper industry to make the product stronger and not porous. In water treatment plants alums are used as clarifiers. The alum surrounds the suspended particles in the water and precipitates them out of solution.

When aluminum salts react with bases, aluminum hydroxide forms as a gelatinous white precipitate. The ionic equation is

$$Al^{3+} + 3\ OH^- \longrightarrow Al(OH)_3(s)$$

This hydroxide is amphoteric and dissolves in either a strong base or a strong acid:

$$Al(OH)_3(s) + NaOH(aq) \longrightarrow NaAlO_2(aq) + 2\ H_2O(l)$$
$$2\ Al(OH)_3(s) + 6\ HCl(aq) \longrightarrow 2\ AlCl_3(aq) + 3\ H_2O(l)$$

Aluminum hydroxide is called a *mordant* in the textile industry. It binds to cloth and also to dye molecules, adhering the dye to the cloth. Some common antacid medications are also preparations of aluminum hydroxide.

Certain aluminum salts exert an astringent or constricting action on the skin and mucous membranes, a property exploited in many commercial antiperspirant preparations. The principal active agents in many of these formulations are aluminum chlorohydrates, actually aluminum hydroxy chlorides having formulas of the type $Al_2(OH)_5Cl \cdot 2\ H_2O$ or $Al(OH)_2Cl$. These substances prevent or retard perspiration by constricting the openings of the sweat glands.

TABLE 20.7 Physical Properties of Group IVA Elements

Element	Atomic number	Atomic mass	Electron configuration	Density (20°C, g/mL)	Melting point (°C)	Boiling point (°C)
C	6	12.01	$[He]2s^2 2p^2$	3.15	3550	Sublimes
Si	14	28.09	$[Ne]3s^2 3p^2$	2.33	1410	2355
Ge	32	77.59	$[Ar]4s^2 3d^{10} 4p^2$	5.32	937	2830
Sn	50	118.7	$[Kr]5s^2 4d^{10} 5p^2$	7.30	232	2270
Pb	82	207.2	$[Xe]6s^2 4f^{14} 5d^{10} 6p^2$	11.4	328	1744

20.8 Group IVA Elements

Physical Properties

The elements of Group IVA exhibit the full range of metallic character. Carbon is distinctly nonmetallic, silicon is classified as a metalloid, and the heavier elements, tin and lead, are most certainly metals. All of these elements are characterized by the half-filled valence shell. Physical properties of these elements are summarized in Table 20.7.

Occurrence and Applications

Carbon is widely distributed in many forms throughout the globe. As an element it commonly occurs in two allotropes, diamond and graphite. The diamond form of carbon has long been of interest to us. The most common naturally occurring diamonds are mined mostly in South Africa. The use of diamonds for abrasives and cutting tools in industry, in addition to their gem status, has led to the production of synthetic diamonds. These synthetic crystals are made by subjecting graphite to intense pressures and temperatures (70,000 atm and 2000°C).

Graphite is by far the more common form of elemental carbon. It melts at extremely high temperatures (3570°C), so it is sometimes used to line electric furnaces. It is remarkably soft and sticks to paper, making graphite the common pencil "lead." Harder "lead" contains more clay mixed with the graphite. This allotrope is also used as a lubricant. There are a variety of other relatively pure forms of carbon, including charcoal and carbon black. Charcoal has remarkable properties as an adsorbant after it has been heated to remove volatile impurities in a process known as activation. The activated charcoal is used to remove colored impurities in the refining of sugar, and can adsorb a variety of contaminants when used in gas masks or a water purification system. Carbon black is used as a reinforcing agent in rubber, and so is the reason that tires are black and not the natural color of rubber. Inks used on nonglossy productions (such as your newspaper) are also made with carbon black. Graphite can also be mixed with plastics to produce highly flexible, strong products such as tennis racquets, golf clubs, and fishing poles.

Carbon is easily oxidized to either carbon monoxide or carbon dioxide, depending upon the amount of oxygen available. For this reason carbon is used industrially for the reduction of metal oxides to metals. Approximately half of the carbon dioxide produced in the United States is used as a refrigerant. When cooled and compressed into solid blocks it is known as dry ice. About one-fourth of the CO_2 produced is used in the production of carbonated beverages. The carbon dioxide released into the atmosphere during combustion of fossil fuels has become a subject of controversy in recent years concerning its effect on the global heat balance ("greenhouse effect").

Carbon monoxide is a toxic gas. It can interact with the iron in the hemoglobin of the blood, preventing proper transport of oxygen. Results of exposure to even a small amount of carbon monoxide can be serious and even fatal.

Carbon exhibits a tremendous ability to form compounds with hydrogen, nitrogen, and oxygen. These compounds are so numerous and diverse that entire branches of chemistry are devoted to their properties and reactions. Chapters 21 through 35 provide a closer look at carbon in the fields of organic chemistry and biochemistry.

Silicon is the second most abundant element in the crust of the earth. Its principal sources are in the form of silicon dioxide, found in quartz, sand, sandstone, agate, and amethyst. It also occurs in a wide variety of silicate minerals ranging from very hard garnet to very soft asbestos. Silicate minerals also include clay, mica, feldspar, talc, and zeolite.

Crystalline quartz is used commercially to control the frequency of virtually all radio and television transmissions. It is used in instruments and timepieces as well. Silica can be dissolved in hot sodium carbonate to produce sodium silicate.

$$Na_2CO_3 + SiO_2 \longrightarrow Na_2SiO_3 + CO_2$$

The sodium silicate produced is known as water glass. It dissolves in water, and the silicate ions can act as buffers to maintain the pH of the solution. They also allow animal and vegetable oils to be dispersed in water. For this reason water glass is frequently used in detergents. Sodium silicate can be treated with a dilute acid to form a gelatinous noncrystalline SiO_2. This substance when dried is known as silica gel. It has many uses as a drying agent or a dehumidifier. It is placed (in small packets) in manufacturers' boxes to protect moisture-susceptible products (like electronics).

Silicon has the ability to form long chains in combination with oxygen in molecules known as silicones. The silicones are a part of a group of compounds called polymers (see Chapter 26). Silicones are nontoxic and resistant to heat, light, and oxygen. They also have some interesting anti-sticking properties. Commercial applications of silicones are truly incredible, ranging from lipstick, lotions, and breast implants to Silly Putty, auto ignition systems, and car polish.

Silicon also exhibits the ability to behave as a semiconductor. This property has made silicon the backbone of the computer and microchip industry. Microchips are used in the myriad of small electronic gadgets and personal computers so prevalent in our society. Silicon is also used in photoelectric cells to generate electricity.

Tin and lead are metals easily recovered from their ores. They have long been used in alloys such as bronze (tin and copper) for making tools and decorative objects. Tin is especially corrosion resistant and is used to plate the surfaces of other metals. The best-known example of this is "tin cans," which are really steel cans coated with a layer of tin. Lead is primarily used in the manufacture of automotive batteries (see Section 18.6). Lead oxide is often used in rust-inhibiting paints and primers. Lead chromate is the brilliant yellow pigment found in highway markings. The toxicity of lead has been known for centuries. Recent concern over the effects of lead poisoning have resulted in the removal of lead from paints and gasoline.

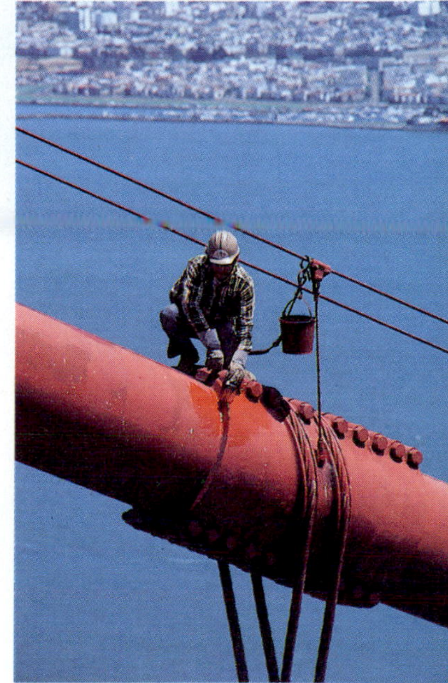

▲ **The Golden Gate Bridge is painted to protect the metal from corrosion.**

20.9 Group VA Elements

Physical Properties

Group VA of the periodic table includes the elements nitrogen, phosphorus, arsenic, antimony, and bismuth. Nitrogen is the most abundant element in the atmosphere, and phosphorus is the 12th most abundant element in the earth's crust. Compounds of nitrogen and phosphorus are essential for all forms of life.

The electron structures of these elements show that they have five electrons in their outer shells, two *s* and three *p* electrons. The elements of this group become increasingly metallic from top to bottom. Nitrogen is a colorless gas with virtually no hint of metallic properties, but bismuth is distinctly metallic in character. Some of the physical properties of Group VA elements are listed in Table 20.8.

Occurrence and Applications of Nitrogen

Nitrogen is found in the atmosphere, living organic matter, decayed matter, ammonia, ammonium salts, and nitrate deposits. Elemental nitrogen, N_2, is useful in many ways. It is extremely inert and highly stable as a result of the triple bond between nitrogen atoms. Nitrogen gas is used to provide an inert atmosphere for sensitive chemical reactions. It also provides a nonoxidizing atmosphere for packaging foods and wine, or preserving articles of sentimental

TABLE 20.8 Physical Properties of Group VA Elements

Element	Atomic number	Atomic mass	Electron configuration	Density (20°C, g/mL)	Melting point (°C)	Boiling point (°C)
N	7	14.01	$[He]2s^22p^3$	0.81[a]	−210.1	−195.8
P	15	30.97	$[Ne]3s^23p^3$	1.82[b]	44.2	280
As	33	74.92	$[Ar]4s^23d^{10}4p^3$	5.7	817[c]	615[d]
Sb	51	121.7	$[Kr]5s^24d^{10}5p^3$	6.6	630.5	1380
Bi	83	209.0	$[Xe]6s^24f^{14}5d^{10}6p^3$	9.8	271.3	1450

[a] liquid at boiling point
[b] white phosphorus
[c] at 36 atm
[d] sublimes

value (such as a wedding dress). Nitrogen is also easily liquefied to produce a coolant used to freeze such things as termites, warts, and biological samples.

In the Haber process, ammonia is manufactured by passing nitrogen and hydrogen gases over a catalyst at elevated temperatures and pressures:

$$N_2(g) + 3\,H_2(g) \rightleftharpoons 2\,NH_3(g)$$

The current production of ammonia in the United States is over 32 billion pounds (1.5×10^{10} kg) annually. By far the greatest consumption of ammonia is in the manufacture of fertilizers or in its direct use as a fertilizer. Other industrial uses of ammonia or its derivatives include the manufacture of nitric acid, plastics, refrigerants, explosives, pulp and paper, and textiles.

Another important hydrogen compound of nitrogen is hydrazine, N_2H_4. It is a powerful reducing agent which has been used in liquid and solid form. The most spectacular use of hydrazine is in rocket fuel. A combination of monomethylhydrazine ($CH_3N_2H_3$) and dinitrogen tetroxide powers the booster rockets of the space shuttle. Hydrazine was also used during lunar landings. Here on earth hydrazine is most commonly found as a blowing agent in the production of plastics. It decomposes to nitrogen gas, producing the porous structure of the plastic.

Nitrogen combines with oxygen in various stoichiometric ratios to form a series of oxides. Several have interesting applications in society. Dinitrogen oxide, N_2O, is odorless and tasteless and is used as an anesthetic for minor surgery. It is commonly known as "laughing gas." Dinitrogen oxide also has the ability to easily dissolve in vegetable fats. This results in its use as a propellant in canned whipped cream. Nitrogen oxide, NO, is a colorless gas that can be produced in the laboratory by reacting copper with concentrated nitric acid. In the presence of air this oxide immediately undergoes oxidation to the brown nitrogen dioxide gas, which is a visible component in the air pollution plaguing our cities.

Nitric acid is an important industrial chemical used in the manufacture of many products. Most nitric acid is made by the Ostwald process. In this process, an ammonia-air mixture is passed over a platinum catalyst at about 800°C. Ammonia is first oxidized to nitrogen oxide, which in turn is further oxidized to nitrogen dioxide. The latter is then absorbed in water to form the acid. The equations are

$$4\,NH_3(g) + 5\,O_2(g) \longrightarrow 4\,NO(g) + 6\,H_2O(g)$$
$$2\,NO(g) + O_2(g) \longrightarrow 2\,NO_2(g)$$
$$3\,NO_2(g) + H_2O \longrightarrow 2\,HNO_3(aq) + NO(g)$$

The acid produced is about 60% HNO_3. The nitrogen oxide formed in the last step is recycled back through the process.

When nitric acid comes into contact with the skin, it reacts with the proteins to cause destruction of the tissues, producing a yellow discoloration of the skin.

The major uses for nitric acid are for manufacturing ammonium nitrate fertilizers, explosives, plastics, pharmaceuticals, dye intermediates, and organic nitro compounds, and for pickling stainless steel in steel refining.

The Nitrogen Cycle

The process by which nitrogen is circulated and recirculated from the atmosphere through living organisms and back to the atmosphere is known as the **nitrogen cycle**.

nitrogen cycle

Nitrogen compounds are required by all living organisms. Despite the fact that the atmosphere is about four-fifths nitrogen, all animals as well as the higher plants are unable to utilize free nitrogen. Higher plants require inorganic nitrogen compounds, and animals must have nitrogen in the form of organic compounds.

Atmospheric nitrogen is *fixed*—that is, converted into chemical compounds that are useful to higher forms of life—by three general routes:

1. **Bacterial action** Certain soil bacteria are capable of converting N_2 into nitrates. Most of these bacteria live in the soil in association with legumes (for example, peas, beans, clover). Some free-living soil bacteria and the blue-green algae, which live in water, are also capable of fixing nitrogen.
2. **Lightning** The high temperature of lightning flashes causes the formation of substantial amounts of nitrogen oxide in the atmosphere. This NO is dissolved in rainwater and is eventually converted to nitrate ions in the soil. The combustion of fuels also provides temperatures high enough to form NO in the atmosphere. The total amount of nitrogen fixed by combustion is relatively insignificant on a worldwide basis. However, NO produced by combustion, especially in automobile engines, is a serious air pollution problem in some areas.
3. **Chemical fixation** Chemical processes have been devised for making nitrogen compounds directly from atmospheric nitrogen. By far the most important of these is the Haber process for making ammonia from nitrogen and hydrogen. This process is the major means of production for the millions of tons of nitrogen fertilizers that are produced synthetically each year.

A schematic diagram of the nitrogen cycle is shown in Figure 20.3. Starting with the atmosphere, the cycle begins with the fixation of nitrogen by any one of the three routes. In the soil, nitrogen or nitrogen compounds are converted to nitrates, taken up by higher plants, and converted to organic compounds. The plants eventually die or are eaten by animals. During the life of the animal, part of the nitrogen from plants in its diet is returned to the soil in the form of fecal and urinary excreta. Eventually, after death, both plants and animals are decomposed by bacterial action. Part of the nitrogen from plant and animal tissues is returned to the atmosphere as free nitrogen and part is retained in the soil. The cycle thus continues.

Other Elements of the Nitrogen Family

Phosphorus is the 12th most abundant element in the earth's crust. Free phosphorus is never found in nature. The element occurs mainly in phosphate minerals. Calcium phosphate is present in the bones and teeth of animals, and small amounts of phosphorus compounds are present in the cells of all plants and animals. The free element exists in several allotropic modifications—white, red, and black being the most common. The white form is tetratomic, P_4; the red and black forms are polymeric chains of phosphorus atoms.

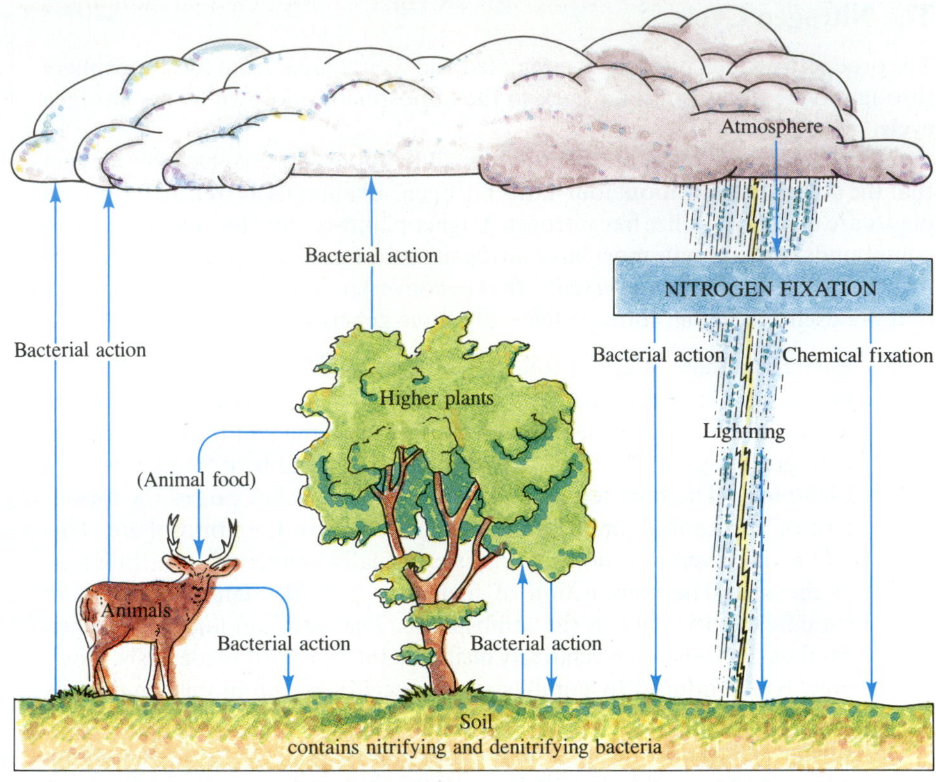

FIGURE 20.3 ► The nitrogen cycle.

Phosphorus is essential for plant growth. It is often present in soil as insoluble minerals. Phosphate rock can be treated with sulfuric or phosphoric acid to make fertilizer high in phosphates.

The sodium salts of phosphates are used in a variety of everyday products. Monosodium phosphate, NaH_2PO_4, forms an acidic solution and is a component in baking powder. Disodium phosphate, Na_2HPO_4, is used to soften boiler water (by precipitating the Ca^{2+} and Mg^{2+} ions). Trisodium phosphate, Na_3PO_4, is used as a common household cleaner (TSP).

In the human body phosphorus makes up about 1% of the body's weight. It is an essential component of the bones, teeth, muscle, brain, and nervous tissue. Phosphorus plays a role in many metabolic reactions occurring in the body. Phosphorus is a component in nucleic acids and the phospholipids in the cell membrane. It is an essential component of ATP and therefore is important in much of the cell's metabolism. Good sources of phosphate in the diet include protein foods, whole-grain cereals, milk, and legumes.

Although nitrogen and phosphorus are nonmetals, arsenic, antimony, and bismuth become progressively more metallic, with bismuth having properties approaching those of a true metal. Arsenic, antimony, and bismuth are mainly used in alloys to increase the hardness of metals. Lead shot is hardened by adding up to 0.5% arsenic; pewter contains 6% bismuth and 1.7% antimony; babbitt metal contains 7% antimony. Low-melting alloys containing over 50% bismuth are used in safety sprinkler heads for fire protection. Arsenic and antimony compounds, although very toxic, have been successfully used as medicinals in controlled dosages.

Table 20.9 Physical Properties of Group VIA Elements

Element	Atomic number	Atomic mass	Electron configuration	Density (20°C, g/mL)	Melting point (°C)	Boiling point (°C)
O	8	16.00	$[He]2s^22p^4$	1.43×10^{-3}	−218	−183
S	16	32.06	$[Ne]3s^23p^4$	2.07	112	444
Se	34	78.96	$[Ar]4s^23d^{10}4p^4$	4.79	217	685
Te	52	127.6	$[Kr]5s^24d^{10}5p^4$	6.24	450	990
Po	84	209	$[Xe]6s^24f^{14}5d^{10}6p^4$	9.32	254	962

20.10 Group VIA Elements

Physical Properties

The elements in Group VIA all have six electrons in their outermost shell. Each forms compounds with metals or hydrogen in which the oxidation number is −2. These elements are mainly nonmetals, yet have the ability to form compounds with more electronegative elements as well. The physical properties of the Group VIA elements are summarized in Table 20.9.

Oxygen

Occurrence and Physical Properties Oxygen is the most abundant element on earth; it occurs both as free oxygen gas and combined with other elements. It forms compounds with all the elements except some noble gases. Water is 88.9% oxygen; the human body is 65% oxygen by mass. The atmosphere contains about 21% oxygen by volume and about 23% oxygen by mass. The oxygen molecule is diatomic.

Oxygen is a colorless, odorless, tasteless gas. It is a pale blue solid or liquid, melting at −218.4°C and boiling at −183°C. The liquid is *paramagnetic*; that is, it is attracted by a magnet. The solubility of the gas is about 0.03 mL per gram of water at 20°C, a fact that is very important for most aquatic life.

Preparation of Oxygen We have seen that oxygen can be prepared by heating mercury(II) oxide (Section 4.3), by electrolysis of water (Section 4.8), and by decomposition of hydrogen peroxide. Many additional methods for the laboratory preparation are known.

Oxygen is obtained commercially from air. Air is first freed of carbon dioxide and moisture and then liquefied to give a mixture that is essentially liquid oxygen and nitrogen. Oxygen is separated from nitrogen by fractional distillation.

One of the outstanding properties of oxygen is its ability to support combustion. By combustion we generally mean the act of burning or, commonly, the union of oxygen with other substances accompanied by the evolution of light and heat. Oxygen is consumed during combustion but does not burn. We are highly dependent on heat, electrical, and mechanical energy obtained from combustion of fossil fuels.

Animal life as we know it is dependent on oxygen. Absorbed in the blood, oxygen is carried by way of the arteries to all tissues of the body, where it is used in metabolism. Metabolism includes the oxidation of carbohydrates, fats, and proteins and provides the energy needed to carry on the normal life processes. The free oxygen in the atmosphere and that dissolved in fresh water and seawater is constantly replenished by photosynthesis carried on by green plants (Section 35.10).

Sulfur

Occurrence and Physical Properties Sulfur has been known for thousands of years; evidence of its use dates back 16 to 20 centuries. It is mentioned in the Bible, where it is called *brimstone*, a name also sometimes used today.

Sulfur occurs both in the free state and in compounds. It is found in various metal sulfides and sulfates, hydrogen sulfide, sulfur dioxide, and in a variety of compounds associated with petroleum and natural gas. Sulfur is an essential element in life processes and is present in all living organisms.

On a commercial basis elemental sulfur is obtained from deposits in Sicily, Texas, Louisiana, and in volcanic regions of Japan and Mexico. Large quantities of sulfur and sulfur compounds are recovered as by-products from petroleum, natural gas, and metallurgical processing. Sulfur is widely used in many phases of agriculture and industry. More than four-fifths of all sulfur produced is made into sulfuric acid.

Sulfur is a bright yellow, odorless solid. In the solid form, sulfur exists in a crown-shaped ring structure containing eight atoms connected by covalent bonds. The molecular aggregate is actually S_8, but in many of its reactions sulfur is represented by a single S. Figure 20.4 shows several ways of representing the S_8 molecule.

The element sulfur exists only as S_2 molecules in the gas phase at high temperatures. It is generally found as one of several allotropes containing the S_8 ring. As sulfur is heated beyond its melting point, a viscous (resistant to flowing) liquid of S_8 rings develops and the color darkens. If the heating process is continued the rings rearrange into long chains, which are, in turn, broken down at high temperatures. If the liquid is cooled quickly a substance called *plastic sulfur* is formed. Eventually this reverts back to the more stable rings.

Chemical Properties of Sulfur Since sulfur is less electronegative than oxygen, its compounds are less ionic and more covalent than those of oxygen. Ionic sulfur compounds are formed with only the most electropositive metals—the alkali metals and the alkaline earth metals (Group IIA). The oxidation states of sulfur range from -2 to $+6$. Sulfur combines directly with many elements, both metals and nonmetals. For example,

$$Zn(s) + S(s) \xrightarrow{\Delta} ZnS(s)$$

$$H_2(g) + S(l) \xrightarrow{200^\circ C} H_2S(g)$$

$$C(s) + 2\ S(g) \xrightarrow{3000^\circ C} CS_2(g)$$

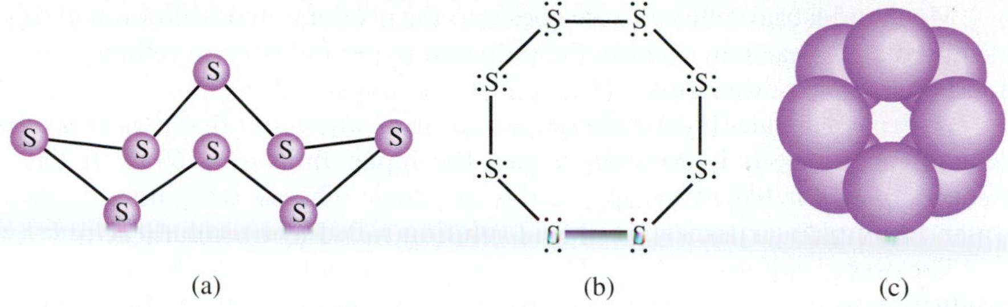

◄ FIGURE 20.4
Diagrams of the S_8 molecule: (a) crown-shaped ring formation with bond angles of 105°; (b) Lewis dot structure; (c) compact molecular orbital structure. The latter most nearly represents the S_8 molecule.

There are at least ten known oxides of sulfur, two of which are well known—sulfur dioxide (SO_2) and sulfur trioxide (SO_3).

The prefix *thio* is used to indicate the presence of sulfur in a compound. More specifically, it means that sulfur is used in place of oxygen in an analogous compound. Two examples of thio compounds are sodium thiosulfate and thiourea. Their formulas and the formulas of their oxygen analogs are shown below.

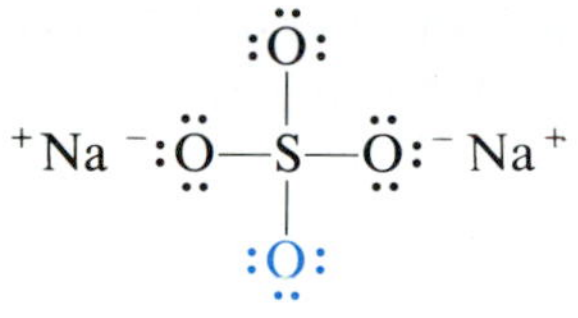

Sodium sulfate, Na_2SO_4

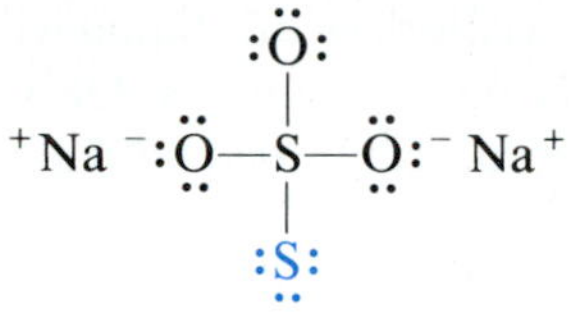

Sodium thiosulfate, $Na_2S_2O_3$

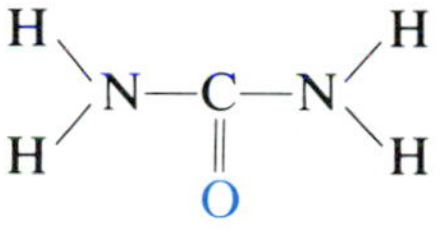

Urea

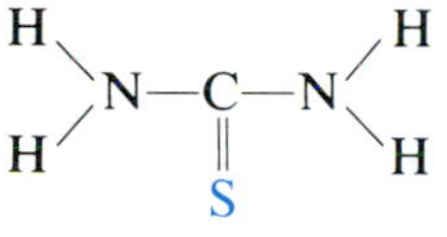

Thiourea

Application of Sulfur Compounds An important use of the thiosulfate ion is in photography. The $S_2O_3^{2-}$ ion dissolves the suspension of silver bromide (AgBr) which coats the photographic film. In areas exposed to light the AgBr is "activated." This silver bromide is reduced to free Ag by the developing solution, and the film is treated with thiosulfate solution to remove the excess AgBr, "fixing" the image.

Sulfur burns in air or oxygen with a bluish flame to form sulfur dioxide:

$$S(l) + O_2(g) \longrightarrow SO_2(g)$$

Sulfur dioxide is also formed by reacting iron pyrites (FeS_2) or other metal sulfides with oxygen (air) at elevated temperatures:

$$4\ FeS_2 + 11\ O_2 \xrightarrow{\Delta} 2\ Fe_2O_3 + 8\ SO_2$$

Sulfur dioxide is a colorless gas that is irritating to the throat, causing choking and coughing. It also causes eye irritation and is an air pollutant known to be damaging to plant life.

Sulfur dioxide is an acid anhydride and forms sulfurous acid (H_2SO_3) when dissolved in water. It is a weak acid, decomposing easily into sulfur dioxide and water.

$$SO_2(g) + H_2O(l) \rightleftharpoons H_2SO_3(aq)$$

Much industrial sulfur dioxide goes into the production of sulfuric acid. It is also used as a bleaching agent in the pulp and paper industry, in refining sugar, and in processing dried fruit.

Hydrogen sulfide (H_2S) is the sulfur analog of water, but that is as far as the resemblance goes. It is normally a gas; the liquid boils at $-60°C$. It has a disagreeable odor, like rotten eggs, and is very toxic. The gas is slightly soluble in water, forming an extremely weak acid solution called hydrosulfuric acid.

The uses of sulfuric acid (H_2SO_4) are numerous and varied. The current U.S. production is greater than 79 billion lb (3.6×10^{10} kg) annually. Large amounts are used to convert insoluble phosphates into soluble phosphates for fertilizers. Sulfuric acid is used in the production of hydrochloric acid, ammonium sulfate, and aluminum sulfate. It is also used in petroleum refining; the manufacture of rayon; the making of explosives (nitroglycerin and TNT), alcohols, detergents, and metal sulfates, textile printing, dyeing, and in scores of other applications. Dilute sulfuric acid is the electrolyte used in millions of ordinary automobile storage batteries.

Sulfuric acid is a colorless, oily liquid with a density of 1.84 g/mL. Its freezing point is 10.4°C; its boiling point is 338°C. The acid decomposes as it boils, liberating its anhydride, sulfur trioxide (SO_3):

$$H_2SO_4(l) \xrightarrow{\Delta} H_2O(g) + SO_3(g)$$

Ordinary concentrated sulfuric acid contains about 96–98% H_2SO_4; the remainder is water.

Other Group VI Elements

Selenium and tellurium are relatively rare elements. In recent years there has been a growing interest in the chemistry of selenium. It is used in the manufacture of red glass, such as that found in traffic lights. It has unusual properties of electrical conduction that make it valuable for use in switches for automatic light timers. The most familiar use of selenium is in xerography. Inside a photocopying machine is an aluminum plate coated with a thin film of selenium. Light coming in discharges the electric charge on the selenium. The black toner sticks only to the remaining charged areas, creating the photocopy as soon as the toner is transferred to paper.

20.11 The Halogens: Group VIIA Elements

Physical Properties

halogens

The **halogens**, Group VIIA of the periodic table, consist of five elements: fluorine (F), chlorine (Cl), bromine (Br), iodine (I), and astatine (At). None of these are found as free elements in nature. The term *halogen* is derived from the Greek words *halos* and *gen*, which mean "salt producer." Chlorine, bromine, and iodine are easily prepared in the laboratory; fluorine is much more difficult to prepare. Astatine is found only in very small amounts as a radioactive decay product of other elements.

TABLE 20.10 Physical Characteristics of the Halogens

Property	Fluorine	Chlorine	Bromine	Iodine
Molecular formula	F_2	Cl_2	Br_2	I_2
Atomic number	9	17	35	53
Atomic mass	19.00	35.45	79.90	126.9
Electron configuration	$[He]2s^2 2p^5$	$[Ne]3s^2 3p^5$	$[Ar]4s^2 3d^{10} 4p^5$	$[Kr]5s^2 4d^{10} 5p^5$
Atomic radius (nm)	0.072	0.099	0.114	0.133
Radius of X^- ion (nm)	0.136	0.181	0.195	0.216
Appearance	Pale yellow gas	Yellow-green gas	Reddish-brown liquid	Bluish-black solid
Density (g/mL)	1.11 (liquid at bp)	1.57 (liquid at bp)	3.14	4.94
Melting point (°C)	−219.6	−102.4	−7.2	113.6
Boiling point (°C)	−187.9	−34.0	58.8	184.5
Relative electronegativity	4.0	3.0	2.8	2.5

All the halogen atoms have seven electrons in their outermost shell and are followed, in each case, in the periodic table by a noble gas. Consequently, their most stable oxidation state is −1. All the halogens except fluorine are known to exist in several oxidation states. A halogen atom has a great tendency to fill its outer-orbital electron vacancy either by gaining an electron, as in the formation of a sodium halide, Na^+X^- (where X = F, Cl, Br, I), or by sharing electrons and forming a covalent bond. In the free state, the halogens exist as diatomic molecules, F_2, Cl_2, Br_2, and I_2:

$$:\ddot{\underset{..}{F}}:\ddot{\underset{..}{F}}: \quad :\ddot{\underset{..}{Cl}}:\ddot{\underset{..}{Cl}}: \quad :\ddot{\underset{..}{Br}}:\ddot{\underset{..}{Br}}: \quad :\ddot{\underset{..}{I}}:\ddot{\underset{..}{I}}:$$

The differences in reactivity of the halogens are due to differences in their electronegativities. Fluorine, as the most electronegative, has the greatest attraction for electrons and is the most reactive, followed by chlorine, bromine, and iodine, respectively (see Table 20.10).

Since fluorine has the highest electronegativity, it is the most characteristically nonmetallic of all the elements. The general trend of the elements to become less electronegative (from top to bottom within a periodic group), and hence more metallic in characteristics, is clearly evident among the halogens. Fluorine and chlorine are gases, bromine is a liquid, and iodine is a solid with a characteristic metallic luster.

Chemical Properties and Applications

The halogen family exhibits a very smooth transition in chemical properties as we progress down the group. Some of the chemical properties of the halogens are summarized in Table 20.11.

Over three-fourths of the fluorine produced in the United States is used for processing uranium for nuclear power plants. The UF_6 formed vaporizes at a fairly low temperature, allowing easy separation from other substances.

TABLE 20.11 Chemical Properties of the Halogens

Characteristic	F_2	Cl_2	Br_2	I_2
General activity	Extreme	Very high	Less active	Least active
Activity with H_2	Violent	Slow (dark) Fast (light)	Requires heat	Slow
Hydrogen halide formula	HF	HCl	HBr	HI
Oxidizing strength	Highest	Second	Third	Lowest
Reducing strength of anion	Lowest	Third	Second	Highest

Fluorine has become an important industrial chemical. It is used in familiar compounds such as Teflon (nonstick coating on cookware) and Freon, used in refrigerants and formerly in aerosol cans as a propellant. Some of these fluorocarbons have recently been implicated in the destruction of the ozone layer.

Fluorine is also present in the bones and teeth of mammals. Some fluoride ions are necessary for the development of sound teeth in children, although too much results in mottled teeth. Controlled amounts of sodium fluoride are sometimes added to water supplies in areas deficient in fluoride.

Over 90% of all commercial chlorine is produced by the electrolysis of molten or aqueous sodium chloride. About half of this inorganic chlorine ends up eventually as vinyl chloride, C_2H_3Cl, used in the manufacture of the plastic polyvinyl chloride (PVC) as well as other chlorine-containing organic solvents. About 20% of the chlorine produced is used to bleach wood pulp for the paper industry and cotton and linen fabrics for the textile industry. Chlorine is used to oxidize and destroy bacteria in virtually all water treatment plants in urban areas in the United States. In a similar manner swimming pools are chlorinated.

Neither bromine nor iodine is as widely found as chlorine and fluorine. Bromine is the only nonmetallic element that is a liquid at ordinary temperatures. It is used in the petroleum, pharmaceutical, and photographic industries. The compound 1,2-dibromoethane, $C_2H_4Br_2$, was used as an additive in leaded gasoline to reduce engine "knock." In unleaded fuels this has been replaced by methyl tertiary butyl ether (MTBE). A variety of drugs and dyes also contain bromine. Bromine compounds are known to have a soothing effect on the nerves. In photography light-sensitive silver bromide is coated on the surface of the film. Precautions must be taken when working with bromine, because its fumes are toxic and it produces painful sores when spilled on the skin.

Iodine is a bluish-black solid that forms violet vapors when heated. Compounds of iodine are found in seawater, where they are selectively absorbed and concentrated by certain types of seaweed (kelp). The first commercial source was from ashes of seaweed. Today iodine is obtained commercially from certain oil well brines and from Chilean sodium nitrate deposits, where it occurs as sodium iodate ($NaIO_3$). The brines are treated with chlorine to liberate iodine:

$$Cl_2(g)\ 2\,I^-(aq) \longrightarrow I_2(s) + 2\,Cl^-(aq)$$

Sodium iodate is separated from other salts found in the nitrate deposits, then reacted with sodium hydrogen sulfite to liberate free iodine:

$$2\ NaIO_3 + 5\ NaHSO_3 \longrightarrow I_2 + 3\ NaHSO_4 + 2\ Na_2SO_4 + H_2O$$

Iodine is only slightly soluble in water, but it is very soluble in organic solvents such as alcohol, carbon tetrachloride, and carbon disulfide. The brown alcohol solution is known as *tincture* of iodine and is used as an antiseptic and disinfectant for cuts and scratches. The carbon tetrachloride and carbon disulfide solutions are violet. Iodine is soluble in aqueous potassium iodide because of the formation of the soluble complex triiodide ion, I_3^-. The solution is labeled "I_2 in KI" and is brown in color.

$$I_2(s) + KI(aq) \longrightarrow KI_3(aq)$$
$$I_2(s) + I^-(aq) \longrightarrow I_3^-(aq)$$

An intense blue complex is formed when iodine is added to a starch suspension. This iodine–starch color is used as a very sensitive test for iodine; it is a useful indicator that allows iodine to be used in the analysis of other substances.

The thyroid gland is richer in iodine than any other part of the human body. A deficiency of iodine causes this gland to enlarge, creating the condition known as goiter. Iodine and iodides have been used for many years as an effective treatment for goiter. Iodine in the regular diet may be supplemented by iodized salt, which contains a small amount of a soluble iodide.

Silver iodide (AgI) is very sensitive to light; it is used in high-speed photographic films.

All of the halogens form stable compounds when reacted with hydrogen. The aqueous solutions of the compounds are all strongly acidic with the exception of hydrogen fluoride.

The melting and boiling points of HF are abnormally high as a direct result of hydrogen bonding. Hydrogen fluoride is dangerous to inhale, since it causes edema of the lungs. The eyes are also quite sensitive to this compound.

A unique property of HF is its ability to etch glass. The art of etching glass with HF was practiced as early as the 16th century. The chemical reaction that takes place is the dissolving of silicon dioxide in the glass, forming silicon tetrafluoride:

$$SiO_2(s) + 4\ HF(g) \longrightarrow SiF_4(g) + 2\ H_2O(l)$$

The area to be etched is first coated with paraffin. The design is then scratched through the wax. The glass is dipped in the HF solution and then washed, removing the paraffin, which is not attacked by hydrogen fluoride. The glass in light bulbs is also "frosted" in the same manner.

Hydrochloric acid—the technical grade is called *muriatic acid*—is one of the most widely used acids. It is essential in the manufacture of textiles, soaps, glue, and for the cleaning of metals. Hydrochloric acid is also used for cleaning mortar from stone or brick structures. The acidity of gastric juice is the result of the presence of HCl. It is necessary for digestion and is also the source of heartburn and acid indigestion.

20.12 Noble Gases

The last column on the right-hand side of the periodic table consists of the group of elements known as the noble or inert gases. These elements are characterized by filled *s* and *p* orbitals in their outermost electron shell. This results in a lack of reactivity as a common characteristic for these gases. No noble gas compounds were known until 1962. More information on these elements can be found in Section 10.10.

20.13 Transition Metals

All the elements we have surveyed to this point have been representative elements, which show similarities in vertical columns and incredible diversity across a period. In contrast, the transition metals exhibit similarities across each period as well as in vertical columns. The last electron in these elements is added to a *d* or an *f* orbital which is not in the outermost shell. This results in more gradual changes than those seen in the representative elements.

Overall, the transition metals exhibit the typical metallic characteristics of luster and high electrical and thermal conductivity. The best heat conductor is silver. You may have practical knowledge of this if you have handled a hot sterling silver gravy ladle. Copper is also a very good conductor of heat and electricity, and is much less expensive than silver. For this reason many electrical systems contain copper.

The transition metals also exhibit considerable diversity in other properties. Mercury is a liquid at room temperature, yet tungsten (used in light bulbs for the filament) does not melt until 3400°C. Some transition metals are relatively soft, such as silver, gold, and copper, while others, like iron, are strong and used for structural materials. Some transition metals react with oxygen to form oxides. These oxides are of two types. In the first type, the oxide coating adheres to the metal surface, protecting it from further corrosion. Examples of this type include chromium and nickel. The second type of oxide is scaly, constantly exposing the metal underneath to further oxidation. The most common example of this type is the rusting of iron. A few of the transition metals, such as gold, silver, and platinum, form oxides only with difficulty.

Ionic transition metal compounds also exhibit specific characteristics. The metals are often found in multiple oxidation states. Many of these compounds are highly colored and quite often they are paramagnetic. Let's consider several transition metals that are important in our daily lives.

Iron

Iron is the fourth most abundant element, following oxygen, silicon, and aluminum. Iron was known in prehistoric times, and its known use dates back almost 8000 years. The first iron to be used was probably obtained from meteorites, but iron was produced from ores as early as 1300 B.C.

Pure iron is a silvery-white, relatively soft, very ductile metal. It has a melt-

TABLE 20.12 Effects of Selected Alloying Elements on the Properties of Steel

Alloying element	Effect on steel
Carbon	Makes tempering possible; increases tensile strength (up to 0.83% C); increases hardness
Cobalt	Imparts high-temperature strength
Chromium	Improves hardness, abrasion resistance, corrosion resistance, and high-temperature properties
Manganese	Imparts hardness, toughness, and resistance to wear and abrasion
Molybdenum	Imparts hardness, shock resistance, and high-temperature strength
Nickel	Imparts corrosion resistance; increases toughness and high-temperature strength; reduces brittleness at very low temperatures
Silicon	Increases strength without affecting ductility; modifies magnetic and electrical properties; high Si concentrations impart resistance to acid corrosion
Tungsten	Imparts hardness, which is retained at high temperatures
Vanadium	Increases strength and toughness; improves heat-treating characteristics

ing point of 1532°C and it boils at 3000°C. One of its most distinguishing characteristics is its magnetism, which it loses when it is heated to 770°C. In the presence of a magnetic field or electric current, iron shows greater magnetic properties than any other element. (Cobalt and nickel are also strongly magnetic.)

Very little iron is used in its pure form; most of it is used in the form of steel, an iron–carbon alloy. Steel may contain small amounts of other elements such as Mn, Si, Cr, Ni, V, W, Mo, Co, and Ti. These alloying elements enhance certain qualities of the steel such as toughness, hardness, corrosion resistance, and so on (see Table 20.12).

Steel is an iron–carbon alloy (0.05–1.7% C) that usually contains small amounts of other elements. Steel making is a basic industry; the quantity of steel produced each year in the world far exceeds the combined quantities of all other metals.

The first step in steel making is the reduction of iron ore to impure iron, commonly called *pig iron*. Reduction occurs in a huge reactor called a *blast furnace*. Furnace operation is continuous; ore, coke, and limestone are charged at the top while steady streams of preheated air are blown in through bottom nozzles (*tuyeres*). Furnace temperatures range from about 200°C near the top to about 1900°C near the bottom (the combustion zone). Coke is both the fuel and the source of the principal reducing agent (CO). The limestone reacts with impurities in the ore and coke to form a slag that coats and protects the reduced iron from reoxidation in the combustion zone. Molten pig iron and slag are removed at intervals from the bottom of the furnace. The following are equations for typical blast furnace reactions.

$$2\,C(s) + O_2(g) \longrightarrow 2\,CO(g)$$
Coke

$$Fe_3O_4(s) + CO(g) \longrightarrow 3\,FeO(s) + CO_2(g)$$
Ore

$$FeO(s) + CO(g) \longrightarrow Fe(l) + CO_2(g)$$

$$CaCO_3(s) \longrightarrow CaO(s) + CO_2(g)$$
Limestone

$$CaO(s) + SiO_2(s) \longrightarrow CaSiO_3(l)$$
Slag

Pig iron contains on the average about 1% Si, 0.03% S, 0.27% P, 2.4% Mn, and 4.6% C, the balance being Fe. These impurities must be removed or lowered to carefully controlled levels to convert pig iron into steel.

Iron is a fairly reactive metal and forms two principal series of compounds: iron(II) or ferrous compounds and iron(III) or ferric compounds. Pure iron reacts with dilute acids or steam (at high temperatures) to form ferrous compounds. These can then be oxidized to ferric compounds.

In addition to its use as a structural material in a diverse array of objects, iron also finds uses based on its magnetic properties. Iron is found in magnets, telephones, electric motors, generators, and many other common devices.

Dipping iron into concentrated nitric acid (a very powerful oxidizing agent) results in the formation of *passive iron*, which is coated with a very thin layer of oxide. Passive iron will not rust unless the film of oxide is broken.

Corrosion of iron, commonly called *rusting*, is a serious economic problem that causes losses of billions of dollars annually in the United States. Rusting occurs on the surface of the iron, where it is exposed to oxygen and moisture in the atmosphere. The iron is transformed into reddish-brown rust, a hydrated iron(III) oxide ($Fe_2O_3 \cdot xH_2O$). As the rust is formed, it flakes off the surface, allowing the corrosion to penetrate deeper into the iron. Both oxygen and water are needed for rusting. The process is summarized in the following equations:

$$2\,Fe + O_2 + 2\,H_2O \longrightarrow 2\,Fe(OH)_2$$
$$4\,Fe(OH)_2 + O_2 + 2\,H_2O \longrightarrow 4\,Fe(OH)_3$$
$$2\,Fe(OH)_3 \longrightarrow Fe_2O_3 \cdot xH_2O$$

▲ **Iron is a principal component in the girders of large buildings.**

The most common general methods for protecting iron from corrosion are (1) protective coatings, (2) alloying, and (3) cathodic protection. Protective coatings of paint, enamel, tar, grease, or metal are commonly used to prevent contact with the atmosphere. The term *galvanized* is applied to iron or steel that is protected by a coating of zinc metal. Highly corrosion resistant iron alloys are produced industrially.

To afford *cathodic protection* to a steel pipeline or tank, magnesium (or zinc) stakes or rods are driven into the earth and connected by wires to the pipeline or tank. This in effect sets up an electrochemical cell, with the more easily oxidized magnesium metal acting as the anode. Electrons flow from the magnesium to the iron, causing the iron to become negatively charged or cathodic with respect to the magnesium. Oxidation does not occur at the iron cathode, and thus no corrosion occurs on the iron pipe or tank. The magnesium or zinc rods or stakes must be replaced periodically, since they are consumed in affording protection to the iron or steel. The iron hulls of ships can be protected by fastening blocks of magnesium metal to them. The zinc coating on galvanized iron actually provides cathodic protection for the iron.

In the human body iron is found primarily in the blood with reserves in the liver, spleen, and bone marrow. Iron is central in the hemoglobin molecule, where it is responsible for the transport of oxygen. Another complex molecule, known as myoglobin, also contains iron. Myoglobin is synthesized in the muscle cells where it serves as a temporary storage depot for oxygen. Iron also is important in many enzymes that catalyze oxidation reactions such as the conversion of β-carotene to vitamin A. Sources of iron in the diet include liver, lean meat, whole-grain cereals, apricots, raisins, and legumes.

Copper

Copper is a soft, extremely ductile and malleable metal with a reddish-brown color. Its excellent electrical and thermal conductivity make it the primary source for wires in electrical systems. Copper is not very active chemically, but when exposed to moist air over a period of time it becomes coated with the green *patina* seen on statues. This patina consists of thin layers of two compounds, $CuSO_4 \cdot 3\,Cu(OH)_2$ and $CuSO_4 \cdot 2\,Cu(OH)_2$. A good deal of chemistry was applied in the restoration of the Statue of Liberty to transplant weathered copper onto the statue and affix the patina to the surface to match the old.

Copper forms alloys with a variety of metals. Table 20.13 indicates the uses of some of these alloys. Copper-containing alloys also constitute material used in coins today. The nickel coin is an alloy of 75% copper and 25% nickel. Most other coins are clad-metal. These coins are now constructed with a core metal, then coated with a thin layer of an alloy. The *dime*, *quarter*, half-dollar, and dollar coins have copper cores coated with a 75% copper, 25% nickel alloy. Other copper-containing alloys include white gold, used in jewelry, and dental gold.

Trace amounts of copper are essential for life. Copper is concentrated in the liver, heart, and brain in human beings. It is instrumental in the synthesis of hemoglobin, bone and connective tissue development, the production of melanin, and for the formation of myelin in the nervous system. Sources of copper in the diet include liver, oysters, crab, nuts, and whole-grain cereal.

Copper in large amounts can have a toxic effect. Salts containing copper are used to kill bacteria, fungi, and algae. Copper sulfate pentahydrate, $CuSO_4 \cdot 5\,H_2O$, commonly known as blue vitriol, is used to treat water in

TABLE 20.13 Uses of Copper Alloys

Alloy metal	Name	Use
Zinc	Brass	Cartridges, musical instruments, hardware
Tin, zinc	Bronze	Statues, medals, primitive tools
Tin	Bell metal	Bells

reservoirs and swimming pools. Paints containing copper are used on the hulls of ships to keep them free from marine organisms.

Zinc

Zinc is a silvery white metal that tarnishes easily. It melts at 420°C and boils at 907°C. The surface film protects zinc from further corrosion. Zinc is a very active metal and is an excellent reducing agent. Approximately 90% of the zinc produced is used in galvanizing steel. Galvanized objects include water pails, gutters, fencing, and nails. Zinc forms the case and the negative electrode in most dry cell batteries. It is also an important component in alloys. In combination with aluminum, the alloy is used in radio and auto parts. The "copper" penny is minted from an alloy of 97.6% zinc and 2.4% copper.

Many chemists do not classify zinc as a transition metal because this element has a filled *d* subshell. In fact, the chemistry of zinc is more like that of the alkaline earth metals than the transition metals.

Compounds of zinc also are important in our lives. Zinc oxide, ZnO, is a white pigment used in paints. It is also used in making tire rubber and as an antiseptic or skin protectant. A paste of zinc oxide and concentrated hydrochloric acid solidifies into a hard cement used in dentistry.

Trace amounts of zinc are found in the human body. It is a constituent of many enzymes in the digestive and respiratory systems. Zinc also plays a role in bone and liver metabolism. It is necessary for healing wounds and plays a key role in skin integrity. Sources of zinc include seafood and meats.

Concepts in Review

1. Indicate the similarities and differences in the physical and chemical properties of metals and nonmetals.
2. Summarize the chemical process that occurs when an ore is converted to a free element.
3. Summarize the important trends on the periodic table.
4. State the definition of an alloy, indicate the types of alloy, and the reason for making them.
5. List the alkali metals (Group IA) and compare their atomic radii, densities, melting points, and first ionization energies.
6. Write balanced chemical equations for the reaction of any alkali metal with (a) water, (b) any halogen, (c) hydrogen.
7. Describe the preparation of magnesium metal from seawater and write equations for the reactions involved.
8. Explain why colors appear when certain elements are heated strongly in a nonluminous flame, and list the specific colors obtained from Li, Na, K, Rb, Ca, Sr, and Ba.
9. Explain why aluminum is more expensive than iron or steel even though it is more abundant than iron in the earth's crust.
10. List the common allotropes of carbon and state two uses for each.

11. Tell which element of Group VA is the most metallic and which is the least metallic.
12. Diagram and discuss the nitrogen cycle.
13. Identify the halogen that has the highest electronegativity (and first ionization energy) and explain why it has such an especially high electronegativity.
14. Arrange the halogens in order of increasing strength as oxidizing agents and the halide ions in order of increasing strength as reducing agents.
15. State the essential difference between pig iron and steel.
16. Explain cathodic protection of metals.
17. Explain how a clad-metal coin is made and list several examples of this type of coin.

EXERCISES

1. Identify each of the following as a metal, nonmetal, or metalloid:
   (a) strontium (e) bromine
   (b) silicon (f) arsenic
   (c) krypton (g) palladium
   (d) tungsten (h) hydrogen
2. Why is the chemistry of the first element in a family often significantly different from the other elements in the family?
3. Which element in Group IVA would have the least metallic character? Give two properties that support these characteristics.
4. Consider the following elements: Na, K, S, O, Ar, Kr. Choose the element with the:
   (a) highest electronegativity
   (b) smallest atomic radius
   (c) smallest ionization energy
   (d) greatest metallic character
5. Why are the alkali metals of higher molar mass more active than those of lower molar mass, while in the halogens the reverse is true?
6. Identify the chemical process used in the conversion of an ore to a free metal.
7. Write an equation for the reduction of a metal oxide by the following methods:
   (a) reaction with carbon (c) heating
   (b) reaction with hydrogen
8. Indicate two methods for the preparation of nonmetals.
9. Differentiate between a solution alloy and an intermetallic compound. Give an example of each.
10. Why are metals combined to form alloys?
11. Write the equations for the production of ammonia by the Haber process. What happens to the majority of ammonia manufactured in this way?
12. Why is hydrogen often considered to be in a family or group of one?
13. Explain in terms of electronic structure why potassium reacts more violently with water than does lithium.
14. How would the ionization energy of francium be expected to compare with that of cesium?
15. Graph the melting points of the alkali metals versus atomic number. Predict the melting point for francium. Will it be a liquid or a solid at room temperature?
16. What is the oxidation number of hydrogen in a hydride? Write an equation for the formation of potassium hydride from its elements.
17. Write equations for the formation of the oxide, peroxide, and superoxide of potassium.
18. Explain how potassium superoxide (KO_2) can be used as a source of oxygen in emergency breathing apparatus.
19. State two biologically significant uses of alkali metals.
20. Write the formulas of an alkali metal oxide, hydroxide, acetate, bicarbonate, and carbonate using a different metal for each compound.
21. On the basis of electron structure, propose an explanation for why a Group IIA metal is less reactive than a Group IA metal that precedes it in the periodic table.
22. Outline the process for isolating magnesium from seawater. Include equations in your answer.

23. Explain the chemical process that occurs as mortar hardens.
24. List two major functions of calcium and magnesium in living organisms.
25. Explain why it is possible to use barium sulfate, but not barium chloride, as an aid to X-ray photography of the gastrointestinal tract.
26. Briefly discuss the nature and origin of the colored flames that are used for the qualitative detection of certain Group IA and Group IIA elements.
27. State three uses for the element boron.
28. What is a mordant and how is it used commercially?
29. What compound is formed if carbon is burned
    (a) in a limited supply of oxygen?
    (b) in an excess of oxygen?
30. What is an allotrope? State an example from Group IVA.
31. State three commercial uses for the element silicon.
32. Does the tendency to become more metallic increase or decrease from top to bottom in Group VA elements? Explain.
33. Draw Lewis-dot diagrams for:
    (a) a nitrogen molecule
    (b) an ammonia molecule
    (c) a nitride ion
    (d) an ammonium ion
    (e) a nitrate ion
34. Outline and discuss the nitrogen cycle in nature.
35. Which Group VA elements are absolutely essential to our health and well-being?
36. Explain why nitrogen is so inert in its free state. What commercial uses result from this property?
37. Explain in terms of electronic structure why the reactivity of Group VIIA elements decreases from top to bottom in the periodic table.
38. How are the compounds of sulfur different from the compounds of oxygen?
39. Use Lewis-dot structures to illustrate how the halogen elements, designated X, can form both covalent bonds (X_2, HX) and ionic bonds (Na^+X^-).
40. Why is the boiling point of HF abnormally high when compared to that of the other hydrogen halides?
41. Arrange the halogens in order of increasing strength as oxidizing agents. Justify the order in terms of electronic structure.
42. What ill effects, if any, would occur from a lack of the following ions in the diet?
    (a) fluoride ions (c) bromide ions
    (b) chloride ions (d) iodide ions
43. How do the transition metals differ from the representative elements?
44. What is the essential chemical difference between pig iron and steel?
45. Explain the process commonly known as the rusting of iron.
46. Bare aluminum and magnesium metal structures are not damaged by prolonged exposure to the atmosphere. Yet bare iron, generally less reactive than either aluminum or magnesium, is severely corroded by long exposure to the atmosphere. Explain.
47. Explain how cathodic protection can prevent oxidation of an iron storage tank or a ship's hull.
48. State four important uses for copper and its compounds in our daily lives.
49. Explain why a metal is galvanized. What element is used in this process?
50. Which of the following statements are correct? Rewrite each incorrect statement to make it correct.
    (a) Most of the common metals are found in the free or uncombined state in nature.
    (b) Metals may be alloyed to increase tensile strength, hardness, and electrical conductivity.
    (c) The chemical reactivity of the alkali metals (Group IA) increases from top to bottom in the periodic table.
    (d) In binary compounds between metals and nonmetals, the metals generally form positive ions.
    (e) Chile saltpeter is another name for sodium nitrate.
    (f) Both calcium and magnesium are essential elements for animals, but only calcium is essential for plants.
    (g) Very pure iron is used in some industrial applications because of its great hardness and strength, especially at high temperatures.
    (h) Copper and aluminum alloys are used for electrical wiring because of their excellent electrical conductivity.
    (i) Galvanized steel is protected from rusting or corrosion by a zinc coating.
    (j) Aluminum is the most abundant metal in the earth's crust.
    (k) The electronegativity of the halogens (Group VIIA) increases from top to bottom in the periodic table.
    (l) Iron plays a central role in bone development in the human body.
    (m) The oxidation number of oxygen in oxygen difluoride (OF_2) is $+2$.

(n) The major use of fluorine in the United States is in the manufacture of toothpaste.
(o) Pure sulfur is a bright yellow, odorless solid.
(p) Sulfur occurs as molecules of S_2 in the gaseous state.
(q) The Ostwald process is used to prepare ammonia on an industrial scale.
(r) Phosphorus is indispensable to plants and animals.
(s) The copper patina on statues protects them from further oxidation.

51. How many liters of SO_2 (at STP) can be obtained by combustion of 150 g of sulfur?

*52. The water coming from a large reservoir is being treated with 0.20 ppm (parts per million) of chlorine. Calculate the number of grams of chlorine required each hour when water is being pumped from the reservoir at a rate of 300,000 L per hour.

53. Calculate the molarity of each of the following concentrated acids:
(a) Sulfuric acid: $d = 1.84$ g/mL, containing 96% H_2SO_4 by mass
(b) Nitric acid: $d = 1.42$ g/mL, containing 70% HNO_3 by mass
(c) Hydrochloric acid: $d = 1.19$ g/mL, containing 37% HCl by mass

21

Organic Chemistry: Saturated Hydrocarbons

◀ **Chapter Opening Photo: Saturated hydrocarbons provide the fuel for us to enjoy many of the wonders of nature.**

Many substances throughout nature contain silicon or carbon within their molecular structures. Silicon is the staple of the geologist—it combines with oxygen in a variety of ways to produce silica and a family of compounds known as the silicates. These compounds form the chemical foundation of most types of sand, rocks, and soil, the essential materials to the construction industry.

In the living world, carbon provides the basis for millions of organic compounds in combination with hydrogen, oxygen, nitrogen, and sulfur. Carbon compounds provide us with energy sources in the form of hydrocarbons and their derivatives that allow us to heat and light our homes, drive our automobiles to work, and fly off to Paris for an elegant dinner. Small substitutions in these carbon molecules can produce chlorofluorocarbons, compounds used in plastics and refrigerants. An understanding of these molecules and their effect upon our global environment is vital in the continuing search to find ways to maintain our lifestyles while preserving the planet.

21.1 Organic Chemistry: History and Scope

During the late 18th and the early 19th centuries, chemists were baffled by the fact that compounds obtained from animal and vegetable sources defied the established rules for inorganic compounds—namely, that compound formation is due to a simple attraction between positively and negatively charged elements. In their experience with inorganic chemistry, only one, or at most a few, compounds were composed of any group of only two or three elements. However, they observed that a group of only four elements—carbon, hydrogen, oxygen, and nitrogen—gave rise to a large number of different compounds that often were remarkably stable.

Because no organic compounds had been synthesized from inorganic substances and because there was no other explanation for the complexities of organic compounds, chemists believed that organic compounds were formed by some "vital force." The **vital force theory** held that organic substances could originate only from living material. In 1828 Friedrich Wöhler (1800–1882), a German chemist, did a simple experiment that eventually proved to be the death blow to this theory. In attempting to prepare ammonium cyanate (NH_4CNO) by heating cyanic acid (HCNO) and ammonia, Wöhler obtained a white crystalline

vital force theory

substance that he identified as urea ($H_2N—CO—NH_2$). Wöhler knew that urea is an authentic organic substance because it is a product of metabolism that had been isolated from urine. Although Wöhler's discovery was not immediately and generally recognized, the vital force theory was overthrown by this simple experiment since *one* organic compound had been made from nonliving materials.

After the work of Wöhler, it was apparent that no vital force other than skill and knowledge was needed to make organic chemicals in the laboratory and that inorganic as well as organic substances could be used as raw materials. Today, **organic chemistry** designates the branch of chemistry that deals with carbon compounds but does not imply that these compounds must originate from some form of life. A few special kinds of carbon compounds (for example, carbon oxides, metal carbides, and metal carbonates) are often excluded from the organic classification because their chemistry is more conveniently related to that of inorganic substances.

organic chemistry

The field of organic chemistry is vast, for it includes not only the composition of all living organisms but also of a great many other materials that we use daily. Examples of organic materials are foodstuffs (fats, proteins, carbohydrates), fuels, fabrics (cotton, wool, rayon, nylon), wood and paper products, paints and varnishes, plastics, dyes, soaps and detergents, cosmetics, medicinals, rubber products, and explosives.

What makes carbon compounds special and different from the other elements in the periodic table? Carbon has the unique ability to bond to itself in long chains and rings of varying size. The greater the number of carbon atoms, the more ways there are to link these atoms in different arrangements. This flexibility in arrangement of atoms produces compounds with the same chemical composition and different structures. There is no theoretical limit on the number of organic compounds that can exist. In addition to this unique bonding property, carbon can form strong covalent bonds with a variety of elements, most often including hydrogen, nitrogen, oxygen, sulfur, phosphorus, and the halogens. These are the elements most commonly found in organic compounds.

21.2 The Carbon Atom: Bonding and Shape

The carbon atom is central to all organic compounds. The atomic number of carbon is 6, and its electron structure is $1s^2 2s^2 2p^2$. Two stable isotopes of carbon exist, ^{12}C and ^{13}C. In addition there are several radioactive isotopes; ^{14}C is the most widely known of these because of its use in radiocarbon dating. Having four electrons in its outer shell, carbon has oxidation numbers ranging from $+4$ to -4 and forms predominantly covalent bonds. Carbon occurs as the free element in diamond, graphite, coal, coke, carbon black, charcoal, and lampblack.

A carbon atom usually forms four covalent bonds, that is, bonds that result from the sharing of one or more pairs of electrons by two atoms. The number of electron pairs shared between atoms determines whether the bond is single or multiple. In a single bond only one pair of electrons is shared by the atoms, with one electron most often contributed by each atom. If both atoms have the same

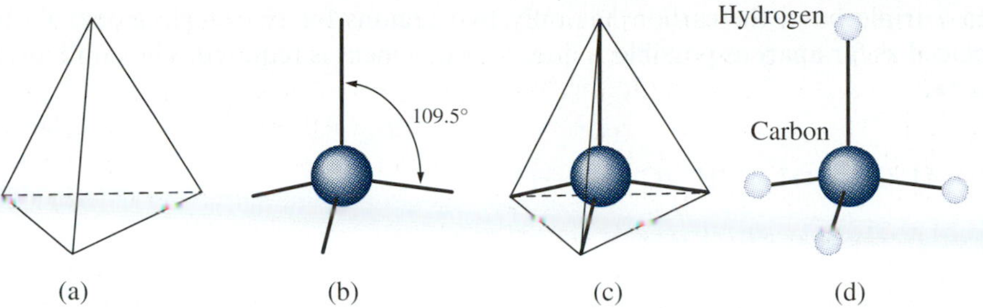

◀ **FIGURE 21.1**
Tetrahedral structure of carbon: (a) a regular tetrahedron; (b) a carbon atom with tetrahedral bonds; (c) a carbon atom within a regular tetrahedron; (d) a methane molecule, CH_4.

electronegativity, the bond is classified as nonpolar. In this type of bond there is no separation of positive and negative charge between the atoms.

Carbon can also form multiple bonds by sharing two or three pairs of electrons between two atoms. The double bond formed by sharing two electron pairs is stronger than a single bond, but not twice as strong. It is also shorter than a single bond. Similarly, a triple bond is formed by the sharing of three electron pairs and is stronger and shorter than a double bond. An organic compound is classified as **saturated** if it contains only single bonds, **unsaturated** if the molecules possess one or more multiple carbon–carbon bonds.

saturated

unsaturated

Lewis structures are useful in representing the bonding between atoms in a molecule, but these representations tell us little about the geometry of the molecules. There are a number of bonding theories that can be used to predict the shape of molecules. One of the common theories is called the valence shell electron pair repulsion (VSEPR) theory. This is a fairly simple, yet accurate method for determining the shape of a molecule.

The basis of the VSEPR theory is the concept that electron pairs repel each other since they have like charges. The electron pairs will therefore try to spread out as far as possible around an atom. In addition, unshared pairs of electrons occupy more space than shared electron pairs.

What does the VSEPR theory tell us about the shapes of carbon-containing compounds? Consider the simplest organic molecule, methane (CH_4). It contains one carbon atom with four single bonds to hydrogen atoms. The four shared pairs of electrons must be placed as far apart as possible in three dimensions. This results in the hydrogen atoms forming the corners of a tetrahedron with the carbon atom in the center (see Figure 21.1). The angle between these tetrahedral bonds is 109.5°.

Double or triple bonds have a significant effect on the shape of the molecule. The additional pairs of electrons in close proximity take up more space than those in a single bond as a result of increased repulsion between the pairs. Consider the Lewis structure for C_2H_4.

H:C::C:H
  ¨   ¨
  H H

Each carbon atom has three separate regions for shared electrons. To place them as far apart as possible requires placing each atom at the corner of a triangle (see Figure 21.2). The bond angles around the carbon atoms are 120°.

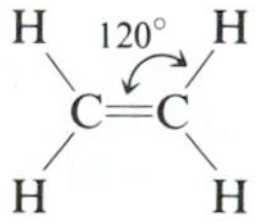

H 120° H
  C=C
H   H

▲
FIGURE 21.2
Shape of a molecule with a carbon–carbon double bond. The hydrogens and carbon form the vertices of a triangle. The bond angles around the carbon are 120°.

In a triple bond the carbon has only two regions for shared electrons. To be placed as far apart as possible, a linear arrangement is required. The bond angle is 180°.

H:C:::C:H   H—C≡C—H (180°)

21.3 Molecular Models

Models are often used in organic chemistry to illustrate molecules. Many representations of molecules can be drawn. Each has its own advantages and disadvantages. Two common methods are: ball-and-stick models and space-filling models. Both types of models are illustrated in Figure 21.3 for the methane molecule.

In the ball-and-stick model different atoms are represented by different colored balls, and bonds are clearly shown with sticks. The space-filling model gives a more accurate representation of the actual molecule but is not as clear in representing the chemical bonds between the atoms. Since most organic molecules contain many atoms, it is often difficult or inconvenient to draw either of these types of model on paper every time we wish to represent a molecule. Chemists often translate a three-dimensional model into a two-dimensional representation (a Lewis structure or bond-line drawing). An example of all four models is shown in Figure 21.4. In both Lewis and bond-line drawings it is important to remember that the bond angle is not 90° the way it appears in the drawing. Rather it is 109.5° since the molecule is actually three-dimensional.

Formulas of organic molecules are also represented differently than those of inorganic compounds. A formula gives information about the composition of a compound. Inorganic compounds contain relatively few groups of atoms and are frequently represented by *empirical formulas*, which give only the simplest ratio of the atoms in a molecule. Larger molecules are often represented by a *molecular formula*, which gives more information. The molecular formula gives the actual number of atoms in a molecule. For inorganic compounds the

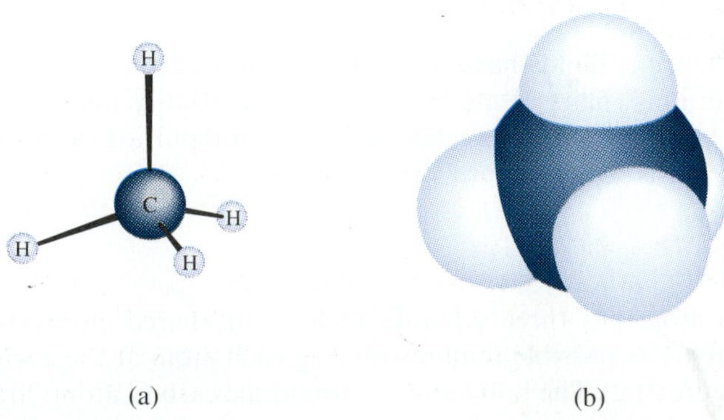

Figure 21.3 ▸ Molecular models of methane. (a) Ball-and-stick model; (b) space-filling model.

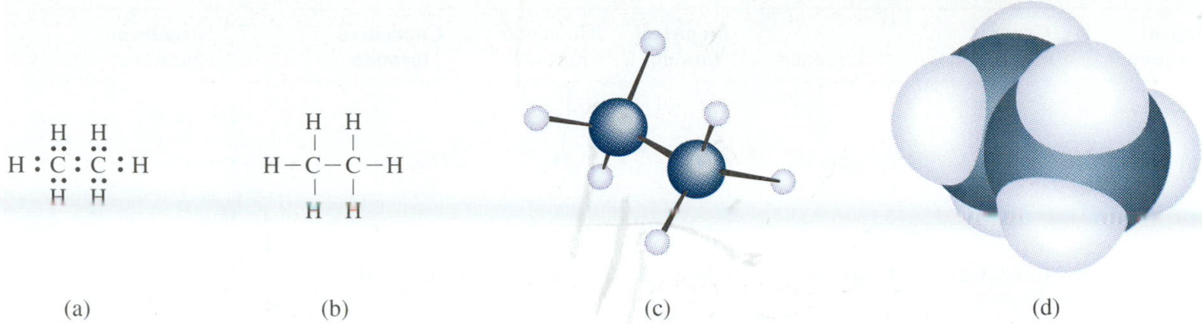

(a) (b) (c) (d)

▲ **FIGURE 21.4**
Models of ethane. (a) Lewis structure; (b) bond-line drawing; (c) ball-and-stick model; (d) space-filling model.

empirical and molecular formulas are often the same. The models we've just been considering represent yet another even more informative type of formula. In a *structural formula* the arrangement of the atoms within the molecule is clearly shown. Organic chemists often shorten these structural formulas into a final type called *condensed structural formulas.* In these formulas each carbon is grouped with adjacent hydrogens and then written as a formula, for example CH_3CH_3 or $CH_3CH_2CH_2CH_3$. The diversity of methods for writing organic formulas is illustrated in Table 21.1, which gives examples of the four types of formulas.

21.4 Classifying Organic Compounds

It is impossible for anyone to study the properties of each of the millions of known organic compounds. Organic compounds with similar structures are grouped into classes as shown in Table 21.1. The members of each class of compounds contain a characteristic atom or group of atoms called a **functional group** shown as the colored portion of the structural formula in Table 21.1.

functional group

Organic compounds from different classes may have the same empirical and molecular formulas, but completely different chemical and physical properties. Look at the compounds in Table 21.1. Notice that the molecular formulas for diethyl ether and ethyl alcohol are C_2H_6O. Compounds that have the same molecular formula but different structural formulas are called **isomers**.

isomers

21.5 Hydrocarbons

Hydrocarbons are compounds that are composed entirely of carbon and hydrogen atoms bonded to each other by covalent bonds. Several classes of hydrocarbons are known. These include the alkanes, alkenes, alkynes, and aromatic hydrocarbons (see Figure 21.5).

hydrocarbons

Fossil fuels—natural gas, petroleum, and coal—are the principal sources of hydrocarbons. Natural gas is primarily methane with small amounts of ethane, propane, and butane. Petroleum is a mixture of hydrocarbons from which gasoline, kerosene, fuel oil, lubricating oil, paraffin wax, and petrolatum (themselves

TABLE 21.1 Classes of Organic Compounds

Class of compound	General formula[a]	Example	Empirical formula	Molecular formula	Condensed formula	Structural formula
Alkane	RH	Ethane	CH_3	C_2H_6	CH_3CH_3	H H │ │ H—C—C—H │ │ H H
Alkene	R—CH=CH$_2$	Ethylene	CH_2	C_2H_4	$H_2C{=}CH_2$	H H ╲ ╱ C=C ╱ ╲ H H
Alkyne	R—C≡C—H	Acetylene	CH	C_2H_2	HC≡CH	H—C≡C—H
Alkyl halide	RX	Ethyl chloride	C_2H_5Cl	C_2H_5Cl	CH_3CH_2Cl	H H │ │ H—C—C—Cl │ │ H H
Alcohol	ROH	Ethyl alcohol	C_2H_6O	C_2H_6O	CH_3CH_2OH	H H │ │ H—C—C—OH │ │ H H
Ether	R—O—R	Diethyl ether	C_2H_6O	C_2H_6O	CH_3OCH_3	H H │ │ H—C—O—C—H │ │ H H
Aldehyde	R—C=O │ H	Acetaldehyde	C_2H_4O	C_2H_4O	CH_3CHO	H │ H—C—C—H │ ‖ H O
Ketone	R—C—R ‖ O	Acetone	C_3H_6O	C_3H_6O	CH_3COCH_3	H H │ │ H—C—C—C—H │ ‖ │ H O H
Carboxylic acid	R—C—OH ‖ O	Acetic acid	CH_2O	$C_2H_4O_2$	CH_3COOH	H │ H—C—C—OH │ ‖ H O
Ester	R—C—OR ‖ O	Methyl acetate	$C_3H_6O_2$	$C_3H_6O_2$	CH_3COOCH_3	H H │ │ H—C—C—O—C—H │ ‖ │ H O H
Amide	R—C—NH$_2$ ‖ O	Acetamide	C_2H_5ON	C_2H_5ON	CH_3CONH_2	H H │ ╱ H—C—C—N │ ‖ ╲ H O H
Amine	R—CH$_2$—NH$_2$	Ethylamine	C_2H_7N	C_2H_7N	$CH_3CH_2NH_2$	H H │ │ H—C—C—N—H │ │ │ H H H

[a] The letter R is used to indicate any of the many possible alkyl groups.

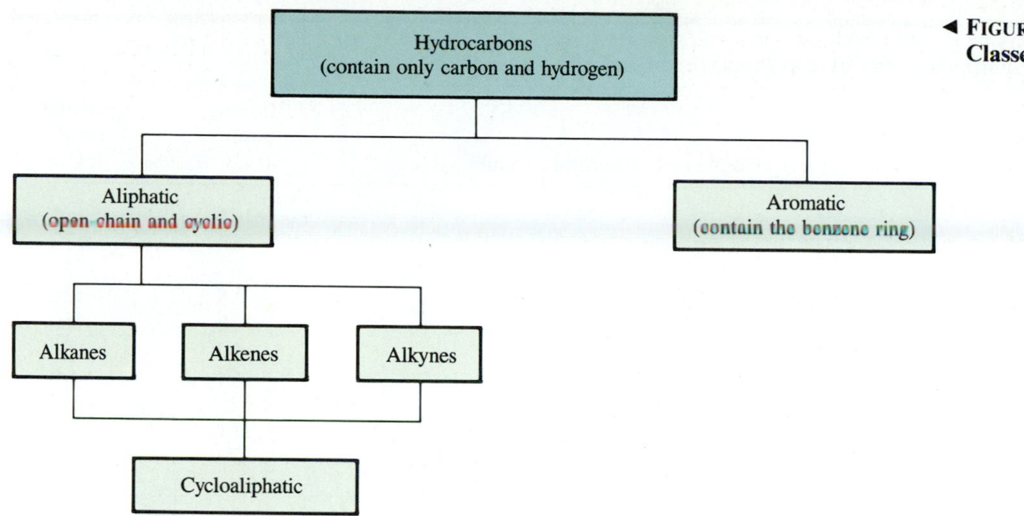

◀ **FIGURE 21.5**
Classes of hydrocarbons.

mixtures of hydrocarbons) are separated. Coal tar, a volatile by-product of the process of making coke from coal for use in the steel industry, is the source of many valuable chemicals including the aromatic hydrocarbons benzene, toluene, and naphthalene.

21.6 SATURATED HYDROCARBONS: ALKANES

alkanes

The **alkanes**, also known as *paraffins* or *saturated hydrocarbons*, are straight- or branched-chain hydrocarbons with only single covalent bonds between the carbon atoms. We shall study the alkanes in some detail because many other classes of organic compounds can be considered as derivatives of these substances. For example, it is necessary to learn the names of the first ten members of the alkane series, because these names are used as a basis for naming other classes of compounds.

Methane (CH_4) is the first member of the alkane series. Members having two, three, and four carbon atoms are ethane, propane, and butane, respectively. The names of the first four alkanes are of common or trivial origin and must be memorized; but the names beginning with the fifth member, pentane, are derived from Greek numbers and are relatively easy to recall. The names and formulas of the first ten members of the series are given in Table 21.2.

homologous series

Adjacent compounds in the alkane series differ from each other in composition by one carbon and two hydrogen atoms. When each member of a series differs from the next member by a CH_2 group, the series is called a **homologous series**. The members of a homologous series are similar in structure but have a regular difference in formula. All common classes of organic compounds exist in homologous series. Each homologous series can be represented by a general formula. For all open-chain alkanes, the general formula is C_nH_{2n+2}, where n corresponds to the number of carbon atoms in the molecule. The formula of any specific alkane is easily determined from this general formula. Thus, for pentane,

TABLE 21.2 Names, Formulas, and Physical Properties of Straight-Chain Alkanes

Name	Molecular formula C_nH_{2n+2}	Condensed structural formula	Boiling point (°C)	Melting point (°C)
Methane	CH_4	CH_4	−161	−183
Ethane	C_2H_6	CH_3CH_3	−88	−172
Propane	C_3H_8	$CH_3CH_2CH_3$	−45	−187
Butane	C_4H_{10}	$CH_3CH_2CH_2CH_3$	−0.5	−138
Pentane	C_5H_{12}	$CH_3CH_2CH_2CH_2CH_3$	36	−130
Hexane	C_6H_{14}	$CH_3CH_2CH_2CH_2CH_2CH_3$	69	−95
Heptane	C_7H_{16}	$CH_3CH_2CH_2CH_2CH_2CH_2CH_3$	98	−90
Octane	C_8H_{18}	$CH_3CH_2CH_2CH_2CH_2CH_2CH_2CH_3$	125	−57
Nonane	C_9H_{20}	$CH_3CH_2CH_2CH_2CH_2CH_2CH_2CH_2CH_3$	151	−54
Decane	$C_{10}H_{22}$	$CH_3CH_2CH_2CH_2CH_2CH_2CH_2CH_2CH_2CH_3$	174	−30

$n = 5$ and $2n + 2 = 12$, so the formula is C_5H_{12}. For hexadecane, the 16-carbon alkane, the formula is $C_{16}H_{34}$.

21.7 CARBON BONDING IN ALKANES

As was pointed out in Section 21.4, a carbon atom is capable of forming single covalent bonds with one, two, three, or four other carbon atoms. To understand this remarkable bonding ability, we must look at the electron structure of carbon. The valence electrons of carbon in their ground state are $2s^2 2p_x^1 2p_y^1$. These are the electrons that enter into bonding. When a carbon atom is bonded to other atoms by single bonds (for example, to four hydrogen atoms in CH_4), it would appear at first that there should be two different types of bonds—bonds involving the 2*s* electrons and bonds involving the 2*p* electrons of the carbon atom. However, this is not the case. All four carbon–hydrogen bonds are found to be identical.

If the carbon atom is to form four equivalent bonds, its electrons in the 2*s* and 2*p* orbitals must rearrange to four equivalent orbitals. To form the four equivalent orbitals, imagine that a 2*s* electron is promoted to a 2*p* orbital, giving carbon an outer-shell electron structure of $2s^1 2p_x^1 2p_y^1 2p_z^1$. The 2*s* orbital and the three 2*p* orbitals then hybridize to form four equivalent hybrid orbitals, which are designated sp^3 orbitals. The orbitals formed (sp^3) are neither *s* orbitals nor *p* orbitals but are a hybrid of those orbitals, having one-fourth *s* orbital character and three-fourths *p* orbital character. This process is illustrated in Figure 21.6. It is these sp^3 orbitals that are directed toward the corners of a regular tetrahedron (see Figure 21.7).

A single bond is formed when one of the sp^3 orbitals overlaps an orbital of another atom. Thus each C—H bond in methane is the result of the overlapping of a carbon sp^3 orbital and a hydrogen *s* orbital [Figure 21.7, part (c)]. Once the bond is formed, the pair of bonding electrons constituting it are said to be in a

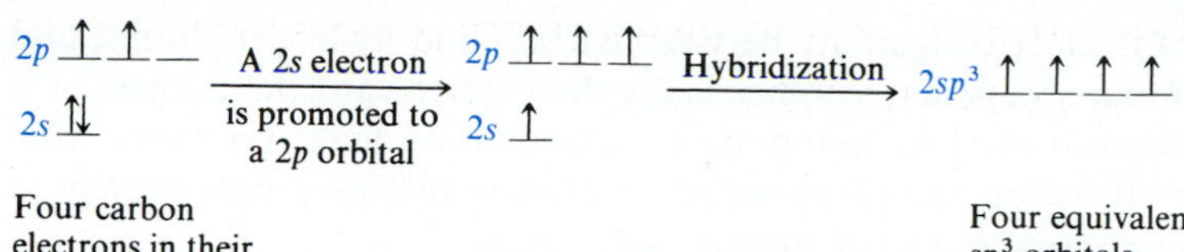

◀ **FIGURE 21.6**
Schematic hybridization of $2s^2 2p_x^1 2p_y^1$ orbitals of carbon to form four sp^3 electron orbitals.

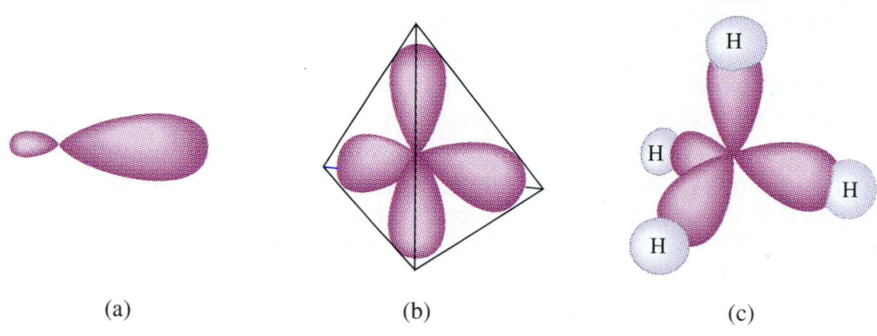

◀ **FIGURE 21.7**
Tetrahedral nature of sp^3 orbitals: (a) a single sp^3 hybridized orbital; (b) four sp^3 hybridized orbitals in tetrahedral arrangement; (c) sp^3 and s orbitals overlapping to form C—H bonds in methane.

molecular orbital. In a similar way a C—C single bond results from the overlap of sp^3 orbitals between two carbon atoms. This type of bond is called a sigma (σ) bond. A **sigma bond** exists if the electron cloud formed by the pair of bonding electrons is symmetrical about a straight line drawn between the nuclei of the bonded atoms.

sigma bond

21.8 Isomerism

The properties of an organic substance are dependent on its molecular structure. The majority of organic compounds are made from relatively few elements—carbon, hydrogen, oxygen, nitrogen, and the halogens. The valence bonds or points of attachment may be represented in structural formulas by a corresponding number of dashes attached to each atom:

—C— (with bonds above and below)   H—   —O—   —N— (with a bond below)   Cl—   Br—   I—   F—

Thus carbon has four bonds to each atom, nitrogen three bonds, oxygen two bonds, and hydrogen and the halogens one bond to each atom.

In an alkane each carbon atom is joined to four other atoms by four single covalent bonds. These bonds are separated by angles of 109.5° (the angles correspond to those formed by lines drawn from the center of a regular tetrahedron to its corners). Alkane molecules contain only carbon–carbon and carbon–hydrogen bonds and are essentially nonpolar. Alkane molecules are nonpolar because (1) carbon–carbon bonds are nonpolar since they are between like atoms; (2) carbon–hydrogen bonds are only slightly polar since there is only a small difference in electronegativity between carbon and hydrogen atoms; and (3) the bonds in an alkane are symmetrically directed toward the corners of a tetrahedron. Because of their low polarity, alkane molecules have very little intermolecular attraction and therefore relatively low boiling points compared with other organic compounds of similar molar mass.

Without the use of models or perspective drawings, the three-dimensional character of atoms and molecules is difficult to portray accurately. However, concepts of structure can be conveyed to some extent by Lewis (or electron-dot) diagrams of structural formulas.

To write the correct structural formula for propane (C_3H_8) we must determine how to place each atom in the molecule. An alkane contains only single bonds. Therefore, each carbon atom must be bonded to four other atoms by either C—C or C—H bonds. Hydrogen must be bonded to only one carbon atom by a C—H bond, since C—H—C bonds do not occur, and an H—H bond would simply represent a hydrogen molecule. Applying this information, we find that the only possible structure for propane is

```
    H  H  H
    |  |  |
 H—C—C—C—H
    |  |  |
    H  H  H
```

Propane

However, it is possible to write two structural formulas corresponding to the molecular formula C_4H_{10}:

```
                                          H
                                       H  |  H
                                        \ | /
                                          C
    H  H  H  H                         H  |  H
    |  |  |  |                         |  |  |
 H—C—C—C—C—H        and            H—C—C—C—H
    |  |  |  |                         |  |  |
    H  H  H  H                         H  H  H
```

Butane   Isobutane

Two C_4H_{10} compounds with the structural formulas shown above actually exist. The butane with the unbranched carbon chain is called *butane* or *normal butane* (abbreviated *n*-butane); it boils at 0.5°C and melts at −138.3°C. The branched-chain butane is called *isobutane*; it boils at −11.7°C and melts at −159.5°C. These differences in physical properties are sufficient to establish that the two compounds, though they have the same molecular formula, are different substances. Models illustrating the structural arrangement of the atoms in methane, ethane, propane, butane, and isobutane are shown in Figure 21.8.

isomerism

isomers

This phenomenon of two or more compounds having the same molecular formula but different structural arrangements of their atoms is called **isomerism**. The various individual compounds are called **isomers.** For example, there are two isomers of butane, C_4H_{10}. Isomerism is very common among organic compounds and is another reason for the large number of known compounds. There are 3 isomers of pentane, 5 isomers of hexane, 9 isomers of heptane, 18 isomers of octane, 35 isomers of nonane, and 75 isomers of decane. The phenomenon of isomerism is a very compelling reason for the use of structural formulas.

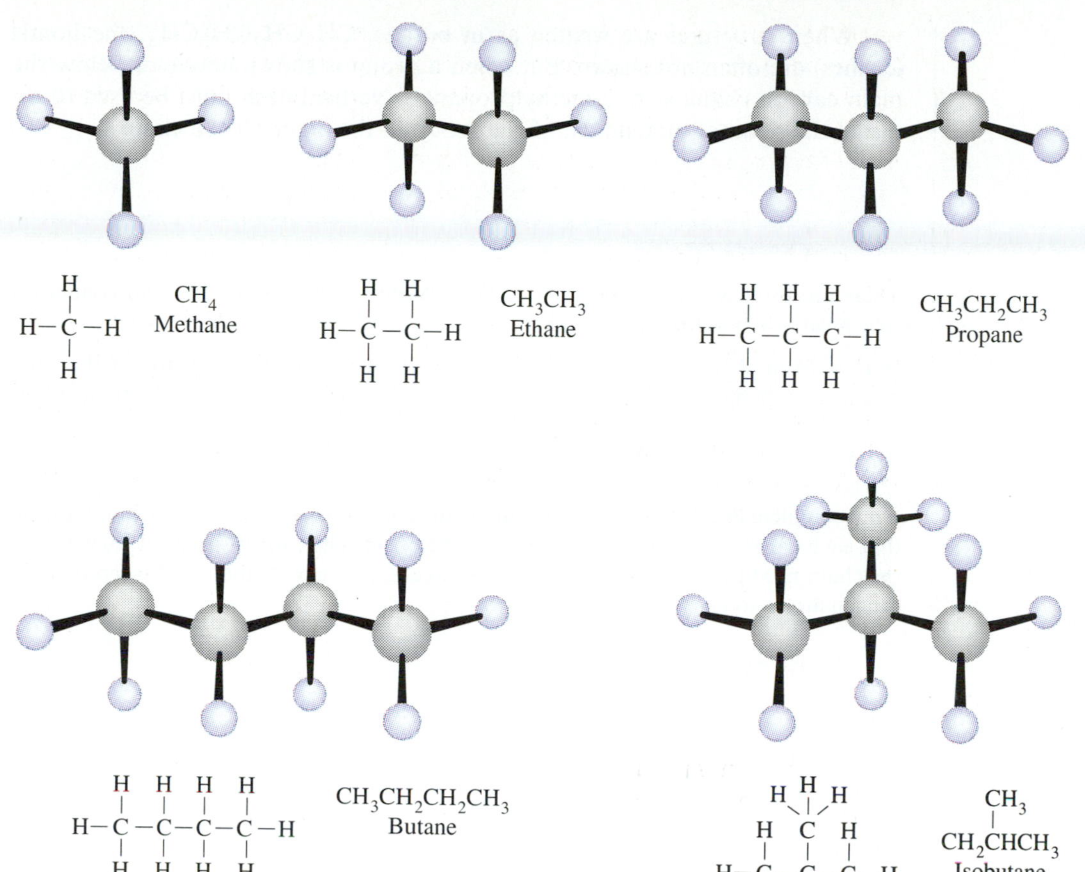

▲
FIGURE 21.8
Ball-and-stick models illustrating structural formulas of methane, ethane, propane, butane, and isobutane.

Isomers are compounds that have the same molecular formula but different structural formulas.

To save time and space in writing, condensed structural formulas are often used. In the condensed structural formulas, atoms and groups that are attached to a carbon atom are generally written to the right of that carbon atom.

Let us interpret the condensed structural formula for propane:

$$\overset{1}{CH_3}—\overset{2}{CH_2}—\overset{3}{CH_3}$$

Carbon number 1 has three hydrogen atoms attached to it and is bonded to carbon number 2, which has two hydrogen atoms on it and is bonded to carbon number 3, which has three hydrogen atoms bonded to it.

When structures are written as in butane, $CH_3CH_2CH_2CH_3$, the bonds (dashes) are often not shown. But when a group is shown above or below the main carbon chain, as in 2-methylpropane, a vertical dash must be used to indicate the point of attachment of the group to the main carbon chain.

EXAMPLE 21.1

There are three isomers of pentane, C_5H_{12}. Write structural formulas and condensed structural formulas for these isomers.

SOLUTION

In a problem of this kind it is best to start by writing the carbon skeleton of the compound containing the longest continuous carbon chain. In this case it is five carbon atoms:

C—C—C—C—C

Now complete the structure by attaching hydrogen atoms around each carbon atom so that each carbon atom has four bonds attached to it. The carbon atoms at each end of the chain need three hydrogen atoms. The three inner carbon atoms each need two hydrogen atoms to give them four bonds.

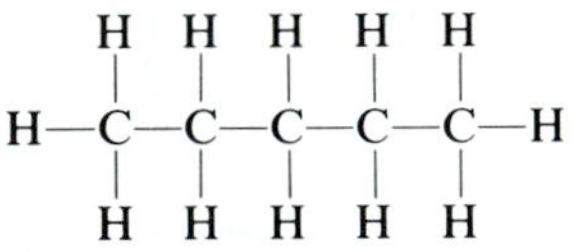

$CH_3CH_2CH_2CH_2CH_3$

For the next isomer, start by writing a four-carbon chain and attach the fifth carbon atom to either of the middle carbon atoms—do not use the end ones.

Both of these structures represent the same compound.

Now add the 12 hydrogen atoms to complete the structure:

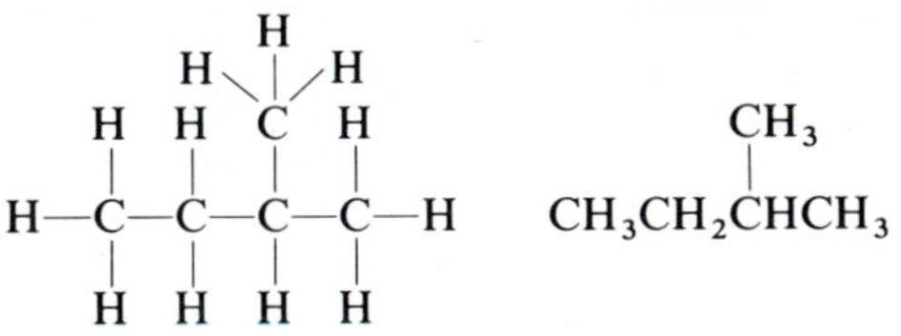

or $CH_3CH_2CH(CH_3)_2$

For the third isomer, write a three-carbon chain, attach the other two carbon atoms to the central carbon atom, and complete the structure by adding the 12 hydrogen atoms:

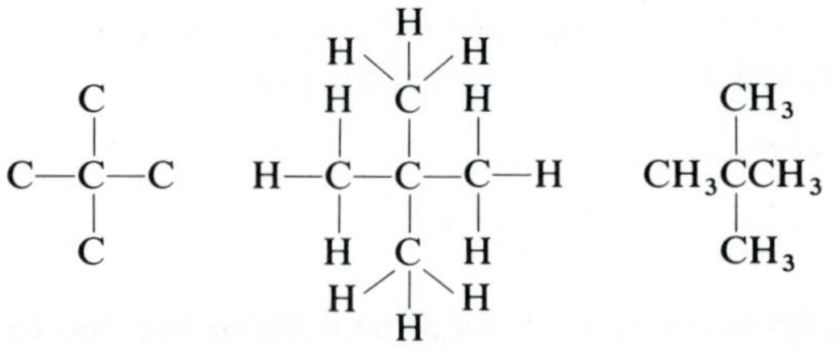

or $C(CH_3)_4$

Practice Write structural and condensed formulas for the isomers of hexane, C_6H_{14}.

Answer:

```
 H  H  H  H  H  H                          H
 |  |  |  |  |  |                          |
H—C——C——C——C——C——C—H                     H—C—H
 |  |  |  |  |  |                 H  H  H  |  H
 H  H  H  H  H  H                 |  |  |  |  |
                               H—C——C——C——C——C—H
                                  |  |  |  |  |
                                  H  H  H  H  H
```

$CH_3CH_2CH_2CH_2CH_2CH_3$

$$\begin{array}{c} CH_3\\ |\\ CH_3CH_2CH_2CHCH_3 \end{array}$$

```
        H                              H
        |                              |
      H—C—H                          H—C—H
 H  H   |   H  H                 H  H  |  H
 |  |   |   |  |                 |  |  |  |
H—C——C——C——C——C—H             H—C——C——C——C—H
 |  |   |   |  |                 |  |  |  |
 H  H   H   H  H                 H  H  |  H
                                       |
                                     H—C—H
                                       |
                                       H
```

$$\begin{array}{c} CH_3\\ |\\ CH_3CH_2CHCH_2CH_3 \end{array}$$

$$\begin{array}{c} CH_3\\ |\\ CH_3CHCHCH_3\\ |\\ CH_3 \end{array}$$

```
        H
        |
      H—C—H
 H  H   |   H
 |  |   |   |
H—C——C——C——C—H
 |  |   |   |
 H  H   |   H
        |
      H—C—H
        |
        H
```

$$\begin{array}{c} CH_3\\ |\\ CH_3CH_2C—CH_3\\ |\\ CH_3 \end{array}$$

21.9 Naming Organic Compounds

In the early development of organic chemistry, each new compound was given a name, usually by the person who had isolated or synthesized it. Names were not systematic but often did carry some information—usually about the origin

TABLE 21.3 Names and Formulas of Selected Alkyl Groups

Formula	Name	Formula	Name
$CH_3—$	Methyl	$CH_3CH(CH_3)—$	Isopropyl
$CH_3CH_2—$	Ethyl	$CH_3CH(CH_3)CH_2—$	Isobutyl
$CH_3CH_2CH_2—$	Propyl	$CH_3CH_2CH(CH_3)—$	*sec*-Butyl (secondary butyl)
$CH_3CH_2CH_2CH_2—$	Butyl	$CH_3C(CH_3)_2—$	*tert*-Butyl or *t*-Butyl (tertiary butyl)
$CH_3(CH_2)_3CH_2—$	Pentyl		
$CH_3(CH_2)_4CH_2—$	Hexyl		
$CH_3(CH_2)_5CH_2—$	Heptyl		
$CH_3(CH_2)_6CH_2—$	Octyl		
$CH_3(CH_2)_7CH_2—$	Nonyl		
$CH_3(CH_2)_8CH_2—$	Decyl		

of the substance. *Wood alcohol* (methanol), for example, was so named because it was obtained by destructive distillation or pyrolysis of wood. Methane, formed during underwater decomposition of vegetable matter in marshes, was originally called *marsh gas.* A single compound was often known by several names. For example, the active ingredient in alcoholic beverages has been called *alcohol, ethyl alcohol, methyl carbinol, grain alcohol, spirit,* and *ethanol.*

Beginning with a meeting in Geneva in 1892, an international system for naming compounds was developed. In its present form the method recommended by the International Union of Pure and Applied Chemistry is systematic, generally unambiguous, and internationally accepted. It is called the **IUPAC System**. Despite the existence of the official IUPAC System, a great many well-established common, or trivial, names and abbreviations (such as TNT and DDT) are used because of their brevity and/or convenience. So it is necessary to have a knowledge of both the IUPAC System and many common names.

IUPAC System

In order to name organic compounds systematically, you must be able to recognize certain common alkyl groups. **Alkyl groups** have the general formula C_nH_{2n+1} (one less hydrogen atom than the corresponding alkane). The missing H atom can be from any carbon in the alkane. The name of the group is formed from the name of the corresponding alkane by simply dropping *-ane* and substituting a *-yl* ending. The names and formulas of selected alkyl groups up to and including four carbon atoms are given in Table 21.3. The letter R is often used in formulas to mean any of the many possible alkyl groups.

alkyl groups

$$R = C_nH_{2n+1} \quad \text{Any alkyl group}$$

The following relatively simple rules are all that are needed to name a great many alkanes according to the IUPAC System. In later sections these rules will be extended to cover other classes of compounds, but advanced texts or references must be consulted for the complete system. The IUPAC rules for naming

NAMING

alkanes are

1. Select the longest continuous chain of carbon atoms as the parent compound, and consider all alkyl groups attached to it as side chains that have replaced hydrogen atoms of the parent hydrocarbon. The name of the alkane consists of the name of the parent compound prefixed by the names of the side-chain alkyl groups attached to it.
2. Number the carbon atoms in the parent carbon chain starting from the end closest to the first carbon atom that has an alkyl or other group substituted for a hydrogen atom.
3. Name each side-chain alkyl group and designate its position on the parent carbon chain by a number (for example, 2-methyl means a methyl group attached to carbon number 2).
4. When the same alkyl-group side chain occurs more than once, indicate this by a prefix (*di-*, *tri-*, *tetra-*, and so forth) written in front of the alkyl-group name (for example, *dimethyl* indicates two methyl groups). The numbers indicating the positions of these alkyl groups are separated by a comma and followed by a hyphen and are placed in front of the name (for example, 2,3-dimethyl).
5. When several different alkyl groups are attached to the parent compound, list them in alphabetical order; for example, *ethyl* before *methyl* in 3-ethyl-4-methyloctane.

The compound shown below is commonly called isopentane. Consider naming it by the IUPAC System:

$$\overset{4}{CH_3}-\overset{3}{CH_2}-\overset{2}{\underset{\displaystyle CH_3}{\underset{|}{CH}}}-\overset{1}{CH_3}$$

2-Methylbutane
(Isopentane)

The longest continuous chain contains four carbon atoms. Therefore, the parent compound is butane and the compound is named as a butane. The methyl group (CH_3—) attached to carbon number 2 is named as a prefix to butane, the "2-" indicating the point of attachment of the methyl group on the butane chain.

How would we write the structural formula for 2-methylpentane? An analysis of its name gives us this information.

1. The parent compound, pentane, contains five carbons. Write and number the five-carbon skeleton of pentane:

$$\overset{5}{C}-\overset{4}{C}-\overset{3}{C}-\overset{2}{C}-\overset{1}{C}$$

2. Put a methyl group on carbon number 2 ("2-methyl" in the name gives this information):

$$\overset{5}{C}-\overset{4}{C}-\overset{3}{C}-\overset{2}{\underset{\displaystyle CH_3}{\underset{|}{C}}}-\overset{1}{C}$$

3. Add hydrogens to give each carbon four bonds. The structural formula is

$$\begin{array}{l} CH_3{-}CH_2{-}CH_2{-}\underset{\displaystyle CH_3}{\underset{|}{CH}}{-}CH_3 \end{array}$$

2-Methylpentane

Should this compound be called 4-methylpentane? No, the IUPAC System specifically states that the parent carbon chain shall be numbered starting from the end nearest to the side or branch chain.

It is very important to understand that it is the sequence of atoms and groups that determines the name of a compound, and not the way the sequence is written. Each of the following formulas represents 2-methylpentane:

$$\overset{1}{CH_3}{-}\underset{\displaystyle CH_3}{\underset{|}{\overset{2}{CH}}}{-}\overset{3}{CH_2}{-}\overset{4}{CH_2}{-}\overset{5}{CH_3} \qquad \overset{5}{CH_3}{-}\overset{4}{CH_2}{-}\overset{3}{CH_2}{-}\overset{\displaystyle CH_3}{\overset{|2}{CH}}{-}\overset{1}{CH_3}$$

$$\begin{array}{l} \qquad\ \ CH_3 \\ \overset{1}{CH_3}{-}\overset{2|}{CH}{-}\overset{3}{CH_2} \\ \qquad\qquad\quad \overset{4|}{CH_2}{-}\overset{5}{CH_3} \end{array} \qquad \begin{array}{c} CH_3 \\ 2| \qquad\quad 4 \\ CH \qquad CH_2 \\ 1\ \ / \backslash 3\ \ / \backslash 5 \\ CH_3 \quad CH_2 \quad CH_3 \end{array}$$

The following formulas and names demonstrate other aspects of the official nomenclature system:

$$\overset{4}{CH_3}{-}\underset{\displaystyle CH_3}{\underset{|}{\overset{3}{CH}}}{-}\underset{\displaystyle CH_3}{\underset{|}{\overset{2}{CH}}}{-}\overset{1}{CH_3}$$

2,3-Dimethylbutane

The name of this compound is 2,3-dimethylbutane. The longest carbon atom chain is four, indicating butane; "dimethyl" indicates two methyl groups; "2,3-" means that one CH_3 is on carbon 2 and one is on carbon 3.

$$\overset{4}{CH_3}{-}\overset{3}{CH_2}{-}\underset{\displaystyle CH_3}{\underset{|}{\overset{\displaystyle CH_3}{\overset{|2}{C}}}}{-}\overset{1}{CH_3}$$

2,2-Dimethylbutane

(Both methyl groups are on the same carbon atom; both numbers are required.)

$$\overset{1}{CH_3}{-}\underset{\displaystyle CH_3}{\underset{|}{\overset{2}{CH}}}{-}\overset{3}{CH_2}{-}\overset{\displaystyle CH_3}{\overset{4|}{CH}}{-}\overset{5}{CH_2}{-}\overset{6}{CH_3}$$

2,4-Dimethylhexane

(The molecule is numbered from left to right.)

$$\begin{array}{l} \quad\;\; ^{3} \qquad\;\; ^{4} \qquad\quad ^{5} \qquad\quad ^{6} \\ CH_3—CH—CH_2—CH_2—CH_3 \\ \qquad\;\; {}_{2}| \\ \qquad\;\; CH_2 \\ \qquad\;\; {}_{1}| \\ \qquad\;\; CH_3 \end{array}$$

3-Methylhexane

(There are six carbons in the longest continuous chain.)

$$\begin{array}{l} \qquad\qquad\qquad\qquad\qquad\qquad\quad CH_2—CH_3 \\ ^{8} \qquad\;\; ^{7} \qquad\quad ^{6} \qquad\quad ^{5} \qquad\quad |^{4} \qquad ^{3} \qquad ^{2} \qquad ^{1} \\ CH_3—CH_2—CH_2—CH_2—C——CH—CH—CH_3 \\ \qquad\qquad\qquad\qquad\qquad\qquad\quad | \qquad\; | \qquad\;\; | \\ \qquad\qquad\qquad\qquad\qquad\qquad\;\; CH_3 \;\; Cl \quad CH_3 \end{array}$$

3-Chloro-4-ethyl-2,4-dimethyloctane

The longest carbon chain is eight. The groups that are attached or substituted for H on the octane chain are named in alphabetical order.

EXAMPLE 21.2

Write the formulas for
(a) 3-Ethylpentane (b) 2,2,4-Trimethylpentane

SOLUTION

(a) The name *pentane* indicates a five-carbon chain. Write five connected carbon atoms and number them:

$$\begin{array}{l} ^{1} \quad ^{2} \quad ^{3} \quad ^{4} \quad ^{5} \\ C—C—C—C—C \end{array}$$

An ethyl group is written as $CH_3CH_2—$. Attach this group at the open bond to carbon number 3:

$$\begin{array}{l} ^{1} \quad ^{2} \quad ^{3} \quad ^{4} \quad ^{5} \\ C—C—C—C—C \\ \qquad\quad\;\, | \\ \qquad\quad CH_2CH_3 \end{array}$$

Now add hydrogen atoms to give each carbon atom four bonds. Carbons 1 and 5 each need three H atoms; carbons 2 and 4 each need two H atoms; and carbon 3 needs one H atom. The formula is complete:

$$\begin{array}{l} CH_3CH_2CHCH_2CH_3 \\ \qquad\qquad | \\ \qquad\quad CH_2CH_3 \end{array}$$

(b) Pentane indicates a five-carbon chain. Write five connected carbon atoms and number them:

$$\begin{array}{l} ^{1} \quad ^{2} \quad ^{3} \quad ^{4} \quad ^{5} \\ C—C—C—C—C \end{array}$$

There are three methyl groups ($CH_3—$) in the compound (*trimethyl*), two attached to carbon 2 and one attached to carbon 4. Attach these three methyl groups to their respective carbon atoms:

$$\begin{array}{ccccccccc} & & CH_3 & & & & CH_3 & & \\ 1 & & 2| & & 3 & & 4| & & 5 \\ C & - & C & - & C & - & C & - & C \\ & & | & & & & & & \\ & & CH_3 & & & & & & \end{array}$$

Now add H atoms to give each carbon atom four bonds. Carbons 1 and 5 each need three H atoms; carbon 2 does not need any H atoms; carbon 3 needs two H atoms; and carbon 4 needs one H atom. The formula is complete:

$$\begin{array}{c} \quad CH_3 \;\; CH_3 \\ \quad | \qquad | \\ CH_3CCH_2CHCH_3 \\ | \qquad\quad \\ CH_3 \qquad \end{array}$$

EXAMPLE 21.3 Name the following compounds:

(a)
$$\begin{array}{c} \qquad\qquad\qquad CH_3 \\ \qquad\qquad\qquad | \\ CH_3CH_2CH_2CH_2CHCH_3 \end{array}$$

(b)
$$\begin{array}{c} \qquad\quad CH_2CH_3 \\ \qquad\quad | \\ CH_3CH_2CH_2CHCH_2CHCH_3 \\ \qquad\qquad\qquad | \\ \qquad\qquad\qquad CH_2CH_3 \end{array}$$

SOLUTION

(a) The longest continuous carbon chain contains six carbon atoms (Rule 1). Thus the parent name of the compound is hexane. Number the carbon chain from right to left so that the methyl group attached to carbon 2 is given the lowest possible number (Rule 2). With a methyl group on carbon 2, the name of the compound is 2-methylhexane (Rule 3).

(b) The longest continuous carbon chain contains eight carbon atoms:

$$\begin{array}{ccccccccccccc} & & & & & & C & - & C & & & & \\ 8 & & 7 & & 6 & & |5 & & 4 & & 3 & & \\ C & - & C & - & C & - & C & - & C & - & C & - & C \\ & & & & & & & & & & |2 & & 1 \\ & & & & & & & & & & C & - & C \end{array}$$

Thus the parent name is octane. As the chain is numbered, there is a methyl group on carbon 3 and an ethyl group on carbon 5. Thus the name of the compound is 5-ethyl-3-methyloctane. Note that ethyl is named before methyl (alphabetical order) (Rule 5).

PRACTICE Write the formula for: (a) 2-methylhexane; (b) 3,4-dimethylheptane; (c) 2-chloro-3-ethylpentane

Answers:

(a)
$$\begin{array}{c} CH_3CHCH_2CH_2CH_2CH_3 \\ | \qquad\qquad\qquad \\ CH_3 \qquad\qquad\quad \end{array}$$

(b)
$$\begin{array}{c} \qquad CH_3 \\ \qquad | \\ CH_3CH_2CHCHCH_2CH_2CH_3 \\ \qquad\quad | \\ \qquad\quad CH_3 \end{array}$$

(c)
$$\begin{array}{c} \quad Cl \\ \quad | \\ CH_3CHCHCH_2CH_3 \\ \qquad | \\ \qquad CH_2CH_3 \end{array}$$

Practice Name the following compounds:

(a) $CH_3CHCH_2CH_3$ (with CH_3 on C-2)

(b) $CH_3CH_2CHCH_2CHCH_3$ (with CH_3 on C-3 and CH_2CH_3 on C-5)

Answers: (a) 2-methylbutane (b) 3,5-dimethylheptane

21.10 Reactions of Alkanes

One single type of reaction of alkanes has inspired people to explore equatorial jungles, endure the heat and sandstorms of the deserts of Africa and the Middle East, mush across the frozen Arctic, and drill holes in the earth more than 30,000 feet deep! These strenuous and expensive activities have been undertaken because alkanes, as well as other hydrocarbons, undergo combustion with oxygen with the evolution of large amounts of heat energy. Methane, for example, reacts with oxygen:

$$CH_4(g) + 2\,O_2(g) \longrightarrow CO_2(g) + 2\,H_2O(g) + 802.5\text{ kJ (191.8 kcal)}$$

The thermal energy can be converted to mechanical and electrical energy. Combustion reactions overshadow all other reactions of alkanes in economic importance. But combustion reactions are not usually of great interest to organic chemists, since carbon dioxide and water are the only chemical products of complete combustion.

Aside from their combustibility, alkanes are relatively sluggish and limited in reactivity. But with proper activation, such as high temperature and/or catalysts, alkanes can be made to react in a variety of ways. Some important noncombustion reactions of alkanes are the following:

1. *Halogenation* (substitution of halogens for hydrogen). *Halogenation* is a general term. When a specific halogen such as chlorine is used, the reaction is called chlorination. RH is an alkane (alkyl group +H atom).

$$RH + X_2 \longrightarrow RX + HX \qquad (X = Cl \text{ or } Br)$$

$$CH_3CH_3 + Cl_2 \longrightarrow \underset{\text{Chloroethane}}{CH_3CH_2Cl} + HCl$$

This reaction yields alkyl halides (RX), which are useful as intermediates for the manufacture of other substances.

2. *Dehydrogenation* (removal of hydrogen)

$$C_nH_{2n+2} \xrightarrow{700\text{–}900^\circ C} C_nH_{2n} + H_2$$

$$CH_3CH_2CH_3 \xrightarrow{\Delta} \underset{\text{Propene}}{CH_3CH{=}CH_2} + H_2$$

This reaction yields alkenes, which, like alkyl halides, are useful chemical intermediates. Hydrogen is a valuable by-product.

▲ **The combustion of alkanes provides the energy for cars and camp stoves.**

3. *Cracking* (breaking up large molecules to form smaller ones)

Example: $\underset{\text{Alkane}}{C_{16}H_{34}} \xrightarrow{\Delta} \underset{\text{Alkane}}{C_8H_{18}} + \underset{\text{Alkene}}{C_8H_{16}}$ (one of several possibilities)

4. *Isomerization* (rearrangement of molecular structures)

Example: $CH_3—CH_2—CH_2—CH_2—CH_3 \longrightarrow CH_3—CH_2—CH(CH_3)—CH_3$

Halogenation is used extensively in the manufacture of petrochemicals (chemicals derived from petroleum and used for purposes other than fuels). The other three reactions—dehydrogenation, cracking, and isomerization—singly or in combination, are of great importance in the production of motor fuels and petrochemicals.

A well-known reaction of methane and chlorine is shown by the equation

$$CH_4 + Cl_2 \longrightarrow \underset{\substack{\text{Chloromethane} \\ \text{(Methyl chloride)}}}{CH_3Cl} + HCl$$

The reaction of methane and chlorine gives a mixture of mono-, di-, tri-, and tetra-substituted chloromethanes.

$$CH_4 \xrightarrow{Cl_2} CH_3Cl \xrightarrow{Cl_2} CH_2Cl_2 \xrightarrow{Cl_2} CHCl_3 \xrightarrow{Cl_2} CCl_4 + 4\ HCl$$

However, if an excess of chlorine is used, the reaction can be controlled to give all tetrachloromethane (carbon tetrachloride). On the other hand, if a large ratio of methane to chlorine is used, the product will be predominantly chloromethane (methyl chloride). Table 21.4 shows the formulas and names for all the chloromethanes. The names for the other halogen-substituted methanes follow the same pattern as for the chloromethanes; for example, CH_3Br is bromomethane, or methyl bromide, and CHI_3 is triiodomethane, or iodoform.

monosubstitution

Chloromethane is a monosubstitution product of methane. The term **monosubstitution** refers to the fact that one hydrogen atom in an organic molecule is substituted by another atom or by a group of atoms. In hydrocarbons, for example, when we substitute one chlorine atom for a hydrogen atom, the new compound is a monosubstitution (monochlorosubstitution) product. In a like manner we can have di-, tri-, tetra-, and so on, substitution products.

TABLE 21.4 Chlorination Products of Methane

Formula	IUPAC name	Common name
CH_3Cl	Chloromethane	Methyl chloride
CH_2Cl_2	Dichloromethane	Methylene chloride
$CHCl_3$	Trichloromethane	Chloroform
CCl_4	Tetrachloromethane	Carbon tetrachloride

This kind of chlorination (or bromination) is general with alkanes. There are nine different chlorination products of ethane. See if you can write the structural formulas for all of them.

When propane is chlorinated, two isomeric monosubstitution products are obtained, because a hydrogen atom may be replaced on either the first or second carbon:

$$CH_3CH_2CH_3 + Cl_2 \xrightarrow[25°C]{Light} \underset{\substack{\text{1-Chloropropane}\\ \text{(}n\text{-Propyl chloride)}}}{CH_3CH_2CH_2Cl} + \underset{\substack{\text{2-Chloropropane}\\ \text{(Isopropyl chloride)}}}{CH_3CHClCH_3} + HCl$$

The letter X is commonly used to indicate a halogen atom. The formula RX indicates a halogen atom attached to an alkyl group and represents the class of compounds known as the **alkyl halides**. When R is CH_3, then CH_3X can be CH_3F, CH_3Cl, CH_3Br, or CH_3I.

alkyl halide

Alkyl halides are named systematically in the same general way as alkanes. Halogen atoms are identified as *fluoro-*, *chloro-*, *bromo-*, or *iodo-* and are named as substituents like side-chain alkyl groups. Study these examples:

$$\underset{\text{2-Chlorobutane}}{CH_3{-}CHCl{-}CH_2{-}CH_3} \qquad \underset{\text{2-Bromo-1-chloropropane}}{CH_2Cl{-}CHBr{-}CH_3}$$

$$\underset{\text{2-Chloro-3-methylpentane}}{CH_3{-}CH_2{-}\underset{\underset{\displaystyle CH_3}{|}}{CH}{-}CHCl{-}CH_3}$$

Example 21.4

How many monochlorosubstitution products can be obtained from pentane?

Solution

First write the formula for pentane:

$$\overset{5}{C}H_3\overset{4}{C}H_2\overset{3}{C}H_2\overset{2}{C}H_2\overset{1}{C}H_3$$

Now rewrite the formula five times substituting a Cl atom for an H atom on each C atom:

I	$CH_3CH_2CH_2CH_2\overset{1}{C}H_2Cl$	Cl on carbon 1
II	$CH_3CH_2CH_2\overset{2}{C}HClCH_3$	Cl on carbon 2
III	$CH_3CH_2\overset{3}{C}HClCH_2CH_3$	Cl on carbon 3
IV	$CH_3\overset{4}{C}HClCH_2CH_2CH_3$	Cl on carbon 4
V	$\overset{5}{C}H_2ClCH_2CH_2CH_2CH_3$	Cl on carbon 5

Compounds I and V are identical. By numbering compound V from left to right, we find that both compounds (I and V) are 1-chloropentane. Compounds II and IV are identical. By numbering compound IV from left to right, we find that both compounds (II and IV) are 2-chloropentane. Thus there are three monochlorosubstitution products of pentane: 1-chloropentane, 2-chloropentane, and 3-chloropentane.

PRACTICE How many dichlorosubstitution products can be obtained from hexane?
Answer: 12

21.11 SOURCES OF ALKANES

The two main sources of alkanes are natural gas and petroleum. Natural gas is formed by the anaerobic decay of plants and animals. The composition of natural gas varies in different locations. Its main component is methane (80–95%), the balance being varying amounts of other hydrocarbons, hydrogen, nitrogen, carbon monoxide, carbon dioxide, and in some locations, hydrogen sulfide. Economically significant amounts of methane are now obtained by the decomposition of sewage, garbage, and other organic waste products.

Petroleum, also called *crude oil*, is a viscous black liquid consisting of a mixture of hydrocarbons with smaller amounts of nitrogen and sulfur-containing organic compounds. Petroleum is formed by the decomposition of plants and animals over millions of years. The composition of petroleum varies widely from one locality to another. Crude oil is refined into many useful products such as gasoline, kerosene, diesel fuel, jet fuel, lubricating oil, heating oil, paraffin wax, petroleum jelly (petrolatum), tars, and asphalt. Petroleum jelly is a jellylike semisolid that is used in preparing medicinal ointments and for lubrication (see Figure 21.9).

A petroleum refinery ▶ processes crude oil into the materials used to produce many consumer products.

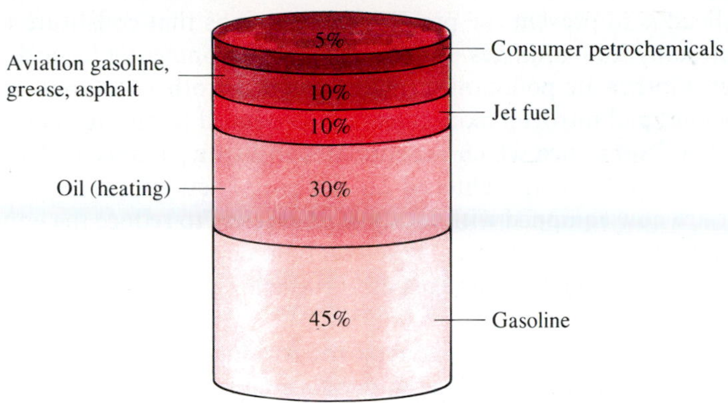

◀ **FIGURE 21.9** **Uses of petroleum.**

At the rate that natural gas and petroleum are being used, these sources of hydrocarbons are destined to be in short supply and virtually exhausted in the not-too-distant future. Alternate sources of fuels must be developed.

21.12 Octane Rating of Gasoline

Gasoline, aside from the additives put into it, consists primarily of hydrocarbons. Gasoline, as it is distilled from crude oil, causes "knocking" when burned in high-compression automobile engines. Knocking, which is due to a too-rapid combustion or detonation of the air–gasoline mixture, is a severe problem in high-compression engines. The knock-resistance of gasolines, a quality that varies widely, is usually expressed in terms of *octane number*, or *octane rating*.

Isooctane (2,2,4-trimethylpentane), because of its highly branched chain structure, is a motor fuel that is resistant to knocking. Mixtures of isooctane and *n*-heptane, a straight-chain alkane that knocks badly, have been used as standards to establish octane ratings of gasolines. Isooctane is arbitrarily assigned an octane number of 100 and *n*-heptane an octane number of 0. To determine the octane rating, a gasoline is compared with mixtures of isooctane and *n*-heptane in a test engine. The octane number of the gasoline corresponds to the percentage of isooctane present in the isooctane and *n*-heptane mixture that matches the knocking characteristics of the gasoline being tested. Thus a 90 octane gasoline has knocking characteristics matching those of a mixture of 90% isooctane and 10% *n*-heptane.

When first used to establish octane numbers, isooctane was the most knock-resistant substance available. Technological advances have resulted in engines with greater power and compression ratios. Higher quality fuels were subsequently necessary. Fuels containing more highly branched hydrocarbons, unsaturated hydrocarbons, or aromatic hydrocarbons may burn more smoothly than isooctane, and have a higher octane rating than 100.

An alternative method to boost octane rating and minimize engine knocking is to add small amounts of an additive to the fuel. One such additive commonly used in gasoline was tetraethyllead, $(C_2H_5)_4Pb$. Adding only 3 mL per gallon of gasoline can increase the octane rating by 15 units. The function of the tetraethyllead is to prevent the premature explosions that constitute knocking. Use of tetraethyllead additives poses a serious environmental hazard. Lead becomes yet another air pollutant, in addition to the others (carbon monoxide, hydrocarbons, and nitrogen oxides) already produced by the automobile. Since lead is a toxic substance which accumulates in living organisms, legal restrictions are constantly being tightened on the use of tetraethyllead. In addition, most cars are now equipped with catalytic converters to reduce the emissions of pollutants into the atmosphere. These catalytic converters are deactivated by lead. These environmental problems are producing changes in the formulation of gasoline.

Major oil companies have changed the formulation of gasoline in recent years to eliminate leaded gasoline and reduce emissions into the atmosphere. Current additives in unleaded gasoline include aromatic compounds such as toluene and xylene (Chapter 22) or methyl *tert*-butyl ether (MTBE):

$$\begin{array}{c} \quad\quad\quad\quad\ CH_3 \\ \quad\quad\quad\quad\ | \\ CH_3{-}O{-}C{-}CH_3 \\ \quad\quad\quad\quad\ | \\ \quad\quad\quad\quad\ CH_3 \end{array}$$

Chemical conversion of straight-chain alkanes into branched or cyclic compounds is also a method for "reformulating" gasoline. Fuels are also changed from season to season as well as from one region of the country to another. These seasonal and regional changes reflect variations in air pollution and environmental standards.

21.13 Cycloalkanes

cycloalkanes

Cyclic, or closed-chain, alkanes also exist. These substances, called **cycloalkanes**, *cycloparaffins*, or *naphthenes*, have the general formula C_nH_{2n}. Note that this series of compounds has two fewer hydrogen atoms than the open-chain alkanes. The bonds for the two missing hydrogen atoms are accounted for by an additional carbon–carbon bond in forming the cyclic ring of carbon atoms. Structures for the four smallest cycloalkanes are shown in Figure 21.10.

With the exception noted below for cyclopropane and cyclobutane, cycloalkanes are generally similar to open-chain alkanes in both physical properties and chemical reactivity. Cycloalkanes are saturated hydrocarbons; they contain only single bonds between carbon atoms.

The reactivity of cyclopropane, and to a lesser degree that of cyclobutane, is greater than that of other alkanes. This greater reactivity exists because the carbon–carbon bond angles in these substances deviate substantially from the normal tetrahedral angle. The carbon atoms form a triangle in cyclopropane, and in cyclobutane they approximate a square. Cyclopropane molecules therefore have carbon–carbon bond angles of 60°, and in cyclobutane the bond angles are about 90°. In the open-chain alkanes and in larger cycloalkanes,

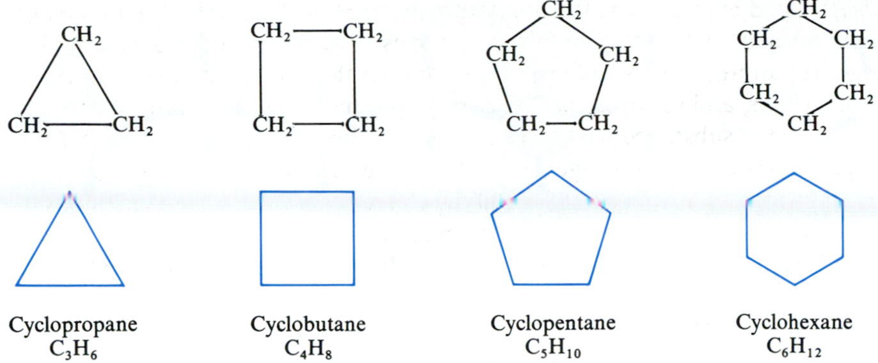

◀ **FIGURE 21.10 Cycloalkanes. In the line representations, each corner of the diagram represents a CH_2 group.**

the carbon atoms are in a three-dimensional zigzag pattern in space and have normal (tetrahedral) bond angles of about 109.5°.

Bromine adds to cyclopropane readily and to cyclobutane to some extent. In this reaction the ring breaks and an open-chain dibromopropane is formed:

$$\underset{\text{Cyclopropane}}{\overset{CH_2}{CH_2{-}CH_2}} + Br_2 \longrightarrow \underset{\text{1,3-Dibromopropane}}{BrCH_2CH_2CH_2Br}$$

Cyclopropane and cyclobutane react in this way because their carbon–carbon bonds are strained and therefore weakened. Cycloalkanes with rings having more than four carbon atoms do not react in this way because their molecules take the shape of nonplanar puckered rings. These rings can be considered to be formed by simply joining the end carbon atoms of the corresponding normal alkanes. The resulting cyclic molecules are nearly strain-free, with carbon atoms arranged in space so that the bond angles are close to 109.5° (Figure 21.11).

Molecular models show that cyclohexane can assume two distinct nonplanar conformations. One form is shaped like a chair, while the other is shaped like a boat (see Figure 21.12). In the chair form the hydrogen atoms are separated as effectively as possible, so this is the more stable conformation. Six of the hydrogens in "chair" cyclohexane lie approximately in the same plane as the carbon ring. These are called **equatorial** hydrogens. The other six hydrogen atoms are approximately at right angles above or below the plane of the ring. These hydrogens are called **axial**. Substituent groups usually are found in equatorial positions where they are furthest from hydrogens and other groups. Five- and six-membered rings are a common occurrence in organic chemistry and biochemistry. These conformations and isomers are evident in carbohydrates as well as in nucleic acids.

equatorial

axial

Cyclopropane is a useful general anesthetic and, along with certain other cycloalkanes, is used as an intermediate in some chemical syntheses. The high reactivity of cyclopropane requires great care in its use as an anesthetic because it is an extreme fire and explosion hazard. The cyclopentane and cyclohexane ring structures are present in many naturally occurring molecules such as prostaglandins, steroids (for example, cholesterol and sex hormones), and some vitamins.

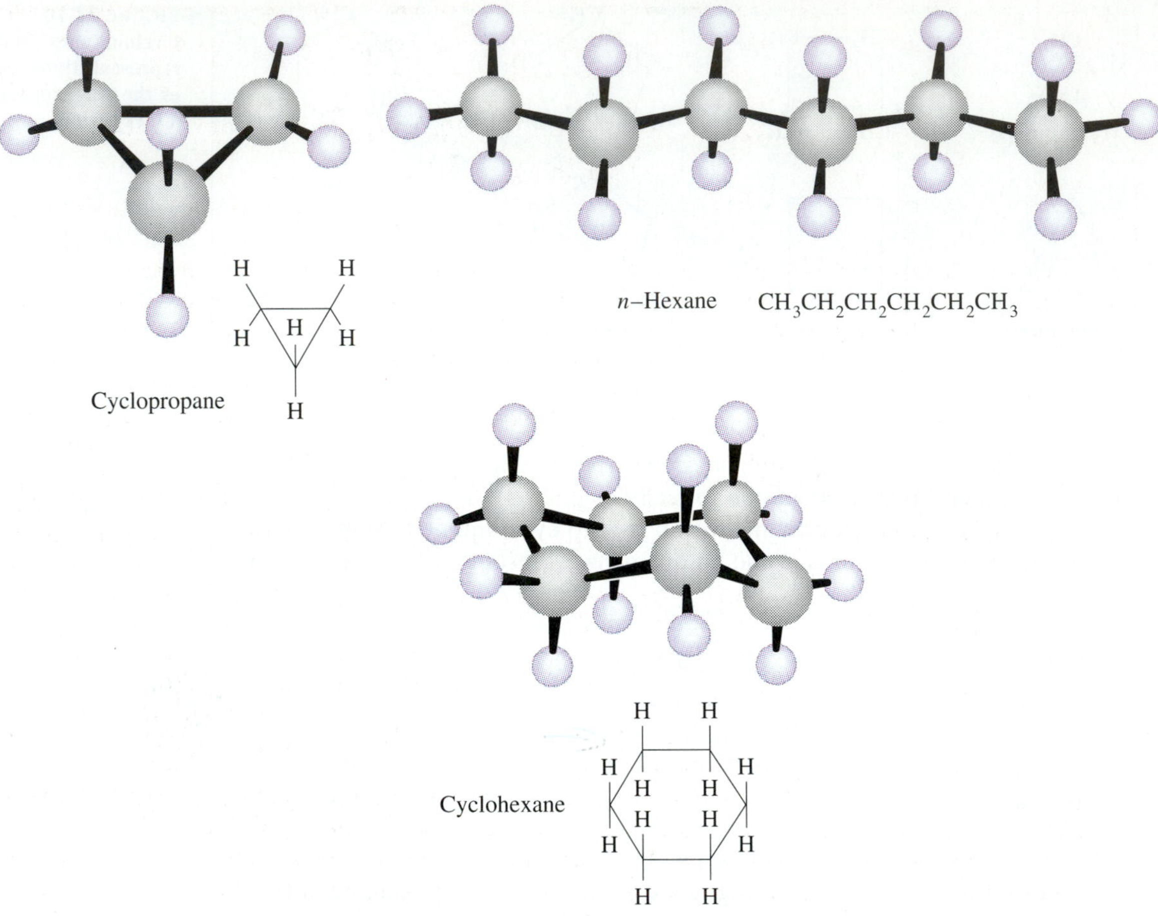

▲
FIGURE 21.11
Ball-and-stick models illustrating cyclopropane, hexane, and cyclohexane: In cyclopropane all the carbon atoms are in one plane. The angle between carbon atoms is 60°, not the usual 109.5°; therefore the cyclopropane ring is strained. In cyclohexane the carbon–carbon bonds are not strained. This is because the molecule is puckered, with carbon–carbon bond angles of 109.5°, as found in normal hexane.

FIGURE 21.12 ► Conformations of cyclohexane: (a) chair conformation; (b) boat conformation. Axial hydrogens are shown in color in the chair conformation.

(a)

(b)

Chemistry in Action

Molecules to Communicate, Refrigerate, and Save Lives

By far the major role of alkanes in our world is as a source of energy for industrial and consumer use. Hydrocarbons do not function as an energy source in the physiological world of living organisms. The compounds that supply fuel necessary for life contain oxygen as well as carbon and hydrogen.

Alkanes do find other significant uses in our lives. High-molar-mass alkanes can be used to soften or moisten the skin (see Section 14.17). Petroleum jelly and mineral oil are both mixtures of hydrocarbons used to protect the skin or as a lubricant.

Insects can use hydrocarbons (as well as other organic compounds) as chemical communication devices. These compounds, called pheromones, are secreted by an insect and recognized by other members of the species as a message. The meaning of the pheromone varies with its composition. It could be a sex attractant, an alarm, or an indication of the path to a source of food. Ants release alarm pheromones when disturbed which have been identified as undecane, $CH_3(CH_2)_9CH_3$, and tridecane, $CH_3(CH_2)_{11}CH_3$. Our growing understanding of these molecules is beginning to lead to their use in insect abatement. Commercial traps are baited with sex pheromone and the insects are captured without the use of pesticides.

Halogenated hydrocarbons serve several significant functions in society. Tetrachloroethylene, C_2Cl_4, and carbon tetrachloride, CCl_4, are both good solvents for nonpolar molecules and have been used as cleaning solvents. Carbon tetrachloride exposure can lead to kidney and liver damage. Both compounds are carcinogens. The dry-cleaning industry is eliminating the use of these compounds.

Freons and other chlorofluorocarbons (CFCs) are useful because they are nontoxic, nonflammable, and noncorrosive. Many of these compounds have low boiling points and make excellent refrigerants. They have also been used as propellants in aerosol sprays, and in the production of some fast food containers. Unfortunately the use of CFCs is a major factor in the destruction of the ozone layer (see Section 13.17). Use of these compounds is currently being discontinued by major industries in an effort to protect our atmosphere.

Some fluorinated hydrocarbons are used as blood substitutes. All the hydrogens in the hydrocarbon are replaced with fluorine, producing a compound known as a fluorocarbon. Dispersions of these compounds in water can absorb nearly 3 times the oxygen per unit as whole blood. Organisms receiving blood substitutes continue to produce whole blood. The substitute permits the tissues to absorb oxygen while it remains chemically inert and is excreted over a period of time.

Concepts in Review

1. Describe the tetrahedral nature of the carbon atom.
2. Explain why the concept of hybridization is used to describe the bonding of carbon in simple compounds such as methane.
3. Explain the bonding in alkanes.
4. Write the Lewis structures for alkanes and halogenated alkanes.

5. Write the names and formulas for the first ten normal alkanes.
6. Understand the concept of isomerization.
7. Write structural formulas and IUPAC names for the isomers of an alkane or a halogenated alkane.
8. Give the IUPAC name of a hydrocarbon or a halogenated hydrocarbon when given the structural formula and vice versa.
9. Write equations for the halogenation of an alkane, giving all possible monohalosubstitution products.
10. Write structural formulas and names for simple cycloalkanes.
11. Draw the two major conformations for cyclohexane.
12. Understand the octane number rating system for gasoline and discuss methods for increasing the octane number.
13. Indicate several biological uses of alkanes.

EXERCISES

1. What are the major reasons for the large number of organic compounds?
2. Why is it believed that a carbon atom must form hybrid electron orbitals when it bonds to hydrogen atoms to form methane?
3. Write Lewis structures for:
   (a) CCl_4 (b) C_2Cl_6 (c) $CH_3CH_2CH_3$
4. Write the names and formulas for the first ten normal alkanes.
5. The name of the compound of formula $C_{11}H_{24}$ is undecane. What is the formula for dodecane, the next higher homologue in the alkane series?
6. How many sigma bonds are in a molecule of
   (a) Ethane (b) Butane (c) Isobutane
7. Which of these formulas represent isomers?
   (a) $CH_3CH_2CH_2CH_3$
   (b) $CH_3CH_2CH_2CH_2CH_3$
   (c)

```
CH3CHCH3
   |
CH3CH2
```

   (d) $CH_3CH_2CH_2CH_2CH_2CH_3$
   (e)

```
CH3       CH3
   \     /
   CH—CH
   /     \
CH3       CH3
```

   (f)

```
CH2—CH2
|    |
CH2  CH2
  \  /
   CH2
```

   (g)

```
CH3CHCH2CH2CH3
   |
   CH3
```

   (h)

```
CH2
|  \
|   CHCH2CH3
|  /
CH2
```

   (i)

```
CH3CH2
     |
     CH2CH3
```

8. (a) How many methyl groups are in each formula in Question 7?
   (b) How many ethyl groups?
9. Which of these formulas represent the same compound?
   (a)

```
CH3CHCH2CHCH3
   |     |
   CH3   CH3
```

   (b)

```
   CH3
   |
CH2CHCH2CHCH3
|       |
CH3     CH3
```

   (c)

```
                CH3
                |
CH3CHCH2CH2CHCH3
   |
   CH3
```

   (d)

```
     CH3
     |
CH3CHCHCH2CH2
   |      |
   CH3    CH3
```

   (e)

```
   CH3
   |
CH3CHCH2
       |
CH3CH2CHCH3
```

   (f)

```
   CH3
   |
CH3CH
   |
   CH2CHCH3
       |
       CH2CH3
```

10. Draw structural formulas for all the isomers of
(a) CH_3Br (b) CH_2Cl_2 (c) C_2H_5Cl (d) C_3H_7Br (e) C_4H_9I (f) $C_3H_6Cl_2$ (g) C_3H_6BrCl (h) $C_4H_8Cl_2$

11. What is the molar mass of an alkane that contains 30 carbon atoms?

12. Write condensed structural formulas for:
(a) The five isomers of hexane
(b) The nine isomers of heptane

13. Give common and IUPAC names for the following:
(a) CH_3CH_2Cl (b) $CH_3CHClCH_3$ (c) $(CH_3)_2CHCH_2Cl$ (d) $CH_3CH_2CH_2Cl$ (e) $(CH_3)_3CCl$ (f) $CH_3CHClCH_2CH_3$

14. The following names are incorrect. Tell why the name is wrong and give the correct name.
(a) 3-Methylbutane
(b) 2-Ethylbutane
(c) 2-Dimethylpentane
(d) 3-Methyl-5-ethyloctane
(e) 3,5,5-Triethylhexane

15. Draw structural formulas of the following compounds:
(a) 2,4-Dimethylpentane
(b) 2,2-Dimethylpentane
(c) 3-Isopropyloctane
(d) 4-Ethyl-2-methylhexane
(e) 4-*t*-Butylheptane
(f) 4-Ethyl-7-isopropyl-2,4,8-trimethyldecane

16. Give the IUPAC name for each of the following compounds:

(a)
```
CH3CH2CHCH3
       |
       CH3
```
(b) $(CH_3)_2CHCH_2CH(CH_3)_2$

(c)
```
CH2
|  \
|   CHCH3
|  /
CH2
```
(d)
```
           CH3CHCH3
              |
CH3CH2CH2CHCH3
```

17. Complete the equations for (a) the monochlorination and (b) complete combustion of butane.

(a) $CH_3CH_2CH_2CH_3 + Cl_2 \xrightarrow{hv}$

(b) $CH_3CH_2CH_2CH_3 + O_2 \xrightarrow{\Delta}$

18. The structure for hexane is

$$CH_3CH_2CH_2CH_2CH_2CH_3$$

Draw structural formulas for all the monochlorohexanes, $C_6H_{13}Cl$, that have the same carbon structures as hexane.

19. Draw structures for the ten dichlorosubstituted isomers ($C_5H_{10}Cl_2$) of 2-methylbutane.

20. A hydrocarbon sample of formula C_4H_{10} was brominated, and four different monobromo compounds of formula C_4H_9Br were isolated. Was the sample a pure compound or a mixture of compounds? Explain your answer.

21. There are two cycloalkanes that have the formula C_4H_8. Draw their structures and name them.

22. Which of these statements are correct? Rewrite the incorrect statements to make them correct.
(a) Alkane hydrocarbons are essentially nonpolar.
(b) The C—H sigma bond in methane is made from an overlap of an *s* electron orbital and an sp^3 electron orbital.
(c) The valence electrons of every carbon atom in an alkane are in sp^3-hybridized orbitals.
(d) The four carbon–hydrogen bonds in methane are equivalent.
(e) Hydrocarbons are composed of carbon and water.
(f) In the alkane homologous series, the formula of each member differs from its preceding member by CH_3.
(g) Carbon atoms can form single, double, and triple bonds with other carbon atoms.
(h) The name for the alkane C_5H_{12} is propane.
(i) There are eight carbon atoms in a molecule of 2,3,3-trimethylpentane.
(j) The general formula for an alkyl halide is RX.
(k) The IUPAC name for $CH_3CH_2CH_2CHClCH_3$ is 4-chloropentane.
(l) Isopropyl chloride is also called 2-chloropropane.
(m) The molecular formula for chlorocyclohexane is $C_6H_{11}Cl$.
(n) Chlorocyclohexane and 1-chlorohexane are isomers.
(o) The products of complete combustion of a hydrocarbon are carbon monoxide and water.
(p) When pentane is chlorinated, three monochlorosubstitution products can be obtained.
(q) Isobutane, 2-methylpropane, and 1,1-dimethylethane are all correct names for the same compound.
(r) Only one monochloro-substituted product results from the chlorination of butane.
(s) Cycloalkanes have the general formula C_nH_{2n}.

22

Unsaturated Hydrocarbons

◀ Chapter Opening Photo: **The brilliant dyes used in the textile industry are unsaturated organic compounds.**

Think of the many images a particular fragrance can evoke. A favorite perfume may provide memories of a romantic evening, the aroma of a light-bodied red wine may remind us of a favorite restaurant, while the smell of cloves forces us to remember the dentist's office. The molecules associated with the essential oils in plants often contain multiple bonds between carbon atoms. These oils are widely used in cosmetics, medicines, flavorings, and perfumes.

Many of these molecules are necessary to enhance our lives. Vitamin A prevents night blindness and maintains the integrity of the eyes. Unsaturated molecules can also link together to form polymers. These large molecules find many uses, such as polypropylene in plastic bottles and carpet fibers, polystyrene in foam cups, Teflon, plastic wrap, and Orlon fiber in clothing.

Hydrocarbons also often form into rings. These ring molecules are the basis for many consumer products such as detergents, insecticides, and dyes. Some aromatic carbon compounds are also found in living organisms.

22.1 Bonding in Unsaturated Hydrocarbons

The unsaturated hydrocarbons consist of three families of homologous compounds which contain multiple bonds between carbon atoms. In each family the compounds contain fewer hydrogens than the corresponding alkanes. One family of compounds contains carbon–carbon double bonds. These compounds are called **alkenes** or olefins. The family of compounds containing carbon–carbon triple bonds is called **alkynes** or acetylenes. The third family of hydrocarbons contains the benzene ring and is known as the **aromatic compounds.**

alkenes

alkynes

aromatic compounds

The double bonds in alkenes are different from the sp^3 hybrid bonds found in the alkanes. The alkene hybridization may be visualized in the following way. One of the $2s$ electrons of carbon is promoted to a $2p$ orbital forming the four half-filled orbitals, $2s^1 2p_x^1 2p_y^1 2p_z^1$. Three of these orbitals ($2s^1 2p_x^1 2p_y^1$) hybridize, forming three equivalent orbitals designated as sp^2. Thus the four orbitals available for bonding are three sp^2 orbitals and one p orbital. This process is illustrated in Figure 22.1.

The three sp^2 hybrid orbitals form angles of 120° with each other and lie in a single plane. The remaining $2p$ orbital is oriented perpendicular to this plane,

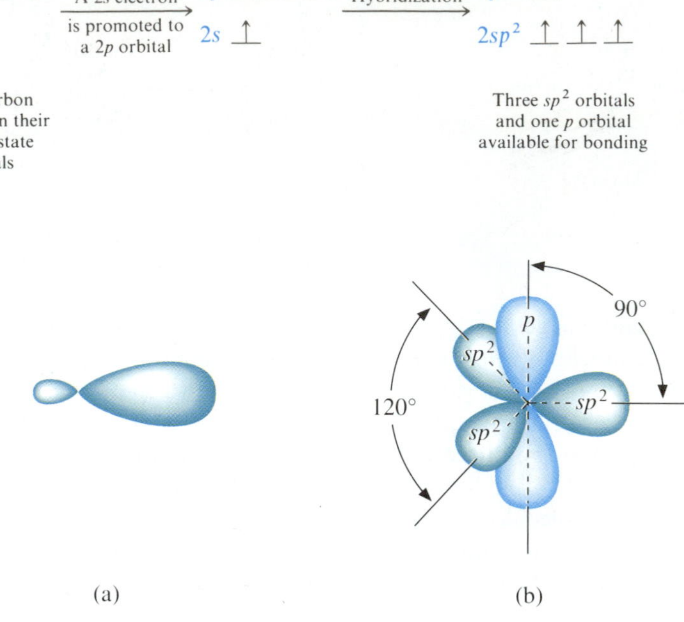

FIGURE 22.1 ▶ Schematic hybridization of $2s^2 2p_x^1 2p_y^1$ orbitals of carbon to form three sp^2 electron orbitals and one p electron orbital.

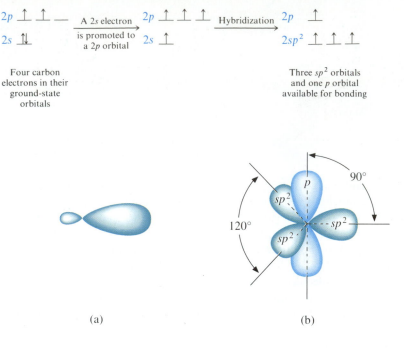

FIGURE 22.2 ▶ (a) A single sp^2 electron orbital and (b) a side view of three sp^2 orbitals all lying in the same plane with a p orbital perpendicular to the three sp^2 orbitals.

with one lobe above and one lobe below the plane (see Figure 22.2). In the formation of a double bond, an sp^2 orbital of one carbon atom overlaps an identical sp^2 orbital of another carbon to form a sigma bond. At the same time the two perpendicular p orbitals (one on each carbon atom) overlap to form a **pi (π) bond** between the two carbon atoms. This pi bond consists of two electron clouds, one above and one below the sigma bond (see Figure 22.3). The $H_2C{=}CH_2$ molecule is completed as the remaining sp^2 orbitals (two on each carbon atom) overlap hydrogen s orbitals to form sigma bonds between the carbon and hydrogen atoms. Thus there are five sigma bonds and one pi bond in an ethylene molecule.

pi bond

In the formula commonly used to represent ethylene ($CH_2{=}CH_2$), no distinction is made between the sigma bond and the pi bond in the carbon–carbon double bond. Each bond is represented by a dash. However, these bonds are actually very different from each other. The sigma bond is formed by the overlap of sp^2 orbitals; the pi bond is formed by the overlap of p orbitals. The sigma bond electron cloud is distributed about a line joining the carbon nuclei, but the pi bond electron cloud is distributed above and below the sigma bond region (see Figure 22.3). The carbon–carbon pi bond is much weaker and, as a consequence, much more reactive than the carbon–carbon sigma bond.

The formation of a triple bond between carbon atoms, as in acetylene, $HC{\equiv}CH$, may be visualized as follows:

1. A carbon atom $2s$ electron is promoted to a $2p$ orbital ($2s^1 2p_x^1 2p_y^1 2p_z^1$).
2. The $2s$ orbital hybridizes with one of the $2p$ orbitals to form two equivalent orbitals known as sp orbitals. These two hybrid orbitals lie on a straight line that passes through the center of the carbon atom. The re-

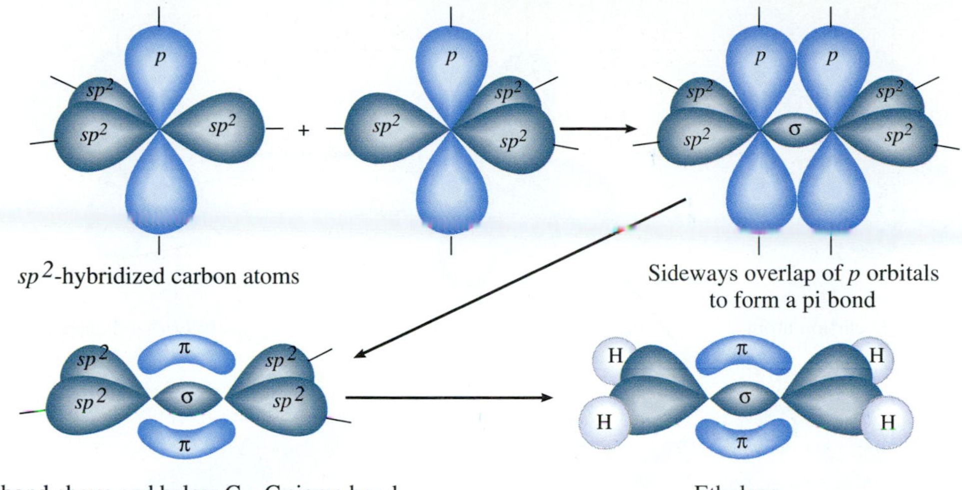

▲ **FIGURE 22.3**
Pi (π) and sigma (σ) bonding in ethylene.

maining two unhybridized $2p$ orbitals are oriented at right angles to these *sp* orbitals and to each other.

3. In forming carbon–carbon bonds, one carbon *sp* orbital overlaps an identical *sp* orbital on another carbon atom to establish a sigma bond between the two carbon atoms.
4. The remaining *sp* orbitals (one on each carbon atom) overlap *s* orbitals on hydrogens to form sigma bonds and establish the H—C—C—H bond sequence. Because all the atoms forming this sequence lie in a straight line, the acetylene molecule is linear.
5. Simultaneously the two $2p$ orbitals on each carbon overlap to form two pi bonds. These two pi bond orbitals occupy sufficient space that they overlap each other to form a continuous tubelike electron cloud surrounding the sigma bond between the carbon atoms (Figure 22.4). These pi bond electrons (as in ethylene) are not as tightly held by the carbon nuclei as are the sigma bond electrons. Acetylene, consequently, is a very reactive substance.

22.2 Nomenclature of Alkenes

The names of alkenes are derived from the names of corresponding alkanes. To name an alkene by the IUPAC System:

1. Select the longest carbon–carbon chain that contains the double bond.
2. Name this parent compound as you would an alkane but change the *-ane* ending to *-ene*; for example, propane is changed to propene.

$CH_3CH_2CH_3$ (Propane) $\qquad$ $CH_3CH{=}CH_2$ (Propene)

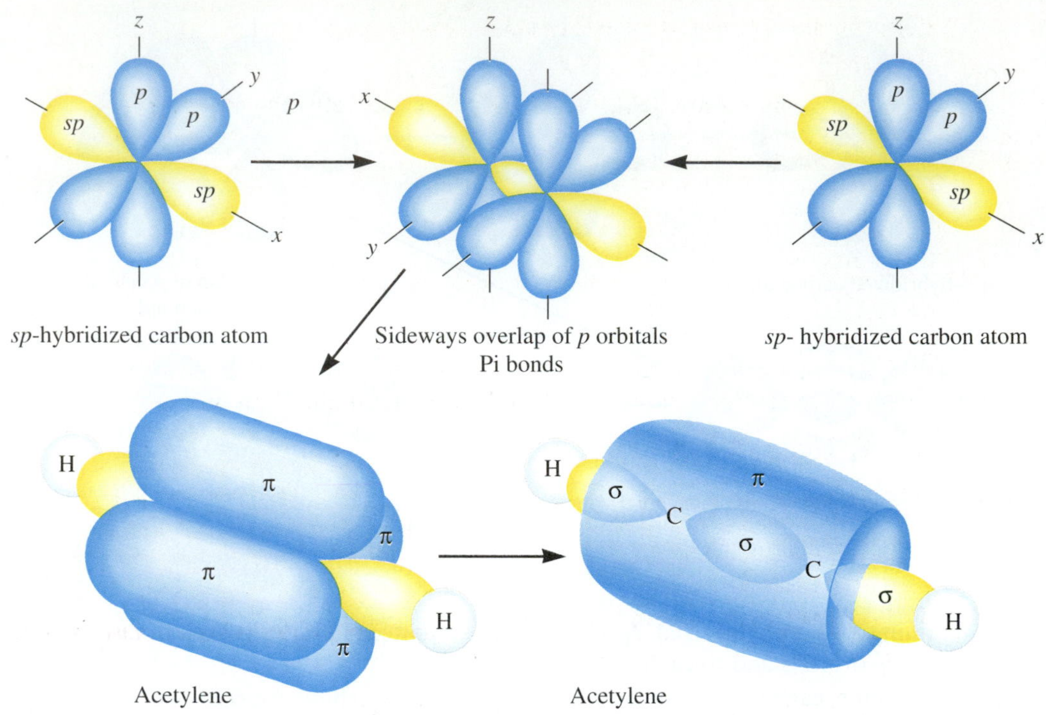

FIGURE 22.4 ▶ Pi (π) and sigma (σ) bonding in acetylene.

3. Number the carbon chain of the parent compound starting with the end nearer to the double bond. Use the smaller of the two numbers on the double-bonded carbon atoms to indicate the position of the double bond. Place this number in front of the alkene name; for example, 2-butene means that the carbon–carbon double bond is between carbon numbers 2 and 3.
4. Side chains and other groups are treated as in naming alkanes, by numbering and assigning them to the carbon atom to which they are bonded.

Study the following examples of named alkenes:

$$\overset{4}{C}H_3\overset{3}{C}H_2\overset{2}{C}H{=}\overset{1}{C}H_2$$
1-Butene

$$\overset{1}{C}H_3\overset{2}{C}H{=}\overset{3}{C}H\overset{4}{C}H_3$$
2-Butene

$$\overset{4}{C}H_3\overset{3}{\underset{}{C}}(CH_3)H\overset{2}{C}H{=}\overset{1}{C}H_2$$
3-Methyl-1-butene

$$(\overset{6}{C}H_3\overset{5}{C}H_2\overset{4}{C}H_2)(CH_3CH_2CH_2)\overset{3}{C}H\overset{2}{C}H{=}\overset{1}{C}H_2$$
3-Propyl-1-hexene

To write a structural formula from a systematic name, the naming process is reversed. For example, how would we write the structural formula for 4-methyl-2-pentene? The name indicates:

1. Five carbons in the longest chain

2. A double bond between carbons 2 and 3
3. A methyl group on carbon 4

Write five carbon atoms in a row. Place a double bond between carbons 2 and 3, and place a methyl group on carbon 4:

```
1  2  3  4  5
C  C─C  C─C
        |
        CH3
```

Carbon skeleton

Now add hydrogen atoms to give each carbon atom four bonds. Carbons 1 and 5 each need three H atoms; carbons 2, 3, and 4 each need one H atom. The complete formula is

```
CH3CH=CHCHCH3
        |
        CH3
```

4-Methyl-2-pentene

EXAMPLE 22.1

Write structural formulas for (a) 2-pentene and (b) 7-methyl-2-octene.

SOLUTION

(a) The stem *pent-* indicates a five-carbon chain; the suffix *-ene* indicates a carbon–carbon double bond; the number 2 locates the double bond between carbons 2 and 3. Write five carbon atoms in a row and place a double bond between carbons 2 and 3:

```
1   2   3   4   5
C─C═C─C─C
```

Add hydrogen atoms to give each carbon atom four bonds. Carbons 1 and 5 each need three H atoms; carbons 2 and 3 each need one H atom; carbon 4 needs two H atoms. The complete formula is

$CH_3CH{=}CHCH_2CH_3$

2-Pentene

(b) Octene, like octane, indicates an eight-carbon chain. The chain contains a double bond between carbons 2 and 3 and a methyl group on carbon 7. Write eight carbon atoms in a row, place a double bond between carbons 2 and 3, and place a methyl group on carbon 7:

```
1   2   3   4   5   6   7   8
C─C═C─C─C─C─C─C
                        |
                        CH3
```

Now add hydrogen atoms to give each carbon atom four bonds. The complete formula is

```
CH3CH=CHCH2CH2CH2CHCH3
                  |
                  CH3
```

7-Methyl-2-octene

EXAMPLE 22.2 Name this compound:

$$\begin{array}{l} CH_3CH_2\underset{\displaystyle \| \atop \displaystyle CH_2}{C}CH_2CH_2CH_3 \end{array}$$

SOLUTION The longest carbon chain contains six carbons. However, since the compound is an alkene, we must include the double bond in the chain. The longest carbon chain containing the double bond has five carbons. Therefore, the compound is named as a pentene.

$$CH_3\overset{2}{C}H_2\underset{\displaystyle \| \atop \displaystyle {}^1CH_2}{\overset{3}{C}}\overset{4}{C}H_2\overset{5}{C}H_2CH_3$$

Attached to carbon 2 is an ethyl group. The name is 2-ethyl-1-pentene.

PRACTICE Write structural formulas for (a) 3-hexene, (b) 4-ethyl-2-heptene, and (c) 3,4-dimethyl-2-pentene.

Answers: (a) $CH_3CH_2CH{=}CHCH_2CH_3$
(b) $CH_3CH{=}CH\underset{\displaystyle | \atop \displaystyle CH_2CH_3}{C}HCH_2CH_2CH_3$
(c) $CH_3CH{=}\overset{\displaystyle CH_3 \atop \displaystyle |}{\underset{\displaystyle | \atop \displaystyle CH_3}{C}}CHCH_3$

PRACTICE Name these compounds.
(a) $CH_3\underset{\displaystyle | \atop \displaystyle CH_2CH_3}{C}{=}CHCH_3$ (b) $CH_3\underset{\displaystyle | \atop \displaystyle CH_3}{C}{=}CH\underset{\displaystyle | \atop \displaystyle CH_3}{C}HCH_3$

Answers: (a) 3-methyl-2-pentene (b) 2,4-dimethyl-2-pentene

22.3 GEOMETRIC ISOMERISM IN ALKENES

There are only two dichloroethanes: 1,1-dichloroethane ($CHCl_2CH_3$) and 1,2-dichloroethane (CH_2ClCH_2Cl). But surprisingly there are three dichloroethenes—namely, 1,1-dichloroethene ($CCl_2{=}CH_2$) and *two* isomers of 1,2-dichloroethene ($CHCl{=}CHCl$). There is only one 1,2-dichloroethane because carbon atoms can rotate freely about a single bond. Thus, the structural formulas I and II that follow represent the same compound. The chlorine atoms are simply shown in different relative positions in the two formulas due to rotation of the CH_2Cl group about the carbon–carbon single bond.

$$\begin{array}{cc} \text{Cl}\quad\text{Cl} & \text{Cl}\quad\text{H} \\ \text{H—C—C—H} & \text{H—C—C—H} \\ \text{H}\quad\text{H} & \text{H}\quad\text{Cl} \\ \text{I} & \text{II} \end{array}$$

1,2-Dichloroethane

Compounds containing a carbon–carbon double bond have restricted rotation about that double bond. Restricted rotation in a molecule gives rise to a type of isomerism known as *geometric isomerism.* Isomers that differ from each other only in the geometry of the molecules and not in the order of their atoms are known as **geometric isomers**. They are also called cis–trans isomers. Two isomers of 1,2-dichloroethene exist because of geometric isomerism.

geometric isomers

For a further explanation, let us look at the geometry of an ethylene molecule. This molecule is planar, or flat, with all six atoms lying in a single plane as in a rectangle:

$$\begin{array}{c} \text{H}\diagdown\quad\diagup\text{H} \\ \text{C=C} \\ \text{H}\diagup\quad\diagdown\text{H} \end{array}$$

Because the hydrogen atoms are identical, only one structural arrangement is possible for ethylene. But if one hydrogen atom on each carbon atom is replaced by chlorine, for example, two different geometric isomers are possible:

$$\begin{array}{c} \text{Cl}\diagdown\quad\diagup\text{Cl} \\ \text{C=C} \\ \text{H}\diagup\quad\diagdown\text{H} \end{array} \quad \text{and} \quad \begin{array}{c} \text{H}\diagdown\quad\diagup\text{Cl} \\ \text{C=C} \\ \text{Cl}\diagup\quad\diagdown\text{H} \end{array}$$

cis-1,2-Dichloroethene (bp = 60.1°C) — *trans*-1,2-Dichloroethene (bp = 48.4°C)

Both of these isomers are known and have been isolated. The fact that they have different boiling points as well as other different physical properties is proof that they are not the same compound. Note that, in naming these geometric isomers, the prefix *cis-* is used to designate the isomer having the substituent groups (chlorine atoms) on the same side of the double bond, and the prefix *trans-* is used to designate the isomer having the substituent groups on opposite sides of the double bond.

Molecules of *cis-* and *trans*-1,2-dichloroethene are not superimposable. That is, we cannot pick up one molecule and place it over the other in such a way that all the atoms in each molecule occupy the same relative positions in space. Nonsuperimposability is a general test for isomerism that all kinds of isomers must meet.

An alkene shows cis–trans isomerism when each of the carbon atoms of the double bond has two different kinds of groups attached to it.

$$\begin{array}{c} \text{a}\diagdown\quad\diagup\text{a} \\ \text{C=C} \\ \text{b}\diagup\quad\diagdown\text{b} \end{array} \qquad \begin{array}{c} \text{a}\diagdown\quad\diagup\text{b} \\ \text{C=C} \\ \text{b}\diagup\quad\diagdown\text{a} \end{array}$$

cis isomer — trans isomer

An alkene does not show cis–trans isomerism if one of the carbon atoms of the double bond has two identical groups attached to it. Thus there are no geometric isomers of ethene or propene.

$$\begin{matrix} H & & H \\ & C{=}C & \\ H & & H \end{matrix}\Big\} \longleftarrow \text{Two groups the same} \longrightarrow \Big\{\begin{matrix} H & & CH_3 \\ & C{=}C & \\ H & & H \end{matrix}$$

Four structural isomers of butene (C_4H_8) are known. Two of these, 1-butene and 2-methylpropene, do not show geometric isomerism.

$$\underset{\text{1-Butene}}{\begin{matrix} CH_3CH_2 & & H \\ & C{=}C & \\ H & & H \end{matrix}\Big\}} \longleftarrow \text{Two groups the same} \longrightarrow \underset{\text{2-Methylpropene}}{\Big\{\begin{matrix} CH_3 & & H \\ & C{=}C & \\ CH_3 & & H \end{matrix}}$$

The other two butenes are the cis–trans isomers shown below.

$$\underset{\textit{cis}\text{-2-Butene}}{\begin{matrix} CH_3 & & CH_3 \\ & C{=}C & \\ H & & H \end{matrix}} \quad \text{and} \quad \underset{\textit{trans}\text{-2-Butene}}{\begin{matrix} CH_3 & & H \\ & C{=}C & \\ H & & CH_3 \end{matrix}}$$

EXAMPLE 22.3

Draw structures for (a) *trans*-3-heptene and (b) *cis*-5-chloro-2-hexene.

SOLUTION

(a) The compound contains seven carbon atoms with a double bond between C 3 and C 4. First draw a C=C double bond in a planar arrangement:

$$\begin{matrix} \diagdown & & \diagup \\ & C{=}C & \\ \diagup & & \diagdown \end{matrix}$$

In the trans positions attach a two-carbon chain to one carbon atom and a three-carbon chain to the other carbon atom:

$$\begin{matrix} \overset{1}{C}{-}\overset{2}{C} & & \diagup \\ & \overset{3}{C}{=}\overset{4}{C} & \\ \diagup & & \overset{5}{C}{-}\overset{6}{C}{-}\overset{7}{C} \end{matrix}$$

Now attach hydrogen atoms to give each carbon atom four bonds. C 3 and C 4 need only one hydrogen atom apiece; these two hydrogen atoms are also trans to each other. The structure is

$$\begin{matrix} CH_3CH_2 & & H \\ & C{=}C & \\ H & & CH_2CH_2CH_3 \end{matrix}$$

(b) The compound contains six carbons, a double bond between C 2 and C 3, and a Cl atom on C 5. Draw a C=C double bond in a planar arrangement:

$$\begin{matrix} \diagdown & & \diagup \\ & C{=}C & \\ \diagup & & \diagdown \end{matrix}$$

In the cis positions attach a $-CH_3$ to one carbon and a three-carbon chain to the other carbon. Place a Cl on C 5.

$$\overset{1}{CH_3}\overset{2}{C}=\overset{3}{C}-\overset{4}{C}-\underset{Cl}{\overset{5}{C}}-\overset{6}{C}$$

Now add H atoms to give each carbon four bonds. C 2, C 3, and C 5 each need only one H atom. The structure is

$$\underset{H}{\overset{CH_3}{}}C=C\underset{H}{\overset{CH_2CHCH_3\ (Cl\ on\ CH)}{}}$$

EXAMPLE 22.4

Draw structural formulas and names for all the isomers of pentene, C_5H_{10}. Identify all geometric isomers.

SOLUTION

Start by drawing the isomers of the five-carbon chain, placing the C=C in all possible positions (there are two possible isomers). Then proceed to the four-carbon chains with a methyl group side chain. Locate the C=C and the CH_3 group in all possible positions. Check for duplications from the names of the compounds.

$CH_2=CHCH_2CH_2CH_3$ 1-Pentene

$CH_3CH=CHCH_2CH_3$ 2-Pentene

$CH_2=C(CH_3)CH_2CH_3$ 2-Methyl-1-butene

$CH_3C(CH_3)=CHCH_3$ 2-Methyl-2-butene

$CH_2=CHCH(CH_3)CH_3$ 3-Methyl-1-butene

Of these five compounds, only 2-pentene can have cis–trans isomers. All of the others have two identical groups on one of the carbon atoms of the double bond. Draw the cis–trans isomers.

$$\underset{CH_3}{\overset{H}{}}C=C\underset{CH_2CH_3}{\overset{H}{}} \qquad \underset{CH_3}{\overset{H}{}}C=C\underset{H}{\overset{CH_2CH_3}{}}$$

cis-2-Pentene *trans*-2-Pentene

EXAMPLE 22.5

Is the compound below the cis or the trans isomer?

$$\underset{H}{\overset{\overset{1}{CH_3}}{}}\overset{2}{C}=\overset{3}{C}\underset{\overset{4}{CH_2}\overset{5}{CH_3}}{\overset{CH_3}{}}$$

SOLUTION

In branched-chain alkenes, the cis–trans designation ordinarily is given to the structure containing the longest carbon chain that includes the carbon–carbon double bond. In this case the longest chain is the 2-pentene, in which the methyl group on carbon 2 and the ethyl group on carbon 3 are trans to each other. Thus the name of the compound shown is *trans*-3-methyl-2-pentene. The cis isomer is

$$\begin{array}{ccc} CH_3 & & CH_2CH_3 \\ & C{=}C & \\ H & & CH_3 \end{array}$$

Draw structures for (a) *cis* -3-methyl-2-pentene and (b) *trans* -2-bromo-2-pentene.

Answers: (a) $\begin{array}{ccc} CH_3CH_2 & & CH_3 \\ & C{=}C & \\ CH_3 & & H_3 \end{array}$ (b) $\begin{array}{ccc} Br & & CH_2CH_3 \\ & C{=}C & \\ CH_3 & & H \end{array}$

PRACTICE Determine whether geometric isomers exist for the following compounds. Draw structures for the *cis* and *trans* isomers.
(a) 1-chloro-2-methyl-2-butene (b) 3-hexene (c) 2,3-dimethyl-2-pentene

Answers: (a) Yes $\begin{array}{ccc} CH_3 & & CH_3 \\ & C{=}C & \\ CH_2Cl & & H \end{array}$ trans $\begin{array}{ccc} CH_3 & & H \\ & C{=}C & \\ CH_2Cl & & CH_3 \end{array}$ cis

(b) Yes $\begin{array}{ccc} CH_3CH_2 & & CH_2CH_3 \\ & C{=}C & \\ H & & H \end{array}$ cis $\begin{array}{ccc} H & & CH_2CH_3 \\ & C{=}C & \\ CH_3CH_2 & & H \end{array}$ trans

(c) No

PRACTICE Identify each of the following as the cis or trans isomer.

(a) $\begin{array}{ccc} H & & CH_3 \\ & C{=}C & \\ CH_3CH_2 & & Cl \end{array}$ (b) $\begin{array}{ccc} CH_3CH_2 & & CH_2CH_3 \\ & C{=}C & \\ CH_3 & & Br \end{array}$

Answers: (a) trans (b) cis

Enzymes (biological catalysts) test molecules for superimposability. In order to react with an enzyme, a molecule must twist and turn in solution until it achieves a specific shape. If this specific shape cannot be attained, the molecule won't react. For example, our bodies commonly use the cis isomer of unsaturated fatty acids but not the trans isomer in building more complex fats. That is why the great majority of unsaturated fatty acids in most cells are cis isomers.

$$\begin{array}{ccc} H & & CH_2CH_2CH_2CH_2CH_2CH_2CH_2COOH \\ & C & \\ & \| & \\ & C & \\ H & & CH_2CH_2CH_2CH_2CH_2CH_2CH_2CH_3 \end{array}$$

cis isomer

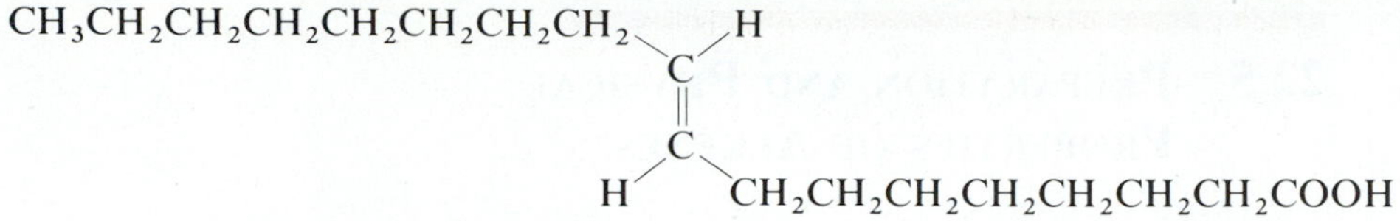

$$\begin{array}{c} CH_3CH_2CH_2CH_2CH_2CH_2CH_2CH_2 \quad\; H \\ \diagdown\; / \\ C \\ \| \\ C \\ / \;\diagdown \\ H \quad\; CH_2CH_2CH_2CH_2CH_2CH_2CH_2COOH \end{array}$$

trans isomer

Enzymes differentiate between the "hairpin" shape of the cis isomer and the more linear trans form. Although these molecules are constantly tumbling and vibrating in solution, they cannot normally attain the same shape due to restricted rotation about the carbon–carbon double bond. Because the two shapes cannot be superimposed, the two isomers are treated differently by the body's enzymes.

22.4 CYCLOALKENES

As the name implies, **cycloalkenes** are cyclic compounds that contain a carbon–carbon double bond in the ring. The two most common cycloalkenes are cyclopentene and cyclohexene. The double bond may be placed between any two carbon atoms in cyclopentene and cyclohexene.

cycloalkenes

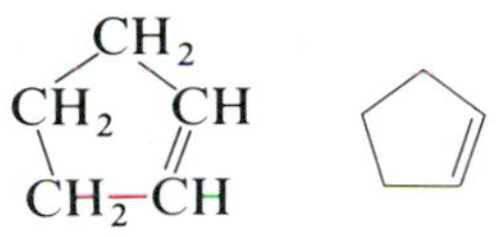

Cyclopentene (C_5H_8)

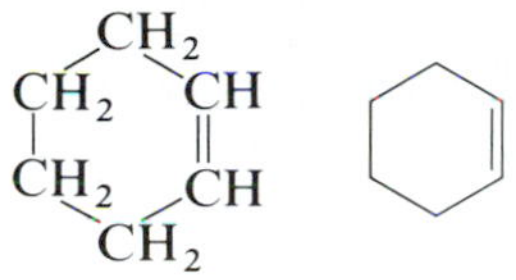

Cyclohexene (C_6H_{10})

In cycloalkenes the carbons of the double bond are assigned numbers 1 and 2. Thus the positions of the double bond need not be indicated in the name of the compound. Other substituents on the ring are named in the usual manner, and their positions on the ring is indicated with the smallest possible numbers.

In the examples below note that a Cl or a CH_3 has replaced one H atom in the molecule. The ring is numbered either clockwise or counterclockwise starting with the carbon–carbon double bond so that the substituted group(s) have the smallest possible numbers.

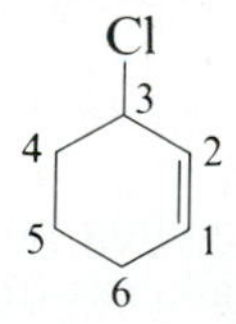

3-Chlorocyclohexene (C_6H_9Cl)

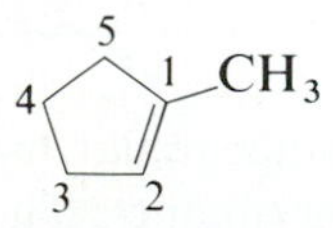

1-Methylcyclopentene ($C_5H_7CH_3$)

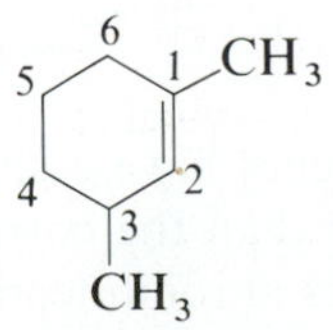

1,3-Dimethylcyclohexene [$C_6H_8(CH_3)_2$]

No cis–trans designation is necessary for cycloalkenes containing up to seven carbon atoms in the ring. Cyclooctene has been shown to exist in both cis and trans forms.

22.5 Preparation and Physical Properties of Alkenes

Cracking

cracking

Ethylene can be produced by the cracking of petroleum. **Cracking**, or pyrolysis, is the process in which saturated hydrocarbons are heated to very high temperatures in the presence of a catalyst (usually silica-alumina). This results in the breaking of large molecules into smaller ones, with the elimination of hydrogen, forming alkenes and small hydrocarbons like methane and ethane. Unfortunately, cracking always results in mixtures of products and is therefore not used often in the laboratory.

Dehydration of Alcohols

dehydration

Dehydration involves the elimination of a molecule of water from a reactant molecule. Dehydration reactions are very common in organic chemistry as well as in biochemistry.

To produce an alkene by dehydration, an alcohol is heated in the presence of concentrated sulfuric acid.

$$CH_3\overset{\overset{\large H}{|}}{\underset{\underset{\large H}{|}}{C}}-\overset{\overset{\large H}{|}}{\underset{\underset{\large OH}{|}}{C}}CH_3 \xrightarrow[\Delta]{\text{con } H_2SO_4} CH_3\overset{\overset{\large H}{|}}{C}=\overset{\overset{\large H}{|}}{C}CH_3 + H_2O$$

Physical Properties

Alkenes have physical properties very similar to the corresponding alkanes. This is not surprising since the difference between an alkane and an alkene is simply two hydrogen atoms.

General formula for alkanes	C_nH_{2n+2}
General formula for alkenes	C_nH_{2n}

Since alkenes have a slightly smaller molar mass, their boiling points are slightly lower than the corresponding alkanes. The smaller alkenes (to 5 carbons) are gases at room temperature. As the chain lengthens (5–17 carbons) the alkenes are liquid, and above 17 carbons they are solid. The alkenes are nonpolar, like the other hydrocarbons, and so are insoluble in water but are soluble in organic solvents. The densities of most alkenes are much less than water. Table 22.1 shows the properties of some alkenes. Notice that isomers (C_4H_8) have similar boiling points although they differ significantly in melting points. The different shapes of the isomers fit into crystals in significantly different ways.

TABLE 22.1 Physical Properties of Alkenes

Molecular formula	Structural formula	IUPAC name	Density (g/mL)	Melting point (°C)	Boiling point (°C)
C_2H_4	$CH_2{=}CH_2$	Ethene	—	−169	−104
C_3H_6	$CH_3CH{=}CH_2$	Propene	—	−185	−47
C_4H_8	$CH_3CH_2CH{=}CH_2$	1-Butene	0.595	−130	−6
C_4H_8	$(CH_3)_2C{=}CH_2$	2-Methylpropene	0.594	−14	−7
C_5H_{10}	$CH_3(CH_2)_2CH{=}CH_2$	1-Pentene	0.641	−138	30

22.6 Chemical Properties of Alkenes

The alkenes are much more reactive than the corresponding alkanes. This greater reactivity is due to the carbon–carbon double bonds.

Addition

A reaction in which two substances join together to produce one compound is called an **addition reaction**. Addition at the carbon–carbon double bond is the most common reaction of alkenes. Hydrogen, halogens (Cl_2 or Br_2), hydrogen halides, sulfuric acid, and water are some of the reagents that can be added to unsaturated hydrocarbons. Ethylene (ethene), for example, reacts in the presence of a platinum catalyst in this fashion:

addition reaction

$$\underset{\text{Ethylene}}{CH_2{=}CH_2} + H_2 \xrightarrow[\text{1 atm}]{\text{Pt, 25°C}} \underset{\text{Ethane}}{CH_3{-}CH_3}$$

$$CH_2{=}CH_2 + Br{-}Br \longrightarrow \underset{\text{1,2-Dibromoethane}}{CH_2Br{-}CH_2Br}$$

The product in the foregoing reaction is colorless. Therefore the disappearance of the reddish-brown bromine color provides visible evidence of reaction. Other reactions of ethylene follow.

$$CH_2{=}CH_2 + HCl \longrightarrow \underset{\text{(Ethyl chloride)}}{\underset{\text{Chloroethane}}{CH_3CH_2Cl}}$$

$$CH_2{=}CH_2 + \underset{\text{Sulfuric acid}}{HOSO_3H(\textit{conc.})} \longrightarrow \underset{\text{Ethyl hydrogen sulfate}}{CH_3CH_2OSO_3H}$$

$$CH_2{=}CH_2 + HOH \xrightarrow{H^+} \underset{\text{(Ethyl alcohol)}}{\underset{\text{Ethanol}}{CH_3CH_2OH}}$$

The H^+ indicates that the reaction is carried out under acidic conditions.

Note that the double bond is broken and the unsaturated alkene molecules become saturated by an addition reaction.

The preceding examples dealt with ethylene, but reactions of this kind can be made to occur on almost any molecule that contains a carbon–carbon double bond. If a symmetrical molecule such as Cl_2 is added to propene, only one product, 1,2-dichloropropane, is formed:

$$CH_2{=}CH{-}CH_3 + Cl_2 \longrightarrow \underset{\text{1,2-Dichloropropane}}{CH_2Cl{-}CHCl{-}CH_3}$$

But if an unsymmetrical molecule such as HCl is added to propene, two products are theoretically possible, depending upon which carbon atom adds the hydrogen. The two possible products are 1-chloropropane and 2-chloropropane. Experimentally we find that 2-chloropropane is formed almost exclusively:

$$CH_3{-}CH{=}CH_2 + HCl \begin{cases} \longrightarrow CH_3CHClCH_3 \ \text{(About 100\%)} \\ \not\longrightarrow CH_3CH_2CH_2Cl \ \text{(Trace)} \end{cases}$$

A single product is obtained because the reaction proceeds stepwise according to the following mechanism:

1. A proton (H^+) from HCl bonds to the number 1 carbon of propene utilizing the pi bond electrons. The intermediate formed is a positively charged alkyl group, or carbocation. The positive charge is localized on the number 2 carbon atom of this carbocation.

$$CH_2{=}CH{-}CH_3 + HCl \longrightarrow \underset{\text{Isopropyl carbocation}}{CH_3{-}\overset{+}{C}H{-}CH_3} + Cl^-$$

2. The chloride ion then adds to the positively charged carbon atom to form a molecule of 2-chloropropane:

$$CH_3{-}\overset{+}{C}H{-}CH_3 + Cl^- \longrightarrow \underset{\text{2-Chloropropane}}{CH_3{-}CHCl{-}CH_3}$$

An ion in which a carbon atom has a positive charge is known as a **carbocation**. There are four types of carbocations: methyl, primary (1°), secondary (2°), and tertiary (3°). Examples of these four types follow:

carbocation

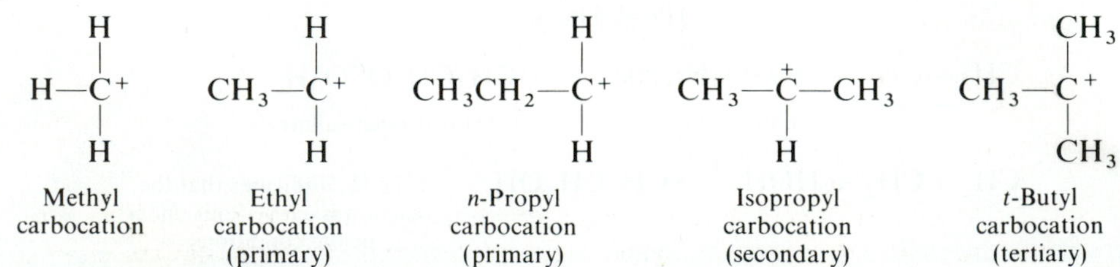

Methyl carbocation | Ethyl carbocation (primary) | *n*-Propyl carbocation (primary) | Isopropyl carbocation (secondary) | *t*-Butyl carbocation (tertiary)

A carbon atom is designated as primary if it is bonded to one carbon atom, secondary if it is bonded to two carbon atoms, and tertiary if it is bonded to three carbon atoms. In a primary carbocation the positive carbon atom is bonded to only one carbon atom. In a secondary carbocation the positive carbon atom is bonded to two carbon atoms. In a tertiary carbocation the positive carbon atom is bonded to three carbon atoms.

The order of stability of carbocations and hence the ease with which they are formed is tertiary > secondary > primary. Thus, in the reaction of propene and HCl, isopropyl carbocation (secondary) is formed as an intermediate in preference to *n*-propyl carbocation (primary).

Stability of carbocations: $3^\circ > 2^\circ > 1^\circ > \overset{+}{C}H_3$

In the middle of the 19th century, a Russian chemist, V. Markovnikov, observed reactions of this kind, and in 1869 he formulated a useful generalization now known as **Markovnikov's rule**. This rule in essence states:

Markovnikov's rule

When an unsymmetrical molecule such as HX(HCl) adds to a carbon–carbon double bond, the hydrogen from HX goes to the carbon atom that has the greater number of hydrogen atoms.

As you can see, the addition of HCl to propene, discussed above, follows Markovnikov's rule. The addition of HI to 2-methylpropene (isobutylene) is another example illustrating this rule:

$$CH_3-\underset{}{\overset{CH_3}{\overset{|}{C}}}=CH_2 + HI \longrightarrow CH_3-\underset{\underset{I}{|}}{\overset{CH_3}{\overset{|}{C}}}-CH_3$$

2-Iodo-2-methylpropane
(*tert*-Butyl iodide)

General rules of this kind are useful in predicting the products of reactions. However, exceptions are known for most such rules.

EXAMPLE 22.6

Write formulas for the organic products formed when 2-methyl-1-butene reacts with (a) H_2, Pt/25°C; (b) Cl_2; (c) HCl; (d) H_2O, H^+.

SOLUTION

First write the formula for 2-methyl-1-butene:

$$\overset{1}{C}H_2=\overset{CH_3}{\overset{|}{\overset{2}{C}}}-\overset{3}{C}H_2-\overset{4}{C}H_3$$

(a) The double bond is broken when a hydrogen molecule adds. One H atom adds to each carbon atom of the double bond. Platinum, Pt, is a necessary catalyst in this reaction. The product is

$$\begin{array}{l} \quad CH_3 \\ \quad | \\ CH_3CHCH_2CH_3 \end{array}$$

2-Methylbutane

(b) The Cl_2 molecule adds to the carbons of the double bond. One Cl atom adds to each carbon atom of the double bond. The product is

$$\begin{array}{l} \qquad CH_3 \\ \qquad | \\ CH_2CCH_2CH_3 \\ | \quad\; | \\ Cl \quad Cl \end{array}$$

1,2-Dichloro-2-methylbutane

(c) HCl adds to the double bond according to Markovnikov's rule. The H^+ goes to C 1 (the more stable 3° carbocation is formed as an intermediate product), and the Cl^- goes to C 2. The product is

$$\begin{array}{l} \qquad CH_3 \\ \qquad | \\ CH_3CCH_2CH_3 \\ \qquad | \\ \qquad Cl \end{array}$$

2-Chloro-2-methylbutane

(d) The net result of this reaction is a molecule of water added across the double bond. The H adds to C 1 (the carbon with the greater number of hydrogen atoms), and the OH adds to C 2 (the carbon of the double bond with the lesser number of hydrogen atoms). The product is

$$\begin{array}{l} \qquad CH_3 \\ \qquad | \\ CH_3CCH_2CH_3 \\ \qquad | \\ \qquad OH \end{array}$$

2-Methyl-2-butanol
(an alcohol)

EXAMPLE 22.7 Write equations for the addition of HCl to (a) 1-pentene and (b) 2-pentene.

SOLUTION

(a) In the case of 1-pentene, $CH_3CH_2CH_2CH{=}CH_2$, the proton from HCl adds to carbon 1 to give the more stable secondary carbocation, followed by the addition of Cl^- to give the product 2-chloropentane. The addition is directly in accordance with Markovnikov's rule.

$$CH_3CH_2CH_2CH{=}CH_2 + HCl \longrightarrow CH_3CH_2CH_2CHClCH_3$$

2-Chloropentane

(b) In 2-pentene, $CH_3CH_2CH{=}CHCH_3$, each carbon of the double bond has one hydrogen atom, and the addition of a proton to either one forms a secondary carbocation. After the addition of Cl^-, the result is two isomeric products that are formed in almost equal quantities:

$$CH_3CH_2CH{=}CHCH_3 + HCl \longrightarrow \underset{\text{2-Chloropentane}}{CH_3CH_2CH_2CHClCH_3} + \underset{\text{3-Chloropentane}}{CH_3CH_2CHClCH_2CH_3}$$

PRACTICE Write formulas for the organic products formed when 3-methyl-2-pentene reacts with (a) H_2/Pt, (b) Br_2, (c) HCl, and (d) H_2O, H^+

Answers:

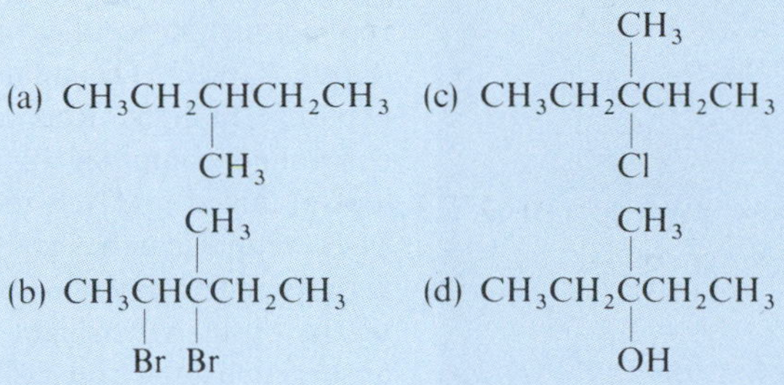

(a) $CH_3CH_2CH(CH_3)CH_2CH_3$ (c) $CH_3CH_2CCl(CH_3)CH_2CH_3$

(b) $CH_3CHBrCBr(CH_3)CH_2CH_3$ (d) $CH_3CH_2C(OH)(CH_3)CH_2CH_3$

PRACTICE Write equations for (a) addition of water to 1-methylcyclopentene and (b) addition of HI to 2-methyl-2-butene.

Answers:

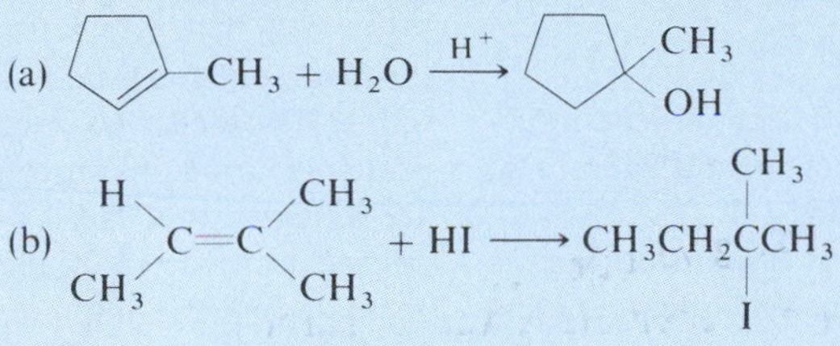

(a) 1-methylcyclopentene $-CH_3 + H_2O \xrightarrow{H^+}$ cyclopentane bearing CH_3 and OH on the same carbon

(b) $(H)(CH_3)C{=}C(CH_3)(CH_3) + HI \longrightarrow CH_3CH_2CI(CH_3)CH_3$

Oxidation

Another typical reaction of alkenes is oxidation at the double bond. For example, when shaken with a cold, dilute solution of potassium permanganate ($KMnO_4$), an alkene is converted to a glycol (glycols are dihydroxy alcohols). Ethylene reacts in this manner:

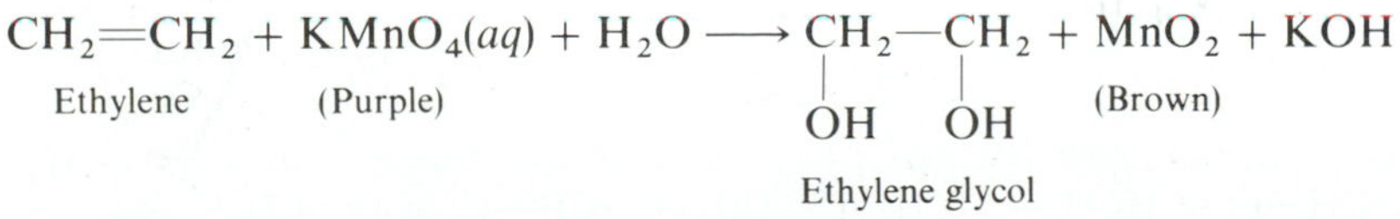

$$\underset{\text{Ethylene}}{CH_2{=}CH_2} + \underset{\text{(Purple)}}{KMnO_4(aq)} + H_2O \longrightarrow \underset{\text{Ethylene glycol}}{\underset{OH\quad OH}{CH_2{-}CH_2}} + \underset{\text{(Brown)}}{MnO_2} + KOH$$

The *Baeyer test* makes use of this reaction to detect or confirm the presence of double (or triple) bonds in hydrocarbons. Evidence of reaction (positive Baeyer test) is the disappearance of the purple color of permanganate ions. The Baeyer test is not specific for detecting unsaturation in hydrocarbons, since other classes of compounds may also give a positive Baeyer test.

Chemistry in Action

Molecules of Sight and Smell

Complex alkenes containing multiple double bonds occur throughout nature. Many of these molecules are built from the molecule isoprene:

$$CH_2{=}C(CH_3){-}CH{=}CH_2$$

Isoprene
(2-Methyl-1,3-butadiene)

Isoprene is a conjugated diene, a molecule in which two double bonds are separated by a single bond. Conjugated double bonds are very stable and so occur in many biochemical molecules. In general isoprene molecules do not appear alone in nature. Compounds generally contain several isoprene units.

Essential oils, substances which give plants their pleasant odors, are frequently extracted from the plant and used in cosmetics, flavorings and perfumes (see Chemistry in Action, Chapter 7). They are all part of a class of compounds known as the terpenes. These molecules are combinations of isoprene and therefore contain a multiple of 5 carbons.

▲ **Scanning electron microscope of rods and cones in the human eye.**

Two of the larger terpene molecules are β-carotene and vitamin A, illustrated below.

β-Carotene is responsible for the color of carrots and tomatoes. The carotenes are an important intermediate in the formation of vitamin A. In animals vitamin A is produced by the splitting of β-carotene in the small intestine. Vitamin A is a precursor to 11-*cis*-retinal, a compound essential to vision.

The retina of the eye contains colored compounds called the visual pigments in a receptor known as a rod. In the dark these pigments are rose-colored, hence their name rhodopsin (Greek). The color fades on exposure to light. Rhodopsin is a compound composed of a protein (opsin) linked to 11-*cis*-retinal. The 11-*cis*-retinal has a shape that fits perfectly into a cavity on the surface of the opsin. When light energy is absorbed by the rod cells, the cis bond at the 11-carbon atom is broken and reformed as the trans isomer. Since the shape of the molecule is now different, it no longer fits into the cavity on the opsin. The 11-*trans*-retinal is now cleaved from the opsin protein. During this process enzymes are activated that cause a change in electrical character of the rod and generate a nerve impulse, which is perceived as light by the brain. The 11-*trans*-retinal is reconverted to 11-*cis*-retinal, which can then recombine with the opsin. The process is summarized in the figure on the facing page.

(Ring with CH_3, CH_3, CH_3) $-CH{+}CH{-}C(CH_3){=}CH{-}CH{+}CH{-}C(CH_3){=}CH{-}CH_2OH$

Vitamin A

(Ring with CH_3, CH_3, CH_3) $-CH{+}CH{-}C(CH_3){=}CHCH{+}CH{-}C(CH_3){=}CHCH{+}CHCH{=}C(CH_3){-}CH{+}CHCH{=}C(CH_3){-}CH{+}CH-$ (Ring with CH_3, CH_3, CH_3)

β-Carotene

▲ **Structure of vitamin A and β-carotene. The colored lines indicate the isoprene units in each molecule.**

(continued)

(continued)

◄ **Chemistry and vision.**

all-*trans*-retinal
+
Opsin
Isomerization
11-*cis*-retinal
Bonding reaction
Opsin (a protein)
$-H_2O$
(11-*cis*)
RHODOPSIN
Opsin
Light
(all-*trans*)
Cleavage reaction
Opsin
Nerve impulse
H_2O

Carbon–carbon double bonds are found in many different kinds of molecules. Most of these substances react with potassium permanganate and undergo somewhat similar reactions with other oxidizing agents including oxygen in the air and, especially, with ozone. Such reactions are frequently troublesome. For example, premature aging and cracking of tires in smoggy atmospheres occur because ozone attacks the double bonds in rubber molecules. Cooking oils and fats sometimes develop disagreeable odors and flavors because the oxygen of the air reacts with the double bonds present in these materials. Potato chips, because of their large surface area, are especially subject to flavor damage caused by oxidation of the unsaturated cooking oils that they contain.

▲ **The cracking on this automobile tire is the result of ozone attacking the rubber molecules.**

22.7 Alkynes: Nomenclature and Preparation

Nomenclature

The procedure for naming alkynes is the same as that for alkenes, but the ending used is *-yne* to indicate the presence of a triple bond. The smaller members of the series are often referred to by their common names. Table 22.2 lists names and formulas for some common alkynes.

TABLE 22.2 Nomenclature for Some Common Alkynes

Molecular formula	Structural formula	Common name	IUPAC name
C_2H_2	$H{-}C{\equiv}C{-}H$	Acetylene	Ethyne
C_3H_4	$CH_3{-}C{\equiv}C{-}H$	Methylacetylene	Propyne
C_4H_6	$CH_3CH_2{-}C{\equiv}C{-}H$	Ethylacetylene	1-Butyne
C_4H_6	$CH_3{-}C{\equiv}C{-}CH_3$	Dimethylacetylene	2-Butyne

Preparation

Although triple bonds are very reactive it is relatively easy to synthesize alkynes. Acetylene can be prepared inexpensively from calcium carbide and water.

$$CaC_2 + 2\,H_2O \longrightarrow HC{\equiv}CH + Ca(OH)_2$$

Acetylene is also prepared by the cracking of methane in an electric arc.

$$2\,CH_4 \xrightarrow{1500^\circ C} HC{\equiv}CH + 3\,H_2$$

22.8 Physical and Chemical Properties of Alkynes

Physical Properties

Acetylene is a colorless gas, with little odor when pure. Its common disagreeable odor is the result of impurities (usually PH_3). Acetylene is insoluble in water and a gas at normal temperature and pressure (bp = -84°C). As a liquid acetylene is very sensitive and may decompose violently (explode) spontaneously or from a slight shock.

$$HC{\equiv}CH \longrightarrow H_2 + 2\,C + 227\text{ kJ }(54.3\text{ kcal})$$

To eliminate the danger of explosions, acetylene is dissolved under pressure in acetone and is packed in cylinders that contain a porous inert material.

Chemical Properties

Acetylene is used mainly (1) as fuel for oxyacetylene cutting and welding torches and (2) as an intermediate in the manufacture of other substances. Both uses are dependent upon the great reactivity of acetylene. Acetylene and oxygen mixtures produce flame temperatures of about 2800°C. Acetylene readily undergoes addition reactions rather similar to those of ethylene. It reacts with chlorine and bromine and decolorizes a permanganate solution (Baeyer's test). Either one or two molecules of bromine or chlorine can be added:

$$HC{\equiv}CH + Br_2 \longrightarrow \underset{\text{1,2-Dibromoethene}}{CHBr{=}CHBr}$$

or

$$HC \equiv CH + 2\,Br_2 \longrightarrow \underset{\text{1,1,2,2-Tetrabromoethane}}{CHBr_2{-}CHBr_2}$$

It is apparent that either unsaturated or saturated compounds can be obtained as addition products of acetylene. Often, unsaturated compounds capable of undergoing further reactions are made from acetylene. For example, vinyl chloride, which is used to make the plastic polyvinyl chloride (PVC), can be made by simple addition of HCl to acetylene:

$$CH \equiv CH + HCl \longrightarrow \underset{\substack{\text{Chloroethene} \\ \text{(Vinyl chloride)}}}{CH_2{=}CHCl}$$

(*Note:* The common name for the $CH_2{=}CH{-}$group is *vinyl.*) If the reaction is not properly controlled, another HCl adds to the chloroethene:

$$CH_2{=}CHCl + HCl \longrightarrow \underset{\text{1,1-Dichloroethane}}{CH_3CHCl_2}$$

Hydrogen chloride reacts with other alkynes in a similar fashion to form substituted alkenes. The addition follows Markovnikov's rule. Alkynes can react with 1 or 2 moles of HCl. Consider the reaction of propyne with HCl:

$$CH_3C{\equiv}CH + HCl \longrightarrow \underset{\text{2-Chloropropene}}{CH_3CCl{=}CH_2}$$

$$CH_3CCl{=}CH_2 + HCl \longrightarrow \underset{\text{2,2-Dichloropropane}}{CH_3CCl_2CH_3}$$

There are certain unique reactions for the alkynes. They are capable of reacting at times when alkenes will not. Acetylene, with certain catalysts, reacts with HCN to form cyanoethylene ($CH_2{=}CHCN$). This chemical is used industrially to manufacture Orlon, a polymer commonly found in clothing. It is also used to form the superabsorbants, which are capable of retaining up to 2000 times their mass of distilled water. These superabsorbants are found in disposable diapers as well as in soil additives to retain water.

The reactions of other alkynes are similar to those of acetylene. Although many other alkynes are known, acetylene is by far the most important industrially.

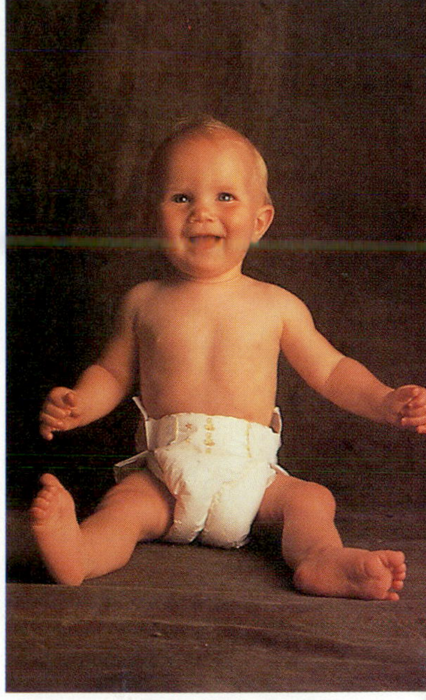

▲ **Superabsorbants are used in disposable diapers to increase the amount of liquid the diaper will absorb without leaking.**

22.9 Aromatic Hydrocarbons: Structure

Benzene and all substances that have structures and chemical properties resembling benzene are classified as **aromatic compounds**. The word *aromatic* originally referred to the rather pleasant odor possessed by many of these substances, but this meaning has been dropped. Benzene, the parent substance of the aromatic hydrocarbons, was first isolated by Michael Faraday in 1825; its correct molecular formula, C_6H_6, was established a few years later. The establishment

aromatic compounds

of a reasonable structural formula that would account for the properties of benzene was a very difficult problem for chemists in the mid-19th century.

Finally, in 1865, August Kekulé proposed that the carbon atoms in a benzene molecule are arranged in a six-membered ring with one hydrogen atom bonded to each carbon atom and with three carbon–carbon double bonds, as shown in the following formula:

```
        H
        |
  H   C   H
    C   C
    ‖   |
    C   C
  H   C   H
        |
        H
```

Kekulé soon realized that there should be two dibromobenzenes, based on double- and single-bond positions relative to the two bromine atoms:

```
        Br                          Br
        |                           |
  H   C   Br                  H   C   Br
    C   C         and           C   C
    ‖   |                       |   ‖
    C   C                       C   C
  H   C   H                   H   C   H
        |                           |
        H                           H
```

Since only one dibromobenzene (with bromine atoms on adjacent carbons) could be produced, Kekulé suggested that the double bonds are in rapid oscillation within the molecule. He therefore proposed that the structure of benzene could be represented in this fashion:

```
        H                           H
        |                           |
  H   C   H                   H   C   H
    C   C                       C   C
    ‖   |          ⇌            |   ‖
    C   C                       C   C
  H   C   H                   H   C   H
        |                           |
        H                           H
```

Kekulé's concepts are a landmark in the history of chemistry. They are the basis of the best representation of the benzene molecule devised in the 19th century, and they mark the beginning of our understanding of structure in aromatic compounds.

Kekulé's formulas have one serious shortcoming: They represent benzene and related substances as highly unsaturated compounds. Yet benzene does not react like a typical alkene (olefin); it does not decolorize bromine solutions rapidly, nor does it destroy the purple color of permanganate ions (Baeyer's test). Instead, the chemical behavior of benzene resembles that of an alkane. Its typical reactions are the substitution type, wherein a hydrogen atom is replaced

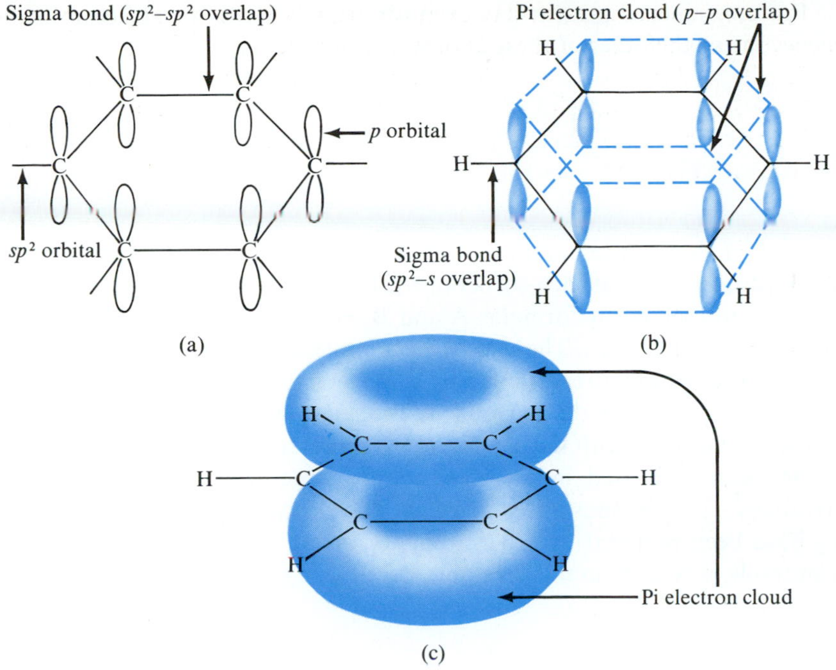

◀ **FIGURE 22.5**
Bonding in a benzene molecule: (a) sp^2–sp^2 orbital overlap to form the carbon ring structure; (b) carbon–hydrogen bonds formed by sp^2–s orbital overlap and overlapping of p orbitals; (c) pi electron clouds above and below the plane of the carbon ring.

by some other group; for example,

$$C_6H_6 + Cl_2 \xrightarrow{Fe} C_6H_5Cl + HCl$$

This problem was not fully resolved until the technique of X-ray diffraction, developed in the years following 1912, permitted us to determine the actual distances between the nuclei of carbon atoms in molecules. The center-to-center distances between carbon atoms in different kinds of hydrocarbon molecules are

Ethane (single bond)	0.154 nm
Ethylene (double bond)	0.134 nm
Benzene	0.139 nm

Because only one carbon–carbon distance (bond length) is found in benzene, it is apparent that alternating single and double bonds do not exist in the benzene molecule.

Modern theory accounts for the structure of the benzene molecule in this way: The orbital hybridization of the carbon atoms is sp^2 (see structure of ethylene in Section 22.1). A planar hexagonal ring is formed by the overlapping of two sp^2 orbitals on each of six carbon atoms. The other sp^2 orbital on each carbon atom overlaps an s orbital of a hydrogen atom, bonding the carbon to the hydrogen by a sigma bond. The remaining six p orbitals, one on each carbon atom, overlap each other and form doughnut-shaped pi electron clouds above and below the plane of the ring (see Figure 22.5). The electrons composing these clouds are not attached to particular carbon atoms but are delocalized and associated with the entire molecule. This electronic structure imparts unusual stability to benzene and is responsible for many of the characteristic properties of aromatic compounds.

For convenience present-day chemists usually write the structure of benzene as one or the other of these abbreviated forms:

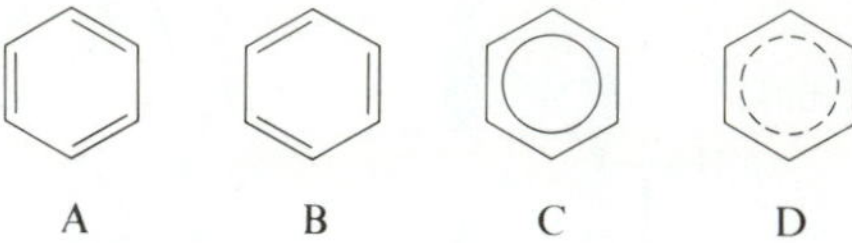

In all of these representations, it is understood that there is a carbon atom and a hydrogen atom at each corner of the hexagon. The classical Kekulé structures are represented by formulas A and B. However, neither of these Kekulé structures actually exists. The real benzene molecule is a hybrid of these structures and is commonly represented by either formula C or D. The circle or the dotted circle represents the pi cloud of electrons, as shown in Figure 22.5. We will use the hexagon with the solid circle to represent a benzene ring.

Hexagons are used in representing the structural formulas of benzene derivatives—that is, substances in which one or more hydrogen atoms in the ring have been replaced by other atoms or groups. Chlorobenzene (C_6H_5Cl), for example, is written in this fashion:

Cl

Chlorobenzene, C_6H_5Cl

This notation indicates that the chlorine atom has replaced a hydrogen atom and is bonded directly to a carbon atom in the ring. Thus, the correct formula for chlorobenzene is C_6H_5Cl, not C_6H_6Cl.

22.10 Naming Aromatic Compounds

A substituted benzene is derived by replacing one or more hydrogen atoms of benzene by another atom or group of atoms. Thus, a monosubstituted benzene has the formula C_6H_5G, where G is the group replacing a hydrogen atom.

Monosubstituted Benzenes

Some monosubstituted benzenes are named by adding the name of the substituent group as a prefix to the word *benzene*. The name is written as one word. Note that the position of the substituent is not important here as all the positions in the hexagon are equivalent. Several examples follow:

O_2N — Nitrobenzene

CH_2CH_3 — Ethylbenzene

Cl — Chlorobenzene

Br — Bromobenzene

Certain monosubstituted benzenes have special names. These are used as parent names for further substituted compounds, so they should be learned.

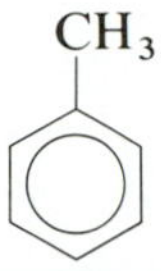

CH_3 OH $CH{=}CH_2$

Toluene (Methylbenzene)

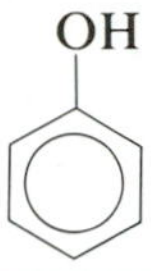

Phenol (Hydroxybenzene)

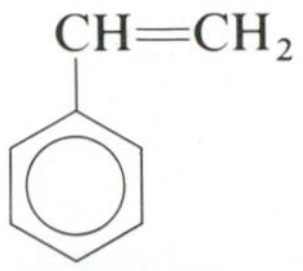

Styrene (Vinylbenzene)

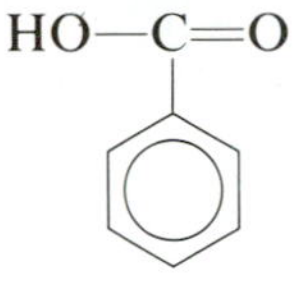

HO—C=O H—C=O NH_2

Benzoic acid (Benzene carboxylic acid)

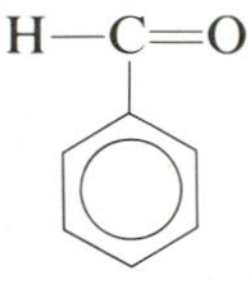

Benzaldehyde (Benzene carboxaldehyde)

Aniline (Aminobenzene)

The C_6H_5— group is known as the phenyl group, and the name *phenyl* is used to name compounds that cannot easily be named as benzene derivatives. For example, the following compounds are named as derivatives of alkanes:

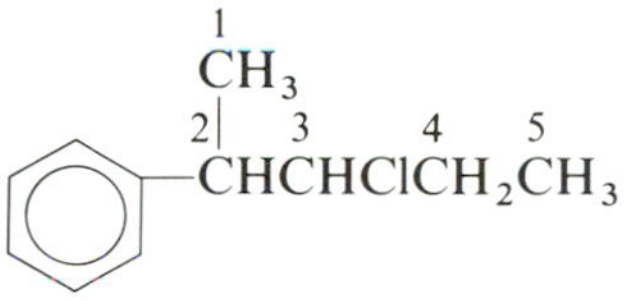

$\overset{1}{CH_3}$ — $\overset{2}{C}H\overset{3}{C}HCl\overset{4}{C}H_2\overset{5}{C}H_3$

$-CH_2-$

3-Chloro-2-phenylpentane

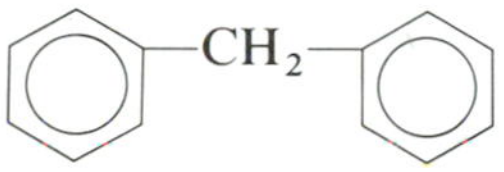

Diphenylmethane

Disubstituted Benzenes

When two substituent groups replace two hydrogen atoms in a benzene molecule, three different isomeric compounds are possible. The prefixes *ortho-*, *meta-*, and *para-* (abbreviated *o-*, *m-*, and *p-*) are used to name these disubstituted benzenes in one nomenclature system. Ortho designates 1,2 disubstitution, meta represents 1,3 disubstitution, and para designates 1,4 disubstitution.

Consider the dichlorobenzenes, $C_6H_4Cl_2$. Note that the three isomers have different physical properties, indicating that they are truly different substances:

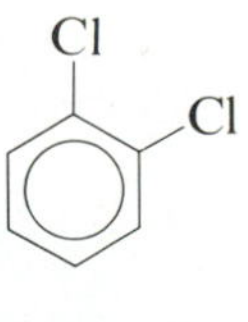

ortho-Dichlorobenzene (1,2-Dichlorobenzene) (mp −17.2°C, bp 180.4°C)

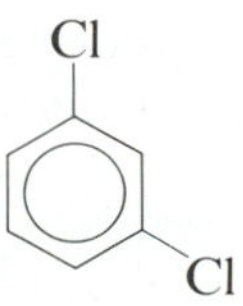

meta-Dichlorobenzene (1,3-Dichlorobenzene) (mp −24.8°C, bp 172°C)

para-Dichlorobenzene (1,4-Dichlorobenzene) (mp 53.1°C, bp 174.4°C)

When the two substituents are different and neither is part of a compound with a special name, the names of the two substituents are given in alphabetical order, followed by the word *benzene*. For example,

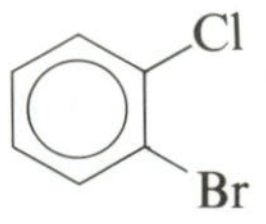

NO_2 / Cl / Br / CH_2CH_3

o-Bromochlorobenzene 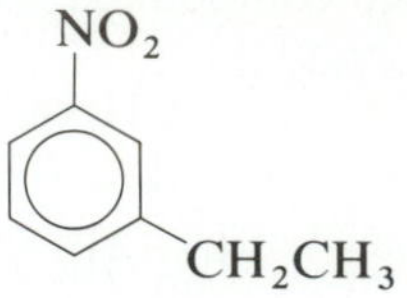 *m*-Ethylnitrobenzene

The dimethylbenzenes have the special name *xylene*.

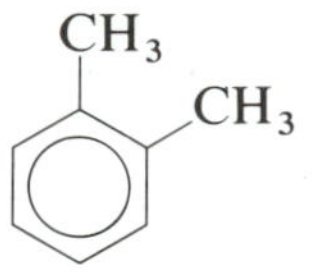

CH_3 CH_3 CH_3 CH_3 CH_3 CH_3

ortho-Xylene 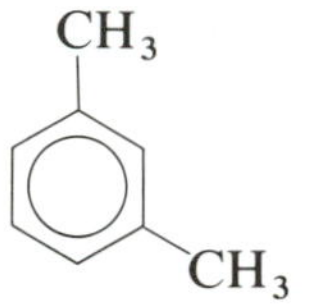 *meta*-Xylene  *para*-Xylene

When one of the substituents corresponds to a monosubstituted benzene that has a special name, the disubstituted compound is named as a derivative of that parent compound. In the following examples the parent compounds are phenol, aniline, and toluene:

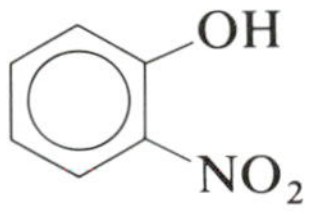

NH_2 CH_3 OH NO_2 Br NO_2

o-Nitrophenol  *p*-Bromoaniline 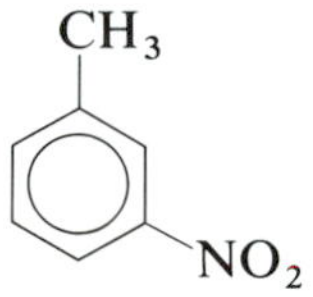 *m*-Nitrotoluene

The numbering system as described under polysubstituted benzenes is also used to name disubstituted benzenes.

Polysubstituted Benzenes

When there are more than two substituents on a benzene ring, the carbon atoms in the ring are numbered starting at one of the substituted groups. Numbering may be either clockwise or counterclockwise but must be done in the direction that gives the lowest possible numbers to the substituent groups. When the compound is named as a derivative of one of the special parent compounds, the substituent of the parent compound is considered to be on carbon 1 of the ring (the CH_3 group is on C 1 in 2,4,6-trinitrotoluene). The following examples illustrate this system:

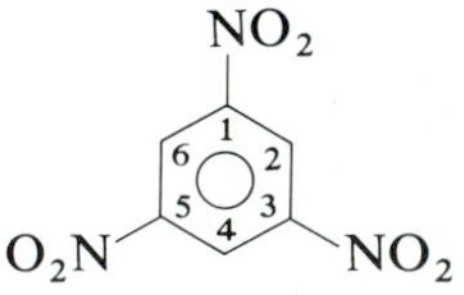

NO_2 / O_2N / NO_2 — Br / Br / Br — CH_3 / O_2N / NO_2 / NO_2 — Br / OH / Cl

1,3,5-Trinitrobenzene

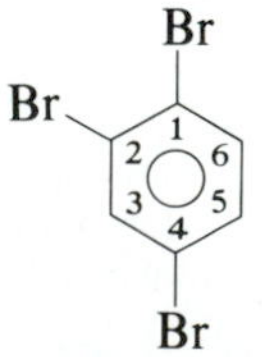

1,2,4-Tribromobenzene (not 1,4,6-)

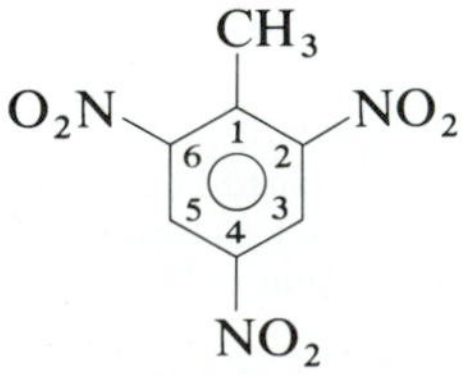

2,4,6-Trinitrotoluene (TNT)

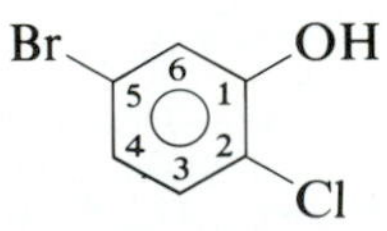

5-Bromo-2-chlorophenol

EXAMPLE 22.8

Write formulas and names for all the possible isomers of (a) chloronitrobenzene, $C_6H_4Cl(NO_2)$, and (b) tribromobenzene, $C_6H_3Br_3$.

SOLUTION

(a) The name and formula indicate a chloro group (Cl) and a nitro group (NO_2) attached to a benzene ring. There are six positions in which to place these two groups. They can be ortho, meta, or para to each other.

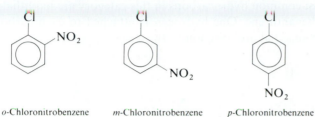

o-Chloronitrobenzene *m*-Chloronitrobenzene *p*-Chloronitrobenzene

(b) For tribromobenzene start by placing the three bromo groups in the 1-, 2-, and 3-positions; then the 1-, 2-, and 4-positions; and so on until all the possible isomers are formed. The name of each isomer will allow you to check that no duplication of formulas has been written.

1,2,3-Tribromobenzene 1,2,4-Tribromobenzene 1,3,5-Tribromobenzene

There are only three isomers of tribromobenzene. If one erroneously writes the 1,2,5- compound, a further check will show that, by numbering the rings as indicated, it is in reality the 1,2,4- isomer:

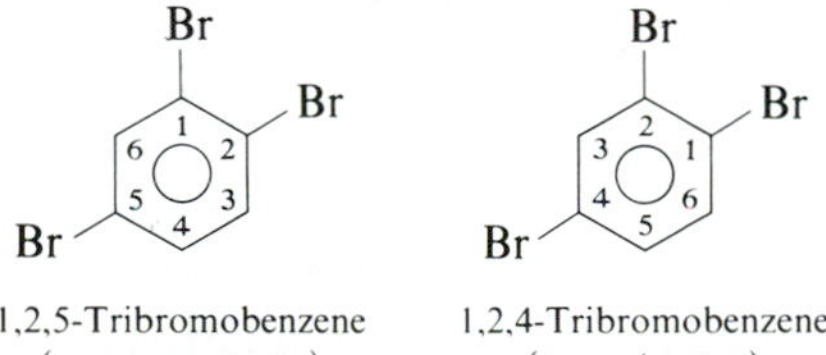

1,2,5-Tribromobenzene (erroneous name) 1,2,4-Tribromobenzene (correct name)

PRACTICE Write formulas and names for all possible isomers of chlorophenol.

Answer:

o-Chlorophenol *m*-Chlorophenol *p*-Chlorophenol

22.11 Polycyclic Aromatic Compounds

polycyclic or fused aromatic ring systems

There are many other aromatic ring systems besides benzene. Their structures consist of two or more rings in which two carbon atoms are common to two rings. These compounds are known as **polycyclic or fused aromatic ring systems**. Three of the most common hydrocarbons in this category are naphthalene, anthracene, and phenanthrene. One hydrogen is attached to each carbon atom except at the carbons that are common to two rings.

Naphthalene, $C_{10}H_8$

Anthracene, $C_{14}H_{10}$ Phenanthrene, $C_{14}H_{10}$

All three of these substances may be obtained from coal tar. Naphthalene is known as moth balls and has been used as a moth repellant for many years. A number of the polycyclic aromatic hydrocarbons (and benzene) have been shown to be carcinogenic (cancer-producing). Formulas for some of the more notable ones, found in coal tar, tar from cigarette smoke, and soot in urban environments follow.

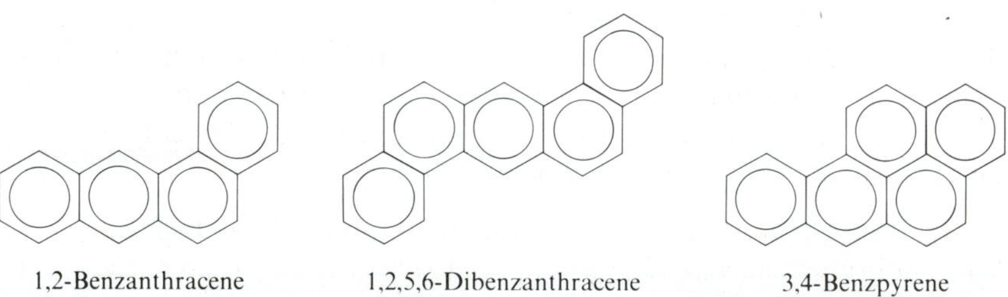

1,2-Benzanthracene 1,2,5,6-Dibenzanthracene 3,4-Benzpyrene

The mechanism by which these compounds cause cancer has not yet been determined. One hypothesis is that the polycyclic hydrocarbons are not active carcinogens; rather, the metabolites (products of chemical transformations in living cells) of these hydrocarbons are currently thought to be the active carcinogens. These fused ring hydrocarbons are formed when organic molecules are heated to high temperatures as they are in the burning of cigarettes.

22.12 Sources and Physical Properties of Aromatic Hydrocarbons

When coal is heated to high temperatures (450–1200°C) in the absence of air to produce coke (C), coal gas and a complex mixture of condensable substances called *coal tar* are driven off:

$$\text{Coal} \xrightarrow{\Delta} \text{Coke} + \text{Coal gas} + \text{Coal tar}$$

The aromatic hydrocarbons, such as benzene, toluene, xylene, naphthalene, and anthracene, were first obtained in quantity from coal tar. Since coal tar itself is a by-product of the manufacture of coke, the total amount of aromatics that can be obtained from this source is limited. The demand for aromatic hydrocarbons, which are used in the production of a vast number of materials such as drugs, dyes, detergents, explosives, insecticides, plastics, and synthetic rubber, became too great to be obtained from coal tar alone. Processes were devised to make aromatic hydrocarbons from the relatively inexpensive alkanes found in petroleum. Currently, about one-third of our benzene supply, and the greater portion of our toluene and xylene supplies, are obtained from petroleum.

Aromatic hydrocarbons are essentially nonpolar substances, insoluble in water but soluble in many organic solvents. They are liquids or solids and usually have densities less than that of water. Aromatic hydrocarbons burn readily, usually with smoky yellow flames as a result of incomplete carbon combustion. Some are good motor fuels with excellent antiknock properties.

22.13 Chemical Properties of Aromatic Hydrocarbons

The most characteristic reactions of aromatic hydrocarbons involve the substitution of some group for a hydrogen on one of the ring carbons. The following are examples of typical aromatic substitution reactions. In each of these reactions, a functional group is substituted for a hydrogen atom.

1. *Halogenation* (chlorination or bromination) When benzene reacts with chlorine or bromine in the presence of a catalyst such as iron(III) chloride or iron(III) bromide, a Cl or a Br atom replaces an H atom to form the products.

$$C_6H_6 + X_2 \xrightarrow{FeX_3} C_6H_5X + HX$$

Benzene, Chlorine or bromine, Chlorobenzene or bromobenzene

2. *Nitration* When benzene reacts with a mixture of concentrated nitric acid and concentrated sulfuric acid at about 50°C, nitrobenzene is formed. In this reaction a nitro group, $—NO_2$, is substituted for an H atom of benzene.

$$C_6H_6 + HO—NO_2 \xrightarrow{H_2SO_4} C_6H_5—NO_2 + H_2O$$

Benzene Nitric acid Nitrobenzene

3. *Alkylation* (Friedel–Crafts reaction) There are many variations of the Friedel–Crafts reaction. In this type the alkyl group from an alkyl halide (RX), in the presence of $AlCl_3$ catalyst, substitutes for an H atom on the benzene ring.

$$C_6H_6 + CH_3CH_2Cl \xrightarrow{AlCl_3} C_6H_5—CH_2CH_3 + HCl$$

Benzene Chloroethane Ethylbenzene

From about 1860 onward, especially in Germany, a great variety of useful substances such as dyes, explosives, and drugs were synthesized from aromatic hydrocarbons by reactions of the types just described. These early syntheses were developed by trial-and-error methods. A good picture of the reaction mechanism, or step-by-step sequence of intermediate stages in the overall reaction, was not obtained until about 1940.

It is now recognized that aromatic substitution reactions usually proceed by a mechanism called *electrophilic substitution*. Three steps are involved: (1) the formation of an electrophile (electron-seeking group), (2) the attachment of the electrophile to the benzene ring forming a positively charged carbocation intermediate, and finally (3) the loss of a hydrogen ion from the carbocation to form the product. This reaction mechanism is illustrated in the chlorination of benzene catalyzed by iron(III) chloride.

Step 1 $FeCl_3 + Cl_2 \longrightarrow FeCl_4^- + Cl^+$

Iron(III) chloride Chloronium ion

Step 2 $C_6H_6 + Cl^+ \longrightarrow [C_6H_6Cl]^+$

A carbocation

Step 3 $[C_6H_6Cl]^+ + FeCl_4^- \longrightarrow C_6H_5Cl + FeCl_3 + HCl$

Chlorobenzene

In Step 1 the electrophile (chloronium ion) is formed. In Step 2 the chloronium ion adds to benzene to form an intermediate carbocation, which loses a

Chemistry in Action

Aromatic Hydrocarbons

Benzene is the starting material for the production of many consumer products. Detergents, insecticides, plastics, dyes, and a multitude of drugs are all synthesized from benzene. It was once widely used as an organic solvent, but now benzene has been classified as a hazardous substance. Prolonged inhalation of benzene can cause nausea and death from cardiac or respiratory failure. Repeated exposure to benzene can result in aplastic anemia, a condition in which both the red and white blood cells decrease. Most laboratories have greatly reduced the use of benzene as a solvent, often substituting toluene or another aromatic compound. Toluene is used in the production of dyes, drugs, lacquers, and in explosives. It can be used to increase the octane rating of fuels and to preserve specimens of urine.

Substances containing the benzene ring are commonly found in living organisms. Plants synthesize benzene rings from carbon dioxide, water, and inorganic materials. Animals are not capable of this synthesis and so must ingest aromatic compounds to survive. Important aromatic compounds necessary to animals include some amino acids and vitamins.

proton (H^+) in Step 3 to form the products, C_6H_5Cl and HCl. The catalyst, $FeCl_3$, is regenerated in Step 3.

This same mechanism is used by living organisms when aromatic rings gain or lose substituents. For example, the thyroxines (thyroid gland hormones) contain aromatic rings that are iodinated by following an electrophilic substitution mechanism. Iodination is a key step in producing these potent hormones. In general, scientists find that most of life's reactions follow mechanisms that have been elucidated in the organic chemist's laboratory.

4. *Oxidation of side chains* Carbon chains attached to an aromatic ring are fairly easy to oxidize. Reagents most commonly used to accomplish this in the laboratory are $KMnO_4$ or $K_2Cr_2O_7 + H_2SO_4$. No matter how long the side chain is, the carbon atom attached to the aromatic ring is oxidized to a carboxylic acid group, —COOH. For example, toluene, ethylbenzene, and propylbenzene are all oxidized to benzoic acid:

$$C_6H_5{-}CH_3 \xrightarrow[\Delta]{KMnO_4/H_2O} C_6H_5{-}COOH$$

Toluene → Benzoic acid

$$C_6H_5{-}CH_2CH_3 \xrightarrow[\Delta]{K_2Cr_2O_7/H_2SO_4} C_6H_5{-}COOH + CO_2$$

Ethylbenzene

$$C_6H_5{-}CH_2CH_2CH_3 \xrightarrow[\Delta]{K_2Cr_2O_7/H_2SO_4} C_6H_5{-}COOH + CH_3COOH$$

Propylbenzene

Concepts in Review

1. Explain the sp^2 and sp hybridization of carbon atoms.
2. Explain the formation of a pi bond.
3. Explain the formation of double and triple bonds.
4. Distinguish, by formulas, the difference between saturated and unsaturated hydrocarbons.
5. Name and write structural formulas of alkenes, alkynes, cycloalkenes, and aromatic compounds.
6. Determine from structural formulas whether a compound can exist as geometric isomers.
7. Name geometric isomers by the cis–trans method.
8. Write equations for the addition reactions of alkenes and alkynes.
9. Explain the formation of carbocations and the role they play in chemical reactions.
10. Apply Markovnikov's rule to the addition of HCl, HBr, HI, and H_2O/H^+ to alkenes and alkynes.
11. Explain the Baeyer test for unsaturation.
12. Distinguish, using simple chemical tests, among alkanes, alkenes, and alkynes.
13. Describe the nature of benzene and how its properties differ from open-chain unsaturated compounds.
14. Explain the role of geometric isomers in vision.
15. Name monosubstituted, disubstituted, and polysubstituted benzene compounds.
16. Draw structural formulas of substituted benzene compounds.
17. Recognize the more common fused aromatic ring compounds.
18. Write equations for the following reactions of benzene and substituted benzenes: halogenation (chlorination or bromination), nitration, alkylation (Friedel–Crafts reaction), and side-chain oxidation.
19. Describe and write equations for the mechanism by which benzene compounds are brominated in the presence of $FeBr_3$ or chlorinated in the presence of $FeCl_3$.

Equations in Review

Alkenes and Alkynes

Addition Reactions

Hydrogenation: $RCH{=}CHR + H_2 \longrightarrow RCH_2CH_2R$

Halogenation: $RCH{=}CHR + X_2 \longrightarrow RCHX{-}CHXR$

Hydrogen halide: $RCH{=}CH_2 + HX \longrightarrow RCH(X){-}CH_3$

Water: $RCH{=}CH_2 + H_2O \xrightarrow{H^+} RCH(OH){-}CH_3$

Oxidation

$$RCH{=}CH_2 \xrightarrow[H_2O]{KMnO_4} RCH(OH){-}CH_2(OH)$$

Alkynes

Same types as alkenes

Aromatic hydrocarbons

Halogenation:

$$C_6H_6 + X_2 \xrightarrow{FeX_3} C_6H_5X + HX$$

Nitration:

$$C_6H_6 + HNO_3 \xrightarrow{H_2SO_4} C_6H_5NO_2 + H_2O$$

Alkylation:

$$C_6H_6 + R{-}Cl \xrightarrow{AlCl_3} C_6H_5R + HCl$$

Oxidation of side chains:

$$C_6H_5CH_2CH_3 \xrightarrow{KMnO_4/H_2O} C_6H_5{-}COOH + CO_2$$

EXERCISES

1. The double bond in ethylene (C_2H_4) is made up of a sigma bond and a pi bond. Explain how the pi bond differs from the sigma bond.
2. Draw Lewis structures to represent the following molecules:
   (a) Ethane (b) Ethene (c) Ethyne
3. (a) Draw structural formulas for the four isomeric chloropropenes, C_3H_5Cl.
   (b) There is another compound with this same molecular formula. What is its structure?
4. There are 17 possible isomeric hexenes including geometric isomers.
   (a) Write the structural formula for each isomer.
   (b) Name each isomer and include the prefix *cis-* or *trans-* where appropriate.

5. Draw structural formulas for the following:
   (a) 2,5-Dimethyl-3-hexene
   (b) 2-Ethyl-3-methyl-1-pentene
   (c) *cis*-4-Methyl-2-pentene
   (d) *cis*-1,2-Diphenylethene
   (e) 3-Pentene-1-yne
   (f) 3-Phenyl-1-butyne
   (g) Vinyl bromide
   (h) Cyclopentene
   (i) *trans*-3-Hexene
   (j) 1-Methylcyclohexene
   (k) 3-Methyl-1-pentyne
   (l) 3-Isopropylcyclopentene
   (m) 3-Methyl-2-phenylhexane
6. Name the following compounds:
   (a) H, H / C=C / CH_3, $CHCH_2CH_3$ (with CH_3 on the CH)
   (b) CH_3, CH_3 / C=C / CH_3, CH_3
   (c) $CH_3CH_2CHCH{=}CH_2$ (with CH on the CH, bearing CH_3 and CH_3)
   (d) CH_3CH_2, CH_3 / C=C / H, CH_2CH_3
   (e) $—CH_2C{\equiv}CH$ (attached to a benzene ring)
   (f) $CH_3CHBrCHBrC{\equiv}CCH_3$
7. Write the structural formulas and IUPAC names for all the (a) pentynes and (b) hexynes.
8. Why is it possible to obtain cis and trans isomers of 1,2-dichloroethene but not of 1,2-dichloroethane?
9. Which of the following molecules have structural formulas that permit cis–trans isomers to exist?
   (a) $(CH_3)_2C{=}CHCH_3$
   (b) $CH_2{=}CHCl$
   (c) $CH_3CH_2C{\equiv}CCH_3$
   (d) $CH_3CH{=}CHCl$
   (e) $CCl_2{=}CBr_2$
   (f) $CH_2ClCH{=}CHCH_2Cl$
10. The following names are incorrect. Tell why each name is wrong and give the correct name.
   (a) 3-Methyl-3-butene
   (b) 3-Pentene
   (c) *cis*-2-Methyl-2-pentene
   (d) 3-Ethyl-1-butene
   (e) 2-Chlorocyclohexene
11. Complete the following equations:
   (a) $CH_3CH_2CH_2CH{=}CH_2 + Br_2 \longrightarrow$
   (b) $CH_3CH_2CH_2CH{=}CH_2 + HCl \longrightarrow$
   (c) $CH_3CH_2\underset{CH_3}{\underset{|}{C}}{=}CHCH_3 + HI \longrightarrow$
   (d) $CH_3CH_2CH{=}CHCH_3 + HBr \longrightarrow$
   (e) $CH_3CH_2CH{=}CH_2 + H_2O \xrightarrow{H^+}$
   (f) (benzene ring)$—CH{=}CH_2 + HCl \longrightarrow$
   (g) (benzene ring)$—CH{=}CH_2 + H_2 \xrightarrow[\text{1 atm}]{\text{Pt, 25°C}}$
   (h) $CH_2{=}CHCl + HBr \longrightarrow$
   (i) $CH_3CH{=}CHCH_3 + KMnO_4 \xrightarrow[\text{Cold}]{H_2O}$
12. Complete the following equations:
   (a) $CH_3C{\equiv}CH + 2\,H_2 \xrightarrow[\text{1 atm}]{\text{Pt, 25°C}}$
   (b) $CH_3C{\equiv}CCH_3 + Br_2\ (1\ \text{mol}) \longrightarrow$
   (c) $CH_3C{\equiv}CCH_3 + Br_2\ (2\ \text{mol}) \longrightarrow$
   (d) The two-step reaction
   $CH{\equiv}CH + HCl \longrightarrow \quad \xrightarrow{HCl}$
   (e) The two-step reaction
   $CH_3C{\equiv}CH + HCl \longrightarrow \quad \xrightarrow{HCl}$
13. Write the formula and name for the product when cyclohexene reacts with
   (a) Br_2 (c) H_2O, H^+
   (b) HI (d) $KMnO_4(aq)$(cold)
14. Write equations to show how 2-butyne can be converted to:
   (a) 2,3-Dibromobutane
   (b) 2,2-Dibromobutane
   (c) 2,2,3,3-Tetrabromobutane
15. Two alkyl bromides are possible when 2-methyl-1-pentene is reacted with HBr. Which one will predominate? Why?
16. Cyclohexane and 2-hexene both have the formula C_6H_{12}. How could you readily distinguish one from the other by chemical tests?
17. Why do many rubber products deteriorate rapidly in smog-ridden areas?
18. Explain the two different kinds of explosion hazards present when acetylene is being handled.
19. Write structural formulas for
   (a) Benzene (f) Phenol
   (b) Toluene (g) *o*-Bromochlorobenzene
   (c) *p*-Xylene (h) 1,3-Dichloro-5-nitrobenzene
   (d) Styrene (i) *m*-Dinitrobenzene
   (e) Aniline

20. Write structural formulas for
    (a) Ethylbenzene
    (b) Benzoic acid
    (c) 1,3,5-Tribromobenzene
    (d) Naphthalene
    (e) Anthracene
    (f) *tert*-Butylbenzene
    (g) 1,1-Diphenylethane
21. Write structural formulas and names for all the isomers of:
    (a) Trichlorobenzene ($C_6H_3Cl_3$)
    (b) Dichlorobromobenzene ($C_6H_3Cl_2Br$)
    (c) The benzene derivatives of formula C_8H_{10}
22. (a) Write the structures for all the isomers that can be written by substituting a third chlorine atom in *o*-dichlorobenzene.
    (b) Write the structures for all the isomers that can be written by substituting an additional chlorine atom in *o*-chlorobromobenzene.
23. Name the following compounds:
    (a) CH_2CH_3 / Cl (para-substituted benzene) (b) benzene–CH=CH$_2$
    (c) benzene–$CH_2CH_2CH_3$
    (d) NH_2 / NO_2 (benzene)
    (e) COOH / Br / Br (benzene) (f) CH_3 / NO_2 (benzene)
    (g) OH / Br / Br / Br (benzene) (h) benzene–CH(CH_3)CH_3
    (i) H–C(benzene)$_3$
24. Explain how the reactions of benzene provide evidence that its structure does not include double bonds like those found in alkenes.
25. Complete the following equations and name the organic products:
    (a) benzene + $Br_2 \xrightarrow{FeBr_3}$
    (b) benzene + CH_3CHCH_3 (with Cl on the central C) $\xrightarrow{AlCl_3}$
    (c) CH_3 / CH_3 (para-xylene) + $HNO_3 \xrightarrow{H_2SO_4}$
    (d) benzene–CH_3 + $KMnO_4 \xrightarrow{H_2O}$
26. Describe the reaction mechanism by which benzene is brominated in the presence of $FeBr_3$ [Question 25, part (a)]. Show equations.
27. In terms of historical events, why did the major source of aromatic hydrocarbons shift from coal tar to petroleum during the 10-year period 1935–1945?
28. Which of the following statements are correct? Rewrite the incorrect statements to make them correct.
    (a) The compound with the formula C_5H_{10} can be either an alkene or a cycloalkane.
    (b) If C_8H_{10} is an open-chain compound with multiple double bonds, it needs an additional ten hydrogen atoms to become a saturated hydrocarbon.
    (c) Propene and propane are isomers.
    (d) The pi bond is formed from two sp^2 electron orbitals.
    (e) A double bond consists of two equivalent bonds called pi bonds.
    (f) A triple bond consists of one sigma bond and two pi bonds.
    (g) The hybridized electron structure of a carbon atom in alkynes is *sp, sp, p, p*.
    (h) The acetylene molecule is linear.
    (i) A molecule of 2,3-dimethyl-1-pentene contains seven carbon atoms.
    (j) When an alkene is reacted with cold $KMnO_4$, a glycol is formed.
    (k) The compound C_6H_{10} can have in its structure: two carbon–carbon double bonds, or

one carbon–carbon triple bond, or one cyclic ring and one carbon–carbon double bond.

(l) The disappearance of the purple color when $KMnO_4$ reacts with an alkene is known as the Markovnikov test for unsaturation.

(m) $CH_3CH_2CH_2^+$ is a primary carbocation.

(n) A secondary carbocation is more stable than a primary carbocation and less stable than a tertiary carbocation.

(o) After bromine has added to an alkene, the product is no longer unsaturated.

(p) Alkynes have the general formula C_nH_{2n-4}.

(q) Cis–trans isomerism occurs in alkenes and alkynes.

(r) Geometric isomers are superimposable on each other.

(s) All six hydrogen atoms in benzene are equivalent.

(t) The chemical behavior of benzene is similar to that of alkenes.

(u) Toluene and benzene are isomers.

(v) Two substituents on a benzene ring that are in the 1,2-position are ortho to each other.

(w) 1,4-Dichlorobenzene and *p*-dichlorobenzene are different names for the same compound.

(x) The oxidation of toluene with hot $KMnO_4$ yields benzoic acid.

(y) The oxidation of ethylbenzene with hot $KMnO_4$ yields benzoic acid.

(z) Toluene and benzene are homologues.

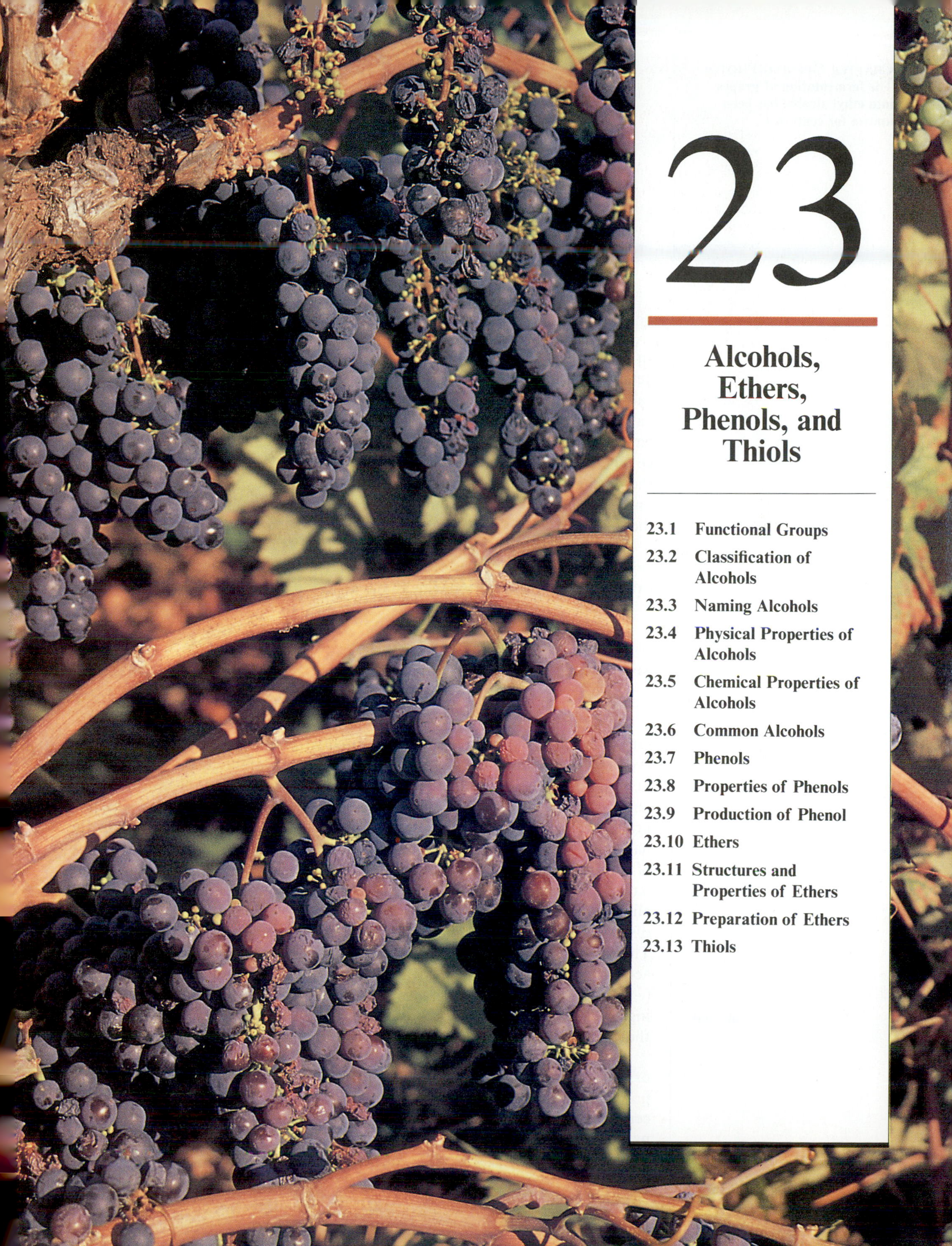

23

Alcohols, Ethers, Phenols, and Thiols

◄ CHAPTER OPENING PHOTO: **The fermentation of grapes into ethyl alcohol has been known for centuries.**

In today's active society we often play as strenuously as we work. A vacation getaway might include an initial car voyage, followed by some vigorous physical activity (hiking, cycling, skiing, or swimming). After a hard day of fun, friends gather to relax and commiserate over sore muscles. During these activities, organic molecules containing an —OH group play a significant role. Ethylene glycol acts as a coolant in the radiator of the car, sugar and carbohydrates provide the biochemical energy for our sport activity, ethyl alcohol is a component in any alcoholic beverage consumed, and phenolic compounds are an active ingredient in the muscle rubs and analgesics that relieve our sore muscles.

Just what changes does the addition of an OH group produce in the physical and chemical properties of an organic molecule? In this chapter we begin to examine the effect of various functional groups on organic molecules.

23.1 Functional Groups

Organic compounds were obtained originally from plants and animals, which, even today, are still the direct sources of many important chemicals. As a case in point, millions of tons of sucrose (table sugar) are obtained from sugar cane and sugar beet juices each year. As chemical knowledge developed, many naturally occurring compounds were synthesized, often at far less cost than the natural products. Of even greater significance than the cheaper manufacture of natural substances was the synthesis of new substances totally unlike any natural product. The synthesis of new substances was aided greatly by the realization that organic chemicals can be divided into a relatively small number of classes and studied on the basis of similar chemical properties (Table 21.1). The various classes of compounds are identified by the presence of certain characteristic groups called functional groups. For example, if a hydroxyl group (—OH) is substituted for a hydrogen atom in an alkane molecule, the resulting compound is an **alcohol**. Thus alcohols are a class of compounds in which the functional group is the hydroxyl group.

alcohol

Through the chemical reactions of functional groups, it is possible to create or synthesize new substances. The synthesis of new and possibly useful compounds or the more economical synthesis of known compounds is a main concern of modern organic chemistry. Most chemicals used today do not occur in nature but are synthesized from naturally occurring materials. The chemical and physical properties of an organic compound depend on (1) the kinds and number of functional groups present and (2) the shape and size of the molecule.

The structures of alcohols, ethers, and phenols may be derived from water by replacing the hydrogen atoms of water with alkyl groups (R) or aromatic rings:

H—O—H	R—O—H	R—O—R	C_6H_5—O—H
Water	Alcohol	Ether	Phenol

The R— groups in ethers can be the same or different and can be alkyl groups or aromatic rings.

23.2 Classification of Alcohols

Structurally, an alcohol is derived from a nonaromatic hydrocarbon by the replacement of at least one hydrogen atom with a hydroxyl group (—OH). Alcohols are represented by the general formula ROH, with methanol (CH_3OH) being the first member of the homologous series. R represents an alkyl or substituted alkyl group. Models illustrating the structural arrangements of the atoms in methanol and ethanol are shown in Figure 23.1.

Alcohols are classified as **primary** (1°), **secondary** (2°), or **tertiary** (3°), depending on whether the carbon atom to which the —OH group is attached is directly bonded to one, two, or three other carbon atoms, respectively. Generalized formulas for 1°, 2°, and 3° alcohols follow.

primary alcohol

secondary alcohol

tertiary alcohol

$R\text{—}\overset{H}{\underset{H}{C}}\text{—}OH$	$R\text{—}\overset{R}{\underset{H}{C}}\text{—}OH$	$R\text{—}\overset{R}{\underset{R}{C}}\text{—}OH$
Primary alcohol	Secondary alcohol	Tertiary alcohol

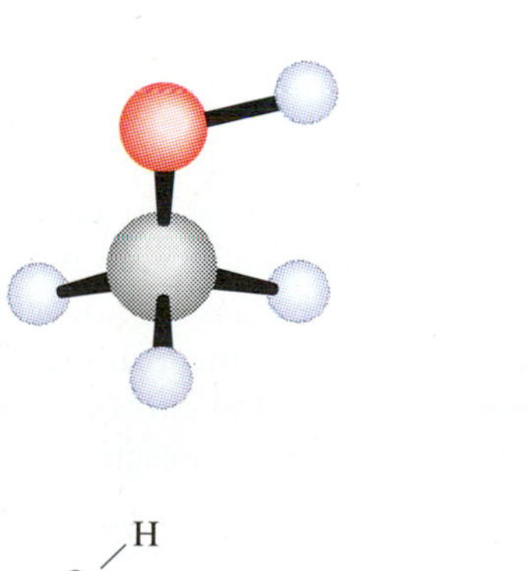

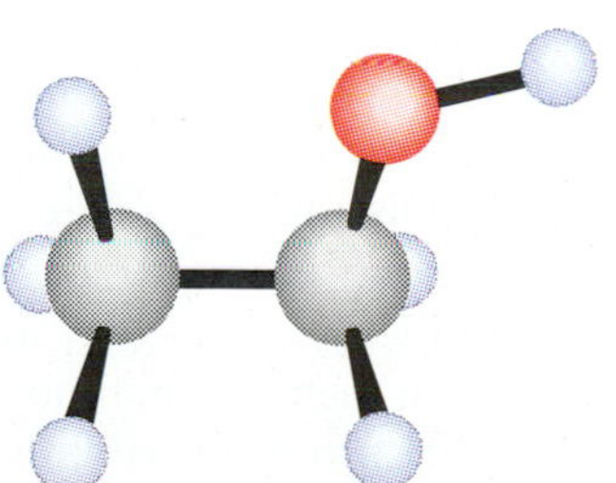

H—C(H)(H)—O—H CH_3OH Methanol

H—C(H)(H)—C(H)(OH)—H CH_3CH_2OH Ethanol

◄ **Figure 23.1** Ball-and-stick models illustrating structural formulas of methanol and ethanol.

TABLE 23.1 Names and Classification of Alcohols

Class	Formula	IUPAC name	Common name[a]	Boiling point (°C)
Primary	CH_3OH	Methanol	Methyl alcohol	65.0
Primary	CH_3CH_2OH	Ethanol	Ethyl alcohol	78.5
Primary	$CH_3CH_2CH_2OH$	1-Propanol	*n*-Propyl alcohol	97.4
Primary	$CH_3CH_2CH_2CH_2OH$	1-Butanol	*n*-Butyl alcohol	118
Primary	$CH_3(CH_2)_3CH_2OH$	1-Pentanol	*n*-Amyl or *n*-pentyl alcohol	138
Primary	$CH_3(CH_2)_6CH_2OH$	1-Octanol	*n*-Octyl alcohol	195
Primary	CH_3CHCH_2OH | CH_3	2-Methyl-1-propanol	Isobutyl alcohol	108
Secondary	CH_3CHCH_3 | OH	2-Propanol	Isopropyl alcohol	82.5
Secondary	$CH_3CH_2CHCH_3$ | OH	2-Butanol	*sec*-Butyl alcohol	91.5
Tertiary	CH_3 | $CH_3—C—OH$ | CH_3	2-Methyl-2-propanol	*t*-Butyl alcohol	82.9
Dihydroxy	$HOCH_2CH_2OH$	1,2-Ethanediol	Ethylene glycol	197
Trihydroxy	$HOCH_2CHCH_2OH$ | OH	1,2,3-Propanetriol	Glycerol or glycerine	290

[a] The abbreviations *n*, *sec*, and *t* stand for normal, secondary, and tertiary, respectively.

Formulas of specific examples of these classes of alcohols are shown in Table 23.1. Methanol (CH_3OH) is grouped with the primary alcohols.

Molecular structures with more than one —OH group attached to a single carbon atom are generally not stable. But an alcohol molecule can contain two or more —OH groups if each —OH is attached to a different carbon atom. Accordingly, alcohols are also classified as monohydroxy, dihydroxy, trihydroxy, and so on, on the basis of the number of hydroxyl groups per molecule. **Polyhydroxy alcohols** and *polyols* are general terms for alcohols that have more than one —OH group per molecule. Polyhydroxy compounds are very important molecules in living cells as they include the carbohydrate class of biochemicals.

polyhydroxy alcohol

An alcohol such as 2-butanol can be written in a single-line formula by putting the —OH group in parentheses and placing it after the carbon to which it is bonded. For example, the following two formulas represent the same compound:

$$\underset{\large OH}{CH_3CH_2\overset{}{C}HCH_3} \qquad CH_3CH_2CH(OH)CH_3$$

23.3 Naming Alcohols

If you know how to name alkanes, it is easy to name alcohols by the IUPAC System. Unfortunately, several of the alcohols are generally known by common or nonsystematic names, so it is often necessary to know more than one name for a given alcohol. The common name is usually formed from the name of the alkyl group that is attached to the —OH group, followed by the word *alcohol*. See examples given below and in Table 23.1. To name an alcohol by the IUPAC System:

1. Select the longest continuous chain of carbon atoms containing the hydroxyl group.
2. Number the carbon atoms in this chain so that the one bearing the —OH group has the lowest possible number.
3. Form the parent alcohol name by replacing the final *e* of the corresponding alkane name by *ol*. When isomers are possible, locate the position of the —OH by placing the number (hyphenated) of the carbon atom to which the —OH is bonded immediately before the parent alcohol name.
4. Name each alkyl side chain (or other group) and designate its position by number.

For example, let us go through the steps above to name the alcohol $CH_3CH_2CH_2CH_2OH$.

Step 1 The longest carbon chain containing the —OH group has four carbons.

Step 2 Number the carbon atoms, giving the carbon bonded to the —OH the number 1.

$$\overset{4}{C}—\overset{3}{C}—\overset{2}{C}—\overset{1}{C}—OH$$

Step 3 The name of the four-carbon alkane is butane. Replace the final *e* in butane with *ol*, forming the name *butanol*. Since the —OH is on carbon 1, place a *1* before butanol to give the complete alcohol name 1-butanol.

Step 4 No groups of atoms other than hydrogen are attached to the butanol chain, so the name of this alcohol is 1-butanol.

Study the application of this naming system to these examples and to those shown in Table 23.1.

$$\overset{3}{CH_3}—\overset{2}{CH_2}—\overset{1}{CH_2OH}$$

1-Propanol
(*n*-Propyl alcohol)

$$\overset{1}{CH_3}—\underset{\displaystyle OH}{\overset{2}{CH}}—\overset{3}{CH_3}$$

2-Propanol
(Isopropyl alcohol)

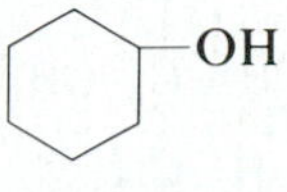

(cyclohexane ring)—OH

Cyclohexanol

$$\overset{4}{CH_3}-\overset{3}{\underset{\underset{CH_3}{|}}{CH}}-\overset{2}{CH_2}-\overset{1}{CH_2OH}$$

3-Methyl-1-butanol
(Isoamyl alcohol or isopentyl alcohol)

$$HO\overset{2}{CH_2}-\overset{1}{CH_2OH}$$

1,2-Ethanediol
(Ethylene glycol)

Example 23.1 Name this alcohol by the IUPAC method.

$$CH_3CH_2\underset{\underset{CH_3}{|}}{C}HCH_2\underset{\underset{OH}{|}}{C}HCH_3$$

SOLUTION

Step 1 The longest continuous carbon chain containing the —OH group has six carbon atoms.

Step 2 This carbon chain is numbered from right to left so that the —OH group has the smallest possible number:

$$\overset{6}{C}-\overset{5}{C}-\overset{4}{\underset{\underset{CH_3}{|}}{C}}-\overset{3}{C}-\overset{2}{\underset{\underset{OH}{|}}{C}}-\overset{1}{C}$$

In this case, the —OH is on carbon 2.

Step 3 The name of the six-carbon alkane is hexane. Replace the final *e* in hexane by *ol*, forming the name *hexanol*. Since the —OH is on carbon 2, place a *2* before hexanol to give the parent alcohol name 2-hexanol.

Step 4 A methyl group (—CH_3) is located on carbon 4. Therefore the full name of the compound is 4-methyl-2-hexanol.

Example 23.2 Write the structural formula of 3,3-dimethyl-2-hexanol.

SOLUTION

Step 1 The "2-hexanol" refers to a six-carbon chain with an —OH group on carbon 2. Write the skeleton structure as follows:

$$\overset{1}{C}-\overset{2}{\underset{\underset{OH}{|}}{C}}-\overset{3}{C}-\overset{4}{C}-\overset{5}{C}-\overset{6}{C}$$

Step 2 Place the two methyl groups ("3,3-dimethyl") on carbon 3:

$$\overset{1}{C}-\underset{\underset{HO}{|}}{\overset{2}{C}}-\overset{\overset{CH_3}{|}}{\underset{\underset{CH_3}{|}}{\overset{3}{C}}}-\overset{4}{C}-\overset{5}{C}-\overset{6}{C}$$

Step 3 Finally, add H atoms to give each carbon atom four bonds:

$$CH_3\underset{\underset{OH}{|}}{C}H-\overset{\overset{CH_3}{|}}{\underset{\underset{CH_3}{|}}{C}}-CH_2CH_2CH_3$$

3,3-Dimethyl-2-hexanol

Practice Write the correct IUPAC name for each of the following:

(a) $CH_3CH_2\underset{\large |}{C}HCH_2CH_2CH_3$ with CH_2OH on the third carbon

(b) $CH_3CH_2CHCH_2CCH_3$ with OH on the third carbon and Br and CH_3 on the fifth carbon

Answers: (a) 2-ethyl-1-pentanol (b) 5-bromo-5-methyl-3-hexanol

Practice Write the structural formula for each of the following:
(a) 3-methylcyclohexanol (b) 4-ethyl-2-methyl-3-heptanol

Answers: (a) cyclohexane ring with OH on carbon 1 and CH_3 on carbon 3

(b) $CH_3CHCHCCH_2CH_2CH_3$ with CH_3 on the second carbon, OH on the third carbon, and CH_2CH_3 on the fourth carbon

23.4 Physical Properties of Alcohols

The physical properties of alcohols are related to those of both water and alkane hydrocarbons. This is easily understandable if we recall certain facts about water and the alkanes. Water molecules are quite polar. The properties of water, such as its high boiling point and its ability to dissolve many polar substances, are largely due to the polarity of its molecules. Alkane molecules possess almost no polarity. The properties of the alkanes reflect this lack of polarity—for example, their relatively low boiling points and inability to dissolve water and other polar substances. An alcohol molecule is made up of a water-like hydroxyl group joined to a hydrocarbon-like alkyl group.

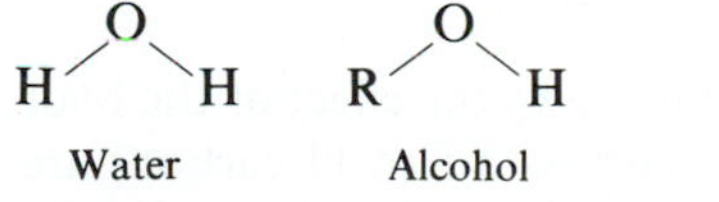

One striking property of alcohols is their relatively high boiling points. The simplest alcohol, methanol, boils at 65°C. But methane, the simplest hydrocarbon, boils at −162°C. The boiling points of the normal alcohols increase in a regular fashion with increasing number of carbon atoms. Branched-chain alcohols have lower boiling points than the corresponding straight-chain alcohols (see Table 23.1).

Alcohols containing up to three carbon atoms are infinitely soluble in water. With one exception (*t*-butyl alcohol), alcohols with four or more carbon atoms have limited solubility in water. In contrast, all hydrocarbons are essentially insoluble in water.

The hydroxyl group on the alcohol molecule is responsible for both the water solubility and the relatively high boiling points of the low molecular-mass alcohols. Hydrogen bonding (see Section 14.11) between water and alcohol

TABLE 23.2 Boiling Points of Alkanes and Monohydroxy Alcohols

Name	Boiling point (°C)	Name	Boiling point (°C)
Hexane	69	1-Hexanol	156
Octane	126	1-Octanol	195
Decane	174	1-Decanol	228

TABLE 23.3 Comparison of the Boiling Points of Ethanol, Ethylene Glycol, and 1-Propanol

Name	Formula	Molar mass	Boiling point (°C)
Ethanol	CH_3CH_2OH	46	78
Ethylene glycol	$CH_2(OH)CH_2OH$	62	197
1-Propanol	$CH_3CH_2CH_2OH$	60	97

molecules accounts for the solubility, and hydrogen bonding between alcohol molecules accounts for their high boiling points.

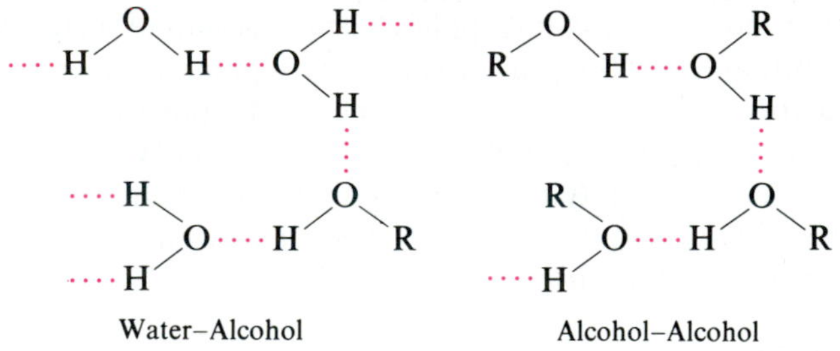

Water–Alcohol

Alcohol–Alcohol

As the length of the hydrocarbon chain increases, the effect of the hydroxyl group becomes relatively less important. Alcohols with 5 to 11 carbons are oily liquids of slight water solubility, and in physical behavior they resemble the corresponding alkane hydrocarbons. However, the effect of the —OH group is still noticeable in that their boiling points are higher than those of alkanes with similar molar masses (see Table 23.2). Alcohols containing 12 or more carbons are waxlike solids that resemble solid alkanes in physical appearance.

In alcohols with two or more hydroxyl groups, the effect of the hydroxyl groups on intermolecular attractive forces is, as we might suspect, even more striking than in monohydroxy alcohols. Ethanol, CH_3CH_2OH, boils at 78°C, but the boiling point of ethylene glycol, or 1,2-ethanediol, $CH_2(OH)CH_2OH$, is 197°C. Comparison of the boiling points of ethanol, 1-propanol, and ethylene glycol shows that increased molar mass does not account for the high boiling point of ethylene glycol (see Table 23.3). The higher boiling point is primarily a result of additional hydrogen bonding due to the two —OH groups in the ethylene glycol molecule.

EXAMPLE 23.3

Glucose is one of the most important carbohydrates in biochemistry. It has six carbons and five alcohol groups (molar mass = 180.2 g). How would you predict the water solubility of glucose will differ from that of hexanol?

SOLUTION

Each polar alcohol group attracts water molecules and increases the solubility of organic compounds in water. Because glucose contains five —OH groups whereas hexanol contains only one, glucose should dissolve to a much greater extent than hexanol. In fact, only about 0.6 g of hexanol will dissolve in 100 g of water (at 20°C). In contrast, about 95 g of glucose will dissolve in 100 g of water. The high water solubility of glucose is important because this molecule is transported in a water solution to the body's cells.

23.5 Chemical Properties of Alcohols

Acidic and Basic Properties

Aliphatic alcohols are similar to water in their acid/basic properties. If an alcohol is placed in a strong acid it will accept a proton (act as a Brønsted–Lowry base) to form a protonated alcohol or **oxonium ion**.

oxonium ion

$$CH_3{-}\ddot{\underset{\cdot\cdot}{O}}H + H_2SO_4 \longrightarrow CH_3{-}\overset{+}{\ddot{O}}\!\begin{smallmatrix}H\\H\end{smallmatrix} + HSO_4^-$$

Methanol and ethanol are approximately the same strength as water as an acid, while the larger alcohols are weaker acids than water. Both water and alcohols react with alkali metals to release hydrogen gas and an anion.

$$2\,H_2O + 2\,Na \longrightarrow \underset{\text{Sodium hydroxide}}{2\,Na^+\,{}^-OH} + H_2\uparrow$$

$$2\,CH_3CH_2OH + 2\,Na \longrightarrow \underset{\text{Sodium ethoxide}}{2\,Na^+\,{}^-OCH_2CH_3} + H_2\uparrow$$

The resulting anion in the alcohol reaction is known as an **alkoxide ion** (RO^-). Alkoxides are strong bases (stronger than hydroxide) and so are used in organic chemistry when a strong base is required in a nonaqueous solution.

alkoxide ion

The order of reactivity of alcohols with sodium or potassium is primary > secondary > tertiary. Alcohols do not react with sodium as vigorously as water. Reactivity decreases with increasing molar mass since the —OH group becomes a relatively smaller, less significant part of the molecule.

Oxidation

We will consider only a few of the many reactions that alcohols are known to undergo. One important reaction is oxidation. We saw in Chapter 18 that the oxidation number of an element increases as a result of oxidation. Carbon can exist in several oxidation states, ranging from −4 to +4. In the −4 oxidation state, such as in methane, the carbon atom is considered to be completely reduced. In carbon dioxide the carbon atom is completely oxidized; that is, it is

TABLE 23.4 Oxidation States of Carbon in One-Carbon Compounds

	Compound	Number of C—O bonds	Oxidation state
CH_4	Methane	0	−4
CH_3OH	Methanol	1	−2
$H_2C{=}O$	Methanal (formaldehyde)	2	0
$HC(=O)OH$	Methanoic acid (formic acid)	3	+2
$O{=}C{=}O$	Carbon dioxide	4	+4

in its highest oxidation state (+4). In many cases, oxidation reactions in organic and biochemistry can be considered in a simple manner without the use of oxidation numbers. Oxidation is the loss of hydrogen or the gain of bonds to oxygen by the organic reactant. Table 23.4 illustrates the progression of oxidation states for various compounds containing one carbon atom.

Carbon atoms exist in progressively higher stages of oxidation in different functional-group compounds:

$$\text{Alkanes} \longrightarrow \text{Alcohols} \longrightarrow \left\{\begin{matrix}\text{Aldehydes}\\ \text{Ketones}\end{matrix}\right\} \longrightarrow \text{Carboxylic acids} \longrightarrow \text{Carbon dioxide}$$

Increasing oxidation state ⟶

The different stages shown do not necessarily indicate direct methods of synthesis. For example, some, but not all alcohols can be converted to aldehydes by oxidation; but it is not practical to convert alkanes directly to alcohols. The following equations represent generalized oxidation reactions in which the oxidizing agent is represented by [O].

$$\underset{\text{Primary alcohol}}{R{-}CH(H){-}OH} \xrightarrow{[O]} \underset{\text{Aldehyde}}{R{-}C(=O){-}H} + H_2O \xrightarrow{[O]} \underset{\text{Carboxylic acid}}{R{-}C(=O){-}OH}$$

$$\underset{\text{Secondary alcohol}}{R{-}C(H)(OH){-}R} \xrightarrow{[O]} \underset{\text{Ketone}}{R{-}C(=O){-}R} + H_2O$$

$$R{-}C(R)(OH){-}R \xrightarrow{[O]} \text{No reaction}$$

Tertiary alcohols do not have a hydrogen on the —OH carbon and so cannot react with oxidizing agents except by such drastic procedures as combustion. Both primary and secondary alcohols are oxidized as shown in the equations by the loss of the colored hydrogen atoms.

Some common oxidizing agents used for specific reactions are potassium permanganate ($KMnO_4$) in an alkaline solution, potassium dichromate ($K_2Cr_2O_7$) in an acid solution, or oxygen of the air. A complete equation for an alcohol oxidation may be fairly complex. Since our main interest is in the changes that occur in the functional groups, we can convey this information in abbreviated form:

$$\underset{\text{Ethanol}}{CH_3CH_2OH} \xrightarrow[\Delta]{K_2Cr_2O_7/H_2SO_4} \underset{\text{Ethanal}}{CH_3\overset{\overset{\displaystyle O}{\|}}{C}{-}H} + H_2O$$

Although the abbreviated equation lacks some of the details, it does show the overall reaction involving the organic compounds. Additional information is provided by notations above and below the arrow, which indicate that this reaction is carried out in heated potassium dichromate–sulfuric acid solution. Abbreviated equations of this kind will be used frequently in the remainder of this book.

EXAMPLE 23.4

What are the products when (a) 1-propanol, (b) 2-propanol, and (c) cyclohexanol are oxidized with $K_2Cr_2O_7/H_2SO_4$?

SOLUTION

(a) The formula for 1-propanol is $CH_3CH_2CH_2OH$. Since it is a primary alcohol, it can be oxidized to an aldehyde or a carboxylic acid. The oxidation occurs at the carbon bonded to the —OH group, and this carbon atom becomes an aldehyde or a carboxylic acid. The rest of the molecule remains the same.

$$CH_3CH_2CH_2OH \nearrow \underset{\text{Propanal}}{CH_3CH_2\overset{\overset{\displaystyle H}{|}}{C}{=}O}$$

$$CH_3CH_2CH_2OH \searrow \underset{\text{Propanoic acid}}{CH_3CH_2COOH}$$

(b) 2-Propanol is a secondary alcohol and is oxidized to a ketone. Ketones resist further oxidation. The oxidation occurs at the carbon bonded to the —OH group.

$$CH_3\underset{\underset{\displaystyle OH}{|}}{C}HCH_3 \longrightarrow \underset{\text{Propanone (Acetone)}}{CH_3\underset{\underset{\displaystyle O}{\|}}{C}CH_3}$$

(c) Cyclohexanol is also a secondary alcohol and is oxidized to cyclohexanone, a ketone.

Cyclohexanol (cyclohexane ring bearing OH) ⟶ cyclohexanone (cyclohexane ring bearing =O)

Cyclohexanone

PRACTICE Write the structure for the products of the oxidation of these alcohols with $K_2Cr_2O_7$ and H_2SO_4:
(a) 1-Butanol (b) 2-Methyl-2-butanol (c) Cyclopentanol

Answers:

(a) $CH_3CH_2CH_2\overset{\overset{\displaystyle O}{\|}}{C}{-}H \xrightarrow{[O]} CH_3CH_2CH_2\overset{\overset{\displaystyle O}{\|}}{C}{-}OH$

(b) No reaction (2-methyl-2-butanol is a tertiary alcohol)

(c) O

Dehydration

The term dehydration implies the elimination of water. Alcohols can be dehydrated to form alkenes (Section 22.5) or ethers. One of the more effective dehydrating agents is sulfuric acid. Whether an ether or an alkene is formed depends on the ratio of alcohol to sulfuric acid, the reaction temperature, and the type of alcohol.

Alkenes: Intramolecular Dehydration The formation of alkenes from an alcohol molecule requires a relatively high temperature. Water is removed from within a *single* alcohol molecule and a new carbon–carbon double bond forms as shown below:

$$H{-}\underset{\underset{\displaystyle H}{|}}{\overset{\overset{\displaystyle H}{|}}{C}}{-}\underset{\underset{\displaystyle OH}{|}}{\overset{\overset{\displaystyle H}{|}}{C}}{-}H \xrightarrow[180^\circ C]{96\%\ H_2SO_4} CH_2{=}CH_2 + H_2O$$

For many alcohols there are a number of ways to remove water. Therefore the double bond can be located in different positions. The major product in these cases is the compound in which the double bond has the greatest number of alkyl substituents (or lesser number of hydrogens).

$$\underset{\text{2-Butanol}}{CH_3CH_2\underset{\underset{\displaystyle OH}{|}}{C}HCH_3} \xrightarrow[100^\circ C]{60\%\ H_2SO_4} \underset{\substack{\text{2-Butene}\\\text{(major product)}}}{CH_3CH{=}CHCH_3} + \underset{\text{1-Butene}}{CH_3CH_2CH{=}CH_2} + H_2O$$

To predict the major product in an intramolecular dehydration, follow these steps: (1) Remove H and OH from adjacent carbons forming a carbon–carbon double bond; (2) if there are choices (multiple hydrogen containing neighboring carbon atoms adjacent to the —OH carbon), remove the hydrogen from the carbon with fewer hydrogens. This is known as **Saytzeff's rule**.

Saytzeff's rule

Ethers: Intermolecular Dehydration A dehydration reaction can take place between two alcohol molecules to produce an ether. However, the dehydration to make an ether is only useful for primary alcohols since secondary and tertiary alcohols predominantly yield alkenes.

$$\begin{array}{l} CH_3CH_2O\boxed{H} \\ CH_3CH_2\boxed{OH} \end{array} \xrightarrow[140^\circ C]{96\%\ H_2SO_4} \underset{\text{Diethyl ether}}{CH_3CH_2OCH_2CH_3} + H_2O$$

A reaction in which *two* molecules are combined by removing a small molecule is known as a **condensation** reaction. There are many examples of other condensation reactions in the formation of biochemical molecules. **condensation**

The type of dehydration which occurs depends on the temperature and the number of reactant molecules. Lower temperatures and *two* alcohol molecules produce ethers while higher temperatures and a *single* alcohol produce alkenes.

Dehydration reactions are often used in industry to make relatively expensive ethers from lower priced alcohols. Alkenes are less expensive than alcohols and so are not produced industrially by this method.

Esterification (Conversion of Alcohols to Esters)

An alcohol can react with a carboxylic acid to form an ester and water. The reaction is represented as follows:

$$\underset{\text{Carboxylic acid}}{R{-}\overset{\overset{\displaystyle O}{\|}}{C}{-}OH} + \underset{\text{Alcohol}}{H{-}OR'} \overset{H^+}{\rightleftharpoons} \underset{\text{Ester}}{R\overset{\overset{\displaystyle O}{\|}}{C}{-}OR'} + HOH$$

$$\underset{\text{Acetic acid}}{CH_3\overset{\overset{\displaystyle O}{\|}}{C}{-}OH} + \underset{\text{Ethanol}}{HOCH_2CH_3} \overset{H^+}{\rightleftharpoons} \underset{\text{Ethyl acetate}}{CH_3\overset{\overset{\displaystyle O}{\|}}{C}{-}OCH_2CH_3} + HOH$$

Esterification is an important reaction of alcohols and is discussed in greater detail in Chapter 25.

23.6 Common Alcohols

Three general methods for making alcohols are

1. *Hydrolysis of an ester*

$$\underset{\text{Ester}}{R\overset{\overset{\displaystyle O}{\|}}{C}{-}OR'} + HOH \xrightarrow[\Delta]{H^+} \underset{\text{Carboxylic acid}}{R\overset{\overset{\displaystyle O}{\|}}{C}{-}OH} + \underset{\text{Alcohol}}{R'OH}$$

2. *Alkaline hydrolysis of an alkyl halide* (1° and 2° alcohols only)

$$\underset{\text{Alkyl halide}}{RX} + NaOH(aq) \longrightarrow \underset{\text{Alcohol}}{ROH} + NaX$$

$$CH_3CH_2Cl + NaOH(aq) \longrightarrow CH_3CH_2OH + NaCl$$

Hydrolysis is a reaction of water with another species in which the water molecule is split (see Section 17.13). The hydrolysis of an ester is the reverse reaction of esterification. A carboxylic acid and an alcohol are formed as products. The reaction can be conducted in an acid or an alkaline medium.

3. *Catalytic reduction of aldehydes and ketones* to produce primary and secondary alcohols. These reactions are discussed in Chapter 24.

In theory, these general methods provide a way to make almost any desired alcohol, but they may not be practical for a specific alcohol because the necessary starting material—ester, alkyl halide, aldehyde, or ketone—cannot be obtained at a reasonable cost. Hence, for economic reasons most of the widely used alcohols are made on an industrial scale by special methods that have been developed for specific alcohols. The preparation and properties of several of these alcohols are described in the following paragraphs.

Methanol

When wood is heated to a high temperature in an atmosphere lacking oxygen, methanol (wood alcohol) and other products are formed and driven off. The process is called *destructive distillation*, and until about 1925 nearly all methanol was obtained in this way. In the early 1920s the synthesis of methanol by high-pressure catalytic hydrogenation of carbon monoxide was developed in Germany. The reaction is

$$CO + 2\,H_2 \xrightarrow[300\text{–}400^\circ C,\ 200\ \text{atm}]{ZnO\text{–}Cr_2O_3} CH_3OH$$

Nearly all methanol is now manufactured by this method.

Methanol is a volatile (bp 65°C), highly flammable liquid. It is poisonous and capable of causing blindness or death if taken internally. Exposure to methanol vapors for even short periods of time is dangerous. Despite this danger, over 3.6×10^9 kg (7.9×10^9 lb) annually is manufactured and used for

1. Conversion to formaldehyde (methanal) primarily for use in the manufacture of polymers
2. Manufacture of other chemicals, especially various kinds of esters
3. Denaturing ethyl alcohol (rendering it unfit as a beverage)
4. An industrial solvent
5. An inexpensive and temporary antifreeze for radiators (it is not a satisfactory permanent antifreeze, because its boiling point is lower than that of water)

The experimental use of 5–30% methanol or ethanol in gasoline (*gasohol*) has shown promising results in reducing the amount of air pollutants emitted in automobile exhausts. Another benefit of using methanol in gasoline is that methanol can be made from nonpetroleum sources. The most economical nonpetroleum source of carbon monoxide for making methanol is coal. In addition to coal, burnable materials such as wood, agricultural wastes, and sewage sludge also are potential sources of methanol.

Ethanol

Ethanol is without doubt the earliest and most widely known alcohol. It is or has been known by a variety of names such as ethyl alcohol, "alcohol," grain alcohol, and spirit. Huge quantities of this substance are prepared by fermentation. Starch and sugar are the raw materials. Starch is first converted to sugar by enzyme- or acid-catalyzed hydrolysis. (An enzyme is a biological catalyst, as discussed in Chapter 32.) Conversion of simple sugars to ethanol is accomplished by yeast:

$$\underset{\text{Glucose}}{C_6H_{12}O_6} \xrightarrow{\text{Yeast}} 2\ \underset{\text{Ethanol}}{CH_3CH_2OH} + 2\ CO_2$$

For legal use in beverages, ethanol is made by fermentation; but a large part of the alcohol for industrial uses (5.9×10^8 kg [about 1.3×10^9 lb] annually) is made from petroleum-derived ethylene. Ethylene is passed into an aqueous acid solution to form ethanol.

$$CH_2{=}CH_2 + H_2O \xrightarrow{H^+} CH_3CH_2OH$$

Some of the economically significant uses of ethanol are the following:

1. Intermediate in the manufacture of other chemicals such as acetaldehyde, acetic acid, ethyl acetate, and diethyl ether
2. Solvent for many organic substances
3. Compounding ingredient for pharmaceuticals, perfumes, flavorings, and so on
4. Essential ingredient of alcoholic beverages

Ethanol acts physiologically as a food, as a drug, and as a poison. It is a food in the limited sense that the body is able to metabolize it to carbon dioxide and water with the production of energy. As a drug, ethanol is often mistakenly considered to be a stimulant, but it is in fact a depressant. In moderate quantities ethanol causes drowsiness and depresses brain functions so that activities requiring skill and judgment (such as automobile driving) are impaired. In larger quantities ethanol causes nausea, vomiting, impaired perception, and incoordination. If a very large amount is consumed, unconsciousness and ultimately death may occur.

Authorities maintain that the effects of ethanol on automobile drivers are a factor in about half of all fatal traffic accidents in the United States. The gravity of this problem can be grasped when you realize that traffic accidents are responsible for about 50,000 deaths each year in the United States.

Heavy taxes are imposed on alcohol in beverages. A gallon of pure alcohol costs only a few dollars to produce, but in a distilled beverage it bears a tax of about 20 dollars or more.

Ethanol for industrial use is often denatured (rendered unfit for drinking) and, thus, is not taxed. Denaturing is done by adding small amounts of methanol and other denaturants that are extremely difficult to remove. Denaturing is required by the federal government to protect the beverage-alcohol tax source. Special tax-free use permits are issued to scientific and industrial users who require pure ethanol for nonbeverage uses.

Isopropyl Alcohol (2-Propanol)

Isopropyl alcohol (2-propanol) is made from propene derived from petroleum. This synthesis is analogous to that used for making ethanol from ethylene:

$$\underset{\text{Propene}}{CH_3CH{=}CH_2} + H_2O \xrightarrow{H^+} \underset{\text{Isopropyl alcohol (2-Propanol)}}{CH_3\underset{\displaystyle OH}{\underset{|}{C}H}CH_3}$$

Note that 2-propanol, not 1-propanol, is produced. This is because, in the first step of the reaction, an H^+ adds to carbon 1 of propene according to Markovnikov's rule (Section 22.6). The —OH group then ends up on carbon 2 to give the final product.

Isopropyl alcohol is a relatively low-cost alcohol that is manufactured in large quantities, about 7.3×10^8 kg (1.6×10^9 lb) annually. It is not a potable alcohol, and even breathing large quantities of the vapor may cause dizziness, headache, nausea, vomiting, mental depression, and coma. Isopropyl alcohol is used (1) to manufacture other chemicals (expecially acetone), (2) as an industrial solvent, and (3) as the principal ingredient in rubbing alcohol formulations.

Ethylene Glycol (1,2-Ethanediol)

Ethylene glycol is the simplest alcohol containing two —OH groups. Like most other relatively cheap, low-molar-mass alcohols, it is commercially derived from petroleum. One industrial synthesis is from ethylene via ethylene oxide (oxirane).

$$\underset{\text{Ethylene}}{2\,CH_2{=}CH_2} + O_2 \xrightarrow[200\text{–}300^\circ C]{\text{Ag catalyst}} \underset{\substack{\text{Oxirane}\\ \text{(Ethylene oxide)}}}{2\,\overset{\diagup O \diagdown}{CH_2{-}CH_2}}$$

$$\overset{\diagup O \diagdown}{CH_2{-}CH_2} + H_2O \xrightarrow{H^+} \underset{\substack{\text{1,2-Ethanediol}\\ \text{(Ethylene glycol)}}}{HOCH_2CH_2OH}$$

Major uses of ethylene glycol are (1) in the preparation of the synthetic polyester fibers (Dacron) and film (Mylar), (2) as a major ingredient in "permanent type" antifreeze for cooling systems, (3) as a solvent in the paint and plastics industries, and (4) in the formulations of printing ink and ink for ballpoint pens.

The low molar mass, complete water solubility, low freezing point, and high boiling point make ethylene glycol a nearly ideal antifreeze. A 58% by mass aqueous solution of ethylene glycol freezes at $-48°C$. Its high boiling point and high heat of vaporization prevent it from being boiled away and permit higher, and therefore more efficient, engine operating temperatures than are possible with water alone. The U.S. production of ethylene glycol amounts to about 1.8×10^8 kg (4.0×10^8 lb) annually. Ethylene glycol is extremely toxic when ingested.

Glycerol

Glycerol, also known as *glycerine* or 1,2,3-propanetriol, is an important trihydroxy alcohol. Glycerol is a syrupy liquid with a sweet, warm taste. It is about 0.6 times as sweet as cane sugar. It is obtained as a by-product of processing animal and vegetable fats to make soap and other products, and it is also synthesized commercially from propene. The major uses of glycerol are (1) as a raw material in the manufacture of polymers and explosives, (2) as an emollient in cosmetics, (3) as a humectant in tobacco products, and (4) as a sweetener. Each use is directly related to the three —OH groups on glycerol.

The —OH groups provide sites through which the glycerol unit may be bonded to other molecules to form a polymer (Chapter 27). The explosive nitroglycerine or glyceryltrinitrate, is made by reacting the —OH groups with nitric acid:

$$\begin{matrix} CH_2OH \\ | \\ CHOH \\ | \\ CH_2OH \\ \text{Glycerol} \end{matrix} + \underset{\text{Nitric acid}}{3\,HONO_2} \longrightarrow \begin{matrix} CH_2ONO_2 \\ | \\ CHONO_2 \\ | \\ CH_2ONO_2 \\ \text{Glyceryltrinitrate} \\ \text{(Nitroglycerine)} \end{matrix} + 3\,H_2O$$

The three polar —OH groups on the glycerol molecule are able to hold water molecules by hydrogen bonding. Consequently, glycerol is a hygroscopic substance; that is, it has the ability to take up water vapor from the air. It is therefore used as a skin moisturizer in cosmetic preparations. Glycerol is also used as an additive in tobacco products; by taking up moisture from the air, it prevents the tobacco from becoming excessively dry and crumbly.

23.7 Phenols

The term **phenol** is used for the class of compounds that have a hydroxy group attached to an aromatic ring. The parent compound is called *phenol* (C_6H_5OH) and is also known as carbolic acid.

phenol

Naming Phenols

Many phenols are named as derivatives of the parent compound using the general methods for naming aromatic compounds. For example,

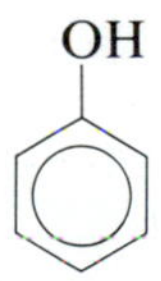

OH | OH, Br | OH, NH_2 | OH, O_2N, NO_2, NO_2

Phenol (Carbolic acid) | *m*-Bromophenol  | *p*-Aminophenol 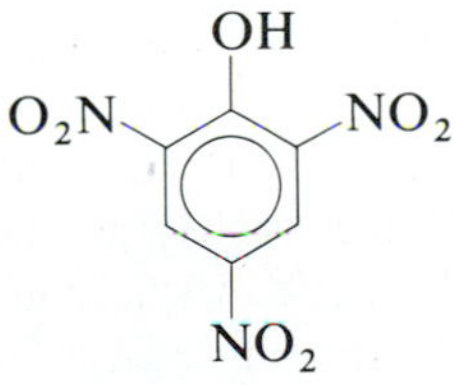 | 2,4,6-Trinitrophenol (Picric acid)

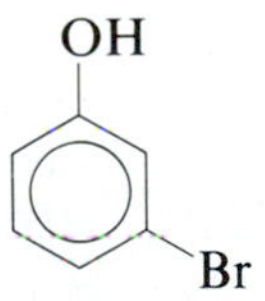

The *ortho*-, *meta*-, and *para*-dihydroxybenzenes have the special names catechol, resorcinol, and hydroquinone, respectively. The catechol structure occurs in many natural substances; and hydroquinone, a manufactured product, is commonly used as a photographic reducer and developer.

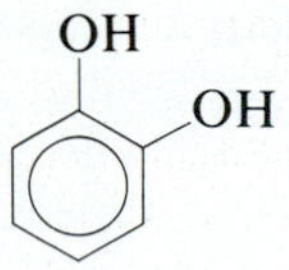

OH, OH | OH, OH | OH, OH

Catechol (1,2-Dihydroxybenzene) | Resorcinol (1,3-Dihydroxybenzene) | Hydroquinone (1,4-Dihydroxybenzene)

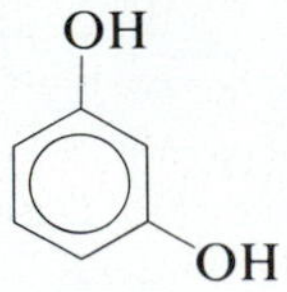

Chemistry in Action

Common Phenols

Many well-known natural substances have phenolic groups in their structures. The formulas and brief descriptions of several examples follow.

Vanillin is the principal odorous component of the vanilla bean. It is one of the most widely used flavorings and is also used for masking undesirable odors in many products such as paints.

Eugenol is the essence of oil of cloves. Two of its uses are as a dental analgesic and for the manufacture of synthetic vanillin. Thymol occurs in the oil of thyme. It has a pleasant odor and flavor and is used as an antiseptic in preparations such as mouthwashes. Thymol is the starting material for the synthesis of menthol, the main constituent of oil of peppermint. Thymol is a widely used flavoring and pharmaceutical.

Butylated hydroxytoluene (BHT) is used in small amounts as an antioxidant preservative for food, synthetic rubber, vegetable oils, soap, and some plastics.

The active irritants in poison ivy and poison oak are called urushiols. They are catechol derivatives with an unbranched 15-carbon side chain in position 3 on the phenol ring.

The widely discussed active principle of marijuana is tetrahydrocannabinol. It is obtained from the dried leaves and flowering tops of the hemp plant and has been used since antiquity for its physiological effects. The common acid–base indicator phenolphthalein is a phenol derivative. Phenolphthalein is also used as a medical cathartic. Epinephrine (adrenalin) is secreted by the adrenal gland in response to stress, fear, anger, or other heightened emotional states. It stimulates the conversion of glycogen to glucose in the body. Phenol is the starting material for the manufacture of aspirin, one of the most widely used drugs for self-medication.

OH, OCH_3, H—C=O

Vanillin

OH, OCH_3, $CH_2CH{=}CH_2$

Eugenol

OH, OH, $C_{15}H_{25-31}$

Urushiols

CH_3, OH, CH, CH_3 CH_3

Thymol

OH, $(CH_3)_3C$, $C(CH_3)_3$, CH_3

2,6-di-*t*-Butyl-4-methylphenol (Butylated hydroxytoluene, BHT)

CH_3, OH, CH_3, O, CH_3, *n*-C_5H_{11}

Tetrahydrocannabinol (from Marijuana)

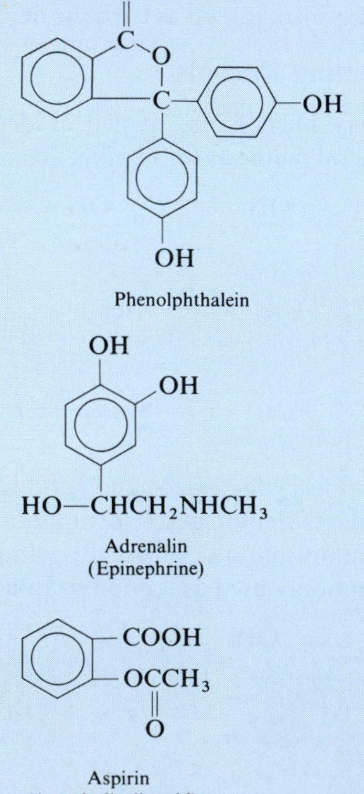

Phenolphthalein

Adrenalin (Epinephrine)

Aspirin (Acetylsalicylic acid)

The *ortho*-, *meta*-, and *para*-methylphenols are present in coal tar and are known as cresols. They are all useful disinfectants.

o-Cresol (*o*-Methylphenol) — OH and CH_3 on adjacent ring carbons

m-Cresol (*m*-Methylphenol) — OH and CH_3 in meta positions

p-Cresol (*p*-Methylphenol) — OH and CH_3 in para positions

23.8 Properties of Phenols

In the pure state, phenol is a colorless crystalline solid with a melting point at about 41°C and a characteristic odor. Phenol is highly poisonous. Ingestion of even small amounts of it may cause nausea, vomiting, circulatory collapse, and death from respiratory failure.

Phenol is a weak acid; it is more acidic than alcohols and water but less acidic than acetic and carbonic acids. The pH values are as follows: 0.1 *M* acetic acid, 2.87; water, 7.0; 0.1 *M* phenol, 5.5. Thus, phenol reacts with sodium hydroxide solution to form a salt but does not react with sodium bicarbonate. The salt formed is called sodium phenoxide or sodium phenolate. Sodium hydroxide does not remove a hydrogen atom from an alcohol because alcohols are weaker acids than water.

$$C_6H_5OH + NaOH \longrightarrow C_6H_5ONa + H_2O$$

Sodium phenoxide (Sodium phenolate)

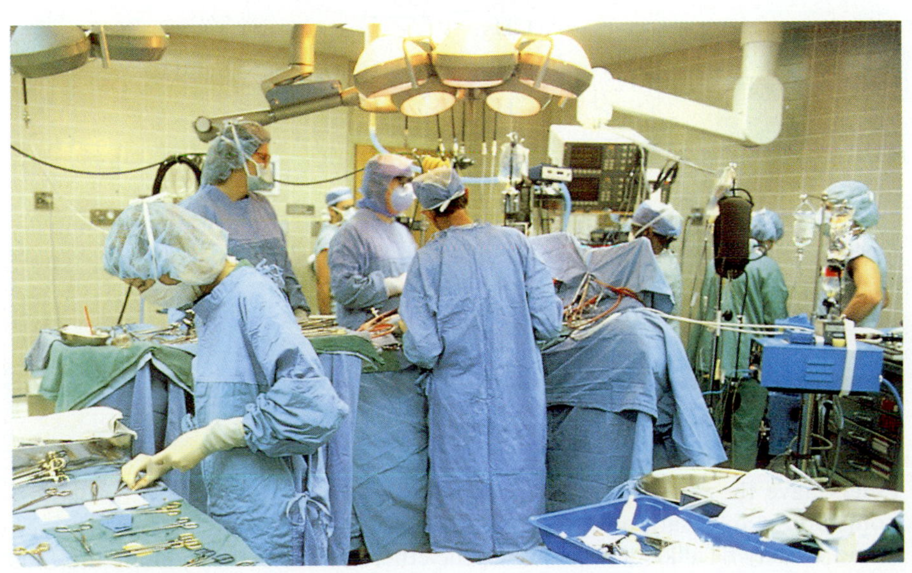

◀ **Phenols were among the first antiseptics to be used in operating rooms to prevent the spread of bacteria. Today other antiseptics such as iodine solutions and germicidal soaps have replaced phenols.**

In general the phenols are toxic to microorganisms. They are widely used as antiseptics and disinfectants. Phenol was the first compound to be used extensively as an operating room disinfectant. Joseph Lister (1827–1912) first used phenol for this purpose in 1867. The antiseptic power of phenols is increased by substituting alkyl groups (up to six carbons) in the benzene ring. For example, 4-hexylresorcinol is used as an antiseptic in numerous pharmaceuticals.

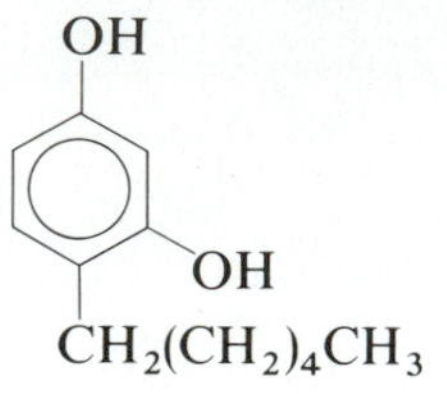

OH

OH

$CH_2(CH_2)_4CH_3$

4-Hexylresorcinol

23.9 Production of Phenol

Phenol is obtained from coal tar. In addition, there are several commercial methods used to produce phenol synthetically. The most economical of these methods starts with benzene and propylene, which react to form cumene (isopropylbenzene). Cumene is then oxidized by air to cumene hydroperoxide, which is treated with dilute sulfuric acid to obtain phenol and acetone. The economic feasibility of the process is due to the fact that two important commercial products are produced. The equations for the reactions are

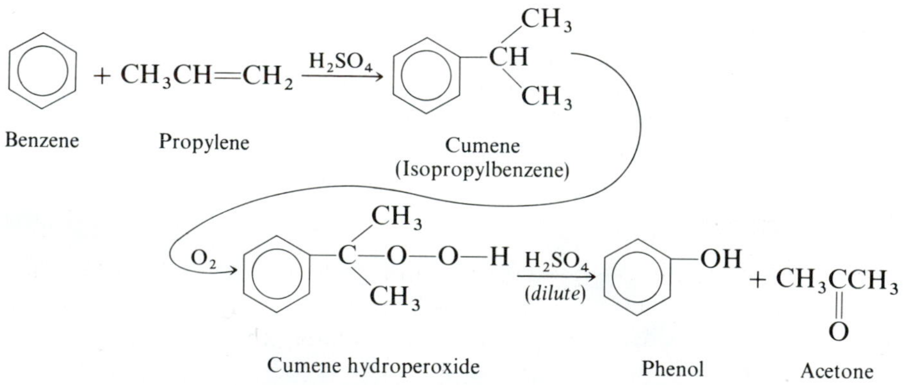

$$C_6H_6 + CH_3CH{=}CH_2 \xrightarrow{H_2SO_4} C_6H_5CH(CH_3)_2$$

Benzene Propylene Cumene (Isopropylbenzene)

$$\xrightarrow{O_2} C_6H_5C(CH_3)_2{-}O{-}O{-}H \xrightarrow[(dilute)]{H_2SO_4} C_6H_5{-}OH + CH_3\overset{\overset{\displaystyle O}{\|}}{C}CH_3$$

Cumene hydroperoxide Phenol Acetone

Over 1.1×10^9 kg (2.4×10^9 lb) of synthetic phenol is produced annually in the United States. The chief use of phenol is for the manufacture of phenol–formaldehyde resins and plastics (see Chapter 27).

23.10 Ethers

ether

Ethers have the general formula ROR′. The groups R and R′ can be derived from saturated, unsaturated, or aromatic hydrocarbons; and, for a given ether, R and R′ may be alike or different. Cyclic ethers are formed by joining the ends

TABLE 23.5 Names and Structural Formulas of Ethers

Name[a]	Formula	Boiling point (°C)
Dimethyl ether (Methoxymethane)	$CH_3—O—CH_3$	−25
Methyl ethyl ether (Methoxyethane)	$CH_3CH_2—O—CH_3$	8
Diethyl ether (Ethoxyethane)	$CH_3CH_2—O—CH_2CH_3$	35
Ethyl isopropyl ether (2-Ethoxypropane)	$CH_3CH_2—O—CH(CH_3)CH_3$	54
Divinyl ether	$CH_2{=}CH—O—CH{=}CH_2$	39
Anisole (Methoxybenzene) Methyl phenyl ether	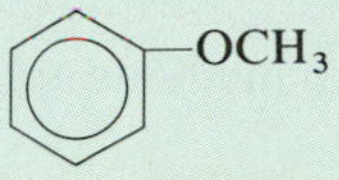 $C_6H_5—OCH_3$	154
Diphenyl ether	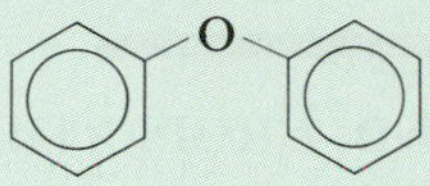 $C_6H_5—O—C_6H_5$	259
Tetrahydrofuran (THF)	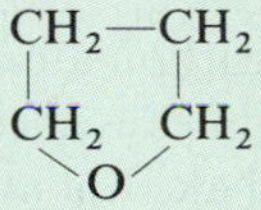 $CH_2—CH_2$, CH_2 CH_2, O (ring)	66

[a] The IUPAC name is in parentheses when given.

of a single hydrocarbon chain through an oxygen atom to form a ring structure. Table 23.5 shows structural formulas and names for some of the different kinds of ethers.

Naming Ethers

Individual ethers, like alcohols, may be known by several names. The ether having the formula $CH_3CH_2—O—CH_2CH_3$ and formerly widely used as an anesthetic is called diethyl ether, ethyl ether, ethoxyethane, or simply ether. Common names of ethers are formed from the names of the groups attached to the oxygen atom, followed by the word *ether*.

$$\underset{\text{Methyl}}{CH_3}—\underset{\text{Ether}}{\boxed{O}}—\underset{\text{Methyl}}{CH_3} \qquad \underset{\text{Methyl}}{CH_3}—\underset{\text{Ether}}{\boxed{O}}—\underset{\text{Ethyl}}{CH_2CH_3}$$

Dimethyl ether　　　　Methyl ethyl ether

In the IUPAC System, ethers are named as alkoxy (RO—) derivatives of the alkane corresponding to the longest carbon–carbon chain in the molecule. To name an ether by this system:

1. Select the longest carbon–carbon chain and label it with the name of the corresponding alkane.

2. Change the *yl* ending of the other hydrocarbon group to *oxy* to obtain the alkoxy group name. For example, CH_3O— is called *methoxy*.
3. Combine the two names from Steps 1 and 2, giving the alkoxy name first, to form the ether name.

CH_3—O—CH_2CH_3

This is the longest C—C chain, so call it *ethane*.

CH_3O— is the other group; modify its name from *methyl* to *methoxy* and combine with *ethane* to obtain the name of the ether, methoxyethane. Thus

CH_3CH_2—O—CH_2CH_3 is ethoxyethane

$CH_3CH_2CH_2$—O—$CH_2CH_2CH_2CH_3$ is propoxybutane

Additional examples named by both methods are found in Table 23.5.

23.11 Structures and Properties of Ethers

An oxygen atom linking together two carbon atoms is the key structural feature of an ether molecule. This oxygen atom causes ether molecules to have a bent shape somewhat like that of water and alcohol molecules:

H—Ö—H (Water) R—Ö—H (Alcohol) R—Ö—R (Ether)

Ethers are somewhat more polar than alkanes, because alkanes lack the oxygen atom with its exposed, nonbonded electrons. But ethers are much less polar than alcohols, since no hydrogen is attached to the oxygen atom in an ether. The solubility and boiling point (vapor pressure) characteristics of ethers are related to the C—O—C structure. Alkanes have virtually no solubility in water or acid. But about 7.5 g of diethyl ether will dissolve in 100 g of water at 20°C. Diethyl ether also dissolves in sulfuric acid. Hydrogen bonding between ether and water molecules and between ether and acid molecules is responsible for this solubility.

Ether····Water: CH_3CH_2—O—CH_2CH_3 ···· H—O—H

Ether····Acid: CH_3CH_2—O—CH_2CH_3 ···· H—O—S(O)(O)—O—H

← Hydrogen bond →

Because no —OH group is present, hydrogen bonding does not occur between ether molecules. This lack of hydrogen bonding can be seen by comparing the boiling points of a hydrocarbon, an ether, and an alcohol of similar molecular mass, as in Table 23.6. The boiling point of the ether is somewhat above that of the hydrocarbon but much lower than that of the more polar alcohol.

TABLE 23.6 Boiling Points of Ethers, Alkanes, and Alcohols

Name	Formula	Molar mass	Boiling point (°C)
Dimethyl ether	CH_3OCH_3	46	−24
Propane	$CH_3CH_2CH_3$	44	−42
Ethanol	CH_3CH_2OH	46	78
Methoxyethane	$CH_3OCH_2CH_3$	60	8
Butane	$CH_3CH_2CH_2CH_3$	58	−0.6
1-Propanol	$CH_3CH_2CH_2OH$	60	97
2-Propanol	$CH_3CH(OH)CH_3$	60	83

Ethers, especially diethyl ether, are exceptionally good solvents for organic compounds. Many polar compounds, including water, acids, alcohols, and other oxygenated organic compounds, dissolve, at least to some extent, in ethers. This solubility is a result of intermolecular attractions between the slightly polar ether molecules and the molecules of the other polar substance. Nonpolar compounds such as hydrocarbons and alkyl halides also dissolve in ethers. These substances dissolve because the ether molecules are not very polar and therefore are not strongly attracted either to one another or to the other kinds of molecules. Thus, ether molecules are able to intermingle freely with the molecules of a nonpolar substance and form a solution by simple mixing.

In summary, ethers are polar enough to dissolve some polar substances; but their polarity is so slight that they act as nonpolar solvents for a great many nonpolar substances.

Ethers have little chemical reactivity, but because a great many organic substances dissolve readily in ethers, they are often used as solvents in laboratory and manufacturing operations. Their use may be dangerous, since low-molar-mass ethers are volatile, and their highly flammable vapors form explosive mixtures with air. Another hazard of ethers is that, despite their generally low chemical reactivity, oxygen of the air slowly reacts with them to form unstable peroxides that are subject to explosive decomposition:

$$CH_3CH_2{-}O{-}CH_2CH_3 + O_2 \longrightarrow \underset{\substack{|\\ O{-}O{-}H}}{CH_3CH}{-}O{-}CH_2CH_3$$

Diethyl ether hydroperoxide

23.12 PREPARATION OF ETHERS

We have seen that ethers can be made by intermolecular dehydration of alcohols by heating in the presence of an acid (see Section 23.5). Ethers are also made from sodium alkoxides or sodium phenoxides and alkyl halides, a reaction called the *Williamson synthesis.*

Chemistry in Action

Ethers

The most widely known use of ethyl ether has been as a general anesthetic for surgery. The introduction of ether for this purpose is one of the great landmarks of medicine. Two Americans, Crawford W. Long and William T. Morton, played important roles in this development. Long, a physician, used ether in a surgical operation as early as 1842 but did not publish his discovery until 1849. Morton, a dentist, used ether as an anesthetic for dental work in 1846. He publicly demonstrated its effectiveness by administering ether to a patient undergoing surgery at the Massachusetts General Hospital on October 16, 1846.

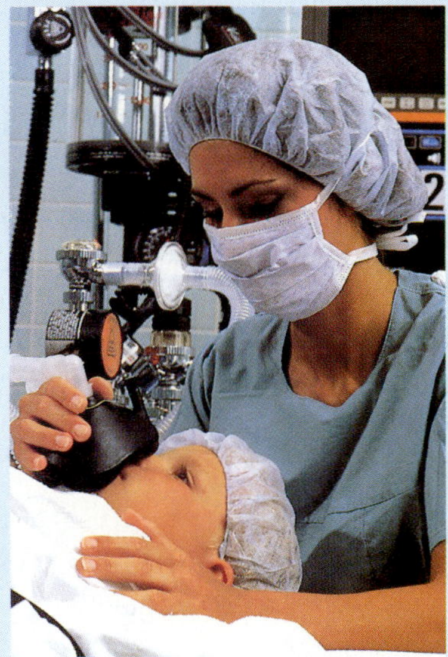

The word *anesthesia* is from the Greek, meaning insensibility, and was suggested to Morton by the poet and physician Oliver Wendell Holmes. A *general anesthetic* is a substance or combination of substances that produces both unconsciousness and insensitivity to pain. Many other substances, including other ethers such as divinyl ether (Vinethene) and methoxyflurane (Penthrane), are general inhalation anesthetics. These substances are superior in some respects, and in recent years have replaced ethyl ether as a general anesthetic.

$$CH_2{=}CH{-}O{-}CH{=}CH_2$$

Divinyl ether

$$CHCl_2CF_2{-}O{-}CH_3$$

Methoxyflurane

Ether produces unconsciousness by depressing the activity of the central nervous system. The major disadvantages of ether include flammability, irritation of respiratory passages, and occurrence of nausea and vomiting after its use. These hazards have resulted in a change toward the use of other substances (such as nitrous oxide, N_2O, or halogenated compounds like halothane, $CF_3CHClBr$) as general anesthetics. These substitute compounds also pose hazards and must be used with caution.

A number of theories explaining the biochemical activity of anesthetics have been proposed. All are based on the nonpolar nature of the compounds. This property leads to high solubility in fats and membranes. One theory suggests that anesthetics interfere with electrical activity in nerve impulses by dissolving in brain cells.

Ether also serves as an excellent extracting medium. It acts as a good solvent for separating lipids (ether soluble) from carbohydrates and proteins (ether insoluble). Ether is the solvent of choice for cocaine users as well. The technique known as "free-basing" involves the separation of cocaine from other substances by extracting it into ether. One method used to capture cocaine manufacturers involves tracking large ether shipments.

$$\underset{\text{Sodium alkoxide}}{RONa} + \underset{\text{Alkyl halide}}{R'X} \longrightarrow \underset{\text{Ether}}{ROR'} + \underset{\text{Sodium halide}}{NaX}$$

The Williamson synthesis is especially useful in the preparation of mixed ethers (where $R \neq R'$) and aromatic ethers:

$$\underset{\substack{\text{Sodium}\\\text{ethoxide}}}{CH_3CH_2ONa} + \underset{\text{Bromomethane}}{CH_3Br} \longrightarrow \underset{\substack{\text{Methoxyethane}\\\text{(Methyl ethyl ether)}}}{CH_3CH_2{-}O{-}CH_3} + NaBr$$

$$C_6H_5ONa + CH_3Br \longrightarrow C_6H_5OCH_3 + NaBr$$

Sodium phenoxide — Bromomethane — Methyl phenyl ether (Anisole)

It is interesting to note that organic chemistry is replete with "name reactions," such as the Williamson synthesis, named after the scientists who invented the reactions.

23.13 Thiols

Sulfur and oxygen are found next to each other in the same family on the periodic table. This indicates some similarity in the formulas of their compounds. Organic compounds which contain the —SH group are analogs of alcohols. The —SH-containing compounds are known as **thiols,** or mercaptans. Examples include:

thiols

CH_3SH
Methanethiol
(Methyl mercaptan)

$$CH_3CH_2\underset{\displaystyle SH}{\underset{|}{C}}HCH_3$$

2-Butanethiol
(*sec*-Butyl mercaptan)

Thiols have a higher molar mass than corresponding alcohols, but boil at lower temperatures (ethanol 78°C, ethanethiol 36°C). The reason for this discrepancy lies in the fact that alcohols form hydrogen bonds while thiols do not.

The major important properties of thiols are summarized below:

1. Foul odors: some of these compounds smell so awful that companies make special labels to warn consumers. The odor given off by a frightened skunk has thiols as the active ingredient. Natural gas is odorized to be detectable by adding small amounts of methanethiol.
2. Oxidation to disulfides

$$2\,RSH \xrightarrow{[O]} R{-}S{-}S{-}R$$

Thiol — Disulfide

This reaction can be accomplished using many oxidizing agents. The sulfur is being oxidized just as the carbon was being oxidized in alcohols. The conversion between thiols and disulfides is important in proteins. When oxidation occurs one part of a protein chain is linked to another, giving the protein a characteristic three-dimensional shape as shown in Figure 23.2.

Coenzyme A (CoA or CoASH) is a complex coenzyme (Chapter 36) which plays a central role in the metabolism of carbohydrates, lipids, and proteins. The acetyl derivative of CoA, called acetyl CoA, occurs in all living organisms. The link between the acetyl group and the coenzyme is a high-energy bond. Acetyl CoA is very reactive and therefore functions as a carrier for acetyl groups in the cell. The biochemical synthesis of acetyl CoA is complex, involving many steps, but may be described overall as a thioesterification.

$$CH_3\overset{\overset{\displaystyle O}{\|}}{C}-OH + H-S-CoA \longrightarrow CH_3-\overset{\overset{\displaystyle O}{\|}}{C}-S-CoA + H_2O$$

Acetic acid Coenzyme A Acetyl CoA

FIGURE 23.2 ▶ Oxidation of thiols to disulfides.

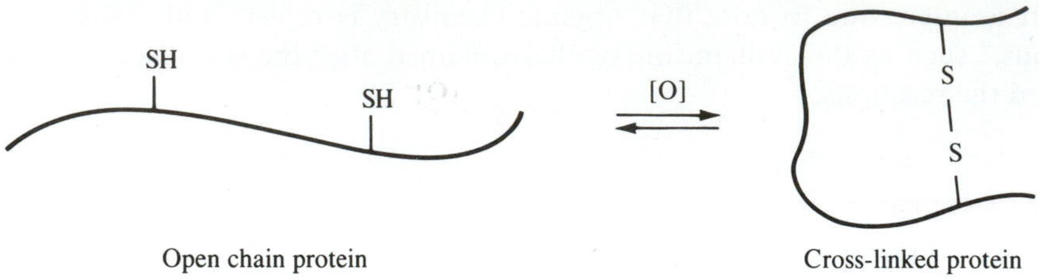

SH SH [O] S S

Open chain protein Cross-linked protein

CONCEPTS IN REVIEW

1. Name alcohols by common and IUPAC methods.
2. Write the structural formula when given the name of an alcohol.
3. Write the structural formulas for all the isomeric alcohols of a given molecular formula.
4. Recognize and identify primary, secondary, and tertiary alcohols.
5. Compare and explain the relative solubilities of alcohols and ethers in water.
6. Indicate a class of biochemicals that has many of the same properties as the polyhydroxy alcohols.
7. Summarize the acid/base properties of alcohols and alkoxides.
8. Write equations for the oxidation of alcohols.
9. Write equations for the dehydration of alcohols to ethers and alkenes.
10. Differentiate between intramolecular and intermolecular dehydration, indicating proper conditions for each.
11. Write equations for the synthesis of alcohols from alkyl halides and from alkenes.
12. Explain the relative reactivities of primary, secondary, and tertiary alcohols.
13. Understand the common methods of preparing methyl alcohol, ethyl alcohol, isopropyl alcohol, ethylene glycol, and glycerol.
14. Name phenols and write their formulas.
15. Understand the differences in properties of the hydroxyl group when bonded to an aromatic ring (a phenol) and to an aliphatic group (an alcohol).
16. Be familiar with the general properties of phenols.
17. Name ethers and write their formulas.
18. Write equations for preparing ethers by the Williamson synthesis.
19. List major properties for thiols.
20. Discuss relative boiling points and water solubilities of comparable hydrocarbons, alcohols, dihydroxy alcohols, ethers, and thiols.
21. Discuss the hazards of using ethers in the laboratory.

Equations in Review

Alcohols

Oxidation reactions

$$R{-}CH_2{-}OH \xrightarrow{[O]} R{-}\overset{\overset{\displaystyle O}{\|}}{C}{-}H + H_2O \xrightarrow{[O]} R{-}\overset{\overset{\displaystyle O}{\|}}{C}{-}OH$$

Primary alcohol — Aldehyde — Carboxylic acid

$$R{-}\underset{\underset{\displaystyle OH}{|}}{\overset{\overset{\displaystyle H}{|}}{C}}{-}R \xrightarrow{[O]} R{-}\overset{\overset{\displaystyle O}{\|}}{C}{-}R + H_2O$$

Secondary alcohol — Ketone

$$R{-}\underset{\underset{\displaystyle OH}{|}}{\overset{\overset{\displaystyle R}{|}}{C}}{-}R \xrightarrow{[O]} \text{No reaction}$$

Tertiary alcohol

Dehydration reactions

Intramolecular

$$H{-}\underset{\underset{\displaystyle H}{|}}{\overset{\overset{\displaystyle H}{|}}{C}}{-}\underset{\underset{\displaystyle OH}{|}}{\overset{\overset{\displaystyle H}{|}}{C}}{-}H \xrightarrow[180^\circ C]{H_2SO_4} CH_2{=}CH_2 + H_2O$$

Intermolecular

$$\begin{matrix} CH_3CH_2OH \\ CH_3CH_2OH \end{matrix} \xrightarrow[140^\circ C]{H_2SO_4} CH_3CH_2OCH_2CH_3 + H_2O$$

Esterification

$$R{-}\overset{\overset{\displaystyle O}{\|}}{C}{-}OH + H{-}OR' \xrightarrow{H^+} R{-}\overset{\overset{\displaystyle O}{\|}}{C}{-}OR' + H_2O$$

Ethers

Williamson synthesis

$$RO^-\ {}^+Na + R'X \longrightarrow ROR' + NaX$$

Thiols

Oxidation to disulfides

$$RSH \xrightarrow{[O]} R{-}S{-}S{-}R$$

Exercises

1. Write structural formulas for and give an example of
   (a) An alkyl halide (e) A ketone
   (b) A phenol (f) A carboxylic acid
   (c) An ether (g) An ester
   (d) An aldehyde (h) A thiol
2. Although it is possible to make alkenes from alcohols, alkenes are seldom, if ever, made in this way on an industrial scale. Why not?
3. Isopropyl alcohol is usually used in rubbing alcohol formulations. Why is this alcohol used in preference to normal propyl alcohol?
4. What classes of compounds can be formed by the oxidation of primary alcohols? Cite examples.
5. What is the molar mass of myricyl alcohol, an open-chain saturated alcohol containing 30 carbon atoms? Myricyl alcohol is present in beeswax as an ester.
6. Why is ethylene glycol (1,2-ethanediol) superior to methyl alcohol (methanol) as an antifreeze for automobile radiators?
7. Briefly outline the physiological effects of
   (a) Methanol (b) Ethanol
8. Explain, in terms of molecular structure, why ethanol (CH_3CH_2OH, molar mass = 46) is a liquid at room temperature and dimethyl ether (CH_3OCH_3, molar mass = 46) is a gas.
9. Write structural formulas for
   (a) Methanol
   (b) 2-Butanol
   (c) 3-Methyl-1-hexanol
   (d) 2-Methyl-2-butanol
   (e) Propylene glycol
   (f) Isopropyl alcohol
   (g) 1-Phenylethanol
   (h) Cyclopentanol
   (i) 2,3-Butanediol
   (j) *sec*-Butyl alcohol
   (k) 2-Propanethiol
   (l) *n*-Pentyl mercaptan
10. Write structures for all the isomers (alcohols and ethers) with the formula
    (a) C_3H_8O
    (b) $C_4H_{10}O$
11. There are eight open-chain isomeric alcohols that have the formula $C_5H_{11}OH$. Write the structural formula and the IUPAC name for each of these alcohols.
12. Which of the isomers in Question 11 are
    (a) Primary alcohols?
    (b) Secondary alcohols?
    (c) Tertiary alcohols?
13. Write the structural formula of a glycol that is both a primary and a secondary alcohol.
14. Name the following compounds:
    (a) CH_3CH_2OH
    (b) $CH_3CH(OH)CH_3$
    (c) C_6H_5—CH_2CH_2OH
    (d) $CH_3CH_2CH(OH)CH_2CH_3$
    (e) $CH_3\underset{\displaystyle OH}{\overset{\displaystyle CH_2CH_3}{C}}CH_2CH_3$
    (f) cyclopentane ring bearing OH and CH_3 on the same carbon
    (g) C_6H_5—$CH_2\underset{\displaystyle CH_3}{CH}CH_2OH$
    (h) CH_2—CH_2 bridged by O (epoxide)
    (i) $CH_3\underset{\displaystyle OH}{CH}$—$\underset{\displaystyle OH}{CH_2}$
    (j) $HOCH_2\underset{}{\overset{\displaystyle CH_3}{CH}}$—$\overset{\displaystyle CH_3}{CH}CH_2OH$
15. Cyclic glycols show cis–trans isomerism because of restricted rotation about the carbon–carbon bonds in the ring. Draw and label the structures for *cis*- and *trans*-cyclopentane-1,2-diol.
16. Write the formula and the name of the chief product when the following alcohols are dehydrated to alkenes:
    (a) $CH_3\underset{\displaystyle OH}{CH}\overset{\displaystyle CH_3}{CH}CH_3$
    (b) $CH_3\underset{\displaystyle OH}{CH}CH_2CH_2CH_3$
    (c) cyclohexane ring—OH
17. When 1-butanol is dehydrated to an alkene, it yields mainly 2-butene rather than 1-butene. This indicates that the dehydration process is at least a two-step reaction. Suggest a mechanism to explain the reaction. (*Hint: n*-Butyl carbocation is formed initially.)

18. Alcohols can be made by reacting alkyl halides with aqueous sodium hydroxide, as follows:

$$RX + NaOH(aq) \longrightarrow ROH + NaX(aq)$$

Give the names and formulas of the alkyl bromides (RBr) needed to prepare the following alcohols by this method:
(a) Isopropyl alcohol (c) Cyclohexanol
(b) 3-Methyl-1-butanol
Write the equation for the preparation of each alcohol.

19. Write the equation for the preparation of an alcohol by reacting each of the following alkenes with sulfuric acid and water:
(a) Propene (d) 1-Pentene
(b) 1-Butene (e) 2-Methyl-2-butene
(c) 2-Butene

20. Write equations to show how each of the following transformations can be accomplished. Some conversions may require more than one step, and some reactions studied in previous chapters may be needed.

(a) $CH_3\underset{|}{\underset{OH}{C}}HCH_3 \longrightarrow CH_3\overset{O}{\overset{\|}{C}}CH_3$

(b) $CH_3CH_2CH_2CH{=}CH_2 \longrightarrow CH_3CH_2CH_2\underset{OH}{\underset{|}{C}}HCH_3$

(c) $CH_3CH_2OH \longrightarrow CH_3CH_2O^-\,Na^+$

(d) $CH_3CH_2CH{=}CH_2 \longrightarrow CH_3CH_2\overset{O}{\overset{\|}{C}}CH_3$

(e) $CH_3CH_2CH_2CH_2OH \longrightarrow CH_3CH_2\underset{Cl}{\underset{|}{C}}HCH_3$

(f) $CH_3CH_2CH_2Cl \longrightarrow CH_3CH_2\underset{H}{\underset{|}{C}}{=}O$

21. Complete the following equations and name the principal organic product formed in each case:

(a) $2\ CH_3CH_2OH \xrightarrow[140^\circ C]{96\%\ H_2SO_4}$

(b) $CH_3CH_2CH_2OH \xrightarrow[180^\circ C]{96\%\ H_2SO_4}$

(c) $CH_3CH(OH)CH_2CH_3 \xrightarrow[\Delta]{K_2Cr_2O_7/H_2SO_4}$

(d) $CH_3CH_2\overset{O}{\overset{\|}{C}}{-}OCH_2CH_3 + H_2O \xrightarrow{H^+}$

22. Benzyl alcohol ($C_6H_5CH_2OH$) is a primary alcohol. Write the formulas of two different organic compounds that can be obtained by oxidizing benzyl alcohol.

23. Name the following compounds:

(a) OH
(b) OH, OH
(c) OH, CH_3
(d) OH, Br, NO_2
(e) OH, OH, $CH_2(CH_2)_4CH_3$
(f) OH, NO_2, NO_2

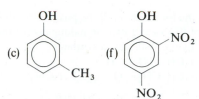

24. Write structural formulas for each of the following:
(a) *o*-Cresol
(b) *p*-Nitrophenol
(c) Resorcinol
(d) 2,6-Dimethylphenol
(e) 4-Hydroxy-3-methoxybenzaldehyde (vanillin)

25. Summarize the general properties of phenols.

26. Write equations for the cumene hydroperoxide synthesis of phenol and acetone.

27. Which of the following compounds would you expect to react with (a) sodium metal and (b) sodium hydroxide solution? Write equations for those that react.

(a) OH
(b) CH_2OH
(c) O

28. Starting with *p*-cresol and ethane, show equations for the synthesis of ethyl-*p*-methylphenyl ether (p-$CH_3C_6H_4OCH_2CH_3$).

29. What two hazards may be present when working with low molar mass ethers?

30. Arrange the following substances in order of increasing solubility in water:
(a) $CH_3CH_2{-}O{-}CH_2CH_2CH_3$
(b) $CH_3CH(OH)CH_2CH_2CH_3$
(c) $CH_3CH_2CH_2CH_2CH_3$
(d) $CH_3CH(OH)CH(OH)CH_2CH_3$

31. Arrange these three compounds in order of increasing acidity:

C_6H_5—CH_2OH (benzyl alcohol)   m-cresol (OH and CH_3 on benzene ring) H_2O

32. There are six isomeric saturated ethers that have the formula $C_5H_{12}O$. Write the structural formula and name for each of these ethers.

33. Write the balanced chemical equation for the complete combustion of diethyl ether,

$CH_3CH_2—O—CH_2CH_3$

34. Write the formulas of all the possible combinations of RONa and RCl for making each of these ethers by the Williamson synthesis:

(a) $CH_3CH_2—O—CH_3$
(b) $CH_3CH_2CH_2—O—CH_2CH_2CH_3$
(c) $C_6H_5—CH_2—O—CH_2CH_3$
(d) $(CH_3)_2HC—O—CH_2CH_2CH_3$

35. Write equations for the following transformations. Name each of the products.

(a) $C_6H_5—CH_2OH \longrightarrow C_6H_5—CH_2—O—CH_2CH_3$

(b) cyclopentyl—Br ⟶ cyclopentyl—O—cyclopentyl

36. Give a simple chemical test that will distinguish between the compounds in each of the following pairs:

(a) Ethanol and dimethyl ether
(b) 1-Pentanol and 1-pentene
(c) *p*-Cresol and methyl phenyl ether

37. Complete the following equations giving only the major organic products:

(a) $CH_3CH_2OH + Na \longrightarrow$

(b) $CH_3CH_2CH_2CH_2OH \xrightarrow[H_2SO_4]{K_2Cr_2O_7}$

(c) $CH_3CH_2OCH_2CH_3 + Na \longrightarrow$

(d) cyclopentyl—$CH{=}CH_2 \xrightarrow{H_2SO_4/H_2O}$

(e) $CH_3CH_2CH_2OH + NaOH \longrightarrow$

38. Which of the following statements are correct? Rewrite the incorrect statements to make them correct.

(a) Another name for isopropyl alcohol is 2-propanol.
(b) Ethanol and dimethyl ether are isomers.
(c) Alcohols and phenols are more acidic than water.
(d) Sodium ethoxide can be prepared by reacting ethyl alcohol and sodium hydroxide solution.
(e) Methyl alcohol is a very poisonous substance that can lead to blindness if ingested.
(f) Tertiary alcohols are easier to oxidize than primary alcohols.
(g) A correct name for $CH_3CH_2CH(OH)CH_3$ is *sec*-butyl alcohol.
(h) $(CH_3)_3CCH_2OH$ is a primary alcohol.
(i) When a secondary alcohol is oxidized, a ketone is formed.
(j) Alcohols have higher boiling points than ethers with comparable molecular masses due to hydrogen bonding between the alcohol molecules.
(k) The product formed when a molecule of water is split out between an alcohol and a carboxylic acid is called an ether.
(l) When 1-butene is reacted with dilute H_2SO_4, the alcohol formed is 1-butanol.
(m) Ethanol used for industrial purposes and rendered unfit for use in beverages is said to be denatured.
(n) The common name for 1,2,3-propanetriol is ethylene glycol.
(o) Although ethyl alcohol is used in beverages, it is still classified physiologically as a depressant and a poison.
(p) Cyclohexanol is a primary alcohol.
(q) Dihydroxy alcohols are more soluble in water than monohydroxy alcohols.
(r) Aldehydes and ketones may be prepared by the oxidation of primary alcohols.
(s) Thiols have a higher molar mass and lower boiling point than alcohols containing an equal number of carbons.
(t) Coenzyme A is an alcohol that has a central role in metabolism.

24

Aldehydes and Ketones

◀ CHAPTER OPENING PHOTO: **Many plastic products begin with molecules known as aldehydes.**

Many organic molecules contain a carbon atom that is connected to oxygen with a double bond. This particular grouping of atoms is particularly reactive and is present in both aldehydes and ketones. Formaldehyde is by far the most common aldehyde molecule. It finds commercial uses as a preservative for animal specimens and in the formation of formaldehyde polymers. These polymers are often used to form plastics, such as Formica and Melmac, and as adhesives in the manufacture of plywood and fiberboard. Other aldehydes are found in nature as spices.

Acetone is the simplest and most common ketone in our lives. It is used in nail polish remover, paints, varnishes, and resins. Acetone is also produced in the body during lipid metabolism. It generally is metabolized, but in diabetic patients more is formed than can be oxidized. The presence of acetone in a urine sample or on the breath is a positive indicator of diabetes. An understanding of aldehydes and ketones forms the basis for a discussion of many of the important organic and biochemical reactions.

24.1 STRUCTURE OF ALDEHYDES AND KETONES

carbonyl group

aldehydes

ketones

The aldehydes and ketones are closely related classes of compounds. Their structures contain the **carbonyl group**, $\rangle C{=}O$, a carbon–oxygen double bond. **Aldehydes** have at least one hydrogen atom bonded to the carbonyl group, whereas **ketones** have only alkyl or aryl (aromatic, denoted Ar) groups bonded to the carbonyl group.

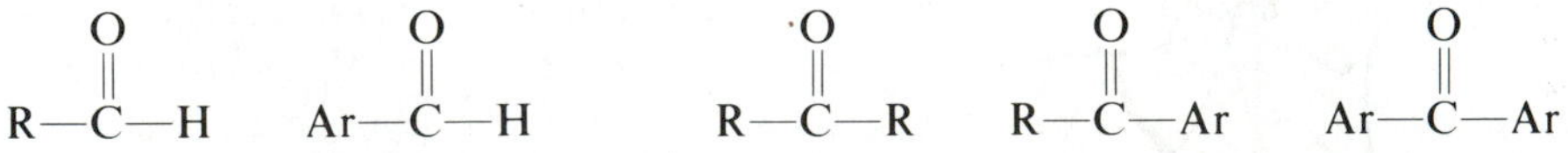

$$\underset{\text{Aldehydes}}{R{-}\overset{O}{\overset{\|}{C}}{-}H \qquad Ar{-}\overset{O}{\overset{\|}{C}}{-}H} \qquad\qquad \underset{\text{Ketones}}{R{-}\overset{O}{\overset{\|}{C}}{-}R \qquad R{-}\overset{O}{\overset{\|}{C}}{-}Ar \qquad Ar{-}\overset{O}{\overset{\|}{C}}{-}Ar}$$

In a linear expression, the aldehyde group is often written as CHO. For example,

$$CH_3CHO \quad \text{is equivalent to} \quad CH_3\overset{O}{\overset{\|}{C}}{-}H$$

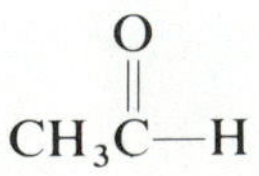

In the linear expression of a ketone, the carbonyl group is written as CO; for example,

$$CH_3COCH_3 \quad \text{is equivalent to} \quad CH_3\overset{\overset{\displaystyle O}{\|}}{C}CH_3$$

The general formula for the saturated homologous series of aldehydes and ketones is $C_nH_{2n}O$.

24.2 Naming Aldehydes and Ketones

Aldehydes

The IUPAC names of aliphatic aldehydes are obtained by dropping the final *e* and adding *al* to the name of the parent hydrocarbon (that is, the longest carbon–carbon chain carrying the —CHO group). The aldehyde carbon is always at the beginning of the carbon chain, is understood to be carbon number 1, and does not need to be numbered. The first member of the homologous series, $H_2C{=}O$, is methanal. The name *methanal* is derived from the hydrocarbon methane, which contains one carbon atom. The second member of the series is ethanal; the third member of the series is propanal; and so on.

CH_4	$H{-}\overset{\overset{\displaystyle O}{\|}}{C}{-}H$	CH_3CH_3	$CH_3\overset{\overset{\displaystyle O}{\|}}{C}{-}H$
Methane	Methanal (from methan~~e~~ + *al*)	Ethane	Ethanal (from ethan~~e~~ + *al*)

The longest carbon chain containing the aldehyde group is the parent compound. Other groups attached to this chain are numbered and named as we have done previously. For example,

$$CH_3CH_2\overset{\overset{\displaystyle CH_3}{|}}{C}HCH_2CH_2\overset{\overset{\displaystyle O}{\|}}{C}{-}H$$

4-Methylhexanal

Common names for some aldehydes are widely used. The common names for the aliphatic aldehydes are derived from the common names of the carboxylic acids (see Table 25.1). The *-ic acid* or *-oic acid* ending of the acid name is dropped and is replaced with the suffix *-aldehyde*. Thus, the name of the one-carbon acid, formic acid, becomes formaldehyde for the one-carbon aldehyde.

$H{-}\overset{\overset{\displaystyle O}{\|}}{C}{-}OH$	$H{-}\overset{\overset{\displaystyle O}{\|}}{C}{-}H$
Formic acid	Formaldehyde

Naming aldehydes by both methods is illustrated in Table 24.1.

TABLE 24.1 IUPAC and Common Names of Selected Aldehydes

Formula	IUPAC name	Common name
$H-\overset{O}{\overset{\|}{C}}-H$	Methanal	Formaldehyde
$CH_3\overset{O}{\overset{\|}{C}}-H$	Ethanal	Acetaldehyde
$CH_3CH_2\overset{O}{\overset{\|}{C}}-H$	Propanal	Propionaldehyde
$CH_3CH_2CH_2\overset{O}{\overset{\|}{C}}-H$	Butanal	Butyraldehyde
$CH_3\underset{CH_3}{\underset{\|}{C}}H\overset{O}{\overset{\|}{C}}-H$	2-Methylpropanal	Isobutyraldehyde

Aromatic aldehydes contain an aldehyde group bonded to an aromatic ring. They are also named after the corresponding carboxylic acids. Thus the name benzaldehyde is derived from benzoic acid, and the name *p*-tolualdehyde is from *p*-toluic acid.

$C_6H_5-\overset{O}{\overset{\|}{C}}-H$ Benzaldehyde

$C_6H_5-\overset{O}{\overset{\|}{C}}-OH$ Benzoic acid

$CH_3-C_6H_4-\overset{O}{\overset{\|}{C}}-H$ *p*-Tolualdehyde

$CH_3-C_6H_4-COOH$ *p*-Toluic acid

In dialdehydes the suffix *dial* is added to the corresponding hydrocarbon name; for example,

$$H-\overset{O}{\overset{\|}{C}}CH_2CH_2\overset{O}{\overset{\|}{C}}-H$$

is named butanedial.

Ketones

The IUPAC name of a ketone is derived from the name of the alkane corresponding to the longest carbon chain that contains the ketone carbonyl group.

The parent name is formed by changing the *e* ending of the alkane to *one*. If the chain is longer than four carbons, it is numbered so that the carbonyl carbon has the smallest number possible, and this number is prefixed to the name of the ketone. Other groups bonded to the parent chain are named and numbered as previously indicated for hydrocarbons and alcohols. See the following examples:

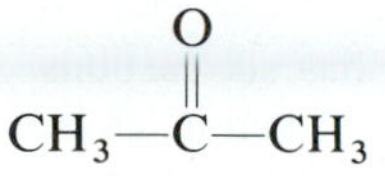

$$CH_3-\overset{\overset{\displaystyle O}{\|}}{C}-CH_3 \qquad \overset{5}{C}H_3\overset{4}{C}H_2\overset{3}{C}H_2-\overset{\overset{\displaystyle O}{\|}}{\underset{2}{C}}-\overset{1}{C}H_3 \qquad \overset{1}{C}H_3\overset{2}{C}H_2-\overset{\overset{\displaystyle O}{\|}}{\underset{3}{C}}-\overset{4}{C}H(CH_3)\overset{5}{C}H_2\overset{6}{C}H_3$$

Propanone 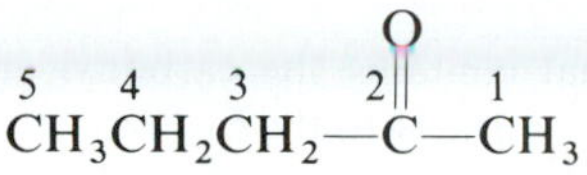 2-Pentanone 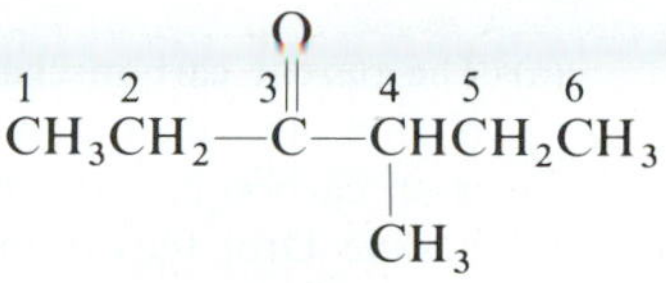 4-Methyl-3-hexanone

Note that in 4-methyl-3-hexanone the carbon chain is numbered from left to right to give the ketone group the lowest possible number.

An alternate non-IUPAC method commonly used to name simple ketones is to list the names of the alkyl or aromatic groups attached to the carbonyl carbon together with the word *ketone*. Thus, butanone ($CH_3COCH_2CH_3$) is methyl ethyl ketone:

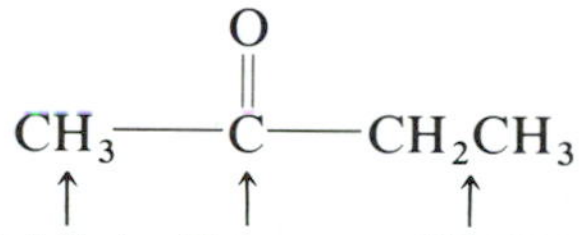

$$\underset{\substack{\uparrow \\ \text{Methyl}}}{CH_3}-\underset{\substack{\uparrow \\ \text{Ketone}}}{\overset{\overset{\displaystyle O}{\|}}{C}}-\underset{\substack{\uparrow \\ \text{Ethyl}}}{CH_2CH_3}$$

Two of the most widely used ketones have special common names: Propanone is called acetone, and butanone is known as methyl ethyl ketone, or MEK.

Aromatic ketones are named in a fashion similar to that for aliphatic ketones and often have special names as well.

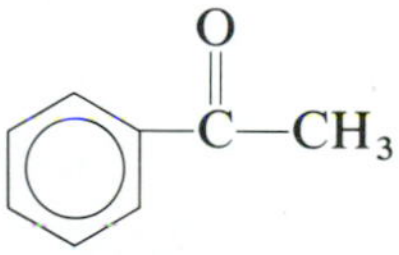

$$C_6H_5-\overset{\overset{\displaystyle O}{\|}}{C}-CH_3 \qquad C_6H_5-\overset{\overset{\displaystyle O}{\|}}{C}-CH_2CH_3$$

Methyl phenyl ketone
Acetophenone
1-Phenylethanone

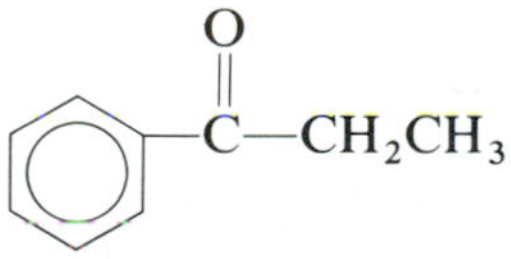

Ethyl phenyl ketone
Propiophenone
1-Phenyl-1-propanone

EXAMPLE 24.1

Write the formulas and the names for the straight-chain five- and six-carbon aldehydes.

SOLUTION

The IUPAC names are based on the five- and six-carbon alkanes. Drop the *e* of the alkane name and add the suffix *al*. Pentane (C_5) becomes pentanal and hexane (C_6) becomes hexanal. (The aldehyde group does not need to be renumbered; it is understood to be on carbon 1.) The common names are derived from valeric acid and caproic acid, respectively.

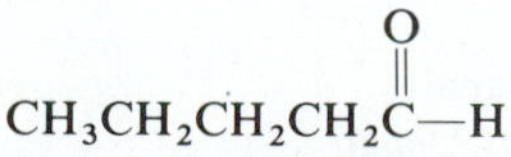

$$CH_3CH_2CH_2CH_2\overset{\overset{\displaystyle O}{\|}}{C}-H \qquad CH_3CH_2CH_2CH_2CH_2\overset{\overset{\displaystyle O}{\|}}{C}-H$$

Pentanal (Valeraldehyde) 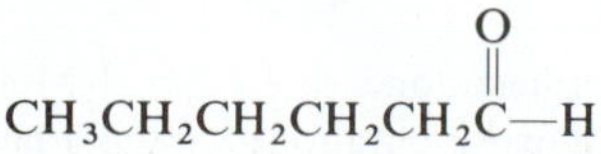 Hexanal (Caproaldehyde)

EXAMPLE 24.2 Give two names for each of the following ketones:

(a) $CH_3CH_2\overset{\overset{\large O}{\|}}{C}CH_2\overset{\overset{\large CH_3}{|}}{C}HCH_3$ (b) $CH_3CH_2CH_2\overset{\overset{\large O}{\|}}{C}-C_6H_5$

SOLUTION

(a) The parent carbon chain that contains the carbonyl group has six carbons. Number this chain from the end nearer to the carbonyl group. The ketone group is on carbon 3, and a methyl group is on carbon 5. The six-carbon alkane is hexane. Drop the *e* from hexane and add *one* to give the parent name, hexanone. Prefix the name hexanone with a 3- to locate the ketone group and with 5-methyl- to locate the methyl group. The name is 5-methyl-3-hexanone. The common name is ethyl isobutyl ketone since the C═O has an ethyl group and an isobutyl group bonded to it.

(b) The longest aliphatic chain has four carbons. The parent ketone name is butanone, derived by dropping the *e* of butane and adding *one*. The butanone has a phenyl group attached to carbon 1. The IUPAC name is therefore 1-phenyl-1-butanone. The common name for this compound is phenyl *n*-propyl ketone, since the C═O group has a phenyl and an *n*-propyl group bonded to it.

PRACTICE Write structures for the following carbonyl compounds:
(a) **4-Bromo-5-hydroxyhexanal** (c) 3-Buten-2-one
(b) Phenylethanal (d) Diphenylmethanone (diphenyl ketone)

Answers:

(a) $CH_3\overset{\overset{\large OH}{|}}{C}H\underset{\underset{\large Br}{|}}{C}HCH_2CH_2\overset{\overset{\large O}{\|}}{C}-H$ (c) $CH_3\overset{\overset{\large O}{\|}}{C}CH{=}CH_2$

(b) $C_6H_5-CH_2\overset{\overset{\large O}{\|}}{C}-H$ (d) $C_6H_5-\overset{\overset{\large O}{\|}}{C}-C_6H_5$

PRACTICE Name each of the following compounds using the IUPAC system:

(a) cyclohexane ring$=$O (c) $CH_3\underset{\underset{\large CH_3}{|}}{C}HCH_2\overset{\overset{\large O}{\|}}{C}-H$

(b) $ClCH_2CH_2\overset{\overset{\large O}{\|}}{C}-CH_3$ (d) $\underset{\underset{\large C_6H_5}{|}}{C}H{=}CH-\overset{\overset{\large O}{\|}}{C}-H$

Answers: (a) Cyclohexanone (c) 3-Methylbutanal
(b) 4-Chloro-2-butanone (d) 3-Phenyl-2-propenal

24.3 Bonding and Physical Properties

The carbon atom of the carbonyl group is sp^2 hybridized and is joined to three other atoms by sigma bonds. The fourth bond is made by overlapping p electrons of carbon and oxygen to form a pi bond between the carbon and oxygen atoms.

Because the oxygen atom is considerably more electronegative than carbon, the carbonyl group is polar, with the electrons shifted toward the oxygen atom. This makes the oxygen atom partially negative (δ^-) and leaves the carbon atom partially positive (δ^+). Many of the chemical reactions of aldehydes and ketones are due to this polarity.

$$\overset{\delta^+}{\text{C}}{=}\overset{\delta^-}{\ddot{\text{O}}}\!:$$

$$\longleftrightarrow$$

Polarity

Unlike alcohols, aldehydes and ketones cannot interact with themselves through hydrogen bonding, because there is no hydrogen atom attached to the oxygen atom of the carbonyl group. Aldehydes and ketones, therefore, have lower boiling points than alcohols of comparable molar mass (Table 24.2).

Low molar mass aldehydes and ketones are soluble in water, but for five or more carbons, the solubility decreases markedly. Ketones are very efficient organic solvents.

The lower molar mass aldehydes have a penetrating, disagreeable odor and are partially responsible for the taste of some rancid and stale foods. As the molar mass increases, the odor of both aldehydes and ketones, especially the aromatic ones, becomes more fragrant. Some are even used in flavorings and perfumes. A few of these and other selected aldehydes and ketones are shown in Figure 24.1.

Table 24.2 Boiling Points of Selected Aldehydes and Ketones and Corresponding Alcohols

Name	Molar mass	Boiling point (°C)
1-Propanol	60	97
Propanal	58	49
Propanone	58	56
1-Butanol	74	118
Butanal	72	76
Butanone	72	80
1-Pentanol	86	138
Pentanal	84	103
2-Pentanone	84	102

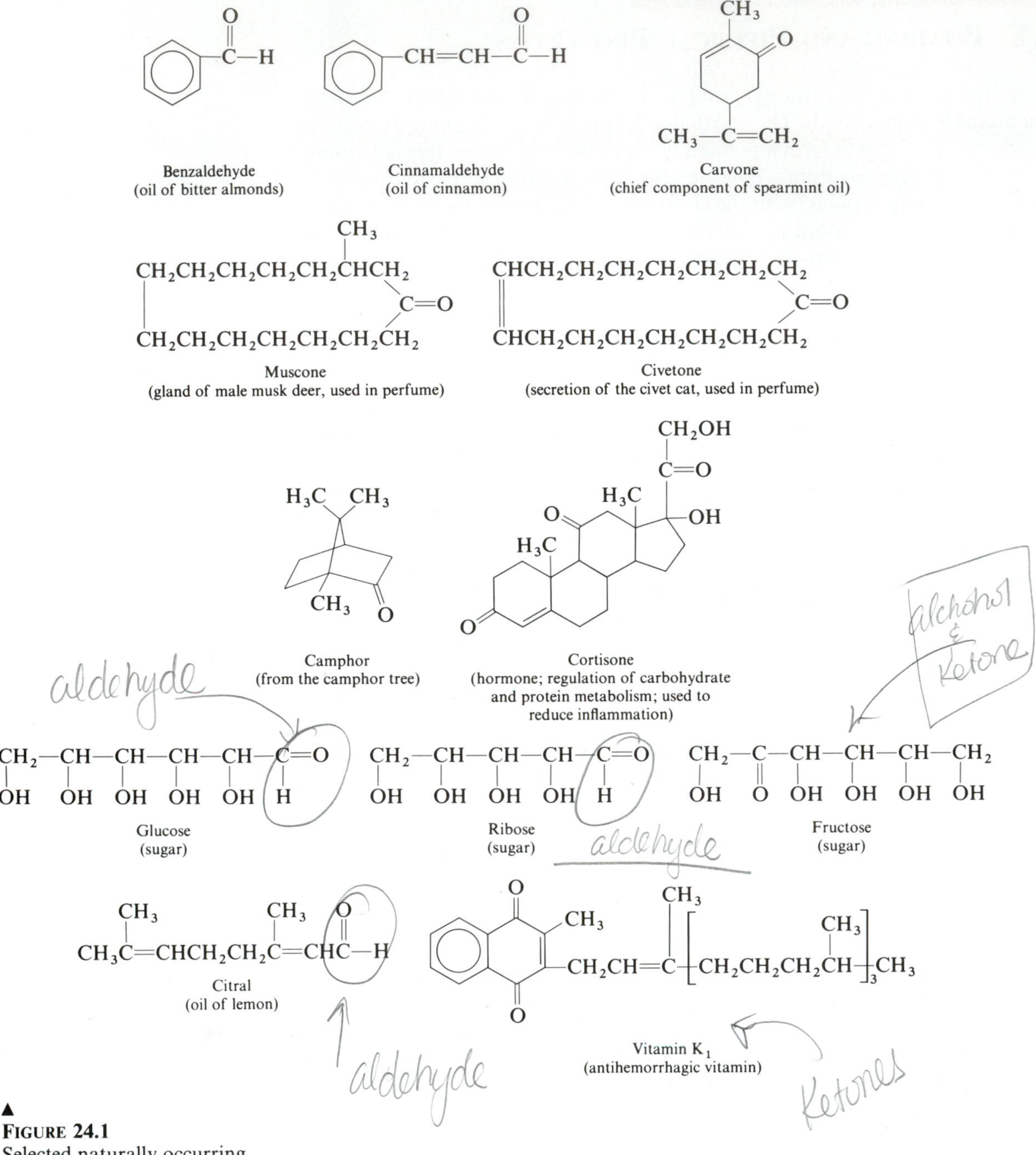

▲
FIGURE 24.1
Selected naturally occurring aldehydes and ketones.

24.4 Chemical Properties of Aldehydes and Ketones

The carbonyl group undergoes a great variety of reactions. Although there are differences, aldehydes and ketones undergo many similar reactions. However, ketones are generally less reactive than aldehydes. Some typical reactions follow.

Oxidation

Aldehydes are easily oxidized to carboxylic acids by a variety of oxidizing agents, including (under some conditions) oxygen of the air. Oxidation is the reaction in which aldehydes differ most from ketones. In fact, aldehydes and ketones may be separated into classes by their relative susceptibilities to oxidation. Aldehydes are easily oxidized to carboxylic acids by $K_2Cr_2O_7 + H_2SO_4$ and by mild oxidizing agents such as Ag^+ and Cu^{2+} ions; ketones are unaffected by such reagents. Ketones can be oxidized under drastic conditions—for example, by treatment with hot potassium permanganate solution. However, under these conditions carbon–carbon bonds are broken, and a variety of products are formed. Equations for the oxidation of aldehydes by dichromate are

$$3\ \mathrm{R}\overset{\overset{\displaystyle \mathrm{O}}{\|}}{\mathrm{C}}\!-\!\mathrm{H} + Cr_2O_7^{2-} + 8\ H^+ \longrightarrow 3\ \underset{\text{Carboxylic acid}}{\mathrm{R}\overset{\overset{\displaystyle \mathrm{O}}{\|}}{\mathrm{C}}\!-\!\mathrm{OH}} + 2\ Cr^{3+} + 4\ H_2O$$

$$3\ \mathrm{CH_3}\overset{\overset{\displaystyle \mathrm{O}}{\|}}{\mathrm{C}}\!-\!\mathrm{H} + Cr_2O_7^{2-} + 8\ H^+ \longrightarrow 3\ \underset{\text{Acetic acid}}{\mathrm{CH_3}\overset{\overset{\displaystyle \mathrm{O}}{\|}}{\mathrm{C}}\!-\!\mathrm{OH}} + 2\ Cr^{3+} + 4\ H_2O$$

Tollens test

The **Tollens test** (silver-mirror test) for aldehydes is based on the ability of silver ions to oxidize aldehydes. The Ag^+ ions are thereby reduced to metallic silver. In practice a little of the suspected aldehyde is added to a solution of silver nitrate and ammonia in a clean test tube. The appearance of a silver mirror on the inner wall of the tube is a positive test for the aldehyde group. The abbreviated equation is

$$\mathrm{R}\overset{\overset{\displaystyle \mathrm{O}}{\|}}{\mathrm{C}}\!-\!\mathrm{H} + 2\ Ag^+ \xrightarrow[H_2O]{NH_3} \mathrm{R}\overset{\overset{\displaystyle \mathrm{O}}{\|}}{\mathrm{C}}\!-\!\mathrm{O^-\ NH_4^+} + 2\ Ag(s) \qquad \text{(general reaction)}$$

$$\mathrm{CH_3}\overset{\overset{\displaystyle \mathrm{O}}{\|}}{\mathrm{C}}\!-\!\mathrm{H} + 2\ Ag^+ \xrightarrow[H_2O]{NH_3} CH_3COO^-\ NH_4^+ + 2\ Ag(s)$$

Fehling and Benedict tests

Fehling's and Benedict's solutions contain Cu^{2+} ions in an alkaline medium. In the **Fehling and the Benedict tests**, the aldehyde group is oxidized to an acid by Cu^{2+} ions. The blue Cu^{2+} ions are reduced and form brick-red copper(I) oxide (Cu_2O), which precipitates during the reaction. These tests can be used for detecting carbohydrates that have an available aldehyde group. The abbreviated equation is

$$\underset{\text{}}{\mathrm{R\overset{\overset{\displaystyle O}{\|}}{C}{-}H}} + \underset{\text{(blue)}}{2\,Cu^{2+}} \xrightarrow[H_2O]{NaOH} RCOO^-\,Na^+ + \underset{\text{(brick-red)}}{Cu_2O(s)}$$

Most ketones do not give a positive test with Tollens', Fehling's, or Benedict's solutions. These tests are used to distinguish between aldehydes and ketones.

$$\mathrm{R{-}\overset{\overset{\displaystyle O}{\|}}{C}{-}R} + Ag^+ \xrightarrow[H_2O]{NH_3} \text{No reaction}$$

$$\mathrm{R{-}\overset{\overset{\displaystyle O}{\|}}{C}{-}R} + Cu^{2+} \xrightarrow[H_2O]{OH^-} \text{No reaction}$$

Aldehydes and ketones are highly combustible, yielding carbon dioxide and water when completely burned. Their vapors, like those of nearly all volatile organic substances, form explosive mixtures with air. Adequate safety precautions must be taken to guard against fire and explosions when working with aldehydes and ketones or other volatile organic compounds, especially hydrocarbons, alcohols, and ethers.

The oxidation of aldehydes is a very important reaction in biochemistry. When our cells "burn" carbohydrates, they take advantage of the aldehyde reactivity. The aldehyde group is oxidized to a carboxylic acid and is eventually converted to carbon dioxide, which is then exhaled. This stepwise oxidation provides some of the energy necessary to sustain life.

Reduction

Aldehydes and ketones are easily reduced to alcohols, either by elemental hydrogen in the presence of a catalyst or by chemical reducing agents such as lithium aluminum hydride ($LiAlH_4$) or sodium borohydride ($NaBH_4$). Aldehydes yield primary alcohols; ketones yield secondary alcohols:

$$\mathrm{R{-}\overset{\overset{\displaystyle O}{\|}}{C}{-}H} \xrightarrow[\Delta]{H_2/Ni} \underset{1^\circ\ \text{alcohol}}{RCH_2OH} \qquad \text{(general reaction)}$$

$$\mathrm{R{-}\overset{\overset{\displaystyle O}{\|}}{C}{-}R} \xrightarrow[\Delta]{H_2/Ni} \underset{2^\circ\ \text{alcohol}}{\mathrm{R{-}\overset{\overset{\displaystyle OH}{|}}{C}H{-}R}} \qquad \text{(general reaction)}$$

$$\mathrm{CH_3\overset{\overset{\displaystyle O}{\|}}{C}{-}H} \xrightarrow[\Delta]{H_2/Ni} CH_3CH_2OH$$

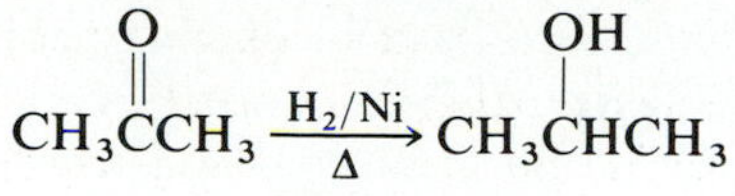

$$CH_3\overset{\overset{O}{\|}}{C}CH_3 \xrightarrow[\Delta]{H_2/Ni} CH_3\overset{\overset{OH}{|}}{C}HCH_3$$

Addition Reactions

Addition of Alcohols Compounds derived from aldehydes and ketones that contain an alkoxy and a hydroxy group on the same carbon atom are known as **hemiacetals** and **hemiketals**. In a like manner, compounds that have two alkoxy groups on the same carbon atom are known as **acetals** and **ketals**.

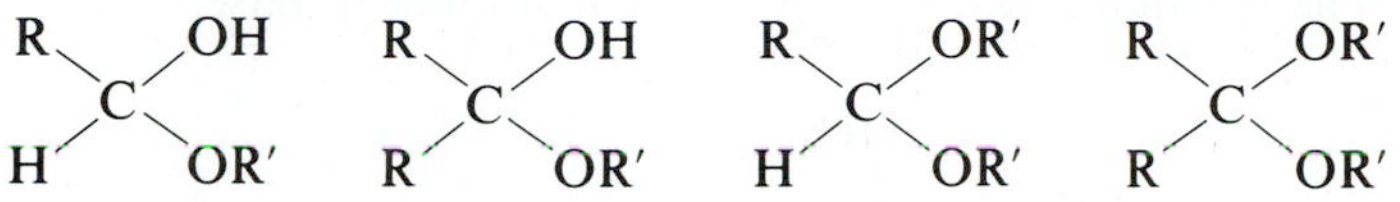

hemiacetal

hemiketal

acetal

ketal

$$\underset{\text{Hemiacetal}}{\overset{R\quad OH}{\underset{H\quad OR'}{C}}} \qquad \underset{\text{Hemiketal}}{\overset{R\quad OH}{\underset{R\quad OR'}{C}}} \qquad \underset{\text{Acetal}}{\overset{R\quad OR'}{\underset{H\quad OR'}{C}}} \qquad \underset{\text{Ketal}}{\overset{R\quad OR'}{\underset{R\quad OR'}{C}}}$$

Most open-chain hemiacetals and hemiketals are so unstable that they cannot be isolated. On the other hand, acetals and ketals are stable in alkaline solutions but are unstable in acid solutions, in which they are hydrolyzed back to the original aldehyde or ketone.

In the reactions below we show only aldehydes in the equations, but keep in mind that ketones behave in a similar fashion, although they are not as reactive.

Aldehydes react with alcohols in the presence of a trace of acid to form hemiacetals:

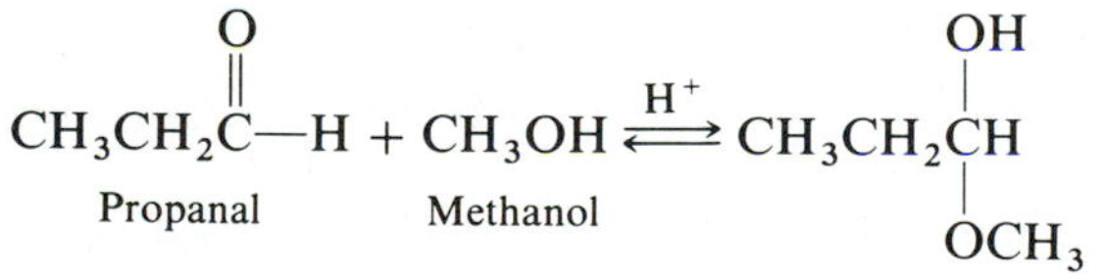

$$\underset{\text{Propanal}}{CH_3CH_2\overset{\overset{O}{\|}}{C}{-}H} + \underset{\text{Methanol}}{CH_3OH} \overset{H^+}{\rightleftharpoons} CH_3CH_2\overset{\overset{OH}{|}}{\underset{\underset{OCH_3}{|}}{C}}H$$

1-Methoxy-1-propanol
(Propionaldehyde methyl hemiacetal)

In the presence of excess alcohol and a strong acid such as dry HCl, aldehydes or hemiacetals react with a second molecule of the alcohol to give an acetal:

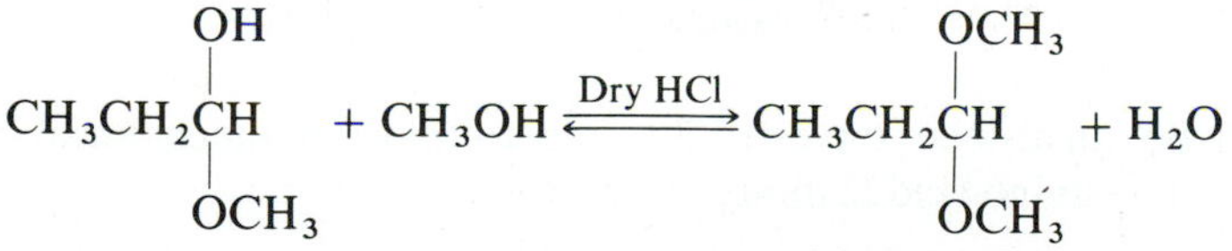

$$CH_3CH_2\overset{\overset{OH}{|}}{\underset{\underset{OCH_3}{|}}{C}}H + CH_3OH \xrightleftharpoons{\text{Dry HCl}} CH_3CH_2\overset{\overset{OCH_3}{|}}{\underset{\underset{OCH_3}{|}}{C}}H + H_2O$$

1,1-Dimethoxypropane
(Propionaldehyde dimethyl acetal)

A hemiacetal has both an alcohol and an ether group attached to the aldehyde carbon. An acetal has two ether groups attached to the aldehyde carbon.

If the alcohol and carbonyl groups are within the same molecule, the result is the formation of a cyclic hemiacetal (or hemiketal). This *intramolecular* cyclization is particularly significant in carbohydrate chemistry during the study of monosaccharides (Chapter 29).

$$HOCH_2CH_2CH_2CH_2CHO \underset{}{\overset{H^+}{\rightleftarrows}} \text{cyclic } CH_2CH_2CH_2CH_2\text{—O—CH(OH)}$$

5-Hydroxypentanal Stable hemiacetal

Addition of Hydrogen Cyanide The addition of hydrogen cyanide, HCN, to aldehydes and ketones forms a class of compounds known as cyanohydrins. **Cyanohydrins** have a cyano (—CN) group and a hydroxyl group on the same carbon atom. The reaction is catalyzed by a small amount of base:

cyanohydrin

$$CH_3\overset{O}{\overset{\|}{C}}\text{—}H + HCN \xrightarrow{OH^-} CH_3\overset{OH}{\overset{|}{C}}HCN$$

Acetaldehyde Acetaldehyde cyanohydrin

$$CH_3\underset{O}{\underset{\|}{C}}CH_3 + HCN \xrightarrow{OH^-} CH_3\overset{CH_3}{\overset{|}{\underset{OH}{\underset{|}{C}}}}\text{—}CN$$

Acetone Acetone cyanohydrin

In the cyanohydrin reaction, the more positive H atom of HCN adds to the oxygen of the carbonyl group, and the —CN group adds to the carbon atom of the carbonyl group. In the aldehyde addition, the length of the carbon chain is increased by one carbon. The ketone addition product also contains an additional carbon atom.

Cyanohydrins are useful intermediates for the synthesis of several important compounds. For example, the hydrolyses of cyanohydrins produce α-hydroxy acids:

$$CH_3\underset{OH}{\underset{|}{C}}H\text{—}CN + H_2O \xrightarrow{H^+} CH_3\underset{OH}{\underset{|}{C}}HCOOH + NH_4^+$$

Lactic acid

Acetaldehyde can also be converted into other important biochemical compounds such as the amino acid alanine:

$$CH_3\text{—}\overset{O}{\overset{\|}{C}}\text{—}H \xrightarrow{HCN} CH_3\text{—}\overset{OH}{\overset{|}{\underset{H}{\underset{|}{C}}}}\text{—}CN \xrightarrow{NH_3} CH_3\text{—}\overset{NH_2}{\overset{|}{\underset{H}{\underset{|}{C}}}}\text{—}CN \xrightarrow[H^+]{H_2O} CH_3\text{—}\overset{H_2N}{\overset{|}{\underset{H}{\underset{|}{C}}}}\text{—}\overset{O}{\overset{\|}{C}}\text{—}OH$$

Alanine

Some commercial reactions also involve the use of cyanohydrins. Acetone cyanohydrin can be converted to methyl methacrylate when refluxed with methanol and a strong acid. The methyl methacrylate can then be polymerized to Lucite or Plexiglas, both transparent plastics. See Chapter 27.

▲ **A sculpture formed from Lucite.**

Aldol Condensation (Self-addition) In a carbonyl compound the carbon atoms are labeled alpha (α), beta (β), gamma (γ), delta (δ), and so on, according to their positions with respect to the carbonyl group. The alpha carbon is adjacent to the carbonyl carbon, the beta carbon is next, the gamma carbon is third, and so forth. The hydrogens attached to the alpha-carbon atom are therefore called alpha hydrogens, and so on, as shown below:

$$\overset{\delta}{-\mathrm{C}}-\overset{\gamma}{\mathrm{C}}-\overset{\beta}{\underset{|}{\mathrm{C}}}-\overset{\alpha}{\underset{|}{\mathrm{C}}}-\mathrm{C}{=}\mathrm{O}$$

Beta H atom → H H ← Alpha H atom

The hydrogen atoms attached to the alpha carbon atom have the unique ability to be more easily released as protons than other hydrogens within the molecule.

An aldehyde or ketone that contains α-hydrogens may add to itself or to another α-hydrogen containing aldehyde or ketone. The product of this reaction contains both a carbonyl group and an alcohol group within the same molecule. The reaction is known as an **aldol condensation** and is catalyzed by dilute base. Remember that a *condensation* reaction is one in which two smaller molecules combine to form a larger molecule, usually with the loss of a small molecule in the process. The aldol condensation is very similar to the other carbonyl addition reactions. An α-hydrogen adds to the carbonyl oxygen, and the remainder of the molecule adds to the carbonyl carbon.

aldol condensation

$$\mathrm{CH_3-\overset{\overset{\displaystyle O}{\|}}{C}-H} \quad + \quad \alpha\,\mathrm{H}:\ \mathrm{H-\overset{\overset{\displaystyle H}{|}}{\underset{\underset{\displaystyle H}{|}}{C}}-\overset{\overset{\displaystyle O}{\|}}{C}-H} \xrightarrow{\text{Dilute NaOH}} \mathrm{CH_3\overset{\overset{\displaystyle OH}{|}}{C}HCH_2\overset{\overset{\displaystyle O}{\|}}{C}-H}$$

Aldol

(3-Hydroxybutanal)

Acetone also undergoes the aldol condensation:

$$\mathrm{CH_3\overset{\overset{\displaystyle O}{\|}}{C}CH_3} + \mathrm{H-CH_2\overset{\overset{\displaystyle O}{\|}}{C}CH_3} \xrightarrow[\text{NaOH}]{\text{Dilute}} \mathrm{CH_3\overset{\overset{\displaystyle OH}{|}}{\underset{\underset{\displaystyle CH_3}{|}}{C}}-CH_2\overset{\overset{\displaystyle O}{\|}}{C}CH_3}$$

Acetone | Acetone | Diacetone alcohol (4-Hydroxy-4-methyl-2-pentanone)

In the foregoing reaction an alpha hydrogen first transfers from one molecule to the oxygen of the other molecule. This breaks the C=O pi bond, leaving a carbon atom of each molecule with three bonds. The two carbon atoms then bond to each other, forming the product, diacetone alcohol.

EXAMPLE 24.3 Write the equation for the aldol condensation of propanal.

SOLUTION First write the structure for propanal and locate the alpha-hydrogen atoms:

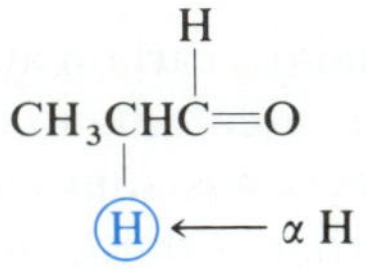

$$\begin{array}{c} \text{H} \\ | \\ \text{CH}_3\text{CHC}{=}\text{O} \\ | \\ \text{(H)} \longleftarrow \alpha\,\text{H} \end{array}$$

Now write two propanal molecules and transfer an alpha hydrogen from one molecule to the oxygen of the second molecule. After the pi bond breaks, the two carbon atoms that are bonded to only three other atoms are attached to each other to form the product.

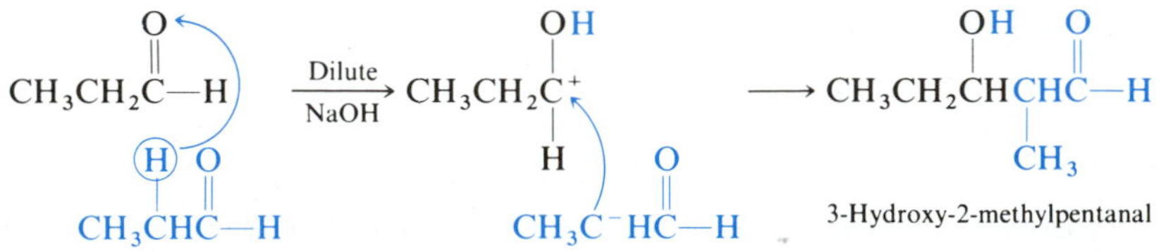

$$\begin{array}{c} \text{O} \\ \| \\ \text{CH}_3\text{CH}_2\text{C{—}H} \\ \text{(H)}\;\text{O} \\ |\quad \| \\ \text{CH}_3\text{CHC{—}H} \end{array} \xrightarrow[\text{NaOH}]{\text{Dilute}} \begin{array}{c} \text{OH} \\ | \\ \text{CH}_3\text{CH}_2\text{C}^+ \\ | \\ \text{H} \qquad \text{O} \\ \qquad \| \\ \text{CH}_3\text{C}^-\text{HC{—}H} \end{array} \longrightarrow \begin{array}{c} \text{OH}\quad\text{O} \\ |\qquad \| \\ \text{CH}_3\text{CH}_2\text{CHCHC{—}H} \\ | \\ \text{CH}_3 \end{array}$$

3-Hydroxy-2-methylpentanal

PRACTICE Write the equation for the aldol condensation of butanal.

Answer:

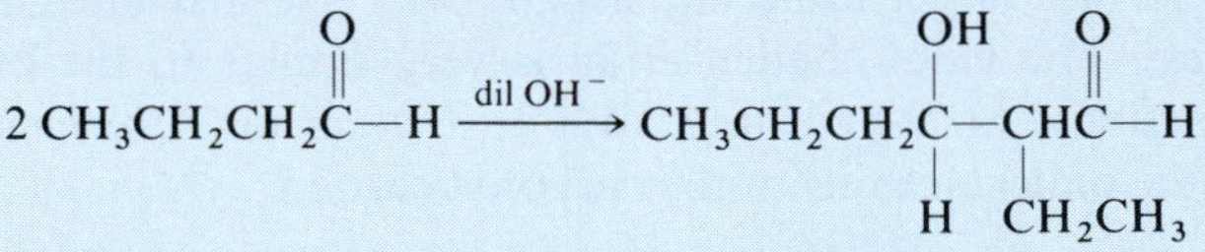

$$2\,\text{CH}_3\text{CH}_2\text{CH}_2\overset{\text{O}}{\overset{\|}{\text{C}}}\text{—H} \xrightarrow{\text{dil OH}^-} \text{CH}_3\text{CH}_2\text{CH}_2\underset{\text{H}}{\underset{|}{\overset{\text{OH}}{\overset{|}{\text{C}}}}}\text{—}\underset{\text{CH}_2\text{CH}_3}{\underset{|}{\text{CH}}}\overset{\text{O}}{\overset{\|}{\text{C}}}\text{—H}$$

PRACTICE Write the equation for the aldol condensation of 3-pentanone.

Answer:

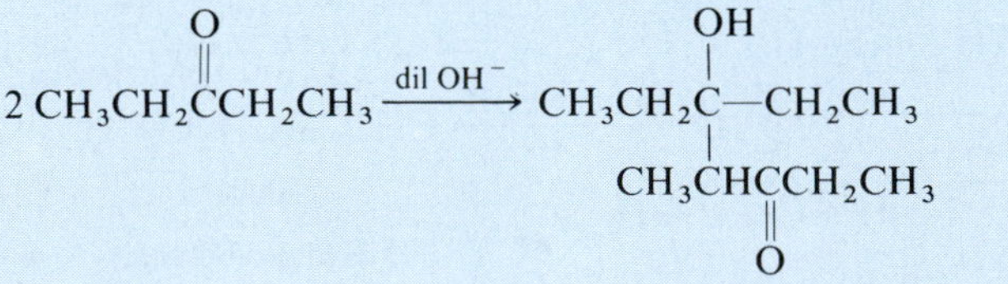

$$2\,\text{CH}_3\text{CH}_2\overset{\text{O}}{\overset{\|}{\text{C}}}\text{CH}_2\text{CH}_3 \xrightarrow{\text{dil OH}^-} \text{CH}_3\text{CH}_2\overset{\text{OH}}{\overset{|}{\text{C}}}\text{—CH}_2\text{CH}_3 \\ \qquad\qquad | \\ \qquad\qquad \text{CH}_3\text{CH}\underset{\text{O}}{\underset{\|}{\text{C}}}\text{CH}_2\text{CH}_3$$

The aldol condensation reaction is used often by living cells. By means of this reaction, smaller biochemicals can be combined to make larger molecules. For example, most cells can join two three-carbon carbohydrates to make one six-carbon sugar using the aldol condensation. Ultimately, this reaction sequence leads to the formation of glucose (see Chapter 36).

24.5 COMMON ALDEHYDES AND KETONES

Numerous methods have been devised for making aldehydes and ketones. The oxidation of alcohols is a very general method. Special methods are often used for the commercial production of individual aldehydes and ketones.

Formaldehyde (Methanal) This aldehyde is made from methanol by reaction with oxygen (air) in the presence of a silver or copper catalyst:

$$2\,CH_3OH + O_2 \xrightarrow[400^\circ C]{\text{Ag or Cu}} \underset{\substack{\text{Formaldehyde}\\ \text{(methanal)}}}{2\,H_2C{=}O} + 2\,H_2O$$

Formaldehyde is a poisonous, irritating gas that is very soluble in water. It is marketed as a 40% aqueous solution called *formalin*. Because formaldehyde is a powerful germicide, it is used in embalming and to preserve biological specimens. Formaldehyde is also used for disinfecting dwellings, ships, and storage houses; for destroying flies; for tanning hides; and as a fungicide for plants and vegetables. But by far the largest use of this chemical is in the manufacture of polymers (Chapter 27). About 2.59×10^9 kg (5.7×10^9 lb) of formaldehyde is manufactured annually in the United States.

Formaldehyde vapors are intensely irritating to the mucous membranes. Ingestion may cause severe abdominal pains, leading to coma and death.

It is of interest that formaldehyde may have had a significant role in chemical evolution. Formaldehyde is believed to have been a component of the primitive atmosphere of the earth. It is theorized that the reactivity of this single-carbon aldehyde enabled it to form more complex organic molecules—molecules that were precursors of the still more complicated substances that today are essential components of every living organism.

Acetaldehyde (Ethanal) Acetaldehyde is a volatile liquid (bp 21°C) with a pungent, irritating odor. It has a general narcotic action and in large doses may cause respiratory paralysis. Its principal use is as an intermediate in the manufacture of other chemicals such as acetic acid, 1-butanol, and paraldehyde.

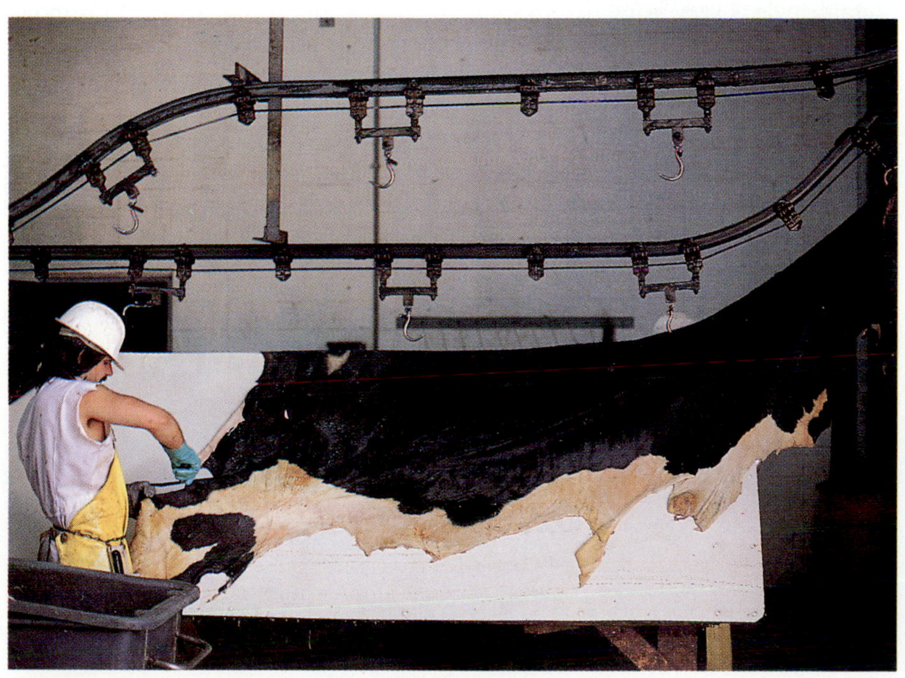

◄ **This Holstein hide is being scraped in preparation for tanning. Formalin can be used during the tanning process to soften and preserve the hide.**

Acetic acid, for example, is made by air oxidation of acetaldehyde:

$$2\,CH_3\overset{\overset{\large O}{\|}}{C}{-}H + O_2 \xrightarrow[\Delta]{Mn^{2+}} 2\,CH_3\overset{\overset{\large O}{\|}}{C}{-}OH$$

Acetaldehyde undergoes reactions in which three or four molecules condense or polymerize to form the cyclic compounds paraldehyde and metaldehyde:

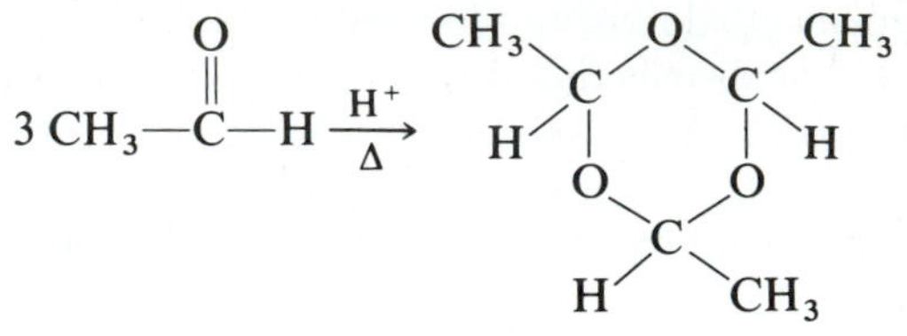

$$3\,CH_3{-}\overset{\overset{\large O}{\|}}{C}{-}H \xrightarrow[\Delta]{H^+} \text{(cyclic trimer: three CH(CH}_3\text{) units joined by O atoms)}$$

Paraldehyde
(bp 125°C)

Paraldehyde is used in medical practice as a hypnotic or sleep-inducing drug.

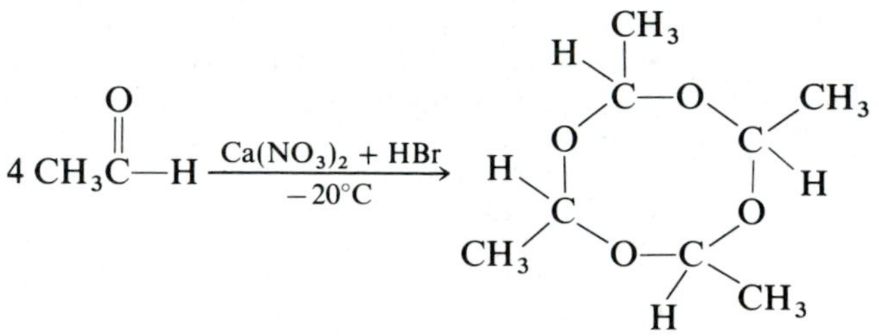

$$4\,CH_3\overset{\overset{\large O}{\|}}{C}{-}H \xrightarrow[-20°C]{Ca(NO_3)_2 + HBr} \text{(cyclic tetramer: four CH(CH}_3\text{) units joined by O atoms)}$$

Metaldehyde
(mp 246°C)

Metaldehyde is very attractive to slugs and snails, and it is also very poisonous to them. For this reason it is an active ingredient in some pesticides that are sold for lawn and garden use. Metaldehyde is also used as a solid fuel.

Benzaldehyde Benzaldehyde (C_6H_5CHO) is known as *oil of bitter almonds.* It is found in almonds and in the seeds of stone fruits—for example, apricots and peaches. Benzaldehyde is made synthetically and used in the manufacture of artificial flavors. The synthesis begins with the free-radical chlorination of toluene, as shown in the following sequence of reactions:

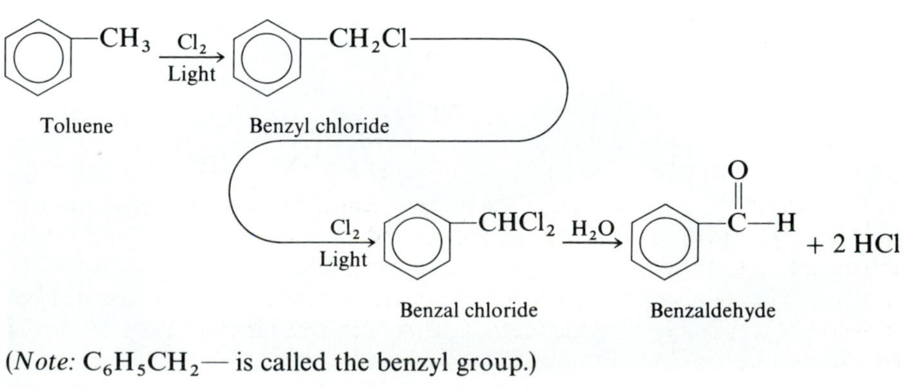

$$C_6H_5{-}CH_3 \xrightarrow[Light]{Cl_2} C_6H_5{-}CH_2Cl \xrightarrow[Light]{Cl_2} C_6H_5{-}CHCl_2 \xrightarrow{H_2O} C_6H_5{-}\overset{\overset{\large O}{\|}}{C}{-}H + 2\,HCl$$

Toluene | Benzyl chloride | Benzal chloride | Benzaldehyde

(*Note:* $C_6H_5CH_2$— is called the benzyl group.)

Chemistry in Action

Familiar Aldehydes and Ketones

Aldehydes are most often used commercially in the production of plastics. The common polymers known as Formica, used on counters and in telephones, Melmac, used in plastic tableware, and Bakelite, found in handles on kitchen utensils, are all derived from formaldehyde.

Spices and flavorings are often complex mixtures of ingredients. The characteristic flavor and aroma of many common flavorings and spices result from the presence of aromatic aldehydes. Examples include vanillin (see Chemistry in Action Chapter 7), cinnamaldehyde, citral, and benzaldehyde (see Table 24.1).

Higher molar mass ketones are also useful because of their unique aromas. Camphor (Table 24.1) is a moth repellant. Muscone, extracted from the musk glands of the male musk deer, is frequently used to give the musky aroma to perfumes. A tiny amount of these chemicals can be used to produce long-lasting aromas. Perfume manufacturers closely guard the secret of their ingredients to keep their formulations unique and selling well.

In living organisms aldehydes and ketones are found in many different compounds. Sugars are often classified as containing an aldehyde group, as in glucose or ribose, or a ketone group, as in fructose. The compounds known as steroids, belonging to the lipid family, also often contain aldehyde or ketone groups. These lipids may function as hormones as in the case of cortisone.

▲ **Perfumes contain aromas attributable to aldehydes or ketones.**

Cinnamaldehyde, $C_6H_5CH{=}CHCHO$, is the principal substance contributing to the flavor of oil of cinnamon. Like benzaldehyde, it is made synthetically and used in the preparation of artificial flavoring agents.

$$C_6H_5\text{—CH=CH—}\overset{\overset{\large O}{\|}}{C}\text{—H}$$

Cinnamaldehyde

Acetone and Methyl Ethyl Ketone Ketones are widely used organic solvents. Acetone, in particular, is used in very large quantities for this purpose. U.S. production of acetone is about 8.2×10^8 kg (1.8×10^9 lb) annually. It is used as a solvent in the manufacture of drugs, chemicals, and explosives; for removal of paints, varnishes, and fingernail polish; and as a solvent in the plastics industry. Methyl ethyl ketone (MEK) is also widely used as a solvent, especially for lacquers. Both acetone and MEK are made by oxidation (dehydrogenation) of secondary alcohols. Acetone is also a coproduct in the manufacture of phenol (see Section 23.11).

$$\underset{\text{2-Propanol}}{CH_3\underset{\underset{OH}{|}}{C}HCH_3} \xrightarrow[250\text{–}300^\circ C]{Cu} \underset{\text{Acetone (Propanone)}}{CH_3\underset{\underset{O}{\|}}{C}CH_3} + H_2$$

$$\underset{\text{2-Butanol}}{CH_3CH_2\underset{\underset{OH}{|}}{C}HCH_3} \xrightarrow[250\text{–}300^\circ C]{Cu} \underset{\text{Methyl ethyl ketone (2-Butanone)}}{CH_3CH_2\underset{\underset{O}{\|}}{C}CH_3} + H_2$$

Acetone is also formed in the human body during lipid metabolism. Usually it is oxidized to carbon dioxide and water. Normal concentrations of acetone in the body are less than 1 mg/100 mL of blood volume. In patients with diabetes mellitus the concentration of acetone may rise, and it is then excreted in the urine where it can be easily detected. Sometimes the odor of acetone may also be detected on the breath of these patients.

Concepts in Review

1. Recognize aldehydes and ketones from their formulas.
2. Give IUPAC and common names of aldehydes and ketones.
3. Write formulas of aldehydes and ketones when given their names.
4. Understand why aldehydes and ketones have lower boiling points than alcohols.
5. Write equations showing the oxidation of alcohols to aldehydes and ketones.
6. Write the structure of the alcohol formed when an aldehyde or ketone is reduced.
7. Discuss the Tollens, Benedict, and Fehling tests, including the reagents used, evidence of a positive test, and the equations of the reactions that occur in positive tests.
8. Use the Tollens, Benedict, and Fehling tests to distinguish between aldehydes and ketones.
9. Recognize whether an aldehyde or ketone undergoes the aldol condensation.
10. Write equations showing the aldol condensation of aldehydes and ketones.
11. Write equations for the formation and hydrolysis of cyanohydrins.
12. Write equations for the formation and decomposition of hemiacetals, hemiketals, acetals, and ketals.

Equations in Review

Oxidation

$$R-\overset{\overset{\displaystyle O}{\|}}{C}-H \xrightarrow{[O]} R-\overset{\overset{\displaystyle O}{\|}}{C}-OH$$

$$R-\overset{\overset{\displaystyle O}{\|}}{C}-R \xrightarrow{[O]} \text{No reaction}$$

Reduction

$$R-\overset{\overset{\displaystyle O}{\|}}{C}-H \xrightarrow{[H]} R-CH_2-OH$$

$$R-\overset{\overset{\displaystyle O}{\|}}{C}-R \xrightarrow{[H]} R-\underset{\underset{\displaystyle OH}{|}}{\overset{\overset{\displaystyle H}{|}}{C}}-R$$

Addition reactions

Alcohol

$$R-\overset{\overset{\displaystyle O}{\|}}{C}-H + R-OH \longrightarrow R-\underset{\underset{\displaystyle OR}{|}}{\overset{\overset{\displaystyle OH}{|}}{C}}-H$$

$$R-\overset{\overset{\displaystyle O}{\|}}{C}-R + R-OH \longrightarrow R-\underset{\underset{\displaystyle OR}{|}}{\overset{\overset{\displaystyle OH}{|}}{C}}-R$$

Hydrogen cyanide

$$R-\overset{\overset{\displaystyle O}{\|}}{C}-H + HCN \xrightarrow{OH^-} R-\underset{\underset{\displaystyle H}{|}}{\overset{\overset{\displaystyle OH}{|}}{C}}-CN$$

$$R-\overset{\overset{\displaystyle O}{\|}}{C}-R + HCN \xrightarrow{OH^-} R-\underset{\underset{\displaystyle OH}{|}}{\overset{\overset{\displaystyle R}{|}}{C}}-CN$$

Aldol condensation

$$2\,R-CH_2-\overset{\overset{\displaystyle O}{\|}}{C}-H \xrightarrow{[\text{dil } OH^-]} R-CH_2-\overset{\overset{\displaystyle OH}{|}}{CH}-\underset{\underset{\displaystyle R}{|}}{CH}-\overset{\overset{\displaystyle O}{\|}}{C}-H$$

EXERCISES

1. Write generalized structures for
   (a) An aldehyde (e) A hemiketal
   (b) A ketone (f) An acetal
   (c) A dialdehyde (g) A ketal
   (d) A hemiacetal (h) A cyanohydrin
2. Name each of these aldehydes:
   (a) $H_2C{=}O$ (two names)
   (b) $CH_3CH_2CH_2C({=}O){-}H$ (two names)
   (c) $CH_3CH(CH_3)CH_2C({=}O){-}H$ (one name)
   (d) $C_6H_5{-}C({=}O){-}H$ (one name)
   (e) $H{-}C({=}O)CH_2CH_2C({=}O){-}H$ (one name)
   (f) benzene ring bearing $-C({=}O){-}H$, with Cl ortho and $CH(CH_3)_2$ para (one name)
   (g) $C_6H_5{-}CH{=}CH{-}C({=}O){-}H$ (one name)
   (h) $CH_3CH(OH)CH_2C({=}O){-}H$ (two names)
   (i) $CH_3(H)C{=}C(H){-}C({=}O){-}H$ (one name)
3. Name each of these ketones:
   (a) CH_3COCH_3 (three names)
   (b) $CH_3CH_2COCH_3$ (two names)
   (c) $C_6H_5{-}C({=}O){-}CH_2CH_3$ (two names)
   (d) $CH_3C({=}O){-}C(CH_3)_2{-}CH_3$ (two names)
   (e) cyclopentane ring $=O$ (one name)
   (f) $CH_3C({=}O)CH_2CH_2C({=}O)CH_3$ (one name)
   (g) $CH_3C(CH_3)(OH){-}CH_2C({=}O)CH_3$ (two names)
   (h) $C_6H_5{-}CH_2C({=}O)CH_3$ (two names)
4. Write structural formulas for
   (a) 1,3-Dichloropropanone
   (b) Phenylacetaldehyde
   (c) 3-Butenal
   (d) 3-Hydroxypropanal
   (e) Diisopropyl ketone
   (f) 4-Methyl-3-hexanone
   (g) Hexanal
   (h) Cyclohexanone
5. Write structural formulas for propanal and propanone. Judging from these formulas, do you think that aldehydes and ketones are isomeric with each other? Show evidence and substantiate your answer by testing with a four-carbon aldehyde and a four-carbon ketone.
6. Explain, in terms of structure, why aldehydes and ketones have lower boiling points than alcohols of similar molar masses.
7. Which compound in each of the following pairs has the higher boiling point? (Try to answer without consulting tables.)
   (a) 1-Pentanol or pentanal
   (b) Pentane or pentanal
   (c) Benzaldehyde or benzyl alcohol
   (d) 2-Pentanone or 2-pentanol
   (e) Propanone or butanone
8. Write equations to show how each of the following is oxidized by (1) $K_2Cr_2O_7 + H_2SO_4$ and (2) air + Cu or Ag + heat:
   (a) 1-Propanol
   (b) 3-Pentanol
   (c) 2,3-Dimethyl-2-butanol

9. Ketones are prepared by oxidation of secondary alcohols. Which alcohol should be used to prepare:
   (a) Diethyl ketone
   (b) Diisopropyl ketone
   (c) 4-Phenyl-2-butanone

10. (a) What functional group is present in a compound that gives a positive Tollens test?
    (b) What is the visible evidence for a positive Tollens test?
    (c) Write an equation showing the reaction involved in a positive Tollens test.
11. (a) What functional group is present in a compound that gives a positive Fehling test?
    (b) What is the visible evidence for a positive Fehling test?
    (c) Write an equation showing the reaction involved in a positive Fehling test.
12. Give the products of the reaction of the following with Tollens' reagent:
    (a) Butanal
    (b) Benzaldehyde
    (c) Methyl ethyl ketone
13. Write equations showing the aldol condensation for the following compounds:
    (a) Acetaldehyde (c) Butanal
    (b) Propanal (d) 3-Pentanone
14. How many aldol condensation products are possible if a mixture of ethanal and propanal is reacted with dilute NaOH?
15. Complete the following equations:

$$\text{(a) } CH_3CH_2\overset{\overset{\displaystyle O}{\|}}{C}{-}H + CH_3CH_2CH_2OH \xrightleftharpoons{\text{Dry HCl}}$$

$$\text{(b) } CH_3\underset{\underset{\displaystyle O}{\|}}{C}CH_3 + \underset{\underset{\displaystyle OH}{|}}{CH_2}\underset{\underset{\displaystyle OH}{|}}{CH_2} \xrightleftharpoons{\text{Dry HCl}}$$

$$\text{(c) } CH_3CH_2\overset{\overset{\displaystyle O}{\|}}{C}{-}H + CH_3CH_2OH \xrightleftharpoons{H^+}$$

(d) (cyclohexane ring)$=O + CH_3OH \xrightleftharpoons{H^+}$

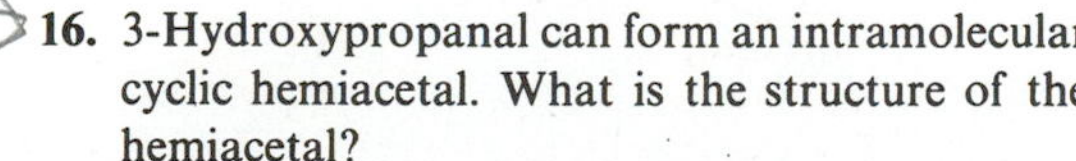

$$\text{(e) } CH_3CH_2CH_2CH(OCH_3)_2 \xrightarrow[H^+]{H_2O}$$

16. 3-Hydroxypropanal can form an intramolecular cyclic hemiacetal. What is the structure of the hemiacetal?
17. Write equations for the following sequence of reactions:
    (a) Benzaldehyde + HCN $\longrightarrow$
    (b) Product from part (a) + $H_2O \longrightarrow$
    (c) Product from part (b) + $K_2Cr_2O_7 + H_2SO_4 \longrightarrow$
18. Write equations to show how you could prepare lactic acid, $CH_3CH(OH)COOH$, from acetaldehyde through a cyanohydrin intermediate.
19. Give a simple visible chemical test that will distinguish between the compounds in each of the following pairs:

$$\text{(a) } CH_3CH_2\overset{\overset{\displaystyle O}{\|}}{C}{-}H \text{ and } CH_3\overset{\overset{\displaystyle O}{\|}}{C}CH_3$$

$$\text{(b) } CH_3CH_2\overset{\overset{\displaystyle O}{\|}}{C}{-}H \text{ and } CH_2{=}CH\overset{\overset{\displaystyle O}{\|}}{C}{-}H$$

(c) $C_6H_5{-}CH_2CH_2OH$ and

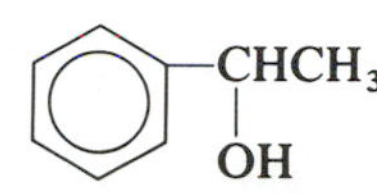

$C_6H_5{-}\underset{\underset{\displaystyle OH}{|}}{CH}CH_3$

20. Which of the following statements are correct? Rewrite the incorrect statements to make them correct.
    (a) The functional group that characterizes aldehydes and ketones is called a carboxyl group.
    (b) The carbonyl group contains a sigma and a pi bond.
    (c) The carbonyl group is polar, with the oxygen atom being more electronegative than the carbon atom.
    (d) The higher molar mass aldehydes and ketones are very soluble in water.
    (e) Ketones, like aldehydes, are easily oxidized to carboxylic acids.
    (f) A compound of formula $C_6H_{12}O$ can be either an aliphatic aldehyde or ketone.
    (g) Diethyl ketone has the same molecular formula as butyraldehyde.
    (h) In aldehydes and ketones, the hydrogen atoms that are bonded to carbon atoms adjacent to the carbonyl group (alpha position) are more reactive than other hydrogen atoms in the molecule.
    (i) In order for an aldehyde or a ketone to undergo the aldol condensation, it must have at least one alpha-hydrogen atom.
    (j) A hemiacetal has an alcohol and an ether group bonded to the same carbon atom.
    (k) Acetals are stable in acid solution but not in alkaline solution.
    (l) Formaldehyde is a gas, but it is usually handled in a solution.
    (m) The major use for formaldehyde is for making plastics.

(n) Ethanal may be distinguished from propanal by use of Tollens' reagent.
(o) The general formula for the saturated homologous series of aldehydes and ketones is $C_nH_{2n}O$.
(p) The compound C_2H_4O can be an aldehyde or a ketone.
(q) The name for O=C(H)—C(H)=O is ethanedial.
(r) The oxidation product of 3-pentanol is diethylketone.
(s) $C_6H_5CH_2CH(OC_2H_5)_2$ is a ketal.
(t) When hydrolyzed, cyanohydrins form α-hydroxy acids.
(u) Of the three compounds ethanal, propanal, and butanal, ethanal has the lowest vapor pressure.
(v) Of the three compounds ethanal, propanal, and butanal, butanal has the highest boiling point.
(w) Reduction of aldehydes yields secondary alcohols.

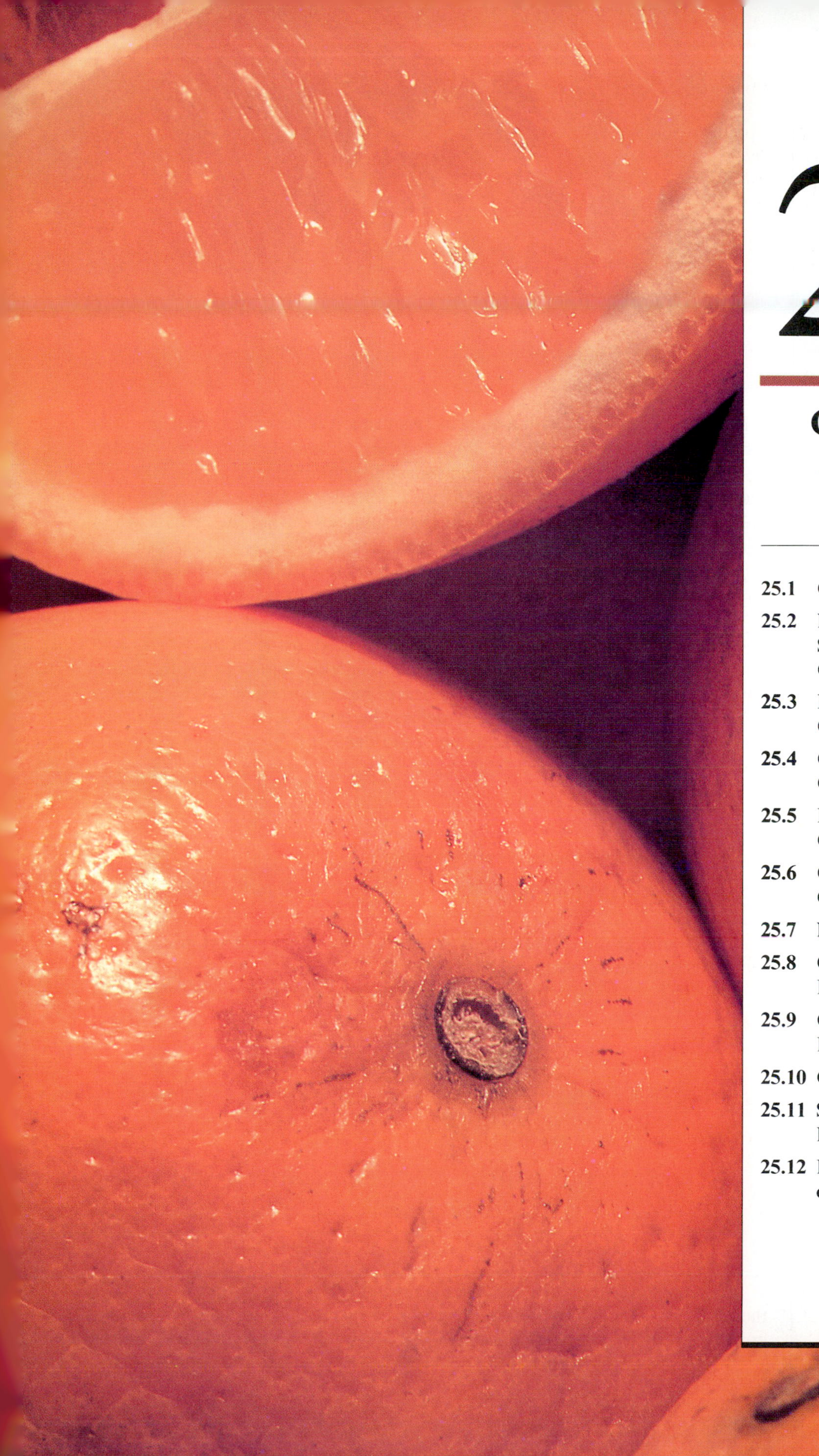

25

Carboxylic Acids and Esters

◀ CHAPTER OPENING PHOTO: **All citrus fruits contain carboxylic acids.**

Whenever we eat foods with a sour or tart taste, it is very likely that the taste results from the presence of at least one carboxylic acid. Lemons contain citric acid, vinegar contains acetic acid, and sour milk contains lactic acid. Carboxylic acids are also important compounds in biochemistry. Citric acid is found in the blood, and lactic acid is produced in the muscles during the breakdown of glucose. Most often in living systems these acids are found in the form of salts or acid derivatives. Carboxylic acid salts are commonly used in our lives as preservatives, especially in cheeses and breads. Some carboxylic acid salts are used to treat skin irritations like diaper rash and athlete's foot.

The sweet and pleasant odors and tastes of food are often the direct result of carboxylic acid derivatives known as esters. These compounds are frequently found as artificial flavors in foods in place of more expensive natural extracts. In biochemistry ester-like molecules act as the energy carriers in many cells. The properties of carboxylic acids are quite distinct from those of aldehydes and alcohols.

25.1 CARBOXYLIC ACIDS

carboxyl group

The functional group of the carboxylic acids is called a **carboxyl group** and is represented in the following ways:

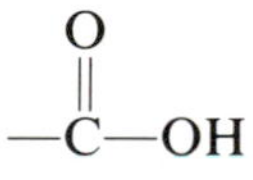

$$-\overset{\overset{\displaystyle O}{\|}}{C}-OH \qquad \text{or} \qquad -COOH \qquad \text{or} \qquad -CO_2H$$

Carboxylic acids can be either aliphatic or aromatic:

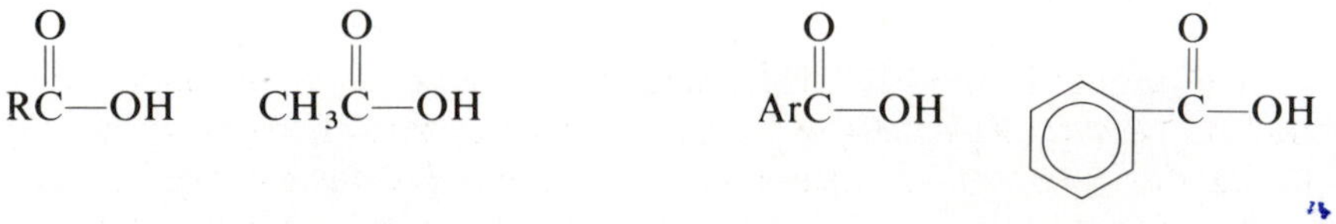

$$R\overset{\overset{\displaystyle O}{\|}}{C}-OH \qquad CH_3\overset{\overset{\displaystyle O}{\|}}{C}-OH \qquad\qquad Ar\overset{\overset{\displaystyle O}{\|}}{C}-OH \qquad C_6H_5-\overset{\overset{\displaystyle O}{\|}}{C}-OH$$

Aliphatic — Aromatic

Carboxylic acids react with many substances to produce derivatives. In these reactions the hydroxyl group is replaced by a halogen (—Cl), an acyloxy group (—OOCR), an alkoxy group (—OR), or an amino group ($-NH_2$). The general reactions are summarized below:

$$R{-}\overset{\overset{\displaystyle O}{\|}}{C}{-}OH \xrightarrow{-Cl} R{-}\overset{\overset{\displaystyle O}{\|}}{C}{-}Cl \qquad \text{(acyl halide)}$$

$$R{-}\overset{\overset{\displaystyle O}{\|}}{C}{-}OH \xrightarrow{-OOCR} R{-}\overset{\overset{\displaystyle O}{\|}}{C}{-}O{-}\overset{\overset{\displaystyle O}{\|}}{C}{-}R \qquad \text{(acid anhydride)}$$

$$R{-}\overset{\overset{\displaystyle O}{\|}}{C}{-}OH \xrightarrow{-OR} R{-}\overset{\overset{\displaystyle O}{\|}}{C}{-}OR \qquad \text{(ester)}$$

$$R{-}\overset{\overset{\displaystyle O}{\|}}{C}{-}OH \xrightarrow{-NH_2} R{-}\overset{\overset{\displaystyle O}{\|}}{C}{-}NH_2 \qquad \text{(amide)}$$

25.2 NOMENCLATURE AND SOURCES OF ALIPHATIC CARBOXYLIC ACIDS

Aliphatic carboxylic acids form a homologous series. The carboxyl group is always at the beginning of a carbon chain, and the C atom in this group is understood to be carbon number 1 in naming the compound.

To name a carboxylic acid by the IUPAC System, first identify the longest carbon chain including the carboxyl group. Then form the acid name by dropping the final *e* from the corresponding parent hydrocarbon name and adding *oic acid.* Thus, the names corresponding to the one-, two-, and three-carbon acids are methanoic acid, ethanoic acid, and propanoic acid. These names are derived from methane, ethane, and propane.

CH_4	Methane	$HCOOH$	Methanoic acid
CH_3CH_3	Ethane	CH_3COOH	Ethanoic acid
$CH_3CH_2CH_3$	Propane	CH_3CH_2COOH	Propanoic acid

Other groups bonded to the parent chain are numbered and named as we have done previously. For example,

$$\overset{5}{C}H_3\overset{4}{C}H_2\overset{3}{\underset{\underset{\displaystyle CH_3}{|}}{C}}H\overset{2}{C}H_2\overset{1}{C}OOH$$

3-Methylpentanoic acid

Unfortunately the IUPAC method is not the only, nor the most used, method of naming acids. Organic acids are usually known by common names. Methanoic, ethanoic, and propanoic acids are called formic, acetic, and propionic acids, respectively. These names usually refer to a natural source of the acid and are not systematic. Formic acid was named from the Latin word *formica,* meaning ant. This acid contributes to the stinging sensation of ant bites. Acetic acid is found in vinegar and is so named from the Latin word for vinegar. The name of butyric acid is derived from the Latin term for butter, since it is

▲ **The sting of an ant bite is caused by formic acid.**

TABLE 25.1 Names, Formulas, and Physical Properties of Saturated Carboxylic Acids

Common name (IUPAC name)	Formula	Melting point (°C)	Boiling point (°C)	Solubility in water[a]
Formic acid (Methanoic acid)	HCOOH	8.4	100.8	∞
Acetic acid (Ethanoic acid)	CH_3COOH	16.6	118	∞
Propionic acid (Propanoic acid)	CH_3CH_2COOH	−21.5	141.4	∞
Butyric acid (Butanoic acid)	$CH_3(CH_2)_2COOH$	−6	164	∞
Valeric acid (Pentanoic acid)	$CH_3(CH_2)_3COOH$	−34.5	186.4	3.3
Caproic acid (Hexanoic acid)	$CH_3(CH_2)_4COOH$	−3.4	205	1.1
Caprylic acid (Octanoic acid)	$CH_3(CH_2)_6COOH$	16.3	239	0.1
Capric acid (Decanoic acid)	$CH_3(CH_2)_8COOH$	31.4	269	Insoluble
Lauric acid (Dodecanoic acid)	$CH_3(CH_2)_{10}COOH$	44.1	225[b]	Insoluble
Myristic acid (Tetradecanoic acid)	$CH_3(CH_2)_{12}COOH$	54.2	251[b]	Insoluble
Palmitic acid (Hexadecanoic acid)	$CH_3(CH_2)_{14}COOH$	63	272[b]	Insoluble
Stearic acid (Octadecanoic acid)	$CH_3(CH_2)_{16}COOH$	69.6	287[b]	Insoluble
Arachidic acid (Eicosanoic acid)	$CH_3(CH_2)_{18}COOH$	77	298[b]	Insoluble

[a] Grams of acid per 100 g of water

[b] Boiling point is given at 100 mm Hg pressure instead of atmospheric pressure because thermal decomposition occurs before this acid reaches its boiling point at atmospheric pressure.

a constituent of butterfat. The 6-, 8-, and 10-carbon acids are found in goat fat and have names derived from the Latin word for goat. These three acids—caproic, caprylic, and capric—along with butyric acid have characteristic and disagreeable odors. In a similar way the names of the 12-, 14-, and 16-carbon acids—lauric, myristic, and palmitic—are from plants from which the corresponding acid has been isolated. The name stearic acid is derived from a Greek word meaning beef fat or tallow, which is a good source of this acid. Many of the carboxylic acids, principally those having even numbers of carbon atoms ranging from 4 to about 20, exist in combined form in plant and animal fats. These are called *fatty acids* (see Chapter 30). Table 25.1 lists the common and IUPAC names, together with some of the physical properties, of the more important saturated aliphatic acids.

Another common nomenclature method using letters of the Greek alphabet ($\alpha, \beta, \gamma, \delta, \ldots$) has traditionally been used in naming certain acid derivatives, especially hydroxy, amino, and halogen acids. When Greek letters are used, the carbon atoms, beginning with the one adjacent to the carboxyl group, are labeled α, β, γ, $\delta, \ldots$. When numbers are used, the numbers begin with the carbon in the —COOH group. Common and IUPAC nomenclature systems should not be intermixed.

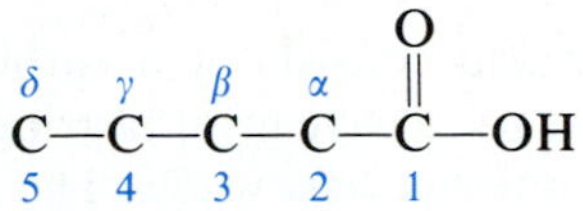

$$\overset{\delta}{\underset{5}{C}}-\overset{\gamma}{\underset{4}{C}}-\overset{\beta}{\underset{3}{C}}-\overset{\alpha}{\underset{2}{C}}-\underset{1}{\overset{O}{\overset{\|}{C}}}-OH$$

	$CH_3CH_2CH(OH)COOH$	$CH_3CH(NH_2)COOH$	CH_2ClCH_2COOH
Common name:	α-Hydroxybutyric acid	α-Aminopropionic acid	β-Chloropropionic acid
IUPAC name:	2-Hydroxybutanoic acid	2-Aminopropanoic acid	3-Chloropropanoic acid

Example 25.1

Write formulas for the following:
(a) 3-Chloropentanoic acid (b) γ-Hydroxybutyric acid (c) Phenylacetic acid

Solution

(a) *Pentanoic* indicates a five-carbon acid. Substituted on carbon 3 is a chlorine atom. Write five carbon atoms in a row. Make carbon 1 a carboxyl group, place a Cl on carbon 3, and add hydrogens to give each carbon four bonds. The formula is

$$CH_3CH_2CHClCH_2COOH$$

(b) *Butyric* indicates a four-carbon acid. The gamma (γ) position is three carbons removed from the carboxyl group. Therefore, the formula is

$$HO-\underbrace{CH_2}_{\gamma\ \text{carbon}}CH_2CH_2COOH$$

(c) *Acetic acid* is the familiar two-carbon acid. There is only one place to substitute the phenyl group and still call the compound an acid–that is, at the CH_3 group. Substitute a phenyl group for one of the three H atoms to give the formula:

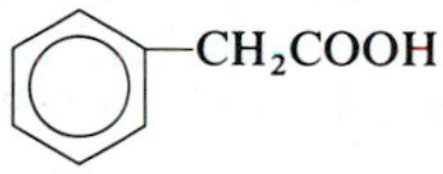

$$C_6H_5-CH_2COOH$$

Practice Write formulas for (a) 2-methylpropanoic acid, (b) β-chlorocaproic acid, and (c) cyclohexanecarboxylic acid.

Answers:

(a) $CH_3CH(CH_3)COOH$

(b) $CH_3CH_2CH_2CH(Cl)CH_2COOH$

(c) 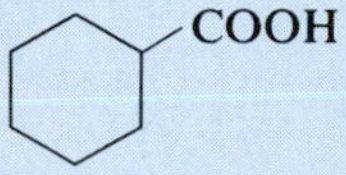 $C_6H_{11}-COOH$

25.3 Physical Properties of Carboxylic Acids

Each aliphatic carboxylic acid molecule is polar and consists of a carboxyl group and a hydrocarbon radical. These two unlike parts have great bearing on the physical, as well as chemical, behavior of the molecule as a whole. The first four acids, formic through butyric, are completely soluble (miscible) in water (Table 25.1). Beginning with pentanoic acid (valeric acid), the water solubility falls sharply and is only about 0.1 g of acid per 100 g of water for octanoic acid (caprylic acid). Acids of this series with more than eight carbons are virtually insoluble in water. The water-solubility characteristics of the first four acids are evidently determined by the highly soluble polar carboxyl group. Thereafter the water solubility of the nonpolar hydrocarbon chain is dominant.

The polarity due to the carboxyl group is evident in the boiling-point data. Formic acid (HCOOH) boils at about 101°C. Carbon dioxide, a nonpolar substance of similar molar mass, remains in the gaseous state until it is cooled to −78°C. In like manner, the boiling point of acetic acid (molar mass 60) is 118°C, whereas nonpolar butane (molar mass 58) boils at −0.6°C. The comparatively high boiling points for carboxylic acids are due to intermolecular attractions resulting from hydrogen bonding. In fact, molar mass determinations on gaseous acetic acid (near its boiling point) show a value of about 120, indicating that two molecules are joined together to form a *dimer*, $(CH_3COOH)_2$.

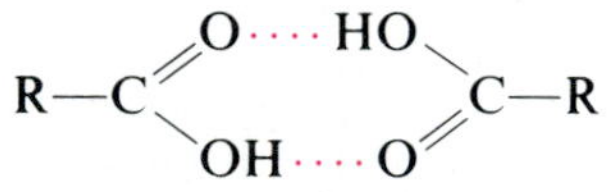

$$R{-}C\begin{matrix}{\nearrow}O\cdots HO{\searrow}\\ {\searrow}OH\cdots O{\nearrow}\end{matrix}C{-}R \qquad CH_3{-}C\begin{matrix}{\nearrow}O\cdots HO{\searrow}\\ {\searrow}OH\cdots O{\nearrow}\end{matrix}C{-}CH_3$$

Hydrogen bonding in carboxylic acids

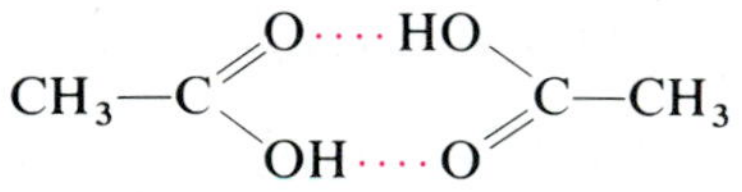

Acetic acid dimer

Saturated monocarboxylic acids that have fewer than ten carbon atoms are liquids at room temperature, whereas those with more than ten carbon atoms are waxlike solids.

Carboxylic acids and phenols, like mineral acids such as HCl, ionize in water to produce hydronium ions and anions (Chapter 16). Carboxylic acids are generally weak; that is, they are only slightly ionized in water. Phenols are, in general, even weaker acids than carboxylic acids. For example, the ionization constant for acetic acid is 1.8×10^{-5} and that for phenol is 1.1×10^{-10}. Equations illustrating these ionizations are given below.

$$\underset{\text{Hydrogen chloride}}{HCl} + H_2O \longrightarrow \underset{\text{Hydronium ion}}{H_3O^+} + \underset{\text{Chloride ion}}{Cl^-}$$

$$\underset{\text{Acetic acid}}{CH_3\overset{\overset{O}{\parallel}}{C}{-}OH} + H_2O \rightleftharpoons \underset{\text{Hydronium ion}}{H_3O^+} + \underset{\text{Acetate ion}}{CH_3\overset{\overset{O}{\parallel}}{C}{-}O^-}$$

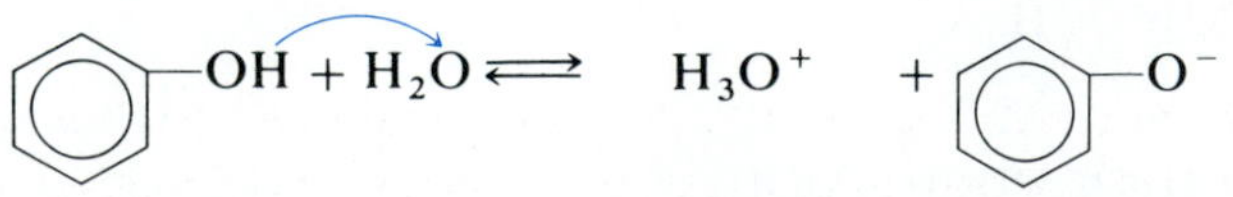

$$\underset{\text{Phenol}}{C_6H_5{-}OH} + H_2O \rightleftharpoons \underset{\text{Hydronium ion}}{H_3O^+} + \underset{\text{Phenoxide ion}}{C_6H_5{-}O^-}$$

Carboxylic acids are very common in biological systems. Their tendency to ionize and form anions means that many biological molecules carry negative charges.

▲ **Rhubarb is one of a number of vegetables containing carboxylic acids.**

25.4 Classification of Carboxylic Acids

Thus far our discussion has dealt mainly with a single type of acid—that is, saturated monocarboxylic acids. But various other kinds of carboxylic acids are known. Some of the more important ones are discussed here.

Unsaturated Acids

An unsaturated acid contains one or more carbon–carbon double bonds. The first member of the homologous series of unsaturated carboxylic acids, containing one carbon–carbon double bond, is acrylic acid, $CH_2{=}CHCOOH$. The IUPAC name for $CH_2{=}CHCOOH$ is propenoic acid. Derivatives of acrylic acid are used to manufacture a class of synthetic polymers known as the acrylates (see Chapter 27). These polymers are widely used as textiles and in paints and lacquers. Unsaturated carboxylic acids undergo the reactions of both an unsaturated hydrocarbon and a carboxylic acid.

Even one carbon–carbon double bond in the molecule exerts a major influence on the physical and chemical properties of an acid. The effect of a double bond can be seen when comparing the two 18-carbon acids, stearic and oleic. Stearic acid, $CH_3(CH_2)_{16}COOH$, a solid that melts at 70°C, shows only the reactions of a carboxylic acid. On the other hand, oleic acid, $CH_3(CH_2)_7CH{=}CH(CH_2)_7COOH$ (mp 16°C), with one double bond, is a liquid at room temperature and shows the reactions of an unsaturated hydrocarbon as well as those of a carboxylic acid.

Aromatic Carboxylic Acids

In an aromatic carboxylic acid, the carbon of the carboxyl group (—COOH) is bonded directly to a carbon in an aromatic nucleus. The parent compound of this series is benzoic acid. Other common examples are the three isomeric toluic acids:

Benzoic acid	*o*-Toluic acid	*m*-Toluic acid	*p*-Toluic acid
C_6H_5—COOH	CH_3 and —COOH on adjacent ring carbons	CH_3 and —COOH in 1,3 positions	CH_3 and COOH in 1,4 positions

Dicarboxylic Acids

Acids of both the aliphatic and aromatic series that contain two or more carboxyl groups are known. These are called dicarboxylic acids. The simplest member of the aliphatic series is oxalic acid. The next member in the homologous

TABLE 25.2 Names and Formulas of Selected Dicarboxylic Acids

Common name	IUPAC name	Formula
Oxalic acid	Ethanedioic acid	HOOCCOOH
Malonic acid	Propanedioic acid	$HOOCCH_2COOH$
Succinic acid	Butanedioic acid	$HOOC(CH_2)_2COOH$
Glutaric acid	Pentanedioic acid	$HOOC(CH_2)_3COOH$
Adipic acid	Hexanedioic acid	$HOOC(CH_2)_4COOH$
Fumaric acid	*trans*-2-Butenedioic acid	HOOCCH=CHCOOH
Maleic acid	*cis*-2-Butenedioic acid	HOOCCH=CHCOOH

series is malonic acid. Several dicarboxylic acids and their names are listed in Table 25.2.

The IUPAC names for dicarboxylic acids are formed by modifying the corresponding hydrocarbon names to end in *dioic acid*. Thus the two-carbon acid is ethanedioic acid (derived from ethane). However, the common names for dicarboxylic acids are frequently used.

Oxalic acid is found in various plants including spinach, cabbage, and rhubarb. Among its many uses are bleaching straw and leather and removing rust and ink stains. Although oxalic acid is poisonous, the amounts present in the above-mentioned vegetables are usually not harmful.

Malonic acid is made synthetically but was originally prepared from malic acid, which is commonly found in apples and many fruit juices. Malonic acid is one of the major compounds used in the manufacture of the class of drugs known as barbiturates. When heated above their melting points, malonic acid and substituted malonic acids lose carbon dioxide to give monocarboxylic acids. Thus, malonic acid yields acetic acid when strongly heated:

$$\begin{matrix} COOH \\ | \\ CH_2 \\ | \\ COOH \end{matrix} \xrightarrow{150^\circ C} CH_3COOH + CO_2\uparrow$$

Malonic acid — Acetic acid

Malonic acid is used as the biological precursor for the synthesis of fatty acids. In living cells, when malonic acid loses carbon dioxide, the acetic acid units are linked together to begin formation of fatty acids.

Succinic acid has been known since the 16th century, when it was obtained as a distillation product of amber. Succinic, fumaric, and citric acids are among the important acids in the energy-producing metabolic pathway known as the citric acid cycle (see Chapter 36). Citric acid is a *tricarboxylic acid* that is widely distributed in plant and animal tissue, especially in citrus fruits (lemon juice contains 5–8%). The formula for citric acid is

$$\begin{matrix} & & CH_2COOH \\ & & | \\ HO & - & C-COOH \\ & & | \\ & & CH_2COOH \end{matrix}$$

Citric acid

When succinic acid is heated, it loses water, forming succinic anhydride, an acid anhydride. Glutaric acid behaves similarly, forming glutaric anhydride.

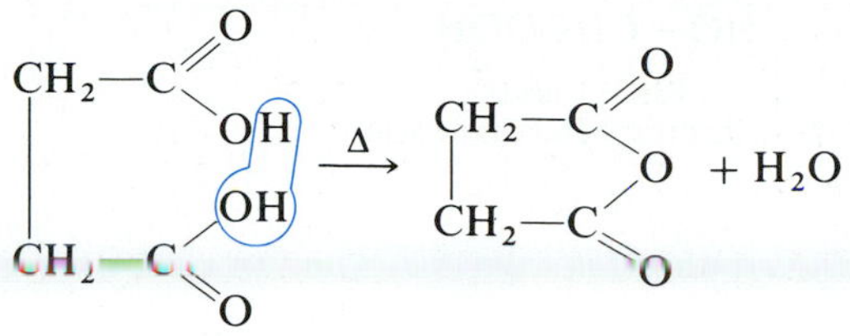

$$\begin{array}{l} CH_2-C(=O)-OH \\ | \\ CH_2-C(=O)-OH \end{array} \xrightarrow{\Delta} \begin{array}{l} CH_2-C(=O) \\ | \quad\quad\quad\; \backslash O \\ CH_2-C(=O) \end{array} + H_2O$$

Succinic acid — Succinic anhydride

Adipic acid is the most important commercial dicarboxylic acid. It is made from benzene by converting it first to cyclohexene and then by oxidation to adipic acid. About 5.9×10^8 kg (1.3×10^9 lb) of adipic acid is produced annually in the United States. Most of the adipic acid is used to produce nylon (Chapter 27). It is also used in polyurethane foams, plasticizers, and lubricating-oil additives.

Aromatic dicarboxylic acids contain two carboxyl groups attached directly to an aromatic nucleus. Examples are the three isomeric phthalic acids, $C_6H_4(COOH)_2$:

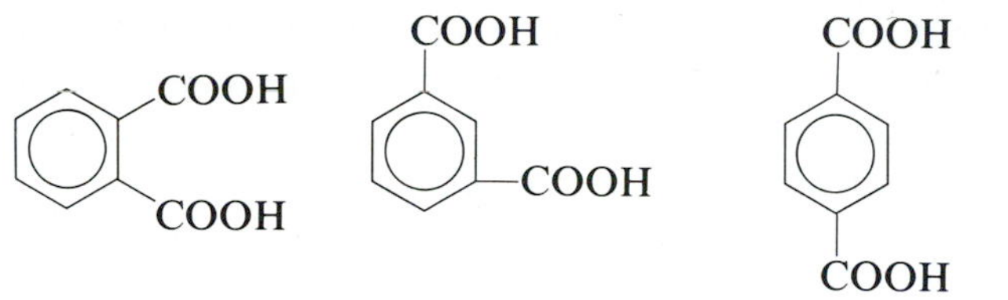

o-Phthalic acid (Phthalic acid) — *m*-Phthalic acid (Isophthalic acid) — *p*-Phthalic acid (Terephthalic acid)

Dicarboxylic acids are *bifunctional*; that is, they have two sites where reactions can occur. Therefore they are often used as monomers in the preparation of synthetic polymers such as Dacron polyester (Section 27.7).

Hydroxy Acids

Lactic acid, found in sour milk, sauerkraut, and dill pickles, has the functional groups of both a carboxylic acid and an alcohol. Lactic acid is the end product when our muscles use glucose for energy in the absence of oxygen, a process called *glycolysis* (see Chapter 36). Salicylic acid is both a carboxylic acid and a phenol. It is of special interest because a family of useful drugs—the salicylates—are derivatives of this acid. The salicylates include aspirin and function as *analgesics* (pain relievers) and as *antipyretics* (fever reducers). The structural formulas of several hydroxy acids follow:

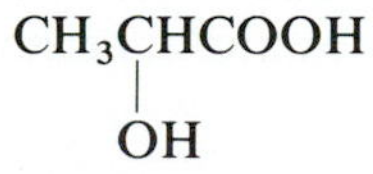

$$CH_3CH(OH)COOH$$

Lactic acid
(α-Hydroxypropionic acid)

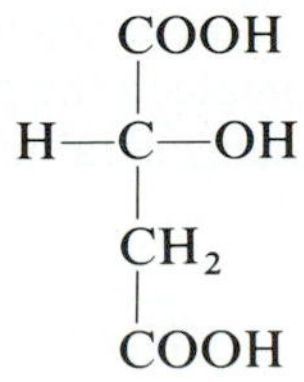

$$HOOC-CH(OH)-CH_2-COOH$$

Malic acid
(α-Hydroxysuccinic acid)

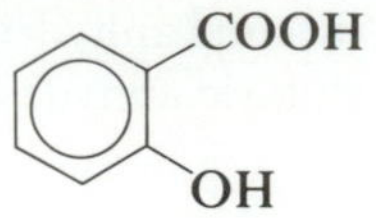

$$o\text{-}C_6H_4(OH)COOH$$

Salicylic acid
(o-Hydroxybenzoic acid)

$$\begin{array}{l} HO\text{—}CHCOOH \\ \quad\;\;| \\ HO\text{—}CHCOOH \end{array}$$

Tartaric acid
(2,3-Dihydroxybutanedioic acid)

Amino Acids

Naturally occurring amino acids have this general formula; the amino group is in the alpha position:

$$R\text{—}\underset{\underset{NH_2}{|}}{CH}\text{—}\overset{\overset{O}{\|}}{C}\text{—}OH$$

Each amino acid molecule has a carboxyl group that acts as an acid and an amino group that acts as a base. About 20 biologically important amino acids, each with a different group represented by R, have been found in nature. (In amino acids, R does not always represent an alkyl group.) The immensely complicated protein molecules, found in every form of life, are built from amino acids. Some protein molecules contain more than 10,000 amino acid units. Amino acids and proteins are discussed in more detail in Chapter 31.

25.5 Preparation of Carboxylic Acids

Many different methods of preparing carboxylic acids are known. We will consider only a few examples.

Oxidation of an Aldehyde or a Primary Alcohol This is a general method that can be used to convert an aldehyde or primary alcohol to the corresponding carboxylic acid:

$$RCH_2OH \xrightarrow{[O]} RCOOH$$

$$R\overset{\overset{O}{\|}}{C}\text{—}H \xrightarrow{[O]} RCOOH$$

Butyric acid (butanoic acid) can be obtained by oxidizing either 1-butanol or butanal with potassium dichromate in the presence of sulfuric acid. Aromatic acids may be prepared by the same general method. For example, benzoic acid is obtained by oxidizing benzyl alcohol.

$$\underset{\text{1-Butanol}}{CH_3CH_2CH_2CH_2OH} \xrightarrow[H_2SO_4,\ \Delta]{Cr_2O_7^{2-}} \underset{\text{Butanoic acid}}{CH_3CH_2CH_2COOH}$$

$$\underset{\text{Butanal}}{CH_3CH_2CH_2\overset{O}{\overset{\|}{C}}—H} \xrightarrow[H_2SO_4 \\ \Delta]{Cr_2O_7^{2-}} \underset{\text{Butanoic acid}}{CH_3CH_2CH_2COOH}$$

$$\underset{\text{Benzyl alcohol}}{C_6H_5—CH_2OH} \xrightarrow[H_2SO_4 \\ \Delta]{Cr_2O_7^{2-}} \underset{\text{Benzoic acid}}{C_6H_5—COOH}$$

Carboxylic acids can also be obtained by the hydrolysis or saponification of esters (see Section 25.9).

Oxidation of Alkyl Groups Attached to Aromatic Rings When reacted with a strong oxidizing agent (alkaline permanganate solution or potassium dichromate and sulfuric acid), alkyl groups bonded to aromatic rings are oxidized to carboxyl groups. Regardless of the size or length of the alkyl group, the carbon atom adjacent to the ring remains bonded to the ring and is oxidized to a carboxyl group. The remainder of the alkyl group goes either to carbon dioxide or to a salt of a carboxylic acid. Thus sodium benzoate is obtained when toluene, ethylbenzene, or propylbenzene is heated with alkaline permanganate solution:

$$\underset{\text{Toluene}}{C_6H_5—CH_3} \xrightarrow[NaOH \\ \Delta]{NaMnO_4} \underset{\text{Sodium benzoate}}{C_6H_5—\overset{O}{\overset{\|}{C}}—O^- \, Na^+}$$

$$\underset{\text{Ethylbenzene}}{C_6H_5—CH_2CH_3} \xrightarrow[NaOH \\ \Delta]{NaMnO_4} \underset{\text{Sodium benzoate}}{C_6H_5—\overset{O}{\overset{\|}{C}}—O^- \, Na^+} + CO_2\uparrow$$

$$\underset{\text{Propylbenzene}}{C_6H_5—CH_2CH_2CH_3} \xrightarrow[NaOH \\ \Delta]{NaMnO_4} \underset{\text{Sodium benzoate}}{C_6H_5—\overset{O}{\overset{\|}{C}}—O^- \, Na^+} + \underset{\text{Sodium acetate}}{CH_3COO^- \, Na^+}$$

Since the reaction is conducted in an alkaline medium, a salt of the carboxylic acid (sodium benzoate) is formed instead of the free acid. To obtain the free carboxylic acid, the reaction mixture is acidified with a strong mineral acid (HCl or H_2SO_4) in a second step.

$$\underset{\text{Sodium benzoate}}{C_6H_5—COONa} + H^+ \longrightarrow \underset{\text{Benzoic acid}}{C_6H_5—COOH} + Na^+$$

Hydrolysis of Nitriles Nitriles, RCN, which can be prepared by adding HCN to aldehydes and ketones (Section 24.4) or by reacting alkyl halides with KCN, can be hydrolyzed to carboxylic acids:

$$\underset{\text{Alkyl halide}}{RX} + KCN \longrightarrow \underset{\text{A nitrile}}{RCN} + KX$$

$$RCN + 2\,H_2O \xrightarrow{H^+} RCOOH + NH_4^+$$

$$CH_3CN + 2\,H_2O \xrightarrow{H^+} CH_3COOH + NH_4^+$$

25.6 Chemical Properties of Carboxylic Acids

Acid–Base Reactions Because of their ability to form hydrogen ions in solution, acids in general have the following properties:

1. Sour taste
2. Change blue litmus to red and affect other suitable indicators
3. Form water solutions with pH values of less than 7
4. Undergo neutralization reactions with bases to form water and a salt

All of the foregoing general properties of an acid are readily seen in low molar mass carboxylic acids such as acetic acid. However, these general acid properties can be greatly influenced by the size of the hydrocarbon chain attached to the carboxyl group. In stearic acid, for example, taste, effect on indicators, and pH are not detectable because the large size of the hydrocarbon chain makes the acid insoluble in water. But stearic acid reacts with a base to form water and a salt. With sodium hydroxide, the equation for the reaction is

$$\underset{\text{Stearic acid}}{C_{17}H_{35}COOH} + NaOH \longrightarrow \underset{\text{Sodium stearate}}{C_{17}H_{35}COONa} + H_2O$$

The salts formed from this neutralization reaction have different properties than the acids. Salts are soluble in water and dissociate completely in solution. These properties assist in the separation of carboxylic acids from other nonpolar compounds. A base, like NaOH, is added to the mixture of compounds. The carboxylic acid reacts, forming its sodium salt and water. The salt dissolves in water while the remaining nonpolar molecules stay in the organic layer. Once the layers are separated, some mineral acid, HCl, can be added to the carboxylic acid salt to recover the acid.

Carboxylic acids generally react with sodium bicarbonate to release carbon dioxide. This reaction can be used to distinguish a carboxylic acid from a phenol (also a weak acid). Phenols do not react with bicarbonate, although they will be neutralized by a strong base.

Acid Chloride Formation Thionyl chloride ($SOCl_2$) reacts with carboxylic acids to form acid chlorides, which are very reactive and can be used to synthesize other substances such as amides and esters:

$$\underset{\text{Acid}}{\mathrm{R}\overset{\overset{\mathrm{O}}{\|}}{\mathrm{C}}\text{—OH}} + \underset{\text{Thionyl chloride}}{\mathrm{SOCl_2}} \longrightarrow \underset{\text{Acid chloride}}{\mathrm{R}\overset{\overset{\mathrm{O}}{\|}}{\mathrm{C}}\text{—Cl}} + \mathrm{SO_2} + \mathrm{HCl}$$

$$\mathrm{CH_3}\overset{\overset{\mathrm{O}}{\|}}{\mathrm{C}}\text{—OH} + \mathrm{SOCl_2} \longrightarrow \underset{\text{Acetyl chloride}}{\mathrm{CH_3}\overset{\overset{\mathrm{O}}{\|}}{\mathrm{C}}\text{—Cl}} + \mathrm{SO_2} + \mathrm{HCl}$$

Acid chlorides are extremely reactive substances. They must be kept away from moisture, or they will hydrolyze back to the acid:

$$\mathrm{R}\overset{\overset{\mathrm{O}}{\|}}{\mathrm{C}}\text{—Cl} + \mathrm{H_2O} \longrightarrow \mathrm{R}\overset{\overset{\mathrm{O}}{\|}}{\mathrm{C}}\text{—OH} + \mathrm{HCl}$$

Acid chlorides are more reactive than acids and can be used in place of acids to prepare esters and amides:

$$\mathrm{CH_3}\overset{\overset{\mathrm{O}}{\|}}{\mathrm{C}}\text{—Cl} + \mathrm{CH_3OH} \longrightarrow \underset{\text{Methyl acetate}}{\mathrm{CH_3}\overset{\overset{\mathrm{O}}{\|}}{\mathrm{C}}\text{—OCH_3}} + \mathrm{HCl}$$

$$\mathrm{CH_3}\overset{\overset{\mathrm{O}}{\|}}{\mathrm{C}}\text{—Cl} + 2\,\mathrm{NH_3} \longrightarrow \underset{\text{Acetamide}}{\mathrm{CH_3}\overset{\overset{\mathrm{O}}{\|}}{\mathrm{C}}\text{—NH_2}} + \mathrm{NH_4Cl}$$

Acid Anhydride Formation Inorganic anhydrides are formed by the elimination of a molecule of water from an acid or a base:

$$\mathrm{H_2SO_3} \longrightarrow \mathrm{SO_2} + \mathrm{H_2O}$$

$$\mathrm{Ba(OH)_2} \longrightarrow \mathrm{BaO} + \mathrm{H_2O}$$

An organic anhydride could be formed by the elimination of a molecule of water from two molecules of acid.

$$\mathrm{R}\text{—}\overset{\overset{\mathrm{O}}{\|}}{\mathrm{C}}\text{—OH} + \text{HO—}\overset{\overset{\mathrm{O}}{\|}}{\mathrm{C}}\text{—R}' \longrightarrow \mathrm{R}\text{—}\overset{\overset{\mathrm{O}}{\|}}{\mathrm{C}}\text{—O—}\overset{\overset{\mathrm{O}}{\|}}{\mathrm{C}}\text{—R}' + \mathrm{H_2O}$$

The most commonly used organic anhydride is acetic anhydride. It can be prepared by the reaction of acetyl chloride with sodium acetate.

$$\mathrm{CH_3}\overset{\overset{\mathrm{O}}{\|}}{\mathrm{C}}\text{—Cl} + \mathrm{Na^+}\ {}^{-}\mathrm{O}\overset{\overset{\mathrm{O}}{\|}}{\mathrm{C}}\text{—CH_3} \longrightarrow \underset{\text{Acetic anhydride}}{\mathrm{CH_3}\overset{\overset{\mathrm{O}}{\|}}{\mathrm{C}}\text{—O—}\overset{\overset{\mathrm{O}}{\|}}{\mathrm{C}}\text{—CH_3}} + \mathrm{NaCl}$$

Acid anhydrides are very reactive and can be used to synthesize amides and esters. The anhydrides are not used as often as the acid chlorides in organic synthesis, however. In living cells acid anhydrides are commonly used to activate carboxylic acids for further reaction.

Ester Formation When an acid and an alcohol are heated in an acidic medium, a condensation reaction occurs. The products are an ester and water.

$$\underset{\text{Carboxylic acid}}{\overset{\overset{\displaystyle O}{\|}}{RC}\!-\!\boxed{OH}} + \underset{\text{Alcohol}}{R'O\boxed{H}} \overset{H^+}{\rightleftharpoons} \underset{\substack{\text{Ester}\\ \text{(R can be H,}\\ \text{but R' cannot be H)}}}{\overset{\overset{\displaystyle O}{\|}}{RC}\!-\!OR'} + H_2O$$

$$\underset{\text{Formic acid}}{H-\overset{\overset{\displaystyle O}{\|}}{C}-OH} + CH_3CH_2OH \overset{H^+}{\rightleftharpoons} \underset{\text{Ethyl formate}}{H-\overset{\overset{\displaystyle O}{\|}}{C}-OCH_2CH_3} + H_2O$$

At first glance this looks like the familiar acid–base neutralization reaction. But this is not the case, because the alcohol does not yield OH^- ions, and the ester, unlike a salt, is a molecular, not an ionic, substance. The forward reaction of an acid and an alcohol is called *esterification*; the reverse reaction of an ester with water is called *hydrolysis*. The work of a chemist may call for manipulating reaction conditions to favor the formation of either esters or their component parts, alcohols and acids.

Esterification is one of the most important reactions of carboxylic acids. Many biologically significant substances are esters.

25.7 Nomenclature of Esters

The general formula for an ester is RCOOR′, where R may be a hydrogen, alkyl group, or aryl group, and R′ may be an alkyl group, or aryl group, but *not* a hydrogen. Esters are found throughout nature. The ester linkage is particularly important in the study of fats and oils, both of which are esters. Esters of phosphoric acid are of vital importance to life as well.

Esters are alcohol derivatives of carboxylic acids. They are named in much the same way as salts. The alcohol part is named first, followed by the name of the acid modified to end in *ate*. The *ic* ending of the organic acid name is replaced by the ending *ate*. Thus in the IUPAC System, *ethanoic acid* becomes *ethanoate*. In the common names, *acetic acid* becomes *acetate*. To name an ester it is necessary to recognize the portion of the ester molecule that comes from the acid and the portion that comes from the alcohol. In the general formula for an ester, the RC=O comes from the acid, and the R′O comes from the alcohol:

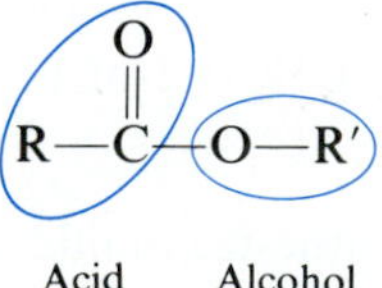

Acid Alcohol

The R′ in R′O is named first, followed by the name of the acid modified by replacing *ic acid* with *ate*. The ester derived from ethyl alcohol and acetic acid

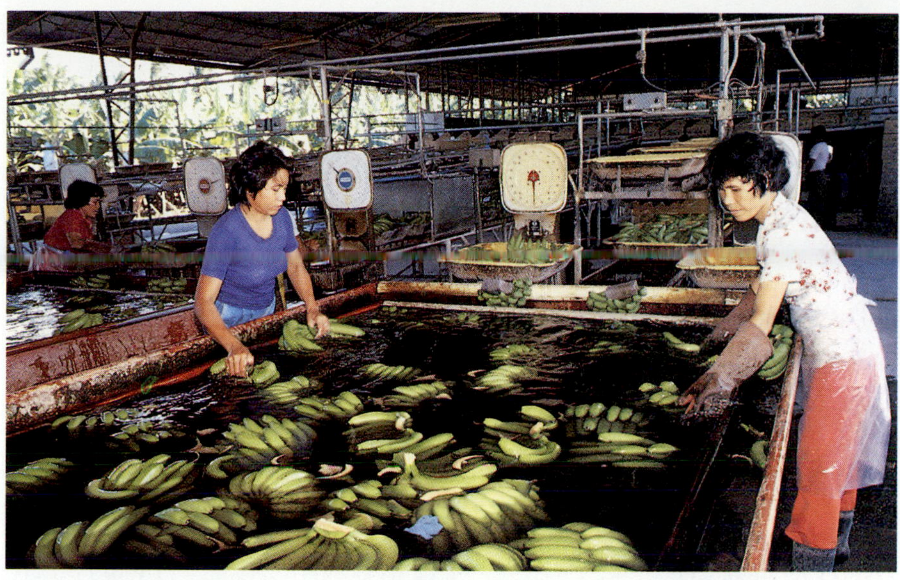

◄ **Bananas during the harvesting process. The odor of ripe bananas in the store or at home is caused by the ester isoamyl acetate.**

is called ethyl acetate or ethyl ethanoate. Consider the ester formed from CH_3CH_2COOH and CH_3OH:

$$\underset{\text{(Propionic acid)}}{\underset{\text{Propanoic acid}}{CH_3CH_2\overset{\overset{\displaystyle O}{\|}}{C}}}\!\!-\!\boxed{OH + H}\!-\!\underset{\text{(Methyl alcohol)}}{\underset{\text{Methanol}}{OCH_3}} \overset{H^+}{\rightleftharpoons} \underset{\text{(Methyl propionate)}}{\underset{\text{Methyl propanoate}}{CH_3CH_2\overset{\overset{\displaystyle O}{\|}}{C}-OCH_3}} + H_2O$$

Esters of aromatic acids are named in the same general way as those of aliphatic acids. For example, the ester of benzoic acid and isopropyl alcohol is

$$C_6H_5-\overset{\overset{\displaystyle O}{\|}}{C}-O-CH(CH_3)_2$$

Isopropyl benzoate

Formulas and names for additional esters are given in Table 25.3.

25.8 Occurrence and Physical Properties of Esters

Since many acids and many alcohols are known, the number of esters theoretically possible is very large. In fact, both natural and man-made esters exist in almost endless variety. Simple esters derived from monocarboxylic acids and

TABLE 25.3 Formulas and Names of Selected Esters

Formula	IUPAC name	Common name	Odor or flavor
$CH_3C(=O){-}OCH_2CH_2CH(CH_3)CH_3$	Isopentyl ethanoate	Isoamyl acetate	Banana, pear
$CH_3CH_2CH_2C(=O){-}OCH_2CH_3$	Ethyl butanoate	Ethyl butyrate	Pineapple
$HC(=O){-}OCH_2CH(CH_3)CH_3$	Isobutyl methanoate	Isobutyl formate	Raspberry
$CH_3C(=O){-}OCH_2(CH_2)_6CH_3$	Octyl ethanoate	*n*-Octyl acetate	Orange
$C_6H_4(OH){-}C(=O){-}OCH_3$ (ortho)	Methyl-2-hydroxybenzoate	Methyl salicylate	Wintergreen

monohydroxy alcohols are colorless, generally nonpolar liquids or solids. The low polarity of ester molecules is substantiated by the fact that both their water solubility and boiling points are lower than those of either acids or alcohols of similar molar masses.

Low and intermediate molar mass esters (from both acids and alcohols up to about ten carbons) are liquids with characteristic, usually fragrant or fruity odors. The distinctive odor and flavor of many fruits are caused by one or more of these esters. The difference in properties between an acid and its esters is remarkable. For example, in contrast to the extremely unpleasant odor of butyric acid, ethyl butyrate has the pleasant odor of pineapple and methyl butyrate the odor of artificial rum. Esters are used in flavoring and scenting agents (see Table 25.3). They are generally good solvents for organic substances, and those having relatively low molar masses are volatile. Therefore esters such as ethyl acetate, butyl acetate, and isoamyl acetate are extensively used in paints, varnishes, and lacquers.

High molar mass esters (formed from acids and alcohols of 16 or more carbons) are waxes and are obtained from various plants. They are used in furniture wax and automobile wax preparations; for example, carnauba wax contains esters of 24- and 28-carbon fatty acids and 32- and 34-carbon alcohols. Polyesters with very high molar masses, such as Dacron, are widely used in the textile industries (see Chapter 27).

EXAMPLE 25.2

Name the following esters:

(a) $H-\overset{\overset{O}{\|}}{C}-OCH_2CH_2CH_3$

(b) $C_6H_5-\overset{\overset{O}{\|}}{C}-OCH_2CH_3$

(c)
$$\begin{array}{c} O{=}C-OCH_2CH_3 \\ | \\ CH_2 \\ | \\ O{=}C-OCH_2CH_3 \end{array}$$

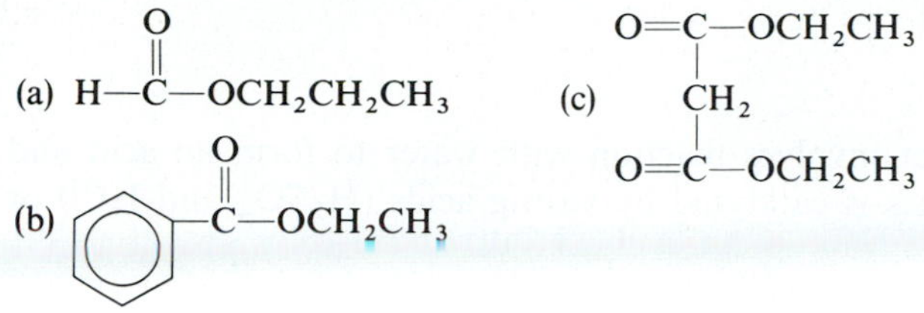

SOLUTION

(a) First identify the acid and alcohol components. The acid contains one carbon and is formic acid. The alcohol is propyl alcohol.

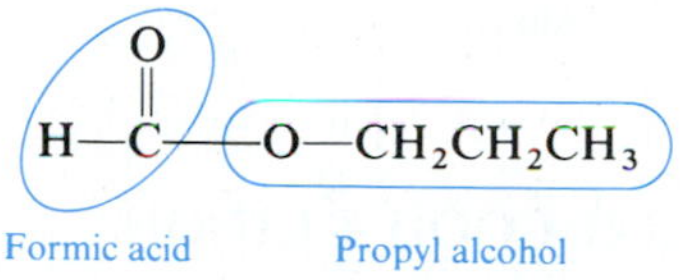

$$\underbrace{H-\overset{\overset{O}{\|}}{C}}_{\text{Formic acid}}-\underbrace{O-CH_2CH_2CH_3}_{\text{Propyl alcohol}}$$

Change the *ic* ending of the acid to *ate*, making the name formate or methanoate. The name of the ester then is propyl formate or propyl methanoate.

(b) The acid is benzoic acid; the alcohol is ethyl alcohol. Using the same procedure as in part (a), the name of the ester is ethyl benzoate.

(c) The acid is the three-carbon dicarboxylic acid, malonic acid. The alcohol is ethyl alcohol. Both acid groups are in the ester form. The name, therefore, is diethyl malonate.

PRACTICE Name the following esters:

(a) $CH_3CH_2\underset{\underset{O}{\|}}{C}-O-CH_3$

(b) $CH_3-\underset{\underset{O}{\|}}{C}-O-C_6H_5$

(c) $CH_3-O-\overset{\overset{O}{\|}}{C}-CH_2CH_2-\overset{\overset{O}{\|}}{C}-O-CH_3$

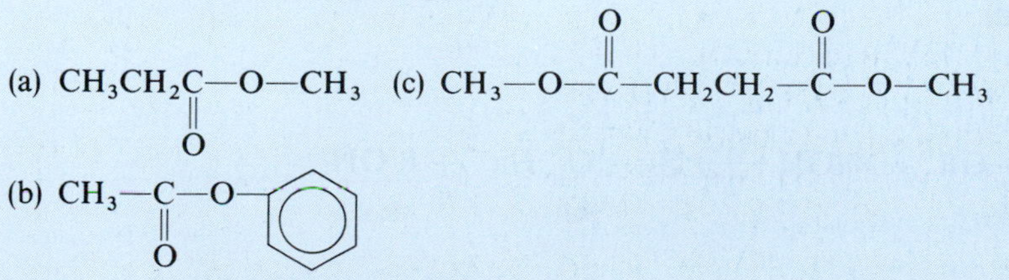

Answers: **(a) Methyl propanoate (b) Phenyl ethanoate (c) Dimethly succinate**

25.9 CHEMICAL PROPERTIES OF ESTERS

The most important reaction of esters is *hydrolysis*. Hydrolysis is the splitting of molecules through the addition of water. The majority of organic and biochemical substances react only very slowly, if at all, with water. In order to increase the rate of these reactions, a catalyst is required. In the laboratory the chemist often employs an acid or base as a catalyst for hydrolysis. In living systems the role of catalyst is filled by enzymes.

Acid Hydrolysis

Hydrolysis of an ester involves reaction with water to form an acid and an alcohol. The hydrolysis is catalyzed by strong acids (H_2SO_4 and HCl) or by certain enzymes.

$$\underset{\text{Ester}}{\mathrm{R\overset{\overset{\displaystyle O}{\|}}{C}{-}OR'}} + H_2O \xrightarrow[\text{or enzyme}]{H^+} \underset{\text{Acid}}{\mathrm{R\overset{\overset{\displaystyle O}{\|}}{C}{-}OH}} + \underset{\text{Alcohol}}{\mathrm{R'OH}}$$

$$\underset{\text{Methyl propanoate}}{\mathrm{CH_3CH_2\overset{\overset{\displaystyle O}{\|}}{C}{-}OCH_3}} + H_2O \xrightarrow{H^+} \underset{\text{Propanoic acid}}{\mathrm{CH_3CH_2COOH}} + \underset{\text{Methanol}}{\mathrm{CH_3OH}}$$

$$\underset{\text{Methyl salicylate}}{\mathrm{C_6H_4(OH)\overset{\overset{\displaystyle O}{\|}}{C}{-}OCH_3}} + H_2O \xrightarrow{H^+} \underset{\text{Salicylic acid}}{\mathrm{C_6H_4(OH)COOH}} + \underset{\text{Methanol}}{\mathrm{CH_3OH}}$$

Alkaline Hydrolysis (Saponification)

saponification

Saponification is the hydrolysis of an ester by a strong base (NaOH or KOH) to produce an alcohol and a salt (or soap if the salt formed is from a high molar mass acid):

$$\underset{\text{Ester}}{\mathrm{R\overset{\overset{\displaystyle O}{\|}}{C}{-}OR'}} + NaOH \xrightarrow[\Delta]{H_2O} \underset{\text{Salt}}{\mathrm{R\overset{\overset{\displaystyle O}{\|}}{C}{-}O^-\,Na^+}} + \underset{\text{Alcohol}}{\mathrm{R'OH}}$$

$$\underset{\text{Ethyl stearate}}{\mathrm{CH_3(CH_2)_{16}\overset{\overset{\displaystyle O}{\|}}{C}{-}OCH_2CH_3}} + NaOH \xrightarrow[\Delta]{H_2O} \underset{\text{Sodium stearate}}{\mathrm{CH_3(CH_2)_{16}\overset{\overset{\displaystyle O}{\|}}{C}ONa}} + \mathrm{CH_3CH_2OH}$$

The carboxylic acid may be obtained by reacting the salt with a strong acid such as HCl:

$$\mathrm{CH_3(CH_2)_{16}COONa} + HCl \longrightarrow \underset{\text{Stearic acid}}{\mathrm{CH_3(CH_2)_{16}COOH}} + NaCl$$

Notice that in saponification the base is a reactant and not a catalyst.

25.10 Glycerol Esters

Fats and **oils** are esters of glycerol and predominantly long-chain fatty acids. Fats and oils are also called **triacylglycerols** or triglycerides, since each molecule is derived from one molecule of glycerol and three molecules of fatty acid.

fats, oils

triacylglycerols

Glycerol portion →

$$\begin{array}{l} CH_2{-}O{-}\overset{O}{\overset{\|}{C}}{-}R \\ R'{-}\overset{O}{\overset{\|}{C}}{-}O{-}CH \\ CH_2{-}O{-}\overset{O}{\overset{\|}{C}}{-}R'' \end{array}$$

General formula for a triacylglycerol

$$\begin{array}{ll} CH_2{-}O{-}\overset{O}{\overset{\|}{C}}{-}C_{17}H_{35} & \text{(Stearic)} \\ C_{15}H_{31}\overset{O}{\overset{\|}{C}}{-}O{-}CH & \text{(Palmitic)} \\ CH_2{-}O{-}\overset{O}{\overset{\|}{C}}{-}C_{11}H_{23} & \text{(Lauric)} \end{array}$$

Typical triacylglycerol containing three different fatty acids

The structural formulas of triacylglycerol molecules vary because

1. The length of the fatty acid chain may vary from 4 to 20 carbons, but the number of carbon atoms in the chain is nearly always even.
2. Each fatty acid may be saturated or may be unsaturated and contain one, two, or three carbon–carbon double bonds.
3. A triacylglycerol may, and frequently does, contain three different fatty acids.

The most abundant saturated fatty acids in fats and oils are lauric, myristic, palmitic, and stearic acids (Table 25.4). The most abundant unsaturated acids in fats and oils contain 18 carbon atoms and have one, two, or three carbon–carbon double bonds. In all of these naturally occurring unsaturated acids, the configuration about the double bond is cis. Their formulas are:

$$CH_3(CH_2)_7CH{=}CH(CH_2)_7COOH$$

Oleic acid

$$CH_3(CH_2)_4CH{=}CHCH_2CH{=}CH(CH_2)_7COOH$$

Linoleic acid

$$CH_3CH_2CH{=}CHCH_2CH{=}CHCH_2CH{=}CH(CH_2)_7COOH$$

Linolenic acid

The major physical difference between fats and oils is that fats are solid and oils are liquid at room temperature (see Section 30.2). Since the glycerol part of the structure is the same for a fat and an oil, the difference must be due to the fatty acid end of the molecule. Fats contain a larger proportion of saturated fatty acids, whereas oils contain greater amounts of unsaturated fatty

▲ Solid fats such as lard contain saturated fatty acid portions. Vegetable oils contain unsaturated fatty acids.

TABLE 25.4 Fatty Acid Composition of Selected Fats and Oils

	Fatty acid (%)				
Fat or oil	**Myristic acid**	**Palmitic acid**	**Stearic acid**	**Oleic acid**	**Linoleic acid**
Animal fat					
Butter[a]	7–10	23–26	10–13	30–40	4–5
Lard	1–2	28–30	12–18	41–48	6–7
Tallow	3–6	24–32	14–32	35–48	2–4
Vegetable oil					
Olive	0–1	5–15	1–4	49–84	4–12
Peanut	—	6–9	2–6	50–70	13–26
Corn	0–2	7–11	3–4	43–49	34–42
Cottonseed	0–2	19–24	1–2	23–33	40–48
Soybean	0–2	6–10	2–4	21–29	50–59
Linseed[b]	—	4–7	2–5	9–38	3–43

[a] Butyric acid, 3–4%
[b] Linolenic acid, 25–58%

acids. The term *polyunsaturated* has been popularized in recent years; it means that each molecule of fat in a particular product contains several double bonds.

Fats and oils are obtained from natural sources. In general, fats come from animal sources and oils from vegetable sources. Thus, lard is obtained from hogs and tallow from cattle and sheep. Olive, cottonseed, corn, peanut, soybean, linseed, and other oils are obtained from the fruit or seed of their respective vegetable sources. Table 25.4 shows the major constituents of several fats and oils.

Triacylglycerols are the principal form in which energy is stored in the body. The caloric value per unit of mass is over twice as great as that for carbohydrates and proteins. As a source of energy, triacylglycerols can be completely replaced by either carbohydrates or proteins. However, some minimum amount of fat is needed in the diet, because fat supplies the nutritionally essential unsaturated fatty acids (linoleic, linolenic, and arachidonic) required by the body (see Section 30.2).

Hydrogenation of Glycerides Addition of hydrogen is a characteristic reaction of the carbon–carbon pi bonds. Industrially, low-cost vegetable oils are partially hydrogenated to obtain solid fats that are useful as shortening in baking or in making margarine. In this process, hydrogen gas is bubbled through hot oil containing a finely dispersed nickel catalyst. The hydrogen adds to the carbon–carbon double bonds of the oil to saturate the double bonds and form fats:

$$H_2 + \underset{\text{In oil or fat}}{—CH{=}CH—} \xrightarrow{Ni} —CH_2—CH_2—$$

In actual practice, only some of the double bonds are allowed to become saturated. The degree of hydrogenation can be controlled to obtain a product of any desired degree of saturation. The products resulting from the partial hydrogenation of oils are marketed as solid shortening (Crisco, Spry, and so on) and are used for cooking and baking. Oils and fats are also partially hydrogenated to improve their keeping qualities. Rancidity in fats and oils results from air oxidation at points of unsaturation, producing low molar mass aldehydes and acids of disagreeable odor and flavor.

Hydrogenolysis Triacylglycerols can be split and reduced in a reaction called hydrogenolysis (splitting by hydrogen). Hydrogenolysis requires higher temperatures and pressures and a different catalyst (copper chromite) than does hydrogenation of double bonds. Each triacylglycerol molecule yields a molecule of glycerol and three primary alcohol molecules. The hydrogenolysis of glyceryl trilaurate is represented as follows:

$$\underset{\text{Glyceryl trilaurate}}{\begin{array}{l} CH_2{-}O{-}\overset{O}{\overset{\|}{C}}{-}C_{11}H_{23} \\ C_{11}H_{23}{-}\overset{O}{\overset{\|}{C}}{-}O{-}CH \\ CH_2{-}O{-}\overset{O}{\overset{\|}{C}}{-}C_{11}H_{23} \end{array}} + 6\,H_2 \xrightarrow[\Delta,\ \text{pressure}]{\text{Copper chromite}} \underset{\substack{\text{Lauryl alcohol} \\ \text{(1-Dodecanol)}}}{3\,CH_3(CH_2)_{10}CH_2OH} + \underset{\text{Glycerol}}{\begin{array}{c} CH_2OH \\ | \\ CHOH \\ | \\ CH_2OH \end{array}}$$

Long-chain primary alcohols obtained by this reaction are important, since they are used to manufacture other products, especially synthetic detergents (see Section 25.11).

Hydrolysis Triacylglycerols can be hydrolyzed, yielding fatty acids and glycerol. The hydrolysis is catalyzed by digestive enzymes at room temperatures and by mineral acids at high temperatures:

$$\underset{\text{Triacylglycerol}}{\begin{array}{l} CH_2{-}O{-}\overset{O}{\overset{\|}{C}}{-}R \\ R'{-}\overset{O}{\overset{\|}{C}}{-}O{-}CH \\ CH_2{-}O{-}\overset{O}{\overset{\|}{C}}{-}R'' \end{array}} + 3\,H_2O \xrightarrow[\text{enzymes}]{H^+ \text{ or}} \underset{\substack{\text{Fatty acids} \\ \text{(3 molecules)}}}{\begin{array}{c} RCOOH \\ R'COOH \\ R''COOH \end{array}} + \underset{\text{Glycerol}}{\begin{array}{c} CH_2OH \\ | \\ CHOH \\ | \\ CH_2OH \end{array}}$$

The enzyme-catalyzed reaction occurs in digestive reactions and in biological degradation (or metabolic) processes. The acid-catalyzed reaction is employed in the commercial preparation of fatty acids and glycerol.

Saponification Saponification of a fat or oil involves the alkaline hydrolysis of a triester. The products formed are glycerol and the alkali metal salts of fatty acids, which are called soaps. As a specific example, glyceryl tripalmitate reacts with sodium hydroxide to produce sodium palmitate and glycerol:

$$\underset{\text{Glyceryl tripalmitate}}{\begin{array}{l} \phantom{C_{15}H_{31}-C-O-}CH_2-O-\overset{\overset{O}{\|}}{C}-C_{15}H_{31} \\ \phantom{C_{15}H_{31}-C-O-}| \\ C_{15}H_{31}-\overset{\overset{O}{\|}}{C}-O-CH \\ \phantom{C_{15}H_{31}-C-O-}| \\ \phantom{C_{15}H_{31}-C-O-}CH_2-O-\overset{\overset{O}{\|}}{C}-C_{15}H_{31} \end{array}} + 3\,NaOH \xrightarrow{\Delta} 3\,\underset{\substack{\text{Sodium palmitate} \\ \text{(a soap)}}}{C_{15}H_{31}COONa} + \underset{\text{Glycerol}}{\begin{array}{c} CH_2OH \\ | \\ CHOH \\ | \\ CH_2OH \end{array}}$$

Glycerol, fatty acids, and soaps are valuable articles of commerce, and the processing of fats and oils to obtain these products is a major industry.

25.11 Soaps and Synthetic Detergents

In the broadest sense possible, a detergent is simply a cleansing agent. Soap has been used as a cleansing agent for at least 2000 years and thereby is classified as a detergent under this definition. Beginning about 1930 a number of new cleansing agents that were superior in many respects to ordinary soap began to appear on the market. Because they were both synthetic organic products and detergents, they were called **synthetic detergents**, or syndets. A soap is distinguished from a synthetic detergent on the basis of chemical composition and not on the basis of function or usage.

synthetic detergents

Soaps

In former times soap-making was a crude operation. Surplus fats were boiled with wood ashes or with some other alkaline material. Today, soap is made in large manufacturing plants under controlled conditions. Salts of long-chain fatty acids are called **soaps**. However, only the sodium and potassium salts of carboxylic acids containing 12 to 18 carbon atoms are of great value as soaps, because of their abundance in fats.

soaps

$$\text{Fat or oil} + NaOH \longrightarrow \text{Soap} + \text{Glycerol}$$

To understand how a soap works as a cleansing agent, let us consider sodium palmitate, $CH_3(CH_2)_{14}COONa$, as an example of a typical soap. In water this substance exists as sodium ions, Na^+, and palmitate ions, $CH_3(CH_2)_{14}COO^-$. The sodium ion is an ordinary hydrated metal ion. The cleansing property, then, must be centered in the palmitate ion. The palmitate ion contains both a **hydrophilic** (water-loving) and a **hydrophobic** (water-fearing) group. The hydrophilic end is the polar, negatively charged carboxylate group. The hydrophobic end is the long hydrocarbon group. The hydrocarbon group is soluble in oils and greases, but is not soluble in water. The hydrophilic carboxylate group is soluble in water.

hydrophilic

hydrophobic

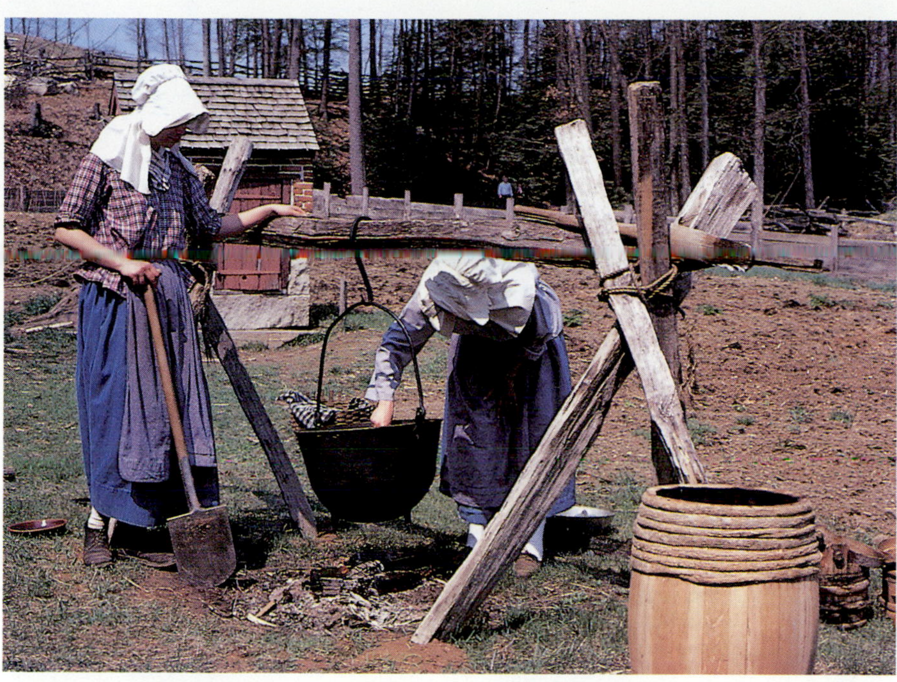

◄ **Old-fashioned soap making at Sturbridge Village. These women are making soap from lard boiled with wood ashes.**

The cleansing action of a soap is explained in this fashion: When the soap comes in contact with grease on a soiled surface, the hydrocarbon end of the soap dissolves in the grease, leaving the negatively charged carboxylate end exposed on the grease surface. Because the negatively charged carboxylate groups are strongly attracted by water, small droplets are formed, and the grease is literally lifted or floated away from the soiled object (see Figure 25.1).

The cleansing property of a soap is due to its ability to act as an emulsifying agent between water and water-insoluble greases and oils. The grease–soap emulsion is stable because the oil droplets repel each other due to the negatively charged carboxyl groups on their surfaces. Some insoluble particulate matter is carried away with the grease; the remainder is wetted and mechanically washed away in the water. Synthetic detergents function in a similar way.

Ordinary soap is a good cleansing agent in soft water, but it is not satisfactory in hard water because insoluble calcium, magnesium, and iron(III) salts are formed. Palmitate ions, for example, are precipitated by calcium ions:

$$\underset{}{Ca^{2+}(aq)} + 2\,\underset{\text{Palmitate ion}}{CH_3(CH_2)_{14}COO^-(aq)} \longrightarrow \underset{\text{Calcium palmitate}}{[CH_3(CH_2)_{14}COO]_2Ca(s)}$$

These precipitates are sticky substances and are responsible for "bathtub ring" and the sticky feel of hair after being shampooed with soap in hard water.

Soaps are ineffective in acidic solutions because water-insoluble molecular fatty acids are formed:

$$CH_3(CH_2)_{14}COO^- + H^+ \longrightarrow \underset{\text{Palmitic acid molecule}}{CH_3(CH_2)_{14}COOH}$$

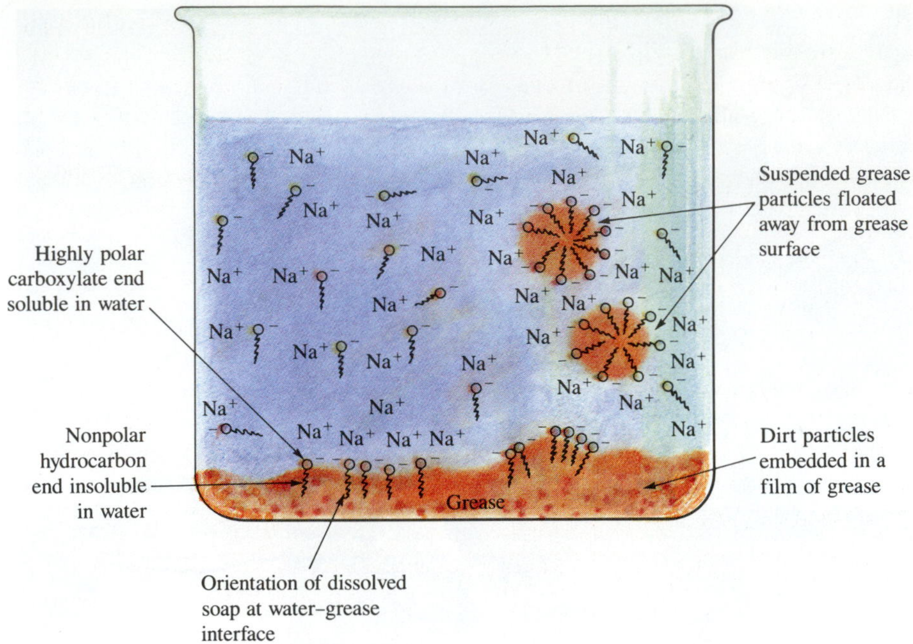

FIGURE 25.1 ▶ Cleansing action of soap: Dirt particles are embedded in a surface film of grease. The hydrocarbon ends of negative soap ions dissolve in the grease film, leaving exposed carboxylate groups. These carboxylate groups are attracted to water, and small droplets of grease-bearing dirt are formed and floated away from the surface.

Synthetic Detergents

Once it was recognized that the insoluble hydrocarbon radical joined to a highly polar group was the key to the detergent action of soaps, chemists set out to make new substances that would have similar properties. About 1930, synthetic detergents (syndets) began to replace soaps, and at present about four pounds of syndets are sold for each pound of soap.

Although hundreds of substances having detergent properties are known, an idea of their general nature can be obtained from consideration of sodium lauryl sulfate and sodium *p*-dodecylbenzene sulfonate.

$$CH_3(CH_2)_{10}CH_2OSO_3^-\ Na^+$$

Sodium lauryl sulfate

$$CH_3(CH_2)_{10}CH_2\text{—}C_6H_4\text{—}SO_3^-\ Na^+$$

Sodium *p*-dodecylbenzene sulfonate

Sodium lauryl sulfate and sodium *p*-dodecylbenzene sulfonate act in water in much the same way as sodium palmitate. The negative ion of each substance is the detergent. For example, like the palmitate ion, the negative lauryl sulfate ion has a long hydrocarbon chain that is soluble in grease and a sulfate group that is attracted to water:

$$\underbrace{CH_3CH_2CH_2CH_2CH_2CH_2CH_2CH_2CH_2CH_2CH_2CH_2}_{\text{Nonpolar hydrophobic end, grease soluble}}\text{—}\underbrace{OSO_3^-}_{\text{Polar hydrophilic end, water soluble}}$$

The one great advantage these synthetic detergents have over soap is that their calcium, magnesium, and iron(III) salts, as well as their sodium salts, are soluble in water. Therefore, they are nearly as effective in hard water as in soft water.

The foregoing are anionic detergents, because the detergent activity is located in negative ions. Other detergents, both cationic and nonionic, have been developed for special purposes. The detergent activity of a cationic detergent is located in a cation that has a long hydrocarbon chain and a positive charge.

$$\underbrace{CH_3(CH_2)_{14}CH_2}_{\text{Grease soluble, hydrophobic}}—\underbrace{\overset{+}{N}(CH_3)_3}_{\text{Water soluble, hydrophilic}}$$

Nonionic detergents are molecular substances. The molecule of a nonionic detergent contains a grease-soluble component and a water-soluble component. Some of these substances are especially useful in automatic washing machines because they have good detergent but low sudsing properties. The structure of a representative nonionic detergent is

$$\underbrace{CH_3(CH_2)_{10}CH_2}_{\text{Grease soluble}}—O—\underbrace{(CH_2CH_2O)_7—CH_2CH_2OH}_{\text{Water soluble}}$$

Biodegradability

Organic substances that are readily decomposed by microorganisms in the environment are said to be **biodegradable**. All naturally occurring organic substances are eventually converted to simple inorganic molecules and ions such as CO_2, H_2O, N_2, Cl^-, and SO_4^{2-}. Most of these conversions are catalyzed by enzymes produced by microorganisms. These enzymes are capable of attacking only certain specific molecular configurations that are found in substances occurring in nature.

biodegradable

A number of years ago a serious environmental pollution problem arose in connection with synthetic detergents. Some of the early syndets containing highly branched-chain hydrocarbons had no counterparts in nature. Therefore, enzymes capable of degrading them did not exist, and the detergents were essentially nonbiodegradable and broke down very, very slowly. As a result these syndets accumulated in water supplies, where they caused severe pollution problems due to excessive foaming and other undesirable effects.

Detergent manufacturers, acting on the recommendations of chemists and biologists, changed from a branched-chain alkyl benzene to a straight-chain alkyl benzene raw material. Detergents containing the straight-chain alkyl groups are biodegradable.

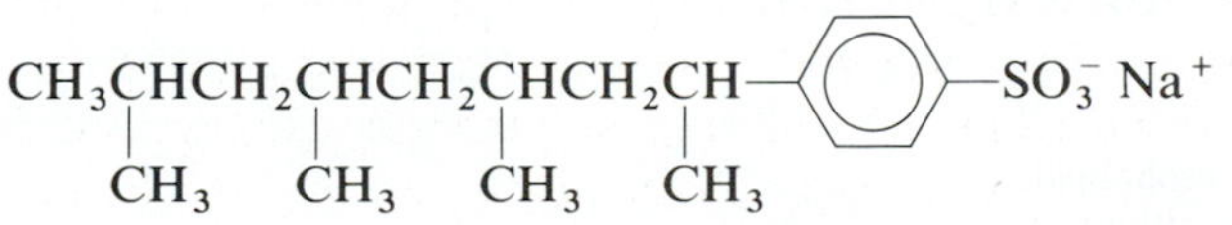

$$CH_3\underset{\displaystyle CH_3}{\underset{|}{C}}HCH_2\underset{\displaystyle CH_3}{\underset{|}{C}}HCH_2\underset{\displaystyle CH_3}{\underset{|}{C}}HCH_2\underset{\displaystyle CH_3}{\underset{|}{C}}H—C_6H_4—SO_3^-\ Na^+$$

A nonbiodegradable detergent

$$CH_3CH_2CH_2CH_2CH_2CH_2CH_2CH_2CH_2CH_2CH_2CH_2—C_6H_4—SO_3^-\ Na^+$$

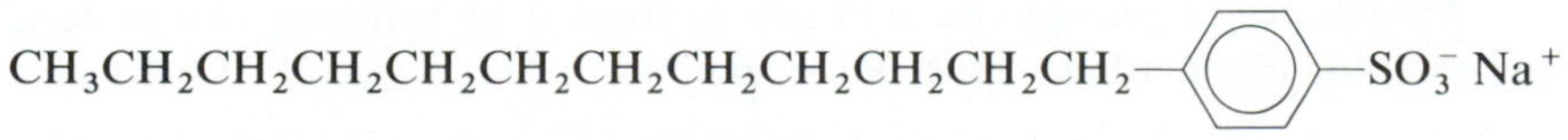

A biodegradable detergent

25.12 Esters and Anhydrides from Phosphoric Acid

Phosphoric acid has a Lewis structure similar to that of a carboxylic acid.

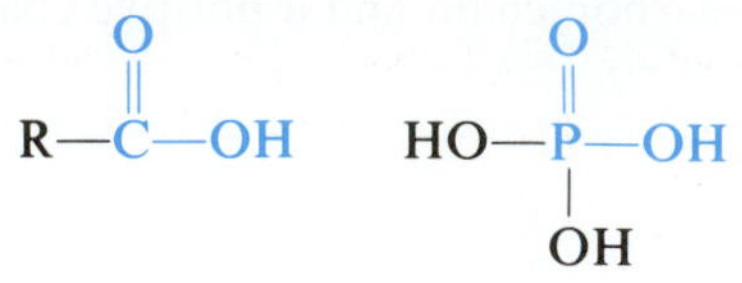

$$\mathrm{R{-}\overset{\overset{\displaystyle O}{\|}}{C}{-}OH} \qquad \mathrm{HO{-}\overset{\overset{\displaystyle O}{\|}}{\underset{\underset{\displaystyle OH}{|}}{P}}{-}OH}$$

Carboxylic acid Phosphoric acid

In both molecules an —OH is attached to an element which is double bonded to an oxygen. In fact, phosphoric acid has three such —OH groups. This similarity in structure permits phosphoric acid to behave as a carboxylic acid in reaction with an alcohol. The product of the esterification reaction is called a phosphate ester.

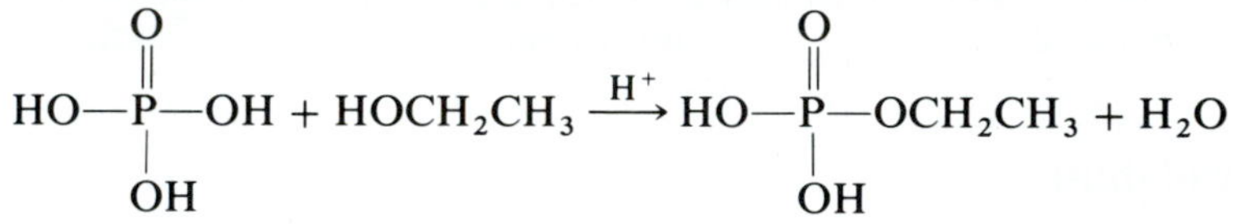

$$\mathrm{HO{-}\overset{\overset{\displaystyle O}{\|}}{\underset{\underset{\displaystyle OH}{|}}{P}}{-}OH} + \mathrm{HOCH_2CH_3} \xrightarrow{H^+} \mathrm{HO{-}\overset{\overset{\displaystyle O}{\|}}{\underset{\underset{\displaystyle OH}{|}}{P}}{-}OCH_2CH_3} + \mathrm{H_2O}$$

Phosphoric acid Ethanol Monoethyl phosphate

phosphorylation

Phosphate esters play a significant role in physiological processes. The esterification of an alcohol by phosphoric acid is one example of a process called **phosphorylation**.

The phosphate ester still has other —OH groups, which allow further phosphorylation to occur. The result of a further condensation reaction is a molecule with two phosphate ester linkages. This is the structure commonly found in phospholipids, molecules that are key components of cell membranes and play significant roles in brain and nerve tissues (Chapter 30).

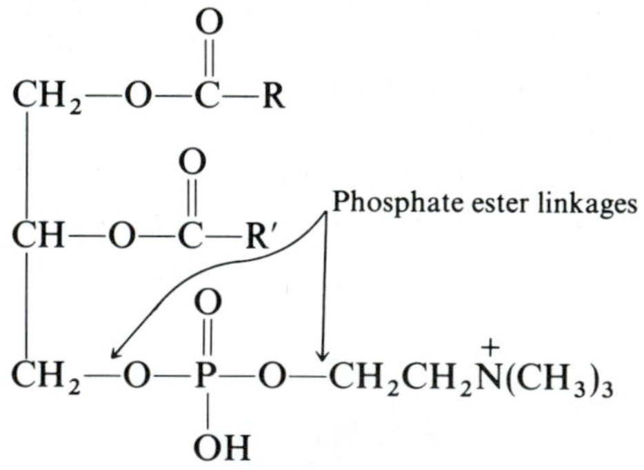

$$\begin{array}{l} \mathrm{CH_2{-}O{-}\overset{\overset{\displaystyle O}{\|}}{C}{-}R} \\ | \\ \mathrm{CH{-}O{-}\overset{\overset{\displaystyle O}{\|}}{C}{-}R'} \\ | \\ \mathrm{CH_2{-}O{-}\overset{\overset{\displaystyle O}{\|}}{\underset{\underset{\displaystyle OH}{|}}{P}}{-}O{-}CH_2CH_2\overset{+}{N}(CH_3)_3} \end{array}$$

Phosphate ester linkages

A phospholipid
(Lecithin)

Anhydrides of phosphoric acid can be formed by bringing two molecules together and eliminating a water molecule.

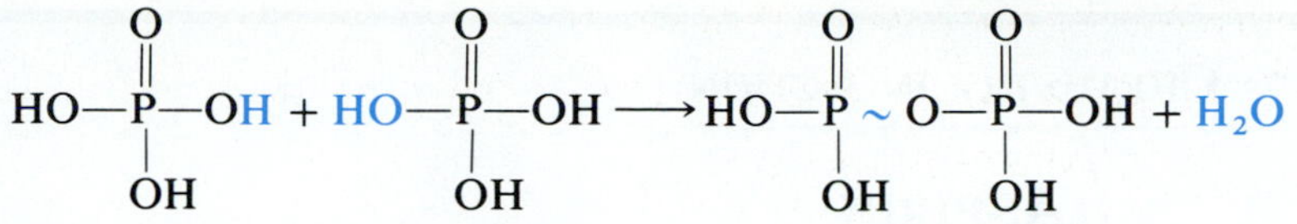

$$\underset{\text{Phosphoric acid}}{\mathrm{HO{-}\overset{\overset{\displaystyle O}{\|}}{\underset{\underset{\displaystyle OH}{|}}{P}}{-}OH}} + \underset{\text{Phosphoric acid}}{\mathrm{HO{-}\overset{\overset{\displaystyle O}{\|}}{\underset{\underset{\displaystyle OH}{|}}{P}}{-}OH}} \longrightarrow \underset{\text{Pyrophosphoric acid}}{\mathrm{HO{-}\overset{\overset{\displaystyle O}{\|}}{\underset{\underset{\displaystyle OH}{|}}{P}} \sim O{-}\overset{\overset{\displaystyle O}{\|}}{\underset{\underset{\displaystyle OH}{|}}{P}}{-}OH}} + H_2O$$

This reaction is similar to a condensation reaction and results in a phosphoric anhydride linkage (shown as ~). The pyrophosphoric acid can react with another phosphoric acid to produce triphosphoric acid.

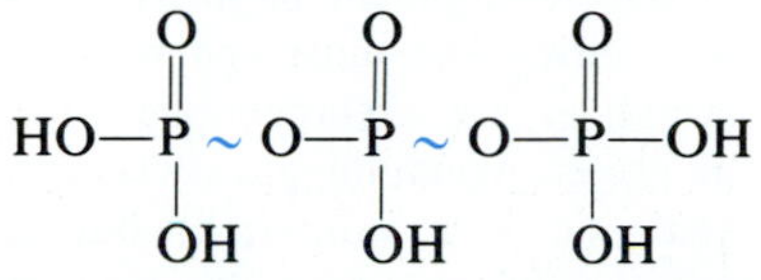

$$\mathrm{HO{-}\overset{\overset{\displaystyle O}{\|}}{\underset{\underset{\displaystyle OH}{|}}{P}} \sim O{-}\overset{\overset{\displaystyle O}{\|}}{\underset{\underset{\displaystyle OH}{|}}{P}} \sim O{-}\overset{\overset{\displaystyle O}{\|}}{\underset{\underset{\displaystyle OH}{|}}{P}}{-}OH}$$

These anhydrides can then react with alcohols to form diphosphates or triphosphates. The most common triphosphate molecules in living systems are adenosine triphosphate (ATP).

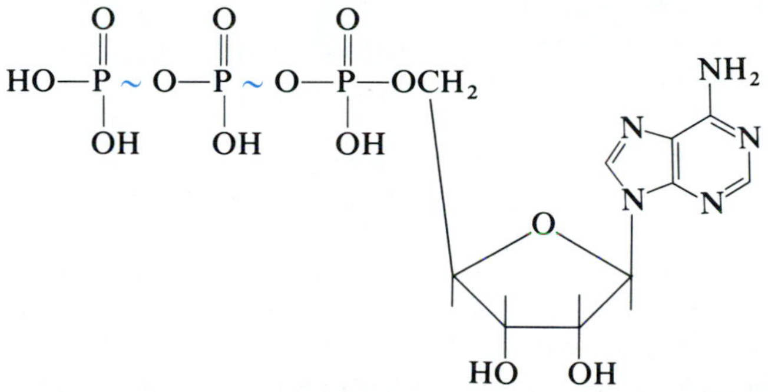

$$\mathrm{HO{-}\overset{\overset{\displaystyle O}{\|}}{\underset{\underset{\displaystyle OH}{|}}{P}} \sim O{-}\overset{\overset{\displaystyle O}{\|}}{\underset{\underset{\displaystyle OH}{|}}{P}} \sim O{-}\overset{\overset{\displaystyle O}{\|}}{\underset{\underset{\displaystyle OH}{|}}{P}}{-}OCH_2}$$

NH$_2$ N N N N O HO OH

Adenosine triphosphate (ATP)

ATP was first isolated from skeletal tissue and is now known to occur in all kinds of plant and animal cells. Adenosine triphosphate is a carrier of energy in cells. The bonds between phosphates are high-energy bonds. These high energy bonds release large amounts of energy (>7000 cal/mol) when they are hydrolyzed, for instance when ATP is hydrolyzed to adenosine diphosphate (ADP):

$$\mathrm{ATP} \underset{}{\overset{H_2O}{\rightleftharpoons}} \mathrm{ADP} + H_2PO_4^- + \text{Energy}$$

An important feature of this reaction is its reversibility. Hydrolysis releases energy while synthesis requires energy. A typical ATP molecule can be recycled within a minute of forming. Although ATP is the principal energy carrier in the cell, there are other phosphate anhydrides that are also high energy compounds in living organisms.

CHEMISTRY IN ACTION

ASPIRIN

From the earliest days of medicine, people have obtained pain relief by chewing willow bark. In 1840 the active compound was isolated from the bark and identified as salicylic acid. Unfortunately, salicylic acid has several undesirable side effects, including a very sour taste and irritation of the stomach lining.

In 1883 an organic chemist reacted salicylic acid with acetic anhydride to form the ester acetylsalicylic acid.

O—C(=O)—CH_3 and COOH on a benzene ring

Acetylsalicylic acid

The chemist happened to work for the Bayer Company, and the name "aspirin" became popularized worldwide.

Aspirin is the most widely used drug in the world. It acts as a fever reducer (antipyretic), a pain reliever (analgesic), and an anti-inflammatory agent. In large doses (lethal is between 30 and 40 g) aspirin is also a poison. Tablets are manufactured by mixing 0.32 g of aspirin with an inert binder (often starch) to hold the tablet together. Approximately half of the aspirin manufactured in the United States (more than 45 million pounds) is made into aspirin tablets. The remainder is used in combination pain relievers and cold remedies.

Until relatively recently it was not understood how aspirin acted within the body. Chemists have now determined that aspirin inhibits an enzyme necessary for the synthesis of prostaglandins (Chapter 30). Functions of prostaglandins include elevation of the blood pressure, tissue inflammation, and activation of the pain receptors in tissues. A reduction in the synthesis of prostaglandins produces both analgesic and anti-inflammatory effects.

Aspirin usage poses undesirable side effects as well. It causes irritation of the stomach lining, can inhibit blood clotting, prolong labor, and is associated with the development of Reyes syndrome, a brain disease which occurs in children recovering from chicken pox or flu. Some people are allergic to aspirin.

Publicity over Reyes syndrome and allergic reactions have resulted in the development of alternatives to aspirin. The most common alternatives are acetaminophen and ibuprofen. Acetaminophen acts as an analgesic and antipyretic, but is not anti-inflammatory. Ibuprofen, a fairly recent addition to the nonprescriptive drug category, acts as a prostaglandin inhibitor in a manner similar to aspirin.

NH—C(=O)—CH_3 and OH on a benzene ring

Acetaminophen

$CH_2CH(CH_3)CH_3$ and $CH(CH_3)$—COOH on a benzene ring

Ibuprofen

Other esters of salicylic acid also show medicinal effects. Methyl salicylate is an oil with the odor of wintergreen. It is commonly found as the active ingredient in pain-relieving liniments. It has the unusual property of penetrating the skin surface where it hydrolyzes, releasing salicylic acid and relieving the pain. Phenyl salicylate is an ester that is not hydrolyzed by acids. It is used to coat pills in order to permit them to pass through the stomach before disintegrating and releasing their contents.

Concepts in Review

1. Give the common and IUPAC names of carboxylic acids.
2. Write structural formulas for saturated carboxylic acids, unsaturated acids, hydroxy acids, aromatic carboxylic acids, and dicarboxylic acids.
3. Tell how water solubility of carboxylic acids varies with increasing molar mass.
4. Relate the boiling points of carboxylic acids to their structure.
5. Write equations for the preparation of carboxylic acids by (a) oxidation of alcohols and aldehydes, (b) hydrolysis or saponification of esters and fats, (c) oxidation of aromatic hydrocarbons, and (d) hydrolysis of nitriles.
6. Write equations showing the effect of heat on malonic, succinic, and glutaric acids.
7. Write equations for the reactions of carboxylic acids to form (a) salts, (b) esters, and (c) acid chlorides.
8. Write an equation to show the formation of an ester from an acid chloride.
9. Write common names, IUPAC names, and formulas of esters.
10. Identify the portion of an ester that is derived from a carboxylic acid and the portion derived from an alcohol.
11. Write the structure of a triacylglycerol (triglyceride) when given the fatty acid composition.
12. Explain the differences between a fat and an oil.
13. Write equations illustrating the (a) hydrogenation, (b) hydrogenolysis, (c) hydrolysis, and (d) saponification of a fat or oil.
14. Write the structural formulas for the three principal unsaturated carboxylic acids found in fats and oils.
15. Explain how a soap or synthetic detergent acts as a cleansing agent.
16. Explain why syndets are effective and soaps are not effective as cleansing agents in hard water.
17. Differentiate among cationic, anionic, and nonionic detergents.
18. Explain the similarities between a carboxylic acid and phosphoric acid.
19. Write an equation showing the formation of a phosphate ester.
20. Write an equation for the formation of a phosphoric anhydride.
21. Indicate the significant role of phosphate esters in living organisms.
22. Indicate the ways in which aspirin acts within the body.
23. Explain how aspirin and ibuprofen differ from acetaminophen in medicinal use within the body.

Equations in Review

Acid–Base Reactions

$$RCOOH + M^{+}OH^{-} \longrightarrow RCOO^{-}M^{+} + H_2O$$

Formation of Acid Chlorides

$$R{-}\overset{\overset{\displaystyle O}{\|}}{C}{-}OH + SOCl_2 \longrightarrow R{-}\overset{\overset{\displaystyle O}{\|}}{C}{-}Cl + SO_2 + HCl$$

Esterification

$$R{-}\overset{\overset{\displaystyle O}{\|}}{C}{-}OH + HOR' \xrightarrow{H^+} R{-}\overset{\overset{\displaystyle O}{\|}}{C}{-}OR' + H_2O$$

Phosphorylation

$$HO{-}\underset{\underset{\displaystyle OH}{|}}{\overset{\overset{\displaystyle O}{\|}}{P}}{-}OH + HOR' \xrightarrow{H^+} HO{-}\underset{\underset{\displaystyle OH}{|}}{\overset{\overset{\displaystyle O}{\|}}{P}}{-}OR' + H_2O$$

Formation of Anhydrides

$$R{-}\overset{\overset{\displaystyle O}{\|}}{C}{-}OH + HO{-}\overset{\overset{\displaystyle O}{\|}}{C}{-}R' \longrightarrow R{-}\overset{\overset{\displaystyle O}{\|}}{C}{-}O{-}\overset{\overset{\displaystyle O}{\|}}{C}{-}R' + H_2O$$

$$HO{-}\underset{\underset{\displaystyle OH}{|}}{\overset{\overset{\displaystyle O}{\|}}{P}}{-}OH + HO{-}\underset{\underset{\displaystyle OH}{|}}{\overset{\overset{\displaystyle O}{\|}}{P}}{-}OH \longrightarrow HO{-}\underset{\underset{\displaystyle OH}{|}}{\overset{\overset{\displaystyle O}{\|}}{P}}{-}O{-}\underset{\underset{\displaystyle OH}{|}}{\overset{\overset{\displaystyle O}{\|}}{P}}{-}OH + H_2O$$

Acid Hydrolysis

$$R{-}\overset{\overset{\displaystyle O}{\|}}{C}{-}OR' + H_2O \xrightarrow{H^+} R{-}\overset{\overset{\displaystyle O}{\|}}{C}{-}OH + HOR'$$

Basic Hydrolysis (Saponification)

$$R{-}\overset{\overset{\displaystyle O}{\|}}{C}{-}OR' + NaOH \longrightarrow R{-}\overset{\overset{\displaystyle O}{\|}}{C}{-}O^{-}\,{}^{+}Na + HOR'$$

EXERCISES

1. Using a specific compound in each case, write structural formulas for the following:
   (a) An aliphatic carboxylic acid
   (b) An aromatic carboxylic acid
   (c) An α-hydroxy acid
   (d) An α-amino acid
   (e) A β-chloro acid
   (f) A dicarboxylic acid
   (g) An unsaturated acid
   (h) An ester
   (i) A nitrile
   (j) A sodium salt of a carboxylic acid
   (k) An acid halide
   (l) A triacylglycerol
2. Name the following compounds:
   (a) $CH_3(CH_2)_5COOH$
   (b) 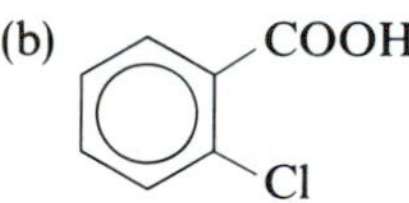

   (c) $CH_3CH{=}CHCOOH$
   (d) $CH_3(CH_2)_7CH{=}CH(CH_2)_7COOH$
   (e) COOH / COOH (benzene ring with two adjacent COOH groups)
   (f) COOH / CH_3 (benzene ring with COOH and meta CH_3)
   (g) $CH_3(CH_2)_{16}COOH$
   (h) $CH_3CH_2\underset{\displaystyle OH}{\underset{|}{C}}HCOOH$
   (i) $CH_3CH_2COO^- \, Na^+$
   (j) $CH_3CH_2COO^- \, NH_4^+$
3. Write structures for the following compounds:
   (a) Caproic acid
   (b) Adipic acid
   (c) *o*-Toluic acid
   (d) Oxalic acid
   (e) β-Hydroxybutyric acid
   (f) *p*-Phthalic acid
   (g) Sodium benzoate
   (h) Linolenic acid
   (i) 2-Chloropropanoic acid
   (j) Ammonium benzoate
   (k) *p*-Aminobenzoic acid
   (l) *m*-Phthalic acid
4. Which of the following would have the more objectionable odor?
   (a) A 1% solution of butyric acid (C_3H_7COOH)
   (b) A 1% solution of sodium butyrate (C_3H_7COONa)
   Cite a satisfactory reason for your answer.
5. Suggest a logical scheme for obtaining reasonably pure stearic acid from a solution containing 2.0% sodium stearate, $CH_3(CH_2)_{16}COONa$, dissolved in water.
6. Assume that you have a 0.01 *M* solution of each of the following substances:
   (a) NH_3 (c) NaCl (e) CH_3COOH
   (b) HCl (d) NaOH (f) (benzene ring)—OH
   Arrange them in order of increasing pH (list the most acidic solution first).
7. Give at least one name, more if you can, for each of the following:
   (a) $HC(=O)—OCH_3$
   (b) 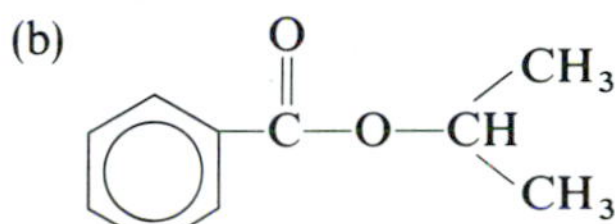

   (c) 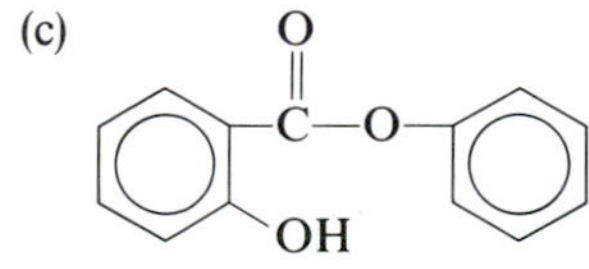

   (d) $CH_3CH_2C(=O)—OCH_2CH_3$
   (e) $CH_3CH_2O—C(=O)—C(=O)—OCH_2CH_3$
   (f) $CH_2{=}CHC(=O)—OCH_3$
8. Write structural formulas for each:
   (a) Methyl formate (c) *n*-Propyl propanoate
   (b) *n*-Octyl acetate (d) Ethyl benzoate
9. Write the structural formula and name of the principal organic product for each of the following reactions:

   (a) $CH_3(CH_2)_7CH{=}CH(CH_2)_7\overset{\displaystyle O}{\overset{\|}{C}}OH + H_2 \xrightarrow{Ni}$

(b) $CH_3CH_2C(=O)—OH + SOCl_2 \longrightarrow$

(c) 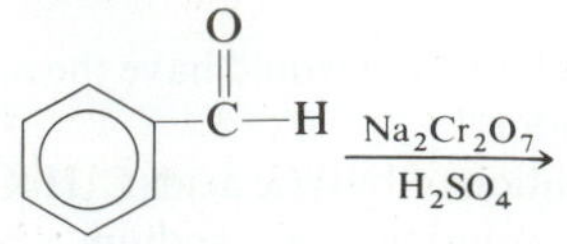 $C_6H_5—C(=O)—H \xrightarrow[H_2SO_4]{Na_2Cr_2O_7}$

(d) 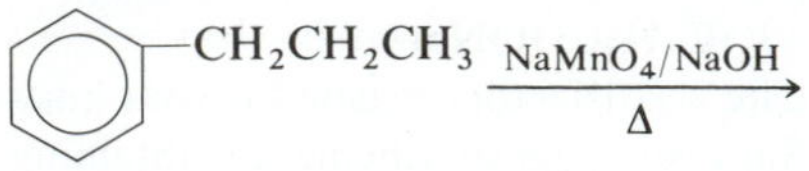 $C_6H_5—CH_2CH_2CH_3 \xrightarrow[\Delta]{NaMnO_4/NaOH}$

(e) 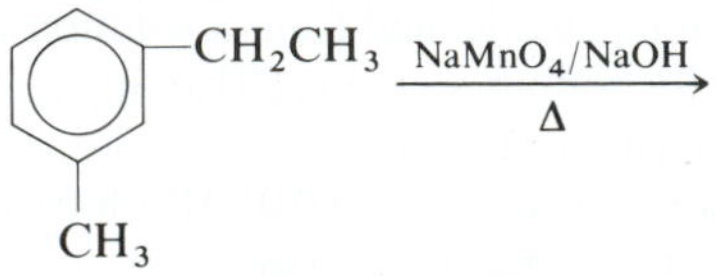 $CH_3—C_6H_4—CH_2CH_3 \xrightarrow[\Delta]{NaMnO_4/NaOH}$

(f) $CH_3CH_2COOH + NaOH \longrightarrow$

(g) $CH_3(CH_2)_3CH_2C(=O)OCH_2CH_2CH_3 + NaOH \xrightarrow{\Delta}$

10. Write structural formulas for the organic products of the following reactions:

(a) 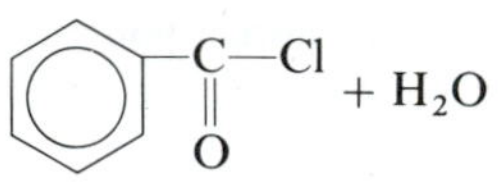 $C_6H_5—C(=O)—Cl + H_2O$

(b) 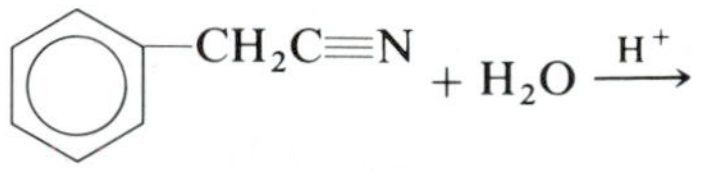 $C_6H_5—CH_2C\equiv N + H_2O \xrightarrow{H^+}$

11. Write structural formulas for the organic products of the following reactions:

(a) 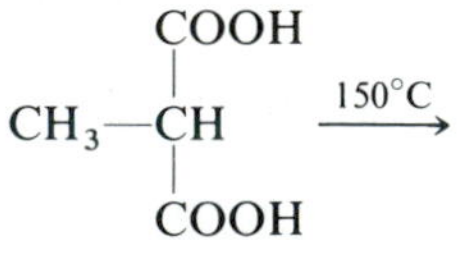 $CH_3—CH(COOH)_2 \xrightarrow{150°C}$

(b) 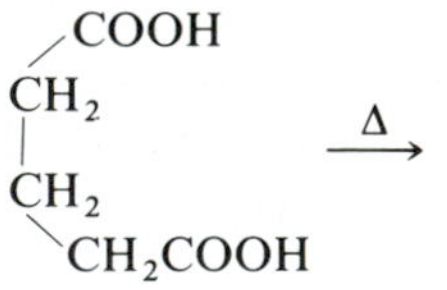 $HOOC—CH_2—CH_2—CH_2COOH \xrightarrow{\Delta}$

Glutaric acid

(c) 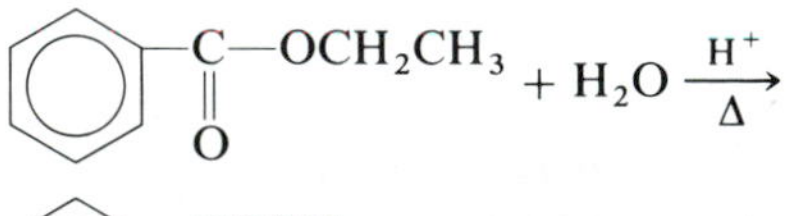 $C_6H_5—C(=O)—OCH_2CH_3 + H_2O \xrightarrow[\Delta]{H^+}$

(d) 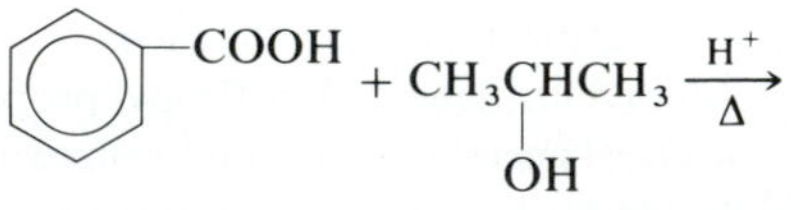 $C_6H_5—COOH + CH_3CH(OH)CH_3 \xrightarrow[\Delta]{H^+}$

(e) 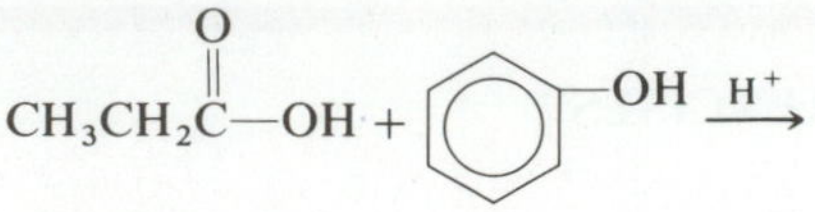 $CH_3CH_2C(=O)—OH + C_6H_5—OH \xrightarrow{H^+}$

(f) 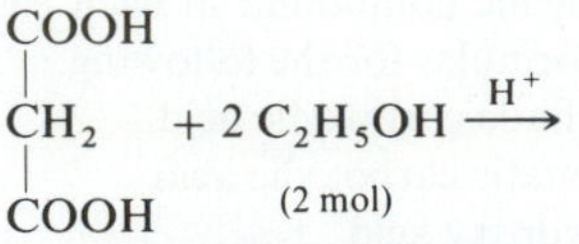 $HOOC—CH_2—COOH + 2\ C_2H_5OH \xrightarrow{H^+}$ (2 mol)

12. Write the structural formulas for the organic products formed in the following reactions:

(a) 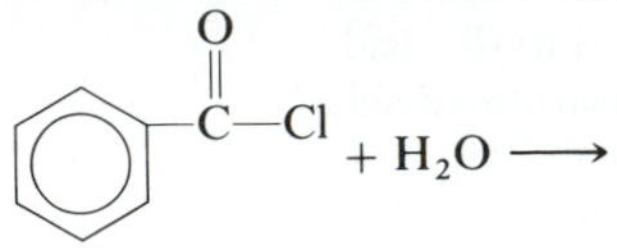 $C_6H_5—C(=O)—Cl + H_2O \longrightarrow$

(b) $CH_2{=}CHCOOH + Br_2 \longrightarrow$

(c) $CH_3COOH + SOCl_2 \longrightarrow$

(d) $CH_3C(=O)—Cl + NH_3 \longrightarrow$

(e) $CH_3C(=O)—Cl + CH_3CH_2OH \longrightarrow$

13. Write the structural formula of the ester that when hydrolyzed would yield
(a) Methanol and acetic acid
(b) Ethanol and formic acid
(c) 1-Octanol and acetic acid
(d) 2-Propanol and benzoic acid
(e) Methanol and salicylic acid

14. Write structural formulas for the following compounds:
(a) Isopropyl formate
(b) Methyl palmitate
(c) Diethyl adipate
(d) Benzyl benzoate
(e) Phenyl propionate
(f) Methyl-2-chloropentanoate

15. What simple tests can be used to distinguish between the following pairs of compounds?
(a) Benzoic acid and sodium benzoate
(b) Maleic acid and malonic acid

16. The geometric configuration of naturally occurring unsaturated 18-carbon acids is all cis. Draw structural formulas for
(a) *cis*-Oleic acid (b) *cis,cis*-Linoleic acid

17. Write the structural formula of a triacylglycerol that contains one unit each of lauric acid, palmitic acid, and stearic acid. How many other triacylglycerols, each containing all three of these acids, are possible?

18. Triolein (glyceryl trioleate) has this structure:

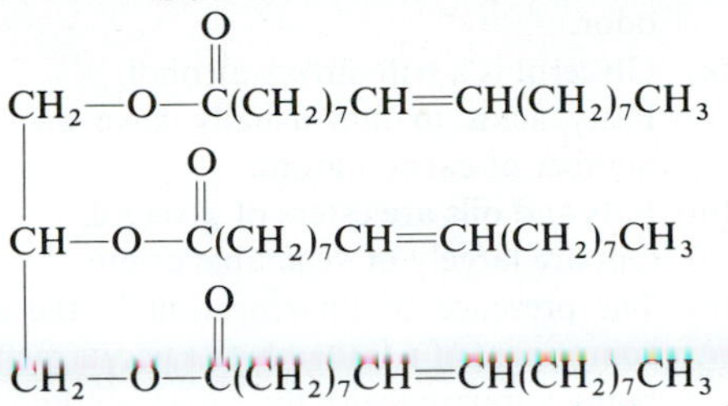

$$\begin{array}{l} CH_2{-}O{-}\overset{\displaystyle O}{\overset{\|}{C}}(CH_2)_7CH{=}CH(CH_2)_7CH_3 \\ | \\ CH{-}O{-}\overset{\displaystyle O}{\overset{\|}{C}}(CH_2)_7CH{=}CH(CH_2)_7CH_3 \\ | \\ CH_2{-}O{-}\overset{\displaystyle O}{\overset{\|}{C}}(CH_2)_7CH{=}CH(CH_2)_7CH_3 \end{array}$$

Write the names and formulas of all products expected when triolein is
(a) Reacted with hydrogen in the presence of Ni
(b) Reacted with water at high temperature and pressure in the presence of mineral acid
(c) Boiled with potassium hydroxide
(d) Reacted with hydrogen at relatively high pressure and temperature in the presence of a copper chromite catalyst

19. Which has the greater solubility in water?
(a) Methyl propanoate or propanoic acid
(b) Sodium palmitate or palmitic acid
(c) Sodium stearate or barium stearate
(d) Phenol or sodium phenoxide

20. Write the structural formulas for each pair of compounds mentioned in Question 19.

21. Explain the difference between
(a) A fat and an oil
(b) A soap and a syndet
(c) Hydrolysis and saponification

22. Cite the principal advantages synthetic detergents (syndets) have over soaps.

23. Would $CH_3(CH_2)_{12}COOH$ or $CH_3(CH_2)_{12}COONa$ be the more useful cleansing agent in soft water? Explain.

24. Would $CH_3(CH_2)_{11}OSO_3Na$ (sodium lauryl sulfate) or $CH_3CH_2CH_2OSO_3Na$ (sodium propyl sulfate) be the more effective detergent in hard water? Explain.

25. Explain the cleansing action of detergents.

26. Which one of the following substances is a good detergent in water? Is this substance a nonionic, anionic, or cationic detergent?

(a) $C_{16}H_{33}N(CH_3)_3^+ \; Cl^-$

Hexadecyltrimethyl ammonium chloride

(b) $C_{16}H_{34}$

Hexadecane

(c) $C_{15}H_{31}COOH$

Palmitic acid (Hexadecanoic acid)

(d) $C_{15}H_{31}\overset{\displaystyle O}{\overset{\|}{C}}{-}O{-}C_{16}H_{33}$

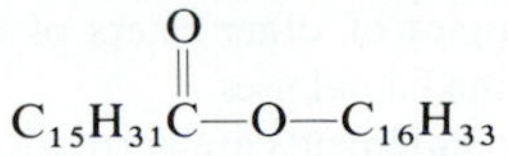

Cetyl palmitate (Hexadecyl hexadecanoate)

27. Margarine consists principally of a relatively small amount of water emulsified in vegetable or animal fats and oils. Monoacylglycerols (monoglycerides) such as glycerol monooleate,

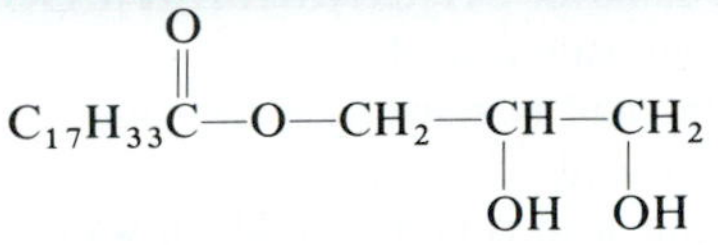

$$C_{17}H_{33}\overset{\displaystyle O}{\overset{\|}{C}}{-}O{-}CH_2{-}\underset{\displaystyle OH}{\underset{|}{CH}}{-}\underset{\displaystyle OH}{\underset{|}{CH_2}}$$

are often used in small quantities in making margarine.
(a) What is the function of the monoacylglycerols?
(b) Explain how this function is achieved.

28. If 1.00 kg of triolein (glyceryl trioleate) is converted to tristearin (glyceryl tristearate) by hydrogenation:
(a) How many liters of hydrogen (at STP) are required?
(b) What is the mass of the tristearin that is produced?

29. Which acid requires more base for neutralization, (a) 1 g of acetic acid or (b) 1 g of propanoic acid? Explain your answer.

30. Starting with ethyl alcohol as the only source of organic material and using any other reagents you desire, write equations to show the synthesis of
(a) Acetic acid
(b) Ethyl acetate
(c) β-Hydroxybutyric acid

31. Upon hydrolysis, an ester of formula $C_6H_{12}O_2$ yields an acid A and an alcohol B. When B is oxidized, it yields a product identical to A. What is the structure of the ester? Explain your answer.

32. Show the products of the reaction of 1 mole of phosphoric acid with
(a) 1 mole of ethanol
(b) 2 moles of ethanol
(c) 3 moles of ethanol

33. What type of chemical reaction is the conversion of ATP into ADP?

34. List the medicinal effects of aspirin within the body.

35. What are the risks associated with the use of aspirin?

36. What substances are commonly substituted for aspirin? Indicate their effects in contrast to those of aspirin.

37. Give two examples of other esters of salicylic acid that have medicinal uses.

38. Which of these statements are correct? Rewrite the incorrect statements to make them correct.

(a) Carboxylic acids can be either aliphatic or aromatic.

(b) The functional group —COOH is known as a carboxyl group.

(c) The name for $CH_3CH_2CHBrCH_2COOH$ is γ-bromovaleric acid.

(d) Acetic acid is a stronger acid than hydrochloric acid.

(e) Benzoic acid is more soluble in water than sodium benzoate.

(f) Carboxylic acids have relatively high boiling points because of hydrogen bonding between molecules.

(g) The formula $C_{17}H_{33}COOH$ represents oleic acid.

(h) Fumaric and maleic acids are cis–trans isomers.

(i) If $CH_3CH_2CH_2COCl$ comes into contact with water, it is hydrolyzed to butyric acid.

(j) Volatile esters generally have a pleasant odor.

(k) Glycerol is a trihydroxy alcohol.

(l) Fatty acids in fats usually have an even number of carbon atoms.

(m) Fats and oils are esters of glycerol.

(n) Oils are largely of vegetable origin.

(o) The presence of unsaturation in the acid component of a fat tends to raise its melting point compared with the corresponding saturated compound.

(p) Alkali metal salts of long-chain fatty acids are called soaps.

(q) The chemical name for aspirin is acetylsalicylic acid.

(r) ATP has one high energy phosphate bond.

(s) The hydrophobic end of a detergent molecule is water soluble.

(t) Saponification is the hydrolysis of a fat or an ester in an acid medium.

(u) Methyl salicylate is a salt.

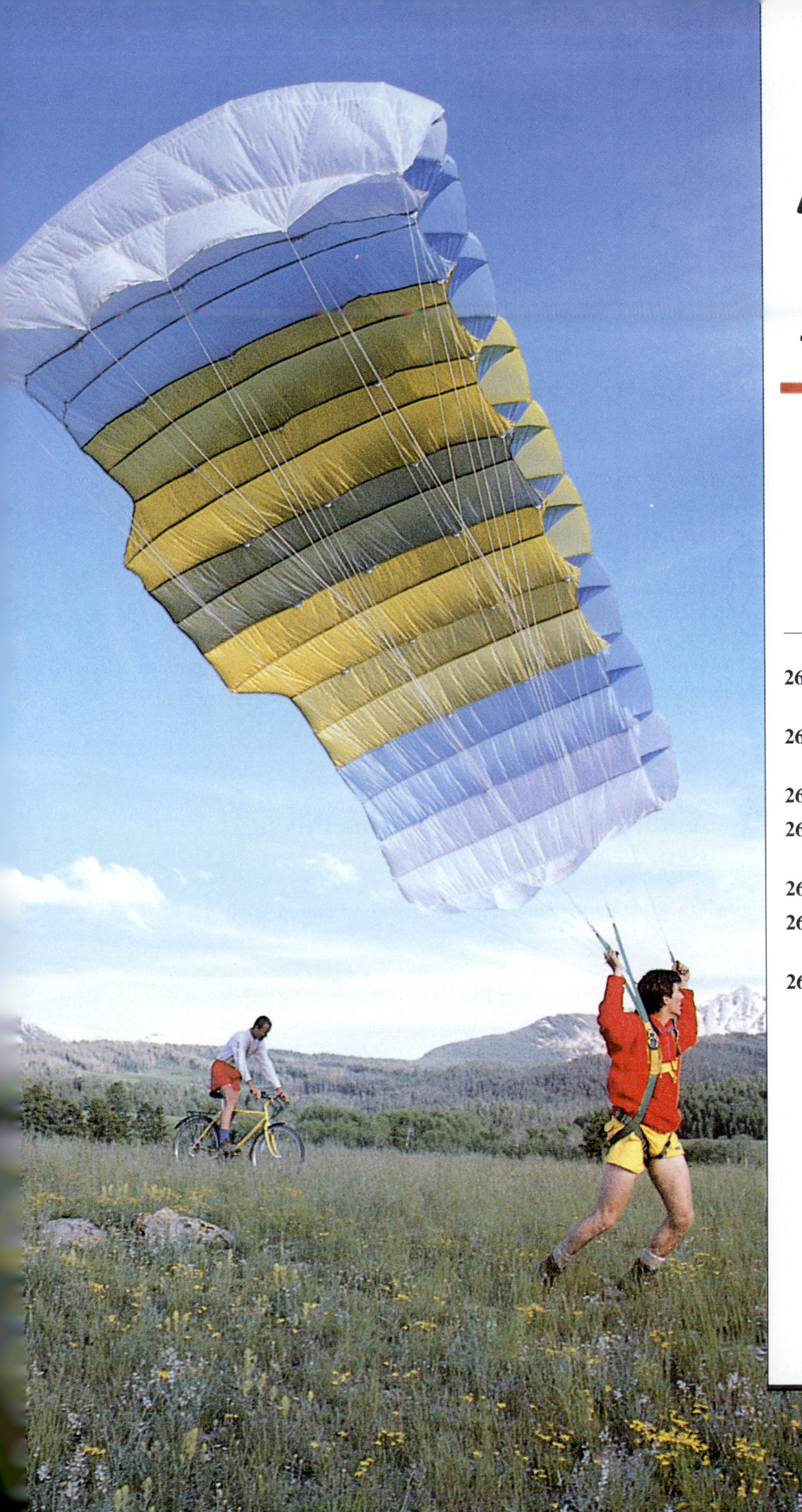

26

Amides and Amines: Organic Nitrogen Compounds

◀ CHAPTER OPENING PHOTO: **The nylon used in parachutes and in sports clothing is formed by linking amines and carboxylic acids into amides.**

Organic compounds can also contain nitrogen. Two major classes of nitrogen-containing compounds are amines and amides. Amines isolated from plants form a group of compounds called alkaloids. Thousands of alkaloids have been isolated. Many of these compounds exhibit physiological activity. Examples of common alkaloid compounds include quinine, used in the treatment of malaria; strychnine, a poison; morphine, a narcotic; and nicotine, a stimulant. Many other drugs are also nitrogen-containing compounds.

Amides are nitrogen derivatives of carboxylic acids. These compounds are found as polymers, both commercially as in nylon, and biologically as in proteins. An understanding of the chemistry of organic nitrogen compounds is the cornerstone of genetics, and is essential to unlocking the chemical secrets of living organisms.

26.1 Amides: Nomenclature and Physical Properties

Carboxylic acids react with ammonia to form ammonium salts:

$$\underset{\text{Carboxylic acid}}{R\overset{\overset{\displaystyle O}{\|}}{C}\text{—}OH} + \underset{\text{Ammonia}}{NH_3} \longrightarrow \underset{\text{Ammonium salt}}{R\overset{\overset{\displaystyle O}{\|}}{C}\text{—}O^-\,NH_4^+}$$

$$\underset{\text{Acetic acid}}{CH_3\overset{\overset{\displaystyle O}{\|}}{C}\text{—}OH} + NH_3 \longrightarrow \underset{\text{Ammonium acetate}}{CH_3\overset{\overset{\displaystyle O}{\|}}{C}\text{—}O^-\,NH_4^+}$$

Ammonium salts of carboxylic acids are ionic substances. Ammonium acetate, for example, is ionized and exists as ammonium ions and acetate ions, both in the crystalline form and when dissolved in water.

When heated, ammonium salts of carboxylic acids are converted to *amides*:

$$\underset{\text{Ammonium salt}}{R\text{—}\overset{\overset{\displaystyle O}{\|}}{C}\text{—}O^-\,NH_4^+} \xrightarrow{\Delta} \underset{\text{Amide}}{R\overset{\overset{\displaystyle O}{\|}}{C}\text{—}NH_2} + H_2O$$

$$CH_3\overset{\overset{\large O}{\|}}{C}—O^-NH_4^+ \xrightarrow{\Delta} CH_3\overset{\overset{\large O}{\|}}{C}—NH_2 + H_2O$$

Ammonium acetate — Acetamide (Ethanamide)

Amides are molecular substances and exist as molecules (not ions) both in the crystalline form and when dissolved in water. An amide contains the following characteristic structure:

$$R—\boxed{\overset{\overset{\large O}{\|}}{C}—N}H_2$$

Amide structure

In amides the carbon atom of a carbonyl group is bonded directly to a nitrogen atom of an $—NH_2$, —NHR, or $—NR_2$ group. The amide structure occurs in numerous substances, including proteins and some synthetic polymers such as nylon.

Two systems are in use for naming amides. The IUPAC name is based on the longest carbon chain that includes the amide group. The amide name is formed by dropping the *oic acid* ending from the corresponding IUPAC acid name and adding the suffix *amide*. Thus,

Methanoic acid	becomes	Methanamide
HCOOH		$HC\begin{smallmatrix}\nearrow O \\ \searrow NH_2\end{smallmatrix}$ (O double-bonded)
Ethanoic acid	becomes	Ethanamide
CH_3COOH		$CH_3C\begin{smallmatrix}\nearrow O \\ \searrow NH_2\end{smallmatrix}$ (O double-bonded)

In a like manner, the common names for amides are formed from the common names of the corresponding carboxylic acids by dropping the *ic* or *oic acid* ending and adding the suffix *amide*. Thus,

Formic acid	becomes	Formamide
HCOOH		$HC\begin{smallmatrix}\nearrow O \\ \searrow NH_2\end{smallmatrix}$ (O double-bonded)
Butyric acid	becomes	Butyramide
$CH_3CH_2CH_2COOH$		$CH_3CH_2CH_2C\begin{smallmatrix}\nearrow O \\ \searrow NH_2\end{smallmatrix}$ (O double-bonded)

Benzoic acid becomes Benzamide

$$C_6H_5\text{—COOH} \qquad C_6H_5\text{—}\overset{\overset{\displaystyle O}{\|}}{C}\text{—}NH_2$$

When the nitrogen of an amide is connected to an alkyl or aryl group, the group is named as a prefix preceded by the letter *N*.

$$CH_3\text{—}\overset{\overset{\displaystyle O}{\|}}{C}\text{—}\underset{\underset{\displaystyle H}{|}}{N}\text{—}CH_3 \quad \text{is} \quad N\text{-methylacetamide}$$

$$CH_3CH_2\text{—}\overset{\overset{\displaystyle O}{\|}}{C}\text{—}\underset{\underset{\displaystyle CH_2CH_3}{|}}{N}\text{—}CH_2CH_3 \quad \text{is} \quad N,N\text{-diethylpropionamide}$$

Formulas and names of selected amides are shown in Table 26.1.

Except for formamide, a liquid, all other unsubstituted amides are solids at room temperature. Many are odorless and colorless. Low molar mass amides are soluble in water, but solubility decreases quickly as molar mass

Table 26.1 Formulas and Names of Selected Amides

Formula	IUPAC name	Common name
$H\overset{\overset{O}{\|}}{C}\text{—}NH_2$	Methanamide	Formamide
$CH_3\overset{\overset{O}{\|}}{C}\text{—}NH_2$	Ethanamide	Acetamide
$CH_3CH_2\overset{\overset{O}{\|}}{C}\text{—}NH_2$	Propanamide	Propionamide
$CH_3\underset{}{\overset{\overset{CH_3}{|}}{C}H}\text{—}\overset{\overset{O}{\|}}{C}\text{—}NH_2$	2-Methylpropanamide	Isobutyramide
$CH_3\text{—}\overset{\overset{O}{\|}}{C}\text{—}\overset{\overset{H}{|}}{N}\text{—}C_6H_5$	N-phenylethanamide	Acetanilide
$CH_3\text{—}\overset{\overset{O}{\|}}{C}\text{—}\underset{\underset{H}{|}}{N}\text{—}C_6H_4\text{—}OH$	N-*p*-hydroxyphenylethanamide	Acetaminophen

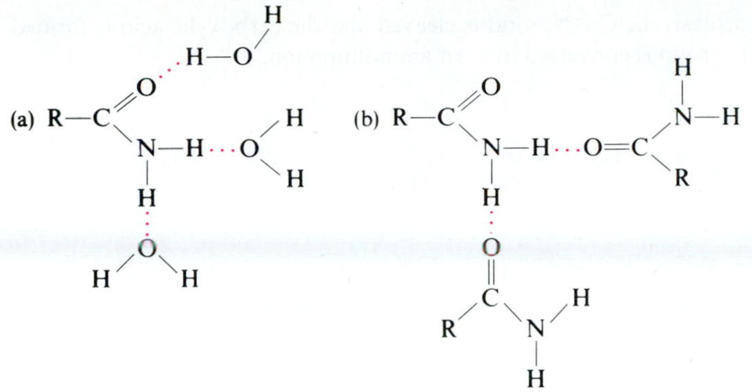

FIGURE 26.1 Hydrogen bonding in amides. (a) Hydrogen bonding between amides and water molecules; (b) intermolecular hydrogen bonding.

increases. The amide functional group is polar, and nitrogen is capable of hydrogen bonding. The solubility of these molecules and their exceptionally high melting points and boiling points are the result of this polarity and hydrogen bonding between molecules, shown in Figure 26.1.

26.2 CHEMICAL PROPERTIES OF AMIDES

One of the more important reactions of amides is hydrolysis. This type of reaction is analogous to hydrolysis of carboxylic esters. The amide is cleaved into two parts, the carboxylic acid portion and the nitrogen-containing portion. As in ester hydrolysis, this reaction requires the presence of a strong acid or a strong base for it to occur in the laboratory. Amide hydrolysis is accomplished in living systems during the degradation of proteins by enzymatic reactions under much milder conditions (Chapter 32). Hydrolysis of an unsubstituted amide in an acid solution produces a carboxylic acid and an ammonium salt.

$$CH_3CH_2CH_2\overset{\overset{\Large O}{\|}}{C}{-}NH_2 + H_2O + HCl \xrightarrow{\Delta} CH_3CH_2CH_2\overset{\overset{\Large O}{\|}}{C}{-}OH + NH_4Cl$$

Basic hydrolysis results in the production of ammonia and the salt of a carboxylic acid.

$$CH_3{-}\overset{\overset{\Large O}{\|}}{C}{-}NH_2 + OH^- \longrightarrow CH_3\overset{\overset{\Large O}{\|}}{C}{-}O^- + NH_3$$

EXAMPLE 26.1

Show the products of acid and basic hydrolysis of $C_6H_5{-}\overset{\overset{\Large O}{\|}}{C}{-}NH_2$

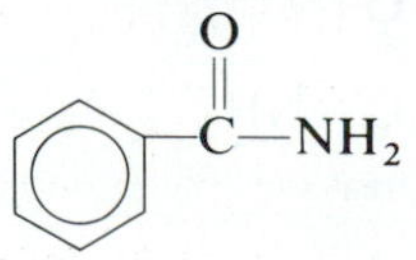

SOLUTION

(a) In acid hydrolysis the C—N bond is cleaved and the carboxylic acid is formed. The —NH_2 group is converted into an ammonium ion.

$$H^+ + C_6H_5\text{—}\overset{\overset{\displaystyle O}{\|}}{C}\text{—}NH_2 + HOH \longrightarrow C_6H_5\text{—}\overset{\overset{\displaystyle O}{\|}}{C}\text{—}OH + NH_4^+$$

(b) In basic solution the C—N bond is also cleaved, but since the solution is basic the salt of the carboxylic acid is formed along with ammonia.

$$C_6H_5\text{—}\overset{\overset{\displaystyle O}{\|}}{C}\text{—}NH_2 + OH^- \longrightarrow C_6H_5\text{—}\overset{\overset{\displaystyle O}{\|}}{C}\text{—}O^- + NH_3$$

PRACTICE Give the products for acid and basic hydrolysis of:

(a) $CH_3CH_2\overset{\overset{\displaystyle O}{\|}}{C}\text{—}NH_2$ (b) $CH_3\overset{\overset{\displaystyle O}{\|}}{C}\text{—}N(CH_3)_2$

Answers: (a) Acid hydrolysis $CH_3CH_2\overset{\overset{\displaystyle O}{\|}}{C}\text{—}OH + NH_4^+$

Basic hydrolysis $CH_3CH_2\overset{\overset{\displaystyle O}{\|}}{C}\text{—}O^- + NH_3$

(b) Acid hydrolysis $CH_3\overset{\overset{\displaystyle O}{\|}}{C}\text{—}OH + \overset{+}{N}H_2(CH_3)_2$

Basic hydrolysis $CH_3\overset{\overset{\displaystyle O}{\|}}{C}\text{—}O^- + NH(CH_3)_2$

26.3 UREA

Hydrolysis of proteins in the digestion process involves cleavage of the amide linkage between adjacent amino acids (Chapter 31). The process begins in the acidic environment of the stomach, and hydrolysis continues in the small intestine.

The body disposes of nitrogen by the formation of a diamide known as urea.

$$H_2N\text{—}\overset{\overset{\displaystyle O}{\|}}{C}\text{—}NH_2$$

Urea

Urea is a white solid that melts at 133°C. It is soluble in water and therefore is excreted from the body as urine, an aqueous solution of urea. The normal adult excretes about 30 g of urea daily in urine.

Urea is a common commercial product as well. It is found most often as a fertilizer to add nitrogen to the soil, or as a starting material in the production of plastics and barbiturates.

26.4 AMINES: NOMENCLATURE AND PHYSICAL PROPERTIES

An **amine** is a substituted ammonia molecule and has the general formula RNH_2, R_2NH, or R_3N, where R is an alkyl or an aryl group. Amines are classified as primary (1°), secondary (2°), or tertiary (3°), depending on the number of hydrocarbon groups attached to the nitrogen atom. Examples are given below.

amine

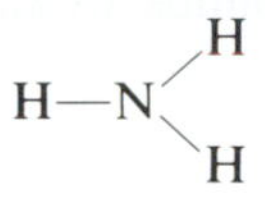

$H—NH_2$
Ammonia

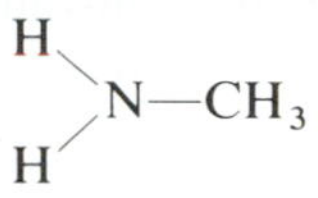

$CH_3—NH_2$
Methylamine (1° amine)

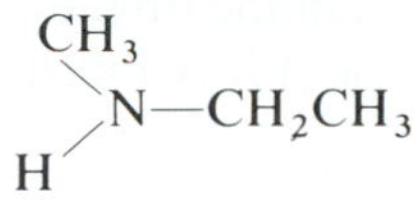

$CH_3—NH—CH_2CH_3$
Methylethylamine (2° amine)

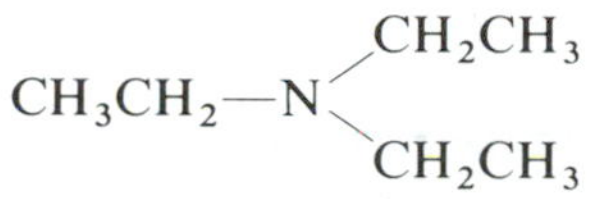

$CH_3CH_2—N(CH_2CH_3)_2$
Triethylamine (3° amine)

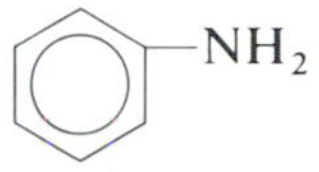

$C_6H_5—NH_2$
Aniline (1° amine)

◀ Mulching and fertilizing are ways of putting nitrogen into the soil.

Common names for aliphatic amines are formed by naming the alkyl group or groups attached to the nitrogen atom followed by the ending *amine*. Thus, CH_3NH_2 is methylamine, $(CH_3)_2NH$ is dimethylamine, and $(CH_3)_3N$ is trimethylamine. A few more examples follow:

$CH_3CH_2NH_2$ — Ethylamine

$CH_3CH_2CH_2NH_2$ — Propylamine

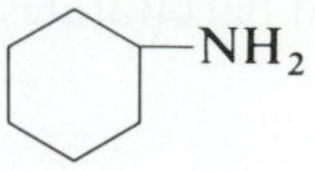

$—NH_2$ — Cyclohexylamine ($C_6H_{11}NH_2$)

In the IUPAC System $—NH_2$ is called an amino group, and amines are named as amino-substituted hydrocarbons using the longest carbon chain as the parent compound (for example, $CH_3CH_2CH_2NH_2$ is called 1-aminopropane).

The most important aromatic amine is aniline ($C_6H_5NH_2$). Derivatives are named as substituted anilines. To identify a substituted aniline in which the substituent group is attached to the nitrogen atom, an *N*- is placed before the group name to indicate that the substituent is bonded to the nitrogen atom and not to a carbon atom in the ring. For example, the following compounds are called *N*-methylaniline and *N*,*N*-dimethylaniline:

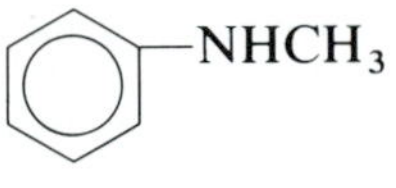

$—NHCH_3$ — *N*-Methylaniline

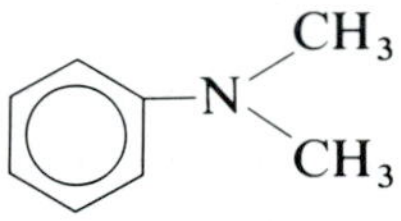

$—N(CH_3)_2$ — *N*,*N*-Dimethylaniline

When a group is substituted for a hydrogen atom in the ring, the resulting ring-substituted aniline is named as we have previously done for naming aromatic compounds. The monomethyl ring-substituted anilines are known as toluidines. Study the names for the following substituted anilines:

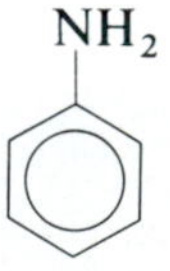

NH_2 — Aniline

NH_2, CH_3 — *p*-Toluidine (*p*-Methylaniline)

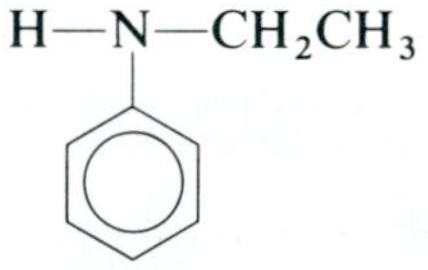

$H—N—CH_2CH_3$ — *N*-Ethylaniline

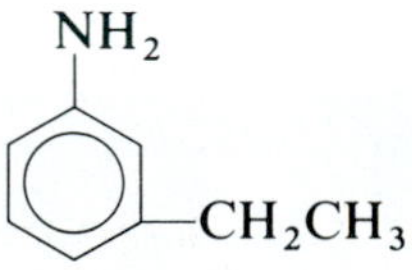

NH_2, CH_2CH_3 — *m*-Ethylaniline

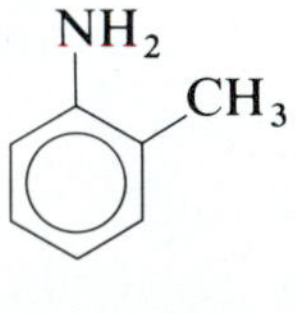

NH_2, CH_3 — *o*-Toluidine (*o*-Methylaniline)

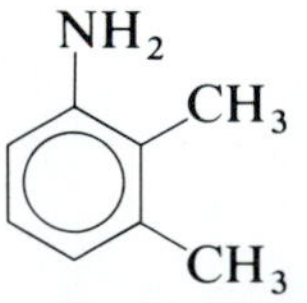

NH_2, CH_3, CH_3 — 2,3-Dimethylaniline

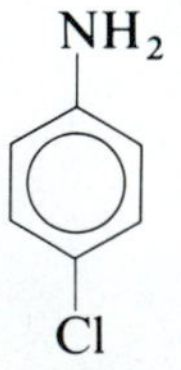

NH_2, Cl — *p*-Chloroaniline

Physiologically, aniline is a toxic substance. It is easily absorbed through the skin and affects both the blood and the nervous system. Aniline reduces the oxygen-carrying capacity of the blood by converting hemoglobin to methemoglobin. Methemoglobin is the oxidized form of hemoglobin in which the iron has gone from a +2 to a +3 oxidation state.

EXAMPLE 26.2

Name the two compounds given:

(a) $CH_3\underset{\displaystyle CH_3}{CH}CH_2—NH_2$ (b) $HOOC—C_6H_4—NH_2$ (*para*)

SOLUTION

(a) The alkyl group attached to NH_2 is an isobutyl group. Thus the common name is isobutylamine. The longest chain containing the NH_2 has three carbons. Therefore the parent carbon chain is propane, and the compound is called 1-amino-2-methylpropane.

(b) The parent compound on which the name is based is benzoic acid. With an amino group in the para position, the name is *p*-aminobenzoic acid. The acronym for *p*-aminobenzoic acid is PABA. In the human body PABA is a growth factor for certain bacteria. The main source of PABA for the human body is folic acid, an essential vitamin in the diet. Esters of PABA are some of the most effective ultraviolet screening agents and are used in suntanning lotions.

PRACTICE Name the following compounds.

(a) $CH_3CH_2CH_2NH_2$ (c) $CH_3\underset{\displaystyle NH_2}{CH}CH_2CH_3$

(b) $NHCH_2CH_3$

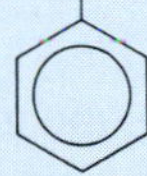

Answers: (a) propylamine (b) *N*-ethylaniline (c) 2-aminobutane

Ring compounds in which all the atoms in the ring are not alike are known as **heterocyclic compounds**. The most common heteroatoms are oxygen, nitrogen, and sulfur. A number of the nitrogen-containing heterocyclic compounds are present in naturally occurring biological substances such as DNA, which controls heredity. The structural formulas of several nitrogen-containing heterocyclics follow:

heterocyclic compounds

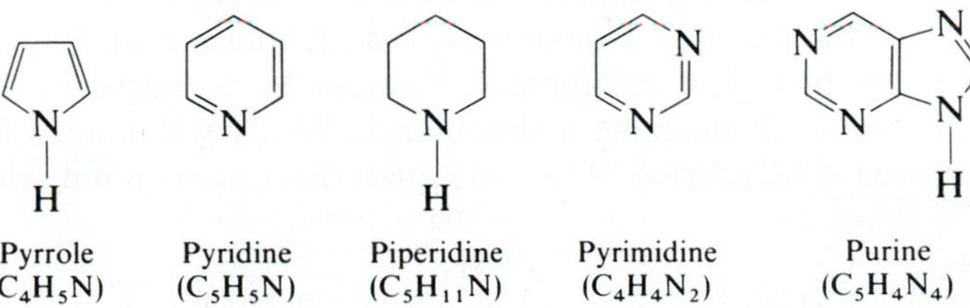

Pyrrole (C_4H_5N) Pyridine (C_5H_5N) Piperidine ($C_5H_{11}N$) Pyrimidine ($C_4H_4N_2$) Purine ($C_5H_4N_4$)

Amines are capable of hydrogen bonding with water. As a result the aliphatic amines with up to six carbons are quite soluble in water. Methylamine and ethylamine are flammable gases with a strong ammoniacal odor.

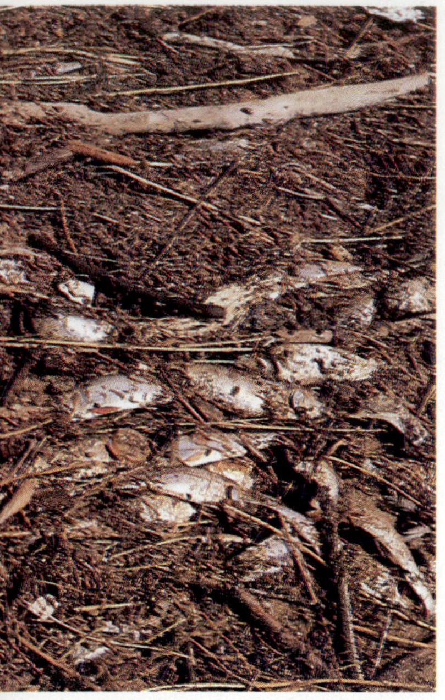

▲ **The odor of dead fish is characteristic of the amines.**

Trimethylamine has a "fishy" odor. Higher molar mass amines have obnoxious odors. The foul odors arising from dead fish and decaying flesh are due to amines released by bacterial decomposition. Two of these compounds are diamines, 1,4-butanediamine and 1,5-pentanediamine. Each compound contains two amino groups.

$$H_2NCH_2CH_2CH_2CH_2NH_2 \qquad H_2NCH_2CH_2CH_2CH_2CH_2NH_2$$

1,4-Butanediamine (Putrescine) — 1,5-Pentanediamine (Cadaverine)

Simple aromatic amines are all liquids or solids. They are colorless or almost colorless when freshly prepared but become dark brown or red when exposed to air and light.

26.5 Preparation of Amines

Alkylation of Ammonia and Amines The substitution of alkyl groups for hydrogen atoms of ammonia can be done by reacting ammonia with alkyl halides. Thus, in successive reactions, a primary, a secondary, and a tertiary amine can be formed.

$$NH_3 \xrightarrow{CH_3Br} \underset{\substack{\text{Methylamine} \\ (1^\circ)}}{CH_3NH_2} \xrightarrow{CH_3Br} \underset{\substack{\text{Dimethylamine} \\ (2^\circ)}}{(CH_3)_2NH} \xrightarrow{CH_3Br} \underset{\substack{\text{Trimethylamine} \\ (3^\circ)}}{(CH_3)_3N}$$

quaternary ammonium salts

Tertiary amines can be further alkylated so that there are four organic groups bonded to the nitrogen atom. These types of compounds are called **quaternary ammonium salts**. For example,

$$CH_3-\ddot{N}(CH_3)-CH_3 + CH_3Br \longrightarrow \underset{\text{Tetramethylammonium bromide}}{CH_3-N^+(CH_3)_2-CH_3Br^-}$$

Quaternary ammonium salts are well known in biologically active compounds and in many popular medicinals. For example, acetylcholine, an active neurotransmitter in the brain, is a quaternary salt. Choline is an important component of many biological membranes. Vitamin B_1 is marketed as the quaternary ammonium salt thiamine hydrochloride. Many well-known fabric softening agents used in laundering clothes are quaternary ammonium salts.

$$\underset{\text{Choline}}{CH_3-N^+(CH_3)_2-CH_2CH_2OH} \qquad \underset{\text{Acetylcholine}}{CH_3-N^+(CH_3)_2-CH_2CH_2O\overset{\underset{\|}{O}}{C}CH_3}$$

Reduction of Amides and Nitriles Amides can be reduced with lithium aluminum hydride to give amines. For example, acetamide can be reduced to ethylamine; and when *N*,*N*-diethylacetamide is reduced, triethylamine is formed:

$$CH_3\overset{\overset{\displaystyle O}{\|}}{C}-NH_2 \xrightarrow{LiAlH_4} \underset{\text{Ethylamine}}{CH_3CH_2NH_2}$$

$$CH_3\overset{\overset{\displaystyle O}{\|}}{C}-N\begin{matrix} CH_2CH_3 \\ CH_2CH_3 \end{matrix} \xrightarrow{LiAlH_4} \underset{\text{Triethylamine}}{(CH_3CH_2)_3N}$$

Nitriles, RCN, are also reducible to amines using hydrogen and a metal catalyst:

$$\underset{\text{Propionitrile}}{CH_3CH_2C{\equiv}N} \xrightarrow{H_2/Ni} \underset{n\text{-Propylamine}}{CH_3CH_2CH_2NH_2}$$

Reduction of Aromatic Nitro Compounds Aniline, the most widely used aromatic amine, is made by reducing nitrobenzene. The nitro group can be reduced by several reagents; Fe and HCl, or Sn and HCl, are commonly used.

$$C_6H_5NO_2 \xrightarrow[HCl]{Sn} C_6H_5NH_3^+Cl^- \xrightarrow{NaOH} C_6H_5NH_2$$

26.6 Chemical Properties of Amines

Alkaline Properties of Amines

In many respects amines resemble ammonia in their reactions. Thus, amines are bases and, like ammonia, ionize in water to produce OH^- ions:

$$\underset{\text{Ammonia molecule}}{\ddot{N}H_3} + (H)OH \rightleftharpoons \underset{\text{Ammonium ion}}{NH_4^+} + \underset{\text{Hydroxide ion}}{OH^-}$$

Methylamine and aniline react in the same manner:

$$\underset{\text{Methylamine molecule}}{CH_3\ddot{N}H_2} + (H)OH \rightleftharpoons \underset{\text{Methylammonium ion}}{CH_3NH_3^+} + \underset{\text{Hydroxide ion}}{OH^-}$$

$$\underset{\text{Aniline molecule}}{C_6H_5\ddot{N}H_2} + (H)OH \rightleftharpoons \underset{\text{Anilinium ion}}{C_6H_5NH_3^+} + \underset{\text{Hydroxide ion}}{OH^-}$$

The ions formed are substituted ammonium ions. They are named by replacing the amine ending by ammonium and, for the aromatic amines, by replacing the aniline name by anilinium.

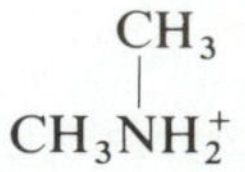

$CH_3NH_2^+$–CH_3 Dimethylammonium ion

$^+NH_3$ on benzene ring with ortho CH_3 *o*-Methylanilinium ion

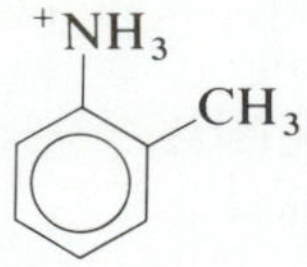

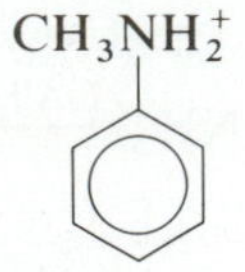

$CH_3NH_2^+$ on benzene ring   *N*-Methylanilinium ion

Like ammonia, amines are weak bases. Methylamine is a slightly stronger base than ammonia, and aniline is considerably weaker than ammonia. The pH values for 0.1 *M* solutions are: methylamine, 11.8; ammonia, 11.1; and aniline, 8.8.

Because amine groups form substituted ammonium ions under physiological conditions, they can provide the positive charge for biological molecules. For example, neurotransmitters are often positively charged. The structures of two such transmitter compounds, dopamine and serotonin, are shown here:

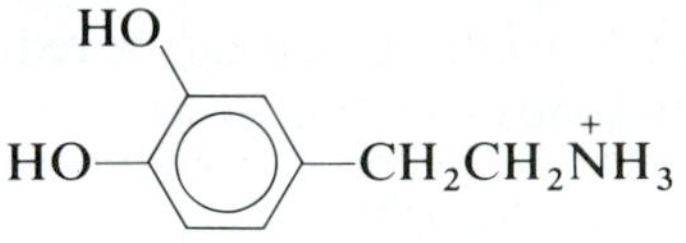

HO, HO–C_6H_3–$CH_2CH_2\overset{+}{N}H_3$

Dopamine

OH, indole ring (N–H)–$CH_2CH_2\overset{+}{N}H_3$

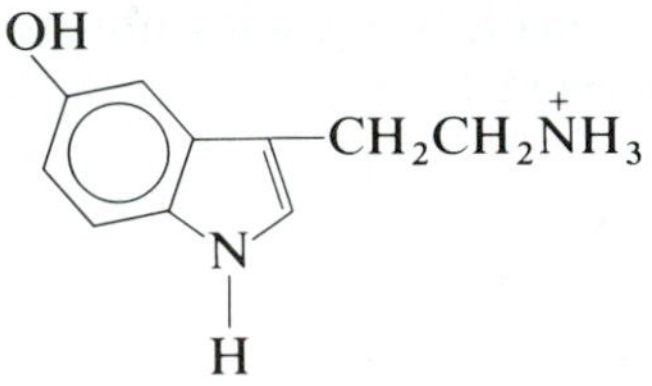

Serotonin

Salt Formation

An amine reacts with a strong acid to form a salt; for example, methylamine and hydrogen chloride react in this fashion:

$$CH_3NH_2(g) + HCl(g) \longrightarrow CH_3NH_3{}^+\,Cl^-$$

Methylamine molecule   Hydrogen chloride molecule   Methylammonium chloride (salt)

Methylammonium chloride is made up of methylammonium ions, $CH_3NH_3^+$, and chloride ions, Cl^-. It is a white crystalline salt that in physical appearance resembles ammonium chloride very closely.

Aniline reacts in a similar manner, forming anilinium chloride:

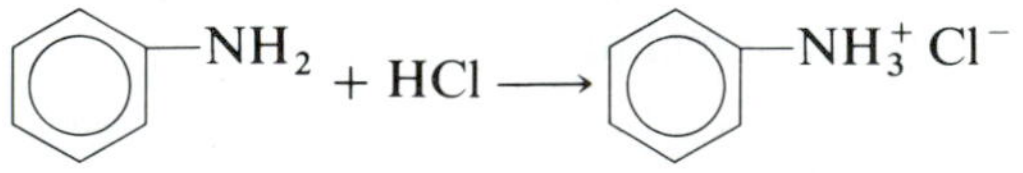

$$C_6H_5NH_2 + HCl \longrightarrow C_6H_5NH_3^+\,Cl^-$$

Anilinium chloride

Many amines or amino compounds are more stable in the form of the hydrochloride salt. When the free amine is wanted, the HCl is neutralized to liberate the free amine. Thus,

$$RNH_3^+\,Cl^-\ (\text{or } RNH_2\cdot HCl) + NaOH \longrightarrow RNH_2 + NaCl + H_2O$$

An amine hydrochloride salt   Free amine

(*continues on page 685*)

Chemistry in Action

Drugs for the Central Nervous System

Drugs that affect the central nervous system, including the brain and spinal cord, have become increasingly prevalent in modern society. Classification of these substances is often complex as a single drug may elicit different responses, depending upon dosage or activity of the individual user. A classification system based upon the major response and pattern of chemical action has been developed by the World Health Organization. Major categories include opiates, barbiturates, cocaine, cannabis, amphetamines, hallucinogens, tranquilizers, and designer drugs.

Basic compounds that are derived from plants and show physiological activity are known as **alkaloids**. These substances are usually amines. The opium alkaloids are often called the opiates and include both compounds derived from the opium poppy and synthetic compounds which have morphine-like activity (sleep-inducing and analgesic properties). These drugs are classified as *narcotics*, producing physical addiction, and are strictly regulated by federal law.

Barbiturates are synthetic drugs classified as sedatives. They are prepared from urea and substituted malonic acid and are usually prescribed as sodium salts to increase their solubility in water. The activity of these compounds depends on the nature of the substituents on the barbituric acid molecule. Seconal and Nembutal act over a period of three to four hours, and so are commonly found as sleeping pills. Pentothal acts within seconds and wears off quickly, explaining its use as a dental anesthetic. The formulas on this page show the structures of some common barbiturates.

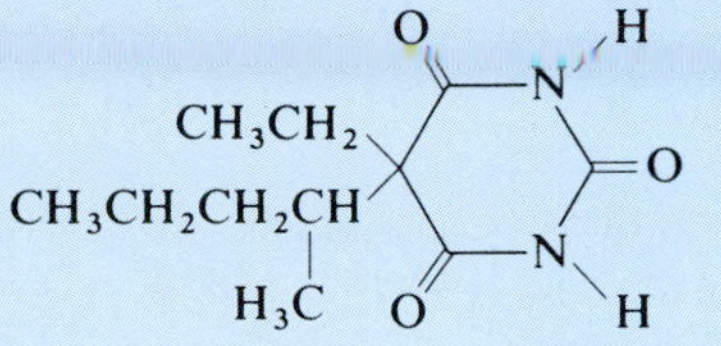

Nembutal

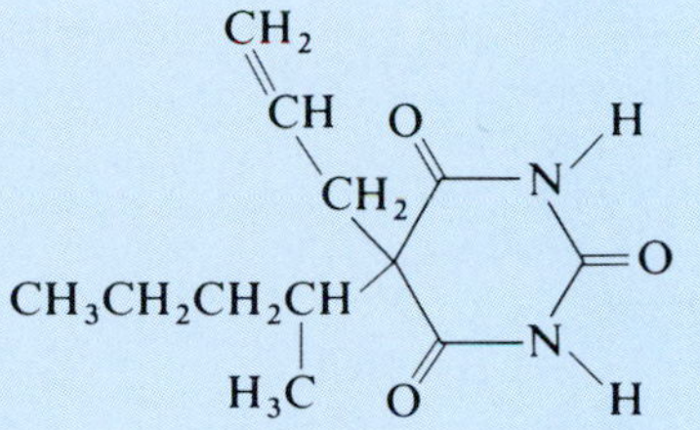

Seconal

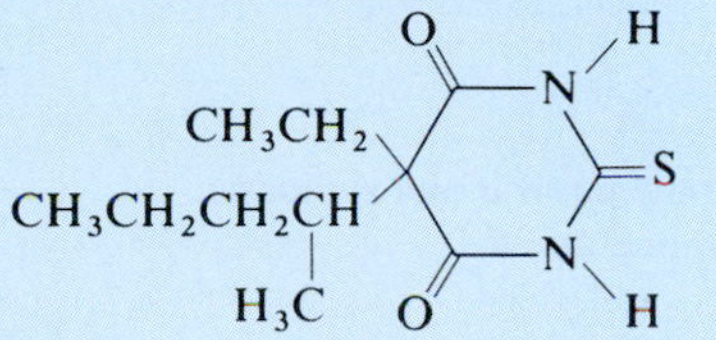

Pentothal

▲ **Some common barbiturates.**

The action of a barbiturate is to depress the activity of brain cells. For this reason they are often called "downers," and they are one of the more widely abused drugs. Barbiturate users require larger and larger doses of the drug to achieve the same effect as the body develops a tolerance to the drug. Dependency on barbiturates is becoming a more serious problem than narcotic addiction. This is because withdrawal from barbiturates is more dangerous and the margin of safety between a lethal dose and that dose necessary to achieve the desired effect decreases as an addict's tolerance increases. Barbiturate overdoses now account for over 20% of all reported poisonings.

Tranquilizers are drugs taken to modify psychotic behavior without inducing sleep, or to reduce anxiety or restlessness. Psychotic behavior is treated with strong tranquilizers such as Thorazine or reserpine. Thorazine (chlorpromazine) is extremely effective in treating schizophrenia and manic depression. Milder tranquilizers are often used to relieve the pressure and anxiety of daily life. These substances generally belong to a group known as benzodiazepines. The most popular tranquilizer is diazepam (Valium), followed closely by Librium, a drug used in treating neuroses and acute alcoholism. The structures of these common tranquilizers are shown on the next page.

▲ **The nicotine in cigarettes, the toxins in rat poison, the quinine in tonic water and the amphetamines in diet pills are all examples of alkaloids.**

(*continued*)

(continued)

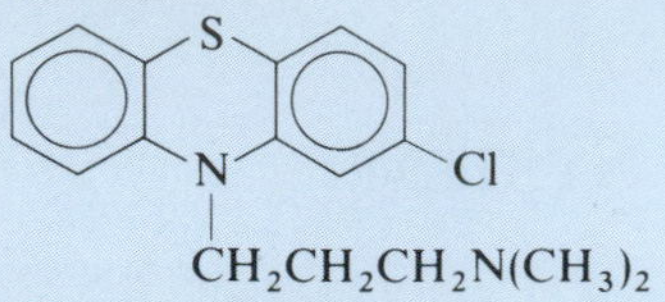

Chlorpromazine
(Thorazine)

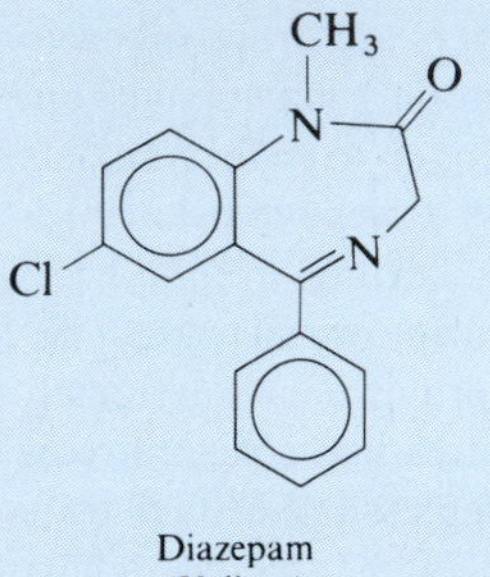

Diazepam
(Valium)

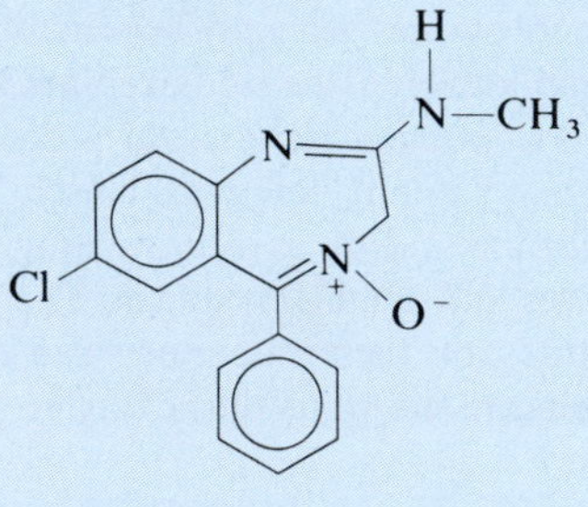

Chlordiazepoxide
(Librium)

▲
Several common tranquilizers.

In contrast, amphetamines are part of a group of drugs that stimulate the central nervous system. These drugs are referred to as *sympathomimetic amines* since they mimic the action of epinephrine and norepinephrine, both substances naturally produced in the body. In fact the parent drug, amphetamine, was first synthesized to simulate epinephrine. Several natural and synthetic amphetamines are shown above.

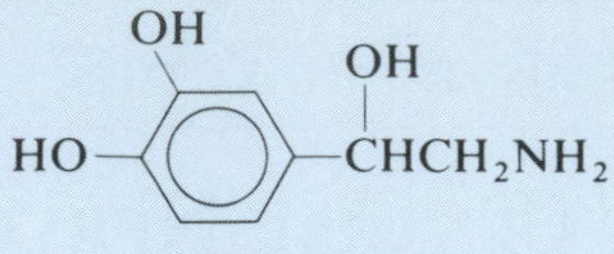

Norepinephrine

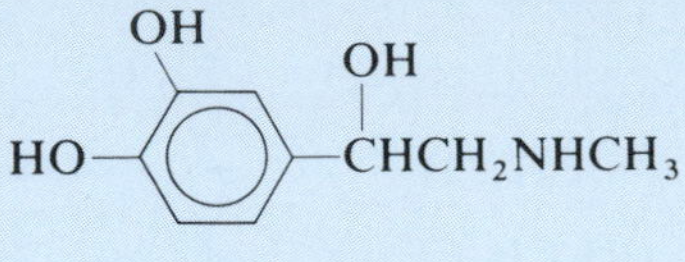

Epinephrine
(Adrenalin)

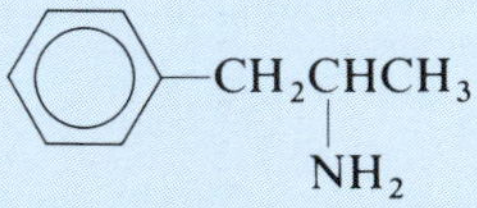

Amphetamine
(Benzedrine)

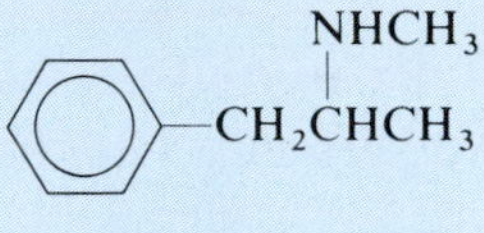

Methamphetamine
(Methedrine)

▲
Some natural and synthetic amphetamines.

These drugs are used medically to treat depression, narcolepsy, and obesity. Use of amphetamines produces a feeling of well-being, loss of fatigue, and increasing alertness. Overuse of amphetamines can result in severe health effects. These compounds tend to concentrate in the central nervous system and are not easily broken down. The most widely abused amphetamine is methamphetamine, commonly called "speed." This drug is used in progressively higher doses producing a state of constant wakefulness followed by exhaustion in an ongoing cycle.

In the 1980s analogs of pharmacological drugs have come to be known as designer drugs. These are manufactured in illegal laboratories and are sold on the street as other well-known drugs. By making slight modifications in the structure of a drug, these underground chemists try to produce more potent or longer lasting drugs. An example of this technology is an analog of Demerol (mepiridine) called MPPP, shown above. MPPP is 25 times as potent as Demerol and is also about 3 times as potent as heroin. Unfortunately, an impurity in the synthesis of MPPP resulted in several cases of irreversible Parkinson's disease. Designer drugs can often be lethal as a result of these impurities and a lack of testing under controlled conditions.

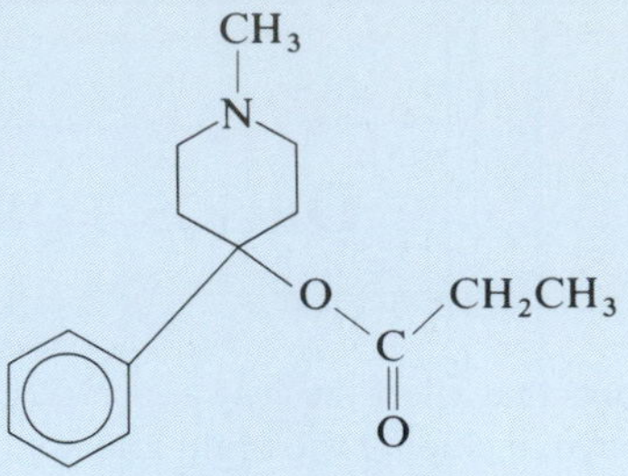

MPPP

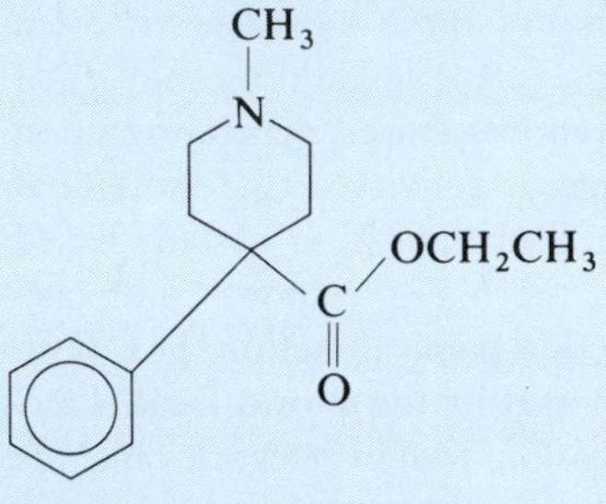

Demerol

▲
The designer drug MPPP is made by a slight change in the structure of a known drug, Demerol.

Formation of Amides

Primary and secondary amines react with acid chlorides to form substituted amides. For example,

$$CH_3\overset{O}{\overset{\|}{C}}-Cl + 2\,(CH_3CH_2)_2NH \longrightarrow CH_3\overset{O}{\overset{\|}{C}}-N\begin{matrix}CH_2CH_3\\CH_2CH_3\end{matrix} + (CH_3CH_2)_2NH_2^+\,Cl^-$$

N,N-Diethylacetamide

26.7 Sources and Uses of Amines

Nitrogen compounds are found throughout the plant and animal kingdoms. Amines, substituted amines, and amides occur in every living cell. Colorful dyes, vitamins, and many medicinals including alkaloids, quinine and morphine, sulfa drugs, and amphetamines are but a few of the classes of substances in which organic nitrogen compounds are found. The formulas for several well-known nitrogen compounds are shown below:

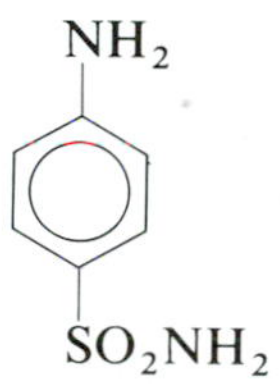

$H_2N-C_6H_4-SO_2NH_2$

Sulfanilamide
(antibacterial agent)

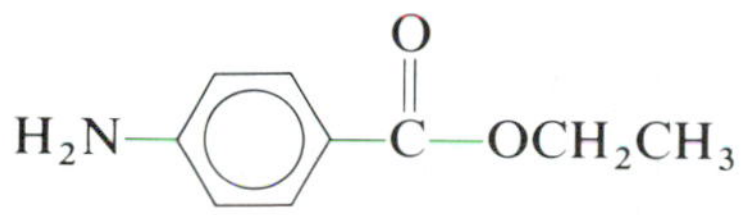

$H_2N-C_6H_4-\overset{O}{\overset{\|}{C}}-OCH_2CH_3$

Benzocaine (Ethyl-*p*-aminobenzoate)
(topical anesthetic)

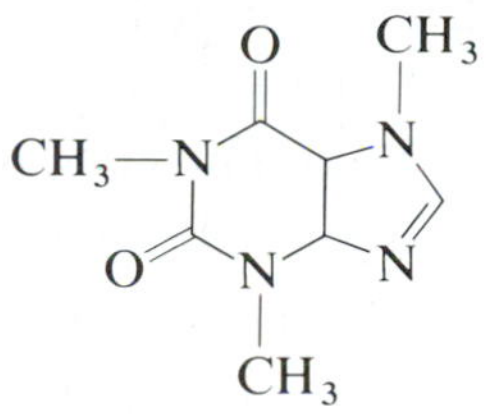

Caffeine
(stimulant and diuretic)

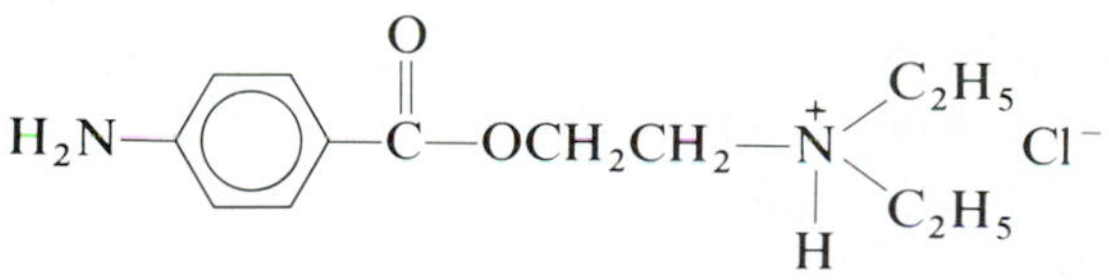

$H_2N-C_6H_4-\overset{O}{\overset{\|}{C}}-OCH_2CH_2-\overset{+}{N}H(C_2H_5)_2\;Cl^-$

Procaine hydrochloride (novocaine)
(local anesthetic)

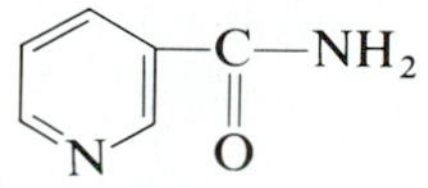

Nicotinamide (niacin)
(antipellagra vitamin)

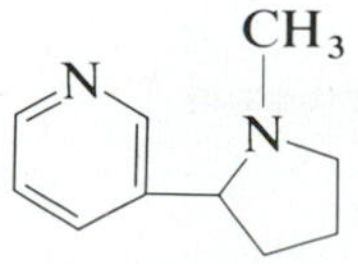

Nicotine
(from tobacco leaves, used as an agricultural insecticide)

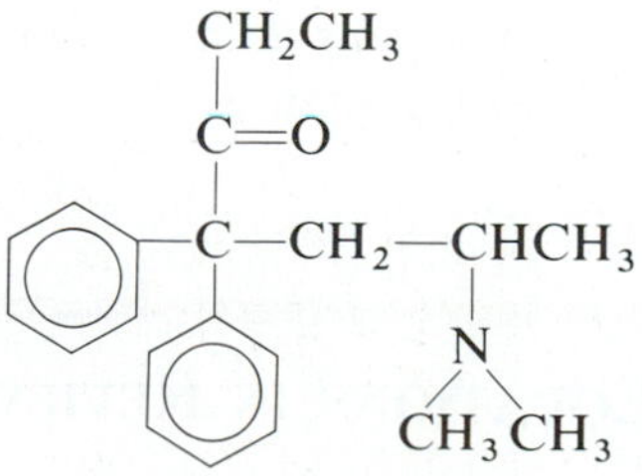

Methadone
(narcotic analgesic, substitute for heroin)

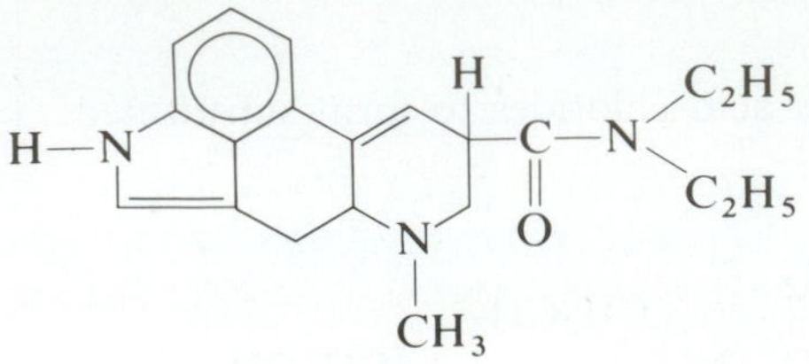

Lysergic acid diethylamide (LSD)
(hallucinogen)

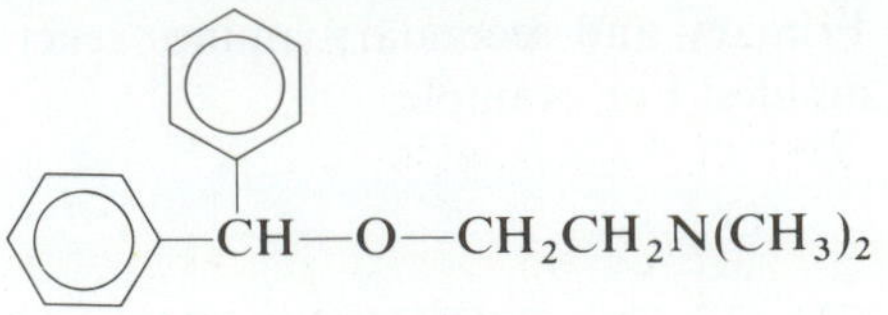

$C_6H_5-CH(C_6H_5)-O-CH_2CH_2N(CH_3)_2$

Benadryl
(antihistamine)

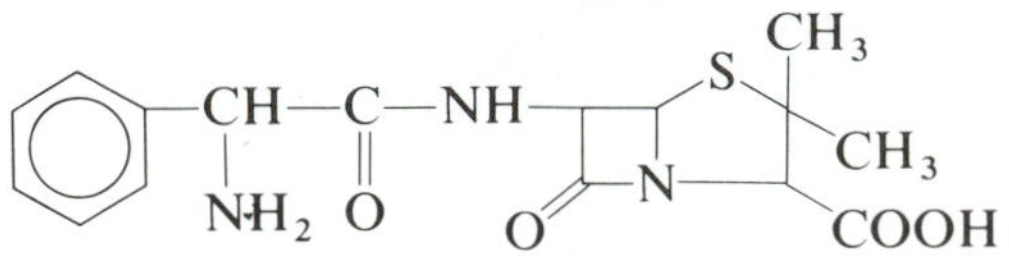

Ampicillin
(antibacterial agent)

Concepts in Review

1. Name and write structural formulas for amides.
2. Explain the high melting and boiling points of the amides, compared to alkanes of similar molar mass.
3. Write equations for both acidic and basic hydrolysis of amides.
4. Name and write structural formulas for amines.
5. Distinguish among primary, secondary, and tertiary amines.
6. Show that amines are bases in their reactions with water and with acids.
7. Write equations for reactions of amines to form substituted amides.
8. Write equations for the formation of amines from (a) alkyl halides plus ammonia, (b) reduction of amides with $LiAlH_4$, (c) reduction of nitriles with H_2 and Ni, and (d) reduction of aromatic nitro compounds.
9. Identify and write equations for the formation of quaternary ammonium salts.
10. Indicate the major physiological responses to barbiturates, tranquilizers, amphetamines, and alkaloids.
11. Explain the term "designer drug" and indicate the hazards associated with these substances.

Equations in Review

Formation of Amides

$$R-\overset{O}{\overset{\|}{C}}-Cl + NH_2R' \xrightarrow{\Delta} R-\overset{O}{\overset{\|}{C}}-NHR'$$

$$R{-}\overset{\displaystyle O}{\overset{\|}{C}}{-}Cl + NHR'_2 \longrightarrow R{-}\overset{\displaystyle O}{\overset{\|}{C}}{-}NR'_2$$

Hydrolysis of Amides

$$R{-}\overset{\displaystyle O}{\overset{\|}{C}}{-}NHR' + H_2O \longrightarrow R{-}\overset{\displaystyle O}{\overset{\|}{C}}{-}OH + H_2NR'$$

Amine Reaction as a Base

$$\underset{\text{Amine}}{R{-}NH_2} + \underset{\text{Acid}}{HX} \longrightarrow \underset{\text{Amine salt}}{R{-}\overset{+}{N}H_3X^-}$$

Exercises

1. Draw structural formulas for:
   (a) Butylamine
   (b) *p*-Methylaniline
   (c) *N*-Methyl-*p*-bromobenzamide
   (d) *N*,*N*-Diethylbenzamide
2. Name the following compounds.
   (a) $CH_3{-}\overset{\displaystyle O}{\overset{\|}{C}}{-}NH_2$
   (b) 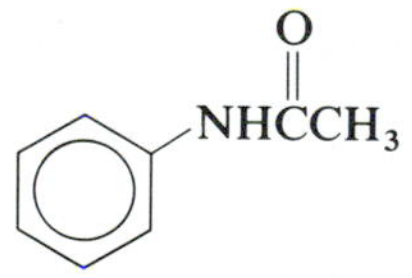
   (c) 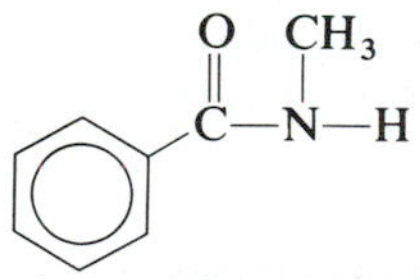
3. Predict which of the four compounds will be the most soluble in water, ethylamine, $CH_3CH_2CH_2CH_2NH_2$, acetamide, or $CH_3CH_2CH_2\overset{\displaystyle O}{\overset{\|}{C}}{-}NH_2$. Explain your answer.
4. Predict the substituted amide products:
   (a) Acetic acid and diethylamine and heat
   (b) Acetic acid and propylamine and heat
   (c) Butanoic acid and methylamine and heat
5. Predict whether each of the following compounds is an acid, a base, or neither.
   (a) CH_3CH_2OH
   (b) $CH_3CH_2NH_2$
   (c) 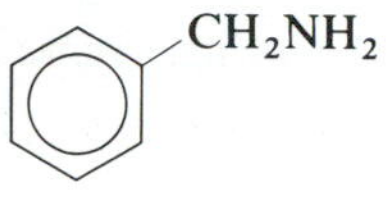
   (d) 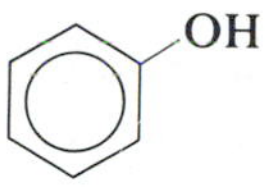
   (e) $CH_3CH_2{-}\overset{\displaystyle O}{\overset{\|}{C}}{-}NH_2$
   (f) $CH_3CH_2{-}\overset{\displaystyle O}{\overset{\|}{C}}{-}OH$
6. Show the hydrolysis products of these compounds.
   (a) 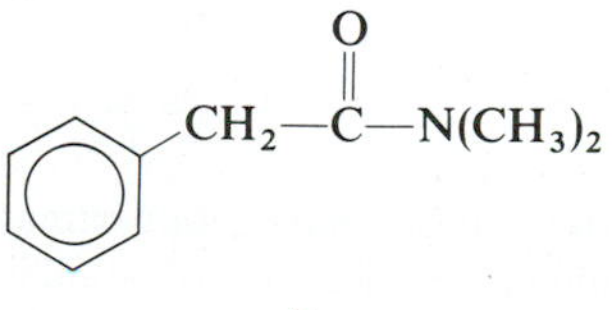
   (b) $CH_3{-}NH{-}\overset{\displaystyle O}{\overset{\|}{C}}{-}CH_3$
   (c) 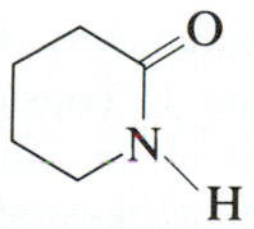
7. Contrast the physical properties of amides with those of amines.
8. Arrange each set of compounds in order of increasing solubility in water.

(a) $CH_3CH_2C(=O)—NHCH_3$

$CH_3CH_2C(=O)—N(CH_3)_2$

$CH_3CH_2C(=O)—NH_2$

(b) $CH_3(CH_2)_4C(=O)—NH_2$ $CH_3C(=O)—NH_2$

$C_6H_5—C(=O)—NH_2$

9. Urea is soluble in water. Show the possible hydrogen bonding that helps to explain this property.
10. Classify each of the following amines as primary, secondary, or tertiary.
   (a) $CH_3CH_2CH_2NH_2$
   (b) CH_3NHCH_3
   (c) $C_6H_5NH_2$ (benzene ring with NH_2)
   (d) cyclohexyl—N(H)—phenyl
   (e) piperidine ring with N—CH_3
11. Explain why amines have approximately the same solubility as alcohols of similar molar mass.
12. Propylamine, $CH_3CH_2CH_2NH_2$, ethylmethylamine, $CH_3CH_2NHCH_3$, and trimethylamine, $(CH_3)_3N$, are all isomers. Explain why the boiling point of trimethylamine is considerably lower than the boiling points of the other two.
13. After cleaning or packing fish, workers often use lemon juice to clean their hands. What is the purpose of the lemon juice?
14. Low molar mass aliphatic amines generally have odors suggestive of ammonia and/or stale fish. Which would have the more objectionable odor? Give a satisfactory reason for your answer.
   (a) A 1% trimethylamine solution in 1.0 *M* H_2SO_4 or
   (b) A 1% trimethylamine solution in 1.0 *M* NaOH
15. Draw structural formulas for
   (a) All the amines of formula C_3H_9N
   (b) All the amines of formula $C_4H_{11}N$
   (c) All the quaternary ammonium salts of formula $C_4H_{12}N^+Cl^-$
16. Name the following compounds.
   (a) benzene ring with CH_2CH_3 and NH_2 (ortho)
   (b) $CH_3CH_2NH_3^+\ Br^+$
   (c) $H_2N—C(=O)—CH_2—CH_2—C(=O)—NH_2$
   (d) $C_6H_5—NHCH_2CH_3$
   (e) pyridine ring (N)
   (f) $CH_3C(=O)—NHC_2H_5$
   (g) pyridine ring with $—C(=O)—NH_2$
   (h) cyclohexane ring with $—NH_2$
   (i) benzene ring with NH_2 and NO_2 (meta)
   (j) $(C_2H_5)_4N^+\ I^-$
   (k) $C_6H_5—NH—C_6H_5$
17. Draw structural formulas for the following compounds.
   (a) Methylethylamine
   (b) Tributylamine
   (c) Aniline
   (d) Ethylammonium chloride
   (e) *N*-methylanilinium chloride
   (f) 1,4-Butanediamine
   (g) *N,N*-dimethylaniline
   (h) Ethylisopropymethylamine
   (i) Pyridine
   (j) 2-Amino-1-pentanol
18. Write equations to show how each conversion may be accomplished. (Some conversions may require more than one step.)
   (a) $CH_3CH_2CH_2Br \longrightarrow CH_3CH_2CH_2NH_2$
   (b) $CH_3CH_2CH_2Br \longrightarrow CH_3CH_2CH_2CH_2NH_2$
   (c) $CH_3—C_6H_4—NO_2 \longrightarrow CH_3—C_6H_4—NH_2$ (para)

19. (a) What is a heterocyclic compound?
(b) How many heterocyclic rings are present in (1) LSD, (2) ampicillin, (3) methadone, and (4) nicotine?

20. Which of the following statements are correct? Rewrite the incorrect statements to make them correct.
(a) The common name for CH_3CONH_2 is methanamide.
(b) The general formula for a primary amine is $\overset{O}{\overset{\|}{RCNH_2}}$.
(c) Uric acid is the primary way that the body loses nitrogen.
(d) Heterocyclic compounds contain more than one ring.
(e) Alkaloids are used to settle upset stomachs.
(f) Barbiturates are stimulants.
(g) Amphetamines are also called sympathetic amines.
(h) Designer drugs are used in the manufacture of psychedelic clothing.
(i) The name for $[(CH_3)_2CH]_2NH$ is isopropyl amine.
(j) Aniline is soluble in dilute HCl because it forms a soluble salt.
(k) Most amines have pleasant odors.
(l) Lactic acid is an α-amino acid.
(m) Sulfanilamide is an aniline derivative.
(n) Isopropyl amine is a secondary amine.
(o) Butylamine and diethylamine are isomers.
(p) Aniline is made by the reduction of nitrobenzene.

21. List three classes of drugs and indicate the major effects of each.

22. Indicate the functional groups present in:
(a) Aspirin
(b) Acetaminophen
(c) Ibuprofen
(d) Adrenalin

23. Show the structure known as an amide linkage. Indicate the important classes of biochemicals that contain this linkage.

27

Polymers: Macromolecules

◀ **Chapter Opening Photo: River rafting through the Grand Canyon is often done on rafts made from polymers.**

What is a "mer"? The terms *polymer* and *monomer* have entered our everyday speech, and we are familiar with the prefixes, *poly* meaning "many" and *mono* meaning "one." Although a "mer" sounds like something from a child's cartoon show, scientists have chosen this small three-letter root to convey an important concept. It is derived from the Greek *meros*, meaning "part." So, a monomer is a "one part" and a polymer is a "many part."

Look around you for a few minutes and see all the objects that could be described in these terms. A brick building could be described as a polymer, with each brick representing a monomer. A string of pearls is an elegant polymer, with each pearl a monomer. Starting with simple chemical monomers, chemists have learned to construct both utilitarian and elegant polymers. As you will see in this chapter, many of the polymers are of great commercial importance.

27.1 Macromolecules

Up to now we have dealt mainly with rather small organic molecules that contain up to 50 atoms and some (fats) that contain up to about 150 atoms. But there exist in nature some very large molecules (macromolecules) containing tens of thousands of atoms. Some of these, such as starch, glycogen, cellulose, proteins, and DNA, have molar masses in the millions and are central to many of our life processes. Synthetic macromolecules touch every phase of modern living. It is hard today to imagine a world without polymers. Textiles for clothing, carpeting, and draperies, shoes, toys, automobile parts, construction materials, synthetic rubber, chemical equipment, medical supplies, cooking utensils, synthetic leather, recreational equipment—the list could go on and on. All these and a host of others that we consider to be essential in our daily lives are wholly or partly synthetic polymers. Most of these polymers were unknown 60 years ago. The vast majority of these polymeric materials are based on petroleum. Because petroleum is a nonreplaceable resource, our dependence on polymers is another good reason for not squandering the limited world supply of petroleum.

Polyethylene is an example of a synthetic polymer. Ethylene, derived from petroleum, is made to react with itself to form polyethylene (or polythene). Polyethylene is a long-chain hydrocarbon made from many ethylene units:

$$\underset{\text{Ethylene}}{n\,CH_2{=}CH_2} \longrightarrow \underset{\text{Polyethylene}}{-CH_2CH_2CH_2[CH_2CH_2]_nCH_2CH_2CH_2-}$$

A typical polyethylene molecule is made up of about 2,500–25,000 ethylene molecules joined in a continuous structure.

polymerization

polymer

monomer

The process of forming very large, high molar mass molecules from smaller units is called **polymerization**. The large molecule, or unit, is called the **polymer** and the small unit, the **monomer**. The term *polymer* is derived from the Greek word *polumerēs*, meaning "having many parts." Ethylene is a monomer, and polyethylene is a polymer. Because of their large size, polymers are often called *macromolecules*. Another commonly used term is *plastics*. The world *plastic* means "to be capable of being molded, or pliable." Although not all polymers are pliable and capable of being remolded, the word *plastics* has gained general use and has come to mean any of a variety of polymeric substances.

27.2 Synthetic Polymers

Some of the early commercial polymers were merely modifications of naturally occurring substances. One chemically modified natural polymer, nitrated cellulose, was made and sold as Celluloid late in the 19th century. But the first commercially successful fully synthetic polymer, Bakelite, was made from phenol and formaldehyde by Leo Baekeland in 1909. This was the beginning of the modern plastics industry. Chemists began to create many synthetic polymers in the late 1920s. Since then, ever-increasing numbers of synthetic macromolecular materials have become articles of commerce. Even greater numbers of polymers have been made and discarded for technical or economic reasons. Polymers are presently used extensively in nearly every industry. For example, the electronics industry uses "plastics" in applications ranging from microchip production to fabrication of heat and impact resistant cases for the assembled products. In the auto industry, huge amounts of polymers are used as body, engine, transmission, and electrical system components, as well as for tires. Vast quantities of polymers with varied and sometimes highly specialized characteristics are used for packaging; for example, packaging for frozen foods must withstand subfreezing temperatures as well as those met in either conventional or microwave ovens.

Although there is a great variety of synthetic polymers on the market, they can be classified into the following general groups based on properties and uses:

1. Rubberlike materials or elastomers
2. Flexible films
3. Synthetic textiles and fibers
4. Resins (or plastics) for casting, molding, and extruding
5. Coating resins for dip, spray, or solvent dispersed applications
6. Miscellaneous, including hydraulic fluids, foamed insulation, ion-exchange resins

27.3 Polymer Types

addition polymer

Two general types of polymers—addition and condensation—are known. An **addition polymer** is one that is produced by the successive addition of repeating

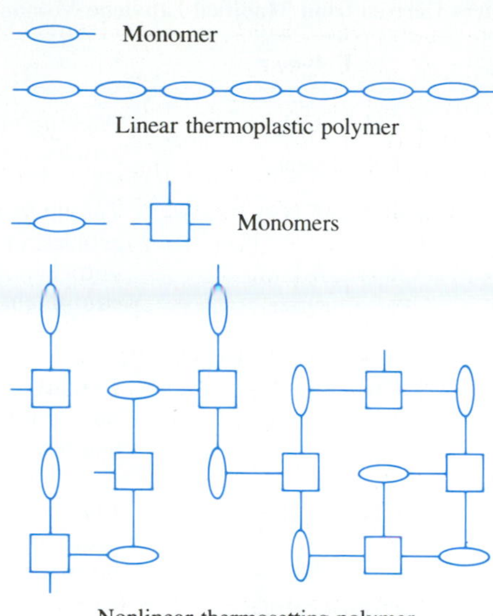

◀ **FIGURE 27.1 Schematic diagrams of thermoplastic and thermosetting polymer structures.**

monomer molecules. Polyethylene is an example of an addition polymer. A **condensation polymer** is one that is formed from monomers that react to split out water or some other simple substance.

condensation polymer

Polymers are also classified as being either thermoplastic or thermosetting. Those that soften on reheating are **thermoplastic polymers**; those that set to an infusible solid and do not soften on reheating are **thermosetting polymers**. Thermoplastic polymers are formed when monomer molecules join end to end in a linear chain with little or no cross-linking between the chains. Thermosetting polymers are macromolecules in which the polymeric chains are cross-linked to form a network structure. The structures of thermoplastic and thermosetting polymers are illustrated in the schematic diagram in Figure 27.1.

thermoplastic polymers

thermosetting polymers

27.4 Addition Polymerization

Ethylene polymerizes by addition to form polyethylene according to the reaction shown in Section 27.1. Polyethylene is the most important and widely used polymer on the market today. It is a tough, inert, but flexible thermoplastic material. Over 9.5×10^9 kg (2.1×10^{10} lb) of polyethylene is produced annually in the United States alone. Polyethylene is made into hundreds of different articles such as bread wrappers, toys, squeeze bottles, containers of all kinds, and electrical insulation.

The double bond is the key structural feature involved in the polymerization of ethylene. Ethylene derivatives, in which one or more hydrogen atoms have been replaced by other atoms or groups, can also be polymerized. This is often called *vinyl polymerization.* Many of our commercial synthetic polymers are made from such modified ethylene monomers. The names, structures, and uses of some of these polymers are given in Table 27.1.

Free radicals catalyze or initiate many addition polymerizations. Organic peroxides, ROOR, are frequently used for this purpose. The steps in the reaction

TABLE 27.1 Polymers Derived from Modified Ethylene Monomers

Monomer	Polymer	Uses
$CH_2{=}CH_2$ Ethylene	$\text{+}CH_2{-}CH_2\text{+}_n$ Polyethylene	Packing material, molded articles, containers, toys
$CH_2{=}CH{-}CH_3$ Propylene	$\left(CH_2{-}CH(CH_3)\right)_n$ Polypropylene	Textile fibers, molded articles, lightweight ropes, autoclavable biological equipment
$CH_2{=}C(CH_3)_2$ Isobutylene	$\left(CH_2{-}C(CH_3)_2\right)_n$ Polyisobutylene	Pressure-sensitive adhesives, butyl rubber (contains some isoprene as copolymer)
$CH_2{=}CH{-}Cl$ Vinyl chloride	$\left(CH_2{-}CH(Cl)\right)_n$ Polyvinyl chloride (PVC)	Phonograph records, garden hoses, pipes, molded articles, floor tile, electrical insulation, vinyl leather
$CH_2{=}CCl_2$ Vinylidene chloride	$\text{+}CH_2{-}CCl_2\text{+}_n$ Saran	Food packaging, textile fibers, pipes, tubing (contains some vinyl chloride as copolymer)
$CH_2{=}CH{-}CN$ Acrylonitrile	$\left(CH_2{-}CH(CN)\right)_n$ Orlon, Acrilan	Textile fibers
$CF_2{=}CF_2$ Tetrafluoroethylene	$\text{+}CF_2{-}CF_2\text{+}_n$ Teflon	Gaskets, valves, insulation, heat-resistant and chemical-resistant coatings, linings for pots and pans
$CH_2{=}CH{-}C_6H_5$ Styrene	$\left(CH_2{-}CH(C_6H_5)\right)_n$ Polystyrene	Molded articles, styrofoam, insulation, toys, disposable food containers
$CH_2{=}CH{-}OC(=O){-}CH_3$ Vinyl acetate	$\left(CH_2{-}CH(OC(=O){-}CH_3)\right)_n$ Polyvinyl acetate	Adhesives, paint, and varnish
$CH_2{=}C(CH_3){-}C(=O){-}O{-}CH_3$ Methylmethacrylate	$\left(CH_2{-}C(CH_3)(C(=O)OCH_3)\right)_n$ Lucite, Plexiglas (acrylic resins)	Contact lenses, clear sheets for windows and optical uses, molded articles, automobile finishes

are as follows.

Step 1 *Free radical formation.* The peroxide splits into free radicals:

$$RO{:}OR \longrightarrow 2\ RO\cdot$$

Step 2 *Propagation of polymeric chain.*

$$RO\cdot + CH_2{=}CH_2 \longrightarrow ROCH_2CH_2\cdot$$
$$ROCH_2CH_2\cdot + CH_2{=}CH_2 \longrightarrow ROCH_2CH_2CH_2CH_2\cdot \quad \text{(and so on)}$$

Step 3 *Termination.* Polymerization stops when the free radicals are used up:

$$RO(CH_2CH_2)_n\cdot + \cdot OR \longrightarrow RO(CH_2CH_2)_nOR$$
$$RO(CH_2CH_2)_n\cdot + \cdot(CH_2CH_2)_nOR \longrightarrow RO(CH_2CH_2)_n(CH_2CH_2)_nOR$$

Addition (or vinyl) polymerization of ethylene and its substituted derivatives yields saturated polymers—that is, polymer chains without carbon–carbon double bonds. The pi bond is eliminated when a free radical adds to an ethylene molecule. One of the electrons of the pi bond pairs with the unpaired electron of the free radical, thus bonding the radical to the ethylene unit. The other pi bond electron remains unpaired, generating a new and larger free radical. This new free radical then adds another ethylene molecule, continuing the building of the polymeric chain. This process is illustrated by the following electron-dot diagram:

```
   H  H              H  H
   |  |              |  |
H—C::C—H  ——→  H—C:C—H      (ROCH2CH2·)
RO·                  R̈O  ·
```

27.5 Butadiene Polymers

A diene is a compound that contains two carbon–carbon double bonds. Another type of addition polymer is based on the compound 1,3-butadiene or its derivatives.

$$\overset{1}{C}H_2{=}\overset{2}{C}H{-}\overset{3}{C}H{=}\overset{4}{C}H_2$$

1,3-Butadiene

Natural rubber is a polymer of isoprene (2-methyl-1,3-butadiene). Many synthetic elastomers or rubberlike materials are polymers of isoprene or of butadiene. Unlike the saturated ethylene polymers, these polymers are unsaturated; that is, they have double bonds in their polymeric structures.

The tires of these cars are ▶ made of rubber that has been vulcanized. This process improves the temperature range for use and resistance to abrasion.

$$n\,CH_2{=}\overset{\displaystyle CH_3}{\overset{|}{C}}{-}CH{=}CH_2 \longrightarrow \left(CH_2{-}\overset{\displaystyle CH_3}{\overset{|}{C}}{=}CH{-}CH_2\right)_n$$

Isoprene — Rubber polymer chain (Polyisoprene)

$$n\,CH_2{=}CH{-}CH{=}CH_2 \longrightarrow \left(CH_2{-}CH{=}CH{-}CH_2\right)_n$$

1,3-Butadiene — Butadiene polymer chain

$$n\,CH_2{=}\overset{\displaystyle Cl}{\overset{|}{C}}{-}CH{=}CH_2 \longrightarrow \left(CH_2{-}\overset{\displaystyle Cl}{\overset{|}{C}}{=}CH{-}CH_2\right)_n$$

2-Chloro-1,3-butadiene (Chloroprene) — Neoprene polymer chain (Polychloroprene)

In this kind of polymerization, the free radical adds to the butadiene monomer at carbon 1 of the carbon–carbon double bond. At the same time, a double bond is formed between carbon 2 and carbon 3, and a new free radical is formed at carbon 4. This process is illustrated in the following diagram:

$$RO\cdot + H{-}\overset{H}{\overset{|_1}{C}}{::}\overset{H}{\overset{|_2}{C}}{:}\overset{H}{\overset{|_3}{C}}{::}\overset{H}{\overset{|_4}{C}}{-}H \longrightarrow RO{:}\underset{H}{\underset{|}{\overset{H}{\overset{|}{C}}}}{:}\overset{H}{\overset{|}{C}}{::}\overset{H}{\overset{|}{C}}{:}\underset{H}{\underset{|}{\overset{H}{\overset{|}{C}}}}\cdot$$

Free radical — 1,3-Butadiene — Radical chain lengthened by four carbon atoms

One of the outstanding synthetic rubbers (styrene–butadiene rubber, SBR) is made from two monomers, styrene and 1,3-butadiene. These substances form a **copolymer**—that is, a polymer containing two different kinds of monomer

copolymer

units. Styrene and butadiene do not necessarily have to combine in a 1:1 ratio. In the actual manufacture of SBR polymers, about three moles of butadiene are used per mole of styrene. Thus, the butadiene and styrene units are intermixed, but in a ratio of about 3:1.

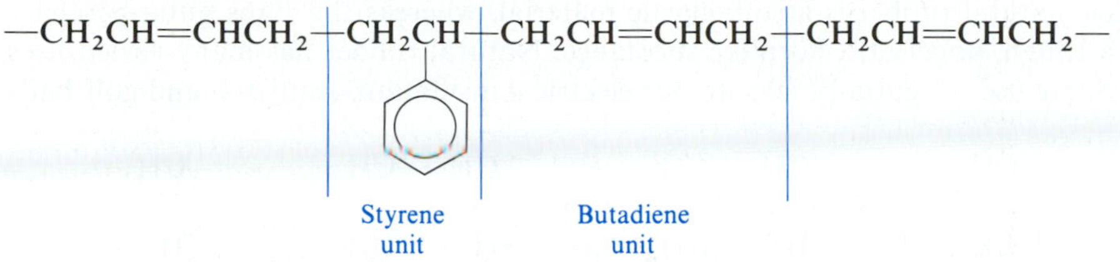

$$-CH_2CH{=}CHCH_2{-}CH_2CH(C_6H_5){-}CH_2CH{=}CHCH_2{-}CH_2CH{=}CHCH_2-$$

Styrene unit   Butadiene unit

Segment of styrene–butadiene rubber (SBR)

The presence of double bonds at intervals along the chains of rubber and rubberlike synthetic polymers designed for use in tires is almost a necessity and, at the same time, a disadvantage. On the positive side, double bonds make vulcanization possible. On the negative side, double bonds afford sites where ozone, present especially in smoggy atmospheres, can attack the rubber, causing "age hardening" and cracking. Vulcanization extends the useful temperature range of rubber products and imparts greater abrasion resistance to them. The vulcanization process is usually accomplished by heating raw rubber with sulfur and other auxiliary agents. It consists of introducing sulfur atoms that connect or cross-link the long strands of polymeric chains. Vulcanization was devised through trial-and-error experimentation by the American inventor Charles Goodyear in 1839, long before any real understanding of the chemistry of the process was known. Goodyear's patent on "Improvement in India Rubber" was issued on June 15, 1844. In the segment of vulcanized rubber shown below, the chains of polymerized isoprene are cross-linked by sulfur–sulfur bonds giving the polymer more strength and elasticity.

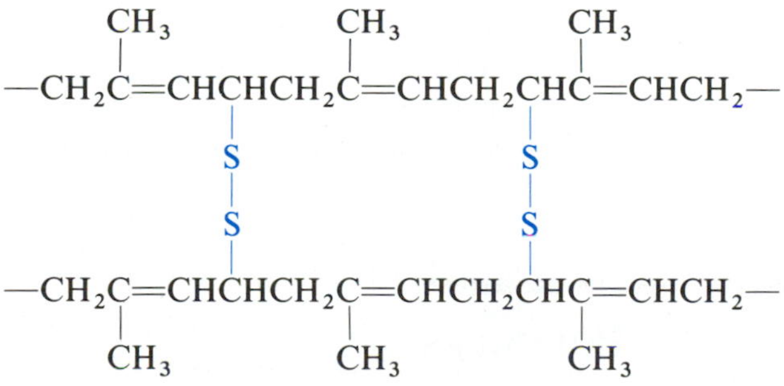

$$\begin{array}{l} \quad\ \ CH_3 \qquad\qquad\quad CH_3 \qquad\qquad\quad CH_3 \\ -CH_2\overset{|}{C}{=}CH\underset{|}{C}HCH_2\overset{|}{C}{=}CHCH_2\underset{|}{C}H\overset{|}{C}{=}CHCH_2- \\ \qquad\qquad\quad S \qquad\qquad\qquad\qquad\ \ S \\ \qquad\qquad\quad | \qquad\qquad\qquad\qquad\ \ | \\ \qquad\qquad\quad S \qquad\qquad\qquad\qquad\ \ S \\ -CH_2\underset{|}{C}{=}CH\overset{|}{C}HCH_2\underset{|}{C}{=}CHCH_2\overset{|}{C}H\underset{|}{C}{=}CHCH_2- \\ \quad\ \ CH_3 \qquad\qquad\quad CH_3 \qquad\qquad\quad CH_3 \end{array}$$

27.6 Geometric Isomerism in Polymers

The recurring double bonds in isoprene and butadiene polymers make it possible to have polymers with specific spatial orientation as a result of cis–trans isomerism. Recall from Section 22.3 that two carbon atoms joined by a double bond are not free to rotate and thus give rise to cis–trans isomerism. An isoprene polymer can have all-cis, all-trans, or a random distribution of cis and trans configurations about the double bonds.

Natural rubber is *cis*-polyisoprene with an all cis configuration about the carbon–carbon double bonds. Gutta-percha, also obtained from plants, is a *trans*-polyisoprene with an all trans configuration. Although these two polymers have the same composition, their properties are radically different. The cis natural rubber is a soft, elastic material, whereas the trans gutta-percha is a tough, nonelastic, hornlike substance. Natural rubber has many varied uses. Some uses of gutta-percha are for electrical insulation, dentistry, and golf balls.

```
—CH2          CH2—CH2           CH2—CH2           CH2—
     \       /       \         /       \         /
      C=C             C=C               C=C
     /       \       /         \       /         \
CH3           H   CH3           H   CH3           H
```

All cis configuration of natural rubber

```
—CH2          H   CH3           CH2—CH2           H
     \       /       \         /       \         /
      C=C             C=C               C=C
     /       \       /         \       /         \
CH3           CH2—CH2           H   CH3           CH2—
```

All trans configuration of gutta-percha

Chicle is another natural substance containing polyisoprenes. It is obtained by concentrating the latex from the sapodilla tree and contains about 5% *cis*-polyisoprene and 12% *trans*-polyisoprene. The chief use of chicle is in chewing gum.

Only random or nonstereospecific polymers are obtained by free radical polymerization. Synthetic polyisoprenes made by free radical polymerization are much inferior to natural rubber, since they contain both the cis and the trans isomers. But in the 1950s Karl Ziegler (1898–1973) of Germany and Giulio Natta (1903–1979) of Italy developed catalysts [for example, $(C_2H_5)_3Al/TiCl_4$] that allowed polymerization to proceed by an ionic mechanism, producing stereochemically controlled polymers. Ziegler–Natta catalysts made possible the synthesis of polyisoprene with an all cis configuration and with properties fully comparable to those of natural rubber. This material is known by the odd but logical name *synthetic natural rubber*. In 1963 Natta and Ziegler were jointly awarded the Nobel prize for their work on stereochemically controlled polymerization reactions.

27.7 Condensation Polymers

Condensation polymers are formed by reactions between functional groups on adjacent monomer molecules. As a rule a smaller molecule, usually water, is eliminated in the reaction. The monomers must be at least bifunctional, and if cross-linking is to occur, there must be more than two functional groups on some monomer molecules.

Most biochemical polymers are of the condensation type. Proteins are polyamides, whereas nucleic acids are polyesters. Just as with synthetic polymers, biological polymers can have varied physical properties and varied functions. The collagen protein is used to make durable structures such as feathers, hair, and hooves. The elastin protein can be used for structures that need to stretch, such as tendons. Other proteins provide the mucous coating that protects our nasal passages. These differences in physical properties are determined by (1) the functional groups that are incorporated in the polymer and (2) the structure into which the atoms have been bonded. This generalization applies to synthetic polymers as well.

Many different condensation polymers have been synthesized. Important classes include the polyesters, polyamides, phenol–formaldehyde polymers, and polyurethanes.

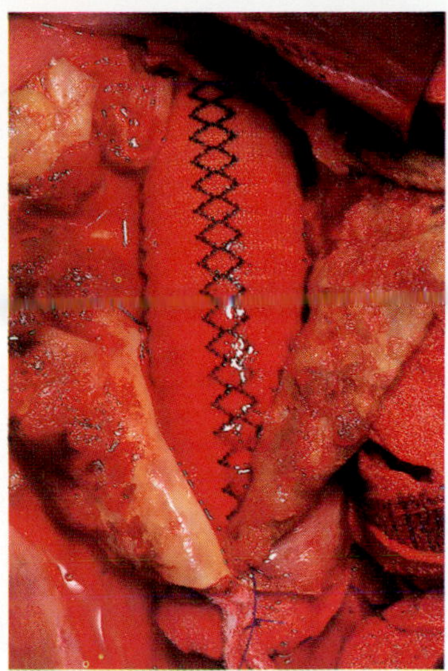

▲ **The synthetic fiber Dacron is used for patching in heart surgery.**

Polyesters

Polyesters are joined by ester linkages between carboxylic acid and alcohol groups; the macromolecule formed may be linear or cross-linked. From the bifunctional monomers terephthalic acid and ethylene glycol, a linear polyester is obtained. Esterification occurs between the alcohol and acid groups on both ends of both monomers, forming long-chain macromolecules:

$$\underset{\text{Terephthalic acid}}{HOOC{-}C_6H_4{-}COOH} \qquad \underset{\text{Ethylene glycol}}{HOCH_2CH_2OH}$$

$$-OCH_2CH_2O{-}\left[\overset{O}{\overset{\|}{C}}{-}C_6H_4{-}\overset{O}{\overset{\|}{C}}{-}OCH_2CH_2O\right]_n{-}\overset{O}{\overset{\|}{C}}{-}C_6H_4{-}\overset{O}{\overset{\|}{C}}-$$

This polymer may be drawn into fibers or formed into transparent films of great strength. Dacron and Terylene synthetic textiles and Mylar films are made from this polyester. In actual practice the dimethyl ester of terephthalic acid is used, and the molecule split out is methyl alcohol instead of water.

When trifunctional acids or alcohols are used as monomers, cross-linked thermosetting polyesters are obtained (see Figure 27.1). One common example is the reaction of glycerol and *o*-phthalic acid. The polymer formed is one of a group of polymers known as alkyd resins. Glycerol has three functional —OH groups, and phthalic acid has two functional —COOH groups:

$$\underset{\text{Glycerol}}{HOCH_2\overset{OH}{\overset{|}{C}}HCH_2OH} \qquad \underset{o\text{-Phthalic acid}}{C_6H_4(COOH)_2}$$

A cross-linked macromolecular structure is formed that, with modifications, has proved to be one of the most outstanding materials used in the coatings industry. Alkyd resins have been used as "baked-on" finishes for automobiles and household appliances. Each year more than 3.6×10^8 kg (7.9×10^8 lb) of these resins are used in paints, varnishes, lacquers, electrical insulation, and so on.

Polyamides

Although there are several nylons, one of the best known and the first commercially successful polyamide is Nylon-66. This polymer was so named because it

Fibers made from polymers ▶ can be used to conduct light to a very specific location. This property is used extensively in the field of fiberoptics.

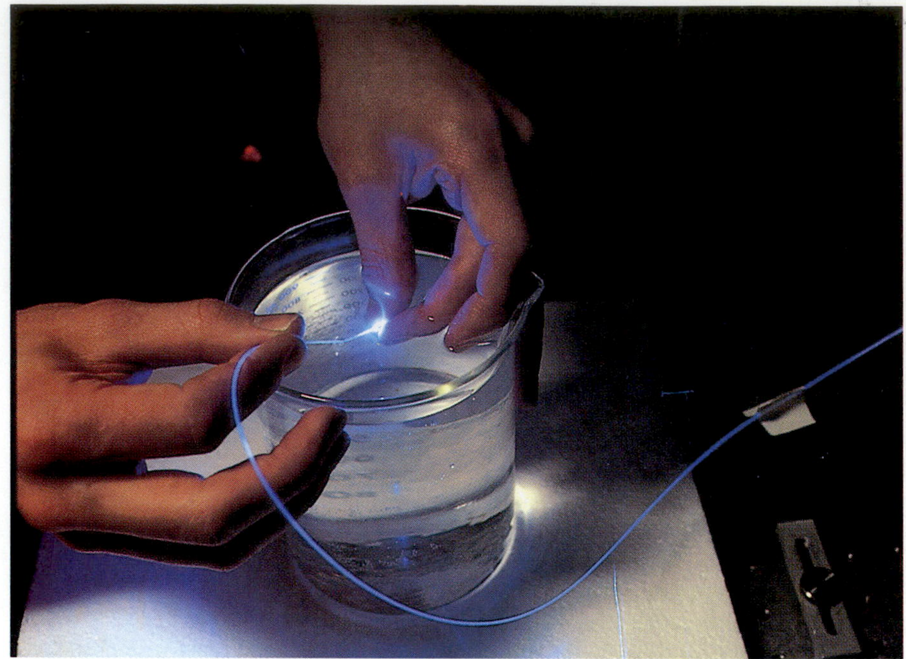

was made from two six-carbon monomers, adipic acid [$HOOC(CH_2)_4COOH$] and 1,6-diaminohexane [$H_2N(CH_2)_6NH_2$]. The polymer chains of polyamides contain recurring amide linkages. The amide linkage can be made by reacting a carboxylic acid group with an amine group:

$$\underset{\text{Carboxylic acid group}}{R-\overset{\overset{\large O}{\|}}{C}-OH} + \underset{\text{Amine group}}{H\overset{\overset{\large H}{|}}{N}-CH_2-R'} \xrightarrow{\Delta} \underset{\text{Amide linkage}}{R-\overset{\overset{\large O}{\|}}{C}-NH}-CH_2-R' + H_2O$$

The repeating structural unit of the Nylon-66 chain consists of one adipic acid unit and one 1,6-diaminohexane unit:

$$\underset{\text{Adipic acid}}{HOOC-(CH_2)_4-COOH} \qquad \underset{\text{1,6-Diaminohexane (Hexamethylenediamine)}}{H_2N-(CH_2)_6-NH_2}$$

$$-NH(CH_2)_6-NH-\left[\overset{\overset{\large O}{\|}}{C}(CH_2)_4-\overset{\overset{\large O}{\|}}{C}-NH(CH_2)_6-NH\right]_n\overset{\overset{\large O}{\|}}{C}-(CH_2)_4-\overset{\overset{\large O}{\|}}{C}-$$

Segment of Nylon-66 polyamide

Nylon was developed as a synthetic fiber for the production of stockings and other wearing apparel. It was introduced to the public at the New York World's Fair in 1939. Nylon is used to make fibers for clothing and carpeting, filaments for fishing lines and ropes, bristles for brushes, and molded objects such as gears and bearings. For the latter application no lubrication is required, because nylon surfaces are inherently slippery.

Phenol–Formaldehyde Polymers

As noted earlier, a phenol–formaldehyde condensation polymer (Bakelite) was first marketed over 75 years ago. Polymers of this type are still widely used, especially in electrical equipment, because of their insulating and fire-resistant properties.

Each phenol molecule can react with formaldehyde to lose an H atom from the para position and from each of the ortho positions (indicated by arrows):

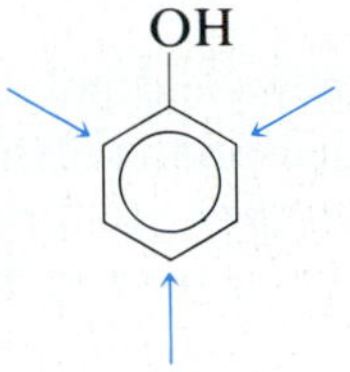

OH

Each formaldehyde molecule reacts with two phenol molecules to eliminate water:

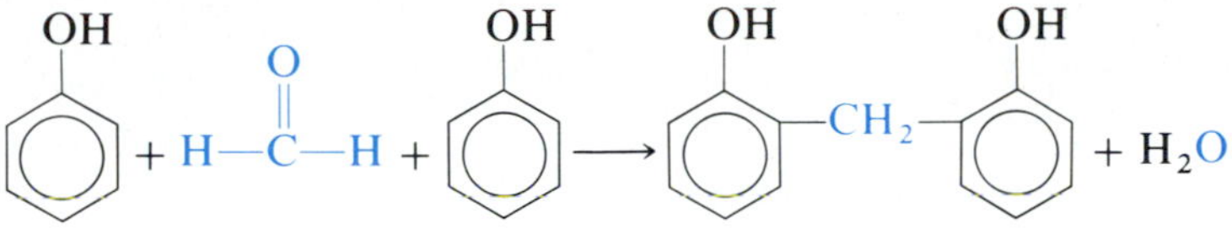

OH + H—C(=O)—H + OH ⟶ OH —CH_2— OH + H_2O

Similar reactions occur at the other two reactive sites on each phenol molecule, leading to the formation of the polymer. This polymer is thermosetting because it has an extensively cross-linked network structure. A typical section of this structure is illustrated as follows:

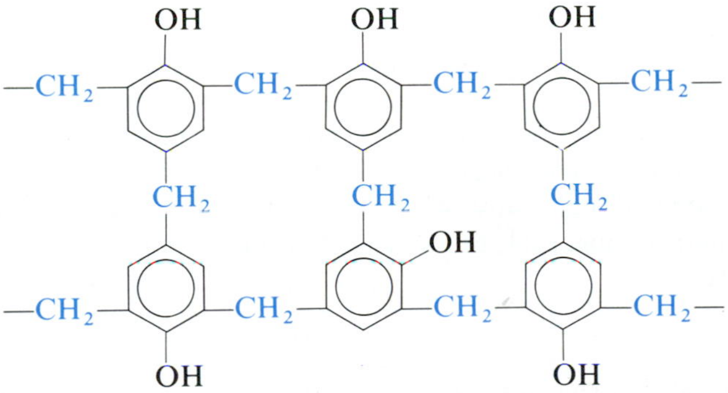

Phenol–formaldehyde polymer

Polyurethanes

The compound urethane has structural features of both an ester and an amide. Its formula is

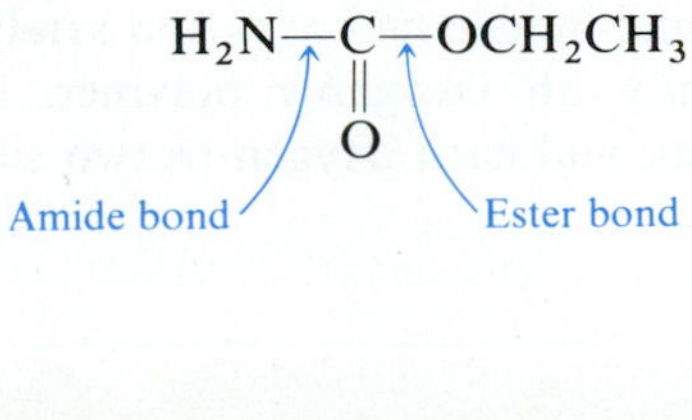

$$H_2N—C(=O)—OCH_2CH_3$$

Amide bond   Ester bond

A substituted urethane can be made by reacting an isocyanate with an alcohol. For example,

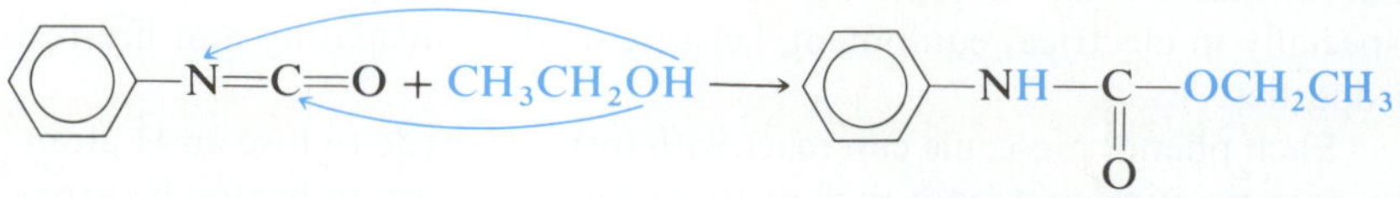

$$C_6H_5{-}N{=}C{=}O + CH_3CH_2OH \longrightarrow C_6H_5{-}NH{-}\overset{\overset{\displaystyle O}{\|}}{C}{-}OCH_2CH_3$$

Phenyl isocyanate → *N*-Phenyl urethane

Diisocyanates and diols are both difunctional; therefore, they yield polymers called *polyurethanes*. The polyurethanes are classified as condensation polymers, although no water or other small molecule is split out when they are formed. From phenylene diisocyanate and ethylene glycol, we obtain a polyurethane that has a structure as shown:

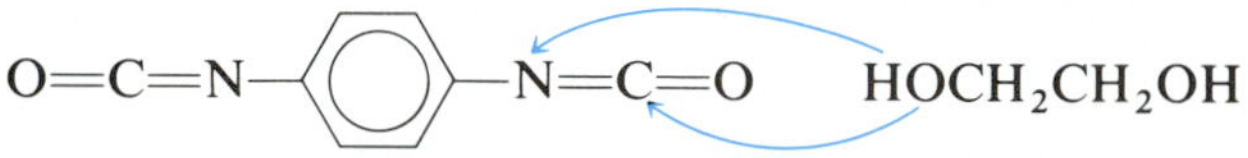

$$O{=}C{=}N{-}C_6H_4{-}N{=}C{=}O \qquad HOCH_2CH_2OH$$

p-Phenylene diisocyanate Ethylene glycol

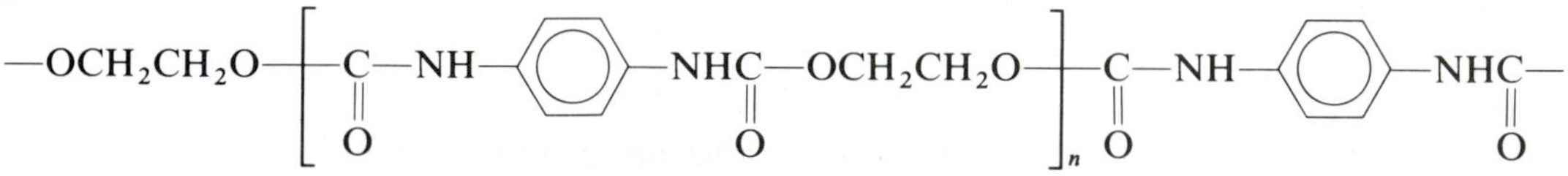

$$-OCH_2CH_2O{-}\left[\underset{\underset{\displaystyle O}{\|}}{C}{-}NH{-}C_6H_4{-}NH\underset{\underset{\displaystyle O}{\|}}{C}{-}OCH_2CH_2O\right]_n{-}\underset{\underset{\displaystyle O}{\|}}{C}{-}NH{-}C_6H_4{-}NH\underset{\underset{\displaystyle O}{\|}}{C}-$$

Segment of a polyurethane

Some polyurethanes are soft elastic materials that are widely used as *foam rubber* in upholstery and similar applications. There are many other applications, including automobile safety padding, insulation, life preservers, elastic fibers, and semirigid or rigid foams. Polyurethane can be made into a foam or spongy polymer by adding water during the polymerization or during the molding process. Water reacts with the isocyanate to produce carbon dioxide, which causes the polymer to foam. The effect is similar to that of baking powder releasing carbon dioxide in dough, causing it to rise and become light. The result is a polymer containing innumerable tiny gas filled cavities that give the product a spongelike quality.

27.8 Silicone Polymers

The silicones are an unusual group of polymers. They include oils and greases, molding resins, rubbers (elastomers), and Silly Putty, the latter being a remarkable material that bounces like a rubber ball when dropped but that can be shaped like putty! Silicones have properties common to both organic and inorganic compounds. The mineral quartz (found in igneous rocks and sand) has the empirical formula SiO_2 and is actually an inorganic polymer. Each silicon atom is bonded to four oxygen atoms, and each oxygen to two silicon atoms to form a three-dimensional structure:

$$\begin{array}{ccccccc} & & | & & | & & \\ & & \mathrm{O} & & \mathrm{O} & & \\ | & & | & & | & & | \\ -\mathrm{Si} & -\mathrm{O}- & \mathrm{Si} & -\mathrm{O}- & \mathrm{Si} & -\mathrm{O}- & \mathrm{Si}- \\ | & & | & & | & & | \\ & & \mathrm{O} & & \mathrm{O} & & \\ & & | & & | & & \\ & & -\mathrm{Si}- & & -\mathrm{Si}- & & \\ & & | & & | & & \end{array}$$

The silicon–oxygen bonds are very strong, and quartz is stable at very high temperatures. But it is also very hard and brittle and therefore difficult to form into useful shapes.

Linear silicone polymers, also called silicones or polysiloxanes, consist of silicon–oxygen chains with two alkyl groups attached to each silicon atom:

$$-\mathrm{O}-\underset{\mathrm{R}}{\overset{\mathrm{R}}{\underset{|}{\overset{|}{\mathrm{Si}}}}}-\mathrm{O}-\left[\underset{\mathrm{R}}{\overset{\mathrm{R}}{\underset{|}{\overset{|}{\mathrm{Si}}}}}-\mathrm{O}\right]_n-\underset{\mathrm{R}}{\overset{\mathrm{R}}{\underset{|}{\overset{|}{\mathrm{Si}}}}}-\mathrm{O}-$$

The physical properties of silicones can be modified by (1) varying the length of the polymer chain, (2) varying the R groups, and (3) introducing cross-linking between the chains.

Because of their special properties, silicones have found a variety of uses despite their high cost. Some of their useful qualities are (1) good insulating properties (used in high temperature applications), (2) little viscosity change over a wide temperature range (therefore they are used as lubricating oils and greases at extreme temperatures), (3) excellent water repellency (used to waterproof many types of surfaces), and (4) good biological compatibility (hence their use in medical and plastic surgery applications). A few of the specific uses of silicone polymers are for coatings on printed electronic circuits, synthetic lubricants, hydraulic systems, brake fluids, electrical insulation, foam shields for nuclear power plants, solar energy, heat transfer systems, hair sprays, body and hand lotions, automobile and furniture polishes, and urethane foams. Because of their stability and tissue compatibility, silicones are frequently used for permanent surgical implants. For example, heart pacemakers are encased in protective casings made of silicone, and silicones are used to replace destroyed nose cartilage.

▲ **The safety of using silicone for breast implants has recently become an issue of controversy.**

Concepts in Review

1. Write formulas for addition polymers when given the monomer(s).
2. Write formulas for condensation polymers when given the monomers.

3. Describe the properties of a thermoplastic polymer and a thermosetting polymer.
4. Explain the free radical mechanism for polymer formation.
5. Identify polymers from their trade names (for example, Dacron, nylon, and Teflon).
6. Explain the effect of cross-linking in polymers.
7. Draw a segment of the structural formula of natural rubber to illustrate the all-cis configuration.
8. Explain how butadiene-type polymers are formed.
9. Explain vulcanization and its effect on rubber.
10. Identify polymers by type (such as vinyl, polyester, polyamide, or polyurethane).
11. Identify several useful qualities of silicones.

Equations in Review

Steps in Addition Polymerization

Free radical formation

$$RO{:}OR \longrightarrow 2\,RO\cdot$$

Propagation of polymeric chain

$$RO\cdot + CH_2{=}CH_2 \longrightarrow ROCH_2CH_2\cdot$$
$$ROCH_2CH_2\cdot + CH_2{=}CH_2 \longrightarrow ROCH_2CH_2CH_2CH_2\cdot$$

Termination

$$RO(CH_2CH_2)_n\cdot + \cdot OR \longrightarrow RO(CH_2CH_2)_nOR$$
$$RO(CH_2CH_2)_n\cdot + \cdot(CH_2CH_2)_nOR \longrightarrow RO(CH_2CH_2)_n(CH_2CH_2)_nOR$$

General Reactions Forming Condensation Polymers

Polyesters

$$HO{-}\overset{\overset{\displaystyle O}{\|}}{C}{-}(CH_2)_n{-}\overset{\overset{\displaystyle O}{\|}}{C}{-}OH + HO{+}CH_2)_n{-}OH \longrightarrow \left(\overset{\overset{\displaystyle O}{\|}}{C}{-}(CH_2)_n{-}\overset{\overset{\displaystyle O}{\|}}{C}{-}O{-}(CH_2)_n{-}O\right)_n$$

Polyamides

$$HO{-}\overset{\overset{\displaystyle O}{\|}}{C}{-}(CH_2)_n{-}\overset{\overset{\displaystyle O}{\|}}{C}{-}OH + H_2N{+}CH_2)_n{-}NH_2 \longrightarrow \left(\overset{\overset{\displaystyle O}{\|}}{C}{-}(CH_2)_n{-}\overset{\overset{\displaystyle O}{\|}}{C}{-}NH{-}(CH_2)_n{-}NH\right)_n$$

Urethanes

$$O{=}C{=}N{-}C_6H_4{-}N{=}C{=}O + HOCH_2CH_2OH \longrightarrow$$

$$\left(\overset{O}{\overset{\|}{C}}{-}NH{-}C_6H_4{-}NH\overset{O}{\overset{\|}{C}}{-}OCH_2CH_2O \right)_n$$

EXERCISES

1. How does condensation polymerization differ from addition polymerization?
2. What property distinguishes a thermoplastic polymer from a thermosetting polymer?
3. Show the free radical mechanism for the polymerization of propylene to polypropylene.
4. How many ethylene units are in a polyethylene molecule that has a molar mass of approximately 25,000?
5. Write a structural formula showing the polymer that can be formed from
   (a) Ethylene (c) 1-Butene
   (b) Propylene (d) 2-Butene
6. Write structural formulas for the following polymers:
   (a) Saran (d) Polystyrene
   (b) Orlon (e) Lucite
   (c) Teflon
7. Write structures showing two possible ways in which vinyl chloride can polymerize to form polyvinyl chloride. Show four units in each structure.
8. (a) Write the structure for a polymer that can be formed from 2,3-dimethyl-1,3-butadiene.
   (b) Can 2,3-dimethyl-1,3-butadiene form cis and trans polymers? Explain.
9. Write the chemical structures for the monomers of
   (a) Natural rubber
   (b) Synthetic (SBR) rubber
   (c) Synthetic natural rubber
10. Why is the useful life of natural rubber and that of many synthetic rubbers shortened in smoggy atmospheres?
11. How are rubber molecules modified by vulcanization?
12. Natural rubber is the all cis polymer of isoprene, and gutta-percha is the all trans isomer. Write the structure for each polymer, showing at least three isoprene units.
13. Nitrile rubber (Buna N) is a copolymer of two parts of 1,3-butadiene to one part of acrylonitrile ($CH_2{=}CHCN$). Write a structure for this synthetic rubber, showing at least two units of the polymer.
14. Ziegler–Natta catalysts can orient the polymerization of propylene to form isotactic polypropylene—that is, polypropylene with all the methyl groups on the same side of the long carbon chain. Write the structure for (a) isotactic polypropylene and (b) another possible geometric form of polypropylene.
15. Write formulas showing the structure of
    (a) Dacron (c) Bakelite
    (b) Nylon-66 (d) Polyurethane
16. Using *p*-cresol ($p\text{-}CH_3{-}C_6H_4{-}OH$) in place of phenol to form a phenol–formaldehyde polymer results in a thermoplastic rather than a thermosetting polymer. Explain why this occurs.
17. Why must at least some monomer molecules be trifunctional to form a thermosetting polyester?
18. Glyptal polyesters are made from glycerol and phthalic acid. Would this kind of polymer more likely be thermoplastic or thermosetting? Explain.
19. How is "foam" introduced into foam or sponge rubber materials?
20. Silicone polymers are more resistant to high temperatures than the usual organic polymers. Suggest an explanation for this property of the silicones.
21. Which of these statements are correct? Rewrite each incorrect statement to make it correct.
    (a) The process of forming macromolecules from small units is called polymerization.
    (b) The monomers in condensation polymerization must be at least bifunctional.
    (c) The monomers of Nylon-66 are a dicarboxylic acid and a diol.
    (d) Dacron and Mylar are both made from the same monomers.

(e) Hexamethylene diamine is a secondary amine.
(f) Teflon is an addition polymer.
(g) Vulcanization was invented by Charles Goodrich.
(h) The monomer for polystyrene is

$$C_6H_5\text{—}CH_2CH{=}CH_2$$

(i) Polyurethanes have both ester and amide bonds in their structure.
(j) Bakelite is a copolymer of phenol and ethylene glycol.
(k) The monomer of natural rubber is 2-methyl-1,3-butadiene.

22. Write structures for the monomers of each of the polymers shown below, and classify each as polyvinyl, polyester, polyamide, or polyurethane.

(a) $$\text{—}CH_2\underset{\substack{|\\ C=O\\ |\\ OC_2H_5}}{CH}CH_2\underset{\substack{|\\ C=O\\ |\\ OC_2H_5}}{CH}CH_2\underset{\substack{|\\ C=O\\ |\\ OC_2H_5}}{CH}CH_2\underset{\substack{|\\ C=O\\ |\\ OC_2H_5}}{CH}\text{—}$$

(b) $$\text{—}\underset{\substack{\|\\ O}}{C}\text{—}C_6H_4\text{—}\underset{\substack{\|\\ O}}{C}\text{—}OCH_2\text{—}C_6H_4\text{—}CH_2O\text{—}\underset{\substack{\|\\ O}}{C}\text{—}C_6H_4\text{—}\underset{\substack{\|\\ O}}{C}\text{—}OCH_2\text{—}C_6H_4\text{—}CH_2O\text{—}$$

(c) $$\text{—}\overset{\substack{O\\ \|}}{C}\text{—}NH\text{—}C_6H_3(CH_3)\text{—}NH\overset{\substack{O\\ \|}}{C}OCH_2CH_2\text{—}O\text{—}\overset{\substack{O\\ \|}}{C}\text{—}NH\text{—}C_6H_3(CH_3)\text{—}NH\text{—}\underset{\substack{\|\\ O}}{C}\text{—}OCH_2CH_2O\text{—}$$

(d) $$\text{—}\underset{\substack{\|\\ O}}{C}\text{—}(CH_2)_8\text{—}\underset{\substack{\|\\ O}}{C}\text{—}NH\text{—}(CH_2)_6\text{—}NH\text{—}\underset{\substack{\|\\ O}}{C}\text{—}(CH_2)_8\text{—}\underset{\substack{\|\\ O}}{C}\text{—}NH\text{—}(CH_2)_6\text{—}NH\text{—}$$

(e) $$\text{—}CH_2\overset{\substack{CH_3\\ |}}{\underset{\substack{|\\ Cl}}{C}}\text{—}CH_2\text{—}\overset{\substack{CH_3\\ |}}{\underset{\substack{|\\ Cl}}{C}}\text{—}CH_2\text{—}\overset{\substack{CH_3\\ |}}{\underset{\substack{|\\ Cl}}{C}}\text{—}$$

(f) $$\left[OCH_2\text{—}C_6H_{10}\text{—}CH_2O\text{—}\underset{\substack{\|\\ O}}{C}\text{—}C_6H_4\text{—}\underset{\substack{\|\\ O}}{C}\right]_n$$

28

Stereoisomerism

◀ Chapter Opening Photo: These crystals of an ester of choline are seen through a polarizing light microscope.

Many of us grew up hearing such comments as, "Can't you tell your left from your right?" Have you watched a small child try to differentiate between a right and left shoe? Not surprisingly, the distinction between right and left is difficult. After all, our bodies are reasonably symmetrical. For example, both hands are made up of the same components (four fingers, a thumb, and a palm) ordered in the same way (from thumb through little finger). Yet, there is a difference if we try to put a left-handed glove on our right hand or a right shoe on our left foot.

Molecules possess similar, subtle structural differences, which can have a major impact on chemical reactivity. For example, although there are two forms of blood sugar, related as closely as our left and right hands, only one of these structures can be used by our bodies for energy. Stereoisomerism is a subject that attempts to define these subtle differences in molecular structure.

28.1 Review of Isomerism

The phenomenon of two or more compounds having the same number and kinds of atoms is isomerism. Thus isomers are different compounds that have the same composition and the same molecular formula.

There are two types of isomerism. In the first type, known as structural isomerism, the difference between isomers is due to a different structural arrangement of the atoms to form different molecules. For example, butane and isobutane, ethanol and dimethyl ether, and 1-chloropropane and 2-chloropropane are structural isomers:

$CH_3CH_2CH_2CH_3$ Butane	CH_3CH_2OH Ethanol	$CH_3CH_2CH_2Cl$ 1-Chloropropane
CH_3CHCH_3 (with CH_3 on central C) Isobutane	CH_3OCH_3 Dimethyl ether	$CH_3CHClCH_3$ 2-Chloropropane

In the second type of isomerism, the isomers have the same structural formulas but differ in the arrangement of the atoms in space. This type of isomerism is known as **stereoisomerism**. Thus, compounds that have the same

stereoisomerism

structural formulas but differ in their spatial arrangement are called **stereoisomers**. There are two types of stereoisomers: cis–trans or geometric isomers, which we have already considered, and optical isomers, the subject of this chapter. The outstanding feature of optical isomers is that they have the ability to rotate plane-polarized light.

stereoisomers

FIGURE 28.1 ▶
(a) Diagram of ordinary light vibrating in all possible directions (planes) and (b) diagram of plane-polarized light vibrating in a single plane. The beam of light is coming toward the reader.

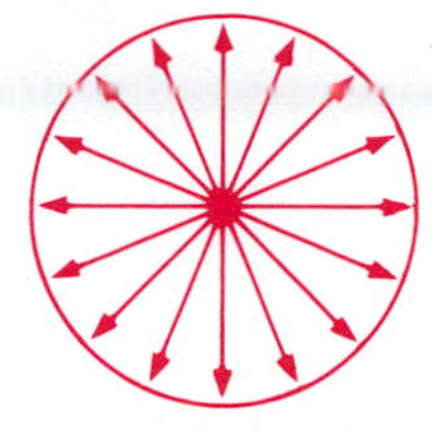

Ordinary (unpolarized) light
(a)

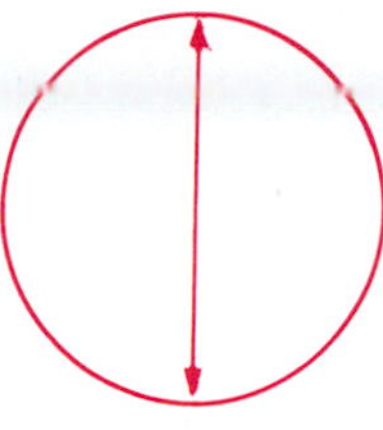

Plane-polarized light
(b)

28.2 Plane-Polarized Light

Plane-polarized light is light that is vibrating in only one plane. Ordinary (unpolarized) light consists of electromagnetic waves vibrating in all directions (planes) perpendicular to the direction in which it is traveling. When ordinary light passes through a polarizer, it emerges vibrating in only one plane and is called plane-polarized light (Figure 28.1).

plane-polarized light

Polarizers can be made from calcite or tourmaline crystals or from a Polaroid filter, which is a transparent plastic containing properly oriented embedded

(a)

(b)

◀ FIGURE 28.2
Two Polaroid filters (a) with axes parallel and (b) with axes at right angles. In (a), light passes through both filters and emerges polarized. In (b), the polarized light that emerges from one filter is blocked and does not pass through the second filter, which is at right angles to the first. With no light emerging, the filters appear black.

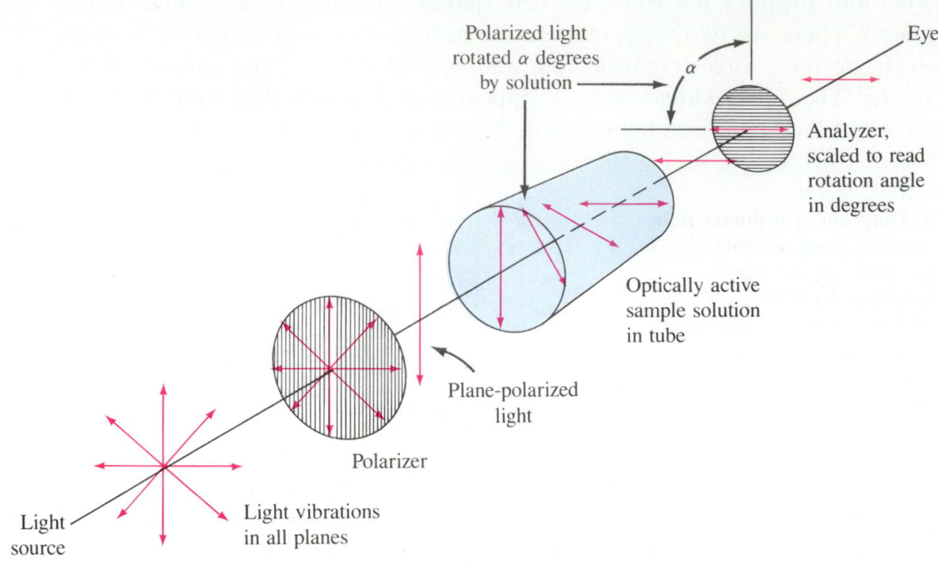

FIGURE 28.3 ► Schematic diagram of a polarimeter. This instrument is used to measure the angle α through which an optically active substance rotates the plane of polarized light.

crystals. Two Polaroid filters with their axes parallel allow the passage of plane-polarized light. But when one filter is placed so that its axis is at right angles to that of the other filter, the passage of light is blocked and the filters appear black (Figure 28.2).

Specific Rotation

The rotation of plane-polarized light is quantitatively measured with a polarimeter. The essential features of this instrument are (1) a light source (usually a sodium lamp), (2) a polarizer, (3) a sample tube, (4) an analyzer (which is another matched polarizer), and (5) a calibrated scale (360°) to measure the number of degrees the plane of polarized light is rotated. The calibrated scale is attached to the analyzer (see Figure 28.3). When the sample tube contains a solution of an optically inactive material, the axes of the polarizer and the analyzer are parallel and the scale is at zero degrees; the light passing through is at maximum intensity. When a solution of an optically active substance is placed in the sample tube, the plane in which the polarized light is vibrating is rotated through an angle (α). The analyzer is then rotated to the position where the emerging light is at maximum intensity. The number of degrees and the direction of rotation by the solution are then read from the scale as the observed rotation.

specific rotation

The **specific rotation**, [α], of a compound is the number of degrees that polarized light would be rotated by passage through 1 decimeter (dm) of a solution of the substance at a concentration of 1 gram per milliliter. The specific rotation of optically active substances is listed in chemical handbooks along with other physical properties. The following formula is used to calculate specific rotation from polarimeter data:

$$[\alpha] = \frac{\text{Observed rotation in degrees}}{\left(\begin{array}{c}\text{Length of}\\ \text{sample tube in decimeters}\end{array}\right)\left(\begin{array}{c}\text{Sample concentration in}\\ \text{grams per milliliter}\end{array}\right)}$$

28.3 Optical Activity

Many naturally occurring substances are able to rotate the plane of polarized light. Because of this ability to rotate polarized light, such substances are said to be **optically active**. When plane-polarized light passes through an optically active substance, the plane of the polarized light is rotated. If the rotation is to the right (clockwise), the substance is said to be **dextrorotatory**; if the rotation is to the left (counterclockwise), the substance is said to be **levorotatory**.

optical activity

dextrorotatory

levorotatory

Some minerals, notably quartz, rotate the plane of polarized light (in fact, optical activity was discovered in minerals). However, when such mineral crystals are melted, the optical activity disappears. This means that the optical activity of these crystals must be due to an ordered arrangement within the crystals.

In 1848 Louis Pasteur (1822–1895) observed that sodium ammonium tartrate, a salt of tartaric acid, exists as a mixture of two kinds of crystals. Pasteur carefully hand separated the two kinds of crystals. Investigating their properties, he found that solutions made from either kind of crystal would rotate the plane of polarized light, but in opposite directions. Since this optical activity was present in a solution, it could not be caused by a specific arrangement within a crystal.

The tetrahedral arrangement of single bonds around a carbon atom makes asymmetry (lack of symmetry) possible in organic molecules. When four different atoms or functional groups are bonded to a carbon atom, the molecule formed is asymmetric, and the carbon atom is called an **asymmetric carbon atom** (see Figure 28.4). In 1874 J. H. van't Hoff (1852–1911) and J. A. Le Bel (1847–1930) concluded that the presence of at least one asymmetric carbon atom in a molecule of an optically active substance is a key factor for optical activity. The first Nobel prize in chemistry was awarded to van't Hoff in 1901.

asymmetric carbon atom

D

A B

C

○ = Carbon atom

A, B, C, D = Four different atoms or groups of atoms

◄ **FIGURE 28.4 Three-dimensional representation of an asymmetric carbon atom with four different groups bonded to it. The carbon atom is a sphere. Bonds to A and B project from the sphere toward the observer. Bonds to C and D project from the sphere away from the observer.**

▲ **The mineral quartz can rotate a plane of polarized light.**

Molecules of optically active substances must have at least one center of dissymmetry. Although optically active compounds are known that do not contain asymmetric carbon atoms (their center of dissymmetry is due to some other structural feature), most optically active organic substances do contain one or more asymmetric carbon atoms.

28.4 Projection Formulas

Molecules of a compound that contain one asymmetric carbon atom occur in two optically active isomeric forms. This is because the four different groups bonded to the asymmetric carbon atom can be oriented in space in two different configurations. It is important to understand how we represent such isomers on paper.

Let us consider the spatial arrangement of a lactic acid molecule, $CH_3CH(OH)COOH$, that contains one asymmetric carbon atom:

$$\begin{array}{c} {}^{1}\text{COOH} \\ | \\ \text{H}-{}^{2}\text{C*}-\text{OH} \\ | \\ {}^{3}\text{CH}_3 \end{array} \qquad \text{C*} = \text{Asymmetric carbon atom}$$

Lactic acid

Three-dimensional models are the best means of representing such a molecule, but by adopting certain conventions and using imagination, we can formulate the images on paper. The geometrical arrangement of the four groups about the asymmetric carbon (carbon 2) is the key to the stereoisomerism of lactic acid. The four bonds attached to carbon 2 are separated by angles of about 109.5°. Diagram I in Figure 28.5 is a three-dimensional representation of lactic acid in which the asymmetric carbon atom is represented as a sphere, with its center in the plane of the paper. The —H and —OH groups are projected forward from the paper (toward the observer), and the —COOH and —CH_3 groups are projected back from the paper (away from the observer).

For convenience of expression, simpler diagrams such as II and III are used. These are much easier and faster to draw. In II, it is understood that the groups (—H and —OH) attached to the horizontal bonds are coming out of the plane of the paper toward the observer, and the groups attached to the vertical bonds are projected back from the paper. The molecule represented by formula III is made by drawing a cross and attaching the four groups in their respective positions, as in formula II. The asymmetric carbon atom is understood to be located where the lines cross. Formulas II and III are called **projection formulas**.

projection formula

COOH / H ◄ ○ ► OH / CH_3 — I

$$\begin{array}{c} \text{COOH} \\ | \\ \text{H}-\text{C}-\text{OH} \\ | \\ \text{CH}_3 \end{array} \qquad \begin{array}{c} \text{COOH} \\ | \\ \text{H}-\!\!\!+\!\!\!-\text{OH} \\ | \\ \text{CH}_3 \end{array}$$

II III

Figure 28.5 ► **Methods of representing three-dimensional formulas of a compound that contains one asymmetric carbon atom. All three structures represent the same molecule.**

It is important to be careful when comparing projection formulas. Two rules apply: (1) projection formulas must not be turned 90°; (2) projection formulas must not be lifted or flipped out of the plane of the paper. Projection formulas may, however, be turned 180° in the plane of the paper without changing the spatial arrangement of the molecule. Consider the following projection formulas:

COOH
II—OII
CH_3

III

CH_3
IIO—II
COOH

IV

CH_3
HO H
COOH

V

Formulas I, III, IV, and V represent identical molecules. Formula IV was obtained by turning formula III 180°. Formula V is formula IV drawn in a three-dimensional representation. If formula III is turned 90°, the other stereoisomer of lactic acid is represented, as shown in formulas VI and VII.

H
CH_3—COOH
OH

VI

H
CH_3 COOH
OH

VII

EXAMPLE 28.1

(a) Redraw the three-dimensional formula (A) into a projection formula. (b) Draw the three-dimensional formula represented by the projection formula (B).

(A)
CH_3
H CH_2NH_2
Br

$CH_3CHBrCH_2NH_2$

(B)
CH_3
H—CH_2CH_3
OH

$CH_3CH(OH)CH_2CH_3$

SOLUTION

(a) Draw a vertical and a horizontal line crossing each other. Place the CH_3 at the top and the Br at the bottom of the vertical line. Place the H at the left and the CH_2NH_2 at the right of the horizontal line to complete the projection formula:

CH_3
H—CH_2NH_2
Br

(b) Draw a small circle to represent the asymmetric carbon atom (C2) that is located where the two lines cross in the projection formula. Draw a short line extending from the top and from the bottom of the circle. Place the CH_3 on the top and the OH on the bottom at the end of these lines. Now draw two short lines from within the circle coming toward your left and right arms. Place the H at the end of the left line and the CH_2CH_3 at the end of the right line. The finished formula should look like this.

CH_3

H — ⟨ ⟩ — CH_2CH_3

OH

▲
FIGURE 28.6
The left hand is the same as the mirror image of the right hand. Right and left hands are not superimposable; hence they are chiral.

PRACTICE Redraw this projection formula as a three-dimensional formula.

Br

Cl —+— CH_3

F

Answer:

Br

Cl — ⟨ ⟩ — CH_3

F

28.5 ENANTIOMERS

Your right and your left hands are mirror images of each other; that is, the left hand is a mirror reflection of the right hand and vice versa. Furthermore, the two hands are not superimposable on each other (see Figure 28.6). **Superimposable** means that, when we lay one object upon another, all parts of both objects coincide exactly.

superimposable

A molecule that is not superimposable on its mirror image is said to be **chiral**. The word *chiral* comes from the Greek word *cheir*, meaning hand. Chiral molecules have the property of "handedness"; that is, they are related to each other in the same manner as the right and left hands. An asymmetric carbon atom is often referred to as a chiral carbon or a chiral center. Molecules, or objects, that are superimposable are **achiral**. Some chiral and achiral objects are shown in Figure 28.7.

chiral

achiral

The formulas developed in Section 28.4 dealt primarily with a single kind of lactic acid molecule. But two stereoisomers of lactic acid are known, one that rotates the plane of polarized light to the right and one that rotates it to the left. These two forms of lactic acid are shown in Figure 28.8. If we examine these two structural formulas carefully, we can see that they are mirror images of each other. The reflection of either molecule in a mirror corresponds to the structure of the other molecule. Even though the two molecules have the same molecular formula and the same four groups attached to the central carbon atom, they are not superimposable. Therefore, the two molecules are not identical but are isomers. One molecule rotates the plane of polarized light to the left and is termed *levorotatory*; the other molecule rotates it to the right and is *dextrorotatory*. A plus (+) or a minus (−) sign written in parentheses and placed in front of a name or formula indicates the direction of rotation of polarized light and becomes part of the name of the compound. Plus (+) indicates rotation to the right and minus (−) to the left. Using projection formulas, we write the two lactic acids as follows:

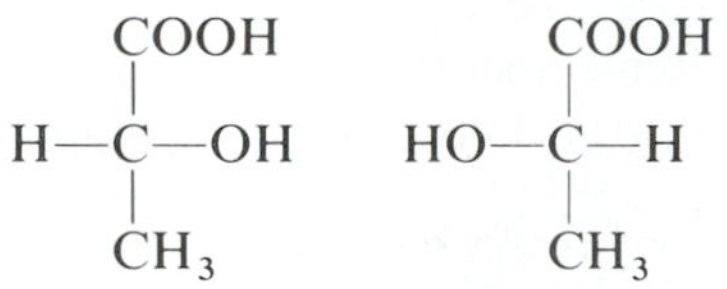

COOH
H—C—OH
CH3
COOH
HO—C—H
CH3

(−)-Lactic acid (+)-Lactic acid

Originally it was not known which lactic acid structure was the (+) or the (−) compound. However, it is now known that they are as shown. Isomers that are

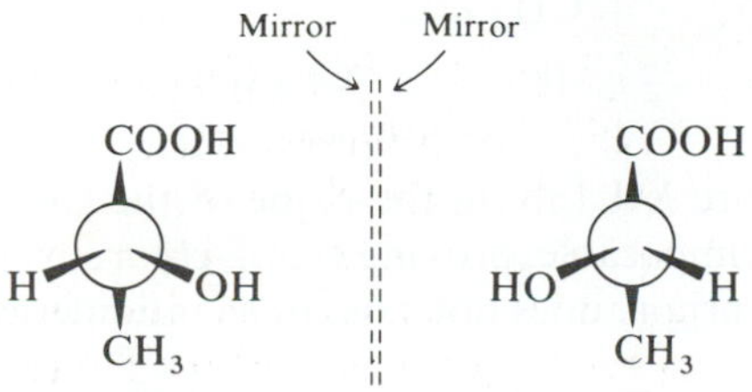

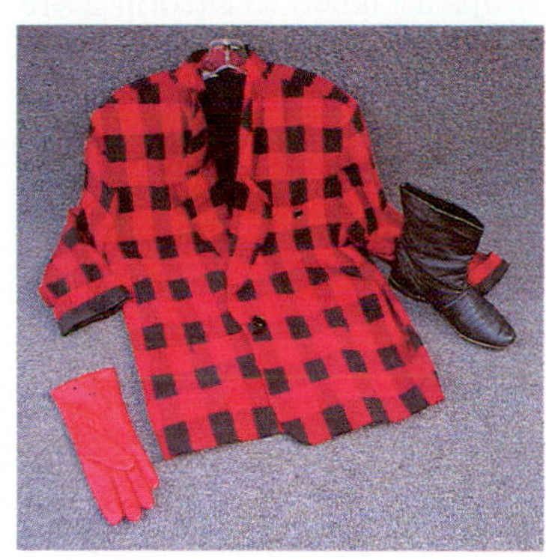

◄ **FIGURE 28.7**
(a) Chiral and (b) achiral objects.

Mirror
Mirror
COOH
H
OH
CH3
COOH
HO
H
CH3

(−)-Lactic acid (+)-Lactic acid

◄ **FIGURE 28.8**
Mirror-image isomers of lactic acid. Each isomer is the mirror reflection of the other. (−)-Lactic acid rotates plane-polarized light to the left, and (+)-lactic acid rotates plane-polarized light to the right.

enantiomer

mirror images of each other are called **enantiomers**. The word *enantiomer* comes from the Greek word *enantios*, which means opposite.

Many, but not all, molecules that contain an asymmetric carbon are chiral. Most of the molecules we shall study that have an asymmetric carbon atom are chiral. To decide whether a molecule is chiral and has an enantiomer, make models of the molecule and of its mirror image and see if they are superimposable. This is the ultimate test, but instead of making models every time, first examine the formula to see if it has an asymmetric carbon atom. If an asymmetric carbon atom is found, draw a cross and attach the four groups on the asymmetric carbon to the four ends of the cross. The asymmetric carbon is understood to be located where the lines cross. Remember that an asymmetric carbon atom has four different groups attached to it. Let us test the compounds 2-butanol and 2-chloropropane:

$$\underset{\text{2-Butanol}}{CH_3CH_2\underset{\displaystyle OH}{\underset{|}{C}}HCH_3} \qquad \underset{\text{2-Chloropropane}}{CH_3\underset{\displaystyle Cl}{\underset{|}{C}}HCH_3}$$

In 2-butanol, carbon 2 is asymmetric. The four groups attached to carbon 2 are H, OH, CH_3, and CH_2CH_3. Draw the structure and its mirror image:

$$\underset{\text{VIII}}{\begin{array}{ccc} & CH_3 & \\ H & + & OH \\ & CH_2 & \\ & | & \\ & CH_3 & \end{array}} \quad \Big| \text{ Mirror} \quad \underset{\text{IX}}{\begin{array}{ccc} & CH_3 & \\ HO & + & H \\ & CH_2 & \\ & | & \\ & CH_3 & \end{array}} \qquad \underset{\text{X (IX turned 180°)}}{\begin{array}{ccc} & CH_3 & \\ & | & \\ & CH_2 & \\ H & + & OH \\ & CH_3 & \end{array}}$$

Enantiomers

Turning structure IX 180° in the plane of the paper allows H and OH to coincide with their position in VIII, but CH_3 and CH_2CH_3 do not coincide. Therefore, we conclude that the mirror-image structures VIII and IX are enantiomers since they are not superimposable.

In 2-chloropropane, the four groups attached to carbon 2 are H, Cl, CH_3, and CH_3. Note that two groups are the same. Draw the structure and its mirror image:

$$\underset{\text{XI}}{\begin{array}{ccc} & CH_3 & \\ H & + & Cl \\ & CH_3 & \end{array}} \quad \Big| \text{ Mirror} \quad \underset{\text{XII}}{\begin{array}{ccc} & CH_3 & \\ Cl & + & H \\ & CH_3 & \end{array}} \qquad \underset{\text{XIII (XII turned 180°)}}{\begin{array}{ccc} & CH_3 & \\ H & + & Cl \\ & CH_3 & \end{array}}$$

When we turn structure XII 180° in the plane of the paper, the two structures XI and XIII are superimposable, proving that 2-chloropropane, which does not have an asymmetric carbon, does not exist in enantiomeric forms.

EXAMPLE 28.2

Draw mirror-image isomers for any of the following compounds that can exist as enantiomers:

(a) $CH_3CH_2CH_2OH$ (c) $CH_3CH_2CHClCH_2CH_2CH_3$

(b) $CH_3CH_2CHClCH_2CH_3$

SOLUTION

First check each formula for asymmetric carbon atoms:

(a) No asymmetric carbon atoms; each carbon has at least two groups that are the same.

(b) No asymmetric carbon atoms; carbon 3 has H, Cl, and two CH_2CH_3 groups.

(c) Carbon 3 is asymmetric; the four groups on carbon 3 are H, Cl, CH_3CH_2, and $CH_3CH_2CH_2$. Draw mirror images:

```
              H            |← Mirror       H
              |            |               |
CH3CH2—C—CH2CH2CH3   |  CH3CH2CH2—C—CH2CH3
              |            |               |
             Cl            |              Cl
```

PRACTICE Draw the mirror-image isomers for the following compounds that can exist as enantiomers:

(a) $NH_2CH_2CHBrCH_2OH$ (c) $NH_2CH_2CH_2CH_2OH$

(b) $NH_2CH_2CHBrCH_2NH_2$

Answers: (a)

```
             H          |← Mirror     H
             |          |             |
NH2CH2—C—CH2OH    |  HOCH2—C—CH2NH2
             |          |             |
            Br          |            Br
```

(b), (c) have no asymmetric carbons

The relationship between enantiomers is such that if we change the positions of any two groups on a compound containing only one asymmetric carbon atom, we obtain the structure of its enantiomer. If we make a second change, the structure of the original isomer is obtained again. In both cases shown below, (+)-lactic acid is formed by interchanging the positions of two groups on (−)-lactic acid:

```
     CH3                               COOH                            COOH
     |                                 |                               |
H————+————OH  ← Change position of  H——+——OH  Change position of →  HO——+——H
     |          CH3 and COOH           |         H and OH              |
     COOH                              CH3                             CH3
(+)-Lactic acid                  (−)-Lactic acid                (+)-Lactic acid
```

To compare two projection formulas to test whether they are the same structure or are enantiomers, (1) turn one structure 180° in the plane of the paper and compare to see if they are superimposable, or (2) make successive group interchanges until the formulas are identical. If an odd number of interchanges are made, the two original formulas represent enantiomers; if an even number of interchanges are made, the two formulas represent the same compound. The following two examples illustrate this method. Are structures XIV and XV the same compound?

CH_3
Br—+—H
Cl
XIV

H
Br—+—Cl
CH_3
XV

H
Br—+—Cl
CH_3
XV

Interchange CH_3 and Cl →

H
Br—+—CH_3
Cl

Interchange CH_3 and H →

CH_3
Br—+—H
Cl
XIV

Two interchanges were needed to make structure XV identical to structure XIV. Therefore, structures XIV and XV represent the same compound.

Do structures XVI and XVII represent the same compound?

CH_3
HO—+—H
CH_2CH_3
XVI

OH
CH_3CH_2—+—CH_3
H
XVII

OH
CH_3CH_2—+—CH_3
H
XVII

Interchange CH_3CH_2 and OH →

CH_2CH_3
HO—+—CH_3
H

Interchange CH_3 and H →

CH_2CH_3
HO—+—H
CH_3

Interchange CH_3 and CH_3CH_2 →

CH_3
HO—+—H
CH_2CH_3
XVI

Three interchanges were needed to make structure XVII identical to structure XVI. Therefore, structures XVI and XVII do not represent the same compound; they are enantiomers.

Enantiomers ordinarily have the same chemical properties and the same physical properties other than optical rotation (see Table 28.1). They rotate plane-polarized light the same number of degrees but in opposite directions. Enantiomers differ in their physiological properties. Enzymes act on only one of a pair of enantiomers, since enzymes are stereoselective. For example, enzyme-catalyzed reduction of pyruvic acid in muscle tissue yields only (+)-lactic acid, but reduction of pyruvic acid catalyzed with H_2/Pt yields both (+)- and (−)-lactic acids (see Section 28.6).

A summary of the key factors of enantiomers and optical isomerism follows:

1. A carbon atom that has four different groups bonded to it is called an asymmetric or a chiral carbon atom.
2. A compound with one asymmetric carbon atom can exist in two isomeric forms called enantiomers.
3. Enantiomers are nonsuperimposable mirror-image isomers.
4. Enantiomers are optically active; that is, they are able to rotate plane-polarized light.
5. One isomer of an enantiomeric pair rotates polarized light to the left (counterclockwise). The other isomer rotates polarized light to the right (clockwise). The degree of rotation is the same but in opposite directions.
6. Rotation of polarized light to the right is indicated by (+), placed in front of the name of the compound. Rotation to the left is indicated by (−), for example, (+)-lactic acid and (−)-lactic acid.

28.6 Racemic Mixtures

racemic mixture

A mixture containing equal amounts of a pair of enantiomers is known as a **racemic mixture**. Such a mixture is optically inactive and shows no rotation of polarized light when tested in a polarimeter. Each of the enantiomers rotates the plane of polarized light by the same amount but in opposite directions. Thus the rotation by each isomer is canceled. The (±) symbol is often used to designate racemic mixtures. For example, a racemic mixture of lactic acid is written as (±)-lactic acid because this mixture contains equal molar amounts of (+)-lactic acid and (−)-lactic acid.

Racemic mixtures are usually obtained in laboratory syntheses of compounds in which an asymmetric carbon atom is formed. Thus, catalytic reduction of pyruvic acid (an achiral compound) to lactic acid produces a racemic mixture containing equal amounts of (+)- and (−)-lactic acid:

$$\underset{\text{Pyruvic acid}}{CH_3\underset{\underset{O}{\|}}{C}COOH} + H_2 \xrightarrow{Ni} \underset{(\pm)\text{-Lactic acid}}{CH_3\underset{\underset{OH}{|}}{C}HCOOH}$$

As a general rule, in the biological synthesis of potentially optically active compounds, only one of the isomers is produced. For example, (+)-lactic acid is produced by reactions occurring in muscle tissue, and (−)-lactic acid is produced by lactic acid bacteria in the souring of milk. These stereospecific reactions occur because biochemical syntheses are enzyme catalyzed. The preferential production of one isomer over another is apparently due to the configuration (shape) of the specific enzyme involved. Returning to the hand analogy, if the "right-handed" enantiomer is produced, then the enzyme responsible for the product can be likened to a right-handed glove.

The mirror-image isomers (enantiomers) of a racemic mixture are alike in all ordinary physical properties except in their action on polarized light. It is possible to separate or resolve racemic mixtures into their optically active components. In fact, Pasteur's original work with sodium ammonium tartrate involved such a separation. But a general consideration of the methods involved in such separations is beyond the scope of our present discussion.

Enantiomers usually differ in their biochemical properties. In fact, most living cells are able to use only one isomer of a mirror-image pair. For example, (+)-glucose ("blood sugar") can be used for metabolic energy, whereas (−)-glucose cannot. Enantiomers are truly different molecules and are treated as such by most organisms.

28.7 Diastereomers and Meso Compounds

The enantiomers discussed in the preceding sections are stereoisomers. That is, they differ only in the spatial arrangement of the atoms and groups within the molecule. The number of stereoisomers increases as the number of asymmetric carbon atoms increases. The maximum number of stereoisomers for a given compound is obtained by the formula 2^n, where n is the number of asymmetric carbon atoms in the compound.

> **2^n = Maximum number of stereoisomers for a given chiral compound**
>
> **n = Number of asymmetric carbon atoms in a molecule**

As we have seen, there are two ($2^1 = 2$) stereoisomers of lactic acid. But for a substance with two nonidentical asymmetric carbon atoms, such as 2-bromo-3-chlorobutane ($CH_3CHBrCHClCH_3$), four stereoisomers are possible ($2^2 = 4$). These four possible stereoisomers are written as projection formulas in this way (carbons 2 and 3 are asymmetric):

```
     CH3           CH3           CH3           CH3
      |             |             |             |
  H --+-- Br   Br --+-- H    H --+-- Br   Br --+-- H
      |             |             |             |
  H --+-- Cl   Cl --+-- H   Cl --+-- H     H --+-- Cl
      |             |             |             |
     CH3           CH3           CH3           CH3
    XVIII          XIX            XX           XXI

         Enantiomers                 Enantiomers
```

Remember that, for comparison, projection formulas may be turned 180° in the plane of the paper, but they cannot be lifted (flipped) out of the plane. Formulas XVIII and XIX, and formulas XX and XXI, represent two pairs of nonsuperimposable mirror-image isomers and are, therefore, two pairs of enantiomers. All four compounds are optically active. But the properties of XVIII and XIX differ from the properties of XX and XXI because they are not mirror-image isomers of each other. Stereoisomers that are not enantiomers (not mirror images of each other) are called **diastereomers**. There are four different pairs of diastereomers of 2-bromo-3-chlorobutane: They are XVIII and XX, XVIII and XXI, XIX and XX, and XIX and XXI.

diastereomer

The 2^n formula indicates that four stereoisomers of tartaric acid are possible. The projection formulas of these four possible stereoisomers are written in this way (carbons 2 and 3 are asymmetric):

```
      COOH              COOH
  HO---+---H        H---+---OH
   H---+---OH      HO---+---H
      COOH              COOH
      XXII              XXIII

      COOH              COOH
   H---+---OH      HO---+---H
   H---+---OH      HO---+---H
      COOH              COOH
      XXIV              XXV
```

Formulas XXII and XXIII represent nonsuperimposable mirror-image isomers and are, therefore, enantiomers. Formulas XXIV and XXV are also mirror images. But by turning XXV 180°, we see that it is exactly superimposable on XXIV. Therefore XXIV and XXV represent the same compound, and only *three* stereoisomers of tartaric acid actually exist. Compound XXIV is achiral and does not rotate polarized light. A plane of symmetry can be passed between carbons 2 and 3 so that the top and bottom halves of the molecule are mirror images:

```
      COOH
       |
   H—C—OH
-------|-------  <— Plane of symmetry
   H—C—OH
       |
      COOH
```

Thus the molecule is internally compensated. The rotation of polarized light in one direction by half of the molecule is exactly compensated by an opposite rotation by the other half. Stereoisomers that contain asymmetric carbon atoms and are superimposable on their own mirror images are called **meso compounds**, or **meso structures**. All meso compounds are optically inactive.

meso compound

meso structure

The term *meso* comes from the Greek word *mesos*, meaning middle. It was first used by Pasteur to name a kind of tartaric acid that was optically inactive and could not be separated into different forms by any means. Pasteur called it *meso*-tartaric acid, because it seemed intermediate between the (+)- and (−)-tartaric acid. The three stereoisomers of tartaric acid are represented and designated in this fashion:

```
       COOH               COOH              COOH
        |                  |                 |
   HO—C—H            H—C—OH            H—C—OH
        |                  |                 |
    H—C—OH          HO—C—H             H—C—OH
        |                  |                 |
       COOH               COOH              COOH
 (−)-Tartaric acid  (+)-Tartaric acid  meso-Tartaric acid
```

Table 28.1 Properties of Tartaric Acid [HOOCCH(OH)CH(OH)COOH]

Name	Specific gravity	Melting point (°C)	Solubility (g/100 g H_2O)	Specific rotation [α]
(+)-Tartaric acid	1.760	170	$147^{20°C}$	+12°
(−)-Tartaric acid	1.760	170	$147^{20°C}$	−12°
(±)-Tartaric acid (racemic mixture)	1.687	206	$20.6^{20°C}$	0°
meso-Tartaric acid	1.666	140	$125^{15°C}$	0°

The physical properties of tartaric acid stereoisomers are given in Table 28.1. Note that the properties of (+)-tartaric acid and (−)-tartaric acid are identical except for opposite rotation of polarized light. However, *meso*-tartaric acid has properties that are entirely different from those of the other isomers. But most surprising is the fact that the racemic mixture, though composed of equal parts of the (+) and (−) enantiomers, differs from them in specific gravity, melting point, and solubility. Why, for example, is the melting point of the racemic mixture higher than that of any of the other forms? The melting point of any substance is largely dependent on the attractive forces holding the ions or molecules together. The melting point of the racemic mixture is higher than that of either enantiomer. Therefore, we can conclude that the attraction between molecules of the (+) and (−) enantiomers in the racemic mixture is greater than the attraction between molecules of the (+) and (+) or the (−) and (−) enantiomers.

Example 28.3

How many stereoisomers can exist for the following compounds? Write their structures and label any pairs of enantiomers and meso compounds. Point out any diastereomers.

(a) $CH_3CHBrCHBrCH_2CH_3$ (b) $CH_2BrCHClCHClCH_2Br$

SOLUTION

(a) Carbons 2 and 3 are asymmetric, so there can be a maximum of four stereoisomers ($2^2 = 4$). Write structures around the asymmetric carbons:

```
    CH3          CH3            CH3          CH3
     |            |              |            |
H ---+--- Br  Br ---+--- H   H ---+--- Br  Br ---+--- H
     |            |              |            |
H ---+--- Br  Br ---+--- H   Br ---+--- H   H ---+--- Br
     |            |              |            |
    CH2          CH2            CH2          CH2
     |            |              |            |
    CH3          CH3            CH3          CH3
     I            II             III          IV
       Enantiomers                 Enantiomers
```

There are four stereoisomers: two pairs of enantiomers (I and II, and III and IV) and no meso compounds. Structures I and III, I and IV, II and III, and II and IV are diastereomers.

(b) Carbons 2 and 3 are asymmetric, so there can be a maximum of four stereoisomers. Write structures around the asymmetric carbons:

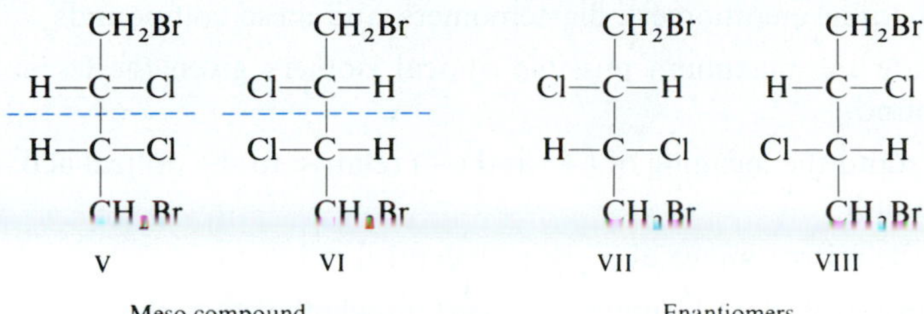

Meso compound Enantiomers

There are three stereoisomers: one pair of enantiomers (VII and VIII) and one meso compound V. Structures V and VI represent the meso compound because there is a plane of symmetry between carbons 2 and 3, and turning VI 180° makes it superimposable on V. Structures V, VII, and VIII are diastereomers.

PRACTICE Write all stereoisomer structures and label any pairs of enantiomers and meso compounds for the following compound. Also, point out any diastereomers.

$HOOCCHClCH_2CH_2CHClCOOH$

Answer:

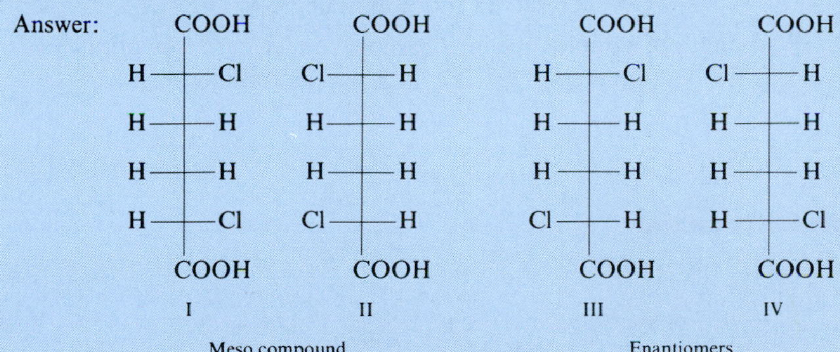

Meso compound Enantiomers

Structures I and III, I and IV, II and III, and II and IV are diasteromers.

Concepts in Review

1. Identify all asymmetric (chiral) carbon atoms in a given formula.
2. Explain the use of a polarimeter.
3. Explain how polarized light is obtained.
4. Calculate the specific rotation of a compound.
5. Explain the phenomenon of optical isomerism.

6. Determine whether a compound is chiral.
7. Draw projection formulas for all possible stereoisomers of a given compound. Label enantiomers, diastereomers, and meso compounds.
8. Calculate the maximum possible optical isomers given the formula of a compound.
9. Understand the meaning of (+) and (−) relative to the optical activity of a compound.
10. Draw the mirror image of a given structure.
11. Compare projection formulas to ascertain whether they represent identical compounds or enantiomers.
12. Compare the physical properties of enantiomers, diastereomers, and racemic mixtures.
13. Explain why meso compounds are optically inactive.
14. Determine whether optical isomers are formed in simple chemical reactions.

Equations in Review

Specific Rotation

$$[\alpha] = \frac{\text{Observed rotation in degrees}}{\left(\begin{array}{c}\text{Length of sample tube in}\\ \text{decimeters}\end{array}\right)\left(\begin{array}{c}\text{Sample concentration in grams}\\ \text{per milliliter}\end{array}\right)}$$

Exercises

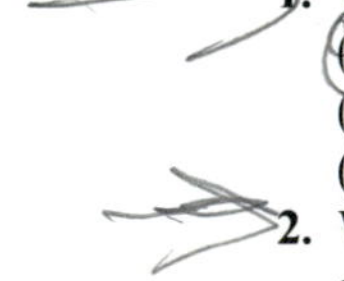

1. Which of these objects are chiral?
   (a) A wood screw
   (b) A pair of pliers
   (c) The letter O
   (d) The letter G
   (e) A coiled spring
   (f) Your ear
2. What is an asymmetric carbon atom? Draw structural formulas of three different compounds that contain one asymmetric carbon atom, and mark the asymmetric carbon in each with an asterisk.
3. How can you tell when the axes of two Polaroid filters are parallel? When one filter has been rotated by 90°?
4. How many asymmetric carbon atoms are present in each of the following?

(a)
```
      H   Cl
      |   |
  Cl—C—C—Br
      |   |
      H   H
```

(b)
```
     Br  Cl
     |   |
  H—C—C—H
     |   |
     H   H
```

(c) $CH_3CH_2CH_2CHClCH_3$

(d)
```
  H—C=O
     |
  H—C—OH
     |
  H—C—OH
     |
  H—C—OH
     |
     H
```

(e)
```
  CH3CH2CHCH2CH3
         |
         OH
```

(f)
```
        H
        |
   H—C—OH
        |
       C=O
        |
  HO—C—H
        |
   H—C—OH
        |
   H—C—OH
        |
   H—C—OH
        |
        H
```

(g)
```
        H   H   O
        |   |   ‖
   HO—C—C—C—OH
        |   |
        H  NH2
```

5. Write the formulas and decide which of the following compounds will show optical activity:
   (a) 1-Chloropentane
   (b) 2-Chloropentane
   (c) 3-Chloropentane
   (d) 1-Chloro-2-methylpentane
   (e) 2-Chloro-2-methylpentane
   (f) 3-Chloro-2-methylpentane
   (g) 4-Chloro-2-methylpentane
   (h) 3-Chloro-3-methylpentane
6. Suppose a carbon atom is located at the center of a square with four different groups attached to the corners in a planar arrangement. Would the compound rotate polarized light? Explain.
7. What is a necessary and sufficient condition for a compound to show enantiomerism?
8. Glucose ($C_6H_{12}O_6$) has four asymmetric carbon atoms. How many stereoisomers of glucose are theoretically possible?
9. Do structures A and B represent enantiomers or the same compound? Justify your answer.

```
           Cl                 Br
           |                  |
(A) Br—C—H      (B) H—C—Cl
           |                  |
           F                  F
```

10. Which of these projection formulas represent (−)-lactic acid and which represent (+)-lactic acid?

```
(a)      CH3            (c)      COOH
     HO——+——H               H——+——CH3
         COOH                     OH

(b)      OH             (d)      CH3
      H——+——COOH            H——+——OH
         CH3                     COOH

(e)      COOH           (f)      H
    CH3——+——H              CH3——+——OH
         OH                      COOH
```

11. Draw projection formulas for all the possible stereoisomers of the following compounds. Label pairs of enantiomers and meso compounds.
    (a) 1,2-Dibromopropane
    (b) 2,3-Dichlorobutane
    (c) 2-Butanol
    (d) 2,4-Dibromopentane
    (e) 3-Chlorohexane
12. Draw projection formulas for all the stereoisomers of 1,2,3-trihydroxybutane. Point out enantiomers, meso compounds, and diastereomers, where present.

*13. Write structures for the four stereoisomers of 3-pentene-2-ol.

*14. Some substituted cycloalkanes are chiral. Draw the structures and enantiomers for any of the following that are chiral:

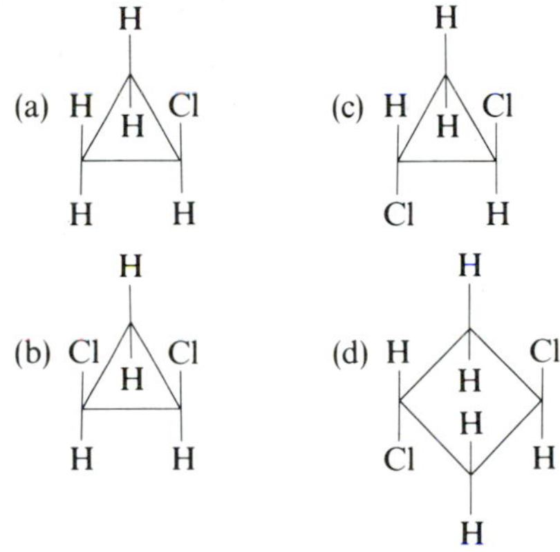

15. The physical properties for (+)-2-methyl-1-butanol are specific rotation, +5.76°; bp, 129°C; density, 0.819 g/mL. What are these same properties for (−)-2-methyl-1-butanol?

*16. Draw projection formulas for all the stereoisomers of 2,3,4-tribromopentane and point out enantiomers and meso compounds, where present.

17. (a) Draw the nine structural isomers of $C_4H_8Cl_2$.
    (b) Identify which structures represent chiral molecules, and draw all possible pairs of enantiomers and meso compounds, if any.
18. (+)-2-Chlorobutane is further chlorinated to give dichlorobutanes ($C_4H_8Cl_2$). Write structures for all the possible isomers formed and indicate which of these isomers will be optically active. [Remember that (+)-2-chlorobutane is optically active.]

19. In the chlorination of butane, 1-chlorobutane and 2-chlorobutane are obtained as products. After separation by distillation, neither product rotates the plane of polarized light. Explain these results.

20. (a) Explain why it is not possible to separate enantiomers by ordinary chemical and physical means.
(b) Explain why diastereomers can usually be separated by ordinary physical and chemical means.

21. The observed rotation of polarized light was 12.5° for a solution of compound X at a concentration of 10.0 g/100. mL. The measurement was made using a polarimeter sample tube 20.0 cm in length. Calculate the specific rotation, $[\alpha]$, for compound X.

22. Which, if any, of the following are meso compounds?

```
        COOH                 CH3
(a)  H ──┼── OH      (c)  H ──┼── Cl
    HO ──┼── COOH        Br ──┼── CH3
         H                    H

        CH3                  CH3
(b)  H ──┼── Cl      (d)  H ──┼── Cl
   CH3 ──┼── Cl           H ──┼── Br
         H                H ──┼── Cl
                              CH3
```

***23.** Substances of the type shown below are known to rotate the plane of polarized light. What inference can be drawn regarding the spatial character of the nitrogen bonds?

```
⌈      R      ⌉+
|      |      |
| R'''—N—R'   |  [X]⁻
|      |      |
⌊      R''    ⌋
```

***24.** (a) A chiral substance was identified as a primary alcohol of formula $C_5H_{12}O$. What is its structure?
(b) The compound $C_6H_{14}O_3$ has three primary alcohol groups and is chiral. What is its structure?

***25.** What is the structure of a substance of formula $C_3H_8O_2$ that is (a) chiral and contains two —OH groups; (b) chiral and contains one —OH group; (c) achiral and contains two —OH groups? (Only one —OH group can be bonded to a carbon atom.)

26. Which of these statements are correct? Rewrite each incorrect statement to make it correct.
(a) The polarizer and the analyzer of a polarimeter are made of the same material.
(b) The specific rotation of a compound is dependent on the number of molecules in the path of the plane-polarized light.
(c) Very few natural products are optically active.
(d) Cis-trans isomers are not considered to be stereoisomers.
(e) J. A. Le Bel received the first Nobel prize in chemistry in 1901.
(f) A compound that rotates plane-polarized light +25° would be at the same position on the polarimeter as one that rotates the light −335°.
(g) Molecules that contain only one asymmetric carbon atom are chiral, but not all chiral molecules contain an asymmetric carbon atom.
(h) The compound $CH_3CHBrCHBrCH_2OH$ has eight optical isomers.
(i) The compounds shown in these projection formulas are enantiomers:

```
      Cl               H
Br ──┼── H      Br ──┼── Cl
      CH3              CH3
```

(j) Diastereomers have identical melting points.
(k) A molecule that contains two asymmetric carbon atoms may not be chiral.
(l) A racemic mixture contains equal amounts of dextrorotatory and levorotatory molecules of a compound.

29

Carbohydrates

◀ Chapter Opening Photo: These varieties of pasta are examples of carbohydrates.

What is the most abundant organic chemical in the world? The answer is not petroleum products, plastics, or drugs. Rather, it is cellulose. An amazing ten billion tons of this carbohydrate are formed daily in the biosphere. Aggregates of this substance allow the California redwoods to stretch hundreds of feet toward the sky and make a Brazil nut a "hard nut to crack." Products as diverse as the paper in this book and cotton in many articles of clothing also derive from cellulose. Perhaps it is not so surprising that this carbohydrate is the most widespread organic chemical in the world.

Carbohydrates are molecules of exceptional utility. These molecules provide a basic diet for many of us (starch and sugar), a roof over our heads, and clothes for our bodies (cellulose). But they also thicken our ice cream, stick our postage stamps to our letters, and provide biodegradable plastic trash sacks. Starting from relatively simple components (using carbon, hydrogen, and oxygen), nature has created one of the premier classes of biochemicals.

29.1 The Role of Carbohydrates

Carbohydrates are among the most widespread and important biochemicals. Most of the matter in plants, except water, consists of these substances. Carbohydrates are one of the three principal classes of energy yielding nutrients; the other two are fats and proteins. Because of their widespread distribution and their role in many vital metabolic processes such as photosynthesis, carbohydrates have been subjected to a great deal of scientific study over the last 150 years.

The name *carbohydrates* was given to this class of compounds many years ago by French scientists who called them *hydrates de carbone*, because their empirical formulas approximated $(C \cdot H_2O)_n$. It was found later that not all substances classified as carbohydrates conform to this formula (for example, rhamnose, $C_6H_{12}O_5$, and deoxyribose, $C_5H_{10}O_4$). It seems clear that carbohydrates are not simply hydrated carbon; they are complex substances containing from three to many thousands of carbon atoms. The general definition is:

carbohydrates

Carbohydrates are polyhydroxy aldehydes or polyhydroxy ketones or substances that yield these compounds when hydrolyzed.

The simplest carbohydrates are glyceraldehyde and dihydroxyacetone:

$$\begin{array}{c} \text{H—C=O} \\ | \\ \text{H—C—OH} \\ | \\ \text{H—C—OH} \\ | \\ \text{H} \end{array} \qquad \begin{array}{c} \text{H} \\ | \\ \text{H—C—OH} \\ | \\ \text{C=O} \\ | \\ \text{H—C—OH} \\ | \\ \text{H} \end{array}$$

Glyceraldehyde Dihydroxyacetone

These substances are "polyhydroxy" because each molecule has more than one hydroxyl group. Glyceraldehyde contains a carbonyl carbon in a terminal position and, therefore, is an aldehyde. The internal carbonyl of dihydroxyacetone identifies it as a ketone. Much of the chemistry and biochemistry of carbohydrates can be understood from a basic knowledge of the chemistry of the hydroxyl and carbonyl functional groups (see Chapters 23 and 24 for review).

Carbohydrates provide an essential base for the world's food chain. Plants use light energy to convert carbon dioxide and water into carbohydrates. Herbivores eat these plants and are, in turn, eaten by carnivores. The energy originally present in the plant carbohydrate is transferred up the food chain. As we begin to study carbohydrates, it is worthwhile to consider this question: What are the essential characteristics that allow carbohydrates to act as foodstuffs and major sources of biological energy? In answer to this question, three relevant biochemical principles will be set forth.

First, a foodstuff must contain carbon. The preceding formulas show that carbohydrates provide much carbon. The average carbohydrate is about 40% carbon by mass.

Second, the carbon atoms in an energy providing molecule must be in a more reduced state than those in carbon dioxide. Reduced carbon atoms can react with atmospheric oxygen to produce carbon dioxide and energy (witness the heat of a wood fire or the energy of a gasoline engine). A useful estimate of a carbon's reduced state is the oxidation number (see Table 29.1). The lower the oxidation number, the more reduced the carbon is and the more energy it can supply. Carbohydrates are energy yielding nutrients partly because they contain carbons in an intermediate state of reduction, with oxidation numbers ranging from -1 to $+1$. As these carbons are oxidized to carbon dioxide during metabolism, much energy is released.

Third, an energy-providing foodstuff not only must contain reduced carbons, but it also must have a chemical reactivity and a physical structure appropriate for use in cells. Carbohydrate metabolism is aided by the presence of reactive hydroxyl and carbonyl functional groups.

29.2 Classification

A carbohydrate is classified as a monosaccharide, a disaccharide, an oligosaccharide, or a polysaccharide, depending on the number of monosaccharide units linked together to form a molecule. A **monosaccharide** is a carbohydrate that

monosaccharide

TABLE 29.1 Relative Potential Energy of Carbon in Different Oxidation States

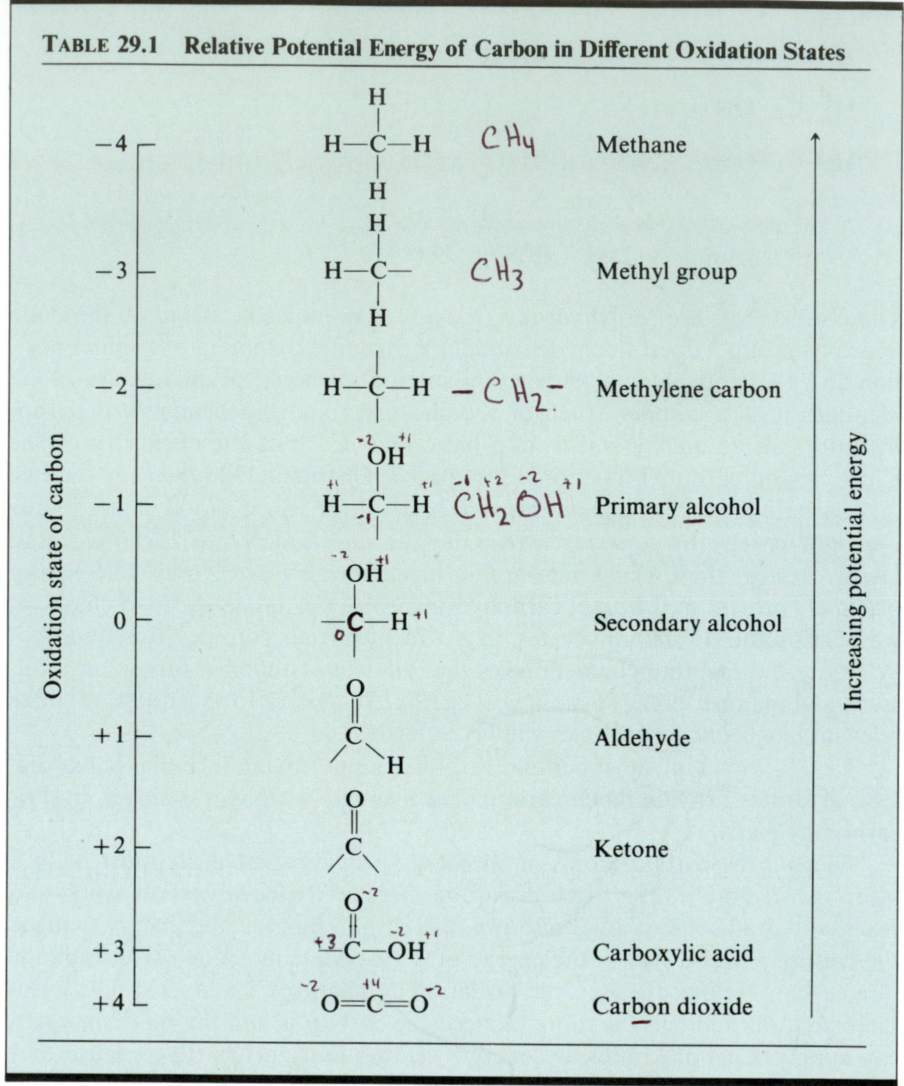

Oxidation state of carbon		
−4	H—C(H)(H)—H	Methane
−3	H—C(H)(H)—	Methyl group
−2	H—C—H	Methylene carbon
−1	H—C(OH)—H	Primary alcohol
0	—C(OH)—H	Secondary alcohol
+1	C(=O)—H	Aldehyde
+2	C(=O)	Ketone
+3	C(=O)—OH	Carboxylic acid
+4	O=C=O	Carbon dioxide

Increasing potential energy ↑ (from carbon dioxide to methane)

cannot be hydrolyzed to simpler carbohydrate units. The monosaccharide is the basic carbohydrate unit of cellular metabolism. A **disaccharide** yields two monosaccharides—either alike or different—when hydrolyzed:

disaccharide

$$\text{Disaccharide} + \text{Water} \xrightarrow{H^+ \text{ or enzyme}} 2 \text{ Monosaccharides}$$

Disaccharides are often used by plants or animals to transport monosaccharides from one cell to another. The monosaccharides and disaccharides generally have names ending in *ose*—for example, glucose, sucrose, and lactose. These water-soluble carbohydrates, which have a characteristically sweet taste, are also called *sugars*.

oligosaccharide

An **oligosaccharide** has two to six monosaccharide units linked together. *Oligo* comes from the Greek word *oligos*, which means small or few. Free oligosaccharides containing more than two monosaccharide units are rarely found in nature.

A **polysaccharide** is a macromolecular substance that can be hydrolyzed to yield many monosaccharide units:

polysaccharide

$$\text{Polysaccharide} + \text{Water} \xrightarrow{H^+ \text{ or enzyme}} \text{Many monosaccharide units}$$

Polysaccharides are important as structural supports (particularly in plants) and also serve as a storage depot for monosaccharides (which cells use for energy).

Carbohydrates also can be classified in other ways. A monosaccharide might be described with respect to several of these categories:

1. As a *triose*, *tetrose*, *pentose*, *hexose*, or *heptose*. Theoretically, a monosaccharide can have any number of carbons greater than three, but only monosaccharides of three to seven carbons are commonly found in the biosphere.

Trioses	$C_3H_6O_3$
Tetroses	$C_4H_8O_4$
Pentoses	$C_5H_{10}O_5$
Hexoses	$C_6H_{12}O_6$
Heptoses	$C_7H_{14}O_7$

2. As an *aldose* or *ketose*, depending on whether an aldehyde group (—CHO) or keto group (>C=O) is present.
3. As a D or L isomer, depending on the spatial orientation of the —H and —OH groups attached to the carbon atom that is adjacent to the terminal primary alcohol group. When the —OH is written to the right of this carbon in the *projection formula*, the D isomer is represented. When the —OH is written to the left, the L isomer is represented. The reference compounds for this classification are the trioses D-glyceraldehyde and L-glyceraldehyde, whose formulas follow. Also shown are two aldohexoses (D- and L-glucose) and a ketohexose (D-fructose).

D-Glyceraldehyde: H—C=O / H—C—OH (D-configuration) / CH_2OH (Terminal 1° ROH)

L-Glyceraldehyde: H—C=O / HO—C—H (L-configuration) / CH_2OH (Terminal 1° ROH)

L-Glucose: H—C=O / HO—C—H / H—C—OH / HO—C—H / HO—C—H (L-configuration) / CH_2OH (Terminal 1° ROH)

D-Glucose: H—C=O / H—C—OH / HO—C—H / H—C—OH / H—C—OH (D-configuration) / CH_2OH (Terminal 1° ROH)

D-Fructose: CH_2OH / C=O / HO—C—H / H—C—OH / H—C—OH (D-configuration) / CH_2OH (Terminal 1° ROH)

The letters D and L do not in any way refer to the direction of optical rotation of a carbohydrate. The D and L forms of any specific compound are *enantiomers* (for example, D- and L-glucose).

4. As a (+) or (−) isomer, depending on whether the monosaccharide rotates the plane of polarized light to the right (+) or to the left (−). (See Section 28.5.)
5. As a *furanose* or a *pyranose*, depending on whether the cyclic structure of the carbohydrate is related to that of the five-membered or six-membered heterocyclic ring compound furan or pyran (a heterocyclic ring contains more than one kind of atom in the ring):

Furan, C_4H_4O
(five-membered ring containing oxygen in the ring)

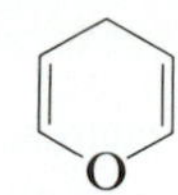

Pyran, C_5H_6O
(six-membered ring containing oxygen in the ring)

6. As having an alpha (α) or beta (β) configuration, based on the orientation of the —H and —OH groups about a specific asymmetric carbon in the cyclic form of the monosaccharide (Section 29.5).

EXAMPLE 29.1

Write projection formulas for (a) an L-aldotriose, (b) a D-ketotetrose, and (c) a D-aldopentose. Determine which carbons can provide the most energy for a cell.

SOLUTION

(a) *Triose* indicates a three-carbon carbohydrate; *aldo* indicates that the compound is an aldehyde; L- indicates that the —OH on carbon 2 (adjacent to the terminal CH_2OH) is on the left. The aldehyde group is carbon 1.

```
   H—C=O
     |
  HO—C—H
     |
     CH₂OH
```

An L-aldotriose

The aldehyde carbon has an oxidation number of +1; the secondary alcohol, 0; the primary alcohol, −1. The primary alcohol carbon is most reduced and should provide the most energy.

(b) *Tetrose* indicates a four-carbon carbohydrate; *keto* indicates a ketone group (on carbon 2); D- indicates that the —OH on carbon 3 (adjacent to the terminal CH_2OH) is on the right. Carbons 1 and 4 are primary alcohols.

```
     CH₂OH
     |
     C=O
     |
   H—C—OH
     |
     CH₂OH
```

A D-ketotetrose

The ketone carbon has an oxidation number of +2; the secondary alcohol carbon, 0; each primary alcohol carbon, −1. The primary alcohol carbons are the most reduced and have the most potential energy.

(c) *Pentose* indicates a five-carbon carbohydrate; *aldo* indicates an aldehyde group (on carbon 1); D- indicates that the —OH on carbon 4 (adjacent to the terminal CH_2OH) is on the right. The orientation of the —OH groups on carbons 2 and 3 is not specified here and therefore can be written in either direction for this problem.

```
H—C=O
  |
H—C—OH
  |
H—C—OH
  |
H—C—OH
  |
  CH2OH
```

A D-aldopentose

The aldehyde carbon has an oxidation number of +1; each secondary alcohol carbon, 0; the primary alcohol carbon, −1. The primary alcohol carbon is the most reduced and has the most potential energy.

Practice Identify the most reduced carbons in a ketoheptose.
Answer: The primary alcohol carbons

Practice In a projection formula for a D-aldotriose, is the —OH of the secondary alcohol carbon written on the right or the left side?
Answer: The right side

29.3 Monosaccharides

Although a great many monosaccharides have been synthesized, only a very few appear to be of much biological significance. One pentose monosaccharide (ribose) and its deoxy derivative are essential components of ribonucleic acid (RNA) and of deoxyribonucleic acid (DNA) (see Chapter 33). However, the hexose monosaccharides are the most important carbohydrate sources of cellular energy. Three hexoses—glucose, galactose, and fructose—are of major significance in nutrition. All three have the same molecular formula, $C_6H_{12}O_6$, and thus contain an equal number of reduced carbons. They differ in structure but are biologically interconvertible. Glucose plays a central role in carbohydrate energy utilization. Other carbohydrates are usually converted to glucose before cellular utilization. The structure of glucose is considered in detail in Section 29.4.

Glucose

Glucose is the most important of the monosaccharides. It is an aldohexose and is found in the free state in plant and animal tissue. Glucose is commonly known as *dextrose* or *grape sugar*. It is a component of the disaccharides sucrose, maltose, and lactose, and is also the monomer of the polysaccharides starch, cellulose,

▲
Sucrose is converted into monosaccharides, fructose and glucose by an enzyme found in bees.

and glycogen. Among the common sugars, glucose is of intermediate sweetness (see Section 29.11).

Glucose is the key sugar of the body and is carried by the bloodstream to all body parts. The concentration of glucose in the blood is normally 80–100 mg per 100 mL of blood. Because glucose is the most abundant carbohydrate in the blood, it is also sometimes known as *blood sugar*. Glucose requires no digestion and therefore may be given intravenously to patients who cannot take food by mouth. Glucose is found in the urine of those who have diabetes mellitus (sugar diabetes). The condition in which glucose is excreted in the urine is called glycosuria.

Galactose

Galactose is also an aldohexose and occurs, along with glucose, in lactose and in many oligo- and polysaccharides such as pectin, gums, and mucilages. Galactose is an isomer of glucose, differing only in the spatial arrangement of the —H and —OH groups around carbon 4 (see Section 29.4). Galactose is synthesized in the mammary glands to make the lactose of milk. It is also a constituent of glycolipids and glycoproteins in many cell membranes such as those in nervous tissue. Galactose is less than half as sweet as glucose.

A severe inherited disease, called galactosemia, is the inability of infants to metabolize galactose. The galactose concentration increases markedly in the blood, and also appears in the urine. Galactosemia causes vomiting, diarrhea, enlargement of the liver, and often mental retardation. If not recognized within a few days after birth, it can lead to death. If diagnosis is made early and lactose is excluded from the diet, the symptoms disappear and normal growth may be resumed.

Fructose

Fructose, also known as *levulose*, is a ketohexose that occurs in fruit juices, honey, and (along with glucose) as a constituent of sucrose. Fructose is the major constituent of the polysaccharide inulin, a starchlike substance present in many plants such as dahlia tubers, chicory roots, and Jerusalem artichokes. Fructose is the sweetest of all the sugars, being about twice as sweet as glucose. This accounts for the sweetness of honey. The enzyme invertase, present in bees, splits sucrose into glucose and fructose. Fructose is metabolized directly but is also readily converted to glucose in the liver.

29.4 Structure of Glucose and Other Aldoses

In one of the classic feats of research in organic chemistry, Emil Fischer (1852–1919), working in Germany, established the structural configuration of glucose along with that of many other sugars. He received the Nobel prize in chemistry in 1902. Fischer devised projection formulas that relate the structure of a sugar to one or the other of the two enantiomeric forms of glyceraldehyde. These projection formulas represent three-dimensional stereoisomers (see Chapter 28) in a two-dimensional plane. (Remember that stereoisomers cannot be interconverted without breaking and reforming covalent bonds. Each carbohydrate isomer has a different shape and thus reacts differently in biological systems.)

In Fischer projection formulas, the molecule is represented with the aldehyde (or ketone) group at the top. The —H and —OH groups attached to interior carbons are written to the right or to the left as they would appear when projected toward the observer. The two glyceraldehydes are represented thus:

H—C=O H ◄ ► OH CH_2OH	H—^{1}C=O H—^{2}C—OH $^3CH_2OH$	H—C=O HO ◄ ► H CH_2OH	H—^{1}C=O HO—^{2}C—H $^3CH_2OH$
D-Glyceraldehyde (three-dimensional representation)	D-Glyceraldehyde (projection formula)	L-Glyceraldehyde (three-dimensional representation)	L-Glyceraldehyde (projection formula)

In the three-dimensional molecules represented by these formulas, the number-2 carbon atoms are in the plane of the paper. The —H and —OH groups project forward (toward the observer); the —CHO and —CH_2OH groups project backward (away from the observer). Any two monosaccharides that differ only in the configuration around a single carbon atom are called **epimers**. Thus D- and L-glyceraldehyde are epimers.

epimer

Fischer recognized that there were two enantiomeric forms of glucose. To these forms he assigned the following structures and names:

H—^{1}C=O H—^{2}C—OH HO—^{3}C—H H—^{4}C—OH H—^{5}C—OH $^6CH_2OH$	H—^{1}C=O HO—^{2}C—H H—^{3}C—OH HO—^{4}C—H HO—^{5}C—H $^6CH_2OH$
D-Glucose	L-Glucose

The structure called D-glucose is so named because the —H and —OH on carbon 5 are in the same configuration as the —H and —OH on carbon 2 in D-glyceraldehyde. The configuration of the —H and —OH on carbon 5 in L-glucose corresponds to the —H and —OH on carbon 2 in L-glyceraldehyde.

Fischer recognized that 16 different aldohexoses, 8 with the D configuration and 8 with the L configuration, were possible. This follows our formula 2^n for optical isomers (see Section 28.7). Glucose has four asymmetric carbon atoms and should have 16 stereoisomers (2^4). The configurations of the D-aldose family are shown in Figure 29.1. In this family, new asymmetric carbon atoms are formed as we go from triose to tetrose to pentose to hexose. Each time a new asymmetric carbon is added, a pair of epimers is formed that differ only in the structure around carbon 2. This sequence continues until eight D-aldohexoses are created. A similar series starting with L-glyceraldehyde is known, making a total of 16 aldohexoses.

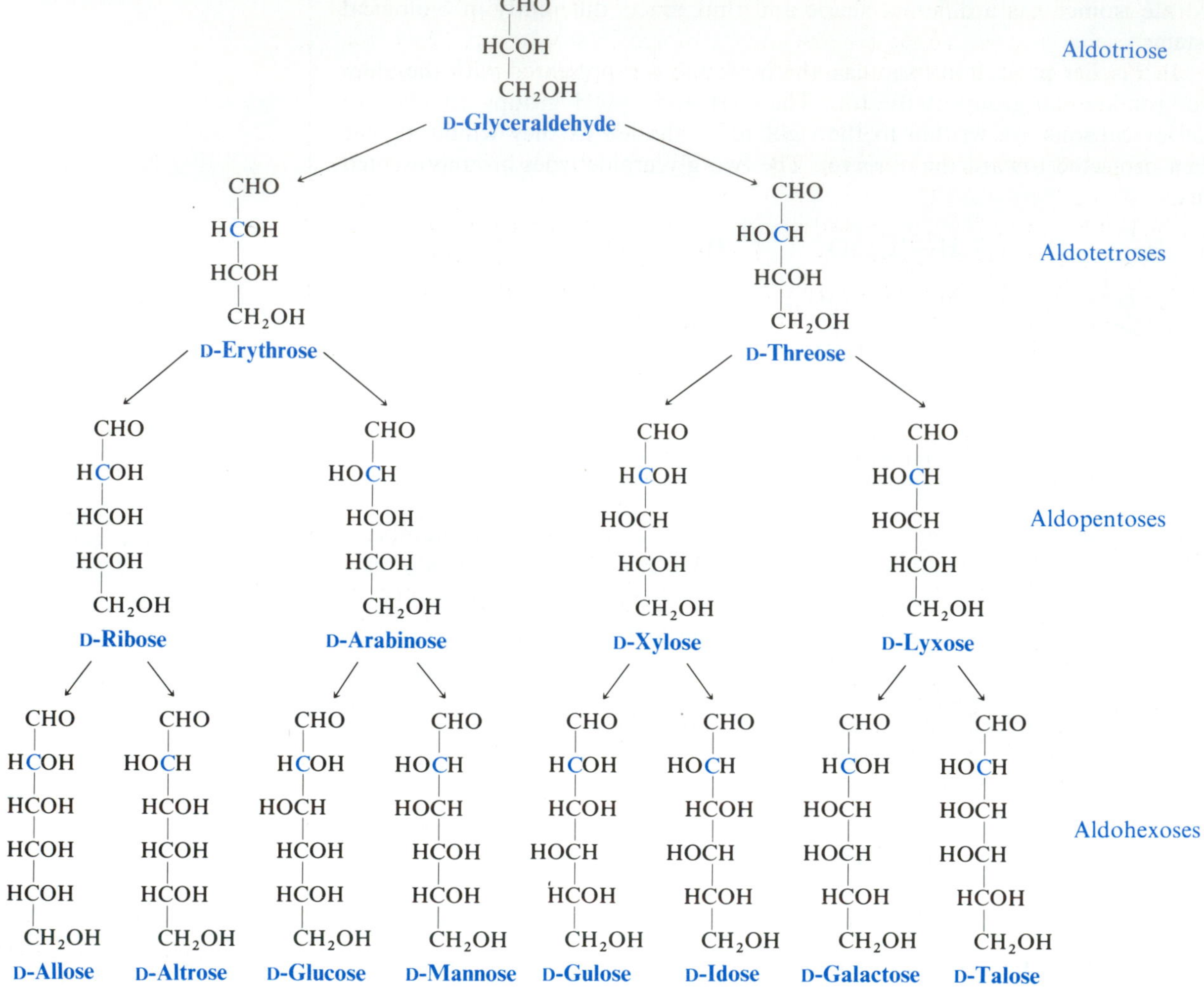

▲
FIGURE 29.1
Configurations of the D family of aldoses. The new asymmetric carbon atom added in going from triose to tetrose to pentose to hexose is marked in color.

FIGURE 29.2 ►
An example of the Kiliani–Fischer synthesis in which two aldotetrose molecules are formed from an aldotriose molecule.

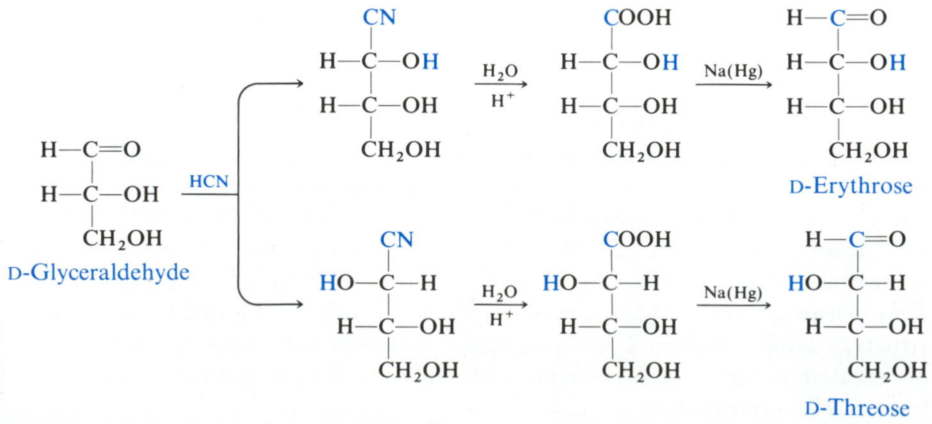

D-Glyceraldehyde
HCN
H_2O, H^+
Na(Hg)
D-Erythrose
D-Threose

All of the 16 aldohexoses have been synthesized, but only D-glucose and D-galactose appear to be of considerable biological importance. Since the metabolism of most living organisms revolves about D-glucose, our discussion will be centered on this substance.

The laboratory conversion of one aldose into another aldose containing one more carbon atom is known as the Kiliani–Fischer synthesis. This synthesis makes use of the aldehyde's ability to bond to an additional group (see Section 24.4). The synthesis involves (1) the addition of HCN to form a cyanohydrin, (2) hydrolysis of the —CN group to —COOH, and (3) reduction with sodium amalgam, Na(Hg), to form the aldehyde. As an example, the formation of two aldotetroses from an aldotriose is shown in Figure 29.2.

29.5 Cyclic Structure of Glucose; Mutarotation

Straight open chain D-glucose is so reactive that almost all molecules quickly rearrange their bonds to form two new structures. These two forms are diastereomers and differ with respect to their rotation of polarized light. One form, labeled α-D-glucopyranose, has a specific rotation, [α], of +112°; the other, labeled β-D-glucopyranose, has a specific rotation of +18.7°. An interesting phenomenon occurs when these two forms of glucose are put into separate solutions and allowed to stand for several hours. The specific rotation of each solution changes to +52.7°. This phenomenon is known as mutarotation. An explanation of mutarotation is that D-glucose exists in solution as an equilibrium mixture of two cyclic forms and the open-chain form (see Figure 29.3). The two

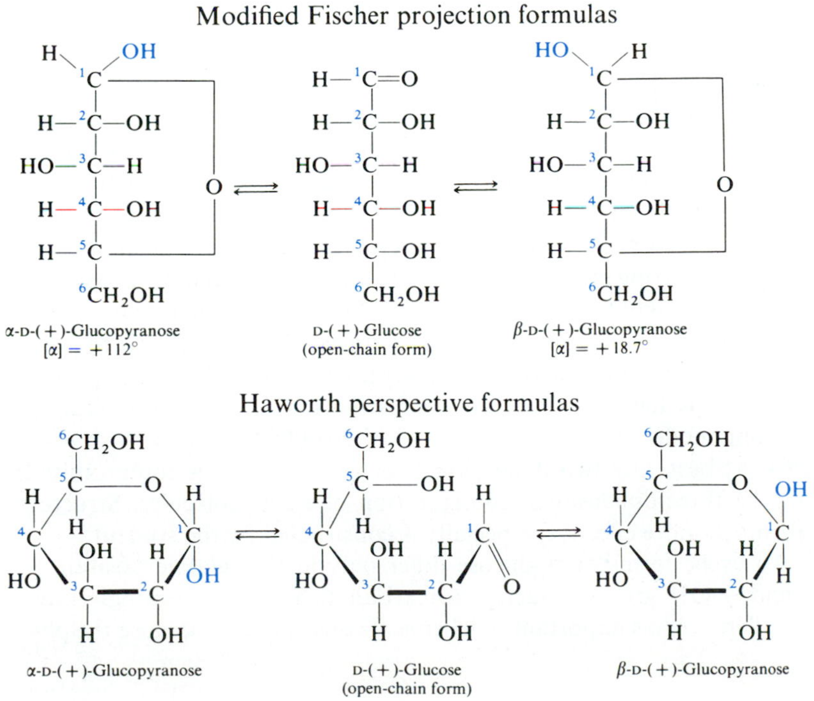

Modified Fischer projection formulas

α-D-(+)-Glucopyranose
[α] = +112°

D-(+)-Glucose
(open-chain form)

β-D-(+)-Glucopyranose
[α] = +18.7°

Haworth perspective formulas

α-D-(+)-Glucopyranose

D-(+)-Glucose
(open-chain form)

β-D-(+)-Glucopyranose

FIGURE 29.3 Mutarotation of D-glucose, shown in both the modified Fischer projection formulas and the Haworth perspective formulas.

cyclic molecules are optical isomers, differing only in the orientation of the —H and —OH groups about carbon 1. When dissolved, some α-D-glucopyranose molecules are transformed into β-D-glucopyranose, and vice versa, until an equilibrium is reached between the α and β forms. (Note that no other chiral centers in D-glucose are altered when this sugar is dissolved.) The equilibrium solution contains about 36% α molecules and 64% β molecules, with a trace of open-chain molecules. When two cyclic isomers differ only in their stereo arrangement about the carbon involved in mutarotation, they are called **anomers**. For example, α- and β-D-glucopyranose are anomers (see Figure 29.3). **Mutarotation** is the process by which anomers are interconverted.

anomer

mutarotation

The cyclic forms of D-glucose may be represented by either Fischer projection formulas or by Haworth perspective formulas. These structures are shown in Figure 29.3. In the cyclic Fischer projection formulas of the D-aldoses, the α form has the —OH on carbon 1 written to the right; in the β form the —OH on carbon 1 is on the left. The Haworth structure represents the molecule as a flat hexagon with the —H and —OH groups above and below the plane of the hexagon. In the α form the —OH on carbon 1 is written below the plane; in the β form the —OH on carbon 1 is above the plane. In converting the projection formula of a D-aldohexose to the Haworth formula, the —OH groups on carbons 2, 3, and 4 are written below the plane if they project to the right and above the plane if they project to the left. Carbon 6 is written above the plane.

Haworth formulas are sometimes shown in abbreviated schematic form. For example, α-D-(+)-glucopyranose is shown in this diagram:

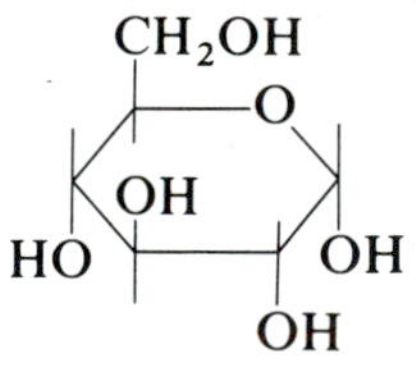

CH_2OH
O
OH
HO OH
OH

α-D-(+)-Glucopyranose

Although both the Fischer projection formula and the Haworth formula provide useful representations of carbohydrate molecules, it is important to understand that these structures only approximate the true molecular shapes. We know, for example, that the pyranose ring is not flat but, rather, can assume either a chair or boat conformation like the cycloalkanes (see Section 21.13). Most naturally occurring monosaccharides are found in the chair form as shown in Figure 29.4 for α-D-glucopyranose. Even this three-dimensional structure does not truly capture how a sugar molecule must appear. For, unlike this representation, we know that atoms move as close together as possible when they form molecules. Perhaps the most accurate representation of a sugar molecule is the space-filling model. A space-filling model of α-D-glucopyranose is shown in Figure 29.4. At best, any two-dimensional representation is a compromise in portraying the three-dimensional configuration of such molecules. Structural models are much more effective, especially if constructed by the student.

The two cyclic forms of D-glucose differ only in the relative positions of the —H and —OH groups attached to carbon 1. Yet, this seemingly minor structural difference has important biochemical consequences because the phys-

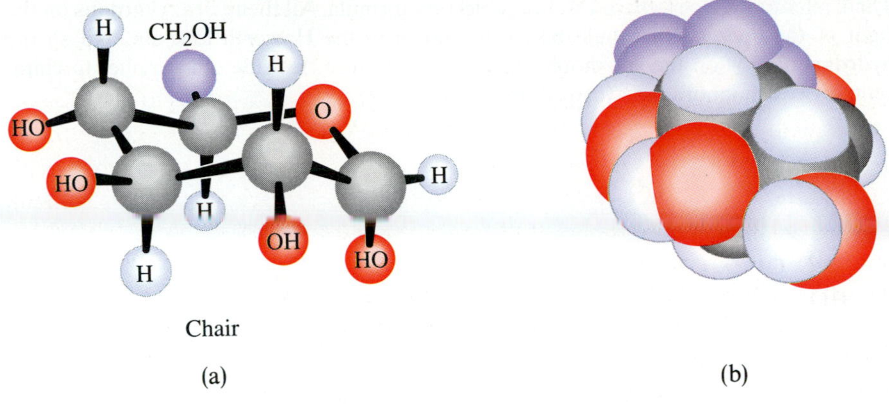

◀ **FIGURE 29.4 Three-dimensional representations of α-D-glycopyranose: (a) ball-and-stick model; (b) space-filling model.**

ical shape of a molecule often determines its biological use. For example, the fundamental structural difference between starch and cellulose is that starch is a polymer of α-D-glucopyranose, whereas cellulose is a polymer of β-D-glucopyranose. As a consequence, starch is a major and easily digested human food, but we are totally unable to digest cellulose.

EXAMPLE 29.2

Write the pyranose Haworth perspective formulas for the two anomers of D-mannose and name these isomers.

SOLUTION

First, write the open-chain Fischer projection formula for D-mannose (you must memorize this structure or know where to find it in the text) and number the carbons from top (the aldehyde group) to bottom (the primary alcohol group).

H—^{1}C=O
HO—^{2}C—H
HO—^{3}C—H
H—^{4}C—OH
H—^{5}C—OH
$^6CH_2OH$

Next, draw the structure of the Haworth pyranose ring. Number the carbons from the right-hand point of the hexagon clockwise around the cyclic form, placing the CH_2OH(C6) group in the up position on the ring.

$^6CH_2OH$
5 O
4 1
3 2

Then, refer to the open-chain Fischer projection formula. All the hydroxyl groups on the right of the open chain should be written down in the Haworth formula, and all the hydroxyl groups on the left should be written up. Since this rule only applies to chiral centers, we can ignore the hydroxyl on carbon 6.

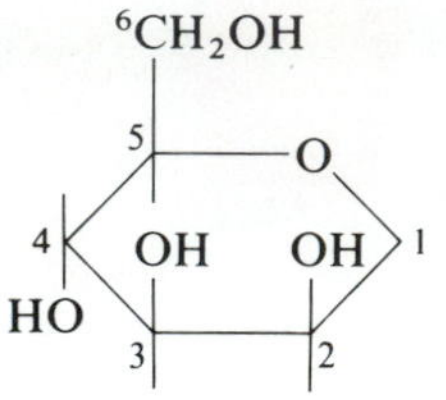

The carbon involved in mutarotation, carbon 1, can have either of two configurations, the α-anomer when the hydroxyl is pointed down or the β-anomer when the hydroxyl is pointed up. The last step in this exercise is to add the hydroxyl group at carbon 1 and name the two anomers.

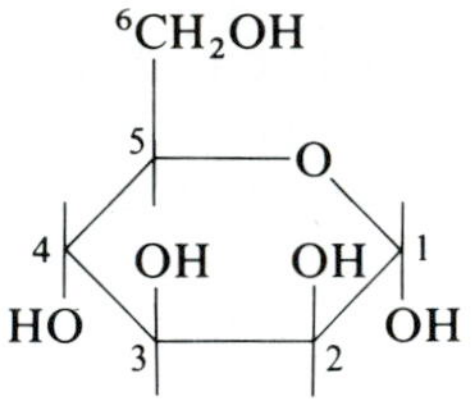

α-D-Mannopyranose

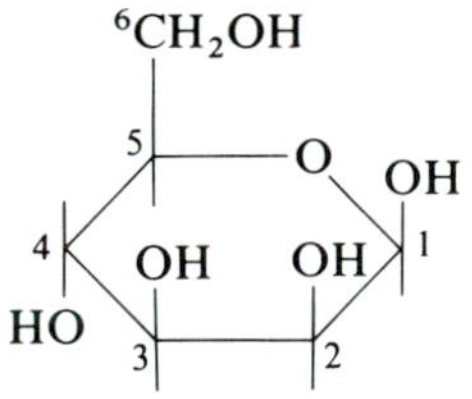

β-D-Mannopyranose

Practice Draw the Haworth formula for α-D-galactopyranose.

Answer:

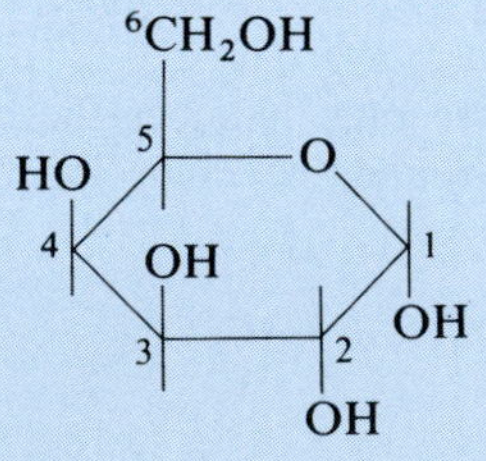

29.6 Hemiacetals and Acetals

In Chapter 24 we studied the reactions of aldehydes (and ketones) to form hemiacetals and acetals. The hemiacetal structure consists of an ether linkage and an alcohol linkage on the same carbon atom (shown in red), whereas the acetal structure has two ether linkages to the same carbon atom:

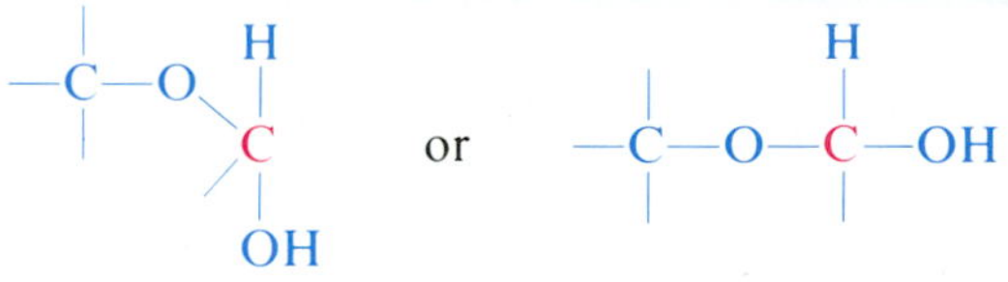

Hemiacetal

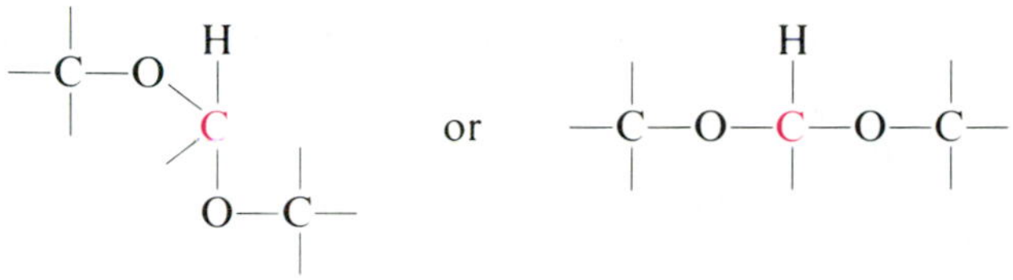

Acetal

Cyclic structures of monosaccharides are intramolecular hemiacetals. Five- or six-membered rings are especially stable.

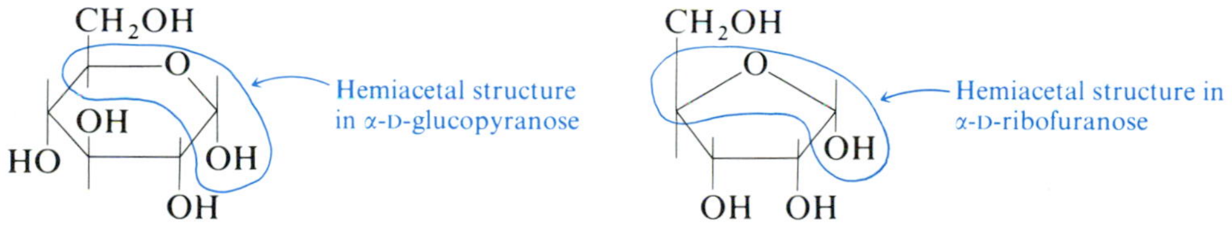

Hemiacetal structure in α-D-glucopyranose

Hemiacetal structure in α-D-ribofuranose

However, in an aqueous solution, the ring often opens and the hemiacetal momentarily reverts to the open-chain aldehyde. When the open chain closes, it forms either the α or the β anomer. Mutarotation results from this opening and closing of the hemiacetal ring (see Figure 29.3).

When an alcohol, ROH, reacts with another alcohol, R′OH to split out H_2O, the product formed can be an ether, ROR′. Carbohydrates are alcohols and behave in a similar manner. When a monosaccharide hemiacetal reacts with an alcohol, the product is an acetal. In carbohydrate terminology this acetal structure is called a **glycoside** (derived from the Greek word *glykys*, meaning sweet). In the case of glucose, it would be a glucoside; if galactose, a galactoside; and so on.

glycoside

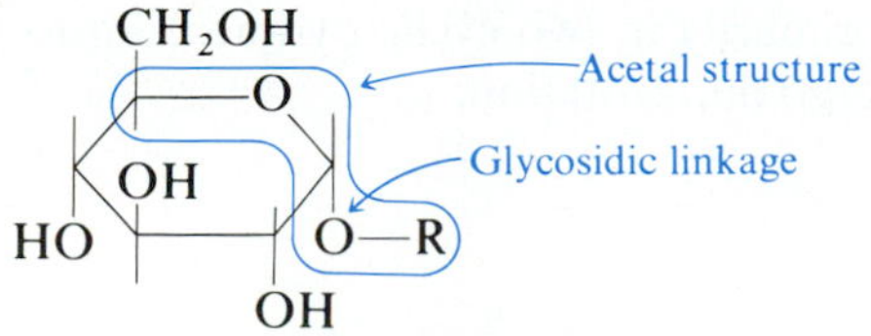

Acetal structure

Glycosidic linkage

An α-glycoside
(R = a variety of groups)

A glycoside differs significantly from a monosaccharide with respect to chemical reactivity.

When α-D-glucopyranose is heated with methyl alcohol and a small quantity of hydrogen chloride is added, two optically active isomers are formed—methyl α-D-glucopyranoside and methyl β-D-glucopyranoside:

$$\alpha\text{-D-Glucopyranose} \underset{HCl}{\overset{CH_3OH}{\rightleftharpoons}} \text{Methyl } \alpha\text{-D-glucopyranoside} + \text{Methyl } \beta\text{-D-glucopyranoside}$$

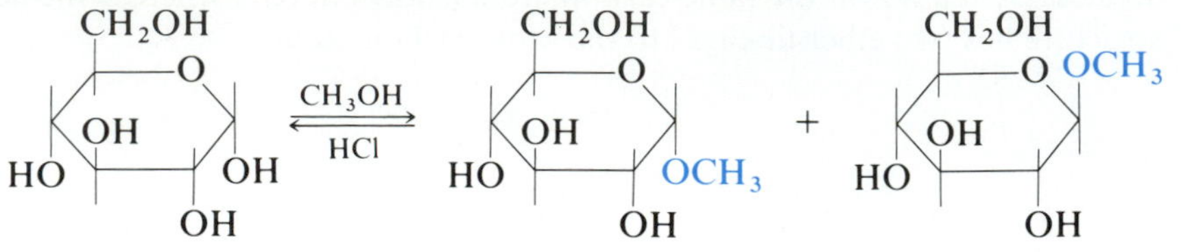

α-D-Glucopyranose

Methyl α-D-glucopyranoside (mp 165°C, [α] = +158°)

Methyl β-D-glucopyranoside (mp 107°C, [α] = −33°)

Unlike D-glucose, the two glycoside products no longer undergo mutarotation. They do not form open-chain compounds in aqueous solution. Acetals tend to be more stable and less reactive than hemiacetals.

The glycosidic linkage occurs in a wide variety of natural substances. All carbohydrates other than monosaccharides are glycosides. Heart stimulants such as digitalis and ouabain are known as heart glycosides. Several antibiotics such as streptomycin and erythromycin are also glycosides.

29.7 Structures of Galactose and Fructose

Galactose, like glucose, is an aldohexose and differs structurally from glucose only in the configuration of the —H and —OH group on the fourth carbon:

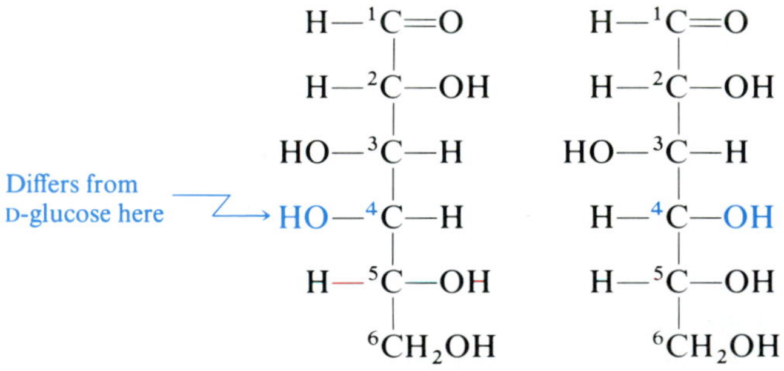

D-Galactose	D-Glucose
$H—{}^1C{=}O$	$H—{}^1C{=}O$
$H—{}^2C—OH$	$H—{}^2C—OH$
$HO—{}^3C—H$	$HO—{}^3C—H$
$HO—{}^4C—H$ (Differs from D-glucose here)	$H—{}^4C—OH$
$H—{}^5C—OH$	$H—{}^5C—OH$
${}^6CH_2OH$	${}^6CH_2OH$

Galactose, like glucose, also exists primarily in two cyclic pyranose forms that have hemiacetal structures and undergo mutarotation:

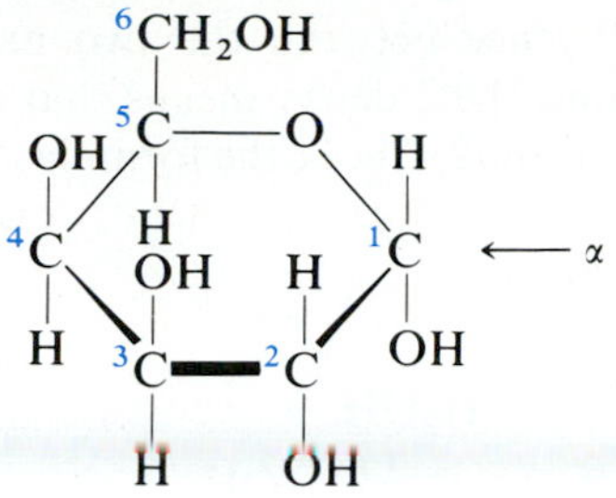

α — α-D-Galactopyranose

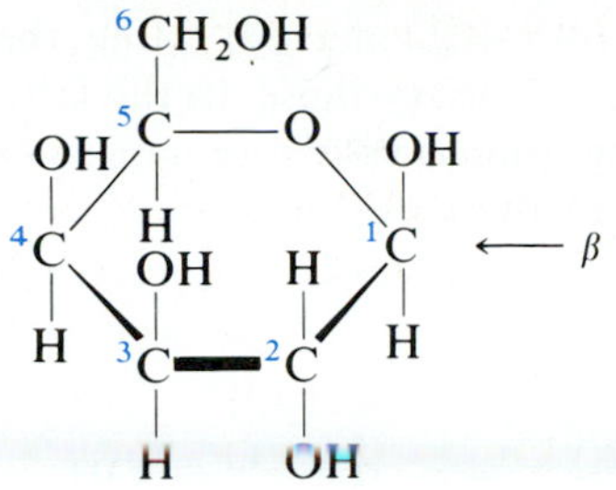

β — β-D-Galactopyranose

Fructose is a ketohexose. The open-chain form may be represented in a Fischer projection formula:

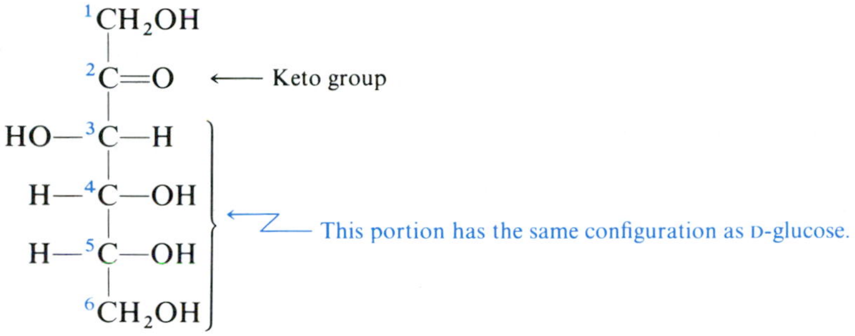

$^{1}CH_2OH$
$^{2}C{=}O$ ← Keto group
$HO{-}^{3}C{-}H$
$H{-}^{4}C{-}OH$
$H{-}^{5}C{-}OH$
$^{6}CH_2OH$

← This portion has the same configuration as D-glucose.

D-Fructose

Like glucose and galactose, fructose exists in both cyclic and open-chain forms. One common cyclic structure is a five-membered furanose ring in the β configuration:

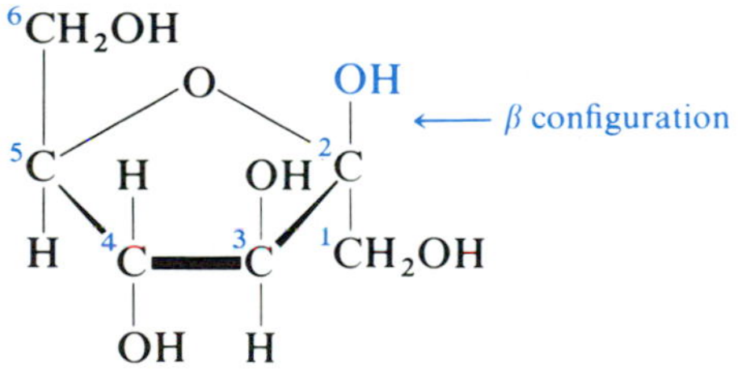

← β configuration

β-D-Fructofuranose

29.8 Pentoses

An open chain aldopentose has three asymmetric carbon atoms. Therefore eight (2^3) isomeric aldopentoses are possible. The four possible D-pentoses are shown on page 736 (Figure 29.1). Arabinose and xylose occur in some plants as polysaccharides called pentosans. D-ribose and its derivative D-2-deoxyribose are the most interesting pentoses because of their relationship to nucleic acids and the genetic code (Chapter 33). Note the difference between the two names D-ribose and D-2-deoxyribose. In the latter name, the 2-deoxy means that oxygen is missing from the D-ribose molecule at carbon 2. Check the formulas that follow to verify this difference.

H—^{1}C=O
H—^{2}C—OH
H—^{3}C—OH
H—^{4}C—OH
$^5CH_2OH$

D-Ribose

$^5CH_2OH$ O OH 4 1 H H 3 H 2 H OH OH or CH_2OH O OH OH OH

β-D-Ribofuranose

H—^{1}C=O
H—^{2}C—H
H—^{3}C—OH
H—^{4}C—OH
$^5CH_2OH$

D-2-Deoxyribose

$^5CH_2OH$ O OH 4 1 H H 3 H 2 H OH H or CH_2OH O OH OH

β-D-2-Deoxyribofuranose

The ketose that is closely related to D-ribose is named D-ribulose. (Ketose names are often derived from the corresponding aldose name by modifying the suffix *-ose* to *-ulose*.) This ketose is an intermediate that allows cells to make many other monosaccharides. In photosynthetic organisms, D-ribulose is used to capture carbon dioxide and thus make new carbohydrates.

CH_2OH
C=O
H—C—OH
H—C—OH
CH_2OH

D-Ribulose

▲ **Sugar cane is a major source for sucrose, or table sugar.**

29.9 Disaccharides

Disaccharides are carbohydrates composed of two monosaccharide residues united by a glycosidic linkage. The two important disaccharides that are found in the free state in nature are sucrose and lactose. Sucrose, commonly known as *table sugar*, exists throughout the plant kingdom. Sugar cane contains 15–20%

sucrose, and sugar beets 10–17%. Maple syrup and sorghum are also good sources of sucrose. Lactose, also known as *milk sugar*, is found free in nature mainly in the milk of mammals. Human milk contains about 6.7% lactose and cow milk about 4.5% of this sugar.

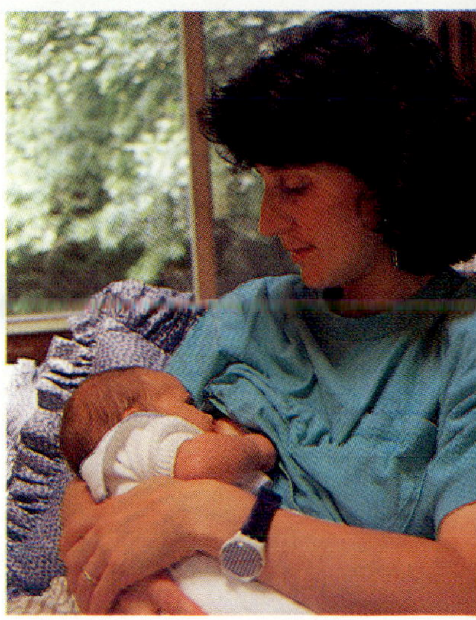

Unlike sucrose and lactose, several other important disaccharides are derived directly from polysaccharides by hydrolysis (see Section 29.14). For example, maltose, isomaltose, and cellobiose are formed when specific polysaccharides are hydrolyzed. Of this group, maltose is the most common and is found as a constituent of sprouting grain.

Upon hydrolysis, disaccharides yield two monosaccharide molecules. The hydrolysis is catalyzed by hydrogen ions (acids), usually at elevated temperatures, or by certain enzymes that act effectively at room or body temperatures. A different enzyme is required for the hydrolysis of each of the three disaccharides:

$$\text{Sucrose} + \text{Water} \xrightarrow{\text{H}^+ \text{ or sucrase}} \text{Glucose} + \text{Fructose}$$

$$\text{Lactose} + \text{Water} \xrightarrow{\text{H}^+ \text{ or lactase}} \text{Galactose} + \text{Glucose}$$

$$\text{Maltose} + \text{Water} \xrightarrow{\text{H}^+ \text{ or maltase}} \text{Glucose} + \text{Glucose}$$

The enzyme lactase is present in the small intestine of infants and allows them to easily digest lactose from their milk diet. Unfortunately, as people mature their intestines often stop producing the lactase enzyme, and they lose the ability to digest lactose. Instead, this sugar is metabolized by common bacteria that live in the large intestine. The gas and intestinal discomfort that results is termed milk or lactose intolerance and is a condition that afflicts many adults. Carefully note the small structural differences between lactose and the other common disaccharides in the next section. Small differences in molecular shape often determine what our bodies do with specific molecules.

29.10 Structures and Properties of Disaccharides

Disaccharides contain an acetal structure (glycosidic linkage), and some also contain a hemiacetal structure. The acetal structure in maltose may be considered as being derived from two glucose molecules by the elimination of a molecule of water between the —OH group on carbon 1 of one glucose unit and the —OH group of carbon 4 on the other glucose unit. This is an α-1,4-glycosidic linkage, since the glucose units have the α configuration and are joined at carbons 1 and 4. In a more systematic nomenclature, this form of maltose is known as α-D-glucopyranosyl-(1-4)-α-D-glucopyranose. If the structure of glucose is known, this name provides a complete description for drawing the maltose formula.

α-D-Glucopyranose unit α-D-Glucopyranose unit

α-1,4-Glycosidic linkage

Acetal structure

$+ H_2O$

Maltose
[α-D-Glucopyranosyl-(1-4)-α-D-glucopyranose]

Common lactose consists of a β-D-galactopyranose unit linked to an α-D-glucopyranose unit. These are joined by a β-1,4-glycosidic linkage from carbon 1 on galactose to carbon 4 on glucose. The more systematic name for lactose is β-D-galactopyranosyl-(1-4)-α-D-glucopyranose. Note that, although glycosidic bonds are straight carbon–oxygen linkages, the structural formula represents them in a bent fashion to provide stereochemical information. Carbon 1 of the galactose is in a β configuration, so its bent bond initially points up. The oxygen on carbon 4 of the glucose is below the carbon in the Haworth formula, and thus the bent bond initially points down.

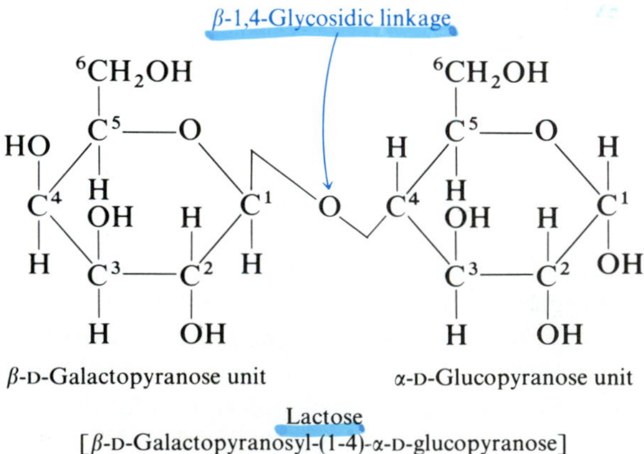

β-D-Galactopyranose unit α-D-Glucopyranose unit

Lactose
[β-D-Galactopyranosyl-(1-4)-α-D-glucopyranose]

Sucrose consists of an α-D-glucopyranose unit and a β-D-fructofuranose unit. These monosaccharides are joined by an oxygen bridge from carbon 1 on glucose to carbon 2 on fructose—that is, by an α-1,2-glycosidic linkage.

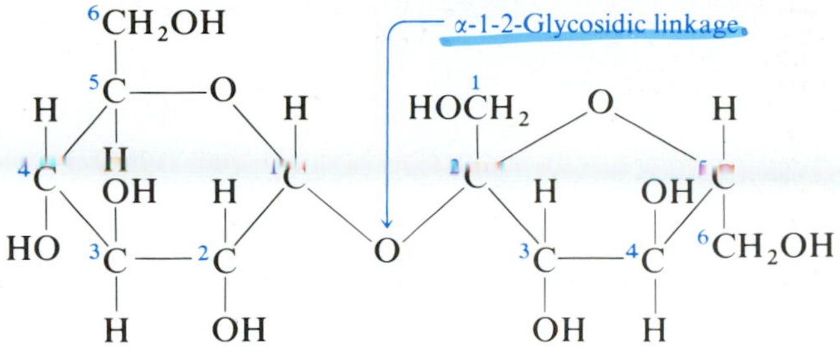

α-1-2-Glycosidic linkage

α-D-Glucopyranose unit β-D-Fructofuranose unit

Sucrose
[α-D-Glucopyranosyl-(1-2)-β-D-fructofuranose]

In this perspective formula, the fructose unit has been flipped to bring its number 2 carbon close to the number 1 carbon on the glucose unit. The groups on the fructose unit are therefore shown reversed from the perspective representation in Section 29.7.

EXAMPLE 29.3

Write the Haworth formula for isomaltose, α-D-glucopyranosyl-(1-6)-α-D-glucopyranose.

SOLUTION

Recognize that this disaccharide is composed of two α-D-glucopyranose units linked between carbon 6 of one sugar and carbon 1 of the other. First, write the Haworth formula for the monosaccharides and number their carbons.

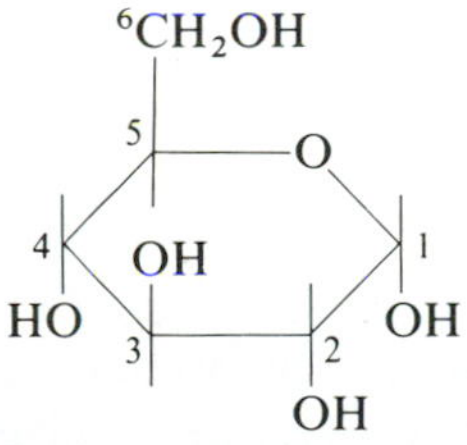

The two α-D-glucopyranose units must be linked in such a way that the stereochemistry at carbon 1 is preserved (carbon 6 is not an asymmetric center). One correct way to write the isomaltose structure is as follows:

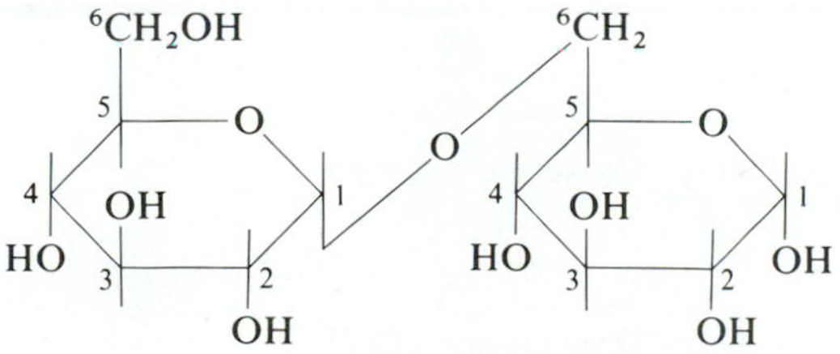

PRACTICE Write the structure for cellobiose (a disaccharide that can be derived from plants), β-D-glucopyranosyl-(1-4)-β-D-glucopyranose.

Answer: $^6CH_2OH$ $^6CH_2OH$

5 O 5 O OH

4 OH 1 O 4 OH 1

HO 3 2 3 2

OH OH

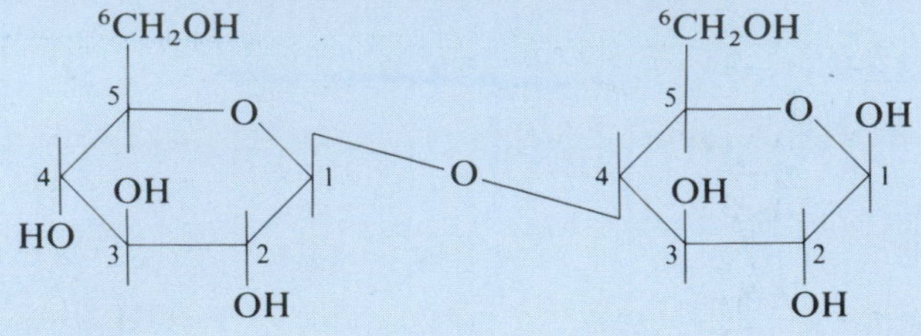

Lactose and maltose both show mutarotation, which indicates that one of the monosaccharide units has a hemiacetal ring that can open and close to interchange anomers. Sucrose has no hemiacetal structure and hence does not mutarotate.

The three disaccharides sucrose, lactose, and maltose have physical properties associated with large polar molecules. All three are crystalline solids and are quite soluble in water; the solubility of sucrose amounts to 179 g per 100 g of water at 0°C. Hydrogen bonding between the polar —OH groups on the sugar molecules and the water molecules is a major factor in this high solubility. These sugars are not easily melted. In fact, lactose is the only one with a clearly defined melting point (201.6°C). Sucrose and maltose begin to decompose when heated to 186°C and 102.5°C, respectively. When sucrose is heated to melting, it darkens and undergoes partial decomposition. The resulting mixture is known as caramel, or burnt sugar, and is used as coloring and as a flavoring agent in foods.

29.11 Sweeteners and Diet

Carbohydrates have long been valued for their ability to sweeten foods. Fructose is the sweetest of the common sugars (a scale of relative sweetness is given in Table 29.2), although sucrose (table sugar) is the most commonly used sweetener.

TABLE 29.2 Relative Sweetness of Sugars

Sugar	Sweetness	Sugar	Sweetness
Fructose	100	Galactose	19
Sucrose	58	Lactose	9.2
Glucose	43	Invert sugar	75
Maltose	19		

Astonishingly large amounts of sucrose are produced from sugar beets and cane: World production is on the order of 90 million tons annually. There are no essential chemical differences between cane and beet sugar. In the United States, approximately 20–30% of the average caloric intake is sucrose (about 150 g/day per person). Low price and sweet taste are the major reasons for high sucrose consumption. Note that sucrose is only 58% as sweet as fructose (Table 29.2). However, because sucrose is inexpensive to produce and is amenable to a variety of food processing techniques, approximately 60–80% of all sweeteners is sucrose.

Sucrose has a tendency to crystallize from concentrated solutions or syrups. Therefore, in commercial food preparations (for example, candies, jellies, and canned fruits) the sucrose is often hydrolyzed:

$$\text{Sucrose} + H_2O \xrightarrow{H^+} \text{Glucose} + \text{Fructose}$$

invert sugar

The resulting mixture of glucose and fructose, usually in solution, is called **invert sugar**. Invert sugar has less tendency to crystallize than sucrose, and it has greater sweetening power than an equivalent amount of sucrose. The nutritive value of the sucrose is not affected in any way by the conversion to invert sugar because the same hydrolysis reaction occurs in normal digestion.

High fructose corn syrups provide an alternate means of sweetening liquids such as soft drinks. These syrups are produced from corn starch and derive their sweet taste from fructose. Of course, starch is not sweet; as discussed in Section 29.14, it is a glucose polymer and contains no fructose. So, to produce this sweetener, the starch polymer must be broken down to yield some glucose monosaccharides, and in turn, glucose must be converted to fructose. Biotechnology has provided the means to accomplish this conversion. The starch polymers are hydrolyzed and some of the glucose is converted into fructose by a relatively new manufacturing process using enzymes. High-fructose corn syrups are economical because corn starch is much cheaper than cane or beet sugar.

Unfortunately, high sugar consumption has presented problems. Sucrose and other common sugars are ready sources of metabolic energy. Thus, for many people, sucrose is a source of too many calories. Oral bacteria also find sucrose easy to metabolize, increasing the incidence of dental caries. Finally, because the monosaccharides, fructose and glucose, are quickly absorbed from the small intestine, sugar consumption leads to a rapid increase in blood sugar. Such a sharp rise can be dangerous for people with impaired carbohydrate metabolism—for example, diabetes mellitus.

Scientists have searched for sugar substitutes. The ideal substitute might be sweeter than sucrose but lack the structural features and chemical reactivity that allow sugar to be absorbed and metabolized. Purely artificial, noncarbohydrate sweeteners have been developed starting with saccharin in 1879. This molecule is about 300 times sweeter than sucrose, cannot be metabolized, and so is nonnutritive. However, the search for other artificial sweeteners has continued because of saccharin's aftertaste and because of its potential health risks. Sodium cyclamate (20 times sweeter than fructose) was introduced in the early 1960s, but this molecule has been shown to cause cancer in laboratory animals. More recently, aspartame has become the artificial sweetener of choice. This molecule is about 200 times sweeter than fructose and poses no known health risks for most individuals, with the notable exception of people who suffer from phenylketonuria (PKU). Aspartame is composed of two amino acids (see Chapter 31) and can be metabolized to yield energy. However, foods sweetened with aspartame have much fewer calories than those containing sucrose because the same sweet taste is achieved with about 200 times less aspartame.

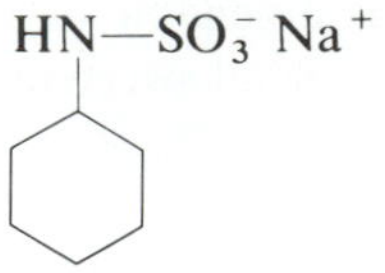

$HN{-}SO_3^- \; Na^+$ (on cyclohexane ring)

Sodium cyclamate

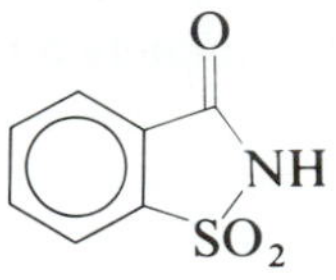

Saccharin

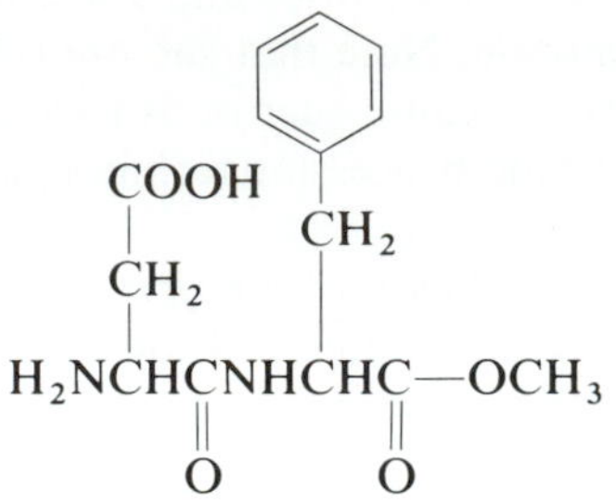

$H_2NCHC(=O)NHCHC(=O){-}OCH_3$ with CH_2COOH and $CH_2C_6H_5$ side chains

Aspartame

29.12 Reducing Sugars

reducing sugar

Some sugars are capable of reducing silver ions to free silver, and copper(II) ions to copper(I) ions, under prescribed conditions. Such sugars are called **reducing sugars**. This reducing ability, which is useful in classifying sugars and in certain clinical tests, is dependent on the presence of (1) aldehydes, (2) α-hydroxyketone groups ($-CH_2COCH_2OH$) such as in fructose, or (3) hemiacetal structures in cyclic molecules such as maltose. These groups are easily oxidized to carboxylic acid (or carboxylate ion) groups; the metal ions are thereby reduced ($Ag^+ \longrightarrow Ag^0; Cu^{2+} \longrightarrow Cu^+$). Several different reagents, including Tollens', Fehling's, Benedict's, and Barfoed's reagents, are used to detect reducing sugars (see Section 24.4). The Benedict, Fehling, and Barfoed tests depend on the formation of copper(I) oxide precipitate to indicate a positive reaction.

$$\underset{\text{Aldehyde group}}{RC(H){=}O} + \underset{\text{(blue)}}{2\,Cu^{2+}} + 5\,OH^- \longrightarrow \underset{\text{Carboxylate ion group}}{RC({=}O){-}O^-} + \underset{\text{Copper(I) oxide (brick red)}}{Cu_2O(s)} + 3\,H_2O$$

Barfoed's reagent contains Cu^{2+} ions in the presence of acetic acid. It is used to distinguish reducing monosaccharides from reducing disaccharides. Under the same reaction conditions, the reagent is reduced more rapidly by monosaccharides.

Glucose and galactose contain aldehyde groups; fructose contains an α-hydroxyketone group. Therefore, all three of these monosaccharides are reducing sugars.

A carbohydrate molecule need not have a free aldehyde or α-hydroxyketone group to be a reducing sugar. A hemiacetal structure (see below) is a potential aldehyde group. Maltose and the cyclic form of glucose are examples of molecules with the hemiacetal structure.

Hemiacetal structure that opens to form an aldehyde group

Glucose Maltose

Under mildly alkaline conditions, the rings open at the points indicated by the arrows to form aldehyde groups:

Ring opens here

$$\xrightarrow[H_2O]{OH^-}$$

Glucose (ring structure) Glucose (open-chain structure)

Any sugar that has the hemiacetal structure is classified as a reducing sugar. Among the disaccharides, lactose and maltose have hemiacetal structures and are therefore reducing sugars. Sucrose is not a reducing sugar because it does not have the hemiacetal structure.

Many clinical tests monitor glucose as a reducing sugar. For example, Benedict's and Fehling's reagents are used to detect the presence of glucose in urine. Initially the reagents are deep blue in color. A positive test is indicated by a color change to greenish-yellow, yellowish-orange, or brick-red, corresponding to an increasing glucose (reducing sugar) concentration. These tests are used for estimating the amount of glucose in the urine of diabetics in order to adjust the amount of insulin needed for proper glucose utilization.

Alternatively, a clinical test (glucose oxidase test) makes use of an enzyme catalyzed oxidation of glucose to test for urine sugar. The inclusion of an enzyme ensures a reaction that is specific for the glucose structure, allowing a more selective test for glucose in the urine.

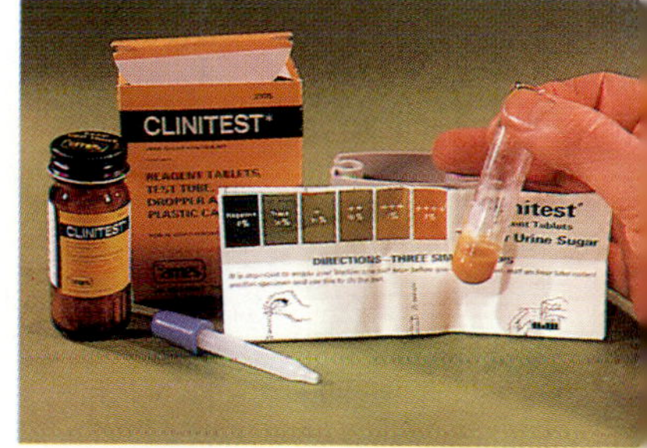

29.13 Reactions of Monosaccharides

Oxidation

The oxidation of monosaccharides by copper ions is described in Section 29.12. The aldehyde groups in monosaccharides are also oxidized to monocarboxylic acids by other mild oxidizing agents such as bromine water. The carboxylic acid group is formed at carbon 1. The name of the resulting acid is formed by changing the *ose* ending to *onic acid*. Glucose yields gluconic acid; galactose, galactonic acid; and so on.

$$
\begin{array}{c}
\text{H—C=O} \\ | \\ \text{H—C—OH} \\ | \\ \text{HO—C—H} \\ | \\ \text{H—C—OH} \\ | \\ \text{H—C—OH} \\ | \\ \text{CH}_2\text{OH} \\ \text{D-Glucose}
\end{array}
+ Br_2 + H_2O \longrightarrow
\begin{array}{c}
\text{COOH} \\ | \\ \text{H—C—OH} \\ | \\ \text{HO—C—H} \\ | \\ \text{H—C—OH} \\ | \\ \text{H—C—OH} \\ | \\ \text{CH}_2\text{OH} \\ \text{D-Gluconic acid}
\end{array}
+ 2\,HBr
$$

Dilute nitric acid, a vigorous oxidizing agent, oxidizes both carbon 1 and carbon 6 of aldohexoses to form dicarboxylic acids. The resulting acid is named by changing the *ose* sugar suffix to *aric acid*. Glucose yields glucaric acid (saccharic acid); galactose, galactaric acid (mucic acid).

$$
\begin{array}{c}
\text{H—C=O} \\ | \\ \text{H—C—OH} \\ | \\ \text{HO—C—H} \\ | \\ \text{H—C—OH} \\ | \\ \text{H—C—OH} \\ | \\ \text{CH}_2\text{OH} \\ \text{D-Glucose}
\end{array}
\xrightarrow{\text{Warm HNO}_3}
\begin{array}{c}
\text{COOH} \\ | \\ \text{H—C—OH} \\ | \\ \text{HO—C—H} \\ | \\ \text{H—C—OH} \\ | \\ \text{H—C—OH} \\ | \\ \text{COOH} \\ \text{Glucaric acid}
\end{array}
\qquad
\begin{array}{c}
\text{COOH} \\ | \\ \text{H—C—OH} \\ | \\ \text{HO—C—H} \\ | \\ \text{HO—C—H} \\ | \\ \text{H—C—OH} \\ | \\ \text{COOH} \\ \text{Galactaric acid}
\end{array}
$$

This reaction serves as the basis of the *galactaric* (*mucic*) *acid test*, which is sometimes used to distinguish glucose from galactose. Galactaric acid is only slightly soluble in dilute nitric acid, whereas glucaric acid is quite soluble. Hence, when oxidized with nitric acid, galactose (and lactose, which hydrolyzes to form glucose and galactose) yields a precipitate of galactaric acid crystals. When glucose is oxidized under the same conditions, the glucaric acid does not precipitate. Note also that galactaric acid is an optically inactive meso compound, whereas glucaric acid is an optically active compound (see Section 28.7).

Osazone Formation

Phenylhydrazine ($C_6H_5NHNH_2$) reacts with carbons 1 and 2 of reducing sugars to form derivatives called osazones. The formation of these distinctive crystalline derivatives is useful for comparing the structures of sugars. Glucose and fructose react as shown in Figure 29.5.

Identical osazones are obtained from D-glucose and D-fructose. This demonstrates that carbons 3 through 6 of D-glucose and D-fructose molecules are identical. The same osazone is also obtained from D-mannose. This indicates that

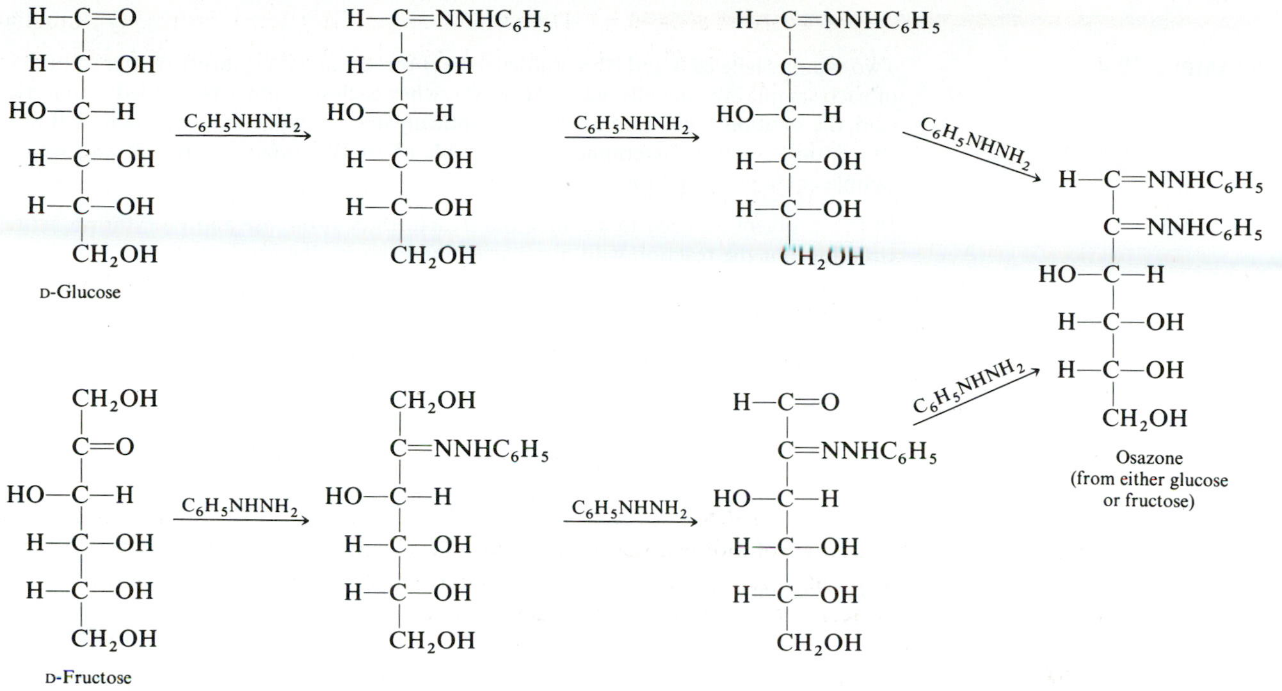

$$\begin{array}{c} H-C=O \\ | \\ H-C-OH \\ | \\ HO-C-H \\ | \\ H-C-OH \\ | \\ H-C-OH \\ | \\ CH_2OH \end{array} \xrightarrow{C_6H_5NHNH_2} \begin{array}{c} H-C=NNHC_6H_5 \\ | \\ H-C-OH \\ | \\ HO-C-H \\ | \\ H-C-OH \\ | \\ H-C-OH \\ | \\ CH_2OH \end{array} \xrightarrow{C_6H_5NHNH_2} \begin{array}{c} H-C=NNHC_6H_5 \\ | \\ C=O \\ | \\ HO-C-H \\ | \\ H-C-OH \\ | \\ H-C-OH \\ | \\ CH_2OH \end{array} \xrightarrow{C_6H_5NHNH_2}$$

D-Glucose

$$\begin{array}{c} CH_2OH \\ | \\ C=O \\ | \\ HO-C-H \\ | \\ H-C-OH \\ | \\ H-C-OH \\ | \\ CH_2OH \end{array} \xrightarrow{C_6H_5NHNH_2} \begin{array}{c} CH_2OH \\ | \\ C=NNHC_6H_5 \\ | \\ HO-C-H \\ | \\ H-C-OH \\ | \\ H-C-OH \\ | \\ CH_2OH \end{array} \xrightarrow{C_6H_5NHNH_2} \begin{array}{c} H-C=O \\ | \\ C=NNHC_6H_5 \\ | \\ HO-C-H \\ | \\ H-C-OH \\ | \\ H-C-OH \\ | \\ CH_2OH \end{array} \xrightarrow{C_6H_5NHNH_2}$$

D-Fructose

$$\begin{array}{c} H-C=NNHC_6H_5 \\ | \\ C=NNHC_6H_5 \\ | \\ HO-C-H \\ | \\ H-C-OH \\ | \\ H-C-OH \\ | \\ CH_2OH \end{array}$$

Osazone
(from either glucose or fructose)

▲
FIGURE 29.5
Reaction of glucose and fructose to form osazones. Common sugars form the same structure at carbons 1 and 2. Since D-glucose and D-fructose are identical at all other positions, these sugars yield the same osazone.

carbons 3 through 6 of the D-mannose molecule are the same as those of D-glucose and D-fructose molecules. In fact, D-mannose differs from D-glucose only in the configuration of the —H and —OH groups on carbon 2.

Reduction

Monosaccharides may be reduced to their corresponding polyhydroxy alcohols by reducing agents such as H_2/Pt or sodium amalgam, Na(Hg). For example, glucose yields sorbitol (glucitol), galactose yields galactitol (dulcitol), and mannose yields mannitol; all of these are hexahydric alcohols (containing six —OH groups).

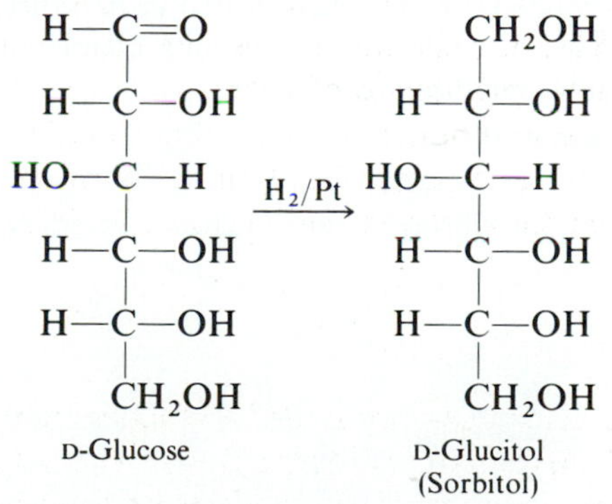

$$\begin{array}{c} H-C=O \\ | \\ H-C-OH \\ | \\ HO-C-H \\ | \\ H-C-OH \\ | \\ H-C-OH \\ | \\ CH_2OH \end{array} \xrightarrow{H_2/Pt} \begin{array}{c} CH_2OH \\ | \\ H-C-OH \\ | \\ HO-C-H \\ | \\ H-C-OH \\ | \\ H-C-OH \\ | \\ CH_2OH \end{array}$$

D-Glucose D-Glucitol (Sorbitol)

Hexahydric alcohols have properties resembling those of glycerol (Section 23.6). Because of their affinity for water, they are used as moisturizing agents in food and cosmetics. Sorbitol, galactitol, and mannitol occur naturally in a variety of plants.

EXAMPLE 29.4

Two samples labeled A and B are known to be D-threose and D-erythrose. Water solutions of each sample are optically active. However, when each solution was warmed with nitric acid, the solution from sample A became optically inactive while that from sample B was still optically active. Determine which sample (A or B) contains D-threose and which sample contains D-erythrose.

SOLUTION

In this problem we need to examine the structures of D-threose and D-erythrose, write equations for the reaction with nitric acid, and examine the products to see why one is optically active and the other optically inactive. Start by writing the formulas for D-threose and D-erythrose:

$$
\begin{array}{c}
\text{H—C=O} \\
| \\
\text{HO—C—H} \\
| \\
\text{H—C—OH} \\
| \\
\text{CH}_2\text{OH} \\
\text{D-Threose}
\end{array}
\qquad
\begin{array}{c}
\text{H—C=O} \\
| \\
\text{H—C—OH} \\
| \\
\text{H—C—OH} \\
| \\
\text{CH}_2\text{OH} \\
\text{D-Erythrose}
\end{array}
$$

The oxidation of these tetroses would yield dicarboxylic acids:

$$
\begin{array}{c}
\text{H—C=O} \\
| \\
\text{HO—C—H} \\
| \\
\text{H—C—OH} \\
| \\
\text{CH}_2\text{OH}
\end{array}
\xrightarrow[\text{HNO}_3]{\text{Warm}}
\begin{array}{c}
\text{COOH} \\
| \\
\text{HO—C—H} \\
| \\
\text{H—C—OH} \\
| \\
\text{COOH} \\
\text{I}
\end{array}
\qquad
\begin{array}{c}
\text{H—C=O} \\
| \\
\text{H—C—OH} \\
| \\
\text{H—C—OH} \\
| \\
\text{CH}_2\text{OH}
\end{array}
\xrightarrow[\text{HNO}_3]{\text{Warm}}
\begin{array}{c}
\text{COOH} \\
| \\
\text{H—C—OH} \\
| \\
\text{H—C—OH} \\
| \\
\text{COOH} \\
\text{II}
\end{array}
$$

Product I is a chiral molecule and is optically active. Product II is a meso compound and is optically inactive. Therefore, sample A is D-erythrose, since oxidation yields the meso acid. Sample B then must be D-threose.

PRACTICE A disaccharide yields no copper(I) oxide when treated in Benedict's test. This carbohydrate is composed of two α-D-galactopyranose units. Identify the carbon from each monosaccharide involved in the acetal linkage.

Answer: **Carbon 1 from each x-D-galactopyranose.**

PRACTICE Write the structure of the product formed when D-galactose is reduced with H_2/Pt. Is this compound optically active?

Answer:

$$
\begin{array}{c}
\text{CH}_2\text{OH} \\
| \\
\text{H—C—OH} \\
| \\
\text{HO—C—H} \\
| \\
\text{HO—C—H} \\
| \\
\text{H—C—OH} \\
| \\
\text{CH}_2\text{OH}
\end{array}
$$

No, it is a meso compound.

29.14 Polysaccharides Derived from Glucose

Although many naturally occurring polysaccharides are known, three—starch, cellulose, and glycogen—are of outstanding importance. All three, when hydrolyzed, yield D-glucose as the only product, according to this approximate general equation:

$$(C_6H_{10}O_5)_n + n\,H_2O \longrightarrow n\,C_6H_{12}O_6$$

Polysaccharide molecule (approximate formula) — D-Glucose

This hydrolysis reaction establishes that all three polysaccharides are polymers made up of glucose monosaccharide units. It also means that the differences in properties among the three polysaccharides must be due to differences in the structures and/or sizes of these molecules.

Many years of research were required to determine the detailed structures for polysaccharide molecules. Consideration of all this work is beyond the scope of our discussion, but an abbreviated summary of the results is given in the following paragraphs.

Starch

Starch is found in plants, mainly in the seeds, roots, or tubers (see Figure 29.6). Corn, wheat, potatoes, rice, and cassava are the chief sources of dietary starch. The two main components of starch are amylose and amylopectin. Amylose molecules are unbranched chains composed of about 25 to 1300 α-D-glucose

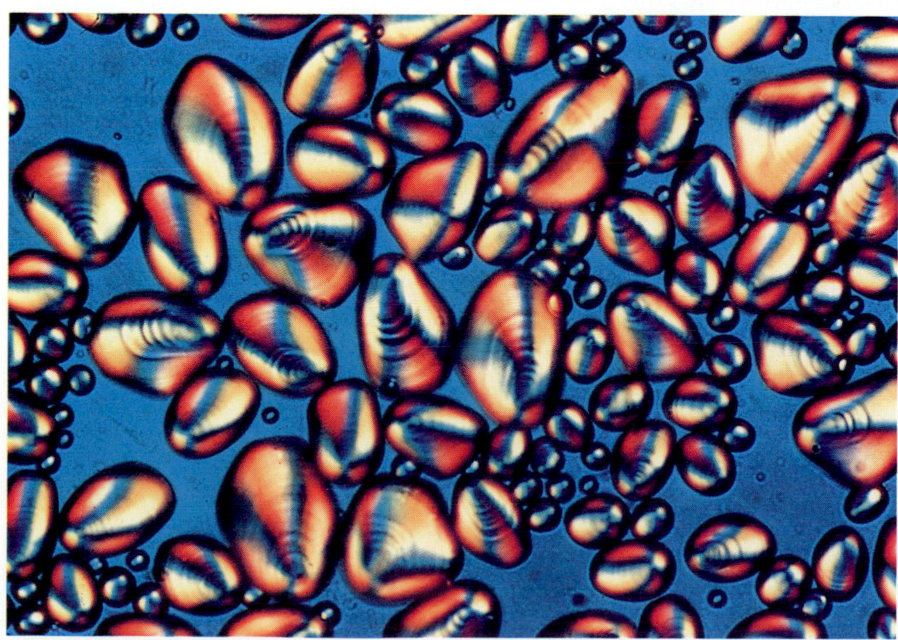

◄ Figure 29.6 **Polarized light micrograph of starch granules in potato tuber cells.**

units joined by α-1,4-glycosidic linkages, as shown in Figure 29.7. The stereochemistry of the α-anomer causes amylose to coil into a helical conformation. Partial hydrolysis of this linear polymer yields the disaccharide maltose.

Amylopectin is a branched-chain polysaccharide with much larger molecules than those of amylose. Amylopectin molecules consist on the average of several thousand α-D-glucose units with molar masses ranging up to 1 million or more. The main chain contains glucose units connected by α-1,4-glycosidic linkages. Branch chains are linked to the main chain through α-1,6-glycosidic linkages about every 25 glucose units, as shown in Figure 29.7. This molecule has a characteristic treelike structure because of its many branch chains. Partial hydrolysis of amylopectin yields both maltose and the related disaccharide isomaltose, α-D-glucopyranosyl-(1-6)-α-D-glucopyranose.

Despite the presence of many polar —OH groups, starch molecules are insoluble in cold water, apparently because of their very large size. Starch readily forms colloidal dispersions in hot water. Such starch "solutions" form an intense blue-black color in the presence of free iodine. Hence, a starch solution can be used to detect free iodine, or a dilute iodine solution can be used to detect starch.

Starch is readily converted to glucose by heating with water and a little acid (for example, hydrochloric or sulfuric acid). It is also readily hydrolyzed at room temperature by certain digestive enzymes. The hydrolysis of starch to maltose and glucose is shown in the following equation:

$$\text{Starch} \xrightarrow[\substack{\text{or salivary and} \\ \text{pancreatic amylase}}]{\text{Acid} + \Delta} \text{Dextrins and Maltose} \xrightarrow[\substack{\text{or maltase and other} \\ \text{intestinal enzymes}}]{\text{Acid} + \Delta} \text{D-Glucose}$$

The hydrolysis of starch can be followed qualitatively by periodically testing samples from a mixture of starch and saliva with a very dilute iodine solution. The change of color sequence is blue-black ⟶ blue ⟶ purple ⟶ pink ⟶ colorless as the starch molecules are broken down into smaller and smaller fragments.

Hydrolysis is a key chemical reaction in the digestion of starchy foods. If these foods are well-chewed, salivary amylase normally decreases the starch polymer chain length from on the order of a thousand glucose units to about eight per chain. In the small intestine, pancreatic amylase continues digestion to form maltose. Enzymes in the small intestine membranes complete the conversion of starch to glucose, which is then absorbed into the bloodstream.

Starch is the most important energy storage carbohydrate of the plant kingdom. In turn, humans and other animals consume huge quantities of starch. This polymer is such an important food source because it has the appropriate structure to be readily broken down to D-glucose. The reduced carbons from starch provide much of our daily energy needs as they are oxidized to carbon dioxide.

Glycogen

Glycogen is the energy storage carbohydrate of the animal kingdom. It is formed in the body by polymerization of glucose and stored in the liver and in muscle tissues. Structurally, it is very similar to the amylopectin fraction of starch except that it is more highly branched. The α-1,6-glycosidic linkages occur on one of every 12 to 18 glucose units.

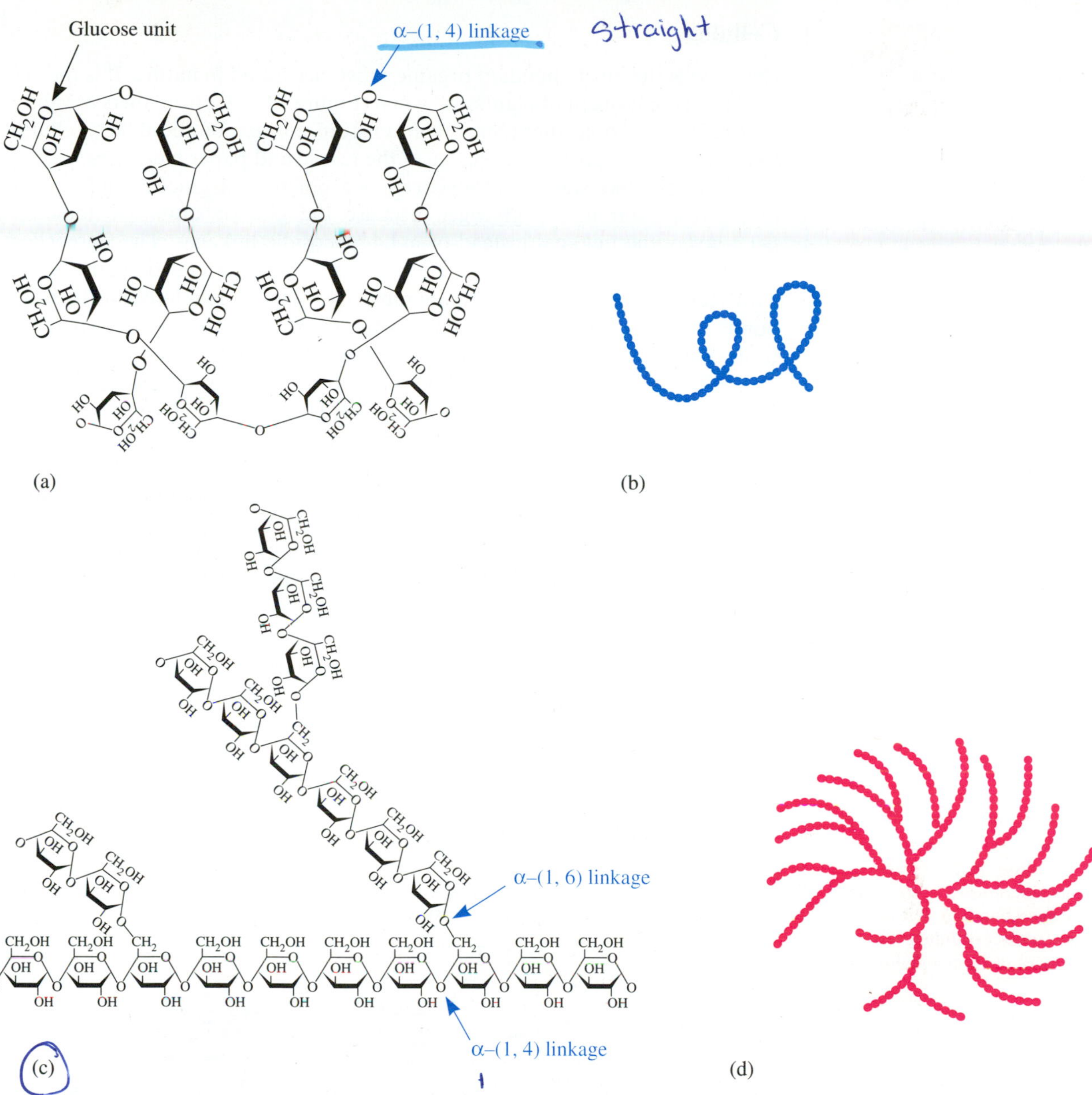

▲
FIGURE 29.7
(a) Molecular structure of amylose chain; (b) array of glucose units (dots) in amylose; (c) molecular structure of amylopectin; (d) branched array of glucose units (dots) in amylopectin.

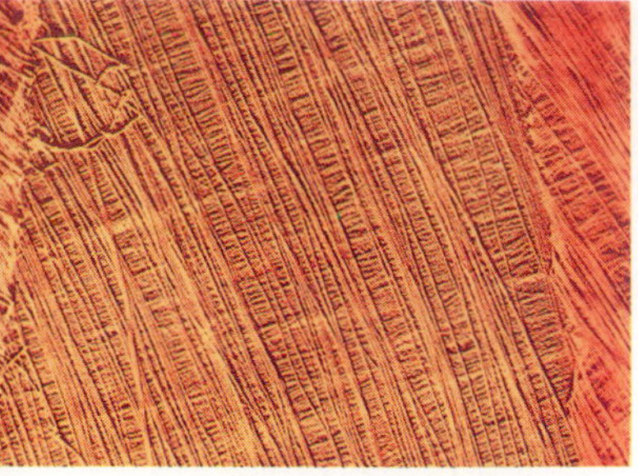

▲
FIGURE 29.8
Scanning electron micrograph of cellulose fibers in a plant cell wall.

Cellulose

Cellulose is the most abundant organic substance found in nature. It is the chief structural component of plants and wood (Figure 29.8). Cotton fibers are almost pure cellulose; wood, after removal of moisture, consists of about 50% cellulose. Cellulose is an important substance in the textile and paper industries.

Cellulose, like starch and glycogen, is a polymer of glucose. But cellulose differs from starch and glycogen because the glucose units are joined by β-1,4-glycosidic linkages instead of α-1,4-glycosidic linkages. The stereochemistry of the β-anomer allows this polymer to form an extended chain that can hydrogen-bond to adjacent cellulose molecules. The large number of hydrogen bonds so formed partially accounts for the strength of the resulting plant cell walls. The cellulose structure is illustrated in Figure 29.9.

A partial hydrolysis of cellulose produces the disaccharide cellobiose, β-D-glucopyranosyl-(1-4)-β-D-glucopyranose. However, cellulose has greater resistance to hydrolysis than either starch or glycogen. It is not appreciably hydrolyzed when boiled in a 1% sulfuric acid solution. It does not show a color reaction with iodine. Humans cannot digest cellulose, because they have no enzymes capable of catalyzing its hydrolysis. Fortunately, some microorganisms found in soil and in the digestive tracts of certain animals produce enzymes that do catalyze the breakdown of cellulose. The presence of these microorganisms explains why cows and other herbivorous animals thrive on grass—and also why termites thrive on wood.

The —OH groups of starch and cellulose can be reacted without destruction of the macromolecular structures. For example, nitric acid converts an —OH

FIGURE 29.9 ►
Haworth and three-dimensional representations of cellulose. In the three-dimensional drawing, note the hydrogen bonding that links the extended cellulose polymers to form cellulose fibers.

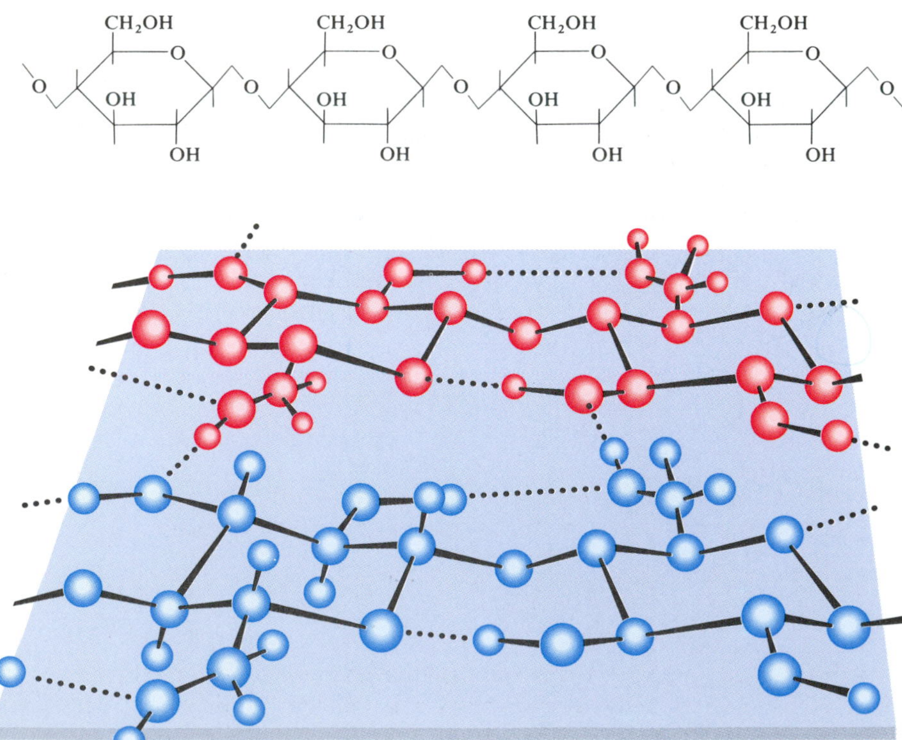

CHEMISTRY IN ACTION

POLYSACCHARIDES

Many carbohydrate polymers are composed of monosaccharides other than glucose and serve a variety of purposes for both plants and animals. Some polysaccharides are made up of only one type of monosaccharide. For example, D-mannose-containing polymers are found in many plants including pine trees and orchid tubers. Polysaccharides composed of the pentose D-xylose are also found widely distributed in wood and vegetable products.

More complex polysaccharides are often found in animals. These carbohydrates are linked together by glycosidic bonds at various positions on the monosaccharides. Thus, a sugar may be bonded to its neighbor from carbon 1 to carbon 3, or carbon 1 to carbon 4, or carbon 1 to carbon 6, and so on. Because of these numerous linkages, the polysaccharides often have a complex, branching structure. Also, it is not uncommon for these molecules to contain many different kinds of monosaccharides. These complex carbohydrates serve a variety of functions in animals.

Mucopolysaccharides (or glycosaminoglycans) make up part of the connective tissue, and are found in the joints and the skin. The slimy, mucuslike consistency of these molecules derives from their special chemical properties. About half of the sugar units are acidic in these carbohydrates. These acid groups become negatively charged under physiological conditions, causing the polymer chains to repel each other.

Water fills the space between the polymer chains and gives the mucopolysaccharide a spongelike consistency. One gram of some mucopolysaccharides can absorb up to 20 liters of water. This natural shock absorber and lubricant is necessary for animal locomotion.

Even more complicated polysaccharides are found on the surfaces of almost all cells. These carbohydrates serve as "labels" or antigens allowing organisms to distinguish their own cells from invading bacteria, for example. Antigen recognition illustrates a very important biochemical principle: *molecular shape carries information which guides the reactions of life.*

In humans, the polysaccharides on the surface of the red blood cells give rise to a number of blood types, often classified by the ABO system. Cells of different blood types have surface polysaccharides with different structures. Cells carrying one carbohydrate structure are commonly not tolerated by an individual of another blood type. Thus, for example, if type AB blood is transfused into someone with type O red blood cells, the new cells will be attacked by the immunosystem. Red cell destruction can lead to serious injury or death.

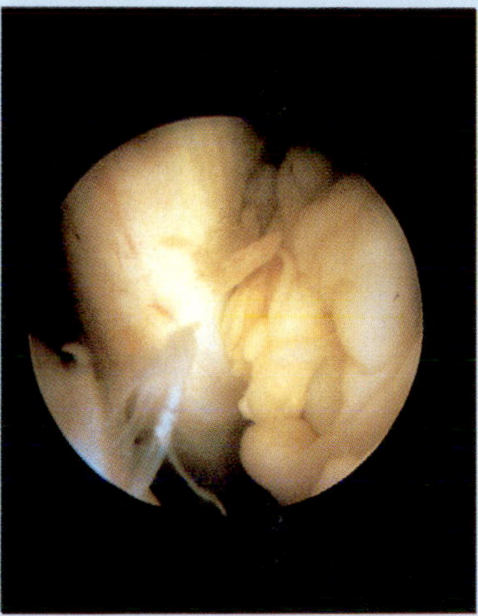

▲ **The connective tissue inside the knee contains mucopolysaccharides.**

group to a nitrate group in this fashion:

$$\underset{\substack{\text{Hydroxyl group}\\\text{on cellulose}\\\text{molecule}}}{\text{Cellulose—}\boxed{\text{OH} + \text{H}}}\underset{\substack{\text{Nitric}\\\text{acid}}}{\text{ONO}_2} \longrightarrow \underset{\substack{\text{Nitrate group}\\\text{on cellulose}\\\text{molecule}}}{\text{Cellulose—ONO}_2} + \underset{\text{Water}}{\text{H}_2\text{O}}$$

If only a portion of the —OH groups on the cellulose molecule are nitrated, a plastic nitrocellulose material known as celluloid or pyroxylin is obtained. This material has been used to make such diverse articles as billiard balls, celluloid shirt collars, and photographic film. By nitration of nearly all —OH groups, a powerful high explosive is obtained. This highly nitrated cellulose, or "guncotton," is the basic ingredient in modern "smokeless" gunpowder.

Another modified cellulose, cellulose acetate, is made by esterification of —OH groups with acetic acid (acetic anhydride). About two-thirds of the —OH groups are esterified:

$$\text{Cellulose—OH} + \text{CH}_3\text{—}\overset{\overset{\displaystyle O}{\|}}{C}\text{—O—}\overset{\overset{\displaystyle O}{\|}}{C}\text{—CH}_3 \longrightarrow \text{Cellulose—O—}\overset{\overset{\displaystyle O}{\|}}{C}\text{—CH}_3 + \text{CH}_3\text{COOH}$$

Hydroxyl group on cellulose molecule

Acetic anhydride

Acetate group on cellulose molecule

Cellulose acetate, unlike the dangerously flammable cellulose nitrate, can be made to burn only with difficulty. For this reason, cellulose acetate has displaced cellulose nitrate in almost all kinds of photographic films. The textile known as acetate rayon is made from cellulose acetate. Cellulose acetate is also used as a clear, transparent packaging film. In another process, cellulose reacts with carbon disulfide in the presence of sodium hydroxide to form a soluble cellulose derivative called cellulose xanthate, from which cellulose can be regenerated. Viscose rayon textiles and cellophane packaging materials are made of regenerated cellulose prepared by this process.

Concepts in Review

1. List three important characteristics of an energy-providing food.
2. Compare the oxidation state of carbon in carbohydrates with that of carbon in carbon dioxide.
3. Classify carbohydrates as mono-, di-, oligo-, or polysaccharides.
4. Explain the use of D, L, (+), and (−) in naming carbohydrates.
5. Identify and write pyranose and furanose ring structures of carbohydrates.
6. Identify and write Fischer projection and Haworth formulas for carbohydrates.
7. Identify the structural feature of a carbohydrate that makes it a reducing sugar.
8. Explain the phenomenon of mutarotation.
9. Distinguish between hemiacetal and acetal structures in a carbohydrate.
10. Understand what a glycoside linkage is.
11. Understand disaccharide composition and the manner in which monosaccharides are linked together in sucrose, lactose, and maltose.

12. Identify monosaccharides that are epimers.
13. List the major sources of glucose, galactose, fructose, sucrose, lactose, and maltose.
14. Describe some disadvantages of a diet high in sucrose and some alternatives to this sweetener.
15. Identify three disaccharides that are derived from polysaccharides.
16. Describe the Benedict test and tell what evidence must be seen to indicate a positive test.
17. Write chemical equations for the oxidation of monosaccharides by bromine and by nitric acid.
18. Write chemical equations for the reduction of monosaccharides.
19. Write chemical equations for the formation of osazones.
20. Identify monosaccharides that give identical osazones.
21. Identify monosaccharides that give optically inactive (meso) products when reduced to polyhydroxy alcohols or when oxidized to dicarboxylic acids.
22. Discuss the structural differences between amylose, amylopectin, and cellulose.
23. Compare the common polymers of glucose with more complex polysaccharides.

Equations in Review

Formation of Sugar Ring Structures from Open-Chain Forms—Hemiacetal/Hemiketal Formation (example)

$^6CH_2OH$ ^5C—O H H ^4C H OH H ^1C HO ^3C—^2C OH H OH ⇌ $^6CH_2OH$ ^5C—OH H H ^4C H OH H ^1C=O HO ^3C—^2C H OH ⇌ $^6CH_2OH$ ^5C—O OH H ^4C H OH H ^1C H HO ^3C—^2C H OH

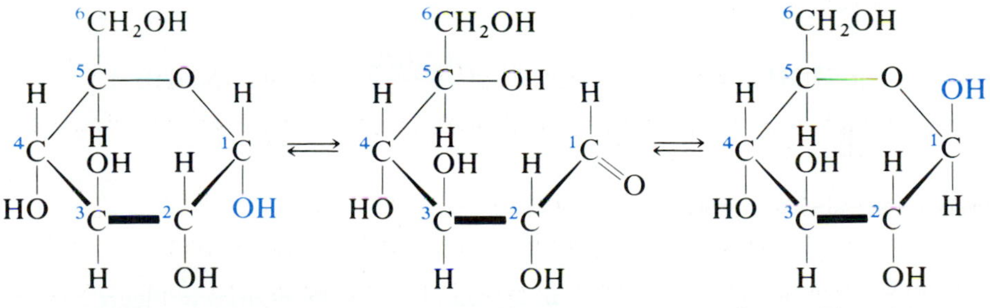

α-D-(+)-Glucopyranose — D-(+)-Glucose (open-chain form) — β-D-(+)-Glucopyranose

Reaction of Monosaccharides with Alcohols—Acetal/Ketal Formation (example)

CH_2OH O OH HO OH OH $\underset{HCl}{\overset{ROH}{\rightleftharpoons}}$ CH_2OH O OH HO OR OH + CH_2OH O OR OH HO OH

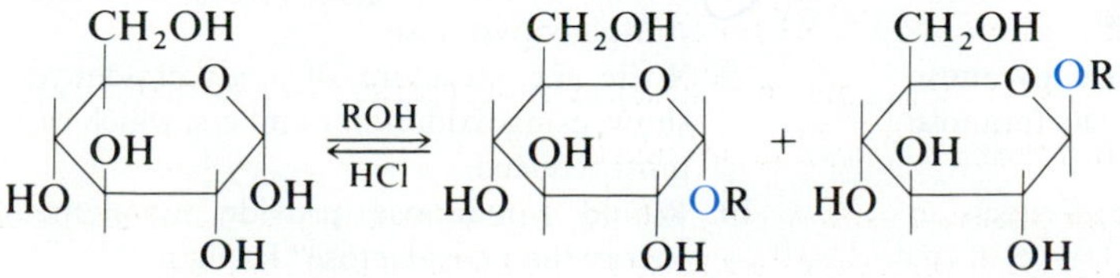

Kiliani–Fischer Synthesis for Increasing the Number of Carbons in an Aldose

$$\underset{\substack{|\\ \text{R}}}{\text{H—C=O}} \xrightarrow{\text{HCN}} \underset{\substack{|\\ \text{CHOH}\\ |\\ \text{R}}}{\text{CN}} \xrightarrow[\text{H}^+]{\text{H}_2\text{O}} \underset{\substack{|\\ \text{CHOH}\\ |\\ \text{R}}}{\text{COOH}} \xrightarrow{\text{Na(Hg)}} \underset{\substack{|\\ \text{CHOH}\\ |\\ \text{R}}}{\text{H—C=O}}$$

Oxidation of Carbohydrates

$$\underset{\substack{|\\ (\text{CHOH})_n\\ |\\ \text{CH}_2\text{OH}}}{\text{H—C=O}} + \text{Br}_2 + \text{H}_2\text{O} \longrightarrow \underset{\substack{|\\ (\text{CHOH})_n\\ |\\ \text{CH}_2\text{OH}}}{\text{COOH}} + 2\,\text{HBr}$$

$$\underset{\substack{|\\ (\text{CHOH})_n\\ |\\ \text{CH}_2\text{OH}}}{\text{H—C=O}} + 2\,\text{Cu}^{2+} + 5\,\text{OH}^- \longrightarrow \underset{\substack{|\\ (\text{CHOH})_n\\ |\\ \text{CH}_2\text{OH}}}{\text{COOH}} + \text{Cu}_2\text{O} + 3\,\text{H}_2\text{O}$$

$$\underset{\substack{|\\ (\text{CHOH})_n\\ |\\ \text{CH}_2\text{OH}}}{\text{H—C=O}} \xrightarrow{\text{Warm HNO}_3} \underset{\substack{|\\ (\text{CHOH})_n\\ |\\ \text{COOH}}}{\text{COOH}}$$

Reduction of Carbohydrates

$$\underset{\substack{|\\ \text{R}}}{\text{H—C=O}} \xrightarrow{\text{H}_2/\text{Pt}} \underset{\substack{|\\ \text{R}}}{\text{CH}_2\text{OH}}$$

Formation of Osazones from Carbohydrates

$$\underset{\substack{|\\ \text{CHOH}\\ |\\ \text{R}}}{\text{H—C=O}} \quad \text{or} \quad \underset{\substack{|\\ \text{C=O}\\ |\\ \text{R}}}{\text{CH}_2\text{OH}} \xrightarrow{2\,\text{C}_6\text{H}_5\text{NHNH}_2} \underset{\substack{|\\ \text{C=NNHC}_6\text{H}_5\\ |\\ \text{R}}}{\text{H—C=NNHC}_6\text{H}_5}$$

EXERCISES

1. Why is the oxidation state of a carbohydrate carbon important in metabolism?
2. What is the significance of the notations D and L in the name of a carbohydrate?
3. What is the significance of the notations (+) and (−) in the name of a carbohydrate?
4. Explain how a carbohydrate with a pyranose structure differs from one with a furanose structure.
5. What is galactosemia and what are its effects on humans?
6. Which of the D-aldohexoses in Figure 29.1 are epimers?
7. Write the cyclic structures for α-D-glucopyranose, β-D-galactopyranose, and α-D-mannopyranose.
8. Explain how α-D-glycopyranose differs from β-D-glucopyranose.
9. Write the structure of L-glyceraldehyde and show, using oxidation numbers, which carbon is most reduced.
10. Would D-mannose provide more metabolic energy than D-galactose? Explain.

11. Starting with the proper D-tetrose, show the steps for the synthesis of D-glucose by the Kiliani–Fischer synthesis.
12. Are the cyclic forms of monosaccharides hemiacetals or glycosides?
13. Explain the phenomenon of mutarotation.
14. Is D-2-deoxyarabinose the same as D-2-deoxyribose? Explain.
15. What is (are) the major source(s) of each?
(a) Sucrose (b) Lactose (c) Maltose
16. What is the monosaccharide composition of each of the following?
(a) Sucrose (c) Lactose (e) Amylose
(b) Maltose (d) Glycogen
17. How does the structure of cellobiose differ from that of isomaltose?
18. Cite two advantages of aspartame, as a sweetener, over sucrose.
19. Explain why invert sugar is sweeter than sucrose.
20. Which of the disaccharides sucrose, maltose, and lactose show mutarotation? Explain why.
21. Draw structural formulas for maltose and sucrose. Point out the portion of the structure that is responsible for one of these disaccharides being classified as a reducing sugar.
22. Give the systematic names for isomaltose and lactose.
23. What visual difference would you expect between a dilute glucose solution and a concentrated glucose solution in the Benedict test?
24. Which of the D-aldopentoses (Figure 29.1) is optically inactive when reduced with H_2/Pt? Write structures for the reduced products and explain why they are optically inactive.
25. Glyceraldehyde, $CH_2(OH)CH(OH)CHO$, is the simplest aldose.
(a) Write projection formulas for and identify the D and L forms of glyceraldehyde.
(b) Write the structural formula of the product obtained by reacting glyceraldehyde with hydrogen in the presence of a platinum catalyst.
26. Consider the eight aldohexoses given in Figure 29.1 and answer the following:
(a) Which of these aldohexoses give the same osazone?
(b) The eight aldohexoses are oxidized by nitric acid to dicarboxylic acids. Which of these give meso (optically inactive) dicarboxylic acids?
(c) Write the structures and names for the enantiomers of D-(+)-mannose and D-(−)-idose.
27. Describe the principal structural differences and similarities between the members of each of the following pairs:
(a) D-Glucose and D-fructose
(b) D-Ribose and D-2-deoxyribose
(c) Maltose and sucrose
(d) Amylose and amylopectin
(e) Cellulose and glycogen
(f) D-(−)-Ribose and D-(+)-glucose
28. When glucose is oxidized with nitric acid, glucaric acid is formed. Write the structural formula and the name of the dicarboxylic acid that is formed when galactose is oxidized with nitric acid.
29. Write the formulas for the four L-aldopentoses. Indicate which pair (or pairs) of the L-aldopentoses give identical osazones with phenylhydrazine.
30. Write the structure of a disaccharide that hydrolyzes to give only D-glucose and (a) is a reducing sugar, shows mutarotation, and forms an osazone; (b) is a nonreducing sugar, does not mutarotate, and does not form an osazone.
31. What are the structural differences between starch and cellulose?
32. What are the two main components of starch? How are they alike and how do they differ?
33. Write the structure for cellobiose, a disaccharide obtained by the hydrolysis of cellulose.
34. Draw the Haworth formulas for
(a) β-D-glucopyranosyl-(1-4)-α-D-galactopyranose
(b) β-D-galactopyranosyl-(1-6)-β-D-mannopyranose
35. A molar mass value of about 325,000 was calculated for a sample of amylopectin. Approximately how many glucose units are present in an average molecule of this amylopectin? (*Hint:* As each glucose is added to a growing amylopectin chain, a molecule of water is removed.)
36. Draw enough of the structural formula of cellulose acetate to show the repeating units and how they are linked.
37. Write three units of the polymeric structure of amylose.
38. How is the acidic nature of the mucopolysaccharides related to their biological function?
39. What changes must take place to convert cornstarch into the sweetener high fructose corn syrup?
40. Glucose units in cellulose are connected by β-1,4-glycosidic linkages. How is the β stereochemistry important in allowing the cellulose polymers to bond together to form cellulose fibers?
41. We sometimes hear the comment, "It is important that you chew your food carefully." How

does this statement apply to the digestion of starchy foods?

42. Which of these statements are correct? Rewrite each incorrect statement to make it correct.

(a) α-D-Glucopyranose and β-D-glucopyranose are enantiomers.
(b) The carbon of a secondary alcohol is more reduced than the carbon of a primary alcohol.
(c) D-Glyceraldehyde and L-glyceraldehyde are epimers.
(d) D-Threose and L-threose are epimers.
(e) There are eight stereoisomers of the aldopentoses.
(f) D-Glucose and L-glucose form identical osazones.
(g) Raffinose, which consists of one unit each of galactose, glucose, and fructose, is an oligosaccharide.
(h) Two aldohexoses that react with phenylhydrazine to yield identical osazones are epimers.
(i) Dextrose is another name for glucose.
(j) D-Mannitol is obtained from D-mannose by oxidation.
(k) Methyl glucosides are capable of reducing Fehling's and Barfoed's reagents.
(l) Humans are incapable of using cellulose directly as a food.
(m) Fructose can be classified as a hexose, a monosaccharide, and an aldose.
(n) The change in the specific rotation of a carbohydrate solution to an equilibrium value is called mutarotation.
(o) The disaccharide found in mammalian milk is galactose.
(p) The reserve carbohydrate of animals is glycogen.
(q) Starch consists of two polysaccharides known as amylose and amylopectin.
(r) Carbohydrates that are capable of reducing copper ions in Benedict's reagent are called reducing sugars.
(s) The polysaccharides cellulose, starch, and glycogen are all composed of glucose units.
(t) Invert sugar is sweeter than fructose.
(u) Sucrose, glucose, galactose, and fructose are reducing sugars.
(v) Methyl glycosides do not undergo mutarotation.
(w) Oxidation of D-erythrose with dilute HNO_3 yields *meso*-tartaric acid.
(x) The amylose polysaccharide coils into a helical shape.
(y) Aspartame is a non-nutritive sweetener.
(z) Mucopolysaccharides have many sugar units that act as bases.

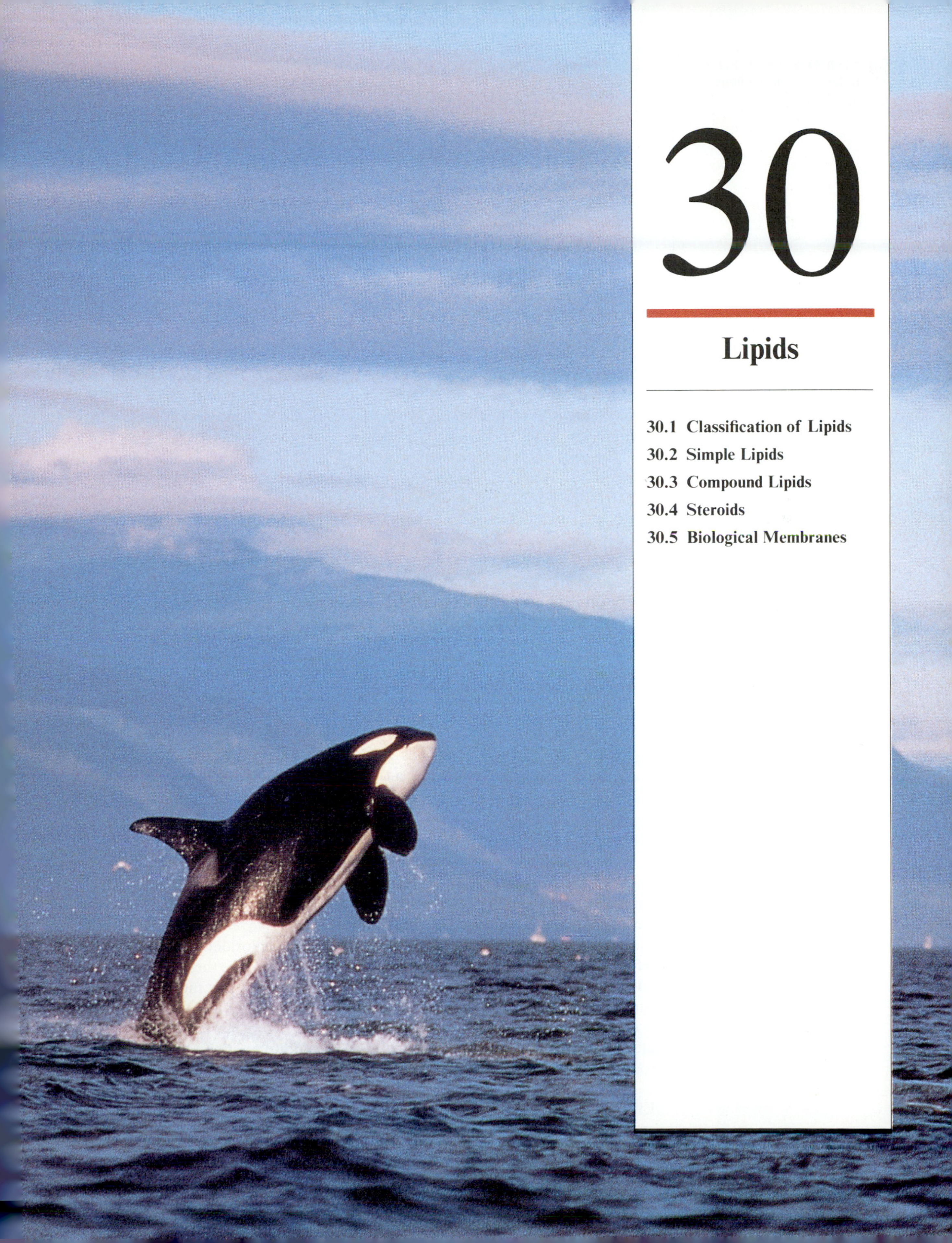

30

Lipids

◀ CHAPTER OPENING PHOTO: **The killer whale has a huge reserve of lipids.**

Lipids provide the bad, the good, and the ugly of the cleaning industry. For example, it can be bad luck to splatter a drop of oil (a lipid) on a new shirt—this spot just won't wash out with water. But, thank goodness for soap, another form of lipid. Even the ugliest stack of greasy dishes is no match for a sudsy basin of hot water and soap.

The way lipids interact with water is a key to their importance in nature. They are insoluble. An oil slick can spread for many square miles on the surface of the ocean partly because oil and water don't mix. Based on this same principle, cells surround themselves with a thin film of lipid, the cell membrane. We protect a fine wood floor with wax, another lipid, because we can depend on this material to adhere to the floor and not dissolve in water. A lipid's stickiness and lack of water solubility also can create diseases such as atherosclerosis, where arteries become partially clogged by lipids. Lipid characteristics bring both benefits and problems—these molecules are truly a mixed blessing.

30.1 CLASSIFICATION OF LIPIDS

lipids

Lipids are water insoluble, oily or greasy biochemical compounds that can be extracted from cells by nonpolar solvents such as ether, chloroform, or benzene. Unlike carbohydrates, lipids share no common chemical structure but are a catch all class. Still, these molecules must possess some structural similarities because of their shared water insolubility.

What makes a molecule, such as a lipid, insoluble in water? To answer this question, we must establish two important principles about water solutions: A compound may dissolve in water (1) if the water molecules bond well to the potential solute and (2) if the water molecules can still move relatively freely around the dissolved compound. For example, salt (sodium chloride) dissolves because it forms ions to which water molecules can bond *and* because these ions are small and do not significantly impede the movement of the water molecules. Sugar (sucrose) dissolves because it can form hydrogen bonds with water and because it is still a relatively small molecule. On the other hand, starch, a collection of glucose polymers, is too large to truly dissolve (although it can hydrogen-bond to water).

Lipid structures and solubilities differ from both salts and carbohydrates. Even though lipid molecules are not as big as a starch polymer, they are big enough to substantially affect the free movement of water molecules. In addition, lipids cannot hydrogen-bond to the extent of carbohydrates, nor do they form the large number of positive and negative charges found in a salt solution. Lipids are large and relatively nonpolar molecules and, thus, are water insoluble.

Compounds such as most carbohydrates and salts are said to be *hydrophilic* ("water loving"). In contrast, lipids are said to be *hydrophobic* ("water fearing").

Consider fatty acids, which are common components of lipids. As shown in Table 30.1, when the number of atoms in a fatty acid molecule increases, the water solubility of the fatty acid decreases dramatically. Water molecules can easily maneuver around smaller compounds like butyric acid, which is infinitely soluble in water. However, these same water molecules run into a huge barrier when they encounter the 18-carbon chain of stearic acid, and so only a little of this fatty acid dissolves in water (0.0003 g/100 g of water).

The hydrophobic nature of lipids contributes significantly to the biological functions of these molecules. Their water insolubility allows lipids to serve as barriers to aqueous solutions. This property, as we shall see later, is of great importance when lipids form cellular membranes. Lipids are classified as follows:

1. **Simple lipids**
   - **(a)** *Fats and oils:* esters of fatty acids and glycerol
   - **(b)** *Waxes:* esters of high molar mass fatty acids and high molar mass alcohols
2. **Compound lipids**
   - **(a)** *Phospholipids:* substances that yield glycerol, phosphoric acid, fatty acids, and nitrogen-containing bases upon hydrolysis
   - **(b)** *Sphingolipids:* substances that yield an unsaturated amino alcohol (sphingosine), a long-chain fatty acid, and either a carbohydrate or phosphate and a nitrogen base upon hydrolysis
   - **(c)** *Glycolipids:* substances that yield sphingosine, fatty acids, a nitrogen-containing base, and a carbohydrate upon hydrolysis
3. **Steroids**
   Substances possessing the steroid nucleus, which is a 17-carbon structure consisting of four fused carbocyclic rings. Cholesterol and several hormones are in this class.
4. **Miscellaneous lipids**
   Substances that do not fit into the preceding classifications; these include the fat-soluble vitamins A, D, E, and K.

The most abundant lipids are the fats and oils. These substances constitute one of the three important classes of foods. The discussion that follows is centered on fats and oils. A more complete consideration of the properties and composition of various fats and oils is given in Section 25.10.

30.2 Simple Lipids

Fatty Acids

Fatty acids, which form a part of most lipids, are carboxylic acids with long, hydrophobic side chains. The formulas for some of the most common fatty acids are shown in Table 30.1. All these acids are straight chain compounds with an even number of carbon atoms. Five of the fatty acids in this table—palmitoleic,

TABLE 30.1 Some Naturally Occurring Fatty Acids

Fatty acid	Number of C atoms	Formula	Solubility (g/100 g water)	Melting point (°C)
Saturated acids				
Butyric acid	4	$CH_3CH_2CH_2COOH$	∞	−4.7
Caproic acid	6	$CH_3(CH_2)_4COOH$	1.08	−1.5
Caprylic acid	8	$CH_3(CH_2)_6COOH$	0.07	16
Capric acid	10	$CH_3(CH_2)_8COOH$	0.015	32
Lauric acid	12	$CH_3(CH_2)_{10}COOH$	0.006	48
Myristic acid	14	$CH_3(CH_2)_{12}COOH$	0.002	57
Palmitic acid	16	$CH_3(CH_2)_{14}COOH$	0.0007	63
Stearic acid	18	$CH_3(CH_2)_{16}COOH$	0.0003	70
Arachidic acid	20	$CH_3(CH_2)_{18}COOH$	—	77
Unsaturated acids				
Palmitoleic acid	16	$CH_3(CH_2)_5CH{=}CH(CH_2)_7COOH$	—	0.5
Oleic acid	18	$CH_3(CH_2)_7CH{=}CH(CH_2)_7COOH$	—	13
Linoleic acid	18	$CH_3(CH_2)_4CH{=}CHCH_2CH{=}CH(CH_2)_7COOH$	—	−5
Linolenic acid	18	$CH_3CH_2CH{=}CHCH_2CH{=}CHCH_2CH{=}CH(CH_2)_7COOH$	—	−11
Arachidonic acid	20	$CH_3(CH_2)_4(CH{=}CHCH_2)_4CH_2CH_2COOH$	—	−50

oleic, linoleic, linolenic, and arachidonic—are unsaturated, having carbon–carbon double bonds in their structures. Animal and higher plant cells produce lipids in which palmitic, oleic, linoleic, and stearic acids predominate. Over one-half of plant and animal fatty acids are unsaturated, plant lipids tending to be more unsaturated than their animal counterparts.

Double bonds impart some special characteristics to the unsaturated fatty acids. Remember that the presence of double bonds raises the possibility of geometric isomerism (Section 22.3). Unsaturated fatty acids may be either cis or trans isomers. To illustrate the effect of these double bonds on fatty acid structure, the following two fatty acids are portrayed in a simplified manner with each of the many $—CH_2—$ groups as an apex at the intersection between two single bonds.

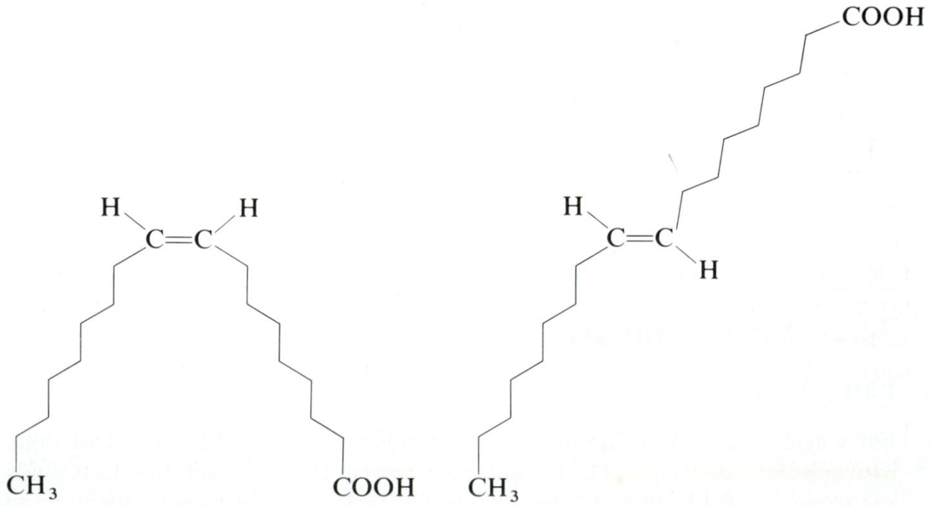

H C=C H
CH_3 COOH

Cis isomer

COOH
H C=C H
CH_3

Trans isomer

Note that the trans isomer is almost a linear molecule while the double bond in the cis isomer introduces a kink in the fatty acid structure. Unsaturated fatty acids found in nature are almost always cis isomers. These kinked fatty acids cannot stack closely together and, hence, they do not solidify easily. As shown in Table 30.1, unsaturated fatty acids have lower melting points than saturated fatty acids of a similar size. Cooking oils purchased from your market are liquids at room temperature because a high percentage of their fatty acids are unsaturated. In like manner, biological membranes are very fluid because of the presence of double bonds in their component fatty acids (see Section 30.5).

Three unsaturated fatty acids—linoleic, linolenic, and arachidonic—are essential for animal nutrition and must be present in the diet. Diets lacking these fatty acids lead to impaired growth and reproduction, and skin disorders such as eczema and dermatitis. A dermatitis disorder can be attributed to an unsaturated fatty acid deficiency if the symptoms clear up when that fatty acid is supplied in the diet.

Selected fatty acids, as well as other lipids, are biochemical precursors of several classes of hormones. The well-known steroid hormones are synthesized from cholesterol and will be discussed later in this chapter. Arachidonic acid and, to a lesser extent, linolenic acid are also used by the body to make hormonelike substances. The biochemicals derived from arachidonic acid are collectively termed eicosanoids using a derivative of the Greek word for twenty, *eikosi*, to indicate that these compounds have 20 carbon atoms. Prostaglandins are perhaps the best known of the eicosanoid class, which also includes the leukotrienes, prostacyclins, and thromboxanes. Cell membranes release arachidonic acid in response to a variety of circumstances, including infection and allergic reactions. In turn, enzymes in the surrounding fluid convert this fatty acid to specific eicosanoids by catalyzing the addition of oxygen to the arachidonic double bonds. Some examples of eicosanoids are shown in Figure 30.1.

Unlike true hormones, eicosanoids are not transported via the bloodstream to their site of action but rather take effect where they are synthesized. Prostaglandins are a primary cause of the swelling, redness, and pain associated with tissue inflammation. Platelets in the bloodstream form the thromboxanes, which act as vasoconstrictors and stimulate platelet aggregation, as an initial step in blood clotting. Leukotrienes are formed by a variety of white blood cells as well as other tissues and cause many of the symptoms associated with an allergy attack. For example, asthma is thought to be mediated by the leukotrienes.

Some relatively common medications affect the formation of the eicosanoids. Aspirin, by blocking the reaction between arachidonic acid and oxygen, stops inflammation caused by the prostaglandins. Recent research indicates that low levels of aspirin may prevent heart attacks and strokes, possibly by blocking synthesis of the thromboxanes that participate in blood clotting. Cortisone acts as an anti-inflammatory drug by decreasing release of arachidonic acid from the cell membranes. It also seems that diet may have a significant effect on eicosanoid formation. Fish oils contain fatty acids that inhibit formation of the thromboxanes and lead to formation of the less potent leukotrienes. It has been suggested that cultures for which fish is a dietary staple (such as the Greenland Eskimos) have a low level of heart disease, possibly because of a decrease in thromboxane formation.

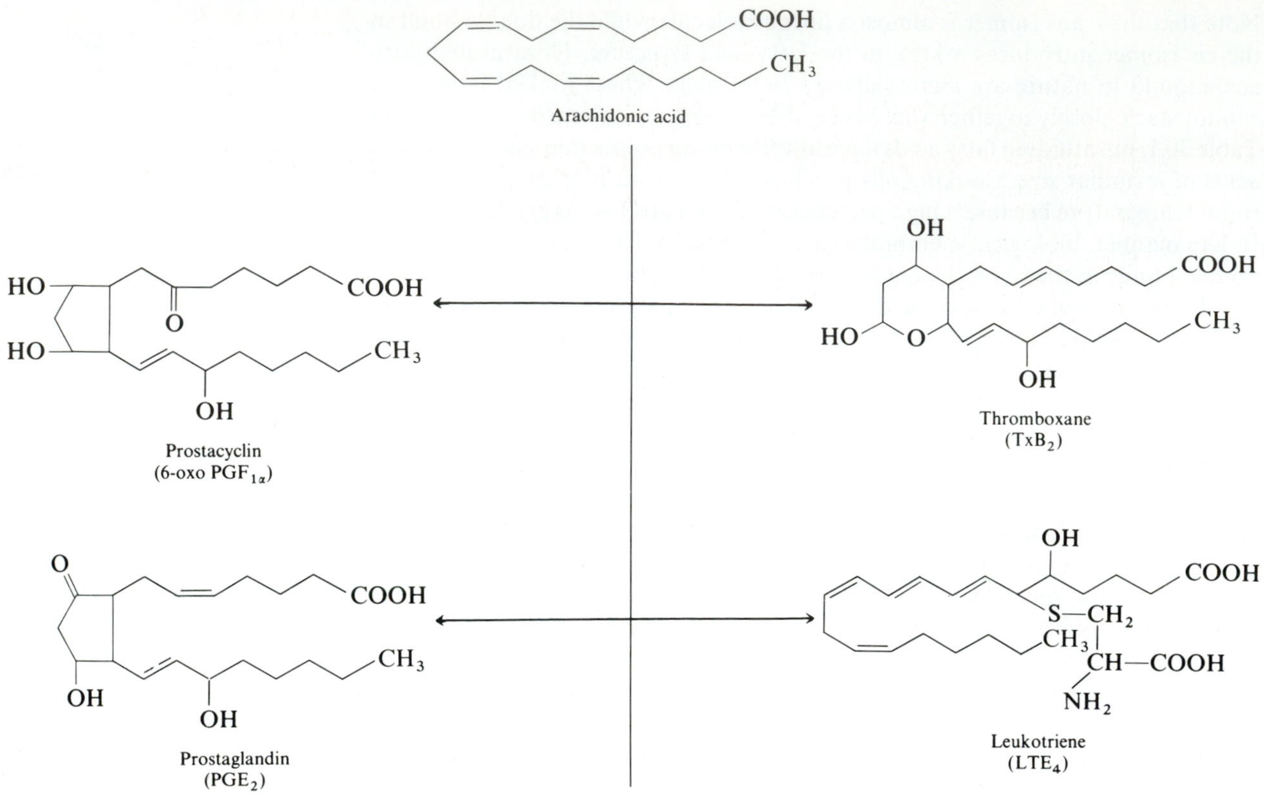

Arachidonic acid

Prostacyclin
(6-oxo $PGF_{1\alpha}$)

Thromboxane
(TxB_2)

Prostaglandin
(PGE_2)

Leukotriene
(LTE_4)

FIGURE 30.1
Several examples of eicosanoids. Each of these molecules is derived from arachidonic acid.

Fats and Oils

Chemically, fats and oils are esters of glycerol and the higher molar mass fatty acids. They have the general formula

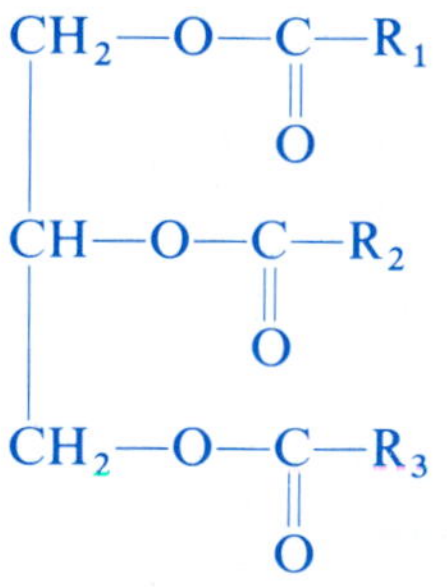

$$\begin{array}{l} CH_2-O-\overset{\displaystyle O}{\overset{\|}{C}}-R_1 \\ | \\ CH-O-\overset{\displaystyle O}{\overset{\|}{C}}-R_2 \\ | \\ CH_2-O-\overset{\displaystyle O}{\overset{\|}{C}}-R_3 \end{array}$$

where the R's can be either long-chain saturated or unsaturated hydrocarbon groups. Figure 30.2 shows a three-dimensional representation of a typical fat.

Fats may be considered to be triesters formed from the trihydroxy alcohol glycerol and three molecules of fatty acids. Most of the fatty acids in these esters have 14 to 18 carbons. Because there are three ester groups per glycerol, these molecules are called triacylglycerols, or triglycerides (an older name that is still commonly used) The three R groups are usually different.

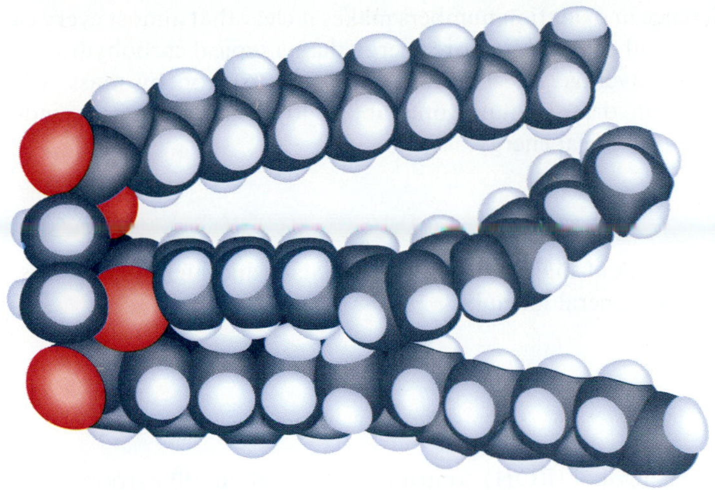

◀ FIGURE 30.2
Space-filling model of a triacylglycerol formed by reacting glycerol with one palmitic acid, one oleic acid, and one stearic acid. Note the kink introduced into oleic acid by the cis double bond.

$$\begin{array}{l} CH_2{-}O{-}\boxed{H + H{-}O}{-}\underset{\|}{\overset{}{C}}{-}R_1 \\ \qquad\qquad\qquad\qquad O \\ CH{-}O{-}\boxed{H + H{-}O}{-}\underset{\|}{C}{-}R_2 \\ \qquad\qquad\qquad\qquad O \\ CH_2{-}O{-}\boxed{H + H{-}O}{-}\underset{\|}{C}{-}R_3 \\ \qquad\qquad\qquad\qquad O \end{array} \longrightarrow \begin{array}{l} CH_2{-}O{-}\underset{\|}{C}{-}R_1 \\ \qquad\qquad O \\ CH{-}O{-}\underset{\|}{C}{-}R_2 + 3\,H_2O \\ \qquad\qquad O \\ CH_2{-}O{-}\underset{\|}{C}{-}R_3 \\ \qquad\qquad O \end{array}$$

Glycerol Fatty acids A triacylglycerol

Fats and oils fit the general description of a lipid. They are large molecules, averaging more than 50 carbon atoms per molecule, with many nonpolar, uncharged groups. Because they contain large numbers of saturated carbons, the triacylglycerols are hydrophobic and water insoluble.

Fats are an important food source for humans and normally account for about 25–50% of their caloric intake. When oxidized to carbon dioxide and water, fats supply about 40 kJ of energy per gram (9.5 kcal/g), which is more than twice the amount obtained from carbohydrates or proteins.

The energy from a fat is released when the reduced carbons are oxidized. In general, the more reduced a carbon, the more energy it contains (see Section 29.1). Triacylglycerol carbons are more reduced than those of most other foods. The typical carbon from a fat has an oxidation number of −2, whereas the typical carbon from a carbohydrate has an oxidation number of 0:

$$\begin{array}{c} OH \\ | \\ C{-}C{-}C \\ | \\ H \end{array} \leftarrow 0 \text{ (Oxidation number)} \qquad \begin{array}{c} H \\ | \\ C{-}C{-}C \\ | \\ H \end{array} \leftarrow -2 \text{ (Oxidation number)}$$

In a carbohydrate In a fat

This difference in oxidation numbers makes it clear that almost every carbon in a fat contains and can release more energy than a typical carbohydrate carbon. In addition, the average fat contains about 75% carbon by mass, whereas the average carbohydrate contains only about 40% carbon. Fats are indeed a rich source of biochemical energy.

Waxes

waxes

Waxes are esters of high molar mass fatty acids and high molar mass alcohols. They have the general formula

$$R'-\overset{\overset{\large O}{\|}}{C}-O-R$$

in which the alcohol (ROH) contributes up to about 30 carbons, and the fatty acid (R′COOH) provides an equivalent number of carbons. Waxes are very large molecules with almost no polar groups. They represent one of the most hydrophobic lipid classes.

Their extreme water insolubility allows waxes to serve a protective function. Leaves, feathers, fruit, and fur are often naturally coated with a wax. Hardwood floors, cars, and leather goods are just a few of the man-made products that can be protected by a wax. Waxes tend to be the hardest of the lipids because their carbon chains are long and have very few double bonds. As with fats and oils, the size of the wax molecule and the number of double bonds contained in its carbon chains determine how solid or liquid the wax will be.

30.3 Compound Lipids

Phospholipids

phospholipids

The **phospholipids** are a group of compounds that yield one or more fatty acid molecules, a phosphate group, and usually a nitrogenous base upon hydrolysis. The phosphate group and the nitrogenous base, which are found at one end of the phospholipid molecule, often have negative and positive charges. Consequently, in contrast to the triacylglycerols, phospholipids have hydrophobic ends that repel water and hydrophilic ends that interact with water.

$$\left.\begin{array}{l} CH_2-O-\boxed{\text{Fatty acid}} \\ \ | \\ CH-O-\boxed{\text{Fatty acid}} \\ \ | \\ CH_2-O-\boxed{\text{Fatty acid}} \end{array}\right\} \text{All hydrophobic}$$

A triacylglycerol

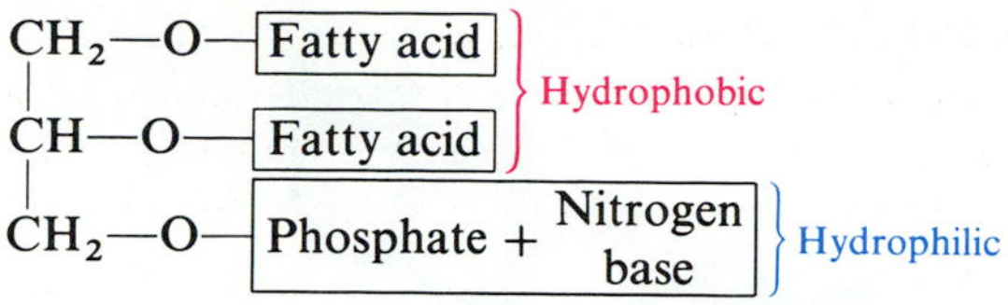

$$\begin{array}{l} CH_2-O-\boxed{\text{Fatty acid}} \\ | \\ CH-O-\boxed{\text{Fatty acid}} \\ | \\ CH_2-O-\boxed{\text{Phosphate} + \text{Nitrogen base}} \end{array} \quad \begin{array}{l} \left.\right\} \text{Hydrophobic} \\ \\ \left.\right\} \text{Hydrophilic} \end{array}$$

A phospholipid

As will be seen later in this chapter, a lipid with both hydrophobic and hydrophilic character is needed to make membranes. It is not surprising that phospholipids are one of the most important membrane components.

Phospholipids are also involved in the metabolism of other lipids and nonlipids. Although they are produced to some extent by almost all cells, most of the phospholipids that enter the bloodstream are formed in the liver. Representative phospholipids are described below.

Phosphatidic Acids Phosphatidic acids are glyceryl esters of fatty acids and phosphoric acid. The phosphatidic acids are important intermediates in the synthesis of triacylglycerols and other phospholipids.

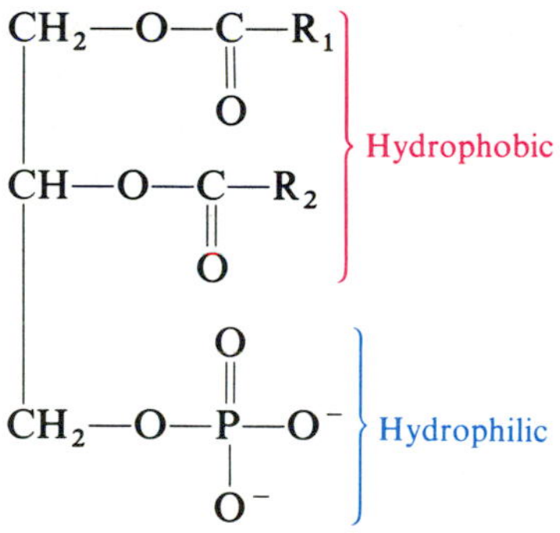

$$\begin{array}{l} CH_2-O-\underset{\underset{O}{\|}}{C}-R_1 \\ | \\ CH-O-\underset{\underset{O}{\|}}{C}-R_2 \\ | \\ CH_2-O-\overset{\overset{O}{\|}}{\underset{\underset{O^-}{|}}{P}}-O^- \end{array} \quad \begin{array}{l} \left.\right\} \text{Hydrophobic} \\ \\ \left.\right\} \text{Hydrophilic} \end{array}$$

A phosphatidic acid

A three-dimensional representation of a typical phosphatidic acid is given in Figure 30.3. As with all common phospholipids, the fatty acid chains are large relative to the rest of this molecule. Other phospholipids are formed from a given phosphatidic acid when specific nitrogen-containing compounds are linked to the phosphate group by an ester bond. Three commonly used nitrogen compounds are choline, ethanolamine, and L-serine:

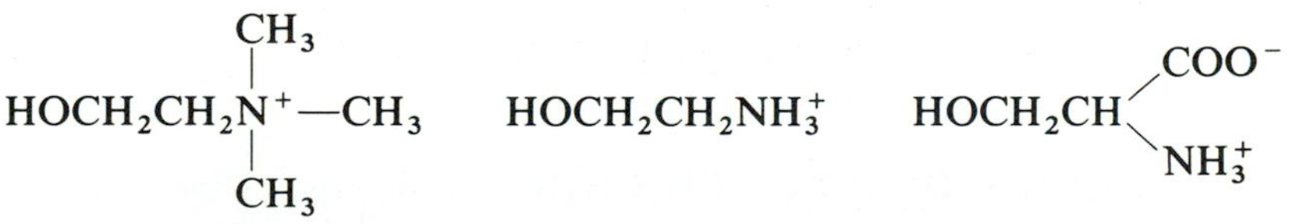

$$HOCH_2CH_2\overset{\overset{CH_3}{|}}{\underset{\underset{CH_3}{|}}{N^+}}-CH_3 \qquad HOCH_2CH_2NH_3^+ \qquad HOCH_2CH\begin{array}{l} ^{\diagup COO^-} \\ _{\diagdown NH_3^+} \end{array}$$

Choline Ethanolamine L-Serine

Because other phospholipids are structurally related to phosphatidic acids, their names are also closely related.

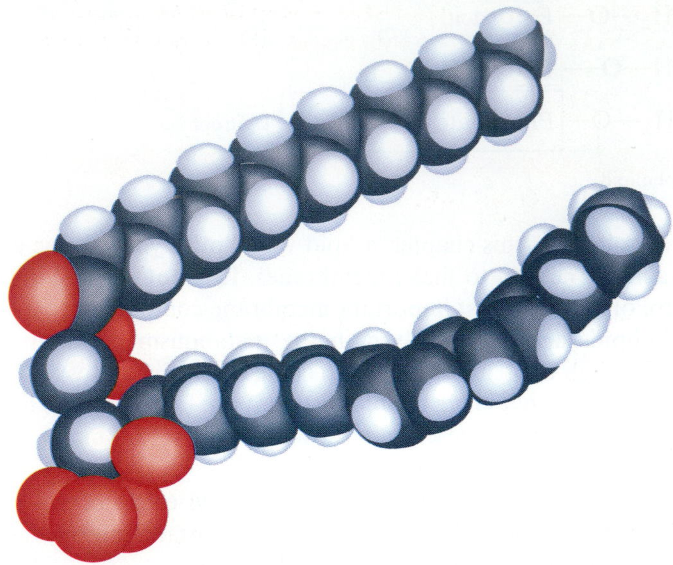

FIGURE 30.3 ▶ **Space-filling model of a phosphatidic acid which is esterified to palmitic acid at the top glycerol carbon and to oleic acid at the middle carbon. Note the kink introduced into the oleic acid by the cis double bonds.**

▲ **Chocolate is emulsified with phosphatidyl choline.**

Phosphatidyl Cholines (Lecithins) Phosphatidyl cholines (lecithins) are glyceryl esters of fatty acids, phosphoric acid, and choline. The synonym *lecithin* is an older term that is still used, particularly in commercial products that contain phosphatidyl choline. Phosphatidyl cholines are synthesized in the liver and are present in considerable amounts in nerve tissue and brain substance. Most commercial phosphatidyl choline is obtained from soybean oil and contains palmitic, stearic, palmitoleic, oleic, linoleic, linolenic, and arachidonic acids. Phosphatidyl choline is an edible and digestible emulsifying agent that is used extensively in the food industry. For example, chocolate and margarine are generally emulsified with phosphatidyl choline. Phosphatidyl choline is also used as an emulsifier in many pharmaceutical preparations.

The single most important biological function for phosphatidyl choline is as a membrane component. This phospholipid makes up between 10 and 20 percent of many membranes.

$$\begin{array}{l} CH_2{-}O{-}\overset{\displaystyle O}{\overset{\|}{C}}{-}R_1 \\ | \\ CH{-}O{-}\overset{\displaystyle O}{\overset{\|}{C}}{-}R_2 \\ | \\ CH_2{-}O{-}\underset{\displaystyle O^-}{\underset{|}{\overset{\displaystyle O}{\overset{\|}{P}}}}{-}O{-}CH_2CH_2\underset{\displaystyle CH_3}{\underset{|}{\overset{\displaystyle CH_3}{\overset{|}{N^+}}}}{-}CH_3 \end{array}$$

Hydrophobic (acyl groups); Choline; Hydrophilic (phosphate–choline head)

A phosphatidyl choline (lecithin) molecule

Phosphatidyl Ethanolamines (Cephalins) Another important constituent of biological membranes is the phosphatidyl ethanolamines (cephalins). These lipids are glyceryl esters of fatty acids, phosphoric acid, and ethanolamine ($HOCH_2CH_2NH_2$). They are found in essentially all living organisms.

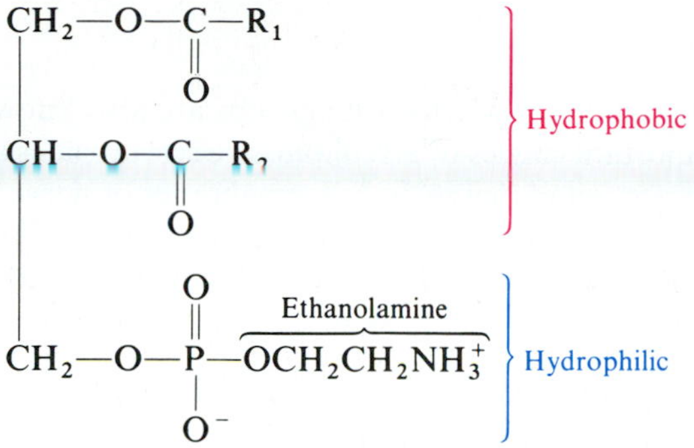

$$\begin{array}{l} CH_2-O-\overset{\displaystyle O}{\overset{\|}{C}}-R_1 \\ | \\ CH-O-\underset{\displaystyle O}{\underset{\|}{C}}-R_2 \\ | \\ CH_2-O-\overset{\displaystyle O}{\overset{\|}{\underset{\displaystyle O^-}{\underset{|}{P}}}}-\overbrace{OCH_2CH_2NH_3^+}^{\text{Ethanolamine}} \end{array} \quad \begin{array}{l} \left.\right\}\text{Hydrophobic} \\ \left.\right\}\text{Hydrophilic} \end{array}$$

A phosphatidyl ethanolamine (cephalin) molecule

Sphingolipids

Sphingolipids are compounds that, when hydrolyzed, yield a long chain fatty acid (18 to 26 carbons), a hydrophilic group (either phosphate and choline or a carbohydrate), and sphingosine (an unsaturated amino alcohol). Sphingosine substitutes for glycerol in these lipids. When drawn as follows, sphingosine can be seen as similar to glycerol esterified to one fatty acid:

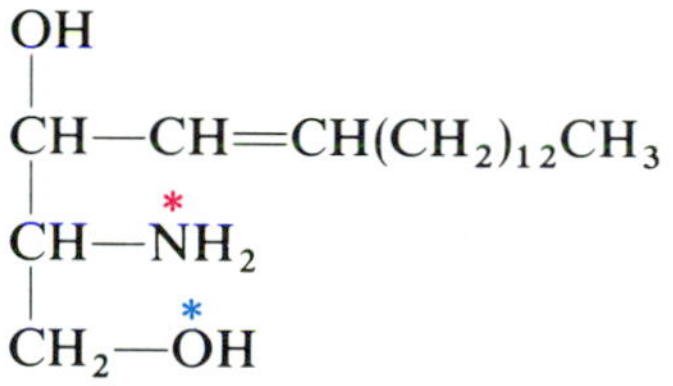

sphingolipids

$$\begin{array}{l} OH \\ | \\ CH-CH{=}CH(CH_2)_{12}CH_3 \\ | \\ CH-\overset{*}{N}H_2 \\ | \\ CH_2-\overset{*}{O}H \end{array} \qquad \begin{array}{l} CH_2-O-\overset{\displaystyle O}{\overset{\|}{C}}(CH_2)_nCH_3 \\ | \\ CH-\overset{*}{O}H \\ | \\ CH_2-\overset{*}{O}H \end{array}$$

Sphingosine 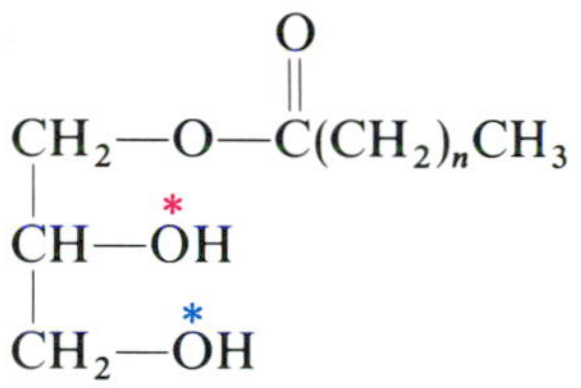 — Glycerol esterified with one fatty acid

The starred atoms on sphingosine react further to make sphingolipids, just as the starred atoms on the glycerol compound react further to give triacylglycerols or phospholipids.

Sphingolipids are common membrane components because they have both hydrophobic and hydrophilic character. For example, sphingomyelins are found in the myelin sheath membranes that surround nerves:

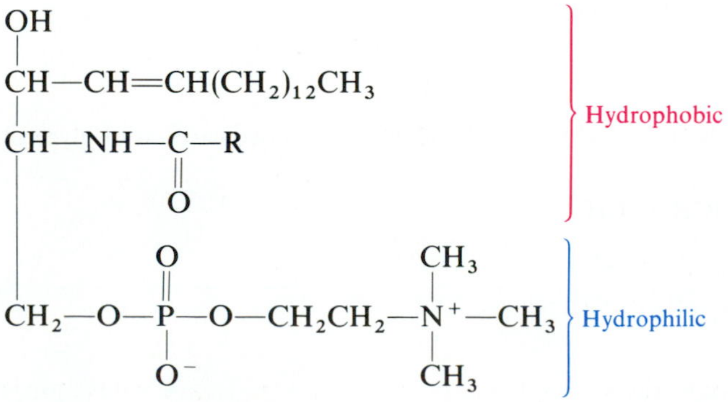

$$\begin{array}{l} OH \\ | \\ CH-CH{=}CH(CH_2)_{12}CH_3 \\ | \\ CH-NH-\underset{\displaystyle O}{\underset{\|}{C}}-R \\ | \\ CH_2-O-\overset{\displaystyle O}{\overset{\|}{\underset{\displaystyle O^-}{\underset{|}{P}}}}-O-CH_2CH_2-\overset{\displaystyle CH_3}{\overset{|}{\underset{\displaystyle CH_3}{\underset{|}{N^+}}}}-CH_3 \end{array} \quad \begin{array}{l} \left.\right\}\text{Hydrophobic} \\ \left.\right\}\text{Hydrophilic} \end{array}$$

A sphingomyelin

Notice the hydrophobic and hydrophilic parts of this molecule. Sphingomyelins can also be classified as phospholipids.

Glycolipids

glycolipids

Sphingolipids that contain carbohydrate groups are also known as **glycolipids**. The two most important classes of glycolipids are cerebrosides and gangliosides. These substances are found mainly in cell membranes of nerve and brain tissue. A cerebroside may contain either D-galactose or D-glucose. The following formula of a galactocerebroside shows the typical structure of cerebrosides:

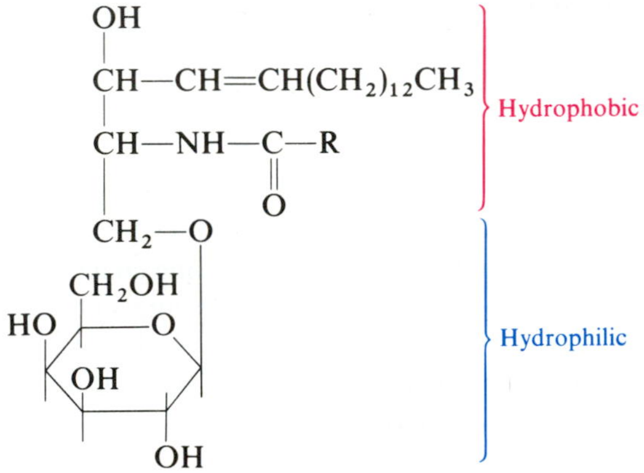

```
OH
|
CH—CH=CH(CH2)12CH3
|                            } Hydrophobic
CH—NH—C—R
|     ||
|     O
CH2—O
  CH2OH |
HO   —O |                    } Hydrophilic
   OH   (ring)
     OH
```

A β-galactocerebroside

Gangliosides are similar to cerebrosides in structure but contain complex oligosaccharides instead of simple monosaccharides.

EXAMPLE 30.1

Write the formula for a phosphatidyl ethanolamine that contains two palmitic acid groups.

SOLUTION

Phosphatidyl ethanolamine is a phospholipid and so contains glycerol, phosphate, fatty acids, and a nitrogen base. First, write the structure for glycerol.

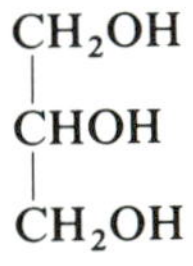

$$\begin{array}{l} CH_2OH \\ | \\ CHOH \\ | \\ CH_2OH \end{array}$$

The two palmitic acids are linked by ester bonds to the top two carbons of glycerol.

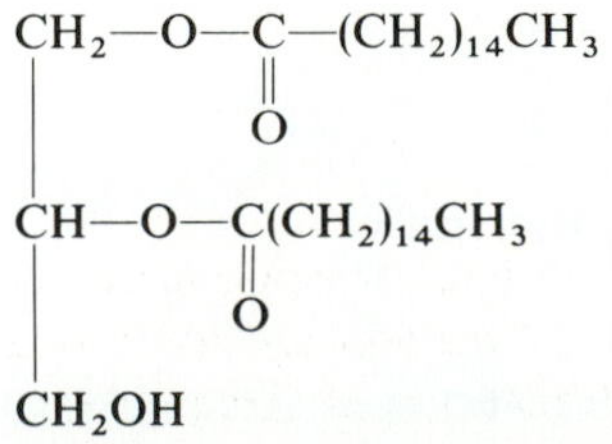

$$\begin{array}{l} CH_2\text{—O—}\underset{\underset{\displaystyle O}{\|}}{C}\text{—}(CH_2)_{14}CH_3 \\ | \\ CH\text{—O—}\underset{\underset{\displaystyle O}{\|}}{C}(CH_2)_{14}CH_3 \\ | \\ CH_2OH \end{array}$$

The phosphate group is connected via an ester bond to the bottom glycerol carbon to form a phosphatidic acid.

$$
\begin{array}{l}
CH_2{-}O{-}\underset{\displaystyle \overset{\|}{O}}{C}{-}(CH_2)_{14}CH_3 \\
|\\
CH{-}O{-}\underset{\displaystyle \overset{\|}{O}}{C}(CH_2)_{14}CH_3 \\
|\\
CH_2{-}O{-}\underset{\displaystyle \overset{|}{O^-}}{\overset{\displaystyle O \atop \|}{P}}{-}O^-
\end{array}
$$

Finally, the ethanolamine is linked to the phosphate group to yield phosphatidyl ethanolamine.

$$
\begin{array}{l}
CH_2{-}O{-}\underset{\displaystyle \overset{\|}{O}}{C}{-}(CH_2)_{14}CH_3 \\
|\\
CH{-}O{-}\underset{\displaystyle \overset{\|}{O}}{C}(CH_2)_{14}CH_3 \\
|\\
CH_2{-}O{-}\underset{\displaystyle \overset{|}{O^-}}{\overset{\displaystyle O \atop \|}{P}}{-}OCH_2CH_2NH_3^+
\end{array}
$$

PRACTICE Give the structure of a sphingomyelin that contains stearic acid.

Answer:

$$
\begin{array}{l}
OH \\
|\\
CH{-}CH{=}CH(CH_2)_{12}CH_3 \\
|\\
CH{-}NH{-}\underset{\displaystyle \overset{\|}{O}}{C}(CH_2)_{16}CH_3 \\
|\\
CH_2{-}O{-}\underset{\displaystyle \overset{|}{O^-}}{\overset{\displaystyle O \atop \|}{P}}{-}OCH_2CH_2{-}\underset{\displaystyle \overset{|}{CH_3}}{\overset{\displaystyle CH_3 \atop |}{N^+}}{-}CH_3
\end{array}
$$

30.4 STEROIDS

Steroids are compounds that have the steroid nucleus, which consists of four fused carbocyclic rings. This nucleus contains 17 carbon atoms in one five-membered and three six-membered rings. Modifications of this nucleus in the

steroids

▲ **Muscle mass can be increased through the use of steriods.**

various steroid compounds include added side chains, hydroxyl groups, carbonyl groups, ring double bonds, and so on.

Steroid ring nucleus

Steroids are closely related in structure but are highly diverse in function. Examples of steroids and steroid containing materials are (1) cholesterol, the most abundant steroid in the body, which is widely distributed in all cells and serves as a major membrane component; (2) bile salts, which aid in the digestion of fats; (3) ergosterol, a yeast steroid, which is converted to vitamin D by ultraviolet radiation; (4) digitalis and related substances called cardiac glycosides, which are potent heart drugs; (5) the adrenal cortex hormones, which are involved in metabolism; and (6) the sex hormones, which control sexual characteristics and reproduction. The formulas for several steroids are given in Figure 30.4.

Cholesterol is the parent compound from which the steroid hormones are synthesized. As we will see, only small changes in steroid structure can lead to large changes in hormonal action. Cholesterol is first converted to progesterone, a compound which helps control the menstrual cycle and pregnancy. This hormone is, in turn, the parent compound from which testosterone and the adrenal corticosteroids are produced. Notice that the long side chain on carbon 17 in cholesterol (Figure 30.4) is smaller in progesterone and is eliminated when testosterone is formed. Interestingly, testosterone is the precursor for the female sex hormones such as estradiol. These sex hormones are produced by the gonads, either the male testes or the female ovaries. The small structural differences between testosterone and estradiol trigger vastly different physiological responses. If the embryonic male gonads are surgically removed and testosterone is no longer available, the embryo develops as a female. In contrast, the female hormones seem to be important in sexual maturation and function but not in embryonic development. It appears that embryonic mammals are programmed to develop as females unless this program is overridden by the action of testosterone.

Cholesterol is also used to build cell membranes, many of which contain about 25% by mass of this steroid. In fact, often there is as much cholesterol as there is phospholipid, sphingolipid, or glycolipid. These latter three lipid classes cause the membrane to be more oily; cholesterol solidifies the membrane. This different behavior arises from an important structural difference. Cholesterol's four fused rings make it a rigid molecule while other membrane lipids are more flexible because of their fatty acid chains. When biological membranes are synthesized, their cholesterol level is adjusted to achieve an appropriate balance between a solid and liquid consistency.

Cholesterol is a steroid of special interest not only because it is the precursor of many other steroids and an important membrane component, but because of its association with atherosclerosis. This metabolic disease leads to the deposition

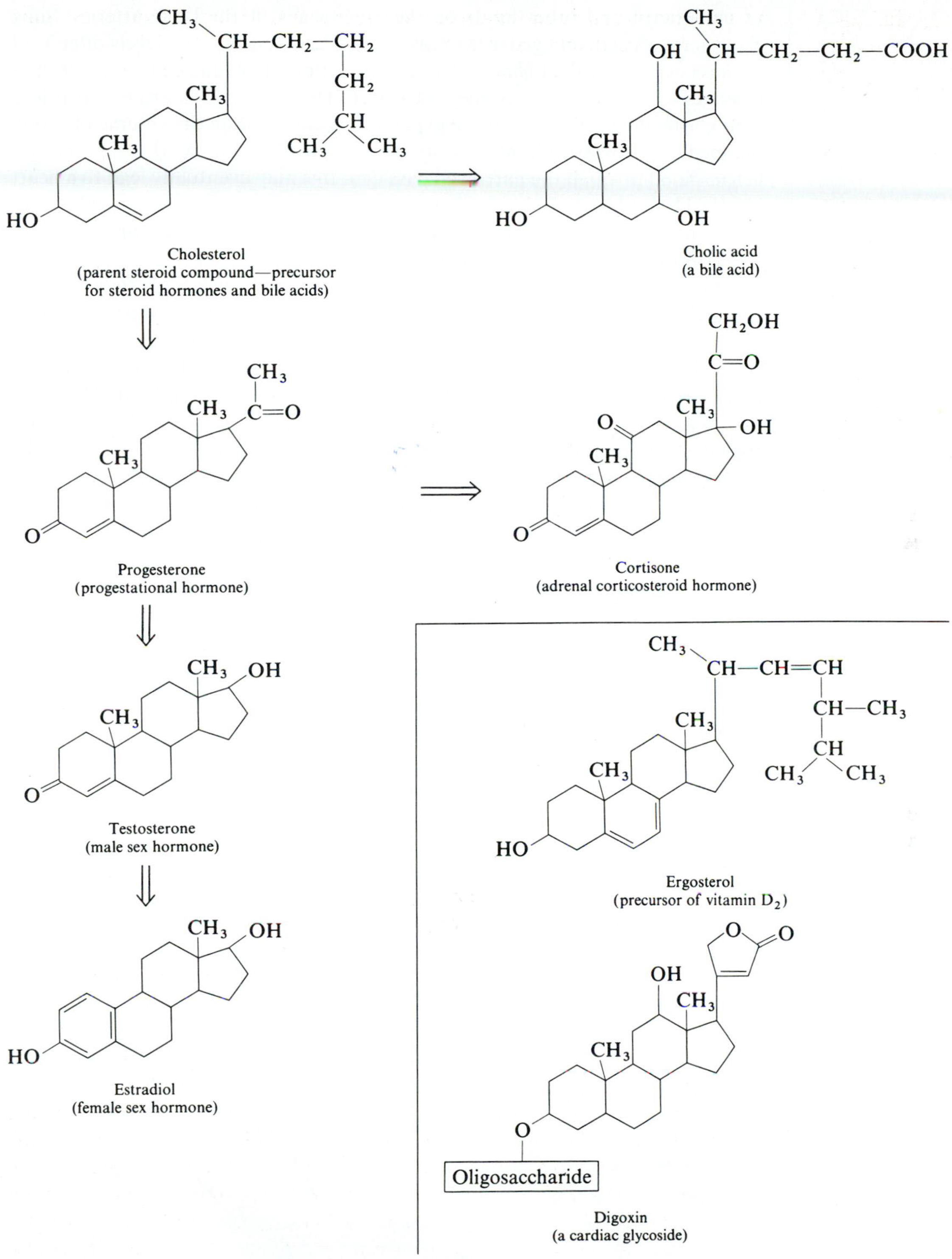

Cholesterol
(parent steroid compound—precursor for steroid hormones and bile acids)

Cholic acid
(a bile acid)

Progesterone
(progestational hormone)

Cortisone
(adrenal corticosteroid hormone)

Testosterone
(male sex hormone)

Ergosterol
(precursor of vitamin D_2)

Estradiol
(female sex hormone)

Digoxin
(a cardiac glycoside)

▲
FIGURE 30.4
Structures of selected steroids. Arrows show the biosynthetic relationship between steroids derived from cholesterol.

of cholesterol and other lipids on the inner walls of the large arteries. Since cholesterol is relatively water insoluble, high serum cholesterol levels often lead to these deposits, called *plaque*. The accumulation of plaque causes the arterial passages to become progressively narrower. The walls of the arteries also lose their elasticity and their ability to expand to accommodate the volume of blood pumped by the heart. Blood pressure increases as the heart works to pump sufficient blood through the narrowed passages; this may eventually lead to a heart attack. The build-up of plaque also causes the inner walls to have a rough rather than a normal smooth surface. This condition is favorable to coronary thrombosis (heart attack due to blood clots).

Physicians are concerned about serum cholesterol levels and often advise us to watch our dietary intake. We ingest about 300 mg of cholesterol per day on the average while our bodies synthesize another 1 g. Note that only a small part of the daily supply of cholesterol comes from the diet. However, the dietary amount is critical because the quantity of cholesterol which is excreted (about 500 mg per day) is also very small. It appears that once cholesterol has been absorbed from the intestines, this steroid is only very gradually eliminated from the body. Thus, an important means for lowering serum cholesterol levels is to severely restrict major dietary sources such as red meat, liver, and eggs.

Because of the variety of both beneficial and dangerous effects of cholesterol, the body has developed a careful distribution system for this steroid (see Figure 30.5). Dietary cholesterol is transferred to the liver, which is the clearinghouse for cholesterol transport to the rest of the body. Cholesterol leaves the liver in association with other lipids and certain proteins in the form of an aggregate known as the *very low density lipoprotein* (VLDL). As the other lipids and proteins are removed from this aggregate, it is converted first to an *intermediate density lipoprotein* (IDL) and then to a *low density lipoprotein* (LDL). Cells needing cholesterol are able to absorb the low density lipoprotein. Proper

FIGURE 30.5 ▶ Schematic representation of the lipoprotein distribution system.

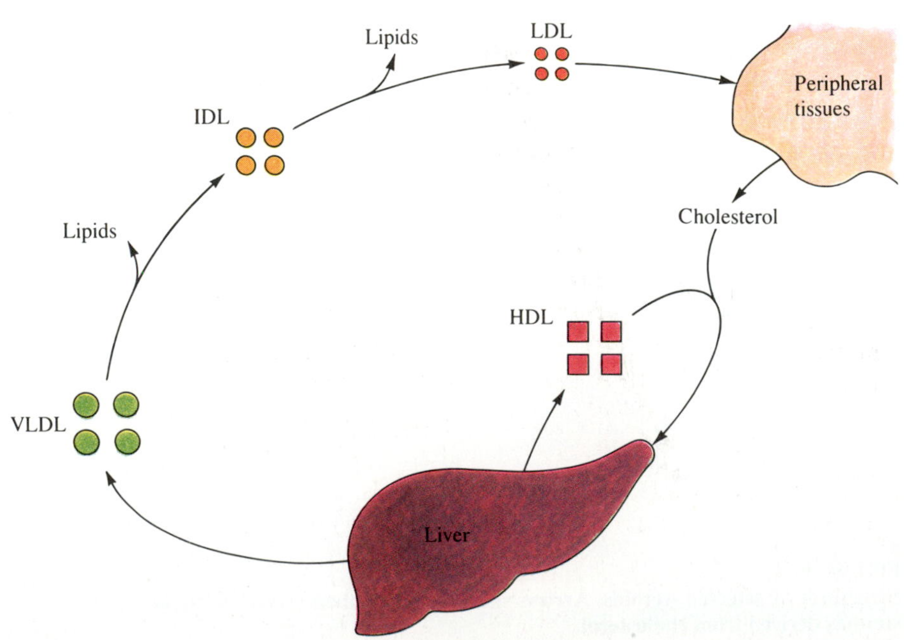

transfer of cholesterol also depends on a fourth lipoprotein, the *high density lipoprotein* (HDL). HDL acts as a cholesterol scavenger by collecting this steroid and returning it to the liver, essentially opposing the action of LDL, which delivers cholesterol to body tissues. A normal cholesterol metabolism must involve a proper balance of these lipoprotein factors.

Atherosclerosis can result from an imbalance in the cholesterol distribution system as well as an increase in serum cholesterol. High cholesterol diets correlate with plaque formation and also lead to an increase in amounts of LDL. However, any defect in the steroid distribution system can also cause plaque formation. People with high plasma LDL concentrations are prone to atherosclerosis, even though they may be on low cholesterol diets. In contrast, large amounts of HDL seem to prevent plaque formation. HDL levels can be increased by strenuous exercise, weight loss, and the female sex hormones. However, the most common drug treatments to prevent atherosclerosis still involve the reduction of serum cholesterol, either by preventing absorption of cholesterol from the intestine or by inhibiting cellular synthesis of this steroid. Treatments such as these can decrease serum cholesterol levels by about 50%.

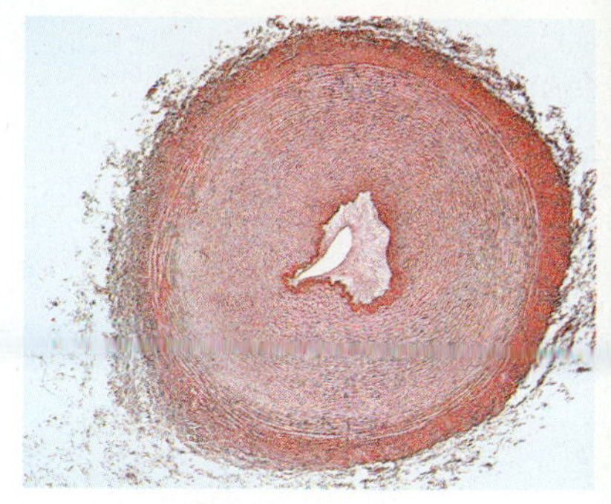

▲ **This clogged artery shows plaque formation on the artery wall.**

30.5 Biological Membranes

Biological membranes are thin, semipermeable cellular barriers. The general function of these barriers is to exclude dangerous chemicals from the cell while allowing nutrients to enter. Membranes also confine special molecules to specific sections of the cell. Because almost all the dangerous chemicals, nutrients, and special molecules are water soluble, the membranes can act as effective barriers only if they impede the movement of hydrophilic (water soluble) molecules.

To act as such a barrier, a membrane must have some special properties. To exclude water and water solutes, the bulk of a membrane must be hydrophobic. But a membrane necessarily touches water both inside and outside the cell. Therefore, the surface of a membrane must be hydrophilic. Thus a membrane can be visualized as being layered much like a piece of laminated plywood.

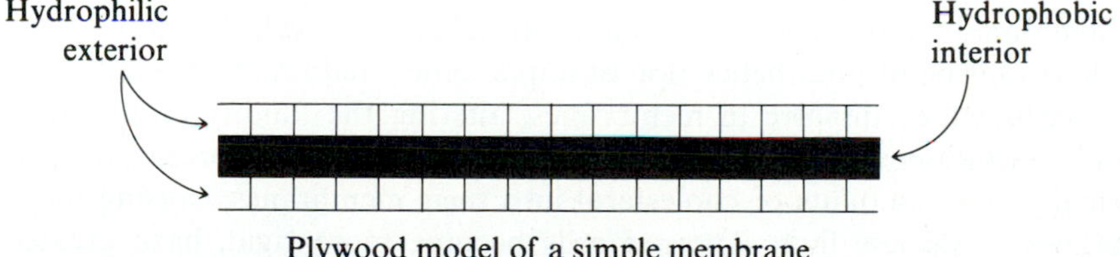

Plywood model of a simple membrane

The interior provides the barrier while the exterior interacts with the aqueous environment.

The cell uses lipids to give the membrane its hydrophobic nature. In fact, by selecting the right lipids, both the hydrophobic and hydrophilic portions of a membrane can be assembled. There are several classes of membrane lipids. The most important of these are the phospholipids and sphingolipids. Remember that these lipids have both a hydrophobic and a hydrophilic section.

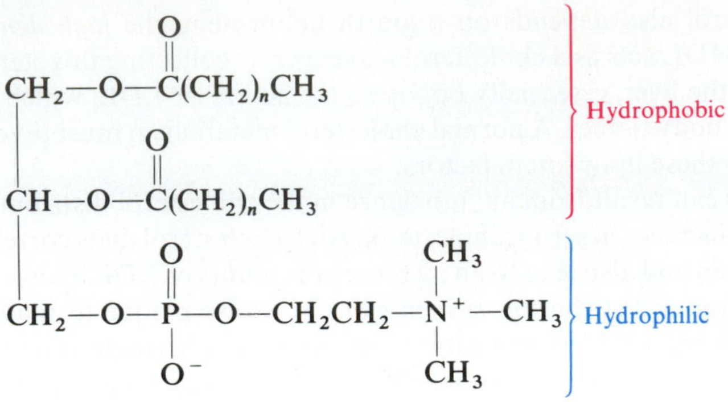

$$
\begin{array}{l}
\quad\quad\;\; O \\
\quad\quad\;\; \| \\
CH_2-O-C(CH_2)_nCH_3 \\
| \quad\quad\;\; O \\
| \quad\quad\;\; \| \\
CH-O-C(CH_2)_nCH_3 \\
| \quad\quad\;\; O \quad\quad\quad\quad\quad\quad CH_3 \\
| \quad\quad\;\; \| \quad\quad\quad\quad\quad\quad\;\; | \\
CH_2-O-P-O-CH_2CH_2-N^+-CH_3 \\
\quad\quad\quad\;\; | \quad\quad\quad\quad\quad\quad\;\; | \\
\quad\quad\quad\; O^- \quad\quad\quad\quad\quad\; CH_3
\end{array}
$$

Hydrophobic (fatty acid chains); Hydrophilic (phosphate and choline)

A phosphatidyl choline (lecithin) molecule

A membrane lipid can be pictured as having two long, hydrophobic tails (the fatty acids in the example given) and a small, hydrophilic end (the phosphate and choline in the example). Often a membrane lipid is drawn schematically as follows:

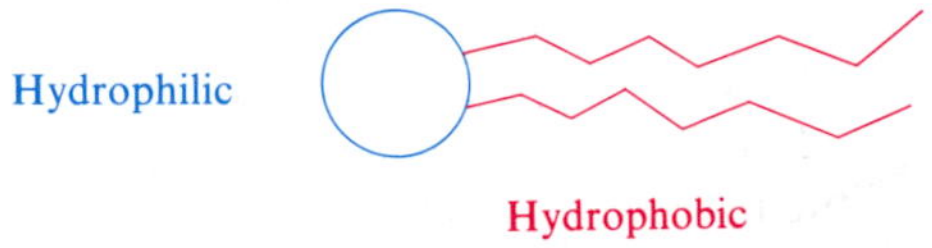

Hydrophilic

Hydrophobic

lipid bilayer

By stacking these lipids together, a basic membrane is formed. This structure is known as a **lipid bilayer**. It has the necessary properties of a membrane—a hydrophobic barrier and a hydrophilic surface.

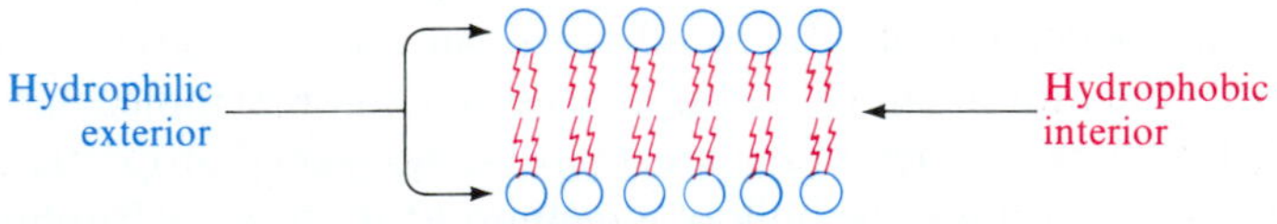

Hydrophilic exterior

Hydrophobic interior

Lipids give this barrier an oily or fluid appearance. The more unsaturated fatty acids in the membrane, the more fluid it will be. Other lipids, most importantly cholesterol, cause the membrane to be less fluid.

Membrane fluidity can have significant effects on cell function. It is thought that general anesthetics (for example, ether, halothane) are effective partly because they dissolve in membranes, altering the fluidity of the lipid bilayer. During severe cirrhosis of the liver, red cells are forced to take abnormally large amounts of cholesterol into their membranes, causing these membranes to be less fluid. The red cells become more rigid, have greater difficulty passing through narrow capillaries, and are destroyed more easily.

liposomes

Artificial lipid bilayers have important medical applications. When membrane lipids like phospholipids are specially mixed with aqueous solutions, small spheres of lipid bilayers form. These **liposomes** enclose an aqueous core into which solutes can be placed. Because of the barrier properties of the bilayer, the

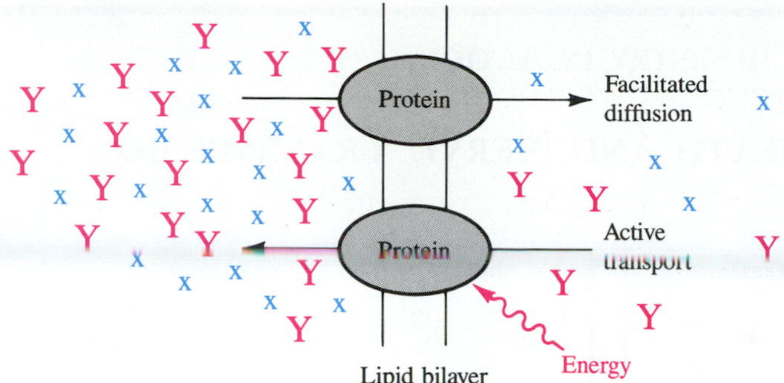

◄ **FIGURE 30.6**
How membrane transport is aided by proteins: Molecules or ions (symbolized by Y and x) can move from high concentration to low concentration without energy (facilitated diffusion), but movement in the reverse direction requires energy (active transport).

interior solute is completely separated from the external solution. Such liposome packages are being developed for drug delivery; even though the patient is injected with the drug, the drug is not exposed until the liposome reaches its particular target cell or organ. (The liposome bilayer is designed to be recognized, bound and absorbed *only* by the target cells.) Ideally, the drug does not affect other parts of the body.

The lipid bilayer acts as a barrier to water and hydrophilic molecules because of its hydrophobic interior. Because a cell is surrounded by a lipid bilayer, most charged or polar molecules have difficulty entering the cell. But the cell must have nutrients, so it is faced with a dilemma. How can it selectively allow some hydrophilic molecules to cross the lipid bilayer while excluding others?

Proteins in the fluid bilayer solve this dilemma. These proteins allow specific molecular transport through the hydrophobic interior. (For a general discussion of proteins, see Chapter 31.) They recognize specific molecules on the exterior of a cell membrane and shuttle these molecules into the cell. This process may be as simple as providing a tunnel through the membrane for selected nutrients. If the protein helps (facilitates) transport without using energy, the process is called **facilitated diffusion**. Some transport requires energy, such as when molecules are moved from areas of low concentration to areas of high concentration (the opposite direction from that of diffusion). This energy requiring transport is termed **active transport** (see Figure 30.6).

facilitated diffusion

active transport

Other proteins are located in the lipid bilayer to allow special reactions to occur. These special reactions, which often could be harmed by water, are enclosed in the hydrophobic interior of the lipid bilayer. Much of the cellular energy production (oxidation–reduction) occurs in this environment.

Thus a complete cellular membrane must have both lipid and protein. A typical membrane includes about 60% protein, 25% phospholipid, 10% cholesterol, and 5% sphingolipid. The fluid lipid bilayer is studded with many solid proteins. The proteins form a random pattern on the outer surface of the oily lipid. This general membrane is called the **fluid-mosaic model** (see Figure 30.7). Proteins that are primarily inside the lipid bilayer are termed *intrinsic membrane proteins*, and those on the surface are named *extrinsic membrane proteins*.

fluid-mosaic model

Chemistry in Action

The Myelin Sheath and Nerve Transmission

Human nerve tissue is a good example of the importance of membranes. Nerves coordinate many of the body processes, allowing life to continue. Their ability to function depends primarily on the characteristics of their membranes.

Examine the human motor neuron shown below. Notice the thickness of the axon (long cylindrical portion). The membranes of the axon are so important that two cells join together to provide them; the outer wrapping from one cell is the myelin sheath, and the thin inner neuronal membrane is from the other cell.

The neuronal membrane transmits electrical nerve impulses. To accomplish this task the neuron, using energy, slowly concentrates potassium ions inside the cell while expelling sodium ions (active transport shown in the diagram above). Throughout life, proteins within the neuronal membranes constantly pump these ions. When a neuron transmits a signal, membrane proteins allow a small portion of the potassium ions to flow rapidly out of the cell while sodium ions flow in. The sodium and potassium ions now flow from areas of high concentration to areas of low concentration without using energy. This is an example of facilitated diffusion. The rapid ion movement is an electrical nerve impulse.

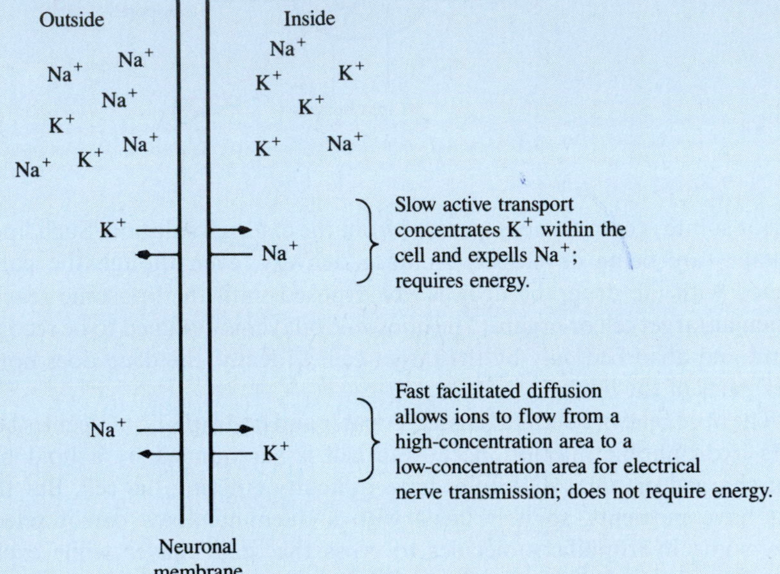

Scanning electron micrograph of motor nerve end plates in muscle. ▼

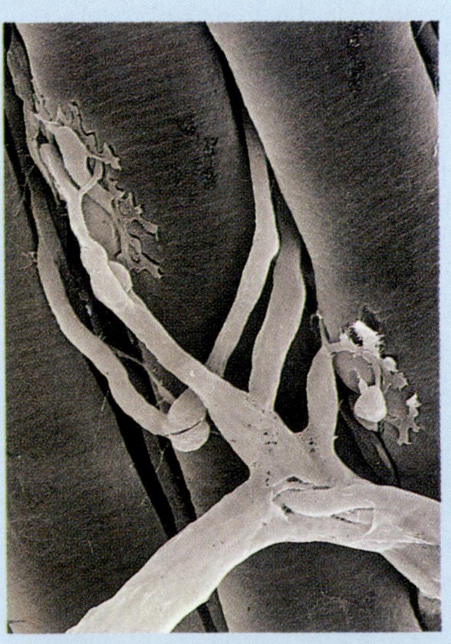

The myelin sheath serves as insulation for electrical nerve transmission (like the insulation around a copper wire). It is made up of concentric wrappings of a membrane that is high in lipid but low in protein (about 30% protein and 70% lipid). Remember that a lipid bilayer impedes movement of hydrophilic groups such as ions. The major function of the myelin sheath is to act as a barrier.

If the myelin sheath is removed, nerve transmission is faulty. Multiple sclerosis is a disease that causes destruction of the myelin sheath in portions of the central nervous system. The cause of this crippling disease is not completely understood. As the myelin disappears, the affected person experiences vision impairment, muscle weakness, and lack of coordination. These nerve-based symptoms arise because of poor nerve electrical transmission.

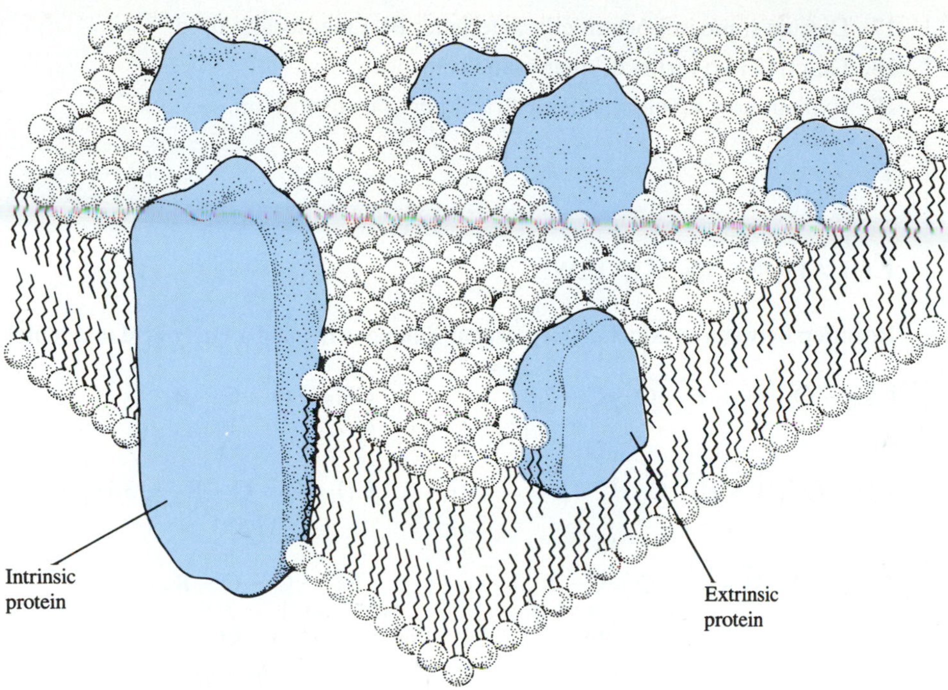

◀ **FIGURE 30.7**
The fluid-mosaic model of a membrane.

CONCEPTS IN REVIEW

1. Describe several general features of lipid structures.
2. Describe the classes of lipids and their functions.
3. Briefly explain why fatty acids are hydrophobic.
4. State which fatty acids commonly occur in fats and oils.
5. State which fatty acids are essential to human diets.
6. Explain why unsaturated fatty acids have lower melting points than the corresponding saturated fatty acids.
7. Briefly discuss the biological importance of eicosanoids.
8. Define the major biological function of waxes.
9. Describe the general structural makeup of phospholipids.
10. Describe the similarities between phospholipids and sphingolipids.
11. Draw the structural feature common to all steroids.
12. Draw the structures for cholesterol and several other common steroids.
13. Describe the biochemical relationships between common steroid hormones.
14. Discuss various components of the steroid distribution system.
15. Discuss atherosclerosis and the factors that affect it.
16. Draw a schematic of a lipid bilayer.

17. Discuss some important characteristics of a membrane lipid.

18. Describe two forms of membrane transport.

19. Explain how myelin helps electrical nerve transmission.

EQUATIONS IN REVIEW

Reaction Between an Alcohol (Glycerol or Sphingosine) and a Fatty acid (example)

$$\begin{array}{l} CH_2-O-\boxed{H + H-O}-\underset{\underset{O}{\|}}{C}-R_1 \\ | \\ CH-O-\boxed{H + H-O}-\underset{\underset{O}{\|}}{C}-R_2 \\ | \\ CH_2-O-\boxed{H + H-O}-\underset{\underset{O}{\|}}{C}-R_3 \end{array} \longrightarrow \begin{array}{l} CH_2-O-\underset{\underset{O}{\|}}{C}-R_1 \\ | \\ CH-O-\underset{\underset{O}{\|}}{C}-R_2 \\ | \\ CH_2-O-\underset{\underset{O}{\|}}{C}-R_3 \end{array} + 3\,H_2O$$

Glycerol   Fatty acids   A triacylglycerol

EXERCISES

1. Why are the lipids, which are dissimilar substances, classified as a group?
2. Briefly explain why caproic acid is more water-soluble than stearic acid.
3. Write the structural formula of a triacylglycerol that contains one unit each of palmitic, stearic, and oleic acids. How many other triacylglycerols are possible, each containing one unit of each of these acids?
4. Draw formulas for the products of a hydrolysis reaction involving a triacylglycerol that contains palmitic acid, oleic acid, and linoleic acid.
5. What are the three essential fatty acids? What are the consequences of their being absent from the diet?
6. Briefly explain why arachidonic acid is of special biological importance.
7. List two reasons why fats contain more biochemical energy than carbohydrates.
8. How is aspirin thought to relieve inflammation?
9. Wax serves a protective function on many types of leaves. How does it do this?
10. What two properties must a membrane lipid possess?
11. In what organ in the body are phospholipids mainly produced?
12. Lecithins and cephalins are both derivatives of phosphatidic acid. Indicate how they differ from each other.
13. Draw the structure of phosphatidyl serine.
14. In what ways is sphingosine similar to a glycerol molecule that has been esterified to one fatty acid?
15. Show the structure of a glucocerebroside.
16. What two structural features allow sphingomyelin to serve as a membrane lipid?
17. What common structural feature is possessed by all steroids? Write the structural formulas of two steroids.
18. List the four classes of eicosanoids.
19. How does a diet that contains a large amount of fish possibly decrease the risk of a heart attack?
20. Why are waxes generally the most solid of the lipids?
21. What is atherosclerosis? How is it produced and what are its symptoms?
22. Why is dietary cholesterol intake potentially critical in controlling high serum cholesterol levels?

23. Briefly describe the body's cholesterol distribution system.
24. Why is HDL a potential aid in controlling serum cholesterol levels?
25. Draw the structure of a membrane lipid.
26. Why can a lipid bilayer be described as a barrier?
27. What advantage does a liposome provide as a vehicle for drug delivery?
28. Why is the myelin sheath known as an insulator?
29. Would you expect an intrinsic protein to be hydrophobic? Explain.
30. Distinguish active transport from facilitated diffusion.
31. Which of these statements are correct? Rewrite each incorrect statement to make it correct.
    (a) A lipid will dissolve in water.
    (b) Lipids are often small molecules.
    (c) A triacylglycerol is high in biochemical energy because it contains many oxidized carbons.
    (d) Linoleic acid has a lower melting point than stearic acid.
    (e) Phosphatidyl choline is also known as lecithin.
    (f) A sphingolipid always contains carbohydrate.
    (g) Cholesterol is often found in membranes.
    (h) Movement of potassium ion into a nerve cell is an example of facilitated diffusion.
    (i) A lipid bilayer has a hydrophilic interior.
    (j) The "fluid" in the fluid-mosaic model refers to the lipid bilayer.
    (k) Estradiol is the precursor of testosterone.
    (l) Cortisone blocks inflammation by limiting the release of arachidonic acid from membranes.
    (m) Cells that need cholesterol are able to absorb the low density lipoprotein.
    (n) Thromboxane formation may be a factor in heart disease.

31

Amino Acids, Polypeptides, and Proteins

◀ **Chapter Opening Photo: This Olympic weight lifter uses muscles composed of long chains of proteins.**

The word "protein" derives from the Greek word *proteios* meaning "first." When these chemicals were discovered, they were placed first in biological importance. Although biochemists now take a more balanced view of the importance of proteins, think of the startling properties provided us by these molecules. Spider-web protein is many times stronger than the toughest steel; hair, feathers, and hooves are all made from one related group of proteins; another protein provides the glass clear lens material needed for vision. If very small quantities (milligram amounts) of some proteins are missing, a person's metabolic processes may be out of control. Juvenile-onset diabetes mellitus results from a lack of the insulin protein. Dwarfism can arise when the growth hormone protein is lacking. A special "antifreeze" blood protein allows Antarctic fish to survive at body temperatures below freezing.

This list could go on and on, but what is perhaps most amazing is that this great variety of proteins is made from the same, relatively small, group of amino acids. By using various amounts of these amino acids in different sequences, nature has created proteins to serve the many tasks needed to sustain life.

31.1 The Structure–Function Connection

Proteins are present in every living cell. Their very name, derived from the Greek word *proteios*, which means holding first place, signifies the importance of these substances. Proteins are one of the three major classes of foods. The other two, carbohydrates and fats, are needed for energy; proteins are needed for growth and maintenance of body tissue. Some common foods with high (over 10%) protein content are fish, beans, nuts, cheese, eggs, poultry, and meat. These foods tend to be scarce and relatively expensive. Proteins are, therefore, the class of foods that is least available to the undernourished people of the world. Hence, the question of how to secure an adequate supply of high-quality protein for an ever-increasing population is one of our more critical problems (see Chapter 34).

Proteins function as structural materials and as enzymes (catalysts) that regulate the countless chemical reactions taking place in every living organism, including the reactions involved in the decomposition and synthesis of proteins.

All proteins are polymeric substances that yield amino acids on hydrolysis. Those that yield only amino acids when hydrolyzed are classified as **simple proteins**; those that yield amino acids and one or more additional products are

simple proteins

conjugated proteins

classified as **conjugated proteins**. There are approximately 200 different known amino acids in nature. Some are found in only one particular species of plant or animal, others in only a few life forms. But 20 of these amino acids are found in almost all proteins. Furthermore, these same 20 amino acids are used by all forms of life in the synthesis of proteins.

All proteins contain carbon, hydrogen, oxygen, and nitrogen. Some proteins contain additional elements, usually sulfur, phosphorus, iron, copper, or zinc. The significant presence of nitrogen in all proteins sets them apart from carbohydrates and lipids. The average nitrogen content is about 16%.

Proteins are highly specific in their functions. The amino acid units in a given protein molecule are arranged in a definite sequence. An amazing fact about proteins is that in some cases if just one of the hundreds or thousands of amino acid units is missing or out of place, the biological function of that protein is seriously damaged or destroyed. The sequence of amino acids in a protein establishes the function of that protein.

This relationship between structure and function contrasts sharply with that for other classes of biochemicals. For example, carbohydrates can provide cellular energy because they contain one particular type of atom, reduced carbon, that is readily oxidizable. This important function does not depend directly on the sequence in which the atoms are arranged. On the other hand, an appropriate sequence of amino acids produces a protein strong enough to form a horse's hoof, a different sequence produces a protein capable of absorbing oxygen in the lungs and releasing it to needy cells; yet another sequence produces a hormone capable of directing carbohydrate metabolism for an entire organism. As will be seen, full understanding of the function of a protein requires that the structure of that protein be understood.

31.2 The Nature of Amino Acids

Each amino acid has two functional groups, an amino group (—NH_2) and a carboxyl group (—COOH). The amino acids that are found in proteins are called alpha (α) amino acids because the amino group is attached to the first or alpha carbon atom adjacent to the carboxyl group. The beta (β) position is the next adjacent carbon; the gamma (γ) position the next carbon; and so on. The following formula represents an alpha (α) amino acid:

$$\overset{\gamma}{C}H_3\overset{\beta}{C}H_2\overset{\alpha}{C}HCOOH \quad (\text{with } NH_2 \text{ on the } \alpha \text{ carbon})$$

α-Amino butyric acid

Alpha amino acids are represented by this general formula:

$$\mathrm{R{-}C(H)(NH_2){-}COOH}$$

Variable group (R) — Carboxyl group (COOH) — Amino group (NH_2)

The portion of the molecule designated R is commonly referred to as the *amino acid side chain.* It is not restricted to alkyl groups and may contain (a) open chain, cyclic, or aromatic hydrocarbon groups; (b) additional amino or carboxyl groups; (c) hydroxyl groups; or (d) sulfur-containing groups.

Amino acids are divided into three groups: neutral, acidic, and basic. They are classified as neutral amino acids when their molecules have the same number of amino and carboxyl groups, as acidic when their molecules have more carboxyl groups than amino groups, or as basic when their molecules have more amino groups than carboxyl groups.

The names, formulas, and abbreviations of the common amino acids are given in Table 31.1. Two of these, aspartic acid and glutamic acid, are classified as acidic; three—lysine, arginine, and histidine—as basic; the remainder are classified as neutral amino acids.

TABLE 31.1 Common Amino Acids Derived from Proteins

Name	Abbreviation	Formula
Alanine	Ala	$CH_3CH(NH_2)COOH$
Arginine	Arg	$NH_2{-}C({=}NH){-}NH{-}CH_2CH_2CH_2CH(NH_2)COOH$
Asparagine	Asn	$NH_2C({=}O){-}CH_2CH(NH_2)COOH$
Aspartic acid	Asp	$HOOCCH_2CH(NH_2)COOH$
Cysteine	Cys	$HSCH_2CH(NH_2)COOH$
Glutamic acid	Glu	$HOOCCH_2CH_2CH(NH_2)COOH$
Glutamine	Gln	$NH_2C({=}O)CH_2CH_2CH(NH_2)COOH$
Glycine	Gly	$HCH(NH_2)COOH$
Histidine	His	$\overline{N{=}CH{-}N(H){-}C{=}}\,\overline{CH}{-}$ ring: $N{-}CH$, $HC{=}N$, $CH{=}C$, $HC{-}N(H){-}C$; $C{-}CH_2CH(NH_2)COOH$
Isoleucine[a]	Ile	$CH_3CH_2CH(CH_3){-}CH(NH_2)COOH$

(continued)

TABLE 31.1 (*continued*)

Name	Abbreviation	Formula
Leucine[a]	Leu	$(CH_3)_2CHCH_2CH(NH_2)COOH$
Lysine[a]	Lys	$NH_2CH_2CH_2CH_2CH_2CH(NH_2)COOH$
Methionine[a]	Met	$CH_3SCH_2CH_2CH(NH_2)COOH$
Phenylalanine[a]	Phe	C_6H_5—$CH_2CH(NH_2)COOH$
Proline	Pro	pyrrolidine ring (N—H)—COOH
Serine	Ser	$HOCH_2CH(NH_2)COOH$
Threonine[a]	Thr	$CH_3CH(OH)$—$CH(NH_2)COOH$
Tryptophan[a]	Trp	indole (N—H) C—$CH_2CH(NH_2)COOH$ (C=CH)
Tyrosine	Tyr	HO—C_6H_4—$CH_2CH(NH_2)COOH$
Valine[a]	Val	$(CH_3)_2CHCH(NH_2)COOH$

[a] Amino acids essential in human nutrition.

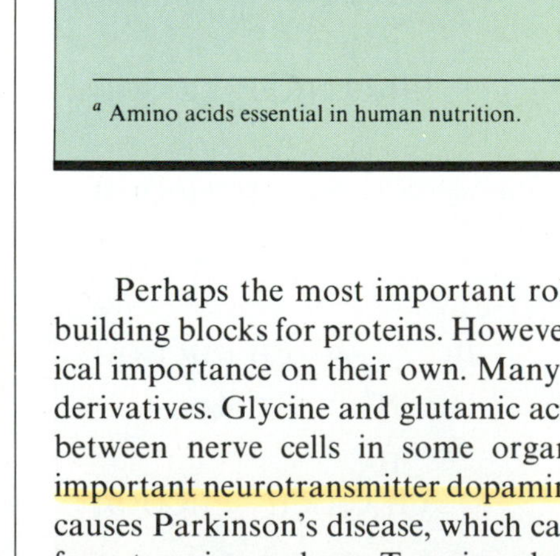

Perhaps the most important role played by amino acids is to serve as the building blocks for proteins. However, selected amino acids also have physiological importance on their own. Many neurotransmitters are amino acids or their derivatives. Glycine and glutamic acid are known to act as chemical messengers between nerve cells in some organisms. Tyrosine is converted to the very important neurotransmitter dopamine. A deficiency of this amino acid derivative causes Parkinson's disease, which can be relieved by another compound formed from tyrosine, L-dopa. Tyrosine also is the parent compound for the "flight or fight" hormone, epinephrine (adrenalin) and the metabolic hormone thyroxine. Still another amino acid with an important physiological role is histidine, which

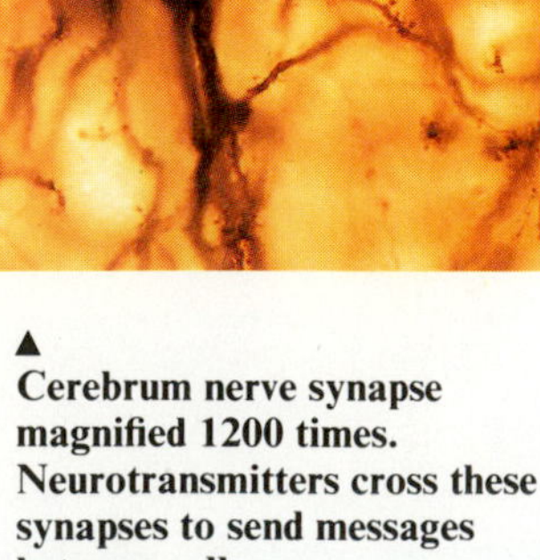

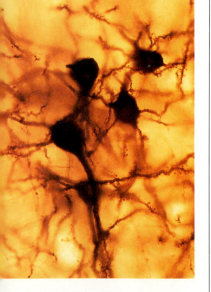

▲ **Cerebrum nerve synapse magnified 1200 times. Neurotransmitters cross these synapses to send messages between cells.**

is converted in the body to histamine. This derivative causes the stomach lining to secrete HCl but probably is best known for causing many of the symptoms associated with tissue inflammation and colds and is the reason antihistamines are such important over-the-counter medications.

31.3 Essential Amino Acids

Dietary protein is broken down during digestion into its constituent amino acids, which supply much of the body's need for amino acids (see Chapter 34). Eight of the amino acids are **essential amino acids**. These amino acids—isoleucine, leucine, lysine, methionine, phenylalanine, threonine, tryptophan, and valine—are essential to the functioning of the human body. Since the body is not capable of synthesizing them, they must be supplied in our diets if we are to enjoy normal health. It is known that some other animals require amino acids in addition to those listed for humans. Rats, for example, require two additional amino acids—arginine and histidine—in their diets.

essential amino acids

On a nutritional basis, proteins are classified as *complete* or *incomplete*. A complete protein supplies all the essential amino acids; an incomplete protein is deficient in one or more essential amino acids. Many proteins, especially those from vegetable sources, are incomplete. For example, protein from corn (maize) is deficient in lysine. The nutritional quality of such vegetable proteins can be greatly improved by supplementing them with the essential amino acids that are lacking, if these can be synthesized at reasonable costs. Lysine, methionine, and tryptophan are being sold at present for enriching human food and livestock feed. This is another way to extend the world's limited supply of high-quality food protein. In still another approach to the problem of obtaining more high-quality protein, plant breeders have developed maize varieties with greatly improved lysine content. Genetic engineering may hold the key to further significant improvements in plant protein quality.

31.4 D-Amino Acids and L-Amino Acids

All amino acids, except glycine, have at least one asymmetric carbon atom. For example, two stereoisomers of alanine are possible:

$$\begin{array}{c} COOH \\ | \\ H-C-NH_2 \\ | \\ CH_3 \end{array} \qquad \begin{array}{c} COOH \\ | \\ H_2N-C-H \\ | \\ CH_3 \end{array}$$

D-(−)-Alanine L-(+)-Alanine

Projection formulas illustrate well the D and L configurations of amino acids in the same way as the configurations of D- and L-glyceraldehyde (Section 29.2). The —COOH group is written at the top of the projection formula, and the

D configuration is indicated by writing the alpha $—NH_2$ to the right of carbon 2. The L configuration is indicated by writing the alpha $—NH_2$ to the left of carbon 2. Although some D-amino acids occur in nature, only L-amino acids occur in proteins. The (+) and (−) signs in the name indicate the direction of rotation of plane-polarized light by the amino acid. Most amino acids have relatively complex structures making use of the projection formula difficult. Thus, unless stereochemical information is explicitly considered, amino acids will be shown using a condensed, structural formula in this chapter.

31.5 AMPHOTERISM

Amino acids are *amphoteric* (or *amphiprotic*); that is, they can react either as an acid or as a base. For example, with a strong base such as sodium hydroxide, alanine reacts as an acid, as shown in equation (1); with a strong acid such as HCl, alanine reacts as a base as shown in equation (2).

$$\underset{\text{Alanine}}{CH_3\underset{NH_2}{\underset{|}{C}}HCOOH} + NaOH \longrightarrow \underset{\text{Sodium alanate}}{CH_3\underset{NH_2}{\underset{|}{C}}HCOO^- Na^+} + H_2O \tag{1}$$

$$CH_3\underset{NH_2}{\underset{|}{C}}HCOOH + HCl \longrightarrow \underset{\text{Alanyl ammonium chloride}}{CH_3\underset{NH_3^+ Cl^-}{\underset{|}{C}}HCOOH} \tag{2}$$

Even in neutral biological solutions, amino acids do not actually exist in the molecular form shown in equations (1) and (2). Instead, they exist mainly as dipolar ions called **zwitterions**. Again using alanine as an example, the proton on the carboxyl group transfers to the amino group, forming a zwitterion by an acid—base reaction within the molecule:

zwitterions

$$\underset{\text{Alanine molecule}}{CH_3\underset{H_2N:}{\underset{|}{C}}HCOO\text{(H)}} \longrightarrow \underset{\text{Alanine zwitterion}}{CH_3\underset{NH_3^+}{\underset{|}{C}}HCOO^-}$$

On an ionic basis, the reaction of alanine with NaOH and HCl is

$$\underset{\text{Alanine zwitterion}}{CH_3\underset{NH_3^+}{\underset{|}{C}}HCOO^-} + OH^- \longrightarrow \underset{\text{Alanate anion}}{CH_3\underset{NH_2}{\underset{|}{C}}HCOO^-} + H_2O \tag{3}$$

$$\underset{\text{Alanine zwitterion}}{CH_3\underset{NH_3^+}{\underset{|}{C}}HCOO^-} + H^+ \longrightarrow \underset{\text{Alanyl ammonium cation}}{CH_3\underset{NH_3^+}{\underset{|}{C}}HCOOH} \tag{4}$$

Other amino acids behave like alanine. Together with protein molecules that contain $—COOH$ and $—NH_2$ groups, they help to buffer or stabilize the pH of the blood at about 7.4. The pH is maintained close to 7.4 because any excess acid or base in the blood is neutralized by reactions such as shown in equations (3) and (4).

$$\underset{\substack{\text{Cation form}\\ \text{II}}}{\underset{\text{NH}_3^+}{\text{RCHCOOH}}} \underset{\text{OH}^-}{\overset{\text{H}^+}{\rightleftharpoons}} \underset{\substack{\text{Zwitterion form}\\ \text{I}}}{\underset{\text{NH}_3^+}{\text{RCHCOO}^-}} \underset{\text{OH}^-}{\overset{\text{H}^+}{\rightleftharpoons}} \underset{\substack{\text{Anion form}\\ \text{III}}}{\underset{\text{NH}_2}{\text{RCHCOO}^-}} \quad (5)$$

When an amino acid in solution has equal positive and negative charges, as in formula I of equation (5), it is electrically neutral and does not migrate toward either the positive or negative electrode when placed in an electrolytic cell. The pH at which there is no migration toward either electrode is called the **isoelectric point** (see Table 31.2). If acid (H^+) is added to an amino acid at its isoelectric point, the equilibrium is shifted toward formula II, and the cation formed migrates toward the negative electrode. When base (OH^-) is added, the anion formed (formula III) migrates toward the positive electrode. Differences in isoelectric points are important in isolating and purifying amino acids and proteins, since their rates and directions of migration can be controlled in an electrolytic cell by adjusting the pH. This method of separation is called *electrophoresis*.

isoelectric point

Amino acids are classified as basic, neutral, or acidic depending on whether the ratio of —NH_2 to —COOH groups in the molecules is greater than 1:1, equal to 1:1, or less than 1:1, respectively. Furthermore, this ratio differs from 1:1 only if the amino acid side chain (R—) contains an additional amino or carboxyl group. For example, if the side chain contains a carboxyl group, the amino acid is considered acidic. Thus, the R— group determines whether an amino acid is classified as basic, neutral, or acidic.

Isoelectric points are found at pH values ranging from 7.8 to 10.8 for basic amino acids, 4.8 to 6.3 for neutral amino acids, and 2.8 to 3.3 for acidic amino acids. It is logical that a molecule such as glutamic acid, with one amino group and two carboxyl groups, would be classified as acidic and that its isoelectric point would be at a pH lower than 7.0. It might also seem that the isoelectric point of an amino acid that is classified as neutral, such as alanine, with one amino and one carboxyl group, would have an isoelectric point of 7.0. However, the isoelectric point of alanine is 6.0, not 7.0. This is because the carboxyl group and the amino group are not equally ionized. The carboxyl group of alanine ionizes to a greater degree as an acid than the amino group ionizes as a base.

TABLE 31.2 Isoelectric Points of Selected Amino Acids

Amino acid	pH at isoelectric point	Amino acid	pH at isoelectric point
Arginine	10.8	Serine	5.7
Lysine	9.7	Glutamic acid	3.2
Alanine	6.0	Aspartic acid	2.9
Glycine	6.0		

EXAMPLE 31.1 Draw the structure of L-serine in a strongly acidic solution.

SOLUTION First, draw the structure for L-serine as the molecule would exist in a neutral solution. The α-amino and carboxylic acid groups form a zwitterion.

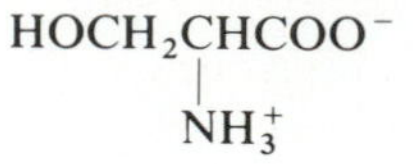

$$\underset{\displaystyle NH_3^+}{HOCH_2\underset{|}{C}HCOO^-}$$

When the solution is made acidic (the concentration of hydrogen ions is increased), the amine group is unaffected because it is already protonated. However, the carboxylate anion bonds to a hydrogen ion, resulting in an L-serine structure with a net positive charge.

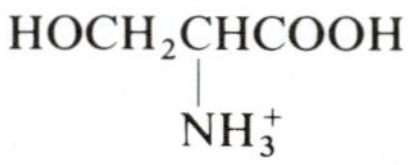

$$\underset{\displaystyle NH_3^+}{HOCH_2\underset{|}{C}HCOOH}$$

PRACTICE Draw the structure of L-valine in a strongly basic solution.
Answer: $\underset{\displaystyle NH_2}{(CH_3)_2CHCH_2\underset{|}{C}HCOO^-}$

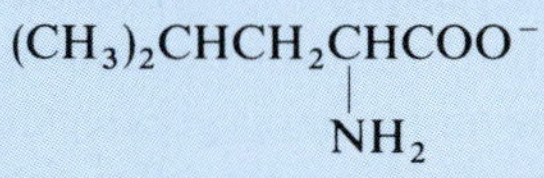

31.6 FORMATION OF POLYPEPTIDES

Proteins are polyamides consisting of amino acid units joined through amide structures. If we react two glycine molecules, with the elimination of a molecule of water, we form a compound containing the amide structure, also called the **peptide linkage**, or peptide bond. The elimination of water occurs between the carboxyl group of one amino acid and the α-amino group of a second amino acid (see Section 26.1). The product formed from two glycine molecules is called glycylglycine (abbreviated Gly-Gly). Because it contains two amino acid units, it is called a dipeptide.

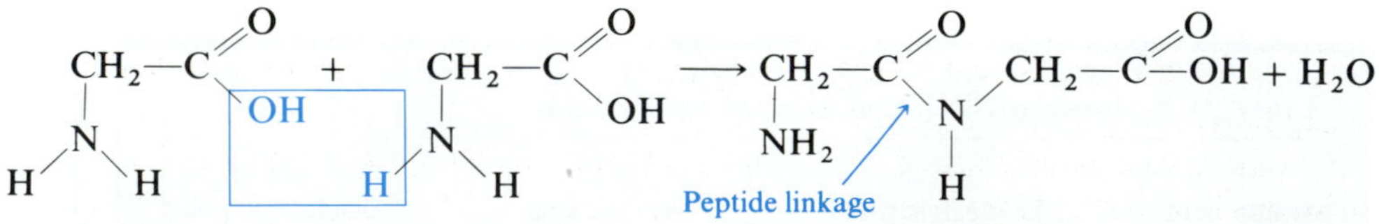

peptide linkage

$$H_2N{-}CH_2{-}\overset{O}{\overset{\|}{C}}{-}OH + H{-}NH{-}CH_2{-}\overset{O}{\overset{\|}{C}}{-}OH \longrightarrow H_2N{-}CH_2{-}\overset{O}{\overset{\|}{C}}{-}\underset{H}{\underset{|}{N}}{-}CH_2{-}\overset{O}{\overset{\|}{C}}{-}OH + H_2O$$

Peptide linkage

Glycylglycine
(Gly-Gly)

If three amino acid units are included in a molecule, it is a tripeptide; if four, a tetrapeptide; if five, a pentapeptide; and so on. Peptides containing up to about 40–50 amino acid units in a chain are called **polypeptides**. The units making up the peptide are amino acids less the elements of water and are referred

polypeptide

to as *amino acid residues* or, simply, residues. Still larger chains of amino acids are known as proteins.

When amino acids form a polypeptide chain, a carboxyl group and an α-amino group are involved in each peptide bond (amide bond). While these groups are joined in peptide bonds, they cannot ionize as acids or bases. Consequently, the physico-chemical properties of a polypeptide/protein are determined to a large extent by the side chains of the amino acid residues.

In linear peptides one end of the chain has a free amino group and the other end a free carboxyl group. The amino-group end is called the *N-terminal residue* and the other end the *C-terminal residue*:

```
  1   2   3   4   5   6   7
Ala-Pro-Tyr-Met-Gly-Lys-Gly
 ↑                       ↑
N-Terminal residue    C-Terminal residue
```

The sequence of amino acids in a chain is numbered starting with the N-terminal residue, which is written to the left with the C-terminal residue at the right. Any segment of the sequence that is not specifically known is placed in parentheses. Thus in the heptapeptide above, if the order of tyrosine and methionine were not known, the structure would be written as

Ala-Pro-(Met, Tyr)-Gly-Lys-Gly

Peptides are named as acyl derivatives of the C-terminal amino acid with the C-terminal unit keeping its complete name. The *ine* ending of all but the C-terminal amino acid is changed to *yl*, and these are listed in the order in which they appear, starting with the N-terminal amino acid.

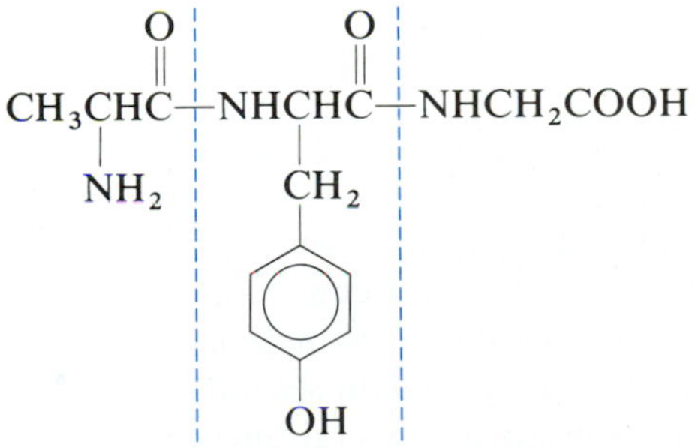

$$\underbrace{CH_3CH(NH_2)C(=O)}_{\text{Alanyl}}-\underbrace{NHCH(CH_2C_6H_4OH)C(=O)}_{\text{Tyrosyl}}-\underbrace{NHCH_2COOH}_{\text{Glycine}}$$

Ala-Tyr-Gly

Thus Ala-Tyr-Gly is called alanyltyrosylglycine. The name of Arg-Gln-His-Ala is arginylglutamylhistidylalanine.

Alanine and glycine can form two different dipeptides, Gly-Ala and Ala-Gly, using each amino acid only once:

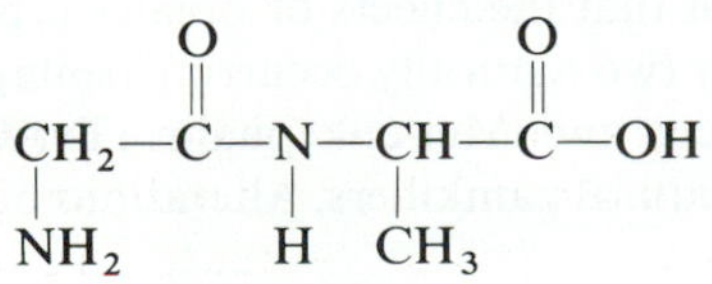

$$CH_2(NH_2)-C(=O)-N(H)-CH(CH_3)-C(=O)-OH \quad \text{and} \quad CH_3CH(NH_2)-C(=O)-N(H)-CH_2-C(=O)-OH$$

Glycylalanine (Gly-Ala) Alanylglycine (Ala-Gly)

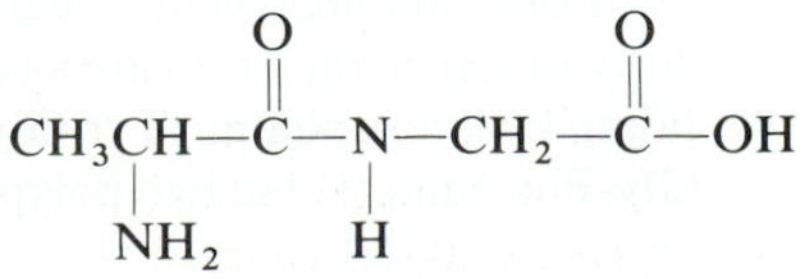

TABLE 31.3 Primary Structures and Functions of Some Biological Polypeptides[a]

Name	Primary structure	General biological function
Substance P	Arg-Pro-Lys-Pro-Gln-Gln-Phe-Phe-Gly-Leu-Met-NH_2	Is a pain-producing agent
Bradykinin	Arg-Pro-Pro-Gly-Phe-Ser-Pro-Phe-Arg	Affects tissue inflammation and blood pressure
Angiotensin II	Asp-Arg-Val-Tyr-Val-His-Pro-Phe	Maintains water balance and blood pressure
Leu-enkephalin Met-enkephalin	Tyr-Gly-Gly-Phe-Leu Tyr-Gly-Gly-Phe-Met	Relieves pain, produces sense of well-being
Vasopressin	Cys-Tyr-Phe-Gln-Asn-Cys-Pro-Arg-Gly-NH_2 (S—S bridge between the two Cys)	Increases blood pressure, decreases kidney water excretion
Oxytocin	Cys-Tyr-Ile-Gln-Asn-Cys-Pro-Leu-Gly-NH_2 (S—S bridge between the two Cys)	Initiates childbirth labor, causes mammary gland milk release, affects kidney excretion of water and sodium

[a] Where —NH_2 is indicated at the end of the sequence, the C-terminal amino acid has an amide structure rather than a free —COOH.

If three different amino acids react—for example, glycine, alanine, and threonine–six tripeptides in which each amino acid appears only once are possible.

Gly-Ala-Thr	Ala-Thr-Gly	Thr-Ala-Gly
Gly-Thr-Ala	Ala-Gly-Thr	Thr-Gly-Ala

The number of possible peptides rises very rapidly as the number of amino acid units increases. For example, there are 120 ($1 \times 2 \times 3 \times 4 \times 5 = 120$) different ways to combine five different amino acids to form a pentapeptide, using each amino acid only once in each molecule. If the same constraints are applied to 15 different amino acids, the number of possible combinations is greater than 1 trillion (10^{12})! Since a protein molecule may contain several hundred amino acid units, with individual amino acids occurring several times, the number of possible combinations from 20 amino acids is simply beyond imagination.

There are a number of small, naturally occurring polypeptides with significant biochemical functions. In general these substances serve as hormones or nerve transmitters. Their functions range from controlling pain and pleasure responses in the brain to controlling smooth muscle contraction or kidney fluid excretion rates (see Table 31.3). The amino acid sequence and chain length give a polypeptide its biological effectiveness and specificity.

For example, recent research has shown that the effects of opiates (opium derivatives) on the brain are also exhibited by two naturally occurring pentapeptides, Leu-enkephalin (Tyr-Gly-Gly-Phe-Leu) and Met-enkephalin (Tyr-Gly-Gly-Phe-Met). These two polypeptides are natural painkillers. Alterations of the amino acid sequence—which alters the side-chain characteristics—cause drastic changes in the analgesic effects of these pentapeptides. The substitution of L-alanine for either of the glycine residues in these compounds (simply changing one side-chain group from —H to —CH_3) causes an approximately 1000-fold decrease in effectiveness as a painkiller! The substitution of L-tyrosine for L-phenylalanine causes a comparable loss of activity. Even the substitution of D-tyrosine for the L-tyrosine residue causes a considerable loss in the analgesic effectiveness of the pentapeptides.

It is clearly evident that a particular sequence of amino acid residues is essential for proper polypeptide function. This sequence aligns the side-chain characteristics (large or small; polar or nonpolar; acidic, basic, or neutral) in the proper positions for a specific polypeptide function.

Oxytocin and vasopressin are similar nonapeptides, differing only at two positions in their primary structure (see Table 31.3). Yet their biological functions differ dramatically. Oxytocin controls uterine contractions during labor in childbirth and also causes contraction of the smooth muscles of the mammary glands, resulting in milk excretion. Vasopressin in high concentration raises the blood pressure and has been used in treatment of surgical shock for this purpose. Vasopressin is also an antidiuretic, regulating the excretion of fluid by the kidneys. The absence of vasopressin leads to diabetes insipidus. This condition is characterized by excretion of up to 30 liters of urine per day, but can be controlled by administration of vasopressin or its derivatives.

The isolation and synthesis of oxytocin and vasopressin was accomplished by Vincent du Vigneaud (1901–1978) and co-workers at Cornell University. Du Vigneaud was awarded the Nobel prize in chemistry in 1955 for this work. Synthetic oxytocin is indistinguishable from the natural material. It is available commercially and is used for the induction of labor in the late stages of pregnancy.

EXAMPLE 31.2

Write the structure of the tripeptide Ser-Gly-Ala.

SOLUTION

First, write the structures of the three amino acids in this tripeptide.

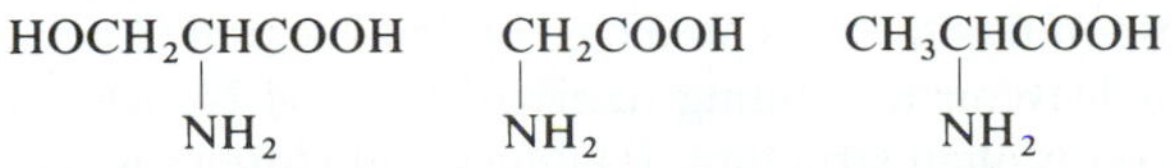

$$\underset{\text{Serine (Ser)}}{\mathrm{HOCH_2\underset{\underset{\displaystyle NH_2}{|}}{C}HCOOH}} \qquad \underset{\text{Glycine (Gly)}}{\mathrm{\underset{\underset{\displaystyle NH_2}{|}}{C}H_2COOH}} \qquad \underset{\text{Alanine (Ala)}}{\mathrm{CH_3\underset{\underset{\displaystyle NH_2}{|}}{C}HCOOH}}$$

By convention, the amino acid residue written at the left end of the tripeptide has a free amino group while the residue at the right end has a free carboxylic acid group. When the amino acids are connected by peptide linkages the following structure results:

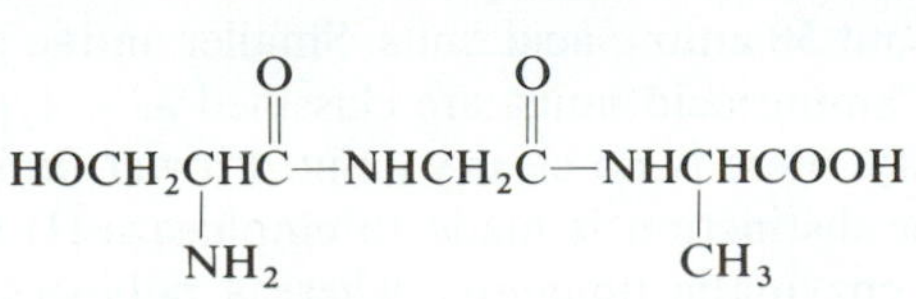

$$\mathrm{HOCH_2\underset{\underset{\displaystyle NH_2}{|}}{C}H\overset{\overset{\displaystyle O}{\|}}{C}{-}NHCH_2\overset{\overset{\displaystyle O}{\|}}{C}{-}NH\underset{\underset{\displaystyle CH_3}{|}}{C}HCOOH}$$

PRACTICE Write the structure of the pentapeptide Gly-Leu-Asp-Ser-Cys.

Answer:

```
      O         O          O         O
      ‖         ‖          ‖         ‖
CH2C—NHCHC—NHCHC—NHCHC—NHCHCOOH
 |        |          |         |         |
NH2     CH2        CH2       CH2       CH2
          |          |         |         |
         CH        COOH       OH        SH
        /  \
     CH3   CH3
```

31.7 PROTEIN STRUCTURE

By 1940 a great deal of information concerning proteins had been assembled. Their elemental composition was known, and they had been carefully classified according to solubility in various solvents. Proteins were known to be polymers of amino acids, and the different amino acids had, for the most part, been isolated and identified. Protein molecules were known to be very large in size, with molar masses ranging from several thousand to several million.

Knowledge of protein structure could help to answer many chemical and biological questions. But for a while the task of determining the actual structure of molecules of such colossal size appeared to be next to impossible. Then Linus Pauling (b. 1901), at the California Institute of Technology, attacked the problem by a new approach. Using X-ray diffraction techniques, Pauling and his collaborators painstakingly determined the bond angles and dimensions of amino acids and of dipeptides and tripeptides. After building accurate scale models of the dipeptides and tripeptides, they determined how these could be fitted into likely polypeptide configurations. Based on this work, Pauling and R. B. Corey proposed in 1951 that two different conformations—the *α-helix* and the *β-pleated-sheet*—were the most probable stable polypeptide chain configurations of protein molecules. These two macromolecular structures are illustrated in Figure 31.1. Within a short time it was established that many proteins do have structures corresponding to those predicted by Pauling and Corey. This work was a very great achievement. Pauling received the 1954 Nobel prize in chemistry for this work on protein structure. Pauling's and Corey's work provided the inspiration for another great biochemical breakthrough—the concept of the double-helix structure for deoxyribonucleic acid, DNA (see Section 33.6).

Proteins are very large molecules. But just how many amino acid units must be present for a substance to be a protein? There is no universally agreed upon answer to this question. Some authorities state that a protein must have a molar mass of at least 6000 or contain about 50 amino acid units. Smaller amino acid polymers, containing from 5 to 50 amino acid units, are classified as polypeptides and are not proteins. In reality, there is no clearly defined lower limit to the molecular size of proteins. The distinction is made to emphasize (1) that proteins often serve structural or enzymatic functions, whereas polypeptides often serve hormone related functions and (2) that the three-dimensional conformation of proteins is directly related to function, whereas the relation is not so clear-cut with polypeptides.

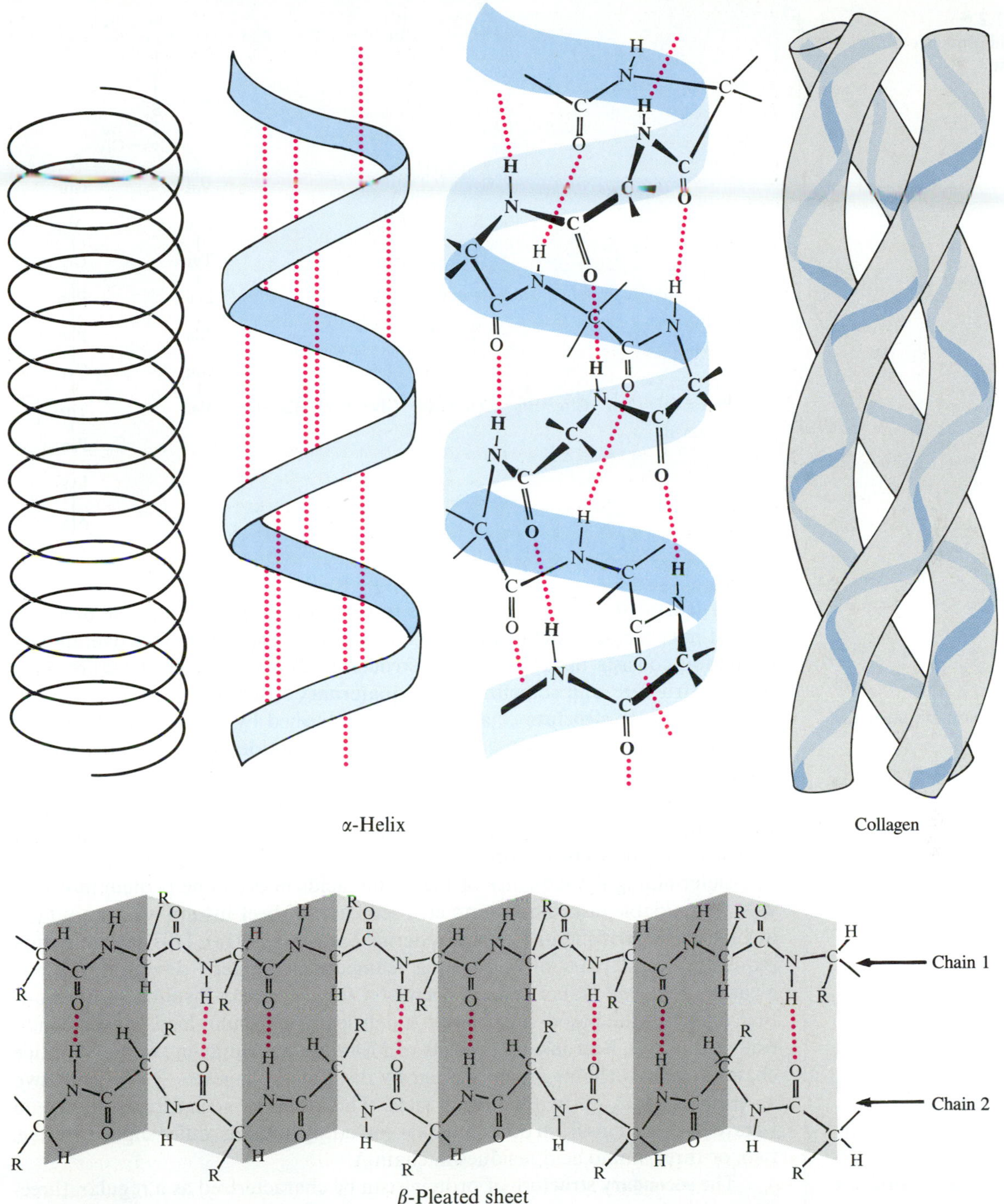

▲

FIGURE 31.1

α-Helix, β-pleated-sheet, and the special collagen protein structures. Collagen is composed of three helical protein chains wound together in a three-stranded helix. As the most abundant protein in the animal world, collagen's function in the body is mainly as connective tissue.

FIGURE 31.2 ▶ Amino acid sequence of beef insulin.

15 Gln—Leu—Glu—Asn—Tyr—Cys 20—Asn 21

Tyr | Leu | Ser

10 Cys—Val

S | S

Ser | Ala

1 Gly—Ile—Val—Glu—Gln 5—Cys—Cys

Chain A

S | S

20 Cys—Gly

Val | Leu | Tyr | 15 Leu | Ala | Glu | Val

Glu | Arg | Gly | Phe | Phe 25 | Tyr | Thr | Pro | Lys | Ala 30

1 Phe—Val—Asn—Gln—His 5—Leu—Cys—Gly—Ser—His 10—Leu—Val

Chain B

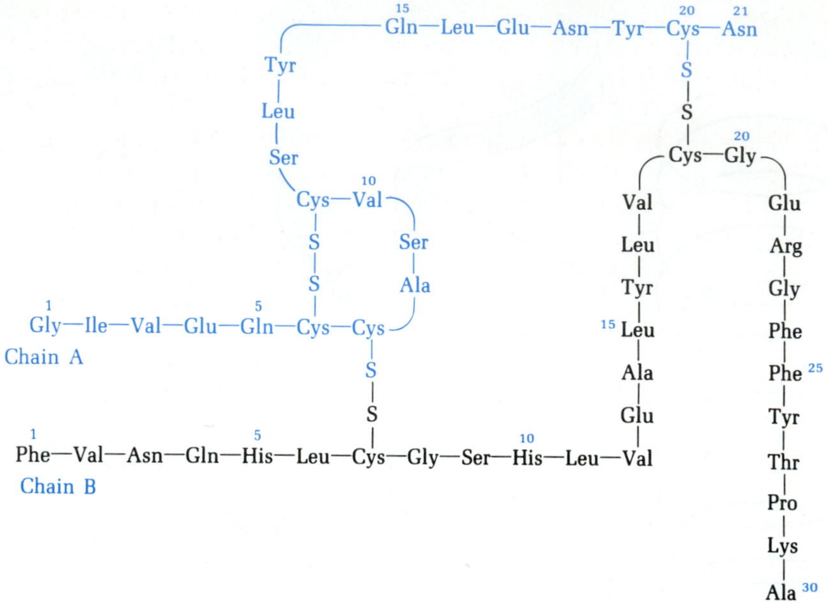

In general, if a protein molecule is to perform a specific biological function, it must have a closely defined overall conformation, or shape. This overall conformation consists of (1) a primary structure, (2) a secondary structure, (3) a tertiary structure, and sometimes (4) a quaternary structure.

primary structure

The **primary structure** of a protein is established by the number, kind, and sequence of amino acid units composing the polypeptide chain or chains making up the molecule. The primary structure determines the alignment of side-chain characteristics, which, in turn, determines the three-dimensional shape into which the protein folds. In this sense the amino acid sequence is of primary importance in establishing protein shape.

Determining the sequence of the amino acids in even one protein molecule was a formidable task. The amino acid sequence of beef insulin was announced in 1955 by the British biochemist Frederick Sanger (b. 1918). This determination required several years of effort by a team under Sanger's direction. He was awarded the 1958 Nobel prize in chemistry for this work. Insulin is a hormone that regulates the blood-sugar level. A deficiency of insulin leads to the condition of diabetes. Beef insulin consists of 51 amino acid units in two polypeptide chains. The two chains are connected by disulfide linkages (—S—S—) of two cysteine residues at two different sites. The primary structure is shown in Figure 31.2. Insulins from other animals, including humans, differ slightly by one, two, or three amino acid residues in chain A.

secondary structure

The **secondary structure** of proteins can be characterized as a regular, three-dimensional structure held together by hydrogen bonding between the oxygen of the >C=O and the hydrogen of the H—N< groups in the polypeptide chains:

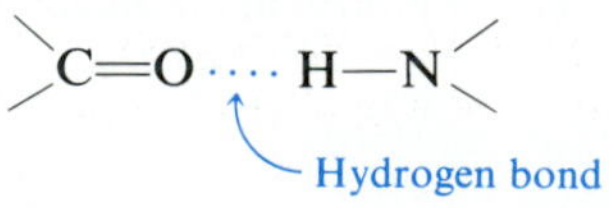

>C=O ···· H—N<

Hydrogen bond

The α-helical and β-pleated-sheet structures of Pauling and Corey are two examples of secondary structure. As shown in Figure 31.1, essentially every peptide bond is involved in at least one hydrogen bond in these structures. Proteins having α-helical or β-pleated-sheet secondary structures are strongly held in particular conformations by virtue of the large number of hydrogen bonds.

The fibrous proteins are an important class of proteins that contain highly developed secondary structures. Because these structures provide strength, these proteins tend to function in support roles. For example, collagen in the form of a three-stranded helix (see Figure 31.1) is the principal protein in connective tissues. The α-keratins depend on the α-helix for strength and to provide support for such diverse structures as hair, fingernails, feathers, and hooves. The silk protein, fibroin, is folded into a β-pleated-sheet structure. In each case the repeating secondary structure with its multitude of hydrogen bonds provides the protein with strength and some degree of rigidity.

tertiary structure

The **tertiary structure** of a protein refers to the distinctive and characteristic conformation, or shape, of a protein molecule. This overall three-dimensional conformation is held together by a variety of interactions between amino acid side chains. These interactions include (1) hydrogen bonding, (2) ionic bonding, and (3) disulfide bonding. For example,

1. Glutamic acid–tyrosine hydrogen bonding

$$\text{Protein}-CH_2CH_2\overset{\displaystyle OH}{\overset{|}{C}}=O\cdots HO-C_6H_4-\text{Protein}$$

2. Glutamic acid–lysine ionic bonding

$$\text{Protein}-CH_2CH_2\overset{\displaystyle O}{\overset{\|}{C}}-O^-\ H_3N^+-CH_2CH_2CH_2CH_2-\text{Protein}$$

3. Cysteine–cysteine disulfide bonding

$$\text{Protein}-CH_2-S-S-CH_2-\text{Protein}$$

The tertiary structure depends on the number and location of these interactions, variables that are fixed when the primary structure is synthesized. Thus, the tertiary structure depends on the primary structure. For example, there are three locations in the insulin molecule (Figure 31.2) where the primary sequence permits disulfide bonding. This, in turn, has an obvious bearing on the shape of insulin.

Hair is especially rich in disulfide bonds. These can be broken by certain reducing agents and restored by an oxidizing agent. This fact is the key to "cold" permanent waving of hair. Some of the disulfide bonds are broken by applying a reducing agent to the hair. The hair is then styled with the desired curls or waves. These are then permanently set by using an oxidizing agent to reestablish the disulfide bonds at different points.

▲ **Frederick Sanger (1918–).**

$$\text{—CH}_2\text{—S—S—CH}_2\text{—} \xrightarrow[\text{agent}]{\text{Reducing}}$$

Bonds in normal hair

$$\text{—CH}_2\text{—S—H H—S—CH}_2\text{—} \xrightarrow[\text{agent}]{\text{Oxidizing}} \text{—CH}_2\text{—S—S—CH}_2\text{—}$$

Broken bonds of "reduced" form

Reestablished bonds of permanently waved hair

If Table 31.1 is examined closely it can be concluded that most of the amino acids have side chains that cannot form hydrogen bonds nor ionic bonds nor disulfide bonds. What then do the majority of amino acids do to hold together the tertiary structure of a protein? This is an important question and leads to an equally important answer. *The uncharged relatively nonpolar amino acids form the center or core of most proteins.* These amino acids have side chains that don't bond very well to water; they are like saturated hydrocarbons or lipids and are hydrophobic. When a protein is synthesized, the uncharged, nonpolar amino acids turn inward toward each other, excluding water and forming the core of the protein structure.

Myoglobin, a small oxygen-carrying protein of the muscle, was the first protein for which a three-dimensional structure was determined (see Figure 31.3). It can be described as having a folded-sausage structure, each section of the sausage representing a segment of the α-helix. Under normal conditions the folds of the myoglobin structure are held firmly in place by interactions between amino acid side chains. The conformation of myoglobin closely accommodates an organically bound iron atom (in the heme group, shown in Figure 31.3 by the rectangular solid), which allows myoglobin to store and transport oxygen in the muscles. The oxygen transport and storage function is determined by myoglobin's specific primary, secondary, and tertiary structures. It is worth noting that myoglobin is a *conjugated* protein—that is, a protein that contains groups other than amino acid residues.

Many proteins, including myoglobin and molecules of a much larger size, have an overall structure which is supported by a basic framework. This framework is composed of pieces of strong, secondary structure. For example, each short α-helical section in myoglobin is relatively rigid because of its many hydrogen bonds. These rigid cylinders then form a cage around the heme group, holding it firmly in place so that it can bind oxygen in the appropriate way.

Like myoglobin, larger proteins also have a skeleton made up of secondary structure, primarily α-helix and β-pleated sheet. Often these bigger molecules are folded into a number of units, each about the same size as myoglobin, which are called **protein domains**. In this way, a larger protein is like several myoglobin-sized proteins connected together. A protein domain has its own secondary skeleton and often serves a discrete task in a protein's overall function. In fact, if selected peptide bonds are broken carefully, domains can actually be isolated and separated.

protein domains

To learn more about these domains and how secondary structure serves as a protein skeleton, scientists have developed a simplified means of visualizing complex proteins. Several proteins are illustrated in Figure 31.4. These simplified structures only sketch the position of the protein chain and leave out the loca-

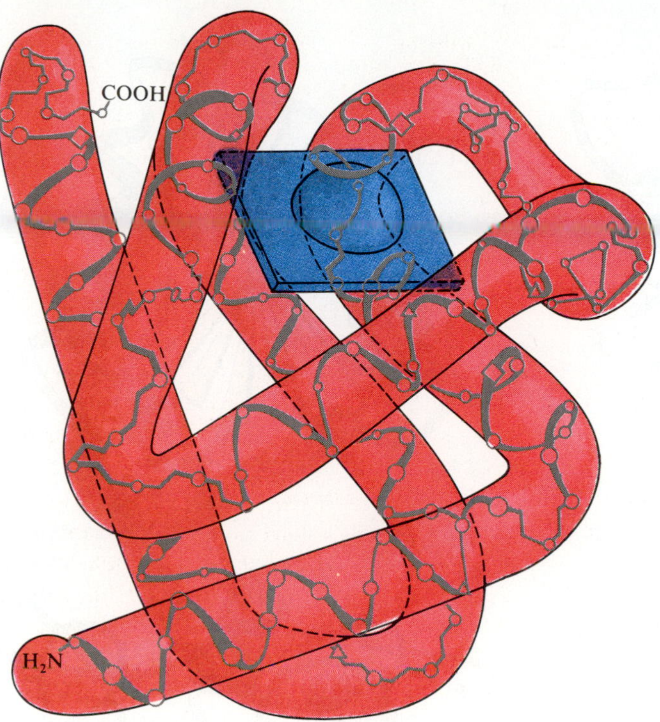

Myoglobin

◄ FIGURE 31.3
Tertiary structure of a protein, represented by myoglobin. The blue portion represents a heme group.

tion of the amino acid side chains. The α-helix is shown with a curling line while each strand of β-pleated sheet is a broader line with an arrowhead at its end to show which way the protein chain is running (from the amino end toward the carboxyl terminus). These pictures are sometimes referred to as "ribbon structures" based on the appearance of the protein strand.

Proteins share common skeletal structures just as various animal species have a common bone structure (see Figure 31.4). Some proteins use the α-helix like myoglobin. Spiral tubes are twisted around to support the bulk of these protein domains. In other proteins the β-pleated sheet is configured to provide a rigid protein framework. Some proteins have a fan of β-pleated sheet (or twisted sheet) while for other molecules the β-pleated sheet has wrapped around on itself to form a barrel. As the structures of more and more proteins are examined, it is apparent that there are at least several common protein skeletons. The twisted sheet and β-barrel are forms taken by the β-pleated sheet while the α-helix often is found on the outside of the β-pleated sheet or in bundles of four (Figure 31.5, page 807).

Note that these ribbon structures make it easy to follow the protein chain and to see how the secondary structure is arranged, *but* they do not provide a true picture of a protein. In reality proteins are solid, as shown with the space-filling model of an enzyme (phosphoglycerate kinase) in Figure 31.6 (page 808). The hydrophobic amino acids tend to form the interior with the charged and polar amino acids on the surface where they can interact with water. Notice that this protein has two domains connected through a relatively narrow center.

quaternary structure

A fourth type of structure, called **quaternary structure**, is found in some complex proteins. These proteins are made up of two or more smaller protein subunits (polypeptide chains). Nonprotein components may also be present. The

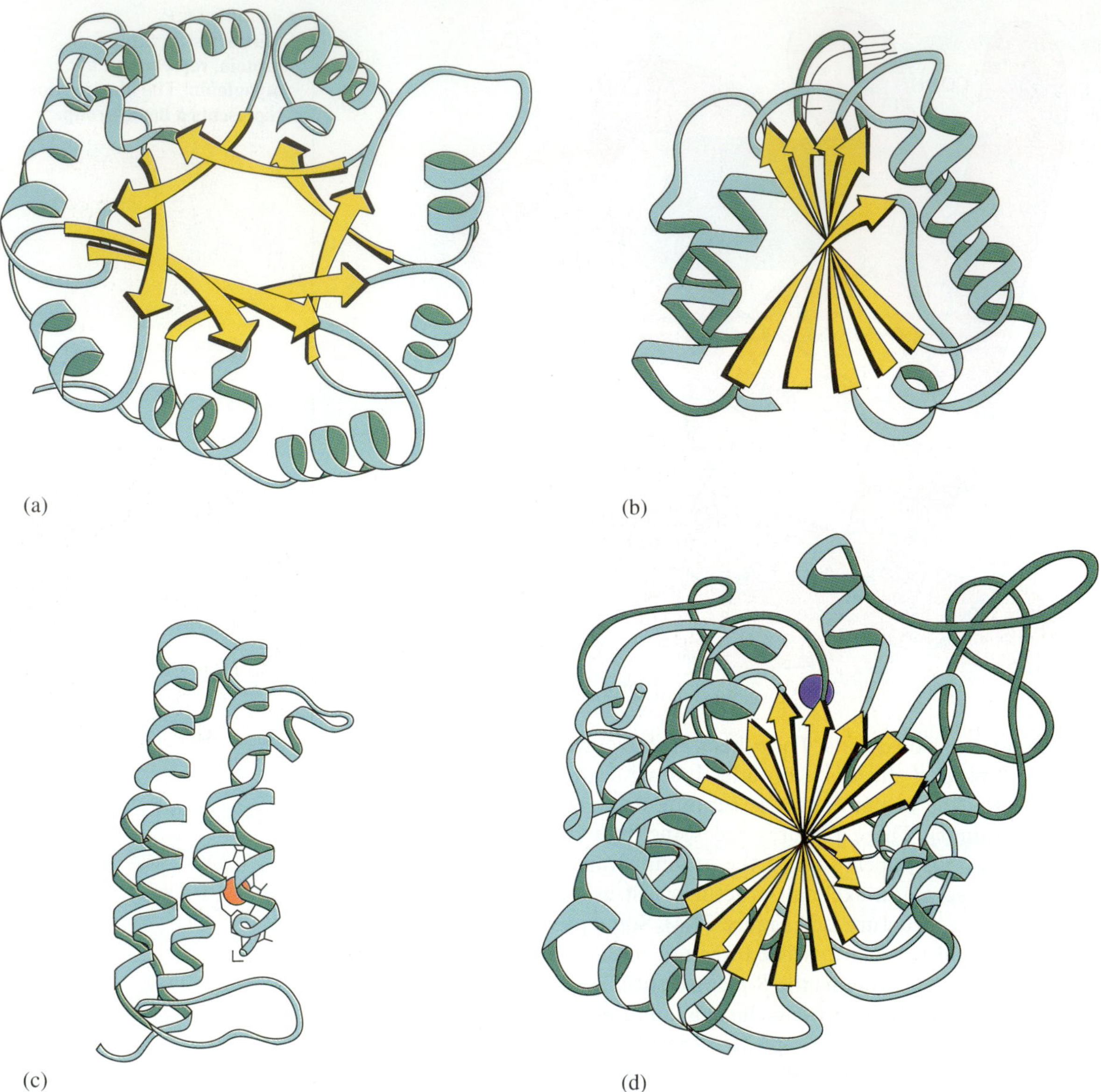

▲
FIGURE 31.4
Ribbon structures of several proteins. (a) Triose phosphate isomerase, an enzyme used in glucose metabolism; (b) flavodoxin, a conjugated protein found in some bacteria; (c) cytochrome c′, a conjugated protein which contains a heme group and is important in cellular energy metabolism; (d) carboxypeptidase, an enzyme used in the digestion of proteins.

quaternary structure refers to the shape of the entire complex molecule and is determined by the way in which the subunits are held together by *noncovalent* bonds—that is, by hydrogen bonding, ionic bonding, and so on.

Quaternary structure is commonly important in proteins that are involved in the control of metabolic processes. For example, hemoglobin, the oxygen transport protein of the blood, is composed of four subunits (see Figure 31.7, page 808). Each subunit is similar to myoglobin in that a set of helical segments surround the oxygen-binding heme group. Of course, the important structural difference is that hemoglobin has a quaternary structure. This leads to an equally important functional difference. Whereas myoglobin always binds and releases oxygen in the same way, hemoglobin changes its oxygen-binding characteristics depending on the available O_2. When hemoglobin binds oxygen on one subunit,

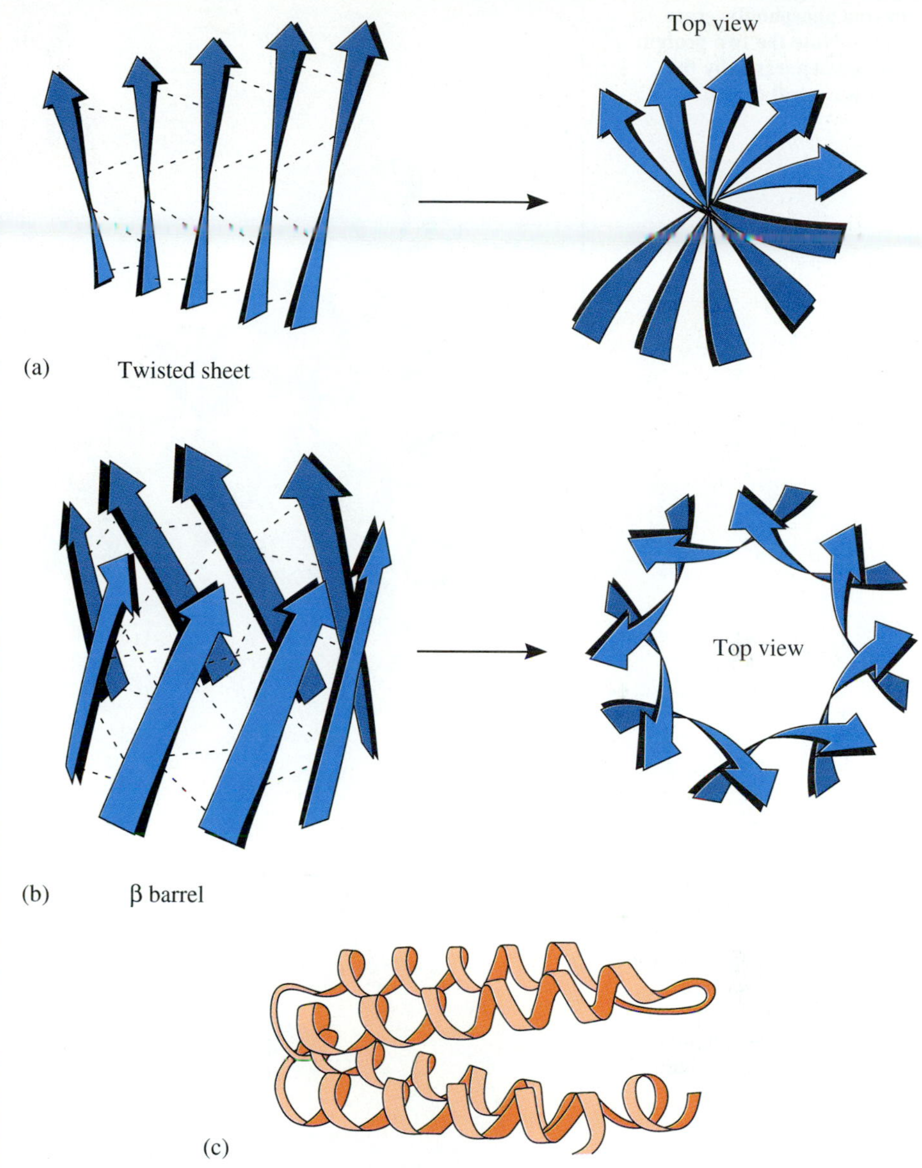

FIGURE 31.5 ▸ Common protein skeletal structures. (a) Twisted sheet of β-pleated sheet; (b) β-barrel of β-pleated sheet; (c) bundle of four α-helices.

the protein's conformation changes to facilitate the binding of three additional oxygen molecules (see Figure 31.8, page 809). The oxygen binding is said to be *cooperative*—that is, the binding of an oxygen at one site promotes oxygen binding on the other three sites. In an oxygen-rich environment, hemoglobin becomes saturated with oxygen.

Conversely, the loss of one oxygen from hemoglobin facilitates the release of oxygen from the other sites. As hemoglobin moves to the oxygen-needy body tissues, its oxygen is cooperatively released. The presence of a quaternary structure allows the binding or removal of one oxygen molecule to control the binding or removal of three other oxygen molecules. Hence, the oxygen transport effectiveness of hemoglobin is multiplied by the quaternary structure.

FIGURE 31.6 ►
Space-filling model of the enzyme phosphoglycerate kinase. Note the two protein domains separated by the narrow central region.

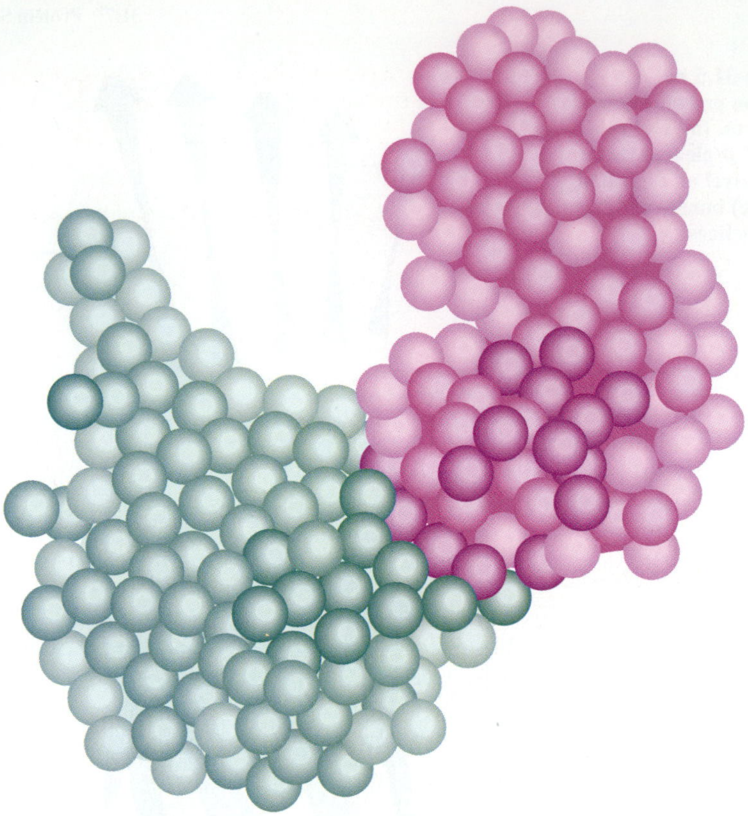

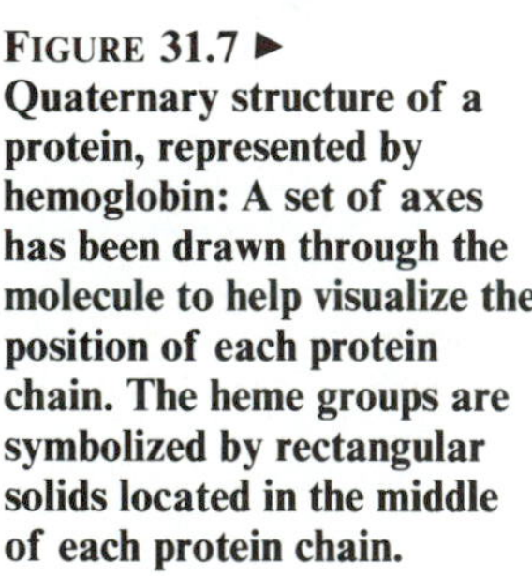

FIGURE 31.7 ►
Quaternary structure of a protein, represented by hemoglobin: A set of axes has been drawn through the molecule to help visualize the position of each protein chain. The heme groups are symbolized by rectangular solids located in the middle of each protein chain.

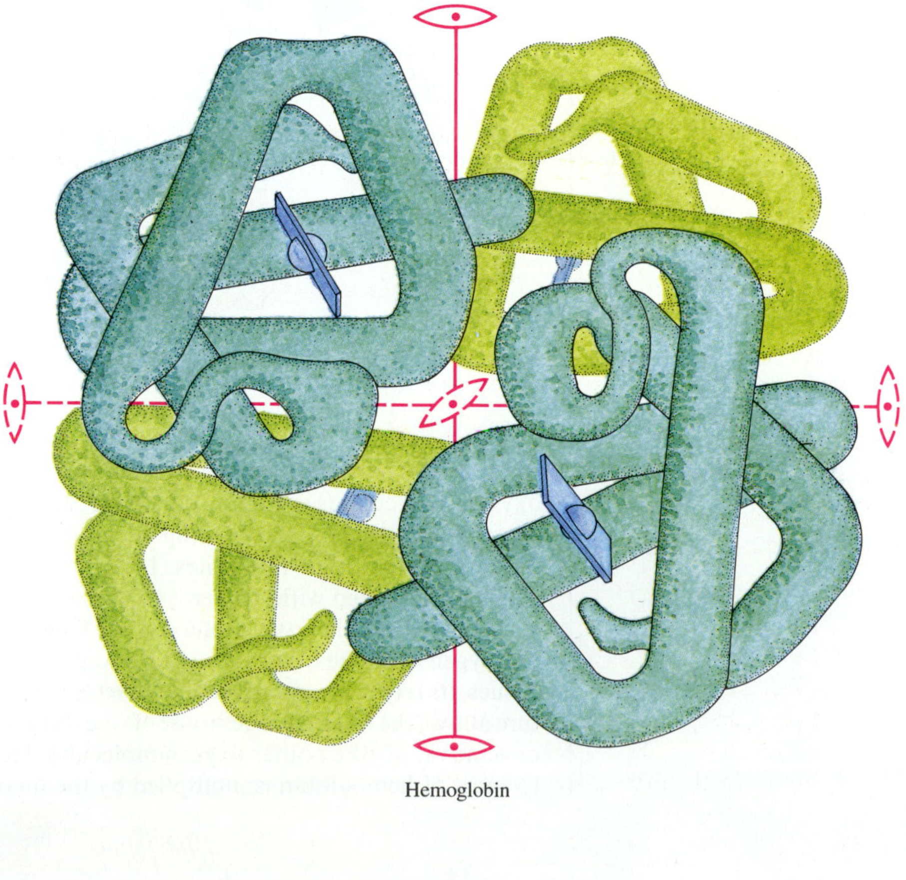

Hemoglobin

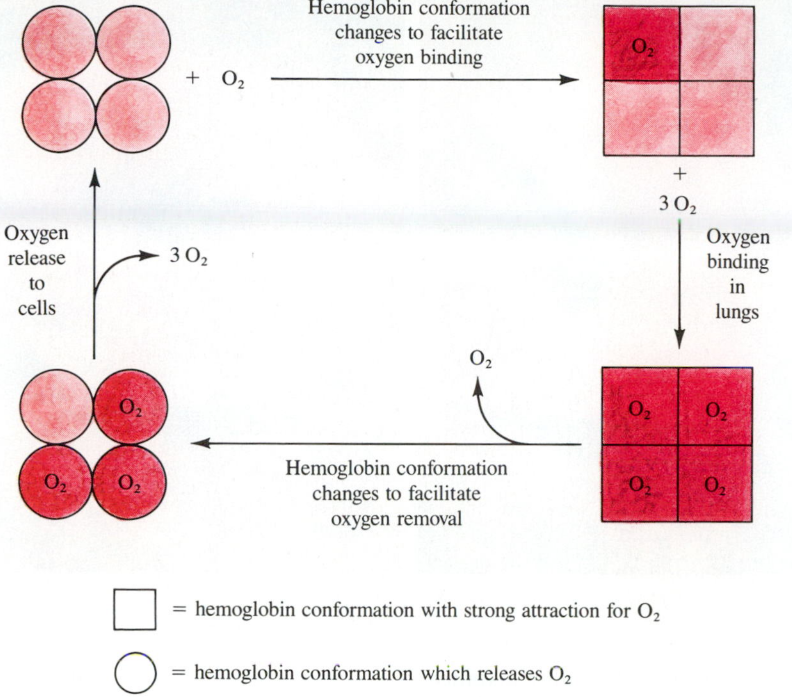

◀ **FIGURE 31.8**
A schematic representation of the oxygen–hemoglobin binding process. The circles and the squares represent two different conformations of the hemoglobin molecule. Oxygen binding or release causes hemoglobin to change its conformation.

Even minor alterations in primary structure may have drastic effects on the three-dimensional structural function of a protein. A graphic example of this fact is provided by sickle-cell anemia. This crippling genetic disease is due to red blood cells assuming a sickle shape. Sickled cells have impaired vitality and function. Sickle-cell anemia has been traced to a small change in the primary structure, or amino acid residue sequence, of hemoglobin. Normal hemoglobin molecules contain four polypeptide subunits—two identical α-polypeptide chains and two identical β-polypeptide chains. Sickle-cell and normal β-polypeptide chains of hemoglobin each contain 146 amino acid residues and differ only in the residue at the sixth position. In the sickle-cell hemoglobin chain, a glutamic acid residue has been replaced by a valine residue in this position. This change is sufficient, under some circumstances, to cause hemoglobin to aggregate into long filaments—a different quaternary structure. Large amounts of hemoglobin are present in red blood cells, and the changed quaternary structure causes sickling of the cell and greatly diminishes cell vitality (see Figure 31.9). This cell affliction has led to the premature deaths of many affected individuals.

To summarize, there are four classes of protein structure, and each has a different function.

1. Primary structure is defined by the amino acid residue sequence of the polypeptide chain. This sequence determines the shapes, or conformations, into which a protein can be arranged.
2. Secondary structure is a regular, repeating three-dimensional conformation held together by hydrogen bonding between components of the amide bonds of the primary chain. Hydrogen bonding can be between bonds in the same chain (for instance, the α-helix) or between bonds in

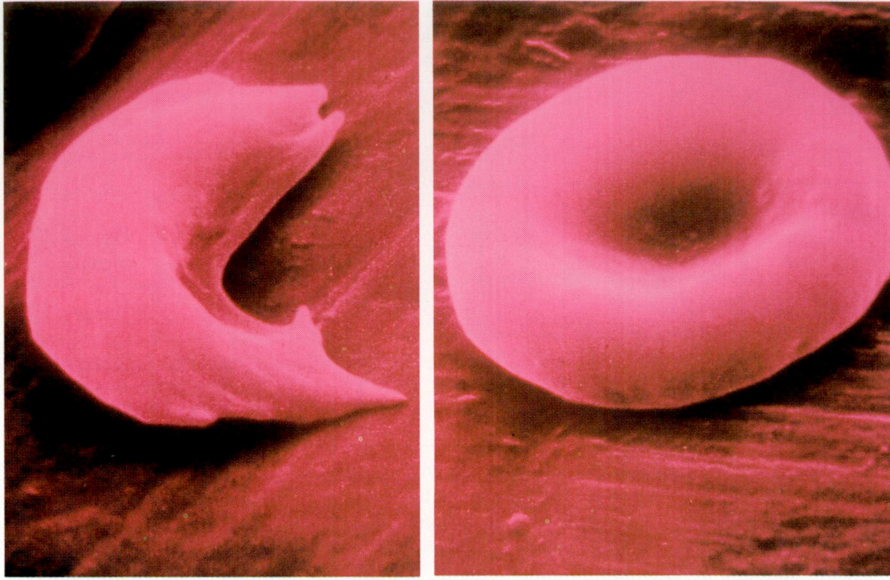

FIGURE 31.9 ▶ Scanning electron micrograph of sickled red blood cell (left) and normal red blood cell (right).

adjacent chains (for instance, the β-pleated sheet). The secondary structure imparts strength to proteins.

3. Tertiary structure is the overall three-dimensional structural conformation of the protein molecule. Various kinds of bonding interactions between components of the amino acid side chains stabilize the tertiary structure. The biological function of a protein is ultimately determined by the tertiary structure.
4. Quaternary structure, present in some proteins, describes the three-dimensional arrangement of subunits linked together by noncovalent bonds. The subunits have their own primary, secondary, and tertiary structures. Quaternary structure is often a prerequisite for proteins involved in the control of metabolic processes.

31.8 Functions of Proteins

Proteins serve a variety of very important biological functions as dictated by their diverse structures. A listing of general protein functions includes (1) structural, (2) catalytic, (3) hormonal, and (4) binding. The catalytic function is perhaps the most important and will be discussed in Chapter 32. However, each of the other functions is also vital to life. In the next several paragraphs, important examples of each function will be discussed. Note how each protein's structure critically determines its function.

Collagen, an example of a structural protein, is the most abundant protein in the body. It forms the bone matrix around which the calcium phosphate mineral can crystallize. Ligaments, tendons, and skin are composed of a large proportion of collagen. The structure of this protein allows it to provide a strong framework for each of these tissues. As shown in Figure 31.1, collagen is a long

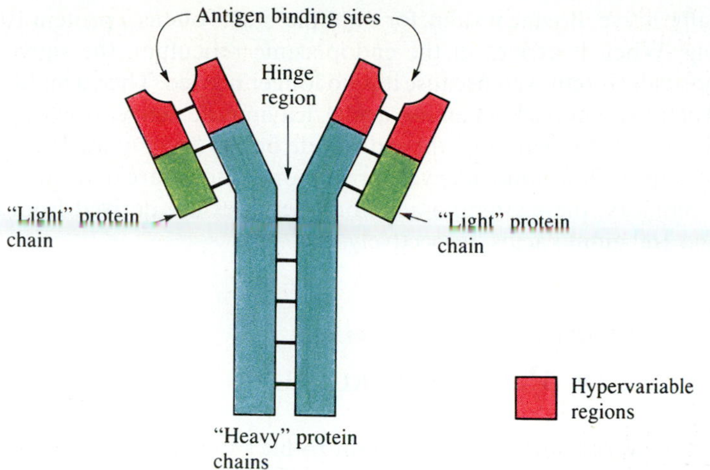

◀ **FIGURE 31.10 Diagrammatic shape of a typical immunoglobulin G protein.**

slender protein whose three strands are wrapped one around another as a rope would be woven, and, like a rope, the finished product is much stronger than a single strand. These triple helices are in turn stacked together like cordwood to form collagen fibers. The resulting tough material can be aligned to form tendons or deposited as a network to form a bone or skin structure. Next time you take a stride, recognize the force exerted on your Achilles tendon and how important the strength of collagen is to life.

Binding proteins serve a multitude of functions from transport of oxygen (by hemoglobin) to transport of vital nutrients across the intestinal membranes to binding of invading bacteria via the immune response. In each case, the protein structure is designed to efficiently accomplish its function. For example, consider the immunoglobulins (antibodies), proteins responsible for binding molecules foreign to the body. A successful immunological response requires that our bodies be prepared with a protein whose binding site is closely complementary to each of millions of different invading molecules (antigens). Just how our body accomplishes this feat can be described starting with the shape of a typical immunoglobulin G (Figure 31.10). Antigen binding regions are located at the ends of two arms. These are hinged together so that the distance between binding sites can vary with the size of the invading particle. The immunoglobulin protein chains are endowed with special hypervariable regions; every different immunoglobulin has a distinct amino acid sequence in its antigen binding sites and, therefore, has sites of unique size and shape. The body ensures that there will be immunoglobulins for many different antigens by producing millions of proteins with different hypervariable regions. Note that two different protein chains make up the antigen binding site. Since each chain has a hypervariable region that can be changed independently, the body has the ability to produce upwards of 10^{10} (ten billion) immunoglobulins with different binding sites, more than enough to handle exposure to invading particles.

Proteins or polypeptides that function as hormones are first synthesized in the cell as larger precursor molecules. Because these proteins are designed to be exported into the bloodstream, their structures share a common polypeptide segment, called a *signal peptide*. This amino acid sequence causes proteins to be moved into the endoplasmic reticulum, where they can be modified to become hormonally active. Bovine insulin, for example, starts out as a protein 107 amino acids long. When it arrives in the endoplasmic reticulum, the signal peptide (23 amino acids) is removed because it is no longer needed. Then a middle section of the protein is removed (33 amino acids), leaving the active form of insulin as two chains bonded together by three disulfide bonds (see Figure 31.2). Even the small polypeptide hormones like vasopressin or oxytocin are derived from much larger precursors; these 9-amino-acid-long peptides are derived from proteins of 215 and 160 amino acids, respectively.

31.9 Loss of Protein Structure

denaturation

Because protein structure is so important to life's functions, the loss of protein structure can be crucial. If a protein loses only its native three-dimensional conformation, the process is referred to as **denaturation**. In contrast, hydrolysis of peptide bonds ultimately converts proteins into their constituent amino acids. Often denaturation precedes hydrolysis.

Denaturation involves alteration or disruption of the secondary, tertiary, or quaternary—but not primary—structure of proteins (see Figure 31.11). Because a protein's function depends on its natural conformation, biological activity is lost with denaturation. This process may involve changes ranging from the subtle and reversible alterations caused by a slight pH shift to the extreme alterations involved in tanning a skin to form leather.

Environmental changes may easily disrupt natural protein structure, which is held together predominantly by noncovalent, relatively weak bonds. As gentle an act as pouring a protein solution can cause denaturation. Purified proteins must often be stored under ice-cold conditions because room temperature denatures them. It is not surprising that a wide variety of chemical and physical agents also can denature proteins. To name a few: strong acids and strong bases, salts (especially those of heavy metals), certain specific reagents such as tannic acid and picric acid, alcohol and other organic solvents, detergents, mechanical action such as whipping, high temperature, and ultraviolet radiation. Denatured proteins are generally less soluble than native proteins and often coagulate or precipitate from solution. Cooks have taken advantage of this for many years. When egg white, which is a concentrated solution of egg albumin protein, is stirred vigorously (as with an egg beater), an unsweetened meringue forms; the albumin denatures and coagulates. A cooked egg solidifies partially because egg proteins including albumin are denatured by heat.

In clinical laboratories the analysis of blood serum for small molecules such as glucose and uric acid is hampered by the presence of serum protein. This problem is resolved by first treating the serum with an acid to denature and precipitate the protein. The precipitate is removed and the protein-free liquid is then analyzed.

Loss of protein structure also occurs with hydrolysis of the peptide bonds to produce free amino acids. This chemical reaction destroys the protein's primary structure. Proteins can be hydrolyzed by boiling in a solution containing a strong acid such as HCl or a strong base such as NaOH. At ordinary temperatures proteins can be hydrolyzed using enzymes (see Chapter 32). These molecules, called proteolytic enzymes, are themselves proteins that function to

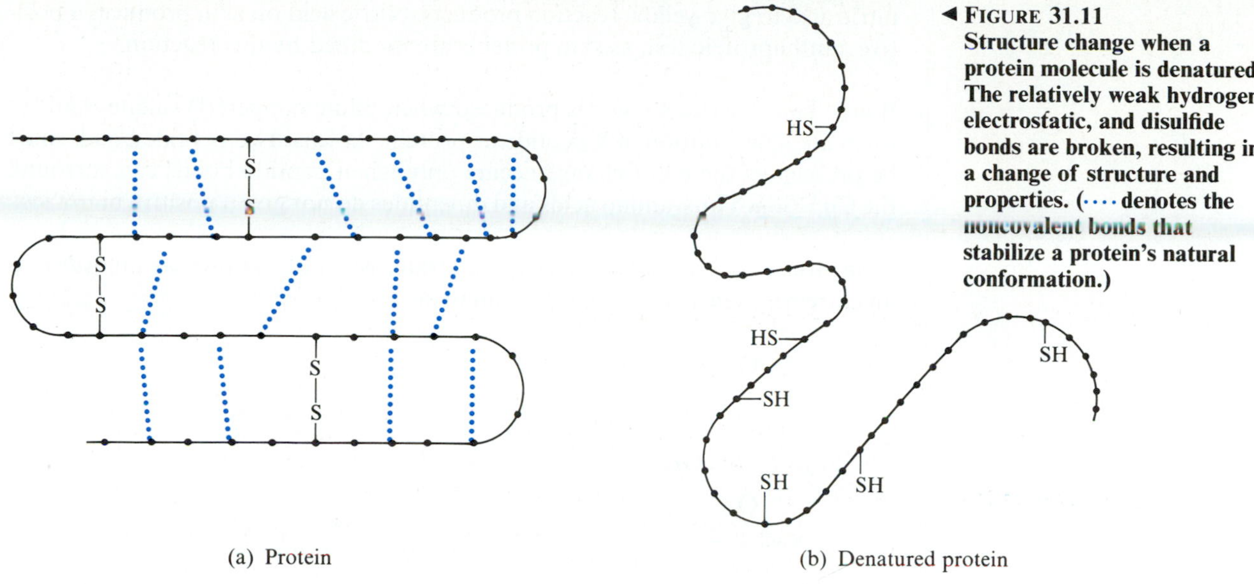

◄ **FIGURE 31.11**
Structure change when a protein molecule is denatured: The relatively weak hydrogen, electrostatic, and disulfide bonds are broken, resulting in a change of structure and properties. (···· denotes the noncovalent bonds that stabilize a protein's natural conformation.)

catalyze or speed up the hydrolysis reaction. The essential reaction of hydrolysis is the breaking of a peptide linkage and the addition of the elements of water:

$$\sim\text{N(H)}-\text{CHR}-\text{C(=O)}-\text{NH}-\text{CHR}-\text{C(=O)}\sim + H_2O \longrightarrow \sim\text{N(H)}-\text{CHR}-\text{C(=O)}-\text{OH} + H_2N-\text{CHR}-\text{C(=O)}\sim$$

Peptide linkage in protein chain (the C—N bond)

(∼ = the rest of the protein chain)

Any molecule with one or more peptide bonds, from the smallest dipeptide to the largest protein molecule, can be hydrolyzed. During hydrolysis, proteins are broken down into smaller and smaller fragments until all component amino acids are liberated.

Dietary protein must have its structure completely destroyed before it can provide nutrition for the body. Thus, digestion involves both the processes of denaturation and hydrolysis. Stomach acid causes most proteins to denature. Then, proteolytic enzymes in the stomach and the small intestine hydrolyze the proteins to smaller and smaller fragments until the free amino acids are formed and can be absorbed through the intestinal membranes into the bloodstream.

31.10 Tests for Proteins and Amino Acids

Many tests have been devised for detecting and distinguishing among amino acids, peptides, and proteins. Some examples are described here.

Xanthoproteic Reaction Proteins containing a benzene ring—for example, the amino acids phenylalanine, tryptophan, and tyrosine—react with concentrated nitric acid to give yellow reaction products. Nitric acid on skin produces a positive xanthoproteic test, as skin proteins are modified by this reaction.

Biuret Test A violet color is produced when dilute copper(II) sulfate is added to an alkaline solution of a peptide or protein. At least two peptide bonds must be present, as the color change occurs only when peptide bonds can surround the Cu^{2+} ion. Thus, amino acids and dipeptides do not give a positive biuret test.

Ninhydrin Test Triketohydrindene hydrate, generally known as *ninhydrin*, is an extremely sensitive reagent for amino acids:

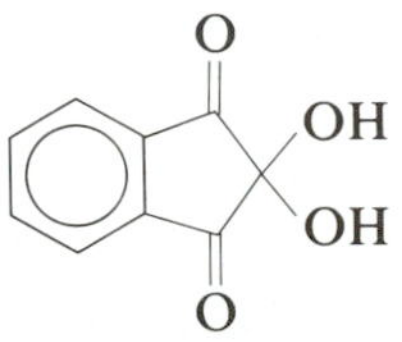

O
OH
OH
O

Ninhydrin

All amino acids, except proline and hydroxyproline, give a blue solution with ninhydrin. Proline and hydroxyproline produce a yellow solution. Less than 1 μg (10^{-6} g) of an amino acid can be detected with ninhydrin.

Chromatographic Separation

Complex mixtures of amino acids are readily separated by thin layer, paper, or column chromatography. In chromatographic methods the components of a mixture are separated by means of differences in their distributions between two phases. Separation depends on the relative tendencies of the components to remain in one phase or the other. In *thin-layer chromatography* (TLC), for example, a liquid and a solid phase are used. The procedure is as follows: A tiny drop of a solution containing a mixture of amino acids (obtained by hydrolyzing a protein) is spotted on a strip (or sheet) coated with a thin layer of dried alumina or some other adsorbant. After the spot has dried, the bottom edge of the strip is put into a suitable solvent. The solvent ascends the strip (by diffusion), carrying the different amino acids upward at different rates. When the solvent front nears the top, the strip is removed from the solvent and dried. The locations of the different amino acids are established by spraying the strip with ninhydrin solution and noting where colored spots appear. The pattern of colored spots is called a chromatogram. The identities of the amino acids in an unknown mixture can be established by comparing the chromatogram of the mixture with a chromatogram produced by known amino acids. A typical chromatogram of amino acids is shown in Figure 31.12.

Since proteins tend to denature during thin-layer chromatography, they are often separated by column chromatography. In this technique a solution containing a mixture of proteins (liquid phase) is allowed to percolate through a column packed with beads of a suitable polymer (solid phase). Separation of the mixture is accomplished as the proteins partition between the solid and liquid phases. The separated proteins move at different rates and are collected as they leave the column. Unlike amino acid chromatography, mild conditions must be

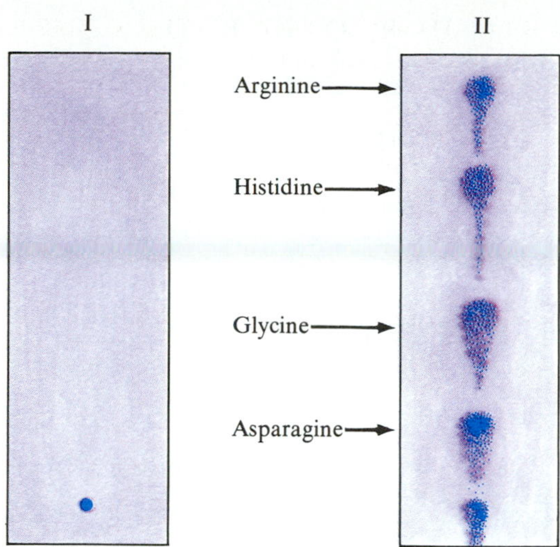

◀ FIGURE 31.12 Chromatogram showing separation of selected amino acids. On the left, spotted on the chromatographic strip, is an amino acid mixture containing arginine, histidine, glycine, and asparagine. On the right is the developed chromatogram showing the separated amino acids after treatment with ninhydrin. The solvent is a 250:60:250 volume ratio of 1-butanol–acetic acid–water.

maintained to limit protein denaturation. In the past it was not uncommon for a protein column chromatography to take from 6 to 12 hours. Recent developments in the field of high-performance liquid chromatography (HPLC) have shortened this time to minutes.

31.11 DETERMINATION OF THE PRIMARY STRUCTURE OF POLYPEPTIDES

Sanger's Reagent The pioneering work of Frederick Sanger gave us the first complete primary structure of a protein, insulin, in 1955. He used specific enzymes to hydrolyze insulin into smaller peptides and amino acids, then separated and identified the hydrolytic products by various chemical reactions.

Sanger's reagent, 2-4-dinitrofluorobenzene (DNFB), reacts with the α-amino group of the N-terminal amino acid of a polypeptide chain. The carbon–nitrogen bond between the amino acid and the benzene ring is more resistant to hydrolysis than are the remaining peptide linkages. Thus, when the substituted polypeptide is hydrolyzed, the terminal amino acid remains with the dinitrobenzene radical and can be isolated and identified. The remaining peptide chain is hydrolyzed to free amino acids in the process. This method marked an important step in the determination of the amino acid sequence in a protein. The reaction of Sanger's reagent is illustrated in the following equations:

$$O_2N\text{-}C_6H_3(NO_2)\text{-}F + H\text{-}N(H)\text{-}CHR\text{-}C(=O)\text{-}N(H)\text{-}CHR\text{-}C(=O)\text{-}$$

DNFB Polypeptide chain

$$\xrightarrow{\text{Alkaline}} O_2N\text{-}C_6H_3(NO_2)\text{-}N(H)\text{-}CHR\text{-}C(=O)\text{-}N(H)\text{-}CHR\text{-}C(=O)\text{-}$$

DNFB polypeptide derivative

$$\xrightarrow{\text{Acid hydrolysis}} O_2N\text{-}C_6H_3(NO_2)\text{-}N(H)\text{-}CH(R)\text{-}COOH + H_3N^+\text{-}CH(R)\text{-}COOH$$

DNFB-terminal amino acid derivative — Mixture of amino acids

From the DNFB hydrolysis, Sanger learned which amino acids were present in insulin. By less drastic hydrolysis, he split the insulin molecule into peptide fragments consisting of two, three, four, or more amino acid residues. After analyzing vast numbers of fragments utilizing the N-terminal method, he pieced them together in the proper sequence by combining fragments with overlapping structures at their ends, finally elucidating the entire insulin structure. As an example, consider the overlap that occurs between the hexapeptide and heptapeptide shown below:

Gly-Glu-Arg-Gly-Phe-Phe	Hexapeptide
Gly-Phe-Phe-Tyr-Thr-Pro-Lys	Heptapeptide
Gly-Glu-Arg-Gly-Phe-Phe-Tyr-Thr-Pro-Lys	Decapeptide

The three residues Gly-Phe-Phe at the end of the hexapeptide match the three residues at the beginning of the heptapeptide. By using these three residues in common, the structure of the decapeptide shown, which occurs in chain B of insulin (residues 20–29), is determined. (See Figure 31.2, which shows the amino acid sequence of insulin.)

Edman Degradation More recently, the Edman degradation method has been developed to split off amino acids one at a time from the N-terminal end of a polypeptide chain. In this procedure the reagent, phenylisothiocyanate, is first added to the N-terminal amino group. The N-terminal amino acid–phenylisothiocyanate addition product is then cleaved from the polypeptide chain. The resultant substituted phenylthiohydantoin is then isolated and identified. The shortened polypeptide chain is then ready to undergo another Edman degradation. By repeating this set of reactions, one can determine directly the amino acid sequence of a polypeptide. A machine, the protein sequenator, which carries out protein sequencing by Edman degradation automatically, is now available.

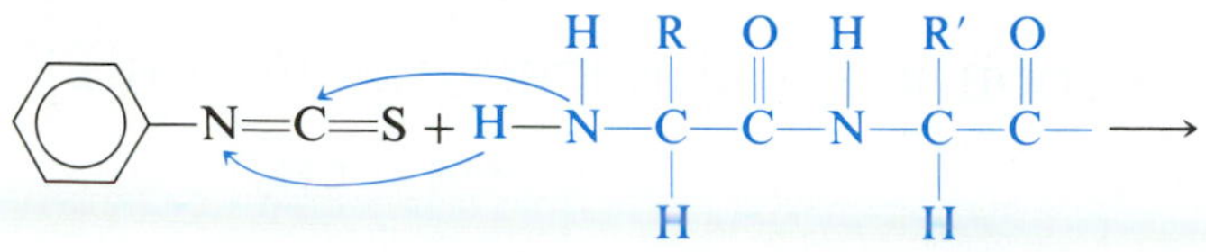

Phenylisothiocyanate Polypeptide chain

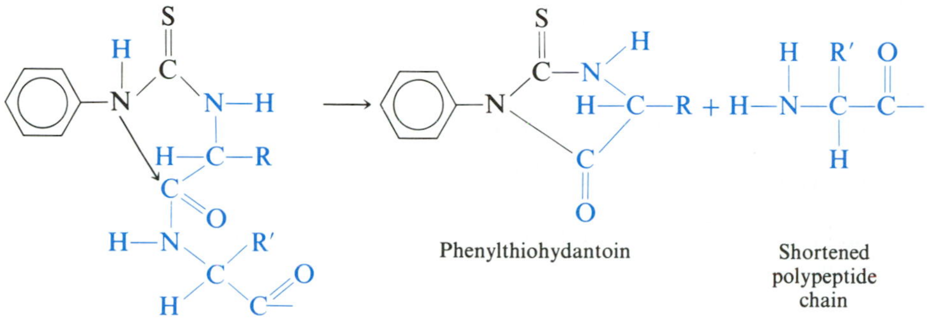

Phenylthiohydantoin

Shortened polypeptide chain

31.12 Synthesis of Peptides and Proteins

Since each amino acid has two functional groups, it is not difficult to form dipeptides or even fairly large polypeptide molecules. As mentioned in Section 31.6, glycine and alanine react to form two different dipeptides, glycylalanine and alanylglycine. If threonine, alanine, and glycine react, six tripeptides, each made up of three different amino acids, are obtained. By reacting a mixture of several amino acids, polypeptides are produced that contain a random arrangement of the amino acid units. This fact makes it clear that even a small protein like insulin could not be produced by simply reacting a mixture of the required amino acids. When such a large polypeptide is synthesized, it is formed by joining amino acids one by one in the proper sequence. Within cells, the proper sequence is maintained by careful genetic control (see Section 33.10).

When a polypeptide chain of known primary structure is synthesized *in vitro* (synthesis in "glass" without the aid of living tissue), the process must be controlled so that only one particular amino acid is added at each stage. Remarkable progress has been made in developing the necessary techniques. Amino acids are used in which either the amino or carboxylic acid group has been inactivated or blocked with a suitable *blocking agent*. The blocked amino or carboxylic acid group is reactivated by removing the blocking agent in the next stage of the synthesis. For example, if the tripeptide Thr-Ala-Gly is to be made, amino-blocked threonine is reacted with carboxyl-blocked alanine to make the doubly blocked Thr-Ala dipeptide. The carboxyl-blocked alanine end of the dipeptide can then be reactivated and reacted with carboxyl-blocked glycine to make the blocked Thr-Ala-Gly tripeptide. The free tripeptide can then be obtained by removing the blocking agents from both ends. Using [B] to designate the blocking agents, this synthesis is represented in schematic form as follows:

$$\boxed{B}\,H_2N\underset{\substack{|\\ H-C-OH\\ |\\ CH_3}}{C}HCOOH + H_2N\underset{\substack{|\\ CH_3}}{C}HCOOH\,\boxed{B} \longrightarrow \boxed{B}\,H_2N\underset{\substack{|\\ H-C-OH\\ |\\ CH_3}}{C}H\overset{\substack{O\\ \|}}{C}-NH\underset{\substack{|\\ CH_3}}{C}HCOOH\,\boxed{B}$$

Amino-blocked threonine — Carboxyl-blocked alanine — Blocked Thr-Ala dipeptide

$$\boxed{B}\,H_2N\underset{\substack{|\\ H-C-OH\\ |\\ CH_3}}{C}H\overset{\substack{O\\ \|}}{C}-NH\underset{\substack{|\\ CH_3}}{C}HCOOH + H_2NCH_2COOH\,\boxed{B} \longrightarrow$$

Amino-blocked Thr-Ala dipeptide — Carboxyl-blocked glycine

$$\boxed{B}\,H_2N\underset{\substack{|\\ H-C-OH\\ |\\ CH_3}}{C}H\overset{\substack{O\\ \|}}{C}-NH\underset{\substack{|\\ CH_3}}{C}H\overset{\substack{O\\ \|}}{C}-NHCH_2COOH\,\boxed{B}$$

Blocked Thr-Ala-Gly tripeptide

Longer polypeptide chains of specified amino acid sequence can be made by this general technique. The procedure is very tedious. The synthesis of insulin required an effort equivalent to one person working steadily for several years.

It is not necessary to prepare an entire synthetic protein polypeptide chain by starting at one end and adding amino acids one at a time. Previously prepared shorter polypeptide chains of known structure can be joined to form a longer polypeptide chain. This method was used in making synthetic insulin. The insulin molecule, shown in Figure 31.2, consists of two polypeptide chains bonded together by two disulfide bonds. A third disulfide bond, between two amino acids in one chain, makes a small loop in that chain. Once the two chains had been assembled from fragments, a seemingly formidable problem remained—how to form the disulfide bonds in the correct positions. As it turned out, it was necessary only to bring the two chains together, with the cysteine side chains in the reduced condition, and to treat them with an oxidizing agent. Disulfide bonds formed at the right places, and biologically active insulin molecules were obtained.

In the mid-1960s a machine capable of automatically synthesizing polypeptide chains of known amino acid sequence was designed by R. B. Merrifield of Rockefeller University. The starting amino acid is bonded to a plastic surface (polystyrene bead) in the reaction chamber of the apparatus. Various reagents needed for building a chain of predetermined structures are automatically delivered to the reaction chamber in a programmed sequence. Twelve reagents and about 100 operations are needed to lengthen the chain by a single amino acid residue. But the machine is capable of adding residues to the chain at the rate of six a day. Such a machine makes it possible to synthesize complex molecules like insulin in a few days. The first large polypeptide (pancreatic ribonuclease), containing 124 amino acid residues, was synthesized by this method in 1969. More recently a human growth hormone (HGH) containing 188 amino acid residues was synthesized using this technique.

Over the past ten years, techniques have been developed to use biological systems to create and produce new proteins. These procedures start with genetic material which codes for the protein of interest. Then this material is properly modified and introduced into rapidly growing cells that are treated to overproduce this foreign protein. The result can be a harvest of much needed protein, such as the human insulin used by diabetics or the human growth hormone used to treat dwarfism. These procedures constitute a part of genetic engineering and will be discussed in more detail in Chapter 33. The advantages of these new techniques can be realized by comparing the rate of biological protein synthesis with that achieved by the machine designed by Merrifield. If a bacterium such as *E. coli* is used to produce a human protein such as growth hormone, one amino acid residue can be added every 0.01 second; a growth hormone molecule can be produced every 10 to 20 seconds. In contrast, the same growth hormone would take days to produce in the chemistry laboratory.

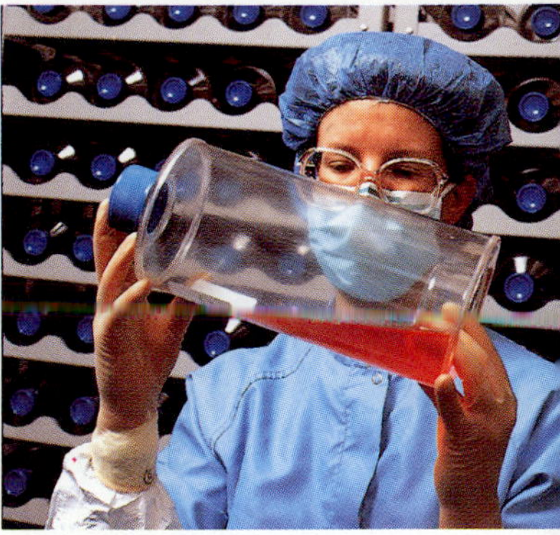

▲
Production of a cloned protein. The technician is holding a container with cultured cells in a nutrient medium.

Concepts in Review

1. List five foods that are major sources of proteins.
2. List the elements that are usually contained in proteins.
3. Write formulas for and know the abbreviations for the common amino acids.
4. Write the zwitterion formula of an amino acid and know how it behaves in a dilute acidic and in a dilute basic solution.
5. Understand why a protein in solution at its isoelectric pH does not migrate in an electrolytic cell.
6. Combine amino acids into polypeptide chains.
7. Understand how peptide chains are named and numbered.
8. Understand what is meant by the N-terminal and C-terminal residues of a peptide chain.
9. Briefly explain the primary, secondary, and tertiary structure of a protein.
10. Explain the meaning of the statement, "The primary structure helps establish a protein's three-dimensional structure."
11. Define quaternary structure and explain its importance in the function of hemoglobin.
12. Understand the bonding in the α-helical and β-pleated-sheet structure of a protein.
13. Give an example of a structural protein.
14. Explain the role cysteine plays in the structure of polypeptides and proteins.
15. Describe how the α-helix and β-pleated sheet are important in the structure of protein domains.

16. List the most important protein functions.
17. Relate the structure of collagen to its biological function.
18. Discuss the role of the immunoglobulins in the body's defense mechanisms and identify important structural elements of these proteins.
19. Explain the importance of peptide bond cleavage in the formation of protein and peptide hormones.
20. Describe Sanger's work on determining the primary structure of proteins.
21. Reconstruct a peptide chain from a knowledge of its hydrolysis products.
22. Explain how Edman degradation allows the primary structure to be determined.
23. Describe chromatographic methods of separating amino acids.
24. State the physical evidence observed in a positive reaction in the (a) xanthoproteic reaction, (b) biuret test, and (c) ninhydrin test.
25. Know what is meant by denaturation and the various methods by which proteins can be denatured.

Equations in Review

Ionization of Amino Acids

$$\underset{\substack{\text{Cation form}\\ \text{II}}}{\underset{NH_3^+}{\underset{|}{RCHCOOH}}} \underset{OH^-}{\overset{H^+}{\rightleftharpoons}} \underset{\substack{\text{Zwitterion form}\\ \text{I}}}{\underset{NH_3^+}{\underset{|}{RCHCOO^-}}} \underset{H^+}{\overset{OH^-}{\rightleftharpoons}} \underset{\substack{\text{Anion form}\\ \text{III}}}{\underset{NH_2}{\underset{|}{RCHCOO^-}}}$$

(Similar reactions may occur on the R groups of specific amino acids.)

Reaction Between Amino Acids to Form a Peptide Linkage

$$\underset{NH_2}{\underset{|}{RCHCOOH}} + \underset{NH_2}{\underset{|}{RCHCOOH}} \longrightarrow \underset{NH_2}{\underset{|}{RCH}}\overset{O}{\overset{\|}{C}}\text{—NH—}\overset{R}{\overset{|}{C}}HCOOH + H_2O$$

Hydrolysis of the Peptide Linkage in Proteins

Peptide linkage in protein chain

$$\sim\underset{H}{\overset{H}{N}}\text{—}\underset{H}{\overset{R}{C}}\text{—}\overset{O}{\overset{\|}{C}}\text{—}\underset{H}{N}\text{—}\underset{R}{\overset{H}{C}}\text{—}\underset{O}{\underset{\|}{C}}\sim + H_2O \longrightarrow \sim\overset{H}{N}\text{—}\underset{H}{\overset{R}{C}}\text{—}\overset{O}{\overset{\|}{C}}\text{—OH} + H_2N\text{—}\underset{R}{\overset{H}{C}}\text{—}\underset{O}{\underset{\|}{C}}\sim$$

($\sim$ = rest of protein chain)

EXERCISES

1. List four foods that are major sources of proteins.
2. Why are the amino acids of proteins called α-amino acids?
3. What elements are present in amino acids and proteins?
4. Why are proteins from some foods of greater nutritional value than others?
5. Write the names of the amino acids that are essential to humans.
6. What two general methods are now available for improving the nutritional value of corn protein?
7. Which amino acids contain a heterocyclic ring?
8. Write the structural formulas for D-serine and L-serine. Which form is found in proteins?
9. Why are the amino acids amphoteric? Why are they optically active?
10. Write the structural formula representing threonine at its isoelectric point.
11. For phenylalanine write:
    (a) The molecular formula
    (b) The zwitterion formula
    (c) The formula in 0.1 M H_2SO_4
    (d) The formula in 0.1 M NaOH
12. Write ionic equations to show how alanine acts as a buffer toward:
    (a) H^+ ion (b) OH^- ion
13. (a) At what pH will arginine not migrate to either electrode in an electrolytic cell?
    (b) In what pH range will it migrate toward the positive electrode?
14. What can you say about the number of positive and negative charges on a protein molecule at its isoelectric point?
15. At what pH will the following amino acids be at their isoelectric points?
    (a) Histidine (c) Glutamic acid
    (b) Phenylalanine
16. Write out the full structural formula of the two dipeptides containing glycine and phenylalanine. Indicate the location of the peptide bonds.
17. Write structures for:
    (a) Glycylglycine
    (b) Glycylglycylalanine
    (c) Leucylmethionylglycylserine
18. Using amino acid abbreviations, write all the possible tripeptides containing one unit each of glycine, phenylalanine, and leucine.
19. How many dipeptides containing glycine can be written using the 20 amino acids from Table 31.1?
20. Explain what is meant by (a) the primary structure, (b) the secondary structure, and (c) the tertiary structure of a protein.
21. What special role does the sulfur-containing amino acid cysteine have in protein structure?
22. How do myoglobin and hemoglobin differ in structure and function?
23. Briefly describe the protein skeletal shapes formed by the β-pleated sheet.
24. How do amino acids like alanine and leucine help stabilize the three-dimensional structure of a protein?
25. Ribonuclease is a protein that is slightly smaller than myoglobin. How many protein domains would you predict for the structure of ribonuclease?
26. List four general functions performed by proteins.
27. How are the hypervariable regions in immunoglobulin chains important to this protein's function?
28. Protein hormones are often modified before they become active. What is a signal peptide, and how is it important in protein modification?
29. Suggest a reason why proteins tend to be easily denatured.
30. Explain how hydrolysis of a protein differs from denaturation.
31. What chemical change occurs when a protein is hydrolyzed to amino acids?
32. What is the visible evidence observed in a positive reaction for the following tests?
    (a) Xanthoproteic reaction
    (b) Biuret test
    (c) Ninhydrin test
33. Would Thr-Ala-Gly react with each of the following?
    (a) Sanger's reagent
    (b) Concentrated HNO_3 to give a positive xanthoproteic test
    (c) Ninhydrin
34. Which amino acids give a positive xanthoproteic test?
35. (a) What is thin-layer chromatography?
    (b) Describe how amino acids are separated using this technique.
    (c) What reagent is used to locate the amino acids in the chromatogram?
36. Briefly describe protein column chromatography.
37. Threonine has two asymmetric centers. Write

Fischer projection formulas for its stereoisomers.

38. Show the reactants and products for one complete cycle of the Edman degradation on the tripeptide Ala-Leu-Gly.

39. What is the amino acid sequence of heptapeptides A and B, which upon hydrolysis yield these tripeptides:
(A) Gly-Phe-Leu, Phe-Ala-Gly, Leu-Ala-Tyr
(B) Phe-Gly-Tyr, Phe-Ala-Ala, Ala-Leu-Phe
Both A and B contain one residue each of Gly, Leu, and Tyr and two residues each of Ala and Phe.

40. A nonapeptide is obtained from blood plasma by treatment with the enzyme trypsin. Analysis showed that both terminal amino acids are arginine. Total hydrolysis of this peptide yielded Gly, Ser, 2 Arg, 2 Phe, 3 Pro. Partial hydrolysis gave Phe-Ser, Phe-Arg, Arg-Pro, Pro-Pro, Pro-Gly-Phe, Ser-Pro-Phe. What is the amino acid sequence of the nonapeptide?

41. One hundred grams (100.0 g) of a food product was analyzed and found to contain 6.0 g of nitrogen. If protein contains an average of 16% nitrogen, what percentage of the food is protein?

42. Human hemoglobin contains 0.33% iron. If each hemoglobin molecule contains four iron atoms, what is the molar mass of hemoglobin?

43. Which of these statements are correct? Rewrite each incorrect statement to make it correct.
(a) Proteins, like fats and carbohydrates, are primarily for supplying heat and energy to the body.
(b) Proteins differ from fats and carbohydrates in that they contain a large amount of nitrogen.
(c) A complete protein is one that contains all the essential amino acids.
(d) Except for glycine, amino acids that are found in proteins have the L-configuration.
(e) All amino acids have an asymmetric carbon atom and are therefore optically active.
(f) The amide linkages by which amino acids are joined together are called peptide linkages.
(g) Two different dipeptides can be formed from the amino acids glycine and phenylalanine.
(h) The compound Ala-Phe-Tyr has two peptide bonds and is therefore known as a dipeptide.
(i) A zwitterion is a dipolar ion form of an amino acid.
(j) The primary structure of a protein is the α-helical or β-pleated-sheet form that it takes.
(k) The tertiary structure of a protein determines that protein's primary structure.
(l) The amino acid residues in a peptide chain are numbered beginning with the C-terminal amino acid.
(m) Insulin contains two polypeptide chains, one with 21 amino acids and the other with 30 amino acids.
(n) Amino acids are often removed from protein hormones to activate them.
(o) Collagen is an example of a binding protein.
(p) α-helices cluster into a "twisted sheet" conformation to form part of many protein skeletons.
(q) A protein structure with hinged arms provides binding flexibility for the immunoglobulins.
(r) When a protein is denatured, the polypeptide bonds are broken, liberating the amino acids.
(s) Irreversible coagulation or precipitation of proteins is called denaturation.
(t) Sanger's reagent reacts with peptide chains, isolating the N-terminal amino acid.

32

Enzymes

◀ CHAPTER OPENING PHOTO: Yeast provides the enzymes necessary to make bread rise.

Perhaps you have seen laundry detergents advertised as "containing enzymes." These molecules are added to detergents in order to clean especially hard-to-remove spots. In life as well, enzymes take on the "tough" metabolic tasks—in fact, life cannot continue in the absence of enzymes.

Enzymes are important because they can accelerate a chemical reaction by one million to one-hundred million times. Imagine what would happen if some of our tasks were accelerated by that amount! A two-hour daily homework assignment would be completed in about one-thousandth of a second. A flight from Los Angeles to New York would take about one-hundredth of a second. The building of Hoover Dam, a monumental task requiring five years of earth moving and complex steel and concrete work, would require only about thirty seconds.

Scientists now understand the general characteristics of enzymes. And, as we will see in this chapter, these molecules achieve almost miraculous results by following some very basic chemical principles.

32.1 MOLECULAR ACCELERATORS

enzymes

Enzymes are the catalysts of biochemical reactions. Enzymes catalyze nearly all of the myriad reactions that occur in living cells. Uncatalyzed reactions that require hours of boiling in the presence of a strong acid or a strong base can occur in a fraction of a second in the presence of the proper enzyme at room temperature and nearly neutral pH. The catalytic functions of enzymes are directly dependent on their three-dimensional structures. It was generally believed until quite recently that all enzymes were proteins. However, research under way since about 1980 has shown that certain ribonucleic acids (RNAs, see Chapter 33) also function as enzymes.

Louis Pasteur was one of the first scientists to study enzyme catalyzed reactions. He believed that living yeasts or bacteria were required for these reactions, which he called *fermentations*—for example, the conversion of glucose to alcohol by yeasts. In 1897 Eduard Büchner (1860–1917) made a cell-free filtrate that contained enzymes prepared by grinding yeast cells with very fine sand. The enzymes in this filtrate converted glucose to alcohol, thus proving that the presence of living cells was not required for enzyme activity. For this work Büchner received the Nobel prize in chemistry in 1907.

It is important to realize that enzymes are essential to life. The critical biochemical reactions go too slowly for life to be maintained in the absence of enzymes. The typical biochemical reaction occurs more than a million times faster when catalyzed by an enzyme. For example, we know that the reduced carbons of carbohydrates can react with oxygen to produce carbon dioxide and energy. Yet the sucrose in the sugar bowl at home never reacts significantly with the oxygen in the air. These sucrose molecules must overcome an energy barrier (activation energy) before reaction can occur (see Figure 32.1). In the sugar bowl, the energy barrier is too large. But in the cell, enzymes lower the activation energy and enable sucrose to react rapidly enough to provide the energy needed for life processes.

▲
Fermentation of glucose.

Each organism contains thousands of enzymes. Some are simple proteins consisting only of amino acid units. Others are conjugated and consist of a protein part, or *apoenzyme*, and a nonprotein part, or *coenzyme*. Both parts are essential, and a functioning enzyme consisting of both the protein and nonprotein parts is called a *holoenzyme*:

$$\text{Apoenzyme} + \text{Coenzyme} = \text{Holoenzyme}$$

Often the coenzyme is derived from a vitamin, and one coenzyme may be associated with many different enzymes.

For some enzymes an inorganic component such as a metal ion (for example, Ca^{2+}, Mg^{2+}, or Zn^{2+}) is required. This inorganic component is an *activator*. From the standpoint of function, an activator is analogous to a coenzyme, but inorganic components are not called coenzymes.

Another remarkable property of enzymes is their specificity of reaction; that is, a certain enzyme catalyzes the reaction of a specific type of substance. For example, the enzyme maltase catalyzes the reaction of maltose and water to form glucose. Maltase has no effect on the other two common disaccharides, sucrose and lactose. Each of these sugars requires a specific enzyme—sucrase to hydrolyze sucrose, lactase to hydrolyze lactose. These reactions are indicated by the following equations:

$$\underset{\text{Maltose}}{C_{12}H_{22}O_{11}} + H_2O \xrightarrow{\text{Maltase}} \underset{\text{Glucose}}{C_6H_{12}O_6} + \underset{\text{Glucose}}{C_6H_{12}O_6}$$

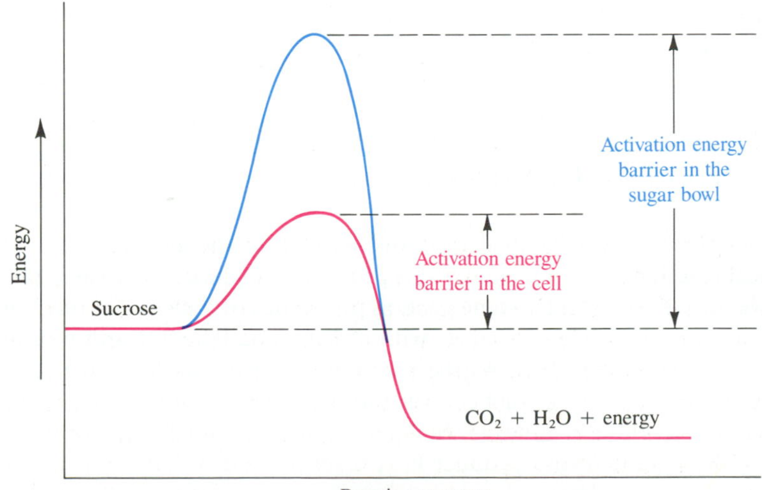

◀ **FIGURE 32.1**
A typical reaction energy profile: The lower activation energy in the cell is due to the catalytic effect of enzymes.

$$\underset{\text{Sucrose}}{C_{12}H_{22}O_{11}} + H_2O \xrightarrow{\text{Sucrase}} \underset{\text{Glucose}}{C_6H_{12}O_6} + \underset{\text{Fructose}}{C_6H_{12}O_6}$$

$$\underset{\text{Lactose}}{C_{12}H_{22}O_{11}} + H_2O \xrightarrow{\text{Lactase}} \underset{\text{Glucose}}{C_6H_{12}O_6} + \underset{\text{Galactose}}{C_6H_{12}O_6}$$

The substance acted on by an enzyme is called the *substrate*. Sucrose is the substrate of the enzyme sucrase. Enzymes have been named by adding the suffix *-ase* to the root of the substrate name. Note the derivations of maltase, sucrase, and lactase from maltose, sucrose, and lactose. Many enzymes, especially digestive enzymes, have common names such as pepsin, rennin, trypsin, and so on. These names have no systematic significance.

In the International Union of Biochemistry (IUB) System, enzymes are assigned to one of six classes, the names of which clearly describe the nature of the reaction they catalyze. Each of the classes has several subclasses. In this system the name of the enzyme has two parts; the first gives the name of the substrate, and the second, ending in *-ase*, indicates the type of reactions catalyzed by all enzymes in the group. The six main classes of enzymes are

1. *Oxidoreductases:* Enzymes that catalyze the oxidation–reduction between two substrates.
2. *Transferases:* Enzymes that catalyze the transfer of a functional group between two substrates.
3. *Hydrolases:* Enzymes that catalyze the hydrolysis of esters, carbohydrates, and proteins (polypeptides).
4. *Lyases:* Enzymes that catalyze the removal of groups from substrates by mechanisms other than hydrolysis.
5. *Isomerases:* Enzymes that catalyze the interconversion of stereoisomers and structural isomers.
6. *Ligases:* Enzymes that catalyze the linking together of two compounds with the breaking of a pyrophosphate bond in adenosine triphosphate (ATP, see Chapter 33).

Because the systematic name is usually long and often complex, working or practical names are used for enzymes. For example, adenosine triphosphate creatine phosphotransferase is called creatine kinase, and acetylcholine acylhydrolase is called acetylcholine esterase.

32.2 Enzyme Catalysis

Enzymes catalyze biochemical reactions and, thus, increase the rate of these chemical reactions—but how does this process take place? To answer this question, we must first consider some general properties of a chemical reaction.

Every chemical reaction starts with at least one reactant and finishes with a minimum of one product. As the reaction proceeds, the reactant concentration decreases and the product concentration increases. Often these changes are plotted as a function of time, as shown in Figure 32.2 for the hypothetical conversion of reactant A into product B. A reaction rate is defined as a change in concentration with time.

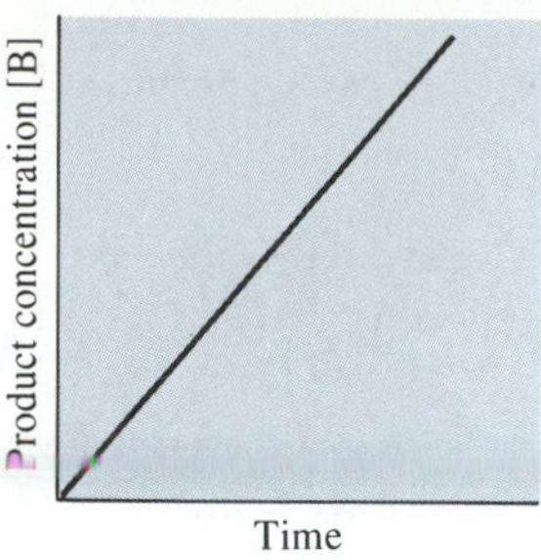

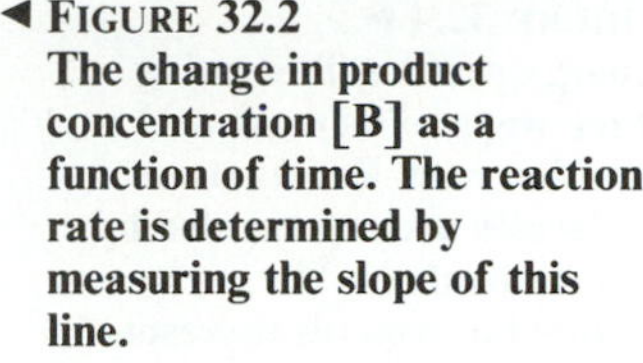
FIGURE 32.2 The change in product concentration [B] as a function of time. The reaction rate is determined by measuring the slope of this line.

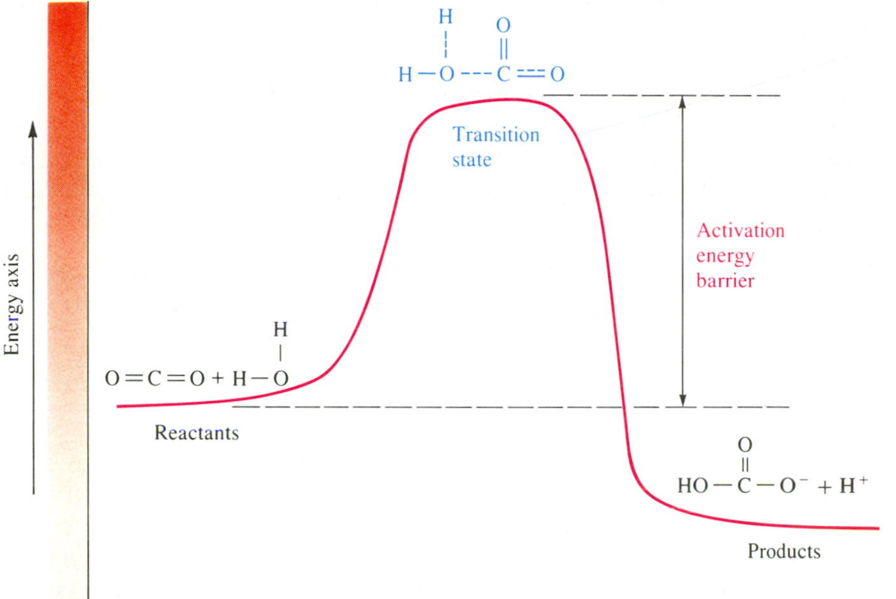

FIGURE 32.3 An energy profile for the reaction between water and carbon dioxide: The dashed lines show bonds that are being formed or broken in the transition state.

transition state

The reactant must pass through a high-energy **transition state** in order to be converted into a product. This transition state is an unstable structure with characteristics of both the reactant and the product. The energy necessary to move a reactant to the transition state is termed the activation energy.

For example, carbon dioxide can be reacted with water to yield bicarbonate. Yet, as shown in Figure 32.3, energy is required to align the water at its appropriate position and to rearrange the bonds within the carbon dioxide. Once the transition state is created, the remainder of the process proceeds easily. As with almost all chemical reactions, reaching the transition state is difficult and limits the rate at which reactants are converted to products.

There are three common ways of increasing a reaction rate (see Figure 32.4):

1. *Increasing the reactant concentration:* When the total reactant concentration is made larger, the number of reactant molecules with the necessary activation energy also increases. For simple reactions in the absence

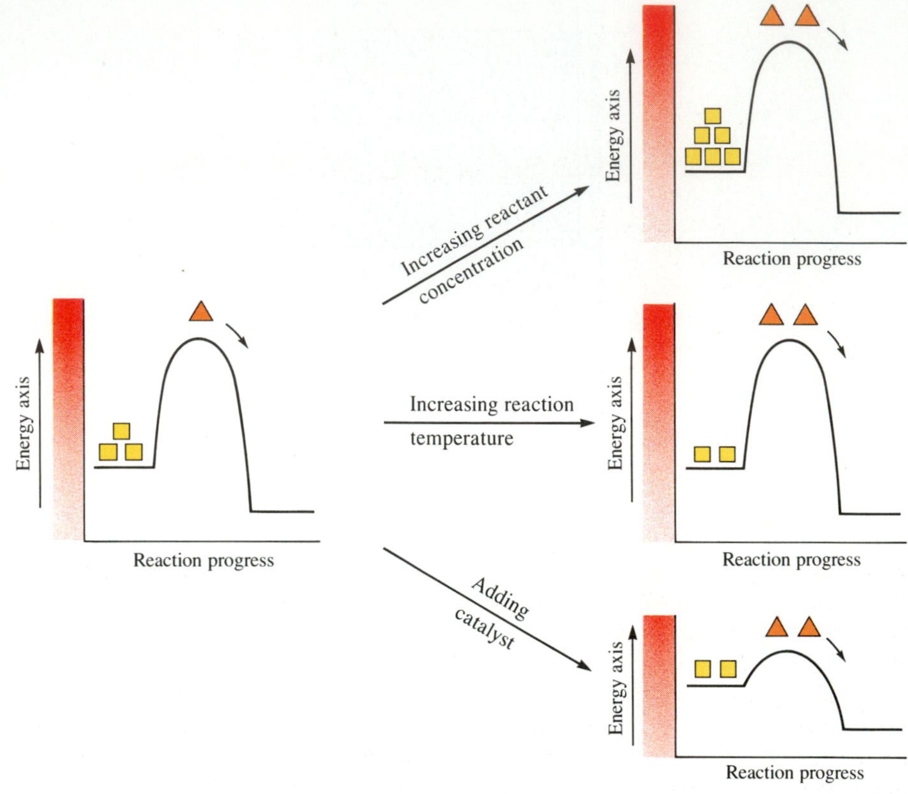

FIGURE 32.4 ▶ Energy profiles illustrating three ways to increase a reaction rate. Reactant molecules are represented by square symbols. The triangular symbols represent the number of transition state molecules that are converted to products.

of a catalyst, the reaction rate increases in a linear way with reactant concentration.

2. *Increasing the reaction temperature:* An increase in temperature generally means that each reactant molecule becomes more energetic. A larger fraction of the reactants have the activation energy necessary to be converted to products, and the reaction rate increases.
3. *Adding a catalyst:* A catalyst lowers the activation energy by allowing a new, lower energy transition state. Since this new process has a lower activation energy, more reactants have the energy to become products. The reaction rate increases.

Biological systems can rarely change reactant concentration or temperature upon demand. Thus, to alter reaction rates, life has evolved a set of superb catalysts—the enzymes. How these protein catalysts work has been a subject of study for many years.

In 1913, two German researchers, Leonor Michaelis (1875–1949) and Maud Menten, measured enzyme-catalyzed reaction rates as a function of substrate (reactant) concentration. They observed that most enzyme-catalyzed reactions show an increasing rate with increasing substrate concentration *but* only to a specific maximum velocity (V_{max}; Figure 32.5). A graph like that in Figure 32.5 is often called a Michaelis–Menten plot.

Scientists have learned much about the nature of enzymes by studying the Michaelis–Menten plot. First, because the rate approaches a maximum, it can be concluded that enzymes have a limited catalytic ability. Once these enzymes are operating at a maximum, a further increase in substrate (reactant) concentration does not change the reaction rate. By analogy, an enzyme shuttles reac-

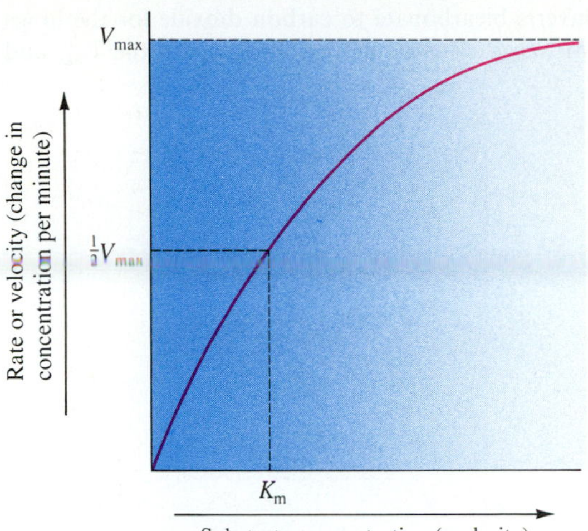

◀ **FIGURE 32.5**
A Michaelis–Menten plot showing the rate of an enzyme-catalyzed reaction as a function of substrate concentration. The shaded portion of the graph approximately marks the area where a change in substrate concentration causes a significant rate change.

tants to the transition state (and on to products) like an usher seating patrons at a theater. No matter how many people arrive at the entrance, a conscientious usher can only lead the ticket holders to their places at a set pace. Similarly, each specific enzyme has a set, maximum catalytic rate.

The V_{max} that can be determined from a Michaelis–Menten plot is one measure of an enzyme's maximum catalytic efficiency. Perhaps a more useful measure, called the **turnover number**, is calculated from the V_{max}. A turnover number measures how many substrate molecules one enzyme can react or "turn over" in a given time span. For example, a particular enzyme might have a turnover number of 1000 s^{-1}, which would mean one enzyme molecule can react 1000 substrate molecules per second.

turnover number

The Michaelis–Menten plot also shows that enzymes are responsive and most useful to the cell only within a specific substrate concentration range. Physiological reactions generally need to change as their environment changes. That is, reaction rates should increase when there are more reactants and decrease when there are less reactants. Figure 32.5 shows such changes only over a limited substrate concentration range. It is within this range that an enzyme is most useful for the cell.

The **Michaelis constant, K_m**, is a numerical measure of the effective substrate concentration range for an enzyme. This constant is defined as the substrate concentration needed for an enzyme to operate at one-half of its maximum rate. Thus, the K_m is poised at the approximate midpoint of the effective concentration range. Perhaps the most direct way to derive K_m is shown in Figure 32.5; after identifying $\frac{1}{2}V_{max}$ a line can be extrapolated to the abscissa (x-axis) and the numerical value for K_m determined.

Michaelis constant, K_m

EXAMPLE 32.1

The enzyme carbonic anhydrase is responsible for catalyzing the reaction between water and carbon dioxide:

$$CO_2 + H_2O \rightleftarrows HCO_3^- + H^+$$

This enzyme is important to the body because it allows carbon dioxide produced by body tissues to dissolve quickly in the bloodstream for transport to the lungs. In addition, this enzyme quickly converts bicarbonate to carbon dioxide for the lungs to exhale. Based on the following Michaelis–Menten plot, estimate both the V_{max} and K_m for carbonic anhydrase.

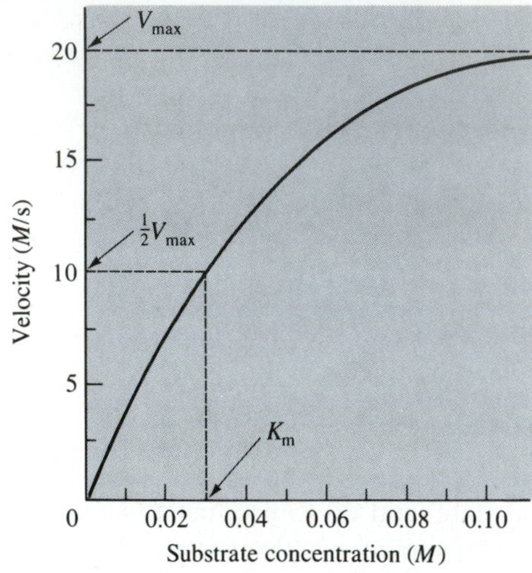

SOLUTION

The V_{max} can be estimated by reading the y-axis where the graphed line becomes horizontal. For this data, carbonic anhydrase has a V_{max} of 20 M/s. The K_m can be estimated by first finding a value on the y-axis that is equal to $\frac{1}{2}$ V_{max} (10.0 M/s). From this value extrapolate over to the graphed line and down to the x-axis. The value on the x-axis is the Michaelis constant and is equal to 0.03 M.

PRACTICE Find the K_m and V_{max} for the digestive enzyme, pepsin, from the following graph:

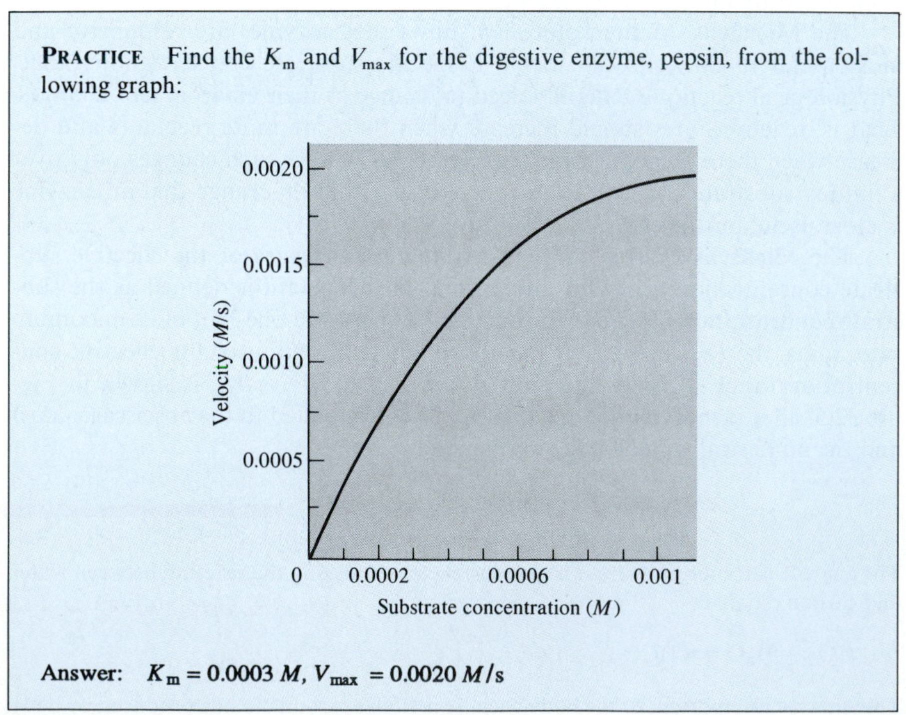

Answer: $K_m = 0.0003\ M$, $V_{max} = 0.0020\ M/s$

Scientists believe that each specific enzyme's catalytic abilities are tailored to fit a specific metabolic need. Some enzymes work well at low substrate concentrations, whereas others require much higher concentrations before they operate efficiently. Some enzymes catalyze reactions very quickly while others fit a biochemical need by reacting much more slowly. Several examples will illustrate these concepts.

There are two common mammalian enzymes that begin glucose metabolism, hexokinase and glucokinase. Both enzymes catalyze the same reaction, an addition of a phosphate group to glucose, but hexokinase has a K_m of about 0.0001 *M* while glucokinase has a K_m of about 0.005 *M*. These two enzymes differ in their effective concentration range; at low glucose concentrations only the hexokinase catalyzes the reaction while at higher glucose concentrations both hexokinase and glucokinase react. Thus, the difference in K_ms means that hexokinase has first priority to use the available glucose. This is important because hexokinase is found in all tissues, where glucose is used for energy, while glucokinase is found only in the liver, where glucose is stored as glycogen. When energy is needed, hexokinase uses the glucose. If there is some excess glucose, the glucokinase enzyme starts converting this sugar to glycogen. The difference in K_ms helps shunt the glucose to where it can best be used.

We have considered how the Michaelis' constant is important; now let us compare the turnover numbers for several enzymes. Catalase is responsible for destroying the cellular toxin, hydrogen peroxide, before it can do biological damage. This enzyme has a large turnover number of 10,000,000 s^{-1}; one catalase enzyme can destroy 10 million hydrogen peroxide molecules per second. This very large turnover number minimizes the danger posed by hydrogen peroxide. In contrast, the enzyme chymotrypsin has a much smaller turnover number of about 0.2 s^{-1} (two reactants are converted to products every 10 seconds by one chymotrypsin molecule). This enzyme digests protein as it moves slowly through the small intestine. Since the digestive process is slow, chymotrypsin is not required to react as quickly as catalase. In many cases it appears that an enzyme's turnover number matches the speed required for a biological process.

EXAMPLE 32.2

Two enzymes are studied under identical conditions. Enzyme A is found to have a turnover number of 1500 s^{-1} and a K_m of 0.01 *M* whereas enzyme B shows a turnover number of 500 s^{-1} and a K_m of 0.005 *M*. Compare (a) the catalytic efficiency of the two enzymes and (b) their respective effective substrate concentration ranges.

SOLUTION

(a) The turnover number is a measure of catalytic efficiency. Under identical conditions, the larger the turnover number, the more efficient an enzyme is as a catalyst. Because enzyme A has a larger turnover number than enzyme B, enzyme A is a more efficient catalyst.

(b) The K_m is a measure of the effective substrate concentration range for an enzyme. A lower K_m value means that an enzyme works well at lower substrate concentrations. Enzyme B is effective at a lower substrate concentration range than enzyme A.

PRACTICE Under optimum conditions, the digestive enzyme pepsin has a turnover number of about 30 min^{-1} while a second digestive enzyme, trypsin, has a turnover number of 12 min^{-1}. Both of these enzymes digest proteins. Which would you judge to be the more efficient?

Answer: Under these conditions, pepsin is the more efficient because it has the larger turnover number, converting 30 reactant molecules to products per minute.

32.3 ENZYME ACTIVE SITE

Catalysis takes place on a small portion of the enzyme structure called the *enzyme active site.* Often this is a crevice or pocket on the enzyme representing only 1–5% of the total surface area. For example, Figure 32.6 shows a space-filling model of the enzyme hexokinase, which catalyzes a first step in the breakdown of glucose to provide metabolic energy. Notice that the active site is located in a crevice. Glucose can enter this site and is bound. The enzyme then must change shape before the reaction takes place. Thus, although catalysis occurs at the small active site, the entire three-dimensional structure of the enzyme is important.

By examining values such as the turnover number and the K_m, scientists have gained a basic understanding of what takes place at an enzyme active site. To function effectively an enzyme must attract and bind the substrate. Once the substrate is bound, a chemical reaction is catalyzed. This two-step process is described by the following general sequence. Enzyme (E) and substrate (S) combine to form an enzyme–substrate intermediate (E–S). This intermediate decomposes to give the product (P) and regenerate the enzyme:

$$E + S \xrightarrow[\text{Binding}]{} E\text{–}S \xrightarrow[\text{Catalysis}]{} E + P$$

For the hydrolysis of maltose by the enzyme maltase, the sequence is

$$\underset{E}{\text{Maltase}} + \underset{S}{\text{Maltose}} \longrightarrow \underset{E\text{–}S}{\text{Maltase–Maltose}}$$

$$\underset{E\text{–}S}{\text{Maltase–Maltose}} + H_2O \longrightarrow \underset{E}{\text{Maltase}} + \underset{P}{\text{2 Glucose}}$$

Each different enzyme has its own unique active site whose shape determines, in part, which substrates can bind. Enzymes are said to be stereospecific; that is, each enzyme catalyzes reactions for only a limited number of different reactant structures. For example, maltase binds to maltose (two glucose units linked by a α-1,4-glycosidic bond) but not to lactose (galactose coupled to glucose). In fact, this enzyme can even distinguish maltose from cellobiose (glucose coupled to glucose by a β-1,4-linkage), in which only the glycosidic linkage is different. Enzyme stereospecificity is a very important means by which the cell controls its biochemistry.

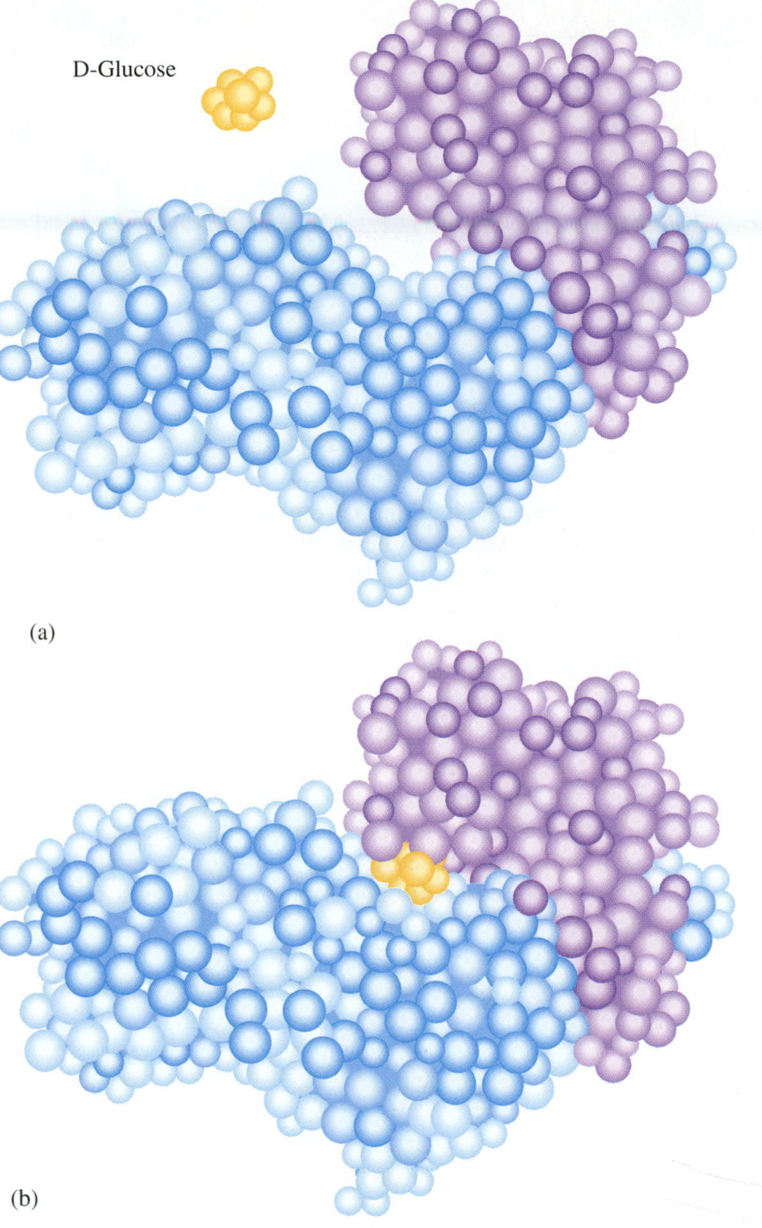

◀ **FIGURE 32.6**
A space-filling model of the enzyme, hexokinase, before (a) and after (b) it binds to the substrate, glucose: Note the two protein domains for this enzyme which are colored differently.

Enzyme specificity is partially due to a complementary relationship between the active site and substrate structures. This **lock-and-key hypothesis** envisions the substrate as a key that fits into the appropriate active site, the lock. Although this hypothesis describes a fundamental property of enzyme–substrate binding, it has been known for some time that enzyme active sites are not rigid as a lock would be. Instead, the active site bends to a certain degree when the appropriate substrate binds. The **induced-fit model** proposes that the active site adjusts its structure in order to prepare the substrate–enzyme complex for catalysis. Enzyme stereospecificity is thus explained in terms of an active site having

lock-and-key hypothesis

induced-fit model

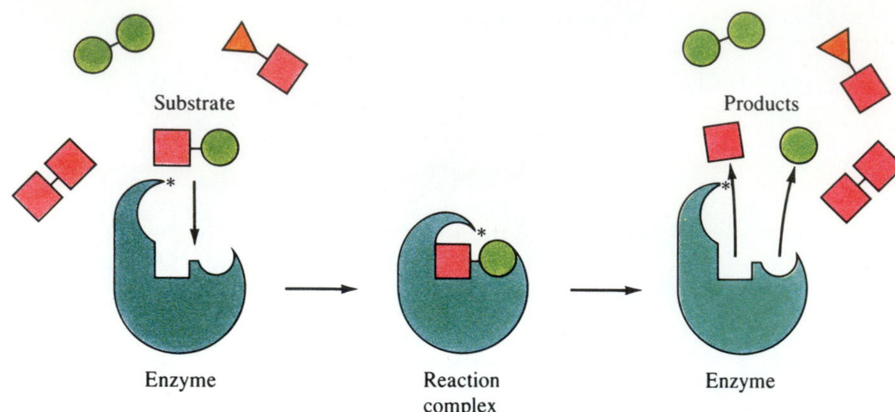

FIGURE 32.7 ▶ Enzyme-substrate interaction illustrating both the lock-and-key hypothesis and the induced-fit model. The correct substrate (■—●) fits the active site (lock-and-key hypothesis). This substrate also causes an enzyme conformation change which positions a catalytic group (*) to cleave the appropriate bond (induced-fit model).

a somewhat flexible shape. This shape is rigid enough to exclude very dissimilar substrates (lock-and-key hypothesis) but flexible enough to accommodate (induced-fit model) and allow catalysis of appropriate substrates (see Figure 32.7).

The fact that an enzyme is flexible can help to explain enzyme binding specificity and also how the enzyme converts reactants into products. *An enzyme is a dynamic catalyst.* As the enzyme attracts the substrate into the active site, the enzyme's shape and the reactant's shape both begin to change. Note the change in hexokinase structure when glucose is bound (Figure 32.6). This alteration of enzyme shape aids the transformation of the reactant into the product. Figure 32.7 schematically depicts an enzyme where a shape change leads to catalysis. This only exemplifies the numerous alterations that can be brought about by enzymes, for example, bonds broken or formed, charges moved, and new molecular substituents added or removed.

An actual metabolic reaction will illustrate some of the basic features of enzyme catalysis. For example, let us take the enzyme hexokinase, which catalyzes the transformation of glucose to glucose-6-phosphate.

$$\text{glucose} \xrightarrow[\text{Mg}^{2+},\ \text{ATP}]{\text{Hexokinase}} \text{glucose-6-phosphate}$$

glucose: $^6CH_2OH$, ring positions 1–5 with O; HO at 4, OH at 3, OH at 2, OH at 1

glucose-6-phosphate: $^6CH_2OPO_3^{2-}$ (drawn as $CH_2O\text{–}P(=O)(O^-)\text{–}O^-$), ring positions 1–5 with O; HO at 4, OH at 3, OH at 2, OH at 1

Like many enzymes, hexokinase requires a second reactant. In this case, adenosine-5′-triphosphate (ATP) supplies the phosphate group that is transferred to the glucose. ATP is a very important molecule within the cell because it often transfers energy as well as phosphate groups from enzyme to enzyme (see Chapter 33). Additionally, the metal ion, Mg^{2+}, is required as an activator

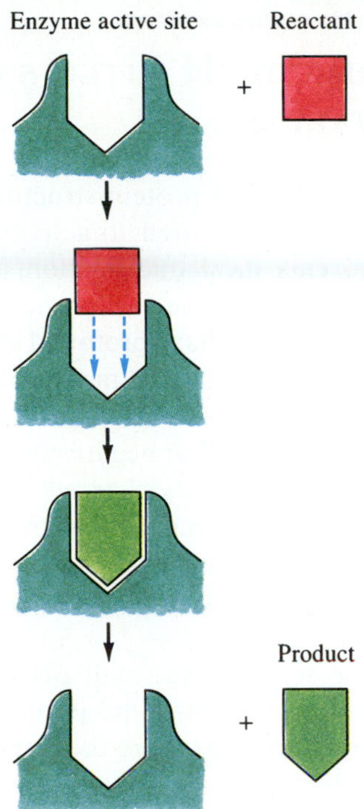

◄ **FIGURE 32.8**
Schematic showing the substrate being forced toward the product shape by enzyme binding.

in this reaction. With all of these components present, the enzyme can start the catalytic process.

When glucose binds, the hexokinase shape changes to bring the ATP close to the C-6 of the glucose, thus forcing the transfer of phosphate to this specific carbon. Hexokinase speeds this reaction in several ways. First, this enzyme acts to bring the reactants close together, a process termed **proximity catalysis**. Second, hexokinase positions the reactants so the proper bonds will form or be broken. (The enzyme ensures that a phosphate is added to C-6 of glucose and not at one of the other carbons.) This is often termed the **productive binding hypothesis**, because reactants are bound/oriented in such a way that products result. Hexokinase is a successful catalyst—glucose reacts with ATP 10 billion (10^{10}) times faster than it would in the absence of this enzyme. Increases in reaction rate of this magnitude are essential to life. In fact, without enzymes, cellular reactions are too slow to keep the cells alive.

proximity catalysis

productive binding hypothesis

Biotechnologists have made use of our understanding of enzyme catalysis to design (with nature's help) completely new enzymes. Antibodies (immunoglobulin proteins) are produced that bind tightly to a molecule like the transition state. Thus, the antibody binding site is not complementary to the shape of the reactant but rather to a shape more like the product. When these antibodies bind reactant molecules, the strong attractive forces "strain" the reactants as illustrated in Figure 32.8. The reactant molecule is impelled to change shape to fit the binding site. Catalysis occurs, and these antibodies act as enzymes! Scientists have termed this mode of catalysis the **strain hypothesis**, and it is thought to be important in many natural enzymes as well as these "antibody enzymes." In the not-too-distant future, such artificial enzymes may serve important industrial and medical applications.

strain hypothesis

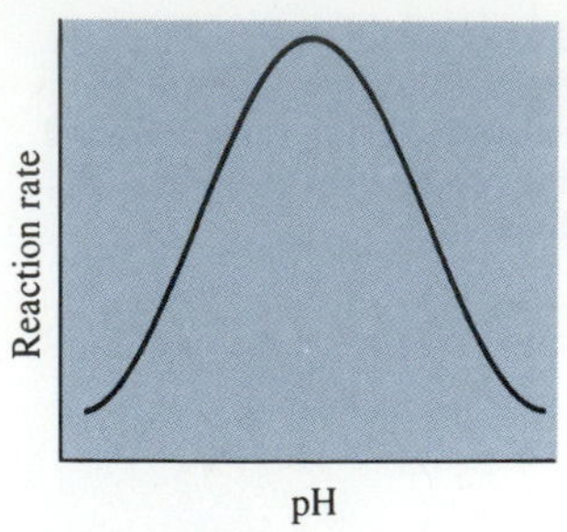

▲
FIGURE 32.9
A typical plot of the enzyme-catalyzed rate as a function of pH.

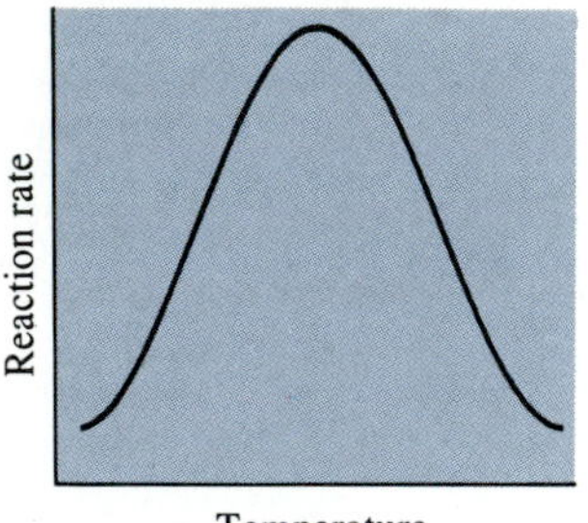

▲
FIGURE 32.10
A plot showing the typical temperature dependence of an enzyme-catalyzed rate.

32.4 TEMPERATURE AND pH EFFECTS ON ENZYME CATALYSIS

Essentially any change that will affect protein structure will affect an enzyme's catalytic function. If an enzyme is denatured, its activity will be lost. Thus, strong acids and bases, organic solvents, mechanical action, and high temperature are examples of treatments that may decrease an enzyme-catalyzed rate of reaction.

Even slight changes in the pH can have profound effects on enzyme catalysis. Remember that some of the amino acids that make up enzymes have side chains whose charge depends on pH. For example, side chains with carboxylic acid functional groups may be either neutral or negatively charged; those with amino groups may be either neutral or positively charged. Thus, as the pH is changed the charges on an enzyme, or more specifically at the active site of an enzyme, also change. Enzyme catalysis is affected. For each enzyme there is an optimum pH; shifts to more acidic or more basic conditions decrease the enzyme activity as shown in Figure 32.9.

Because enzymes are so sensitive to small pH changes, our bodies have developed elaborate mechanisms to control the amount of acid and base present in cellular fluids. The kidneys and lungs share the responsibility for maintaining blood acid and base levels. If the blood pH shifts outside of the narrow range 6.8–7.8, death often results. The body's metabolic processes can no longer maintain life partially because of a pH-induced change in enzyme activities.

Body temperature is also carefully controlled partially because enzyme activities are particularly temperature-sensitive. A typical plot of temperature versus enzyme activity is shown in Figure 32.10. At very low temperatures, few reactant molecules have enough energy to overcome the activation energy barrier and the reaction occurs slowly. As the temperature increases, more reactants have the necessary activation energy and the rate increases. But a point is reached where the high temperature causes enzyme denaturation, and the rate of reaction starts to decrease. Thus, both low and high temperatures cause slow reaction rates, and, as with pH, there is an optimum temperature for each enzyme. Not surprisingly, the optimum temperature for many human enzymes is approximately body temperature (about 37°C).

32.5 ENZYME REGULATION

Since enzymes are vital to life, enzyme catalysis is under careful cellular control. The cell uses a variety of mechanisms to change the rate of substrate conversion to product. Sometimes a new group of atoms is covalently joined to the enzyme in a process called *covalent modification*. In other cases, another molecule is noncovalently bound to the enzyme. The protein structural change that results may cause a decrease in enzyme activity, **enzyme inhibition**, or an increase in activity, **enzyme activation**.

enzyme inhibition

enzyme activation

To learn about enzyme control, scientists have carefully studied both inhibition and activation. They have found that an enzyme's catalytic abilities may be changed by changing its K_m or its turnover number. For example, an enzyme may be inhibited by making it bind to its substrate more weakly (increasing its K_m) or by decreasing its catalytic efficiency (decreasing its turnover number or V_{max}). Reversing these changes causes enzyme activation. A Michaelis–Menten plot illustrates this point for inhibition (see Figure 32.11). Note that a decrease in V_{max} causes the curve to rise less while an increase in K_m results in a curve with a shallower slope.

Hexokinase is an example of an enzyme that is under cellular control. Because this enzyme begins the breakdown of glucose to yield cellular energy, the hexokinase-catalyzed reaction is often not needed when cellular energy levels are high. Control occurs in the following way: (1) if glucose no longer is needed for cellular energy, the glucose-6-phosphate concentration increases; (2) this compound binds to the hexokinase enzyme and inhibits the enzyme activity. Because glucose-6-phosphate is a product of the reaction, this form of enzyme control is called **product inhibition**. Hexokinase responds to the overall cellular energy state and does not use more glucose than needed.

product inhibition

Feedback inhibition and **feedforward activation** are two other common forms of enzyme control. To understand these mechanisms, visualize the various cellular processes as assembly lines with a different enzyme at each step from raw material to finished product. Feedback inhibition affects enzymes at the beginning of the molecular assembly line. A final product acts as "feedback" and inhibits an enzyme from using too many molecular starting materials. In contrast, feedforward activation often controls enzymes at the end of the molecular assembly line. If there is an excess of starting materials, these molecules will "feedforward" and activate enzymes, which in turn cause the whole process to move faster. As with other control mechanisms, both feedback inhibition and feedforward activation serve to coordinate enzyme processes within the cell.

feedback inhibition

feedforward activation

A variety of drug therapies make use of enzyme control to selectively affect target cells. Careful drug design can create a molecule which binds to only one type of enzyme. By binding, the drug blocks normal catalysis and causes enzyme inhibition. In each of the following examples, a specific enzyme is inhibited, which in turn causes a selective metabolic change.

1. Methotrexate is an anticancer drug because it is similar in structure to, but cannot function like, the coenzyme dihydrofolate. This coenzyme is needed to reproduce cellular genetic material. When methotrexate replaces dihydrofolate, an enzyme is inhibited and genetic replication is slowed. Since rapid cell growth requires genetic replication, rapidly growing cancer cells are selectively impacted by methotrexate treatment.
2. AZT (3′-azido-3′-deoxythymidine) was the first drug approved in the United States for treatment of the viral disease, acquired immune deficiency syndrome (AIDS). This drug is structurally similar to a reactant needed to make genetic material for the AIDS virus. When AZT binds to a specific enzyme (reverse transcriptase), the formation of new viral genetic material is inhibited. Virus reproduction is impeded and the progression of AIDS is slowed.

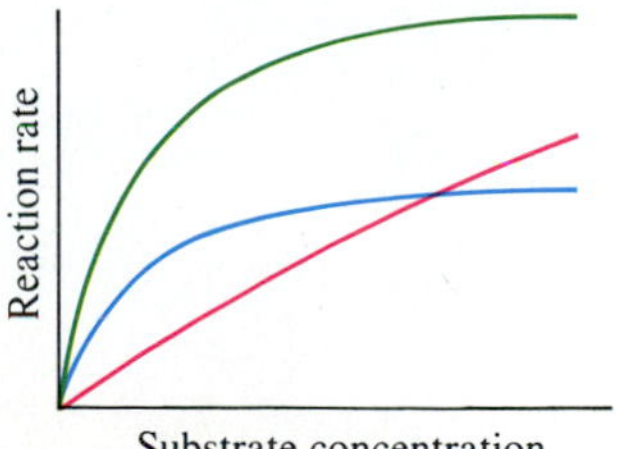

▲
FIGURE 32.11
A Michaelis–Menton plot for an unregulated enzyme (green), the same enzyme inhibited by a decrease in V_{max} (blue) and the enzyme inhibited by an increase in K_m (red)

These examples of drug therapy illustrate a very important principle: Because biological processes depend on enzymes, enzyme control often has a major impact on life.

Chemistry in Action

Industrial Strength Enzymes

Not only are enzymes important in biology, but these proteins also are becoming increasingly important in industry. Enzymes offer two major advantages to manufacturing processes and in commercial products: first, enzymes cause very large increases in reaction rates even at room temperature; second, enzymes are relatively specific and can be used to target selected reactants. Perhaps the biggest disadvantage to industrial enzymes is their relative short supply (and, therefore, higher cost as compared to traditional chemical treatments). Recent developments in biotechnology offer supplies of less expensive enzymes through genetic engineering.

Enzymes have long been used in food processing. The citrus industry has recently perfected a process to remove peel from oranges or grapefruits by using the enzyme pectinase. The pectinase penetrates the peel in a vacuum infusion process. There it dissolves the albedo (the white stringy material) that attaches the peel to the fruit. When the fruit is removed from the solution, the skin can be peeled easily by machine or hand. The industry is now marketing pre-peeled citrus to hospitals, airlines, and restaurants (see photo below).

About 25% of all industrial enzymes are used to convert cornstarch into syrups that are equivalent in sweetness and in calories to ordinary table sugar. More than 5 billion pounds of such syrups are produced annually. The process uses three enzymes: the first, α-amylase, catalyzes the liquefaction of starch to dextrins; the second, a glucoamylase, catalyzes the breakdown of dextrins to glucose; the third, glucose isomerase, converts glucose to fructose.

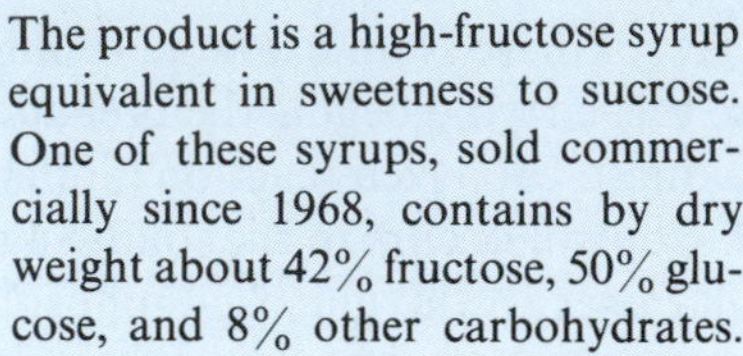

$$\text{Starch} \xrightarrow{\alpha\text{-Amylase}} \text{Dextrins}$$

$$\text{Dextrins} \xrightarrow{\text{Glucoamylase}} \text{Glucose}$$

$$\text{Glucose} \xrightarrow[\text{isomerase}]{\text{Glucose}} \text{Fructose}$$

The product is a high-fructose syrup equivalent in sweetness to sucrose. One of these syrups, sold commercially since 1968, contains by dry weight about 42% fructose, 50% glucose, and 8% other carbohydrates.

Industrial enzymes offer solutions to environmental pollution problems for some manufacturers. For example, the paper industry, like other industries that use chemicals, is concerned with minimizing processes that produce potentially hazardous waste. Paper is produced from wood chips by first digesting the cellulose structure with calcium sulfite and then bleaching the pulp with chlorine to obtain a bright white paper. An excess of chlorine must be used because the pulp is not completely broken down. This excess creates a significant disposal problem as chlorine is environmentally hazardous. Recent developments in biotechnology offer a potential solution. The enzymes needed to complete wood fiber digestion (cellulase and hemicellulase) have been produced in larger quantities via genetic engineering. With such enzymes to finish degrading the wood pulp, paper manufacturers may be able to markedly decrease the amount of chlorine used as bleach.

Consumer goods are increasingly impacted by enzyme technology. Many detergents are better cleansing agents because they contain enzymes; fully 40% of all industrially produced enzymes are used in this way. Meat tenderizers often contain papain, an enzyme which breaks down protein molecules. Even clothing manufacturers are finding uses for newly available enzymes. More and more denim products are en-

(*continued*)

(*continued*)
zyme-treated to replace stonewashing, a process in which the material is washed with pumice to soften the fabric's appearance and remove some of the dye. Because this abrasion may weaken the fabric as well, some manufacturers now use "biostoning." The denim is treated with the enzyme, cellulase, which changes the fabric's appearance without weakening the fabric structure.

The ability to produce large quantities of purified enzymes has raised a number of potential medical applications. For genetic diseases characterized by the loss of a specific enzyme, a treatment known as enzyme-replacement therapy may be useful. For example, a lysosomal storage disease, such as Tay–Sachs disease, leads to the accumulation of excess intracellular polysaccharides because specific digestive enzymes are unavailable. Polysaccharide buildup can lead to mental retardation, paralysis, blindness, and death. Current research is aimed at developing an appropriate microcapsule package that will transport additional digestive enzymes to the affected cells. Such enzyme-replacement therapy has also been proposed for removing toxic substances from the bloodstream.

Concepts in Review

1. Briefly explain why living cells need catalysts.
2. List the six main classes of enzymes and their functions.
3. Compare a simple, uncatalyzed reaction with an enzyme-catalyzed process.
4. Define the symbols K_m and V_{max} and relate them to the Michaelis–Menten plot.
5. Discuss the relationship between the reaction rate and the enzyme concentration.
6. Describe an enzyme active site.
7. Tell what is meant by the specificity of an enzyme.
8. List several ways that an enzyme can facilitate the conversion of substrate to product.
9. Summarize the effects of pH and temperature on an enzyme-catalyzed reaction.
10. Describe how inhibition and activation are important in enzyme control.
11. Discuss the potential role of enzymes in medicine and industry.

Equations in Review

Enzyme-Catalyzed Conversion of Substrate to Product

$$\mathrm{E + S} \xrightarrow[\text{Binding}]{} \mathrm{E\text{–}S} \xrightarrow[\text{Catalysis}]{} \mathrm{E + P}$$

EXERCISES

1. What is the activation energy and how is this energy affected by enzymes?
2. What is the general role of enzymes in the body?
3. Distinguish between a coenzyme and an apoenzyme.
4. Give the names of enzymes that catalyze the hydrolysis of (a) sucrose, (b) lactose, and (c) maltose.
5. What are six general classes of enzymes?
6. By drawing a Michaelis–Menten plot, define the V_{max}.
7. Why are K_m s reported in units of concentration, for example, molarity?
8. The digestive enzyme chymotrypsin breaks peptide bonds adjacent to the amino acids L-tryptophan and L-phenylalanine. This enzyme has a K_m for a specific L-tryptophan–containing substrate of 0.08 millimolar (mM) and a K_m for a specific L-phenylalanine–containing substrate of 1.30 mM. Which substrate do you think binds more tightly to the enzyme? Briefly explain.
9. Chymotrypsin has a turnover number for a glycine-containing substrate of 0.05 s^{-1} and a turnover number for an L-tyrosine–containing substrate of 200 s^{-1}. For which substrate is chymotrypsin a more efficient catalyst? Briefly explain.
10. A catalyst increases the rate of a chemical reaction. List two other means of increasing reaction rates.
11. An enzyme reacts 0.02 moles per liter of substrate every 8 minutes. What is the reaction rate in units of molar per minute?
12. Pepsin, a digestive enzyme found in the stomach, has a turnover number of 0.5 s^{-1} for a specific protein substrate. How many proteins can be digested by one pepsin molecule in 5 minutes?
13. A scientist studies two enzymes which catalyze the same reaction. Enzyme A has a turnover number of 225 s^{-1} and a K_m of 0.03 M whereas enzyme B has a turnover number of 120 s^{-1} and a K_m of 0.15 M. The scientist concludes that enzyme B is more effective than enzyme A. Do you agree? Briefly explain.
14. Differentiate between the lock-and-key hypothesis and the induced-fit model for enzyme function.
15. Two enzymes are being studied; the first enzyme uses *n*-butyl alcohol as a substrate while the second enzyme uses *t*-butyl alcohol as a substrate. Based on the lock-and-key hypothesis, how might the shapes for the enzyme active sites differ?
16. List one important way in which the induced-fit model is important in the hexokinase catalyzed reaction.
17. If the K_m for an enzyme is decreased, would you predict activation or inhibition? Briefly explain.
18. Why does an enzyme catalyzed reaction rate decrease at high temperatures?
19. Feedback inhibition is an important form of enzyme regulation. Based on what you know about this control mechanism, how might a different process, "feedback activation," cause regulatory *problems* for a cell?
20. List three ways that substrate binding to the active site helps the reactants convert to products.
21. How does an enzyme inhibitor differ from an enzyme substrate?
22. Describe briefly how an artificial enzyme that is produced from an immunoglobulin helps substrates react to form products.
23. Which of the following statements are correct? Rewrite the incorrect ones to make them correct.
    (a) Enzymes are proteins.
    (b) Enzymes increase the activation energy for a reaction.
    (c) An enzyme active site is where catalysis occurs.
    (d) The product of an enzyme catalyzed reaction is called the substrate.
    (e) Doubling the reactant concentration always doubles the rate of an enzyme catalyzed reaction.
    (f) An enzyme with a small K_m can bind at low substrate concentrations.
    (g) The substrate concentration that gives a rate of $\frac{1}{4}$ V_{max} is equal to the K_m.
    (h) The turnover number is a measure of an enzyme's catalytic ability.
    (i) The lock-and-key hypothesis requires a flexible enzyme.
    (j) The productive binding hypothesis states that an enzyme binds to and orients the reactants so that products can form most easily.
    (k) Inhibition occurs if an enzyme's K_m is increased.
    (l) Feedforward activation increases the rate for enzymes at or close to the end of a metabolic process ("assembly line").
    (m) Denaturation decreases an enzyme catalyzed reaction rate.
    (n) Enzyme treatment can allow manufacturers to avoid more harsh chemical and physical processes.

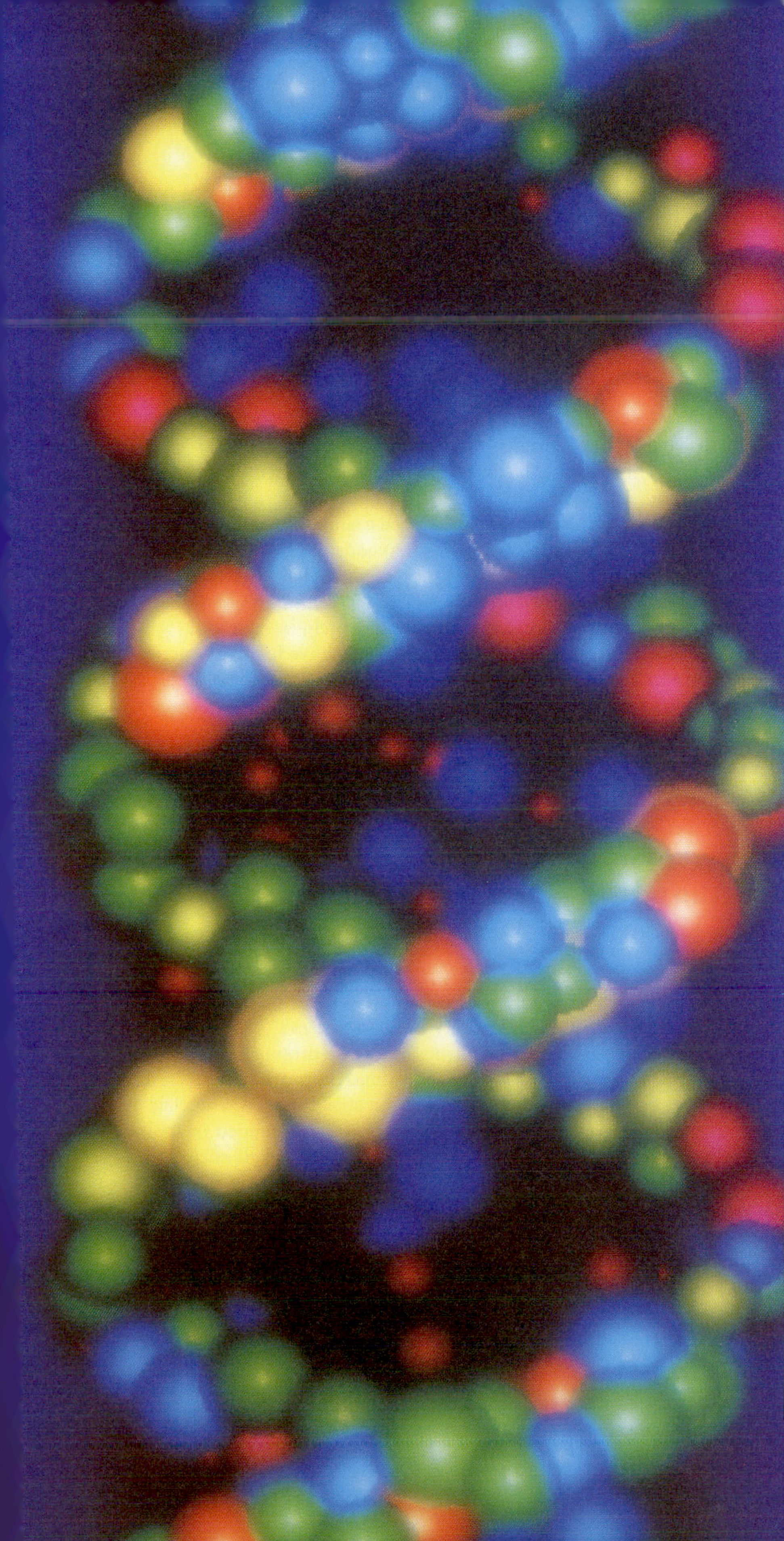

33

Nucleic Acids and Heredity

◀ Chapter Opening Photo: This model shows the double helix structure of DNA.

The plight of Doctor Frankenstein's monster touches a chord in all of us. An evil scientist has given this monster life but cannot control his creation. The monster is "unnatural" and, as such, the story leads to tragedy.

The advent of genetic engineering has raised a similar fear of a tragic ending. Genetic engineers work with the molecules that code life—the nucleic acids. By changing the code, new life forms can be produced. Already bacteria have been altered to make needed human proteins. Recently, both cows and goats have been genetically engineered to produce a human protein in their milk.

The scientists involved in these initial programs have followed careful protocols and have produced valued medicines. However, the day may soon be with us when we must decide whether humans should be made smarter, stronger, etc., via genetic engineering. How this decision will be made is open to discussion. However, an understanding of the biochemistry of nucleic acids has given us the power to consider such decisions.

33.1 Molecules of Heredity—A Link

The question of how hereditary material duplicates itself was for a long time one of the most baffling problems of biology. For many years biologists attempted in vain to solve this problem and also to find an answer to the question "Why are the offspring of a species undeniably of that species?" Many thought the chemical basis for heredity lay in the structure of the proteins. But no one was able to provide evidence showing how protein could reproduce itself. The answer to the heredity problem was finally found in the structure of the nucleic acids.

nucleoprotein

nucleic acids

The unit structure of all living things is the cell. Suspended in the nuclei of cells are chromosomes, which consist largely of proteins and nucleic acids. A simple protein bonded to a nucleic acid is called a **nucleoprotein**. There are two types of **nucleic acids**—those that contain the sugar deoxyribose and those that contain the sugar ribose. Accordingly, they are called deoxyribonucleic acid (DNA) and ribonucleic acid (RNA). DNA was discovered in 1869 by the Swiss physiologist Friedrich Miescher (1844–1895), who extracted it from the nuclei of cells.

33.2 Bases and Nucleosides

Nucleic acids are complex chemicals that combine several different classes of smaller molecules. As with many complex structures, it is easier to understand the whole by first studying its component parts. We will start the examination of nucleic acids by learning about a critical part of these molecules, two classes of heterocyclic bases, the purines and the pyrimidines:

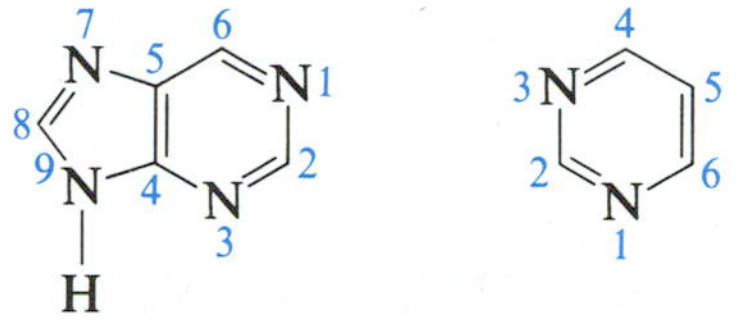

Purine, $C_5H_4N_4$ Pyrimidine, $C_4H_4N_2$

These parent compounds are related in structure, the pyrimidine being a six-membered heterocyclic ring while the purine contains both a five- and six-membered ring. The nitrogen atoms cause these compounds to be known as heterocycles (the rings are made up of more than just carbon atoms) and also as bases. Like the ammonia nitrogen, these heterocycles react with hydrogen ions to make a solution more basic.

There are five major bases commonly found in nucleic acids—two purine bases (adenine and guanine) and three pyrimidine bases (cytosine, thymine, and uracil). Figure 33.1 gives one stable form for each compound. Note that the bases differ one from another in their ring substituents. Each base has a lowermost nitrogen, which is bonded to a hydrogen as well as two carbons. This specific >N—H shares chemical similarities with an alcohol (—OH) group. Just as two sugars can be linked together when an alcohol of one monosaccharide reacts with a second monosaccharide (see Chapter 29), so a purine or pyrimidine can be bonded to a sugar by a reaction with the —NH group.

A **nucleoside** is formed when either a purine or pyrimidine base is linked to a sugar molecule, usually D-ribose or D-2′-deoxyribose.

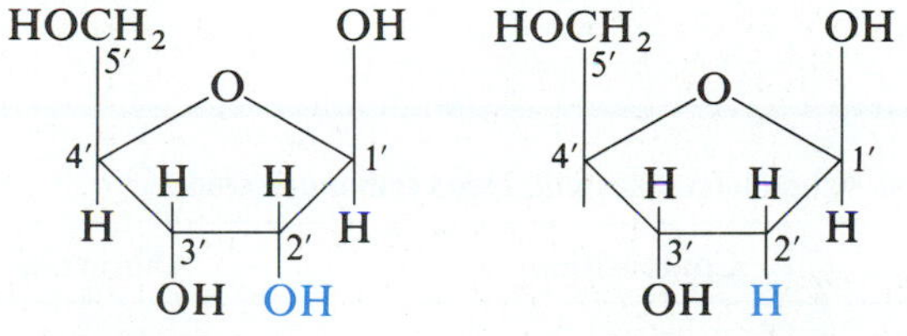

nucleoside

D-ribose D-2′-deoxyribose

The base and sugar are bonded together between carbon 1′ of the sugar and either the purine nitrogen at position 9 or the pyrimidine nitrogen at position 1. Typical structures of nucleosides are shown in Figure 33.2. A prime is added to the position number to differentiate the sugar numbering system from the purine or pyrimidine numbering system.

The name of each nucleoside emphasizes the importance of the base to the chemistry of the molecule. Thus, adenine and D-ribose react to yield adenosine, whereas cytosine and D-2′-deoxyribose yield deoxycytidine. The root of the nucleoside name derives from the purine or pyrimidine name. The compositions of the common ribonucleosides and the deoxyribonucleosides are given in Table 33.1.

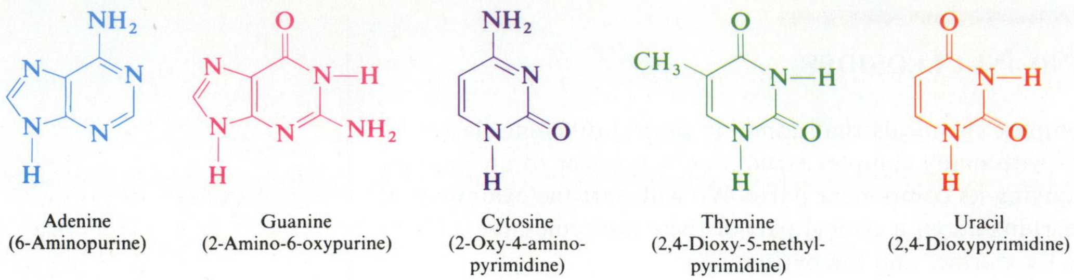

Adenine (6-Aminopurine) Guanine (2-Amino-6-oxypurine) Cytosine (2-Oxy-4-amino-pyrimidine) Thymine (2,4-Dioxy-5-methyl-pyrimidine) Uracil (2,4-Dioxypyrimidine)

▲ **FIGURE 33.1 Purine and pyrimidine bases found in living matter.**

Adenosine Deoxyadenosine Cytidine Deoxycytidine

FIGURE 33.2 ► Typical structures of ribonucleosides and deoxyribonucleosides.

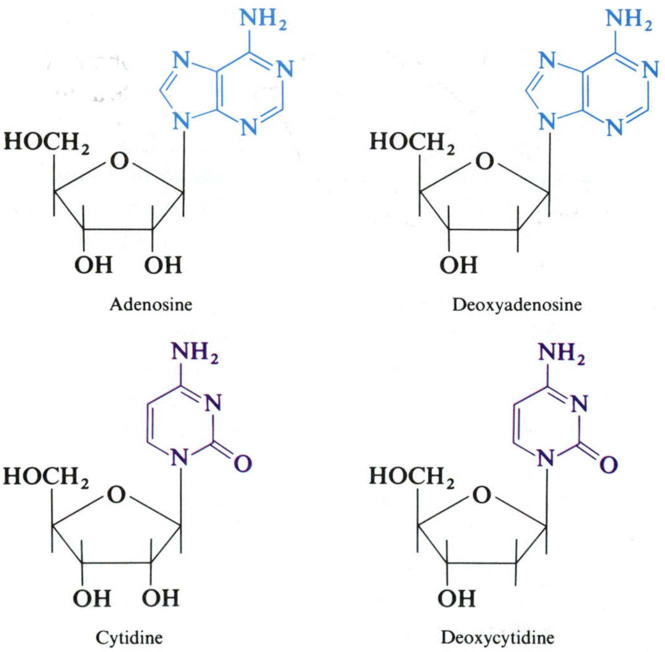

TABLE 33.1 Composition of Ribonucleosides and Deoxyribonucleosides

Name	Composition	Abbreviation
Adenosine	Adenine–ribose	A
Deoxyadenosine	Adenine–deoxyribose	dA
Guanosine	Guanine–ribose	G
Deoxyguanosine	Guanine–deoxyribose	dG
Cytidine	Cytosine–ribose	C
Deoxycytidine	Cytosine–deoxyribose	dC
Thymidine	Thymine–ribose	T
Deoxythymidine	Thymine–deoxyribose	dT
Uridine	Uracil–ribose	U
Deoxyuridine	Uracil–deoxyribose	dU

33.3 Nucleotides: Phosphate Esters

A more complex set of biological molecules is formed by linking phosphate groups to nucleosides. Phosphate esters of nucleosides are termed **nucleotides**. These molecules consist of a purine or a pyrimidine base linked to a sugar, which in turn is bonded to at least one phosphate group.

nucleotides

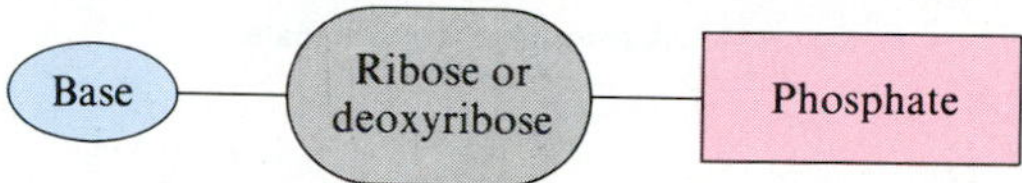

The ester may be a monophosphate, a diphosphate, or a triphosphate. When two or more phosphates are linked together, a high-energy phosphate anhydride bond is formed (see Section 33.4). The ester linkage may be to the hydroxyl group of position 2′, 3′, or 5′ of ribose or of position 3′ or 5′ of deoxyribose. Examples of nucleotide structures are shown in Figure 33.3. The abbreviations for the five nucleosides containing D-ribose are

Adenosine	A
Guanosine	G
Cytidine	C
Thymidine	T
Uridine	U

The letters MP (monophosphate) can be added to any of these to designate the corresponding nucleotide. Thus GMP is guanosine monophosphate. A lowercase d is placed in front of GMP if the nucleotide contains the deoxyribose sugar (dGMP). When the letters such as AMP or GMP are given, it is generally understood that the phosphate group is attached to position 5′ of the ribose unit (5′-AMP). If attachment is elsewhere, it will be designated, for example, 3′-AMP.

Two other important adenosine phosphate esters are adenosine diphosphate (ADP) and adenosine triphosphate (ATP). Note that the letters DP are used for diphosphate and TP for triphosphate. In these molecules the phosphate groups

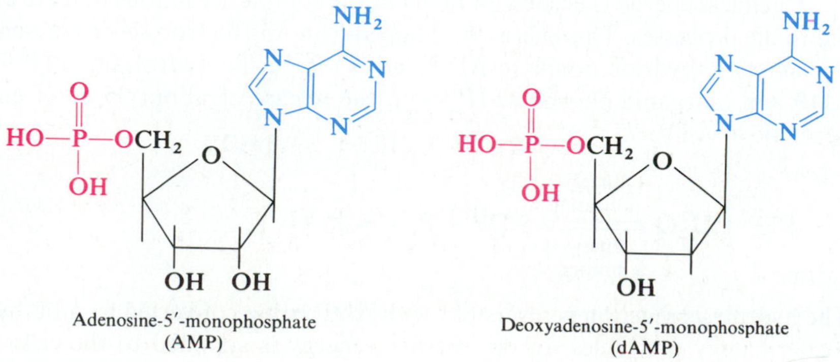

Adenosine-5′-monophosphate (AMP)

Deoxyadenosine-5′-monophosphate (dAMP)

◂ **Figure 33.3** Examples of nucleotides.

FIGURE 33.4 ▶
Structures of ADP and ATP.

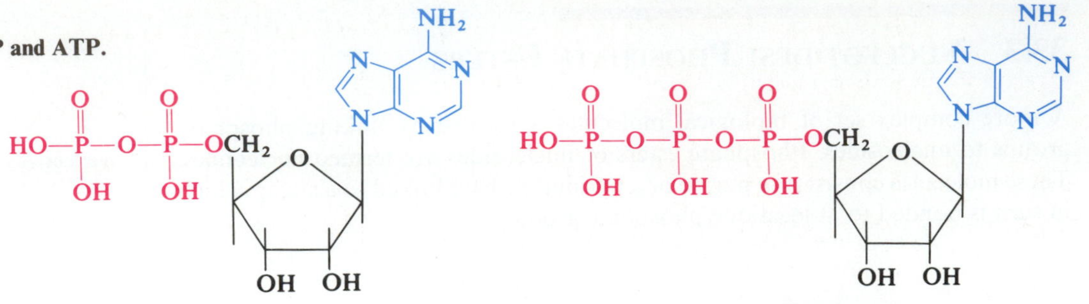

Adenosine-5′-diphosphate (ADP)

Adenosine-5′-triphosphate (ATP)

are linked together. The structures are similar to AMP except that they contain two and three phosphate residues, respectively (see Figure 33.4). All the nucleosides form mono-, di-, and triphosphate nucleotides.

33.4 HIGH-ENERGY NUCLEOTIDES

The nucleotides have a central role in the energy transfers in many metabolic processes. ATP and ADP are especially important in these processes. The role of these two nucleotides is to store and release energy to the cells and tissues. The source of energy is the foods we eat, particularly carbohydrates and fats. Energy is released as the carbons from these foods are oxidized (see Chapter 35). Part of this energy is used to maintain body temperature, and part is stored in the phosphate anhydride bonds of such molecules as ADP and ATP. Because a relatively large amount of energy is stored in these bonds, they are known as high-energy phosphate anhydride bonds.

High-energy phosphate anhydride bonds (ATP)

High-energy anhydride bond (ADP)

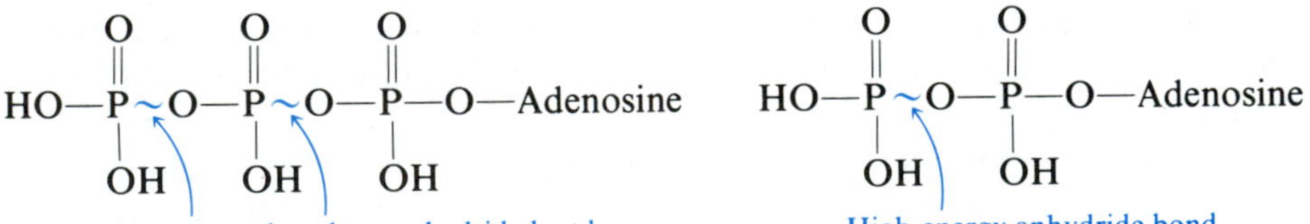

Chemical energy is needed for many of the complex reactions that are essential to life processes. This energy is obtained from the hydrolysis of high-energy phosphate anhydride bonds in ADP and ATP. In the hydrolysis, ATP forms ADP and inorganic phosphate (P_i) with the release of about 35 kJ of energy per mole of ATP:

$$\text{ATP} + \text{H}_2\text{O} \underset{\text{Energy storage}}{\overset{\text{Energy utilization}}{\rightleftharpoons}} \text{ADP} + \text{P}_i + \sim 35\ \text{kJ}$$

The hydrolysis reaction is reversible, with ADP being converted to ATP by still higher energy molecules. In this manner, energy is supplied to the cells from

ATP, and energy is stored by the synthesis of ATP from ADP and AMP. Processes such as muscle movement, nerve sensations, vision, and even the maintenance of our heartbeats are all dependent on energy from ATP.

33.5 Polynucleotides; Nucleic Acids

Starting with two nucleotides, a dinucleotide is formed by splitting out a molecule of water between the —OH of the phosphate group of one nucleotide and the —OH on C-3′ of the ribose or deoxyribose of the other nucleotide. Then another and another nucleotide can be added in the same manner until a polynucleotide chain is formed. Each nucleotide is linked to its neighbors by phosphate ester bonds (see ester formation in Section 25.6).

Two series of polynucleotide chains are known, one containing D-ribose and the other D-2′-deoxyribose. One polymeric chain consists of the monomers AMP, GMP, CMP, and UMP and is known as a polyribonucleotide. The other chain contains the monomers dAMP, dGMP, dCMP, and dTMP is known as a polydeoxyribonucleotide:

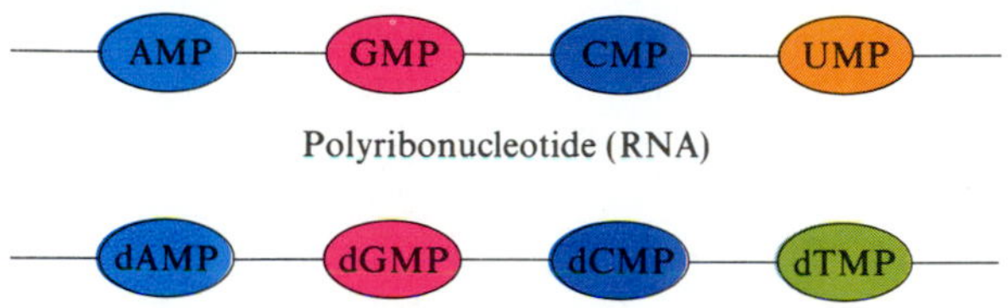

Polyribonucleotide (RNA)

Polydeoxyribonucleotide (DNA)

ribonucleic acid (RNA)

deoxyribonucleic acid (DNA)

The nucleic acids DNA and RNA are polynucleotides. **Ribonucleic acid (RNA)** is a polynucleotide that upon hydrolysis yields ribose, phosphoric acid, and the four purine and pyrimidine bases adenine, guanine, cytosine, and uracil. **Deoxyribonucleic acid (DNA)** is a polynucleotide that yields D-2′-deoxyribose, phosphoric acid, and the four bases adenine, guanine, cytosine, and thymine. Note that RNA and DNA contain one different pyrimidine nucleotide. RNA contains uridine, whereas DNA contains thymidine. A segment of a ribonucleic acid chain is shown in Figure 33.5. As will be described later, RNA and DNA also commonly differ in function: DNA serves as the storehouse for genetic information; RNA aids in expressing genetic characteristics.

33.6 Structure of DNA

Deoxyribonucleic acid (DNA) is a polymeric substance made up of the four nucleotides dAMP, dGMP, dCMP, and dTMP. The size of the DNA polymer varies with the complexity of the organism; more complex organisms tend to have larger DNAs. For example, simple bacteria like *Escherichia coli* have about 8 million nucleotides in their DNA while a human DNA contains up to about 500 million nucleotides. The order in which these nucleotides occur varies in

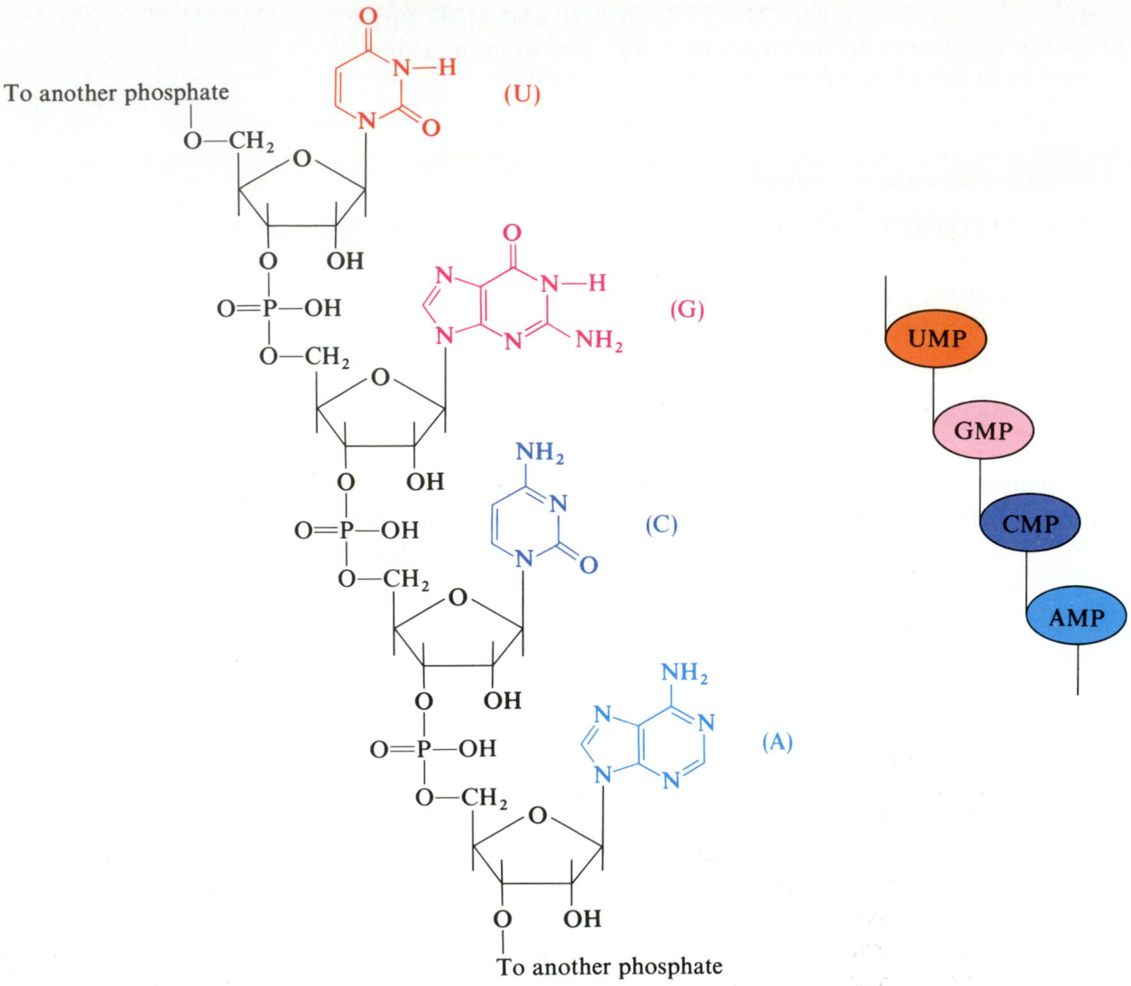

To another phosphate

(U)

(G)

(C)

(A)

To another phosphate

UMP

GMP

CMP

AMP

▲
FIGURE 33.5
A segment of ribonucleic acid (RNA) consisting of the four nucleotides adenosine monophosphate, cytidine monophosphate, guanosine monophosphate, and uridine monophosphate.

different DNA molecules, and it is within this order that the genetic information in a cell is stored.

Scientists have tried to understand DNA structure by determining the nucleotide composition of this molecule from many different sources. For a long time it was thought that the four nucleotides occurred in equal amounts in DNA. However, more refined analyses showed that the amounts of purine and pyrimidine bases varied in different DNA molecules. Surprisingly, careful consideration of these data also showed that the ratios of adenine to thymine and guanine to cytosine were always essentially 1:1. This observation served as an important key to unraveling the structure of DNA. The analyses of DNA from several species are shown in Table 33.2.

A second important clue to the special configuration and structure of DNA came from X-ray diffraction studies. Most significant was the work of Maurice H. F. Wilkins (b. 1916) of Kings College of London. Wilkins's X-ray pictures implied that the nucleotide bases were stacked one on top of another like a stack

Table 33.2 Relative Amounts of Purines and Pyrimidines in Samples of DNA

Source	Adenine	Thymine	Ratio A/T	Guanine	Cytosine	Ratio G/C
Beef thymus	29.0	28.5	1.02	21.2	21.2	1.00
Beef liver	28.8	29.0	0.99	21.0	21.1	1.00
Beef sperm	28.7	27.2	1.06	22.2	22.0	1.01
Human thymus	30.9	29.4	1.05	19.9	19.8	1.00
Human liver	30.3	30.3	1.00	19.5	19.9	0.98
Human sperm	30.9	31.6	0.98	19.1	18.4	1.04
Hen red blood cells	28.8	29.2	0.99	20.5	21.5	0.96
Herring sperm	27.8	27.5	1.01	22.2	22.6	0.98
Wheat germ	26.5	27.0	0.98	23.5	23.0	1.02
Yeast	31.7	32.6	0.97	18.3	17.4	1.05
Vaccinia virus	29.5	29.9	0.99	20.6	20.0	1.03
Bacteriophage T_2	32.5	32.6	1.00	18.2	18.6	0.98

of saucers. From his work as well as that of others, the American biologist James D. Watson (b. 1928) and British physicist Francis H. C. Crick (b. 1916), working at Cambridge University, designed and built a scale model of a DNA molecule. In 1953 Watson and Crick announced their now famous double-stranded helical structure for DNA. This was a milestone in the history of biology, and in 1962 Watson, Crick, and Wilkins were awarded the Nobel prize in medicine and physiology for their studies of DNA.

▲ **Francis Crick (1916–) and James Watson (1928–).**

The structure of DNA, according to Watson and Crick, consists of two polymeric strands of nucleotides in the form of a double helix, with both nucleotide strands coiled around the same axis (see Figure 33.6). Along each strand are alternate phosphate and deoxyribose units with one of the four bases adenine, guanine, cytosine, or thymine attached to deoxyribose as a side group. The double helix is held together by hydrogen bonds extending from the base on one strand of the double helix to a complementary base on the other strand. The structure of DNA has been likened to a ladder that has been twisted into a double helix, with the rungs of the ladder kept perpendicular to the twisted railings. The phosphate and deoxyribose units alternate along the two railings of the ladder, and two nitrogen bases form each rung of the ladder.

In the Watson–Crick model of DNA, the two polynucleotide strands fit together best when a purine base is adjacent to a pyrimidine base. Although this allows for four possible base pairings, A–T, A–C, G–C, and G–T, only the adenine–thymine and guanine–cytosine base pairs can effectively hydrogen-bond together. Under normal conditions, **an adenine on one polynucleotide strand is paired with a thymine on the other strand; a guanine is paired with a cytosine** (see Figure 33.7). This pairing results in A:T and G:C ratios of 1:1, as substantiated by the data in Table 33.2. The hydrogen bonding of complementary base pairs is shown in Figure 33.8. Note that if the sequence of one strand is known, the sequence of the other strand can be determined. The two DNA polymers are said to be *complementary* to each other. As will be discussed in Section 33.7, the cell can chemically "read" one strand in order to synthesize its complementary partner.

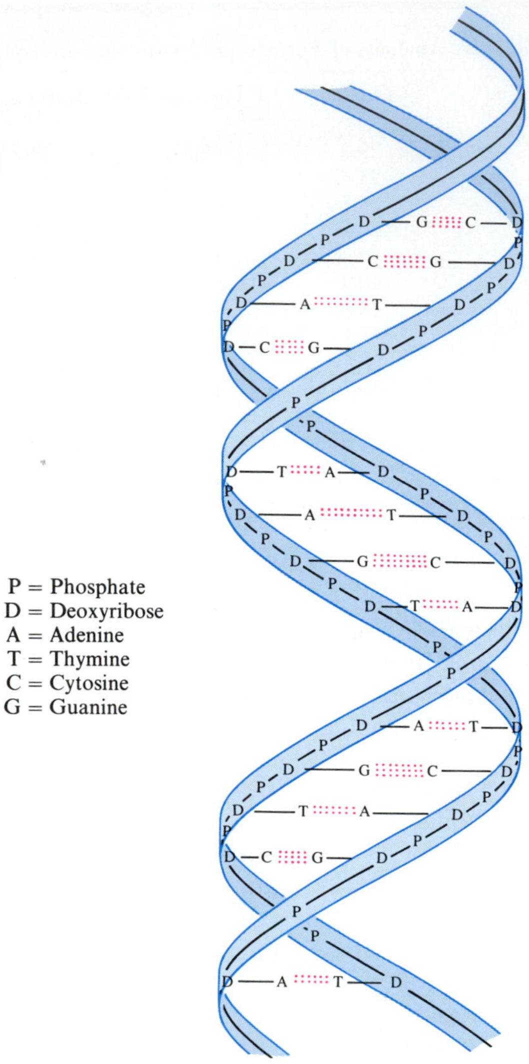

FIGURE 33.6 ► Double-stranded helical structure of DNA. (···· denotes a hydrogen bond between adjoining bases.)

The double helix is an important part of DNA structure. It does not explain how the very large DNA can be packed into a cell or the even smaller cell nucleus. For example, a human DNA molecule can be extended to almost 10 centimeters in length and yet is contained in a nucleus with a diameter about a hundred thousand times smaller. To begin the necessary packing, the DNA is looped around small aggregates of positively charged histone proteins. Hundreds of these aggregates are associated with each DNA molecule so that the DNA is foreshortened and has the appearance of a string of pearls. Further condensation is achieved by wrapping this structure into a tight coil called a solenoid, as shown in Figure 33.9. Finally, the solenoid nucleoprotein complex is wound around a protein scaffold within the nucleus. By following this complete procedure, human DNA can be taken to a length of about 10 μm in condensed chromosomes. Like a high-density computer disk, DNA takes up only a little space relative to the large amount of genetic information it contains.

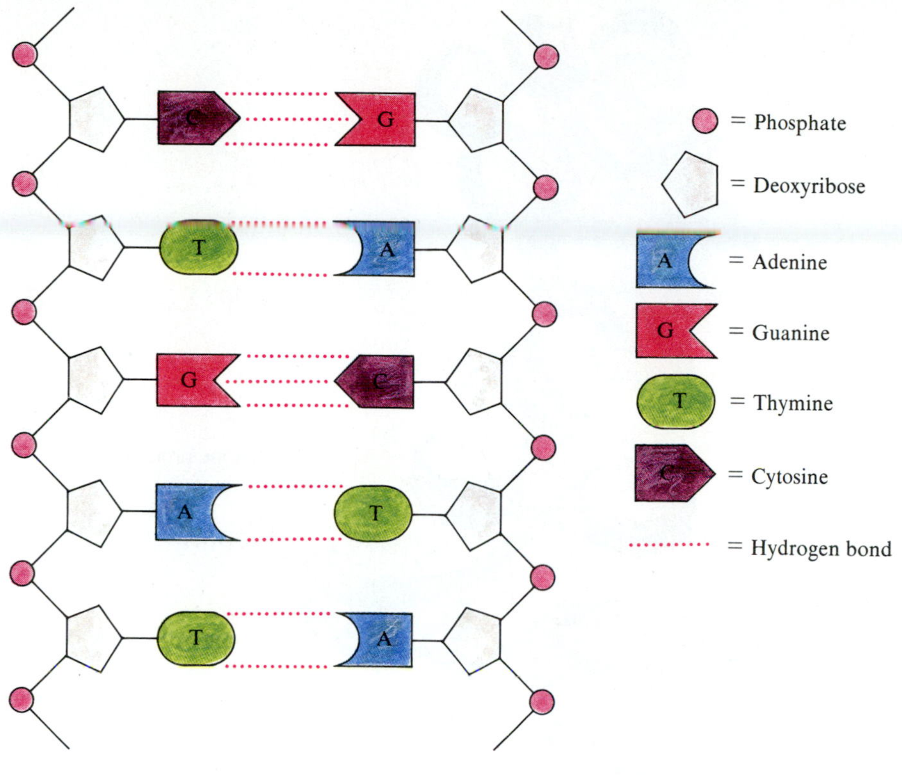

◀ **FIGURE 33.7**
Schematic diagram of a DNA segment showing phosphate, deoxyribose, and complementary base pairings held together by hydrogen bonds.

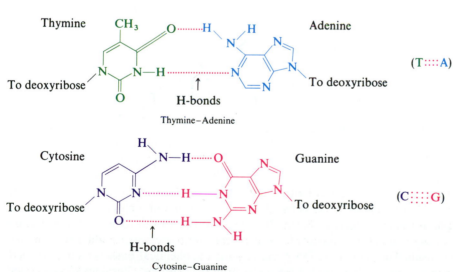

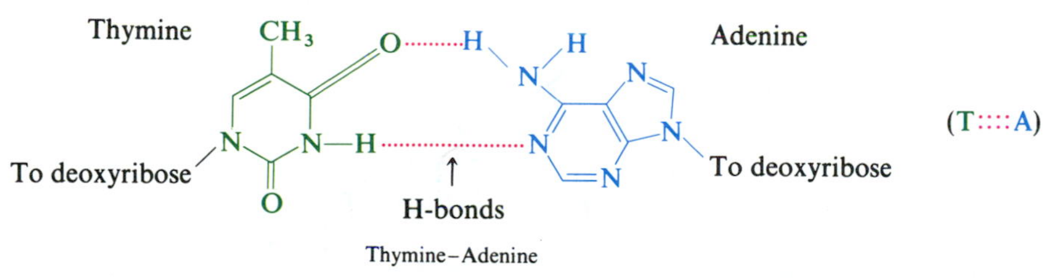

◀ **FIGURE 33.8**
Hydrogen bonding between the complementary bases thymine and adenine (T::::A) and cytosine and guanine (C::::G). Note that one pair of bases has two hydrogen bonds and the other pair has three hydrogen bonds.

33.7 DNA REPLICATION

The foundations of our present concepts of heredity and evolution were laid within the span of a decade. Charles Darwin (1809–1882), in *The Origin of Species* (1859), presented evidence supporting the concept of organic evolution

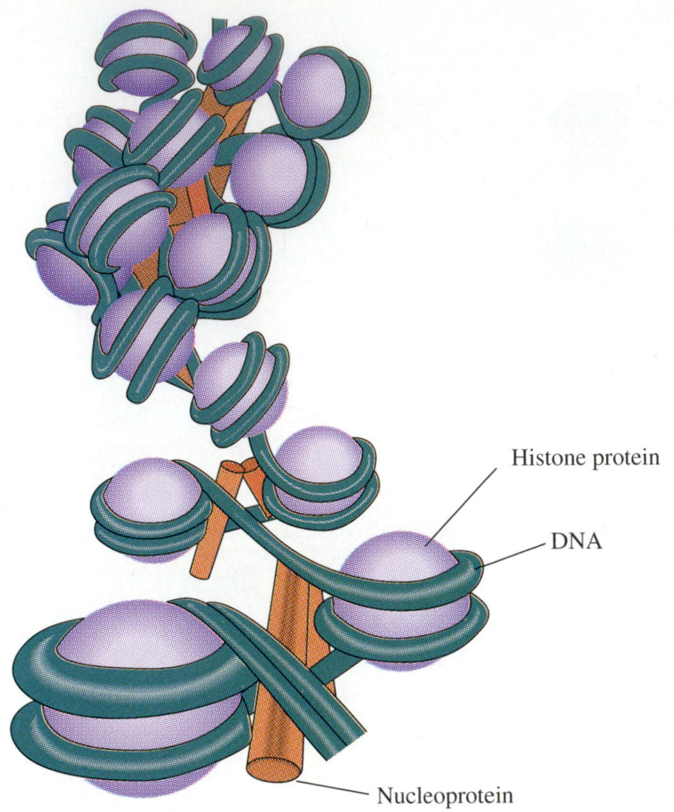

FIGURE 33.9 ▶ Condensed form of DNA: After the DNA polymer is wrapped twice around the histone protein aggregates (shown as round balls), it is coiled around a central axis to form a solenoid structure.

and his theory of natural selection. Gregor Johann Mendel (1822–1884) discovered the basic laws of heredity in 1866, and Friedrich Miescher discovered nucleic acid in 1869. Although Darwin's views were widely discussed and generally accepted by biologists within a few years, Mendel's and Miescher's work went unnoticed for many years.

Mendel's laws were rediscovered about 1900 and led to our present understanding of heredity and the science of genetics. Interest in nucleic acids lagged until nearly the 1950s, when chemical and X-ray data provided the basis for the suggestion by Watson and Crick that DNA exists in a double helix and that DNA has the possible copying mechanism for genetic material.

Heredity is the process by which the physical and mental characteristics of parents are transferred to their offspring. In order for this to occur, it is necessary for the material responsible for genetic transfer to be able to make exact copies of itself. The polymeric DNA molecule is the chemical basis for heredity. The genetic information needed for transmittal of a species' characteristics is coded along the polymeric chain. Although the chain is made from only four different nucleotides, the information content of DNA resides in the sequence of these nucleotides.

genome

The **genome** is the sum of all hereditary material contained in a cell. Within the eucaryotic genome are chromosomes, which are long, threadlike bodies composed of nucleic acids and proteins that contain the fundamental units of heredity, called genes. A **gene** is a segment of the DNA chain that controls formation of a molecule of RNA. In turn, many RNAs determine the amino acid

gene

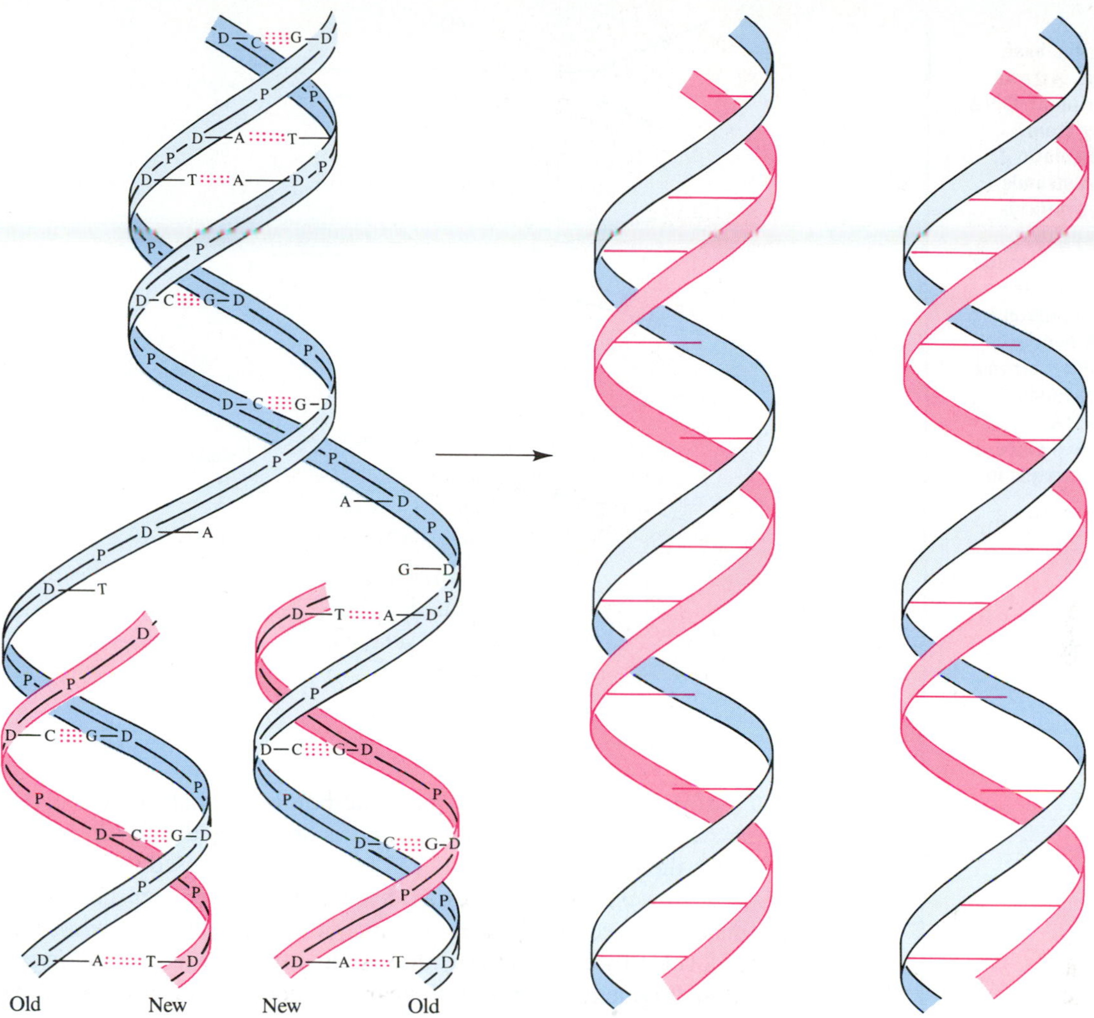

▲ **FIGURE 33.10**
Method of replication of DNA: The two helices unwind, separating at the hydrogen bonds. Each strand then serves as a template, recombining with the proper nucleotides to form a new double-stranded helix. The newly synthesized DNA strands are shown in color.

sequences for specific polypeptides or proteins. One gene commonly directs the synthesis of only one polypeptide or protein molecule. The cell has the capability of producing a multitude of different proteins because each DNA molecule contains a large number of different genes.

For life to continue relatively unchanged, genetic information must be reproduced exactly each time a cell divides. **Replication**, as the name implies, is the biological process for duplicating the DNA molecule. The DNA structure of Watson and Crick holds the key to replication; because of the complementary nature of DNA's nitrogen bases, adenine bonds only to thymine and guanine only to cytosine. Nucleotides with complementary bases can hydrogen-bond to each single strand of DNA and hence be incorporated into a new DNA double helix. Every double-stranded DNA molecule that is produced contains one template strand and one newly formed, complementary strand. This form of DNA synthesis, known as *semiconservative* replication, is illustrated in Figure 33.10.

replication

Replication is one of the most complicated enzyme-catalyzed processes in life. Enzymes are required to unwind the DNA before replication and to repackage the DNA after synthesis. The two template strands are copied differently.

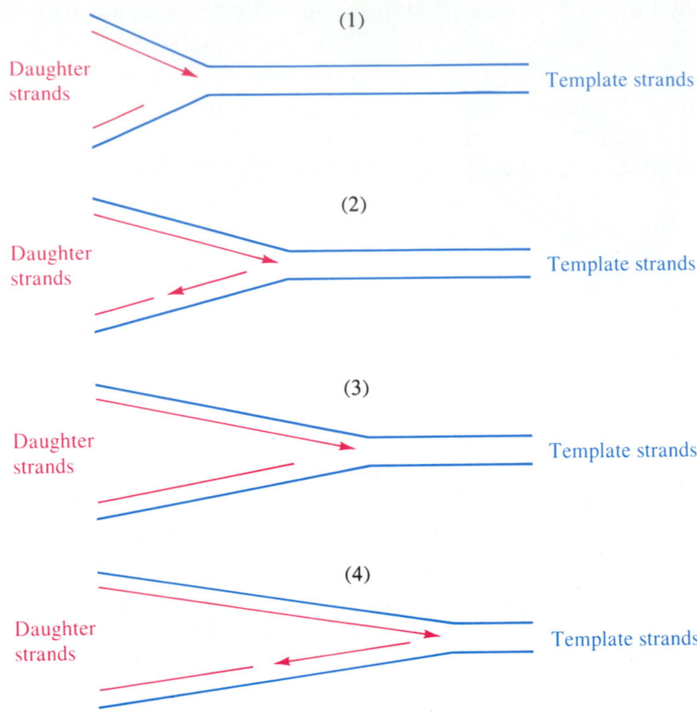

FIGURE 33.11 ▶
Diagram showing the basic replication process. Arrows indicate the direction of DNA synthesis. In going from schematic 1 to schematic 2, the template DNA strands unwind and some synthesis occurs for both daughter strands. Moving to schematic 3, the DNA fragments are connected on one daughter strand while DNA synthesis continues on the other strand and the template strands unwind further. DNA synthesis again takes place for both daughter strands in schematic 4.

One daughter strand grows directly toward the point at which the templates are unwinding while the other daughter strand is synthesized away from this point. In this way, the same enzyme catalyzed reaction can be used for synthesis of both strands. However, the one daughter strand that has been created in small fragments must be connected before replication is complete. So, while one DNA strand is formed by continuous synthesis, the other new strand is formed by a repetition of fragment synthesis followed by a coupling reaction (see Figure 33.11).

Other processes "proofread" the new polymers to check for errors. Amazingly, replication is so carefully coordinated that a mistake is passed on to the new strands only one in about 10^9 times. Of course this important fact means that new cells retain the genetic characteristics of their parents.

The DNA content of cells doubles just before the cell divides, and one-half of the DNA goes to each daughter cell. After cell division is completed, each daughter cell contains DNA and the full genetic code that was present in the original cell. This process of ordinary cell division is known as **mitosis** and occurs in all the cells of the body except the reproductive cells.

mitosis

As we have indicated before, DNA is an integral part of the chromosomes. Each species carries a specific number of chromosomes in the nucleus of each of its cells. The number of chromosomes varies with different species. Humans have 23 pairs, or 46 chromosomes. The fruit fly has 4 pairs, or 8 chromosomes. Each chromosome contains DNA molecules. Mitosis produces cells with the same chromosomal content as the parent cell. However, in the sexual reproductive cycle, cell division occurs by a different process known as meiosis.

In sexual reproduction, two cells, the sperm cell from the male and the egg cell (or ovum) from the female, unite to form the cell of the new individual.

If reproduction took place with mitotic cells, the normal chromosome content would double when two cells united. However, in **meiosis** the cell splits in such a way as to reduce the number of chromosomes to one-half of the number normally present (23 in humans). The sperm cell carries half of the chromosomes from its original cell, and the egg cell also carries half of the chromosomes from its original cell. When the sperm and the egg cells unite during fertilization, the cell once again contains the correct number of chromosomes and the hereditary characteristics of the species. Thus, the offspring derives half its genetic characteristics from the father and half from the mother.

meiosis

33.8 RNA: Genetic Transcription

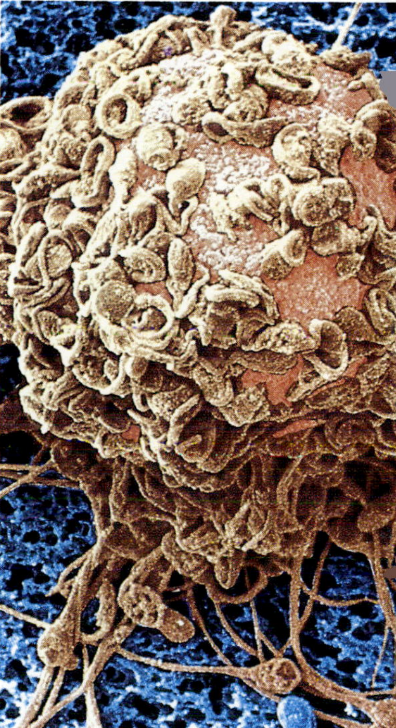

▲ The sperm and egg each provide half the chromosomes as they join during the fertilization process.

One of the main functions of DNA is to direct the synthesis of ribonucleic acids (RNAs). RNA differs from DNA in the following ways: (1) it consists of a single polymeric strand of nucleotides rather than a double helix; (2) it contains the pentose D-ribose instead of D-2′-deoxyribose; (3) it contains the pyrimidine base uracil instead of thymine; and (4) some types of RNA have a significant fraction of unusual bases in addition to the common four. RNA also differs functionally from DNA. Whereas DNA serves as the storehouse of genetic information, RNA is used to process this information into proteins. Three types of RNA are needed to produce proteins: ribosomal RNA (rRNA), messenger RNA (mRNA), and transfer RNA (tRNA).

More than 80% of the cellular RNA is ribosomal RNA. It is found in the ribosomes, where it is associated with protein in proportions of about 60–65% protein to 30–35% rRNA. Ribosomes are the sites for protein synthesis.

Messenger RNA carries genetic information from DNA to the ribosomes. It is a template made from DNA and carries the code that directs the synthesis of proteins. The size of mRNA varies according to the length of the polypeptide chain it will encode.

The primary function of tRNA is to bring amino acids to the ribosomes for incorporation into protein molecules. Consequently there exists at least one tRNA for each of the 20 amino acids required for proteins. Transfer RNA molecules have a number of structural features in common. The end of the chain of all tRNA molecules terminates in a CCA nucleotide sequence to which is attached the amino acid to be transferred to a protein chain. The primary structure of tRNA allows extensive folding of the molecule such that complementary bases are hydrogen-bonded to each other to form a structure that appears like a cloverleaf. The cloverleaf model of tRNA has an anticodon loop consisting of seven unpaired nucleotides. Three of these nucleotides make up an anticodon (see Figure 33.12). The anticodon is complementary to, and hydrogen-bonds with, three bases on an mRNA. The other two loops in the cloverleaf structure enable the tRNA to bind to the ribosome and other specific enzymes during protein synthesis (see Section 33.10.)

The making of RNA from DNA is called **transcription**. The verb *transcribe* literally means to copy, often into a different format. When the nucleotide sequence of one strand of DNA is transcribed into a single strand of RNA, genetic information is copied from DNA to RNA. This transcription occurs in a complementary fashion and depends upon hydrogen bonded pairing between

transcription

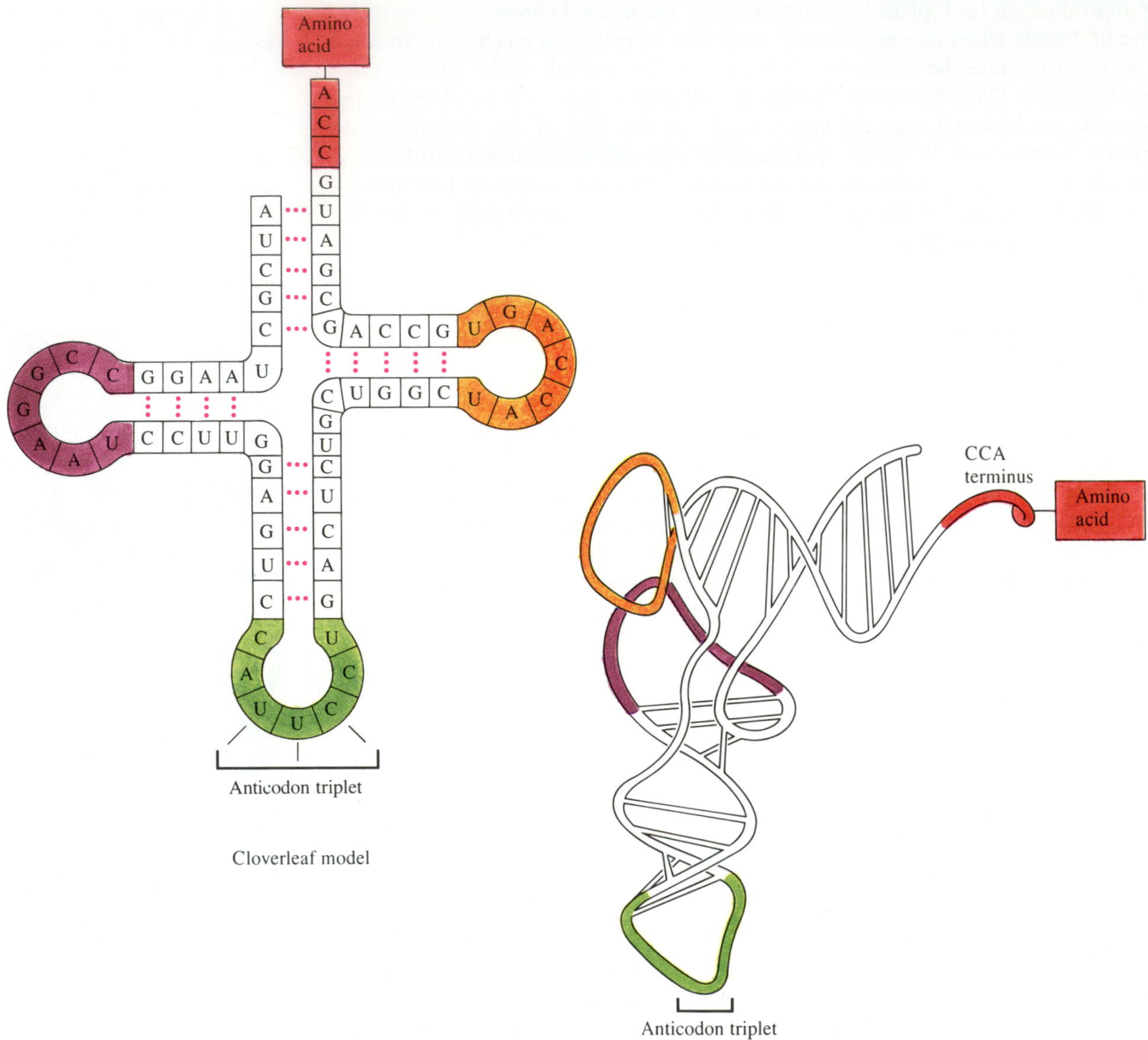

▲
FIGURE 33.12

Cloverleaf model and three dimensional representations of tRNA. The anticodon triplet (CUU) located at the lower loop is complementary to GGA (which is the code for glycine) on mRNA.

appropriate bases (see Figure 33.13). Where there is a guanine base in DNA, a cytosine base will occur in RNA. Cytosine is transcribed to guanine, thymine to adenine, and adenine to uracil (the thymine-like base which is found in RNA).

Because transcription is the initial step in the expression of genetic information, this process is under stringent cellular control. Only a small fraction of the total information stored in DNA is used at any one time. In procaryotic cells (see Section 35.5), related genes are often located together so that they can be transcribed in concert. The control of eucaryotic gene expression is more complex.

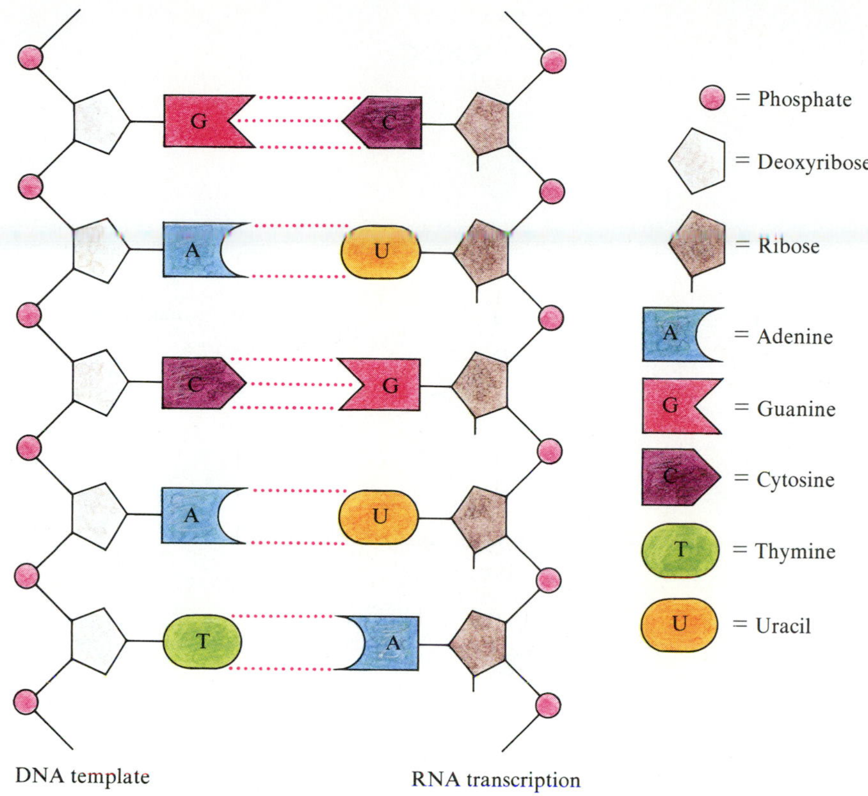

◄ **FIGURE 33.13 Transcription of RNA from DNA. The sugar in RNA is ribose. The complementary base of adenine is uracil. After transcription is complete, the new RNA separates from its DNA template and travels to another location for further use. (.... denotes a hydrogen bond between adjoining bases.)**

The importance of transcription control is illustrated by the recently discovered oncogenes. These genes, which cause cells to become malignant or cancerous, were originally discovered as part of the genome of cancer causing viruses. To the surprise of many investigators, it was also found that these same oncogenes are present in many normal mammalian cells. The cells carry on their normal and necessary functions, and are not cancerous, as long as the oncogenes are under control. But control can be lost due to, for example, a virus infection or exposure to a chemical carcinogen. Then the oncogenes are transcribed too often, and the cell becomes cancerous.

Following transcription, the RNA molecules that are produced from DNA often are modified before they are put to use. Ribosomal RNA and tRNA molecules are formed as larger precursors and are then trimmed to the correct size. After transcription, many of the bases in tRNA are modified by methylation, saturation of a double bond, or by isomerization of the ribose-base linkage. This post-transcriptional modification or processing changes the information content of the RNA.

Eucaryotic mRNA may undergo considerable alteration before the message that it carries is ready to guide protein synthesis. These changes may include the elimination of portions of the mRNA molecule, the splicing together of two or more mRNA molecules, or the addition of new bases to one end of the mRNA. During this processing, the message encoded in the mRNA molecule is apparently refined so that it can be correctly read by the ribosome.

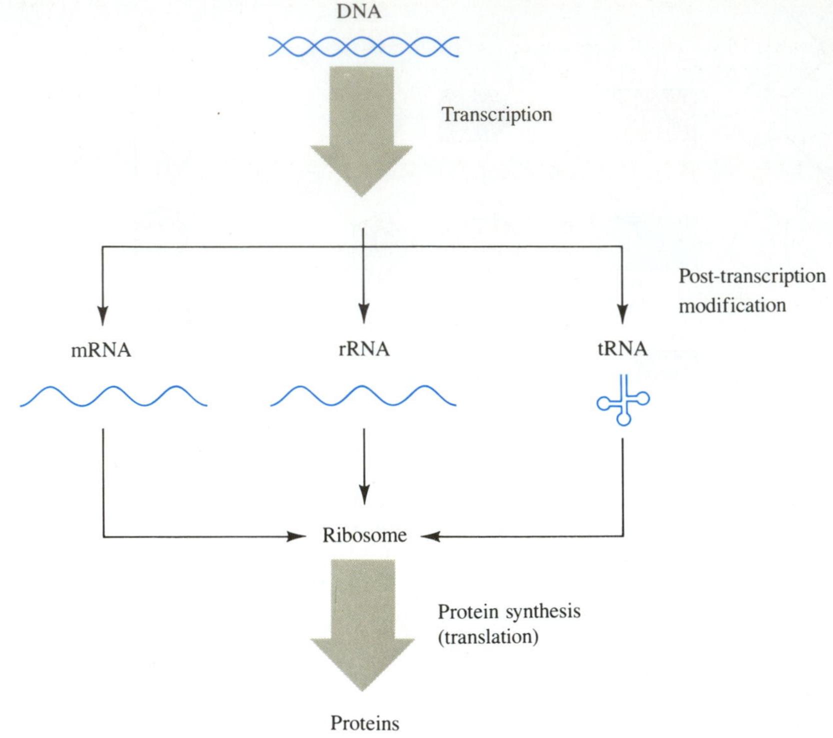

FIGURE 33.14 ▶ A flow diagram representing the processing of cellular genetic information.

Figure 33.14 summarizes the processes by which genetic information is transferred within a cell. These steps culminate in the conversion of a coded nucleotide message into a protein amino acid sequence.

33.9 The Genetic Code

For a long time after the structure of DNA was elucidated, scientists struggled with the problem of how the information stored in DNA could specify the synthesis of so many different proteins. Since the backbone of the DNA molecule contains a regular structure of repeated and identical phosphate and deoxyribose units, the key to the code had to lie with the four bases—adenine, guanine, cytosine, and thymine.

The code, using only the four nucleotides A, G, C, and T, must be capable of coding at least for the 20 amino acids that occur in proteins. If each nucleotide coded one amino acid, only four amino acids could be represented. If the code used two nucleotides to specify an amino acid, 16 (4×4) combinations would be possible—still not enough. Using three nucleotides we can have 64 ($4 \times 4 \times 4$) possible combinations—which is more than enough to specify the 20 common amino acids in proteins. It has now been determined that each code word requires a sequence of three nucleotides. The code is therefore a triplet code. Each triplet of three nucleotides is called a **codon**, and, in general, each codon specifies one

codon

TABLE 33.3 The Genetic Code for Messenger RNA

First nucleotide	Second nucleotide: U	C	A	G	Third nucleotide
U	Phe	Ser	Tyr	Cys	U
	Phe	Ser	Tyr	Cys	C
	Leu	Ser	TC[a]	TC[a]	A
	Leu	Ser	TC[a]	Trp	G
C	Leu	Pro	His	Arg	U
	Leu	Pro	His	Arg	C
	Leu	Pro	Gln	Arg	A
	Leu	Pro	Gln	Arg	G
A	Ile	Thr	Asn	Ser	U
	Ile	Thr	Asn	Ser	C
	Ile	Thr	Lys	Arg	A
	Met	Thr	Lys	Arg	G
G	Val	Ala	Asp	Gly	U
	Val	Ala	Asp	Gly	C
	Val	Ala	Glu	Gly	A
	Val	Ala	Glu	Gly	G

[a] Termination or nonsense codons.

◀ **This table shows the sequence of nucleotides in the triplet codons of messenger RNA that specify a given amino acid. For example, UUU or UUC is the codon for Phe, UCU is the codon for Ser; CAU or CAC is the codon for His.**

amino acid. Thus, to describe a protein containing 200 amino acid units, a gene containing at least 200 codons, or 600 nucleotides, is required.

In the sequence of biological events, the code from a gene in DNA is first transcribed to a coded RNA, which, in turn, is used to direct the synthesis of a protein. The 64 possible codons for mRNA are given in Table 33.3. In this table a three-letter sequence (first nucleotide–second nucleotide–third nucleotide) specifies a particular amino acid. For example, the codon CAC (cytosine–adenine–cytosine) is the code for the amino acid histidine (His). You will note that three codons in the table, marked TC, do not encode any amino acids. These are called *nonsense* or *termination codons*. They act as signals to indicate where the synthesis of a protein molecule is to end. The other 61 codons identify 20 amino acids. Methionine and tryptophan have only one codon each. For the other amino acids, the code is redundant; that is, each amino acid is specified by at least two, and sometimes by as many as six, codons.

It is believed that the genetic code is a universal code for all living organisms; that is, the same nucleotide triplet specifies a given amino acid regardless of whether that amino acid is synthesized by a bacterial cell, a pine tree, or a human being.

Recently scientists have developed a machine that automates the sequencing of DNA. Whereas a researcher might manually sequence about 140 bases per day, the machine can chemically "read" about 10,000 bases per day. This rapid decoding of genetic information is essential if scientists finally hope to

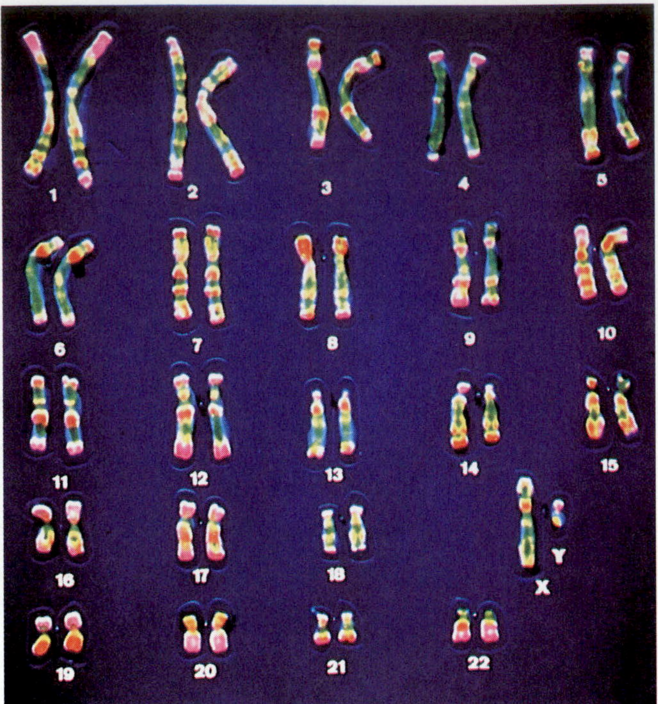

A karyotype of a normal ▶ male showing all the pairs of chromosomes. This information can be used in genetic counseling to identify possible genetic abnormalities in offspring.

understand the human genome, which contains about 3 billion bases. Just such a major collaborative research project, called the Human Genome Project, is in progress. Because one laboratory would take about 1000 years to decode the human genome even with a sequencing machine, the Human Genome Project involves many research laboratories across the United States. The first few years of this project have resulted in identification of the chromosomal location of the genes causing muscular dystrophy, Huntington's disease, and cystic fibrosis. Scientists project that within 20 years they will have the information to diagnose and treat the large number of genetic disorders which afflict humankind.

33.10 Biosynthesis of Proteins

translation

The biosynthesis of proteins is extremely complex, and the following is only a cursory description of the overall process. The production of a polypeptide using an mRNA template is called **translation**. This term is used because it literally means a change from one language to another. The genetic code is *translated* into the primary structure of a polypeptide or a protein.

The biosynthesis of proteins begins when messenger RNA leaves the cellular nucleus and travels to the cytoplasm. Each mRNA is then bound to five or more ribosomes, the bodies responsible for protein synthesis. Amino acids must also be transferred to the ribosomes. To accomplish this step, cellular energy in the form of ATP is used to couple amino acids to tRNAs to form aminoacyl–tRNA complexes that can bind to the ribosomes.

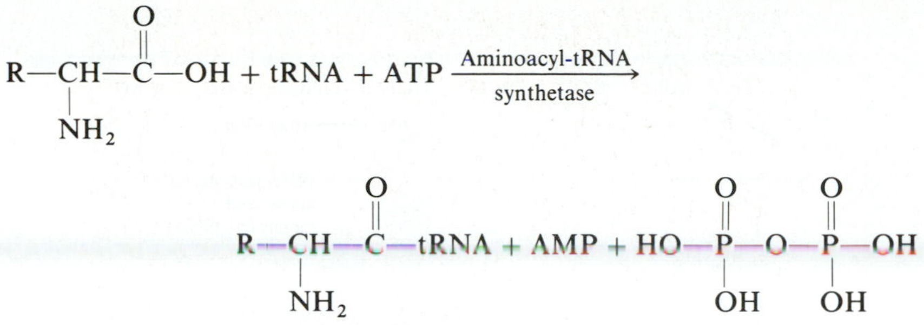

$$\mathrm{R{-}\underset{\displaystyle NH_2}{\underset{|}{CH}}{-}\overset{\displaystyle O}{\overset{\|}{C}}{-}OH} + \mathrm{tRNA} + \mathrm{ATP} \xrightarrow[\text{synthetase}]{\text{Aminoacyl-tRNA}}$$

$$\underset{\text{Aminoacyl-tRNA}}{\mathrm{R{-}\underset{\displaystyle NH_2}{\underset{|}{CH}}{-}\overset{\displaystyle O}{\overset{\|}{C}}{-}tRNA}} + \mathrm{AMP} + \mathrm{HO{-}\underset{\displaystyle OH}{\underset{|}{\overset{\displaystyle O}{\overset{\|}{P}}}}{-}O{-}\underset{\displaystyle OH}{\underset{|}{\overset{\displaystyle O}{\overset{\|}{P}}}}{-}OH}$$

A different specific enzyme is utilized for binding each of the 20 amino acids to a corresponding tRNA. Although only 20 amino acids are involved, there are about 60 different tRNA molecules in the cells. In summary the function of tRNA is to bring amino acids to the ribosome synthesis site.

Initiation The next step in the process is the initiation of polypeptide synthesis. Two codons signal for the start of protein synthesis, AUG and GUG (AUG being the more common). As shown in Table 33.3, these codons also code for incorporation of the amino acids methionine and valine. Most mRNAs have more than one AUG and GUG, and the ribosome must choose at which codon to begin. It appears that the ribosome uses information in addition to the AUG or GUG triplet to choose the correct starting codon for protein synthesis.

Once the correct starting point has been identified, a special initiator tRNA binds to the ribosome. This specific tRNA carries an *N*-formyl methionine in procaryotic cells:

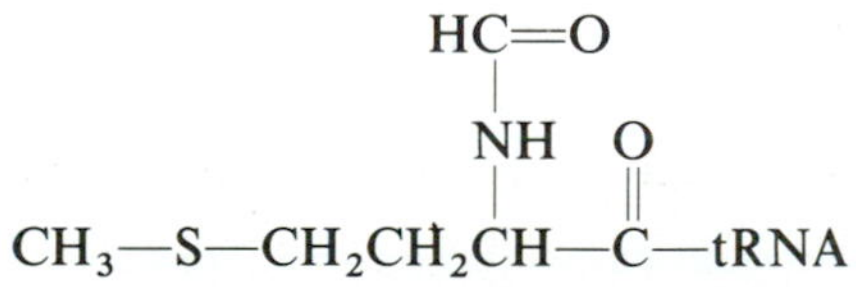

$$\underset{N\text{-formyl methionine-tRNA}}{\mathrm{CH_3{-}S{-}CH_2CH_2\overset{\displaystyle HC{=}O}{\overset{|}{\overset{\displaystyle NH}{\overset{|}{CH}}}}{-}\overset{\displaystyle O}{\overset{\|}{C}}{-}tRNA}}$$

Because of the attached formyl group (CHO), a second amino acid will react only with the carboxyl group of methionine, the correct direction for protein synthesis. In eucaryotic cells, the initiator tRNA carries a methionine group.

Elongation The next stage involves the elongation or growth of the peptide chain, which is assembled one amino acid at a time. After the initiator tRNA is attached to the mRNA codon, the elongation of the polypeptide chain involves the following steps (see Figure 33.15):

1. The next aminoacyl-tRNA enters the ribosome and becomes attached to mRNA through the hydrogen bonding of the tRNA anticodon to the mRNA codon.
2. The peptide bond between the two amino acids is formed by the transfer of the amino acid from the initial aminoacyl-tRNA to the incoming aminoacyl-tRNA. In this step, which is catalyzed by the enzyme peptidyl

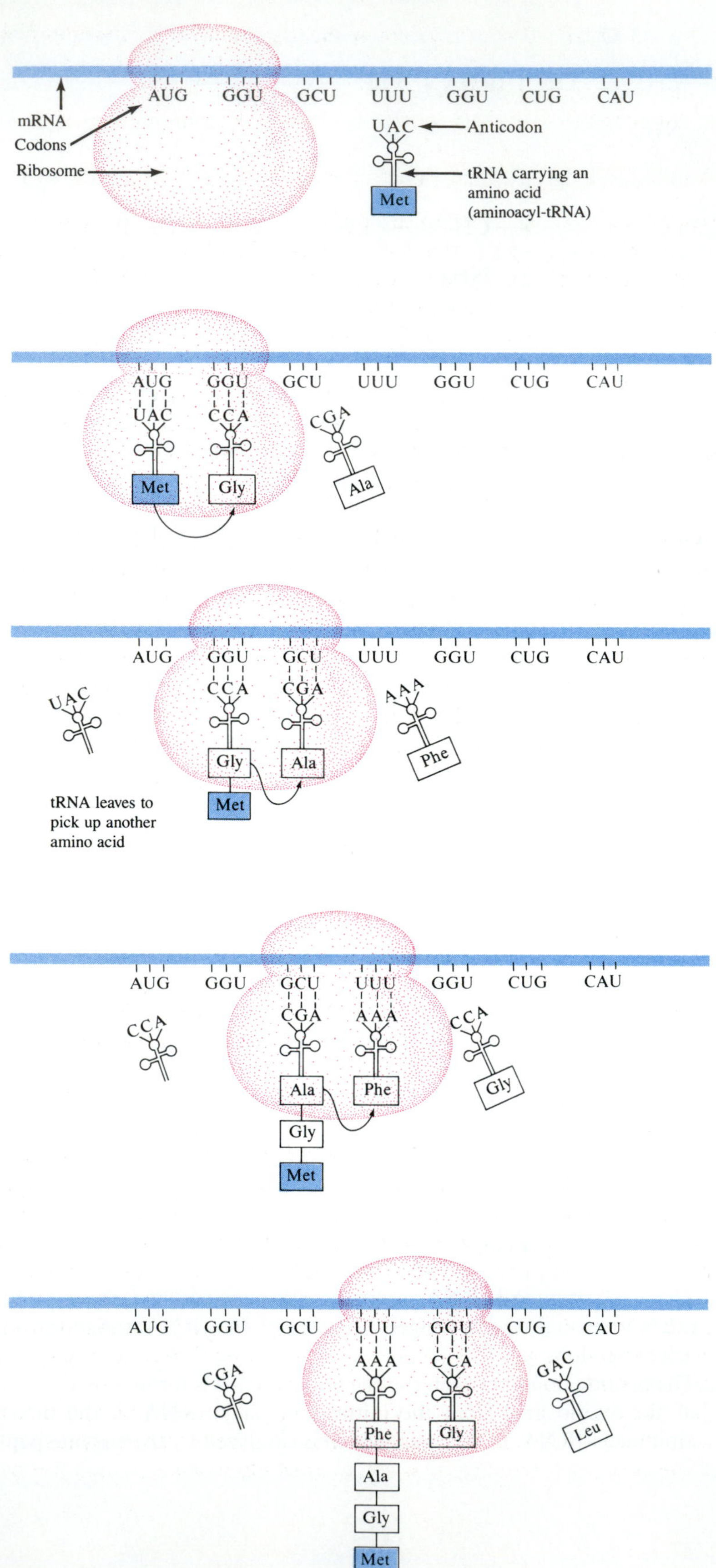

FIGURE 33.15 ▶
Biosynthesis of proteins: mRNA from DNA enters and complexes with the ribosomes. tRNA carrying an amino acid (aminoacyl-tRNA) enters the ribosome and attaches to the mRNA at its complementary anticodon. The peptide chain elongates when another aminoacyl-tRNA enters the ribosome and attaches to the mRNA. The peptide bond is then formed by the transfer of the peptide chain from the initial to the incoming aminoacyl-tRNA. The sequence is repeated until a termination codon appears in mRNA.

transferase, the carboxyl group of the first amino acid separates from its tRNA and forms a peptide bond with the free amino group of the incoming aminoacyl-tRNA.

3. The tRNA carrying the peptide chain (now known as a peptidyl-tRNA) moves over in the ribosome, the free tRNA is ejected, and the next aminoacyl-tRNA enters the ribosome.

The peptide chain is transferred to the incoming amino acid and the sequence is repeated over and over again as the mRNA moves through the ribosome, just like a tape delivering its message. In each step the entire peptide chain is transferred to the incoming amino acid.

The initiation and elongation of polypeptide chains requires cellular energy. The primary source of this energy is the nucleotide guanosine-5′-triphosphate (GTP). At various steps in the growth of the protein chain, a high-energy GTP phosphate anhydride bond is hydrolyzed, yielding GDP and a phosphate group. The energy that is released drives protein synthesis. This reaction is analogous to that involving ATP (see Section 33.4).

Termination The termination of the polypeptide chain occurs when a nonsense or termination codon appears. In normal cells there are no tRNAs that have complementary anticodons to the termination codons. Because there is no new aminoacyl-tRNA to bind to the ribosome, the peptidyl-tRNA is hydrolyzed, and the free polypeptide (protein) is released. All of these amazing, coordinated steps are accomplished at a high rate of speed—about 1 minute for a 150 amino acid chain in hemoglobin and 10–20 seconds for a 300–500 amino acid chain in the bacterium *Escherichia coli* (*E. coli*). This mechanism of protein synthesis is illustrated in Figure 33.15.

33.11 Mutations

It is known that from time to time a new trait appears in an individual that is not present in either parents or ancestors. These traits, which are generally the result of genetic or chromosomal changes, are called **mutations**. Some mutations are beneficial, but most are harmful. Because mutations are genetic, they may be passed on to the next or future generations.

mutation

Mutations can occur spontaneously or can be caused by chemical agents or various types of radiation such as X rays, cosmic rays, and ultraviolet rays. The agent that causes the mutation is called a **mutagen**. Exposure to mutagens may produce changes in the DNA of the sperm or ova. The likelihood of such changes is increased by the intensity and length of exposure to the mutagen. Mutations may then show up as birth defects in the next generation. Common types of genetic alterations include the substitution of one purine or pyrimidine for another during DNA replication. Such a substitution is a change in the genetic code and causes misinformation to be transcribed from the DNA. A mutagen also may alter genetic material by causing a chromosome or chromosome fragment to be added or removed.

mutagen

There is clear evidence that many mutagens are also carcinogens. When these chemicals alter the genome, they may also change genetic controls, causing a cell to become cancerous. Bruce Ames, an American biochemist, made use of this

(continued on page 866)

CHEMISTRY IN ACTION

GENETIC ENGINEERING

Almost since life first evolved, genetic material has been manipulated and changed to achieve specific goals. Nature has its own genetic engineers, the viruses. These small particles commonly carry only a few genes surrounded by a protein coat. Viruses are not living cells but exist as parasites within host cells. The virus invades a cell, adds its genetic material to that of the cell, and forces the cell to produce viral genes and proteins.

For example, the virus that causes the acquired immune deficiency syndrome (AIDS) acts by invading and altering the transcription of cells that are part of the body's immune system (lymphocytes). Within the cell the virus releases viral RNA and the enzyme *reverse transcriptase*. The viral RNA acts as a template and the enzyme synthesizes viral DNA, which then inserts into the cell DNA. New viral RNAs and proteins are produced. The cell has now lost control of its normal transcription and reproductive ability. Progressive damage to the immune system occurs as immune cells produce new viruses and then die.

In recent years laboratory techniques for the controlled production of genetic changes have been developed. These techniques are known as *genetic engineering* and have already been responsible for considerable progress in medicine and biology. Genetic engineering holds the promise of revolutionary advances in medicine, agriculture, the manufacture of pharmaceuticals, and

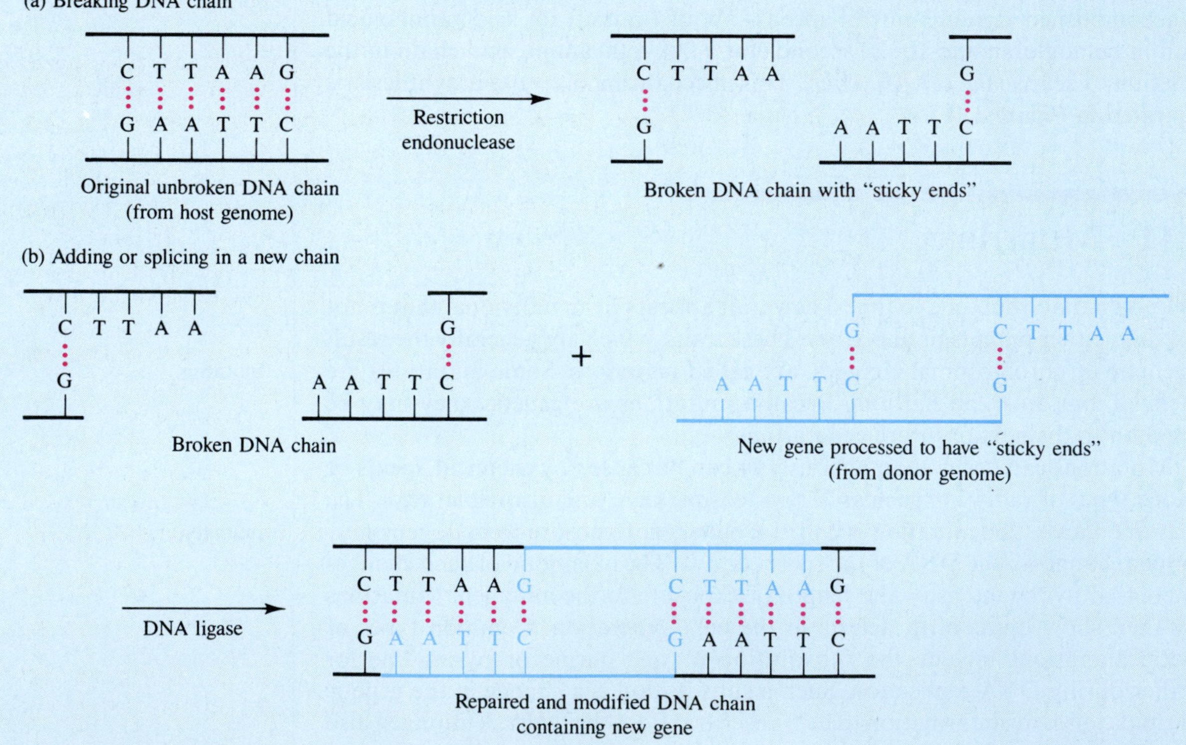

▲
Schematic diagram showing (a) breaking of double-stranded DNA chain to obtain "sticky ends" and (b) adding or splicing in a new gene previously processed to have matching sticky ends. Only the nitrogen bases involved in the sticky ends are shown.

(continued)

(*continued*)
other fields. Using techniques similar to those used by the viruses, the genetic engineer inserts specific genes into the genome of a host cell. The host cell is thus programmed to produce new or different proteins that may benefit humankind.

Genetic engineering has been made possible by several basic advances in nucleic acid biochemistry. First, scientists have gained the ability to isolate, identify, and then synthesize multiple copies of specific genetic messages. Often the gene of interest is present in only one copy in every million genes—this first step is difficult and yet is of critical importance for the genetic engineer.

A recent discovery, the DNA polymerase chain reaction, has considerably improved the success rate for this first process. DNA polymerase is an enzyme that replicates DNA when supplied a starting fragment (the primer) and a DNA strand to copy (the template). By supplying the appropriate primer, the polymerase can be induced to synthesize a specific gene, even in the presence of a wide variety of other genetic material. What makes this reaction especially valuable is that the polymerase can use the newly synthesized DNA as a template for further replication. Thus, after 20 synthetic cycles, almost a million copies of the original gene can be produced. Like a radioactive chain reaction, the polymerase chain reaction leads to an explosion in the number of copied genes.

Once the number of copies of a specific gene has been increased, this gene may be identified, studied, and ultimately perhaps transferred to another DNA molecule. Gene identification has proved valuable in prenatal diagnosis as well as forensic medicine. The polymerase chain reaction allows diagnosis in less than one day of such genetic diseases as phenylketonuria, muscular dystrophy, and cystic fibrosis. Because the DNA from one cell can be analyzed with this technique, fertilized eggs can be grown through several cell divisions and tested for potential fatal genetic diseases before being implanted in the womb. Again the sensitivity of the polymerase chain reaction is important in allowing forensic chemists to compare a suspect's DNA with a small amount of cellular material found at the scene of a crime.

A second important process in genetic engineering is the insertion of genetic material into a "foreign" genome. Special enzymes, the *restriction endonucleases*, have provided a key step during gene insertion. These enzymes split double-stranded DNA at very specific locations. Often the newly formed break has what is termed "sticky ends"; that is, one end of the break can stick to the end of another break via hydrogen bonding between complementary bases (see diagram on page 864). When both a gene and a "foreign" genome are processed in this way, they bond together. Then, with the aid of other enzymes (ligases), the gene can be covalently bonded into place. The resultant modified and repaired "foreign" genome now contains a new gene. The result of this process is a form of recombinant DNA. The term *recombinant DNA* refers to DNA whose genes have been rearranged to contain new or different hereditary information. The process of gene addition to DNA is shown in the diagram on page 864.

Genetic engineering is progressing in a variety of areas. Bacteria have been modified to produce proteins that are relatively scarce. Human insulin and human growth

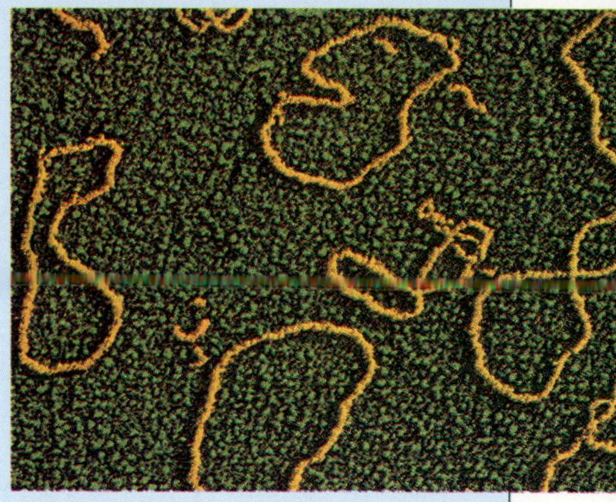

▲
Electron micrograph of plasmids of bacterial DNA from *E. Coli* bacterium.

hormone are now synthesized by microorganisms. Bovine rennin, an enzyme needed in large amounts during cheese production, is now partially supplied by genetically modified bacteria.

Presently, natural proteins can be altered to improve specific properties by genetic engineering. For example, the proteolytic enzymes used in detergents are being modified to improve their stability when exposed to bleach and high pH (conditions that often exist during a laundry cycle). Lipid digestive enzymes (lipases) are being modified to convert inexpensive oils into the more expensive triacylglycerols.

Genetic engineering is also used to alter the properties of living organisms. An immunity to specific weed killers is being incorporated into plants of commercial importance in order to aid growers. Organisms are being developed that can metabolize and therefore clean up toxic wastes. Thus, in the future it is likely that genetic engineering will have a major impact on our way of life.

information to develop a sensitive and easy test for potential carcinogens. The Ames test uses specific strains of bacteria that lack some necessary enzymes due to mutation. These bacteria are exposed to a possible cancer causing substance. If the bacteria regain their missing enzymes, the substance tested has been proved to be a mutagen. By quickly identifying mutagens, this test helps scientists recognize potential carcinogens.

The average person is exposed to many mutagens every day. Yet the frequency of mutation is relatively low. This seeming contradiction can be understood with the knowledge that each cell contains enzymes for repairing damaged DNA. For example, a simple walk on a sunny day exposes the skin cells to ultraviolet radiation, which chemically alters a few of the thymine bases in the DNA. If these changes were allowed to remain, the DNA could not be replicated correctly and a mutation would occur. Fortunately, cellular enzymes locate the damaged DNA, excise the altered thymines, and rebuild the correct nucleotide sequence. Similar repair processes are occurring continuously for most normal cells.

In some cases mutations have weakened the cellular DNA repair machinery. Patients suffering from *xeroderma pigmentosum* lack the enzyme that recognizes chemically altered thymines. For these people sunlight becomes a real danger. Exposure to only small amounts of ultraviolet radiation causes skin ulceration and eventual formation of multiple skin cancers. Although this disease occurs only rarely, it has been estimated that about 1% of all humans are carriers of these mutations.

Concepts in Review

1. Write the structural formulas for the two purine and three pyrimidine bases found in nucleotides.
2. Distinguish between ribonucleotides and deoxyribonucleotides.
3. List the compositions, abbreviations, and structures for the ten nucleotides in DNA and RNA.
4. Write the structural formulas for ADP and ATP.
5. Identify where energy is stored in ADP and ATP.
6. Write a structural formula of a segment of a polynucleotide that contains four nucleotides.
7. Describe the double-helix structure of DNA according to Watson and Crick.
8. Explain the concept of complementary bases.
9. Describe and illustrate the replication process of DNA.
10. Understand how heredity factors are stored in DNA molecules.
11. Distinguish between mitosis and meiosis.
12. Understand how the genetic code is used in the synthesis of proteins.
13. Describe a potential cause of cancer involving an oncogene.
14. Explain the genetic code.
15. State the functions of the three different kinds of RNA.

16. Describe the transcription of the genetic code from DNA to RNA.
17. Describe the biosynthesis of proteins.
18. Understand how mutations are caused.
19. Describe the Ames test.
20. Briefly describe genetic engineering.

Equations in Review

Reaction Between a High-Energy Phosphate Bond of ATP and Water

$$\text{ATP} + \text{H}_2\text{O} \underset{\text{Energy storage}}{\overset{\text{Energy utilization}}{\rightleftharpoons}} \text{ADP} + \text{P}_i + \sim 35 \text{ kJ}$$

Use of ATP Energy to Bond an Amino Acid to a Transfer RNA

$$\text{R—CH(NH}_2\text{)—C(=O)—OH} + \text{tRNA} + \text{ATP} \xrightarrow[\text{synthetase}]{\text{Aminoacyl-tRNA}}$$

$$\text{R—CH(NH}_2\text{)—C(=O)—tRNA} + \text{AMP} + \text{HO—P(=O)(OH)—O—P(=O)(OH)—OH}$$

Exercises

1. Write the names and structural formulas for the five nitrogen bases found in nucleotides.
2. What is the difference between a nucleoside and a nucleotide?
3. Identify the compounds represented by the following letters:
   (a) A, AMP, ADP, ATP
   (b) G, GMP, GDP, GTP
4. What are the three units that make up a nucleotide?
5. Write structural formulas for the substances represented by
   (a) A (c) ADP (e) dGTP
   (b) AMP (d) ATP (f) CDP
6. What is the major function of ATP in the body?
7. What are the principal structural differences between DNA and RNA?
8. Draw the structure for a three nucleotide segment of RNA.
9. Show by structural formulas the hydrogen bonding between adenine and uracil.
10. Briefly describe the structure of DNA as proposed by Watson and Crick.
11. What is meant by the term *complementary bases*?
12. Explain why the ratio of thymine to adenine in DNA is 1:1, but the ratio of thymine to guanine is not necessarily 1:1.
13. Why is DNA considered to be the genetic substance of life?
14. Briefly describe the process of DNA replication.
15. What is the role of DNA in the genetic process?
16. What is the genetic code?
17. Why are at least three nucleotides needed for one unit of the genetic code?
18. List the three kinds of RNA and identify the role of each.
19. What is an oncogene?
20. What is a codon? An anticodon?

21. Explain the relationship between codons and anticodons.

22. There are 146 amino acid residues in the β-polypeptide chain of hemoglobin. How many nucleotides in mRNA are needed to designate this chain? (See Figure 31.7 to review the β-chain.)

23. In RNA does the guanine content have to be equal to the cytosine content? Explain. Do they have to be equal in DNA? Explain.

24. Starting with DNA, briefly outline the biosynthesis of proteins.

25. Explain the role of *N*-formylmethionine in procaryotic protein synthesis.

26. A segment of a DNA strand consists of GCTTAGACCTGA.
   (a) What is the nucleotide order in the complementary mRNA?
   (b) What is the anticodon order in tRNA?
   (c) What is the sequence of amino acids coded by the DNA?

27. What will the anticodon be in tRNA if the codon in mRNA is the following?
   (a) GUC (d) UUU
   (b) ACC (e) CCA
   (c) CGA

28. Complete hydrolysis of RNA would yield what compounds?

29. What is mutation?

30. Why do mutations occur?

31. Briefly describe the Ames test.

32. Does DNA damage always result in a mutation? Explain.

33. Briefly list the basic steps that are often employed in genetic engineering.

34. Which of these statements are correct? Rewrite each incorrect statement to make it correct.
   (a) Adenine and guanine are both purine bases and are found in both DNA and RNA.
   (b) The ratio of adenine to thymine and guanine to cytosine in DNA is about 1:1.
   (c) DNA and RNA are responsible for transmitting genetic information from parent to daughter cells.
   (d) DNA is a polymer made from nucleotides.
   (e) Codons are combinations of the base units in a tRNA molecule.
   (f) The double-helix structure of DNA is held together by peptide linkages.
   (g) Amino acids are linked to tRNA by an ester bond.
   (h) The nucleotide adenosine monophosphate contains adenine, D-ribose, and a phosphate group.
   (i) The letters ATP stand for adenine triphosphate.
   (j) Thymine and uracil are both complementary bases to adenine.
   (k) Messenger RNA is a transcribed section of DNA.
   (l) The ratio of adenine to guanine and thymine to cytosine is 1:1 in DNA.
   (m) Genetic information is based on the nucleotide sequence in DNA.
   (n) Humans have 46 pairs of chromosomes.
   (o) In mitosis the sperm cell and the egg cell, each with 23 chromosomes, unite to give a new cell containing 46 chromosomes.
   (p) The genetic code consists of triplets of nucleotides; each triplet codes an amino acid.
   (q) Transfer RNA carries the code for the synthesis of proteins.
   (r) ADP and ATP contain high energy phosphate anhydride bonds.
   (s) On hydrolysis, DNA yields ribose, phosphoric acid, and the bases adenine, guanine, cytosine, and thymine.
   (t) DNA has a double-helix conformation, whereas all RNAs have a single-helix conformation.

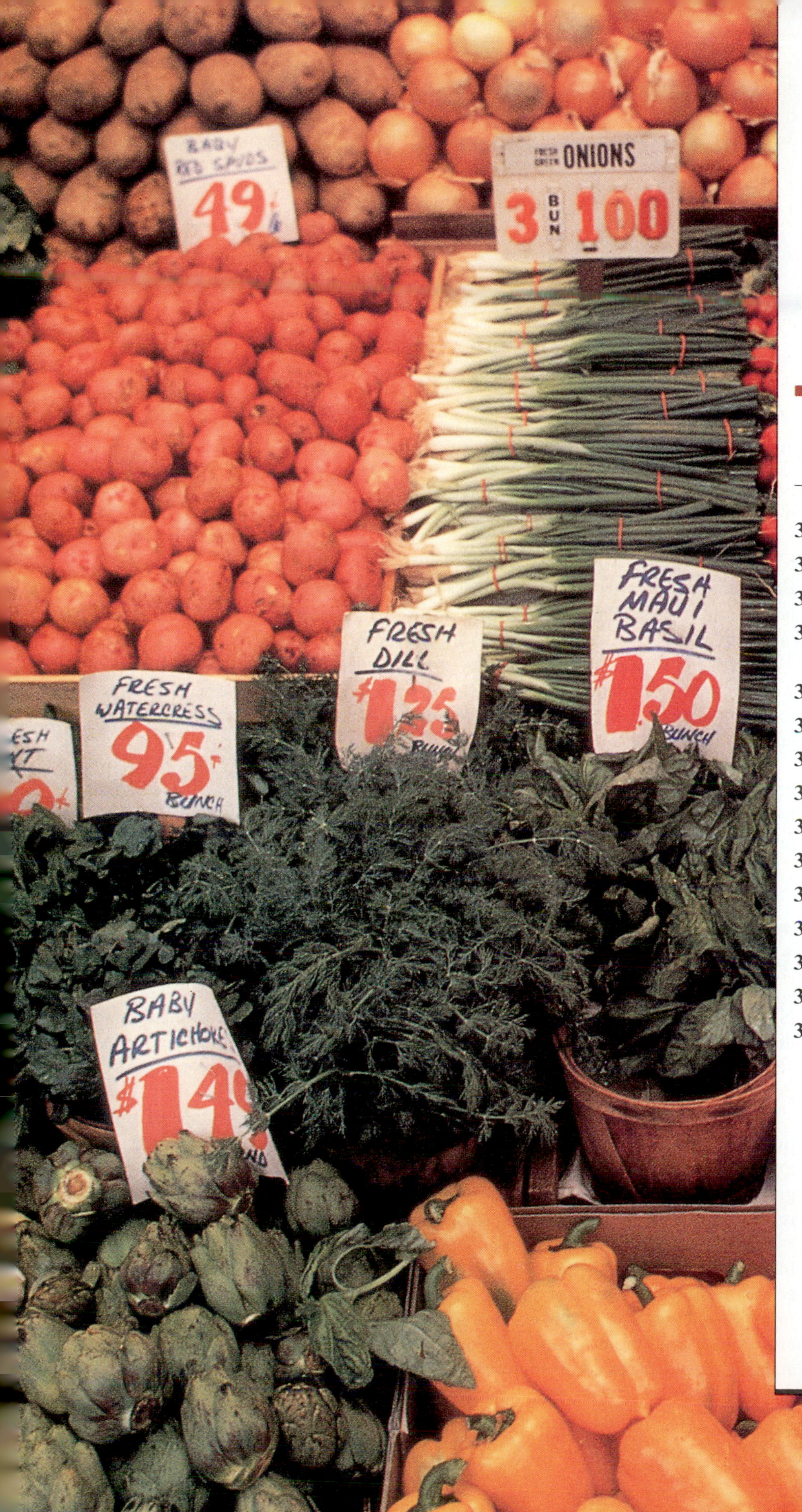

34

Nutrition

34.1 Nutrients
34.2 Diet
34.3 Energy in the Diet
34.4 Carbohydrates in the Diet
34.5 Fats in the Diet
34.6 Proteins in the Diet
34.7 Vitamins
34.8 Minerals
34.9 Water
34.10 Processed Foods
34.11 Food Additives
34.12 A Balanced Diet
34.13 Human Digestion
34.14 Absorption
34.15 Liver Function

◀ **Chapter Opening Photo: A healthy assortment of fresh vegetables is necessary to provide adequate nutrition.**

Jack Sprat could eat no fat,
his wife could eat no lean,
yet twixt the two of them,
they licked the platter clean.
—*Mother Goose*

For hundreds of years it has been known that diet is important to good health and that diets may need to vary from one individual to another. Yet only in recent years have we been able to understand nutritional needs on a molecular level. In the nursery rhyme, Jack Sprat was clearly on a low-fat diet. Today, we can point to evidence that fat intake (especially saturated fat) may be related to heart disease. Fat molecules accumulate on the blood vessel walls, which leads to hardening of the arteries.

An understanding of nutrition and digestion is closely coupled with an understanding of biochemistry. As more is learned about how diet impacts health, we as consumers can make better choices concerning which foods to purchase and eat.

34.1 Nutrients

Nutrition is a science of practical importance. Every person is concerned with diet. As a child you may have heard "Finish your milk" or "Please eat your vegetables." Do you remember being told that chocolate will cause complexion problems? How many of your friends are watching their weight?

The need for choosing the right diet has given rise to the science of nutrition, the study of nutrients—how they are digested, absorbed, metabolized, and excreted. **Nutrients** are components of the food we eat that provide for body growth, maintenance, and repair. Milk is a food, but the calcium from that milk is a nutrient. We eat meat, a food, for protein, a nutrient.

nutrient

As you study biochemistry, you learn about molecules that are required for life. In future chapters, you will learn how the cell uses and produces these molecules. Unfortunately, our cells are not self-sufficient. They require a constant input of energy, a source of carbon and nitrogen, as well as a variety of minerals and special molecules. Cells can synthesize many molecules, but they need starting materials. In addition, most cells require some molecules that they cannot synthesize. Nutrients, when digested, provide the building blocks that allow cells to make carbohydrates, lipids, proteins, and nucleic acids, as well as providing the energy for life.

Nutritionists divide nutrients into six broad classes: (1) carbohydrates, (2) lipids, (3) proteins, (4) vitamins, (5) minerals, and (6) water. The first three classes—carbohydrates, lipids, and proteins—are the major sources of the building materials, replacement parts, and energy needs of the cells. The next three classes of nutrients—water, minerals, and vitamins—have functions other than as sources of energy or building materials. Nevertheless they have vital roles in nutrition. Water is absolutely essential to every diet, because most biochemical reactions occur in aqueous solution. Minerals are required to maintain specific concentrations of certain inorganic ions in the cellular and extracellular fluids. They are also utilized in a variety of other ways—for example, calcium and phosphorus in bones and teeth and iron in hemoglobin. Vitamins are components of some enzyme systems that are vital to the cells.

Nutrients supply the needs of the body's many different cells. For example, red blood cells must produce hemoglobin and thus need iron. Muscle cells produce muscle fibers and need much protein. As they do their jobs, cells wear out and must be replaced. Red blood cells, on an average, are renewed every six weeks, whereas the cells lining the digestive tract must be replaced every three days.

The six classes of nutrients provide the basis for a healthy diet. This chapter will describe briefly (1) the relationship between diet and nutrition, (2) some characteristics of each nutrient category, and (3) the processes by which these nutrients are digested.

▲
"Fast foods" make up a large portion of many peoples daily diet.

34.2 Diet

Diet is the food and drink that we consume. Our choice of meals determines the nutrients available to our bodies. Thus, our health depends directly on our diet.

diet

Unfortunately, some dietary choices are not clear. Controversy surrounds many food selections. Are "fast foods" unhealthy? Should we avoid food additives? What are "natural" foods? To make an intelligent decision, we must understand the role of the food nutrients in maintaining good health.

Many nutrients are contained in each foodstuff. For example, a McDonalds' Big Mac contains 13% protein, 16% fat, and 20% carbohydrate. An order of french fries adds primarily carbohydrate (38%) with some fat (18%) and protein (4%). Such a meal might be supplemented with a chocolate shake (23% carbohydrate, 3% fat, 3% protein). These foods provide a variety of nutrients, but does this represent a healthful or an unhealthful diet?

Nutritionists face this difficult problem of deciding the kinds of nutrients that should be in a diet. They often establish dietary need by correlating physical well-being with nutrient consumption. James Lind (1716–1794), a physician in the British Navy during the middle of the 18th century, was one of the first to use this approach. In a study of scurvy, a disease that afflicted sailors on long voyages, he placed seamen who suffered from scurvy on various diets, some of which contained citrus fruits. By observing changes in the conditions of the seamen, Lind was able to conclude that citrus fruits provide a nutrient that prevents scurvy. This and later work eventually led to the requirement of limes and lemons in the diets of the British Merchant Marine (1765) and the British Navy (1795). Scurvy is now recognized as a deficiency disease caused by a lack of vitamin C, ascorbic acid.

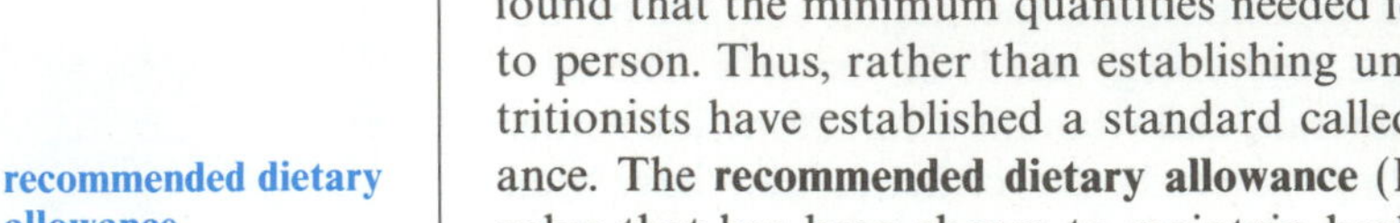

recommended dietary allowance

As scientists like Lind discovered and studied more nutrients, they also found that the minimum quantities needed for good health varied from person to person. Thus, rather than establishing uniform minimum requirements, nutritionists have established a standard called the recommended dietary allowance. The **recommended dietary allowance** (RDA) for a nutrient is an average value that has been shown to maintain health for large groups of people. Although there is considerable variation in the dietary needs of individuals, RDAs are useful yardsticks in judging healthful diets. RDAs for a variety of nutrients are shown in Table 34.1.

34.3 Energy in the Diet

An important component of every diet is the energy allowance which derives primarily from the energy-containing nutrients, the carbohydrates, lipids, and proteins. These molecules have in common that they are a rich source of reduced carbons. As we will see in Chapter 35, almost all of the energy for life is derived from reactions in which cells oxidize carbon compounds. Energy is released as each carbon is ultimately converted to carbon dioxide. Nutrients that supply energy contain carbons that can be oxidized to release energy.

The dietary energy allowance varies with activity, body size, age, and sex (see Table 34.1). Thus a 65 kg male office worker requires a diet that furnishes 2700 kcal/day, whereas a man of similar weight working as a carpenter requires about 3000 kcal/day. Women generally use less energy than men, and energy use decreases with age. [Many nutritionists still use the kilocalorie unit or its synonym, the large Calorie (Cal). Remember, there are 4.184 kilojoules per kilocalorie (or Calorie).]

▲ **A "balanced" diet includes food from each of the major groups in proper proportion.**

The balance between energy needs and the energy allowance is of vital importance. Calorie deficiency leads to a condition called *marasmus*, which affects many of the world's poor, particularly children. Marasmus is a wasting disease due to starvation. People suffering from this disease have limited diets that often consist of bulky, carbohydrate-containing foods—for example, sweet potatoes (32% carbohydrate, 1% fat, 2% protein). These foods are not energy-rich (about 160 kcal for a medium-size sweet potato). Children, with their small stomachs, have difficulty consuming enough to satisfy their energy requirements. It is estimated that 15–20% of the people in underdeveloped countries suffer from malnutrition due to insufficient calories.

In contrast, many people in developed countries consume far more calories than they need for health and well-being. Food is available in abundance, much of it rich in energy. For example, the lunch described previously consisting of a Big Mac (563 kcal), french fries (220 kcal), and a chocolate shake (383 kcal) would supply nearly one-half of the daily energy allowance for an adult male. Excess, accumulated calories lead to a condition called *obesity*, which is characterized by an overabundance of fatty tissue and by many attendant health problems.

TABLE 34.1 Recommended Daily Dietary Allowances[a] (revised 1989)

	Age (years)	Weight (kg)	Weight (lb)	Height (cm)	Height (in.)	Protein (g)	Fat-soluble vitamins: Vitamin *A* (μg RE)[b]	Vitamin *D* (μg)[c]	Vitamin *E* (mg α-TE)[d]	Water-soluble vitamins: Vitamin *C* (mg)	Thiamin (mg)	Riboflavin (mg)	Niacin (mg)	Vitamin B_6 (mg)	Folic acid (μg)	Vitamin B_{12} (μg)	Minerals: *Ca* (mg)	*P* (mg)	*Mg* (mg)	*Fe* (mg)	*Zn* (mg)	*I* (μg)
Infants	0.0–0.5	6	13	61	24	13	375	7.5	3	30	0.3	0.4	5	0.3	25	0.3	400	300	40	6	5	40
	0.5–1.0	9	20	71	28	14	375	10	4	35	0.4	0.5	6	0.6	35	0.5	600	500	60	10	5	50
Children	1–3	13	29	90	35	16	400	10	6	40	0.7	0.8	9	1.0	50	0.7	800	800	80	10	10	70
	4–6	20	44	112	44	24	500	10	7	45	0.9	1.1	12	1.1	75	1.0	800	800	120	10	10	90
	7–10	28	62	132	52	28	700	10	7	45	1.0	1.2	13	1.1	100	1.4	800	800	120	10	10	120
Males	11–14	45	99	157	62	45	1000	10	10	50	1.3	1.5	17	1.7	150	2.0	1200	1200	270	12	15	150
	15–18	66	145	176	69	59	1000	10	10	60	1.5	1.8	20	2.0	200	2.0	1200	1200	400	12	15	150
	19–24	73	160	177	70	58	1000	10	10	60	1.5	1.7	19	2.0	200	2.0	1200	1200	350	10	15	150
	25–50	79	174	176	70	63	1000	5	10	60	1.5	1.7	19	2.0	200	2.0	800	800	350	10	15	150
	51+	77	170	173	68	63	1000	5	10	60	1.2	1.4	15	2.0	200	2.0	800	800	350	10	15	150
Females	11–14	46	101	157	62	46	800	10	8	50	1.1	1.3	15	1.4	150	2.0	1200	1200	280	15	12	150
	15–18	55	120	163	64	49	800	10	8	60	1.1	1.3	15	1.5	180	2.0	1200	1200	300	15	12	150
	19–24	58	128	164	65	46	800	10	8	60	1.1	1.3	15	1.6	180	2.0	1200	1200	280	15	12	150
	25–50	63	138	163	64	50	800	5	8	60	1.0	1.3	15	1.6	180	2.0	800	800	280	15	12	150
	51+	65	143	160	63	50	800	5	8	60	1.0	1.2	13	1.6	180	2.0	800	800	280	10	12	150
Pregnant						50	800	10	10	70	1.5	1.6	17	2.2	400	2.2	1200	1200	320	30	15	175
Lactating						55	1200	10	11	90	1.6	1.8	20	2.1	260	2.6	1200	1200	350	15	17	200

Reference: *Recommended Dietary Allowances*, 10th ed. Food and Nutrition Board, National Research Council–National Academy of Sciences.

[a] The allowances are intended to provide for individual variations among most normal persons as they live in the United States under usual environmental stresses. Diets should be based on a variety of common foods in order to provide other nutrients for which human requirements have been less well defined.

[b] Retinol equivalents. 1 retinol equivalent = 1 μg retinol or 6 μg β-carotene.

[c] As cholecaliciferol. 10 μg cholecalciferol = 400 IU of vitamin D.

[d] α-Tocopherol equivalents. 1 mg α-tocopherol = 1 α-TE.

Each pound of body fat contains about 3500 kcal of energy, or between one and two days' total RDA for energy. Because fat contains so much energy, it is normally accumulated only slowly. Unfortunately, for the same reason, fat is also very difficult to lose. To lose 1 lb of fat, a 65-kg man would have to swim nonstop for about 10 hours, play tennis continuously for about 8 hours, or run for 4 hours. Exercise alone is generally not sufficient to cure obesity without a change in diet as well.

Thus, many people in the affluent, developed countries find it necessary to choose a restricted-calorie diet (to "diet"). The difficulties associated with selecting a restricted-calorie diet have led to the creation of numerous fad diets that fail to provide sound nutrition. Yet nutritionists counsel that successful and nutritionally sound dieting can be accomplished by following these simple guidelines: (1) Reduce calorie intake by only a moderate amount (a 500 kcal/day reduction causes a loss of 1 lb of fat per week) and be prepared to continue the diet for a long time; (2) carefully select foods so that the diet contains adequate amounts of all nutrients. The primary function of any diet is to maintain good health.

34.4 Carbohydrates in the Diet

Carbohydrates are a good source of usable reduced carbon atoms, which makes them important sources of dietary energy. About half of the average daily calorie requirement derives from carbohydrates. These molecules are perhaps the most easily metabolized (see Chapter 36) of the energy-supplying nutrients and can be used for energy under both aerobic and anaerobic conditions. Additionally, carbons from carbohydrates are used to build other cellular molecules—amino acids, nucleic acids, as well as other carbohydrates. Excess dietary carbohydrate is most often converted to fat.

Dietary carbohydrates are primarily the polysaccharides starch and cellulose, the disaccharides lactose and sucrose, and the monosaccharides glucose and fructose. Seeds are the most common source of starch, grains being about 70% by mass starch, whereas dried peas and beans contain about 40%. A second major source of starch is tuber and root crops such as potatoes, yams, and cassava. The disaccharide lactose is an important component of milk, and sucrose is usually consumed as refined sugar (derived from sugar beets or sugar cane). The monosaccharides are often found in fruits.

The polysaccharides, also termed *complex carbohydrates*, are difficult to digest because of their complex structures (see Section 29.14). Starch is digested only slowly, enabling the body to control distribution of this energy nutrient. Cellulose is not digested by humans; however, it is a major source of *dietary fiber*. As cellulose passes through the digestive tract, it absorbs water and provides dietary bulk. This bulk acts to prevent constipation and diverticulosis, a weakening of the intestinal walls. Many nutritionists recommend a daily fiber intake of 15–30 g, supplied by such foods as whole-wheat bread (1.8 g/slice) and bran cereals (7.5 g/0.5 cup).

Although no RDA has been set for carbohydrate, many nutritionists recommend that a minimum of about 500 kcal/day be derived from this source. In the

American diet most of these calories are derived from starch, although in recent years an increasing amount has been coming from sucrose.

The high sucrose content of many modern diets is due primarily to the large amounts of sucrose in commercially prepared foods. In 1900 an American consumed an average of about 20 lb of sucrose annually in prepared foods and beverages; by 1971 this figure had risen to about 70 lb annually. This large increase was caused mainly by (1) increased consumption of prepared foods and (2) an increased percentage of sucrose added to these foods by the manufacturers in attempts to gain larger shares of the market. (It is well known that many people, especially children, have a preference for sweet foods.) Some breakfast cereals on the market today contain more than 40% sucrose!

The increase in sucrose consumption has troubled many nutritionists for several reasons. First, sucrose is a prime factor in the incidence of dental caries. Because it is readily used by oral bacteria, sucrose promotes growth of the microorganisms that cause tooth decay. Second, ingested sucrose is rapidly hydrolyzed to monosaccharides, which are promptly absorbed from the intestine, leading to a rapid increase in blood-sugar levels. Wide variations in the blood-sugar level may stress the body's hormonal system for controlling blood sugar. Finally, sucrose is said to provide "empty calories." This means that sucrose supplies metabolic energy (calories) but lacks other nutrients. Nutritionists often recommend starch over sucrose as a major source of dietary carbohydrate.

34.5 Fats in the Diet

Fat is a more concentrated source of dietary energy than carbohydrate. Not only does fat contain more carbons per unit mass, but the carbons in fats are more reduced. As an energy source fats provide about 9 kcal/g, whereas carbohydrates provide only about 4 kcal/g. Energy can only be obtained from fat when oxygen is present as we will see in Chapter 37; fat metabolism is strictly aerobic in humans. The carbons from fats can be used to synthesize amino acids, nucleic acids, and other fats, but our bodies cannot achieve a net synthesis of carbohydrate from fat. Thus, fats tend not to be as versatile a nutrient as the carbohydrates, although fats are a more concentrated source of energy.

Fats contribute much of the dietary energy of many foods. For example, french fries contain about 18% fat, yet this fat provides about 40% of the calories in this food. A cup of whole milk has 170 kcal, but only 80 kcal comes from a cup of skim (nonfat) milk. Many nutritionists counsel that the best way to reduce calorie intake is to eat foods containing less fat.

In a diet, both the kind and the amount of fat are important. Fatty acids from meat and dairy products are relatively saturated, whereas those from plant sources are generally more unsaturated. Because there is a probable link between high consumption of saturated fats and atherosclerosis, the U. S. Department of Agriculture and other agencies concerned with nutrition have recommended that American diets should contain about equal portions of polyunsaturated and saturated fatty acids. Two ways of increasing polyunsaturated fats in the diets are to (1) cook with vegetable oils such as corn or peanut oil and (2) use soft margarine, which usually contains more unsaturated fats than hard or stick margarine or butter.

Polyunsaturated fats generally also contain three essential fatty acids—linoleic acid, linolenic acid, and arachidonic acid:

$$CH_3(CH_2)_4CH{=}CHCH_2CH{=}CH(CH_2)_7CO_2H$$

Linoleic acid

$$CH_3CH_2CH{=}CHCH_2CH{=}CHCH_2CH{=}CH(CH_2)_7CO_2H$$

Linolenic acid

$$CH_3(CH_2)_4CH{=}CHCH_2CH{=}CHCH_2CH{=}CHCH_2CH{=}CH(CH_2)_3CO_2H$$

Arachidonic acid

Each of these nutrients has been shown to relieve the deleterious physiological changes, such as poor growth, skin lesions, kidney damage, and impaired fertility, that result from a totally fat-free diet. One essential biochemical function of these fatty acids is as precursors for prostaglandin synthesis.

34.6 Proteins in the Diet

The third energy supplying nutrient is protein with an average energy yield of about 4.2 kcal/g. This energy is derived primarily from reduced carbons. However, recall that proteins also contain a large number of nitrogen atoms; in fact, protein is the primary source of nitrogen in our diets. As discussed in Chapter 37, perhaps the most important function of dietary protein is to provide for synthesis of nitrogen containing molecules such as nucleic acids, enzymes and other proteins, nerve transmitters, and many hormones. Because these materials are critical for human growth, protein malnutrition, or *kwashiorkor*, is a particularly serious problem. Unlike carbohydrates and fats (excluding the essential fatty acids), a specific RDA has been established for dietary protein (see Table 34.1).

Kwashiorkor can occur even when calorie intake is sufficient and, thus, is an especially insidious form of malnutrition. In many poverty ridden areas of the world, the only reliable source of protein for children is mother's milk. After weaning, the children eat a protein-poor grain diet. Although the calorie intake is sufficient, these children show the stunted growth, poor disease resistance, and general body wasting that are characteristic of kwashiorkor.

Proteins are obtained from animal sources such as meat, milk, cheese, and eggs, and from plant sources such as cereals, nuts, and legumes (peas, beans, and soybeans). Animal proteins have nutritive values that are generally superior to those of vegetable proteins in that they supply all of the 20 amino acids that the body uses. In contrast, a single vegetable source often lacks several amino acids. This deficiency is critical, because humans cannot synthesize the group of amino acids called the *essential amino acids* (see Section 31.3; Table 34.2). The eight essential amino acids must be obtained from the diet.

Nutritionists often speak of a dietary source (or sources) of complete protein. A **complete protein** is one that supplies all of the essential amino acids. An-

complete protein

Table 34.2 Common Dietary Amino Acids

Essential	Essential	Nonessential	Nonessential
Isoleucine	Tryptophan	Alanine	Glutamine
Leucine	Valine	Arginine	Glycine
Lysine		Asparagine	Histidine
Methionine		Aspartic acid	Proline
Phenylalanine		Cysteine	Serine
Threonine		Glutamic acid	Tyrosine

imal products are, in general, sources of complete proteins. However, by either choice or necessity, animal protein is seldom consumed by a significant fraction of the world's population. Besides the moral, ethical, and religious reasons given for limiting consumption of animal protein, these proteins are relatively expensive sources of amino acids. Thus, most of the people of the underdeveloped countries subsist primarily on vegetable protein. A constant danger inherent in a vegetarian diet is that the source of vegetable protein may be deficient in several of the essential amino acids; that is, the protein may be incomplete. For this reason nutritionists recommend that several sources of vegetable protein be included with each meal in a vegetarian diet. As an example, soybeans, which are rich in lysine, might supplement wheat, which is lysine deficient.

34.7 Vitamins

Vitamins are a group of naturally occurring organic compounds that are essential for good nutrition and must be supplied in the diet. Whereas the energy supplying nutrients are digested and metabolized extensively, vitamins are often used after only minimal modification. Some of the vitamins necessary for humans are listed in Table 34.3. Note that vitamins are often classified according to their solubility, those that are fat soluble and those that are water soluble. The structural formulas of several vitamins are shown in Figure 34.1.

vitamin

A prolonged lack of vitamins in the diet leads to vitamin deficiency diseases such as beriberi, pellagra, pernicious anemia, rickets, and scurvy. Left uncorrected, a vitamin deficiency ultimately results in death. Even when supplementary amounts of vitamins are provided, impaired growth due to a vitamin deficiency may be irreversible. For example, it is difficult to correct the distorted bone structures resulting from a childhood lack of vitamin D. Thus, it is especially important that children receive sufficient vitamins for proper growth and development.

Although vitamins are required in only small amounts (see the RDAs in Table 34.1), the biochemistry of life cannot continue without them. Each vitamin serves at least one specific purpose for an organism. The water soluble compounds are generally involved in cellular metabolism of the energy supplying nutrients. For example, thiamin is required to achieve a maximum energy yield

TABLE 34.3 Some of the Most Important Vitamins

Vitamin	Important dietary sources	Some deficiency symptoms
FAT SOLUBLE		
Vitamin A (Retinol)	Green and yellow vegetables, butter, eggs, nuts, cheese, fish liver oil	Poor teeth and gums, night blindness
Vitamin D (Ergocalciferol, D_2; cholecalciferol, D_3)	Egg yolk, milk, fish liver oils; formed from provitamin in the skin when exposed to sunlight	Rickets (low blood-calcium level, soft bones, distorted skeletal structure)
Vitamin E (α-Tocopherol)	Meat, egg yolk, wheat germ oil, green vegetables; widely distributed in foods	Not definitely known in humans
Vitamin K (Phylloquinone, K_1; menaquinone, K_2)	Eggs, liver, green vegetables; produced in the intestines by bacterial reactions	Blood is slow to clot (antihemorrhagic vitamin)
WATER SOLUBLE		
Vitamin B_1 (Thiamin)	Meat, whole-grain cereals, liver yeast, nuts	Beriberi (nervous system disorders, heart disease, fatigue)
Vitamin B_2 (Riboflavin)	Meat, cheese, eggs, fish, meat products, liver	Sores on the tongue and lips, bloodshot eyes, anemia
Vitamin B_6 (Pyridoxine)	Cereals, liver, meat, fresh vegetables	Skin disorders (dermatitis)
Vitamin B_{12} (Cyanocobalamin)	Meat, eggs, liver, milk	Pernicious anemia
Vitamin C (Ascorbic acid)	Citrus fruits, tomatoes, green vegetables	Scurvy (bleeding gums, loose teeth, swollen joints, slow healing of wounds, weight loss)
Niacin (Nicotinic acid and amide)	Meat, yeast, whole wheat	Pellagra (dermatitis, diarrhea, mental disorders)
Biotin (Vitamin H)	Liver, yeast, egg yolk	Skin disorders (dermatitis)
Folic acid	Liver extract, wheat germ, yeast, green leaves	Macrocytic anemia, gastrointestinal disorders

from carbohydrates, and pyridoxine is of central importance in protein metabolism (see Chapters 36 and 37). Niacin and riboflavin are key components in almost all cellular redox reactions. The fat soluble vitamins often serve very specialized functions. Vitamin D acts as a regulator of calcium metabolism. One function of vitamin A is to furnish the pigment that makes vision possible, while vitamin K enables blood clotting to occur normally.

Because some vitamin functions are not well understood, miraculous properties have been ascribed to these substances, such as vitamins C and E. In the absence of conclusive scientific studies, it is difficult to judge the merits of some of these claims.

FIGURE 34.1 **The structures of selected vitamins.**

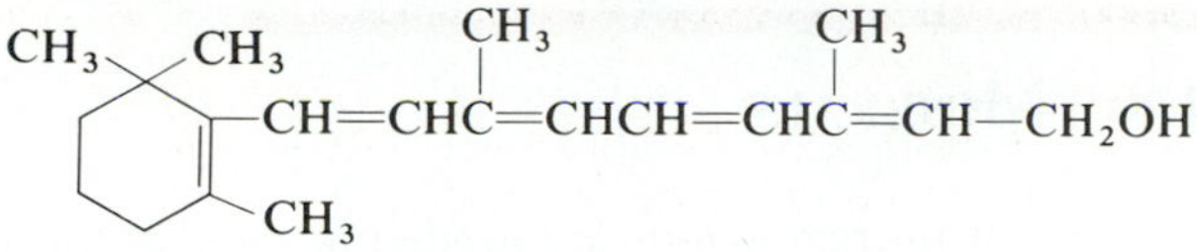

Vitamin A (Retinol)

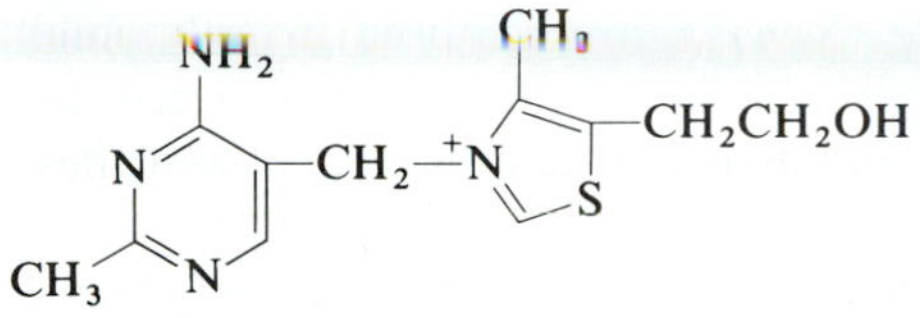

Vitamin B_1 (Thiamin)

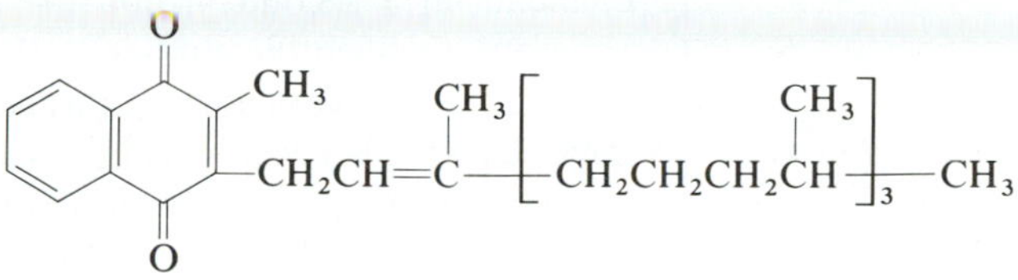

Vitamin K_1 (Phylloquinone)

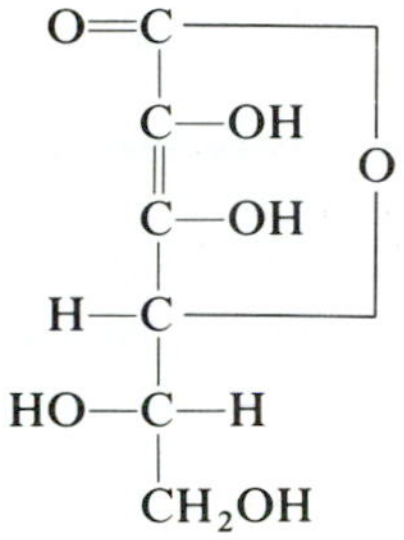

Vitamin C (Ascorbic acid)

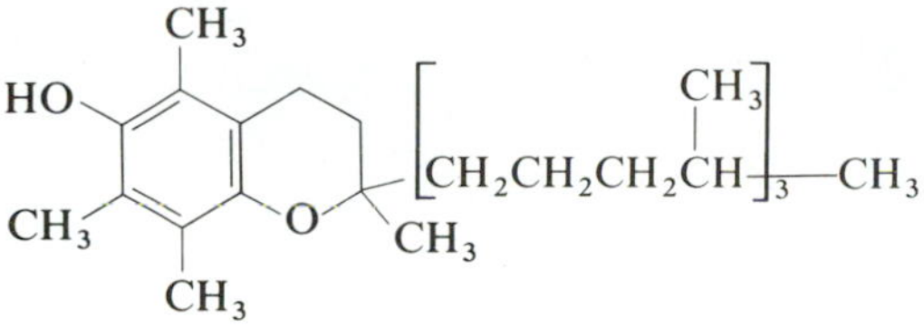

Vitamin E (α-Tocopherol)

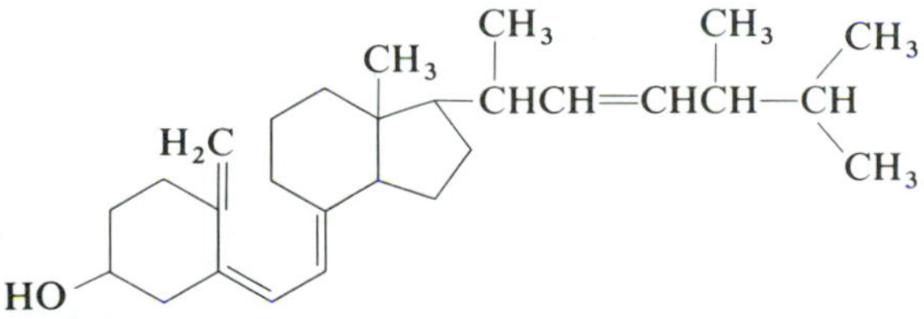

Vitamin D (Ergocalciferol, D_2)

Although vitamins are required only in small quantities, their natural availability is also low. Nutritionists caution that a diet should be balanced to include adequate sources of vitamins. As you can see from Table 34.3, the dietary sources of vitamins are varied. In general, fruits, vegetables, and meats are rich sources of the water soluble vitamins; and eggs, milk products, and liver are good sources of the fat soluble vitamins.

A seemingly balanced diet may be deficient in vitamins due to losses incurred in food processing, storing, and cooking. As much as 50–60% of the water soluble vitamins in vegetables can be lost during cooking. Vitamin C, for example, can be destroyed by exposure to air. Removal of the outside hull from grains drastically decreases their B vitamin content. Thus, when polished rice became a dietary staple in the Orient, beriberi (the thiamin-deficiency disease) grew to epidemic proportions.

34.8 Minerals

A number of inorganic ions are needed in the diet for good health. Those that must be ingested in relatively large amounts, the *major elements*, include sodium, potassium, chloride, calcium, magnesium, and phosphate (phosphorus). Elements in a second group, the *trace elements*, are required in only small amounts. As scientific studies of the body's mineral requirements proceed, the list of needed trace elements constantly lengthens. A recent compilation of these trace elements is given in Table 34.4.

Mineral nutrients differ from organic nutrients in that the body, in general, uses minerals in the ionic form in which they are absorbed. Although these elements are required for good health, they can also be toxic if ingested in quan-

Table 34.4 Essential Trace Elements

Element	Function	Human deficiency signs
Fluorine	Structure of teeth, possibly of bones; possible growth effect	Increased incidence of dental caries; possibly risk factor for osteoporosis
Silicon	Calcification; possible function in connective tissue	Not known
Vanadium	Not known	Not known
Chromium	Efficient use of insulin	Relative insulin resistance, impaired glucose tolerance, elevated serum lipids
Manganese	Mucopolysaccharide metabolism, superoxide dismutase enzyme	Not known
Iron	Oxygen and electron transport	Anemia
Cobalt	Part of vitamin B_{12}	Only as vitamin B_{12} deficiency
Nickel	Interaction with iron absorption	Not known
Copper	Oxidative enzymes; interaction with iron; cross-linking of elastin connective protein	Anemia, changes of ossification; possibly elevated serum cholesterol
Zinc	Numerous enzymes involved in energy metabolism and in transcription and translation	Growth depression, sexual immaturity, skin lesions, depression of immunocompetence, change of taste acuity
Arsenic	Not known	Not known
Selenium	Glutathione peroxidase; interaction with heavy metals	Endemic heart problems conditioned by selenium deficiency
Molybdenum	Xanthine, aldehyde and sulfide oxidase enzymes	Not known
Iodine	Constituent of thyroid hormones	Goiter, depression of thyroid function, cretinism

tities that are too large. Nutritionists have established RDAs for some of these minerals and warn against excess intake.

The major minerals—sodium, potassium, and chloride—are responsible for maintaining the appropriate salt levels in body fluids. Many enzyme reactions require an optimum salt concentration of 0.1 to 0.3 *M*. In addition, individual elements serve specific functions; nerve transmission, for example, requires a supply of extracellular sodium and intracellular potassium. Although there is no RDA for these two elements, 2–4 g/day represents an average NaCl consumption. But because high levels of NaCl can contribute to high blood pressure, many nutritionists advise using salt in moderation.

Calcium and magnesium serve many roles in the body, including being required by some enzymes. Fully 90% of all body calcium and an important percentage of the magnesium are found in the bones and teeth. Calcium also is needed for nerve transmission and blood clotting.

Trace elements are similar to vitamins in that (1) they are required in small amounts and (2) food contains only minute quantities of them. Usually a normal diet contains adequate quantities of all trace elements. The exact function of many trace elements remains unknown.

The most notable exception to the foregoing generalizations is iron. This element is part of hemoglobin, the oxygen-binding protein of the blood. (Iron is also a critical component of some enzymes.) Relatively large amounts of iron are required for hemoglobin replenishment. Unfortunately, many foods are not rich enough in iron to provide the necessary RDA. This is especially true for the higher iron RDA for women. Therefore nutritionists sometimes recommend a daily iron dietary supplement.

34.9 WATER

Water is the solvent of life. As such it carries nutrients to the cells, allows biochemical reactions to proceed, and carries waste from the cells. Water makes up approximately 55–60% of the total body mass.

Our bodies are constantly losing water via urine (700–1400 mL/day), feces (150 mL/day), sweat (500–900 mL/day), and expired air (400 mL/day). If dehydration is to be prevented, this water output must be offset by water intake. In the normal diet, liquids provide about 1200–1500 mL of water per day. The remaining increment is derived from food (700–1000 mL/day) and water formed by biochemical reactions (metabolic water; 200–300 mL/day). Water losses vary and depend on the intake and the activity of the individual.

A proper water balance is maintained by control of water intake and water excretion. The kidney is the center for control of water output. Water intake is controlled by the thirst sensation. We feel thirsty when one or both of the two following events occur: (1) As the body's water stores are depleted, the salt concentration rises. Specific brain cells monitor this salinity and initiate the thirst feeling. (2) As water is drawn away from the salivary glands, the mouth feels dry; this event also leads to the sensation of thirst.

34.10 Processed Foods

Nearly all of the foods available in a supermarket today have been processed to various degrees. The processing may be very extensive with items such as luncheon meats, TV dinners, breakfast cereals, and so on. Loss of nutrients is a common risk associated with processing.

The short history of flour milling is an apt illustration of the risks of too much processing: In 1840 bread had a rough consistency and the color characteristic of stone-ground wheat flour. Over the years vast improvements in milling technology were made. In 1940 bread had a soft, smooth consistency and the white color characteristic of flour made from only the starchy portion of wheat. Consumers generally preferred this bread over the coarse and sometimes gritty product of previous centuries. It was discovered about this time (1940) that the bread had a very low vitamin content and needed vitamin enrichment. Furthermore, in the early 1970s, it was found that the amount of dietary fiber in bread can be greatly increased by less selective milling of the wheat. Excessive processing or milling of the wheat had decreased the nutritive value of bread. The most nutritious bread being baked today is very similar to that made from the minimally processed flour of 1840.

Advocates of "natural," "organic," or "health" foods argue that foods should be minimally affected by modern technology to ensure maximum nutrition. They correctly point out the dangers of nutrient loss during food processing. Yet in the eyes of many consumers in today's busy world, the convenience of ready-prepared foods greatly outweighs any possible nutritional risks associated with their use.

Nutritionists have faced this dilemma by establishing procedures for monitoring nutritional values of processed foods. If a food contains an added nutrient (as do most processed foods) or if a food makes a nutritional claim, then the package must contain a nutritional information panel using the following format:

—Serving or portion size
—Servings or portions per container
—Calorie content per serving
—Protein, grams per serving
—Carbohydrate, grams per serving
—Fat, grams per serving
—Proteins, vitamins, and minerals as percentages of the USRDA.

By using this information, together with a basic knowledge of the nutrients needed for good health, the consumer can make educated choices among processed foods.

34.11 Food Additives

Various substances or chemicals may be added to foods during processing. In fact, more than 3000 of these *food additives* have been given the "generally recognized as safe" (GRAS) rating by the U.S. Food and Drug Administration.

Table 34.5 Some Common Food Additives

Food additive	Purpose
Sodium benzoate Calcium lactate Sorbic acid	Prevention of food spoilage (antimicrobials)
BHA (Butylated hydroxyanisole) BHT (Butylated hydroxytoluene) EDTA (Ethylenediaminetetraacetic acid)	Prevention of changes in color and flavor (antioxidants)
Calcium silicate Silicon dioxide Sodium silicoaluminate	Keeps powders and salt free-flowing (anticaking agents)
Carrageenan Lecithin	Aids even distribution of suspended particles (emulsifiers)
Pectin Propylene glycol	Imparts body and texture (thickeners)
MSG (Monosodium glutamate) Hydrolyzed vegetable protein	Supplements or modifies taste (flavor enhancers)

The purpose of these food additives varies. Some additives enhance the nutritional value of a food, such as when a food is vitamin-enriched. Other additives serve as preservatives. For example, sodium benzoate is used to inhibit bacterial growth. BHA (butylated hydroxyanisole) and BHT (butylated hydroxytoluene) are used as antioxidants. Still other additives may be used to improve the appearance and flavor of a food—for example, emulsifiers, thickeners, anticaking agents, flavors, flavor enhancers, nonsugar sweeteners, and colors (see Table 34.5).

There has been and continues to be a great deal of controversy concerning the use of food additives. Due to the nature and complexity of the subject, there is no doubt that this debate will continue into the foreseeable future. Controversy centers around the problem of balancing the benefit derived from an additive against the risk to consumers. The use of at least some additives is a necessity in the preparation of many foods. To discover that there is a risk and to assess the degree of risk for an additive requires long, difficult, and expensive research. As a case in point, salting and smoking were used in curing meats for centuries before there was any knowledge that these processes might involve a hazard. Research has shown that both processes involve risks to at least some consumers: Salt aggravates certain cardiovascular conditions, and smoke produces carcinogens in the meat. Yet in the eyes of many consumers, these risks are outweighed by the benefits to be had from salted and smoked meats. On the other hand, there would be few consumers, indeed, who would wish to have a compound known to be very carcinogenic used as an additive in their ice cream, even though that compound could improve the flavor of the ice cream remarkably!

Owing to the technical nature of the task, the consumer must rely largely on the judgment and integrity of the professional people who have been assigned the responsibility of protecting our food supply. Consumers also should (1) inform themselves as fully as possible concerning the nature, purpose, and possible hazards of the additives that are used or proposed for us in our foods and (2) as responsible citizens make sure that our government provides adequate support to professionals charged with protecting our food.

34.12 A Balanced Diet

To summarize the preceding sections, carbohydrates, fats, and proteins are a major group of nutrients in our diets. These nutrients supply the molecules that are needed for energy, growth, and maintenance. A typical adult diet includes about 100–200 g each of carbohydrate, fat, and protein daily.

A second group of nutrients—vitamins, minerals, and water—is not used for energy but is nevertheless essential to our existence. Vitamins provide organic molecules that cannot be made in the body; and minerals provide the inorganic ions needed for life. Water is the solvent in which most of the chemical reactions essential to life occur. Vitamins and minerals are required only in small amounts, from a few micrograms to a few milligrams per day. But water must be consumed in large quantities, 2 to 3 liters per day.

For health and well-being, each of six groups of nutrients must be in our diet to satisfy the RDAs given in Table 34.1. With the variety of foods that are available, how can we make sure that our diets contain enough of all the needed nutrients? The answer "Eat a balanced diet" leads to another question, "How can we make sure that our diet is balanced?"

To answer this question, nutritionists have divided foods into six groups: (1) milk products, (2) vegetables, (3) fruits, (4) cereal products, (5) meats, and (6) fatty foods. Each group is a good source of one or more nutrients. To obtain a balanced mixture of carbohydrate, fat, and protein, a diet should contain food from several classes. A balanced diet also assures an adequate supply of essential minerals and vitamins.

A balanced diet must include several food groups, even though each food may contain many nutrients. For example, compare the three breakfasts in Figure 34.2, each of which supplies about 600 kcal of food energy. By choosing only doughnuts (cereal food group) and coffee, the consumer would gain some nutrients but only in small quantities. A breakfast of cold cereal (cereal group) and milk (milk group), toast (cereal group) with margarine (fatty food group) and jelly, and coffee includes more food groups. Still more food groups are found in a breakfast of orange juice (fruit group), a fried egg (meat group), pancakes (cereal group) with margarine (fatty food group) and syrup, milk (milk group) and coffee. Many other breakfasts could be chosen instead, yet an important generalization can be drawn from these three examples: The nutritional value of a meal improves if at least several food groups are included. The third breakfast in this illustration is the most balanced and provides the best nutrition. Although the consumer often does not know the nutrient content of a specific food, overall nutrition can be ensured by choosing a balanced diet.

After the food composing a balanced diet is eaten, it must be digested, absorbed, and transported in order for the proper nutrients to reach the cells.

◀ FIGURE 34.2
A comparison of nutritional values for three breakfast. (No RDA has been established for carbohydrates and fats.)

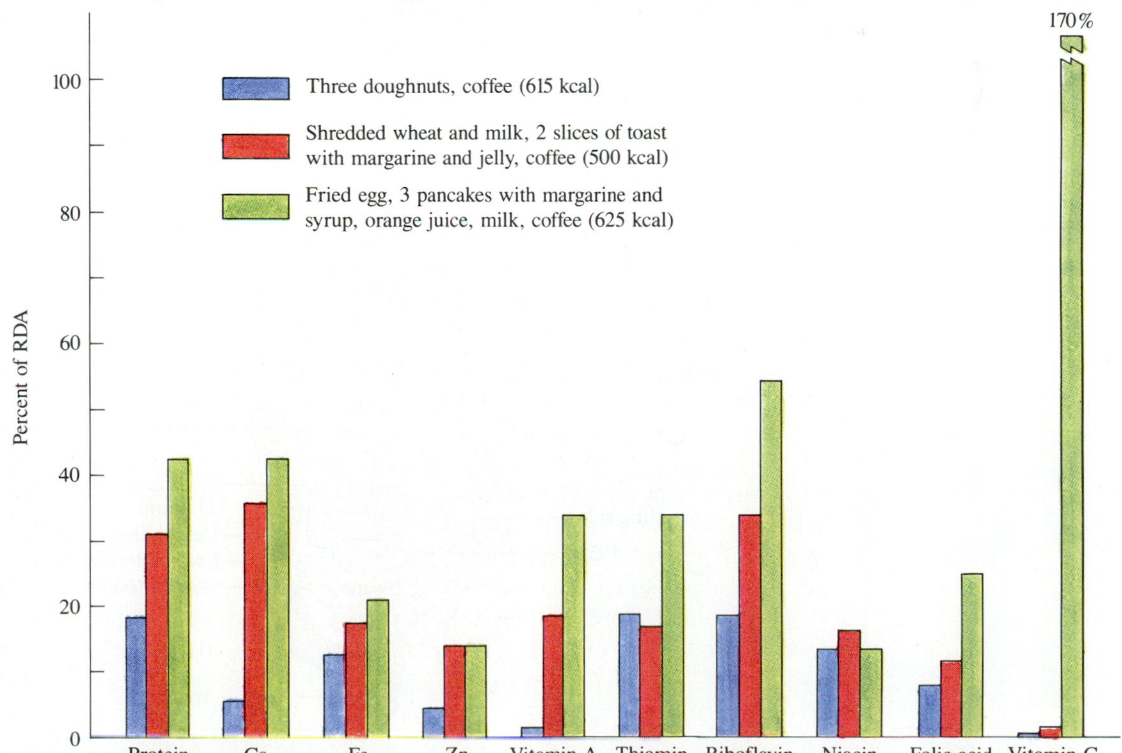

Eating puts food into the alimentary canal (which includes the mouth, esophagus, stomach, small intestine, and large intestine). In a sense this canal is an extension of our external environment; that is, until food is broken down and processed, it cannot enter our internal environment to reach the cells. After digestion and absorption, the nutrients are transported to the cells by the blood and lymph systems. The liver has a vital role in controlling nutrient levels in the blood and neutralizing toxic substances.

34.13 Human Digestion

The human digestive tract is shown diagrammatically in Figure 34.3. Although food is broken up mechanically by chewing and by a churning action in the stomach, digestion is a chemical process. **Digestion** is a series of enzyme-catalyzed reactions by which large molecules are hydrolyzed to molecules small enough to be absorbed through the intestinal membranes. Foods are digested to smaller

digestion

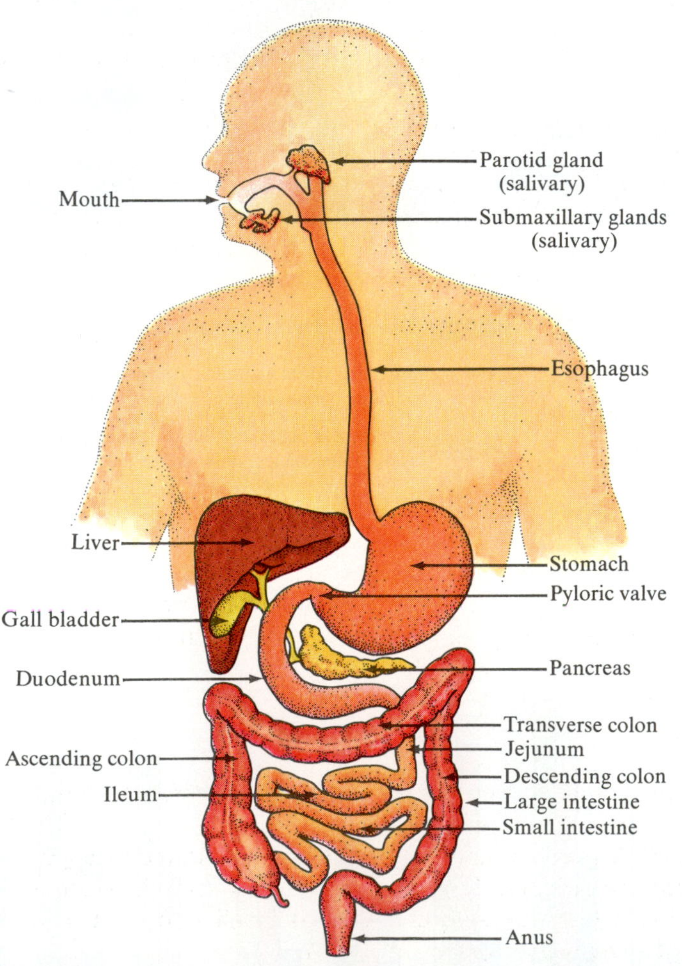

Figure 34.3 ▶ The human digestive tract.

molecules according to this general scheme:

Carbohydrates ⟶ Monosaccharides

Fats ⟶ { Fatty acids; Glycerol; Mono- and diesters of glycerol }

Proteins ⟶ { Amino acids; Dipeptides; Tripeptides }

Food passes through the human digestive tract in this sequence:

Mouth ⟶ esophagus ⟶ stomach ⟶ small intestine (duodenum, jejunum, and ileum) ⟶ large intestine (see Figure 34.3).

Five principal digestive juices (or fluids) enter the digestive tract at various points:

1. Saliva from three pairs of salivary glands in the mouth
2. Gastric juice from glands in the walls of the stomach
3. Pancreatic juice, which is secreted by the pancreas and enters the duodenum through the pancreatic duct
4. Bile, which is secreted by the liver and enters the duodenum via a duct from the gall bladder
5. Intestinal juice from intestinal mucosal cells

The main functions and principal enzymes found in each of these fluids are summarized in Table 34.6. The important digestive enzymes occur in gastric, pancreatic, and intestinal juices. An outline of the digestive process follows. (Detailed accounts of the various stages of digestion are to be found in biochemistry and physiology texts.)

Salivary Digestion Food is chewed (masticated) and mixed with saliva in the mouth, and the hydrolysis of starch begins. The composition of saliva depends on many factors—age, diet, condition of teeth, time of day, and so on. Normal saliva is about 99.5% water. Saliva also contains mucin (a glycoprotein); a number of mineral ions such as K^+, Ca^{2+}, Cl^-, PO_4^{3-}, SCN^-; and one enzyme, salivary amylase (ptyalin). The pH of saliva ranges from slightly acidic to slightly basic, with the optimum pH about 6.6–6.8.

Mucin acts as a lubricant and facilitates the chewing and swallowing of food. The enzyme salivary amylase catalyzes the hydrolysis of starch to maltose:

$$\text{Starch} + \text{Water} \xrightarrow{\text{Salivary amylase}} \text{Maltose}$$

Salivary amylase is inactivated at a pH of 4.0, so it has very little time to act before the food reaches the highly acidic stomach juices.

Saliva is secreted continuously, but the rate of secretion is greatly increased by the sight and odor, or even the thought, of many foods. The mouth-watering effect of the sight or thought of pickles is familiar to most of us. This is an example of a *conditioned reflex*.

TABLE 34.6 Digestive Fluids

Fluid (volume produced daily)	Source	Principal enzymes and/or function
Saliva (1000–1500 mL)	Salivary glands	Lubricant, aids chewing and swallowing; also contains salivary amylase (ptyalin), which begins the digestion of starch
Gastric juice (2000–3000 mL)	Glands in stomach wall	Pepsin, gastric lipase; pepsin catalyzes partial hydrolysis of proteins to proteoses and peptones in the stomach
Pancreatic juice (500–800 mL)	Pancreas	Trypsinogen, chymotrypsinogen, procarboxypeptidase (converted after secretion to trypsin, chymotrypsin, and carboxypeptidase, respectively, which continue protein digestion); amylopsin (α-amylase, a starch-digestive enzyme), steapsin (a lipase)
Bile (500–1000 mL)	Liver	Contains no enzymes; but contains bile salts, which aid digestion by emulsifying lipids; serves to excrete cholesterol and bile pigments derived from hemoglobin
Intestinal juice	Intestinal mucosal cells	Contains a variety of finishing enzymes: sucrase, maltase, and lactase for carbohydrates; aminopolypeptidase and dipeptidase for final protein breakdown; intestinal lipase; nucleases and phosphatase for hydrolysis of nucleic acids

Gastric Digestion When food is swallowed, it passes through the esophagus to the stomach. In the stomach, mechanical action continues; food particles are reduced in size and are mixed with gastric juices until a material of liquid consistency, known as *chyme*, is obtained.

Gastric juice is a clear, pale yellow, acidic fluid having a pH of about 1.5–2.5. It contains hydrochloric acid; the mineral ions Na^+, K^+, Cl^-, and some phosphates; and the digestive enzymes pepsin and lipase. The flow of gastric juice is accelerated by conditioned reflexes and by the presence of food in the stomach. The secretion of the hormone *gastrin* is triggered by food entering the stomach. This hormone, which is produced by the gastric glands, is absorbed into the bloodstream and returned to the stomach wall where it stimulates the secretion of additional gastric juice. Control of gastric secretion by this hormone is an example of one of the many chemical control systems that exist in the body.

The chief digestive function of the stomach is the partial digestion of protein. The principal enzyme of gastric juice is pepsin, which digests protein. The enzyme is secreted in an inactive form called pepsinogen, which is activated by hydrochloric acid to pepsin. Pepsin catalyzes the hydrolysis of proteins to fragments called proteoses and peptones, which are still fairly large molecules. Pepsin splits the peptide bonds adjacent to only a few amino acid residues, particularly tyrosine and phenylalanine:

$$\text{Protein} + \text{Water} \xrightarrow{\text{Pepsin}} \text{Proteoses} + \text{Peptones}$$

The second enzyme in the stomach, gastric lipase, is a fat digesting enzyme. Its action in the stomach is slight because the acidity is too high for lipase activity. Food may be retained in the stomach for as long as 6 hours. It then passes through the pyloric valve into the duodenum.

Intestinal Digestion The next section of the digestive tract, the small intestine, is where most of the digestion occurs. The stomach contents are first made alkaline by secretions from the pancreatic and bile ducts. The pH of the pancreatic juice is 7.5–8.0, and the pH of bile is 7.1–7.7. The shift in pH is necessary because the enzymes of the pancreatic and intestinal juices are active only in an alkaline medium. Enzymes that digest all three kinds of food—carbohydrates, fats, and proteins—are secreted by the pancreas. Pancreatic secretion is stimulated by hormones that are secreted into the bloodstream by the duodenum and the jejunum.

The enzymes occurring in the small intestine include pancreatic amylases (diastase) that hydrolyze most of the starch to maltose, and carbohydrases (α-amylase, maltase, sucrase, and lactase) that complete the hydrolysis of disaccharides to monosaccharides. The proteolytic enzymes trypsin and chymotrypsin attack proteins, proteoses, and peptones, hydrolyzing them to dipeptides. Then the peptidases—carboxypeptidase, aminopeptidase, and dipeptidase—complete the hydrolysis of proteins to amino acids. Pancreatic lipases catalyze the hydrolysis of almost all fats. Fats are split into fatty acids, glycerol, and mono- and diesters of glycerol by these enzymes.

The liver is another important organ in the digestive system. A fluid known as bile is produced by the liver and stored in the gall bladder, a small organ located on the surface of the liver. When food enters the duodenum, the gall bladder contracts, and the bile enters the duodenum through a duct that is also used by the pancreatic juice. In addition to water, the major constituents of the bile are bile acids (as salts), bile pigments, inorganic salts, and cholesterol. The bile acids are steroid monocarboxylic acids, two of which are shown below:

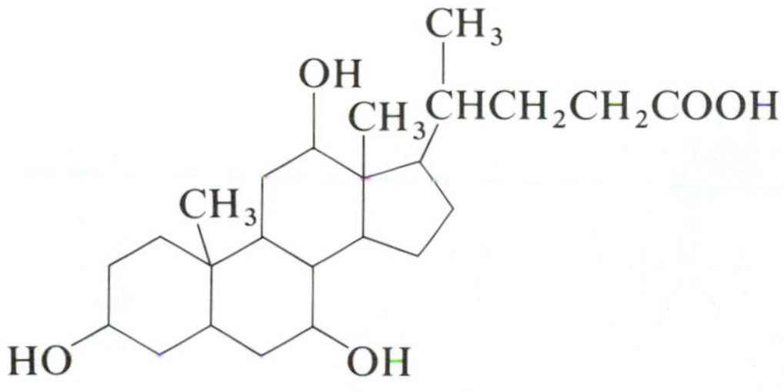

CH_3, OH, $CH_3CHCH_2CH_2COOH$, CH_3, HO, OH

Cholic acid

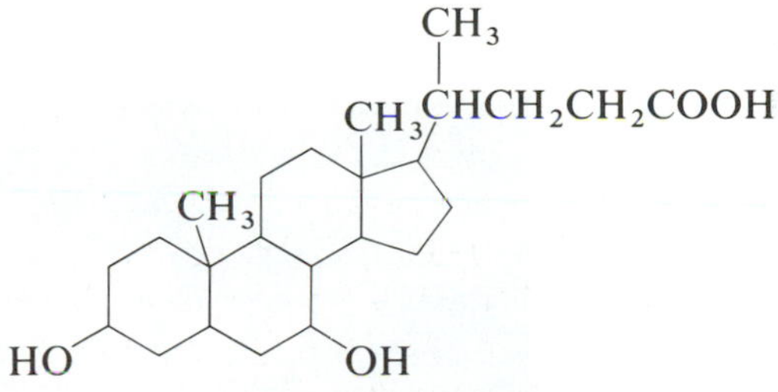

CH_3, $CH_3CHCH_2CH_2COOH$, CH_3, HO, OH

Chenodeoxycholic acid

The bile acids are synthesized in the liver from cholesterol, which is also synthesized in the liver. The presence of bile in the intestine is important for the digestion and absorption of fats. When released into the duodenum, the bile acids emulsify the fats, allowing them to be hydrolyzed by the pancreatic lipases. About 90% of the bile salts are reabsorbed in the lower part of the small intestine and are transported back to the liver and used again.

Most of the digested food is absorbed from the small intestine. Undigested and indigestible material passes from the small intestine to the large intestine, where it is retained for varying periods of time before final elimination as feces. Additional chemical breakdown, sometimes with the production of considerable amounts of gases, is brought about by bacteria (or rather by bacterial enzymes) in the large intestine. For a healthy person this additional breakdown is not important from the standpoint of nutrition since absorption of nutrients does not occur from the large intestine. However, large amounts of water, partly from digestive juices, are absorbed from the large intestine so that the contents become more solid before elimination.

34.14 Absorption

absorption

For digested food to be utilized in the body, it must pass from the intestine into the blood and lymph systems. The process by which digested foods pass through the membrane linings of the small intestine and enter the blood and lymph is called **absorption**. Absorption is complicated, and we shall consider only an overview of it.

After the food you have eaten is digested, the body must absorb billions upon billions of nutrient molecules into the bloodstream. The absorption system is in the membranes of the small intestine, which, upon microscopic inspection, are seen to be wrinkled into hundreds of folds. These folds are covered with thousands of small projections called *villi.* Each projection is itself covered by many minute folds called *microvilli.* The small intestine's wrinkles, folds, and projections increase the surface area available for absorption. The average small intestine is about 4 meters (13.3 ft) long and is estimated to contain about 8360 square meters (90,000 sq ft) of absorbing surface.

The inner surface of the small intestine is composed of mucosal cells that produce many enzymes, such as disaccharidases, aminopeptidases, and dipepti-

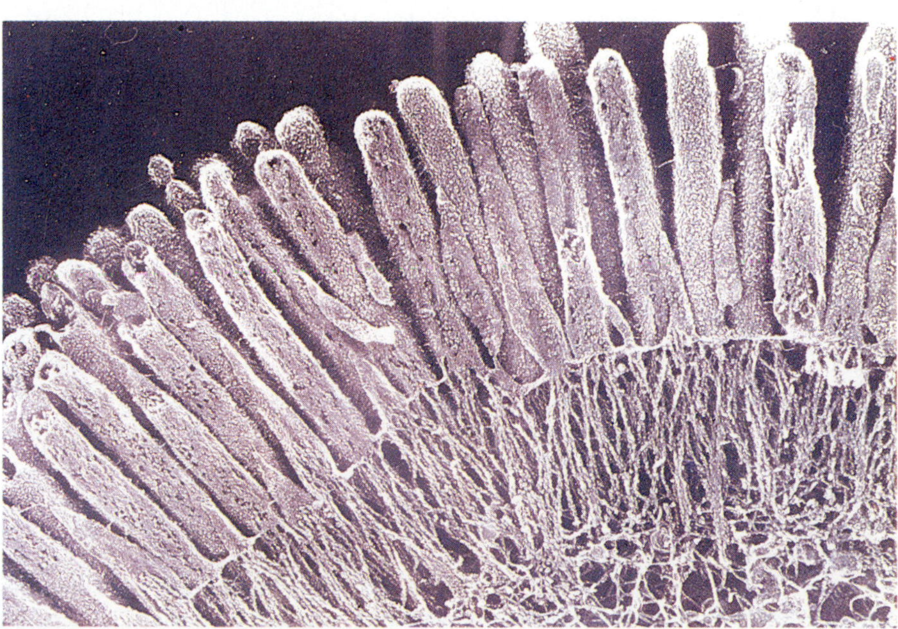

Absorption of digested food ► occurs in the small intestine through the villi and microvilli shown here.

dases, needed to complete the digestive process. As digestion is completed, the resulting nutrient molecules are absorbed by the mucosal cells and transferred to the blood and lymph systems. Water soluble nutrients such as monosaccharides, glycerol, short chain fatty acids, amino acids, and minerals enter directly into the bloodstream. Fat soluble nutrients such as long chain fatty acids and monoacylglycerols first enter the lymph fluid and then enter the bloodstream where the two fluids come together.

An important factor in the absorption process is that the membranes of the intestine are selectively permeable; that is, they prevent the passage of most large molecules but allow the passage of smaller molecules. For example, polysaccharides, disaccharides, and proteins are not ordinarily absorbed, but generally monosaccharides and amino acids are.

34.15 Liver Function

The liver is the largest organ in the body and performs several vital functions. Two of these functions are (1) the regulation of the concentrations of organic nutrients in the blood and (2) the removal of toxic substances from the blood.

The concentration of blood sugar (glucose) is controlled and maintained by processes that occur in the liver. After absorption, excess glucose and other monosaccharides are removed from the blood and converted to glycogen in the liver. The liver is the principal storage organ for glycogen. As glucose is used in other cells, the stored liver glycogen is gradually hydrolyzed to maintain the appropriate blood-glucose concentration. Liver function is under sensitive hormonal control, as are most vital body functions. This regulation will be discussed further in Chapter 36.

A second major function of the liver is the detoxification of harmful and potentially harmful substances. This function apparently developed as higher vertebrates appeared in the evolutionary time scale. The liver is able to deal with most of the toxic molecules that occur in nature. For example, ethanol is oxidized in the liver and nitrogenous metabolic waste products are converted to urea for excretion.

Organic chemists have learned to synthesize new substances that have no counterparts in the biological world. They are used as industrial chemicals, insecticides, drugs, and food additives. When these potentially toxic substances are ingested, even in small amounts, the body is faced with the difficult challenge of metabolizing or destroying substances that are unlike any found in nature.

The liver is actually able to meet this challenge and deal with most of these foreign molecules through an oxidation system located on the endoplasmic reticulum. Bound to these intracellular membranes are enzymes that can catalyze reactions between oxygen and the foreign molecules. As these molecules are oxidized, they become more polar and water soluble. Finally, the oxidation products of the potential toxins are excreted in the urine or bile fluid.

For example, most automobile antifreeze solutions contain ethylene glycol, a toxic substance. Even though this compound does not occur naturally, the liver can metabolize ethylene glycol. When small amounts are ingested accidentally, the following chemical changes occur:

$$\underset{\text{Ethylene glycol}}{\mathrm{H{-}C(OH)(H){-}C(OH)(H){-}H}} \xrightarrow{\text{Oxidation}} \underset{\text{Glyoxal}}{\mathrm{H{-}C(OH)(H){-}C({=}O){-}H}} \xrightarrow{\text{Oxidation}} \underset{\text{Glycolate}}{\mathrm{H{-}C(OH)(H){-}C({=}O){-}O^-}} \xrightarrow{\text{Oxidation}}$$

$$\underset{\text{Glyoxalate}}{\mathrm{H{-}C({=}O){-}C({=}O){-}O^-}} \xrightarrow{\text{Oxidation}} \underset{\text{Formate}}{\mathrm{H{-}C({=}O){-}O^-}} + \underset{\text{Bicarbonate}}{\mathrm{HO{-}C({=}O){-}O^-}}$$

Thus ethylene glycol is converted by oxidation to two more polar (charged) acid anions that are easily eliminated from the body.

Unfortunately oxidation is not effective for some compounds. Halogenated hydrocarbons, which are particularly inert to oxidation, accumulate in fatty tissue or in the liver itself. Examples of halogenated hydrocarbons are carbon tetrachloride, hexachlorobenzene, DDT, dioxins, and polychlorinated biphenyls (PCBs). Some compounds become more toxic after oxidation. For example, polycyclic hydrocarbons (which can be formed when food is barbecued) become carcinogenic upon partial oxidation. Methanol becomes particularly toxic because it is converted by oxidation to formaldehyde. One of the most serious dangers of environmental pollution lies in the introduction of compounds that the liver cannot detoxify.

In a sense the liver is the final guardian along the pathway by which nutrients pass to the cells. This pathway starts with a balanced diet. Once foods are digested and absorbed, the liver adjusts nutrient levels and removes potential toxins. The blood can then provide nutrients for the cellular biochemistry that constitutes life.

Concepts in Review

1. Distinguish a nutrient from a food.
2. Describe the difference between a minimum dietary requirement and a recommended dietary allowance.
3. Summarize the importance of nutrients in metabolism.
4. Briefly describe the general differences between carbohydrate and fat metabolism.
5. Briefly discuss the major functions of dietary protein.
6. Discuss the dangers of kwashiorkor.
7. Explain why fats are a more concentrated source of metabolic energy than carbohydrates.

8. List some of the dangers of a high-sucrose diet.
9. Briefly describe the importance of dietary fiber.
10. List the essential fatty acids and essential amino acids.
11. Discuss a major danger inherent in a vegetarian diet.
12. Summarize some of the functions of the water soluble vitamins.
13. Discuss some of the functions of the fat soluble vitamins.
14. State, in a general way, the results of vitamin deficiencies.
15. List the major mineral elements.
16. Compare the similarities between trace mineral elements and vitamins.
17. Briefly discuss the importance of water in the diet.
18. List four purposes served by food additives.
19. Briefly discuss food additives and their effect on nutrition.
20. List the five principal digestive juices and where they originate in the body.
21. List the principal enzymes of the various digestive juices.
22. Give the main digestive functions of each of the five principal digestive juices.
23. List the classes of products formed when carbohydrates, fats, and proteins are digested.
24. Briefly describe the absorption of nutrients from the digestive tract into the lymph and blood.
25. Discuss the role of the liver in maintaining blood glucose levels.
26. Explain how the liver metabolizes potentially toxic compounds.

Equations in Review

Hydrolysis of Carbohydrates During Digestion (example)

$$\text{Starch} + \text{Water} \xrightarrow{\text{amylase}} \text{Maltose}$$
$$\text{Maltose} + \text{Water} \xrightarrow{\text{maltase}} \text{Glucose}$$

Hydrolysis of Proteins During Digestion

$$\text{Protein} + \text{Water} \xrightarrow{\text{enzyme}} \text{Proteoses} + \text{Peptones}$$
$$\text{Proteoses} + \text{Peptones} + \text{Water} \xrightarrow{\text{enzyme}} \text{Dipeptides}$$
$$\text{Dipeptides} + \text{Water} \xrightarrow{\text{enzyme}} \text{Amino acids}$$

Hydrolysis of Fats During Digestion (example)

$$\text{Triacylglycerols} + \text{Water} \longrightarrow \text{Fatty acids} + \text{Glycerol} + \text{Monoacylglycerols} + \text{Diacylglycerols}$$

Exercises

1. Milk is a food. List four nutrients that can be obtained from milk.
2. List the three classes of nutrients that do *not* commonly supply energy for the cell.
3. What is meant by the term *energy allowance*?
4. How does marasmus differ from kwashiorkor?
5. You are told that a new diet will cause you to lose 9 kg (20 lb) of fat in one week. Is this reasonable? Explain.
6. What is meant by the statement that candy provides empty calories?
7. Why is cellulose important in the diet?
8. Explain why a tablespoon of butter (a fatty food) approximately doubles the calorie content of a medium-size baked potato (a carbohydrate food).
9. What is meant by the term *essential fatty acid*?
10. What chemical characteristics do the essential fatty acids have in common?
11. List the essential amino acids.
12. How do animal proteins differ from vegetable proteins?
13. Which vitamins are water soluble? Which are fat soluble?
14. Why does cooking affect the B-complex vitamin content of vegetables?
15. List three different functions that can be attributed to vitamins.
16. Distinguish the major elements from the trace elements.
17. List two major biological functions for calcium.
18. In what ways are the trace elements similar to vitamins? How do they differ?
19. What percentage of the average water consumption comes from solid foods? How much from liquids?
20. A fruit drink label claims that the juice contains 100% of the RDA for vitamin C. Because of this nutrition claim, what information must be printed on the label by law?
21. List five common categories of food additives.
22. Which federal agency is responsible for regulating the use of food additives?
23. Why must the food of higher animals be digested before it can be utilized?
24. What are the five principal digestive juices?
25. What is chyme?
26. What is the approximate pH of each of the following?
    (a) Saliva (c) Pancreatic juice
    (b) Gastric juice (d) Bile
27. What enzymes are present in each of the digestive juices?
28. In what parts of the digestive system are each of the following digested?
    (a) Carbohydrates (b) Fats (c) Proteins
29. In what part of the digestive tract does the absorption of food occur?
30. What are the end products of carbohydrate digestion? List the specific compounds that are formed.
31. What is the digestive function of the liver?
32. List the principal digestive enzymes that act on proteins.
33. How do the intestinal mucosal cells aid digestion?
34. How does the liver metabolize toxic compounds and remove them from the bloodstream?
35. Which of these statements are correct? Rewrite each incorrect statement to make it correct.
    (a) Water is an essential nutrient.
    (b) Most vegetable proteins are complete proteins.
    (c) Oleic acid is an essential fatty acid.
    (d) A dietary supply of glycine is not needed.
    (e) Of the three classes of energy nutrients, an RDA has been established only for proteins.
    (f) Vitamins are inorganic nutrients.
    (g) Most foods contain a variety of nutrients.
    (h) Calcium is an important trace element.
    (i) The functions of most trace elements are not well understood.
    (j) The main purpose of digestion is to hydrolyze large molecules to smaller ones that can be absorbed through the intestinal membranes.
    (k) Most of the digestion of food occurs in the stomach.
    (l) Gastric juice contains hydrochloric acid and has a pH of 1.5–2.5.
    (m) Digestion in the small intestine occurs in an alkaline medium.
    (n) The function of bile acids is to emulsify carbohydrates; this allows them to be hydrolyzed to monosaccharides.
    (o) The liver functions to reduce potentially toxic compounds.

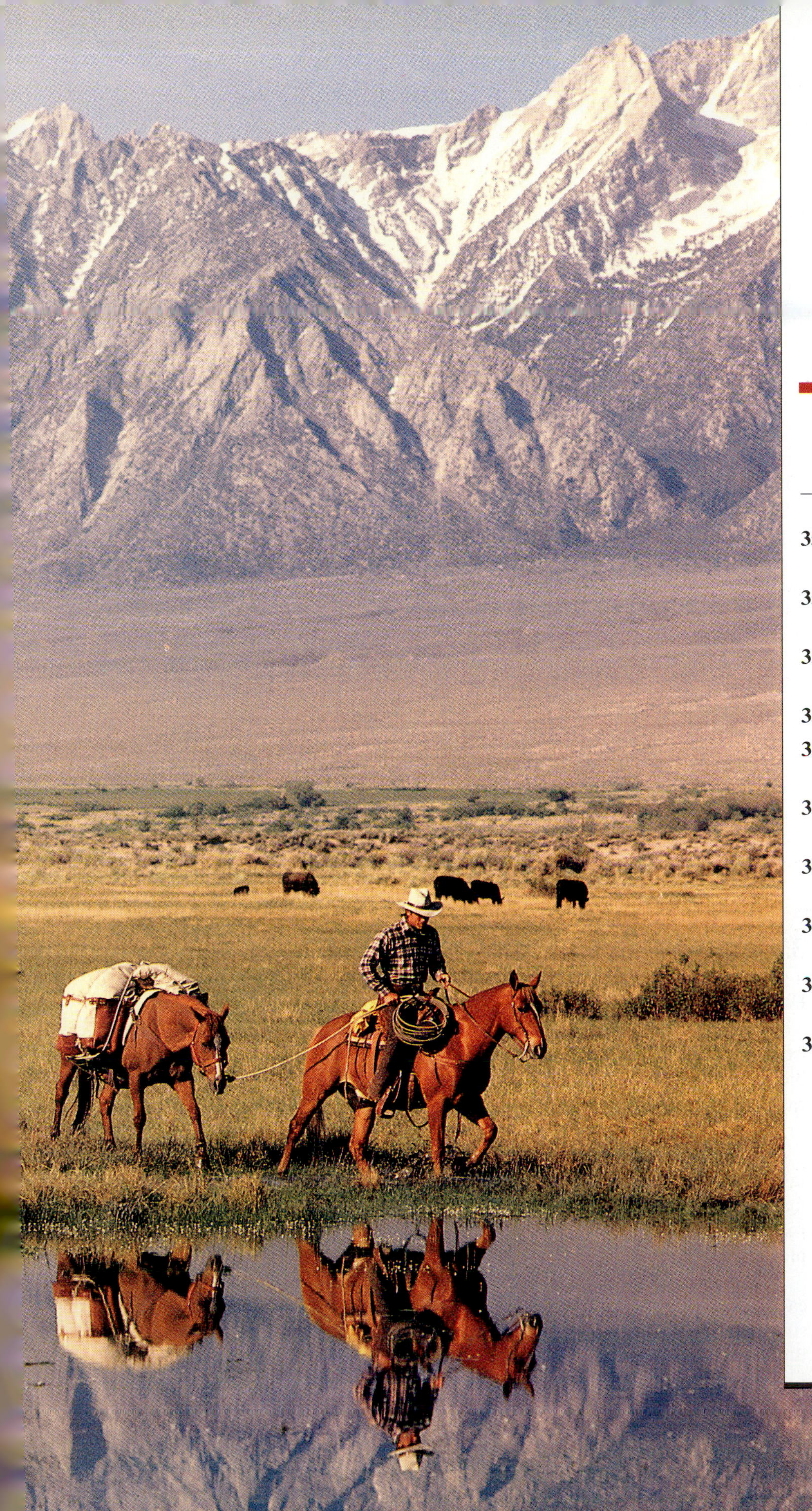

35

Bioenergetics

◀ **Chapter Opening Photo: The energy for life begins with the sun, flows through the grass, the cattle, and continues up the food chain.**

If you are traveling through the midwestern United States on a warm night in early summer, you might notice small yellow lights dancing through the fields. This beautiful sight is evidence that the fireflies, or lightning bugs, have returned. Their flickering light does not arise from a small fire or electrical discharge but rather from a chemical reaction. Bioenergetics are at work.

The monarch butterfly is a very striking insect with a wing span of several inches. As it flies, it is buffeted by the wind, first one direction and then another. Clearly, the monarch is light and not a powerful flier. Yet, this small creature makes an annual migration from the San Francisco Bay area to Mexico and Central America, a distance of about 2000 miles. Chemical reactions in each of the insects' cells make energy available for this arduous trip. Again, bioenergetics are at work.

While you have been reading these paragraphs, fully 20 percent of the oxygen you breathed has been used by your brain. Chemical processes have reacted this oxygen with blood glucose, yielding the energy needed to comprehend your reading. This is another example of bioenergetics—the chemical processes directly related to energy needs and uses in life.

35.1 Energy Changes in Living Organisms

bioenergetics

For life to exist, energy must be available. The part of biochemistry that deals with the transformation, distribution, and utilization of energy by living organism is called **bioenergetics**. Thus, when our muscles undergo physical activity, bioenergetics must have an important role. Let us follow some of the biochemical changes during fairly intense work—for example, cutting grass with a hand-propelled mower on a warm day—as a means of introduction to bioenergetics.

For the most part, bioenergetics occurs inside cells. Consequently, cells must take in needed chemicals from and return waste products to the surrounding fluids. Because of this exchange process, the chemical changes in the fluid surrounding working muscle cells provide important clues concerning the principles of bioenergetics.

As lawn mowing progresses, breathing accelerates; oxygen is taken in and carbon dioxide expelled more rapidly. These gases are transported by the bloodstream between the muscles, where bioenergetics is active, and the lungs. In fact, the blood is an important intermediary in bioenergetics.

◀ **In order to move and live we must transform and use chemical energy. The field of bioenergetics involves the study of these energy interrelationships at a cellular level.**

35.2 Oxygen/Carbon Dioxide Exchange

Muscles must have an adequate oxygen supply, especially when a person is pushing a lawn mower or doing any other work. About 5 g of oxygen (4 L at 25°C and 1 atm) is required for every minute of strenuous activity by a young adult. The necessary oxygen is supplied to the muscle cells by oxygenated arterial blood. Waste carbon dioxide from the muscle cells is removed by oxygen-deficient venous blood.

Normally the partial pressure of oxygen is about 100 torr in the lungs and 35 torr in the cells. Because gases move spontaneously from regions of higher pressure to regions of lower pressure, oxygen moves toward the cells. But the body must move oxygen at rates far faster than can be attained by simple diffusion. Red blood cells, which contain the oxygen transport protein hemoglobin (Hb) (see Section 31.7), increase the oxygen transport rate by 80–90 times that obtained by simple diffusion.

The amount of oxygen bound by hemoglobin changes as the partial pressure of oxygen changes. Figure 35.1 graphically shows the oxygen-binding characteristics of hemoglobin in the form of a steeply sloped **S**-shaped curve, known as a *sigmoidal curve*. The solid line represents the behavior of hemoglobin when the body is resting. The steep portion of the curve is in the 35–40 torr partial pressure range of oxygen found in the fluid surrounding the cell. In this steep portion of the curve, a small change in the partial pressure of oxygen causes a relatively large change in the oxygen bound to hemoglobin. The lowering of the partial pressure of oxygen as hemoglobin moves from the lungs to resting muscle tissue causes the release of about 70% of the hemoglobin-bound oxygen.

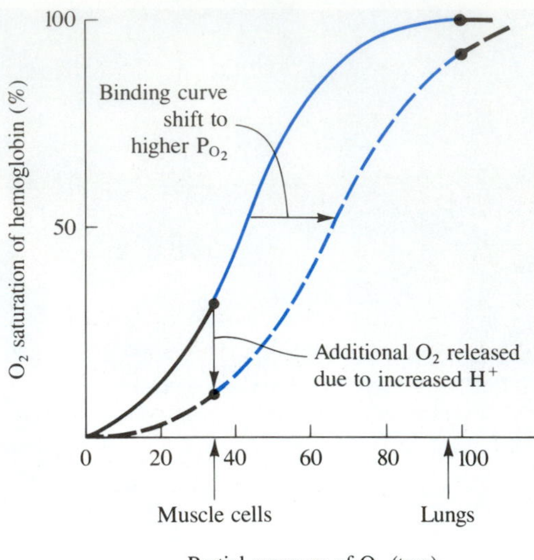

FIGURE 35.1 ▸ The oxygen binding curve for hemoglobin under approximate physiological conditions: As the partial pressure of oxygen is increased, more O_2 binds to hemoglobin. The colored portion of each curve highlights changes in O_2 saturation of hemoglobin as the blood circulates between the lungs and the muscle cells.

When the strenuous work of lawn mowing begins, the muscle cells must have more oxygen to generate the needed energy. The rate of oxygen flow to the cells is speeded up by an increase in the pulse and respiration rates of the person doing the mowing. But more subtle, and very effective, biochemical changes also occur.

First, in hard-working muscle tissue where bioenergetics proceeds rapidly, the partial pressure of oxygen may drop to 10 torr. This decrease in P_{O_2} from 35 torr to 10 torr is small. But because the hemoglobin–oxygen binding curve is steep, the amount of additional oxygen released is large. Hemoglobin may release up to 90% of its oxygen under these conditions.

Second, hydrogen ions are formed as the muscles work (see Section 35.3). These ions bind to oxygenated hemoglobin (HbO_2) and stimulate the release of additional oxygen.

$$H^+ + HbO_2 \longrightarrow HbH^+ + O_2 \tag{1}$$

This acid effect is represented graphically by a shift of the oxygen binding curve toward higher partial pressures of oxygen, as shown in Figure 35.1. This shift, shown by the dashed line, means that more O_2 is released in the muscles. When muscle cells increase bioenergetics and need more oxygen, the acid produced by this activity triggers the release of more oxygen from HbO_2.

When oxygen is used in the muscle cells, carbon dioxide is produced. This metabolic product must be removed from the cells. The partial pressure of carbon dioxide is highest in the muscle cells, decreases in the venous blood, and is lowest in the lungs. Thus, carbon dioxide tends to move from the cells toward the lungs. But as with oxygen, carbon dioxide must be moved at a faster rate than can be attained by simple diffusion.

For the most part the rapid, efficient removal of carbon dioxide is effected by two methods. In the first method hemoglobin acts as a carrier molecule. Carbon dioxide reacts with amino groups on the hemoglobin molecule to form carbamino ion groups and H^+ ions:

$$Hb{-}NH_2 + CO_2 \longrightarrow \underset{\text{Carbamino ion}}{Hb{-}NH{-}CO_2^-} + H^+ \tag{2}$$

In the second method of removal, carbon dioxide reacts with water to form highly soluble bicarbonate and hydrogen ions:

$$CO_2 + H_2O \rightleftharpoons H_2CO_3 \rightleftharpoons H^+ + HCO_3^- \tag{3}$$

Reactions (2) and (3) occur in the fluid surrounding the muscles. The hydrogen ions produced by these reactions stimulate the further release of oxygen in the muscle tissue by the reaction shown in equation (1).

The hemoglobin carbamino ions, bicarbonate ions, and excess hydrogen ions are transported to the lungs via the venous blood. There the exhalation of carbon dioxide enables the reaction of equations (2) and (3) to reverse, as shown in these equations:

$$Hb{-}NH{-}CO_2^- + H^+ \longrightarrow Hb{-}NH_2 + CO_2 \tag{4}$$

$$HCO_3^- + H^+ \rightleftharpoons H_2CO_3 \rightleftharpoons H_2O + CO_2 \tag{5}$$

These reversible reactions serve not only to eliminate CO_2 but also to consume the excess H^+ ions that were formed in the muscle tissue.

As shown in equations (1) to (5), acid (H^+) has an important role in both the delivery of O_2 and the removal of CO_2. In muscle, CO_2 and H^+ are formed. This acid (H^+) causes HbO_2 to release the additional O_2 needed to continue the bioenergetics. In the lungs, CO_2 is exhaled and the blood becomes less acidic, causing Hb to bind more O_2. The overall interrelationships of oxygen, carbon dioxide, and acid can be summarized in this way:

Energy use at the muscle	**Gas balance restored at the lungs**
O_2 used	O_2 inhaled
CO_2 formed	CO_2 exhaled
Acidity increases	Acidity decreases
More O_2 released from HbO_2	More O_2 bound to Hb

35.3 Hydrogen Ion Production and Control

Acid (H^+) is produced as the muscles perform work. Two products of cellular bioenergetics—carbonic acid and lactic acid—are the primary sources of this acidity. Both substances produce hydrogen ions in solution:

$$CO_2 + H_2O \rightleftharpoons H_2CO_3 \rightleftharpoons H^+ + HCO_3^- \tag{6}$$

$$\underset{\text{Lactic acid}}{CH_3CH(OH)COOH} \rightleftharpoons H^+ + \underset{\text{Lactate ion}}{CH_3CH(OH)COO^-} \tag{7}$$

To prevent a toxic buildup of acid (abnormally low pH), the excess H^+ ions must be removed from the cells. The lungs provide a short-term partial solution to the problem of increased acidity in the cells and surrounding fluid. When CO_2 is exhaled from the lungs, reaction (6) is reversed and the H^+ ion concentration is reduced in the bloodstream. During hard work, the rate of removal of carbon dioxide from the lungs is accelerated by an increase in the respiration and pulse rate. Thus, as the breathing rate increases, the blood acidity decreases.

The liver and kidneys also have an important role in controlling the pH of body fluids. The liver converts lactic acid to glucose and thereby helps to prevent toxic acid buildup. The kidneys transfer water and selected ions from the bloodstream to the urine. Hydrogen ions and bicarbonate ions are among those transferred. Because this is a relatively slow process, the kidneys are mainly concerned with the long-term maintenance of correct body fluid pH.

35.4 Metabolism

As we have seen, to do work (for example, to mow a lawn), the muscles must take in extra O_2. This oxygen reacts with nutrients such as carbohydrates or fats. Energy is released as the reduced carbons of the nutrients are converted by a series of reactions into the oxidized carbons of carbon dioxide and lactic acid. These reactions comprise an important fraction of all the chemical reactions that occur in muscle cells.

metabolism

anabolism

catabolism

The sum of all chemical reactions that occur within a living organism is defined as **metabolism**. Many hundreds of different chemical reactions occur in a typical cell. To help make sense of this myriad of reactions, biochemists have subdivided metabolism into two contrasting categories—*anabolism* and *catabolism*. **Anabolism** is the process by which simple substances are synthesized (built up) into complex substances. **Catabolism** is the process by which complex substances are broken down into simpler substances. Anabolic reactions usually involve reduction and consume cellular energy, whereas catabolic reactions usually involve oxidation and produce energy for the cell. The energy needed to mow the lawn is generated from catabolic reactions that occur within muscle cells.

35.5 Metabolism and Cell Structure

procaryote

Cells segregate many of their metabolic reactions into specific, subcellular locations. The simple, procaryotic cells (**procaryotes**)—those without internal membrane-bound bodies—have a minimum amount of spatial organization (see Figure 35.2). The anabolic processes of DNA and RNA synthesis in these cells are localized in the nuclear material, whereas most other metabolic reactions are spread throughout the cytoplasm.

eucaryote

organelle

In contrast, metabolic reactions in the cells of higher plants and animals are often segregated into specialized compartments. These cells, the **eucaryotes**, contain internal, membrane bound bodies called **organelles** (see Figure 35.2). It is within the organelles that many specific metabolic processes occur.

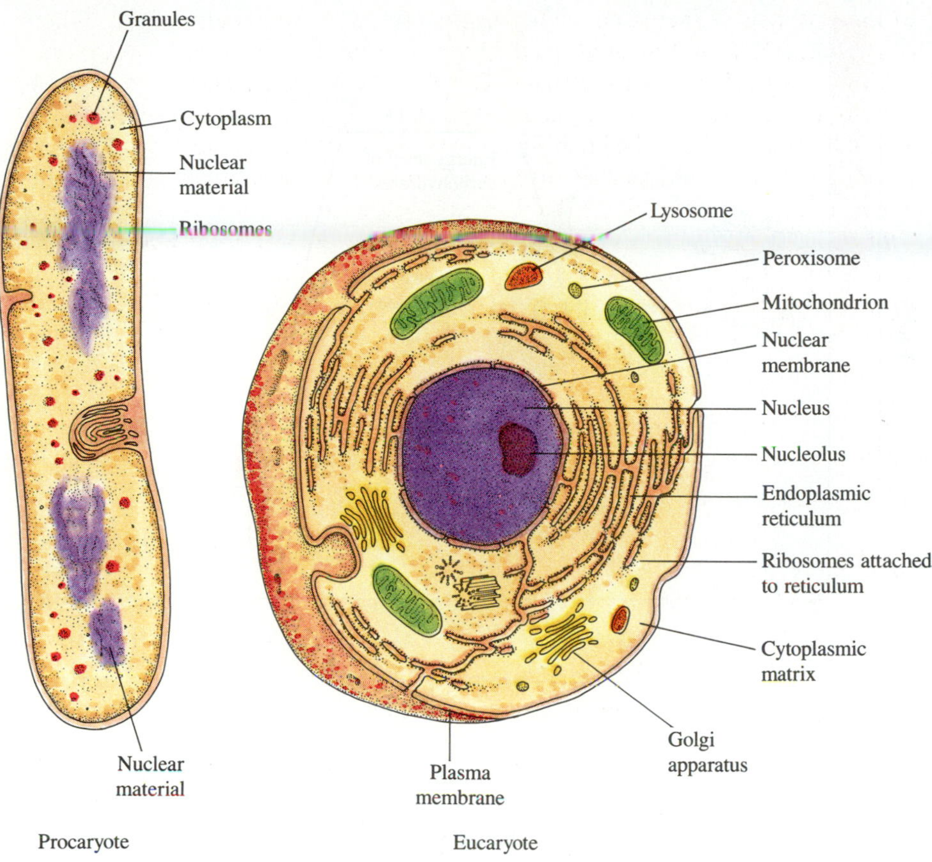

◄ **FIGURE 35.2**
A schematic representation of a procaryotic and a eucaryotic cell. The procaryote lacks much of the organized structure and the numerous organelles found in the eucaryote.

In the eucaryotic cell, most of the DNA and RNA synthesis are localized in the nucleus. Anabolism of proteins takes place on the ribosomes, whereas that of carbohydrates and lipids occurs primarily in the cytoplasm.

There are a variety of specialized catabolic organelles within a eucaryotic cell. The lysosome contains the cell's digestive enzymes, and the peroxisome is the site of oxidative reactions that form hydrogen peroxide. Perhaps the most important catabolic organelle is the mitochondrion (plural, mitochondria). This membrane-bound body provides most of the energy for a typical cell. The energy is released by catabolic processes, which oxidize carbon containing nutrients. Mitochondria (1) consume most of the O_2 that is inhaled and (2) produce most of the CO_2 that is exhaled by the lungs.

35.6 Biochemical Energy Sources

The ultimate source of biological energy on earth is sunlight. Plants capture light energy and transform it to chemical energy by a process called *photosynthesis* (see Section 35.10). This chemical energy is stored in the form of reduced carbon atoms in carbohydrate molecules. It is important to understand that the energy contained in carbohydrates, lipids, and proteins originally came from

FIGURE 35.3 ▶ Energy flow through metabolism using the important energy nutrients, carbohydrates. Reduction of carbon stores energy while oxidation of carbon releases energy

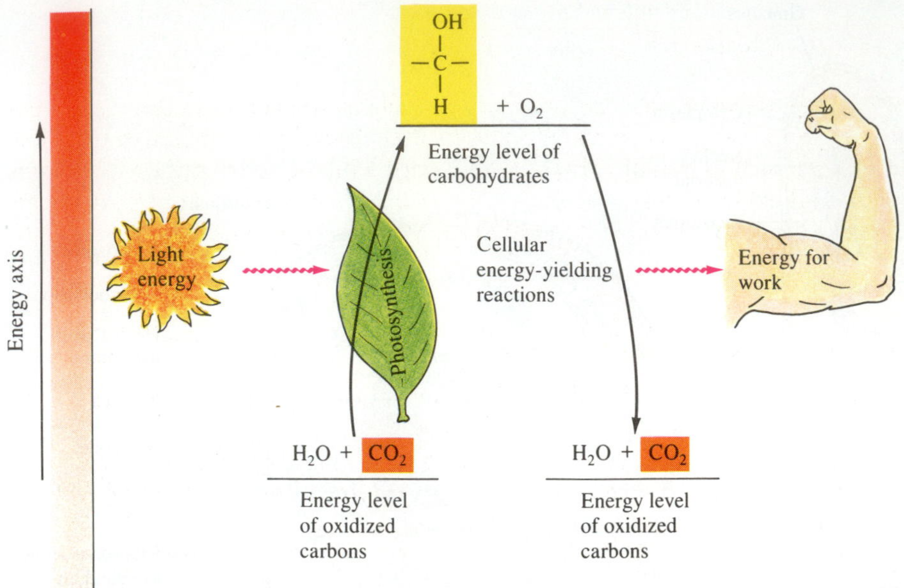

sunlight. As the reduced carbon atoms are oxidized, energy is released and used by the cell.

Figure 35.3 is an energy diagram summarizing the energy flow through metabolism. Several times in this chapter an energy diagram will be used to summarize an important concept. The black arrows trace progress from one stage to the next for the specific process under consideration. Because each stage has its own special energy level, it is presented at a particular height with respect to the energy axis (the higher the level, the more energy is in a stage). When energy levels change, energy must be released or absorbed as shown by the red arrows. Thus, in Figure 35.3, photosynthesis causes oxidized carbons to move to a higher energy level as they are reduced to carbohydrate carbons. The carbohydrate carbons are then oxidized to a lower energy level with the release of energy to do work.

Humans, as well as other animals, draw most of their energy from foodstuffs that contain reduced carbons (see Chapter 34). For example, after we eat a meal, nutrients such as carbohydrates and fats are transported to our cells. The carbons in these compounds are in a reduced state and thus contain stored energy.

H | —C— | H	OH | —C— | H
Typical fatty acid carbon (oxidation number = −2)	Typical carbohydrate carbon (oxidation number = 0)

Through the cell's metabolism these reduced carbons are oxidized, step by step, and are eventually converted to carbon dioxide.

$$O{=}C{=}O$$

Carbon dioxide carbon (oxidation number = +4)

When we work at a strenuous task, the energy needed to do the work is obtained from this general catabolic process that occurs in our muscle cells. Stored nutrients, such as carbohydrates, are oxidized with the consumption of oxygen. Carbon dioxide and lactic acid are produced. To summarize, an oxidative reaction sequence releases the energy that ultimately powers muscle action.

35.7 Biological Oxidation–Reduction

Biological oxidation–reduction, as with all redox reactions, can be separated into two half-reactions (see Section 18.3). Oxidation, the loss of electrons from a reactant,

$$A \xrightarrow{\text{Oxidation}} A^+ + e^- \tag{8}$$

is coupled with a second process, reduction, the gain of electrons by a reactant.

$$B^+ + e^- \xrightarrow{\text{Reduction}} B \tag{9}$$

Reactions (8) and (9) can be combined to give the complete redox reaction if electron transfer is possible:

$$A + B^+ \longrightarrow A^+ + B \tag{10}$$

Oxidation–reduction reactions are vital to cellular bioenergetics. In eucaryotic cells, specific organelles are present that specialize in redox reactions. For example, the *mitochondria* (see Figure 35.4) (often called the powerhouses of the cell) are the sites for most of the catabolic redox reactions. In the mitochondria an electron transport system completes the transfer of electrons to oxygen, forming water. *Chloroplasts* (see Figure 35.4) are organelles found in higher plants and contain an electron transport system that is responsible for the anabolic redox reactions in photosynthesis (see Section 35.10).

The cells maintain close control over these important electron transport systems by isolating them within organelles. Both the mitochondria and the chloroplasts are packed with membranes within which the electron transport systems are located. These lipid-rich membranes act as insulators and force electrons into pathways that are useful to the cell.

To move electrons from one place to another (often outside of the mitochondrion or chloroplast), the cell uses a set of oxidation–reduction coenzymes. Recall from Section 32.1 that a coenzyme is an organic compound that is used and reused to help an enzyme catalyzed reaction. The redox coenzymes facilitate reactions by acting as temporary storage places for electrons. The three most common oxidation–reduction coenzymes (nicotinamide adenine dinucleotide, NAD^+; nicotinamide adenine dinucleotide phosphate, $NADP^+$; and flavin adenine dinucleotide, FAD) are shown in Figure 35.5. Humans synthesize

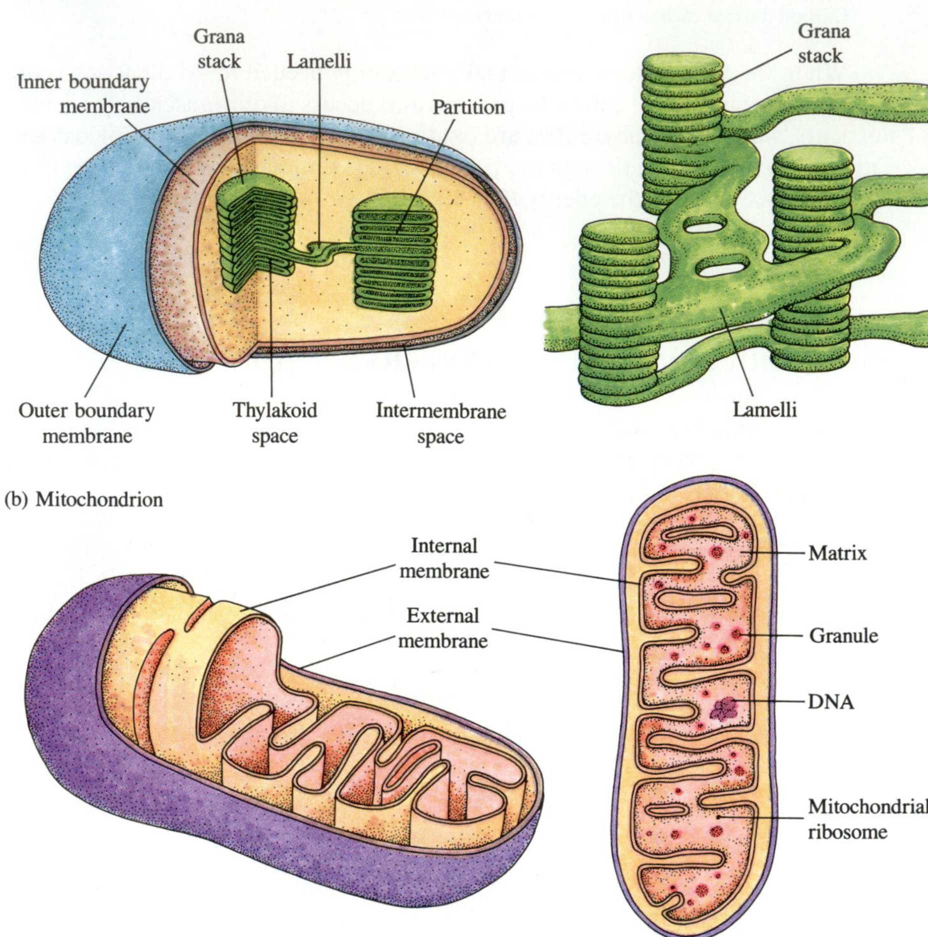

FIGURE 35.4 ▶ Schematic representation of (a) a chloroplast and (b) a mitochondrion.

NAD^+ and $NADP^+$ from the vitamin niacin while FAD is made from the vitamin riboflavin. In each case, the vitamin provides the reaction center of the coenzyme as defined in more detail in the next paragraph.

The addition or removal of electrons occurs in only one portion of each of these complex molecules. The nicotinamide ring is the reactive component within NAD^+ or $NADP^+$:

$$\text{(pyridinium ring, } N^+\text{–R)–}\overset{O}{\overset{\|}{C}}NH_2 + 2\,e^- + H^+ \rightleftharpoons \text{(dihydropyridine ring, N–R, } CH_2\text{)–}\overset{O}{\overset{\|}{C}}NH_2$$

NAD^+ or $NADP^+$ (oxidized form) — NADH or NADPH (reduced form)

(R represents the remainder of each molecule.)

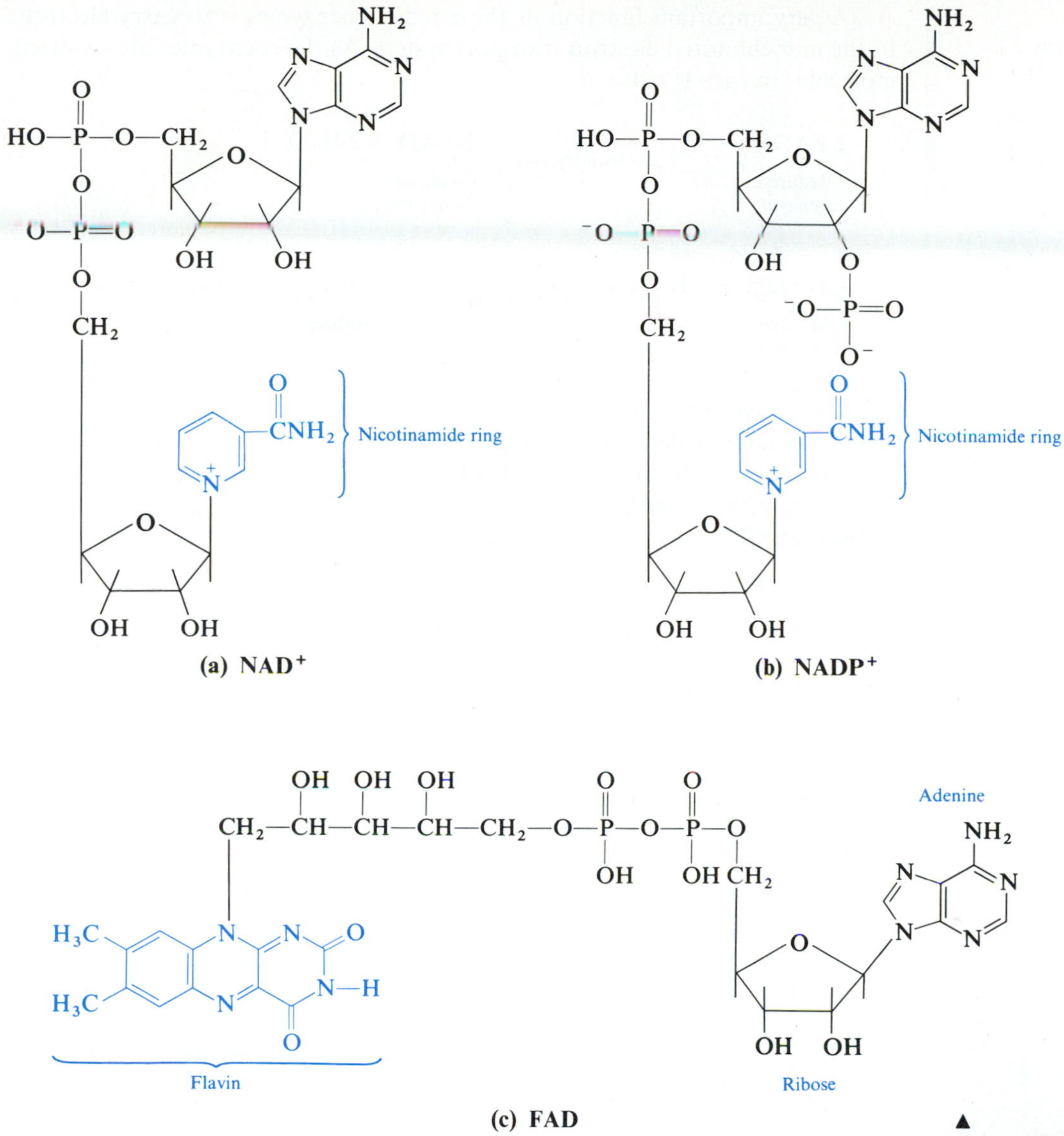

(a) NAD^+

(b) $NADP^+$

Nicotinamide ring

Flavin

Adenine

Ribose

(c) FAD

▲
FIGURE 35.5
Structures of the oxidation–reduction coenzymes (a) nicotinamide adenine dinucleotide (NAD^+), (b) nicotinamide adenine dinucleotide phosphate ($NADP^+$), and (c) flavin adenine dinucleotide (FAD).

For FAD, the flavin ring is the reactive component:

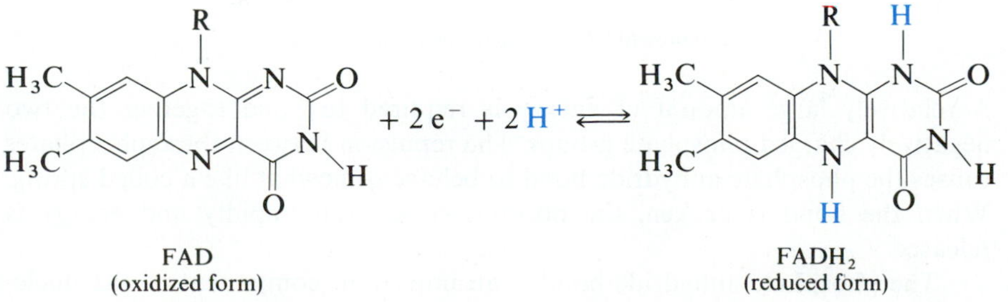

$$\text{FAD (oxidized form)} + 2\,e^- + 2\,H^+ \rightleftharpoons \text{FADH}_2 \text{ (reduced form)}$$

FAD
(oxidized form)

$FADH_2$
(reduced form)

(R represents the remainder of each molecule.)

A very important function of these redox coenzymes is to carry electrons to the mitochondrial electron transport system. As the coenzymes are oxidized, molecular oxygen is reduced:

$$\underset{\text{Reduced coenzyme}}{2\ \text{FADH}_2} + \text{O}_2 \xrightarrow[\text{electron transport}]{\text{Mitochondrial}} \underset{\text{Oxidized coenzyme}}{2\ \text{FAD}} + 2\ \text{H}_2\text{O}$$

$$\underset{\text{Reduced coenzyme}}{2\ \text{NADH}} + 2\ \text{H}^+ + \text{O}_2 \xrightarrow[\text{electron transport}]{\text{Mitochondrial}} \underset{\text{Oxidized coenzyme}}{2\ \text{NAD}^+} + 2\ \text{H}_2\text{O}$$

With the movement of electrons, energy is released. In fact, over 85% of a typical cell's energy is derived from this redox process. (Remember the mitochondria are the powerhouses of the cell!) However, the released energy is not used immediately by the cell but stored, usually in high-energy phosphate bonds such as those in ATP.

35.8 High-Energy Phosphate Bonds

The cell needs an energy delivery system. Most cellular energy is produced in the mitochondria, but this energy must be transported throughout the cell. Such a delivery system is required to carry relatively large amounts of energy and to be easily accessible to cellular reactions. Molecules which contain high-energy phosphate bonds meet this need.

The most common high-energy phosphate bond within the cell is the phosphate anhydride bond (or phosphoanhydride bond; see Section 25.12):

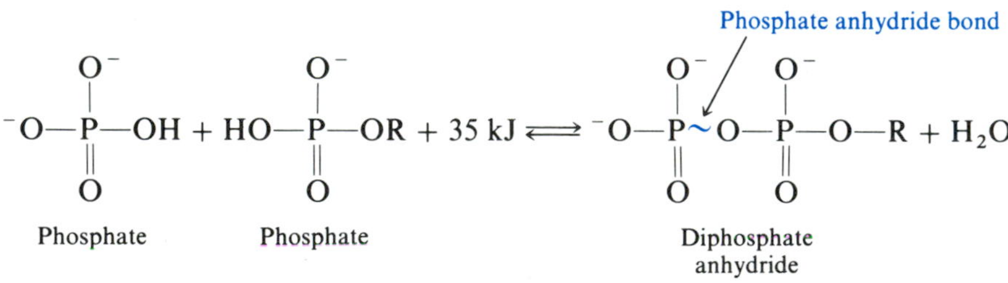

(R represents the remainder of each molecule.)

A relatively large amount of energy is required to bond together the two negatively charged phosphate groups. The repulsion between these phosphates causes the phosphate anhydride bond to behave somewhat like a coiled spring. When the bond is broken, the phosphates separate rapidly and energy is released.

The phosphate anhydride bond is an important component of the nucleotide triphosphates, the most important of which is adenosine triphosphate (ATP) (see Section 33.4):

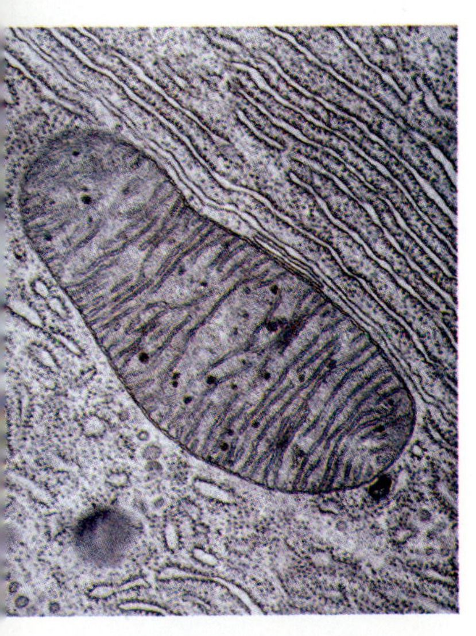

▲ Photomicrograph of a mitochondrion.

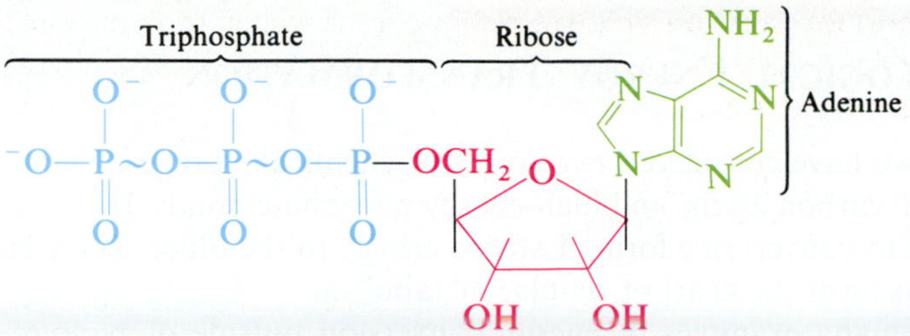

$$\underbrace{^-O-\overset{O^-}{\underset{O}{P}}\sim O-\overset{O^-}{\underset{O}{P}}\sim O-\overset{O^-}{\underset{O}{P}}}_{\text{Triphosphate}}-\underbrace{OCH_2\,(\text{ribose ring with OH, OH})}_{\text{Ribose}}-\text{Adenine}$$

Adenosine triphosphate (ATP)

Adenosine triphosphate plays an important role in all cells, from the simplest to the most complex. ATP functions by storing and transporting the energy in its high-energy phosphate bonds to the places in the cell where energy is needed. ATP is the common intermediary in energy metabolism.

The cell realizes several advantages by storing energy in ATP. First, the stored energy is easily accessible to the cell; it is readily released by a simple hydrolysis reaction yielding adenosine diphosphate (ADP) and an inorganic phosphate ion (P_i):

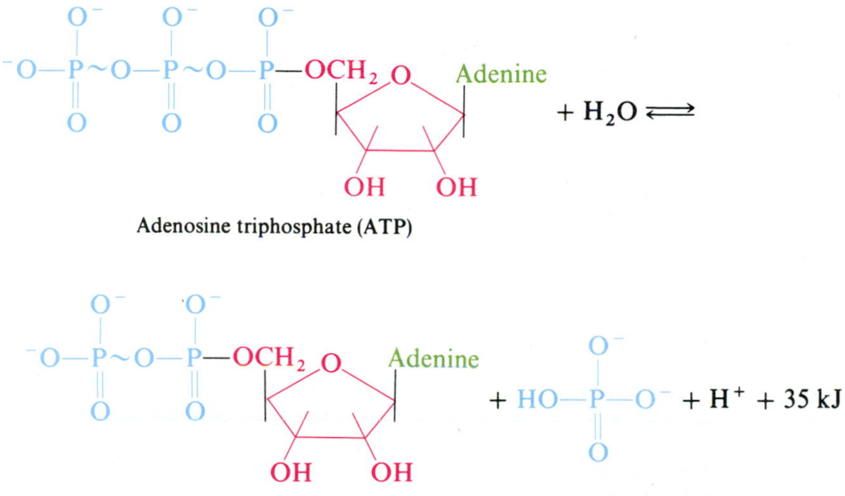

$$\underset{\text{Adenosine triphosphate (ATP)}}{^-O-\overset{O^-}{\underset{O}{P}}\sim O-\overset{O^-}{\underset{O}{P}}\sim O-\overset{O^-}{\underset{O}{P}}-OCH_2(\text{ribose})\text{Adenine}} + H_2O \rightleftharpoons$$

$$\underset{\text{Adenosine diphosphate (ADP)}}{^-O-\overset{O^-}{\underset{O}{P}}\sim O-\overset{O^-}{\underset{O}{P}}-OCH_2(\text{ribose})\text{Adenine}} + \underset{\text{Phosphate ion}}{HO-\overset{O^-}{\underset{O}{P}}-O^-} + H^+ + 35\text{ kJ}$$

Second, ATP serves as the *common energy currency* for the cell. Energy from catabolism of many different kinds of molecules is stored in ATP. For example, the energy obtained from oxidation of carbohydrates, lipids, and proteins is stored in ATP. The oxidation of each of these nutrients requires many different enzyme-catalyzed reactions. To make *direct* use of this energy, the cell would need a separate series of reactions for each different energy source. Instead, the cell channels most of the energy derived from oxidation–reduction reactions into the high-energy bonds of ATP. This process is analogous to an economic system that values all goods and services in terms of a common currency, such as the dollar. Buying and selling within the system is thus greatly simplified. In the cell, energy utilization is greatly simplified by converting all stored energy to ATP, the common energy currency.

35.9 Biological Energy Transformation

To this point we have considered two forms of chemical storage of biological energy: reduced carbon atoms and high-energy phosphate bonds. It is vital that the cell be able to convert one form of stored energy to the other. These energy transformations lie at the heart of cellular metabolism.

There are many examples of energy conversion outside of biology. The chemical energy of gasoline is converted to motion (kinetic energy) of a car. The kinetic energy of a river can be converted to electrical energy. Electrical energy is converted to the heat energy that cooks a meal or heats a room. In each case we can point to a machine or process that transforms energy. For example, an automobile engine converts chemical energy to kinetic energy, and an electric stove transforms electrical energy to heat energy. Similarly, biochemists can point to cellular processes that accomplish energy transformations.

Figure 35.6 summarizes biological energy transformation. In the first process, a variety of cellular reactions oxidize the carbons of energy-supplying nutrients, converting these molecules from a high energy level to a lower energy state. The energy released (red arrow) is used to create high-energy phosphate bonds. Finally, when a cellular reaction needs energy or work must be accomplished, molecules with high-energy phosphate bonds are recycled to low-energy molecules plus inorganic phosphate.

The cell converts energy stored in reduced carbon atoms into energy stored in phosphate anhydride bonds through two biological processes—*substrate-level phosphorylation* and *oxidative phosphorylation.* **Substrate-level phosphorylation** is the process whereby energy derived from the oxidation of reduced carbon atoms is used to form high-energy phosphate bonds on various biochemical molecules (substrates) (Figure 35.7).

substrate-level phosphorylation

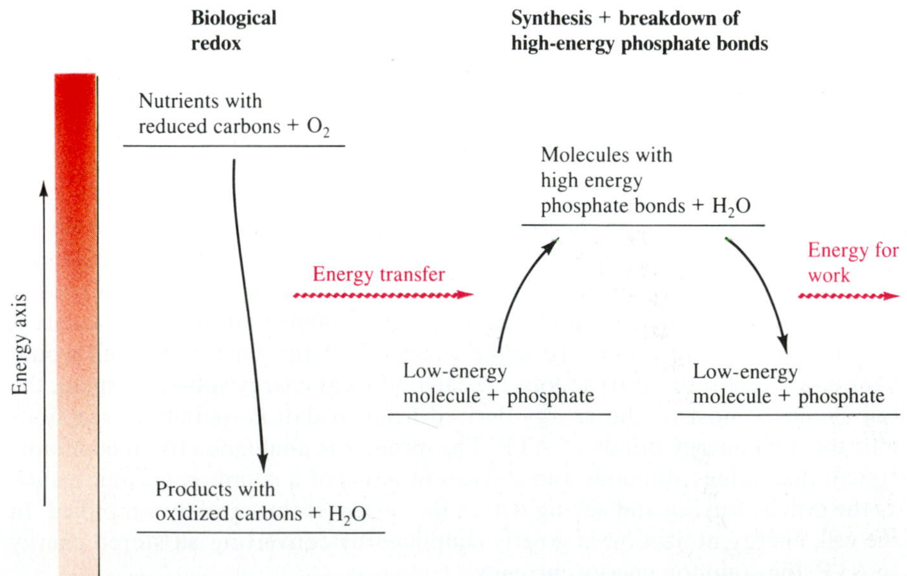

Figure 35.6 ► **Energy flow from nutrients with reduced carbons (energy-yielding nutrients) to high-energy phosphate bonds that are used to do work.**

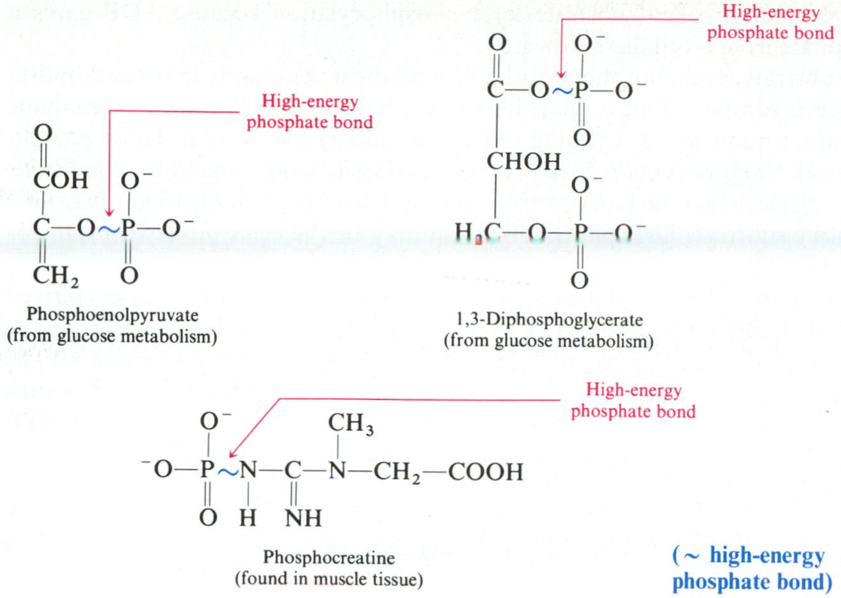

◀ **FIGURE 35.7** **Some biological molecules that contain high-energy phosphate bonds (phosphorylated substrates).**

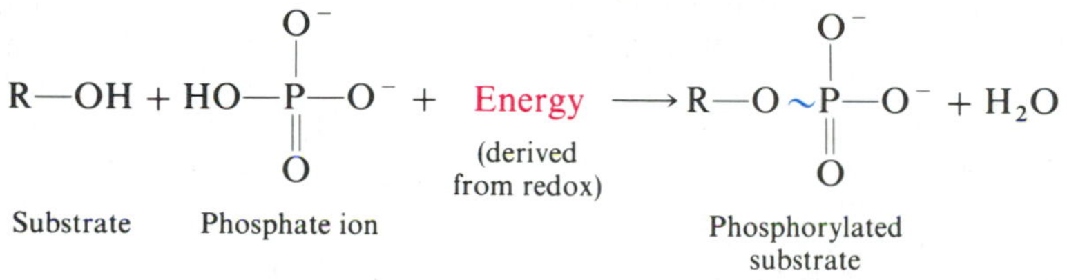

High-energy phosphate bond

Phosphoenolpyruvate (from glucose metabolism)

High-energy phosphate bond

1,3-Diphosphoglycerate (from glucose metabolism)

High-energy phosphate bond

$^{-}O-P(=O)(O^{-})\sim N(H)-C(=NH)-N(CH_3)-CH_2-COOH$

Phosphocreatine (found in muscle tissue)

(~ high-energy phosphate bond)

$$R{-}OH + HO{-}P(=O)(O^{-}){-}O^{-} + \text{Energy} \longrightarrow R{-}O\sim P(=O)(O^{-}){-}O^{-} + H_2O$$

Substrate — Phosphate ion — Energy (derived from redox) — Phosphorylated substrate

(R represents the remainder of the substrate molecule.)

In a succeeding reaction, the phosphorylated substrate transfers the phosphate to ADP and forms ATP.

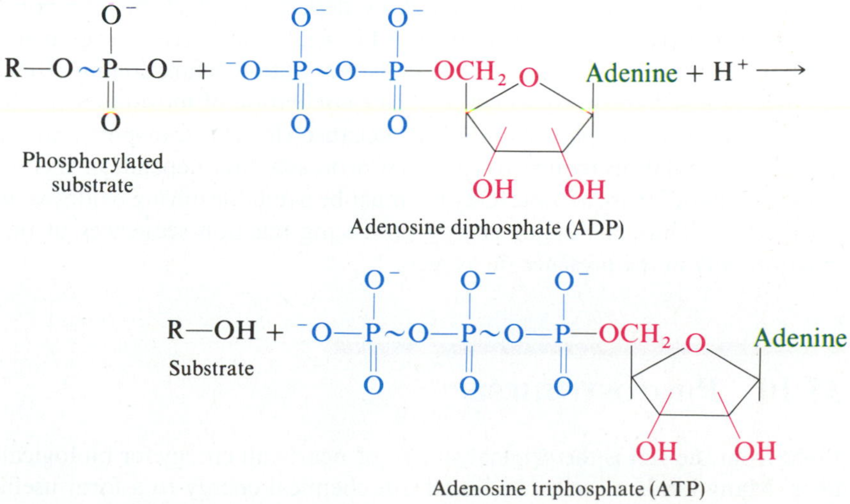

$$R{-}O\sim P(=O)(O^{-}){-}O^{-} + {}^{-}O{-}P(=O)(O^{-})\sim O{-}P(=O)(O^{-}){-}OCH_2\text{-ribose(OH)(OH)-Adenine} + H^{+} \longrightarrow$$

Phosphorylated substrate — Adenosine diphosphate (ADP)

$$R{-}OH + {}^{-}O{-}P(=O)(O^{-})\sim O{-}P(=O)(O^{-})\sim O{-}P(=O)(O^{-}){-}OCH_2\text{-ribose(OH)(OH)-Adenine}$$

Substrate — Adenosine triphosphate (ATP)

This process is called substrate level phosphorylation because ADP gains a phosphate from a cellular substrate.

Substrate level phosphorylation is found most commonly in the catabolism of carbohydrates—that is, glycolysis (see Chapter 36). This process accounts for only a minority of a resting cell's total energy production. However, this means of ATP production does not require oxygen. Under anaerobic conditions (for example, when the blood stream cannot deliver enough O_2 to hardworking muscles), substrate level phosphorylation may be the cell's principal means of forming ATP.

oxidative phosphorylation

Oxidative phosphorylation is a process that directly uses energy derived from oxidation–reduction reactions to form ATP. This process occurs in the mitochondria and depends on the mitochondrial electron transport system. The enzyme-catalyzed oxidation and phosphorylation reactions are coupled in such a way that energy released by the oxidation of the coenzyme is used to form ATP via phosphorylation. The overall process is indicated by these equations:

$$FADH_2 + \tfrac{1}{2}O_2 \xrightarrow[\text{}]{\text{Mitochondrial electron transport}} FAD + H_2O$$

Energy

$$2\ ADP + 2\ P_i \xrightarrow[\text{Oxidative phosphorylation}]{} 2\ ATP + 2\ H_2O$$

$$NADH + H^+ + \tfrac{1}{2}O_2 \xrightarrow{\text{Mitochondrial electron transport}} NAD^+ + H_2O$$

Energy

$$3\ ADP + 3\ P_i \xrightarrow[\text{Oxidative phosphorylation}]{} 3\ ATP + 3\ H_2O$$

Note that, for each $FADH_2$ oxidized, two ATPs are formed; for each NADH oxidized, three ATPs are produced. This combination of mitochondrial electron transport and oxidative phosphorylation creates most cellular ATP.

Several important reaction sequences depend on electron transport and oxidative phosphorylation to produce ATP. Cells can derive energy from fats (see Chapter 37) only when oxidative phosphorylation is functioning. The citric acid cycle (see Chapter 36), which completes oxidation of most nutrients, forms ATP using oxidative phosphorylation. Because electron transport and oxidative phosphorylation require oxygen, the processes that depend on this means of producing ATP are aerobic. Oxygen must be available during oxidative phosphorylation. Thus, the major energy-producing reaction sequences in the cell function only in the presence of oxygen.

35.10 Photosynthesis

Light from the sun is the original source of nearly all energy for biological systems. Many kinds of cells can transform chemical energy to a form useful for doing work. However, there must also be cells that can transform sunlight into

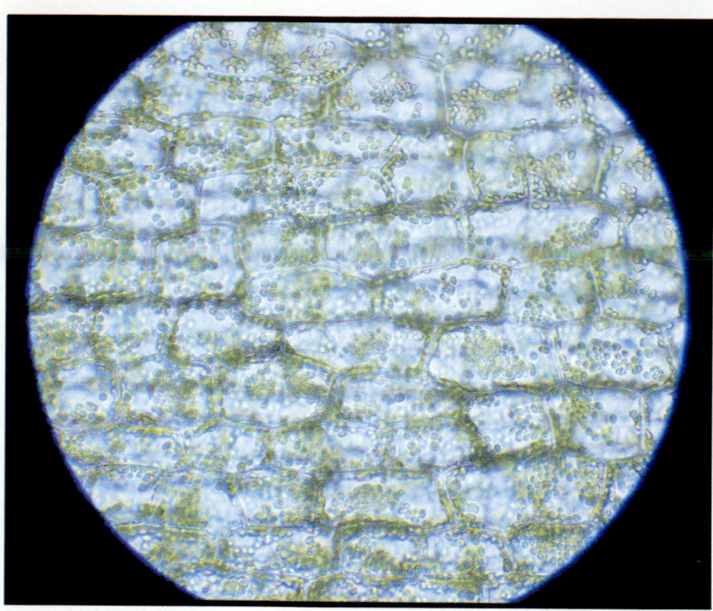

◀ Elodea cells. The chloroplasts are visible within the cytoplasm in the cells.

chemical energy. Such cells use **photosynthesis**, a process by which energy from the sun is converted to chemical energy that is stored in chemical bonds.

photosynthesis

Photosynthesis is performed by a wide variety of organisms, both eucaryotic and procaryotic. Besides the higher plants, photosynthetic eucaryotes include multicellular green, brown, and red algae and unicellular organisms such as euglena. Photosynthetic procaryotes include the green and purple bacteria and the blue-green algae. Although the photosynthetic importance of higher plants is usually emphasized, it has been estimated that more than half of the world's photosynthesis is carried out by unicellular organisms.

Photosynthesis in higher plants is a complex series of reactions in which carbohydrates are synthesized from atmospheric carbon dioxide and water:

$$6\ CO_2 + 6\ H_2O + 2820\ \text{kJ}\ (673\ \text{kcal}) \longrightarrow \underset{\text{Glucose}}{C_6H_{12}O_6} + 6\ O_2$$

Sunlight provides the large energy requirement for this process. An important side benefit of photosynthesis is the generation of oxygen, which is crucial to all aerobic metabolism.

The 1961 Nobel prize in chemistry was awarded to the American chemist Melvin Calvin (1911-1992), of the University of California at Berkeley, for his work on photosynthesis. Calvin and his co-workers used radioactive-carbon tracer techniques to discover the details of the complicated sequence of chemical reactions that occur in the overall process of photosynthesis.

▲ Melvin Calvin (1911-1992).

Photosynthesis traps light energy by reducing carbons. For eucaryotes the necessary electron transfer reactions are segregated in the chloroplast (see Figure 35.4). Like the mitochondrion, the chloroplast contains an electron transport system within its internal membranes. Unlike the mitochondrial system, which oxidizes coenzymes to liberate energy, the chloroplast electron transport system reduces coenzymes with an input of energy:

$$NADP^+ + H_2O + \text{Energy} \xrightarrow[\text{electron transport}]{\text{Chloroplast}} NADPH + H^+ + \tfrac{1}{2}\,O_2$$

$$\text{NADH} + \text{H}^+ + \tfrac{1}{2}\,\text{O}_2 \xrightarrow[\text{electron transport}]{\text{Mitochondrial}} \text{NAD}^+ + \text{H}_2\text{O} + \text{Energy}$$

$$\text{FADH}_2 + \tfrac{1}{2}\,\text{O}_2 \xrightarrow[\text{electron transport}]{\text{Mitochondrial}} \text{FAD} + \text{H}_2\text{O} + \text{Energy}$$

The chloroplasts capture light energy and place it in chemical storage.

The photosynthetic mechanism is complex, but it can be divided into two general components—the *dark reactions* and the *light reactions*. The dark reactions produce glucose from carbon dioxide, reduced coenzymes, and ATP. No light is needed and, in nature, these reactions continue during the night.

The light reactions of photosynthesis form the ATP and NADPH needed to produce glucose. The mechanism for capturing light energy is unique to the photosynthetic process. Although much research has been devoted to this topic, not all of the details are clear. In general, light is absorbed by colored compounds (pigments) located in the chloroplasts. The most abundant of these pigments is chlorophyll. Once the light energy is absorbed, it is transferred to specific molecules (probably special chlorophylls) which lose electrons. These energized electrons travel through the chloroplast electron transport system as shown in Figure 35.8. Two events follow in quick succession: First, the electrons lost by these special chlorophylls are moved to higher energy levels until they can reduce molecules of the coenzyme, $NADP^+$. Second, the special chlorophylls that lost electrons now regain them. Water is the electron donor, giving up electrons and producing oxygen gas (and hydrogen ions) in the process. The overall oxidation–reduction reaction moves four electrons from two water molecules to produce two molecules of NADPH.

$$2\,\text{H}_2\text{O} + 2\,\text{NADP}^+ \xrightarrow{\text{Light}} 2\,\text{NADPH} + \text{O}_2 + 2\,\text{H}^+$$

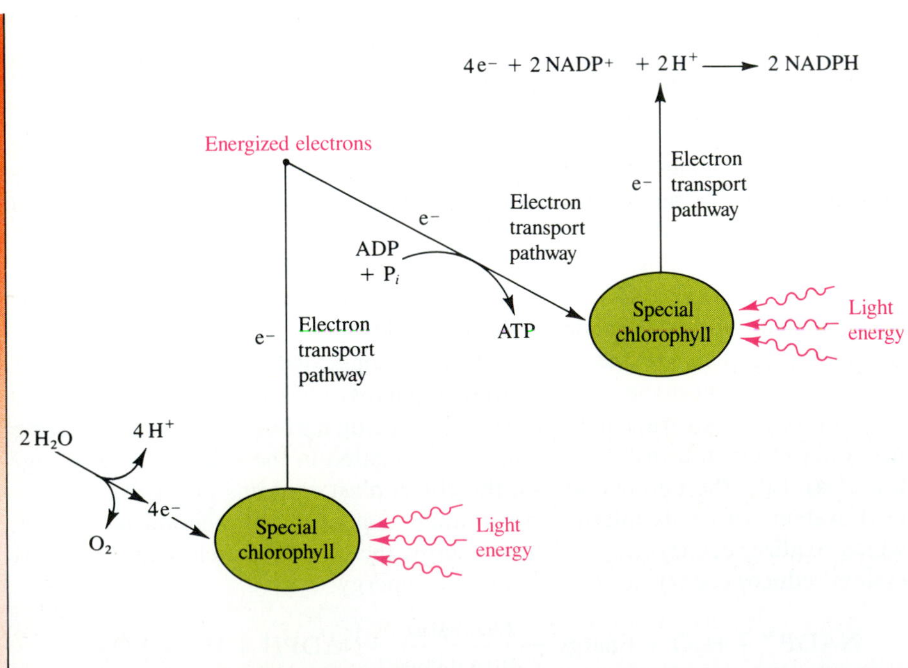

FIGURE 35.8 ▶ A schematic showing the movement of electrons from water to $NADP^+$ in the photosynthetic electron transport pathway: Note that light energy causes the electrons to become more energetic so that they can reduce $NADP^+$.

Photosynthesis uses light energy to force electrons to higher energy levels, as shown in Figure 35.8. As noted in the preceding paragraph, these energetic electrons are used to reduce $NADP^+$. But also notice that, when the electron loses energy in the middle of this electron transport process, the released energy is used to make ATP. Thus, the light reactions of photosynthesis supply both the NADPH and ATP needed to make glucose.

Concepts in Review

1. Describe the importance of hemoglobin in oxygen transport.
2. Explain how oxygen transport is coordinated with carbon dioxide production.
3. Explain how an anabolic process differs from a catabolic process.
4. List three major classes of biochemical substances that provide metabolic energy.
5. Explain how reduced carbon atoms are important in the production of cellular energy.
6. Explain the difference between a procaryote and a eucaryote.
7. List four subcellular organelles and their functions.
8. Explain the importance of membrane lipid in the mitochondrial function.
9. Define the role of the mitochondria in metabolism.
10. Briefly describe electron transport.
11. Discuss the function of NAD^+, $NADP^+$, and FAD in biological processes.
12. Explain the importance of the high-energy phosphate bond in metabolism.
13. Give two advantages for using ATP as a common energy currency for the cell.
14. Contrast substrate level phosphorylation with oxidative phosphorylation.
15. Compare the role of the mitochondrion with that of the chloroplast.
16. Outline the principal steps in the overall process of photosynthesis.

Equations in Review

Binding of Oxygen and Hydrogen Ions to Hemoglobin (Hb)

$$H^+ + HbO_2 \rightleftharpoons HbH^+ + O_2$$

Binding of Carbon Dioxide to Hemoglobin (Hb)

$$Hb{-}NH_2 + CO_2 \rightleftharpoons \underset{\text{Carbamino ion}}{Hb{-}NH{-}CO_2^-} + H^+$$

Important Hydrogen Ion-Forming Reactions in the Bloodstream

$$CO_2 + H_2O \rightleftharpoons H_2CO_3 \rightleftharpoons H^+ + HCO_3^-$$

$$\underset{\text{Lactic acid}}{CH_3CH(OH)COOH} \rightleftharpoons H^+ + \underset{\text{Lactate ion}}{CH_3CH(OH)COO^-}$$

Formation of Phosphate Anhydride Bond

$$\underset{\text{Phosphate}}{{}^-O\text{—}P(=O)(O^-)\text{—}OH} + \underset{\text{Phosphate}}{HO\text{—}P(=O)(O^-)\text{—}OR} + 35\ kJ \rightleftharpoons \underset{\text{Diphosphate anhydride}}{{}^-O\text{—}P(=O)(O^-)\sim O\text{—}P(=O)(O^-)\text{—}O\text{—}R} + H_2O$$

Substrate-Level Phosphorylation

$$\underset{\text{Substrate}}{R\text{—}OH} + \underset{\text{Phosphate ion}}{HO\text{—}P(=O)(O^-)\text{—}O^-} + \underset{\text{(derived from redox)}}{\text{Energy}} \longrightarrow \underset{\text{Phosphorylated substrate}}{R\text{—}O\text{—}P(=O)(O^-)\text{—}O^-} + H_2O$$

(R represents the remainder of the substrate molecule.)

Mitochondrial Electron Transport and Oxidative Phosphorylation

$$FADH_2 + \tfrac{1}{2}O_2 \xrightarrow[\ ]{\text{Mitochondrial electron transport}} FAD + H_2O$$

Energy

$$2\ ADP + 2\ P_i \xrightarrow[\text{Oxidative phosphorylation}]{} 2\ ATP + 2\ H_2O$$

$$NADH + H^+ + \tfrac{1}{2}O_2 \xrightarrow{\text{Mitochondrial electron transport}} NAD^+ + H_2O$$

Energy

$$3\ ADP + 3\ P_i \xrightarrow[\text{Oxidative phosphorylation}]{} 3\ ATP + 3\ H_2O$$

Photosynthetic Electron Transport

$$2\ H_2O + 2\ NADP^+ \xrightarrow{\text{Light}} 2\ NADPH + O_2 + 2\ H^+$$

Exercises

1. Explain how increased carbon dioxide production can create a more acidic environment in muscle tissue.
2. How does rapid breathing (hyperventilation) change the blood pH? Explain.
3. Show a chemical equation for the reaction that produces a bond between hemoglobin and carbon dioxide.
4. What general characteristics are associated with a catabolic pathway? How does this process differ from an anabolic pathway?
5. Why are fats and carbohydrates good sources of cellular energy?
6. Describe the function of the mitochondria in catabolism.
7. Based on the function of the mitochondria, explain why these organelles are composed of about 90% membrane by mass.
8. Could a higher plant cell survive with chloroplasts but no mitochondria? Explain.
9. Give an example of the most common high-energy phosphate bond found in the cell. With what compound is it generally associated?
10. Why is ATP known as the "common energy currency" of the cell?
11. How much energy would be released if 3 mol of ATP was converted to 3 mol of ADP?
12. Define the term *oxidation–reduction coenzyme.*
13. Draw the ring-structure portion of NAD^+ that becomes reduced during metabolism.
14. How does oxidative phosphorylation differ from substrate-level phosphorylation?
15. In what part of the cell does oxidative phosphorylation occur?
16. How many ATPs would be formed from 2 NADH and 3 $FADH_2$ using mitochondrial electron transport and oxidative phosphorylation?
17. The following compound is formed during glucose catabolism:

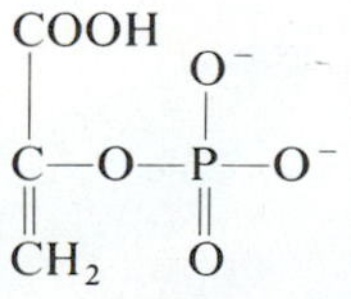

```
COOH
 |     O⁻
 |     |
 C—O—P—O⁻
 ||    ||
CH₂    O
```

Phosphoenolpyruvate

How many ATPs can be formed from one molecule of this compound during substrate level phosphorylation?
18. Compare the structural similarities between the chloroplasts and the mitochondria.
19. Explain how the function of the chloroplast differs from that of the mitochondrion.
20. What role do chloroplast pigments serve in photosynthesis?
21. Give the overall reaction for photosynthesis in higher plants.
22. In photosynthesis, electrons reduce $NADP^+$. From what compound are these electrons ultimately obtained?
23. Which of these statements are correct? Rewrite each incorrect statement to make it correct.
    (a) Slowed breathing, induced by a drug overdose, might cause the blood pH to become more acidic.
    (b) Anabolic processes are those in which complex biological substances are broken down into simpler substances.
    (c) Oxidation of reduced carbon atoms provides energy for the cell.
    (d) Typical carbohydrate carbons supply more energy to the cell than typical fatty acid carbons.
    (e) Many reduced carbons are oxidized to carbon dioxide in the mitochondria.
    (f) NAD^+ and $NADP^+$ are coenzymes that react as oxidizing agents in some metabolic processes.
    (g) Carbon dioxide provides much cellular energy.
    (h) A phosphate anhydride bond is considered to be a high-energy phosphate bond.
    (i) Mitochondrial electron transport serves to oxidize NADH.
    (j) Oxidative phosphorylation occurs in the mitochondria.
    (k) The chloroplast produces carbohydrates.
    (l) Chlorophyll absorbs light during photosynthesis.
    (m) NAD^+ is reduced during photosynthesis.

36

Carbohydrate Metabolism

◄ **Chapter Opening Photo: An aerobics class is one way to increase the metabolism of carbohydrates.**

Have you stopped to think about why carbohydrates are a staple of our world? After all, most of the average diet is carbohydrate, much of our building material derives from this source, and most of the plant kingdom depends on carbohydrates. Yet, why is this class of molecules so prevalent? Biochemists might answer this question in a number of ways, but one of the most important answers is in the air around us. Carbohydrates can be made from carbon dioxide, water, and sunlight using a process called *photosynthesis.*

Of course, a solid carbohydrate like sugar, which is sweet to the taste, has very little in common with gaseous carbon dioxide, which is tasteless and can be hazardous in high concentrations. But a number of chemical reactions can convert carbon dioxide to sugar. Photosynthesis is an example of carbohydrate metabolism and illustrates the fact that metabolism can achieve remarkable conversions. Our world depends on such metabolic processes as photosynthesis.

36.1 Metabolic Pathways

Carbohydrates are rich sources of energy for most organisms. The energy is released when a cell subjects a carbohydrate to a series of chemical reactions, called a *metabolic pathway*. Despite the wide diversity of cell types, carbohydrate metabolic pathways vary little from cell to cell and from organism to organism.

Sunlight provides the energy that is stored in carbohydrates:

$$6\ CO_2 + 6\ H_2O + 2820\ kJ \xrightarrow[\text{Light}]{\text{Enzymes/Chlorophyll}} C_6H_{12}O_6 + 6\ O_2 \qquad (1)$$

Equation (1) summarizes the production of glucose (a carbohydrate) and oxygen by the endothermic process *photosynthesis.* Note that, in the overall transformation, carbon atoms from carbon dioxide are reduced and oxygen atoms from water are oxidized. The light energy is used to cause a net movement of electrons from water to carbon dioxide. As the carbons become more reduced, more energy is stored. Photosynthesis also supplies free oxygen to the atmosphere.

Equation (2) represents the oxidation of glucose and corresponds to the reversal of the overall photosynthesis reaction. Electrons are moved from carbohydrate carbons to oxygen. Energy stored in the reduced carbon atoms in glucose is released and can then be used by the cell to do work.

$$C_6H_{12}O_6 + 6\ O_2 \xrightarrow{\text{Enzymes}} 6\ CO_2 + 6\ H_2O + 2820\ \text{kJ}\ (673\ \text{kcal}) \qquad (2)$$

Glucose can be burned in oxygen in the laboratory to produce carbon dioxide, water, and heat. But in the living cell, the oxidation of glucose does not proceed directly to carbon dioxide and water. Instead, like photosynthesis, the overall process proceeds by a series of enzyme-catalyzed intermediate reactions. These intermediate steps channel some of the liberated energy into uses other than heat production. Specifically, a portion of the energy is stored in the chemical bonds of ATP.

36.2 Enzymes in Metabolism

Specific enzymes are required for essentially all biochemical reactions. Only when enzymes are present do the reactions of metabolic pathways proceed fast enough to keep a cell alive. In a sense, enzymes can be thought of as the valves of a large pumping plant. As long as the valves are open, metabolic chemicals can be processed. If some valves are closed, or missing, biochemicals are shunted to other reactions within the cell.

Enzymes are used to control metabolism just as valves are used to control pumping. Specific enzymes may be either created or destroyed during hormone-induced metabolic changes (see Section 36.8). For example, during a long fast our bodies must break down muscle proteins for energy. The proteins yield amino acids, which need to be converted to carbohydrates and fats by the liver. Normally this organ does not have the necessary enzyme complement to handle a large influx of amino acids. But under fasting conditions, the adrenal glands release *cortisone.* This hormone signals the liver to synthesize the extra enzymes needed to catabolize amino acids.

Metabolism is also controlled by altering the catalytic activity of enzymes. Changes in the three-dimensional structure of an enzyme can cause the rate of the reaction it catalyzes to increase or decrease (see Section 32.5). For example, calcium ions are known to initiate a process that leads to changes in the structure of the enzyme that converts glycogen to glucose. These changes activate the enzyme and increase the rate of glycogen breakdown. When muscles contract, calcium is released, and the conversion rate of glycogen to glucose is increased. Thus more glucose (and therefore energy) is provided automatically to working muscles.

36.3 Overview of Human Carbohydrate Metabolism

Glucose is the key monosaccharide in carbohydrate metabolism. Most carbohydrates that are consumed in a meal are converted to glucose in the liver. Glucose circulates in the blood, and the concentration is maintained within well defined limits in a healthy person. When blood glucose is in excess, it is converted to

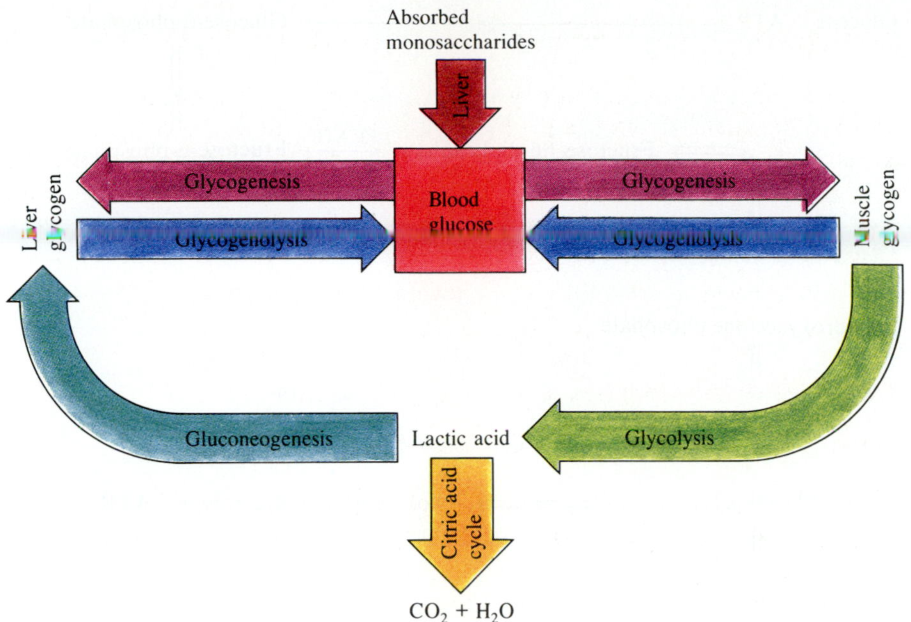

◄ **FIGURE 36.1 Overview of carbohydrate metabolism.**

glycogen in the liver and muscle tissue. Glycogen is a storage polysaccharide; it can be quickly hydrolyzed to replace depleted glucose supplies in the blood. The synthesis of glycogen from glucose is called **glycogenesis**; the hydrolysis, or breakdown, of glycogen to glucose is known as **glycogenolysis** (see Figure 36.1).

glycogenesis

glycogenolysis

Energy (stored as ATP) is obtained from glucose via several common metabolic pathways. Under *anaerobic* (oxygen not required) conditions, a general sequence of reactions known as the *Embden–Meyerhof pathway* oxidizes glucose to produce energy. When the final product of this process is lactic acid, the pathway is known as glycolysis. This reaction sequence is especially important for humans (and other animals) during strenuous activity when muscle tissue generates much lactic acid. Part of the lactic acid produced is sent to the liver and is converted to glycogen in the *gluconeogenesis* pathway (see Section 36.6). The rest is converted to carbon dioxide and water via the citric acid cycle, an *aerobic* (free or respiratory oxygen required) sequence. The *citric acid cycle*, also known as the Krebs cycle or the tricarboxylic acid cycle, is much more efficient at ATP production than is the glycolysis pathway. Consequently, the citric acid cycle is the major vehicle for obtaining metabolic energy from carbohydrates.

The diagram in Figure 36.1 presents a general overview of carbohydrate metabolism. The entire process requires specific enzymes at each reaction stage and is controlled by chemical regulatory compounds.

36.4 ANAEROBIC SEQUENCE

In the absence of oxygen, glucose in living cells can be converted to a variety of end products including lactic acid (in muscle) and alcohol (in yeast). The sequence of reactions involved is similar in different kinds of cells. At least a dozen

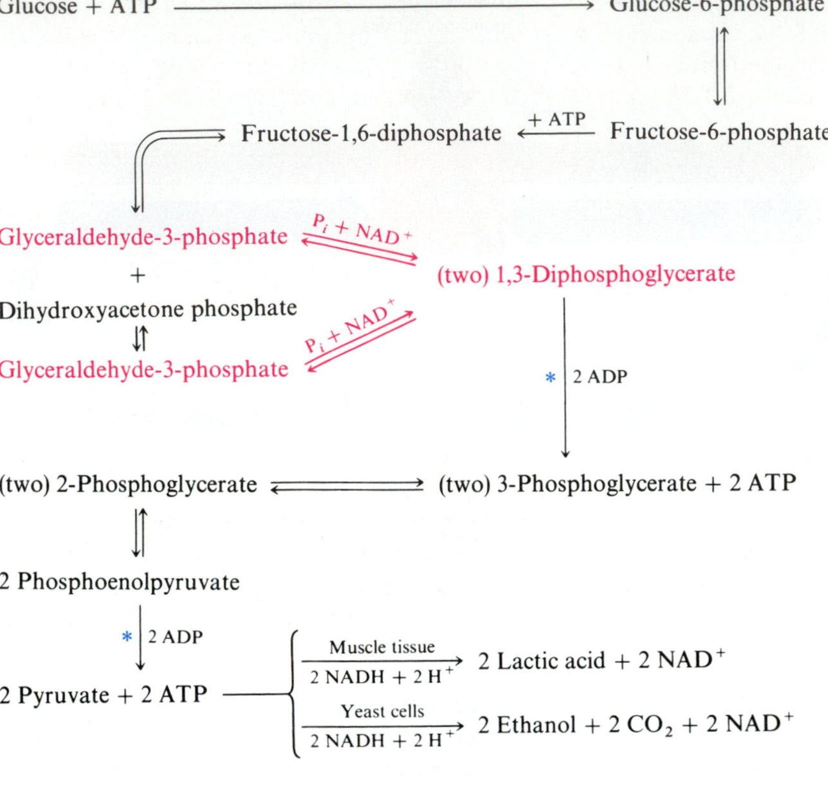

FIGURE 36.2 ▶ Conversion of glucose to pyruvate via the Embden–Meyerhof pathway (anaerobic sequence). Glycolysis proceeds further, forming lactic acid. Formation of ATP is denoted by asterisks.

reactions, many different enzymes, ATP, and inorganic phosphate (P_i) are required. Such a sequence of reactions from a particular reactant to end products is called a *metabolic pathway.*

Embden–Meyerhof pathway

The anaerobic conversion of glucose to pyruvic acid is known as the **Embden–Meyerhof pathway** (see Figure 36.2). The sequence is a catabolic one in which glucose is oxidatively degraded:

$$\underset{\substack{\text{D-Glucose}\\ \text{(sum of oxidation numbers}\\ \text{for carbon = 0)}}}{C_6H_{12}O_6} \xrightarrow{\text{Embden–Meyerhof pathway}} \underset{\substack{\text{Pyruvic acid}\\ \text{(sum of oxidation numbers}\\ \text{for carbon = +4)}}}{2\,CH_3\overset{\overset{\displaystyle O}{\|}}{C}{-}COOH}$$

In glucose the sum of the oxidation numbers of the six carbon atoms is zero. After processing one molecule of glucose to two molecules of pyruvic acid by the pathway, the sum of the oxidation numbers of the six carbon atoms is +4. This makes it evident that an overall oxidation of carbon must occur in the Embden–Meyerhof pathway.

As with most catabolic processes, the Embden–Meyerhof pathway produces energy for the cell. Carbon atoms are oxidized; energy is released and stored in the form of ATP. The pathway uses the process of *substrate-level phosphorylation* (see Section 35.9)—the energy released from carbon oxidation is used to form a high-energy substrate–phosphate bond, which in turn is used to form ATP. It is interesting to note that there is only one oxidation–reduction reaction in the Embden–Meyerhof pathway—the conversion of glyceraldehyde-3-phosphate to 1,3-diphosphoglycerate (marked in red on Figure 36.2):

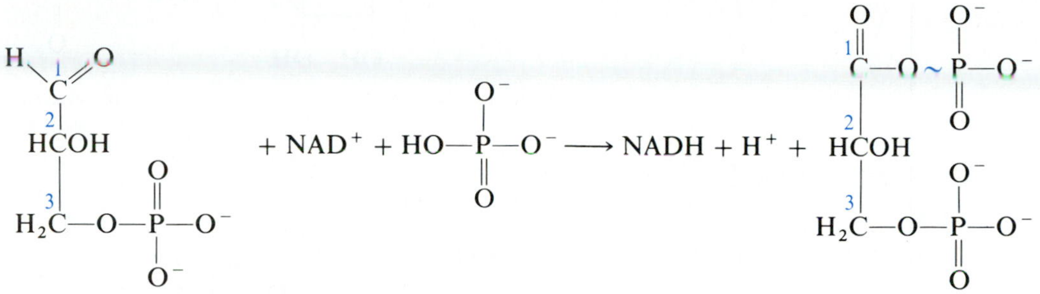

$$\text{Glyceraldehyde-3-phosphate} + NAD^+ + HO{-}P(=O)(O^-){-}O^- \longrightarrow NADH + H^+ + \text{1,3-Diphosphoglycerate}$$

Glyceraldehyde-3-phosphate

1,3-Diphosphoglycerate

Carbon 1 is oxidized from the aldehyde to the carboxylate oxidation state. Simultaneously, the oxidation–reduction coenzyme, NAD^+, is reduced. This single oxidation–reduction supplies most of the energy that is generated by the Embden–Meyerhof pathway. This energy is used to make the following two different high-energy phosphate bonds:

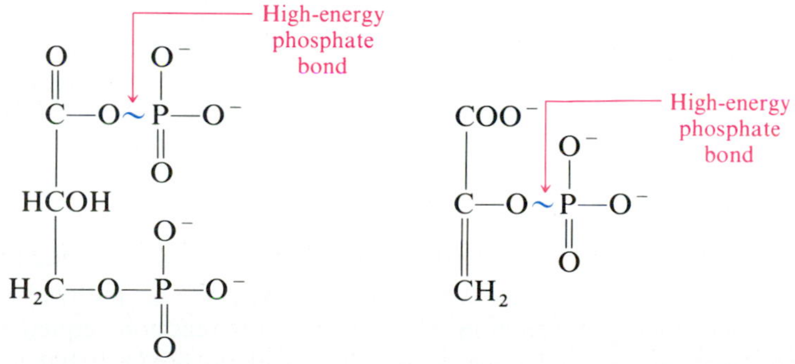

High-energy phosphate bond

1,3-Diphosphoglycerate

High-energy phosphate bond

Phosphoenolpyruvate

These high-energy bonds are found on intermediate compounds or *pathway substrates.* Substrate-level phosphorylation is complete when these substrates transfer their high-energy phosphate bonds to ADP, forming ATP (the reactions marked with an asterisk on Figure 36.2).

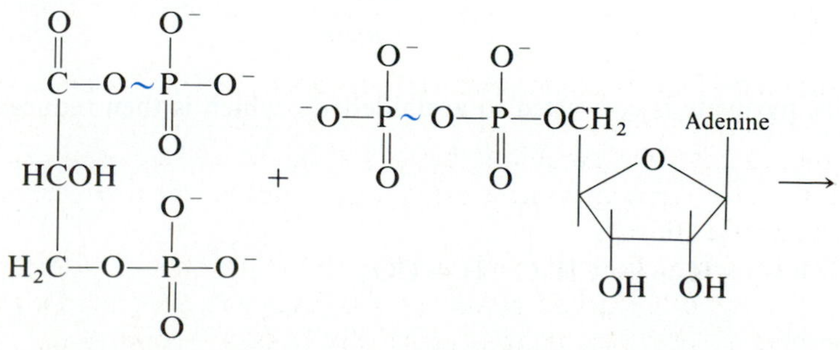

$$\text{1,3-Diphosphoglycerate} + \text{Adenosine diphosphate (ADP)} \longrightarrow$$

1,3-Diphosphoglycerate

Adenine

Adenosine diphosphate (ADP)

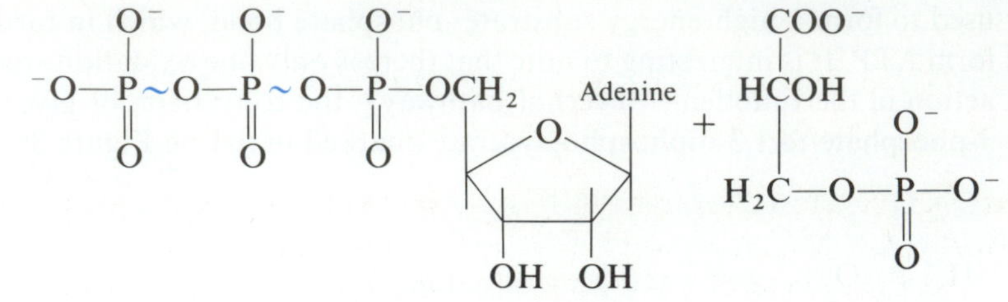

$$\text{Adenosine triphosphate (ATP)} + \text{3-Phosphoglycerate}$$

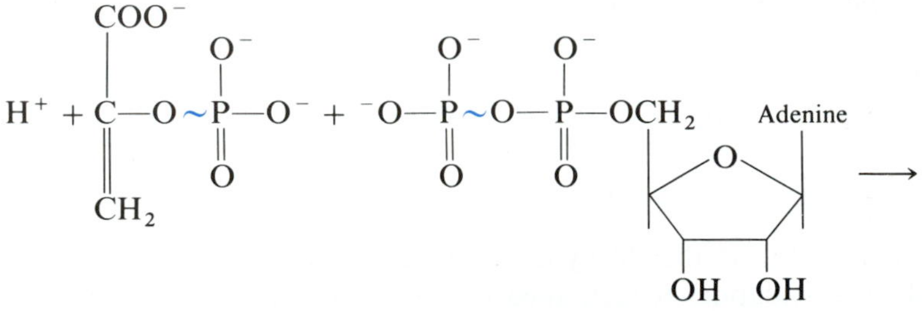

$$H^+ + \text{Phosphoenolpyruvate} + \text{Adenosine diphosphate (ADP)} \longrightarrow \text{Adenosine triphosphate (ATP)} + \text{Pyruvate}$$

Phosphoenolpyruvate Adenosine diphosphate (ADP)

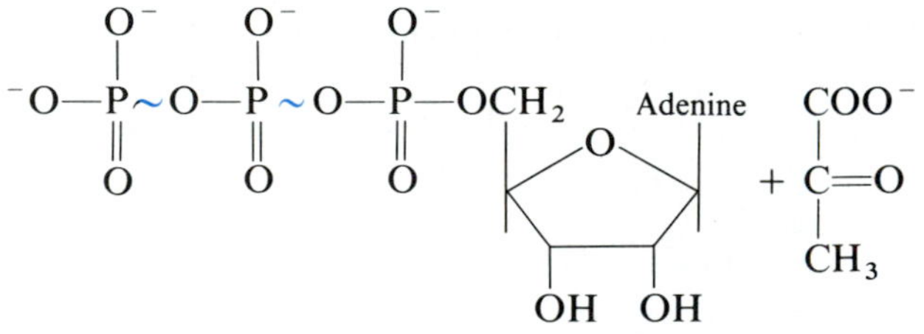

Adenosine triphosphate (ATP) Pyruvate

It is important to note that the Embden–Meyerhof pathway oxidizes carbon and produces ATP in the absence of molecular oxygen. This pathway provides for *anaerobic* energy production. In order for this reaction sequence to continue, the coenzyme NADH must be recycled back to NAD^+; that is, an additional oxidation–reduction reaction is needed to remove electrons from NADH. In the absence of oxygen, pyruvate is used as an electron acceptor. In human muscle cells pyruvate is reduced directly to lactic acid:

$$CH_3C(=O)COO^- + NADH + H^+ \rightleftharpoons CH_3CH(OH)COO^- + NAD^+$$

Pyruvate Lactate

In yeast cells, pyruvate is converted to acetaldehyde, which is then reduced to ethanol:

$$CH_3C(=O)COO^- + H^+ \longrightarrow CH_3C(=O){-}H + CO_2$$

Pyruvate Acetaldehyde

$$\underset{\text{Acetaldehyde}}{CH_3\overset{\overset{O}{\|}}{C}{-}H} + NADH + H^+ \rightleftharpoons \underset{\text{Ethanol}}{CH_3CH_2OH} + NAD^+$$

In each case NADH is reoxidized to NAD^+, and a carbon atom from pyruvate is reduced. When lactic acid is the final product of anaerobic glucose catabolism, the pathway is termed **glycolysis**. As the Embden–Meyerhof pathway produces equal amounts of pyruvate and NADH, there is just enough pyruvate to recycle all the NADH. This is a good example of an important general characteristic of metabolism: Chemical reactions in the cell are precisely balanced so that there is never a large surplus or a large deficit of any metabolic product. If such a situation does occur, the cell may die.

glycolysis

What glycolysis does for the cell can be summarized with the following net chemical equation:

$$C_6H_{12}O_6 + 2\,ADP + 2\,P_i \longrightarrow 2\,CH_3CH(OH)COOH + 2\,ATP + (150\text{ kJ or }36\text{ kcal})$$

This anaerobic process produces cellular energy. However, it is not very efficient because only two moles of ATP are formed per mole of glucose. In fact, if the energy stored in the two ATPs is summed with the energy released as heat during the Embden–Meyerhof pathway, a total of only about 209 kJ/mol (50 kcal/mol) is found to have been removed from glucose. Compare this result with the complete oxidation of glucose:

$$C_6H_{12}O_6 + 6\,O_2 \longrightarrow 6\,CO_2 + 6\,H_2O + (2820\text{ kJ or }673\text{ kcal})$$

We see that the Embden–Meyerhof pathway releases less than one-tenth of the total energy available in glucose. Thus, lactic acid must still contain much energy in its reduced carbon atoms.

36.5 Citric Acid Cycle (Aerobic Sequence)

As discussed previously, only a small fraction of the energy that is potentially available from glucose is liberated during the anaerobic conversion to lactic acid (glycolysis). Consequently, this acid remains valuable to the cells because of its stored energy. The lactic acid formed may be (1) circulated back to the liver and converted to glycogen at the expense of some ATP or (2) converted back to pyruvic acid in order to enter the citric acid cycle.

$$\underset{\text{Lactic acid}}{CH_3\overset{\overset{OH}{|}}{C}HCOOH} + NAD^+ \rightleftharpoons \underset{\text{Pyruvic acid}}{CH_3\overset{\overset{O}{\|}}{C}COOH} + NADH + H^+$$

Pyruvic acid is the link between the anaerobic sequence (Embden–Meyerhof pathway) and the aerobic sequence (citric acid cycle). Pyruvic acid itself does not enter into the citric acid cycle. It is converted to acetyl coenzyme A (acetyl-CoA), a complex substance that like ATP is of great importance in metabolism:

$$\mathrm{CH_3\overset{\overset{\displaystyle O}{\|}}{C}{-}COOH} + \underset{\text{(Coenzyme A)}}{\mathrm{CoASH}} + \mathrm{NAD^+} \longrightarrow \underset{\text{Acetyl-CoA}}{\mathrm{CH_3\overset{\overset{\displaystyle O}{\|}}{C}{-}SCoA}} + \mathrm{NADH} + \mathrm{H^+} + \mathrm{CO_2}$$

This important reaction depends on the availability of several vitamins. In addition to niacin (needed for synthesis of NAD^+), the enzyme that catalyzes this reaction requires riboflavin and thiamine. This is one of the few metabolic reactions that requires thiamine, but because this reaction is key to obtaining large amounts of energy from carbohydrate, thiamine is a vitamin of major importance. Also, our bodies need the vitamin pantothenic acid to synthesize coenzyme A.

Acetyl coenzyme A consists of an acetyl group bonded to a coenzyme A group. Coenzyme A contains the following units: adenine, ribose, diphosphate, pantothenic acid, and thioethanolamine. Coenzyme A is abbreviated as CoASH or CoA. Acetyl coenzyme A is abbreviated as acetyl-CoA or acetyl-SCoA.

Acetyl—Thioethanolamine—Pantothenic acid—Diphosphate—Ribose—Adenine

Acetyl-CoA

The acetyl group is the group that is actually oxidized in the citric acid cycle. This group is attached to the large carrier molecule as a thio ester—that is, by an ester linkage in which oxygen is replaced by sulfur:

$$\mathrm{CH_3{-}\overset{\overset{\displaystyle O}{\|}}{C}{-}S{-}CoA}$$

Acetyl group (CH$_3$—C); Thioester linkage (S); Carrier group (CoA)

Acetyl-CoA not only is a key component of carbohydrate metabolism but also, as we shall see in Chapter 37, serves to tie the metabolism of fats and that of certain amino acids to the citric acid cycle.

▲ **Hans Krebs (1900–1981).**

The citric acid cycle was elucidated by Hans A. Krebs (1900–1981), a British biochemist; thus it is also called the Krebs cycle. For his studies in intermediary metabolism, Krebs shared the 1953 Nobel prize in medicine and physiology with Fritz A. Lipmann (1899–1986), an American biochemist who discovered coenzyme A.

Krebs showed the citric acid cycle to be a series of eight reactions in which the acetyl group of acetyl-CoA is oxidized to carbon dioxide and water, and where many reduced coenzymes (both NADH and $FADH_2$) are formed. These reduced coenzymes then pass electrons through the electron transport system and, in the process, ATP is produced. The citric acid cycle is found in the mitochondria, which are the primary sites for the production of cellular energy.

The sequence of reactions involved in the citric acid cycle is shown in Figure 36.3. It is important to note that this cycle produces little usable cellular energy directly (only one GTP, or guanosine triphosphate, convertible to ATP). However, many ATPs are produced from the oxidation of acetyl-CoA (or from pyruvic acid) because of two other processes—*electron transport* and *oxidative phosphorylation* (see Section 35.9). Electron transport recycles the large number

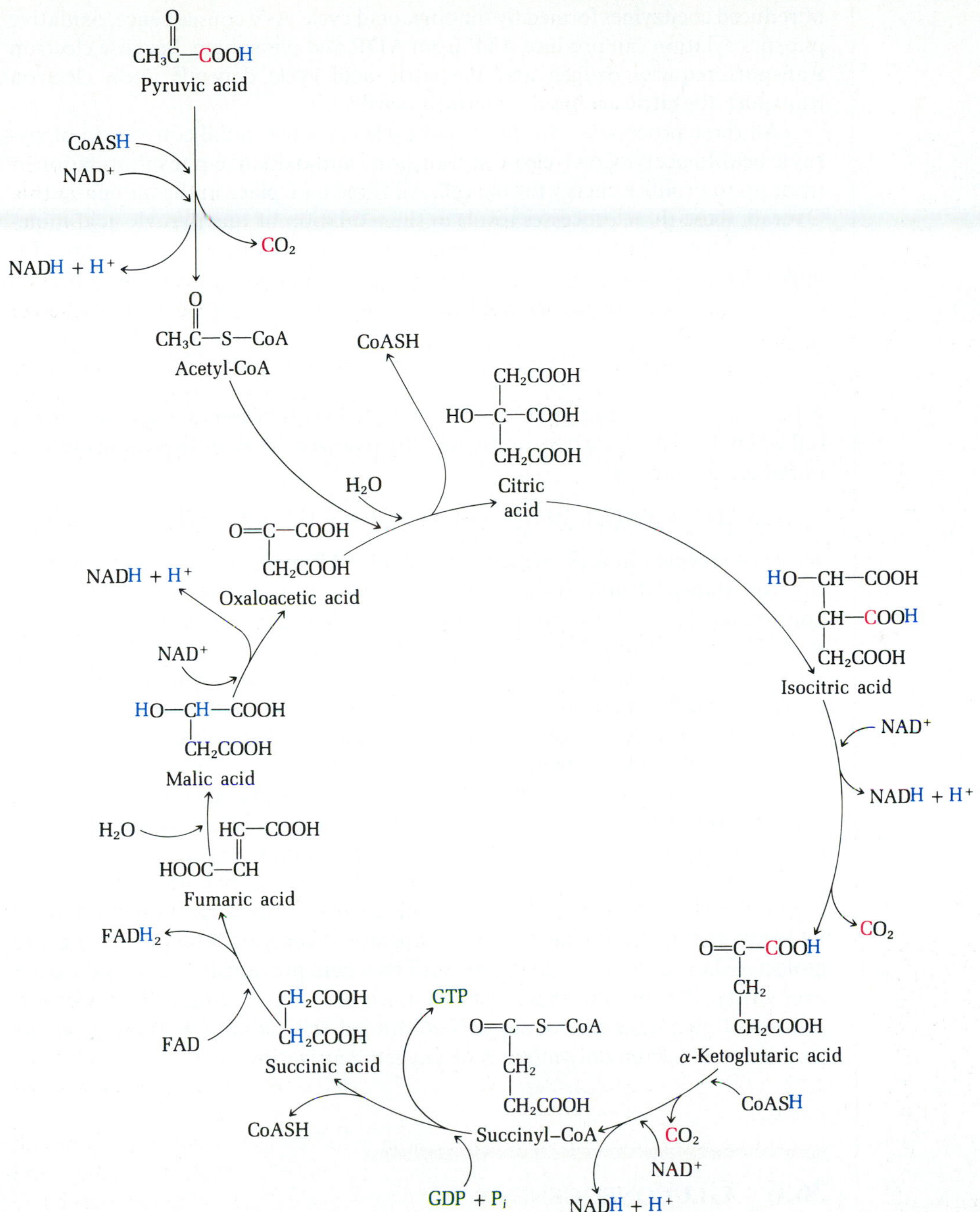

▲
FIGURE 36.3
The citric acid cycle (Krebs cycle).

of reduced coenzymes formed by the citric acid cycle. As a consequence, oxidative phosphorylation can produce ATP from ADP and phosphate. Because electron transport requires oxygen and the citric acid cycle depends upon electron transport, the citric acid cycle is termed *aerobic*.

All three processes—the citric acid cycle (plus the initial conversion of pyruvic acid to acetyl-CoA), electron transport, and oxidative phosphorylation—team up to produce energy for the cell. All three take place in the mitochondria. Overall, these three processes result in the oxidation of one pyruvic acid molecule to three carbon dioxide molecules and the formation of about 15 ATP molecules.

$$CH_3\overset{\overset{\displaystyle O}{\|}}{C}\text{—}COOH \longrightarrow 3\ CO_2$$

A large quantity of energy (1260 kJ; about 300 kcal) becomes available for the cell when these three carbon atoms are fully oxidized. Most of this energy is used to reduce coenzymes:

$$4\ NAD^+ + FAD + 10\ H \longrightarrow 4\ NADH + 4\ H^+ + FADH_2$$

These coenzymes, in turn, yield a total of 14 ATP molecules via mitochondrial electron transport and oxidative phosphorylation. Each mole of ATP stores approximately 35 kJ (8 kcal) of energy. If we include the single GTP (convertible to ATP) that is formed in the citric acid cycle, the cell obtains about 462 kJ (110 kcal) from each mole of pyruvic acid that is oxidized. By using an aerobic process, the cell can produce about 30 ATP molecules from two lactic acids after gaining only two ATPs using the anaerobic pathway, glycolysis. The presence of oxygen yields a large energy bonus for the cell.

Myocardial infarction and stroke are two injuries that are especially serious because they deprive rapidly metabolizing tissue of oxygen. When the heart muscle loses at least part of its normal blood supply, myocardial infarction results; a similar occurrence in the brain results in a stroke. In both cases, a lack of blood circulation means at least part of the tissue loses its normal oxygen supply. Suddenly these cells lose most of their capacity to generate ATP; from a glucose molecule these cells can only gain two ATPs where previously they could garner over thirty. Because the heart and brain are very active tissues, the anaerobic process of glycolysis cannot support continued cell viability. Permanent tissue damage results from only minutes of oxygen deprivation.

36.6 GLUCONEOGENESIS

A continuous supply of glucose is needed by the body, especially for the brain and the nervous system. But the amount of glucose and glycogen present in the body is sufficient to last for only about 4 hours at normal metabolic rates. Thus, there is a metabolic need for a pathway that produces glucose from noncarbohydrate sources. The formation of glucose from noncarbohydrate sources is called **gluconeogenesis**.

gluconeogenesis

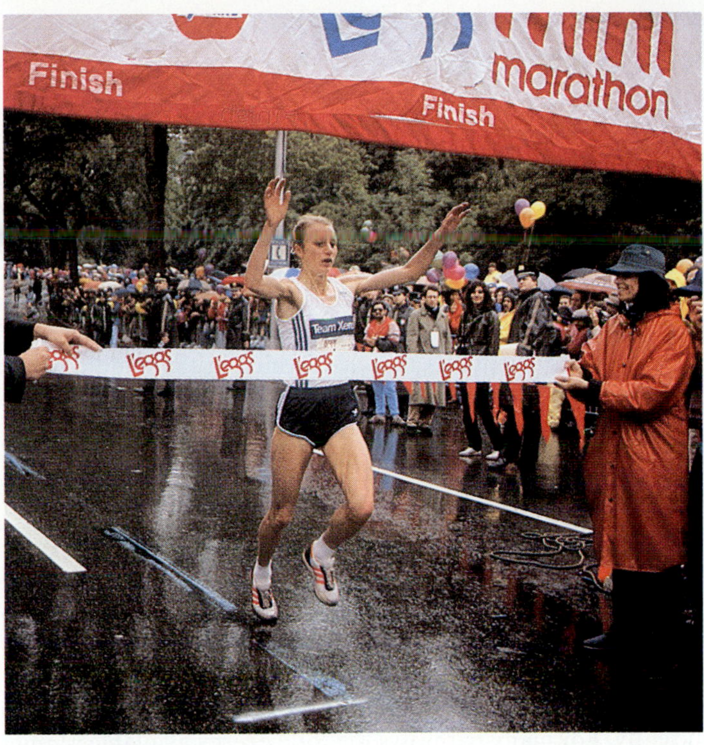

◄ **Grete Waitz winning the New York City Marathon. Long distance runners use gluconeogenesis to provide a source of glucose when more readily available supplies are exhausted.**

Most of the glucose formed during gluconeogenesis comes from lactic acid, certain amino acids, and the glycerol of fats. In each case, these molecules are converted to Embden–Meyerhof pathway intermediates (see Figure 36.2). As shown previously, lactic acid is converted to pyruvic acid. Amino acids first undergo *deamination* (loss of amino groups) and then are converted to phosphoenolpyruvic acid. Glycerol is converted to glyceraldehyde-3-phosphate (see Figure 36.4). Most of the steps in the Embden–Meyerhof pathway are reversed to transform these pathway intermediates into glucose.

Gluconeogenesis takes place primarily in the liver and also in the kidneys. These organs have the enzymes that catalyze reversal of the Embden–Meyerhof pathway. Because of this capability the liver is the organ primarily responsible for maintaining normal blood-sugar levels.

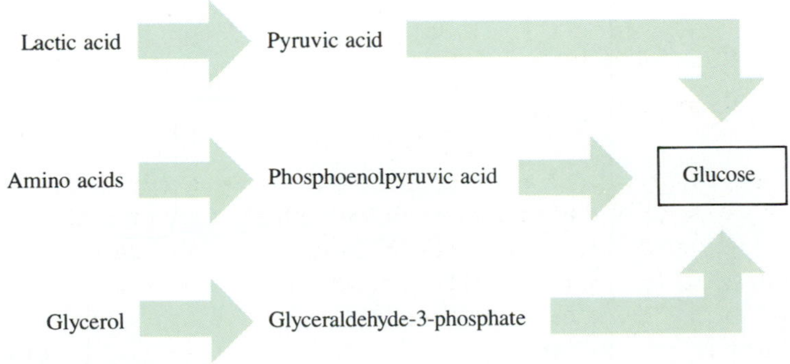

◄ **FIGURE 36.4**
An overview of gluconeogenesis. All transformations except lactic acid to pyruvic acid require several reactions.

36.7 Overview of Complex Metabolic Pathways

When we examine a metabolic pathway such as the Embden–Meyerhof pathway (Figure 36.2) or the citric acid cycle (Figure 36.3), the sheer complexity may be puzzling and overwhelming. We are tempted to ask why so many reactions are used by a cell to achieve its goal. This is a question that biochemists have asked for many years. It appears that a physiological design that includes a number of reactions per pathway achieves several vital objectives for the cell.

First, many of the chemicals that are formed in the middle of a pathway, *pathway intermediates*, are used in other metabolic processes within the cell.

$A \rightleftharpoons B \rightleftharpoons C \rightleftharpoons D$ Pathway I (B, C = pathway intermediates)

$B \rightleftharpoons X \rightleftharpoons Y \rightleftharpoons Z$ Pathway II

$C \rightleftharpoons I \rightleftharpoons J \rightleftharpoons K$ Pathway III

These intermediates are like interchangeable machine parts. Once such machine parts are made, they can be used in more than one machine. And once pathway intermediates are formed, they can be used in several pathways and serve a variety of metabolic functions in the cell.

Second, having multiple-step pathways helps the cell to handle metabolic energy efficiently. For many pathways the total energy available is much greater than the cell can handle in a single reaction. As an example, a one-step complete oxidation of glucose yields enough energy to cook a cell, figuratively speaking. To avoid such disasters the cell extracts only a little energy from glucose at each chemical reaction (Figure 36.5). The quantity of energy released in each step is small enough to be handled by the cell. Thus, in order for a cell to extract the maximum energy, a metabolic pathway must have a number of steps.

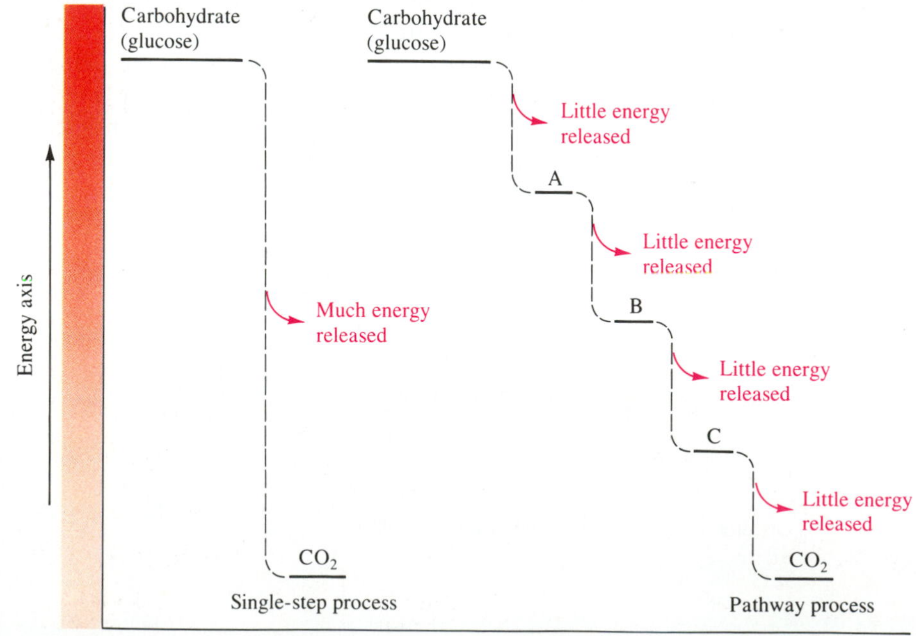

Figure 36.5 ▶ A single-step oxidation process compared with a multiple-step process: In the pathway, A, B, and C represent hypothetical pathway intermediates.

Metabolic reaction sequences seem needlessly complex because we see only a small part of the total cellular function. Although the processes must be studied separately, it is important to remember that they are all interrelated.

36.8 Hormones

Hormones are chemical substances that act as control agents in the body, often regulating metabolic pathways. They help to adjust physiological processes such as digestion, metabolism, growth, and reproduction. For example, the concentration of glucose in the blood is maintained within definite limits by the actions of hormones. Hormones are secreted by the endocrine, or ductless, glands directly into the bloodstream and are transported to various parts of the body to exert specific control functions. The endocrine glands include the thyroid, parathyroid, pancreas, adrenal, pituitary, ovaries, testes, placenta, and certain portions of the gastrointestinal tract. A hormone produced by one species is usually active in some other species. For example, the insulin used to treat diabetes mellitus in humans may be obtained from the pancreas of animals slaughtered in meat packing plants.

hormones

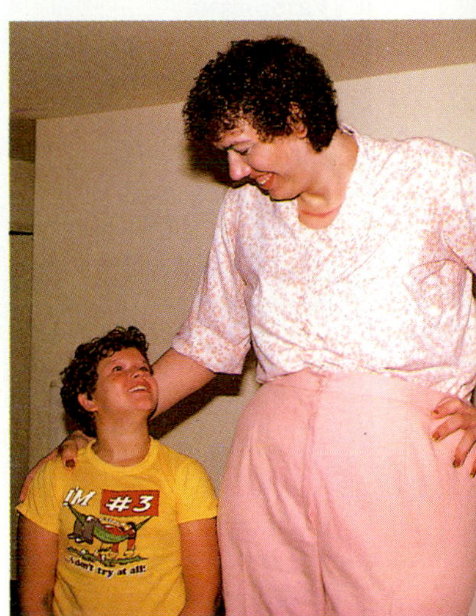

▲ **The height of this woman (7 feet, 7 inches) is due in large part to a constant hypersecretion of the growth hormone during the formative years.**

Hormones are often referred to as the chemical messengers of the body. They do not fit into any single chemical structural classification. Many are proteins or polypeptides, some are steroids, others are phenol or amino acid derivatives; examples are shown in Figure 36.6. Because a lack of any hormone can produce serious physiological disorders, many hormones are produced synthetically or are extracted from their natural sources and made available for medical use.

Like the vitamins, hormones are generally needed in only minute amounts. Concentrations range from 10^{-6} *M* to 10^{-12} *M*. Unlike vitamins, which must be supplied in the diet, the necessary hormones are produced in the body of a healthy person. A number of hormones and their functions are listed in Table 36.1.

HO–(I, I)–O–(I, I)–$CH_2CHCOOH$ (NH_2)

Thyroxin

Cys-Tyr-Ile-Gln-Asn-Cys-Pro-Leu-Gly-NH_2 (S—S bridge between the two Cys)

Oxytocin

1 ... 10 ... 15
His-Ser-Gln-Gly-Thr-Phe-Thr-Ser-Asp-Tyr-Ser-Lys-Tyr-Leu-Asp-Ser-Arg-Arg-
20 ... 29
Ala-Gln-Asp-Phe-Val-Gln-Tyr-Leu-Met-Asn-Thr

Glucagon

CH_3 OH, CH_3, O

Testosterone

CH_3 OH, HO

Estradiol (Estrogen)

OH, OH, $HOCHCH_2NHCH_3$

Epinephrine (Adrenalin)

◄ **Figure 36.6** Structure of selected hormones. Thyroxin is produced in the thyroid gland; oxytocin is a polypeptide produced in the posterior lobe of the pituitary gland; glucagon is a polypeptide produced in the pancreas; testosterone and estradiol are steroid hormones produced in the testes and the ovaries, respectively; epinephrine is produced in the adrenal glands.

TABLE 36.1 Selected Hormones and Their Functions

Hormone	Source	Principal functions
Insulin	Pancreas	Controls blood-sugar level and storage of glycogen
Glucagon	Pancreas	Stimulates conversion of glycogen to glucose; raises blood-sugar level
Oxytocin	Pituitary gland	Stimulates contraction of the uterine muscles and secretion of milk by the mammary glands
Vasopressin	Pituitary gland	Controls water excretion by the kidneys; stimulates constriction of the blood vessels
Growth hormone	Pituitary gland	Stimulates growth
Adrenocorticotrophic hormone (ACTH)	Pituitary gland	Stimulates the adrenal cortex, which, in turn, releases several steroid hormones
Prolactin	Pituitary gland	Stimulates milk production by mammary glands after birth of a baby
Epinephrine (adrenalin)	Adrenal glands	Stimulates rise in blood pressure, acceleration of heartbeat, decreased secretion of insulin, and increased blood sugar
Cortisone	Adrenal glands	Helps control carbohydrate metabolism, salt and water balance, formation and storage of glycogen
Thyroxine and triiodothyronine	Thyroid gland	Increases the metabolic rate of carbohydrates and proteins
Calcitonin	Thyroid gland	Prevents the rise of calcium in the blood above the required level
Parathyroid hormone	Parathyroid gland	Regulates the metabolism of calcium and phosphate in the body
Gastrin	Stomach	Stimulates secretion of gastric juices
Secretin	Duodenum	Stimulates secretion of pancreatic juice
Estrogen	Ovaries	Stimulates development and maintenance of female sexual characteristics
Progesterone	Ovaries	Stimulates female sexual characteristics and maintains pregnancy
Testosterone	Testes	Stimulates development and maintenance of male sexual characteristics

CHEMISTRY IN ACTION

GLUCOSE CONCENTRATION IN THE BLOOD

An adequate blood-glucose level must be maintained to ensure good health. To achieve this goal, hormones regulate and coordinate metabolism in specific organs. The hormones control selected enzymes, which, in turn, regulate the rates of reaction with the appropriate metabolic pathways. In the following paragraphs, we will examine some of the physiological mechanisms that maintain proper blood-sugar levels.

Glucose concentrations average about 70 to 90 mg/100 mL of blood under normal fasting conditions—that is, when no nourishment has been taken for several hours. Most people are in a normal fasting condition before eating breakfast. After the ingestion of carbohydrates, the glucose concentration rises above the normal level, and a condition of *hyperglycemia* exists. If the concentration of glucose rises still further, the renal threshold for glucose is eventually reached. The *renal threshold* is the concentration of a substance in the blood above which the kidneys begin to excrete that substance into the urine. The renal threshold for glucose is about 140 to 170 mg/100 mL of blood. Glucose excreted by the kidneys can be detected in the urine by a test for reducing sugars (for example, the Benedict test; see Section 29.12). When the glucose concentration of the blood is below the normal fasting level, *hypoglycemia* exists (see figure).

Glucose concentration (mg/100 mL of blood)	Condition
140–170	Renal threshold
90–140	Hyperglycemia
70–90	Normal fasting level
50–70	Hypoglycemia

▲ **Conditions related to concentration of glucose in the blood.**

Glucose concentration in the blood is under the control of various hormones. These hormones act as checks on one another and establish an equilibrium condition called *homeostasis*—that is, self-regulated equilibrium. Three hormones—insulin, epinephrine (adrenalin), and glucagon—are of special significance in maintaining glucose concentration within the proper limits. Insulin, secreted by the islets of Langerhans in the pancreas, acts to reduce blood-glucose levels by increasing the rate of glycogen formation. Epinephrine from the adrenal glands and glucagon from the pancreas increase the rate of glycogen breakdown (glycogenolysis) and thereby increase blood-glucose levels. These opposing effects are summarized as follows:

$$\text{Blood glucose} \underset{\text{Glycogenolysis}}{\overset{\text{Glycogenesis}}{\rightleftharpoons}} \text{Glycogen}$$

Glycogenesis: Stimulated by insulin
Glycogenolysis: Stimulated by epinephrine and glucagon

During the digestion of a meal rich in carbohydrates, the blood-glucose level of a healthy person rises into the hyperglycemic range. This stimulates insulin secretion, and the excess glucose is converted to glycogen, thereby returning the glucose level to normal. A large amount of ingested carbohydrates can overstimulate insulin production and thereby produce a condition of mild hypoglycemia. This in turn triggers the secretion of additional epinephrine and glucagon, and the blood-glucose levels are again restored to normal. The body is able to maintain the normal fasting level of blood glucose for long periods of time without food by drawing on liver glycogen, muscle glycogen, and finally on body fat as glucose replacement sources. Thus, in a normal person neither hyperglycemia nor mild hypoglycemia has serious consequences, since the body is able to correct these conditions. However, either condition, if not corrected, can have very serious consequences. Since the brain is heavily dependent on blood glucose for energy, hypoglycemia affects the brain and the central nervous system. Mild hypoglycemia may result in impaired vision, dizziness, and fainting spells. Severe hypoglycemia produces convulsions and unconsciousness; if prolonged, it may result in permanent brain damage and death.

Hyperglycemia may be induced by fear or anger, because the rate

(*continued*)

(*continued*)

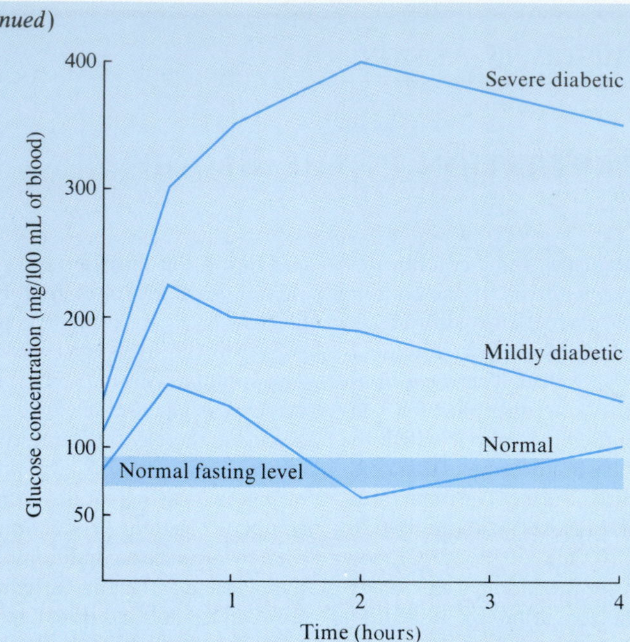

Typical responses to a glucose-tolerance test.

of epinephrine secretion is increased under emotional stress. Glycogen hydrolysis is thereby speeded up and glucose concentration levels rise sharply. This whole sequence readies the individual for the strenuous effort to either fighting or fleeing as the situation demands.

Diabetes mellitus is a serious metabolic disorder characterized by hyperglycemia, glycosuria (glucose in the urine), frequent urination, thirst, weakness and loss of weight. Prior to 1921 diabetes often resulted in death. In that year Frederick Banting and Charles Best, working at the University of Toronto, discovered insulin and devised methods for extracting the hormone from animal pancreases. For his work on insulin, Banting, with J. J. MacLeod, received the Nobel prize in medicine and physiology in 1923. Insulin is very effective in controlling diabetes. It must be given by injection, because, like any other protein, it would be hydrolyzed to amino acids in the gastrointestinal tract.

People with mild or borderline diabetes may show normal fasting blood glucose levels, but they are unable to produce sufficient insulin for prompt control of ingested carbohydrates. As a result, their blood glucose rises to an abnormally high level and does not return to normal for a long period of time. Such a person has a decreased tolerance for glucose, which may be diagnosed by a glucose tolerance test. After fasting for at least 12 hours blood and urine specimens are taken to establish a beginning reference level. The person then drinks a solution containing 100 g of glucose (amount for adults). Blood and urine specimens are then collected at 0.5, 1, 2, and 3 hour intervals and tested for glucose content. In a normal situation the blood glucose level returns to normal after about 3 hours. Individuals with mild diabetes show a slower drop in glucose levels, but the glucose level in a severe diabetic remains high for the entire 3 hours. Responses to a glucose tolerance test are shown in the diagram.

The chemical structure of insulin has been determined and biologically active insulin has been synthesized in the laboratory. In 1978 scientists at the City of Hope Medical Center in Duarte, California, announced the production of insulin identical in structure to that made in the human pancreas. Genes containing the codons required to produce the A and B polypeptide chains of insulin were made and attached to the bacteria *Escherichia coli*, which then synthesized the two chains. The A and B chains were extracted and brought together; insulin was formed when the two chains linked together through the two disulfide groups.

In 1982, this insulin became the first genetically engineered human protein to receive FDA approval for sale. Today many diabetics have the option of choosing to use human insulin rather than the animal protein obtained as a by-product of meat packing.

Concepts in Review

1. Briefly explain why carbohydrates are known as energy-storage molecules.
2. List two general ways that hormones can control metabolic pathways.
3. Briefly explain why enzymes are of vital importance in metabolism.
4. Explain why the Embden–Meyerhof pathway is catabolic.
5. State the purpose of the final reactions in anaerobic glucose catabolism.
6. Explain why the Embden–Meyerhof pathway is termed anaerobic.
7. Explain the role of glycogen in maintenance of blood-glucose levels.
8. Explain the role of epinephrine, insulin, and glucagon in the control of blood-glucose concentration.
9. Briefly describe how substrate level phosphorylation is involved in the Embden–Meyerhof pathway.
10. Give an overall description of the Embden–Meyerhof pathway (anaerobic) for metabolism of glucose.
11. Give an overall description of the citric acid cycle (aerobic).
12. Discuss the relationship between electron transport, oxidative phosphorylation, and the citric acid cycle.
13. Compare the amounts of energy formed in the anaerobic and aerobic metabolic pathways of glucose.
14. Describe the function of acetyl-CoA in carbohydrate metabolism.
15. Describe the formation of glucose by gluconeogenesis.
16. Explain the function of hormones in the body.
17. List the blood glucose levels that are considered to be normal, hyperglycemic, and hypoglycemic.
18. Explain the renal threshold.
19. Predict what might happen to the blood glucose level if a large overdose of insulin is taken.
20. Describe the glucose tolerance test and how it ties in with the condition of diabetes mellitus.

Equations in Review

Interconversion of Glycogen and Glucose

$$\text{Glycogen} \underset{\text{Glycogenesis}}{\overset{\text{Glycogenolysis}}{\rightleftharpoons}} \text{Glucose}$$

Embden–Meyerhof Pathway

$$C_6H_{12}O_6 + 2\,NAD^+ + 2\,ADP + 2\,P_i \longrightarrow 2\,CH_3\overset{\overset{\large O}{\|}}{C}{-}COOH + 2\,ATP + 2\,NADH + 2\,H^+$$

Citric Acid Cycle

$$CH_3{-}\overset{\overset{\large O}{\|}}{C}{-}S{-}CoA + FAD + 2\,H_2O + 3\,NAD^+ + GDP + P_i \longrightarrow 2\,CO_2 + FADH_2 + 3\,NADH + 3\,H^+ + GTP + CoASH$$

Exercises

1. Why are carbohydrates considered to be energy storage molecules?
2. List two general ways in which glucose catabolism differs from photosynthesis.
3. How much energy would be released if three moles of glucose was converted to carbon dioxide?
4. Why are enzymes important in metabolism?
5. What major storage carbohydrate is found in the liver?
6. Draw the structure of the carbohydrate that is produced during glycogenolysis.
7. Define the type of ATP production that occurs in the Embden–Meyerhof pathway.
8. Explain why the Embden–Meyerhof pathway is considered to be anaerobic even though the oxidation of glucose occurs.
9. What are the end products of the anaerobic catabolism of glucose in (a) muscle tissue and (b) yeast cells?
10. What is the purpose of the final reactions in anaerobic glucose catabolism?
11. Explain why the Embden–Meyerhof pathway is considered to be catabolic.
12. How many ATPs would be produced if three glucose molecules were processed through the Embden–Meyerhof pathway?
13. Write the structure of lactic acid. Would you predict that this molecule could provide cellular energy? Briefly explain.
14. How many high-energy phosphate bonds are directly formed in the citric acid cycle? List the number and types of reduced coenzymes that are produced in this cycle.
15. Why are electron transport and oxidative phosphorylation needed when the cell uses the citric acid cycle to produce energy?
16. Compare the amounts of metabolic energy (in the form of ATP) produced per glucose unit via the anaerobic sequence and via the aerobic sequence.
17. Why was the citric acid cycle not involved in the metabolism of life forms that existed before the evolution of photosynthesis?
18. What is acetyl-CoA and what is its function in metabolism?
19. What are the functions of hormones and where are they produced?

20. How does the function of hormones in the body differ from that of enzymes?

21. (a) What is the range of glucose concentration in blood under normal fasting conditions?
(b) What blood-glucose concentrations are considered to be hyperglycemic? Hypoglycemic?

22. What is meant by the renal threshold?

23. What is meant by normal fasting blood sugar level?

24. Does the presence of glucose in the urine establish that the condition of diabetes mellitus is present? Explain.

25. Explain how the body maintains blood-glucose concentrations within certain definite limits despite wide variations in the rates of glucose intake and utilization.

26. Why is epinephrine sometimes called the emergency or crisis hormone?

27. Explain how epinephrine and insulin maintain blood glucose within a definite concentration range.

28. Predict what might happen to blood glucose concentrations if a large overdose of insulin were taken by accident.

29. Why is insulin not effective when taken orally?

30. Describe the glucose tolerance test.

31. Which of these statements are correct? Rewrite each incorrect statement to make it correct.
(a) Carbon dioxide is a final product of glucose catabolism.
(b) In general, only very few metabolic reactions require enzymes.
(c) Hormonal control often changes an enzyme's function.
(d) Glycolysis refers to the process of forming glycogen.
(e) The Embden–Meyerhof pathway uses oxidative phosphorylation to form ATP.
(f) Lactic acid is one product of gluconeogenesis.
(g) ATP is produced directly in the citric acid cycle.
(h) The Embden–Meyerhof metabolic pathway is an anaerobic sequence of reactions.
(i) In the Embden–Meyerhof metabolic pathway, one mole of glucose is converted to two moles of lactic acid in muscle tissue.
(j) The citric acid cycle is much more efficient in energy production than is the Embden–Meyerhof pathway.
(k) Hormones are regulatory agents that are secreted into the stomach and intestine to control metabolism.
(l) A person with diabetes mellitus suffers from hypoglycemia.
(m) Hypoglycemia can affect the brain due to low blood sugar level.
(n) When the blood glucose level exceeds the renal threshold, glucose is eliminated through the kidneys into the urine.

37

Metabolism of Lipids and Proteins

◄ **Chapter Opening Photo: These New York Marathon runners will metabolize lipids and proteins as well as carbohydrates by the end of the race.**

One of the world's most pervasive nutritional problems is protein deficiency, known as kwashiorkor. It especially afflicts children and, when untreated, has a mortality rate of between 30 and 90 percent. These young people suffer from growth retardation, anemia, liver damage, and often appear bloated because of excess water absorption.

In more affluent societies, nutritional problems are often associated with a high intake of saturated fat—stroke and heart disease are closely correlated to lipid intake.

These disparate nutritional problems point toward an important similarity between protein and lipid metabolism. Our biochemical processes have needs for specific amino acids (proteins) and lipids. No matter how much food is available for our diet, we must also be concerned with meeting requirements for selected nutrients.

37.1 Metabolic Energy Sources

The ability to produce cellular energy is a vital characteristic of every cell's metabolism. As we have seen, carbohydrates are one of the major sources of cellular energy. The other two are lipids and proteins.

Of all the lipids, fatty acids are the most commonly used for cellular energy. Each fatty acid contains a long chain of reduced carbon atoms that can be oxidized to yield energy.

$$CH_3CH_2CH_2CH_2CH_2CH_2CH_2CH_2CH_2CH_2CH_2CH_2CH_2CH_2CH_2COOH$$

Palmitic acid

The average oxidation number of the carbon atoms in fatty acids is about -2 compared with 0 in carbohydrates. Thus, when catabolized (oxidized), fatty acids yield more energy per carbon atom than do carbohydrates.

Proteins (amino acids) are also a source of reduced carbon atoms that can be catabolized to provide cellular energy. In addition, amino acids provide the major pool of usable nitrogen for cells. Proteins and amino acids also perform diverse other functions. Some of these functions will be considered later in this chapter.

37.2 Fatty Acid Oxidation (Beta Oxidation)

Fats are the most energy rich class of nutrients. Most of the energy from fats is derived from their constituent fatty acids. Palmitic acid derived from fat yields 39.1 kJ (9.36 kcal) per gram when burned to form carbon dioxide and water. By contrast, glucose yields only 15.6 kJ (3.74 kcal) per gram. Of course, fats are not actually burned in the body simply to produce heat. They are broken down in a series of enzyme catalyzed reactions that also produce useful potential chemical energy in the form of ATP. In complete biochemical oxidation, the carbon and hydrogen of a fat ultimately are combined with oxygen (from respiration) to form carbon dioxide and water.

In 1904 Franz Knoop, a German biochemist, established that the catabolism of fatty acids involved a process whereby their carbon chains are shortened by two carbon atoms at a time. Knoop knew that animals do not metabolize benzene groups to carbon dioxide and water. Instead the benzene nucleus remains attached to at least one carbon atom and is eliminated in the urine as a derivative of either benzoic acid or phenylacetic acid.

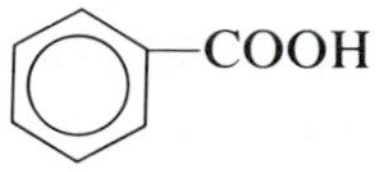

C_6H_5—COOH $\qquad$ C_6H_5—CH_2COOH

Benzoic acid 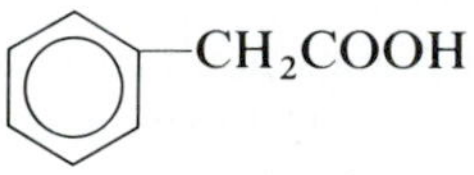 $\qquad$ Phenylacetic acid

Accordingly Knoop prepared a homologous series of straight-chain fatty acids with a phenyl group at one end and a carboxyl group at the other end. He then fed these benzene-tagged acids to test animals. Phenylaceturic acid was identified in the urine of the animals that had eaten acids with an even number of carbon atoms; hippuric acid was present in the urine of the animals that had consumed acids with an odd number of carbon atoms:

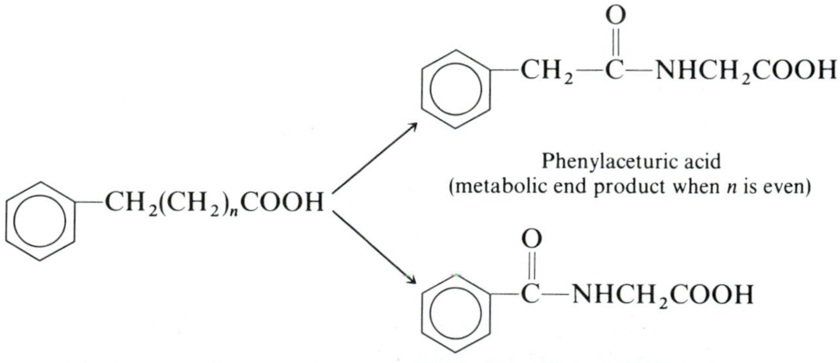

$$C_6H_5\text{—}CH_2(CH_2)_nCOOH \nearrow C_6H_5\text{—}CH_2\text{—}\overset{\overset{\large O}{\|}}{C}\text{—}NHCH_2COOH$$

Phenylaceturic acid
(metabolic end product when n is even)

$$C_6H_5\text{—}CH_2(CH_2)_nCOOH \searrow C_6H_5\text{—}\overset{\overset{\large O}{\|}}{C}\text{—}NHCH_2COOH$$

Hippuric acid
(metabolic end product when n is odd)

These results indicated a metabolic pathway for fatty acids in which the carbon chain is shortened by two carbon atoms at each stage.

Knoop's experiments were remarkable for their time. They involved the use of tagged molecules and served as prototypes for modern research that utilizes isotopes to tag molecules.

Knoop postulated that the carbon chain of a fatty acid is shortened by successive removals of acetic acid units. The process involves the oxidation of the beta-carbon atom and cleavage of the chain between the alpha and beta carbons. A six-carbon fatty acid would produce three molecules of acetic acid, thus:

First reaction sequence

$$\underset{\text{Caproic acid}}{CH_3CH_2CH_2\overset{\beta}{C}H_2\overset{\alpha}{C}H_2COOH} \longrightarrow \underset{\text{Butyric acid}}{CH_3CH_2CH_2COOH} + \underset{\text{Acetic acid}}{CH_3COOH}$$

(β carbon: This C is oxidized. Between β and α: Chain is cleaved here)

Second reaction sequence

$$CH_3\overset{\beta}{C}H_2\overset{\alpha}{C}H_2COOH \longrightarrow CH_3COOH + CH_3COOH$$

(β carbon: This C is oxidized. Between β and α: Chain is cleaved here)

The general validity of Knoop's theory of beta-carbon atom oxidation has been confirmed. However, the detailed pathway for fatty acid oxidation was not established until about 50 years after his original work. The sequence of reactions involved, like those of the Embden–Meyerhof and citric acid pathways, is another fundamental metabolic pathway. Beta oxidation, or the *two-carbon chop*, is accomplished in a series of reactions whereby the first two carbon atoms of the fatty acid chain become the acetyl group in a molecule of acetyl-CoA.

The catabolism proceeds in this manner: A fatty acid reacts with coenzyme A (CoASH) to form an activated thioester. The energy needed for this step of the catabolism is obtained from ATP.

Step 1 Activation: Formation of thioester with CoA

$$\underset{\text{Fatty acid}}{RCH_2CH_2CH_2\overset{\overset{O}{\|}}{C}OH} + CoASH + ATP \longrightarrow$$

$$\underset{\text{CoA thioester of a fatty acid}}{RCH_2CH_2CH_2\overset{\overset{O}{\|}}{C}{-}SCoA} + AMP + \underset{\text{Inorganic phosphate}}{2\,P_i}$$

The activated thioester next undergoes four more steps in the reaction sequence involving *oxidation*, *hydration*, *oxidation*, and *cleavage* to produce acetyl-CoA and an activated thioester shortened by two carbon atoms. The cleavage reaction requires an additional molecule of CoA.

Step 2 Oxidation: Dehydrogenation at carbons 2 and 3 (alpha and beta carbons)

$$RCH_2CH_2CH_2\overset{\overset{O}{\|}}{C}{-}SCoA + FAD \longrightarrow RCH_2CH{=}CH\overset{\overset{O}{\|}}{C}{-}SCoA + FADH_2$$

Step 3 Hydration: Conversion to secondary alcohol

$$\mathrm{RCH_2CH{=}CH\overset{\overset{\displaystyle O}{\|}}{C}{-}SCoA + H_2O \longrightarrow RCH_2\overset{\overset{\displaystyle OH}{|}}{C}HCH_2\overset{\overset{\displaystyle O}{\|}}{C}{-}SCoA}$$

Step 4 Oxidation: Dehydrogenation of carbon 3 (beta carbon) to a keto group

$$\mathrm{RCH_2\overset{\overset{\displaystyle OH}{|}}{C}HCH_2\overset{\overset{\displaystyle O}{\|}}{C}{-}SCoA + NAD^+ \longrightarrow RCH_2\overset{\overset{\displaystyle O}{\|}}{C}CH_2\overset{\overset{\displaystyle O}{\|}}{C}{-}SCoA + NADH + H^+}$$

Step 5 Carbon-chain cleavage: Reaction with CoA to produce acetyl-CoA and activated thioester of a fatty acid shortened by two carbons

$$\mathrm{RCH_2\overset{\overset{\displaystyle O}{\|}}{C}CH_2\overset{\overset{\displaystyle O}{\|}}{C}{-}SCoA + CoASH \longrightarrow RCH_2\overset{\overset{\displaystyle O}{\|}}{C}{-}SCoA + \underset{\text{Acetyl-CoA}}{CH_3\overset{\overset{\displaystyle O}{\|}}{C}{-}SCoA}}$$

The shortened chain thioester repeats the reaction sequence of oxidation, hydration, oxidation, and cleavage to shorten the carbon chain further and produce another molecule of acetyl-CoA. Thus, for example, eight molecules of acetyl-CoA can be produced from one molecule of palmitic acid.

As in the metabolic pathways for glucose, each reaction in the fatty acid oxidation pathway is enzyme catalyzed. No ATP is directly produced during fatty acid catabolism. Instead, ATP is formed when the reduced coenzymes, $FADH_2$ and NADH, are oxidized by the mitochondrial electron transport system in concert with oxidative phosphorylation. Fatty acid oxidation is aerobic because the products, $FADH_2$ and NADH, can only be reoxidized when oxygen is present.

In general, fatty acid catabolism yields more energy than can be derived from the breakdown of glucose. For example, the reduced coenzymes derived from palmitic acid will yield 123 ATP via electron transport and oxidative phosphorylation. Eight additional ATPs can be obtained from the eight GTPs formed in the citric acid cycle while one ATP is used to start the beta oxidation process. Thus, the 16 carbons of palmitic acid yield a total of 130 ATP or about 8.1 ATP per carbon atom. In contrast, glucose can yield between 36 and 38 ATPs as its six carbons are completely oxidized to carbon dioxide. About six ATPs per carbon atom are gained from glucose as compared with about eight ATPs per carbon atom from a fatty acid. Because fatty acid carbons are, in general, more reduced than glucose carbons, fatty acids are a more potent source of energy and yield more ATP molecules during metabolism.

Not surprisingly, the energy storage molecule of choice in the human body is the fatty acid. On the average, a 70-kg male adult carries about 15 kg of fat (as triacylglycerols) but only about 0.22 kg of carbohydrate (as glycogen). Fat is such a good energy storage that many obese people could exist for about one year without food. Unfortunately, fat is not the best energy supply molecule for all tissues. For example, the brain normally derives all of its energy needs from glucose, using about 60% of all glucose metabolized by an adult at rest. Thus, fatty acids are not a universal energy source although they are our most concentrated supply of energy.

37.3 Fat Storage and Utilization

Fats (triacylglycerols) are stored primarily in adipose tissue, which is widely distributed in the body. Fat tends to accumulate under the skin (subcutaneous fat), in the abdominal region, and around some internal organs, especially the kidneys. Fat is deposited around internal organs as a shock absorber, or cushion. Subcutaneous fat acts as an insulating blanket. It is developed to an extreme degree in mammals such as seals, walruses, and whales that live in cold water.

Fat is the major reserve of potential energy. It is metabolized continuously. Stored fat does not remain in the body unchanged; there is a rapid exchange between the triacylglycerols of the plasma lipoproteins and the triacylglycerols in the adipose tissue. The plasma lipoprotein–bound triacylglycerols are broken down by an enzyme (lipoprotein lipase) that is found on the walls of all capillaries, and the resulting free fatty acids are transported into the adipose cells. When the body needs energy from fat, adipose cell enzymes hydrolyze triacylglycerols and the fatty acids are exported to other body tissues. This vital process is under careful hormonal control. For example, a part of the "fight or flight" response caused by the hormone epinephrine (adrenalin) is an increased fatty acid output from the adipose tissue. Conversely, when there is more energy available in the diet than the body needs, the excess energy is used to make body fat. Continued eating of more food than the body can use results in obesity.

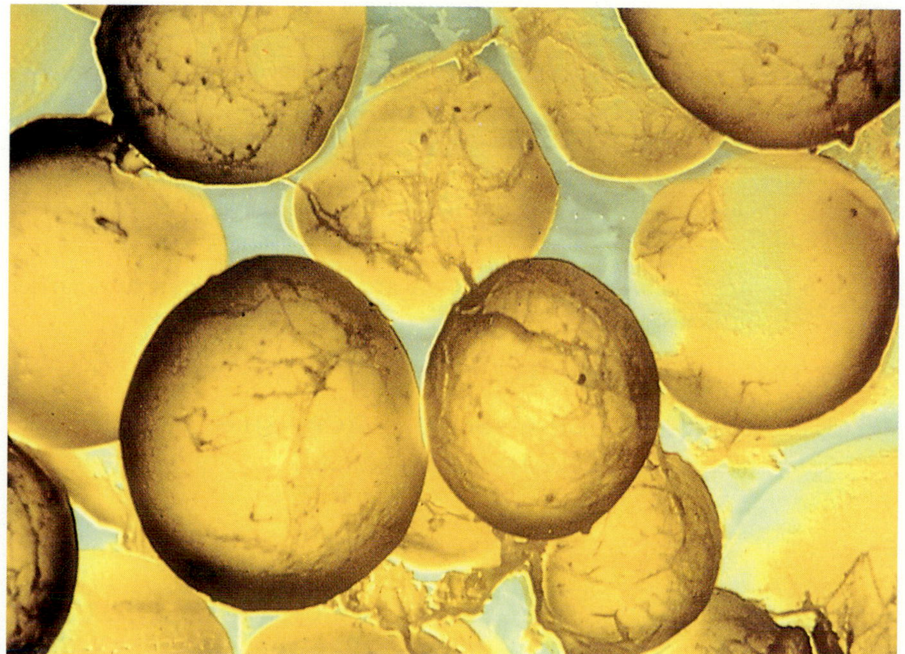

◀ **Photomicrograph of fat cells. These fat cells provide a major reserve of potential energy.**

37.4 Biosynthesis of Fatty Acids (Lipogenesis)

lipogenesis

The biosynthesis of fatty acids from acetyl-CoA is called **lipogenesis**. Acetyl-CoA may be obtained from the catabolism of carbohydrates, fats, or proteins. Fatty acids, in turn, may be combined with glycerol to form triacylglycerols, which are stored in adipose tissue. Consequently, lipogenesis is the pathway by which all three of the major classes of nutrients may ultimately be converted to body fat.

Is lipogenesis just the reverse of fatty acid oxidation (beta oxidation)? By analogy with carbohydrate metabolism (compare glycolysis with gluconeogenesis, see Chapter 36), we might expect this to be the case. However, fatty acid biosynthesis is not simply a reversal of fatty acid oxidation. The following are the major differences between the two pathways:

1. Fatty acid catabolism occurs in the mitochondria, but fatty acid anabolism (lipogenesis) occurs in the cytoplasm.
2. Lipogenesis requires a set of enzymes that are different from the enzymes used in the catabolism.
3. In the anabolic pathway (lipogenesis), the growing fatty acid chain is linked to a special acyl carrier protein (ACP—SH). ACP—SH acts as a handle to transfer the growing chain from one enzyme to another through the series of enzyme-catalyzed reactions in the pathway. Coenzyme A is the carrier in fatty acid catabolism.
4. A preliminary set of reactions, involving malonyl-CoA, occurs for each two-carbon addition cycle in the synthesis. Malonyl is a three-carbon group and has no counterpart in the catabolic pathway. Malonyl-CoA is synthesized from acetyl-CoA and carbon dioxide in the presence of the enzyme acetyl-CoA carboxylase, ATP, and the vitamin biotin.

$$\underset{\text{Acetyl-CoA}}{CH_3\overset{O}{\overset{\|}{C}}-SCoA} + CO_2 \xrightarrow[\text{Acetyl-CoA carboxylase}]{\text{ATP, biotin}} \underset{\text{Malonyl-CoA}}{HO\overset{O}{\overset{\|}{C}}CH_2\overset{O}{\overset{\|}{C}}-SCoA}$$

The biosynthesis of a fatty acid occurs by addition of successive two-carbon-atom increments to a lengthening chain starting with acetyl-CoA. Each incremental addition follows this five-step pathway or reaction sequence.

Step 1 Acetyl-CoA and malonyl-CoA are linked to separate acyl carrier proteins:

$$\underset{\text{Acetyl-CoA}}{CH_3\overset{O}{\overset{\|}{C}}-SCoA} + \underset{\text{Malonyl-CoA}}{HO\overset{O}{\overset{\|}{C}}CH_2\overset{O}{\overset{\|}{C}}-SCoA} + \underset{\text{Acyl protein carrier}}{2\ ACP-SH} \longrightarrow$$

$$\underset{\text{Acetyl-ACP}}{CH_3\overset{O}{\overset{\|}{C}}-SACP} + \underset{\text{Malonyl-ACP}}{HO\overset{O}{\overset{\|}{C}}CH_2\overset{O}{\overset{\|}{C}}-SACP} + 2\ CoASH$$

Step 2 Acetyl-ACP and malonyl-ACP condense, with loss of carbon dioxide (decarboxylation):

$$\underset{\text{Acetyl-ACP}}{CH_3\overset{O}{\overset{\|}{C}}—SACP} + \underset{\text{Malonyl-ACP}}{HO\overset{O}{\overset{\|}{C}}CH_2\overset{O}{\overset{\|}{C}}—SACP} \longrightarrow$$

$$\underset{\text{Acetoacetyl-ACP}}{CH_3\overset{O}{\overset{\|}{C}}CH_2\overset{O}{\overset{\|}{C}}—SACP} + CO_2 + ACP—SH$$

The three steps that follow are approximate reversals of three steps in fatty acid beta oxidation (Section 37.2).

Step 3 Reduction: Hydrogenation of carbon 3 (β-keto group)

$$CH_3\overset{O}{\overset{\|}{C}}CH_2\overset{O}{\overset{\|}{C}}—SACP + NADPH + H^+ \longrightarrow$$

$$\underset{\beta\text{-Hydroxybutyryl-ACP}}{CH_3\overset{OH}{\overset{|}{C}}HCH_2\overset{O}{\overset{\|}{C}}—SACP} + NADP^+$$

Step 4 Dehydration: Formation of a double bond between carbons 2 and 3

$$CH_3\overset{OH}{\overset{|}{C}}HCH_2\overset{O}{\overset{\|}{C}}—SACP \longrightarrow \underset{\text{Crotonyl-ACP}}{CH_3CH{=}CH\overset{O}{\overset{\|}{C}}—SACP} + H_2O$$

Step 5 Reduction: Hydrogenation of carbons 2 and 3

$$CH_3CH{=}CH\overset{O}{\overset{\|}{C}}—SACP + NADPH + H^+ \longrightarrow$$

$$\underset{\text{Butyryl-ACP}}{CH_3CH_2CH_2\overset{O}{\overset{\|}{C}}—SACP} + NADP^+$$

This completes the first cycle of the synthesis; the chain has been lengthened by two carbon atoms. Biosynthesis of longer-chain fatty acids proceeds by a series of such cycles, each lengthening the carbon chain by an increment of two carbon atoms. The next cycle would begin with the reaction of butyryl-ACP and malonyl-ACP, leading to a six-carbon chain, and so on. This synthesis commonly produces palmitic acid (16 carbons) as its end product. The synthesis of palmitic acid from acetyl-CoA and malonyl-CoA requires cycling through the series of steps seven times. The condensed equation for the formation of palmitic acid is

$$\underset{\text{Acetyl-CoA}}{CH_3\overset{\overset{\displaystyle O}{\|}}{C}{-}SCoA} + 7\ \underset{\text{Malonyl-CoA}}{HO\overset{\overset{\displaystyle O}{\|}}{C}CH_2\overset{\overset{\displaystyle O}{\|}}{C}{-}SCoA} + 14\ NADPH + 14\ H^+ \longrightarrow$$

$$\underset{\text{Palmitic acid}}{CH_3(CH_2)_{14}COOH} + 7\ CO_2 + 6\ H_2O + 8\ CoASH + 14\ NADP^+$$

Nearly all naturally occurring fatty acids have even numbers of carbon atoms. A sound reason for this fact is that both the catabolism and the synthesis proceed by two-carbon increments.

In conclusion, it should be noted that the metabolism of fats has some features in common with that of carbohydrates. The acetyl-CoA produced in the catabolism of both carbohydrates and fatty acids can be used as a raw material for making other substances and as an energy source. When acetyl-CoA is oxidized via the citric acid cycle, more potential energy can be trapped in ATP. The ATP in turn serves as the source of energy needed for the production of other substances, including the synthesis of carbohydrates and fats.

37.5 Amino Acid Metabolism

Amino acids serve an important and unique role in cellular metabolism; they are the building blocks of proteins and also provide most of the nitrogen for other nitrogen containing compounds.

Amino acid metabolism differs markedly from the biochemistry of carbohydrates or fatty acids. Amino acids always contain nitrogen, and the chemistry of this element presents unique problems for the cell. In addition, there is no structure common to all the carbon skeletons of amino acids. The carbohydrates can share a common metabolic pathway, as can fatty acids, because they share common structures. But the carbon skeletons of amino acids vary widely, and the cell must use a different metabolic pathway for almost every amino acid. Thus the metabolism of the carbon structures of amino acids is complex. In the sections that follow, a brief overview of amino acid metabolism will be presented together with a more detailed examination of nitrogen metabolism.

37.6 Metabolic Nitrogen Acquisition

Nitrogen is an important component of many biochemicals. In addition to being a component of amino acids, nitrogen is found in nucleic acids, hemoglobin, and many vitamins. Every cell must have a continuous nitrogen supply. This supply might seem easy to obtain because the atmosphere contains about 78% free (elemental) nitrogen. Unfortunately, most cells cannot use elemental nitrogen.

Most higher plant and animal cells require nitrogen that is bonded to other elements before it is biochemically useful. However, most nitrogen on earth exists as N_2 molecules, the two atoms being bonded together by a strong and unreactive triple bond. On an industrial scale, high temperature (400–500°C), high pressure (several hundred atmospheres), and a catalyst are required to react nitrogen gas with the reducing agent, hydrogen, to form ammonia. In the biosphere only a few organisms have the metabolic machinery necessary to use the abundant atmospheric nitrogen. These are procaryotes including *Azobacter* species, *Clostridium pasteurianum*, and *Rhizobium* species. By converting nitrogen gas to compounds, these organisms make nitrogen available to the rest of the biological world.

The conversion of diatomic nitrogen to a biochemically useful form is termed **nitrogen fixation**. Nitrogen is fixed by several methods, including (1) soil bacteria, (2) lightning, (3) Haber synthesis of ammonia, and (4) high-temperature processes such as combustion reactions. Of the procaryotes that are able to catalyze nitrogen fixation, *Rhizobium* bacteria deserve special attention. These bacteria flourish on the roots of legumes, such as clover and soybeans, in a symbiotic relationship. The bacteria have degenerated essentially into nitrogen fixing machines that are maintained by the plants in root nodules. The legumes even provide a special hemoglobin protein to assist the bacteria in their task.

nitrogen fixation

▲ Soybean root nodules containing nitrogen-fixing bacteria.

Currently, in many university and industrial genetic-engineering laboratories, scientists are attempting to transfer the nitrogen fixation capability of procaryotes to higher plants. To accomplish this transfer, the appropriate genes must be removed from procaryotic cells and spliced into the DNA of higher plants (see Chemistry in Action, Chapter 33). This research is particularly challenging and, if successful, would have potentially great rewards. Success would mean that grains and other commercially important crops could be grown in poor soils using little or no nitrogen fertilizer. This achievement would create a vast increase in the world's food production capability.

Higher plants use nitrogen compounds primarily to produce proteins, which, in turn, enter the animal food chain. Both plant and animal proteins are important human nutrients.

Protein is digested and absorbed to provide the amino acid dietary requirements (see Chapter 34). Once absorbed, an amino acid can be used in one of the following ways:

1. Be incorporated into a protein
2. Be utilized in the synthesis of other nitrogenous compounds such as nucleic acids
3. Be deaminated to a keto acid, which either can be used to synthesize other compounds or can be oxidized to carbon dioxide and water to provide energy

Absorbed amino acids enter the amino acid pool. The **amino acid pool** is the total supply of amino acids available for use throughout the body. The amount of amino acids in the pool is maintained in balance with other cellular nitrogen pools (see Figure 37.1).

amino acid pool

One particularly important nitrogen pool is composed of the proteins in all the body's tissues. Amino acids continually move back and forth between the amino acid pool and the tissue proteins. In other words, our body proteins are constantly being broken down and resynthesized. The rate of turnover varies

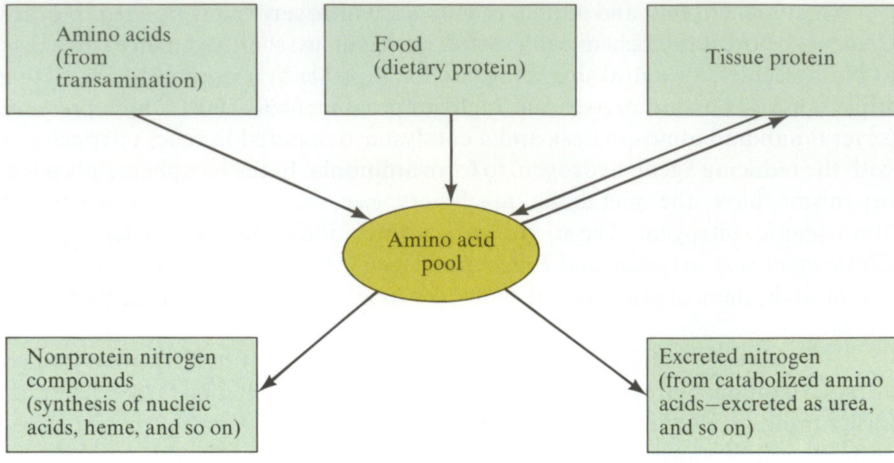

FIGURE 37.1 ▶ Major biological nitrogen pools showing the central amino acid pool.

with different proteins, but research with tagged (isotopically labeled) amino acids has shown that some proteins from liver and other active tissues have a half-life of less than a week, whereas the half-life of some muscle proteins is about six months.

When there is a dietary deficiency of protein (or amino acids), the tissue protein acts as an emergency source to maintain the amino acid pool. For example, one symptom of chronic undernourishment is a wasting of muscle tissue protein. The tissue protein is broken down to yield amino acids. The amino acids may then be used to supply other nitrogen pools or simply may be converted to cellular energy.

nitrogen balance

positive nitrogen balance

negative nitrogen balance

In a healthy, well-nourished adult, the amount of nitrogen excreted is equal to the amount of nitrogen ingested. Such a person is said to be in nitrogen equilibrium, or **nitrogen balance**. In a growing child, the amount of nitrogen excreted is less than that consumed, and the child is in **positive nitrogen balance**. A fasting or starving person or one suffering from certain diseases excretes more nitrogen than is ingested; such a person is in **negative nitrogen balance**. A person on a diet that lacks one or more essential amino acids is in negative nitrogen balance. Tissue proteins can be broken down, but resynthesis is blocked if an amino acid that the body cannot synthesize is missing from the diet. More nitrogen is then excreted than consumed, because the body cannot synthesize specific tissue proteins. The nitrogen balance depends on both the amount and the nature of the nitrogen in the diet.

37.7 Amino Acid Utilization

Amino acids are important in metabolism as carriers of usable nitrogen. If an amino acid is not directly incorporated into tissue proteins, its nitrogen may be incorporated into a variety of other molecules such as amino acids, nucleic acid bases, the heme of hemoglobin, and some lipids. In general, when an amino acid is to be used for some purpose other than protein synthesis, the amino acid carbon skeleton is separated from the amino acid nitrogen.

A process called transamination is responsible for most of the nitrogen transfer to and from amino acids. **Transamination** is the exchange of an oxygen atom for an amino group between an α-keto acid and an α-amino acid:

transamination

$$\underset{\alpha\text{-Amino acid}}{RCH(NH_2)-COOH} + \underset{\alpha\text{-Keto acid}}{R'C(=O)-COOH} \rightleftharpoons RC(=O)-COOH + R'CH(NH_2)-COOH$$

Transamination may involve many different molecules, with each different transamination requiring a different enzyme (transaminase). For example, one enzyme (glutamic-pyruvic transaminase) catalyzes the conversion of pyruvic acid to L-alanine:

$$\underset{\text{Pyruvic acid}}{CH_3C(=O)-COOH} + \underset{\text{L-Glutamic acid}}{NH_2-CH(CH_2CH_2COOH)-COOH} \rightleftharpoons \underset{\text{L-Alanine}}{CH_3CH(NH_2)-COOH} + \underset{\alpha\text{-Ketoglutaric acid}}{O{=}C(CH_2CH_2COOH)-COOH}$$

A different enzyme catalyzes the production of L-aspartic acid from oxaloacetic acid:

$$\underset{\text{Oxaloacetic acid}}{O{=}C(CH_2COOH)-COOH} + \underset{\text{L-Glutamic acid}}{NH_2-CH(CH_2CH_2COOH)COOH} \rightleftharpoons \underset{\text{L-Aspartic acid}}{NH_2-CH(CH_2COOH)COOH} + \underset{\alpha\text{-Ketoglutaric acid}}{O{=}C(CH_2CH_2COOH)-COOH}$$

Note that both of these reactions involve the conversion of L-glutamic acid to α-ketoglutaric acid. In fact, most transaminations use L-glutamic acid. Thus, this amino acid plays a central role in cellular nitrogen transfer.

Transamination is the first step in the conversion of the carbon skeletons of amino acids to energy-storage compounds. Amino acids that are used to produce glucose and glycogen are termed **glucogenic amino acids**. Most amino acids are glucogenic (see Table 37.1). But some amino acids are converted to acetyl-CoA. When fed to starving animals, these amino acids cause an increase in the rate of ketone-body formation and are, therefore, called **ketogenic amino acids** (see Section 37.10). Only leucine is completely ketogenic, but a few amino acids can be converted to either glucose or acetyl-CoA and are both ketogenic and glucogenic.

glucogenic amino acids

ketogenic amino acids

TABLE 37.1 Classification of Amino Acids as Sources of Energy-Storage Molecules

Glucogenic	Ketogenic and glucogenic	Ketogenic
Alanine	Isoleucine	Leucine
Arginine	Lysine	
Aspartic acid	Phenylalanine	
Asparagine	Tyrosine	
Cysteine		
Glutamic acid		
Glutamine		
Glycine		
Histidine		
Methionine		
Proline		
Serine		
Threonine		
Tryptophan		
Valine		

The amino acid pool of Figure 37.1 can now be described in more detail. At the center of this pool is L-glutamic acid, and other amino acids can either add or remove nitrogen from this central compound. Given the important role of L-glutamic acid, it is not surprising that there are other cellular reactions by which nitrogen is interchanged with this amino acid. L-Glutamic acid can accept a second nitrogen atom and form L-glutamine:

$$NH_3 + \underset{\text{L-Glutamic acid}}{NH_2{-}CH(CH_2CH_2C(=O){-}OH)COOH} \rightleftharpoons \underset{\text{L-Glutamine}}{NH_2{-}CH(CH_2CH_2C(=O){-}NH_2)COOH} + H_2O$$

Although this reaction can be considered as the simple addition of ammonia to yield an amide, the actual cellular reactions are more complex. The product, L-glutamine, also serves in biological nitrogen transfer, the amide nitrogen being transferable in a number of cellular reactions.

It is worthwhile to consider the synthesis of L-glutamine in more detail. Ammonia is a base and therefore is toxic to the cell. When ammonia forms an amide bond to L-glutamic acid, the nitrogen becomes less basic and also nontoxic. Thus L-glutamine serves as a safe package for transporting nitrogen. In the human body, L-glutamine is the major compound for transferring nitrogen from one cell to another via the bloodstream.

37.8 Nitrogen Excretion and the Urea Cycle

Nitrogen—unlike carbon, hydrogen, and oxygen—is often conserved for reuse by the cell. But the cell excretes nitrogen when it has an excess of this element or when the carbon skeletons of nitrogen containing compounds are needed for other purposes. Two examples from human nutrition arise: (1) when we consume more protein than is needed (an excess of nitrogen containing molecules) or (2) when we experience starvation (protein is destroyed to provide reduced carbons for energy). Under normal conditions, adult humans excrete about 20 g of urea nitrogen per day.

The nitrogen elimination process poses a major problem for the cell. The simplest excretion product is ammonia, but ammonia is basic and, therefore, toxic to the cell. Fish can excrete ammonia through their gills because ammonia is soluble and is swept away by water passing through the gills. Land animals and birds excrete nitrogen in less toxic forms. Birds and reptiles excrete nitrogen as the white solid uric acid, a derivative of the purine bases. Mammals excrete the water-soluble compound urea. Both of these compounds contain a high percentage of nitrogen in a nontoxic form.

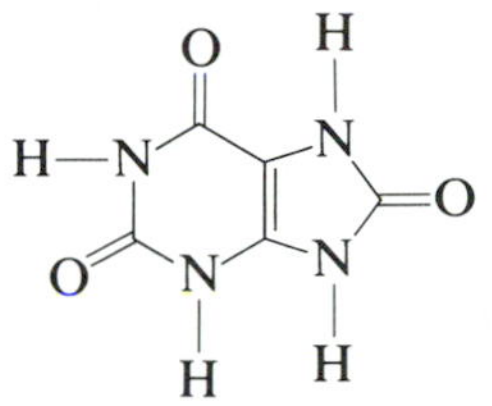

Uric acid

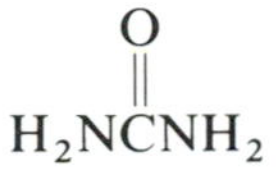

$$H_2N\overset{\overset{O}{\|}}{C}NH_2$$

Urea

Urea synthesis in mammals follows a pathway called the urea cycle (Figure 37.2), which takes place in the liver. Ammonia is first produced from L-glutamic acid in an oxidation–reduction reaction:

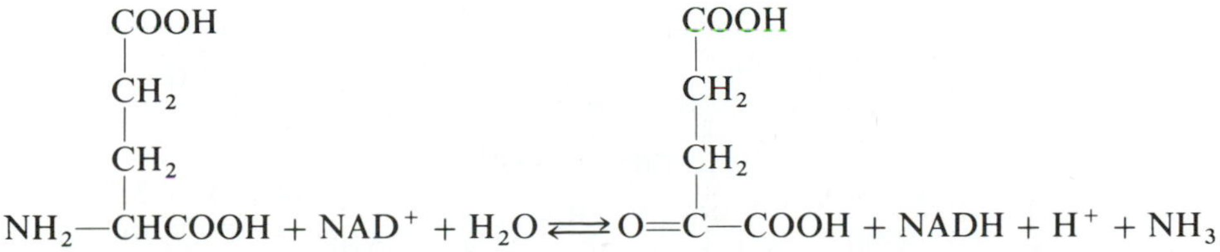

$$\underset{\text{L-Glutamic acid}}{NH_2{-}CH(CH_2CH_2COOH)COOH} + NAD^+ + H_2O \rightleftharpoons \underset{\alpha\text{-Ketoglutaric acid}}{O{=}C(CH_2CH_2COOH){-}COOH} + NADH + H^+ + NH_3$$

The ammonia quickly reacts with bicarbonate and ATP to form carbamoyl phosphate:

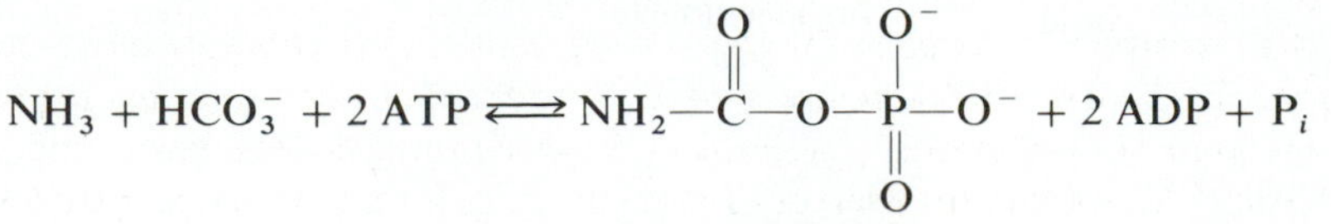

$$NH_3 + HCO_3^- + 2\,ATP \rightleftharpoons \underset{\text{Carbamoyl phosphate}}{NH_2{-}\overset{\overset{O}{\|}}{C}{-}O{-}\overset{\overset{O^-}{|}}{\underset{\underset{O}{\|}}{P}}{-}O^-} + 2\,ADP + P_i$$

FIGURE 37.2 ▶ Urea cycle.

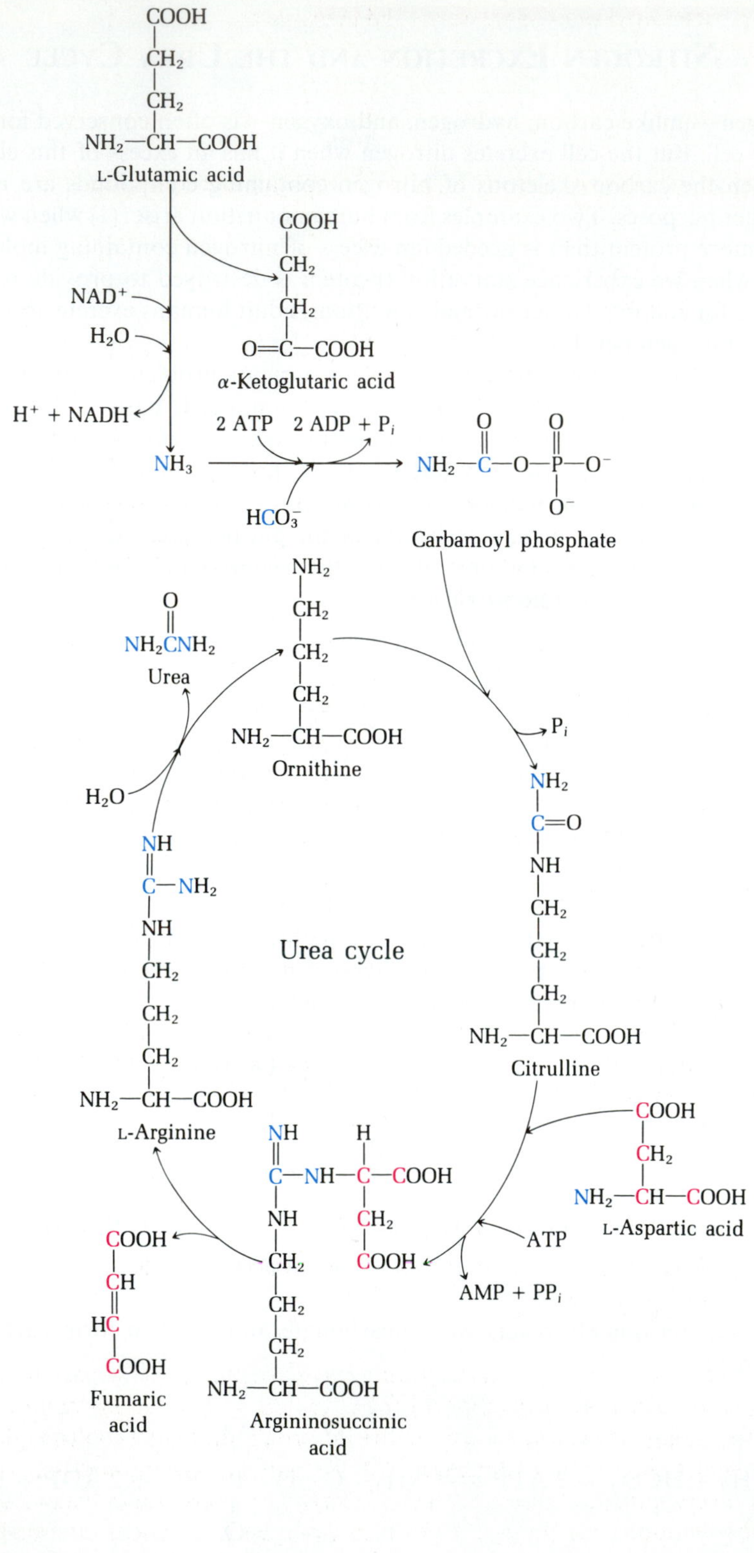

COOH
CH2
CH2
NH2—CH—COOH
L-Glutamic acid
NAD+
H2O
H+ + NADH
NH3
COOH
CH2
CH2
O=C—COOH
α-Ketoglutaric acid
2 ATP
2 ADP + Pi
HCO3-
NH2—C—O—P—O-
Carbamoyl phosphate
NH2
CH2
CH2
CH2
NH2—CH—COOH
Ornithine
O
NH2CNH2
Urea
H2O
Pi
NH2
C=O
NH
CH2
CH2
CH2
NH2—CH—COOH
Citrulline
Urea cycle
NH
C—NH2
NH
CH2
CH2
CH2
NH2—CH—COOH
L-Arginine
COOH
CH2
NH2—CH—COOH
L-Aspartic acid
ATP
AMP + PPi
NH
C—NH—C—COOH
H
CH2
COOH
NH
CH2
CH2
CH2
NH2—CH—COOH
Argininosuccinic acid
COOH
CH
HC
COOH
Fumaric acid

Finally, carbamoyl phosphate enters the urea cycle:

$$\underset{\text{Carbamoyl phosphate}}{NH_2{-}\overset{\overset{\displaystyle O}{\|}}{C}{-}O{-}\overset{\overset{\displaystyle O^-}{|}}{\underset{\underset{\displaystyle O}{\|}}{P}}{-}O^-} + \underset{\text{L-Aspartic acid}}{NH_2{-}\overset{\overset{\displaystyle COOH}{|}}{\overset{\displaystyle CH_2}{|}}\hspace{-0.3em}\underset{\underset{\displaystyle H}{|}}{C}{-}COOH} + ATP \xrightarrow{\text{Urea cycle}}$$

$$\underset{\text{Urea}}{NH_2{-}\overset{\overset{\displaystyle O}{\|}}{C}{-}NH_2} + \underset{\text{Fumaric acid}}{\overset{\overset{\displaystyle COOH}{|}}{\overset{\displaystyle CH}{\|}}\hspace{-0.3em}\underset{\underset{\displaystyle COOH}{|}}{HC}} + PP_i + P_i + AMP$$

Like the intermediate compounds of the citric acid cycle, the compounds intermediate in the urea cycle do not appear in the overall reactions and serve only a catalytic function in the production of urea. Note that the urea cycle intermediates are α-amino acids. These are amino acids that are rarely used in protein synthesis. Their primary role is in the formation of urea.

Also it is important to recognize that the cell must expend energy, ATP, to produce urea. Formation of a nontoxic nitrogen excretion product is essential. In fact, impairment of the urea cycle is one of the major problems of liver cirrhosis caused by alcoholism. As liver function is impaired, more nitrogen is excreted as ammonia, leading to toxic effects.

Finally, let us look at the sources of the nitrogen that is excreted as urea. One nitrogen atom in each urea molecule comes from L-glutamic acid. The other nitrogen atom comes from L-aspartic acid, which may have gained its nitrogen from L-glutamic acid by transamination. Thus, the amino acid that is central to nitrogen transfer reactions is also the major contributor to nitrogen excretion.

37.9 Acetyl-CoA, a Central Molecule in Metabolism

As we think back through metabolism, we can identify some especially important compounds, molecules that are central to that portion of biochemistry. For example, glucose is the central compound in carbohydrate metabolism; glutamic acid is central in amino acid metabolism. There is one compound that is at the hub of all common metabolic processes; acetyl-CoA is central in the metabolism of carbon compounds (Figure 37.3). This molecule is a critical intermediate in the processes that form and break down both fats and amino acids. In addition, essentially all compounds that enter the citric acid cycle first must be catabolized

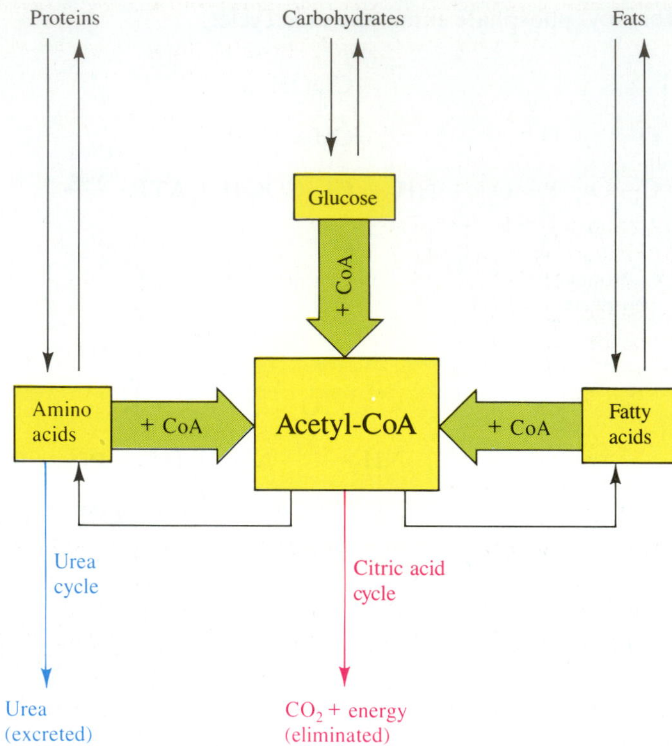

FIGURE 37.3 ► Simplified diagram showing acetyl-CoA at the hub of protein, carbohydrate, and fat metabolism.

to acetyl-CoA. This section will examine the characteristics of acetyl-CoA that make it a central metabolic compound and the advantages of a centralized metabolism for the cell.

Recall that acetyl-CoA consists of a small two-carbon unit (acetyl group) bonded by thioester linkage to a large organic coenzyme molecule, coenzyme A (CoA):

$$\underbrace{CH_3-\overset{\overset{\displaystyle O}{\|}}{C}}_{\text{Acetyl group}}-\underset{\uparrow \text{Thioester linkage}}{S}-\boxed{\text{Coenzyme A}}$$

This structure makes for an almost ideal central metabolic molecule with the following major advantages:

1. *Potential Use in a Wide Variety of Syntheses.* The small size and simple structure of the two-carbon acetyl fragment enables this molecule to be used to build a variety of diverse structures. Complex molecules with very different shapes and functions such as long chain fatty acids, amino acids, and steroid hormones are synthesized from acetyl-CoA.
2. *Special Reactivity of Bonds.* The thioester causes both carbons in the acetyl fragment to be specially reactive. These carbons are "primed" to form new bonds as the acetyl-CoA enters various metabolic pathways.

Chemistry in Action

Ketone Bodies—A Stress Response

Extreme circumstances such as long-term, continuous exercise, starvation, and untreated diabetes mellitus cause a change in metabolism. In such cases the energy obtainable from glucose is severely limited. In response the body produces small molecules from fatty acids, called *ketone bodies*, as glucose substitutes. These molecules include acetoacetic acid and two of its derivatives, β-hydroxybutyric acid and acetone.

$$CH_3\overset{\overset{\displaystyle O}{\|}}{C}CH_2COOH$$

Acetoacetic acid

$$CH_3\overset{\overset{\displaystyle OH}{|}}{C}HCH_2COOH$$

β-Hydroxybutyric acid

$$CH_3\overset{\overset{\displaystyle O}{\|}}{C}CH_3$$

Acetone

Ketone bodies are highly soluble in blood and are able to provide cellular energy via beta oxidation and the citric acid cycle. When blood glucose becomes less available, some body tissues can adapt to using ketone bodies as substitute energy-supply molecules. For example, the brain, which depends upon glucose more than other human tissues, shifts its metabolism to use about 70% ketone bodies after 40 days of minimal food intake.

An increase in ketone-body concentration is an emergency response to an extreme circumstance. This response can have dangerous side effects. For example, most ketone bodies are carboxylic acids. When their concentration increases in the blood, the acidity rises, causing a condition called *ketosis*, or *ketoacidosis*. Ketosis can affect respiration and cause a general deterioration of normal body function.

3. *Structure That Is Recognizable by a Wide Variety of Enzymes.* Coenzyme A (CoA) acts as a kind of handle for the various enzymes that catalyze reactions of the acetyl group. Because many enzymes bind tightly to CoA, the acetyl group of acetyl-CoA can be involved in a great number of diverse reactions.

Acetyl-CoA can be compared with ATP as a central metabolic molecule. Remember that ATP (1) has a potential use in a wide variety of syntheses, (2) has a special reactivity in its phosphate anhydride bonds, and (3) has a structure that is recognizable and used by a wide variety of enzymes. For these important reasons, ATP serves as the "common energy currency" for the cell. In analogous fashion, acetyl-CoA might be termed the "common carbon currency" for the cell.

A consideration of these central metabolic compounds raises an important question: How does centralization aid the cell? We have seen numerous examples of centralization in biochemistry. Not only are there important central metabolites such as ATP and acetyl-CoA, but pathways such as the citric acid cycle centralize the cellular metabolic machinery. A general answer to this question is that centralization improves the efficiency of metabolism and ensures that biochemistry can be under careful metabolic control.

Greater efficiency results from designing central metabolic pathways to handle a variety of different nutrients. For example, the citric acid cycle completes carbon-oxidation for all energy supplying nutrients that are first converted to the central metabolite, acetyl-CoA.

Greater control results from the dependence of many "feeder" pathways on a single central process. Control of a central path like the citric acid cycle will affect in a coordinated way the metabolism of a variety of nutrients. As scientists have learned more about metabolism it has become clear that centralization is an important attribute of the chemistry of life.

Concepts in Review

1. Briefly explain what characteristic of fatty acids allows them to provide large amounts of metabolic energy.
2. Briefly describe Knoop's experiments on fatty acid oxidation and degradation and the conclusions derived from them.
3. Explain what is meant by beta oxidation and beta cleavage in relation to the metabolism of fatty acids.
4. Tell the purpose of ketone-body production and indicate a dangerous side effect.
5. Briefly describe the biosynthesis of fatty acids using palmitic acid as an example.
6. List three major differences between beta oxidation and fatty acid synthesis.
7. Briefly describe how amino acid metabolism differs from carbohydrate and fatty acid metabolism.
8. List the possible metabolic fates of amino acids in humans.
9. Describe the major purpose of transamination.
10. Briefly explain the importance of L-glutamic acid and L-glutamine in nitrogen metabolism.
11. Explain how metabolism of proteins (amino acids) is tied into that of carbohydrates and fats.
12. Explain how a lack of essential amino acids in the diet affects the nitrogen balance.
13. Explain the purpose of the urea cycle.
14. Summarize the importance of acetyl-CoA in metabolism.
15. Discuss the major advantages of a centralized metabolism.

Equations in Review

One Cycle of Beta Oxidation

$$RCH_2CH_2CH_2\overset{\overset{\displaystyle O}{\|}}{C}{-}SCoA + FAD + NAD^+ + H_2O + CoASH \longrightarrow$$

$$RCH_2\overset{\overset{\displaystyle O}{\|}}{C}{-}SCoA + FADH_2 + NADH + H^+ + CH_3\overset{\overset{\displaystyle O}{\|}}{C}{-}SCoA$$

One Cycle of Fatty Acid Synthesis

$$CH_3\overset{\overset{\displaystyle O}{\|}}{C}\text{—SCoA} + HO\overset{\overset{\displaystyle O}{\|}}{C}CH_2\overset{\overset{\displaystyle O}{\|}}{C}\text{—SCoA} + 2\,NADPH + 2\,H^+ + \text{ACP—SH} \longrightarrow$$

$$CH_3CH_2CH_2\overset{\overset{\displaystyle O}{\|}}{C}\text{—SACP} + 2\,NADP^+ + H_2O + 2\,CoASH + CO_2$$

Transamination

$$\underset{\substack{|\\ NH_2 \\ \alpha\text{-Amino}\\ \text{acid}}}{RCH}\text{—COOH} + \underset{\substack{\alpha\text{-Keto}\\ \text{acid}}}{R'\overset{\overset{\displaystyle O}{\|}}{C}\text{—COOH}} \rightleftharpoons \underset{\substack{\alpha\text{-Amino}\\ \text{acid}}}{R\overset{\overset{\displaystyle O}{\|}}{C}\text{—COOH}} + \underset{\substack{|\\ NH_2 \\ \alpha\text{-Keto}\\ \text{acid}}}{R'CH}\text{—COOH}$$

Urea Cycle (overall reaction)

$$HOOC\underset{\substack{|\\ NH_2}}{C}HCH_2COOH + ATP + H_2O + NH_2\text{—}\overset{\overset{\displaystyle O}{\|}}{C}\text{—O—}\underset{\substack{|\\ O^-}}{\overset{\overset{\displaystyle O}{\|}}{P}}\text{—}O^- \longrightarrow$$

$$HOOCCH{=}CHCOOH + P_i + PP_i + AMP + NH_2\overset{\overset{\displaystyle O}{\|}}{C}NH_2$$

EXERCISES

1. What major characteristic of a fatty acid allows it to serve as an energy storage molecule?
2. How is it possible to become obese even though very little fat is included in the diet?
3. Briefly describe Knoop's experiment on fatty acid oxidation and catabolism.
4. What is meant by beta oxidation and beta cleavage in relation to the biochemistry of fatty acids?
5. By means of a diagram, outline how caproic acid is catabolized to butyric acid.
6. Why is the acyl carrier protein (ACP) of importance in fatty acid synthesis?
7. Aside from being the source of certain fatty acids, are fats essential in our diet? Explain your answer.
8. Beta oxidation is used by the cell to produce energy, yet no ATPs are formed in this pathway. Can you explain this seeming contradiction?
9. Give the name and structure of a ketone body that does *not* contain a ketone functional group.
10. How are ketone bodies important in energy metabolism?
11. Define *ketosis*.
12. Is fatty acid synthesis simply the reverse of beta oxidation? Briefly explain.
13. Aside from being a food reserve, what are the two principal functions of body fat?
14. In what way is the citric acid cycle involved in obtaining energy from fats?
15. Why is the ATP yield from a six-carbon fatty acid greater than the ATP yield from a six-carbon hexose (glucose)?
16. Outline the parts played by malonyl-CoA and acetyl-CoA in the biosynthesis of fatty acids.
17. List several reasons why acetyl-CoA is considered to be an important, central intermediate in metabolism.
18. Prepare a diagram showing the principal steps in the conversion of starch to body fat.

19. Briefly describe why soybeans are a crop that enriches the soil.

20. What are the possible metabolic fates of amino acids?

21. How might a low-protein diet cause a negative nitrogen balance?

22. Give an example of a reaction that is central to metabolic nitrogen transfer.

23. Briefly describe why L-glutamic acid is considered to be the central amino acid of the amino acid pool.

24. Why is L-glutamine a better nitrogen transport molecule than ammonia?

25. Is L-aspartic acid a ketogenic amino acid? Explain.

26. Write the structural formulas of the compounds produced by transamination of the following acids:

(a) L-Alanine (c) L-Phenylalanine
(b) L-Aspartic acid (d) L-Serine

27. Give the structure of urea.

28. How many ATPs are used to produce urea? Why is the production of urea of metabolic importance?

29. Draw the structures of the two common amino acids that contribute nitrogen to the urea cycle.

30. Which of these statements are correct? Rewrite each incorrect statement to make it correct.

(a) The average carbon in palmitic acid is relatively oxidized, and, thus, this fatty acid contains little stored metabolic energy.
(b) Carbohydrates provide more metabolic energy than fatty acids on a per gram basis.
(c) Based on his experiments, Knoop postulated that fatty acids are broken down via beta oxidation—that is, by a "two-carbon chop."
(d) Fatty acid oxidation is an anaerobic reaction sequence.
(e) The beta oxidation pathway directly produces 38 ATPs.
(f) Acetyl-CoA is a very important product of beta oxidation.
(g) When energy from glucose is severely limited, the body produces and uses ketone bodies to provide cellular energy.
(h) Triacylglycerols stored in adipose tissue are continuously exchanged with triacylglycerols of the plasma lipoproteins.
(i) The catabolism of fatty acids is known as lipogenesis.
(j) Fatty acid anabolism occurs in the cytoplasm.
(k) An acyl carrier protein is used in both beta oxidation and fatty acid synthesis.
(l) Malonyl-CoA is synthesized from acetyl-CoA.
(m) The biosynthesis of fatty acids occurs by successive additions of two-carbon increments.
(n) During a dietary deficiency of protein, tissue protein acts as an emergency source to maintain a balanced amino acid pool.
(o) Most amino acids are glucogenic.
(p) Nitrogen fixation occurs in higher plants.
(q) A positive nitrogen balance means more nitrogen is excreted than is consumed.
(r) A ketogenic amino acid can cause an increase in the rate of ketone body formation.
(s) Transamination allows formation of amino acids from α-keto acids.
(t) Acetyl-CoA can be produced from carbohydrates, fatty acids, and proteins.
(u) Acetyl-CoA consists of an acetyl group and a coenzyme A molecule linked through a phosphate ester bond.
(v) Mammals excrete nitrogen in the form of urea.
(w) Compared with other nitrogen excretion compounds, urea is especially toxic.

Mathematical Review

1. Multiplication Multiplication is a process of adding any given number or quantity to itself a certain number of times. Thus, 4 times 2 means 4 added two times, or 2 added together four times, to give the product 8. Various ways of expressing multiplication are

$$ab \qquad a \times b \qquad a \cdot b \qquad a(b) \qquad (a)(b)$$

All mean a times b, or a multiplied by b, or b times a.

When $a = 16$ and $b = 24$, we have $16 \times 24 = 384$.

The expression $°F = (1.8 \times °C) + 32$ means that we are to multiply 1.8 times °C and add 32 to the product. When °C equal 50,

$$°F = (1.8 \times 50) + 32 = 90 + 32 = 122°F$$

The result of multiplying two or more numbers together is known as the *product*.

2. Division The word *division* has several meanings. As a mathematical expression, it is the process of finding how many times one number or quantity is contained in another. Various ways of expressing division are:

$$a \div b \qquad \frac{a}{b} \qquad a/b$$

All mean a divided by b.

When $a = 15$ and $b = 3$, $\frac{15}{3} = 5$.

The number above the line is called the *numerator*; the number below the line is the *denominator*. Both the horizontal and the slanted (/) division signs also mean "per." For example, in the expression for density, the mass per unit volume:

$$\text{density} = \text{mass/volume} = \frac{\text{mass}}{\text{volume}} = \text{g/mL}$$

The diagonal line still refers to a division of grams by the number of milliliters occupied by that mass. The result of dividing one number into another is called the *quotient*.

3. Fractions and Decimals A fraction is an expression of division, showing that the numerator is divided by the denominator. A *proper fraction* is one in which the numerator is smaller than the denominator. In an *improper fraction*, the numerator is the larger number. A decimal or a decimal fraction is a proper fraction in which the denominator is some power of 10. The decimal fraction is determined by carrying out the division of the proper fraction. Examples of proper fractions and their decimal fraction equivalents are shown in the following table.

Proper fraction		Decimal fraction		Proper fraction
$\frac{1}{8}$	=	0.125	=	$\frac{125}{1000}$
$\frac{1}{10}$	=	0.1	=	$\frac{1}{10}$
$\frac{3}{4}$	=	0.75	=	$\frac{75}{100}$
$\frac{1}{100}$	=	0.01	=	$\frac{1}{100}$
$\frac{1}{4}$	=	0.25	=	$\frac{25}{100}$

4. Addition of Numbers with Decimals To add numbers with decimals we use the same procedure as that used when adding whole numbers, but we always line up the decimal points in the same column. For example, add 8.21 + 143.1 + 0.325

$$\begin{array}{r} 8.21 \\ +143.1 \\ +\quad 0.325 \\ \hline 151.635 \end{array}$$

When adding numbers that express units of measurement, we must be certain that the numbers added together represent the same units. For example, what is the total length of three pieces of glass tubing: 10.0 cm, 125 mm, and 8.4 cm? If we simply add the numbers, we obtain a value of 143.4, but we are not certain what the unit of measurement is. To add these lengths correctly, first change 125 mm to 12.5 cm. Now all the lengths are expressed in the same units and can be added.

$$\begin{array}{r} 10.0\text{ cm} \\ 12.5\text{ cm} \\ 8.4\text{ cm} \\ \hline 30.9\text{ cm} \end{array}$$

5. Subtraction of Numbers with Decimals To subtract numbers containing decimals, we use the same procedure as for subtracting whole numbers, but we always line up the decimal points in the same column. For example, subtract 20.60 from 182.49.

$$\begin{array}{r} 182.49 \\ -\ 20.60 \\ \hline 161.89 \end{array}$$

6. Multiplication of Numbers with Decimals To multiply two or more numbers together that contain decimals, we first multiply as if they were whole numbers. Then, to locate the decimal point in the product, we add together the number of digits to the right of the decimal in all the numbers multiplied together. The product should have this same number of digits to the right of the decimal point.

Multiply 2.05 × 2.05 (total of four digits to the right of the decimal):

$$\begin{array}{r} 2.05 \\ \times\, 2.05 \\ \hline 1025 \\ 4100 \\ \hline 4.2025 \end{array} \quad \text{(four digits to the right of the decimal)}$$

Here are more examples:

$14.25 \times 6.01 \times 0.75 = 64.231875$ (six digits to the right of the decimal)

$39.26 \times 60 = 2355.60$ (two digits to the right of the decimal)

[*Note*: When at least one of the numbers that is multiplied is a measurement, the answer must be adjusted to contain the correct number of significant figures. (See Section 2.5 on significant figures.)]

7. Division of Numbers with Decimals To divide numbers containing decimals, we first relocate the decimal points of the numerator and denominator by moving them to the right as many places as needed to make the denominator a whole number. (Move the decimal of both the numerator and the denominator the same amount and in the same direction.) For example,

$$\frac{136.94}{4.1} = \frac{1369.4}{41}$$

The decimal point adjustment in this example is equivalent to multiplying both numerator and denominator by 10. Now we carry out the division normally, locating the decimal point immediately above its position in the dividend.

$$\begin{array}{r} 33.4 \\ 41\overline{)1369.4} \\ \underline{123} \\ 139 \\ \underline{123} \\ 164 \\ \underline{164} \end{array} \qquad \frac{0.441}{26.25} = \frac{44.1}{2625} = \begin{array}{r} 0.0168 \\ 2625\overline{)44.1000} \\ \underline{2625} \\ 17850 \\ \underline{15750} \\ 21000 \\ \underline{21000} \end{array}$$

[*Note*: When at least one of the numbers in the division is a measurement, the answer must be adjusted to contain the correct number of significant figures. (See Section 2.5 on significant figures.)]

The foregoing examples are merely guides to the principles used in performing the various mathematical operations illustrated. There are, no doubt, shortcuts and other methods, and the student will discover these with experience. Every student of chemistry should learn to use a scientific electronic calculator for solving mathematical problems. The use of a calculator will save many hours of doing tedious longhand calculations. After solving a problem, the student should check for errors and evaluate the answer to see if it is logical and consistent with the data given.

8. Algebraic Equations Many mathematical problems that are encountered in chemistry fall into the following algebraic forms. Solutions to these problems are simplified by first isolating the desired term on one side of the equation. This rearrangement is accomplished by treating both sides of the equation in an identical manner (so as not to destroy the equality) until the desired term is isolated.

(a) $a = \frac{b}{c}$

To solve for a, divide b by c.
To solve for b, multiply both sides of the equation by c.

$$a \times c = \frac{b}{\cancel{c}} \times \cancel{c}$$

$$b = a \times c$$

To solve for c, multiply both sides of the equation by $\frac{c}{a}$.

$$\cancel{a} \times \frac{c}{\cancel{a}} = \frac{b}{\cancel{c}} \times \frac{\cancel{c}}{a}$$

$$c = \frac{b}{a}$$

(b) $\frac{a}{b} = \frac{c}{d}$

To solve for a, multiply both sides of the equation by b.

$$\frac{a}{\cancel{b}} \times \cancel{b} = \frac{c}{d} \times b$$

$$a = \frac{c \times b}{d}$$

To solve for b, multiply both sides of the equation by $\frac{b \times d}{c}$.

$$\frac{a}{\cancel{b}} \times \frac{\cancel{b} \times d}{c} = \frac{\cancel{c}}{\cancel{d}} \times \frac{b \times \cancel{d}}{\cancel{c}}$$

$$b = \frac{a \times d}{c}$$

(c) $a \times b = c \times d$

To solve for a, divide both sides of the equation by b.

$$\frac{a \times \cancel{b}}{\cancel{b}} = \frac{c \times d}{b}$$

$$a = \frac{c \times d}{b}$$

(d) $\dfrac{(b - c)}{a} = d$

To solve for b, first multiply both sides of the equation by a.

$$\frac{\cancel{a}(b - c)}{\cancel{a}} = d \times a$$

$$b - c = d \times a$$

Then add c to both sides of the equation.

$$b - \cancel{c} + \cancel{c} = d \times a + c$$

$$b = (d \times a) + c$$

When $a = 1.8$, $c = 32$, and $d = 35$,

$$b = (35 \times 1.8) + 32 = 63 + 32 = 95$$

9. Exponents, Powers of 10, Expression of Large and Small Numbers In scientific measurements and calculations, we often encounter very large and very small numbers—for example, 0.00000384 and 602,000,000,000,000,000,000,000. These numbers are troublesome to write and awkward to work with, especially in calculations. A convenient method of expressing these large and small numbers in a simplified form is by means of exponents or powers of 10. This method of expressing numbers is known as **scientific** or **exponential notation**.

An *exponent* is a number written as a superscript following another number; it is also called a *power* of that number, and it indicates how many times the number is used as a factor. In the number 10^2, 2 is the exponent, and the number means 10 squared, or 10 to the second power, or $10 \times 10 = 100$. Three other examples are

$$3^2 = 3 \times 3 = 9$$

$$3^4 = 3 \times 3 \times 3 \times 3 = 81$$

$$10^3 = 10 \times 10 \times 10 = 1000$$

For ease of handling, large and small numbers are expressed in powers of 10. Powers of 10 are used because multiplying or dividing by 10 coincides with moving the decimal point in a number by one place. Thus, a number multiplied by 10^1 would move the decimal point one place to the right; 10^2, two places to the right; 10^{-2}, two places to the left. To express a number in powers of 10, we move the decimal point in the original number to a new position, placing it so that the number is a value between 1 and 10. This new decimal number is multiplied by 10 raised to the proper power. For example, to write the number 42,389 in exponential form (powers of 10), the decimal point is placed between the 4 and the 2 (4.2389), and the number is multiplied by 10^4; thus, the number is 4.2380×10^4.

$$42{,}389 = 4.2389 \times 10^4$$

4 3 2 1

The exponent (power) of 10 (4) tells us the number of places that the decimal point has been moved from its original position. If the decimal point is moved to the left, the exponent is a positive number; if it is moved to the right, the exponent is a negative number. To express the number 0.00248 in exponential notation (as a power of 10), the decimal point is moved three places to the right; the exponent of 10 is -3, and the number is 2.48×10^{-3}.

$$0.00248 = 2.48 \times 10^{-3}$$

1 2 3

Study the following examples.

$$1237 = 1.237 \times 10^{3}$$
$$988 = 9.88 \times 10^{2}$$
$$147.2 = 1.472 \times 10^{2}$$
$$2{,}200{,}000 = 2.2 \times 10^{6}$$
$$0.0123 = 1.23 \times 10^{-2}$$
$$0.00005 = 5 \times 10^{-5}$$
$$0.000368 = 3.68 \times 10^{-4}$$

Exponents in multiplication and division. The use of powers of 10 in multiplication and division greatly simplifies locating the decimal point in the answer. In multiplication, first change all numbers to powers of 10, then multiply the numerical portion in the usual manner, and finally add the exponents of 10 algebraically, expressing them as a power of 10 in the product. In multiplication, the exponents (powers of 10) are added algebraically.

$$10^{2} \times 10^{3} = 10^{(2+3)} = 10^{5}$$
$$10^{2} \times 10^{2} \times 10^{-1} = 10^{(2+2-1)} = 10^{3}$$

Multiply:	$40{,}000 \times 4200$
Change to powers of 10:	$4 \times 10^{4} \times 4.2 \times 10^{3}$
Rearrange:	$4 \times 4.2 \times 10^{4} \times 10^{3}$
	$16.8 \times 10^{(4+3)}$
	16.8×10^{7} or 1.68×10^{8} (Answer)

Multiply: 380×0.00020

$3.80 \times 10^{2} \times 2.0 \times 10^{-4}$

$3.80 \times 2.0 \times 10^{2} \times 10^{-4}$

$7.6 \times 10^{(2-4)}$

7.6×10^{-2} or 0.076 (Answer)

Multiply: $125 \times 284 \times 0.150$

$1.25 \times 10^{2} \times 2.84 \times 10^{2} \times 1.50 \times 10^{-1}$

$1.25 \times 2.84 \times 1.50 \times 10^{2} \times 10^{2} \times 10^{-1}$

$5.325 \times 10^{(2+2-1)}$

5.33×10^{3} (Answer)

In division, after changing the numbers to powers of 10, move the 10 and its exponent from the denominator to the numerator, changing the sign of the exponent. Carry out the division in the usual manner and evaluate the power of 10. The following is a proof of the equality of moving the power of 10 from the denominator to the numerator.

$$1 \times 10^{-2} = 0.01 = \frac{1}{100} = \frac{1}{10^2} = 1 \times 10^{-2}$$

In division, change the sign(s) of the exponent(s) of 10 in the denominator and move the 10 and its exponent(s) to the numerator. Then add all the exponents of 10 together. For example,

$$\frac{10^5}{10^3} = 10^5 \times 10^{-3} = 10^{(5-3)} = 10^2$$

$$\frac{10^3 \times 10^4}{10^{-2}} = 10^3 \times 10^4 \times 10^2 = 10^{(3+4+2)} = 10^9$$

Divide: $\dfrac{2871}{0.0165}$

Change to powers of 10: $\dfrac{2.871 \times 10^3}{1.65 \times 10^{-2}}$

Move 10^{-2} to the numerator, changing the sign of the exponent. This is mathematically equivalent to multiplying both numerator and denominator by 10^2.

$$\frac{2.871 \times 10^3 \times 10^2}{1.65}$$

$$\frac{2.871 \times 10^{(3+2)}}{1.65} = 1.74 \times 10^5 \quad \text{(Answer)}$$

Divide: $\dfrac{0.000585}{0.00300}$

$$\frac{5.85 \times 10^{-4}}{3.00 \times 10^{-3}}$$

$$\frac{5.85 \times 10^{-4} \times 10^3}{3.00} = \frac{5.85 \times 10^{(-4+3)}}{3.00}$$

$$= 1.95 \times 10^{-1} \quad \text{or} \quad 0.195 \quad \text{(Answer)}$$

Calculate: $\dfrac{760. \times 300. \times 40.0}{700. \times 273}$

$$\frac{(7.60 \times 10^2) \times (3.00 \times 10^2) \times (4.00 \times 10^1)}{(7.00 \times 10^2) \times (2.73 \times 10^2)}$$

$$\frac{7.60 \times 3.00 \times 4.00 \times 10^2 \times 10^2 \times 10^1}{7.00 \times 2.73 \times 10^2 \times 10^2}$$

$$= 4.77 \times 10^1 \quad \text{or} \quad 47.7 \quad \text{(Answer)}$$

10. Significant Figures in Calculations The result of a calculation based on experimental measurements cannot be more precise than the measurement that has the greatest uncertainty. (See Section 2.5 for additional discussion.)

Addition and subtraction. The result of an addition or subtraction should contain no more digits to the right of the decimal point than are contained in the quantity that has the least number of digits to the right of the decimal point.

Perform the operation indicated and then round off the number to the proper number of significant figures.

142.8 g	
18.843 g	93.45 mL
36.42 g	− 18.0 mL
198.063 g	75.45 mL
198.1 g (Answer)	75.5 mL (Answer)

Multiplication and division. In calculations involving multiplication or division, the answer should contain the same number of significant figures as the measurement that has the least number of significant figures. In multiplication or division the position of the decimal point has nothing to do with the number of significant figures in the answer. Study the following examples:

	Round off to
$2.05 \times 2.05 = 4.2025$	4.20
$18.48 \times 5.2 = 96.096$	96
$0.0126 \times 0.020 = 0.000252$ or	
$1.26 \times 10^{-2} \times 2.0 \times 10^{-2} = 2.520 \times 10^{-4}$	2.5×10^{-4}
$\frac{1369.4}{41} = 33.4$	33
$\frac{2268}{4.20} = 540.$	540.

11. Dimensional Analysis Many problems of chemistry can be solved readily by dimensional analysis using the factor-label or conversion-factor method. Dimensional analysis involves the use of proper units of dimensions for all factors that are multiplied, divided, added, or subtracted in setting up and solving a problem. Dimensions are physical quantities such as length, mass, and time, which are expressed in such units as centimeters, grams, and seconds, respectively. In solving a problem, we treat these units mathematically just as though they were numbers, which gives us an answer that contains the correct dimensional units.

A measurement or quantity given in one kind of unit can be converted to any other kind of unit having the same dimension. To convert from one kind of unit to another, the original quantity or measurement is multiplied or divided by a conversion factor. The key to success lies in choosing the correct conversion factor. This general method of calculation is illustrated in the following examples.

Suppose we want to change 24 ft to inches. We need to multiply 24 ft by a conversion factor containing feet and inches. Two such conversion factors can be written relating inches to feet.

$$\frac{12 \text{ in.}}{1 \text{ ft}} \quad \text{or} \quad \frac{1 \text{ ft}}{12 \text{ in.}}$$

We choose the factor that will mathematically cancel feet and leave the answer in inches. Note that the units are treated in the same way we treat numbers, multiplying or dividing as required. Two possibilities then arise to change 24 ft to inches:

$$24\ \cancel{\text{ft}} \times \frac{12\text{ in.}}{1\ \cancel{\text{ft}}} \qquad \text{or} \qquad 24\text{ ft} \times \frac{1\text{ ft}}{12\text{ in.}}$$

In the first case (the correct method), feet in the numerator and the denominator cancel, giving us an answer of 288 in. In the second case, the units of the answer are $\text{ft}^2/\text{in.}$, the answer being 2.0 $\text{ft}^2/\text{in.}$ In the first case, the answer is reasonable since it is expressed in units having the proper dimensions. That is, the dimension of length expressed in feet has been converted to length in inches according to the mathematical expression

$$\cancel{\text{ft}} \times \frac{\text{in.}}{\cancel{\text{ft}}} = \text{in.}$$

In the second case, the answer is not reasonable since the units ($\text{ft}^2/\text{in.}$) do not correspond to units of length. The answer is therefore incorrect. The units are the guiding factor for the proper conversion.

The reason we can multiply 24 ft times 12 in./ft and not change the value of the measurement is that the conversion factor is derived from two equivalent quantities. Therefore, the conversion factor 12 in./ft is equal to unity. And when you multiply any factor by 1, it does not change the value.

$$12\text{ in.} = 1\text{ ft} \qquad \text{and} \qquad \frac{12\text{ in.}}{1\text{ ft}} = 1$$

Convert 16 kg to milligrams. In this problem it is best to proceed in this fashion:

$$\text{kg} \longrightarrow \text{g} \longrightarrow \text{mg}$$

The possible conversion factors are

$$\frac{1000\text{ g}}{1\text{ kg}} \quad \text{or} \quad \frac{1\text{ kg}}{1000\text{ g}} \qquad\qquad \frac{1000\text{ mg}}{1\text{ g}} \quad \text{or} \quad \frac{1\text{ g}}{1000\text{ mg}}$$

We use the conversion factor that leaves the proper unit at each step for the next conversion. The calculation is

$$16\ \cancel{\text{kg}} \times \frac{1000\ \cancel{\text{g}}}{1\ \cancel{\text{kg}}} \times \frac{1000\text{ mg}}{1\ \cancel{\text{g}}} = 1.6 \times 10^7\text{ mg}$$

Many problems can be solved by a sequence of steps involving unit conversion factors. This sound, basic approach to problem solving, together with neat and orderly setting up of data, will lead to correct answers having the right units, fewer errors, and considerable saving of time.

12. Graphical Representation of Data A graph is often the most convenient way to present or display a set of data. Various kinds of graphs have been

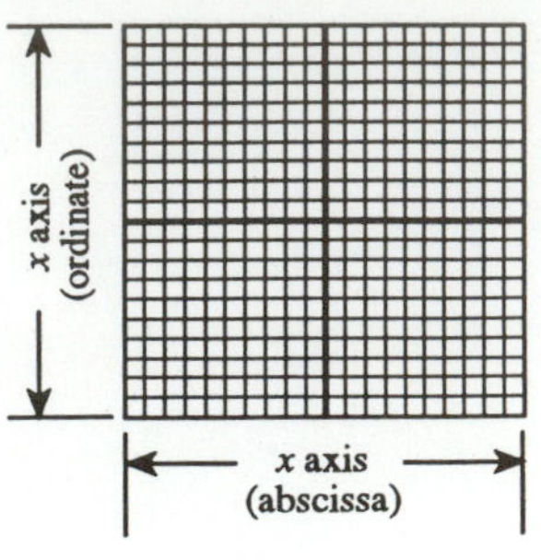

FIGURE I.1

devised, but the most common type uses a set of horizontal and vertical coordinates to show the relationship of two variables. It is called an $x-y$ graph because the data of one variable are represented on the horizontal or x axis (abscissa) and the data of the other variable are represented on the vertical or y axis (ordinate). See Figure I.1.

As a specific example of a simple graph, let us graph the relationship between Celsius and Fahrenheit temperature scales. Assume that initially we have only the information in the following table.

°C	°F
0	32
50	122
100	212

On a set of horizontal and vertical coordinates (graph paper), scale off at least 100 Celsius degrees on the x axis and at least 212 Fahrenheit degrees on the y axis. Locate and mark the three points corresponding to the three temperatures given and draw a line connecting these points (see Figure I.2).

Here is how a point is located on the graph: Using the 50°C–122°F data, trace a vertical line up from 50°C on the x axis and a horizontal line across from 122°F on the y axis and mark the point where the two lines intersect. This process is called *plotting*. The other two points are plotted on the graph in the same way. [*Note*: The number of degrees per scale division was chosen to give a graph of convenient size. In this case there are 5 Fahrenheit degrees per scale division and 2 Celsius degrees per scale division.]

FIGURE I.2 ▶

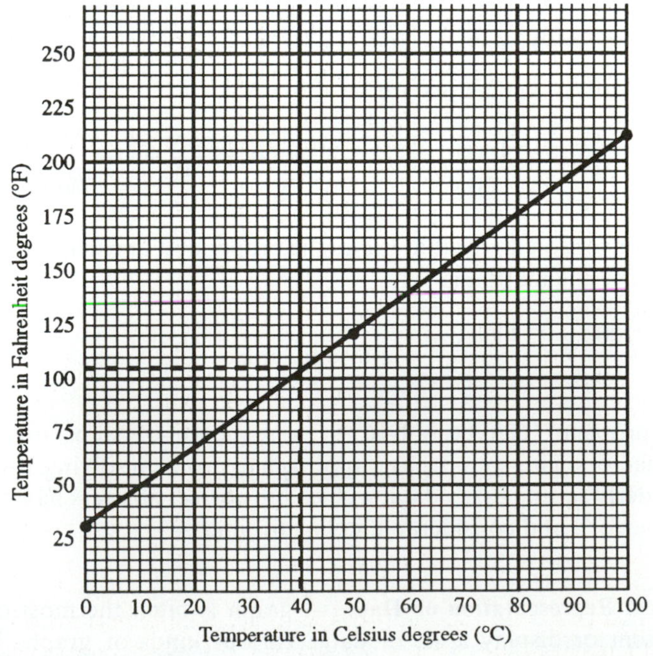

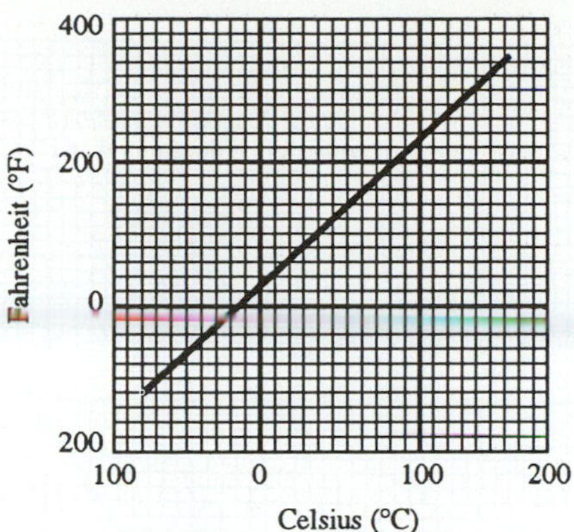

◄ **FIGURE I.3**

The graph in Figure I.2 shows that the relationship between Celsius and Fahrenheit temperature is that of a straight line. The Fahrenheit temperature corresponding to any given Celsius temperature between 0 and 100° can be determined from the graph. For example, to find the Fahrenheit temperature corresponding to 40°C, trace a perpendicular line from 40°C on the *x* axis to the line plotted on the graph. Now trace a horizontal line from this point on the plotted line to the *y* axis and read the corresponding Fahrenheit temperature (104°F). See the dashed lines in Figure I.2. In turn, the Celsius temperature corresponding to any Fahrenheit temperature between 32 and 212° can be determined from the graph by tracing a horizontal line from the Fahrenheit temperature to the plotted line and reading the corresponding temperature on the Celsius scale directly below the point of intersection.

The mathematical relationship of Fahrenheit and Celsius temperatures is expressed by the equation $°F = 1.8 \times °C + 32$. Figure I.2 is a graph of this equation. Because the graph is a straight line, it can be extended indefinitely at either end. Any desired Celsius temperature can be plotted against the corresponding Fahrenheit temperature by extending the scales along both axes as necessary. Negative, as well as positive, values can be plotted on the graph (see Figure I.3).

Figure I.4 is a graph showing the solubility of potassium chlorate in water at various temperatures. The solubility curve on this graph was plotted from the data in the following table.

Temperature (°C)	Solubility (g $KClO_3$/100 g water)
10	5.0
20	7.4
30	10.5
50	19.3
60	24.5
80	38.5

In contrast to the Celsius–Fahrenheit temperature relationship, there is no known mathematical equation that describes the exact relationship between

FIGURE I.4 ►

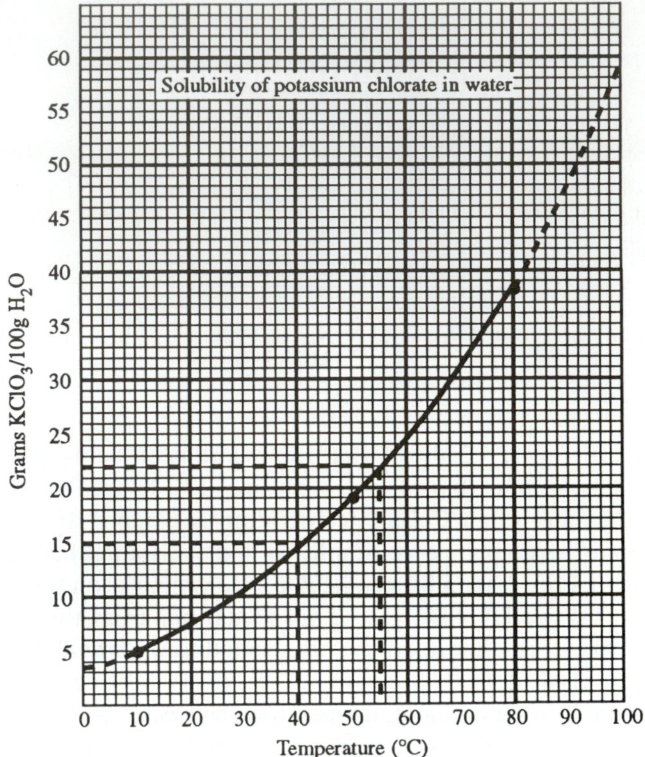

temperature and the solubility of potassium chlorate. The graph in Figure I.4 was constructed from experimentally determined solubilities at the six temperatures shown. These experimentally determined solubilities are all located on the smooth curve traced by the unbroken line portion of the graph. We are therefore confident that the unbroken line represents a very good approximation of the solubility data for potassium chlorate over the temperature range from 10 to 80°C. All points on the plotted curve represent the composition of saturated solutions. Any point below the curve represents an unsaturated solution.

The dashed-line portions of the curve are *extrapolations*; that is, they extend the curve above and below the temperature range actually covered by the plotted solubility data. Curves such as this one are often extrapolated a short distance beyond the range of the known data, although the extrapolated portions may not be highly accurate. Extrapolation is justified only in the absence of more reliable information.

The graph in Figure I.4 can be used with confidence to obtain the solubility of $KClO_3$ at any temperature between 10 and 80°C, but the solubilities between 0 and 10°C and between 80 and 100°C are less reliable. For example, what is the solubility of $KClO_3$ at 55°C, at 40°C, and at 100°C?

First draw a perpendicular line from each temperature to the plotted solubility curve. Now trace a horizontal line to the solubility axis from each point on the curve and read the corresponding solubilities. The values that we read from the graph are

55°	22.0 g $KClO_3$/100 g water
40°	14.2 g $KClO_3$/100 g water
100°	59 g $KClO_3$/100 g water

Of these solubilities, the one at 55°C is probably the most reliable because experimental points are plotted at 50°C and at 60°C. The 40°C solubility value is a bit less reliable because the nearest plotted points are at 30°C and 50°C. The 100°C solubility is the least reliable of the three values because it was taken from the extrapolated part of the curve, and the nearest plotted point is 80°C. Actual handbook solubility values are 14.0 and 57.0 g of $KClO_3$/100 g of water at 40°C and 100°C, respectively.

The graph in Figure I.4 can also be used to determine whether a solution is saturated or unsaturated. For example, a solution contains 15 g of $KClO_3$/100 g of water and is at a temperature of 55°C. Is the solution saturated or unsaturated? *Answer*: The solution is unsaturated because the point corresponding to 15 g and 55°C on the graph is below the solubility curve; all points below the curve represent unsaturated solutions.

Vapor Pressure of Water at Various Temperatures

Temperature (°C)	Vapor pressure (torr)	Temperature (°C)	Vapor pressure (torr)
0	4.6	26	25.2
5	6.5	27	26.7
10	9.2	28	28.3
15	12.8	29	30.0
16	13.6	30	31.8
17	14.5	40	55.3
18	15.5	50	92.5
19	16.5	60	149.4
20	17.5	70	233.7
21	18.6	80	355.1
22	19.8	90	525.8
23	21.2	100	760.0
24	22.4	110	1074.6
25	23.8	—	—

APPENDIX III

Units of Measurement

Physical Constants

Constant	Symbol	Value
Atomic mass unit	amu	1.6606×10^{-27} kg
Avogadro's number	N	6.022×10^{23}/mol
Gas constant	R (at STP)	0.0821 L atm/K mol
Mass of an electron	m_e	9.11×10^{-31} kg 5.5×10^{-4} amu
Mass of a neutron	m_n	1.675×10^{-27} kg 1.00866 amu
Mass of a proton	m_p	1.673×10^{-27} kg 1.00728 amu
Speed of light	c	2.997925×10^8 m/s

SI Units and Conversion Factors

Length		Mass	
SI unit: meter (m)		SI unit: kilogram (kg)	
1 meter	= 1.0936 yards	1 kilogram	= 1000 grams
1 centimeter	= 0.3937 inch		= 2.20 pounds
1 inch	= 2.54 centimeters (exactly)	1 pound	= 453.59 grams
			= 0.45359 kilogram
1 kilometer	= 0.62137 mile		= 16 ounces
1 mile	= 5280 feet	1 ton	= 2000 pounds
	= 1.609 kilometers		= 907.185 kilograms
1 angstrom	= 10^{-10} meter	1 ounce	= 28.3 g
		1 atomic mass unit	= 1.6606×10^{-27} kilograms

Volume		Temperature	
SI unit: cubic meter (m^3)		SI unit: kelvin (K)	
1 liter	= 10^{-3} m^3	0 K	= −273.15°C
	= 1 dm^3		= −459.67°F
	= 1.0567 quarts	K	= °C + 273.15
1 gallon	= 4 quarts	°C	$= \dfrac{(°F - 32)}{1.8}$
	= 8 pints	°F	= 1.8(°C) + 32
	= 3.785 liters	°F	= 1.8(°C + 40) − 40
1 quart	= 32 fluid ounces		
	= 0.946 liter		
1 fluid ounce	= 29.6 mL		

Energy		Pressure	
SI unit: joule (J)		SI unit: pascal (Pa)	
1 joule	= 1 kg m^2/s^2	1 pascal	= 1 kg/m^1 s^2
	= 0.23901 calorie	1 atmosphere	= 101.325 kilopascals
1 calorie	= 4.184 joules		= 760 torr (mmHg)
			= 14.70 pounds per square inch (psi)

Solubility Table

	F^-	Cl^-	Br^-	I^-	O^{2-}	S^{2-}	OH^-	NO_3^-	CO_3^{2-}	SO_4^{2-}	$C_2H_3O_2^-$
H^+	S	S	S	S	S	s	S	S	s	S	S
Na^+	S	S	S	S	S	S	S	S	S	S	S
K^+	S	S	S	S	S	S	S	S	S	S	S
NH_4^+	S	S	S	S	—	S	S	S	S	S	S
Ag^+	S	I	I	I	I	I	—	S	I	I	I
Mg^{2+}	I	S	S	S	I	d	I	S	I	S	S
Ca^{2+}	I	S	S	S	I	d	I	S	I	I	S
Ba^{2+}	I	S	S	S	s	d	s	S	I	I	S
Fe^{2+}	s	S	S	S	I	I	I	S	s	S	S
Fe^{3+}	I	S	S	—	I	I	I	S	I	S	I
Co^{2+}	S	S	S	S	I	I	I	S	I	S	S
Ni^{2+}	s	S	S	S	I	I	I	S	I	S	S
Cu^{2+}	s	S	S	—	I	I	I	S	I	S	S
Zn^{2+}	s	S	S	S	I	I	I	S	I	S	S
Hg^{2+}	d	S	I	I	I	I	I	S	I	d	S
Cd^{2+}	s	S	S	S	I	I	I	S	I	S	S
Sn^{2+}	S	S	S	s	I	I	I	S	I	S	S
Pb^{2+}	I	I	I	I	I	I	I	S	I	I	S
Mn^{2+}	s	S	S	S	I	I	I	S	I	S	S
Al^{3+}	I	S	S	S	I	d	I	S	—	S	S

Key: *S* = soluble in water
s = slightly soluble in water
I = insoluble in water (less than 1 g/100 g H_2O)
d = decomposes in water

Appendix V

Answers to Selected Exercises

Chapter 1

5. The following statements are correct: a, b, d, f, g, h

Chapter 2

2. 135°F
3. 7.6 cm
18. The following statements are correct: a, c, d, e, g, h, i, j, l, n, p, q
19. (a) 2 (b) 3 (c) 3 (d) 2 (e) 3 (f) 6 (g) 3 (h) 4
20. (a) 93.2 (b) 8.87 (c) 0.0285 (d) 21.3 (e) 4.64 (f) 130. (g) 34.3 (h) 2.00×10^6
21. (a) 2.9×10^6 (b) 4.56×10^{-2} (c) 5.8×10^{-1} (d) 4.0822×10^3 (e) 8.40×10^{-3} (f) 4.030×10^1 (g) 1.2×10^7 (h) 5.5×10^{-6}
22. (a) 14.4 (b) 58.5 (c) 1.08×10^8 (d) 4.0×10^1 (e) 2.0×10^2 (f) 2009 (g) 0.504 (h) 1.79×10^3 (i) 7.18×10^{-3} (j) 2.49×10^{-4}
23. (a) 0.833 (b) 0.429 (c) 0.750 (d) 0.500
24. (a) 1.9 (b) 86.7 (c) 2.1 (d) 51.3 (e) 8.93 (f) 0.030
25. (a) 100°C (b) 72°F (c) 298 K (d) 4.6 mL
26. (a) 0.280 m (b) 1.000 km (c) 92.8 mm (d) 1.5×10^{-4} km (e) 6.06×10^{-6} km (f) 4.5×10^8 Å (g) 6.5×10^3 Å (h) 1.21×10^3 cm (i) 8.0×10^3 m (j) 31.5 cm (k) 2.5×10^7 mm (l) 1.2×10^{-6} cm (m) 5.20×10^4 cm (n) 0.3884 nm (o) 107 cm (p) 4.1×10^4 in. (q) 811 km (r) 12.9 cm^2 (s) 117 ft (t) 10.2 mi (u) 7.4×10^4 mm^3 (v) 1.25×10^{19} mm^3
27. (a) 1.068×10^4 mg (b) 6.8×10^{-2} kg (c) 0.00854 kg (d) 94.2 lb (e) 0.164 g (f) 6.5×10^5 mg (g) 5.5×10^3 g (h) 4.3×10^4 g
28. (a) 2.50×10^{-2} L (b) 2.24×10^4 mL (c) 3.3×10^3 mL (d) 1.3×10^3 m^3 (e) 468 mL (f) 9.41 gal (g) 9.0×10^{-3} mL (h) 75.7 L
29. (a) 89 km/hr (b) 81 ft/s
30. (a) 37 ft/s (b) 25 mi/hr
31. 297 g NaCl remaining
32. (a) 7.5 mi/s (b) 12 km/s
33. 77.2 kg
34. 0.32 g
35. 5.0×10^{-1} s
36. 3×10^4 mg
37. \$0.39/kg
38. \$2520
39. 3.0×10^3 times heavier
40. \$21
41. 56 L
42. 230 L
43. 7.6×10^4 drops
44. 160 L
45. 5.0×10^{-3} mL
46. 2.83×10^4 mL
47. 4×10^5 m^2
48. 37.0°C
49. −100°C is 10°F colder than −138°F
50. (a) 72°C (b) −17.8°C (c) 255.2 K (d) −0.40°F (e) 90.°F (f) −22.6°C (g) 546 K (h) −61°C
51. (a) −40°F (b) −11.4°C
52. 1.565 g/mL
53. 3.12 g/mL
54. 7.1 g/mL
55. 595 g
56. 1.28 g/mL
57. 3.40×10^2 g
58. 27.6 g ethyl alcohol, 76.9 g = (mass of cylinder + alcohol)
59. 1.74 g/mL Mg; 2.70 g/mL Al; 10.5 g/mL Ag
60. 3.57×10^3 g
61. 0.965 g/mL
62. 49.8 mL
63. H_2O = 50 mL; alcohol = 60 mL
64. Gold bar is not pure.
65. \$1,033,718
66. d(slug) = 2.7 g/mL; d(liq) = 0.842 g/mL

Chapter 3

44. The following statements are correct: c, f, g, j, m, o, q, t, u, w, x, y, aa, dd
45. 2.67 g/mL
46. 54.5 mL
47. 18 carats
48. 75% C
49. 7.6×10^3 g Au
50. 1570 kg of alloy

Chapter 4

21. The following statements are correct: a, d, f, h, i
22. (a) 391 K (b) 244.4°F
23. 30.3% Fe
24. 2.25 g copper(II)oxide
25. 21.9 g Cu
26. 60.2 g Hg
27. 8.50% fat
28. (a) 6.9 g oxygen (b) 60.3% magnesium
29. 21.7 g chlorine
30. 1.7×10^4 J
31. 1.9×10^3 J
32. 0.301 J/g °C
33. 0.480 J/g °C
34. 17°C
35. 41.6°C
36. 3.0×10^3 cal
37. 29.1°C
38. 5.03×10^{-2} J/g °C
39. The copper pan heats fastest
40. 5°C
41. 6.4 g coal
42. (a) 1.2×10^{14} cal (b) 3.9×10^8 gal

Chapter 5

9. The following statements are correct: a, b, c, f, g
14. The following statements are correct: b, c, f, g, h, k
34. 131 amu
35. Silver = 107.9 amu
36. Magnesium = 24.31 amu
37. 6.03×10^{24} atoms
38. 69.7 amu = gallium

Chapter 6

31. The following statements are correct: a, d, f, g, i, j, l, m, o, p, r, u, w, x, y, aa

Chapter 7

9. The following statements are correct: a, b, c, d, h
15. The following statements are correct: a, d, f, g, j, k, m
16. (a) 119.0 (b) 142.0 (c) 331.2 (d) 46.07 (e) 60.05 (f) 231.6 (g) 342.3 (h) 342.1 (i) 132.1
17. (a) 40.00 (b) 275.8 (c) 152.0 (d) 96.09 (e) 146.3 (f) 122.1 (g) 180.2 (h) 368.4 (i) 244.2
18. (a) 0.344 mol Zn (b) 2.83×10^{-2} mol Mg (c) 7.5×10^{-2} mol Cu (d) 6.48 mol Co (e) 4.6×10^{-4} mol Sn (f) 28 mol N atoms
19. (a) 0.625 mol NaOH (b) 0.275 mol Br_2 (c) 7.18×10^{-3} mol $MgCl_2$ (d) 0.462 mol CH_3OH (e) 2.03×10^{-2} mol Na_2SO_4 (f) 5.97 mol ZnI_2
20. (a) 108 g Au (b) 285 g H_2O (c) 886 g Cl_2 (d) 252 g NH_4NO_3 (e) 0.0417 g H_2SO_4 (f) 11 g CCl_4 (g) 0.122 g Ti (h) 8.0×10^{-7} g S
21. (a) 7.59×10^{23} molecules O_2 (b) 3.4×10^{23} molecules C_6H_6 (c) 6.02×10^{23} molecules CH_4 (d) 1.652×10^{25} molecules HCl
22. (a) 3.441×10^{-22} g Pb (b) 1.791×10^{-22} g Ag (c) 2.992×10^{-23} g H_2O (d) 3.771×10^{-22} g $C_3H_5(NO_3)_3$
23. (a) 550. g Cu (b) 24.6 kg Au (c) 1.7×10^{-23} mol C (d) 8.3×10^{-21} mol CO_2 (e) 0.886 mol S (f) 42.8 mol NaCl (g) 1.05×10^{24} atoms Mg (h) 9.47 mol Br_2
24. (a) 6.022×10^{23} molecules CS_2 (b) 6.022×10^{23} C atoms (c) 1.204×10^{24} S atoms (d) 1.806×10^{24} atoms
25. 8.43×10^{23} atoms P
26. 5.88 g Na
27. 10.8 g/atomic mass
28. 5.54×10^{19} m
29. 1.2×10^{14} dollars/person
30. (a) 8.3×10^{16} drops (b) 7.3×10^6 mi^3
31. (a) 10.3 cm^3 (b) 2.18 cm
32. (a) 6.02×10^{23} atoms 0 (b) 3.75×10^{23} atoms 0 (c) 3.60×10^{23} atoms 0 (d) 6.0×10^{24} atoms 0 (e) 5.4×10^{24} atoms 0 (f) 5.0×10^{18} atoms 0
33. (a) 14.4 g Ag (b) 1.27 g Cl (c) 266 g N (d) 6.74 g O (e) 6.04 g H
34. 10.3 mol H_2SO_4
35. 1.62 mol HNO_3
36. (a) H_2O will contain the most molecules (b) CH_3OH will contain the most atoms
37. 41.58 g Fe_2S_3
38. (a) 22.34% Na; 77.65% Br (b) 39.06% K; 1.007% H; 12.00% C; 47.95% O (c) 34.41% Fe; 65.56% Cl (d) 16.53% Si; 83.46% Cl (e) 15.77% Al; 28.11% S; 56.12% O (f) 63.51% Ag; 8.246% N; 28.25% O
39. (a) 47.97% Zn; 52.02% Cl (b) 18.17% N; 9.153% H; 31.16% C; 41.51% O (c) 12.26% Mg; 31.24% P; 56.48% O (d) 21.21% N; 6.104% H; 24.27% S; 48.45% O

(e) 23.09% Fe; 17.37% N; 59.53% O
(f) 54.39% I; 45.61% Cl

40. (a) 77.73% (b) 69.94% (c) 72.37% (d) 15.16%
41. (a) 47.62% (b) 34.05% (c) 83.46%
(d) 83.63%; Highest %Cl is LiCl; lowest %Cl is $BaCl_2$
42. 43.7% P; 56.3% O
43. 24.2% C; 4.04% H; 71.72% Cl
44. 8.66 g Li
45. (a) 76.98% Hg (b) 46.38% O (c) 17.27% N
(d) 2.721% Mg
46. (a) H_2O (b) N_2O_3 (c) equal (d) $KClO_3$
(e) $KHSO_4$ (f) Na_2CrO_4
47. (a) N_2O (b) NO (c) N_2O_5 (d) Na_2CO_3
(e) $NaClO_4$ (f) Mn_3O_4
48. (a) CuCl (b) $CuCl_2$ (c) Cr_2S_3 (d) K_3PO_4
(e) $BaCr_2O_7$ (f) PBr_8Cl_3
49. SnO_2
50. V_2O_5
51. There is insufficient S
52. $C_6H_6O_2$
53. $C_6H_{12}O_6$
54. $C_9H_8O_4$
55. 4.77 g O
56. GaAs
57. (a) CCl_4 (b) C_2Cl_6 (c) C_6Cl_6 (d) C_3Cl_8

Chapter 8

21. The following statements are correct: a, d, e, f, h, i, j, l, n, o, p, q

Chapter 9

1. (a) 0.247 mol KNO_3
(b) 0.0658 mol $Ca(NO_3)_2$
(c) 4.4 mol $(NH_4)_2C_2O_4$
(d) 25.0 mol $NaHCO_3$
(e) 3.85×10^{-3} mol $ZnCl_2$
(f) 0.056 mol NaOH (g) 16 mol CO_2
(h) 4.3 mol C_2H_5OH (i) 0.237 mol H_2SO_4
2. (a) 273 g $Fe(OH)_3$ (b) 1.31 g $NiSO_4$
(c) 1.25×10^5 g $CaCO_3$ (d) 3.60 g $HC_2H_3O_2$
(e) 179 g NH_3 (f) 373 g Bi_2S_3 (g) 2.6 g HCl
(h) 1.35 g $C_6H_{12}O_6$ (i) 2×10^3 g Br_2
(j) 18 g K_2CrO_4
3. (a) 10.0 g H_2O (b) 25.0 g HCl
6. 2.80 mol Cl_2
7. 5×10^2 g NaOH
8. 0.800 mol $Al(OH)_3$
9. 19.7 g $Zn_3(PO_4)_2$
10. (a) 0.500 mol Fe_2O_3 (b) 12.4 mol O_2
(c) 6.20 mol SO_2 (d) 65.6 g SO_2
(e) 0.871 mol O_2 (f) 332 g FeS_2
11. 4.20 mol HCl
12. (a) 2.22 mol H_2 (b) 162 g HCl
13. 87.4 kg Fe
14. 117 g H_2O, 271 g Fe
15. (a) 52.5 mol O_2 (b) 13.0 g CO_2
(c) 2.20×10^2 g CO_2
16. (a) HNO_3 is limiting; KOH in excess
(b) H_2SO_4 is limiting, NaOH in excess
(c) H_2S is limiting; $Bi(NO_3)_3$ in excess
(d) H_2O is limiting; Fe in excess
17. (a) 2.0 mol Cu; 2.0 mol $FeSO_4$; 1.0 mol $CuSO_4$ (unreacted)
(b) 15.9 g Cu; 38.1 g $FeSO_4$; 6.0 g Fe (unreacted)
18. (a) 3.0 mol CO_2 (b) 9.0 mol CO_2
(c) 1.8 mol CO_2
(d) 6.0 mol CO_2; 8.0 mol H_2O; 4.0 mol O_2
(e) 16.5 g CO_2 (f) 59.9 g CO_2
(g) 59.9 g CO_2
19. 57.8%
20. 45.8 g CH_3OH; 4.24 g H_2 unreacted
21. 95.0% yield of Cu
22. (a) 3.2×10^2 g C_2H_5OH
(b) 1.10×10^3 g $C_6H_{12}O_6$
23. 5.2×10^2 g C
24. 77.8% CaC_2
25. $MgCl_2$ produces more AgCl than $CaCl_2$
26. 3.7×10^2 kg Li_2O
27. 14 kg concentrated H_2SO_4
28. 13 tablets
29. 67.2% $KClO_3$
30. The following statements are correct: a, c, d, f
31. The following statemets are correct: a, c, e

Chapter 10

37. The following statements are correct: a, c, d, e, f, h, i, j, k, l, n, p, q, r, u

Chapter 11

47. The following statements are correct: a, d, e, g, h, i, k, m, o, p, t, v

Chapter 12

37. The following statements are correct: b, c, e, h, i, j, l, o, q, r, s, t, u, v, w, x, z, bb, cc, gg, hh, jj, kk, mm

Chapter 13

27. The following statements are correct: b, d, f1, f4, h, i, n, o, p, q, t
28. (a) 0.941 atm (b) 28.1 in. (c) 13.8 lb/in^2.
(d) 715 torr (e) 953 mbar (f) 95.3 kPa

29. (a) 0.037 atm (b) 78.95 atm (c) 1.05 atm (d) 0.0493 atm
30. (a) 2.6×10^2 mL (b) 8.0×10^2 mL (c) 132 mL
31. (a) 374 mm Hg (b) 7.1×10^2 mm Hg
32. 1.0×10^2 atm
33. (a) 6.60 L (b) 6.17 L (c) 2.42 L (d) 8.35 L
34. 600. K (327°C)
35. 1.5×10^3 torr
36. −62°C
37. 65 atm
38. 320°F
39. (a) 3.6×10^2 mL (b) 7.8×10^2 mL
40. 6.1 L
41. 3.55 atm or 2.70×10^3 torr
42. 2.4×10^5 L
43. 33.4 L
44. 7.39×10^{21} molecules; 2.22×10^{22} atoms; 275 mL at STP
45. 703 torr
46. 1.10×10^3 torr
47. 2.29 L
48. 3.70×10^2 torr
49. 56 L
50. (a) 34 mol (b) 1.2×10^2 g H_2
51. 4.9 g CO_2
52. 43 g/mol
53. (a) 22.4 L (b) 8.301 L (c) 44.6 L
54. 2.69×10^{22} molecules
55. 44.6 mol Cl_2
56. 39.9 g/mol
57. (a) 3.74 g/L (b) 0.18 g/L (c) 3.58 g/L (d) 2.504 g/L
58. (a) 1.70 g/L F_2; (b) 1.54 g/L F_2
59. −78°C
60. (a) 8.4 L H_2 (b) 8.23 g CH_4 (c) 8.48 g/L (d) 63.5 g/mol
61. 1.64×10^2 g/mol
62. 57 L Ne
63. 123 L
64. 2.81 K
65. 0.13 mol N_2
66. 279 L H_2 STP
67. (a) 5.5 mol NH_3 (b) 5.6 mol NH_3 (c) 9.6 L NO (d) 0.640 L NO (e) 2.4 L NO (f) 1.1×10^2 g O_2 (g) 9.7 g NH_3
68. (a) 268 L O_2 (b) 153 L SO_2
69. 9.0 atm
70. He effuses 2.646 times as fast as N_2
71. (a) He/CH_4 = 2:1 (b) 67 cm from He end
72. C_4H_8
73. (a) 10.0 mol CO_2; 3.0 mol O_2; no CO (b) 29 atm
74. 76% $KClO_3$
76. (a) 250 mm Hg (b) 1060 mm Hg
77. (a) 1.1×10^2 torr CO_2; 1.3×10^1 torr H_2 (b) 120 torr
78. 0.72 g/L
79. Air enters the room
80. 60.4 atm

Chapter 14

59. The following statements are correct: a, b, c, f, h, l, m, o, p, s, t, u, w, y
64. (a) 0.420 mol $CoCl_2 \cdot 6H_2O$ (b) 0.262 mol $FeI_2 \cdot 4H_2O$
65. (a) 2.52 mol H_2O (b) 1.05 mol H_2O
66. 51.2% H_2O
67. 48.7% H_2O
68. $Pb(C_2H_3O_2)_2 \cdot 3\,H_2O$
69. $FePO_4 \cdot 4\,H_2O$
70. 3.11×10^5 J
71. 5.5×10^4 J
72. 1.62×10^5 cal
73. 40.7 kJ/mol
74. Yes, sufficient ice
75. 68°C
76. 3.1×10^3 cal
77. Mixture of ice and water, 0°C
78. 82 g ice remains
79. No; insufficient steam
80. 2.29×10^6 J
81. 43.9 kJ
82. 40.2 g H_2O
83. 18.0 mL with volume 22.4 L
84. (a) 18.0 g H_2O (b) 36.0 g H_2O (c) 0.784 g H_2O (d) 18.0 g H_2O (e) 0.447 g H_2O (f) 0.167 g H_2O
85. 6.97×10^{18} molecules/sec
86. 54 g H_2SO_4
87. (a) yes, O_2 remains (b) 20.0 mL O_2

Chapter 15

35. The following statements are correct: a, b, f, h, j, k, l, n, p, r, s, t, v, x, y, z
51. (a) 20.0% NaBr (b) 10.7% K_2SO_4 (c) 7.41% $Mg(NO_3)_2$
52. (a) 240. g solution (b) 544 g solution
53. (a) 23.1% NaCl (b) 22% $HC_2H_3O_2$ (c) 15% $C_6H_{12}O_6$
54. (a) 3.3 g KCl (b) 37.5 g K_2CrO_4 (c) 6.4 g $NaHCO_3$
55. 455 g solution
56. (a) 4.5 g NaCl (b) 449 g H_2O
57. 19 g H_2O needs to evaporate
58. (a) 22.0% CH_3OH (b) 33.6% NaCl
59. (a) 25.0% CH_3OH (b) 22% CCl_4
60. $x = 2.1 \times 10^2$ mL solution

61. (a) 424 g HNO_3 (b) 1.18 L
62. (a) 0.40 *M* (b) 3.8 *M* NaCl (c) 2.5 *M* HCl
(d) 0.59 *M* $BaCl_2 \cdot 2H_2O$
63. (a) 0.327 *M* Na_2CrO_4 (b) 1.8 *M* $C_6H_{12}O_6$
(c) 2.19×10^{-3} *M* $Al_2(SO_4)_3$
(d) 0.172 *M* $Ca(NO_3)_2$
64. (a) 40. mol LiCl (b) 0.0750 mol H_2SO_4
(c) 3.5×10^{-4} mol NaOH (d) 16 mol $CoCl_2$
65. (a) 8.8×10^3 g NaCl (b) 13 g HCl
(c) 4.6×10^2 g H_2SO_4 (d) 8.58 g $Na_2C_2O_4$
66. (a) 1.68×10^3 mL (b) 3.91×10^4 mL
(c) 1.05×10^3 mL (d) 7.82×10^3 mL
67. 6.72 *M* HNO_3
68. (a) 6.0 *M* HCl (b) 0.064 *M* $ZnSO_4$
(c) 1.6 *M* HCl
69. (a) 2.0×10^2 mL 12 *M* HCl
(b) 20. mL 15 *M* NH_3 (c) 16 mL 16 *M* HNO_3
(d) 69 mL 18 *M* H_2SO_4
70. 8.2 L
71. (a) 0.48 *M* H_2SO_4 (b) 0.74 *M* H_2SO_4
(c) 1.83 *M* H_2SO_4
72. 540. mL H_2O to be added
73. 0.32 *M* HNO_3
74. Take 31 mL of 5.00 *M* KOH and dilute with water to a volume of 250 mL
75. (a) 7.60 g $BaCrO_4$ (b) 15 mL of 1.0 *M* $BaCl_2$
76. (a) 33.3 mL 0.250 *M* Na_3PO_4
(b) 1.10 g $Mg_3(PO_4)_2$
77. (a) 0.300 mol H_2 (b) 7.80 L H_2
78. 2.08 *M* HCl
79. (a) 0.67 mol KCl (b) 0.33 mol $CrCl_3$
(c) 0.30 mol $FeCl_2$
(d) 69 mL of 0.060 *M* $K_2Cr_2O_7$
(e) 35 mL of 6.0 *M* HCl
80. (a) 1.3 mol Cl_2 (b) 16 mol HCl
(c) 1.3×10^2 mL 6.0 *M* HCl (d) 3.2 L Cl_2
81. (a) HCl: 36.46 g/eq; NaOH: 40.00 g/eq
(b) HCl: 36.46 g/eq; $Ba(OH)_2$: 85.65 g/eq
(c) H_2SO_4: 49.04 g/eq; $Ca(OH)_2$: 37.05 g/eq
(d) H_2SO_4: 49.04 g/eq; KOH 56.11 g/eq
(e) H_3PO_4: 49.00 g/eq; LiOH 23.95 g/eq
82. (a) 4.0 *N* HCl (b) 0.243 *N* HNO_3
(c) 6.0 *N* H_2SO_4 (d) 5.55 *N* H_3PO_4
(e) 0.250 *N* $HC_2H_3O_2$
83. 1.86 *N* H_2SO_4
84. $Mg(OH)_2$ tablet
85. (a) 9.943 mL NaOH (b) 7.261 mL NaOH
(c) 20.19 mL NaOH
86. (a) 4.37 *m* CH_3OH (b) 10 *m* C_6H_6
(c) 5.5 *m* $C_6H_{12}O_6$
87. (a) CH_3OH (b) Both the same
88. 6.2 *m* H_2SO_4, 5.0 *M* H_2SO_4
89. Solution will freeze in 20°F temperature
90. (a) 101.5°C; (b) 2.9 *m*
91. (a) 10.74 *m* (b) 105.50°C (c) −20.0°C
92. (a) 0.545 *m* (b) 2.7°C (c) 81.5°C
93. 153 g/mol
94. 163 g/mol
95. 500. g H_2O
96. (a) 8.04×10^3 g $C_2H_6O_2$
(b) 7.24×10^3 mL $C_2H_6O_2$ (c) −4.0°F
97. 60. g 10% NaOH
98. 97 g NaOH
99. (a) 1.6×10^2 g sugar (b) 0.47 *M*
(c) 0.516 *m*
100. Molecular formula is $C_8H_4N_2$
101. 16.2 *m* HCl
102. 0.14 *M*; 6.4 g KNO_3

Chapter 16

28. The following statements are correct:
a, c, f, i, j, k, l, n, p, r, t, u
38. (a) 0.015 *M* Na^+; 0.015 *M* Cl^-
(b) 4.25 *M* Na^+; 4.25 *M* K^+; 4.25 *M* SO_4^{2-}
(c) 0.75 *M* Zn^{2+}; 1.5 *M* Br^-
(d) 4.95 *M* SO_4^{2-}; 3.30 *M* Al^{3+}
(e) 0.20 *M* Ca^{2+}; 0.40 *M* Cl^-
(f) 0.265 *M* K^+; 0.265 *M* I^-
(g) 0.682 *M* NH_4^+; 0.341 *M* SO_4^{2-}
(h) 0.0628 *M* Mg^{2+}; 0.126 *M* ClO_3^-
39. (a) 0.034 g Na^+; 0.053 g Cl^-
(b) 9.77 g Na^+; 16.6 g K^+; 40.8 g SO_4^{2-}
(c) 4.9 g Zn^{2+}; 12 g Br^-
(d) 8.90 g Al^{3+}; 47.5 g SO_4^{2-}
(e) 0.80 g Ca^{2+}; 1.4 g Cl^-
(f) 1.04 g K^+; 3.36 g I^-
(g) 1.23 g NH_4^+; 3.28 g SO_4^{2-}
(h) 0.153 g Mg^{2+}; 1.05 g ClO_3^-
40. 0.260 M Ca^{2+}
41. (a) 1.0 *M* Na^+; 1.0 *M* Cl^-
(b) 0.50 *M* Na^+; 0.50 *M* Cl^-
(c) 1.0 *M* K^+; 0.5 *M* Ca^{2+}; 2.0 *M* Cl^-
(d) 0.20 *M* H^+; 0.20 *M* K^+; 0.40 *M* Cl^-
(e) No ions present in solution
(f) 0.67 *M* Na^+; 0.67 *M* NO_3^-
42. 3.0×10^3 mL
43. (a) 0.684 *M* HCl (b) 0.542 *M* HCl
(c) 0.367 *M* HCl (d) 0.147 *M* NaOH
(e) 0.964 *M* NaOH (f) 0.4750 *M* NaOH
44. 0.309 M $Ba(OH)_2$
53. (a) 40.8 mL of 0.245 *M* HCl
(b) 1.57×10^3 mL of 0.245 *M* HCl
54. 0.201 *M* HCl
55. 437 mL of 0.1234 *M* HCl
56. 0.673 g KOH
57. 88.0% NaOH
58. 7.0% NaCl in sample
59. (a) 0.468 L H_2 (b) 0.936 L H_2

60. (a) 2 (b) 0 (c) 8.19 (d) 7 (e) 0.30 (f) 4
61. (a) 3.4 (b) 2.6 (c) 4.3 (d) 10.5
62. 4
63. 13.9 mL of 18.0 M H_2SO_4
64. 0.025 mol NaOH remains; basic solution
65. 0.3586 N NaOH
66. 22.7 mL of 0.325 M HNO_3
67. 22.7 mL of 0.325 M H_2SO_4
68. 0.4536 N H_3PO_4; 0.1512 M H_3PO_4
69. 1.2×10^2 mL
70. (a) 0.32 N H_2SO_4 (b) 0.39 g H_2SO_4
71. 0.0500 N HCl; 0.100 N NaOH
72. 1.2×10^2 g/equiv
73. 57 g/equiv

Chapter 17

38. The following statements are correct: c, d, g, i, j, k, l, n, p, q, s, t, u, v, x, z
39. 4.20 mol HI
40. (a) 3.2 mol HI
(b) 3.4 mol HI; 0.30 mol H_2; 0.57 mol I_2
(c) 57
41. 2.73 mol H_2; 0.538 mol I_2; 0.500 mol HI
42. $K_{eq} = 29$
43. 128 times as fast
44. HOCl, 3.5×10^{-8}; $HC_3H_5O_2$, 1.3×10^{-5}; HCN, 4.0×10^{-10}
45. (a) $[H^+] = 2.1 \times 10^{-3}$ M (b) pH = 2.7
(c) 0.84% ionized
46. $K_a = 2.7 \times 10^{-5}$
47. $K_a = 7 \times 10^{-10}$
48. (a) 0.42% ionized; pH = 2.4
(b) 1.3% ionized; pH = 2.9
(c) 4.2% ionized; pH = 3.4
49. $K_a = 1 \times 10^{-7}$
50. $K_a = 7.2 \times 10^{-6}$
51. $[OH^-] = 1.7 \times 10^{-15}$
52. (a) pH = 4.0; pOH = 10.0
(b) pOH = 2.0; pH = 12.0
(c) pOH = 2.6; pH = 11.4
(d) pH = 4.2; pOH = 9.8
(e) pOH = 4.9; pH = 9.1
53. (a) $[OH^-] = 1.0 \times 10^{-10}$
(b) $[OH^-] = 3.6 \times 10^{-9}$
(c) $[OH^-] = 2.5 \times 10^{-6}$
54. (a) $[H^+] = 1.7 \times 10^{-8}$ (b) $[H^+] = 1 \times 10^{-6}$
(c) $[H^+] = 2.2 \times 10^{-9}$
55. (a) 1.5×10^{-9} (b) 1.9×10^{-12}
(c) 1.2×10^{-23} (d) 2.6×10^{-13}
(e) 3.1×10^{-70} (f) 1.7×10^{-10}
(g) 2.4×10^{-5} (h) 5.13×10^{-17}
(i) 1.81×10^{-18}
56. (a) 4.5×10^{-5} M (b) 7.6×10^{-10} M
(c) 1.6×10^{-2} M (d) 1.2×10^{-4} M
57. (a) 8.9×10^{-4} g $BaCO_3$
(b) 9.3×10^{-9} g $AlPO_4$
(c) 0.50 g Ag_2SO_4 (d) 7.0×10^{-4} g $Mg(OH)_2$
58. (a) 2.1×10^{-4} mol Ca^{2+}/L; 4.2×10^{-4} mol F^-/L
(b) 8.2×10^{-3} g CaF_2
59. (a) Precipitate formed (b) Precipitate formed
(c) No precipitate formed
60. (a) 3.0×10^{-8} M SO_4^{2-}
(b) 7.0×10^{-7} g $BaSO_4$
61. $BaSO_4$ will precipitate first
62. (a) 5.0×10^{-12} mol AgBr
(b) 2.5×10^{-12} mol AgBr
63. No precipitate of $PbCl_2$ will form
64. $K_{eq} = 1 \times 10^4$
65. $K_{sp} = 2.3 \times 10^{-5}$
66. $[NH_3] = 8.0$ M
67. No precipitate should form
68. $K_{sp} = 4.00 \times 10^{-28}$
69. (a) $[H^+] = 3.6 \times 10^{-5}$; pH = 4.4
(b) $[H^+] = 1.8 \times 10^{-5}$; pH = 4.7
70. (a) Change in pH = 5.3 units
(b) Change in pH = 0.02 units

Chapter 18

29. The following statements are correct: a, c, e, g, j, k, m, p, q, r, s
30. 0.0772 mol NO
31. 20.2 L Cl_2
32. 17 g $KMnO_4$
33. 66.2 mL $K_2Cr_2O_7$
34. 10.0 mL $K_2Cr_2O_7$
35. 91.3% KI
36. 149 g Cu
37. 4.22 L NO
38. 5.56 mol H_2

Chapter 19

39. 0.0625 mg Sr-90 remains
40. $381
41. 2064 AD; one-eighth of the starting amount remains
42. 1.1×10^4 years old
43. 18 minutes/half-life
44. (a) 0.0424 g/mol (b) 3.8×10^{12} J/mol
45. (a) 2.9×10^{-11} J/atom U-235
(b) 1.7×10^{13} J/mol (c) 0.08185% mass loss
46. (a) 5.3×10^4 J/mol (b) 0.195% mass loss
47. 7 alpha and 4 beta particles lost
49. 0.16 mg remains
51. The following statements are correct: c, e, f, g, j, k, n, o, q, r, t

Chapter 20

50. The following statements are correct: b, c, d, e, h, i, j, m, o, r, s.
51. 1.1×10^2 L SO_2
52. 60 g Cl_2/hr
53. (a) 18 M H_2SO_4 (b) 16 M HNO_3 (c) 12 M HCl

Chapter 21

11. 422.8 g/mol
22. The following statements are correct: a, b, c, d, g, i, j, l, m, p, s.

Chapter 22

28. The following statements are correct: a, f, g, h, i, j, k, m, n, o, s, v, w, x, y, z.

Chapter 23

5. 438.8 g/mol
38. The following statements are correct: a, b, e, g, h, i, j, m, o, q, s.

Chapter 24

20. The following statements are correct: b, c, f, h, i, j, l, m, o, q, r, t, v.

Chapter 25

28. (a) 75.9 L H_2 (b) 1.01 kg triolein
38. The following statements are correct: a, b, f, g, h, i, j, k, l, m, n, p, q.

Chapter 26

20. The following statements are correct: g, j, m, o, p.

Chapter 27

4. 891 ethylene units
22. The following statements are correct: a, b, d, f, i, k.

Chapter 28

21. 62.5°
26. The following statements are correct: a, f, g, i, k, l.

Chapter 29

42. The following statements are correct: c, e, g, h, i, l, n, p, q, r, s, v, x, w.

Chapter 30

31. The following statements are correct: d, e, g, j, l, m, n.

Chapter 31

41. 38% protein
42. 6.8×10^4 (molar mass)
43. The following statements are correct: b, c, d, f, g, i, m, n, q, s, t.

Chapter 32

11. 0.003 M/min.
12. 150 proteins.
23. The following statements are correct: a, c, f, h, j, k, l, m, n.

Chapter 33

34. The following statements are correct: a, b, d, g, h, j, k, m, p, r.

Chapter 34

35. The following statements are correct: a, d, e, g, i, j, l, m.

Chapter 35

23. The following statements are correct: a, c, e, f, h, i, j, k, l.

Chapter 36

31. The following statements are correct: a, c, h, i, j, m, n.

Chapter 37

30. The following statements are correct: c, f, g, h, j, l, m, n, o, r, s, t, v.

INDEX/GLOSSARY

Entries and page numbers that are in boldface type refer to definitions of key terms in the text.

Photo Credits

These pages are an extension of the copyright page.

Chapter 1

Page 1: Comstock; Page 3: (left) Baltimore Spice Company, (right) Genentech; Page 5: The Granger Collection; Page 7: Rita Amaya; Page 11: Michael Melford/Tony Stone Worldwide.

Chapter 2

Page 14: Murray Alcosser /The Image Bank; Page 19: (top) David F. Hughes/Stock Boston, (bottom) David F. Scharf/Peter Arnold; Page 23: Sara Hunsaker; Page 31: (four photos) Rita Amaya.

Chapter 3

Page 46: Tom Doody/The Picture Cube; Page 48: Dallas and John Heaton/Stock Boston; Page 54: The Granger Collection; Page 55: (two photos) Rita Amaya; Page 61: (three photos) Rita Amaya; Page 62: General Electric Corporation.

Chapter 4

Page 66: Dawson Jones/Stock Boston; Page 72: (two photos) Rita Amaya; Page 73: R. M. Collins, III/The Image Works; Page 76: Paul Conklin/Monkmeyer Press; Page 77: Topham/The Image Works; Page 78: Bob Daemmrich/The Image Works.

Chapter 5

Page 81: J. C. Revy/Phototake NYC; Page 83: The Granger Collection; Page 84: Historical Pictures/Stock Montage; Page 86: IBM Corporation; Page 87: (left) Rita Amaya, (bottom) The Granger Collection; Page 88: The Granger Collection; Page 94: Linda M. Sweeting and Michael M. Rosenblattt/Towson State University.

Chapter 6

Page 99: Bjorn Bolstad/Peter Arnold; Page 106: Sargent-Welch Scientific Company, a VWR Company; Page 119: Comstock.

Chapter 7

Page 124: Comstock; Page 126: The Granger Collection; Page 130: Rita Amaya; Page 142: Sara Hunsaker.

Chapter 8

Pages 147, 151, and 158: Richard Megna/Fundamental Photographs; Page 159: E. R. Degginger; Page 165: Rita Amaya; Page 166: Comstock; Page 167: Cary Wolinsky/Stock Boston.

Chapter 9

Page 171: Michael J. Howell/ProFiles West; Page 178: NASA; Page 182: Comstock; Page 186: R. S. Muller/Berkeley Sensor and Actuator Center.

Chapter 10

Page 191: Lawrence Berkeley Laboratories, Science Photo Library/ Photo Researchers; Page 193: (top) Photo Researchers; (bottom) Mike Greenlar/The Image Works; Page 194: (left and center) The Granger Collection, (right) Topham/The Image Works.

Chapter 11

Page 211: Jack Finch/Science Photo Library; Page 212: Historical Pictures/Stock Montage; Page 224: American Telephone and Telegraph.

Chapter 12

Page 228: Geoff Tompkinson/TSW-Click/Chicago Limited; Page 237: UPI/Bettman; Page 239: Wide World Photos.

Chapter 13

Page 255: Comstock; Page 263: The Granger Collection; Page 268: Rita Amaya; Page 270: The Granger Collection; Page 289: NASA/ Science Source, Photo Researchers; Page 290: Karl I. Wallin/FPG International.

Chapter 14

Page 296: NASA; Page 306: Rita Amaya; Page 314: Sara Hotchkiss; Page 317: Gamma Liaison International; Page 318: Scott Kranhold.

Chapter 15

Page 325: Mike Kirkpatrick/ProFiles West; Page 329: Rita Amaya; Page 334: Bill Varie/Image Bank; Page 352: Runk/Schoenberger/ Grant Heilman.

Chapter 16

Page 362: Tony Stone Worldwide; Page 369: Roy Bishop/Stock Boston; Pages 370, 379, and 380: Rita Amaya; Page 383: Tom Stack and Associates; Page 388: Rita Amaya; Page 390 Sara Hunsaker.

Chapter 17

Page 396: Harald Sund/The Image Bank; Page 406: Rita Amaya; Page 420: Ian Miles/The Image Bank; Page 424: Ed Reschke/Peter Arnold.

Chapter l8

Page 430: Comstock; Page 443: Rita Amaya; Page 446: Stephen Frisch/Stock Boston; Page 450: (2 photos) Corning Glass.

Chapter 19

Page 455: John Mazziotta et al./Neurology/Science Photo Library/ Photo Researchers; Page 457: The Granger Collection; Page 465: NASA, Grant Heilmann; Page 467: (a) Biomedical Communications/Photo Researchers; (b) Hank Morgan/Photo Researchers; (c) Packard Instrument Company.

Chapter 20

Page 483: Allen Russell/ProFiles West; Page 487: AP/Wide World Photos; Page 495: Borax Corporation; Page 496: Grant Le Duc/ Monkmeyer Press; Page 499: Karen R. Preuss/The Image Works; Page 512: Comstock.

Chapter 21

Page 518: Marcella Pedone/The Image Bank; Page 537: Alan Carey/The Image Works; Page 540: Kirby Harrison/The Image Works; Page 545: Russell Wood/Taurus Photos.

Chapter 22

Page 548: John Yurka/The Picutre Cube; Page 566: Ralph Eagle, Jr./Photo Researchers; Page 569: Michael Nelson/FPG International.

Chapter 23

Page 585: Leslye Borden/Photo Edit; Page 603: Visuals Unlimited; Page 608: SIU/Photo Researchers.

Chapter 24

Page 6l5: Comstock; Page 627: Robert Brenner/PhotoEdit; Page 629: David Wells/The Image Works; Page 631: Bonnie Kamin.

Chapter 25

Page 637: W. Marc Bernsau/The Image Works; Page 639: Hans Pfletschinger/Peter Arnold; Page 643: Hanson Carroll/FPG International; Page 651: Holt Studios Limited/Earth Scenes/ Animals Animals; Page 655: Sara Hunsaker; Page 659: Robert S. Arnold, Old Sturbridge Village; Page 664: Sara Hunsaker.

Chapter 26

Page 671: Todd Powell/ProFiles West; Page 676: Sara Hunsaker; Page 677: Gary Bumgarner/ Tony Stone Worldwide; Page 680 G. L. Twiest/Visuals Unlimited; Page 683: Sara Hunsaker.

Chapter 27

Page 690: Wiley Wales/ProFiles West; Page 696: Thomas Zimmermann/FPG International; Page 699: Custom Medical Stock Photo; Page 700: Peter A. Simon/Phototake; Page 703: NCI/Science Source/Photo Researchers.

Chapter 28

Page 707: Dennis Kunkel/Phototake NYC; Page 709: Rita Amaya; Page 712: Runk/Schoenberger/Grant Heilman; Pages 714 and Pages 715: Sara Hunsaker.

Chapter 29

Page 727: Comstock; Page 734: Scott Camazine/Photo Researchers; Page 744: ProPix/Monkmeyer Press; Page 745: Brady/Monkmeyer Press; Page 751: James Webb/Phototake; Page 755: Eric V. Gravè/ Photo Researchers; Page 758: Science Source/Photo Researchers; Page 759: Visuals Unlimited.

Chapter 30

Page 765: Ocean Images/The Image Bank; Page 775: Liane Enkelis/Stock Boston; Page 779: Donna Jernigan/Monkmeyer Press; Page 781: Alfred Pasieka/Peter Arnold; Page 784: D. W. Fawcett/Photo Researchers.

Chapter 31

Page 788: Bob Daemmrich/The Image Works; Page 792: Manfred Kage/Peter Arnold; Page 803: AP/Wide World Photos; Page 810: Stanley Flegler/Visuals Unlimited; Page 819: James Holmes/ Science Photo Library/Photo Researchers.

Chapter 32

Page 823: Simon Yea/Tony Stone Worldwide; Page 825: Rita Amaya; Page 838: Sunkist Corporation.

Chapter 33

Pages 841: Character House/Photo Researchers; 849: Photo Researchers; Page 855: David Scharf/Peter Arnold; Page 860: Science Source/Photo Researchers; Page 865: Science Photo Library/Photo Researchers.

Chapter 34

Page 869: Jeffrey W. Myers/Stock Boston; Page 871: Comstock; Page 872: Art Attack/Monkmeyer Press; Page 882: Paul Conklin/ Monkmeyer Press; Page 885: (3 photos) Sara Hunsaker; Page 890: Fawcett/Hirokawa/Henses/Science Source/Photo Researchers.

Chapter 35

Page 895: John Lawlor/TSN-Click/Chicago Limited; Page 897: Comstock; Page 906: K. R. Porter/Photo Researchers; Page 911: (top) Runk/Schoenberger/Grant Heilman, (bottom) Lawrence Berkeley Laboratory/University of California.

Chapter 36

Page 916: Steve Dunwell/The Image Bank; Page 924: AP/Wide World Photos; Page 927: Ken Levinson/Monkmeyer Press; Page 929: Bettina Cirone/Photo Researchers.

Chapter 37

Page 936: Allen Russell/ProFiles West; Page 941: CNRI/ Phototake.

NAMES, FORMULAS AND CHARGES OF COMMON IONS

Positive Ions (Cations)

1+		2+		3+		4+	
Ammonium	NH_4^+	Calcium	Ca^{2+}	Aluminum	Al^{3+}	Tin(IV)	Sn^{4+}
Cesium	Cs^+	Chromium(II)	Cr^{2+}	Antimony(III)	Sb^{3+}	(Stannic)	
Copper(I)	Cu^+	Cobalt(II)	Co^{2+}	Arsenic(III)	As^{3+}	Lead(IV)	Pb^{4+}
(Cuprous)		Copper(II)	Cu^{2+}	Bismuth(III)	Bi^{3+}	(Plumbic)	
Francium	Fr^+	(Cupric)		Boron	B^{3+}	Manganese(IV)	Mn^{4+}
Gold	Au^+	Iron(II)	Fe^{2+}	Chromium(III)	Cr^{3+}		
Lithium	Li^+	(Ferrous)		Cobalt(III)	Co^{3+}		
Potassium	K^+	Lead(II)	Pb^{2+}	Iron(III)	Fe^{3+}		
Rubidium	Rb^+	Manganese(II)	Mn^{2+}	(Ferric)			
Silver	Ag^+	Mercury(I)	Hg_2^{2+}	Titanium(III)	Ti^{3+}		
Sodium	Na^+	Mercury(II)	Hg^{2+}				
		Nickel(II)	Ni^{2+}				
		Tin(II)	Sn^{2+}				
		(Stannous)					
		Zinc	Zn^{2+}				

Negative Ions (Anions)

1−		2−		3−	
Acetate	$C_2H_3O_2^-$	Carbonate	CO_3^{2-}	Arsenate	AsO_4^{3-}
Bromide	Br^-	Chromate	CrO_4^{2-}	Borate	BO_3^{3-}
Chlorate	ClO_3^-	Dichromate	$Cr_2O_7^{2-}$	Phosphate	PO_4^{3-}
Chloride	Cl^-	Hydrogen Phosphate	HPO_4^{2-}	Phosphite	PO_3^{3-}
Chlorite	ClO_2^-	(Biphosphate)		Phosphide	P^{3-}
Cyanide	CN^-	Oxalate	$C_2O_4^{2-}$		
Dihydrogen phosphate	$H_2PO_4^-$	Oxide	O^{2-}		
Fluoride	F^-	Peroxide	O_2^{2-}		
Hydride	H^-	Sulfate	SO_4^{2-}		
Hydrogen carbonate	HCO_3^-	Sulfide	S^{2-}		
(Bicarbonate)		Sulfite	SO_3^{2-}		
Hydrogen oxalate	$HC_2O_4^-$	Silicate	SiO_3^{2-}		
Hydrogen sulfate	HSO_4^-				
(Bisulfate)					
Hydrogen sulfide	HS^-				
Hydrogen sulfite	HSO_3^-				
(Bisulfite)					
Hydroxide	OH^-				
Hypochlorite	$HClO^-$				
Iodide	I^-				
Nitrate	NO_3^-				
Nitrite	NO_2^-				
Perchlorate	ClO_4^-				
Permanganate	MnO_4^-				
Thiocyanate	SCN^-				